Lecture Notes in Computer Science 5200

Commenced Publication in 1973
Founding and Former Series Editors:
Gerhard Goos, Juris Hartmanis, and Jan van Leeuwen

Teresa Vazão Mário M. Freire
Ilyoung Chong (Eds.)

Information Networking

Towards Ubiquitous Networking and Services

International Conference, ICOIN 2007
Estoril, Portugal, January 23-25, 2007
Revised Selected Papers

Volume Editors

Teresa Vazão
INESC ID Lisboa/IST, TAGUSPARK
Av. Prof. Cavaco Silva, 2744-016 Porto Salvo, Portugal
E-mail: teresa.vazao@tagus.ist.utl.pt

Mário M. Freire
University of Beira Interior
Department of Computer Science
Rua Marques d'Avila e Bolama, 6201-001 Covilhã, Portugal
E-mail: mario@di.ubi.pt

Ilyoung Chong
Hankuk University of Foreign Studies
Department of Information and Communications Engineering
Imun-dong, Dongdaemun-Gu, Seoul 130-790, Korea
E-mail: iychong@hufs.ac.kr

Library of Congress Control Number: 2008939364

CR Subject Classification (1998): C.2, H.4, H.3, D.2.12, D.4, H.5

LNCS Sublibrary: SL 5 – Computer Communication Networks
and Telecommunications

ISSN 0302-9743
ISBN-10 3-540-89523-X Springer Berlin Heidelberg New York
ISBN-13 978-3-540-89523-7 Springer Berlin Heidelberg New York

Springer is a part of Springer Science+Business Media

springer.com

Printed in Germany

Typesetting: Camera-ready by author, data conversion by Scientific Publishing Services, Chennai, India
Printed on acid-free paper SPIN: 12535934 06/3180 5 4 3 2 1 0

Preface

This volume contains the set of revised selected papers presented at the 21st International Conference on Information Networking (ICOIN 2007), which was held in Estoril, Portugal, January 23–25, 2007. The conference series started under the name of Joint Workshop on Computer Communications, in 1986. At that time, it constituted a technical meeting for researchers and engineers on Internet technologies in East Asian countries, where several technical networking issues were discussed. In 1993, the meeting was reorganized as an international conference known as ICOIN. Recent conferences were held in Sendai, Japan (2006), Jeju, Korea (2005), Pusan, Korea (2004), Jeju, Korea (2003), Jeju, Korea (2002), Beppu City, Japan (2001), Hsin-chu, Taiwan (2000), and Tokyo, Japan (1999). In 2007, for the first time since its creation, ICOIN took place outside Asia, and we were very pleased to host it in Portugal. ICOIN 2007 was organized by INESC-ID and IST/Technical University of Lisbon (Portugal) with the technical co-sponsorship of IEEE Communications Society and IEEE Portugal Section-Computer Society Chapter, in cooperation with the Order of Engineers College of Informatics Engineering (Portugal), IPSJ (Information Processing Society of Japan), KISS (Korea Information Science Society), and Lecture Notes in Computer Science (LNCS), Springer, Germany.

The papers presented in this volume were selected in two stages: 1) reviewing and selection for the ICOIN program and 2) on-site presentation review by session chairs or by program committee chairs. Regarding the first step, in response to the Call for Papers, a total of 302 papers were submitted, of which 100 were accepted, after careful assessment, for oral presentation in 24 technical sessions. Each paper was reviewed by at least three members of the Technical Program Committee or by external peer reviewers. A second review round was performed based on the presentation review by session chairs or by program committee chairs and, after a careful assessment by program chairs, 82 revised papers were selected for this volume. The set of revised selected papers covers a wide range of networking-related topics, including sensor networks; ad-hoc, mobile, and wireless networks; optical networks; peer-to-peer networks and systems; routing; transport protocols; quality of service; network design and capacity planning; resource management; performance monitoring; network management; next generation Internet; and networked applications and services.

We believe that this set of revised selected papers will make a significant contribution to the solution of key problems in the field of information networking. We would like to take this opportunity to warmly thank all of the authors who submitted their valuable papers to the conference. We are grateful to the members of the Technical Program Committee and to the numerous reviewers. Without their support, the organization of a high-quality conference program and the selection of this set of revised papers would not have been possible. We are also indebted to the many individuals and organizations that made this event happen, namely to our staff, who handled the logistics and worked to make this meeting a success, and to the sponsors.

We hope that you will find this volume, containing the revised selected papers of the International Conference on Information Networking (ICOIN 2007), a useful and timely document for presenting new ideas, results and recent findings in information networking towards ubiquitous networking and services.

February 2008

Teresa Vazão
Mário Freire
Ilyoung Chong

Conference Committees

Organizing Committee

Conference Chairman

Mário Serafim Nunes	INESC ID/IST, Portugal

Chair

Teresa Vazão	INESC ID/IST, Portugal

Co-chairs

Mário M. Freire	University of Beira Interior, Portugal
Yanghee Choi	Seoul National University, Korea

International Cooperation Committee

Jong Kwon Kim	Seoul National University, Korea
João Picoito	Siemens SA, Portugal

Technical Program Committee

Chair

Mário M. Freire	University of Beira Interior, Portugal

Co-chairs

Ilyoung Chong	HUFS, Korea
Nelson Fonseca	University of Campinas, Brazil
Pascal Lorenz	University of Haute Alsace, France
Teresa Vazão	INESC ID/IST, Portugal

Vice-Chairs

Algirdas Pakstas	University of Sunderland, UK
Fabrizio Granelli	University of Trento, Italy
Hyunkook Kahng	Korea University, Korea
Kenji Kawahara	Kyushu Institute of Technology, Japan
Nikola Rozic	Croatian Academy of Engineering, Croatia

Members

Table of Contents

Part I

Part II

Part III

Signaling-Embedded Short Preamble MAC for Multihop Wireless Sensor Networks

Kyuho Han[1], Sangsoon Lim[1], Sangbin Lee[1], Jin Wook Lee[2], and Sunshin An[1]

[1] Department of Electronics and Computer Engineering,
Korea University, Seoul, Korea
{garget,lssgood,kulsbin,sunshin}@dsys.korea.ac.kr
[2] Networking Technology Lab.,
Samsung Advanced Institute of Technology (SAIT), Seoul, Korea
thetruth.lee@samsung.com

Abstract. Applications of WSN usually generate low-rate traffic so the communication channel is expected to be idle most of the time. Therefore idle listening is very critical source of energy dissipation in wireless sensor networks. To reduce idle listening, we propose preamble sampling MAC that is named SESP-MAC. The main idea of SESP-MAC is to add control information into the short preamble frame. So the stream of short preambles is used not only for preamble sampling but also for avoiding overhearing, decreasing control packet overhead and reducing the listening of the redundant message, caused by message-flooding.

Keywords: duty-cycle, preamble sampling, MAC protocol, energy efficiency.

1 Introduction

The advances which have been made in wireless and micro-machine technologies have made it possible to develop wireless sensor networks. A wireless sensor network (WSN) is generally composed of a large number of sensor nodes and a few data collectors, which are called sink node. Sensor nodes are responsible for generating sensory data and reporting them to a previously specified sink node for prolonged duration.

Because sensor nodes may be densely deployed in remote location, it is likely that replacing their batteries will not be possible. But generally, many applications of WSN are required the long life time of sensor nodes to collect sensory data. Therefore power efficient protocols at each layer of communications are very important for wireless sensor networks [1]. There are several major sources of energy waste in wireless sensor networks [2].

- **Collision** occurs when two nodes transmit at the same time and interfere with each others transmission. Hence, retransmissions increase energy consumption.
- **Control packet overhead** such as RTS/CTS/ACK can be significant for wireless sensor networks that use small data packets.

T. Vazão, M.M. Freire, and I. Chong (Eds.): ICOIN 2007, LNCS 5200, pp. 1–10, 2008.

- **Overhearing** occurs when there is no meaningful activity when nodes receive packets or a part of packets that are destined to other nodes.
- **Idle listening** is the cost of actively listening for potential packets. Because nodes must keep their radio in receive mode, this source causes inefficient use of energy.

Applications of WSN usually generate low-rate traffic so that communication channel is expected to be idle most of the time. Therefore idle listening is very critical source of energy dissipation in WSN. To reduce idle listening, a sensor node decreases the duty cycle of a RF transceiver. It means that the receiver sleeps for long period of time, and the node periodically wakes up to check for activity of channel. If the channel is idle, the receiver goes back to sleep until the next channel-check period. Otherwise the receiver receives the message. But in this mechanism, any specific method is needed to prevent message-loss. Transmissions that are destined to the node that is sleeping make the message-loss.

There are two approaches for reliable message transmission in duty cycle controlled MAC (Medium Access Control) protocols. The first way is that nodes exchange the schedule of sleep/wakeup to synchronize on the wakeup-timing of other nodes. This approach used in S-MAC [2] protocol. S-MAC is a low power RTS-CTS protocol for wireless sensor networks inspired by PAMAS [3] and 802.11. By using RTS-CTS scheme, S-MAC can avoid the energy waste from idle listening. S-MAC uses explicit control message to synchronize the wake-up timing among neighboring sensor nodes. But S-MAC is very complex so it is hard to implement. And as the size of the network increase, S-MAC must maintain an increasing number of neighbor's schedules so additional overhead for synchronization is incurred. The second way is that a sender transmits a packet with long preamble to match the channel-check period of a receiver. This kind of approach is called as a preamble sampling [4]. In Fig.1, we illustrate the operation of preamble sampling. Each node has low duty cycle of the RF transceiver,

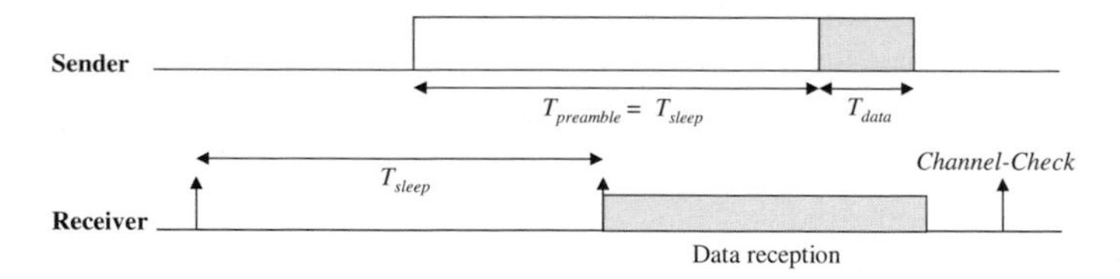

Fig. 1. The operation of preamble sampling

similar to the first approach. But there is no explicit message exchange for synchronizing on the wakeup-timing of other node. Instead, a node sends a data frame with the long preamble as long as the duration of the wakeup-timing. So a receiver can sample the long preamble and prepare the RF transceiver to receive the data frame. We call this kind of MAC protocol as PS-MAC for convenience. B-MAC [5] is one of the famous PS-MAC. However, B-MAC suffers from the waste in energy consumed by a long fixed preamble. Moreover, increasing the sample rate or neighborhood size increases the amount of energy consumption by overhearing of long preamble. WiseMAC[6] that is another duty cycle controlled MAC protocol with preamble sampling. Nodes exchange the information of wakeup-timing through ACK frame. By this synchronization scheme, it is possible to decrease the length of preamble between pre-synchronized pairs. But there is the drift between the clocks at the pairs. So if the duration of periodic sensing is enough long, this synchronization scheme can't reduce the energy waste caused by overhearing and long preamble. This limitation is also commented in WiseMAC. We also proposed new MAC protocol to reduce negative effects of long fixed preamble in DPS-MAC [7]. In this ways, an address of destination is repeatedly inserted in preamble. So a receiver can find the destination address of message by reading preamble. If the incoming message is destined to this node, the node processes the remained message. Otherwise the node stops from receiving the remained message. DPS-MAC successfully reduces the energy waste by idle listening and overhearing. But DPS-MAC protocol can't reduce the energy dissipation by receiving the redundant data frame that is caused by message-flooding. Almost network protocols of multihop wireless environments use message-flooding for finding route paths. So any effort to reduce energy leakage, caused by message-flooding, is necessary to WSN. And single long preamble schemes are hard to implement in latest RF chip. Chipcon CC1000 [8] that is used in famous Mica2 [9] has an interface for transmitting raw bit stream. But most of latest RF chips are only have an interface for transmitting data frame. It means that a RF chip has HWs to generate physical header such as preamble and CRC, and appends these to MAC frame. For example, Chipcon CC2420 [10] that is used in MicaZ [11] and TelosB [12] has HWs for generating and striping physical header, automatically.

So we use the stream of short preambles instead of one long preamble. And we insert additional information into the short preamble frame for energy saving. So we can decrease control message overhead, avoid overhearing and reduce the energy waste by message-flooding. In this paper we named our new proposal SESP-MAC (Signaling-embedded Short Preamble MAC) for convenience.

The remainder of the paper is organized as follows. Section 2 elaborates on the design of SESP-MAC. Section 3 evaluates performance of SESP-MAC through mathematical analysis. Section 4 provides conclusions and future work.

2 Design of SESP-MAC

The main idea of SESP-MAC is to add control information into the short preamble frame. So the stream of short preambles is used not only for preamble sampling but also for avoiding overhearing, decreasing control packet overhead and reducing the listening of the redundant message, caused by message-flooding.

2.1 Basic Operation of SESP-MAC

The message format of our short preamble frame is illustrated in Fig.2. And we named it SESP-frame. This format is based on Chipcon CC2420 and IEEE 802.15.4. We use frame control field as a type indicator. And RPI(Remained

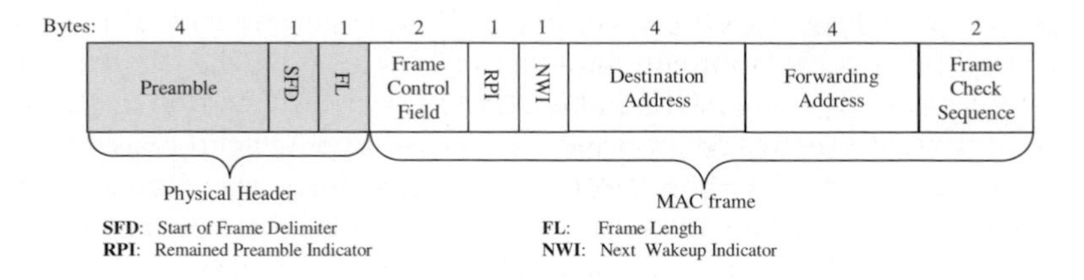

Fig. 2. The format of SESP frame

Preamble Indicator) field means the remained number of preamble to be sent. NWI(Next Wakeup Indicator) is the remained time to the next wakeup-timing of the node that made SESP frames. NWI of each SESP frame is updated before the transmission of SESP frame. Destination address field equals the destination address field of the data frame. Finally, forwarding address field support a cross-layer function. When a node receives the message that will be forwarded, the node copies the source address of the data frame and pastes it to the forward address of SESP frames to be forwarded. If a node is the originator of message-flooding, forward address field is filled with zeros. In the case of unicast, forward address field equals to the stream of zeros, too.

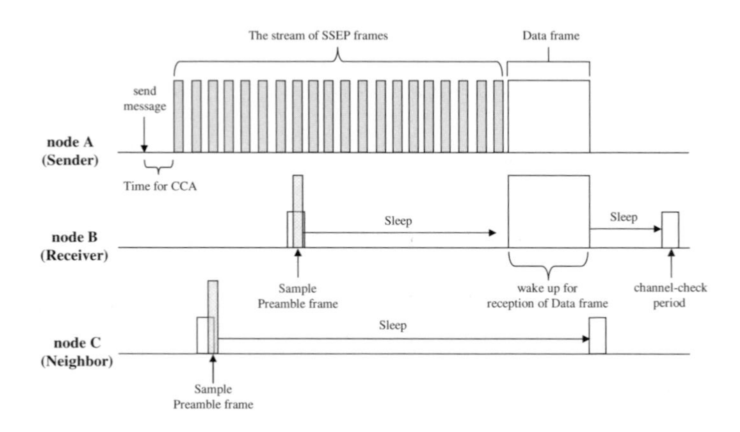

Fig. 3. The operation of SESP-MAC

Fig.3 shows the basic operation of SESP-MAC. When there is a message to be sent, SESP-MAC does CCA (Clear Channel Assessment). If there is no traffic at that time, SESP-MAC sends data frame with the stream of SESP frames. The length of the stream of SESP frames is longer then the length between adjacent channel-check periods in Fig.3. So all node that are located within the radio propagation range of node A can receive one SESP frame. Because each SESP frame has destination address field, the nodes, that receives SESP frame, can decide whether this transmission is irrelevant or not. Therefore node C, that is the neighbor of node A, recognizes that this transmission is not for it. So it sleeps until the next channel-check period for avoiding overhearing. But node B, that is the receiver, continues on communication for receiving the data frame. Also, node B can know the remained number of SESP frames from reading RPI field of the received SESP frame. So to reduce the listening of meaningless SESP frame, node B goes to sleep and after $RPI(T_p + T_{ip})$ time later, it wakes up to prepare the receiving of that data frame. There is the guard time for compensating clock accuracy, $T_{ip} - T_{process}$. $T_{process}$ is small amount of time for processing SESP frame.

2.2 Avoiding Receiving Redundant Flooded-Messages

If a node broadcasts a message for flooding, it will receive the redundant data frames with long preamble that are transmitted by its neighbors, too. The energy waste of this kind of irrelevant reception is increased with increasing the number of neighbors. To avoid the energy waste that is caused by redundant flooded-messages, SESP-MAC examines two fields of SESP frame: destination address field and forwarding address field. If the destination address field of received SESP frame is broadcast address and the forwarding address field of the SESP fame equals to source address of this node, the node decides that this transmission is redundant. It means that the received SESP frame is for the data frame that was transmitted by this node. So it sleeps until the next channel-check period.

2.3 Reducing Control Packet Overhead

For reliable MAC layer transmission, using acknowledgement is very general scheme. But this scheme makes additional control packet overhead. To reduce the number of control packet overhead: the number of ACK frames. SESP-MAC uses both implicit-ACK and generic ACK mechanism. For implicit-ACK, we use two properties of multihop wireless sensor networks. One is overhearing and the other is mutli-hop path. A node can confirm its successful transmission by overhearing of flooded message. If a node sent a flooding message and it receives the SESP frame which forward address field equals to its address, it thinks that previous flooded message was received successfully and forwarded. Because this implicit-ACK scheme is dependent with network layer protocols, this scheme is cross layer function.

2.4 Other Advantages

When a node receives a SESP frame, the node can know the schedule of preamble sampling of the sender that sent the received SESP frame by reading LWI field of the SESP frame. So SESP-MAC can support data transmission without the stream of SESP frames. Because we add LWI in not ACK frame but SESP frame, the schedule of preamble sampling of the sender is distributed to all nodes that are located within the radio propagation range of the sender. This is advance feature compare with WiseMAC. But this scheme has the same problem with WiseMAC. That is problem of the drift between the clocks.

2.5 Considerations for Implementation

To design SESP-MAC, we must seriously consider some parameters. At first we consider T_{ip} in Fig.4. T_{ip} means the inter SESP frame duration. When MAC protocols send a frame, there is some delay θ between the time of transmission of MAC layer and the time of transmission of physical energy. And the value of θ is varies with HW and SW architecture. So to minimize the effect of variation of θ, T_{ip} must be enough long compare with θ. And also T_{ip} must to be enough long to prevent overflow of the receiver buffer. In this paper, we can't find the optimal value of T_{ip} yet. In the case of Chipcon CC2420, the logical time for transmitting one SESP frame (20bytes) is about 0.6 msec. So we assign long time, 1.4 msec this is longer then twice of the logical time for transmitting one SESP frame, to T_{ip}. We will determine the optimal value of T_{ip} in future work. Because of the long period of T_{ip}, the instantaneous CCA(Clear Channel Assessment) is not sufficient for SESP-MAC. For example, if a node does CCA at the right after one SESP frame is propagated, the node determines that channel is idle. This operation will make collusion. To prevent this situation, SESP-MAC does continuous CCA during the interval T_{CCA}. And T_{CCA} must be longer than $T_{ip} + T_p$ to check at lest one SESP frame. The last parameter is T_{CC}. T_{CC} is channel-check period. During T_{CC}, a node must receive at least one SESP frame. So T_{CC} must be longer than $T_{ip} + 2T_p$, too. Fig.4 shows the relation of these parameters.

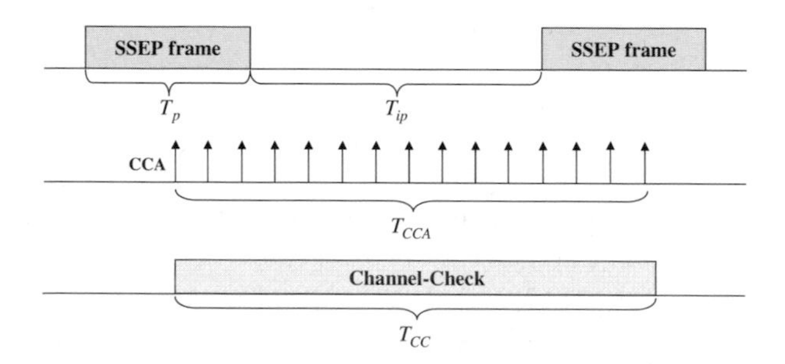

Fig. 4. The CCA and CC

3 Performance Analysis

In this section, we derive analytical expressions for the energy consumption. The main idea of SESP-MAC is to add control information into the short preamble frame. So the stream of short preambles is used not only for preamble sampling but also for avoiding overhearing, decreasing control packet overhead and reducing the listening of the redundant message, caused by message-flooding.

SESP-MAC uses forwarding address field of SESP frame for reducing control packet overhead. For example, we assume that a message is transmitted along with the routing path that is consisted with m hops. SESP-MAC only needs 1 ACK frame that is generated by the destination node. But simple PS-MAC needs m ACK frames. So it seemed straightforward enough that SESP-MAC reduces energy waste that is caused by packet overhead. To verify the other advantages of SESP-MAC, we make the equations that show the energy consumption for reception of one data frame. Fig.5 shows the energy consumption of SESP-MAC. The parameters in Fig.5 are described at Table. 1. All parameters that are used in this section are also described in Table.1.

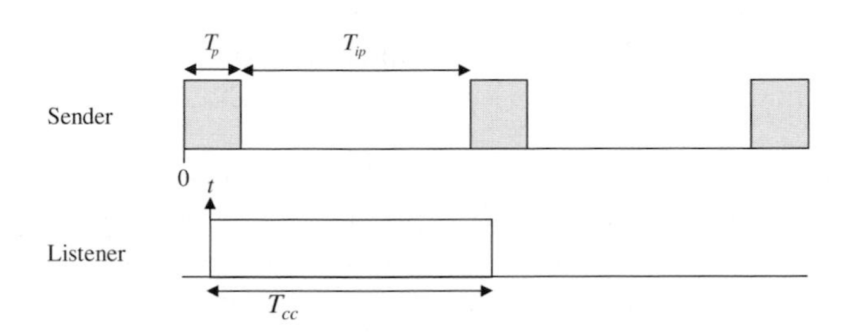

Fig. 5. The reception of SESP frame

Table 1. The parameters. These are based on Chipcon CC2420.

Parameter	Description	Values used in simulation
N	Number of SESP frames	50, 100, 150, 200, 250
B	Data rate	250 kbps
C_{sleep}	Current used in sleep mode	0.426 mA
C_{idle}	Current used in idle listeing mode	19.8 mA
C_{rx}	Current used in rx mode	19.8 mA
V	Voltage	3 V
T_p	The time of SESP frame	0.6 ms (20 bytes)
T_{ip}	The time of inter SESP frame	1.4 ms
T_{sleep}	The time of Sleep	$= N(T_p + T_{ip})$
$T_{preamble}$	The time of PS-MAC preamble	$= T_{sleep}$
T_{data}	The time of data frame	3.2 ms (100 bytes)
t	Uniformly distributed random variable	$0 < t < T_p + T_{ip}$
x	Uniformly distributed random variable	$0 < x < T_{sleep}$

There are two cases. One is that listener hears the some portion of SESP frame and one perfect SESP frame at the channel-check period. The other is that listener hears only one perfect SESP frame. If listener is the receiver of the SESP frame, the necessary energy for data reception is (1).

$$E_{receiver} = \begin{cases} E_{recv}\{(T_p - t) + T_p + T_{data}\} + 2E_{idle}T_{ip} \\ +E_{sleep}\{(N-2)(T_p + T_{ip}) + t\} & ,0 < t \leq T_p \\ E_{recv}\{(T_p - T_{data}) + E_{idle}(T_p + 2T_{ip} - t) \\ +E_{sleep}\{(N-1)(T_p + T_{ip}) - (T_p + T_{ip} - t)\} & ,T_p < t \leq T_p + T_{ip} \end{cases} \quad (1)$$

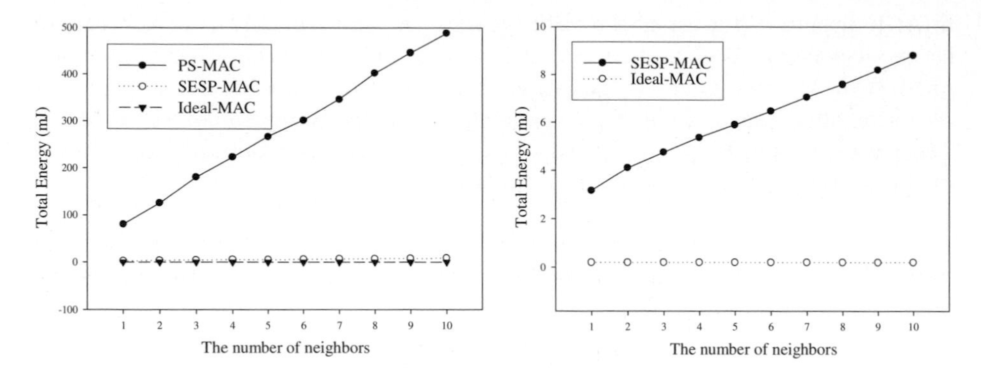

Fig. 6. Energy consumption of ideal, PS-MAC and SESP-MAC as a function of the number of neighbors. The right graph is enlarged-version of the left graph.

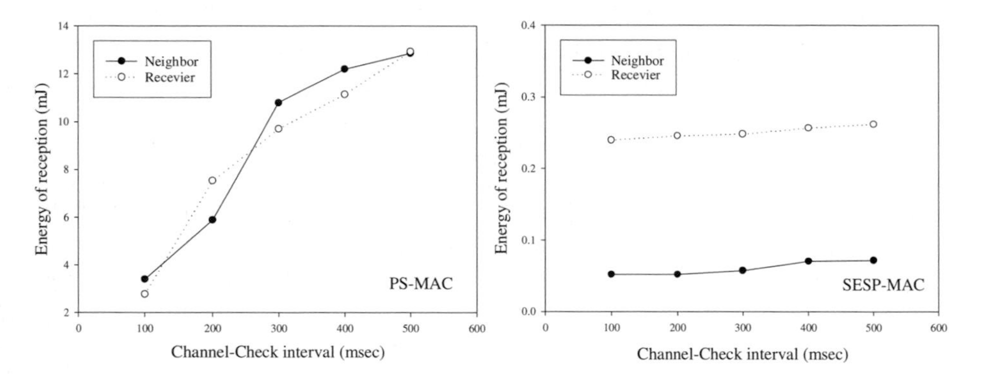

Fig. 7. Energy consumption of PS-MAC and SESP-MAC as a function of channel-check interval

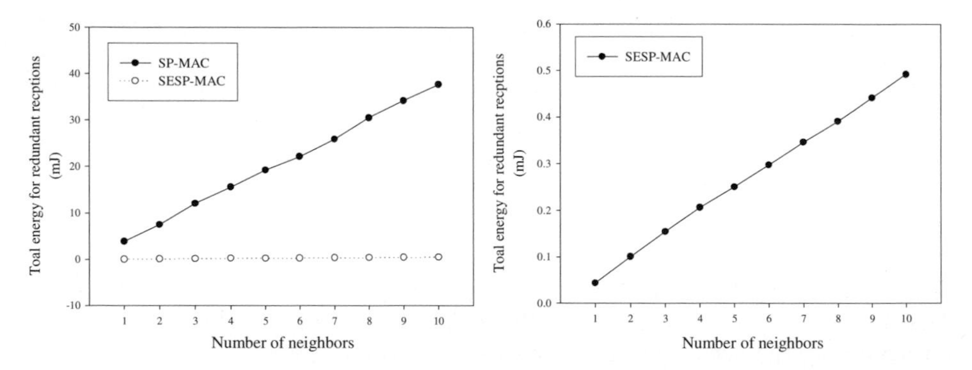

Fig. 8. Energy consumption of SESP-MAC by receiving redundant flooding message as a function of the number of neighbors. The right graph is enlarge-version of the left graph.

If listener is not a receiver of the SESP frame, the necessary energy that is caused by overhearing is (2).

$$E_{overhearing} = \begin{cases} E_{recv}\{(T_p - t) + T_p\} + 2E_{idle}T_{ip} + E(N_{rp}, T_{data}) \\ +E_{sleep}\{(N-2)(T_p + T_{ip}) + t + T_{data}\} & , 0 < t \leq T_p \\ E_{recv}T_p + E_{idle}(T_p + 2T_{ip} - t) + E(N_{rp}, T_{data}) \\ +E_{sleep}\{(N-)(T_p + T_{ip}) - (T_p + T_{ip} - t) + T_{ip} + T_{data}\} & , T_p < t \leq T_p + T_{ip} \end{cases} \quad (2)$$

$E(N_{rp}, T_{data})$ is the energy consumption for listening the portion of the data frame. For convenience, we ignore the value of $E(N_{rp}, T_{data})$. (3) shows the boundary condition of $E(N_{rp}, T_{data})$.

$$0 < E(N_{rp}, T_{data}) \leq E_{recv} \cdot T_p \quad (3)$$

In the case of simple PS-MAC, there is no mechanism to avoid overhearing and receiving meaningless preamble. So the necessary energy for data reception is (4).

$$E_{receiver} = E_{recv}(T_{data} + x) + E_{sleep}(T_{preamble} - x) = E_{overhearing}, 0 < x \leq T_{sleep} \quad (4)$$

In Fig.6, we compute the total energy consumption for transmitting one data frame with 100msec. channel-check interval. Total energy means the summation of energy consumption of a receiver and neighbors. Ideal-MAC means there is no overhearing and no preamble sampling. SESP-MAC has the mechanism to avoid overhearing and reception of useless SESP fames. So SESP-MAC consumes lower energy than simple PS-MAC. In Fig.6 and Fig.7, we can see that SESP-MAC is not only more energy efficient but also more independent of both the density of nodes and the interval of channel-check than simple PS-MAC. In Fig.7, SESP-MAC reduces the reception of redundant message that is caused by message floodinga.

4 Conclusion

An energy-efficiency MAC protocol in WSN is an open research area in which we are conducting further studies. To solve the problem of inefficient operations and reach our goal in WSN, we have proposed SESP-MAC. The main idea of SESP-MAC is to add control information into the short preamble frame. So the stream of short preambles is used not only for preamble sampling but also for avoiding overhearing, reducing control packet overhead and avoiding the listening of the redundant message, caused by message-flooding. The performance results have shown SESP-MAC is more suitable for general WSN and can achieve much conserving energy in reception compared to simple PS-MAC protocol. Fig.6 shows the energy gain that is accepted by avoiding overhearing. Fig.7 shows the energy saving for reducing meaningless SESP frames. In case of SESP-MAC, a neighboring node of the receiver consums much less energy than energy consumption of the receiver. But the case of simple SP-MAC, the energy consumption of both recevier and neighbor is similar. In Fig.7, we solve the energy waste, caused by reception of redundant flooding message.

This novel protocol is the subject of an ongoing study, and we plan to implement SESP-MAC protocol on the node that we have created.

References

1. Akyildiz, W.S., Sankarasubramaniam, Y., Cayirci, E.: A survey on sensor networks. IEEE Communications Magazine 40(8), 102–114 (2002)
2. Ye, W., Heidemann, J., Estrin, D.: An energy-efficient mac protocol for wireless sensor networks. In: Proc. of the 21st Int. Annual Joint Conf. of the IEEE Computer and Communications Societies (INFOCOM 2002), New York, NY (June 2002)
3. Singh, S., Raghavendra, C.S.: PAMAS: Power Aware Multi-Access protocol with Signalling for Ad Hoc Networks. ACM ComputerCommunications Review (1999)
4. El-Hoiydi, A.: Aloha with Preamble Sampling for Sporadic Traffic in Ad Hoc Wireless Sensor Networks. In: Proc. IEEE Int. Conf. on Communications, New York, USA, April 2002, pp. 3418–3423 (2002)
5. Polastre, J., Hill, J., Culler, D.: Versatile low power media access for wireless sensor networks. In: Proc. of the Second ACM Conf. on Embedded Networked Sensor Systems (SenSys), Baltimore, MD (November 2004)
6. El-Hoiyi, A., Decotignie, J.-D., Hernandez, J.: Low power MAC protocols for infrastructure wireless sensor networks. In: Proc. of the Fifth European Wireless Conference (February 2004)
7. Lim, S., Ji, Y., Cho, J., An, S.: An Ultra Low Power Medium Access Control Protocol with the Divided Preamble Sampling. In: Youn, H.Y., Kim, M., Morikawa, H. (eds.) UCS 2006. LNCS, vol. 4239, pp. 210–224. Springer, Heidelberg (2006)
8. CC 1000, http://focus.ti.com/docs/prod/folders/print/cc1000.html
9. Mica2, http://www.xbow.com/Products/productdetails.aspx?sid=174
10. CC2420, http://focus.ti.com/docs/prod/folders/print/cc2420.html
11. MicaZ, http://www.xbow.com/Products/productdetails.aspx?sid=164
12. TelosB, http://www.xbow.com/Products/productdetails.aspx?sid=252

Data Diffusion Considering Target Mobility in Large-Scale Wireless Sensor Network

Yuhwa Suh and Yongtae Shin

Dept. of Computer Science, Soongsil University, Sangdo-Dong, Dongjak-Gu, Seoul, Korea, 156-764
{zzarara,shin}@cherry.ssu.ac.kr

Abstract. In this paper, we propose DDTM, Data Diffusion considering Target Mobility for large-scale WSN. Most of existing routing protocols did not consider mobile target and sink. So it requires frequently flooding and path update whenever targets or sinks move. This can lead to drain battery of sensors excessively and decrease lifetime of WSN. We proposed DDTM for Target Mobility. DDTM decreases a consumption of energy as reusing the existing grid structure of existing routing protocol TTDD, when the target moves in local cell. We evaluate performance of DDTM and TTDD in target mobility and scalability through measuring the mathematical cost. Our results show that DDTM supports target mobility efficiently in WSN.

Keywords: WSN, Target Mobility, DDTM, TTDD, Grid Structure.

1 Introduction

Wireless Sensor Network (WSN) is a network consisted of large number, small size of sensor nodes. WSNs act similarly with Ad-hoc networks, but it is different from Ad-hoc networks. Ad-hocs are constructed for providing communication services between nodes without communication infrastructure. On the other hand, WSNs are constructed for collecting information about environment to sink. In WSNs, energy efficiency for lifetime of network is more important than performance of network, because WSNs are restricted to low hardware resources of terminal devices such as memory, computation power and to frequent network partition caused by exhaustion of battery, failure of sensors during communication between the sensors. End-point mobility is more important issue than node mobility in WSNs. Although several data diffusion protocols have been proposed for sensor networks, such as Directed Diffusion [7], Rumor Routing [8], TTDD [5], SPIN [6], etc., few existing approaches provide a scalable and efficient method to solve these problems. They are required flooding and route update whenever targets or sinks move and it can lead to drain battery of sensors excessively and decrease lifetime of WSN.

In this paper, we proposed DDTM, Data Diffusion consider Target Mobility to address the multiple, mobile target problem. DDTM is based on TTDD (Two-tier Data Dissemination in Large-scale Wireless Sensor Networks)[5] to address sink mobility problem. TTDD handles only a sink mobility efficiently compared with

T. Vazão, M.M. Freire, and I. Chong (Eds.): ICOIN 2007, LNCS 5200, pp. 11–20, 2008.

other existing network protocols, However, it do not consider a target mobility, it requires very high energy resources by reconstructing the grid structure whenever targets move. DDTM decrease a consumption of energy when a target moves in local cells as reusing the existing grid of TTDD.

The rest of this paper is organized as follows. Section 2 describes the operation and problem of TTDD. Section 3 introduces the design of DDTM, Section 4 evaluates performance of DDTM and TTDD in target's mobility through measuring the mathematical cost. Finally, we conclude this paper in Section 5.

2 TTDD

In TTDD, sensors are assumed to know their locations and be stationary. Existing routing protocols suggest that each mobile sink need to continuously propagate its location information throughout the sensor field, so that all sensor nodes are informed of the direction of sending future data reports. However, frequent location updates from multiple sinks can lead to both increased collisions in wireless transmissions and rapid power consumption of the sensor's limited battery supply [5].

As soon as a source generates data, it starts preparing for data dissemination by building a grid structure. The source sends data announcement message to each of its four adjacent crossing points. The node sent this message becomes dissemination node and propagates recursively its four adjacent crossing points. Once a grid for the specified source is built, a sink can flood its queries within a local cell to receive data. The query will be received by the nearest dissemination node on the grid, which then propagates the query upstream through other dissemination nodes toward the source. Requested data will flow down in the reverse direction to the sink [5].

TTDD build the grid on a per-source basis, so that different sources recruit different sets of dissemination node. TTDD considers efficiently sink mobility by using a grid structure, but do not handle the target mobility. If a target moves, a source changes to a current sensor node sensing target. The source may build frequently the new grid, if sensor network has multiple targets and moves frequently, the sensor network may be exhausted rapidly by very high energy consumption.

3 DDTM

This paper studies the problem of efficient data diffusion in a large-scale network having multiple, mobile target. We consider a network made of stationary sensor nodes, sinks and targets may change their locations dynamically.

In this work, a source refers to a sensor node that generates sensing data to report target. Once a stimulus appears, the sensors surrounding it collectively process the signal and one of them becomes the source to generate data reports. A target is a phenomenon or a material object that sensor nodes like to generate information. It is potentially multiple, mobile and each has an identifier. Sinks query the network to collect sensing data. There can be multiple sinks moving around in the sensor field.

3.1 Terminology

Grd_RZ (Grid Reusing Zone)
A new source in this zone reuses the grid built by a previous source. A previous source which constructed the grid stores a cell size R (a maximum distance from grid construction source) in cache table. If the target leaves from this zone, the new source builds the new grid structure and the new Grd_RZ.

Tgt_MZ (Target Mobility Zone)
As soon as a source senses a target, it notifies neighbor sensor nodes by flooding MZ_ADV (Mobility Zone Advertisement) message in a local area. We refer to this local area as Tgt_MZ. This zone is built for setting a route between a new source and a previous source. One of sensor nodes within Tgt_MZ may become the new source if the target moves. We refer to sensor nodes within Tgt_MZ as candidate sources.

3.2 Message and Cache Table

As soon as a source generates data, it starts setting Tgt_MZ by flooding MZ_ADV in a local cell to notify neighbor sensor node appearance of target. Once neighbor sensor nodes receive this message, they stores information in the message to a cache table. MZ_ADV message is composed as follows:

Tgt_ID : identifier of target
Seq : sequence number
Tgt_MZ_ID : ID of source's own belonging Target mobility zone
Grd_src : location, ID of grid construction source
Tgt_MZ_Src : location, ID of source building Tgt_MZ
Grd_RZ cell size: range of grid reusing area (radius R)

The Cache Table is composed as follows:

Tgt_ID (Target ID) : identifier of target
Prev_N (Previous Node) : ID of neighbor sensor node sending first MZ_ADV message
Tgt_MZ_ID (Target Mobility Zone ID): ID of Tgt_MZ that sensor node's own belongs (Increase one by one whenever a source generates Tgt_MZ)
Grd_Src (Grid construction source) : location, ID of grid construction source
Grd_RZ (Grid Reusing Zone) : range of grid reusing area (radius R)

3.3 Operation of DDTM

DDTM is composed tour steps. One step is target sensing and the Grd_RZ setting, second step is the Tgt_MZ building and third step is the route setting, fourth step is data forwarding.

Target Sensing and Grd_RZ Setting
Once the source generates data about target, it decides whether it generates the new grid or reuses the existing grid through the cache table. The source generates the Grd_RZ following two cases. One is that information of the target does not exist in

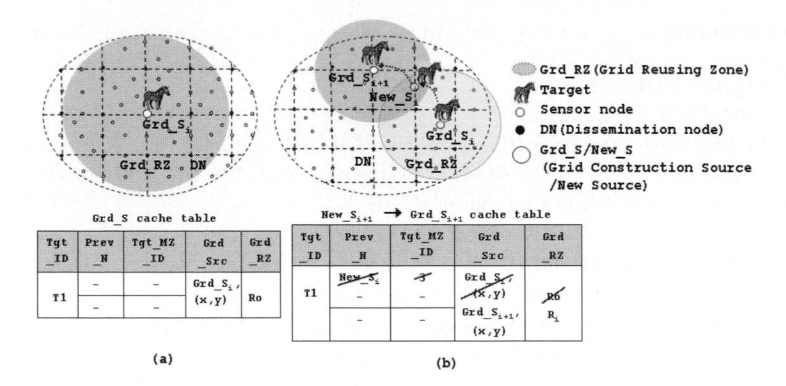

Fig. 1. Grd_RZ building

Variables :

T_j : target , S_i : source node
R_0 : initial cell size of a Grid Reusing Zone
R : cell size of a Grid Reusing Zone
x_{s_j}, y_{s_j} : location of a current source node
x_{G_s}, y_{G_s} : location of a grid construction source node
Max_Tgt_MZ *: The maximum number of a Target Mobility Zone that the source node can include*
Num_of_Tgt_MZ *: the number of a Target Mobility Zone for the Target*

Functions :

Cache_init (Target T, Grd_src S_i, Grd_RZ R)
: initialize a cache table, fill each parameter value in each field of the cache table

Grd_RZ_Construct1 (Target T) {

$$R = \frac{d}{Num\_of\_Tgt\_MZ} \times R_{old} \times \quad (d: \text{constant value})$$

Cache_init (T, S_i, R) }|

Grd_RZ_Construct2 (Target T) {
Cache_init (T, S_i, R_0) }

Operation :

Search T_j in Cache Table of S_i
if exist T_j
then

if ($\sqrt{(x_{s_j} - x_{G\_s_j})^2 + (y_{s_j} - y_{G\_s_j})^2} < R$) AND (Num_of_Tgt_MZ(T_j) < Max_Tgt_MZ)
then Grid Reusing
else Grid Construct
Grd_RZ_Construct1(T_j)
else Grid Construct
Grd_RZ_Construct2(T_j)

Fig. 2. Algorithm 1 - Grd_MZ Setting by $node_i$

cache table. The other is that the source leaves from Grd_RZ range although exist information of target in cache table. The former one sets Grd_RZ by initializing the Grd_RZ field of cache table with cell size R_0(see Fig. 1 (a)). The latter one regenerates a new grid and new Grd_RZ (see Fig. 1 (b)). R changes according to target moving pattern. If the cache table of source has many Tgt_MZ, target's moving range may be small and the number of target's movements may be quite a few. In that case, the source sets up cell size small, in opposite case, one sets up cell size large.

But R_{i+1} is similar to R_i, because the target moving pattern may be regular usually. So R_i has influence on R_{i+1}.

Tgt_MZ building

Once the source generates data, it builds Tgt_MZ after setting Grd_RZ. Before forwarding data to sink, the source floods MZ_ADV message to neighbor sensor nodes in local area. The source retrieves in the cache table whether or not its own Tgt_MZ exists. If not, it is grid construction source and fills Tgt_MZ field with 1 and Prev_N field with NULL. it generates MZ_ADV message with information in cache table, and floods within radius r range. (See Fig. 4 (a))

Sensor nodes received this message are included within Tgt_MZ of the current source. Fig. 4 (a) shows that Cdt_S_1 and Cdt_S_2 is included Tgt_MZ of Grd_S. They caches MZ_ADV message received first from the neighbor sensor. Sensor nodes within Tgt_MZ are candidate source nodes which can become the new source when the target moves.

Variables :

T_i : *target* S_i : *source node*

r_0 : *initial cell size of a Target Mobility Zone*

$r = \dfrac{c}{Num\_of\_Tgt\_MZ} \times r_{old}$: *cell size of a Target Mobility Zone*

Max_Tgt_MZ_ID : *maximum Tgt_MZ_ID in the cache table*

Operation :

Search Tgt_MZ info of T_i in Cache Table of S_i
if not exist
  then *Initiate cache table (Prev_N : NULL, Tgt_MZ_ID : 1)*
    Generate MZ_ADV
    Flooding by r_0 cell size
  else *Cache (Prev_N : NULL, Tgt_MZ_ID: Max_Tgt_MZ_ID + 1)*
    Generate MZ_ADV
    Flooding by r(in cache table) cell size

Fig. 3. Algorithm 2 - Tgt_MZ Constructing by $node_i$

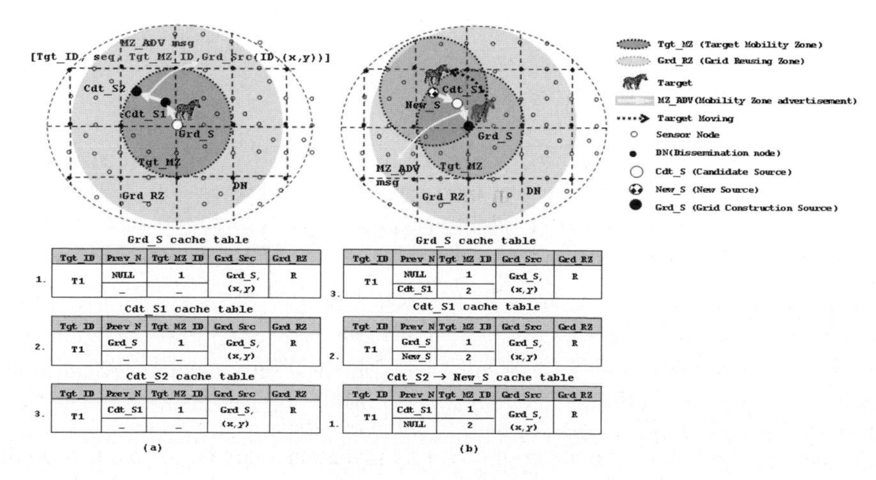

(a)

1. Grd_S cache table

Tgt_ID	Prev_N	Tgt_MZ_ID	Grd_Src	Grd_RZ
T1	NULL	1	Grd_S, (x,y)	R
	_	_		

2. Cdt_S1 cache table

Tgt_ID	Prev_N	Tgt_MZ_ID	Grd_Src	Grd_RZ
T1	Grd_S	1	Grd_S, (x,y)	R
	_	_		

3. Cdt_S2 cache table

Tgt_ID	Prev_N	Tgt_MZ_ID	Grd_Src	Grd_RZ
T1	Cdt_S1	1	Grd_S, (x,y)	R
	_	_		

(b)

3. Grd_S cache table

Tgt_ID	Prev_N	Tgt_MZ_ID	Grd_Src	Grd_RZ
T1	NULL	1	Grd_S, (x,y)	R
	Cdt_S1	2		

2. Cdt_S1 cache table

Tgt_ID	Prev_N	Tgt_MZ_ID	Grd_Src	Grd_RZ
T1	Grd_S	1	Grd_S, (x,y)	R
	New_S	2		

1. Cdt_S2 → New_S cache table

Tgt_ID	Prev_N	Tgt_MZ_ID	Grd_Src	Grd_RZ
T1	Cdt_S1	1	Grd_S, (x,y)	R
	NULL	2		

Fig. 4. Tgt_MZ building

Route Setting

It establishes a route between the new source and the old source that Tgt_MZ setting is repeated. The cache table is filled with information as Tgt_MZ setting is repeated. Fig. 4 (b) shows that the candidate source Cdt_S$_2$ within Tgt_MZ of Grd_S becomes the new source New_S as the target moves (see Fig. 4 (a)). The new source adds to the largest Tgt_MZ_ID + 1 in Tgt_MZ_ID field of cache table, NULL in Prev_N field. And the source generates MZ_ADV message with new values and floods with radius r range. If this is repeated, Tgt_MZ is crossed as Fig. 4 (b). So the previous source and the new source receives MZ_ADV message of each other and is aware of information of each other.

Data Forwarding

Fig. 5 show that the new source forwards the data to the grid construction source, grid construction source forwards the data to dissemination node on a grid. The dissemination node forwards the data to sink. The source generating data forwards data to the neighbor sensor node in Prev_N field which has the smallest Tgt_MZ_ID in cache table. That defends loop of a route, in case the target wander in same the area and performs data for forwarding the shortest path.

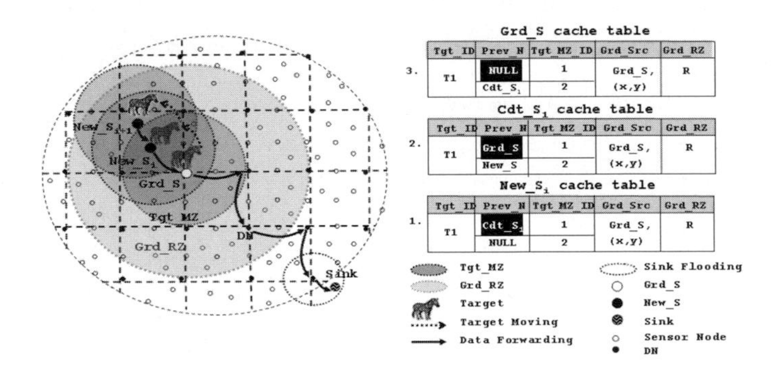

Fig. 5. Data Forwarding

4 Performance Evaluation

In this section, we evaluate the worst-case communication overhead and compare the performance on TTDD and DDTM in the scenarios of mobile target. We use model and notations of TTDD and derive additionally parameters under cover of model of TTDD. We consider a square of area in which sensor nodes are uniformly distributed.

4.1 Model and Notations

We consider a square sensor field of area A in which N sensor nodes are uniformly distributed so that on each side there are approximately $\sqrt{N}$ sensor nodes. There are k sinks and k_t targets in the sensor field. Sinks move at an average speed v, targets move at an average speed v_t, while receiving d data packets from a source during a time period of T. Each data packet has a unit size and both the query and data announcement messages have a comparable size l. The communication overhead to flood an area is proportional to the number of sensor nodes in it.

The source divides the sensor field into cells; each has an area α^2. There are $n = N\alpha^2 / A$ sensor nodes in each cell and $\sqrt{n}$ sensor nodes on each side of a cell. Each sink traverses m cells and m is upper bounded by $1 + vT / \alpha$. For stationary sink, m=1[5]. There are average e neighbor sensor nodes around one sensor nose and $p\sqrt{2N}$ sensor nodes on the route between the previous source and the new source. There are sensor nodes nr^2 / α^2 in per Tgt_MZ. Unit size of Tgt_MZ is r and unit size of Grd_RZ is R. Each target traverses M Tgt_MZs and M is upper bounded by $1 + v_t T / r$. For stationary target M=1. When target moves in sensor field, Grd_RZ is generated G and G is upper bound by $1 + v_t T / R$. We assumes same parameters that TTDD shows comparable performance with existing Protocol as A=2000×2000m^2, N=10000, n=100, k=4, c=1, d=100, l=1, m=1, e=10, p=5, $r = \alpha$, G=1, k_t=1. We consider the performance on a per-Grd_RZ.

4.2 Target Mobility

We analyze the worst-case message overhead of TTDD and DDTM. The overhead for the query to reach the source is $enl + \sqrt{2}(c\sqrt{N})l$ $(0 < c \le \sqrt{2})$, the overhead to deliver d/m data packets from a source to a sink is $\sqrt{2}(c\sqrt{N})\frac{d}{m}$. For k mobile sinks, the overhead to receive d packets about stationary target in m cells is:

$$U = \frac{4N}{\sqrt{n}} l + km\left\{(enl + \sqrt{2}(c\sqrt{N})l + \sqrt{2}(c\sqrt{N})\frac{d}{m}\right\} = \frac{4N}{\sqrt{n}} l + kmenl + kc(ml + d)\sqrt{2N} \quad (1)$$

When k_t targets move M times, the overhead to deliver data packets in M Tgt_MZ is:

$$CO_{TTDD} = k_t G\left[\left\{\frac{4N}{\sqrt{n}} l + kc(ml + d)\sqrt{2N}\right\} M + kmenl\right] \quad (2)$$

When k_t targets move M times, the overhead to build M Tgt_MZ is $e(r^2 / \alpha^2)nlM$ There are $(pr / \alpha)\sqrt{n}$ sensor nodes on route between the new source and the previous source. p is a factor increasing the number of nodes on route. The data message overhead to MZ_ADV message overhead is $e(r^2 / \alpha^2)lM$. The overhead to deliver from the present source to the sink is:

$$CO_{DDTM} = k_t G\left[U + \left\{e\frac{nr^2}{\alpha^2} lM + \frac{pr}{\alpha}\sqrt{n}(M-1)\right\}\right] = k_t G\left[U + \left\{enlM + p\sqrt{n}(M-1)\right\}\right] \quad (r = \alpha) \quad (3)$$

To compare TTDD and DDTM, we have:

$$\frac{CO_{DDTM}}{CO_{TTDD}} = \frac{k_t G\left[U + \left\{enlM + p\sqrt{n}(M-1)\right\}\right]}{k_t G\left[\left\{\frac{4N}{\sqrt{n}} l + kc(ml + d)\sqrt{2N}\right\} M + kmenl\right]} \quad (4)$$

Fig.6 (a) shows the overhead of DDTM/TTDD. Once the target moves, it decreases rapidly the number of target movements to 36. It means that the overhead of the grid reconstruction on TTDD is much more than the one of the local flooding on DDTM. In case of over 36 decreases fluently, this is because high target mobility leads to frequent the local flooding on DDTM. Although the target's movement increases continuously, the curve is flat. That presents that the increase of the local flooding does not influence greatly on overall the message overhead for the sensing data to reach the sink. Also the DDTM/TTDD overhead is more than 1, in case of the target is stationary (M=1). This is because DDTM performs the same operation with TTDD and floods in local cell to build Tgt_MZ. But DDTM/TTDD overhead is close to 1, it shows that DDTM has the performance comparable with TTDD in stationary target scenarios. Consequently, DDTM has the message overhead smaller than TTDD in mobile target scenarios and shows the comparable performance with TTDD even than stationary target scenarios.

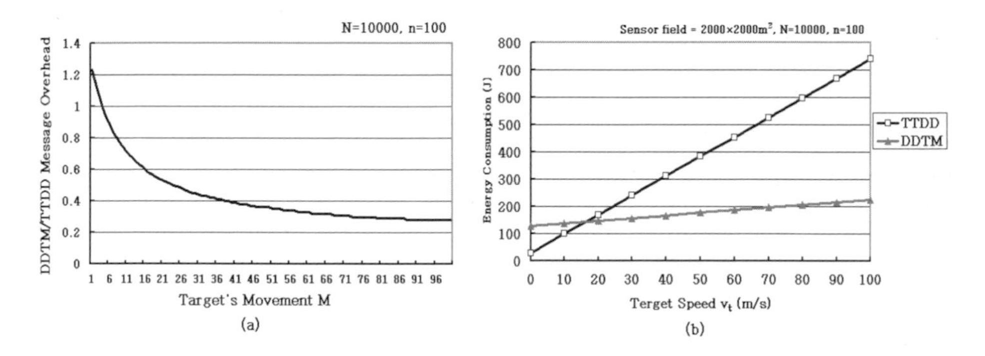

Fig. 6. Message Overhead v.s Target's Mobility v.s. Energy Consumption

We analyze the energy consumption of TTDD and DDTM. We assume same WSN environmental parameters that show the best performance on TTDD. So we assume parameters as A=2000×2000m^2, α=200m, v=4, n=N/100, m=2, M=1+0.3v_t, the other parameters are same with section 4.1. Also the MZ_ADV packet is l=36byte, the data packet is D=64byte. The initial energy is 26kJ, the energy for sending data is E_t=0.47 $\mu J/bit$, energy for receiving data is E_r=0.47 $\mu J/bit$ [9]. As target is stationary, the energy consumption to deliver once data to sink is:

$$E_U = \left[\frac{4N}{\sqrt{n}}l + km\left\{enl + \sqrt{2}(c\sqrt{N})l + \sqrt{2}(c\sqrt{N})\frac{d}{m}D\right\}\right](E_t + E_r) \qquad (M=1)$$
$$= \left\{\frac{4N}{\sqrt{n}}l + kmenl + kc(ml + Dd)\sqrt{2N}\right\}(E_t + E_r) \qquad (5)$$

$$E_{DDTM} = k_t G\left[E_U + \left\{enlM + p\sqrt{n}(M-1)Dd\right\}(E_t + E_r)\right]$$
$$= k_t G\left[E_U + \left\{enl(1+0.3v_t) + 0.3v_t pDd\sqrt{n}\right\}(E_t + E_r)\right] \qquad (6)$$

$$E_{TTDD} = k_t G\left[\left(\frac{4N}{\sqrt{n}}l + kc(ml+Dd)\sqrt{2N}\right)(1+0.3v_t) + kmenl\right](E_t + E_r) \quad (7)$$

Fig. 6 (b) shows that the energy consumption of DDTM is higher than one of TTDD when the target's moving speed is less than 18*m/s*. This is because DDTM performs the grid construction and building Tgt_MZ for reusing the grid. However, the higher speed a target moves at, the more rapidly the energy consumption of TTDD increases to construct a new grid. So the energy consumption of TTDD increases over 3 times than DDTM when it is to 100*m/s*. Consequently, the high target mobility make the source construct frequently a grid, this causes so much energy consumption.

4.3 Scalability

Equation (8) is derived by substituting M=1+0.3v_t , n=N/100 on Equation (3). Equation (8) shows the impact of the number of sensor nodes overhead compared with a TTDD.

$$\frac{CO_{DDTM}}{CO_{TTDD}} = \frac{k_t\left[(617+0.21v_t)\sqrt{N}(0.09+0.003v_t)10N\right]}{k_t\left[(617+173v_t)\sqrt{N}+0.8N\right]} \quad (8)$$

Fig. 7 (a) shows the TTDD communication overhead as a function of N with different target moving speeds for k_t = 1 case. Though the number of nodes highly increases, DDTM/TTDD <1. In this work setup, DDTM has consistently lower overhead compared with TTDD in mobile target scenario. Fig. 7 (b) shows the TTDD and DDTM communication overhead in multiple targets. In TTDD, building grid has higher overhead compared with local flooding for grid reusing in DDTM.

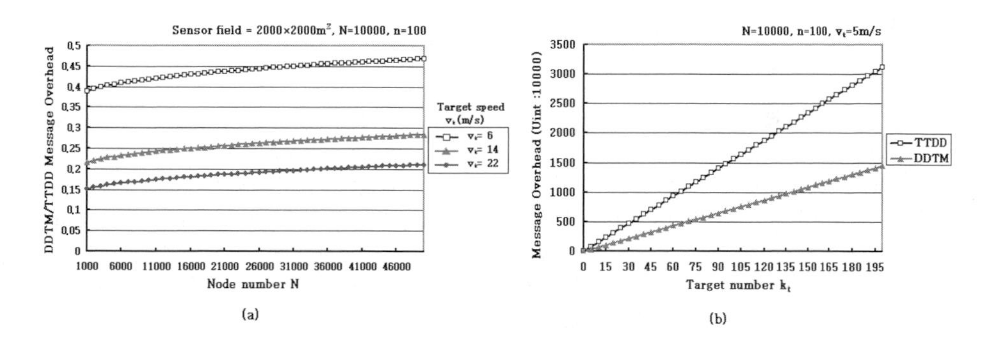

Fig. 7. Node Number v.s Message Overhead v.s. Target Number

5 Conclusion

In a large scale sensor network, previous works do not consider efficiently end-point mobility, so it should frequently flood throughout network or update the route whenever targets and sinks move. TTDD worked up to solve the problem by utilizing a grid structure, but this brings the excessive energy consumption in WSN environment which has high target mobility. Because it addresses only sink mobility.

Such the excessive message by updating the route, reconstruction a grid structure and the flooding requires the high energy consumption. It reduces the lifetime of WSN.

This paper introduces DDTM, data diffusion for the target mobility. DDTM considers the energy efficiency as addressing the end-point mobility. It is based on TTDD handling the sink mobility and addresses the target mobility as reusing a grid structure on TTDD by building the Target Mobility zone and the Grid reusing zone.

We confirmed that DDTM can effectively deliver data from mobile targets to mobile sinks with performance comparable with TTDD.

References

1. Akyildiz, I., Su, W., Sankarasubramaniam, Y., Cayirci, E.: A survey on Sensor Networks. IEEE Communications Magazine 40(8), 102–114 (2002)
2. Jiang, Q., Manivannan, D.: Routing Protocols for Sensor Networks. In: CCNC 2004 (2004); Tubaishat, M. Madria, S.: Sensor Networks: An Overview. IEEE Potentials (April/May 2003)
3. Al-Karaki, J.N., Kamal, A.E.: Routing Techniques in Wireless Sensor Networks: A Survey. IEEE Wireless Communications (December 2004)
4. Tubaishat, M., Madria, S.: Sensor Networks: An Overview. IEEE Potentials (April/May 2003)
5. Ye, F., Luo, H., Cheng, J., Lu, S., Zhang, L.: A Two-tier Data Dissemination Model for Large-scale Wireless Sensor Networks. In: Proc. of the 8th Annual International Conf. on Mobile computing and networking, pp. 148–159 (September 2002)
6. Heinzelman, W.R., Kulik, J., Balakrishnan, H.: Adaptive Protocols for Information Dissemination in Wireless Sensor Networks. In: Fifth ACM/IEEE MOBICOM Conference, Seattle, WA (August 1999)
7. Intanagonwiwat, C., Govindan, R., Estrin, D., Heidemann, J., Silva, F.: Directed Diffusion for Wireless Sensor Networking. IEEE/ACM Transactions on Networking 11, 2–16 (2003)
8. Braginsky, D., Estrin, D.: Rumor Routing Algorithm for Sensor Networks. In: Proc. of the 1st ACM International Workshop on Wireless Sensor Networks and Applications (2002)
9. Carman, D.W., Kruus, P.S., Matt, B.J.: Constraints and Approaches for Distributed Sensor Network Security (Final), NAI Labs Technical Report #00-010 (September 2000)

Improving the Performance of Optical Burst-Switched Networks with Limited-Range Wavelength Conversion through Traffic Engineering in the Wavelength Domain

João Pedro[1,2], Paulo Monteiro[1,3], and João Pires[2]

[1] Nokia Siemens Networks S.A., R. Irmãos Siemens 1, 2720-093 Amadora, Portugal
{joao.pedro,paulo.monteiro}@siemens.com
[2] Instituto de Telecomunicações, Instituto Superior Técnico, Av. Rovisco Pais 1, 1049-001 Lisboa, Portugal
jpires@lx.it.pt
[3] Instituto de Telecomunicações, Universidade de Aveiro, Campus Universitário de Santiago, 3810-193 Aveiro, Portugal

Abstract. Optical Burst Switching (OBS) is a promising switching paradigm to efficiently support Internet Protocol (IP) packets over optical networks, under current and foreseeable limitations of optical technology. The prospects of OBS networks would greatly benefit, in terms of cost and ease of implementation, from limiting the wavelength conversion capabilities at the network nodes. This paper describes and assesses the performance of a traffic engineering strategy for optimizing the wavelength assignment in OBS networks with limited-range wavelength conversion capabilities. Simulation results show that this strategy significantly improves the network performance when the network nodes have wavelength converters with small conversion ranges.

Keywords: Limited-range wavelength conversion, optical burst switching, traffic engineering, wavelength assignment.

1 Introduction

The increasing bandwidth demand of IP-based services is already being supported by transport networks which exploit the huge transmission capacity offered by optical fibres and Wavelength Division Multiplexing (WDM) technology [1]. Despite the widespread use of optical technology in transmission, current optical networks mostly provide static wavelength routing. Hence, switching of data is still performed in the electrical domain, requiring optical-electrical-optical (O-E-O) conversion equipment per each wavelength channel and high speed IP routers. Not only thus the amount of O-E-O conversion equipment grows with the number of wavelengths per link, but also the electrical switching equipment becomes more expensive as channel bit rate increases [2]. As a result, current IP over WDM solutions will not be able to support the increasing bandwidth demands in a cost-effective way. In view of this, optical switching technologies and architectures have attracted considerable interest, as they are expected to reduce switching costs in next-generation transport networks.

T. Vazão, M.M. Freire, and I. Chong (Eds.): ICOIN 2007, LNCS 5200, pp. 21–30, 2008.

The three main optical switching paradigms are Optical Circuit Switching (OCS), Optical Burst Switching (OBS), and Optical Packet Switching (OPS). The former paradigm is easily implemented with existing optical technology, but it is inefficient in supporting bursty IP traffic, mainly due to its coarse wavelength granularity. On the other hand, the latter paradigm provides statistical multiplexing at the packet level, but requires optical buffering and optical processing capabilities, which are still too immature for a near term deployment. The OBS paradigm aims at a compromise between bandwidth utilization efficiency and optical technology requirements [3]. It achieves sub-wavelength granularity by assembling multiple IP packets into bursts, which are the elemental traffic units routed and switched inside the OBS network, while avoiding optical buffering and processing by reserving bandwidth for burst transmission in advance and using out-of-band signalling.

OBS networks use one-way resource reservation mechanisms [4], which allocate wavelengths for burst transmission on a link by link basis, ignoring the wavelengths availability in the downstream links of the burst path. Thus, wavelength contention occurs when two or more bursts, overlapping in time, arrive at a transit node on the same wavelength and are directed to the same output link. Given that unresolved contention leads to burst loss, degrading the network performance, most studies on OBS networks assume that wavelength contention is resolved using wavelength converters to convert the wavelength assigned to the contending bursts to wavelengths that are available on the output link. However, while wavelength conversion devices remain relatively immature and expensive [5], optical networks are expected to only benefit from limited or sparse wavelength conversion capabilities [6].

In this paper, we study the performance improvements achieved when using traffic engineering in the wavelength domain to optimize the wavelength assignment in OBS networks with limited-range wavelength conversion capabilities. Previous works [7], [8] have proposed traffic engineering strategies to improve the performance of OBS networks without wavelength converters. Here, we adapt the strategy proposed in [8] for optimizing the use of wavelength converters with only limited conversion ranges. Simulation results show that this strategy improves the network performance, in some cases by several orders of magnitude in terms of average burst blocking probability, when the wavelength converters have small conversion ranges.

The remainder of the paper is organized as follows. Section 2 outlines the OBS network architecture and enabling optical technologies. Section 3 describes a traffic engineering strategy that aims at minimizing contention for the same sub-set of wavelengths by overlapping burst paths and, thus, improving the performance of OBS networks with limited-range wavelength conversion. The performance of this strategy is evaluated through network simulation in section 4. Finally, section 5 presents the concluding remarks.

2 OBS Network Architecture and Enabling Technologies

Transport of IP packets by an OBS network complies with the following principles. Firstly, the IP packets arriving at an OBS edge node are sorted out based both on their destination node and Quality-of-Service (QoS) requirements, and are assembled into bursts according to a burst assembly strategy. Upon assembling a burst, the edge node sends a Burst Header Packet (BHP) message through a separate wavelength channel to signal the forthcoming burst transmission. The BHP message, preceding the data burst by a short offset time, is electronically processed at every core node along the burst path, whereas the burst is switched in the optical domain. The node's control unit processes the bandwidth reservation request carried by the BHP and tries to both allocate a wavelength on the appropriate output fibre and to configure the node's optical switch matrix for the upcoming burst. If bandwidth reservation is successful in the entire burst path, the burst reaches its egress node, where it is disassembled and the data are delivered to the higher layer protocol.

The separation of control signal and data burst in both the wavelength and time domains relaxes the optical technology requirements. Essentially, the OBS paradigm demands the deployment, at the network nodes, of optical switch matrices with fast switching times [9]. All-optical wavelength converters, although not necessary from a conceptual perspective, are assumed to be available in the majority of proposals for OBS networks, given their ability to efficiently resolve contention, greatly improving network performance [10]. However, these devices are still undergoing research and development and experimental results with some of the most promising techniques for wavelength conversion have shown that their performance strongly depends on the combination of input and output wavelengths [11]. Consequently, in the near future, all-optical wavelength converters will likely only be able to efficiently convert an input wavelength to a limited range of output wavelengths. In an OBS network, this limits the ability of resolving wavelength contention at the network nodes, degrading the network performance as compared to that achieved when any input wavelength can be converted to any output wavelength.

In view of the current limitations of wavelength conversion technology, research efforts should be directed to improving the performance of OBS networks in which the nodes have practical Limited-Range Wavelength Converters (LRWCs), instead of ideal, but complex/expensive, Full-Range Wavelength Converters (FRWCs).

3 Traffic Engineering in the Wavelength Domain for Minimizing Wavelength Contention in OBS Networks with LRWCs

The performance of OBS networks with wavelength conversion restrictions, such as no wavelength conversion, and limited-range and/or sparse wavelength conversion, will eventually be improved if chances of wavelength contention at the network nodes are minimized. The principle underlying the strategies for minimizing wavelength contention in OBS networks is the following [12]: if two or more burst paths share one or more network links, contention on those links will be reduced if each burst path gives preference to using wavelengths different from those preferred by the other overlapping burst paths. That is, the chances of wavelength contention can be reduced by isolating, as much as possible, burst traffic of overlapping burst paths on different wavelengths. In practice, each ingress node maintains a priority-based ordering of the wavelengths per each egress node and uses it to search an available wavelength (in the first link of the burst path) for transmitting bursts towards the egress node.

The work in [7] introduced the use of traffic engineering in the wavelength domain to minimize wavelength contention in OBS networks without wavelength converters at the network nodes. The proposed strategy uses network and traffic conditions that are expected to remain unchanged over relatively long time scales, such as network topology, routing paths, and average offered traffic load between the network nodes, to determine the wavelength orderings that reduce the probability of using the same wavelength in overlapping burst paths. Recently, the authors proposed a new traffic engineering strategy to attain the same goal [8]. This new strategy was shown to significantly outperform that of [7] in improving the performance of OBS networks without wavelength converters. In the following, we extend the use of the strategy in [8] for OBS networks with limited-range wavelength converters at the network nodes.

Consider an OBS network with N nodes, L unidirectional links, and W wavelengths per link. Let Π denote the set of paths used to transmit bursts in the network, and let E_i denote the set of links traversed by path $\pi_i \in \Pi$. Let also γ_i denote the average traffic load offered to path π_i. Define the wavelength conversion range R as the number of output wavelengths each input wavelength can be converted to. Note that if R=1, the output wavelength used by a burst must be the same as its input wavelength (no wavelength conversion), whereas if R=W, each input wavelength can be converted to any of the W output wavelengths (full-range wavelength conversion). Moreover, let $\Lambda=\{\lambda_j: 1 \leq j \leq W\}$ denote the set of wavelength channels in each link and assume that each wavelength $\lambda_j \in \Lambda$ can only be converted to wavelengths of a sub-set of Λ with size R. More precisely, assume Λ is partitioned into W/R disjoint sub-sets, hereafter called wavebands, $\Lambda_i=\{\lambda_j: (i-1)R \leq j \leq iR\}$, $1 \leq i \leq W/R$, and that input wavelengths can only be converted to output wavelengths inside the same waveband.

Under these assumptions, we can generalize the concept of wavelength contention minimization to that of waveband contention minimization, where the objective is then to use the described inputs to determine the waveband orderings of each burst path that minimize the probability that overlapping burst paths will use the same waveband. Therefore, when assigning a wavelength to a burst generated locally, the ingress node will first search for an available wavelength on the waveband with highest priority. If none is available, it searches for an available wavelength on the following ordered waveband and so on. However, the resulting optimization problem inherits the shortcomings of the wavelength contention minimization problem [8], noticeably, the fact that it is not possible to express any relevant performance metric as a function of the known inputs in an analytical closed-form manner.

Waveband contention can only occur between bursts using paths that share at least one network link. Moreover, the chances of waveband contention are expected to increase with both the average traffic load offered to the paths and the number of common links [7]. Hence, define the interference level of path π_i on path π_j as

$$I(\pi_i,\pi_j) = \gamma_i \mid E_i \cap E_j \mid, \; i \neq j, \tag{1}$$

where $|E_i \cap E_j|$ denotes the number of links shared by both paths, and define the combined interference level between paths π_i and π_j as

$$I^{c}(\pi_i,\pi_j) = I(\pi_i,\pi_j) + I(\pi_j,\pi_i) = (\gamma_i + \gamma_j) \mid E_i \cap E_j \mid, \; i \neq j. \tag{2}$$

Basically, the higher the combined interference level of two paths, the higher the likelihood that bursts on those paths will try to go through the same network link at the same time, thus contending for the same resources.

In general, waveband contention between bursts of two paths that share at least one network link is minimized if they use opposite waveband orderings. However, in most practical network scenarios, each path shares network links with many other paths and, therefore, it is not possible to have opposite waveband orderings for each two overlapping paths. Consequently, waveband contention minimization strategies have to use some estimate of the likelihood of waveband contention, such as the combined interference level, to determine which burst paths should have waveband orderings as opposed as possible, in the sense that one burst path gives higher priority to wavebands that have low priority on the other path.

Let $1 \leq P(\pi_i,\Lambda_j) \leq W/R$ denote the priority assigned to waveband Λ_j on the burst path π_i. Thus, the wavelengths of Λ_j will be used by π_i with priority $P(\pi_i,\Lambda_j)$ in the network links E_i. The proposed Heuristic Minimum Priority Interference (HMPI) strategy assigns a priority to each waveband on each burst path as to minimize the *priority interference* on the links. That is, it minimizes, as possible, the amount of burst paths that use with high priorities the same waveband on a link, thus reducing the chances of wavelength contention in the link. The strategy comprises two separate stages. In the first stage the primary (highest priority) waveband used by each burst path is selected, while the second stage orders the non-primary wavebands for each burst path. The primary waveband selection stage consists of the following steps.

(S1) Reorder the paths of Π such that if $i < j$ one of the following conditions holds

$$\sum_{\pi_k \in \Pi} I(\pi_i,\pi_k) > \sum_{\pi_k \in \Pi} I(\pi_j,\pi_k); \tag{3}$$

$$\sum_{\pi_k \in \Pi} I(\pi_i,\pi_k) = \sum_{\pi_k \in \Pi} I(\pi_j,\pi_k) \quad \text{and} \quad |E_i| > |E_j|. \tag{4}$$

(S2) Consider $M=W/R$ sub-sets, one per waveband, initially empty, that is, $|\Pi_j|=0$ for $j=1,\ldots,M$. Following the path ordering of Π, include path π_i in sub-set Π_j such that one of the subsequent conditions holds

$$\forall_{k \neq j} \quad \sum_{\pi_l \in \Pi_j} I^c(\pi_i,\pi_l) < \sum_{\pi_l \in \Pi_k} I^c(\pi_i,\pi_l); \tag{5}$$

$$\forall_{k \neq j} \quad \sum_{\pi_l \in \Pi_j} I^c(\pi_i,\pi_l) = \sum_{\pi_l \in \Pi_k} I^c(\pi_i,\pi_l) \quad \text{and} \quad |\Pi_j| > |\Pi_k|. \tag{6}$$

(S3) Select the waveband Λ_j as the primary waveband of all the paths in sub-set Π_j, that is

$$P(\pi_i,\Lambda_j) = \begin{cases} M & \text{if } \pi_i \in \Pi_j \\ 0 & \text{otherwise} \end{cases}. \tag{7}$$

The non-primary waveband ordering stage comprises the following steps, which are executed for all priorities $1 \leq p < M$ in decreasing order and for all paths $\pi_i \in \Pi$ by the ordering defined in the first stage of the HMPI strategy.

(S1) Let $\Lambda^* = \{\Lambda_j : P(\pi_i, \Lambda_j) = 0,\ 1 \leq j \leq M\}$ denote the initial set of candidate wavebands, containing all wavebands that were not assigned a priority on π_i. If $|\Lambda^*|=1$ go to (**S7**).

(S2) Let $P_{\Lambda^*} = \{p_k : \exists \pi_l, l \neq i, P(\pi_l, \Lambda_j) = p_k, |E_l \cap E_i| > 0, \Lambda_j \in \Lambda^*\}$ denote the set of priorities already assigned to candidate wavebands on paths that overlap with π_i.

(S3) Let $\rho = \min_{\Lambda_j \in \Lambda^*} \{\max\{P(\pi_l, \Lambda_j) : l \neq i, |E_l \cap E_i| > 0\}\}$ be the lowest priority from the set of the highest priorities assigned to candidate wavebands on the paths that use links of π_i. Update the set of candidate wavebands as follows

$$\Lambda^* \leftarrow \Lambda^* \setminus \{\Lambda_j : \exists \pi_l, l \neq i, P(\pi_l, \Lambda_j) > \rho, |E_l \cap E_i| > 0, \Lambda_j \in \Lambda^*\}. \tag{8}$$

If $|\Lambda^*|=1$ go to (**S7**).

(S4) Let $C(e_m, \Lambda_j) = \sum \{\gamma_l : E_l \supset e_m, |E_l \cap E_i| > 0, P(\pi_l, \Lambda_j) = \rho\}$ be the cost associated with waveband Λ_j on link $e_m \in E_i$. Thus, the minimum cost among the highest costs associated with the candidate wavebands on the links of path π_i is $\alpha_e = \min_{\Lambda_j \in \Lambda^*} \{\max\{C(e_m, \Lambda_j) : e_m \in E_i\}\}$. Update the set of candidate wavebands as follows

$$\Lambda^* \leftarrow \Lambda^* \setminus \{\Lambda_j : \exists e_m\ C(e_m, \Lambda_j) > \alpha_e, e_m \in E_i, \Lambda_j \in \Lambda^*\}. \tag{9}$$

If $|\Lambda^*|=1$ go to (**S7**).

(S5) Let $C(\pi_i, \Lambda_j) = \sum_{e_m \in E_i} C(e_m, \Lambda_j)$ be the cost associated with waveband Λ_j on π_i. Thus, $\alpha_\pi = \min_{\Lambda_j \in \Lambda^*} C(\pi_i, \Lambda_j)$ is the minimum cost among the costs associated with the candidate wavebands on π_i. Update the set of candidate wavebands as follows

$$\Lambda^* \leftarrow \Lambda^* \setminus \{\Lambda_j : C(\pi_i, \Lambda_j) > \alpha_\pi, \Lambda_j \in \Lambda^*\}. \tag{10}$$

If $|\Lambda^*|=1$ go to (**S7**).

(S6) Update the set of priorities assigned to the candidate wavebands as follows

$$P_{\Lambda^*} \leftarrow P_{\Lambda^*} \setminus \{p_k : p_k \geq \rho, p_k \in P_{\Lambda^*}\}. \tag{11}$$

If $|P_{\Lambda^*}|>0$ go to (**S3**). Else, randomly select a candidate waveband $\Lambda_j \in \Lambda^*$.

(S7) Assign priority p to the candidate waveband $\Lambda_j \in \Lambda^*$ on path π_i, that is

$$P(\pi_i, \Lambda_j) = p. \tag{12}$$

4 Results and Discussion

The performance of OBS networks with LRWCs is evaluated here using network simulation [10]. The performance metric is the average burst blocking probability, which measures the average fraction of traffic (in bursts) discarded by the network. In the network simulations performed, the OBS network employs Just Enough Time (JET) resource reservation [3], albeit waveband contention minimization strategies can be used with any other one-way resource reservation mechanism [4].

A regular 10-node ring topology and the irregular 14-node NSF network topology [10] are used in this study. In both cases, a uniform traffic pattern is assumed and the network has 60 wavelengths per link, a wavelength capacity of 10 Gb/s, a switch fabric configuration time of 10 μs, and an average burst size of 100 kB. A negative exponential distribution is used for both burst size and burst interarrival time. The burst paths are computed with shortest path routing for the 10-node ring and with the load balancing algorithm of [13] for the NSF network.

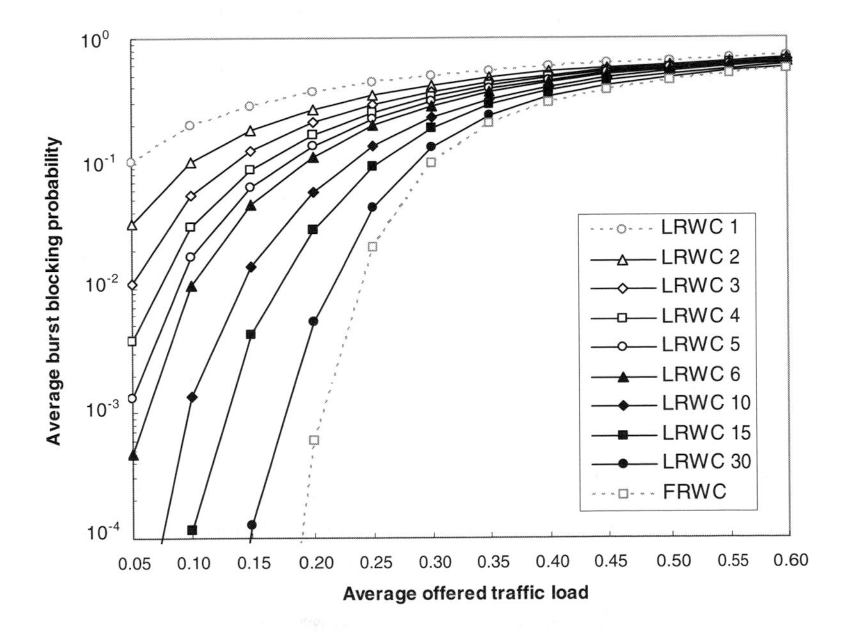

Fig. 1. Performance of the 10-node ring network without waveband contention minimization

Fig. 1 plots the average burst blocking probability as a function of the average offered traffic load normalized to the network capacity for the 10-node ring network with different wavelength conversion ranges (LRWC *R*), but without using waveband contention minimization. Ingress nodes use Random Assignment (RA) to allocate a wavelength to a burst generated locally. The average burst blocking probability and a 95% confidence interval on this value were obtained by simulating 20 separate burst traces per average offered traffic load value. However, the confidence intervals are so narrow that are omitted for improving readability.

These curves illustrate the performance degradation observed in an OBS network using LRWCs with small wavelength conversion ranges, which is of such magnitude that causes the network to become impractical. Thus, it is of paramount importance to improve the performance of OBS networks that employ these realistic LRWCs.

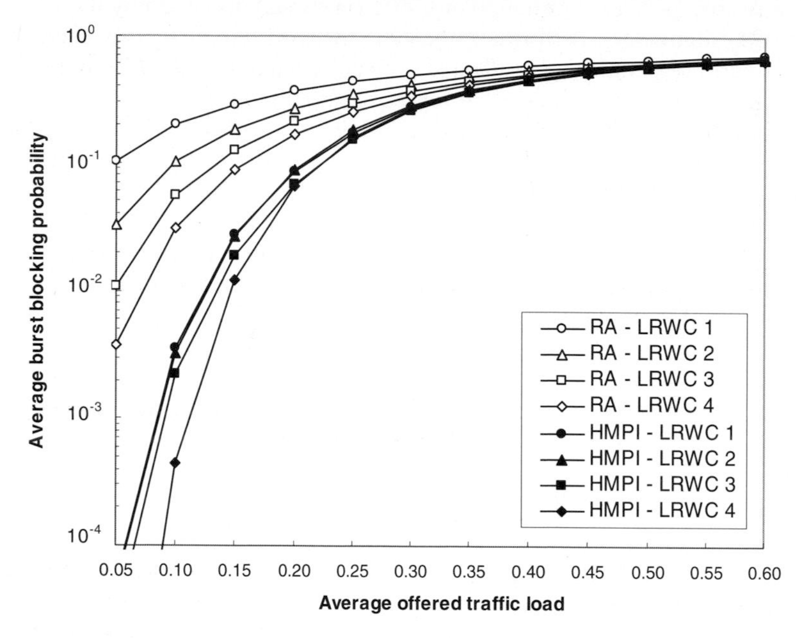

Fig. 2. Performance of the 10-node ring network for small wavelength conversion ranges

Fig. 2 shows the performance of the 10-node ring network employing LRWCs with small wavelength conversion ranges and for both the HMPI and RA strategies, that is, with and without the use of waveband contention minimization strategies.

These results show that the HMPI strategy reduces, in some cases by several orders of magnitude, the average burst blocking probability of an OBS network employing LRWCs with small conversion ranges. This performance improvement is due to the ability of the HMPI strategy to isolate, as possible, the traffic of overlapping burst paths on different wavebands, thus significantly reducing the chances of waveband contention between bursts on these paths.

Noticeably, these results also show that an OBS network with no wavelength conversion (R=1) using the HMPI strategy outperforms the same network with R=4 using RA. Moreover, comparing the curves of Fig. 1 and Fig. 2 it is interesting to observe that with waveband contention minimization a network with R=4 almost achieves the performance of the same network without using waveband contention minimization, but with a much wider wavelength conversion range of R=15.

In order to gain further insight on the impact of the wavelength conversion range on the effectiveness of the HMPI strategy, Fig. 3 plots the average burst blocking probability as a function of the wavelength conversion range for an average offered traffic load of 0.10 and 0.15.

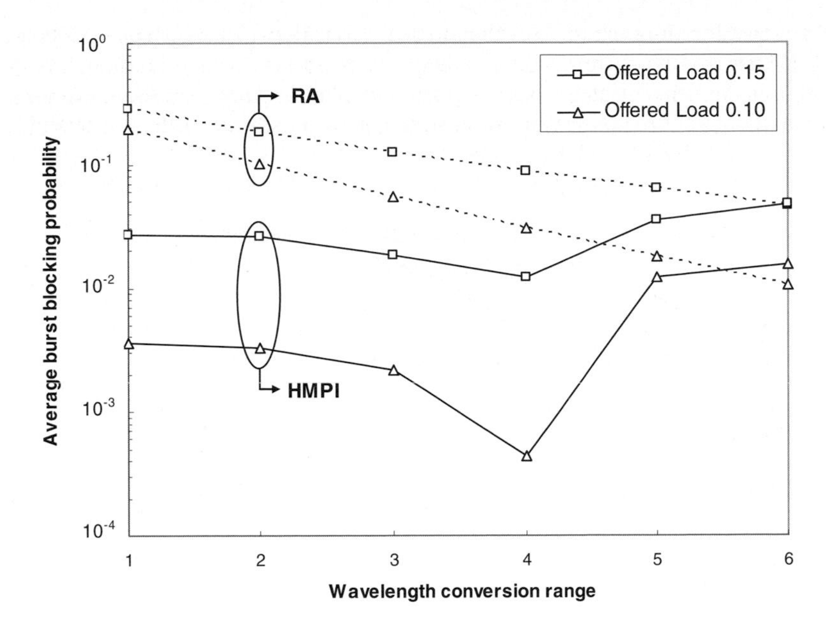

Fig. 3. Performance of the 10-node ring network for an offered traffic load of 0.10 and 0.15

The curves plotted in Fig. 3 show that without waveband contention minimization the average burst blocking probability decreases monotonically as the wavelength conversion range is increased. However, they suggest that a similar behaviour may not be observed when using waveband contention minimization, since for *R*=5 and *R*=6 the performance of the OBS network using the HMPI strategy is worst than that for *R*=4. Moreover, it was also observed that for *R*>6, the HMPI strategy does not outperform the RA strategy. The reason for this behaviour is that, for the same number of wavelengths per link, a wider wavelength conversion range corresponds to a smaller number of wavebands. Hence, when the number of wavebands becomes smaller than the number of burst paths traversing each link, the HMPI strategy can no longer efficiently isolate the traffic of overlapping burst paths on different wavebands and, as a result, does not reduce the chances of waveband contention.

Due to lack of space in the paper the simulation results for the NSF network are not plotted here. However, they follow the same trends as those observed and discussed for the 10-node ring network.

5 Conclusions

In view of the current limitations of wavelength conversion technology and the reduced performance observed in OBS networks employing limited-range wavelength converters at the network nodes, this paper has proposed the use of traffic engineering in the wavelength domain for improving the performance of OBS networks with small wavelength conversion ranges.

A strategy for minimizing waveband contention between overlapping burst paths was described and its performance evaluated through network simulation. The results have shown that this strategy can improve the performance of OBS networks with small wavelength conversion ranges, in some cases by several orders of magnitude in terms of average burst blocking probability. Therefore, it contributes to the feasibility of OBS networks employing practical limited-range wavelength converters.

Acknowledgments. The authors acknowledge the financial support from Siemens Networks S.A. and Fundação para a Ciência e a Tecnologia (FCT), Portugal, through research grant SFRH/BDE/15584/2006. J. Pedro also acknowledges his colleagues S. Pato, R. Morais, and M. Pinho for greatly increasing the computational capacity available for network simulation and J. Castro for helpful discussions.

References

1. Ramaswami, R., Sivarajan, K.: Optical Networks: A Practical Perspective. Morgan Kaufmann, San Francisco (2002)
2. Korotky, S.: Network global expectation model: A statistical formalism for quickly quantifying network needs and costs. IEEE/OSA Journal of Lightwave Technology 22(3), 703–722 (2004)
3. Qiao, C., Yoo, M.: Optical Burst Switching (OBS) – A new paradigm for an optical Internet. Journal of High Speed Networks 8(1), 69–84 (1999)
4. Chen, Y., Qiao, C., Yu, X.: Optical Burst Switching: A new area in optical network research. IEEE Network 18(3), 16–23 (2004)
5. Sartorius, B., Nolting, H.-P.: Techniques and technologies for all-optical processing in communication systems. In: ECOC 2004, paper Mo3.5.1 (2004)
6. Zang, H., Jue, J., Mukherjee, B.: A review of routing and wavelength assignment approaches for wavelength-routed optical WDM networks. Optical Networks Magazine 1(1), 47–60 (2000)
7. Teng, J., Rouskas, G.: Wavelength selection in OBS networks using traffic engineering and priority-based concepts. IEEE Journal of Selected Areas in Communications 23(8), 1658–1669 (2005)
8. Pedro, J., Monteiro, P., Pires, J.: Wavelength contention minimization strategies for optical burst-switched networks. In: IEEE GLOBECOM 2006, paper OPNp1-5. IEEE Press, Los Alamitos (2006)
9. Sun, Y., Hashiguchi, T., Minh, V., Wang, X., Morikawa, H., Aoyama, T.: Design and implementation of an optical burst-switched network testbed. IEEE Communications Magazine 43(11), s48–s55 (2005)
10. Pedro, J., Castro, J., Monteiro, P., Pires, J.: On the modelling and performance evaluation of optical burst-switched networks. In: IEEE CAMAD 2006, pp. 30–37. IEEE Press, Los Alamitos (2006)
11. Eramo, V., Listanti, M., Di Donato, M.: Performance evaluation of a bufferless optical switch with limited-range wavelength converters. IEEE Photonics Technology Letters 16(2), 644–646 (2004)
12. Wang, X., Morikawa, H., Aoyama, T.: Priority-based wavelength assignment algorithm for burst switched WDM optical networks. IEICE Transactions on Communications E86-B(5), 1508–1514 (2003)
13. Castro, J., Pedro, J., Monteiro, P.: Burst loss reduction in OBS networks by minimizing network congestion. In: ConfTele 2005, session FriAmPO1 (2005)

On the Early Release of Burst-Control Packets in Optical Burst-Switched Networks*

José Alberto Hernández and Javier Aracil

Universidad Autónoma de Madrid
Ctra. Colmenar Viejo, km. 15, 28049 Madrid, Spain
{Jose.Hernandez,Javier.Aracil}@uam.es
http://www.ii.uam.es/~networking

Abstract. In Optical Burst-Switched networks, the so-called Burst-Control Packet is sent a given offset-time ahead of the optical data burst to advertise the imminent burst arrival, and reserve a time-slot at each intermediate node to allocate it. This work proposes a methodology to estimate the number of packets to arrive in a given amount of time, in order to make it possible to send the BCP packet straightafter the first packet arrival and reduce the latency experienced during the burst-assembly process.

The following studies the impact of a wrong guess in terms of over-reservation of resources and waiting-time at the assembler, providing a detailed characterisation of their probability density functions. Additionally, a case example in a scenario with non-homogeneous Poisson arrivals is analysed and it is shown how to choose the appropriate burst-assembly algorithm values to never exceed a given over-reservation amount.

1 Introduction

Dense Wavelength Division Multiplexing [1] has been proposed as a promising physical layer technology for the forthcoming next-generation Internet, due to the huge amount of raw bandwidth provided, in the order of gigabits per wavelength, with more than one hundred wavelengths per optical fibre [2,3]. In this light, the Optical Burst Switching (or just OBS) paradigm over DWDM physical layers arises as a cost-effective solution for the high utilisation and multiplexing of such tremendous amount of raw bandwidth with relatively low switching complexity involved [4,5].

In a typical OBS network, ingress nodes aggregate incoming packets into larger-size data bursts, which are transmitted all-optically through the network core. Such optical bursts do not suffer from optical/electrical/optical conversion at the intermediate nodes, leading to a fast and efficient transmission of large volumes of data.

Each optical burst has an associated Burst-Control Packet (or BCP), this is, a small-size packet which carries the control information to get its associated data

* This work was funded by the European Union *e-Photon/ONe+* project and by the Spanish Ministry of Science and Education under the DIOR project.

T. Vazão, M.M. Freire, and I. Chong (Eds.): ICOIN 2007, LNCS 5200, pp. 31–40, 2008.

burst delivered at the other end of the optical network. To do so, the BCP is sent a given offset time in advance of its associated data burst, and is processed electronically (it suffers O/E/O conversion) at each intermediate node along the path. Its main role is to advertise each intermediate node of the size and expected arriving time of its associated data burst [4,6]. With this information, the core node can find and reserve a time-slot at the appropriate output wavelength, and consequently, can immediately switch the data burst in the optical domain as soon as it arrives. This way, the need for temporal buffering of optical data is removed, in contrast to electrical switches.

Typically, the BCP is generated and transmitted straightafter the data burst is assembled at the border node, since it must know the exact burst size and release time to inform the intermediate nodes' scheduler, under *Just-Enough-Time* (JET) scheduling [4,6]. Hence, in addition to the delay suffered by the data packets during the burst assembly process, the packets suffer an extra delay given by the offset-time between the BCP and the data burst. In certain situations, such delay may be excessive.

To alleviate such long delay, this work proposes a mechanism to overlap the burst-assembly delay and the offset delay suffered by the the data packets. Essentially, after the first packet has arrived at the burst assembler, our algorithm generates and sends off the BCP to the next hop in the path. Such early BCP carries out a given burst-release time (which is equal to the offset time) and a rough estimation of the final size of the optical burst. The following studies how to make such estimation, and analyses its impact on the global network performance.

The remainder of this work is organised as follows: Section 2 studies the statistics of the burst generation time and hints how to estimate the final burst size. Section 3 validates the equations derived in the analysis section and further proposes a scenario to evaluate the benefits of the early BCP release mechanism. Finally, section 4 brings the main findings, conclusions and merits of this work.

2 Statistical Analysis of the Burst-Release Time

2.1 Problem Statement

As previously stated, ingress OBS nodes aggregate packets together into the so-called bursts, which are converted to the optical plane. Throughout this work, packet arrivals shall be assumed to follow a Poissonian basis with rate λ packets/seg. This assumption is gaining in importance among the network research community, especially after the recent measurement-based studies in core Internet links [7,8].

Let λ refer to the average rate of incoming packets per unit of time at the burst assembler, and let n refer to the number of packets in a burst. Without loss of generality, incoming packets are assume to have constant size. As shown in figure 1, packet interarrivals x_i, $i = 1, \ldots$ are exponentially distributed with parameter $\lambda = \frac{1}{EX}$.

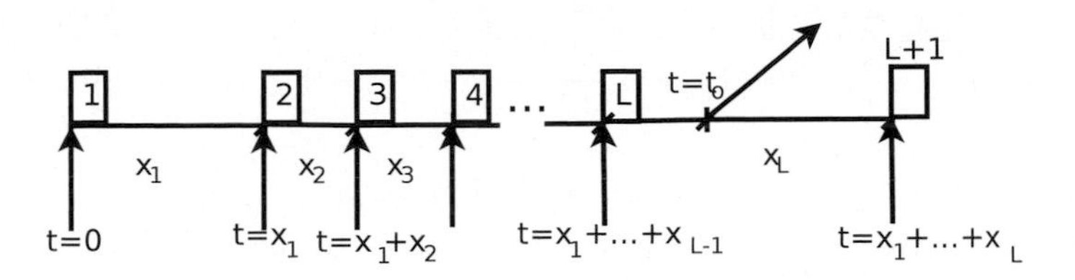

Fig. 1. Notation

Under these assumptions, the burst-assembler proceeds as follows: When the first packet arrives at the border node (packet no. 1 in the figure), the BCP is generated. Essentially, each BCP contains the information of (1) the time at which the burst shall be released, namely t_o, and (2) the number of packets in the optical burst, namely $\hat{L}$. Typically, t_o must not be smaller than the offset time, which is set by the network topology, and represents the amount of time that a BCP needs to configure all the intermediate nodes along the path. Hence, since t_o is fixed, the role of the burst-assembler is to guess the appropriate value of $\hat{L}$, taking into account that:

- It may well happen that the actual number of packets arriving within time t_o, say L, is smaller than the estimated burst-size $\hat{L}$. In this case, the optical burst must be released anyway at time t_o, and cannot wait for the $\hat{L} - L$ packets remaining to fulfill the optical burst. Hence, the BCP has reserved at intermediate nodes for $\hat{L}$ packets, whereas only $L < \hat{L}$ will actually occupy such scheduled time. Thus, the amount of over-reservation is $\hat{L} - L$ packets.
- On the other hand, it may well happen that the optical burst-size reaches the total of L packets before t_o, say at time $t < t_o$. However, the optical burst cannot be released before time t_o, because it is only guaranteed that the BCP has allocated space at time t_o. Therefore, the data burst must wait in queue at the intermediate node during $t_o - t$ units of time, and no resource over-reservation occurs.

The above clearly brings a trade-off when guessing/estimating the value of $\hat{L}$ packet arrivals before time t_o. A conservative estimation ($\hat{L}$ small enough) would lead to the over-reservation of resources at the intermediate nodes, whereas a tight estimation ($\hat{L}$ large) may produce buffer overloading at the burst-assembler. The following sections analysis the impact of choosing $\hat{L}$ small or large by means of, firstly the over-dimensioning probability distribution (when L small), secondly the distribution of waiting time in queue (when L large).

2.2 Probability Distribution of the Burst-Release Time

Under the assumption of Poissonian packet arrivals, the assembly time t for a L-sized burst follows a Gamma distribution with $L-1$ degrees of freedom and parameter λ, as noted in [9,10]. The Probability Density Function (pdf) for such assembly time is given by

$$\Gamma_t(L-1,\lambda) = \frac{\lambda^{L-1}t^{L-2}}{(L-2)!}e^{-\lambda t}, \quad t \geq 0 \tag{1}$$

with mean $E[t] = \frac{L-1}{\lambda}$ and standard deviation $Std[t] = \sqrt{\frac{L-1}{\lambda^2}}$.

In this light, since the BCP is released after the first packet arrival with information t_o and L, the probability to actually have $\hat{L}-1$ additional packet arrivals before release time t_o is given by:

$$P(t < t_o) = \int_0^{t_o} \frac{\lambda^{L-1}t^{L-2}}{(L-2)!}e^{-\lambda t}dt = \frac{\gamma_{\text{inc}}(L-1,\lambda t_o)}{(L-2)!} \tag{2}$$

where γ_{inc} refers to the incomplete gamma function[1].

It is worth noticing here that such probability depends not only on the choice of t_o, but also on the value of L. Clearly, it is easier to complete L_1 packets within time $[0, t_o]$ than $L_2 > L_1$ within the same amount of time. This effect is shown in fig. 2.

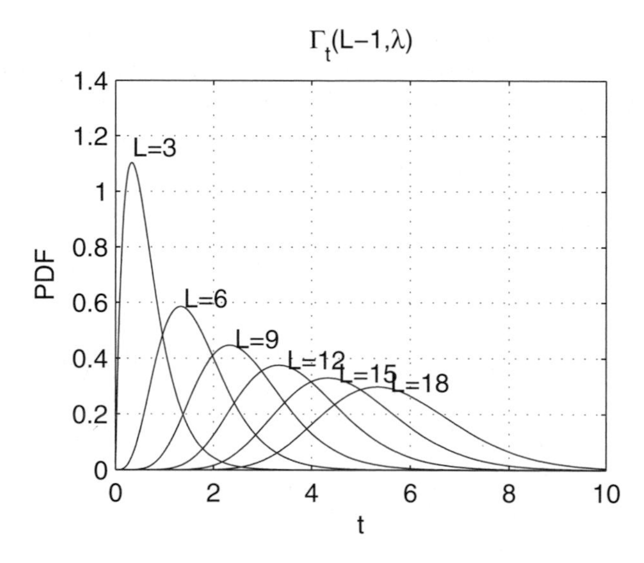

Fig. 2. Burst-release time distribution for various values of L

In this figure, the value of $\lambda = 3$ is fixed, but L takes values in the range $L \in \{3, 6, 9, 12, 15, 18\}$. As shown, given $t_o = 4$ fixed, the probability to complete L packets before t_o decreases the larger the value of L (see table 1).

Hence, the choice of L relatively small is a conservative estimation, that is, high probability to complete L packets before time t_o. Moreover, the choice of small L would probably lead to a situation at which data bursts are completed in a time t much earlier than t_o, thus requiring to allocate them in memory for time $t_o - t$. However, the opposite (relatively large L values) leads to high

[1] Note that $\gamma_{inc}(n,x) = \int_0^x t^{n-1}e^{-t}dt$.

Table 1. A few values of $\gamma_{\text{inc}}(L-1, \lambda t_o)$

L	$P(t < 4)$
3	0.9999
6	0.9924
9	0.9105
12	0.6528
15	0.3185
18	0.1013

probability of transmitting data bursts with less packets than predicted, thus over-loading the network.

The following analyses in detail the amount of over-reserved resources due to large values of L, and the burst waiting-time in queue due to early completion when small L.

2.3 Case 1: Over-Reservation of Resources

This section studies the first situation described above: The case at which the BCP reserves for a $\hat{L}$-sized optical burst, whereas the actual optical burst is of size $n < \hat{L}$ size. Let Y refer to the random variable that represents the excess reservation at intermediate nodes, that is, $Y = \hat{L} - n$. The probability mass function of Y, conditioned to $\hat{L}$, is given by:

$$\begin{aligned}\mathbb{P}(Y = m) &= \mathbb{P}(n = \hat{L} - m \text{ Poisson arrivals in } [0, t_o)) = \\ &= \frac{(\lambda t_o)^{\hat{L}-m}}{(\hat{L}-m)!} e^{-\lambda t_o}\end{aligned} \tag{3}$$

where $0 \leq n \leq \hat{L}-1$. As shown, the random variable Y is distributed as a shifted Poisson distribution.

Finally, the average over-reservation (in packets) is given by:

$$\mathbb{E}[Y] = \sum_{n=1}^{\hat{L}-1} (\hat{L}-n) \frac{(\lambda t_o)^{n-1}}{(n-1)!} e^{-\lambda t_o} \tag{4}$$

2.4 Case 2: Waiting Time Distribution

This section examines the second situation described above: The case at which the $\hat{L}$-th packet arrives at time $t < t_o$, thus fulfilling the data burst, and forcing the completed burst to be buffered for time $t_o - t$. Let Z refer to the random variable that represents the waiting-time in buffer, that is, $Z = t_o - t$. Then, it is clear that the probability density function of Z is the shifted gamma distribution:

$$\begin{aligned}f_Z(t) &= \Gamma_{t_o - t}(\hat{L}-1, \lambda) = \\ &= \frac{\lambda^{\hat{L}-1}(t_o - t)^{\hat{L}-2}}{(\hat{L}-2)!} e^{-\lambda(t_o - t)}, \quad 0 \leq t \leq t_O\end{aligned} \tag{5}$$

The average waiting time can be easily obtained by:

$$
\begin{aligned}
E[t_o - t] &= \int_0^{t_o} (t_o - t) \frac{\lambda^{\hat{L}-1} t^{\hat{L}-2}}{(\hat{L}-2)!} e^{-\lambda t} dt = \\
&= t_o \int_0^{t_o} \frac{\lambda^{\hat{L}-1} t^{\hat{L}-2}}{(\hat{L}-2)!} e^{-\lambda t} dt - \int_0^{t_o} \frac{\lambda^{\hat{L}-1} t^{\hat{L}-1}}{(\hat{L}-2)!} e^{-\lambda t} dt = \\
&= t_o \frac{\gamma_{\text{inc}}(\hat{L}-1, \lambda t_o)}{(\hat{L}-2)!} - \frac{1}{\lambda} \frac{\gamma_{\text{inc}}(\hat{L}, \lambda t_o)}{(\hat{L}-2)!}
\end{aligned}
\quad (6)
$$

3 Experiments

The following experiments are focused on first, demonstrating the validity of eq. 3 and 5 above; and secondly, propose an algorithm to obtain the adequate value of $\hat{L}$ to meet a set of requirements, in an environment with non-homogeneous Poisson arrivals.

3.1 Validation

This experiment aims to show the validity of eq. 3 and 5 above. To do so, we have simulated a burst-assembler receiving $N = 10^6$ incoming packets on a Poissonian basis with parameter $\lambda = 100000$ packets/sec. The estimated size of outgoing optical bursts has been chosen as $\hat{L} = 50$ packets, and the burst-release time is $t_o = 53.52$ms. With these values, an amount of 25% of cases are not able to complete a 50-sized data burst before time t_o, as given by:

$$
P(t > t_o) = \frac{\gamma_{\text{inc}}(\hat{L}-1, \lambda t_o)}{(\hat{L}-2)!} = 0.25
$$

whereas the other 75% do manage to fulfill the data bursts within time.

Figure 3 (top) shows the histogram of the waiting-time in queue of the 75% of cases (around $0.75 \times 10^6/50 \approx 15000$ simulated bursts) that achieve burst-completion within time together with the theoretical values given by equation 5. Similarly, figure 3 (bottom) shows the amount of over-reserved packets at intermediate nodes for the 25% of cases (around $0.25 \times 10^6/50 \approx 5000$ simulated bursts) that do not reach $\hat{L}$ within time. Again, the theoretical values, given by equation 3, are plotted together with the simulated results.

Obviously, the bottom figure is less accurate than the top figure, since in the bottom figure only 25% of the total cases constitute the histogram, whereas the top figure has been computed with 75% of the cases. Overall, it turns out that the analytical expressions match the simulation results very well.

3.2 Numerical Example

This experiment presents a mechanism to obtain the appropriate value of $\hat{L}$ in a sample scenario with non-homogeneous Poisson arrivals. Indeed, incoming

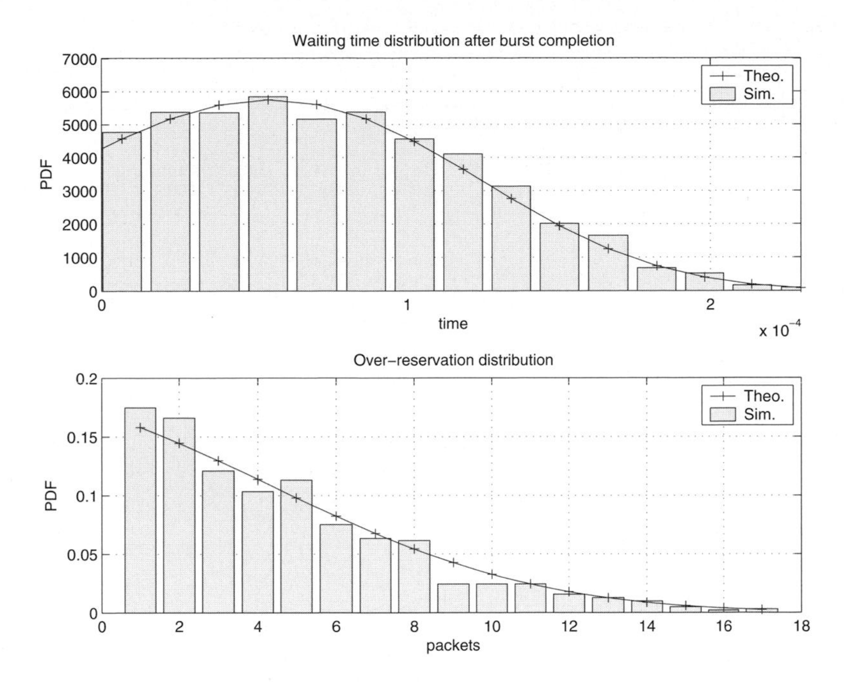

Fig. 3. Waiting-time (top) and over-reservation (bottom) distributions for $\hat{L} = 50$ packets

packets have been assumed to follow a Poissonian distribution with changing λ. For simplicity, we have assumed:

$$\lambda(n) = 10^5 + 2 \cdot 10^5 \cos^2(\frac{2\pi}{1000}n) \quad \text{packets/sec}$$

This is a value of λ ranging from 100000 to 300000 packets/sec with period 500 samples.

Obviously, the burstifier does not know the real value of λ at each instant and has to estimate it. In the experiment we have considered the well-known Exponential-Weighted Moving Average algorithm to estimate λ. Essentially, such algorithm proceeds as follows: For every new packet arrival with interarrival time x_n from the previous one, $n \geq 1$, we estimate the average interarrival time as:

$$\hat{\bar{x}}_n = \frac{W}{W+1}\hat{\bar{x}}_{n-1} + \frac{1}{W+1}x_n$$

for some value of W. With this value, we compute the estimated $\hat{\lambda}$ as $\hat{\lambda}_n = \left(\hat{\bar{x}}_n\right)^{-1}$ since $\lambda = 1/EX$. The choice of parameter W is a measure of the memory of the estimation. That is, W small gives more weight to new samples than W large. However, W provides a smoother estimate of λ. For highly changing environments, a small value of W is preferred. In our case, figure 4 shows the evolution of the estimated $\hat{\lambda}$ for several values of W. Clearly, $W = 25$ shows the

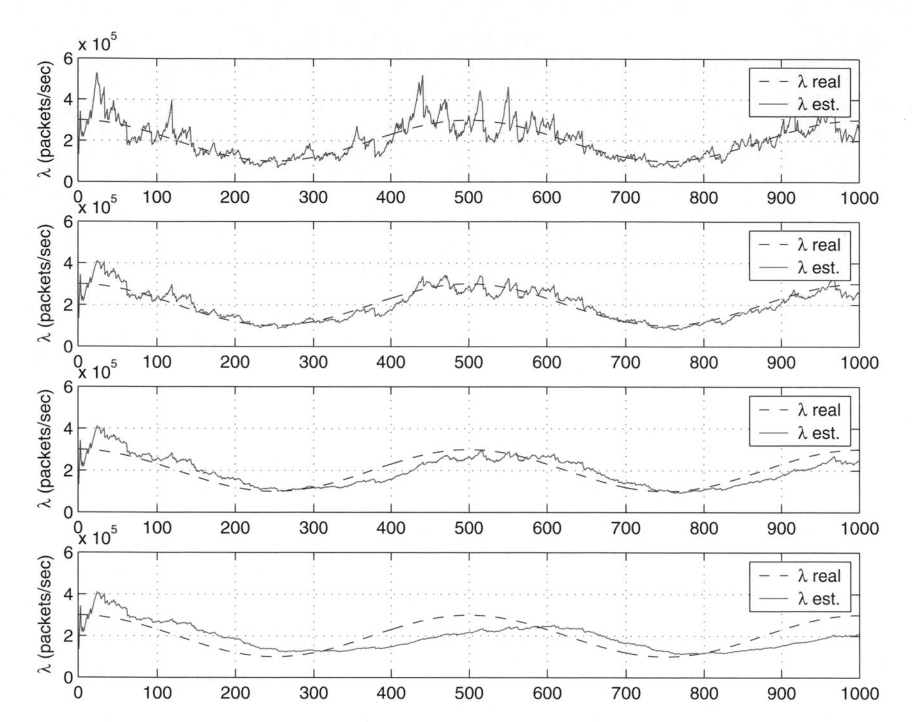

Fig. 4. Example of EMWA for $W = 10$ (top), $W = 25$ (middle-top), $W = 50$ (middle-bottom) and $W = 100$ (bottom) for the case $\lambda(n) = 10^5 + 2 \cdot 10^5 \cos^2(\frac{2\pi}{1000}n)$ packets/sec.

best behaviour in terms of high accuracy in the estimation with fast tracking of the changes in λ.

In our case, we have chosen the value $W = 25$. Figure 5 (top) shows the real value of λ along with the estimated $\hat{\lambda}$ using the EWMA algorithm with $W = 25$.

Concerning the remaining experiment parameters, we have chosen the value $t_o = 10.19$ ms, which is fixed by the network topology, and a strategy of designing $\hat{L}$ on attempts to have bursts than exceed this value no more than 10% of the times. That is:

$$\text{Find } \hat{L} \text{ such that } \frac{\gamma_{\text{inc}}(\hat{L}-1, \hat{\lambda}t_o)}{(\hat{L}-2)!} = 0.9$$

as pointed out in equation 2.

With this parameter set, figure 5 shows the evolution of $\hat{\lambda}_n$ (fig. 5 top), the predicted size (fig. 5 middle) and waiting-time in queue (fig. 5 bottom) with both real λ and estimated $\hat{\lambda}$.

As shown, when the estimation $\hat{\lambda}$ and the real value of λ are close, both the predicted size and waiting-time in queue are close too. Additionally, it is worth remarking that, in those cases where there is an excess of packet reservation, the waiting-time in queue is null (see for instance the inverval $n \in [100, 200]$), and vice versa.

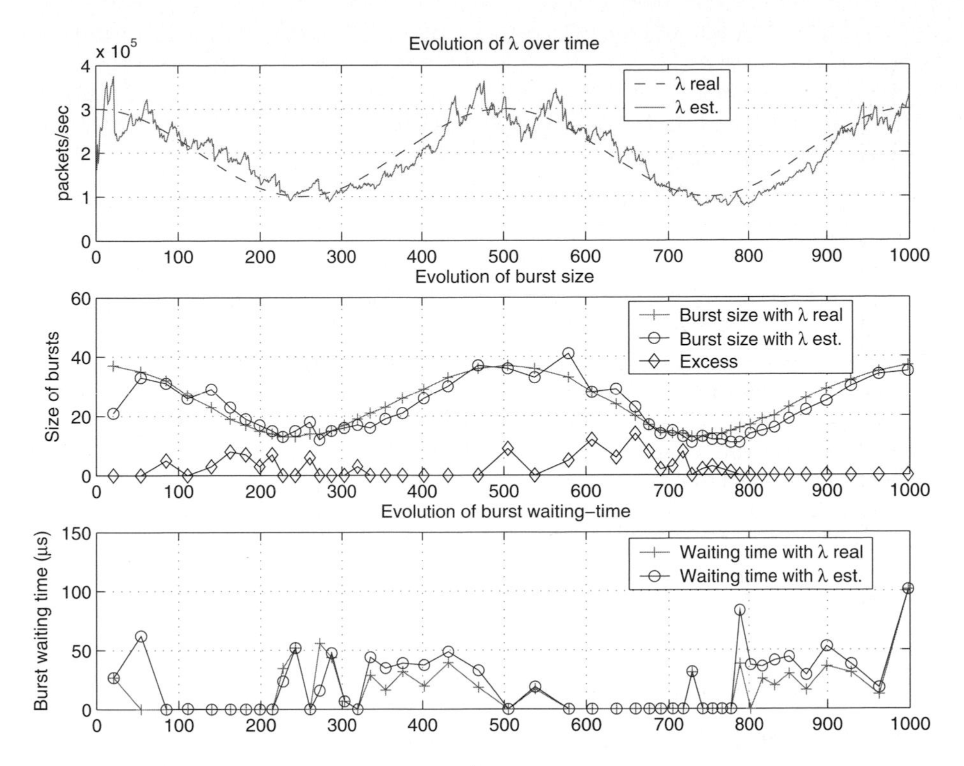

Fig. 5. Numerical example: evolution of $\hat{\lambda}_n$ (top); estimated optical burst size along with their optimal values (middle); and, burst-waiting time in queue along with the optimal values (bottom)

Finally, it is worth emphasising that the algorithm finds the appropriate $\hat{L}$ which produces over-reservation only 10% of the times. However, sometimes the algorithm finds such value based on a wrong estimate of λ, thus leading to situations with higher over-reservation of resources than the designed 10% of the cases.

4 Summary and Conclusions

This work proposes to send the Burst Control Packet of a given optical burst as soon as the first packet comprising the burst has arrived at the burst-assembler. The burst-release time information is set by the network topology, and the final size of the optical burst is thus estimated as the amount of expected packets arriving until burst-release. Furthermore, such number must not be chosen as the expected number of arrivals, but it can be set to any particular value such that the probability to have a smaller number of incoming packets is small.

The impact of choosing a small or large value is further analysed and concluded that, the larger its value, the more likely to over-reserve resources in the network which, indeed, has a clear impact in the global network performance. On the other hand, the smaller the estimated value, the more likely to not exceed it, thus reducing the amount of unnecessary time-slot reservation at intermediate nodes, at the expense of having to temporarily allocate the completed burst until its release. Nevertheless, in a case or another, the burst-assembly delay and the offset-time delay are overlapped, thus reducing the former and providing a more efficient early burst release.

References

1. Brackett, C.: Dense Wavelength Division Multiplexing networks: Principles and applications. IEEE JSAC 8(6), 948–964 (1990)
2. Turner, J.: Terabit burst switching. J. High Speed Networks 8, 3–16 (1999)
3. Verma, S., Chaskar, H., Ravikanth, R.: Optical Burst Switching: A viable solution for Terabit IP backbone. IEEE Network, 48–53 (November/December 2000)
4. Qiao, C., Yoo, M.: Optical Burst Switching (OBS) – A new paradigm for an optical Internet. J. High Speed Networks 8, 69–84 (1999)
5. Chen, Y., Qiao, C., Yu, X.: Optical Burst Switching: A new area in optical networking research. IEEE Network, 16–23 (May/June 2004)
6. Yoo, M., Qiao, C.: Just-Enough Time (JET): A high-speed protocol for bursty traffic in optical networks. In: Proc. IEEE/LEOS Conf. Tech. Global Info. Infrastructure, pp. 26–27 (1997)
7. Karagiannis, T., Molle, M., Faloutsos, M., Broido, A.: A nonstationary Poisson view of Internet traffic. In: IEEE INFOCOM (2004)
8. Haga, P., Diriczi, K., Vattay, G., Csabai, I.: Understanding packet pair separation beyond the fluid model: The key role of traffic. In: IEEE INFOCOM (2006)
9. Yu, X., Li, J., Chen, Y., Qiao, C.: Traffic statistics and performance evaluation in Optical Burst Switching networks. IEEE/OSA Journal of Lightwave Technology 22(12), 2722–2738 (2004)
10. Hernández, J.A., Aracil, J., López, V., López de Vergara, J.E.: On the analysis of burst-assembly delay in OBS networks and applications in delay-based service differentiation. Phot. Network Communications 14(1), 49–62 (2007)

1+X: A Novel Protection Scheme for Optical Burst Switched Networks

Ho-Jeong Yu, Kyoung-Min Yoo, Kyeong-Eun Han, Won-Hyuk Yang, Sang-Yeol Lee, and Young-Chon Kim*

Department of Computer Engineering, Chonbuk National University
664-14, Duckjin-Dong, Duckjin-Gu, Jeonju, Chonbuk 561-756, Korea
Tel.: +82-63-270-2413; Fax: +82-63-270-2394
{gutira,yckim}@chonbuk.ac.kr

Abstract. As a link failure may lead to severe burst loss in optical burst switched networks, survivability has emerged as one of the most important issues in the design of the optical networks. In this paper, we propose the 1+X protection scheme, which is a novel protection scheme that not only reduces the loss of burst caused by a link failure but also improves the efficiency of resources in OBS networks. Through numerical analysis, we confirm that the proposed 1+X protection scheme has better performance than conventional schemes in terms of reservation redundancy, end-to-end delay and resource efficiency.

1 Introduction

Optical Burst Switching (OBS) has attracted considerable much attention, as it has been classed as a very promising technology for next generation optical networks. OBS takes advantage of both high capacity of optical fibers and sophisticated control of electronics simultaneously [1]. In OBS networks, a number of data units (e.g. IP packets or ATM cells) are assembled for transmission as a burst. A burst control packet (BCP) is sent ahead and the data burst is sent after a certain offset time. The BCP is processed electronically at intermediate nodes to reserve wavelength resources and then the data burst is all optically switched without optical/electronic/optical (O/E/O) conversion.

However, as OBS employs a one-way reservation protocol, and a burst cannot be buffered at any intermediate node due to the lack of optical memory (a fiber delay line, if available at all, can only provide limited delay and contention resolution capability), burst loss is of major concern.

There are two major reasons that cause data loss in OBS networks. The first reason is that of burst contention – contention can only occur when bursts compete for the same wavelength on the same output port simultaneously. In an effort to reduce burst loss by contention, many studies, such as deflection routing [2], burst segmentation [3] and wavelength conversion [4] have been conducted. The second reason is that of link/node failure. The work in [5] discussed the possibility of providing the 1+1

* Corresponding Author.

T. Vazão, M.M. Freire, and I. Chong (Eds.): ICOIN 2007, LNCS 5200, pp. 41–49, 2008.

protection scheme for OBS networks. The 1+1 protection scheme has the advantage of fast recovery, but suffers from the disadvantage that it employs inefficient resource utilization, which wastes 50% of the resources of entire networks. To supplement these shortcomings, the double Reservation scheme was proposed [1]. However, this scheme takes a longer period of time to protect against link failure.

Therefore, a reliable protection scheme that can complement the afore-mentioned drawbacks is required. In order to achieve these objectives, we propose a novel 1+X protection scheme that can increase resource utilization while maintaining reasonable end-to-end delay. This, the 1+X protection scheme has better resource utilization than the 1+1 protection scheme and DR scheme and exhibits lower end-to-end delay than that of the DR scheme.

This paper is organized as follows. Section 2 discusses related work. In section 3, we present a novel protection scheme for OBS networks. Numerical analysis is reported in Section 4. Section 5 concludes this paper and outlines the direction for future work.

2 Related Work

OBS network protection schemes can be divided into two groups, path protection and link protection schemes. In path protection schemes, the traffic is rerouted through a backup path when a link failure occurs on the primary path. In link protection schemes, traffic is rerouted only around the failed link. In comparison of the two schemes, path protection is shown to be more efficient in resource utilization of the backup path and exhibits lower end-to-end delay for the recovered route.

These path and link protections schemes can also be dedicated or shared. In the dedicated path protection scheme, such as that employing 1+1 or 1:1 protection, resources along a backup path are dedicated for only one connection and are not shared with the backup paths for other connections. In particular, the ingress node sends two copies of data bursts through two disjoint paths towards the egress node in the 1+1 protection scheme. If failure occurs, the egress node only needs to perform switching to receive the data burst from the alternative data path. In the latter shared protection, the resources along a backup path may be shared with that of other backup paths. Thus, if failure occurs, the relative backup path will be activated and other primary paths will need to seek new backup paths. As a result, backup channels are multiplexed among different failure scenarios, which are not expected to occur simultaneously. The optical cross connects (OXCs) on backup paths cannot be configured until failure occurs if shared protection is used. The recovery time in shared protection is longer, though shared protection is more capacity-efficient when compared with dedicated protection.

To overcome the resource inefficiency of the 1+1 protection scheme and the long recovery time of shared protection, the DR scheme was proposed [1]. In the DR scheme, two disjoint paths (i.e. primary path and backup path) between the ingress node and the egress node are initially reserved. If the reservation on the primary path is successful and the burst reaches the egress node, the egress node will send out a RELEASE packet along the reserved backup path. Upon receiving the RELEASE packet, the intermediate node on the backup path will release the reserved channel to other traffic.

However, as the DR scheme uses a longer offset time on the backup path, if failure occurs, end-to-end delay is increased accordingly. Additionally, since the backup path is not released until the RELEASE packet arrives at the ingress node, resource utilization is not so much improved though it is better than that of the 1+1 protection scheme.

To overcome the shortcomings of the DR scheme, we propose the novel 1+X protection scheme. We focus on improving resource utilization as well as maintaining acceptable end-to-end delay. Therefore, the main challenge is to find the right tradeoff among existing protection schemes in terms of resource usage.

In this paper, we compare the 1+X protection scheme with the 1+1 protection scheme and DR scheme, in terms of reservation redundancy, end-to-end delay and resource utilization.

3 1+X Protection

In this section, we describe in detail the proposed 1+X protection scheme. The main idea is to improve the resource efficiency and reliability according to the partial reservation on the backup path. A process of the 1+X protection scheme is composed of the following steps.

Step 1: Partial duplex reservation step for the primary and backup path.
Step 2: Releasing step for the backup path when burst transmission is succeeded.
Step 3: Burst transmission step through the backup path when link failure occurs.

In step 1, it reserves two disjoint paths, called the primary path and backup path. Each node uses the primary path to transmit a burst. If failure occurs, it uses the backup path. These two paths can be the shortest disjoint paths. Offset time in the backup path is longer than that in the primary path. Unlike the existing protection schemes in OBS, 1+X protection only reserves X portions of the backup path, where X means a ratio of the number of reserved links to total number of links on the backup path. If and only if failure occurs, the scheme additionally it reserves the rest portions of the backup path.

Specifically, two BCPs called BCP1 and BCP2 are sent through the primary path and backup path before transmitting a burst respectively. The two transmitted BCPs will reserve the available channels for their corresponding burst. One reservation through the primary path from source node to destination node is called the primary reservation. The other reservation through the backup path from source node to specific node is called the backup reservation. The actual node located in X portions of the backup path is defined *X Node*.

For example shown in Fig.1, path 1-2-3-4 is the primary path (P1) and path 1-5-6-7-4 is the backup path (P2). As mentioned above, the source node sends BCP1 along the P1 with offset time τ, and sends BCP2 along the P2 with offset time $\tau + \delta$, simultaneously. It is noticeable that BCP2 waits for RELEASE packet or Backward Reservation Message (BRM)/Forward Reservation Message (FRM) at *X Node* without being sent to a destination node. In Fig.1, it is assumed that *X* is equal to 0.5. Therefore, if the distance between two adjacent nodes is the same, the distance between the source node and *X Node* is half that of the distance between the source node and destination node. Due to this assumption, BCP2 is only sent to node 6, which is on the half location of the P2.

Step 2 works as follows. If the reservation on the primary path succeeds, a RELEASE packet will be sent from destination node to *X Node* through the P2. Afterward, *X Node* will send the RELEASE packet towards the source node along the P2 and all intermediate nodes release the reserved resource for the corresponding burst.

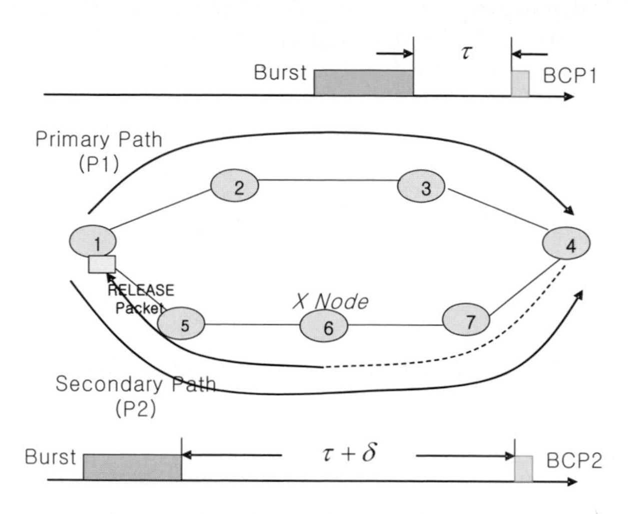

Fig. 1. Primary path reservation succeeds in 1+X(X=0.5)

As shown in the above figure, when the BCP1 along the P1 arrives at the destination node (which means the reservation on the P1 is succeeded), a RELEASE packet is generated at the destination node. We assume that every node is aware of the primary path and backup path for all source-destination pairs. Accordingly, the RELEASE packet is sent towards the *X Node,* along the P2, as soon as it is generated. The resource from destination node to *X Node* is not reserved, therefore the RELEASE packet releases only the reserved resource from the *X Node* to the source node. As a result, resource efficiency will be improved since other bursts can share the rest of the backup path by releasing the reserved resource.

The last step performs burst transmission through the backup path, when link failure occurs. As shown in Fig.2, if link failure occurs, both end nodes of that link detect the failure and then generate the Fault Information Message (FIM). Finally, the FIM is broadcast to adjacent nodes. If the source node receives the FIM, it generates the FRM and sends it to *X Node*. Accordingly, if the destination node receives the FIM, it generates the BRM and sends it to *X Node*. The FRM and BRM notify *X Node* to reserve the rest of the backup path. Thus, the BCP2 waiting at *X Node* begins to reserve the wavelength resource from *X Node* to destination node, and the burst is transmitted along the P2 simultaneously. The BCP2 has a longer offset time than the BCP1, and it is more likely that the reservation on the P2 will be successful. If the reservations on both paths fail, the burst has to be dropped.

As we can predict, although the 1+X protection scheme can not provide absolute guarantee of burst delivery, the chance that a burst is delivered successfully is higher than in conventional OBS. In addition, 1+X protection can not only improve performance but also support QoS according to changing the value of X.

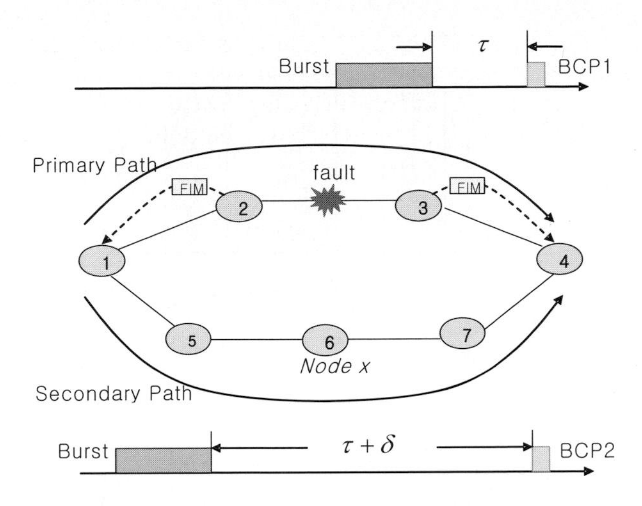

Fig. 2. Link fault occurs on primary path(X=0.5)

4 Numerical Analysis

In order to show the performance of the proposed 1+X protection scheme, we compare the performance of different OBS protection schemes in terms of reservation redundancy, end-to-end delay and resource efficiency.

Firstly, we calculate the reservation redundancy to evaluate the efficiency of resource utilization. The reservation redundancy, which is the ratio of reserved links to total links required for the primary path, is calculated as follows.

$$\text{Reservation redundancy} = \frac{\text{Total capacity of reserved resource}}{\text{Total capacity of primary path}} \quad (1)$$

Fig.3 shows the reservation redundancy of three kinds of protection schemes. It is observed that the actual amount of resources employed is smaller than the total of reserved resources if the reservation redundancy is high. As a result of calculating the reservation redundancy, the 1+1 protection and DR schemes show the worst performance because they reserve wavelength resource on the primary path as well as on the backup path. Whereas, 1+X protection partially reserves wavelength resource on the backup path, according to the value of X, i.e. reservation redundancy depends on the value of X. As shown in Fig.3, the 1+X protection scheme exhibits lowest redundancy when the value of X is equal to 0.25 and shows highest redundancy when the value of X is equal to 0.75. The above results can be summarized by the following rule. The 1+X protection always shows lower redundancy than that of the 1+1 protection scheme and DR scheme.

Secondly, we calculate the end-to-end delay versus the value of X. In order to calculate end-to-end delay, we define the notations as shown in Table 1.

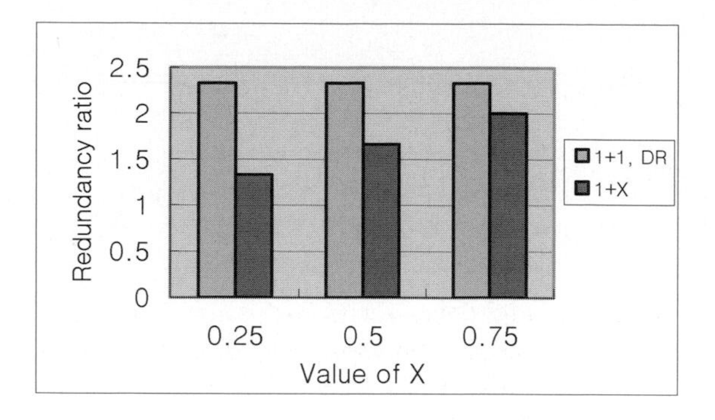

Fig. 3. Reservation redundancy

Table 1. Notations

Notation	Comments
D	Delay time
h	Number of hops
P/B	Primary/ Backup path
(s,d)	Link from source to destination node
(d,x)	Link from destination node to X Node
(s,x)	Link from source node to X Node
R	Channel capacity
Lb	Average burst size
Offset1	Offset time of primary path
Offset2	Offset time of backup path
T_p(scheme)	End to end delay time of primary path
T_b(scheme)	End to end delay time of backup path

Under the non-failure state, the end-to-end delay of the three kinds of protection schemes is the same. However, when link failure occurs, the end-to-end delay of each protection scheme is commonly calculated as the sum of offset time(of BCP2), burst size in time and burst propagation delay. In the case of 1+1 protection and the DR scheme, the burst will arrive after the offset time of BCP2. However, as the offset time of the DR scheme is longer than that of the 1+1 protection scheme, and the end-to-end delay of the DR scheme is longer than that of the 1+1 protection. The maximum end-to-end delay in the 1+X protection scheme is calculated by Eq. (2). Eq. (2) contains the delay (= processing delay + propagation delay) of BCP2, BCP2 holding time at *X Node* and burst size in time. Here, the holding time of BCP2 at *X Node* is caused by delaying BCP2 in order to provide sufficient time to recover from the failure. It contains the detection time of link failure, the broadcasting time of FIM towards source and destination nodes and the sending time of the reservation message (BRM/FRM) towards *X Node*. BCP2 holding time at *X Node* is also preset as the minimum required time period to maintain the lowest end-to-end delay.

$$D^h_{B(s,d)} + \text{BCP2 holding time} + \frac{R}{L_b} \tag{2}$$

BCP2 holding time can be calculated as follows:

$$D^h_{P(s,d)} + D^h_{B(d,x)} - D^h_{B(s,x)} \tag{3}$$

Fig. 4 shows the end-to-end delay versus the value of X. The scenario of Fig. 2 is assumed. The 1+1 protection scheme shows the shortest delay and the DR scheme shows the longest delay, since it has longer offset time on the backup path. We confirm that the performance of the proposed 1+X protection scheme is changed according to the value of X. The Performance of our scheme is worse than that of the 1+1 protection scheme but better than that of the DR scheme. Although 1+1 protection the best performance in terms of end-to-end delay, it has worst performance in terms of reservation redundancy (as shown in Fig.3) because about 50 percent of network resources are wasted.

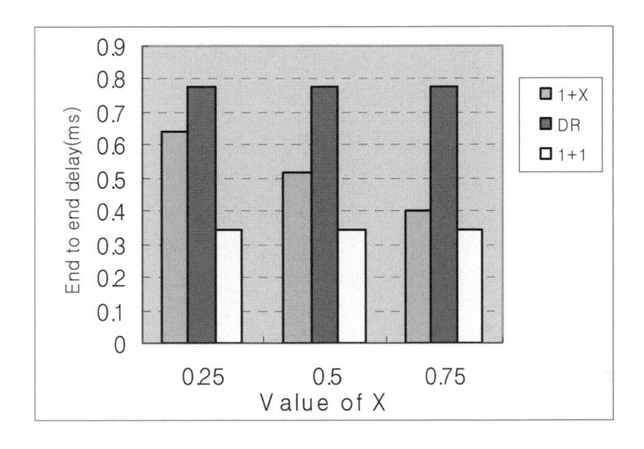

Fig. 4. End-to-end delay

Lastly, in order to observe the variation of resource reservation, we show the reservation redundancy in conjunction with end-to-end delay. Fig.5 shows the resource reservation status when link failure occurs. Three kinds of schemes reserve the same amount of resources because the burst is sent through the backup path. However, the burst arrival time differs according to the scheme. As shown in the figure, the burst of the 1+1 protection scheme is the most rapidly arrived since the gap between offset1 and offset2 is short. Offset1 and offset2 are an offset time of the primary path and backup path, respectively. The DR scheme shows the latest arrival time because the gap between offset1 and offset2 is long. The 1+X protection scheme can obtain different values of arrival time depending on *X Node*. However, it always has shorter arrival time than that of the DR scheme.

Fig. 6 shows the resource reservation status when the burst succeeds in transmitting through the primary path. Because the burst is transmitted through the primary path and the backup path simultaneously, the 1+1 protection scheme reserves

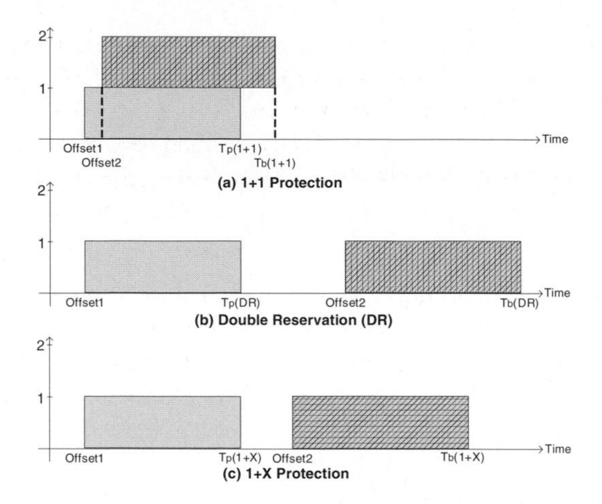

Fig. 5. Resource reservation (failure occurrence, X=0.5)

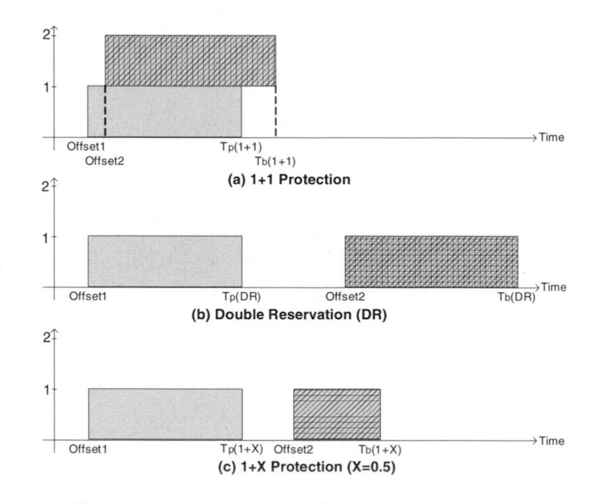

Fig. 6. Resource reservation (non-failure state)

the same amount of resources as link failure. DR scheme also reserves the primary path and the backup path simultaneously. However, due to releasing the backup path previously, the burst is not transmitted through the backup path. The main difference between the DR scheme and our proposed scheme is that our scheme reserves resources from source node to *X Node* on the backup path. Therefore, in our scheme, the amount of resources for reservation is fewer than that of the other schemes. Accordingly, the value of X is an important factor in terms of resource redundancy.

However, although the DR scheme releases the backup path previously, the releasing time is later than that of the 1+X protection scheme. Therefore, competition with the other connections on the backup path frequently occurs. That is, the reserved resource on the backup path is actually not used but occupies the available resources. Thus, it results in inefficient resource allocation.

In summary, the 1+1 protection and DR schemes waste considerable resources in the non-failure state. In contrast, our proposed scheme partially reserves resources from source to *X Node* on the backup path and then releases resources immediately when the burst transmission succeeds through the primary path. Thus, the 1+X protection scheme is a resource-efficient method not only to recover from link failure but also to maintain the lowest level of resource redundancy.

5 Conclusions

In this paper, we proposed a novel protection scheme, the 1+X protection scheme, against single link failure in optical burst switched networks. The main objective is to increase resource utilization while maintaining reasonable end-to-end delay.

The proposed 1+X protection scheme is compared with the 1+1 protection and DR schemes in terms of reservation redundancy, end-to-end delay and resource efficiency, through numerical analysis. The results showed that the 1+X protection scheme has the best performance in terms of resource efficiency, while supporting reasonable end-to-end delay. In particular, the reservation redundancy of the 1+X protection scheme is the lowest among protection schemes under the non-failure state. It is important to decrease the resource reservation redundancy under the non-failure state because the probability of link failure is very low in optical networks.

In the future, we will seek to find an appropriate reservation ratio of the backup path, X, depending on the required quality of services.

Acknowledgement

"This work was supported by KOSEF through OIRC project (No. R11-2000-074-02006-0),"

References

1. Li, J., Cao, X., Xin, C.: Double Reservation in Optical Burst Switching Networks. Proceeding of IEEE of Advances in Wired and Wireless Communication, 180–183 (April 2005)
2. Chen, Y., Wu, H., Xu, D., Qiao, C.: Performance Analysis of Optical Burst Switched Node with Deflection Routing. In: Proceedings of IEEE ICC 2003, pp. 1355–1359 (May 2003)
3. Vokkarane, V., Jue, J.P.: Prioritized Routing and Burst Segmentation for QoS in Optical Burst-Switched Networks. In: Proceedings of Optical Fiber Communication Conference, pp. 221–222 (March 2002)
4. Ramamirthan, J., Turner, J.: Design of Wavelength Converting Switches for Optical Burst Switching. In: Proceeding of INFOCOMM 2002, pp. 362–370 (June 2002)
5. Griffith, D., Lee, S.: A 1+1 Protection Architecture for Optical Burst Switched Networks. IEEE Journal on Selected Areas in Communications 21, 1384–1398 (2003)

Satisfaction-Based Handover Control Algorithm for Multimedia Services in Heterogeneous Wireless Networks

Jong Min Lee[1], Ok Sik Yang[1], Seong Gon Choi[2], and Jun Kyun Choi[1]

[1] Information and Communications University (ICU),
119 Munji-Dong, Yuseong-Gu, Daejeon 305-732, Republic of Korea
{jmlee,yos,jkchoi}@icu.ac.kr
[2] Chungbuk National University (CBNU),
12 Gaeshin-Dong, Heungduk-Gu, Chungbuk 361-763, Republic of Korea
sgchoi@chungbuk.ac.kr

Abstract. In this paper, we propose satisfaction-based Handover Control algorithm to reduce handover blocking probability as well as increase the revenue of service providers over WLAN and WAAN (Wide Area Access Network). This algorithm utilizes dynamic resource allocation to decrease the blocking probability of vertical handover connections within the limited capacity of system. Based on this algorithm, we derive the handover blocking probability as new traffic load and handover traffic load increase. In order to evaluate the performance, we compare proposed algorithm against traditional non-bounded and fixed bound schemes. Numerical results show that the proposed scheme improves handover blocking probability and increase the revenue of service providers.

1 Introduction

Over the past ten years, various wireless communication technologies (e.g. 3G cellular, IEEE 802.11 WLAN, Bluetooth) have been developed. As users who demand various multimedia applications increase, current mobile networks need to provide efficient resource management with different quality of service (QoS) requirements in the presence of heterogeneous wireless networks. In such environment, a users or network will be able to decide where to handover among the different access technologies based on the bandwidth, cost, and user preferences, application requirements and so on. Therefore, efficient radio resource management and connection admission control (CAC) strategies will be key components in such a heterogeneous wireless system supporting multiple types of applications with different QoS requirements [1].

Many admission control schemes have been proposed to enable the network to provide the desired QoS requirements by limiting the number of admitted connections to the networks to reduce or avoid connection dropping and blocking [2], [3]. However, in heterogeneous wireless networks, other aspects of admission control need to be considered due to the handover. If the wireless network is unable to assign a new channel for handover due to the lack of resources, an ongoing connection may be dropped before it finishes services as a result of the mobile user moving from its

T. Vazão, M.M. Freire, and I. Chong (Eds.): ICOIN 2007, LNCS 5200, pp. 50–59, 2008.

current place to another during handover. Since dropping an ongoing connection is generally more sensitive to a mobile user than blocking a new connection request, handover connections should have a higher priority over the new connections in order to minimize the handover blocking probability. On the other hand, reducing the blocking of handover connection by channel reservation or other means could increase blocking for the new connections. There is therefore a trade off between these two QoS measures [4]. The problem of maintaining the service continuity and QoS guarantees to the multimedia applications during handover is deteriorated by the increase of vertical handover in heterogeneous wireless networks.

In the heterogeneous wireless networks, vertical handover considers cost, user preferences, traffic characteristic, user mobility range, and so on. Generally the mobile users, who want to use two or more networks for the wide area mobility, pay more than single network users. Therefore, vertical handover should have higher priority to support QoS requirement of wide area network users because it considers more various factors (e.g. cost, bandwidth, velocity, etc.) than horizontal handover. So we proposed a flexible bound admission control scheme for vertical handover connections in heterogeneous wireless networks. By adopting this scheme, we expect the improvement of handover blocking probability and higher revenue of service providers.

This paper is organized as follows. In the next section, we describe the architecture of proposed algorithm. In section 3, we propose a dynamic bound admission control algorithm using softness profile. Numerical results obtained using the traditional methods are presented and compared in Section 4. Finally, we conclude the paper in section 5.

2 The Architecture of Proposed Algorithm

2.1 Network Architecture

Fig. 1 shows the network architecture for mobility support in heterogeneous wireless networks. There are two kinds of handover that can occur under this architecture. One is between WLAN and WAAN, the other is between WLANs: vertical handover and horizontal handover.

We assume that a user has multi-interface terminal [5]. As shown in Fig. 1, the connection initiated in WAAN and mobile node is moving to right side. When the mobile node jumps into another access area, it requires vertical or horizontal handover to continue their ongoing services. The point A, B and C indicate handover points respectively. For the implementation, Mobile IP [6] mechanism can be installed in all equipment including MNs (Mobile Nodes). Moreover, this approach should support more than one IP address for one mobile user so that one user can access more than two wireless systems simultaneously. Finally, on the top of a network, a suitable resource allocation mechanism is required to control the traffic and system load. Flexible bound resource allocation algorithm is exploited here in this architecture [7].

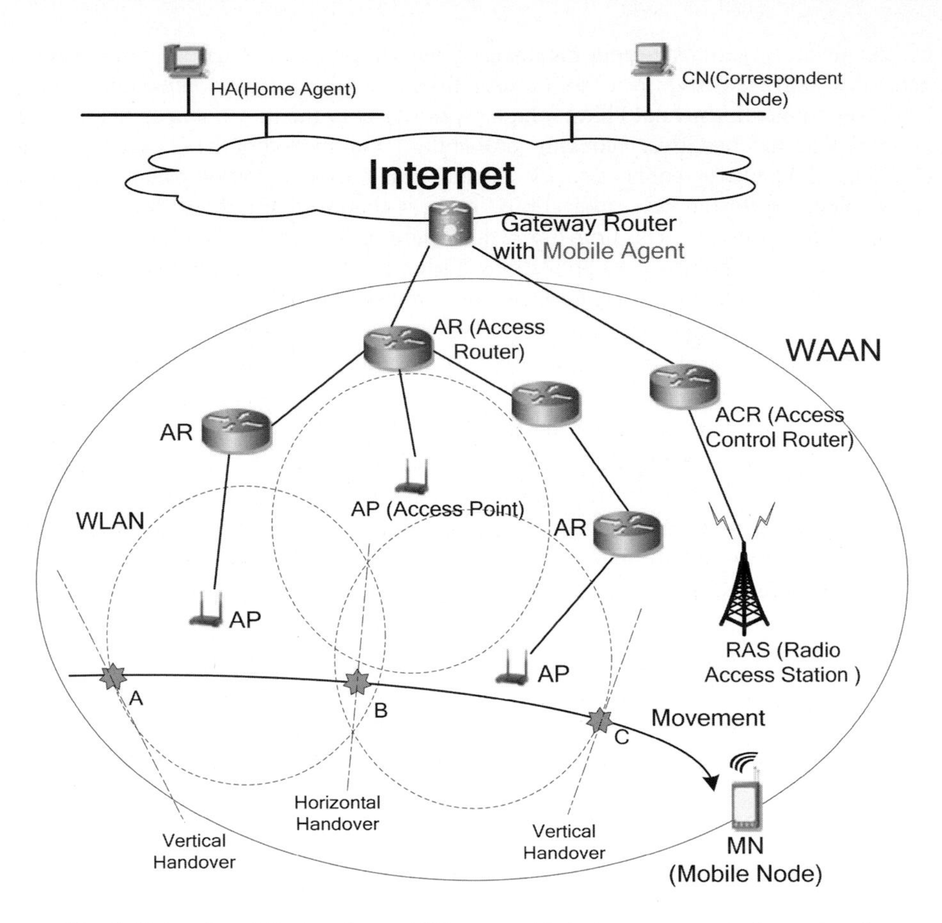

Fig. 1. Network architecture for mobility support in heterogeneous wireless networks

3 Flexible Bound Admission Control Algorithm

3.1 Softness Profile

As shown in Fig. 2, the softness profile is defined on the scales of two parameters: satisfaction index and bandwidth ratio [8]. The satisfaction index is a mean-opinion-based (MOS) value graded from 1 to 5, which is divided by two regions: the acceptable satisfaction region and low satisfaction region.

Bandwidth ratio graded from 0 to 1 can be separated by 3 regions. In the region from 1 to A, it has low degradation of satisfaction index. It means users are not sensitive in this region. However, it has large degradation of satisfaction index in the region from A to B.

The point indicated as B is called the critical bandwidth ratio (ξ) used in proposed algorithm. Since this value is the minimum acceptable satisfaction index, it can be threshold of bandwidth ratio. In the region from B to 0, users do not satisfy their

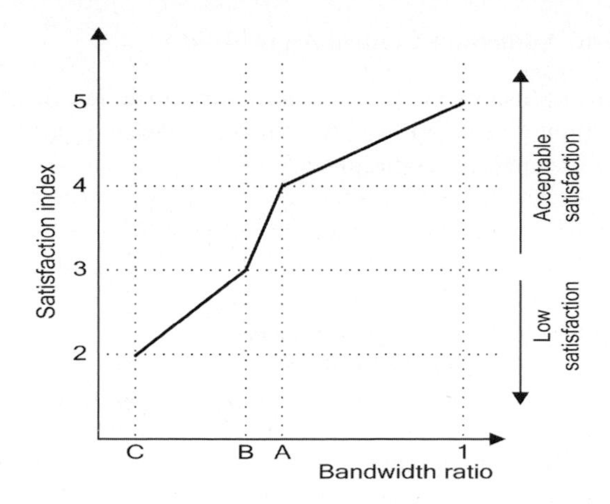

Fig. 2. Softness profile

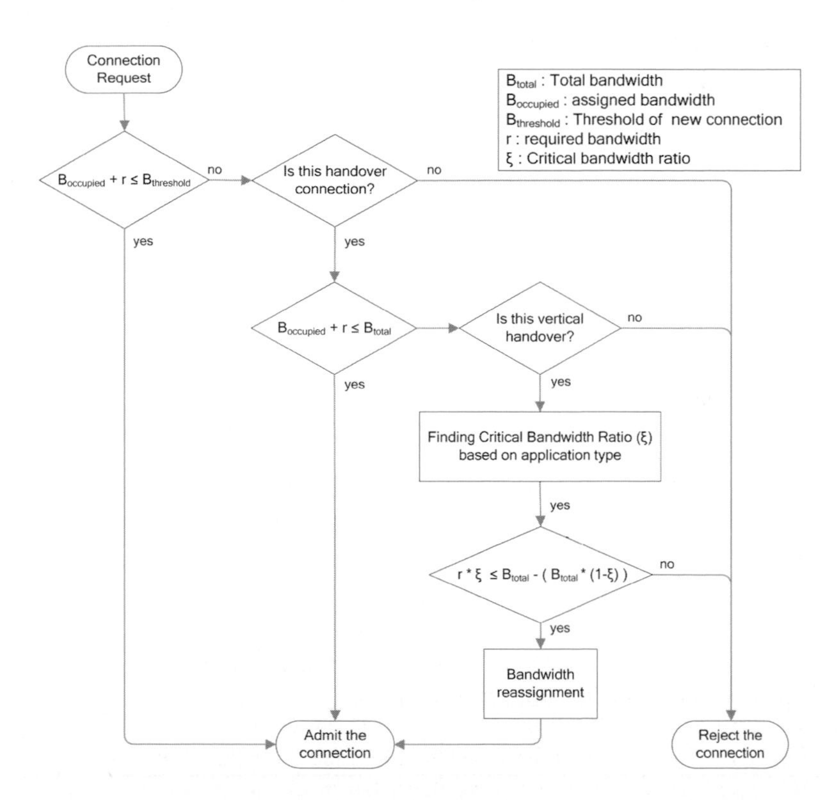

Fig. 3. Flexible Bound Admission Control Algorithm

services. Therefore, this region should not be assigned to any users. Generally, the critical bandwidth ratio (ξ) of Video On Demand (VOD) is 0.6 ~ 0.8 and back ground traffic is 0.2~0.6 [9].

3.2 Flexible Bound Admission Control Algorithm

Proposed algorithm is illustrated in Fig. 3. This algorithm shows dynamic bound handover procedure within given total bandwidth B_{total}. When a mobile node requires bandwidth for new or handover connections, mobile agent checks the available bandwidth within some threshold to decide connection admission or rejection. In this point, handover connections should be treated differently in terms of resource allocation. Since users tend to be much more sensitive to connection dropping (e.g. disconnection during VOD service) than to connection blocking (e.g. fail to initiate connection), handover connections should assign higher priority than the new connection. Especially, the vertical handover connection needs to have higher priority than horizontal handover connection because it considers more various factors (e.g. cost, bandwidth, velocity, etc.) than horizontal handover connection. In this time, if there is no available bandwidth to accept vertical handover connection, mobile agent negotiates and reassigns bandwidth by choosing critical bandwidth ratio (ξ). As a result, the vertical handover connections can be accepted more than horizontal handover connections.

3.3 System Model and Numerical Analysis

Fig. 4 indicates the transition diagram for the proposed scheme. The detailed notations used in this diagram are shown in Table. 1.

Table 1. Notations

Notation	explanation
λ_n	Arrival rate of new connection
λ_h	Arrival rate of horizontal handover connection
λ_v	Arrival rate of vertical handover connection
$1/\mu_h$	Average channel holding time for handover connections
$1/\mu_n$	Average channel holding time for new connections
C	Maximum number of server capacity
T	Threshold (bound of the total bandwidth of all accepted new connections)
ξ	Critical bandwidth ratio
α	$\lfloor (1-\xi)*C \rfloor$
n_n	Number of new connections initiated in the coverage
n_{hv}	Number of handover connections in the coverage

In order to analyze the blocking probability of each connection, we use the two-dimensional Markov chain model with the state space S and M/M/C+α/C+α [10] model is utilized.

$$S = \{(n_n, n_{hv}) | 0 \le n_n \le T, (n_n + n_{hv}) \le C + \alpha\} \tag{1}$$

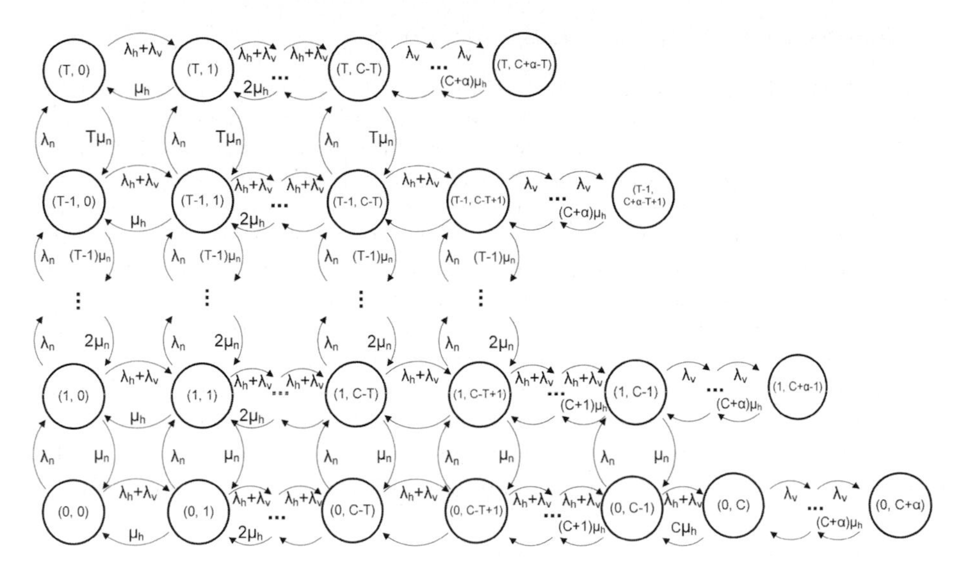

Fig. 4. Transition diagram for the proposed algorithm

Let $q(n_n, n_{hv}; \overline{n_n}, \overline{n_{hv}})$ denote the probability transition rate from state (n_n, n_{hv}) to $(\overline{n_n}, \overline{n_{hv}})$. Then, we obtain the below.

$$
\begin{aligned}
&q(n_n, n_{hv}; n_n + 1, n_{hv}) = \lambda_n && (0 \le n_n < T, 0 \le n_{hv} \le C) \\
&q(n_n, n_{hv}; n_n - 1, n_{hv}) = n_n \mu_n && (0 < n_n \le T, 0 \le n_{hv} \le C) \\
&q(n_n, n_{hv}; n_n, n_{hv} + 1) = \lambda_h + \lambda_v && (0 \le n_n \le T, 0 \le n_{hv} < C) \\
&q(n_n, n_{hv}; n_n, n_{hv} - 1) = n_{hv} \mu_h && (0 \le n_n \le T, 0 < n_{hv} \le C) \\
&q(n_n, n_{hv}; n_n, n_{hv} + 1) = \lambda_v && (0 \le n_n \le T, C \le n_{hv} < C + \alpha) \\
&q(n_n, n_{hv}; n_n, n_{hv} - 1) = n_{hv} \mu_h && (0 \le n_n \le T, C < n_{hv} \le C + \alpha)
\end{aligned}
\tag{2}
$$

Let $p(n_n, n_{hv})$ denote the steady-state probability that there are new connections (n_n) and vertical and horizontal handover connections (n_{hv}). By using the local balance equation [10], we can obtain

$$
\begin{aligned}
&p(n_n, n_{hv}) = \frac{\rho_n{}^{n} \cdot \rho_h{}^{hv} \cdot p(0,0)}{n_n! \cdot n_{hv}!} \\
&\qquad where\ (0 \le n_n \le T, 0 \le n_{hv} < C), \rho_n = \frac{\lambda_n}{\mu_n}, \rho_h = \frac{\lambda_h + \lambda_v}{\mu_h} \\
&p(n_n, n_{hv}) = \frac{\rho_n{}^{n} \cdot \rho_h{}^{C} \cdot \rho^{hv-C}\ p(0,0)}{n_n! \cdot n_{hv}!} \\
&\qquad where\ (0 \le n_n \le T, C \le n_{hv} \le C + \alpha), \rho = \frac{\lambda_n}{\mu_n}, \rho = \frac{\lambda_v}{\mu_h}
\end{aligned}
\tag{3}
$$

From the normalization equation, we also obtain

$$p(0,0)=\left[\sum_{0\le n_n\le T,n_n+n_{hv}\le C+\alpha}\frac{\rho_n^{\;n_n}}{n_n!}\cdot\frac{\rho_h^{\;n_{hv}}}{n_{hv}!}\right]^{-1}$$
$$=\left[\sum_{n_n=0}^{T}\frac{\rho_n^{\;n_n}}{n_n!}h\cdot\sum_{n_{hv}=0}^{C-n_n}\frac{\rho_h^{\;n_{hv}}}{n_{hv}!}+\sum_{n_n=0}^{T}\frac{\rho_n^{\;n_n}}{n_n!}\cdot\sum_{n_{hv}=C}^{(C+\alpha)-n_n}\frac{\rho_h^{\;n_{hv}}\cdot\rho_h^{\;(n_{hv}-C)}}{n_{hv}!}\right]^{-1} \tag{4}$$

From this, we obtain the formulas for new connection blocking probability and handover connection blocking probability as follows:

$$P_{nb}=\frac{\sum_{n_{hv}=0}^{C-T}\frac{\rho^T}{T!}\cdot\frac{\rho^{n_{hv}}}{n_{hv}!}+\sum_{n_n=0}^{T-1}\frac{\rho^{n_n}}{n_n!}\cdot\frac{\rho_h^{\;C-n_n}}{(C-n_n)!}}{P(0,0)}$$
$$P_{hb}=\frac{\sum_{n_n=0}^{T}\frac{\rho^{n_n}}{n_n!}\cdot\frac{\rho_h^{\;C-n_n}}{(C-n_n)!}}{P(0,0)} \tag{5}$$
$$P_{vb}=\frac{\sum_{n_n=0}^{T}\frac{\rho^{n_n}}{n_n!}\cdot\frac{\rho_h^{\;(C+\alpha)-n_n}}{((C+\alpha)-n_n)!}}{P(0,0)}$$

Next, we introduce the cost model defined by the 'reward' and 'penalty' value per each connection to evaluate the revenue [11].

$$\text{revenue }=\sum_{i=1}^{C}i\mu\times R\times\frac{\frac{\rho^i}{i!}}{\sum_{j=0}^{C}\frac{\rho^j}{j!}}-L\times\lambda\times\frac{\frac{\rho^C}{C!}}{\sum_{j=0}^{C}\frac{\rho^j}{j!}} \tag{6}$$

where R : reward of a connection served successfully

L :loss (Penalty) of a connection blocked.

4 Numerical Results

In this section, we present the numerical results for the comparison of performance. We compared three bounding schemes: non-bound, fixed bound and proposed algorithms. Fig. 5 shows handover blocking probability under the following parameters: C=30, T=15, λ_n = 1/30, λ_h =1/60, λ_v =1/60, μ_h =1/450, μ_n is varying from 1/800 to 1/100, and ξ =0.9.

In Fig. 5 handover connection traffic load is given as ρ_h =15. It is observed that when traffic load of the handover connection is higher than the new connection traffic load (e.g. $\rho_h > \rho$), non-bound scheme and fixed bound scheme are not much different. On the other hand, when traffic load of the handover connection is lower than the new connection traffic (e.g. $\rho_h < \rho$), their blocking probabilities become different. This is the case when the new connections arrive in burst (say after finishing a class) [12]. Our proposed algorithm can offer the seamless handover for ongoing connections with lower blocking probability.

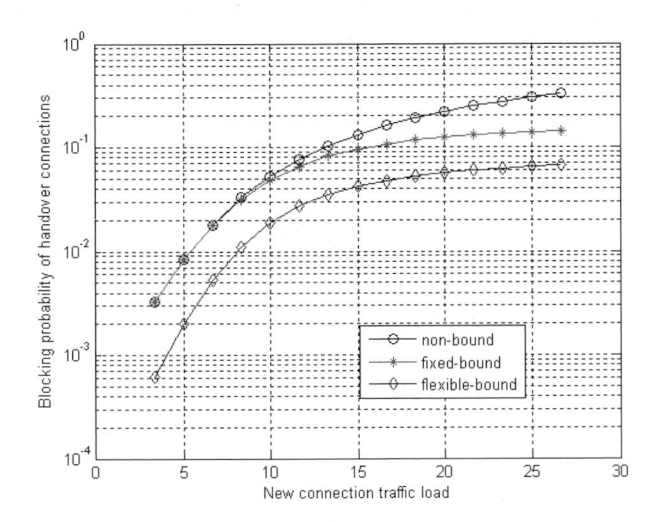

Fig. 5. Blocking probability of handover connections vs. new connection load

In Fig. 6, we increase the handover connection traffic load (ρ_h). This graph shows the blocking probability of handover connection under the following parameters: C=30, T=20, λ_n = 1/20, λ_h =1/60, λ_v =1/60, μ_n =1/300, μ_h is varying from 1/100 to 1/1200, and ξ =0.9.

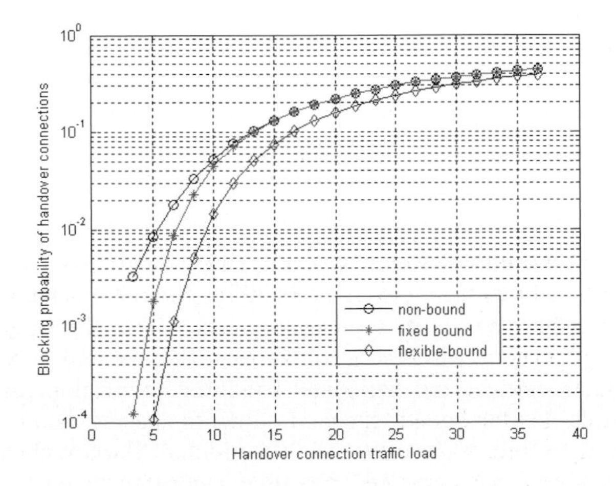

Fig. 6. Blocking probability of handover connections vs. handover connection load

In this case, traffic load of handover connections (ρ_h) are increasing from 0 to 40 and traffic load (ρ_n) of the new connection is 15. Since handover connections are not bounded, as increasing the handover traffic load, the differences among three schemes become similar.

Now we evaluate the revenue of each connection in the aspect of service provider. Fig. 7shows the numerical results obtained by equation (7). This result shows the revenue of each connection under the following parameters: C=30, T=20, λ_h =1/60, λ_v =1/60, μ_n =1/300, μ_h =1/450, λ_n is varying from 1/25 to 1/10, ξ =0.9 and (R : L = 2 : 1).

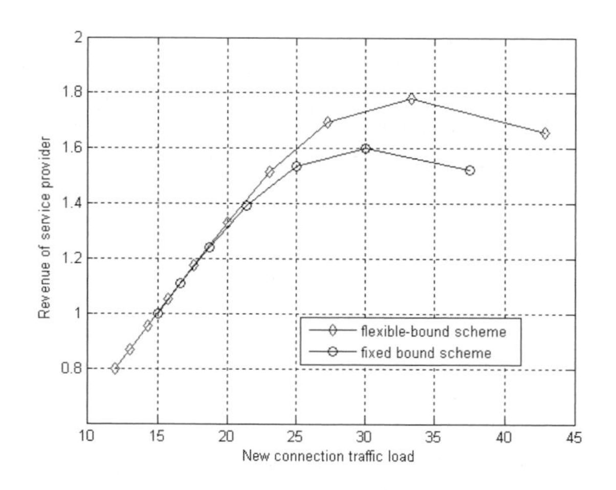

Fig. 7. Revenue of service provider

Fig. 7 shows that when traffic load is light, these schemes are not much different and linearly increase. However, as the traffic load increases the revenue of proposed resource allocation algorithm is higher than fixed bound scheme. As mentioned before, this is the case that the new connections arrive in burst (say after finishing a class). In such case, our proposed algorithm can offer the higher revenue.

5 Conclusion

In this paper, we proposed satisfaction based flexible bound resource allocation algorithm that not only reduces the blocking probability of vertical handover connections, but also increases the revenue of service providers within the limited capacity of system over heterogeneous wireless networks (e.g. WLAN and WAAN). Proposed algorithm considers vertical and horizontal handover connections that have higher priority by utilizing flexible bound scheme. In order to analyze the blocking probability of proposed algorithm, we use the two-dimensional Markov chain model. From the numerical analysis, we compared proposed algorithm against traditional non-bound and fixed bound scheme. As a result, proposed scheme is able to improve handover blocking probability in heterogeneous wireless networks. In addition, by reducing the handover blocking probability, we can also achieve the higher revenue of providers. Future work needs to analyze the optimized critical bandwidth ratio based on the number of user arrivals and revenue.

Acknowledgement. This work was supported in part by the MIC, Korea under the ITRC program (ITAC1090060300350001000100100) supervised by the IITA and the KOSEF under the ERC program.

References

1. Niyato, D., Hossain, E.: Call Admission Control for QoS Provisioning in 4G Wireless Networks: Issues and Approaches. IEEE Computer Society Press, Los Alamitos (2005)
2. Katzela, I., Naghshineh, M.: Channel assignment schemes for cellular mobile telecommunication systems: A comprehensive survey. IEEE Personal Commun. 3, 10–31 (1996)
3. Fang, Y., Zhang, Y.: Call Admission Control Schemes and Performance Analysis in Wireless Mobile Networks. IEEE Transactions on Vehicular Technology 51(2), 371–382 (2002)
4. Yavuz, E.A., Leung, V.C.M.: A Practical Method for Estimating Performance Metrics of Call Admission Control Schemes in Wireless Mobile Networks. IEEE WCNC, 1254–1259 (March 2005)
5. Buddhikot, M., et al.: Design and Implementation of a WLAN/CDMA 2000 Interworking Architecture. IEEE Communications Magazine (November 2003)
6. Johnson, D., Perkins, C., Arkko, J.: Mobility Support for IPv6, RFC 3775 (June 2004)
7. Cheng, S.-T., Lin, J.-L.: IPv6-Based Dynamic Coordinated Call Admission Control Mechanism Over Integrated Wireless Networks. IEEE Journal on Selected Areas in Communications 23(11), 2093–2103 (2005)
8. Reininger, D., Izmailov, R.: Soft quality-of-service control for multimedia traffic on ATM networks. In: Proceedings of IEEE ATM Workshop, pp. 234–241 (1998)
9. Kim, S.H., Jang, Y.M.: Soft QoS-Based Vertical Handover Scheme for WLAN and WCDMA Networks Using Dynamic Programming Approach. In: Lee, J.-Y., Kang, C.-H. (eds.) CIC 2002. LNCS, vol. 2524, pp. 707–716. Springer, Heidelberg (2003)
10. Kleinrock, L.: Queueing System, Theory, vol. 1. John Wiley and Sons, New York (1975)
11. Chen, I.R., His, T.-H.: Performance analysis of admission control algorithms based on reward optimization for real-time multimedia servers. In: The International Journal Performance Evaluation, vol. 33, pp. 89–112 (March 1998)
12. Hou, J., Fang, Y.: Mobility-based call admission control schemes for wireless mobile networks, Wireless Communications and Mobile Computing. Wirel. Commun. Mob. Comput. 1, 269–282 (2001)

Adaptive Qos-Aware Wireless Packet Scheduling in OFDMA Broadband Wireless Systems

Kwangsik Myung[1], Seungwan Ryu[1,2], Byunghan Ryu[2], Seungkwon Hong[3], and Myungsik Yoo[4]

[1] Deaprtment of Information Systems, Chung-Ang University, Korea
[2] Broadband Mobile MAC team, ETRI, Korea
[3] Department of Industrial and Management Engineering, ChungJu Univ.,
[4] Department of Telecom. and Electronic Engineering, Soongsil Univ.

Abstract. In this paper, we propose the *urgency and efficiency based wireless packet scheduling (UEPS)* algorithm designed not only to support multiple users simultaneously but also to serve real time (RT) and non-real time (NRT) traffics for a user at the same time. The design goal of the UEPS algorithm is to maximize throughput of NRT traffics with satisfying QoS requirements of RT traffics. We also developed two packet loading methods, *an optimized* and *a fast implementation-based methods*. Simulation study shows that there is no significant difference on performance of the UEPS algorithm under two packet loading methods.

1 Introduction

Challenges on delivering QoS to users in packet based wireless networks have been watched with keen interest, and the packet scheduler operates at the medium access control (MAC) layer is considered as the key component for QoS provisioning to users. There are many existing packet scheduling algorithms designed to support data traffics. For example, Proportional Fair (PF) [1] and Modified-Largest weighted delay first (M-LWDF) [2] algorithms are designed mainly to support NRT data services in CDMA-1x-EVDO (HDR) system.

In general, QoS requirements of RT and NRT traffics are different each other. RT traffics such as the voice and the video streaming traffics require a low and bounded delay but can tolerate some information loss. Thus, it is imperative for RT traffics to meet delay and loss requirements. In contrast, NRT data traffics require low information loss but less stringent delay requirements compared to the RT traffics. In addition, since the amount of NRT data traffics to be transmitted is much larger than that of RT traffic data, throughput maximization is the main performance measure for NRT data traffic. As a result, performance objectives of RT and NRT traffics to be achieved within a scheduler are conflicting each other.

In this paper, we propose *an urgency and efficiency based packet scheduling (UEPS)* algorithm that allows multiple users to receive packets at any given scheduling time instant. The idea behind UEPS algorithm is to maximize throughput of NRT traffics as long as QoS requirements of RT traffics such

T. Vazão, M.M. Freire, and I. Chong (Eds.): ICOIN 2007, LNCS 5200, pp. 60–69, 2008.

as the packet delay and the loss rate requirements are satisfied. In addition, two packet loading methods, *an optimized* and *a fast implementation-based methods*, on the sub-channel of the selected users are proposed.

This paper is organized as follows. In the next section, we introduce the OFDMA wireless system model and the structure of the UEPS algorithm. In section 3, we discuss concepts of the urgency of scheduling and the efficiency of radio resource usage. In section 4, we proposed the UEPS algorithm followed by two packet loading methods. In section 5, we evaluate performance of the UEPS algorithm via simulation study. Finally, we summarize this study.

2 System Model

We consider an OFDMA system with 20MHz of bandwidth and $100\mu s$ of OFDM symbol duration. The frame and slot periods are assumed to be 12ms and 1ms respectively. It is assumed that there are 1,536 subcarriers, and all subcarriers are shared by all users in a cell in terms of sub-channels, a subset of the subcarriers. We assume that there are 12 sub-channels and each sub-channel is a group of 128 subcarriers. It is also assumed that all subcarriers are used for data transmission for simplification, and subcarriers in each sub-channel are selected by a pre-determined random pattern. The modulation and coding scheme is determined by the prescribed adaptive modulation code (AMC) table based on the instantaneous signal-interference-ratio (SIR) of each sub-channel.

The proposed packet scheduling system in a base station (BS) consists of three blocks: a packet classifier (PC), a buffer management block (BMB), and a packet scheduler (PS). The packet classifier classifies incoming packets according to their userID, traffic types and QoS profiles, and sends them to a user's sub-BMB in BMB. The BMB maintains 128 sub-BMBs to support up to 128 users, and each sub-BMB maintains 4 buffers to store packets of 4 different traffic types seperately. Each sub-BMB maintains QoS statistics such as the arrival time and delay deadline of each packet, the number of packets, and the head-of-line (HOL) delay in each buffer. Finally, the PS transmits packets to users according to the scheduling priority obtained using QoS statistics and channel status reported by user equipments.

3 Scheduling with Time Constraints

3.1 The Urgency of Scheduling

A time-utility function (TUF) of a delay-sensitive RT traffic can be expressed as a hard time-utility, in that utility of an RT traffic drops abruptly to zero when the delay passes its deadline. On the other hand, TUF of an NRT traffic is a continuously decreasing function in delay, in that utility of an NRT traffic decreases slowly as delay increases. Among NRT traffics some has a (soft) deadline like WWW traffics. On the other hand, some NRT traffics such as email and FTP traffics have much longer deadline or no deadline.

The unit change of TUF value at any time instant indicates the urgency of scheduling of packets as time passes by. Let $U_i(t)$ be the TUF of a HOL packet of traffic i at time t. Then the unit change of TUF value of the packet at time t is the absolute value of the first derivative of $U_i(t)$, i.e., $|U_i'(t)|$, at time t. A possible packet scheduling rule is to select a packet among HOL packets based on $|U_i'(t)|$, $\forall i \in I$.

Since the downlink between a BS and UEs is the last link to users, the end-to-end delay can be met as long as packets are delivered to UEs within the deadline. Hence the time interval of an RT traffic packet from its arrival time to its deadline, $[a_i, D_i] = [a_i, a_i + d_i]$, can be divided into two sub-intervals, $[a_i, D_i - j_i)$ and $[D_i - j_i, D_i]$ (so called *the marginal scheduling time interval (MSTI)*), by introducing a negative jitter from its deadline, where a_i, D_i, d_i and $0 \leq j_i < d_i$ are the arrival time, the delay deadline, the maximum allowable delay margin of the packet of an RT traffic i, and the delay jitter respectively. Then, packet of an RT traffic i is transmitted only during the time interval $[D_i - j_i, D_i]$, and NRT packets are transmitted during the remaining time interval, $[a_i, D_i - j_i)$.

To schedule the RT traffic packet during MSTI, a non-zero value is assigned to $|U_i'(t)|$ for this time interval and 0 for the remaining time interval. However, since the TUF of an RT traffic is a hard and discontinuous function in delay, the unit change of the utility, $|U_i'(t)|$, can not be obtained directly at its delay deadline. To address this problem, the TUF of an RT traffic can be relaxed into a continuous *z-shaped* function which has properties similar to the original hard discontinuous function. A z-shaped function relaxation of the TUF of an RT traffic can be easily obtained analytically using an s-shaped function having close relation with z-shaped function. For example, a z-shaped function can be obtained using the s-shaped sigmoid function, $f_{Sigmoid}(t, a, c) = 1/(1+e^{-a(t-c)})$, where a and c are parameters that determine slope and location of the inflection point of the function. Then, the relaxed z-shaped TUF function is $U_{RT}(t) = 1 - f_{Sigmoid}(t, a, c) = e^{-a(t-c)}/(1 + e^{-a(t-c)})$, and the unit change of utility of a RT traffic at the inflection point $(t = c)$ is $|U_{RT}'(t = c)| = a/4$. This value is assigned as the urgency factor of an RT traffic packet during MSTI.

Since TUFs of NRT traffics are monotonic decreasing functions in time (delay), an analytic model can be easily obtained using related monotonic increasing functions. For example, a truncated exponential function, $f(a_i, t, D_i) = exp(a_i t)$, can be used, where a_i is an arbitrary parameter and $D_i \geq t \geq 0$ is the delay deadline of an NRT traffic i. Then a possible TUF of an NRT traffic i is $f_{NRT_i}(t) = 1 - f(a_i, t, D_i) = 1 - exp(a_i t)/exp(D_i)$[1], and the urgency is $|U_{NRT_i}'(t)| = a_i exp(a_i t)/exp(D_i)$.

The urgency factor, $|U_i'(t)|$, of each traffic type is used to determine scheduling precedence among HOL packets, and choice of these values for each traffic type is dependent on designer's preference. A rule of thumb is to give RT traffics a higher scheduling precedence over NRT traffics. In this paper, we set the urgency factors of RT voice, RT video, and NRT traffics as follow.

$$|U_{RT-Voice}'(t)| > |U_{RT-Video}'(t)| > |U_{NRT-Data1}'(t)| > |U_{NRT-Data2}'(t)| \quad (1)$$

[1] It is normalized by the maximum time, D_i, so that it can have smoother slope.

3.2 Efficiency of Radio Resource Usage

Efficiency in wireless communications is related to usage of the limited radio resources, i.e., the limited number of radio channels or limited bandwidth. Thus the channel state of available radio channels can be used as an efficiency indicator. For example, the current channel state ($R_i(t)$), the average channel state ($\overline{R_i}(t)$) or the ratio of the current channel state to the average ($R_i(t)/\overline{R_i}(t)$) can be used as an efficiency indicator. In this study, a moving average of the channel state of each user $i \in M$ in past W timeslots, $\overline{R_i}(t) = (1-1/W)\overline{R_i}+(1/W)R_i(t)$, is used for the average channel state, where W is the time window used in calculation of the moving average of the channel state. Note that $\overline{R_i}(t)$ used in our paper is different from the average throughput of user i, $T_i(t)$, in past t_c timeslots used in PF algorithm [1]. Therefore the higher the user's instantaneous channel quality relative to its average value, the higher the chance of a user to transmit data with a rate near to its peak value.

4 UEPS Algorithm

4.1 The Proposed UEPS Algorithm

The UEPS scheduler transmits NRT traffics during the time interval $[0, d_i - j_i)$, assuming that the packet is arrived at time 0 and channel states of all users are the same. In contrast, the scheduler gives an RT traffic a higher scheduling priority over NRT traffics during MSTI, $[d_i - j_i, d_i]$.

The UEPS algorithm operates at a BS in three steps, STEP 0 for packet arrival events, STEP 1 for scheduling priority of each user and STEP 2 for scheduling and transmission of packets.

- In **STEP 0**, the arrived packet is sent to a user's BMB by the packet classifier based on its userID. There are four sub-buffers in a user's BMB where the arrived packet is stored in a sub-buffer of a user's BMB according to its traffic type. QoS profiles of the arrived packet such as the arrival time, the deadline, the packet type, and the packet size are maintained user i's BMB.
- In **STEP 1**, at each scheduling instant the urgency factor of HOL packets of each sub-buffer, $|U'_{ij_k}(t)|$, is calculated to get a descending ordered index set of each packet type, $J_i = \{j_1, j_2, j_3, j_4\}$ where $j_k, k = 1, 2, 3, 4$ is the index of each packet type. Then the highest urgency factor becomes the representative urgency factor of user i, i.e., $|U'_i(t)| = |U'_{ij_1}(t)|$. In addition, the efficiency factor of the user i, $\bar{R}_i(t) = \bar{R}_i(t-1)(1-1/W)+R_i(t)/W$, is obtained. Finally, the scheduling priority value of the user i is $p_i(t) = |U'_i(t)| * (R_i(t)/\bar{R}_i(t))$.
- In **STEP 2**, at each scheduling time instant, multiple users are selected based on their scheduling priority value obtained as follow

$$i^* = \arg\ \max_{i \in I} |U'_i(t)|(R_i(t)/\overline{R_i}(t)) \quad (2)$$

  Then, a sub-channel is allocated to each selected user i^*. The OFDMA system considered in this study is designed to support up to 12 users simultaneously at each scheduling time instant by allocating one of 12 sub-channels to each of selected users. The capacity of each allocated sub-channel is determined from the AMC option. Finally, the scheduler loads user i^*'s packets on the sub-channel as much as possible when there is room. There are packets of 4 different traffic types stored in 4 sub-buffers in user i^*'s BMB, and these packet are loaded on the sub-channel based on the ordered index set of their urgency factors, $J_i = \{j_1, j_2, j_3, j_4\}$, obtained in STEP 1. In this study, we consider two packet loading methods, an optimized packet loading and a fast implementation-based packet loading.

4.2 An Optimized and A Fast Implementation-Based Packet Loading Methods

Suppose that user i is selected by the scheduler for packet transmission. Then a sub-channel is allocated to the user i, and its capacity, C_i, is determined from AMC option. How to load user i's packets on the allocated sub-channels with the limited channel capacity C_i? Since there are up to 4 different traffic types in user i's BMB, the packet loading problem is how to select packets from 4 sub-buffers in user i's BMB. One possible packet loading method is to use the urgency factor of HOL packets of each sub-buffers in user i's BMB. In this study, we investigate two packet loading methods, an optimized and a fast implementation-based packet loading methods, which use the ordered index set of the urgency factor of each traffic type, $J_i = \{j_1, j_2, j_3, j_4\}$, obtained in STEP1.

An Optimized Packet Loading Method. In this study, we assume that there are 4 different traffic types, RT voice, RT video, NRT email, and NRT WWW traffics, and packets of each traffic type are stored in separate sub-buffers in user i's BMB. In the optimized loading, a packet is selected among four HOL packets of the sub-buffers based on the urgency factors of them. When a packet having the highest urgency factor is picked and loaded on the sub-channel, this packet is removed from the HOL, and the next packet waiting in this sub-buffer moves to the HOL position. Then the scheduler picks a packet by comparing urgency factors of 4 HOL packets, and load it on the sub-channel next to the previously loaded packet. This procedure is repeated until the sub-channel is filled. With this packet loading procedure packets of different traffic type are loaded optimally according to the order of the urgency factor.

As shown in figure 1(a), an initial set of HOL packets of sub-buffers is $H_0 = \{p_{j_1}(1), p_{j_2}(1), p_{j_3}(1), p_{j_4}(1)\}$. If $p_{j_1}(1)$ is picked by the scheduler and loaded on the sub-channel, then it is removed from the sub-buffer of type j_1. The resulting new set of HOL packets is $H_1 = \{p_{j_1}(2), p_{j_2}(1), p_{j_3}(1), p_{j_4}(1)\}$. If $p_{j_2}(1)$ has the highest urgency factor in H, then it is loaded on the sub-channel next to $p_{j_1}(1)$ and removed from the HOL of the sub-buffer of type j_2. Then the new set of HOL packets is $H_2 = \{p_{j_1}(2), p_{j_2}(2), p_{j_3}(1), p_{j_4}(1)\}$. This procedure is repeated until the sub-channel is filled or all packets in the user i's BMB are loaded. Figure 1(b) describes an example of the optimized packet loading procedure.

A Fast Implementation-Based Loading. In the optimized loading, packets are loaded on the sub-channel optimally based on the order of the urgency factors

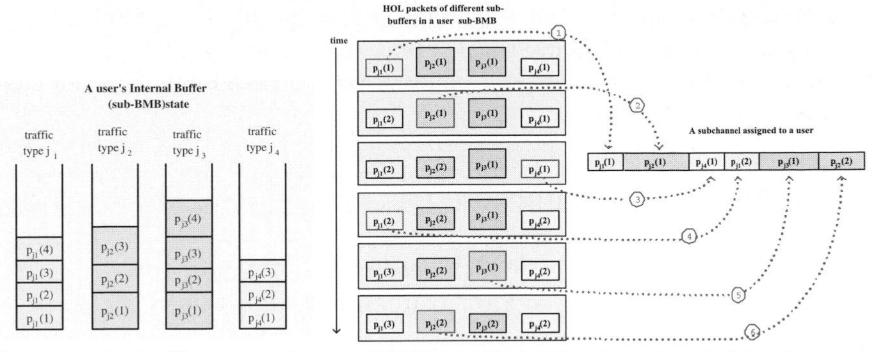

(a) A generic buffer state of a user

(b) The optimization based packet loading

Fig. 1. A generic buffer state of a user having 4 sub-buffers for each traffic type (left) and An example of the optimization based packet loading on a sub-channel

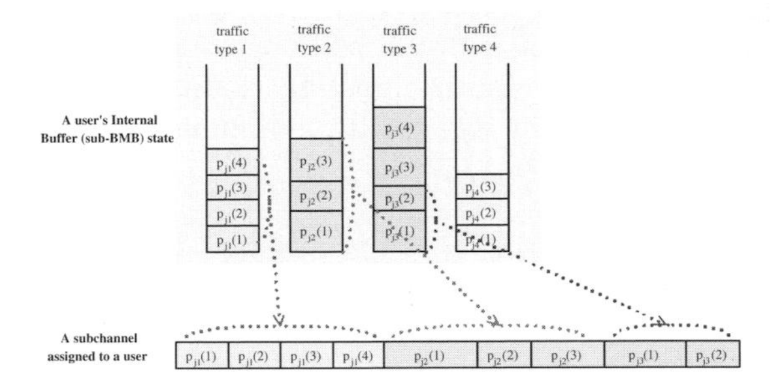

Fig. 2. An example of the Implementation based packet loading on a sub-channel

regardless of traffic types. In this case, to pick a packet and load it on the sub-channel the scheduler must compare urgency factors of 4 HOL packets. And this procedure is repeated until the sub-channel is filled or all packets in the user i's BMB are loaded. In fact, it is impossible to finish this procedure for all 12 selected users within 1ms which is the unit scheduling time interval.

Therefore, we develop *a fast implementation-based packet loading method* to load packets of 12 users on the 12 sub-channels within 1ms. In this method the ordered index set J_i obtained in STEP 1 is used to select a traffic type having the highest urgency factor of the HOL packet. Then packets of the selected traffic type waiting in a sub-buffer are loaded together as many as possible. This procedure is repeated until the sub-channel is filled or all packets in the user i's BMB are loaded.

Figure 2 shows an example of the fast implementation-based packet loading. Since the ordered index set $J_i = \{j_1, j_2, j_3, j_4\}$ means that the HOL packet of traffic type j_1, $p_{j_1}(1)$, has the highest urgency factor among all 4 HOL packets $H_0 = \{p_{j_1}(1), p_{j_2}(1), p_{j_3}(1), p_{j_4}(1)\}$, packets in the sub-buffer of type j_1, i.e., $p_{j_1}(1)$, $p_{j_2}(2)$, $p_{j_3}(3)$, and $p_{j_4}(4)$, are loaded on the sub-channel (see figure 1(a)).

Then 3 packets waiting in the sub-buffer of type j_2, $p_{j_2}(1)$, $p_{j_2}(2)$, and $p_{j_2}(3)$, are loaded on the sub-channel. Finally, only two packets, $p_{j_3}(1)$ and $p_{j_3}(2)$, are loaded on the channel from the sub-buffer of type j_3 because there is not enough room on the sub-channel.

5 Performance Evaluation

5.1 Traffic Types and System Parameters

In the simulation study, it is assumed that there are four different traffic types, and each user generates one of four traffics. *RT voice* is assumed to be the voice over IP (VoIP) traffic modeled as a 2-state Markov (ON/OFF) model. The length of the ON and OFF periods follow the exponential distribution with mean of one second and 1.35 seconds respectively. *RT video* is assumed to be the RT video steaming service that periodically generate packets of variable sizes. We uses 3GPP streaming video traffic for this type of traffic[4]. For *NRT data service type 1*, The WWW model is used, in that a session is assumed to be consisted of several web pages containing multiple packets or datagrams. Characteristics of WWW traffic model are summarized in table 1. Best effort such as emailing traffic is used for the *NRT data type 2* with assuming that messages arrival to the mailboxes is modelled by Poisson process.

Table 1. A summary of characteristics of WWW traffic model

Variables	Levels (mm)
Width of the target (W_t)	2, 4, 6, 8
Width of path (W_p)	4, 6, 10
Length of path (A)	29, 58, 116

We consider a hexagonal cell structure consisting of a reference cell and 6 surrounding cells with 1 km of radius. We assume that all cells use omni-directional antenna. Mobile stations are uniformly distributed in a cell, and move with velocity of an uniform distribution between 3 and 100 km/second in a random direction. The BS transmission power is 12W which evenly distributed to all 12 sub-channels. We also use 3GPP path loss model $L = 128.1 + 37.6log_{10}R$ [4].

5.2 Performance Evaluation

Performance of the UEPS algorithm under various traffic loads is evaluated via extensive simulation study. We generate 4 packets having the same userID simultaneously, and each packet corresponds to one of 4 traffic types, voice, video, WWW and email. As a result, when a userID is generated, 4 packets of different traffic type are generated together. Then these packets are sent to the same user's BMB. Since the proposed UEPS scheduler selects 12 users in each timeslot, we define the number of arrived userID in each timeslot as the offered traffic load. To evaluate performance of the UEPS algorithm under various traffic loads, the number of generated userID varies from 2 to 20 userIDs/timeslot which corresponds to the offered traffic load (λ) of $2/12 = 0.167$ $20/12 = 1.67$.

Performance of the proposed UEPS algorithm is evaluated in terms of three different performance metrics such as the packet loss rate, the average packet delay, and the average throughput. The average throughput under various traffic loads is evaluated for the loss-sensitive NRT traffics, where as the average packet delay is mainly used for the delay-sensitive RT traffics. Since RT traffics have maximum allowable packet loss rate even though they are tolerant to packet loss, performance of RT traffics is also evaluated in terms of the packet loss rate. Delay and loss requirements for RT voice and video streaming traffics at the downlink are [5]

- RT Voice: delay < 40ms, loss rate < 3%
- RT Video: delay <150 ms, loss rate < 1%

Since the length of MSTI of RT traffic is one of important design factors, performance of the UEPS algorithm has been evaluated extensively via simulation study under different sets of MSTI values. In this paper, because of page limitation we set the length of MSTI to 10, 20 and 30 for RT voice traffic and 30 for RT video traffic.

Packet Loss Rates. Figure 3(a) shows the average packet loss rates of RT traffics in the context of the optimized and the fast implementation-based packet loading methods under various traffic loads. The UEPS algorithm with the optimized packet loading method shows slightly lower packet loss rates than the fast implementation-based one under various traffic loads and MSTI values.

For the voice traffic, in terms of the packet loss rate, there is no significant difference between two packet loading methods under light and medium traffic loads, i.e., $\lambda = 2 \sim 12$ when the same MSTI values are used. However, as the traffic load increases, the optimized method shows better performance than the fast implementation-based method. For the video traffic, in case of the same MSTI value, there is no significant difference between two methods under various traffic loads.

Throughput of NRT Traffics. Figure 3(b) shows throughput of NRT traffics such as WWW and emailing traffic with two packet loading methods under various traffic loads when $MSTI_{voice} = 10$ and $MSTI_{video} = 30$. As shown in figure 3(b), there is no significant difference between two packet loading methods in terms of throughput of NRT traffics.

For the WWW traffic, the optimized method gives better throughput under the light traffic load than the fast implementation based method. However, as the traffic load increases, difference in throughput diminishes, and thus there is no difference between two methods under the medium and the heavy traffic loads. In contrast, for the email traffic, the fast implementation-based method gives better throughput performance than the optimized one under the light

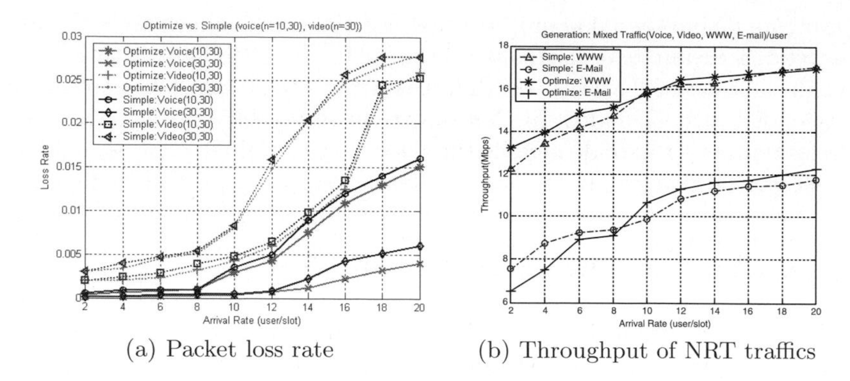

(a) Packet loss rate (b) Throughput of NRT traffics

Fig. 3. Packet loss rates of real-time traffics (left) and throughput of WWW and email traffics (right) under optimization and implementation based UEPS algorithm

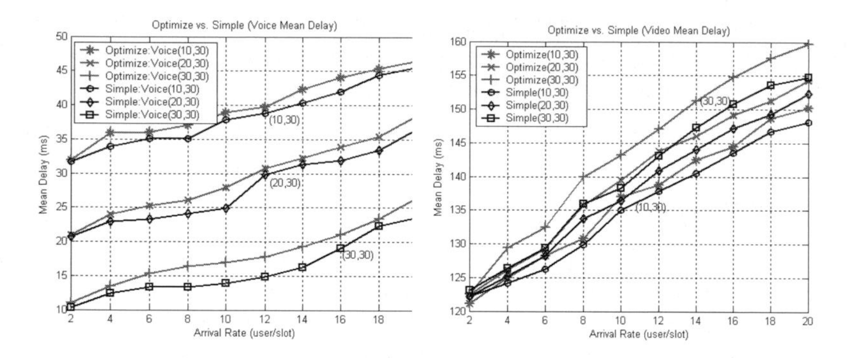

Fig. 4. Average packet delay of real-time voice and video traffics with optimization and implementation based UEPS algorithm

traffic load. However, as the traffic load increases, the optimized method shows better throughput performance than the fast implementation-based one. The reason is that with the optimized method only the HOL packet is picked for loading on the sub-channel when the email traffic is selected for packet loading. In contrast, packets of email traffic are loaded as many as possible when there is room on the sub-channel under the fast implementation-based loading. As a result, throughput of the email traffic with the fast implementation-based method is better than with the optimized one.

Average Packet Delay. Figure 4 shows the average delay of voice and video traffics with two packet loading methods under various traffic loads. For voice traffic, the fast implementation-based packet loading method gives lower packet delay than the optimized one under all traffic loads and all $MSTI_{voice}$ values. Since the fast implementation-based method is designed for fast packet loading, and thus takes only the urgency factor of the HOL packets of each traffic type into account for packet loading, it gives lower packet delay than the optimized method as a result.

For the average delay of video traffics, because of the same reason of the results for the voice traffic, the fast implementation-based packet loading method gives lower packet delay for video traffic than the optimized one under all traffic loads and all $MSTI_{voice}$ values.

6 Conclusions and Further Study Issues

In this paper, we deigned a novel wireless downlink packet scheduling algorithm, *the UEPS algorithm*, designed not only to support multiple users simultaneously, but to serve RT and NRT traffics at the same time for a user. We also developed two packet loading methods for the selected users, an optimized and a fast implementation-based methods. The former method loads packets on the sub-channel packet by packet based on the urgency factor of each packet. The latter method is devised for fast real time implementation of the former method within a small scheduling time interval. Simulation study shows that there is no significant difference on performance of the UEPS algorithm under two packet loading methods.

References

1. Jalali, A., et al.: Data Throughput of CDMA HDR a High Efficiency-High Data Rate Personal Communication Wireless System. In: Proc. of VTC 2000-Spring, pp. 1854–1858 (2000)
2. Andrews, M., et al.: Providing Quality of Service over a Shared Wireless Link. IEEE Communications 39(2), 150–154 (2001)
3. Holma, H., Toskala, A.: WCDMA for UMTS, 2nd edn. John Wiley and Sons, Ltd, Chichester (2002)
4. 3GPP: Physical Layer Aspects of UTRA High Speed Downlink Packet Access (Release 2000), 3G TR25.848 V4.0.0 (March 2001)
5. Janevski, T.: Traffic Analysis and Design of Wireless IP Networks. MA, Artech House (2003)

Scheduled Uplink Packet Transmission Schemes for Voice Traffic in OFDMA-Based Wireless Communication Systems

Seokjoo Shin[1], Kanghee Kim[2], and Sangwook Na[1]

[1] Dept. of Computer Engineering, Chosun University, Gwangju, Korea
[2] Electronics and Telecommunications Research Institute, Daejeon, Korea

Abstract. In this paper diverse uplink packet scheduling algorithms for voice traffic which can be assumed as CBR were proposed and their performance analyzed. Based on voice traffic model analysis, four types of uplink packet scheduling algorithms, BUS, G-PUS, PUS-PLF and PUS-AMC, were proposed. From simulation results, it was verified that the PUS-AMC algorithm can increase channel capacity up to about 5.7 times compared to other algorithms. Superior performance of the PUS-AMC algorithm was also demonstrated by average packet delay time performance analysis.

1 Introduction

In a packet-based wireless communication system, a packet scheduling algorithm is essential to maximize frequency efficiency [1]-[5]. In order to effectively provide finite wireless resource in a packet based system, it can be more reliable the base station to control the uplink resource as well as downlink resource. Most existing research up to now for uplink packet transmission has been done based on dedicated channel assignment or contention based random access, but there has been almost no research on scheduled uplink packet transmission on radio.

Uplink in a cellular network is described as a multipoint-to-point network. Existing commercialized cellular systems for voice traffic are all circuit based resource allocation types and when the terminal establishes call setup with the base station through random access, each terminal is allocated a unique uplink channel by the base station. However, in a packet based uplink transmission for voice traffic, centralized control uplink packet transmission can be applied for the better channel utilization. In a scheduled uplink packet transmission, the terminal is just included as a target of the uplink packet scheduler located in the base station through an initial random access. The terminal can transmit it's packet by using the corresponding resource after the base station maps the specific ID assigned to the terminal with a specific resource (or allocates available resource for a specific ID in time) and notifies the terminal.

In this paper, characteristics of the voice traffic are analyzed and several uplink packet scheduling algorithms are proposed in order to provide good channel utilization in OFDMA (Orthogonal Frequency Division Multiple Access) based

T. Vazão, M.M. Freire, and I. Chong (Eds.): ICOIN 2007, LNCS 5200, pp. 70–79, 2008.

mobile communication system. In advance, three types of scheduling classes can be defined by the amount of terminal state information (transmission buffer state, delay time, packet loss, etc.) possessed by the base station when the base station carries out uplink packet scheduling. Based on the class definition, four types of scheduling algorithms for voice traffic are proposed and their performances are analyzed with considerations of the AMC (Adaptive modulation and coding) method, which is a link adaptation method, and the PLF (Packet Loss Fair)[5] algorithm.

2 System Model

We consider an OFDMA cellular system with packetized transmission. One central base station and multiple distributed users are set to be a cell. In MAC frame structure, time axis is divided into fixed size frames and each frame is partitioned into a certain number of slots. Frequency axis is organized into M sub-channels into which many sub-carriers are grouped and they become the basic unit for data transmission [6]. An example of a proposed frame is shown in Figure 1. In Figure 1, area corresponding to one sub-channel for one slot is defined as a BU (Basic Unit) and this is the basic unit for packet transmission. Control channel and arbitrary approach channel are assumed to be included in a BU.

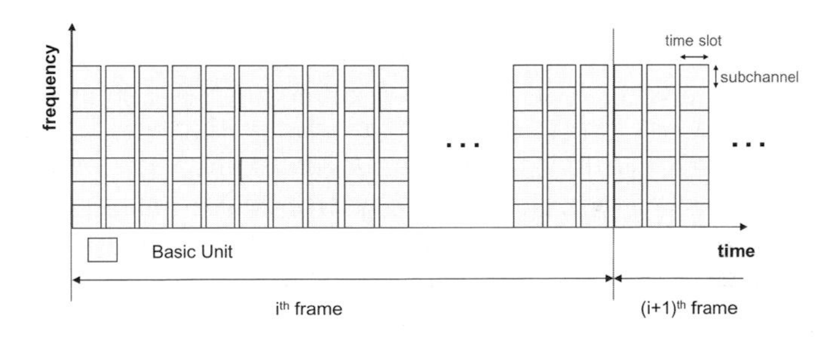

Fig. 1. Proposed frame structure for OFDMA based uplink

For network structure, we considered uplink traffic transmission in a single cell environment. Uplink packet scheduler for a base station selects multiple terminals that require transmitting packets for each slot and notifies the corresponding terminals through the downlink control channel. The terminals that verified this through the control channel transmit packets using the allocated slot and sub-channel information. Size of each packet stored in the terminal buffer is assumed to be identical and this is defined to be the basic packet. If the AMC method is applied, the number of packets that can be transmitted in one BU is determined by the AMC options. More detailed explanation will be given in Secton 3 and Section 4.

Unlike downlink, the uplink packet scheduler in the base station must receive status information about all connections for uplink from the terminals in each instant in order to schedule effectively. The following three types of uplink packet scheduler classes can be defined depending on the availability/non-availability of terminal status information the base station can receive.

- **BUS (Blind Uplink Scheduler):** Uplink packet scheduler for the base station gets actively involved in resource allocation during initial call setup for specific terminals. Subsequent resource allocations, proportional to the assigned band, are provided on a periodic basis and no further status information is received from the terminals.
- **PUS (Passive Uplink Scheduler):** Uplink packet scheduler for the base station gets actively involved in resource allocation during initial call setup for specific terminals. Subsequently, allocated resources are increased or decreased in steps, using a manual method, through 1-2 bit indicators (generated by terminal) based on resources that have been determined.
- **AUS (Active Uplink Scheduler):** Uplink packet scheduler for the base station receives all status information from specific terminals through the uplink control channel that has already been allocated or through additional control packet transmissions and resource allocation for the terminals is carried out effectively, based on this information.

3 Uplink Packet Scheduling Algorithm

3.1 Voice Traffic Model Analysis

Voice source generates patterns for statistically independent voice activated interval and non-activated interval that is determined by an exponential function. Average interval for each is 1 second and 1.35 second respectively. Voice packets are generated only during the voice activated interval and has a bit generation rate of 16kbps. One voice packet is composed of $320bit$ ($40bytes$ w/o header) units and if it is assumed that there are 50 frames in one second, it means that it is possible to maintain a transmission rate that will result in a satisfactory voice sound quality by transmitting one packet per frame.

Figure 2 shows a simple state transition diagram for voice traffic. Length of the ON-OFF interval is assumed to follow an exponential function. Average interval for voice activated interval is t_1 and the probability γ that the active interval ends when time becomes $T$is given by the following equation.

$$\gamma = 1 - e^{-\frac{T}{t_1}} \quad (1)$$

Also, average interval for a non-activated interval is t_2 and the probability σ that it ends at time T is as follows.

$$\sigma = 1 - e^{-\frac{T}{t_2}} \quad (2)$$

For voice traffic, since it can be assumed that a packet is generated once every $20ms$ when ON interval is started, for uplink, the base station only needs to periodically allocate resource for the corresponding terminals once every $20ms$. Figure 3 shows the resource allocation timing relationship for voice packet generation and the base station.

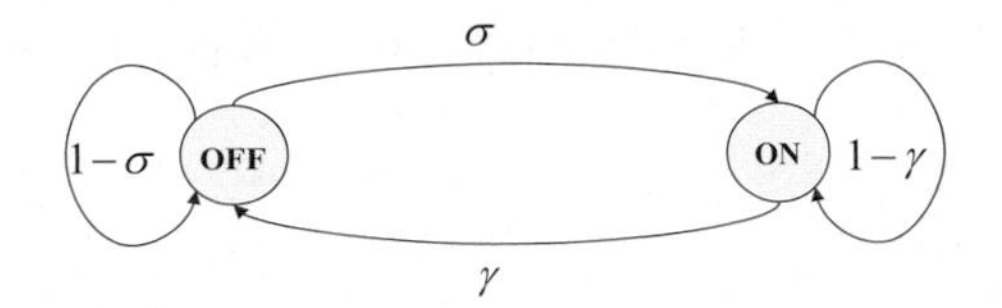

Fig. 2. State transition diagram for the voice traffic model

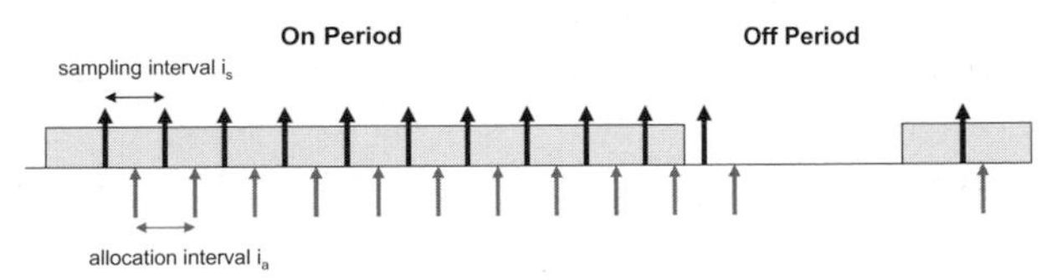

Fig. 3. Timing relationship between voice packet generation period and resource allocation period

Even though $i_s = i_a$ is the ideal case as shown in Figure 3, since packet transmission delay of up to the maximum voice packet delay time (D_{max}) is allowed, we can assume that a certain level of jitter between allocation interval is allowed. Therefore, if the condition $i_a \leq D_{max}$ is satisfied, required QoS can be guaranteed even for the case $i_s = \bar{i_a}$ ($\bar{a}$: mean of a) condition is satisfied. If we assume that the amount of the resource allocated by the base station for a certain user can be α times the size of the generated packet, resource allocation timing relationship of $\bar{i_a} = \alpha i_s$ is possible as well. For example, since the advanced mobile communication systems use the AMC (Adaptive Modulation and Coding) method, the number of basic packets that can be transmitted in a BU is different depending on the channel quality of the user. If the system has been defined to transmit one voice packet, which is the minimum unit generated for the AMC option, it is possible to increase the allocation interval by considering the D_{max}, as the AMC option is improved.

In general, maximum packet delay time and maximum allowed packet loss rate are defined with the QoS parameter for the voice traffic model.

- Maximum packet delay time: D_{max}.
- Maximum allowed packet loss rate: $P_{loss} = P_{drop} + P_{error}$.

Here, P_{drop} is defined to be the packet dropping rate for packets that exceed the maximum transmission delay and P_{error} can be defined to be the packet error probability when the receiving end fails to decode the packet due to the noise that occurs during transmission or when an error occurs during the CRC verification process.

3.2 Proposed Uplink Packet Scheduling Technique for Voice Traffic

For traffic with a periodic characteristic like voice, since the base station only needs to allocate resource periodically or within a certain amount of jitter, on a regular basis, it is sufficient to consider just the BUS, PUS class that were introduced in Section 2. However, by considering the ON-OFF interval, the

AMC option and, P_{loss} a variety of scheduling algorithms shown below can be considered.

1. ON-OFF Interval Not Considered for Voice Traffic (Blind and Deaf After Call Set-up: BUS): • After call setup, ON-OFF interval occurring in terminals is not considered (After initial resource allocation, no further information is collected from the terminals). • Base station PS allocates resource periodically every $20ms$ to the corresponding terminal. • Up to $40ms$ is allowed for jitter between the allocation interval. • AMC option is not considered: Because CQI (Channel Quality Information) information is not needed, uplink control channel use is suppressed and a decrease in inter-channel interference results.

2. ON-OFF Interval Considered (Generic PUS): • Allocate a flag bit in the voice packet header that can express whether ON interval can be sustained or the end: flag(0) indicates continuation of the ON interval and flag(1) indicates the last packet in the current ON interval. In OFF interval, the base station does not allocate resource to the corresponding terminal. • Terminal that enters the ON interval again after OFF notifies the base station by transmitting a 1-2bit information through the allocated control channel. • Base station PS allocates resource periodically every $20ms$ to the corresponding terminal only in the ON interval. • Up to $40ms$ is allowed for jitter between the allocation interval. • AMC option is not considered.

3. Process Packet Loss Based Scheduling with ON-OFF Interval Considered (PUS with PLF (Packet Loss Fair)): • Include content of proposal 2. • Execute priority based scheduling so that P_{loss} is fairly distributed between users. Because the base station knows whether the ON interval will be maintained and since it can find out the packet transmission result by periodically allocating resource and finding out whether they were received, calculation of P_{loss} for each terminal is possible without additional help from the terminal. • Two types of parameters (P_{loss}, D_{curr}: time interval from packet generation to scheduling) can be used for the priority queue mechanism. • Priority order calculation: $j = argmaxP^{i}_{loss}(t)$. Here $P^{i}_{loss}(t)$ represents the current packet loss rate for the i-th terminal.

4. PUS with AMC: • Include content of proposal 3 (Consider the AMC option). • However, the base station determines the AMC option that can be used by the terminal from CQI control channel information for each terminal. • When it is possible to transmit α packets in a BU from AMC option of a terminal, scheduling period can be changed from T_a to $\beta \cdot T_a$ where $\beta = min\left(\alpha, \frac{D_{max}}{i_s} - 1\right)$ for all α. • Decision interval for the scheduling period ($\beta \cdot T_a$) is the same as the uplink pilot channel transmission period because the AMC option or uplink channel state changes in real time. • **AMC option compensation:** If resource is allocated with period $\beta \cdot T_a$ and the number of packets that can be transmitted per BU due to the AMC option of the corresponding terminal has been degraded to $\gamma(\gamma < \beta)$, $\lceil \beta/\gamma \rceil$ BUs are allocated for compensation and the period is changed to $\gamma \cdot T_a$. • **Throughput acceleration:** If resource is allocated with period $\beta \cdot T_a$ and the number of packets that can be transmitted per BU due to the AMC option of the corresponding terminal has been improved to $\gamma(\gamma > \beta)$, resource allocation is delayed by $\left(min(\gamma, \frac{D_{max}}{i_s} - 1) - \beta\right) \cdot T_a$ and scheduling period due to the AMC option is changed to $\left(min(\gamma, \frac{D_{max}}{i_s} - 1)\right) \cdot T_a$.

4 Simulation Environments

Under the wireless environment, we adapt the AMC technique to the proposed packet scheduling algorithm by assigning K MCS levels depending on uplink channel quality of each user. In the frequency selective fading environments, uplink channel quality of different subchannels have different values. BS collects SNR_k values of user k, $SNR_k^0, ..., SNR_k^{N_{sub}-1}$, from predetermined pilot signals corresponding to each subchannel. The SNR is measured as the ratio of pilot signal power to noise power caused within a cell when we assume that there is no other cell interference at all. More specifically, the SNR of the $n\text{-}th$ subchannel received from the $k\text{-}th$ user can be represented as

$$SNR_k^n = \frac{P_p h_{k,n}^2}{N_0 B / N_{sub}}, \tag{3}$$

where $h_{k,n}$ is a random variable representing the fading of the k user and n subchannel. P_p is the transmitted power of the pilot signal. N_0 is the noise power spectral density and B is the total bandwidth of the system. The channel gain, $h_{k,n}^2$, of subchannel n of user k is given by

$$h_{k,n}^2 = |\alpha_{k,n}|^2 \cdot PL_k. \tag{4}$$

Here, PL_k is the path loss for user k and is defined by

$$PL_k = PL(d_0) + 10\beta log(d_k/d_0) + X_\sigma, \tag{5}$$

and $a_{k,n}$ is the short scale fading for user k and subchannel n. The reference distance and distance from BS to user k are d_0 and d_k, respectively. The path loss component is β, while X_σ represents a Gaussian random variable for shadowing with standard deviation σ.

The base station estimates each mobile's SIR using pilot signals transmitted through CQI control channel, and based on which it determines AMC option. The MCS level is classified by the required SNR strength, SNR_{req}, and maps to the number of packets in a BU. The mapping between the MCS level and the number of packets is shown in Table 1, where we assume that all subchannels are allocated with equal power, i.e., 1W.

In this paper, frame structure was simplified to carry out performance evaluation more easily. Length of one frame was defined to be $10ms$ and a frame can be partitioned into 10 slots. Each slot becomes an interval of $1ms$. The number of sub-carriers is 1536 and for uplink, 256 sub-carriers are grouped and assumed to be six sub-channels. MAC stores SDUs of fixed size in the buffer and the size of MAC SDU(MSDU) is assumed to be 40bytes. In addition, MAC header is assumed to be $4bytes$. Parameters used in simulation have been arranged in Table 2.

Table 1. AMC options

Index	$SNR_{req}(dB)$	**Pkts/BU**	**Mod**	**Code rate**
AMC_1	1.5	1	BPSK	1/2
AMC_2	4.0	2	QPSK	1/2
AMC_3	7.0	3	QPSK	3/4
AMC_4	10.5	4	16QAM	1/2
AMC_5	13.5	6	16QAM	3/4
AMC_6	18.5	9	64QAM	3/4

Table 2. System parameters used in simulation

Parameters	**Value**
No. of subchannels	6
No. of slots per frame	10
Length of one frame	$10ms$
Average data rate for voice	$339kbps$
Max. allowed packet loss rate	10^{-2}
MSDU and MPDU size	40, $44bytes$
Max. packet transmission delay	variable
Cell radius (km)	1
User distribution	Uniform
BS transmission power (W)	12
Path loss model (α and δ(dB))	4, 8

5 Simulation Results

Uplink packet scheduling performance evaluation for voice traffic was carried out by considering four schemes as described in Section 3. Figure 4, Figure 5 and Figure 6 show performance evaluation for the BUS algorithm, the Generic PUS and the PUS-PLF algorithm and the PUS-AMC algorithm, respectively. Comparison result for average packet transmission delay when PUS-PLF and PUS-AMC are applied is shown in Figure 7.

It can be seen from Figure 4 that P_{loss} increases abruptly when the number of users exceed 120 if the BUS algorithm is applied. Because this is a method which allocates resource once every $20ms$ without considering ON-OFF intervals, it is easy to see theoretically that 120 persons ($= 6slot * 20ms$) is the maximum number of users that can be used. Intuitively, BUS method is identical to time division multiplexed circuit based scheduling method.

Results for the generic PUS (G-PUS) algorithm and the PUS-PLF algorithm are displayed in Figure 5. Because simulation results were drawn for average P_{loss} of all users, average performance of G-PUS and PUS-PLF are identical. However, unlike the PUS-PLF algorithm in which scheduling is carried out so that P_{loss} for all users is distributed fairly, for G-PUS a probability for occurrence of a call drop (Generation when P_{loss} is greater than the required QoS parameter

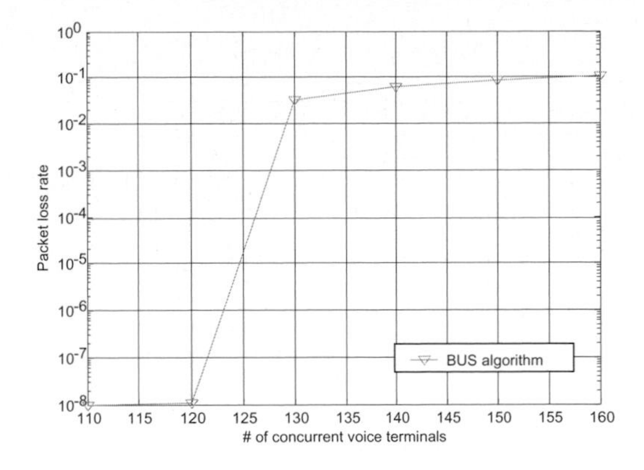

Fig. 4. Packet loss rate for number of users in BUS algorithm (D_{max}=100ms)

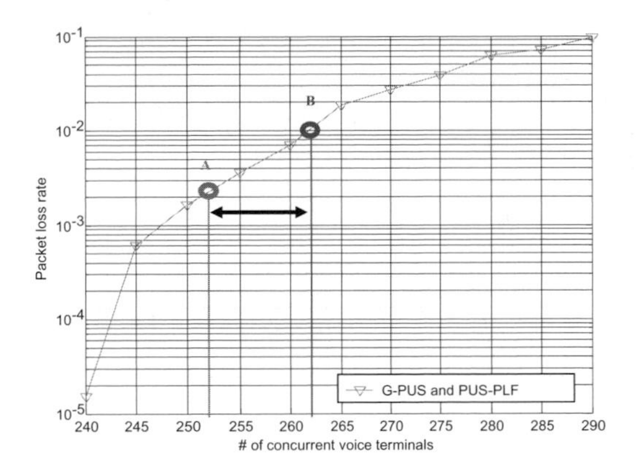

Fig. 5. Packet loss rate for number of users in G-PUS, PUS-PLF algorithm (D_{max}=100ms)

is assumed) between interval A-B in the figure for an arbitrary user exists. In other words, for the G-PUS algorithm, there may be instances when arbitrary users cannot satisfy the QoS they demand between interval A-B in the figure.

Figure 6 shows the results for the algorithm defined as PUS-AMC which was obtained by applying AMC for the user's channel status to the PUS-PLF algorithm. This is an algorithm which can delay packet transmission, according to AMC options that can be used by each user, and transmit many packets simultaneously. As shown in the results, maximum number of simultaneous users that satisfy maximum P_{loss} is about 690. Selection of AMC option is carried out by the base station. AMC option for the terminal can be determined by using the control channel for CQI that is allocated for each terminal through uplink.

Table 3 displays the maximum number of users when P_{loss}^{max} for the proposed algorithm for four types of voice traffic. With BUS algorithm as the reference, capacity increase of 2.2X for the PUS-PLF algorithm and 5.75X for the PUS-AMC algorithm was verified.

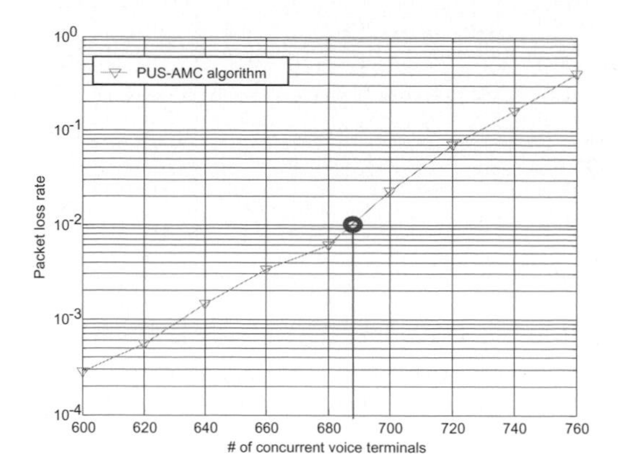

Fig. 6. Packet loss rate for number of users in PUS-AMC algorithm (D_{max}=100ms)

Table 3. Maximum number of simultaneous users for the proposed algorithm when $P_{loss}^{max} = 0.01$, D_{max}=100ms

	BUS	G-PUS	PUS-PLF	PUS-AMC
Max. users (BUS growth rate)	120 (-)	252-262 (2.1-2.18)	262 (2.18)	690 (5.75)

Figure 7 shows the average transmission delay per packet for the PUS-PLF algorithm and the PUS-AMC algorithm. As shown by the results, while the PUS-PLF algorithm has a very low time delay compared to the PUS-AMC algorithm when the number of simultaneous users is less than 260, the time delay increases abruptly when the number of simultaneous users is greater than 260. In addition, even though the PUS-AMC algorithm delays packet transmission by

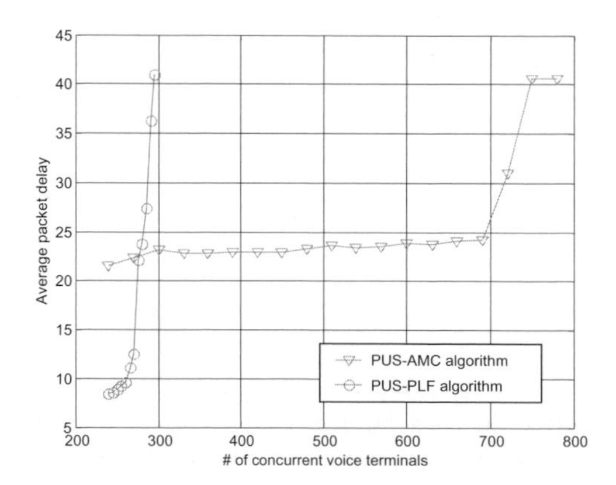

Fig. 7. Average packet transmission delay for number of users in PUS-AMC algorithm and PUS-PLF algorithm (D_{max}=100ms)

force, depending on the AMC option, it can be seen that a packet transmission delay that is a very small value compared to D_{max} is maintained for almost all intervals. In conclusion, this research proved that the PUS-AMC algorithm is an optimum uplink algorithm for voice traffic that can maintain significant capacity increases and packet transmission delay at a certain level.

6 Conclusion

In this paper four uplink packet scheduling algorithms (BUS, G-PUS, PUS-PLF, PUS-AMC) for voice traffic were proposed and their performance analyzed. Because voice traffic not only has a fixed transmission rate but also a periodic packet generation pattern, the base station can predict the terminal status. It was verified that the PUS-AMC algorithm which uses the AMC method can increase channel capacity by about 5.7X compared to other methods by delaying packet transmission by force within the maximum allowed delay and sending many packets simultaneously when channel state is satisfactory. PUS-AMC algorithm also showed excellent average packet transmission delay characteristics. Based on this research, we verified that the PUS-AMC algorithm is the optimum algorithm for voice traffic when an uplink packet transmission system is implemented through base station control.

Acknowledgment

"This research was supported by the MIC, Korea, under the ITRC support program supervised by the IITA" (IITA-2006-C1090-0603-0007)

References

1. Ofuji, Y., Abeta, S., Sawahashi, M.: Fast packet scheduling algorithm based on instantaneous SIR with constraint condition assuring minimum throughput in forward link. In: WCNC 2003, March 2003, vol. 2, pp. 860–865 (2003)
2. Huang, V., Zhuang, W.: Fair packet loss sharing (FPLS) bandwidth allocation in wireless multimedia CDMA communications. In: Proc. Int'l Conf. 3G Wireless and Beyond, pp. 198–203 (May 2001)
3. Elwalid, A., Mitra, D.: Design of generalized processor sharing schedulers which statistically multiplex heterogeneous QoS classes. In: Proc. IEEE INFOCOM 1999, vol. 3, pp. 1220–1230 (1999)
4. Ryu, S., Ryu, B., Seo, H., Shi, M.: Urgency and efficiency based wireless downlink packet scheduling algorithm in OFDMA system. VTC 2005 3, 1456–1462 (2005)
5. Shin, S., Ryu, B.H.: Packet loss fair scheduling scheme for real-time traffic in OFDMA systems. ETRI J. 26(5), 391–396 (2004)
6. Ryu, S.W., Ryu, B.H.: Media approach control structure for next generation mobile communication. Journal of Korean Institute of Communication and Sciences 22(9), 51–62 (2005)

Cognitive Radio MAC Protocol for Hidden Incumbent System Detection*,**

Hyun-Ju Kim[1], Kyoung-Jin Jo[1], Tae-In Hyon[2], Jae-Moung Kim[1], and Sang-Jo Yoo[1]

[1] Graduate School of information Technology & Telecommunication, Inha University, 253 Yonghyun-dong, Nam-gu, Incheon 420-751, Korea
[2] Samsung Advanced Institute of Technology, San 14, Nongseo-Ri, Giheung-Eup, Yongin-Si, Gyeonggi-Do, Korea
{multi,jkjsoul}@inhaian.net, taein.hyon@samsung.com, jaekim@inha.ac.kr, sjyoo@inha.ac.kr

Abstract. In this paper, we propose an inband and outband broadcast method for hidden incumbent system detection of MAC layer for WRAN systems using cognitive radio technology. In order to make sure that cognitive radio WRAN system must be operated with no interference to incumbent system, WRAN system should sense channel during its communication and change channel according to the appearance of incumbent system. To detect hidden incumbent system, we use some extra candidate channels to broadcast current channel list and to report CPE's sensing results.

Keywords: CR, hidden incumbent detection, WRAN.

1 Introduction

Increasing demand of wireless communication services resulted in insufficient frequency assigned to each licensed user or service under the control of government. To resolve the problem, Joseph Mitola [1] [2] [3] [4] suggested the concept of cognitive radio (CR) being able to do self-cognition, user-cognition and radio-cognition. CR is adapted for use of frequency without interference to primary user. After FCC (Federal Communications Commission) [5] referred to the availability of overlapping frequency use in a NPRM (Notice of Proposed Rule Making) [6], IEEE 802.22 WRANs (Wireless Regional Area Networks) working group [7] is constructed for CR technology progress that will define standard interface to PHY and MAC layer so as to use TV spectrum by unlicensed device users. In order to make sure that WRAN system must be operated with no interference to incumbent system, WRAN system should sense channel during its communication and change channel according to the appearance of incumbent system. WRAN base station (BS) and customer promise equipment (CPE) devices assumed that they are able to differentiate between

* This work was supported in part by Samsung Advanced Institute of Technology.
** This work was supported in part by grant No. (R01-2006-000-10266-0) from the Basic Research Program of the Korea Science & Engineering Foundation.

T. Vazão, M.M. Freire, and I. Chong (Eds.): ICOIN 2007, LNCS 5200, pp. 80–89, 2008.

incumbent signal and other unlicensed signal using various sensing methods. In the case that incumbent system appears in the same channel that WRAN system is now using, WRAN system stops using the channel and starts to change service channel to avoid interference to incumbent system. But that includes the problem that incumbent system has interference as WRAN system do not cognize the incumbent system.

This paper suggests a CR MAC protocol that is able to coexist with primary users by means of an inband/outband signaling for hidden incumbent system detection. Inband (defined control signal on current band) and outband signaling (defined control signal on the band other than current band) help to change the current channel to a new channel without interference to incumbent system. CPEs on the overlapping areas can report the incumbent system appearance to the CR BS. CR BS can inform the CPEs of the new channel list to continue the communication (seamless spectrum handover). For initial CR network entry of the CPEs inside the overlapping areas, CPEs can easily find out the current service channel information. CR that takes advantage of the utmost of frequency has the high probability of mutual coexistence with new spectrum utilization technology such as home network market [8] based on WPAN (Wireless Personal Area Network) and UWB (Ultra Wide Band) [9].

The rest of this paper is organized as follows. In Section 2, we present channel sensing and frequency changing procedure. The proposed MAC protocol for hidden incumbent system detection is presented in Section 3. In Section 4 simulation results are shown. Finally, we conclude in Section 5.

2 Channel Sensing and Frequency Changing Procedure

WRAN BS that obtains the sensing results should perform proper reactions to protect incumbent systems. WRAN system should leave the channel for operating of incumbent system without interference to the incumbent system and should be able to provide seamless-service to WRAN CPEs. The channel sensing and changing procedure is as follows:

1. WRAN BS allows WRAN CPEs to start to sense channels by sending some parameters including the value of period and duration into downstream.
2. WRAN CPEs could sense channels after stopping sending data during quiet period.
3. WRAN CPEs transmits the sensing results including some factors over channels to WRAN BS by using upstream.
4. WRAN CPEs which use the channel that is about to using by incumbent system send sensing report to WRAN BS and wait for band change message including the target channel to move and candidate channel list(a set of channels available to communicate with WRAN BS) from WRAN BS.
5. WRAN CPEs which receive the band change message should stop using the channel and response to WRAN BS.
6. WRAN CPEs perform ranging procedure in respect of the new channel and start to communicate with WRAN BS since changing the parameter in related to channel environment.

3 Hidden Incumbent System Detection MAC

3.1 Hidden Incumbent System

WRAN BS may not be able to detect or be informed the existence of incumbent signal. This situation possibly can be happened when the incumbent system signal does not reach to the WRAN BS so that the BS cannot sense it. And CPEs, which are inside both of the WRAN BS coverage and incumbent system coverage, cannot report this overlapping to the WRAN BS because it is not able to decode the WRAN DS signal due to the strong interference. This very serious case is called as "hidden incumbent system case"

For WRAN service, assume WRAN BS had sensed some channels and it recognized channel X was available (it may use a geographical channel usage database information for that area). When an incumbent system begins service with the same channel X, the BS does not know the existence of incumbent system. So, as shown in Figure 1, WRAN CPEs that are located in overlapped area of WRAN and incumbent system and they are not able to decode the WRAN DS signal because of strong incumbent system interference. BS will keep its service on channel X and it will cause strong interference to incumbent system users possibly for a long time.

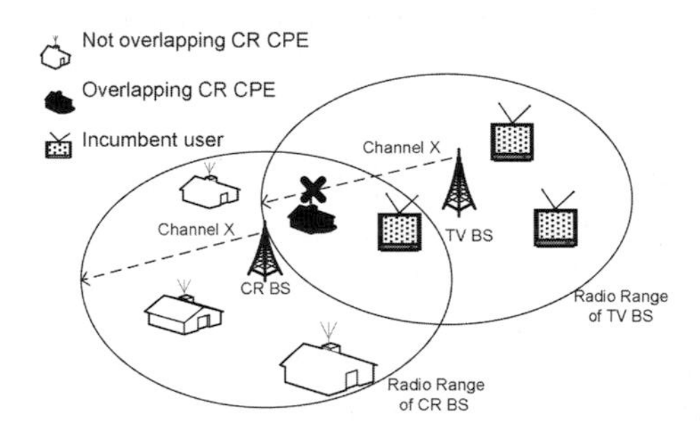

Fig. 1. Hidden incumbent system

The hidden incumbent system case causes the following problems: Harmful interference to the incumbent system users. Some WRAN CPEs cannot communicate with WRAN BS. This hidden incumbent system case may occur not only during the any service period but also during the WRAN BS initialization period (service start time). Incumbent systems can start their service at any time without any notification or may initiate their service disregarding the service pattern as stored in WRAN database. Therefore, WRAN BS can mistakenly change its service channel to the channel that is already being used by an incumbent system because of its unreliable sensing results.

3.2 Hidden Incumbent System Detection Protocol: Inband Signaling

In the situation of hidden incumbent system, it needs a concrete method for detection of hidden incumbent system. WRAN system avoids using same channel that used by incumbent system. This paper suggests the inband (using channel by WRAN system) signaling that allows WRAN CPEs to report the advent of incumbent system through current channel and the outband channels (which are not used by WRAN or incumbent system) signaling method that allows WRAN CPEs to report the advent of incumbent system through candidate channels lists.

It is defined as the inband signaling that WRAN CPEs report a channel occupation of incumbent system through the other synchronous channels currently used by WRAN BS in case of the advent of incumbent system at the certain channel used by WRAN system. WRAN BS knows about the channel table that the CPEs are using. That is good for WRAN CPEs to promptly report the advent of incumbent system in the channel to WRAN BS and to stop the using of the channel and to change from the channel to a new channel. In the situation of that some WRAN CPEs are not able to decode signal from WRAN BS, the method shows now to resolve the problem.

3.3 Hidden Incumbent System Detection Protocol: Outband Signaling

If the WRAN BS only uses a single channel due to the not enough empty bands or other reasons and the incumbent system appears at the current channel, we are not able to solve the hidden incumbent system problem with the inband signaling method. WRAN CR BS cannot recognize hidden incumbent system because of no information. Also, some incumbent users have experienced interference from the WRAN system.

CR BS periodically broadcasts the information for the current channel in some of other unoccupied channels (e.g., candidate channels). The outband channel list is broadcasted in the in-band frame headraces that are not able to decode the frame of the BS's current service channel try to sense the indicated (candidate) channels to locate the BS signal. If CPEs receive the explicit outband broadcast signal, The CPE sends a report to the BS using the up link of the outband channel. After noticing the existence of the hidden incumbent, The CR BS changes its service channel to other available band. The new channel information is also broadcasted with outband signaling.

A Outband signaling procedure for hidden incumbent system detection

Procedures of explicit outband signaling for hidden incumbent system detection is shown in Figure 2.

1. WRAN system provides services to WRAN CPEs in channel X and WRAN CPEs that are using channel X enter the blocking state due to the sudden advent of incumbent system in channel X.
2. WRAN BS periodically broadcasts explicit outband signaling message into the outband channel set {B, C, D}.
3. WRAN CPEs that abruptly do not decode the signal from WRAN BS in channel X due to the appearance of incumbent system start to sense the other channels to communicate with WRAN BS.
4. WRAN CPEs that receive the explicit outband signaling message during the period of sensing channel (in this case channel C) recognize the availability of channel C and report the appearance of hidden incumbent system to WRAN BS.

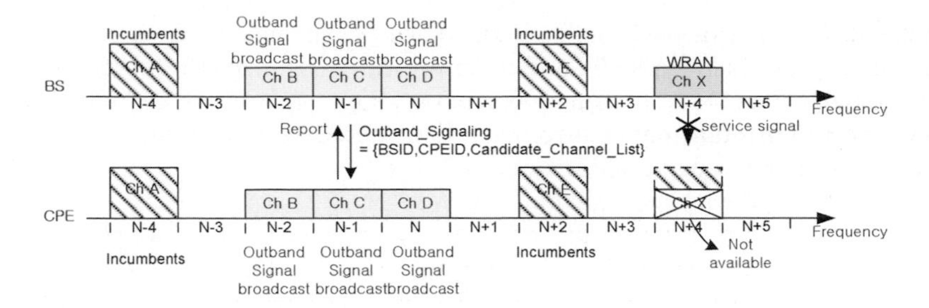

Fig. 2. Explicit outband signaling example

B Broadcast method of explicit outband signaling

The basic broadcast scheme requiring multiple PHY transceivers at the BS is that the outband signal is broadcast in candidate channel B, channel C and channel D at the same time, the BS may choose to broadcast outband signal sequentially in time for simplification of system organization, as shown Figure 3-(a). On the other hand, the sequential outband signal transmission as described in Figure 3-(b) only requires one transceiver for outband signaling. But, this sequential method has a drawback that the probability with which the CPEs can detect the outband signal decreases. Basically, outband signal does not include the CPE list that should respond to the outband signal with Hidden Incumbent System Report because the BS does not know which CPEs are located in the overlapped area in advance. But, optionally BS may choose to include a CPE list to poll specific CPEs. However, in this case the polled CPE should respond even if the current service channel is available to the CPE.

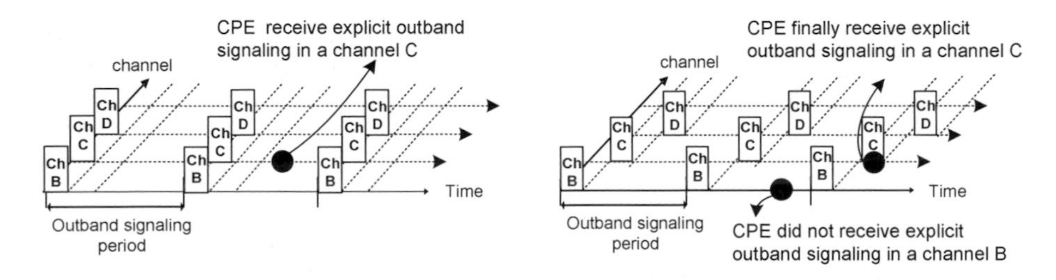

Fig. 3. (a) Simultaneous broadcast method and (b) sequential broadcast method

C Frame structure of explicit outband signaling

This broadcast of outband signal follows the similar PHY and MAC frame architecture to IEEE 802.16. Explicit outband signal DS-Burst includes service channel information, such as current BS's service channels and candidate channels. Because the BS does not know which CPEs will send the report in advance, BS generally cannot allocate up stream (US) resource to each CPE that will send the report. Therefore collisions of reports may be occurred. To increase the probability of successful report delivery to the BS, during one explicit outband signaling time k (2 ~ 3) number of MAC (including DS and US) frames are transmitted as shown in Figure 4. In SCH (Super frame Control Header) of DS, the information for k should be included. For the first US1 all CPEs that detected hidden incumbent systems will try to send reports. In DS-2 and DS-3 BS gives acknowledgements for the previous reports to indicate successful report receptions. Only CPEs that did not receive acknowledgements for the previous reports will try sending reports again.

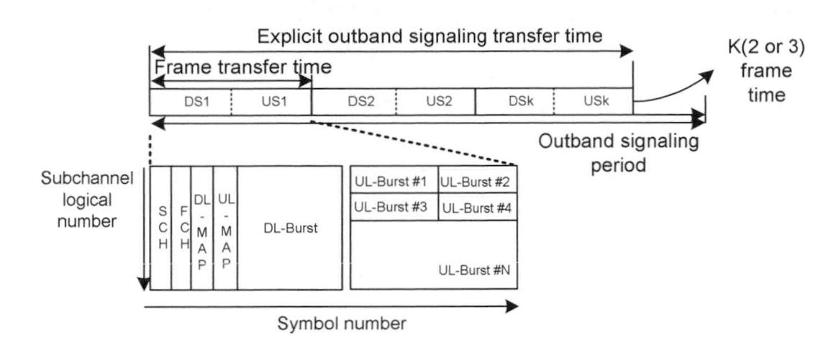

Fig. 4. Frame structures of explicit outband signaling

D CPE upstream bandwidth allocation for hidden incumbent report

For the CPE hidden incumbent system report message, there needs to be a process or method for allocation of upstream resource to CPEs. To reduce this overhead, the BS can allocate upstream resource for CPEs' hidden incumbent system report. First method is that the BS divides US resource into US-Burst slots for all unknown CPEs. When the BS transmits outband signal frame, it divides upstream resource according to maximum hidden incumbent report size and allocates each upstream burst. The second method is to use the CDMA code. The PHY supports the usage of a CDMA mechanism for the purpose of hidden incumbent system report. When the BS transmits outband signal, it allocates resource to transmit the CDMA code in upstream burst. A CPE transmits the selected random CDMA code which is assigned upstream resource to transmit the hidden incumbent system report message.

E Co-existence between WRAN BSs for explicit outband signaling

Transmitting multiple outband signals has a possibility to collide with outband signals of other co-located WRAN BSs. If communication between BSs is possible, the BSs can coexist by sending outband signal in candidate channels at different time after exchanging broadcasting time schedule with each other. But if communication between BSs for exchanging their time schedule is not available, to avoid outband signal collision, BSs randomly select their outband signal broadcasting time within the pre-defined explicit outband signaling period as shown in Figure 5.

Figure 5-(a) illustrates the case when BS broadcasts outband signaling messages at multiple channels simultaneously. And Figure 5-(b) is for the case that BS broadcasts outband signaling message sequentially in time at the different channels. To guarantee that CPE decodes the outband signal within a short time interval without sensing time discordance, CPEs should try to find the signal at one channel at least during one explicit outband signaling period.

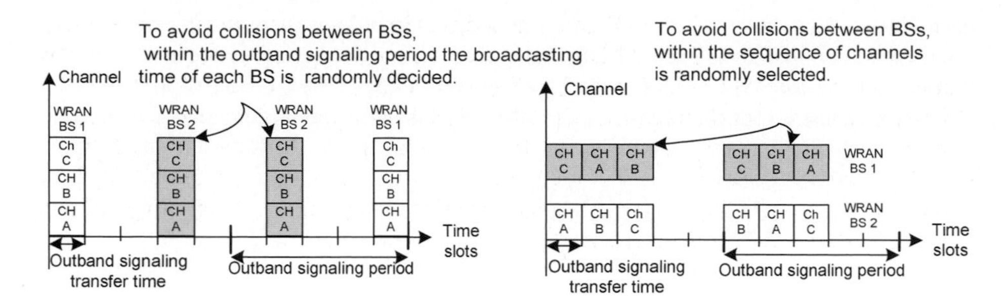

Fig. 5. (a) Simultaneous broadcasting method (b) sequential broadcasting method

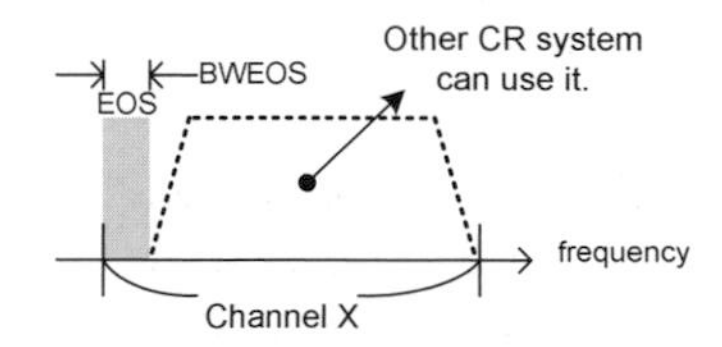

Fig. 6. Fractional bandwidth

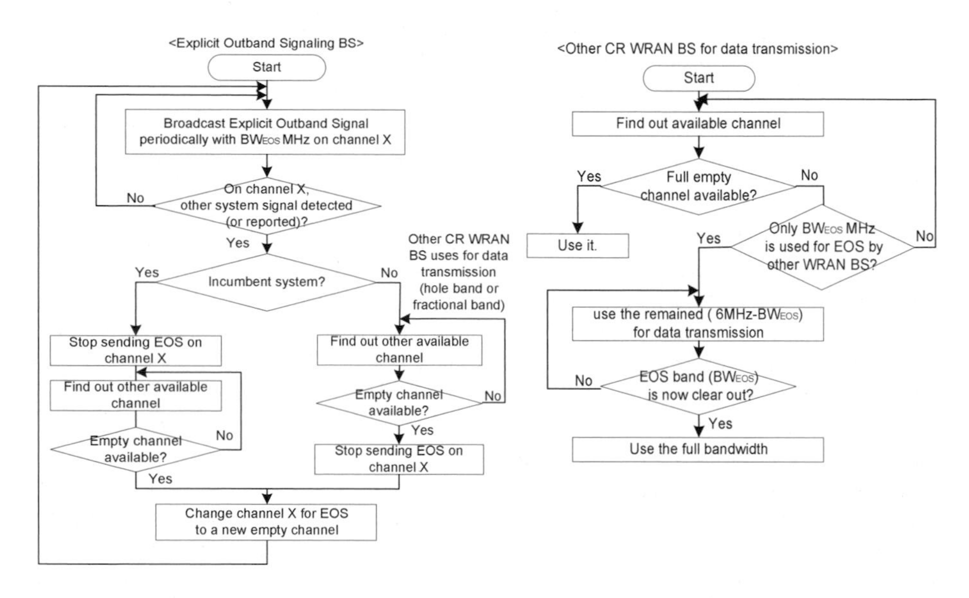

Fig. 7. (a) EOS sending BS operation and (b) other WRAN BS operation

F Efficient channel use for explicit outband signaling with fractional bandwidth usage method

Explicit outband signaling does not need a full bandwidth of a channel. Broadcasting some channel information through an explicit outband signal message requires only small bandwidth as shown in Figure 6. Thus, some bandwidth of a outband frame is used in broadcasting explicit outband signaling and the other is used in transmitting data by other WRAN BSs. Fractional bandwidth usage uses predefined narrow bandwidth to send explicit outband signaling and to receiving sensing report. WRAN BS broadcasts explicit outband signals with only BWEOS bandwidth at the beginning part of a channel bandwidth. If a BS that is broadcasting explicit outband signal detects that other BS uses the remained part, then it will try to change the channel for explicit outband signaling. Figure 7 illustrates how explicit outband signaling BS and other CR WRAN BS for data transmission operate together.

4 Simulation Results

This section describes the results of experiments for hidden incumbent system detection method. In Table 1 some simulation parameters are shown.

Table 1. Experiment parameters

Parameters	Meaning
N_{cpe}	The number of CPEs that detect incumbent system
N_c	The number of candidate channels to send explicit outband signaling onto outband
N_{ch}	The number of total channel inside radio range of WRAN BS = 10
N_s	The number of slots in one frame to send report
k	The number of frames during outband signaling period
P_s	The probability that CPE sense one of candidate channels during explicit outband signaling period. $P_s = \alpha + \beta(N_c / N_{ch}), \alpha = 0.8, \beta = 0.2$

The CPE that detects incumbent system can sense one of candidate channels during explicit outband broadcasting time with P_s. Because during the normal communication time CPEs can know the candidate channel list that is included in BS's DS MAC header, CPEs can access one of the candidate channels at the hidden incumbent system condition. However some CPEs (e.g., CPEs that powered off and just power on or CPEs that did not receive several MAC frames from BS) do not know the exact candidate channel list so that they may not access the candidate channel directly. In this simulation we assume that P_s is larger than 0.8.

First experiment shows accumulated successful rate of sensing report according to increasing k during sending explicit outband signaling. We assume that in WRAN system usually supports from tens to few hundred users at the same time and therefore the incumbent detection CPEs at the edge of system boundary are not many. The more frames we have to send incumbent system detection reports, WRAN BS can detect the hidden incumbent system condition with the higher probability. In this simulation, N_{cpe} varies from 2 to 6 and N_s equals to 10. As we can see in Figure 8 (a) and (b), if we have the more candidate channels, incumbent system detection reports form the CPEs can be distributed to different channels so that the report collision rate will be the smaller. For the given environments, we can see that when k equals to 2 about 90% incumbent detection reports can be received by BS successfully. And if we set k to 3, almost 100% successful rate we can achieve.

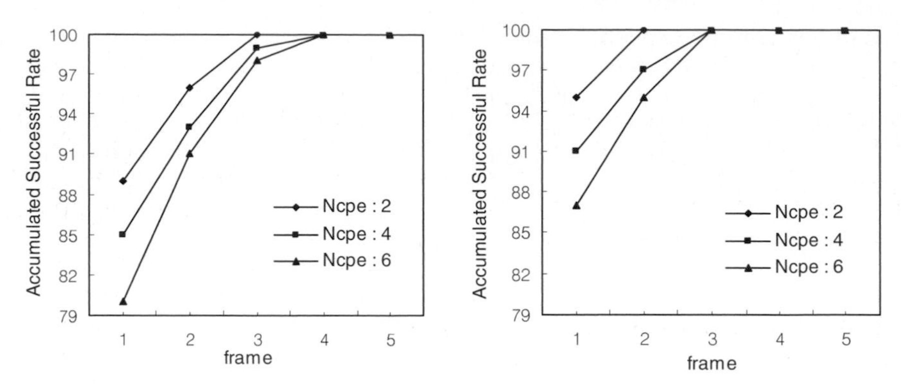

(a) *Ns* (slot) =10, *Nc* (candidate channel) =2 (b) *Ns* (slot) =10, *Nc* (candidate channel) =6

Fig. 8. Comparisons of accumulated successful rate in reporting under different environment

Second experiment measure the explicit out reporting delay in terms of explicit outband signaling transmission time, which means that how much time WRAN BS takes to send explicit outband signaling and receives sensing reports. We compared two outband signaling methods: frequency simultaneous method and time sequential method. As shown in Figure 9, frequency simultaneous method send outband messages at the same time with different frequency bands so that a CPE can sense one of outband channels within one frame time. However, in the time sequential method, if a CPE senses a different channel from the channel that the BS now broadcasts, then the CPE should wait next BS transmission. Therefore, as in Figure 9 reporting delay is linearly increasing as the number of candidate channels for time sequential method.

As the third experiment, we measure the collision probability of sending explicit outband signal messages from different co-located WRAN base stations as increasing the number of candidate channels and the number of co-located base stations. As shown in Figure 10, if the number of candidate channels is increasing, then outband message collisions are decreasing. And if we have the more co-located WRAN base stations, the higher collision probability we get.

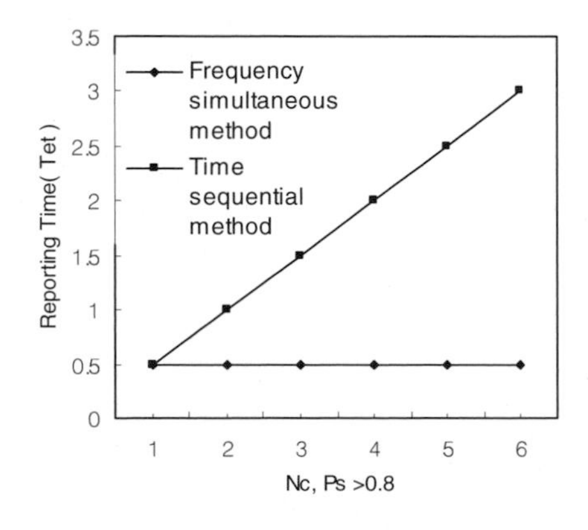

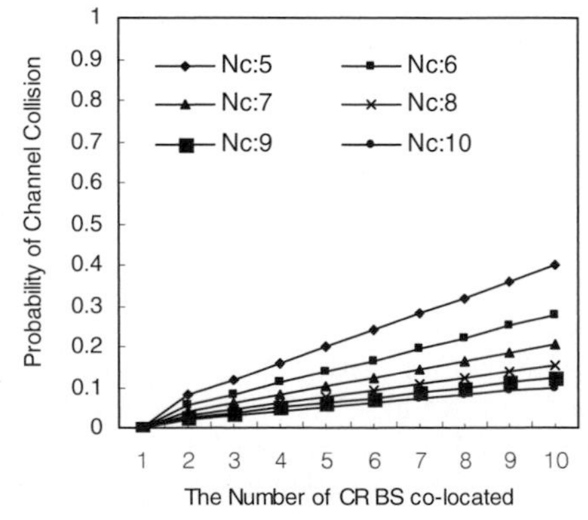

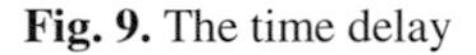

Fig. 9. The time delay

Fig. 10. The probability of channel collision

5 Conclusion

In this paper we have proposed cognitive radio MAC protocol for hidden incumbent system detection in WRAN. To solve hidden incumbent system problem, we proposed two inband and outband reporting procedures. In inband signaling when BS uses multiple channels, CPEs that detect primary system report it with one of other current channels. For outband signaling, BS periodically broadcast outband message through candidate channels. With the proposed hidden incumbent system detection methods, we can support reliable cognitive radio communication and protect incumbent system securely. CR BS can inform the CPEs of the new channel list to continue the communication. For initial CR network entry of the CPEs inside the overlapping areas, CPEs can easily find out the current service channel information.

In simulation study, we can see that for the given network conditions CPEs require 2-3 frames to send the hidden incumbent detection reports with 90% success ratio. For broadcasting methods, even though frequency simultaneous method requires multiple transceivers at BS, it can support short report delay. The other hand, time sequential method requires relatively long delay depend on the number of candidate channels.

References

1. Mitola III, J.: Software radios: Survey, critical evaluation and future directions. IEEE Aerospace and Electronic System Magazine 8(4), 25–36 (1993)
2. Mitola III, J.: Cognitive Radios: Making Software Radios More Personal. IEEE Personal Communications 6(4), 13–18 (1999)
3. Mitola III, J.: Cognitive radio for flexible mobile multimedia communications. In: IEEE International Workshop on Mobile Multimedia Communications, pp. 3–10 (November 1999)
4. Mitola III, J.: Cognitive radio: An integrated agent architecture for software defined radio (2004)
5. FCC, Spectrum policy task force report, No. 02-155 (November 2002)
6. FCC, Notice of rule making and order, No. 03-322 (December 2003)
7. IEEE 802.22-05/0007r47, Functional requirements for the 802.22 WRAN standard (2006)
8. Ball, Ferguso, Rondeau, T.W.: Consumer applications of cognitive radio defined networks. In: 2005 First IEEE International Symposium on New Frontiers in Dynamic Spectrum Access Networks, pp. 518–525 (November 2005)
9. Neel, J.O., Reed, J.H., Gilles, R.P.: Convergence of cognitive radio networks. In: IEEE Wireless Communications and Networking Conference, 4, pp. 2250–2255 (March 2004)

De-triangulation Optimal Solutions for Mobility Scenarios with Asymmetric Links

Pedro Vale Estrela[1,2], Teresa Maria Vazão[1,2], and Mário Serafim Nunes[1,2]

[1] Instituto Superior Técnico-TagusPark / TU Lisbon, Av. Prof Cavaco Silva, 2744-016 Oeiras, Portugal
[2] INESC-ID, Rua Alves Redol, 9, 1000-029, Lisboa, Portugal
{pedro.estrela,teresa.vazao}@tagus.ist.utl.pt,
{mario.nunes}@inov.pt

Abstract. In the field of mobility support, several mobility protocols resort to the use of triangulation mechanisms as a means of supporting fast handovers or basic connectivity. In order to reduce the maximum end-to-end delay of the packets, such triangulations can be later removed to enable direct routing of the data packets. However, using a simple update of this routing entry can cause the reception of out-of-order packets at the mobile node receiver, as the direct packets can arrive earlier than the triangulated packets.

This paper proposes a generic optimal de-triangulation mechanism that neither causes out-of-order packets nor increases the packet delay, for any combination of asymmetric links delays. After an analytic framework analysis contribution, the efficiency of the proposed algorithm is evaluated using simulation studies, in which the packet losses, packet delay, handover latency and control/data load metrics are measured.

1 Introduction

In the field of mobility support, several mobility protocols resort to the use of triangulation mechanisms as a means of supporting fast handovers or basic connectivity. By using these mechanisms, the routing of data packets destinated to mobile nodes is aided by a certain number of intermediate nodes, leading to longer paths and higher end-to-end delays.

Several mobility protocols with micro-mobility capabilities, which the *enhanced Terminal Mobility for IP [1] with Route Optimization (eTIMIP-RO) [2], Fast Handovers for Mobile IP* (FMIP) [3], *BRAIN Candidate for Mobility Management Protocol* (BCMP) [4] or *Micro-mobility support with Efficient Handoff and Route Optimization Mechanism* (MEHROOM) [5] are examples of, can provide a faster handover operation by forwarding the in-flight packets received at the previous Mobile Node (MN) location to the new location, using a simple triangulation scheme. In the same vein, the MIP *Route Optimization* (RO) option can be used to forward data between the involved Foreign Agents in an handover [6]. Also, macro-mobility protocols like MIPv4 [7] can permanently use the triangulation via the Home Agent (HA) as a means of achieving transparency to fixed Correspondent Nodes (CN), or of performing temporary data forwarding before the route optimization operation in MIPv6 [8].

T. Vazão, M.M. Freire, and I. Chong (Eds.): ICOIN 2007, LNCS 5200, pp. 90–102, 2008.

In all cases, in order to reduce the maximum end-to-end delay, such triangulations can be later removed by a ***de-triangulation operation***, in which the sender node is updated with the most current MN location, enabling it to send the packets directly to the MN. In all the previously mentioned protocols, the triangulation is simply removed by updating the sender node with the newest MN location, using a simple update message. However, this simple update can cause the reception of out-of-order packets at the MN. As long as the direct path will typically impose lower latency than the triangulated path, the subsequent directly sent data packets can actually be received *earlier* than the last in-flight packets sent via the triangulated path.

Depending on the type of receiver, these out-of-order packets due to such de-triangulation operations can have a major impact on it, as additional actions may be needed to recover from such phenomena [9]. These out-of-order packets can either be considered as lost packets by UDP receivers which expect ordered arrival [11], or can trigger the retransmit mechanism and needlessly reduce the contention window of the majority of the TCP senders, as no packets have actually been dropped [12].

This problem was addressed by previous research work. The *Celular IP semi-soft handover* [13] solves it by delaying the packets sent through the direct path at a specific node, known as crossover node, for a constant amount of time, in order to allow an earlier reception of the triangulated packets. In spite of avoiding out-of-order packets, this mechanisms forces a constant delay, which is independent of actual de-triangulation phenomena, and results in excessive majorated delays to ensure an ordered reception. Alternatively, the *seamless MIP* [14] marks the packets and reorders them at the destination. This is an inefficient and of limited applicability process, as it requires packet marking in the IP header of all data packets.

This paper proposes a generic optimal de-triangulation mechanism that neither causes out-of-order packets *nor* increases the packet delay. The proposed mechanism achieves its goals by delaying the packets, which will be sent via the direct path, for the exact minimum period of time that causes them to arrive at the destination immediately after the last in-flight packets sent via the triangulated path. Simpler versions of the mechanism that either remain optimal, but assume the presence of symmetric links, or that are non-optimal but continue to guarantee packet ordering, are also presented.

To support this contribution, the paper presents an illustrative example followed by an analytical framework in section 2, which is used to study the problem in section 3. The proposed solutions are described and analyzed in section 4 using the analytical framework, and are additionally validated by NS2 simulation studies in section 5. The paper ends, in section 6, with some conclusions and future work.

2 Analytical Framework and Problem Definition

2.1 Problem Statement

To illustrate the problem, let us consider the triangulation situation depicted in Fig 1a. The figure comprises the crossover node, Node A, which is the node where the triangulated and the direct paths diverge; the old destination node, Node B, which is the one that was previously used to transfer information with the MN and will be used as a triangulated node; and the new destination node, Node C, which is the one that should be used to access the MN after it roams from Node B to Node C. For the sake of simplicity, the MN is omitted in the figure.

At first, when Node A receives data to the MN, it uses Node B the reach the MN, although it is now accessible through Node C; then, Node C will send an update message to Node A, so that data may be routed directly to it, removing the triangulation effect. Even though such operations do not typically incur in the drop of in-transit data packets, it can result in their reordering, as the direct path will typically have lower latency than the triangulated one (otherwise, the routing would benefit from being triangulated in the first place, and no de-triangulation mechanism should be performed). The exact latency difference will be the result of the actual link delays and the amount of data packets queued in the routers, among other factors.

This problem is illustrated in Fig. 1b, where packets sent through the direct path (Data packet 2, A→C) can be received earlier than packets sent through the triangulated path (Data packet 1, A→B→C).

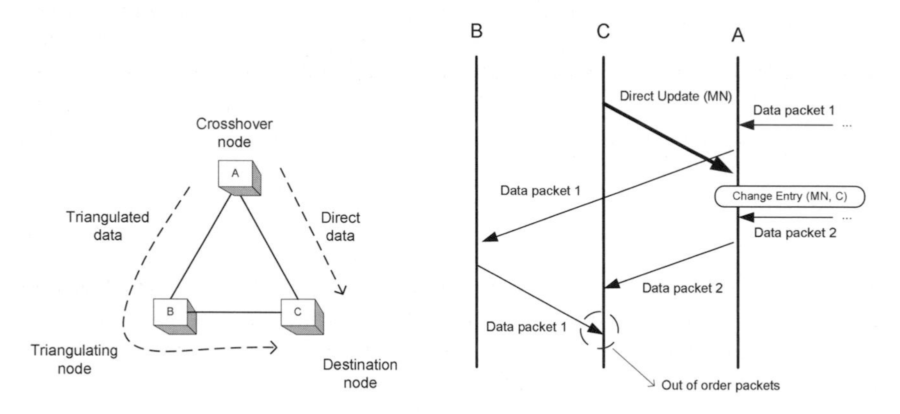

Fig. 1. De-triangulation problem illustration. a) overall problem b) messages exchanged

2.1 De-triangulation Mechanism Description

In this section, we introduce a new framework for analysing the de-triangulation problem.

Let us considered the situation represented in Fig. 2a where the data packet transmission and reception instants are depicted and the time instants represent:

- $t_{0:}$ the time when the de-triangulation process will start at the destination node (Node C);
- $t_x > t_{0:}$ the time when the ***last*** data packet is sent by the crossover node (Node A), through the ***triangulated*** path;
- $t_{x'} > t_x$ the time when this packet is received by the destination node;
- $t_y > t_{x:}$ the time when the ***first*** data packet is sent by the crossover node (Node A) through the ***direct*** path;
- $t_{y'} > t_y$ the time when this packet is received by the destination node.

Considering also the mean delays that occur between the involved nodes for the transmission of packets, depicted in Fig. 2b, which represent, separately, the downstream ($d_{(A,B)}$, $d_{(B,C)}$ and $d_{(A,C)}$) and the upstream ($d_{(B,A)}$, $d_{(C,B)}$ and $d_{(C,A)}$) paths. Each mentioned delay will contain the time needed to support all the operations involved to transmit a packet and to receive it, of which the propagation and queuing delays will be the main drivers.

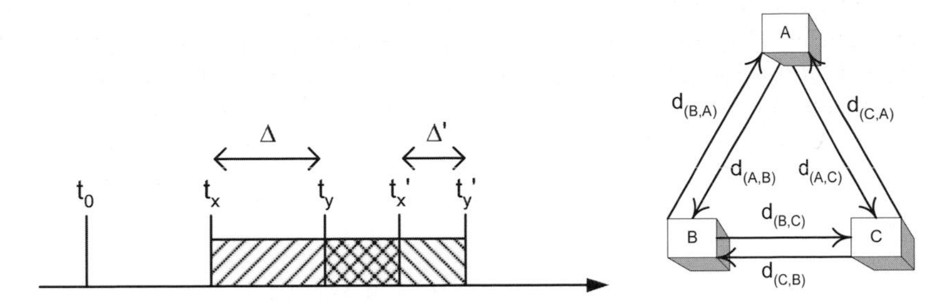

Fig. 2. a) Time instants of data packet transmission and reception b) asymmetric link delays

Using this notation, the following relations occur:

$$t_{x'} = t_x + d_{(A,B)} + d_{(B,C)} \qquad \text{(Eq. 1)}$$

$$t_{y'} = t_y + d_{(A,C)} \qquad \text{(Eq. 2)}$$

Considering now, the time Δ since the transmission of the last triangulated packet until the transmission of the first direct packet at the crossover node; and the time Δ' since the arrival of the first direct packet until the arrival of the last triangulated packet a the destination node, which are given by:

$$\Delta = t_y - t_x \qquad \text{(Eq. 3)}$$

$$\Delta' = t_{y'} - t_{x'} \qquad \text{(Eq. 4)}$$

A positive value of Δ results in buffering being applied to the packets at the crossover node, as incoming packets need to wait before they are forwarded. The value is directly proportional to the maximum buffering capacity required. A value of zero results in no buffering at all. Finally, a negative value of Δ is not used in the context of de-triangulation.

A negative value of Δ' results in the occurrence of out-of-order packets, but no additional delay, as triangulated packets are received later than the direct ones. On other hand, a positive value results in ordered delivery, but with an additional delay being experienced by the involved packets which is equal to Δ'. Finally, a value of zero will result in an optimal de-triangulation mechanism – no out-of-order packets and no delay increase.

Thus, the objective of an ordered de-triangulation mechanism is defined as the operation where the packets will be delayed at the crossover node for a certain period of time Δ, so that Δ' will be positive; additionally, the objective of an optimal ordered de-triangulation mechanism is to ensure that Δ' will be zero.

3 Analysing the Direct De-triangulation Problem

To illustrate the generic solution, let us consider the regular de-triangulation situation depicted in Fig.3.

Firstly, the destination Node C sends an Update Message directly to the crossover node (step 1). Then, the crossover node will update its own Routing Table, being able to start using the direct path (step 2). From now on, incoming packets will be sent directly to Node C, while already in-transit packets continue to use Node B (step 3). Every in-flight packet received by Node B is sent to Node C (triangulated path) (step 4), being received later than the direct paths, unless an ordered de-triangulation mechanism is used.

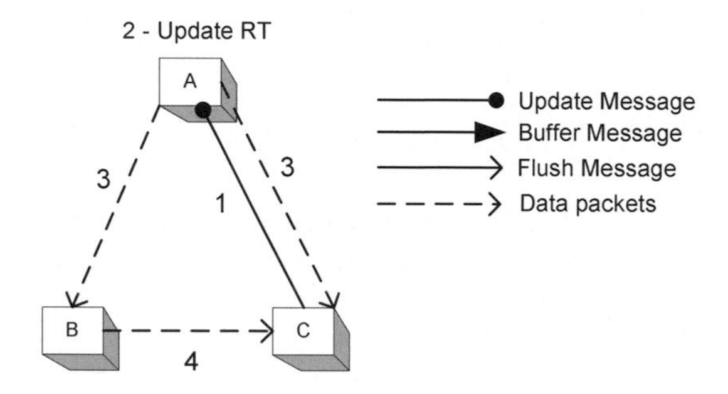

Fig. 3. Direct de-triangulation algorithm

Using the same notation as before:

$$t_x = t_0 + d_{(C,A)} \quad \text{(Eq. 5)}$$

$$t_y = t_x \quad \text{(Eq. 6)}$$

$$\Delta' = (t_y + d_{(A,C)}) - (t_x + d_{(A,B)} + d_{(B,C)}) \quad \text{(Eq. 7)}$$

$$\Delta' = d_{(A,C)} - d_{(A,B)} - d_{(B,C)} \quad \text{(Eq. 8)}$$

In order to avoid out-of-order packets, Δ' must be positive or equal to zero:

$$\Delta' >= 0 \quad \text{(Eq. 9)}$$

$$d_{(A,C)} >= d_{(A,B)} + d_{(B,C)} \quad \text{(Eq. 10)}$$

As this last relation is not verified in the typical situations, as it would negate the major benefit of removing the de-triangulation, this results in out-of-order packets phenomena illustrated in section 2.

4 Proposed Solutions

4.1 Conservative Algorithm

The first presented algorithm, named **conservative algorithm** (Fig. 4a), will ensure that no out-of-order packets will occur, by buffering the received packets at the crossover node, while waiting for the reception of the last packet via the triangulated path. Only when this happens, the buffered packets will be released by the use of a new message called ***Flush***.

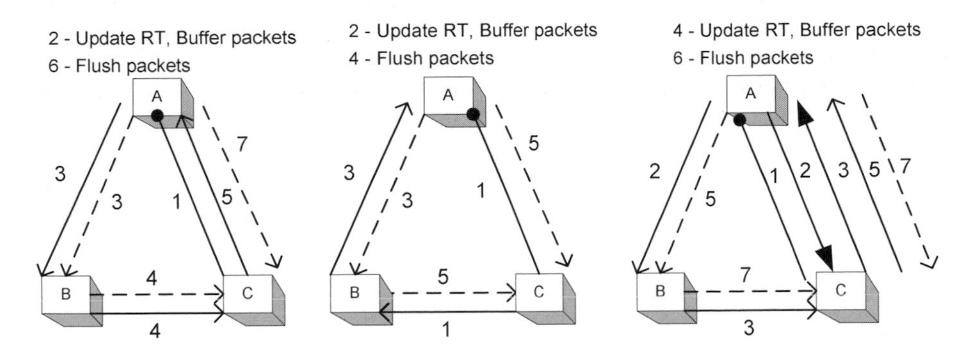

Fig. 4. Proposed solutions: a) conservative b) symmetric c) asymmetric

Firstly, the destination Node C sends an Update Message directly to the crossover node, as in the previous case (step 1), but now it starts buffering all incoming packets to a certain MN (step 2). Then, node A will send a Flush message that will pass through the triangulated and the destination node before returning to itself (step 3, 4 and 5). This forces the control packet to pass through the same paths as the in-transit data packets in the triangulated path (also steps 3 and 4, dashed lines). When node A receives the Flush Message, it transmits the buffered packets via the direct path, and stops the buffering of additional packets (steps 6 and 7).

Using the same notation as above, the last packet forwarded using the triangulated path will guarantedly be received at node C *earlier* than the first packet received via the direct path. However, this operation increases the previous handover latency for an additional amount of time, as shown by the following analysis:

$$t_x = t_0 + d_{(C,A)} \quad \text{(Eq. 11)}$$

$$t_y = t_x + d_{(A,B)} + d_{(B,C)} + d_{(C,A)} \quad \text{(Eq. 12)}$$

$$\Delta' = (t_y + d_{(A,C)}) - (t_x + d_{(A,B)} + d_{(B,C)}) \quad \text{(Eq. 13)}$$

$$\Delta' = (t_x + d_{(A,B)} + d_{(B,C)} + d_{(C,A)} + d_{(A,C)}) - (t_x + d_{(A,B)} + d_{(B,C)}) \quad \text{(Eq. 14)}$$

$$\Delta' = d_{(C,A)} + d_{(A,C)} \quad \text{(Eq. 15)}$$

Thus, out-of-order packets are avoided ($\Delta' > 0$), but the data packets are always unnecessarily delayed for ($d_{(C,A)} + d_{(A,C)}$) time units.

4.2 Optimal Algorithm for Symmetric Links

If the link delays can be assumed to be symmetric, then the previous algorithm can be refined to not incur in any extra delay, besides the imposed by triangulation itself. This has the advantages of solving the latency increase of the previous mechanism, and of reducing the buffering requirements of the **symmetric algorithm** (Fig. 4b).

Again, the destination node sends an Update Message directly to the crossover node (step 1) to start the buffering of all data packets (step 2). However, at the same the, the destination node also sends the Flush Message in parallel to the crossover node, which now passes through the triangulated node in the opposite direction of the data packets (step 1 and 3). As before, when the crossover node receives this message, it stops buffering additional packets (step 4) and transmits the buffered packets via the direct path (step 5).

Using the same notation as before:

$$t_x = t_0 + d_{(C,A)} \quad \text{(Eq. 16)}$$

$$t_y = t_0+d_{(C,B)}+d_{(B,A)} \quad \text{(Eq. 17)}$$

$$\Delta' = (t_y+d_{(A,C)})-(t_x+d_{(A,B)}+d_{(B,C)}) \quad \text{(Eq. 18)}$$

$$\Delta' = (t_0+d_{(C,B)}+d_{(B,A)}+d_{(A,C)})-(t_0+d_{(C,A)}+d_{(A,B)}+d_{(B,C)}) \quad \text{Eq. 19)}$$

$$\Delta' = (d_{(C,B)}+d_{(B,A)}+d_{(A,C)})-(d_{(B,C)}+d_{(A,B)}+d_{(C,A)}) \quad \text{(Eq. 20)}$$

As symmetric links are assumed, $d_{(C,B)}=d_{(B,C)}$; $d_{(B,A)}=d_{(A,B)}$; and $d_{(A,C)}=d_{(C,A)}$; thus:

$$\Delta' = 0 \quad \text{(Eq. 21)}$$

Thus, the algorithm is optimal: out-of-order packets are avoided, and no extra delay is incurred in the data packets.

4.3 Optimal Algorithm for Asymmetric Links

If the link delays are asymmetric, e.g. due to different propagation delays or queuing, then the conservative algorithm can again result in out-of-order packets, in the situations where Δ' is negative. Such situation might happen when the control messages (Update and Flush messages) travel faster than the data packets and the delay imposed to data packets at the crossover node is smaller than needed. Although this could be solved by the addition of an extra small delay after the reception of the Flush Message, this would not result in an optimal solution. In contrast, the final **asymmetric algorithm** presented in this section is able to maintain optimality through slightly higher control and data loads (Fig. 4c).

Firstly, the destination node sends an Update Message directly to the crossover node (step 1). When it receives the message, it sends a Flush Message as in the conservative algorithm via nodes B (step 2) and C (step 3 and 5), but also a new message to itself via node C (steps 2 and 3), called ***Buffer Message***. These pair of messages are used to measure the actual delays experienced by data packets, taking account the asymmetric nature of the links. When the crossover node receives the Buffer Message (step 3), it updates the Routing Table and starts buffering data packets (step 4). The reception of the Flush Message is dealt with as before (step 6 and 7).

Using the same notation as above:

$$t_x=t_0+d_{(C,A)}+d_{(A,C)}+d_{(C,A)} \quad \text{(Eq. 22)}$$

$$t_y=t_0+d_{(C,A)}+d_{(A,B)}+d_{(B,C)}+d_{(C,A)} \quad \text{(Eq. 23)}$$

$$\Delta'=(t_y+d_{(A,C)})-(t_x+d_{(A,B)}+d_{(B,C)}) \quad \text{(Eq. 24)}$$

$$\Delta' = (t_0 + d_{(C,A)} + d_{(A,B)} + d_{(B,C)} + d_{(C,A)} + d_{(A,C)}) - (t_0 + d_{(C,A)} + d_{(A,C)} + d_{(C,A)} + d_{(A,B)} + d_{(B,C)}) \quad \text{(Eq. 25)}$$

$$\Delta' = (d_{(A,B)} + d_{(B,C)} + d_{(C,A)} + d_{(A,C)}) - (d_{(A,B)} + d_{(B,C)} + d_{(C,A)} + d_{(A,C)}) \quad \text{(Eq. 26)}$$

Without the need to assume symmetric links, equation 26 simplifies to:

$$\Delta' = 0 \quad \text{(Eq. 27)}$$

Thus, the algorithm is always optimal regardless of the combination of asymmetric links: out-of-order packets are avoided, and no extra delay is incurred.

5 Simulation Studies

5.1 Simulation Scenario

To evaluate the proposed smooth de-triangulation algorithms, their behaviour was compared with the standard direct de-triangulation using simulation studies. The simulations were carried out using Network Simulator (NS) v2.26, where a simple mobility protocol which uses local temporary triangulations at each handover was modelled. A key addition to the simulator was the modification of the existing UDP LossMonitor object to evaluate and count the received out-of-order packets, besides the dropped packets that are already considered by the base object.

The simulation scenario is illustrated in Fig. 5. Here, a MN will be the receiver of a continuous stream of Constant Bit Rate (CBR) UDP probe packets sent by a CN located outside the domain, that generates 200 packets per second of 100 bytes each. The MN will then make a series of roaming operations between the two existing Access Points, AP1 and AP2, being connected to the network by each one sequentially at a speed of 30 hand/min. The domain has a mobility anchor point (MAP) which receives the packets from the CN, and redirects them to the MN's current AP. Inside the domain the nodes are connected by wired links that feature sufficient bandwidth for the test probes.

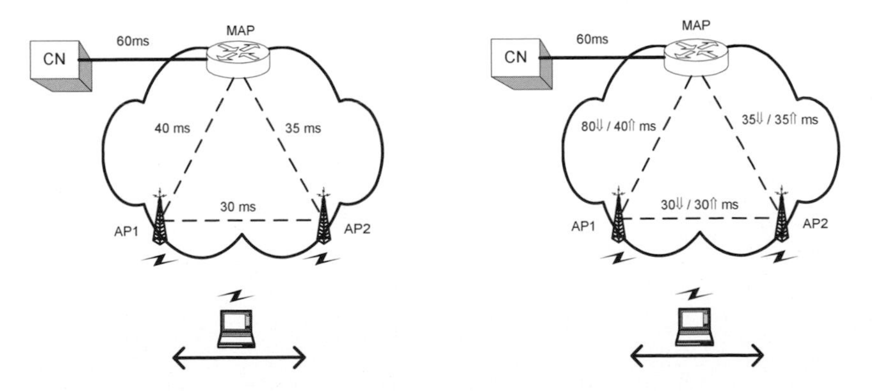

Fig. 5. Simulation Scenario: a) Symmetric case b) Asymmetric case

Two separate scenarios are considered; in the former, all links, with the purposely different delays values as depicted in Fig. 5a, feature symmetric upstream and downstream propagation delays; an asymmetric scenario was also investigated, illustrated in Fig. 5b, where the crossover→AP1 downstream link delay is asymmetric from the upstream delay, in order to simulate local congestion.

At each handover operation, the MN first creates a local tunnel between the involved APs, by sending an Update Message through the new AP destinated to the old AP. To avoid the existence of dropped packets at the old AP, the APs have sufficient transmission power to ensure a continuous coverage to the MN. Then, the handover operation is finalized by updating the MN's entry at the crossover node; in the *direct* model, this is achieved by sending an Update Message directly to the crossover node via the new AP; the other studied models (*smooth_conservative*, *smooth_symmetric* and *smooth_asymmetric*) additionally use the sequence of messages and buffering operations previously described.

In this scenario, each model was tested for a series of 100 handover operations, where the experienced packet losses, delay and buffer usage for the data and control packets were evaluated.

5.2 Loss and Delay Results

Fig. 6 and Fig. 7 show the loss and delay results for all studied alternatives for both symmetric and asymmetric links. The loss graph divides the experienced losses into drops, late packets and out-of-order packets. The delay graph divides the experienced one-way delays into minimum, average and maximum values. As expected from the simulation scenario definitions, no packets are dropped in all cases.

Regarding the regular **direct mechanism**, a large number of packets are received out-of-order, confirming the de-triangulation problem previously defined. On the other hand, the data packets experience different minimum and maximum delays, depending on the stationary and handover situations. The minimum delay (~95 ms) is experienced when the MN is stationary at AP2, as the packets are routed directly from the crossover node to the AP2; the maximum delay is experienced for the triangulated packets, which are sent to AP2 triangulated via AP1 (~130 ms for symmetric test / ~170 ms for asymmetric test).

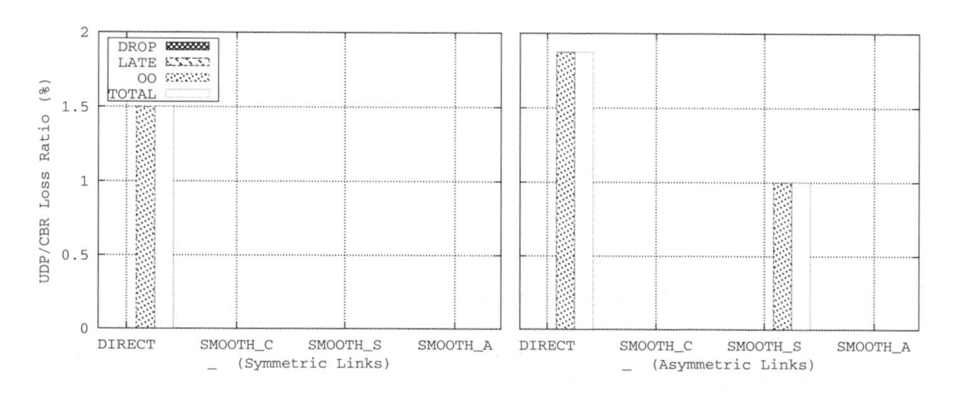

Fig. 6. Loss ratio as a result of Drops, Late and Out-of-Order packets

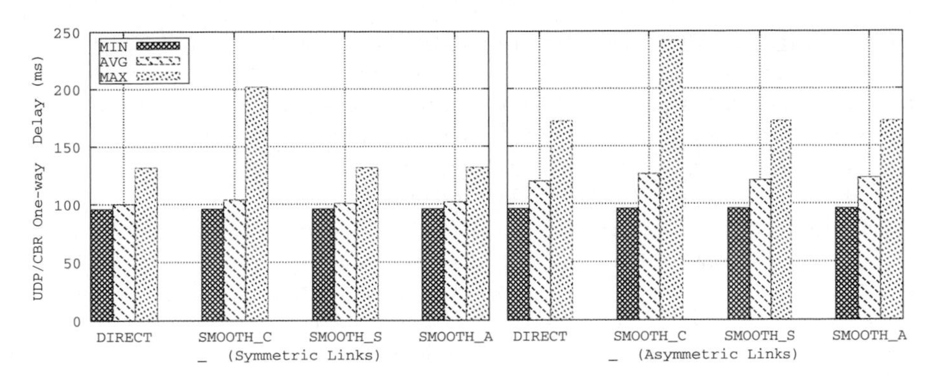

Fig. 7. One-way packet delay (minimum, average and maximum values)

The **conservative mechanism** is able to solve the out-of-order packets phenomena, for both symmetric and asymmetric links. However, the mechanism is not optimal, as it greatly increases the maximum delay of the involved packets (~200 ms for symmetric test / ~240 ms for the asymmetric case).

The **symmetric mechanism** solves this problem, but only if the links are symmetric. In these conditions, no out-of-order packets occur, and the maximum delay is similar to the one experienced in the base direct mechanism (~130 ms). With asymmetric links, even though the maximum delay is not increased (~170 ms), a small number of out-of-order packets is experienced, being this value related to the sum of the delay asymmetries of the involved links, as defined by eq.20.

Finally, the **asymmetric mechanism** solves this last problem, for any combination of asymmetric links. In these conditions, no out-of-order packets occur, and the maximum delay is similar to the one experienced in the base direct mechanism (~130 ms for symmetric test / ~170 ms for asymmetric test).

5.3 Buffer Utilization, Control Load and Data Load Results

Fig. 8, Fig. 9 and Fig. 10 show the buffer utilization, the signalling load and the data load for all studied alternatives, for both symmetric and asymmetric links. The buffer graph shows the average number of queued packets at the crossover node per handover; the control load shows the average number of forwarded control packets per handover, for all the used control packets (Update, Buffer and Flush messages); lastly, data load shows the ratio of data packets that are locally triangulated between the two APs.

Regarding the regular **direct mechanism**, the buffer queues are not used, as no packets are delayed at the crossover node. On the other hand, by only using a single Update message, it has the lowest control load requirements from all the alternatives. Finally, the data packets subject to triangulation for a certain time period, ending when the Update message reaches the crossover node.

The **conservative mechanism** requires the largest buffers at the crossover, which fully corresponds to the maximum delay studied in Fig. 7, needed to hold all the packets during the de-triangulation procedure. Additionally, by propagating its Flush Message through the whole triangle, this scheme has higher control load requirements than the previous case. However, it stops the triangulation effect at the same time as the direct case, by buffering the data packets at the crossover node; thus it has an equal data load as the previous case.

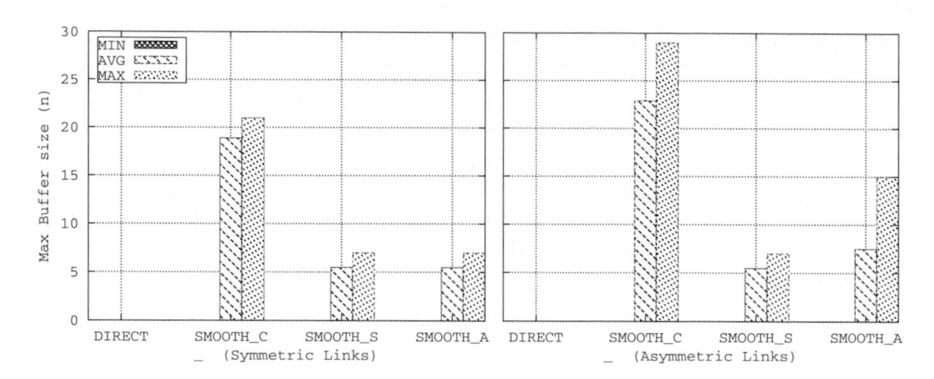

Fig. 8. Average Buffer size per Handover

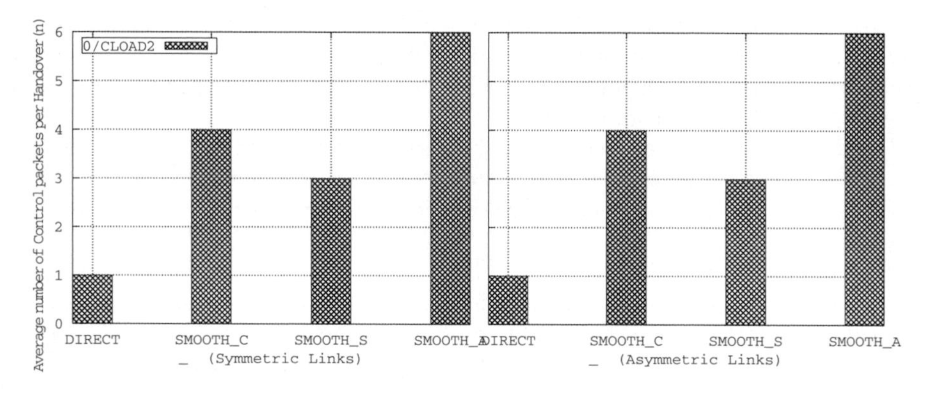

Fig. 9. Average Control load per handover

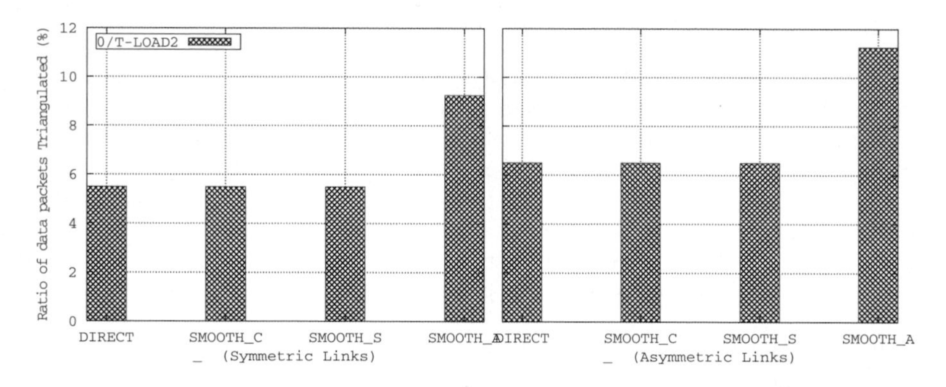

Fig. 10. Ratio of triangulated data packets

By imposing a much lower maximum delay, the **symmetric mechanism** requires a corresponding lower amount of buffer space at the crossover node. Additionally, the Flush message is simplified, which results in lower control load requirements. Again, the symmetric mechanism has a similar data load requirements than the base case, for the same reasons explained previously.

Finally, the **asymmetric mechanism** has the same low buffer usage for symmetric cases as the previous mechanism; however, it now correctly handles the asymmetric cases as well, by buffering the data packets for the necessary time. The downside of these capabilities is an increased control and data load of all the studied mechanisms. The former is related to the usage of additional messages to probe the delay differences; the latter is related to the maintenance of the triangulation effect for an additional small time period, in order to wait for the delay differences's measurements. However, it should be noted that all these control and data messages are always limited to the wired part of the network; thus, these mechanisms have the same efficiency as the base case regarding the usage of the limited wireless resources.

6 Conclusions

While the original MIP protocol keeps the proposed triangulation effect at all times, newer macro and micro mobility protocols solve the problem of average long delays, but suffer from a de-triangulation problem, where out-of-order packets are experienced at each de-triangulation event.

In this paper, we have identified the problem of de-triangulation, introduced a new analytical framework, and then used the framework to study the problem and to propose three increasingly complex de-triangulation schemes that feature different assumptions, results and signalling requirements.

The conservative approach guarantees packet ordering, but increases the maximum delay of the involved de-triangulation data packets; the symmetric approach maintains packet ordering without increasing the maximum delay, for any combination of link delays, but only if these are symmetric; the asymmetric approach guarantees packet ordering without maximum delay increases, for any combination of asymmetric link delays.

The proposed algorithm's efficiency is evaluated using both analytic and NS2 simulation studies which measured the packet losses, packet delay, handover latency and control/data load; besides the ordering and delay results already mentioned, it was found that the buffer requirements are directly co-related to the maximum delay increase, and that the most complex mechanisms have increasing control load requirements, which are nonetheless limited to the wired part of the network.

All these generic algorithms can be easily applied to all mobility protocols that feature a de-triangulation component. An initial state-of-the-art review of this area showed that this can include, at least, the eTIMIP-RO [2], fMIP [3], MIPv4-RO [6], MIPv6-RO [8], BCMP [4], MEHROM [5] and the MIP-SB [15] protocols.

References

[1] Estrela, P., Vazão, T., Nunes, M.: Design and Evaluation of eTIMIP - Overlay Micro-Mobility Architecture based on TIMIP. In: International Conference on Wireless and Mobile Communications ICWMC 2006, Romania (July 2006)

[2] Estrela, P., Vazão, T., Nunes, M.: A Route Optimization Scheme for the eTIMIP micro-mobility protocol. In: PIMRC 2007, Greece (September 2007) (accepted for publication)

[3] Koodli Ed., R.: Fast Handovers for Mobile IPv6. RFC 4068 (July 2005)

[4] Keszei, C., Georganopoulos, N., Tyranyi, Z., Valko, A.: Evaluation of the BRAIN candidate mobility management protocol. In: Proceedings of the IST Mobile Summit (September 2001)
[5] Peters, L., Moerman, I., Dhoedt, B., Demeester, P.: MEHROM: Micromobility support with Efficient Handoff and Route Optimization Mechanism. In: 16th ITC Specialist Seminar on Performance Evaluation of Wireless and Mobile Systems, Antwerp, Belgium, August 31 - September 2, pp. 269–278 (2004)
[6] Perkins, C., Johnson, D.: Route Optimization in Mobile IP, draft-ietf-mobileip-optim-11.txt, IETF (September 2001)
[7] Perkins Ed., C.: IP Mobility Support for IPv4, RFC-3320, IETF (January 2002)
[8] Johnson, D., Perkins, C., Arkko, J.: Mobility Support in IPv6, RFC-3775 (June 2004)
[9] Morton, A., Ciavattone, L., Ramachandran, G., Shalunov, S., Perser, J.: Packet Reordering Metric for IPPM, draft-ietf-ippm-reordering-13.txt (May 2006)
[10] Bennet, J., Partridge, C., Shectman, N.: Packet reordering is not pathological behaviour. IEEE/ACM Transactions on Networking 7(6) (December 1999)
[11] Laor, M., Gendel, L.: The effect of packet reordering in a backbone link on application throughput. IEEE Network 7(6) (September 2002)
[12] Zhou, X., Mieghem, P.: Reordering of IP packets in Internet. In: Proc. Passive and Active Measurement (April 2004)
[13] Campbell, A., et al.: Design, Implementation and Evaluation of Cellular IP. IEEE Personal Communications 7(4) (August 2000)
[14] Hsieh, R., Seneviratne, A.: A comparison of mechanisms for improving mobile IP handoff latency for end-to-end TCP. In: MOBICOM 2003, pp. 29–41 (September 2003)
[15] El Malki, K., Soliman, H.: Simultaneous Bindings for Mobile IPv6 Fast Handovers, draft-elmalki-mobileip-bicasting-v6-06 (work in progress) (July 2005)

Analysis of Hierarchical Paging

Mate Szalay and Sandor Imre

Budapest University of Technology and Economics, Department of Telecommunications

Abstract. In this paper a new location management algorithm called *Hierarchical Paging* is introduced. A mathematical model for the proposed scheme is also presented, which makes it possible to compare it to other solutions and to examine how various parameters affect its performance. It is shown that our solution can be more efficient than traditional paging.

1 Introduction

Over the past few years there has been extreme growth in wireless communications. There is a fierce competition among service providers, and efficiency is very important factor in this competition.

Any mobility protocol has to solve two separate problems: location management (sometimes called reachability) and session continuity (sometimes referred to as handover management). Location management means keeping track of the positions of the mobile nodes in the mobile network, session continuity means to make it possible for the mobile node to continue its sessions (e.g. phone calls) when the mobile node moves to another cell and changes its service access point.

Location management has to address the following questions[2]:

- When should the mobile terminal update its location to the network?
- When a call arrives, how should the exact location of the called mobile equipment be determined?
- How should user location information be stored and disseminated throughout the network?

These problems are solved in two stages: *location registration (or update or tracking) and call delivery (or searching)*[1].

Because of the growth of mobile communications and the limitations of resources (especially frequency), more and more efficient algorithms are needed for routing, call management and location management.

In this paper we present a new efficient location management algorithm.

This paper is structured as follows:

After defining our terminology, explaining the principles of hierarchical location management and paging, related work in the field is presented in Section 3. Then, in Section 4.1, our proposed solution, *hierarchical paging* is introduced. Section 5 presents our mathematical model for the mobile network, the mobility model, the cost model, and the model of the mechanisms of our location

T. Vazão, M.M. Freire, and I. Chong (Eds.): ICOIN 2007, LNCS 5200, pp. 103–112, 2008.

management algorithm. Section 6 presents the comparative analysis of the proposed solution, the effect of changes in various parameters is examined in detail. Finally, in Section 7 the conclusions are drawn.

2 Terminology

The node that is moving around in the mobile network is called *mobile node* or *mobile equipment. Mobile nodes* are connected to the network via *base stations*, each of the base stations covers one *cell.*

The event, when the mobile equipment moves to a new base station (changes its point of connection) is called *handover* or *handoff.*

In a mobile network the terms *location* and *position* usually do not mean geographical position, but they refer the location of the mobile node within the network, the service access point (base station) it is connected to at a specific moment of time.

2.1 Hierarchical Mobility Management

The area where mobility is provided can vary widely from network to network. A satellite-based system may cover a whole continent (or even a whole planet), while a small-scale, targeted mobile network may cover only a building or a part of a building.

Even if the covered area is not too large, mobility management is often handled in a *hierarchical* way for *efficiency* and *scalability* reasons[1]. The area that is covered is divided into smaller areas, the whole network is divided into subnetworks. The point here is that all mobility within a subnetwork can be handled within the boundaries of that subnetwork in a completely transparent way for the other parts of the network. The subnetwork then can be again divided into sub-subnetworks. A solution that is not hierarchical is called *flat*, or *single-layer.*

A hierarchical solution can be more efficient than a flat one by using less resources for mobility management. And at the same time, hierarchical solutions usually scale much better with the size of the network and the number of mobile nodes than flat ones.

2.2 Paging

It is possible to define a single-layer scheme where the location of the mobile node is not always known to the network. In a paging-enabled network the mobile node can switch to *idle state* (sometimes referred to as *standby state*).

The advantage of this scheme is that in idle mode, the mobile node does not have to notify the network every time a handover takes place. Fewer location update messages imply smaller signalling load on the network, and longer battery life for the mobile node.

On the other hand, if the exact location is not known, then the mobile node has to be found in some way, in case it has to be contacted (for example an incoming call or packet arrives), this is called *paging.*

There are various strategies for paging, for references see Section 3.

3 Related Work

The paper[1] of Akyldiz et al. is a good general survey on various existing and proposed mobility management algorithms.

Several solutions exist to both location management and handover management in mobile networks. For some IP (Internet Protocol) based solutions see [2,3,4,5].

Several studies have addressed paging, various paging methods are presented, usually addressing cost minimizations. Almost all of these papers focus on the paging algorithm itself, but the paging is always just at a single (usually the bottom) level (*flat*). Ramjee et al. examine and compare three different paging architectures and protocols, all flat[11]. Hajek, Mitzel and Yang show algorithms to optimize registration and paging together. They examine serial and parallel paging, but both of them are "flat".

In their paper[10], Woo et al. optimize location management for a special types of networks, namely *Two-Way Messaging* networks, using flat paging.

Usually, paging is used at the bottom level of a hierarchical mobility management solution. A real-life example where paging is used is the GSM network architecture [8,9].

4 Hierarchical Paging

4.1 The Approach

Consider a two-layer mobility network: there is a top layer (or top level) and a bottom layer (or bottom level). Both of the layers are implemented using a single-layer paging solution.

4.2 Mechanisms

Consider a network where mobility management is handled in this fashion. The network is divided into subnetworks, and there is an entity in each subnetwork that is responsible for the mobility management. This entity is going to be referred as the *root node*. It either knows the exact location of the mobile node within the subnetwork, or it can determine it by using a (single-layer) paging algorithm.

When an incoming call or packet arrives destined to the mobile node, the old root node is contacted first, because top level mobility management still "thinks" that the mobile node is in the old subnetwork. But the old root node knows that the mobile node is not there, so a top level paging takes place. This top level paging means that the gateway of the network (or the top-level root node) "pages" all the (bottom level) root nodes of the subnetworks, and the root node of the new subnetwork is going to give a positive reply. This top level paging still requires much less signalling than a single layer paging over the whole network would, so it is much more efficient.

The spot, where hierarchical paging is less efficient than single-layer paging is when the mobile node crosses a subnetwork boundary. In that case signalling traffic (location update messages) is generated in case of hierarchical paging, but no messages are sent at all if flat paging is used.

Although hierarchical paging might be much more efficient than single-level paging in certain situations, its scalability might be a problem. If some smart paging algorithm is used at the top-level, this solution might become feasible. To decide wether it is profitable to use hierarchical paging instead of single layer paging various parameters such as mobility intensity, incoming call intensity, various signalling and processing costs have to be taken into account.

We know of no proposal or implementation of this scheme.

5 Analytical Model

In this Section our model is presented for a mobility network, where two-level Hierarchical Paging mobility management algorithm is used. The levels are going to be references as *top* (or *upper*) level and *bottom* (or *lower*) level.

5.1 Network Model

This section describes the network model of a two-layer Hierarchical Paging mobility network. The whole network consists of L cells. The network is divided into k subnetworks, each containing L/k cells. In this case each subnetwork contains L/k cells. Note that this is as approximation, as k might not be a divisor of L.

A handover between two cells of the same subnetwork is called *intra-subnetwork handover*, and a handover between the cells of two neighboring subnetworks is called *inter-subnetwork handover*.

5.2 Mobility Model

The mobility model that is going to be used for our analysis is very similar to the model used in [10].

A mobile node moves to another cell with a constant rate. This is modelled as a Poisson process as in [6,7]. Let λ denote the parameter of the Poisson process and so denote the rate of handovers of the mobile node.

If the cell that the mobile node is staying in has n neighboring cells, then the mobile node is moving to each one of then with a probability of $1/n$.

We assume that voice calls also arrive following a Poisson process. The rate of the call arrival process is denoted by μ.

At any given time a handover takes place *before a call is received* with a probability of $\lambda/(\lambda+\mu)$ and a call is received *before a handover takes place* with a probability of $\mu/(\lambda+\mu)$.

There is one more parameter that has to be defined before our model for Hierarchical Paging can be introduced. If the network is divided into k subnetworks as described in Subsection 5.1, and a handover takes place, the probability that it is an inter-subnetwork handover is $1/\sqrt{(L/k)} = \sqrt{(k/L)}$. It is a good estimate because if the subnetworks are built up of L/k cells, than about one out of $\sqrt{(L/k)}$ cell-borders is a border between different subnetworks.

5.3 Markov-Model

States. One specific mobile node in a mobility network that uses Hierarchical Paging can be modelled using a discrete-time Markov-chain. The Markov-chain has four states: P_0, P_1, P_2 and P_3. The meaning of the states is the following (for reference see Section 4.2):

- P_0: At the top-level it is known which subnetwork the mobile node is currently staying in, and at the bottom level it is known which cell the mobile node is currently staying in. In other words the exact location of the mobile node is currently known.
- P_1: At the top-level it is known which subnetwork the mobile node is currently staying in, but it is not known which cell the mobile node is in. In other words the top-level mobility management has subnetwork-level location information.
- P_2: The subnetwork, which the mobile node is staying in is not known at the top level, but (at the bottom level) the root node of the subnetwork knows the cell, which the mobile node is staying in.
- P_3: The subnetwork which the mobile node is staying in is not known at the top level, and the cell that the mobile node is staying in is not known at the bottom level. In other words, the only information the network has is that the root node of the subnetwork that the mobile node is staying is knows that the mobile is in that specific subnetwork.

Whenever there is a handover or an incoming call, there is a transition in the Markov-chain. Let p denote the probability that the next event is a handover and not a received call. According to the considerations in Subsection 5.2:

$$p = \frac{\lambda}{\lambda + \mu} \tag{1}$$

p is a parameter of our model, we have no way of affecting it by repartitioning the subnetwork or by changing the mobility management in any other way. This parameter is sometimes referred to as "mobility ratio" or "call-to-mobility ratio" [6].

Transition Probabilities. Let's start from P_0; the exact position of the mobile is known. While there are no handovers just incoming calls, the position remains known, thus the states remains P_0. If there is a handover (with the probability of p) is is an inter-subnetwork handover (going to P_2) with the probability of $\sqrt{(k/L)}$, and is an intra-subnetwork handover (to state P_1) with a probability of $1 - \sqrt{(k/L)}$. Figure 1 shows the probabilities of all the possible transitions (on the left of the vertical separator "pipe"). On the right of the separator there are costs that are going to be introduced in the next Subsection. Generally, we move to P_2 with a probability of $p(\sqrt{(k/L)})$, which means that no matter what kind of position information the network had before; after an inter-subnetwork handover the top-level will have no exact position information, and the bottom-level will (the mobile node sends the location update message from a cell, and the bottom-level will know this cell, so it will have exact location information until the mobile node moves away.

Costs. The costs of the various mechanisms are defined in this section. These costs are of course not absolute costs, but relative ones that are used for comparison of different solutions or the same solution with different values of parameters.

In a hierarchical paging scheme an intra-subnetwork handover has a cost of zero. No location update messages are sent, no databases are updated, no entity is notified.

In case of an inter-subnetwork handover, the root nodes of the old and new subnetworks have to be notified about the handover, thus a constant cost of 2 will be assigned to this process.

The cost of receiving an incoming call of packet is made up of the cost of determining the exact location of the mobile node and the cost of receiving the call itself. We have no way of affecting the cost of receiving the call, and it is always there, so we will only consider the cost of determining the exact location of the mobile node.

The cost of determining the location depends on what kind of location information the network had when the call (or packet) arrived. Note that when the exact position is determined, these are transitions to P_0. If we have exact location information an all levels, then the cost is zero (loop from P_0 to P_0). When

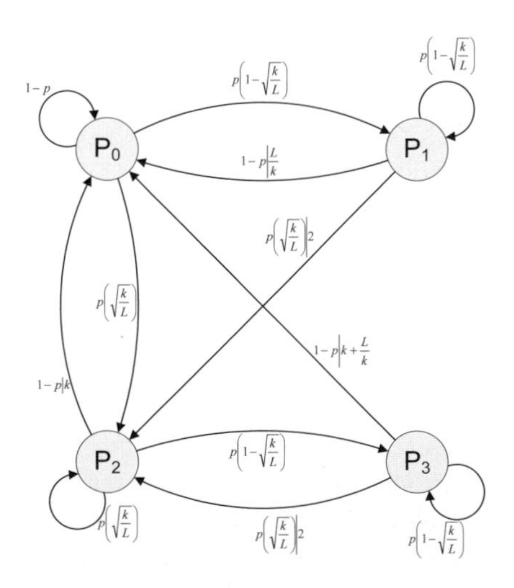

Fig. 1. Hierarchical paging Markov-model transition probabilities and costs

the subnetwork is known, but not the exact cell, a subnetwork-wide paging has to be carried out (from P_1 to P_0), which involves L/k cells, thus a cost of L/k is assigned. When the cell within the subnetwork is know, but the subnetwork is not (form P_2) a top-level paging takes place, k subnetworks are involved, thus a cost of k is assigned. It neither the subnetwork nor the exact cell within the subnetwork is known, then both top-level and bottom-level paging operations are carried out, thus the cost is $k + L/k$.

Figure 1 shows all the probabilities and non-zero costs of both handovers and the determination of exact locations. The cost is always on the right of the "pipe".

Stable State. As the Markov-chain is finite, aperiodic and irreducible, it is stable. The stable state can be computed easily:

$$P_0 = 1 - p$$

$$P_1 = -\frac{-p + \sqrt{\frac{k}{L}}p + p^2 - \sqrt{\frac{k}{L}}p^2}{1 - p + \sqrt{\frac{k}{L}}p}$$

$$P_2 = \sqrt{\frac{k}{L}}p$$

$$P_3 = -\frac{(-1 + \sqrt{\frac{k}{L}})\sqrt{\frac{k}{L}}p^2}{1 - p + \sqrt{\frac{k}{L}}p}$$

where $P_0 ... P_3$ denotes the probability of being in state $P_0 ... P_3$ in the stable state of the Markov-chain.

Average Cost of an Event. Events in our model are handovers and incoming calls. As explained in Section 5.3, we have no way to affect what events occur and when, it is an input parameter. So, the average cost of an event describes how efficient out solution is. The cost of an average event can be determined using the probabilities of the stable state and the costs of transitions. It is a rather complicated, but closed form solution.

6 Analysis

6.1 The Effect of k

What might be the most interesting for us, is the effect of parameter k on the cost. This parameter denotes how many subnetworks our network is divided to, see Section 5.1.

The total number of cells (L) was fixed to 1000, and the mobility ratio (p) was set to 0.9. Figure 2 shows a plot of the *average cost of an event* as a function of k.

The values of k above 500 do not make too much sense as subnetworks with only one cell start to appear in the network. It can be read from Figure 2 that there is an optimal number of subnetworks (somewhere around 25-35).

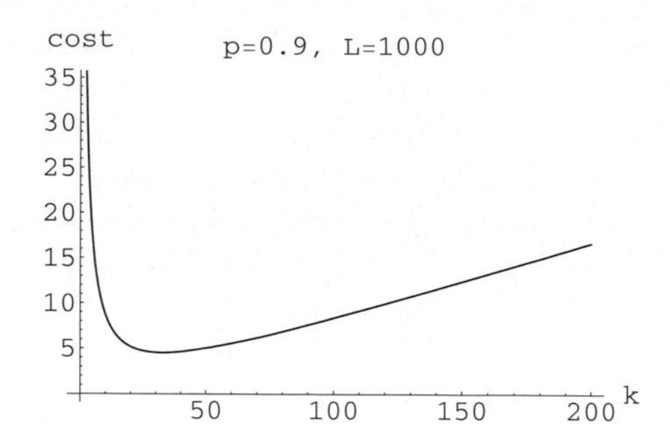

Fig. 2. Cost as a function of the no. of subnetworks, all other parameters are fixed

6.2 Flat vs. Hierarchical Paging

Using the network and mobility models introduced in the previous sections, we can construct a similar markov-model for the single-layer paging solution, see Figure 3. This is a simple, two-state markov-chain, it is easy to see that the average cost of an event in this case is

$$Cost_{flat}(L, p) = p(1-p)L. \tag{2}$$

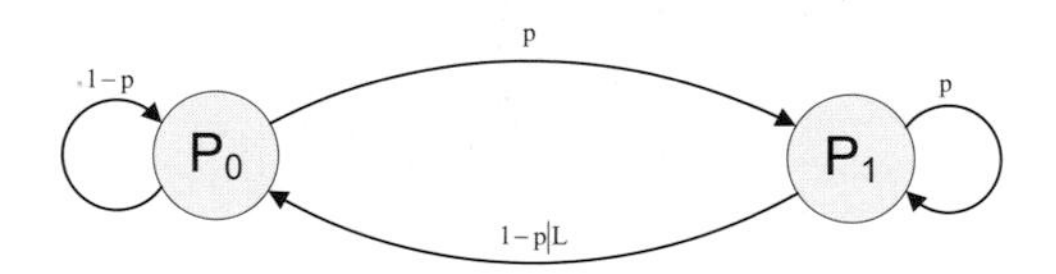

Fig. 3. Markov-model with costs for single-layer paging

Now let's fix the parameters to the same values as when we examined the effect of k on the performance of hierarchical paging. If $L = 1000$ and $p = 0.9$, then the average cost of an event is 90. Figure 2 clearly shows that the performance of hierarchical paging is much better than that of single-layer paging for almost all the k values.

6.3 The Effect of *L*

To see how the cost vs. k plot (Figure 2) is affected by the size of the network, a 3D plot can be used. The same plot can be made for $L = 200 \ldots 1000$. This is depicted in Figure 4. At $L = 1000$ (far back) it is the same as Figure 2.

6.4 The Effect of *p*

Up to now the *mobility ratio* (see Section 5.3 was fixed to 0.9, which means that there is one incoming call per ten *events* on average. We can also examine what happens if the mobility ratio changes. Figure 5 shows how the cost is affected by changes in p, when L is fixed to 1000.

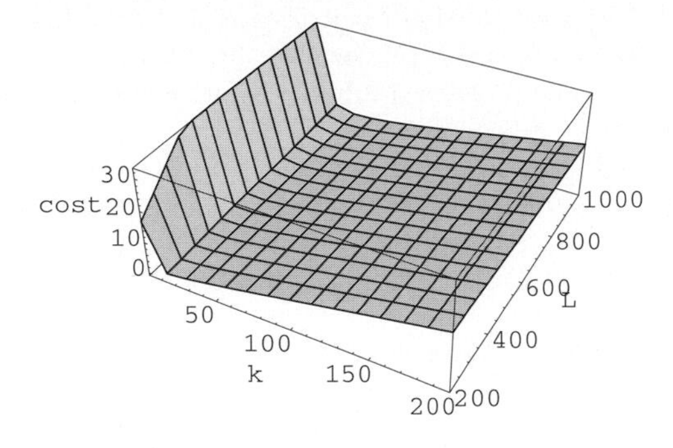

Fig. 4. Cost as a function of the no. of subnetworks and the no. of cells

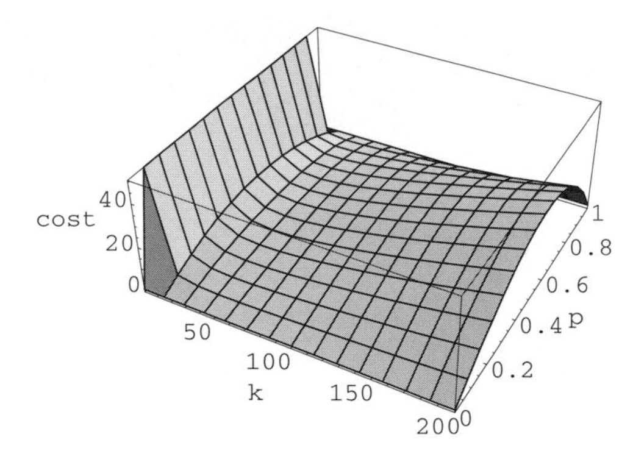

Fig. 5. Cost as a function of the mobility ratio and the no. of cells

7 Conclusions and Future Work

In this paper a novel mobility management approach has been introduced. An analytical model was presented. It has been shown that hierarchical paging is more efficient than simple non-hierarchical paging in the sense of signalling overload.

Hierarchical paging has several parameters to tune, the most important is the number of subnetworks a network is divided into (denoted by k). To minimize the signalling overload an optimal k can be used, which optimum of course depends on various parameters of the network.

Future work shall include the use of more sophisticated mobility models, and further examination of multi-layer hierarchical paging solutions.

References

1. Akyildiz, I., McNair, J., Ho, J., Uzunalioglu, H., Wang, W.: Mobility Management in Next-Generation Wireless Systems. Proceedings of the IEEE 87, 1347–1384 (1999)

2. Castelluccia, C.: HMIPv6: A Hierarchical Mobile IPv6 Proposal, ACM Mobile Computing and Communication Review (MC2R) (April 2000)
3. Gloss, B., Hauser, C.: The IP Micro Mobility Approach. In: EUNICE 2000, September 2000, Eschende, pp. 195–202 (2000)
4. Schulzrinne, H., Rosenberg, J.: The Session Initiation Protocol: Internet-Centric Signaling. IEEE Communications Magazine, 134–141 (October 2000)
5. Turányi, Z., Szabó, Cs., Kail, E., Valkó, A.G.: Global Internet Roaming with ROAMIP. ACM SIGMOBILE Mobile Computer and Communication Review (MC2R) 4(3) (July 2000)
6. Fang, Y., Lin, Y.: Portable Movement Modeling for PCS Networks. IEEE Transactions on Vehicular Technology 49, 1356–1362 (2000)
7. Ma, W., Fang, Y.: Dynamic Hierarchical Mobility Management Strategy for Mobile IP Networks. IEEE Journal of Selected Areas In Communications, 22 (2004)
8. Gibson, J.D. (ed.): The Mobile Communications Handbook. CRC Press LLC and Springer, Heidelberg (1999) ISBN: 3-540-64836-4, 27-3
9. Mohan, S., Jain, R.: Two use location strategies for personal communications services. IEEE Personal Communications 1, 42–50 (1994)
10. Woo, T.Y.C., Porta, T.F.L., Golestani, J., Agarwal, N.: Update and Search Algorithms for Wireless Two-Way Messaging: Design and Performance. In: INFOCOM 1998, pp. 737–747 (1998)
11. Ramjee, R., Li, L., La Porta, T.: IP Paging Service for Mobile Hosts. In: Proceedings of MOBICOM 2001, pp. 332–345 (2001)
12. Hajek, B., Mitzel, K., Yang, S.: Paging and Registration in Cellular Networks: Jointly Optimal Policies and an Iterative Algorithm. In: Proceedings of IEEE INFOCOM (2003)

Wireless Loss Detection for TCP Friendly Rate Control Algorithm in Wireless Networks

Jinyao Yan[1,2], Xuan Zheng[1], Jianbo Liu[1], and Jianzeng Li[1]

[1] Communication University of China, Beijing, P.R.China, 100024
[2] Swiss Federal Institute of Technology (ETH) Zurich, CH-8092
{jyan,zhengxuan,ljb,jzli}@cuc.edu.cn

Abstract. The performance of TFRC algorithm is low in the unreliable and dynamic wireless environment. In particular, the packet loss used in TFRC can be caused either by the wireless error or congestion in wireless networks. Based on the analysis of loss differentiate algorithms such as Biaz and ZigZag, we propose the WLD algorithm aiming to differentiate more precisely between the wireless error and congestion loss. The simulation results show WLD algorithm improves remarkably the throughput of streaming in wireless networks, thus improves the performance of congestion control algorithm such as TFRC for media streaming in wireless networks.

1 Introduction

Research in the field of multimedia communication for applications, like Video on Demand, real-time video and Internet live broadcast has been ongoing for years. Rate control is primarily necessary for multimedia communication applications to deal with the diverse and constantly changing conditions of the Internet. Several congestion control protocols [2], [3], [11], typically taking the form of rate-based congestion control (rate control) protocols, tailored to the rate acceleration and variability requirements of media applications have been developed while handling competing TCP flows in a fair manner. Without congestion control, non-TCP traffic can cause starvation or even congestion collapse to the dominant TCP traffic, if both types of traffic compete for resources at a congested FIFO queue [1]. A representative example of the TCP-friendly congestion control protocols is the TCP-Friendly Rate Control (TFRC) protocol [2]. TFRC is an equation-based congestion control mechanism that uses the TCP throughput function presented in [4] to calculate the actual available rate for a media stream. Previous work [5], [6] on TCP-friendly congestion control protocols that TFRC offer better performance than other TCP-friendly congestion control protocols.

However, the performance of TFRC is very low in wireless networks and MANET, where the network is far more unreliable and dynamic than wired networks. TFRC relies on the recent history of loss event intervals to estimate the current loss event interval, equivalently the loss rate, using a weighted average equation. However, the estimation in predicting loss event interval may not be accurate, due to a number of reasons: 1) highly varying packet losses due to the dynamic wireless link bandwidth; and 2) loss process shows little autocorrelation in wireless network and MANET. The

T. Vazão, M.M. Freire, and I. Chong (Eds.): ICOIN 2007, LNCS 5200, pp. 113–122, 2008.

loss event intervals should possess significant auto-correlation in order for TFRC to have an accurate prediction; otherwise it is impossible to do so no matter how the weights are chosen; the most important reason 3) some packet losses are not caused by congestion but wireless-medium or hand over disruption. Therefore, the throughput of TFRC algorithm is often much lower than the bandwidth that wireless networks provide [15].

Recently, rate control algorithms for media streaming like TFRC have been extended for better performance in wireless environments. There have been two classes of solutions: cross-layer and end-to-end. Cross-layer approaches typically determine a IP packet loss caused by wireless loss based on the information provided by the link layer, and then retransmission the error frame if necessary, thus hide end hosts from packet loss caused by wireless channel error [17]. These cross-layer approaches requires modifications to the network infrastructure, thus end-to-end solutions are the preferred choice.

MULTFRC [13] has been proposed to create multiple simultaneous TFRC connections on the same path when a single connection cannot fully utilize the bandwidth. It is an end-to-end approach and requires no modifications to the Internet infrastructure. On the other hand, MULTFRC requires more resource to manage multiple connections. The sender must also split the data across multiple connections and the receiver must then reassemble the data chunks. This adds overhead to the scheme [18].

Cen et al. [9] have added an end-to-end different loss discrimination (LDA) algorithm to TFRC to improve its efficiency in presence of wireless errors. In the original TFRC throughput equation at end host, packet loss event rate includes both congestion and wireless loss as TFRC does not distinguish between them. Lack of loss discrimination makes the performance of TFRC drop sharply when wireless error is frequent. LDA is to discriminate wireless errors from congestion losses at end hosts. If the loss is induced by wireless errors, it is discounted in the calculation of packet loss event rate by LDA. TFRC equipped with the LDA operates exactly the same way as the original TFRC mechanism.

Paper [9] evaluated LDA algorithms such as Biaz [7],Spike [8] and the new proposed LDA called ZigZag under some various situations. The authors also proposed a hybrid LDA algorithm which is much more complex one. Spike uses relative one-way trip times (*ROTT*). If *ROTT* is close to minimum *ROTT*, the bottleneck is unlikely to be congested, on the contrary, if *ROTT* is close to maximum *ROTT*, the path is heavily loaded and congestion is about to occur. ZigZag uses *ROTT* as a function of loss count while Biaz uses packet inter-arrival times. They will be introduced in detail in section 2.

The video transport protocol (VTP) [14] uses an Achieved Rate estimation scheme to avoid drastic rate fluctuations. In the mean time VTP is equipped also with a variant of Spike to distinguish between congestion loss and error.

The end-to-end solutions usually provide end hosts the ability to differentiate between packet loss caused by congestion and that caused by wireless error. The disadvantage of end-to-end statistics-based approaches is that congestion detection schemes based on statistics are not sufficiently accurate, thus the performance of TFRC in wireless network is still not satisfied.

In this paper, we will propose a hybrid but simple LDA algorithm called WLD (wireless loss detection) algorithm for TFRC in wireless environments, which is aiming to differentiate more precisely between the wireless error and congestion loss. The simulation results show WLD algorithm improves remarkably the throughput of streaming in wireless networks, thus improves the performance of congestion control algorithm for media streaming in wireless networks.

The rest of this paper is organized as follows. Firstly we briefly take look at the related basic algorithms, and then we present our proposed WLD algorithm in section 2. Section 3 reports the simulation results and analysis. We conclude the paper in section 4.

2 Wireless Loss Detection Algorithms

In this section, we introduce the related basic algorithms Biaz and ZigZag first, and then present our WLD algorithms.

2.1 Biaz Algorithm

The Biaz algorithm uses packet inter-arrival time to distinguish congestion losses from wireless corruption losses. Let T_{min} denote minimum inter-arrival time observed so far by the receiver. Let P_o denote an out of order packet received by the receiver. Let P_i denote last in-sequence packet received before P_o. Let T_i denote the time between arrival of packets P_o and P_i. Let n denote the number of packets missing between P_i and P_o (assuming that all packets are of the same size). If $(n+1)T_{min} <= T_i < (n+2)T_{min}$, then n missing packets are lost due to wireless transmission errors. Otherwise, the n missing packets are assumed to be lost due to congestion.

Biaz assumes the last hop to the receiver is a wireless link but also is the bottleneck of the connection, and then the packets tend to queue up at the base station. Therefore, most of the packets are sent back to back on the wireless link. The Biaz algorithm works well when the wireless bottleneck link is not shared.

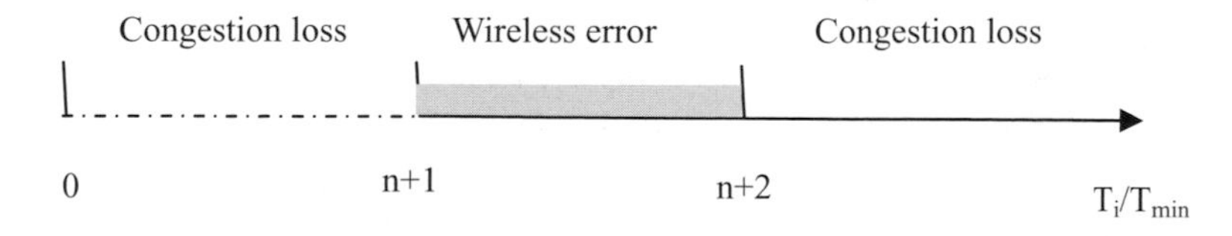

Fig. 1. Biaz Scheme

2.2 ZigZag Algorithm

Both Spike and ZigZag use the Relative One-way Trip Time (*ROTT*) to measure the time a packet takes to travel from the sender to the receiver, and to distinguish the packet losses. ZigZag classifies losses as wireless based on the number of losses, and on the difference $d= rott_i - rott_{mean}$. $rott_i$ is the *rott* of i_{th} packet while $rott_{mean}$ is the calculated by equation (1).

As shown in figure 2, the ZigZag classifies losses with the following criteria:

If $((n=1 \ \& \ d<-rott_{dev}) \ or \ (n=2 \ \& \ d<-rott_{dev}/2) \ or \ (n=3 \ \& \ d<0) \ or \ (n>3 \ \& \ d<rott_{dev}/2))$, then the packet loss is classified as wireless error, else the packet loss is a congestion loss.

The calculation of $rott_{mean}$ and $rott_{dev}$ is as follows, where $\alpha = 1/32$:

$$rott_{mean} = (1-\alpha) * rott_{mean} + \alpha * rott_i \tag{1}$$

$$rott_{dev} = (1-2\alpha) * rott_{dev} + 2\alpha * \left| rott_i - rott_{mean} \right| \tag{2}$$

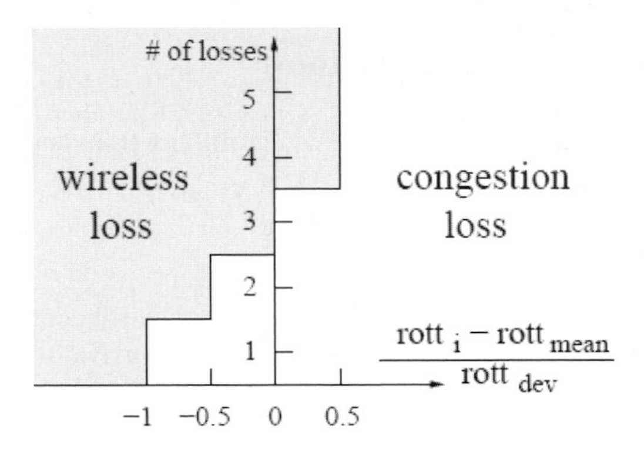

Fig. 2. ZigZag Scheme

As one packet loss is the most common loss pattern in a wired network, and congestion loss usually comes with higher delay, the threshold of $rott > rott_{mean} - rott_{dev}$ intuitively would classify most of the congestion loss correctly. The reasoning behind increasing the threshold with the number of losses encountered is that a more severe loss is associated with higher congestion, and with higher *ROTT*. This way, a loss event containing 4 or more packets is more likely to be classified as wireless loss, as wireless losses often display bursts of correlated errors.

2.3 WLD

From the experiments done in [9], Biaz works ideal when wireless link is the last link with the lowest bandwidth and it is not shared with other flows. The higher the upper limit of T_i, the more likely a loss will be classified as a wireless loss, i.e., the scheme trades off higher accuracy for classifying wireless loss but lower accuracy for congestion loss. The high congestion loss observed with the Biaz scheme indicates that the window is probably too large. The authors then suggest modified Biaz with $(n+1)T_{\min} \le T_i < (n+1.2)T_{\min}$ provides a good tradeoff between low congestion loss misclassification and high throughput when competing with other traffic in the wireless last hop topology. When the bottleneck link bandwidth is shared by a large number of flows in wireless last hop topology, ZigZag performs well and stable while the original Biaz and Spike schemes both have an unacceptably high misclassification rate of congestion loss.

We conclude that none of the base algorithms performs consistently well in the face of competition from other flows in wireless last hop topology. We would combine the advantage of the ZigZag and that of Biaz into one algorithm called WLD algorithm.

Now we present the description of WLD. The average packet inter-arrival time T_{avg} of two sequential packets is computed by equation (3).

$$T_{avg\_i+1} = (1-\beta) * T_{avg\_i} + \beta * T_i/n \tag{3}$$

Where T_i is the instantaneous inter-arrival time; n is the number of packets that between the arrived packets. We set $\beta = 0.2$ during the simulations.

When the lowest bandwidth link is shared by N flows (N>>1), the average packet inter-arrival time would be much greater than T_{min}, where T_{min} is the minimum inter-arrival time. If the slowest link is shared with a small number of flows, or when some sequential packets in the competitive queue belong to the same flow, then T_{avg} should be close T_{min}. We apply modified Biaz algorithm when T_{avg} is close to T_{min}. Otherwise, we apply the ZiZag which performs relatively stable in all cases. We describe the WLD algorithm in Table 1.

Table 1. Description of WLD

```
If (0.95< (T_avg / T_min) <1.3) {
  If (n+1)T_min ≤ T_i < (n+1.2)T_min    wireless error
  Else    congestion loss
}
Else {
  If ((n=1 & d < -rott_dev) or (n=2 & d < -rott_dev /2) or (n=3 &
d<0) or (n>3 & d< rott_dev /2))         wireless error
  Else    congestion loss
}
```

3 Experiments and Analysis

3.1 Configuration of Experiments

In this paper, we consider only the scenario where the sender is on the wired network and the receiver is connected via the wireless link, which is true often in WLAN and cellular environments.

We use ns-2 [12] for our simulation and performance analysis of our proposed algorithm. The simulation topology used in the experiments is depicted in figure 3 as the WLH (wireless last hop) topology in [9], In the simulation topology senders of flows are on the left side and the sender of flow *i* denoted by S_i, while receivers are on the right side and the receiver of flow *i* denoted by K_i. N streams compete for the bandwidth of the bottle neck between router *R1* and router *R2*, which is with the bandwidth of 130*N Kbps and delay of 20ms. The last links to receivers are wireless links with bandwidth of 150Kps and delay of 10ms. We apply the error model for CDMA channels [10] for the wireless link to simulate the wireless error. The packet size was used in the simulation is 762 bytes as the specified packet sizes in the CDMA-2000.

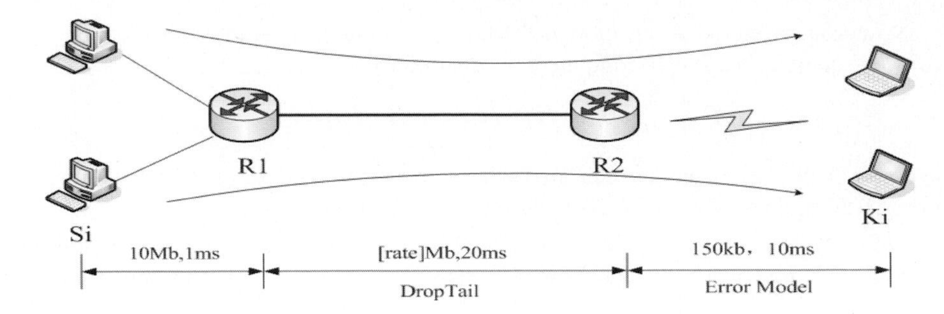

Fig. 3. Simulation Topology

We use the same parameters and random seeds for the compared algorithms. With different random seeds, the same sets of experiments were repeated more than 10 times and results were averaged.

To evaluate the performance of the congestion control algorithms with LDAs, we should compare the metrics of throughput, intra protocol and TCP friendliness, the smoothness and so on. The most important measure is the throughput. As long as the algorithm is opportunistically friendly to TCP, the higher throughput, the better performance of the algorithm is. We adopt the notion of opportunistic friendliness defined in [16]. A new protocol NP is said to be opportunistically friendly to TCP if TCP flows coexisting with NP obtain no less throughput than what they would achieve if all flows were TCP.

3.2 Experiments Results and Analysis

We denote the TFRC in wireless network equipped with wireless loss detection algorithm as TFRC_WLD. In this section, we will compare TFRC_WLD algorithm not only to TCP in figure 4 and the original TFRC in figure 5, but also some LDAs such as Biaz,Spike and Zigzag algorithm in figure 6 for wireless network.

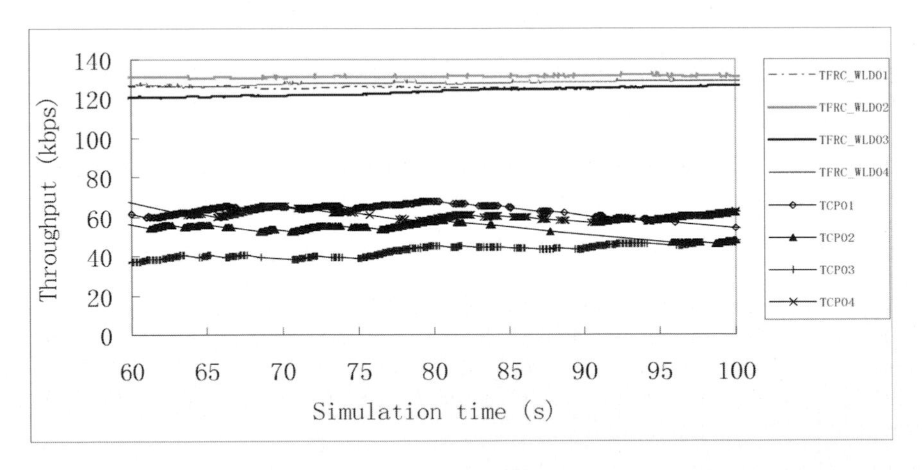

Fig. 4. Comparison of the average throughput and friendliness between TFRC_WLD and TCP (4 flows for each algorithm, wireless error rate =0.1)

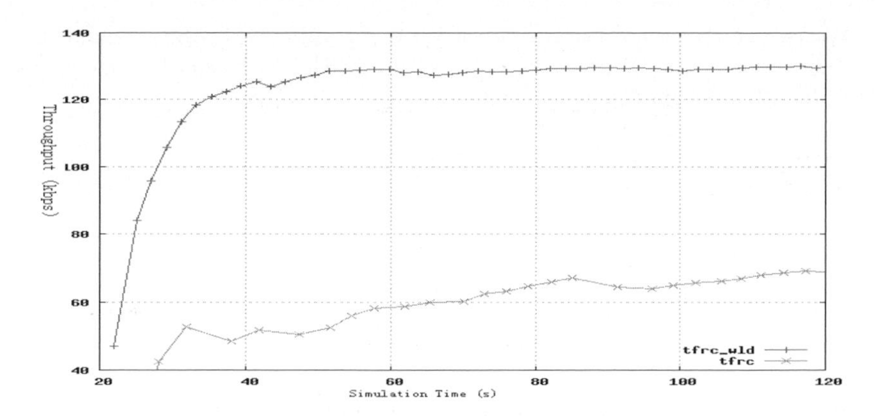

Fig. 5. Comparison of throughput between TFRC_WLD and TFRC flows (wireless error rate =0.1)

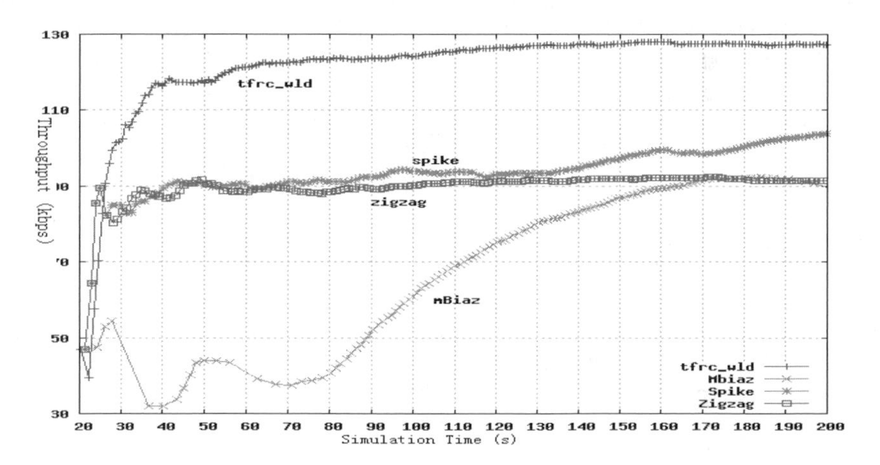

Fig. 6. Comparison of throughput between TFRC_WLD and other LDAs flows (wireless error rate =0.1)

From the experimental results presented above, we conclude the performance of TFRC_WLD algorithm as follows:

1. Good throughput: TFRC_WLD makes good use of the bandwidth in wireless environments.
2. Opportunistic friendliness to TCP and intra protocol friendliness: Each TFRC_WLD flow occupies no more than its bandwidth share in the bottleneck; therefore TFRC_WLD does not steal the bandwidth from TCP. Each flow of TFRC_WLD algorithm shares the comparable bandwidth with other TFRC_WLD flows under the same condition.

Indeed, we have evaluated the performance of TFRC_WLD algorithm under various levels of wireless error rates. We have got the similar results as that under wireless error rate equals 0.1.

We also compare the performance of WLD algorithm with ZBS, which is a hybrid but more complicate algorithm presented in [9]. ZBS1, ZBS2 and ZBS3 are three forms of ZBS with different parameters and schemes.

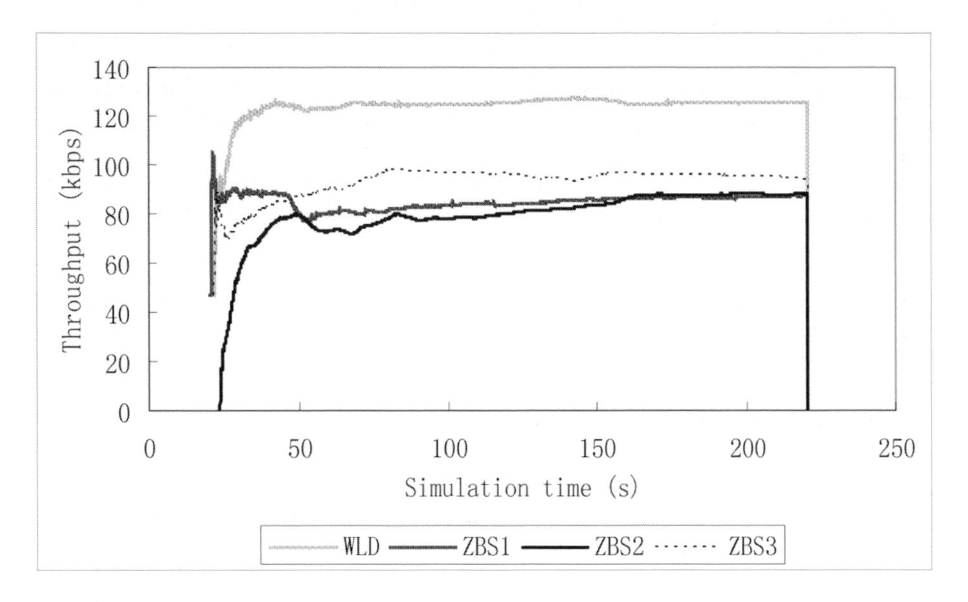

Fig. 7. Comparison of throughput between TFRC_WLD and ZBS flows (wireless error rate =0.1)

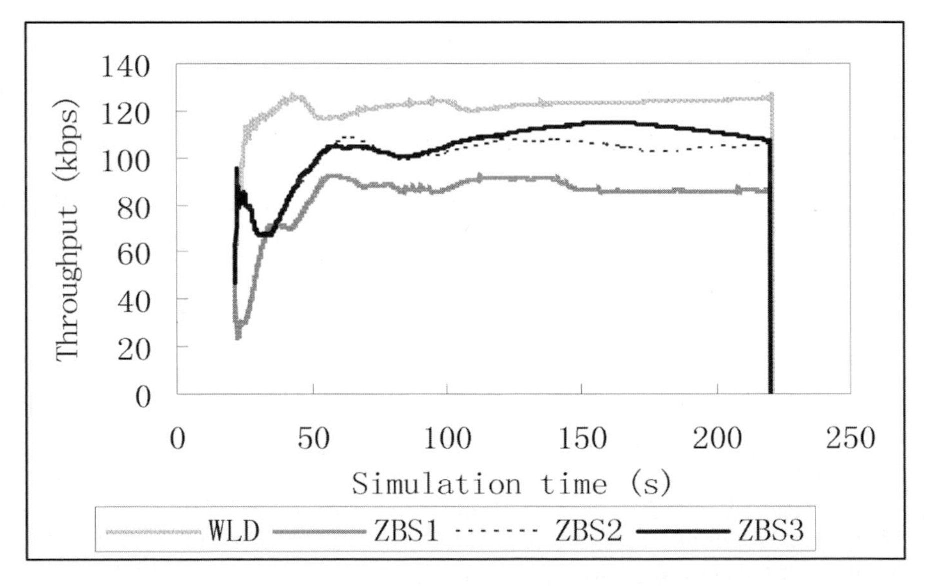

Fig. 8. Comparison of throughput between TFRC_WLD and ZBS flows (wireless error rate =0.01)

We find the throughput of WLD algorithm is higher than ZBS schemes under conditions both that wireless error rate equals 0.1 and 0.01.

4 Conclusions and Future Work

In this paper, we present our proposed WLD algorithm to differentiate the wireless loss and congestion loss for TFRC in wireless networks. WLD is an end-to-end algorithm, which requires no modification of Internet infrastructure.

From the experiments results, we draw the conclusion that TFRC_WLD is superior to TFRC and other LDAs in terms of its higher throughput in various wireless error rates, while keeping the opportunistic friendliness to TCP and the intra protocol friendliness.

We will continue to evaluate TFRC_WLD performance further in more complex topologies and real networks as our future work.

Acknowledgements

This work was partially supported by National Natural Science Foundation of China under Grant 60502015. The first author was partially supported by Swiss National Science Foundation under Grant 200021-112121.

References

1. Floyd, S., Fall, K.: Promoting the Use of End-to-End Congestion Control in the Internet. IEEE\ACM Transactions on Networking (1999)
2. Floyd, S., Handley, M., Padhye, J., et al.: Equation-based Congestion Control for Unicast Applications. In: Proc. ACM SIGCOMM Conference, Sweden (2000-2008)
3. Bansal, D., Balakrishnan, H.: Binomial congestion control algorithms. In: Proc. of IEEE INFOCOM 2001, pp. 631–640 (2001)
4. Padhye, J., Firoiu, V., Towsley, D., Kurose, J.: Modeling TCP Throughput: A Simple Model and its Empirical Validation. In: SIGCOMM Symposium on Communications Architectures and Protocols (August 1998)
5. Bansal, D., Balakrishnan, H., Floyd, S., Shenker, S.: Dynamic behavior of slowly-responsive congestion control algorithms. In: Proc. of ACM SIGCOMM 2001 (August 2001)
6. Yang, R., Kim, M.S., Lam, S.S.: Transient behaviors of TCP-friendly congestion control protocols. In: Proc. of IEEE INFOCOM 2001, pp. 1716–1725 (2001)
7. Biaz, S., Vaidya, N.: Discriminating Congestion Losses from Wireless Losses Using Inter-arrival Times at the Receiver. In: Proc. 1999 IEEE Symposium on Application-Specific Systems and Software Engr. and Techn., Richardson, TX, pp. 10–17 (1999)
8. Tobe, Y., Tamura, Y., Molano, A., Ghosh, S., Tokuda, H.: Achieving moderate fairness for UDP flows by pathstatus classification. In: Proc. 25th Annual IEEE Conf. on Local Computer Networks (LCN 2000), Tampa, FL, November 2000, pp. 252–261 (2000)
9. Cen, S., Cosman, P., Voelker, G.: End-to-end differentiation of congestion and wireless losses. IEEE/ACM Transactions on Networking 11(5), 703–717 (2003)

10. Zhao, Q., Cosman, P., Milstein, L.: Tradeoffs of source coding, channel coding and spreading in CDMA systems. In: Proc. Milcom 2000, Los Angeles, CA (October 2000)
11. Yan, J., May, M., Katrinis, K., Plattner, B.: Media- and TCP-Friendly Congestion Control Algorithm for Scalable Video Streams. IEEE Transactions on Multimedia (April 2006)
12. ns-2 network simulator, `http://www.isi.edu/nsnam/ns/`
13. Chen, M., Zakhor, A.: Rate control for streaming video over wireless. In: Proceedings of the IEEE Infocom. IEEE, Hong Kong (2004)
14. Yang, G., Chen, L.-J., Sun, T., Gerla, M., Sanadidi, M.Y.: Smooth and Efficient Real-time Video Transport in Presence of Wireless Errors. ACM Transactions on Multimedia Computing, Communications and Applications (ACM TOMCCAP) (accepted)
15. Chen, K., Nahrstedt, K.: Limitations of Equation-based Congestion Control in Mobile Ad hoc Networks. In: Proc. of International Workshop on Wireless Ad Hoc Networking (WWAN 2004) in conjuction with ICDCS 2004, Tokyo, Japan (March 2004)
16. Tcp westwood home page, `http://www.cs.ucla.edu/NRL/hpi/tcpw/`
17. Yang, F., Zhang, Q., Zhu, W., Zhang, Y.-Q.: End-to-End TCP-Friendly Streaming Protocol and Bit Allocation for Scalable Video over Wireless Internet. IEEE JSAC special issue on all-IP wireless networks (2004)
18. Yang, G., Chen, L., Sun, T., Gerla, M., Sanadidi, M.Y.: Real-time streaming over wireless links: A comparative study. In: Proceedings of the IEEE ISCC. IEEE, Cartagena (2005)

Rate-Adaptive TCP Spoofing with Segment Aggregation over Asymmetric Long Delay Links*

Jihyun Lee, Hyungyu Park, and Kyungshik Lim

School of EECS, Kyungpook National University
1370 Sankyuk-Dong, Buk-Gu, Daegu, 702-701, Korea
{hyuny,hgpark}@ccmc.knu.ac.kr, kslim@knu.ac.kr

Abstract. To improve the TCP performance over long delay links, TCP spoofing has been mainly proposed as a particular solution. Even if it can reduce the duration of the slow-start, it provides the limited maximum throughput of approximately 1 Mbps with the maximum window size of 64KB. In this paper, we propose a rate control algorithm, called *Rate-Adaptive spoofing*(RA-spoofing) that adjusts *virtual window* dynamically depending on the number of other TCP connections sharing the link, thereby fully utilizing the available bandwidth and maximizing throughput rates. Tightly coupled with RA-spoofing, we also propose a *Segment Aggregation*(SA) mechanism to overcome the bandwidth asymmetry problem. SA is a multiplexing process that a number of TCP segments are transformed into the corresponding number of IP fragments directly, based on *segment aggregation factor*, which results in one true ACK from the destination. The results of the performance analysis provide a good evidence to demonstrate the efficiency of our mechanisms under different assumptions of bandwidth asymmetric ratio and error rate.

Keywords: Rate-Adaptive TCP Spoofing, Segment Aggregation, Asymmetry Long Delay Link, TCP, PEP.

1 Introduction

In a next-generation heterogeneous network environment, satellite system will be an excellent candidate for providing high data-rate Internet access and global connectivity. This stems from its fundamental characteristics of ubiquitous coverage, broadcasting capability, and bandwidth flexibility. However, the environment reveals different link characteristics on an end-to-end basis of the connection path, which may cause a severe degradation of the TCP performance. To achieve high data throughput, TCP should be optimized to accommodate two major satellite link characteristics of long propagation delay and bandwidth asymmetry[1].

* This work was supported in part by MIC, Korea under the ITRC program(C1090-0603-0036) supervised by IITA.

T. Vazão, M.M. Freire, and I. Chong (Eds.): ICOIN 2007, LNCS 5200, pp. 123–132, 2008.

Including new versions of TCP and new transport protocols, a large number of solutions have been extensively proposed. However, most TCP enhancement solutions are impractical for use in the Internet. Because they require changes to be made in all end or intermediate nodes, so in some cases, violating scalability and economical goals. As a good trade-off between the technical model and the business model of satellite Internet, Performance Enhancing Proxy(PEP)[2] has been introduced. The representative mechanisms used in PEPs to improve the TCP performance are as follows; To mitigate the effect of long propagation delay and speed up the slow start phase, TCP spoofing has been mainly proposed in spite of some drawbacks such as the lack of the end-to-end semantics[3]; To overcome highly asymmetric bandwidth, some PEPs implement ACKs filtering and reconstruction so that ACKs are being filtered on the receiver side not to congest the low-speed link and are reconstructed on the other side of the link[4]; To alleviate bit error rate, TCP snooping is usually employed to locally recover the lost segments in response to duplicate ACKs from the receiver[5].

Among those mechanisms and their variants, TCP spoofing can be considered as a particular solution to the long Round Trip Time(RTT) problem in a two-segment satellite environment(e.g., direct broadcast satellite) where the long delay link is the last hop to the destination. TCP spoofing, however, cannot solve the problem completely since it provides the limited maximum throughput of approximately 1 Mbps with the maximum window size of 64KB. This raises a necessity of a new mechanism for maximizing throughput, which should be based on TCP spoofing and dynamically adaptive to the available bandwidth of the link at a safe and fair transmission rate. It should be noted that the mechanism will be much more effective if the bandwidth asymmetry problem is alleviated simultaneously. In our two-segment satellite Internet, however, existing ACKs filtering and reconstruction mechanisms cannot be applied to and then a new mechanism implemented at PEP is needed.

In this paper, we propose two new PEP mechanisms: *Rate-Adaptive spoofing*(RA-spoofing) and *Segment Aggregation*(SA). RA-spoofing is a rate control process that starts operating after the slow start phase, sets the window size based on the link bandwidth and RTT, and adjusts it dynamically depending on the number of other TCP connections sharing the link. Thus, RA-spoofing solves limited maximum throughput and slow startup speed simultaneously on the long delay link. SA is a multiplexing and fragmentation process that assembles an appropriate number of TCP segments based on bit error rate into one IP packet and then divides it into the corresponding number of IP fragments, thereby reducing the number of ACKs on the uplink to 1 for the whole TCP segments. It should be noted that TCP throughput is to be maximized when RA-spoofing and SA are tightly coupled and harmonized.

The remainder of the paper includes Section 2, which presents RA-spoofing and SA in greater detail. Section 3 outlines the simulation work and analyzes its results. Finally, Section 4 concludes our works.

2 New PEP Mechanisms

Our new PEP mechanisms to improve the TCP performance over asymmetric long delay links consider a two-segment network model as shown in Figure 1, where typically the downlink bandwidth is 10 to 1000 times the uplink bandwidth and the one-way propagation delay is about 250 to 280 ms. In the model, we assume that a TCP-aware ARQ protocol is used at the link layer between PEP and the destination. The link layer suppresses duplicate or true ACKs at PEP so that they do not reach the source. The last ACK for a virtual window, described in the next subsection, however, is relayed to the source. If the link layer fails to retransmit the packet, the source will be timeout and retransmit the packet itself. Note that the two-segment network model requires in practice that new PEP mechanisms should work with the existing Internet protocols and infrastructures and be application-independent.

2.1 Rate-Adaptive Spoofing

The long delay link increases the duration of the slow start phase and results in a decrease in initial performance[3]. To solve the problem, TCP spoofing is usually used to shield long delay from the source. In TCP spoofing, PEP near the source sends back ACKs for TCP segments to speed up the sender's data transmission. It then suppresses the true ACKs from the destination and takes responsibility for sending any missing data. Even if TCP spoofing can reduce the duration of the slow start phase, it cannot solve the negative effect of long delay on the TCP performance completely. In a loss-free network, it is well known that the TCP throughput is limited by the formula, *Maximum Throughput* = *Maximum Window Size* / *RTT*. This means that after the slow start phase a maximum throughput of about 1 Mbps can be achieved with maximum window size of 64

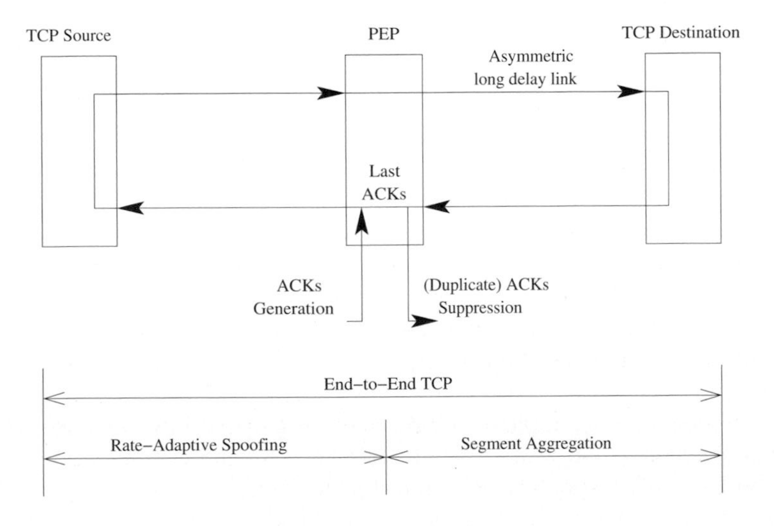

Fig. 1. The Two-Segemnt Network Model with Asymmetric Long Delay Links

KB and RTT of 520 ms regardless of the available bandwidth[6]. It naturally raises the necessity of adding a rate control algorithm to TCP spoofing so that the available bandwidth is fully utilized and the throughput is maximized. Note that the use of the window scale option violates our fundamental requirement of the independence of the existing applications.

procedure. SetVirtualWindowSize()
// BTOTAL is the known total bandwidth of the downlink
// BUFSIZ is the amount of traffic in transit measured at PEP
1 Initialize $W_{virtual}$ to WDFLT
2 **for** every ACK received from the destination
3 **begin**
4 Measure RTT and compute SRTT
5 **if** ACK is not the last one of a virtual window
6 **continue**
7 Compute BW_{avail} = BTOTAL - BUFSIZ/SRTT
8 **if** (BW_{avail} > TDFLT)
9 **if** (measured RTT $\leq$ SRTT + mdev)
10 $W_{virtual} \leftarrow W_{virtual}+$ WDFLT
11 **else**
12 **if** (measured RTT > SRTT + mdev)
13 $W_{virtual} \leftarrow W_{virtual}/2$
14 **end**

To solve the problem of the limited maximum throughput, we propose a rate control algorithm coupled with TCP spoofing, called *Rate-Adaptive spoofing*(RA-spoofing). RA-spoofing operates immediately after the slow start phase and calculates the available bandwidth, BW_{avail} and a *virtual window size*, $W_{virtual}$. The main idea is to dynamically adjust $W_{virtual}$ to a multiple of WDFLT depending on BW_{avail}, where WDFLT is the default window size of 64KB. Let TDFLT be the default maximum throughput of approximately 1 Mbps. If BW_{avail} is greater than TDFLT and measured RTT is less than smoothed RTT(SRTT), $W_{virtual}$ is increased by WDFLT additively. If BW_{avail} is less than TDFLT and measured RTT is greater than smoothed RTT(SRTT), $W_{virtual}$ is dropped to half multiplicatively. In the other cases, $W_{virtual}$ remains unchanged. Using the additive-increase and multiplicative-decrease strategy to adjust $W_{virtual}$ enables RA-spoofing to be fair if multiple competing TCP connections share BW_{avail}, as in the normal TCP congestion control behavior. Given the known total bandwidth of the link, the following procedure determines the size of a virtual window dynamically and fairly based on measured RTT and total amount of traffic in transit on the link.

Figure 2 depicts an example of the RA-spoofing operation when the additive increase is applied to $W_{virtual}$. Assume that the current virtual window is two times WDFLT. In RA-spoofing, PEP nearby the source sends back a cumulative ACK for the first window and then the source speeds up and transmits segments in the second window size of WDFLT. At this time, PEP delays TCP spoofing

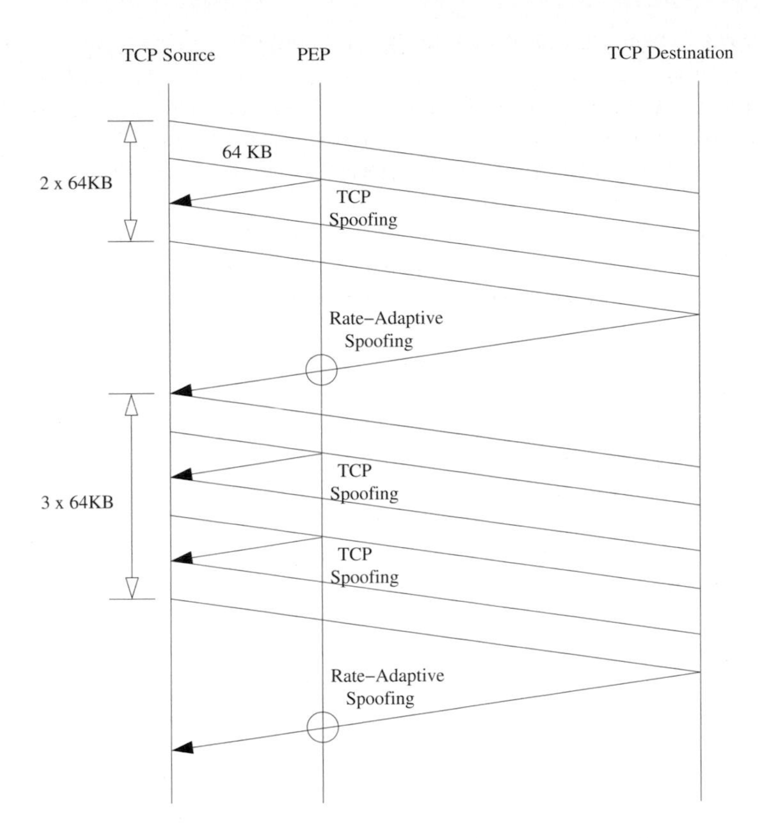

Fig. 2. Example of RA-spoofing Operation When $W_{virtual} = 2\times$WDFLT and Additive Increase Is Applied

for a while until it receives the true ACK for the last segment of the second maximum window from the destination. As soon as PEP receives the true ACK, it determines a new value of $W_{virtual}$. It implies the maximum amount of data the source can send out, without violating the fairness among other competing TCP connections that share BW_{avail}. At this point, PEP refers to BW_{avail} and measured RTT of the ACK, decides how much portion of BW_{avail} should be allocated, and relays the ACK to the source. In this example, $W_{virtual}$ is additively increased in case that $BW_{avail} >$ TDFLT and measured RTT $\leq$ SRTT + mdev.

2.2 Segment Aggregation

When we try to increase the maximum throughput of a connection using RA-spoofing, the bandwidth asymmetry problem raises the issue how to reduce the bandwidth usage of the uplink. Some existing ACKs filtering and reconstruction schemes, however, are not feasible for our two-segment network model. Thus, we propose a *Segment Aggregation*(SA) mechanism implemented at PEP. SA is a kind of multiplexing process where a number of TCP segments are transformed into the corresponding number of IP fragments, which results in one true ACK from the destination.

Let k_a be the *normalized asymmetry ratio*. Then, we define that $k_a = \lceil k_b/d \rceil$, where k_b is the normalized bandwidth ratio as defined in [6] and d is set to 2 for delayed ACKs. The normalized bandwidth ratio between the downstream and upstream paths is the ratio of the raw bandwidths divided by the ratio of the packet sizes used in the two directions. For example, for a 10 Mbps downstream channel and 64 Kbps upstream channel, the raw bandwidth ratio is about 156.3. With 1500-byte data packets and 40-byte ACKs, the ratio of the packet sizes is 37.5. So, k_b is $156.3/37.5 \cong 4.2$ and then $k_a = \lceil 4.2/2 \rceil = 3$. This normalized asymmetry ratio, k_a, represents that only one ACK should be generated for k_a TCP segments in order to avoid congestion on the uplink. In practice, since we have to consider relatively high error rate for occasional fades of satellite links, we introduce a *segment aggregation factor*, k_s, where $k_s = 1$ if BER $\geq 10^{-6}$ and $k_s = k_a$ otherwise. This means that SA is not performed at relatively high bit error rate for occasional fades of satellite links.

Given the segment aggregation factor, k_s, PEP transforms k_s TCP segments into the corresponding number of IP fragments in sequence directly and transmits them to the destination. The transformation process is similar to the normal IP fragmentation process. The difference between them is that all TCP headers of k_s segments except the first are removed and then each of them is encapsulated within an IP fragment separately. Thus, the identification fields of the resulting k_s IP fragments have the same value, their flag fields are set to 1 except that of the last IP fragment which is set to 0, and their offset fields are appropriately set to make the concatenation of the resulting k_s segments. When the destination receives k_s IP fragments, it performs the reassembly task, creates an IP packet, and delivers a TCP segment with the size of $1480 + (k_s - 1) \times 1460$ bytes to the TCP layer, resulting in the generation of one ACK for the aggregated TCP segment. Note that we use a Go-back-N scheme as a TCP-aware ARQ protocol. The Go-back-N scheme shows experimentally much better average performance than the selective-repeat scheme since it can greatly reduce error recovery time on long delay links at the cost of reasonable waste of the downlink bandwidth.

3 Performance Analysis

The simulations in this paper make use of the OPNET simulator, version 11.5. The overall architecture of the simulation software follows Figure 1, where the downlink bandwidth is set to 10 Mbps and RTT to 500 ms. Without any modification of TCP source and destination protocol stacks(TCP NewReno), all mechanisms proposed in this paper was implemented at PEP. PEP provides RA-spoofing, SA, and TCP-aware Go-back-N ARQ mechanisms.

We performs various simulation works under different assumptions of bandwidth asymmetric ratio and error rate. Table 1 shows the calculation of k_b, k_a, and finally the segment aggregation factor, k_s, used in our simulation works. Since the ratio of packet sizes used in the two directions is $1500/40 = 37.5$, the

Table 1. The calculation of k_b, k_a and k_s

Uplink Bandwidth	k_b	k_a	k_s BER $< 10^{-6}$	k_s BER $\geq 10^{-6}$
256 Kbps	1.0	1	1	1
128 Kbps	2.1	2	2	1
64 Kbps	4.2	3	3	1
32 Kbps	8.3	5	5	1

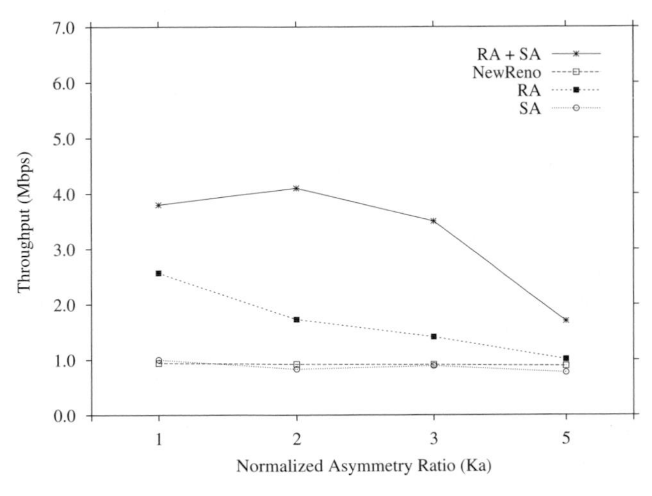

Fig. 3. The effect of each of RA-spoofing and SA when BER $= 10^{-8}$

uplink bandwidth for TCP to achieve 1 Mbps throughput without congestion over the uplink should be at least $1/37.5$ Mbps $= 26.7$ Kbps in an ideal case with no error.

As depicted in Figure 3, TCP NewReno gives the maximum throughput of about 940 Kbps, regardless of the variation of k_a from 1 to 5 when BER $= 10^{-8}$. Rather, SA achieves a slightly degraded performance due to the retransmission of the whole k_s IP fragments for one or more lost fragments in case of the occurrence of intermittent errors.

On the other hand, RA-spoofing achieves about 2.6 Mbps maximum throughput when $k_a = 1$ since no uplink congestion happens. The performance of RA-spoofing is, however, dropped to about 1 Mbps when $k_a = 5$ since RA-spoofing without segment aggregation results in congestion over the uplink. Nevertheless, this has still higher performance than that of the TCP NewReno. As denoted by RA+SA in Figure 3, When SA is to be coupled with RA-spoofing, the performance of RA+SA is greatly improved to about 3.8 Mbps at $k_a = 1$ and about 1.7 Mbps at $k_a = 5$. It is interesting that the throughput of RA+SA is further improved at $k_a = 2$ than at $k_a = 1$. This obviously represents that the segment aggregation feature greatly reduce the usage of the uplink bandwidth and accelerate the ACK-clocking behavior of the conventional TCP.

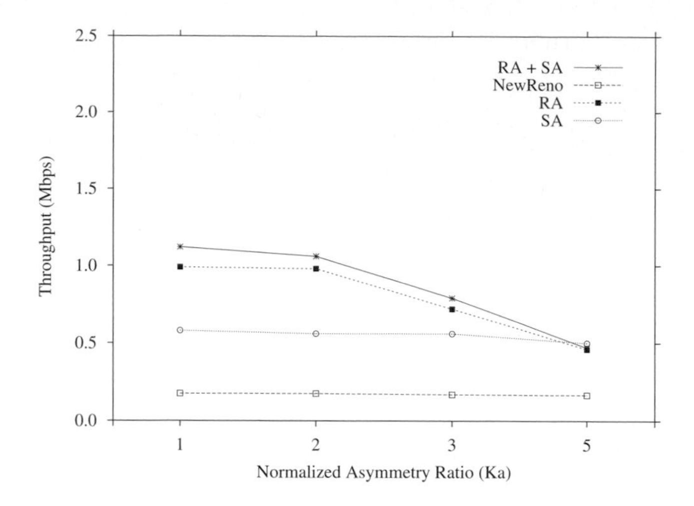

Fig. 4. The effect of each of RA-spoofing and SA when BER = 10^{-6}

Figure 4 also shows the effect of each of RA-spoofing and SA when BER = 10^{-6}. Though k_a varies from 1 to 5, k_s is always set to 1, as calculated in Table 1, which means no segment aggregation function at a high error rate. When $k_a = 1$, TCP NewReno reveals a poor performance of 0.2 Mbps but SA raises it to 0.6 Mbps which comes from the Go-back-N ARQ mechanism used in the link layer. RA and RA-spoofing greatly improve the performance further even at a high error rate when $k_a = 1$ and $k_a = 2$. Note that RA+SA achieves a slightly higher performance than that of RA because the segment aggregation mechanism works occasionally under the random error environment.

Figure 5 summarizes the throughput of RA+SA in a single connection environment. When $k_a = 1$, the throughput of RA+SA reaches 5 to 6 times that of TCP NewReno regardless of various error rates. This mainly owes the RA-spoofing mechanism. As k_a is increased to 2 or 3, SA as well as RA-spoofing raises its throughput to 3 or 4 times that of TCP NewReno. When $k_a = 5$, its throughput is improved to 2 or 3 times that of TCP NewReno even in low and high error rates. Thus, we conclude that the RA+SA mechanism implemented at PEP could greatly improve the performance of the conventional TCP without any modifications of the TCP source and destination. Note that the maximum throughput of RA+SA is limited to approximately 5 Mbps in error free environment. This comes from the fact that the upper limit of $W_{virtual}$ is set to 10 when the downlink bandwidth is 10 Mbps, the transmission time for $W_{virtual}$ blocks is 64 KB $\times W_{virtual}/10$ Mbps $\cong$ 524 ms, and then the achievable maximum throughput of RA+SA is theoretically 64 KB $\times W_{virtual}/(524 + 500)$ ms $\cong$ 5 Mbps.

Figure 6 shows the throughput of RA+SA in an environment that five connections share the downlink of 10 Mbps simultaneously. As k_a increases, the performance of TCP NewReno drops sharply in case of no error and BER = 10^{-8}. When $k_a = 1$ and $k_a = 2$, the throughput of RA+SA reaches 1.7 times that of

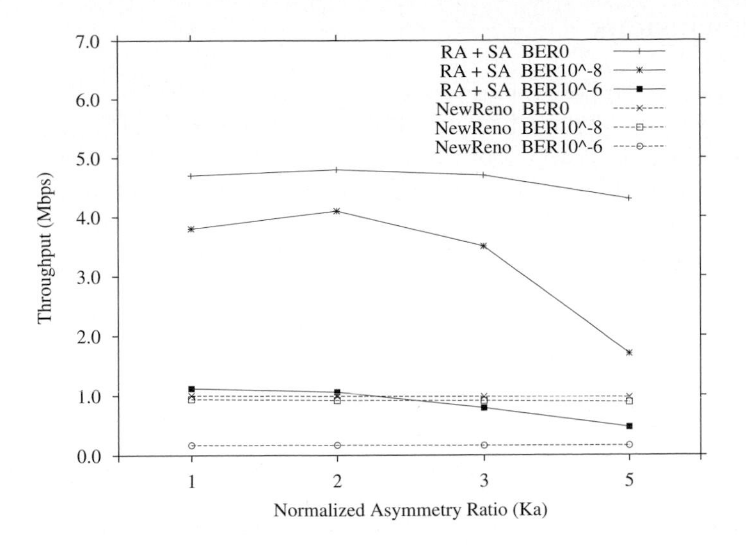

Fig. 5. The throughput of RA-spoofing with SA in a single connection environment

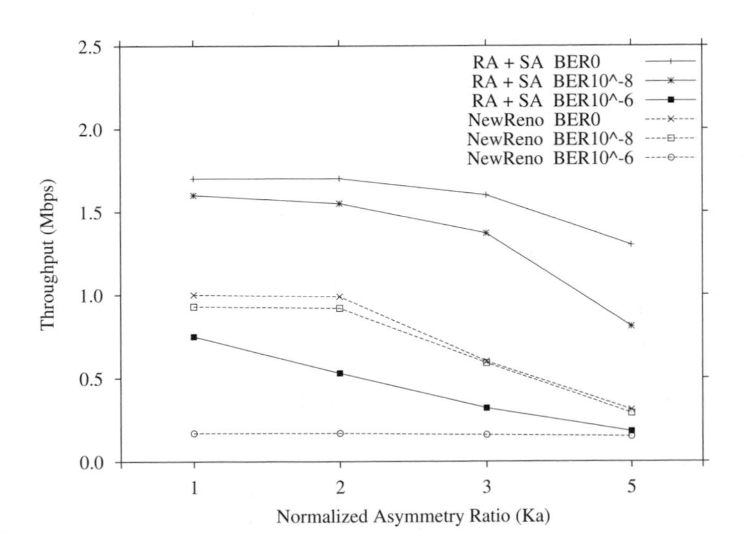

Fig. 6. The throughput of RA-spoofing with SA in a multiple connections environment

NewReno. Furthermore, when $k_a = 3$ and $k_a = 5$, the throughput of RA+SA achieves more than 2 times up to 4 times that of NewReno. In case of $k_a = 5$, the RA+SA throughput reaches 4.2 times that of NewReno when no error occurs and 2.8 times when BER $= 10^{-8}$. This implies that the greater performance is obtained by SA in case of less error, which accelerates RA-spoofing again.

4 Conclusions

In this paper, we propose the two PEP mechanisms to improve TCP performance in a two-segment satellite Internet: Rate-Adaptive spoofing(RA-spoofing) and Segment Aggregation(SA). Both of them are implemented at PEP so that the backward compatibility of the TCP/IP protocol suits is maintained at the TCP source and destination. RA-spoofing is proposed to solve the problem that the maximum throughput of the TCP spoofing is limited to approximately 1 Mbps due to the maximum window size of 64 KB and RTT of 520 ms regardless of the available bandwidth. Given the known total bandwidth of the link, RA-spoofing determines the size of a virtual window dynamically and fairly based on measured RTT and total amount of traffic in transit on the link. With the virtual window, PEP can fully utilize the available bandwidth and maximize the throughput rate. Tightly coupled with RA-spoofing, SA is also proposed to alleviate the effect of bandwidth asymmetry and then accelerates the ACK-clocking of the TCP. The performance analysis reveals that the throughput of RA-spoofing with SA roughly achieves 2 to 4 times that of the conventional TCP under different assumptions of bandwidth asymmetric ratio and error rate.

References

1. Barakat, C., Altman, E., Dabbous, W.: On TCP performance in a heterogeneous network: a survey. IEEE Communications Magazine, 40–46 (2000)
2. Border, J., et al.: Performance enhancing proxies intended to mitigate link-related degradations, RFC 3135, IETF (2001)
3. Ishac, J., Allman, M.: On the performance of TCP spoofing in satellite networks. In: Proc. of IEEE MILCOM, vol. 1, pp. 700–704 (2001)
4. Balakrishnan, H., Padmanabhan, V.N., Katz, R.H.: The effects of asymmetry on TCP performance. ACM Mobile Networks and Applications 4, 219–241 (1999)
5. Balakrishnan, H., Seshan, S., Amir, E., Katz, R.H.: Improving TCP/IP performance over wireless Networks. In: Proc. of ACM MOBICOM, pp. 2–11 (1995)
6. Lakshman, T.V., Madhow, U., Suter, B.: Window-based error recovery and flow control with a slow acknowledgement channel: A study of TCP/IP performance. In: Proc. of IEEE INFOCOM, vol. 3, pp. 1199–1209 (1997)
7. Tian, Y., Xu, K., Ansari, N.: TCP in wireless environments: problems and solutions. IEEE Radio Communications, s27–s32 (2005)

Effects of the Number of Hops on TCP Performance in Multi-hop Cellular Systems

Bongjhin Shin, Youngwan So, Daehyoung Hong, and Jeongmin Yi

Dept. of Electronic Engineering, Sogang University
C.P.O. Box 1142, Seoul, 100-611, Korea
{bjshin,lefteye,dhong,jeongmin}@sogang.ac.kr

Abstract. This paper analyzes performance of Transmission Control Protocol (TCP) in multi-hop cellular networks. Multi-hop transmission shortens communication distance and can make channel quality at each hop better. However it needs channel resources at every hop for packet transmission, and thus effective throughput on end-to-end links can be decreased. Related researches have shown that multi-hop transmission may reduce TCP performance in the systems. However, it may be because improvement of channel quality derived by multi-hop transmission has not been considered fairly. We analyze TCP performance with considering channel quality improvement on a multi-hop system. By simulation, we derived the throughput gain by multi-hop transmission for various values of the number of hops. We also have analyzed the gain according to position of active users in a cell. From the results, we show that multi-hop transmission can improve TCP performance, especially for the users near the cell edge. We also show a proper number of hops that gives maximum TCP performance.

Keywords: TCP, Throughput, Throughput gains, Wireless Multi-Hop Systems.

1 Introduction

Recently, wireless multi-hop systems are gaining increasing attention for their coverage extension and capacity enhancement [1, 2]. These systems aim to provide higher data rates and to improve fairness. These aims can be achieved depending on short communication distance at each hop in multi-hop systems. In this system, Mobile Stations (MS) can connect to a Base Station (BS) directly or through stations for relaying. Increasing commercial interests in wireless multi-hop systems have prompted the IEEE to setup a new task group (802.16j) for formalizing standards on the Physical (PHY) and the Multiple Access Control (MAC) layers. Some simple scenarios using multi-hop transmission are shown in figure 1. This figure contains three possible transmission types: single hop transmission, multi-hop transmission with MSs as relaying stations, and multi-hop transmission with fixed exclusive relay stations.

The multi-hop cellular system is a kind of communication infrastructure that is build from combining conventional single hop cellular systems and fixed relay

T. Vazão, M.M. Freire, and I. Chong (Eds.): ICOIN 2007, LNCS 5200, pp. 133–141, 2008.

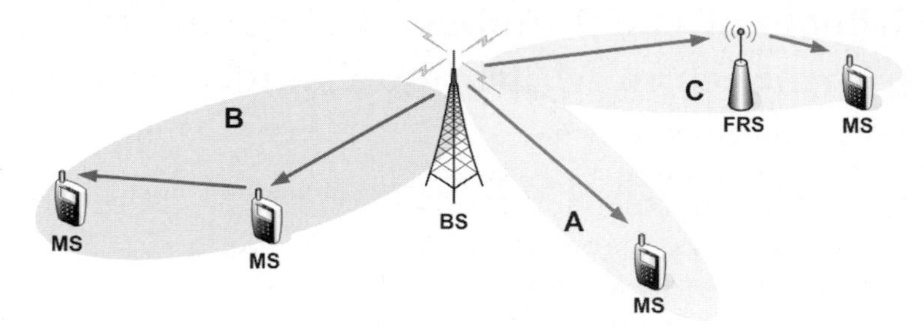

Fig. 1. Simple transmission scenarios on a multi-hop system

stations [3]. The idea of multi-hop transmission is similar to ad hoc networks. In these networks, MSs act as the relay station and forward packets between stations located out of transmission range of each other. In ad hoc networks, both nodes on end of the end-to-end links are always MSs. However, multi-hop cellular systems always have a BS as one of these end nodes.

In single hop cellular network has wide area cell coverage, but serves low data rates, especially near the cell boundary. In multi-hop systems, distance of each hop shortens, and thus channel states and transmission rates can be improved. Since each hop in multi-hop systems requires resources for carrying packets, total resources used on end-to-end multi-hop links can be increased as comparing to single hop systems.

Users' demands that connect to the internet through wireless links have been enlarged. In order to make this connection, methods for stable packet transmission are required. The Transmission Control Protocol (TCP) is the most typical one for reliable packet transmission in the internet. Therefore TCP is also the most potential method in integrated wired-wireless systems. However, conventional TCP have been optimized based on what they are used on wired networks, and thus TCP assumes that the reason of packet losses is network congestion. Therefore most researches about TCP have been concentrated on revising their congestion control algorithm. However, packet losses are frequently occurred from link failure based on low channel quality in wireless networks, and TCP cannot care these losses. This problem is more serious in wireless multi-hop systems, because link failures on a hop may induce the failures of end-to-end multi-hop links. In order to use TCP in wireless multi-hop systems, we should to solve this problem. However before looking for solutions, we should to know about performance of conventional TCP in wireless multi-hop systems.

In order to study on TCP performance over wireless multi-hop systems, several researchers have been reported [4, 5, 6, 7, 8]. Most of them made conclusions that TCP has lower performance in multi-hop systems than in single hop systems. In [4, 5, 6, 7, 8], they showed that TCP throughput decreases to half on 4-hop links. In addition, movement of high speed MSs reduces TCP throughput in [5]. In [8], they proposed a rate control algorithm and showed the algorithm can improve TCP performance. These researches concentrated on the MAC and the

TCP layer solutions. In [4], they revised a routing scheme. A new TCP algorithm is proposed in [5]. In these researches, TCP performance decreases in multi-hop systems despite of applying proposed solutions. This is the why they do not consider effects of improved channel quality in multi-hop systems.

In this paper, we analyze TCP performance with considering channel gain improvement and show that multi-hop transmission can have higher TCP performance than one on single hop systems at near the cell edge. Error and data rates on Link Layers (LL) make effects on TCP performance, because these rates are changed by the number of hops in multi-hop systems. The more number of hops makes shorter communications distance and better channel quality, so a data rate at a hop can be increased. However an effective data rate on end-to-end links can be decreased by using additional resources at each hop. Similarly, better channel quality induces a lower error rate at a hop, but a higher error rate at the end-to-end link. We analyze effects of the number of hops on TCP performance and suggest the numbers of hops can make the maximum TCP performance. We also produce TCP performance in a wireless multi-hop system and compare it with the results in a wireless single hop system. We turn out throughput gain and an error rate for various numbers of maximum using hops and MSs, and distance between the BS and MS in the LL and TCP layer.

The rest of this paper is organized as follows. In Section 2, we describe some models for constructing a multi-hop cellular system. In Section 3, we explain some system models used for simulation. Simulation results of performance are presented in Section 4. Finally, we summarize our work and derive conclusions in Section 5.

2 The Wireless Multi-hop Cellular System Model

In this section, we explain implemented models and define parameters for our wireless multi-hop cellular system. Three models are shown in this section: the resource allocation model for each hop relationship between LL packets and a TCP segment, and process of transmission on a hybrid wired and wireless network. The radio resources for hops and users are divided by time dimension such as a time division multiple access (TDMA) method. The time resource allocation on single and multi-hop is shown in figure 2. A TCP segment is composed of LL packets such as figure 3. We assume that there is no overhead for fragmenting and reassembling of LL packets to a TCP segment.

Figure 4 shows the process that TCP segments are transmitted from a server on the internet to a target MS connected to the wireless multi-hop cellular system. The Base Station Controller (BSC) fragments a TCP segment to LL packets, and then LL packets are sent through the multi-hop system from the BSC to a target MS. The MS receives these LL packets and recomposes to a TCP segment. We assume that our system use the conventional packet transmission method from the server to the BSC. To minimize an effect of packet transmission on the wired internet, we assume the server connected directly to the BSC using wired line.

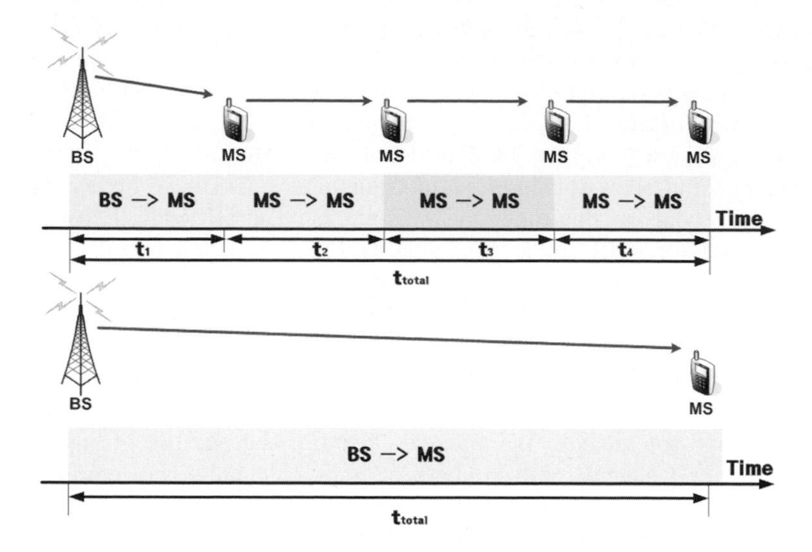

Fig. 2. Packet transmission in single and multi-hop systems

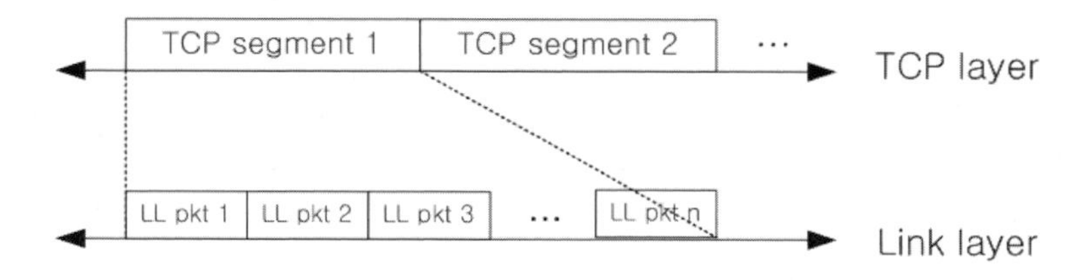

Fig. 3. The relationship between a TCP segment and LL packets

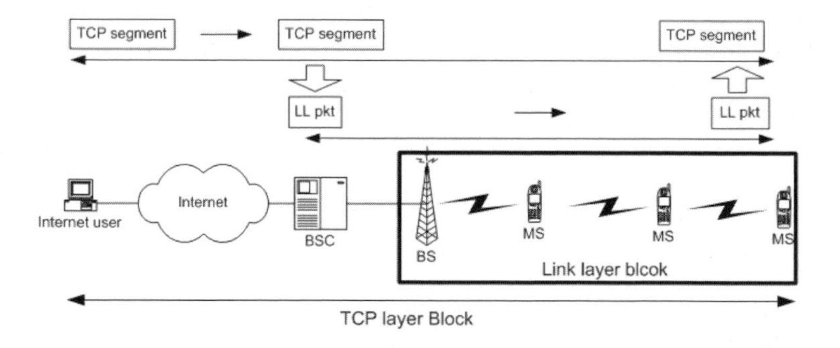

Fig. 4. The process of transmitting TCP segments in the multi-hop cellular system

We define four performance measures; the error rate, the data rate, a cell throughput gain, and a user throughput. All measures are defined both on the LL and TCP layer. An error rate of a kth packet at a jth hop on the LL denotes

$e_{j,k}$, so we can be represented as $E_{s,LL} = e_{1,k}$, and the rate at a multi-hop is represented as

$$E_{m,LL} = 1 - \prod_{j=1}^{J}(1 - e_{j,k}). \tag{1}$$

An error rate of TCP segments at a single hop is represented as

$$E_{s,TCP} = 1 - \prod_{k=1}^{K}(1 - e_{j,k}) \tag{2}$$

and the rate at multi-hop is represented as

$$E_{m,TCP} = 1 - \prod_{k=1}^{K}(1 - \prod_{j=1}^{J}(1 - e_{j,k})). \tag{3}$$

where K denotes the number of LL packets that compose a TCP segment.

To calculate the data rate, we calculate time for data transmission and then produce data rates. Time for packet transmission at jth hop such as in figure 4,

$$t_j = \frac{D}{R_j} \tag{4}$$

and it is also represented as $T_{t,s} = t_1$ where R_j denotes a data rate [bps] on jth hop and D denotes a size of a packet [bits]. Packet transmission time at end-to-end multi-hop is calculated as

$$T_{t,m} = \sum_{j=1}^{J} t_j \tag{5}$$

where J means the number of hops for packet transmission to the target MS. The data rate is calculated by packet size to transmission time. We define that user throughput is user's received bits during a unit time. Cell throughput is defined the sum of user throughput in a cell during a unit time. Throughput gain is defined by throughput of multi-hop to single hop system ratio.

3 Simulation Model

In this section, we show our simulation models and environments of the system. We produce performance of TCP Reno on downlink in wireless multi-hop systems. The system model for performance evaluation is comprised of 19 cells and MSs distribute uniformly in the cells. We assume that all cells can reuse radio resources and MSs on a center cell receive interference from BSs on other cells. We assume that there is no interference within the same cell and thermal noise. The received signal to interference ratio when the ith packet is retransmitted k times at the jth hop is calculated as

$$\left(\frac{C}{I}\right)_j = \frac{L_j P_j}{\sum_{i=1}^{B} L_i P_i} \tag{6}$$

where B denotes the number of interfering BSs. L_j means pathloss from the BS on the center cell to the receiver of the jth hop. P_j denotes transmission power at the jth hop. Transmission power of the BS is 4 times larger than power of the MS.

We consider standard propagation models for performance evaluation that takes into account Rayleigh fast fading, lognormal shadowing, and γth order free space pathloss. The Shannon's capacity formula is used in calculation of the data rate at a hop. U denotes throughput that means the number of received bits during total transmission time. It is represented as

$$U = \frac{D}{T_{t,m}}. \tag{7}$$

The Throughput gain G can be also calculated as the ratio of throughput of most favorable case to that of the direct link. Therefore it is represented as

$$G = \frac{U_r}{U_s}. \tag{8}$$

A Monte Carlo computer simulation has been developed to evaluate performance of our multi-hop system. The shadowing variables are generated with the auto-correlation and the cross correlation, and then we calculate SIR, the data rate, and the metrics. The simulation collects statistical data on the center cell.

4 Numerical Results

In this section, we show and analyze our simulation results. We produce three kind of results; the error rate, cell throughput gain, and user throughput. All results are gathered both on the LL and TCP layer. We present error rates and the cell throughput gain as the number of hops and distance between the BS and MS on the LL and TCP layer. In addition we also yield cell throughput gain and user throughput as the number of MSs and hops on the layers.

The error rate is yielded to show reasons of change of performance. Figure 5 shows the error rate for various values of the transmission hop number and distance between the BS and MS. As a MS goes away from the BS, regardless of the number of hops, C/I received by user reduces, but the error rate are increased. The error rate goes down with increasing the hop number near the cell edge. In figure 5 the error rate on LL is 0.12 near a cell edge in the single hop system, but it decreases to 0.01 on 5-hop transmission. The error rate on TCP layer is higher than the rate on LL, because the error on the LL is enlarged to the error for a TCP segment. That is, a few errors on the LL induce a large number of errors on the TCP layer.

Cell throughput gains are throughput improvement by multi-hop transmission. Figure 6 describes cell throughput gain for various values of the transmission hop

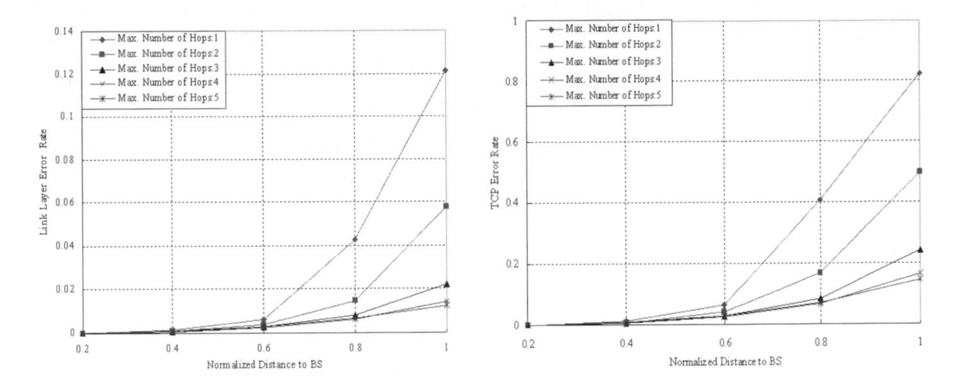

Fig. 5. LL and TCP error rate as MS locations and the number of hops

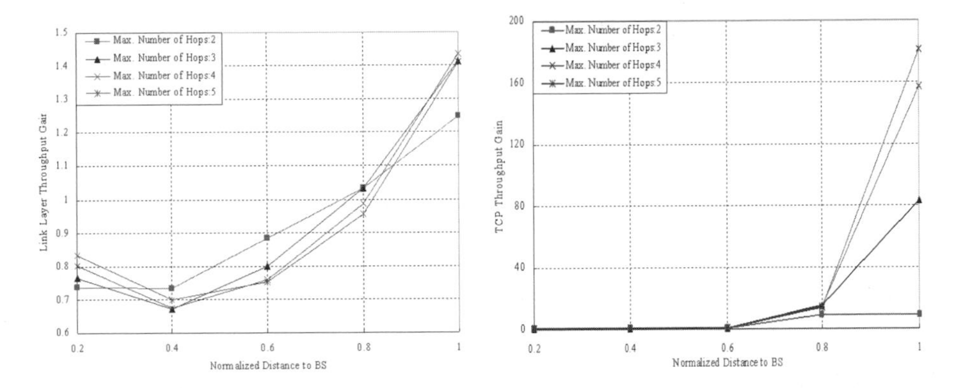

Fig. 6. LL and TCP cell throughput gain as MS locations and the number of hops

number and distance between the BS and MS. In center of the cell, LL and TCP throughput gains in the multi-hop system are less than 1. It means that performance of the single hop system is better than the multi-hop system in this region. In the cell boundary, these gains in the multi-hop system are larger than 1 and improves with increasing hop number. In figure 6 the LL throughput gain is about 1.4 and the TCP throughput gain is about 180. Because we assume that if there is an error on one of the hops contained to multi-hop link, the multi-hop link get an error. Therefore a little difference of error rates at a hop enlarges enormously in total multi-hop links. In figure 6, the LL error rate in the cell boundary has difference about 0.11, but the TCP error rate has difference about 0.6. Therefore improvement of the TCP throughput gain is larger than on the LL.

Figure 7 represents the cell throughput gain with various numbers of MSs that can be used as relay node in the cell. The cell throughput gain multiplies with increasing number of MSs. The LL cell throughput gain increase from 1.2 to 1.4. Figure 8 shows user throughput with various numbers of MSs. The user

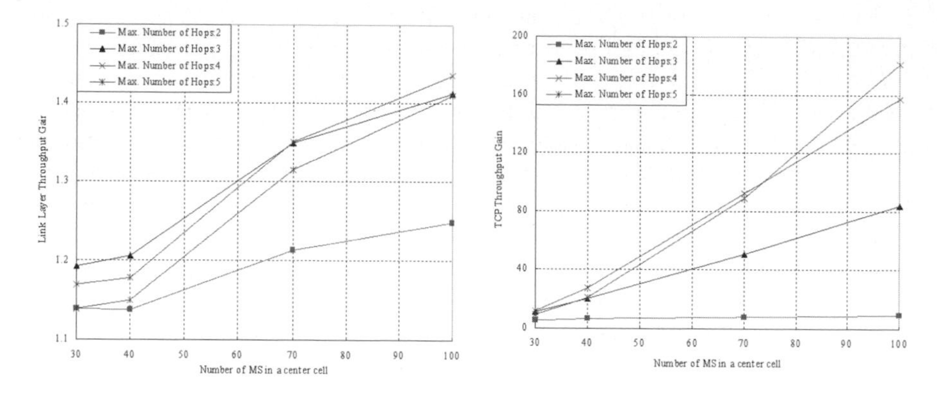

Fig. 7. LL and TCP cell throughput gain as the number of transmitting hops and relaying MSs

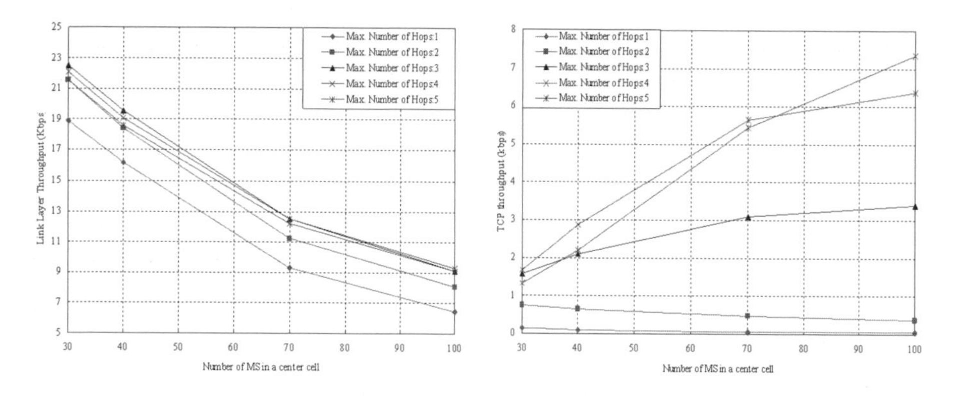

Fig. 8. LL and TCP user throughput as the number of transmitting hops and relaying MSs

throughput decreases with increasing number of MSs, because users should to share same amount of resources.

The results that we obtained in this simulation show that TCP performance in multi-hop cellular systems can be dramatically enhanced. In the multi-hop system, we can get 1.4 times improvement on the LL and 160 times improvement on TCP layer than in the single hop system.

5 Conclusion

This paper analyzes effects of the number of hops on TCP performance in multi-hop cellular systems. As performance measures, we use three parameters on the LL and the TCP layers: error rates, throughput, and throughput gains. These measures are derived as functions of the number of hops, distance between the BS and MSs, and the number of MSs in a cell.

The results in our study show that TCP performance of cellular system can be improved through multi-hop transmission. Our results represent that the controlled number of hops can maximize throughput performance. The number of hops for maximum performance differs with location of target MSs. In case of MSs near the BS, transmission on links consisted of single-hop or fewer numbers of hops enhances TCP performance. In the other case, MSs at the cell boundary, higher number of hops makes better TCP throughput gains. This paper also show that TCP throughput of a cell increases with the number of MSs. It is because those more efficient routes can be established with a lot more MSs.

The results presented in this paper suggest that careful selection of the number of hops is required for maximizing TCP performance on wireless multi-hop networks. The results in this paper can be utilized for planning of wireless multi-hop cellular networks.

References

[1] Huining, H., Yanikomeroglu, H., Falconer, D.D., Periyalwar, S.: Range extension without capacity penalty in cellular networks with digital fixed relays. In: Proc. IEEE GLOBECOM, pp. 3053–3057 (2004)

[2] Esseling, N., Walke, B.H., Pabst, R.: Performance evaluation of a fixed relay concept for next generation wireless systems. In: Proc. IEEE International Symposium on PIMRC, vol. 2, pp. 744–751 (2004)

[3] Cho, J., Haas, Z.J.: On the throughput enhancement of the downstream channel in cellular radio networks through multihop relaying. IEEE Journal on Selected Areas in Communications 22, 1206–1219 (2004)

[4] Gupta, A., Wormsbecker, I., Wilhainson, C.: Experimental evaluation of TCP performance in multi-hop wireless ad hoc networks. In: Proc. International Symposium on MASCOTS, pp. 3–11 (2004)

[5] Zhenghua, F., Xiaoqiao, M., Songwu, L.: How bad TCP can perform in mobile ad hoc networks. In: Proc. International Symposium on ISCC, pp. 298–303 (2002)

[6] Fu, Z., Zerfos, P., Luo, H., Lu, S., Zhang, L., Gerla, M.: The impact of multihop wireless channel on TCP throughput and loss. In: Proc. IEEE INFOCOM, vol. 3, pp. 1744–1753 (2003)

[7] Yumei, W., Ke, Y., Yu, L., Huimin, Z.: Effects of MAC retransmission on TCP performance in IEEE 802.11-based ad-hoc networks. In: Proc. IEEE 59th VTC, vol. 4, pp. 2205–2209 (2004)

[8] Bansal, S., Shorey, R., Kherani, A.A.: Performance of TCP and UDP protocols in multi-hop multi-rate wireless networks. In: IEEE WCNC, vol. 1, pp. 231–236 (2004)

Traffic-Aware MAC Protocol Using Adaptive Duty Cycle for Wireless Sensor Networks*

Seungkyu Bac, Dongho Kwak, and Cheeha Kim

Department of Computer Science and Engineering, POSTECH, Korea
{ktnslt,whitecap,chkim}@postech.ac.kr
http://nds.postech.ac.kr

Abstract. Most of sensor MAC protocols have a fixed duty cycle, which performs poorly under the dynamic traffic condition observed in event-driven sensor applications such as surveillance, fire detection, and object-tracking system. This paper proposes a traffic-aware MAC protocol which dynamically adjusts the duty cycle adapting to the traffic load. Our adaptive scheme operates on a tree topology, and nodes wake up only for the time measured for the successful transmissions. By adjusting the duty cycle, it can prevent packet drops and save energy. The simulation results show that our scheme outperforms a fixed duty cycle scheme [5] and B-MAC [9], hence, it can achieve high packet fidelity and save energy.

Keywords: adaptive duty cycle, MAC, sensor networks.

1 Introduction

In wireless sensor networks, battery-powered nodes operate in an unattended manner after their deployment, which mostly concerns the network lifetime. The network lifetime can be defined in various ways according to different scenarios and is generally categorized into one of network connectivity, the fraction of failure nodes, and sensing coverage [7]. To prolong the network lifetime, each node should take a periodic sleep as long as it can, since the energy consumed in sleep state is much less than in other states up to three orders of magnitude [6]. S-MAC [5] is a representative of MAC protocols which employs periodic sleep with a fixed duty cycle. With S-MAC, nodes awake for a fixed listen interval, and go into sleep for another fixed interval, repeatedly.

In some sensor network applications such as surveillance, fire detection, and object-tracking system, networks are idle most of the time, but a sudden event causes a large amount of packets to yield network congestion. In such circumstances, MAC protocols which have a fixed duty cycle suffer from data loss because they are inadaptable to the heavy traffic load. MAC protocols with a

* "This research was supported by the MIC(Ministry of Information and Communication), Korea, under the ITRC(Information Technology Research Center) support program supervised by the IITA(Institute for Information Technology Advancement)" (IITA-2008-C1090-0801-0045).

T. Vazão, M.M. Freire, and I. Chong (Eds.): ICOIN 2007, LNCS 5200, pp. 142–150, 2008.

small fixed duty cycle can save energy with no events, but packets can be dropped due to queue overflow whenever an event occurs. To reduce the packet drop, the fixed duty cycle should be increased, but energy is largely wasted for frequent listening whenever no events occur. Hence, it is desirable to increase duty cycle to prevent packet drop under heavy traffic, and to decrease duty cycle for nodes that are unaffected by the traffic or are under the light traffic to save energy.

Recently, MAC protocols with adaptive duty cycle have been studied, such as PMAC [3], T-MAC [8], and B-MAC [9]. B-MAC is widely used in sensor networks. It uses a wakeup signal called preamble that lasts during the listen interval. Before transmitting a packet, a node should wake up all neighbors using a long preamble. B-MAC causes a big overhead on networks resulting in low throughput and energy waste. Conceptually, the duty cycle can be adaptive to the traffic condition, but realistically, it has not been discussed upon.

In this paper, an adaptive MAC protocol is proposed which operates on a tree topology where packets are only delivered from a child node to the parent node. In this case, a parent only needs to wake up until all packets from its children are successfully received. To determine the wakeup duration, each child node notifies its parent of the number of packets to transmit and then, we use the Markov chain method [10] which is used to measure the average time for a successful transmission. Our adaptive MAC protocol prevents packet drops by increasing duty cycle under heavy traffic, and saves energy by decreasing duty cycle under light traffic. After presenting our scheme, we conduct simulations to compare our scheme to a fixed duty cycle scheme [5] and B-MAC [9]. The simulation results show that our scheme can achieve higher packet fidelity and consume less energy than other schemes.

2 Proposed Adaptive MAC Protocol

The MAC protocol proposed in this paper adaptively adjusts the duty cycle of nodes adapting to the traffic load. For our work, several assumptions are made. First, tree topology is constructed at network initialization using any protocol such as [12], and each node knows its parent, and child nodes. Second, all data packets generated by nodes have an equal size, and no data aggregation is performed. Third, 802.11 DCF is used for channel access control.

2.1 Tree Topology in Sensor Networks

In data gathering networks such as sensor networks, information reported by each node is collected at the sink node. For easy data gathering and aggregation in sensor networks, tree topology was proposed in [12]. In tree topology, each node has an *id* which is unique in networks and a *level* which measures the distance from the root in hop count. The sink node takes the place of the root and a communication is only allowed between a parent node and its corresponding child node. Using *level*, communication in tree can be regulated to avoid collision and hidden terminal problems between different levels. By assuming superframe

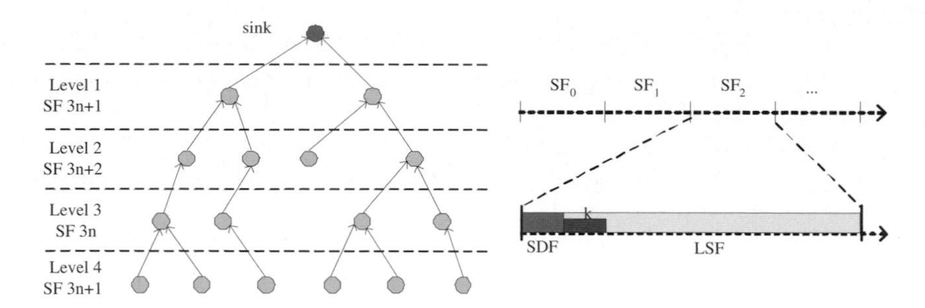

Fig. 1. Tree topology and superframe assignment on each level

structure, superframe $SF_i, i = 3n + (m \bmod 3), n \in \{0, 1, 2, \ldots\}$ is assigned to nodes with level m and each node is only allowed to transmit to its parent in the assigned superframe. Fig 1 shows the tree topology and superframe assignment on each node. The details on a forming superframe will be explained in 2.3.

2.2 Basic Idea

In tree topology, packets are only delivered from children to parent following the edges of tree. Hence, a parent called N_p only needs to be in the listen state at time t for the duration k that all packets from children can be successfully received, where t is decided by the level of N_p's children, and k is determined by the amount of packets that children want to transmit. To decide t, suppose the level of N_p's children is m, then t is the start of every superframe assigned to N_p's children. In this section, we analytically derive the duration k for which N_p should be in the listen state according to the number of packets that its children have.

Suppose that a parent node N_p has c children, $N_1, N_2, \ldots, N_c$, and a child N_i has Q_i packets in its queue for transmission. Let $E[T_{pkt}]$ be the average time between two successful packet transmissions experienced by networks, then the average total time to successfully receive all packets from children can be written as $E[T_{pkt}] \sum_{i=1}^{c} Q_i$. Hence, N_p needs to be in the listen state for k, which is defined by

$$k = E[T_{pkt}] \sum_{i=1}^{c} Q_i. \tag{1}$$

To calculate $E[T_{pkt}]$, Markov chain model is used. In [10], Bianchi provides an extremely accurate and analytical model for 802.11 DCF to compute the normalized system throughput S, defined as the fraction of time the channel is used to successfully transmit a packet, which can be written as

$$S = \frac{P_s P_{tr} E}{(1 - P_{tr})\sigma + P_{tr} P_s T_s + P_{tr}(1 - P_s) T_c}, \tag{2}$$

(see [10] for the definition of each term). Since the packet length E is measured in time unit, E/S represents the average total time to successfully transmit a E-length packet, which is bigger than E due to the backoff, the contention, and the collision. Hence, $E[T_{pkt}]$ can be written as

$$E[T_{pkt}] = \frac{(1 - P_{tr})\sigma + P_{tr} P_s T_s + P_{tr}(1 - P_s)T_c}{P_s P_{tr}}. \tag{3}$$

Now (1) is complete if N_p can obtain the information Q_i, which will be explained in the next section.

2.3 Superframe Format

A superframe whose length is T_{sf}, is formed as shown in Fig 1. In each superframe SF_i which is synchronized over networks, there are two subframes called schedule decision frame (SDF) and listen schedule frame (LSF). The SDF length of T_{sdf} begins at the start of each SF_i, and LSF length of T_{lsf} follows SDF ($T_{sf} = T_{sdf} + T_{lsf}$). In SDF, child nodes N_i assigned to SF_i and the corresponding parent node N_p should wake up, and exchange information to compute k by (1). Each N_i unicasts a packet called SDF-packet which contains Q_i, and N_p broadcasts a SDF-packet containing the computed k after receiving Q_i from all its children. In LSF, which begins after SDF, N_p should be in the listen state for k, and in sleep state for $T_{lsf} - k$. While N_p is in the listen state for k in LSF, each N_i can transmit data packets to N_p.

During SDF and LSF period, every packet transmission is contention-resolved by using 802.11 DCF. Due to an unreliable wireless link, packets can be corrupted or collided, resulting in the loss of Q_i and k sent by N_i and N_p, respectively. The loss of SDF-packets containing Q_i unables N_p to be unable to calculate the correct k, and the loss of SDF-packet having k leaves some children to be unaware of the duty cycle k of N_p. The ACK mechanism of 802.11 DCF can be adopted for reliable unicasts of SDF-packets from N_i to N_p, however, the SDF-packet containing k requires additional mechanism due to the broadcast nature. If N_i has data packets and does not receive k in SDF *(note that a child does not need to know k if it has no data packet)*, then N_i sets a D-flag in the data packets sent in LSF. Upon receiving a data packet set D-flag, N_p piggybacks k in ACK. Using this approach, N_i which does not know k can learn it at the first data transmission.

To synchronize the start and the end of superframe over networks, networks have to agree with a common clock, which means that all nodes should be synchronized to the sink node. For this purpose, a timestamp is inserted into SDF-packet sent by each parent. Using FTSP [2], a timestamped SDF-packet that is transmitted by a parent can synchronize all child nodes with high accuracy. The time synchronization starts at the sink, propagates through networks, and finishes at $h \times SF_i$ where h is the network size measured in hop count. The error of time synchronization will be analyzed in 3.2.

3 Analysis

3.1 Schedule Decision Frame Overhead

The proposed adaptive MAC scheme adjusts the duty cycle k adapting to the traffic intensity measured by queue length Q_i of child nodes N_i. When $Q_i = 0$ in SF_i, N_i does not wake up in LSF of the assigned SF_i regardless of what k of the corresponding parent N_p is, while N_p should be in listen state for a proper duration at k, whenever any Q_i of children is nonzero. Although this adaptive scheme saves energy due to the adaptive duty cycle, it should pay additional overhead called SDF overhead.

Suppose that there is no traffic in networks, then each node wakes up for the minimum amount of time needed to exchange Q_i and k. Since every node has a chance to transmit data packets once in every three SFs, an arbitrary node N_i should minimally wake up two SDFs per three SFs, one for transmitting Q_i in N_i's transmission turn and another for broadcasting k in the turn of its children. Hence, the SDF overhead can be calculated as

$$OH_{sdf} = \frac{2T_{sdf}}{3T_{sf}}. \tag{4}$$

The length of superframe T_{sf} should be properly determined, and it will be explained in 3.2. T_{sdf} is set to $75ms$ throughout this paper since this amount of time is enough for less than 20 nodes to successfully transmit one 10-byte long SDF-packet.

3.2 Superframe Size Decision

In addition to deciding T_{sdf}, the length of a superframe T_{sf} is a very important factor related to performance. When T_{sf} is set inappropriately, sensor networks would show low edge performance. If event-sensing nodes are assumed to generate data packets with a fixed interval λ, principles for nice T_{sf} can be made.

Let Q^{max} be the maximum queue length of a node and packets arrived beyond Q^{max} are to be dropped due to queue overflow. To transmit data packets, an event-sensing node has to wait for the assigned superframe which comes around every three SFs. Hence, packets generated in three superframes should be less than Q^{max} *(constraint 1)*, and all packets generated in three superframes should be transmitted during one assigned superframe *(constraint 2)*. If *constraints 1 and 2* are not confirmed, then packet drop occurs, and each constraint can be written as

$$\frac{3T_{sf}}{\lambda} \leq Q^{max} \qquad (constraint\ 1), \tag{5}$$

$$\frac{3T_{sf}}{\lambda} \leq \frac{T_{sf} - T_{sdf}}{nE[T_{pkt}]} \qquad (constraint\ 2), \tag{6}$$

where n is the number of contending nodes, and $E[T_{pkt}]$ is the average time between two successful packet transmissions experienced by networks. (5) is

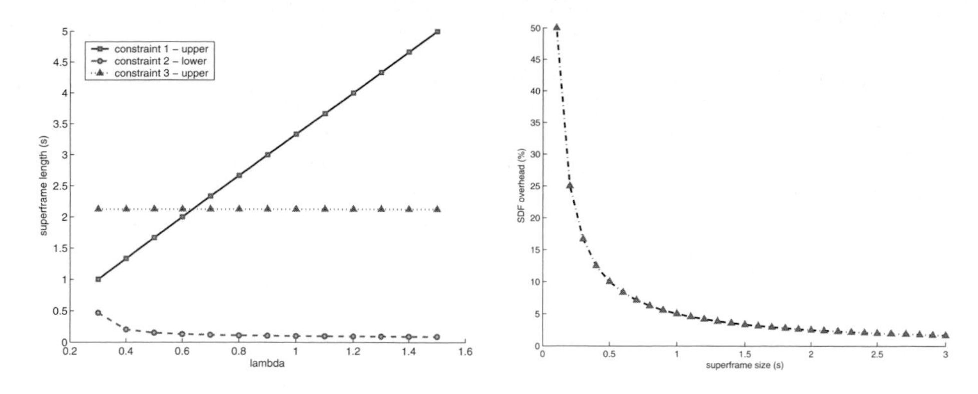

Fig. 2. Three constraints for superframe size, and SDF overhead

trivial and the right hand side of (6) represents the average number of packets transmitted by a node for the time T_{lsf}. Since 802.11 DCF has a long term fairness in transmission opportunity, $nE[T_{pkt}]$ is the average time between two successful packet transmissions experienced by a node making (6) sensible.

In addition to *constraint 1 and 2* for preventing queue overflow, time synchronization burdens another constraint on T_{sf}. The maximum clock error ϵ_{max} in tree topology can be written as $\epsilon_{max} = h\epsilon + hT_{sf}\epsilon_{drift}$, where ϵ, h, and ϵ_{drift} are the maximum single hop error, the height of tree topology, and the maximum clock drift error, respectively. To keep ϵ_{max} below a specific threshold $\bar{\epsilon}_{max}$, T_{sf} should satisfy the constraint below:

$$T_{sf} \leq \frac{\bar{\epsilon}_{max} - h\epsilon}{h\epsilon_{drift}} \quad (constraint\ 3). \tag{7}$$

Fig. 2 shows the three constraints, and *constraint 1, and 3* provide upper bounds and *constraint 2* provides a lower bound for T_{sf}. To calculate $E[T_{pkt}]$, data and ack packet size is set to 100 and 10 bytes long, respectively, n and Q^{max} are both set to 10, and others are set to the same with [10]. For *constraint 3*, $h, \epsilon, \epsilon_{drift}, \bar{\epsilon}_{max}$ is set to $7, 4.2\mu s$[2], $4.75\mu s/s$[11], $100\mu s$, respectively. When choosing the right T_{sf}, SDF overhead should be considered for superframe efficiency also shown in Fig. 2. Using the above analysis, T_{sf} is set to $1.5s$ throughout our work.

4 Simulation Result

Our adaptive MAC protocol is evaluated through simulations of being compared to B-MAC [9] and the fixed duty cycle MAC. The used network topology is shown in Fig. 3, in which an event occurs at time $1.5min$ continuing until $4.5min$, and the simulation finishes at $25min$. When sensing the event, four nodes generate data packets with a fixed interval λ and packets are routed via lines. Since the

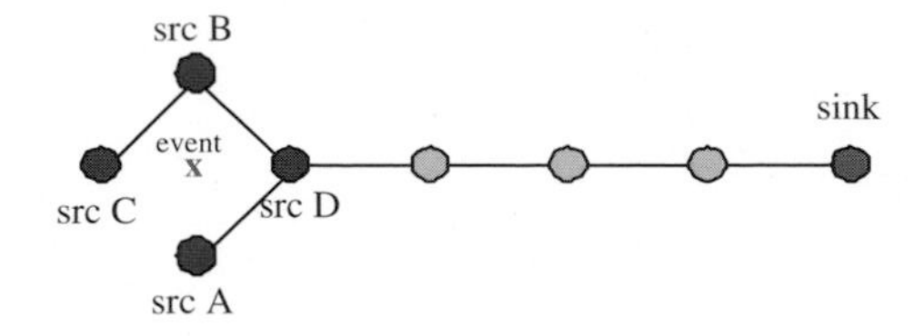

Fig. 3. The network topology used in simulation

packet drop caused by queue overflow is the main concern, no packet corruption and loss in wireless channel are assumed and the Q^{max} is set to 20.

The fixed duty cycle MAC is configured so that each node wakes up at the start of every superframe with a fixed interval k_{fix}, and exchanges data packets during k_{fix}. Three values for k_{fix} are experimented, $100ms, 200ms$, and $300ms$. B-MAC has two parameters *(low power listening, radio sampling interval)*, and the first one is set to $10ms$, and the second one is set to $100ms, 250ms$.

To compare three MAC protocols, the packet fidelity and energy consumption are employed as two performance factors. The packet fidelity is the ratio of the received packets at sink to all packets sent by four sources. The energy consumption is the average energy consumption of all nodes including sink, and the power consumption of radio module is the same with Mica2 described in [6]. The simulation results are shown in Fig. 4 and 5.

Fig. 4 shows the packet fidelity and the energy consumption of the fixed and the adaptive schemes according to the varying λ. The traffic capacity of the fixed scheme is only determined by duty cycle k_{fix}, and large k_{fix} achieves higher packet fidelity when λ becomes low. However, adaptive scheme achieves higher packet fidelity than the fixed one since the duty cycle of the adaptive scheme changes according to λ as indicated in Fig. 4. The energy consumption shows that the adaptive scheme spends less energy than the fixed one. It highly depends on the idle duty cycle defined as the effective duty cycle when networks are idle. k_{fix} determines the idle duty cycle of the fixed scheme, and the idle

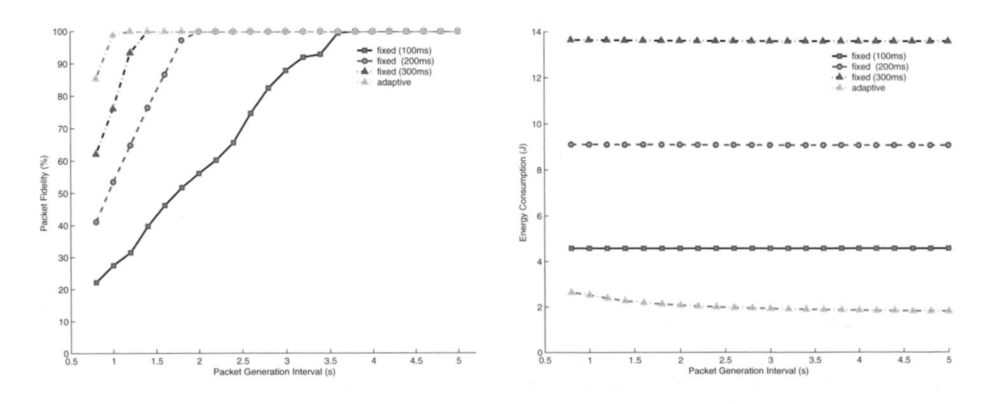

Fig. 4. The comparison results of the adaptive scheme and the fixed scheme

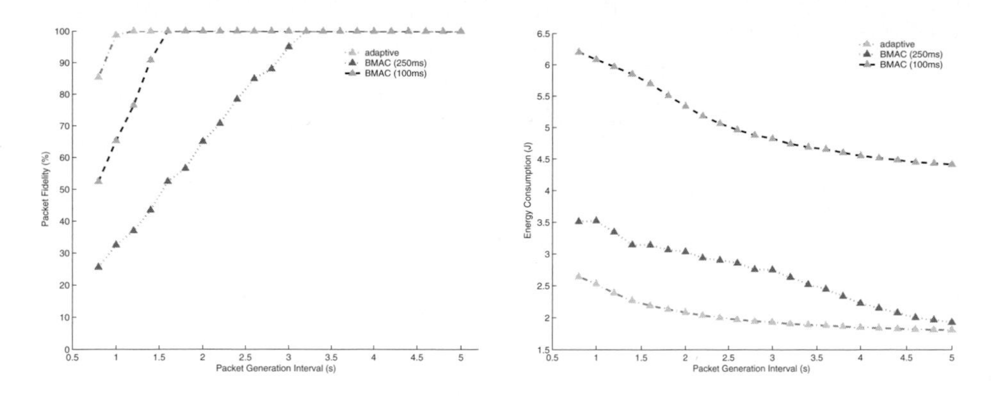

Fig. 5. The comparison results of the adaptive scheme and B-MAC

duty cycle is $100ms/1.5s \times 100 = 6.6\%, 13.3\%, 20\%$ when k_{fix} is $100ms, 200ms$, and $300ms$, respectively, while the idle duty cycle of the adaptive scheme is the same as the SDF overhead (3.3%) defined by (4). Hence, the adaptive scheme is highly superior than the fixed scheme.

Fig. 5 shows the performance comparison between the adaptive scheme and the B-MAC. The long preamble of B-MAC spends a large amount of energy, and also decreases the throughput of B-MAC. In B-MAC, a node should transmit a preamble which is as long as the radio sampling interval, whenever it wish to transmit a packet. Because a node cannot send more than one packet during *(radio sampling interval + data packet transmission time)*, B-MAC suffers from problems such as low throughput, packet drops and low packet fidelity. As shown in Fig. 5, radio sampling interval of B-MAC is highly related to the packet fidelity.

The energy spent by B-MAC and the adaptive scheme also largely depend on the idle duty cycle of each scheme. As mentioned before, the idle duty cycle of the adaptive scheme is 3.3% while B-MAC $(100ms, 250ms)$ is $10\%, 4\%$, respectively. Hence, B-MAC $(250ms)$ consumes almost the same amount of energy as the adaptive scheme when the traffic load is very low.

5 Conclusion

In this paper, an adaptive scheme is proposed which dynamically adjusts the duty cycle adapting to the traffic intensity. Our protocol can prevent packet drops and achieve high packet fidelity, but it keeps idle duty cycle as low as 3.3% by increasing the duty cycle for the high traffic load. To validate this scheme, the superframe structure is formed, which includes SDF for exchanging duty cycle information, and the conditions for proper superframe length are analyzed.

Our scheme is evaluated through simulation, and is compared to the fixed duty cycle scheme and B-MAC. The results indicate that the adaptive scheme achieves high packet fidelity while keeping the energy consumption low.

References

1. Wan, C.Y., Eisenman, S.B., Campbell, A.T.: CODA: Congestion Detection and Avoidance in Sensor Networks. ACM SenSys (2003)
2. Maroti, M., Kusy, B., Simon, G., Ledeczi, A.: The Flooding Time Synchronization Protocol. ACM SenSys (2004)
3. Zheng, T., Radhakrishnan, S., Sarangan, V.: PMAC: An adaptive energy-efficient MAC Protocol for Wireless Sensor Networks. IEEE IPDPS (2005)
4. Bao, L., Garcia-Luna-Aceves, J.J.: Hybrid Channel Access Scheduling in Ad Hoc Networks. In: ICNP (2002)
5. Ye, W., Heidemann, J., Estrin, D.: Medium Access Control With Coordinated Adaptive Sleeping for Wireless Sensor Networks. IEEE transactions on Networking (2004)
6. Anastasi, G., Falchi, A., Passarella, A., Conti, M., Gregori, E.: Performance Measurement of Motes Sensor Networks. ACM MSWiM (2004)
7. Dong, Q.: Maximizing System Lifetime in Wireless Sensor Networks. In: Fourth International Symposium on Information Processing in Sensor Networks (2005)
8. Dam, T., Langendoen, K.: An Adaptive Energy-Efficient MAC Protocol for Wireless Sensor Networks. ACM SenSys (2003)
9. Polastre, J., Hill, J., Culler, D.: Versatile Low Power Media Access for Wireless Sensor Networks. ACM SenSys (2004)
10. Bianchi, G.: Performance analysis of the IEEE 802.11 Distributed Coordination Function. IEEE JSAC (2000)
11. Ganeriwal, S., Kumar, R., Srivastava, M.B.: Timing-sync Protocol for Sensor Networks. ACM SenSys (2003)
12. Krishnamachari, B., Estrin, D., Wicker, S.: Modelling Data-Centric Routing in Wireless Sensor Networks. IEEE INFOCOM (2002)

*m*TBCP-Based Overlay Construction and Evaluation for Broadcasting Mini-system

Mi-Young Kang, Omar F. Hamad, Jin-Han Jeon, and Ji-Seung Nam

Department of Computer Engineering, Chonnam National University
Buk-gu, Gwangju, 500-757, Korea
kmy2221@yahoo.co.kr, omarfh@gmail.com, jhjeon23@naver.com, jsnam@chonnam.ac.kr

Abstract. For better performance and to avoid member service annoyance that results due to joining-clients' waiting durations and time-outs when there are more than one clients wanting to join concurrently for Broadcasting Mini-system's service, this paper proposes a more efficient and better performing Overlay Tree Building Control Protocol by modifying and extending the basic mechanisms building the conventional TBCP. The modified-TBCP (*m*TBCP) proposed is performance-effective mechanism since it considers the case of how fast will children, concurrently, find and join new parents when paths to existing parents are broken. Besides utilizing partial topology information, *m*TBCP also does a LAN-out-degree-check. If the selected child-parent-pair falls under the same LAN, that selected parent does not change the out-degree status. The performance comparison, in terms of Overlay-Connection-Throughput and Latency against Group-Size-Growth, between *m*TBCP, the HMTP, and the traditional TBCP is done through simulations and the results conclude in favour of the proposed *m*TBCP.

Keywords: *m*TBCP, Connection Throughput, OBS, and Overlay Multicast.

1 Introduction

In a conventional TBCP mechanism, the clients being served by a Broadcasting Mini-System may experience annoying and unpleasant services when concurrent members request to join the session or/and when there exist various broken paths at a given time. This is mainly motivated by the fact that the conventional TBCP does not process well more than one join requests concurrently. In TBCP mechanism, more members' join requests are inclined to wait while the protocol is serving fewer join requests. As a result, there might be other waiting members who may experience a disconnected session which may lead to unpleasant reception of the service. With *m*TBCP-based overlay multicast replacing the traditional TBCP mechanism operating the Personal Broadcasting Stations, the problem of joining-time-out can be at large reduced, if not completely terminated. In *m*TBCP mechanism, the source-root maintains the list of existing parents - Potential Parents List (PPL) and keeps updating time after time. Any new client wanting to join sends a request to the source-root. The requests can be sent concurrently from more than one member wanting to join. The

T. Vazão, M.M. Freire, and I. Chong (Eds.): ICOIN 2007, LNCS 5200, pp. 151–160, 2008.

source-root responds the requests by sending the prospective existing PPLs back to respective new join members. The new join member then, establishes a routine for RTT-check for all the parents from the communicated PPL. Consequently, the respective best parent selection is recursively done and respectively adopted. For the purpose of updating the PPL, the same-LAN-out-degree-check is conducted such that: - (a) if a new join member is the out-degree member, register that to the PPL; (b) if a selected parent is the out-degree member, decrease that from the PPL, and (c) if the out-degree measure is full, then delete that parent from the PPL.

The Broadcasting Mini-System is, essentially, comprised of (i) Channel Management Module with two sub-modules – Broadcast Station and Client Access Point; (ii) Personal Broadcast Stations which request to the Channel Management Module for respective channel registration; and (iii) the mTBCP Overlay Multicast Modules which facilitate the client management task including response to client join, leave, failure, broken paths, and fast join/re-join to the sessions. When concurrent clients intend to join a session to the Broadcasting Mini-System, they independently send their Broadcasting Requests (B_REQs) to respective Personal Broadcasting Stations (PBSs). Each PBS, in turn, responds by sending an existing PPL with respect to that particular client. On receiving the PPL, the new client applies the RTT-check Module to calculate the best RTT to the prospective parent with the best RTT.

The notable characteristic of the proposed *m*TBCP compared to the classical TBCP is; in the case of TBCP, if concurrent new members press their B_REQs, the TBCP overlay multicast mechanism processes a fewer B_REQs at a time. That means the waiting time for members to be attended and assigned new parents becomes high. That leads to low overlay connection throughput and a less performing Broadcasting Mini-System. In the case of the proposed *m*TBCP overlay multicast mechanism concurrent new members' B_REQs are attended with negligible waiting time. Therefore, the overall Broadcasting Mini-System experiences higher overlay multicast connection throughput which results into a better performing system with pleasant services.

In sub-section 1.1, the selected research work related to the proposed *m*TBCP has been discussed. Sub-section 1.2 gives an overview of an Overlay Broadcasting System while sub-section 1.3 describes the Session Architecture in Overlay Broadcasting System. Section 2 is devoted to Member Join Requirements while Limitation of Traditional TBCP and Problem with Concurrent Member Join are, respectively, discussed in sub-sections 2.1and 2.2. The details of the Modified-TBCP (*m*TBCP) Scheme is presented in Section 3 while a Simplified Architecture of *m*TBCP and Concurrent Member Join under mTBCP are elaborated in sub-sections 3.1 and 3.2, respectively. Section 4 deals with the Simulation Setup and Results. Finally, Section 5 gives Discussion and presenting Further Scope.

1.1 Related Work

Mathy, Canonico, and Hutchison in [1] have deeply established the founding mechanism for the traditional generic TBCP. They have shown that their main strategy is to reduce convergence time by building the best tree possible early and in advance. However, the fact that the traditional TBCP does not do well when the scenario of concurrent member joining comes into request. Comparatively, our proposed *m*TBCP, efficiently and successfully, intends to overcome this limitation, by introducing a mechanism that allows member joining service concurrently. The idea presented by Farinacci, et al in [3] suggests construction of control trees based on the reverse path concept. However, the control trees proposed here are not overlay trees as they depend on the routers. The limitation of scalability disqualifies the idea that Pendarakis, et al in [5] have described. That is a sort of centralized mechanism where in, for building a distributed topology spanning tree, a session controller maintains, and it must maintain, the total knowledge of membership of a group and also must have the knowledge of the mesh topology that connects the members. To support larger groups, this demands for distributed techniques since no scalability supported in the case of centralized control. Francis, et al in [6] discuss about a similar protocol as the one described in Yoid. However, there is a need for independent mesh for robustness and that the convergence time to optimality is slow. Chu, et al in [7] have introduced the famous Narada protocol where the full knowledge of the group membership needs to be maintained at each node. Therefore, all the protocols that have tried to address the control of the tree building has shown considerable limitations requiring a better performing and efficient proposed *m*TBCP for Broadcasting Mini-System.

1.2 Overlay Broadcasting System

An Overlay Multicast Broadcasting System, here simply referred as Overlay Broadcasting System (OBS), can be efficiently designed and implemented taking the advantages and cross-eliminating the limitations of the three main emerging technologies – the overlay technology, the multicast technology, and the personal broadcasting technology. Fig. 1 is a simplified framework of OBS with a few essential modules shown.

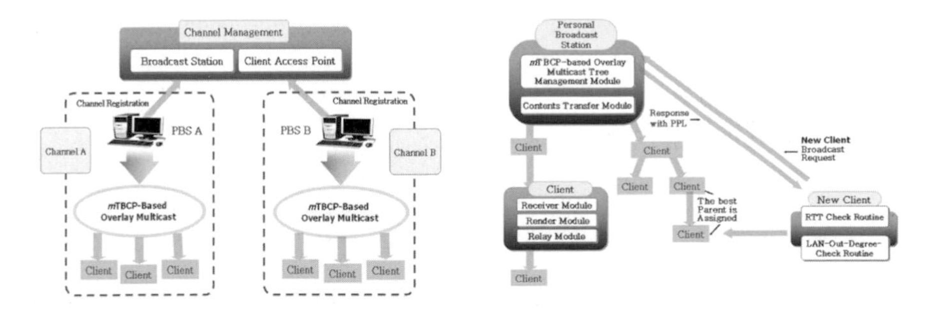

Fig. 1. Overlay Broadcasting System's Framework

Fig. 2. OBS's Session Architecture in

In the Fig. 1, a construction of OBS has been made by, essentially, including the Channel Management Module, Personal Broadcasting Sections, and the *m*TBCP-Based Overlay Multicasting Module. The clients intending to join the broadcasting session sends their requests via an *m*TBCP Overlay Multicast Module which is responsible for creation, controlling, and managing the group trees. The *m*TBCP Overlay Multicast Module then communicate with the Personal Broadcasting Station for channel registration and service assignment. The Broadcasting Station and the Client Access Point are responsible for channel registration and service assignment. The *m*TBCP Overlay Multicast Module enables a given group's members to join and participate for the session, concurrently. The module does not have to maintain all the information and knowledge of the group membership to facilitate the tree building and control. It rather needs to store partial information, say a list of Potential Parent List (PPL). Every client, on session joining requests, is responded with a respective PPL so that that given client can, after routinely checking the respective RTT and the LAN-out-degree-check choose a potential parent to join.

1.3 Session Architecture in Overlay Broadcasting System

An abstract of session architecture in Overlay Broadcasting System is illustrated in Fig 2. A new client, intending to participate in the broadcasting session, sends a "New Client Broadcast Request" (New_B_REQ) to a respective PBS where the *m*TBCP-Based Overlay Multicast Tree Management Module is a part of the PBS. Upon receiving the New_B_REQ, the PBS using the *m*TBCP responds the New_B_REQ by sending appropriate PPL to that particular new Client. This New Client, upon receiving and according to the status o the PPL, applies the two essential routines – the RTT-check Routine and the LAN-Out-Degree-Check Routine – to efficiently decides on the best parent that this New Client can successfully join. Therefore, according to the outcome of the routines, the *best* parent is assigned to the client and hence allowed to participate in a session after sending the response to the PBS which considers the New Client's selected parent for PPL update.

In Fig. 2, the PBS is shown to be constituted, among other parts, by Content Transfer Module which is responsible for the content distribution, including streaming in our Overlay Broadcasting Mini-System. A client is, in principle, equipped with a Receiver Module which accepts content from a Content Transfer Module or other relaying clients, a Render Module for the content bestowing, and a Relay Module which acts as a source of the content source whenever available. The *m*TBCP-Based Overlay Multicast Tree Management Module is responsible for tree building, controlling, and maintaining with partial information and knowledge about the group and the topology with concurrent member-join being served in parallel.

2 Requirements for Member Join

As outlined by Mathy in [1], it gives a significant challenge to build an overlay spanning tree among hosts. It is, by all means, to allow the end-hosts gain knowledge and information about the group through host-to-host metrics. The *m*TBCP- Based protocol must also facilitate member joining such that the status of tree-first, distributive overlay spanning tree building is maintained. Member join requires that the new joining members are assigned with an optimal parent within a good joining time. Traditional TBCP does all these requirements, but the tree convergence time in TBCP, as in Yoid and Narada, is not faster when the concurrent members want to join a particular group at a given time. In our proposed *m*TBCP, we require that many members can join the group independently so that the waiting time to be served can reduce or totally eliminated. We also require that our control protocol results into a low latency, higher Connection-Throughput with respect to group size.

2.1 Limitation of Traditional TBCP

When the New Client joins the session in TBCP, the respective tree can be identified by pairs of only two advertised parameters; the address of the tree root and the port number used by the root for signaling operations of TBCP. The fanout, the number of children a client or a root can accommodate, is fixed and this controls the load traffic on the TBCP tree. Fig. 3 illustrates the limitation of traditional TBCP.

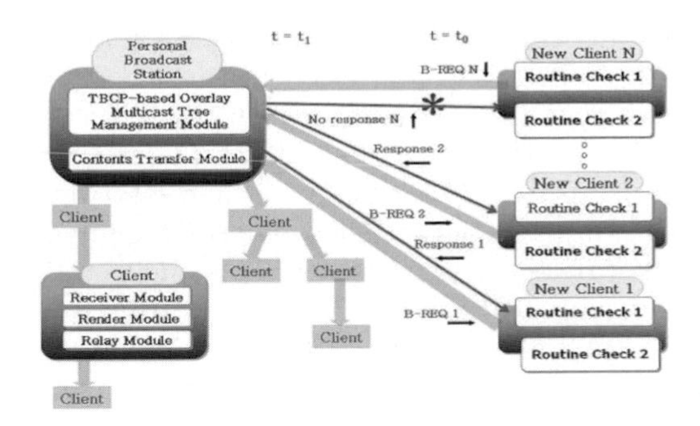

Fig. 3. Limitation of TBCP when Concurrent Members Join a Session

2.2 Problem with Concurrent Member Join

Even though there is a recursive process in the TBCP client join mechanism, each new client must wait for the other (may be privileged members) to join before it gets attended when more than one new clients intend to access the session at the same time. From Fig. 3, it can be realized that at time $t = t_0$, there are *N* New Clients intending to join the same session where the tree building is controlled by traditional TBCP-based overlay. At this time, *N* B_REQs are sent to the PBS at the same time instance. However, assuming that there is no more clients sending their B_REQs during this time period $(t_1 - t_0)$ and since TBCP processes one B_REQ at a time, while serving the first new client, $(N - 1)$ other clients have to wait; while serving the second new client, $(N - 2)$ clients have to wait; and so forth. Therefore, at a certain time t_1, there may be only $(N - k)$ clients responded and the remaining *k* clients un-attended. These un-attended clients may be forced to face unpleasant service reception or sometimes none. Among the remedies that have been thought to overcome this limitation is the modification of the existing TBCP in such a way that the Overlay Multicast module will be able to attend and serve as many New clients intending to join a particular session at a given time. Section 3 is devoted to the proposed mTBCP scheme – its architecture and the way it can handle concurrent members join efficiently.

3 Modified-TBCP (*m*TBCP) Scheme

The modified-TBCP (*m*TBCP) has been proposed being an effort to overcome the limitation of the traditional TBCP especially when concurrent members wish to simultaneously join for the session in an Overlay Broadcasting Mini-System. In principle, this scheme guarantees a better performance as well as overall system efficiency in addition to a pleasant and robust servicing manner. The mTBCP-Based Overlay Multicast Module involves the scheme to maintain a list of potential parents – Potential Parents List (PPL) – at the source root where every entity wishing to join the session sends the B_REQ to report its intention. Upon the reception of the B_REQ, the module cross-checks among its existing groups' members to find out the possible best members who can suffice as a parent of that particular new client. There are cases where the new client is assigned the source root as its potential parent. There are cases where the new client is assigned the other members to be its potential parent, and there are cases where soon after the member has been designated as a parent, it is deleted from the PPL at the source root as the out-degree status does not allow it to act as a parent any further. PPL update is an important recursive routine while operating the *m*TBCP. All the new clients are entitled to report their status soon after securing their new best parents. Among the fields included while reporting include their LAN-relationship with their new parents, their out-degree status (and may be in-degree status), and their RTT status with respect to the other PPL members proposed to them by the source root. The source recursively updates and maintains the recent PPL for any acceptable number of B_REQs that might be addressed to the source root.

3.1 A Simplified Architecture of *m*TBCP

Fig. 4 shows a simplified architecture of *m*TBCP mechanism and concurrent member join and also it illustrates the mechanism that is performed while operating the *m*TBCP. In ①, since it falls within the outdegree unit, the client is registered to PLL itself. When the New Member wants to join for a session, as in ②, it sends its join request to the source root and the source root, as in ③, responds by sending the respective PPL existing at that instant of time. That new member, on receiving the PPL, it does a routine check for RTTs with respect to the members of the PPL it has been proposed to it as well as the RTT from that new member to the source root. In ④, there are two proposed potential parents that can be assigned to the new member and hence, the new member checks for three RTTs, including the one from itself to the source root. After being satisfied with the status of each of the proposed

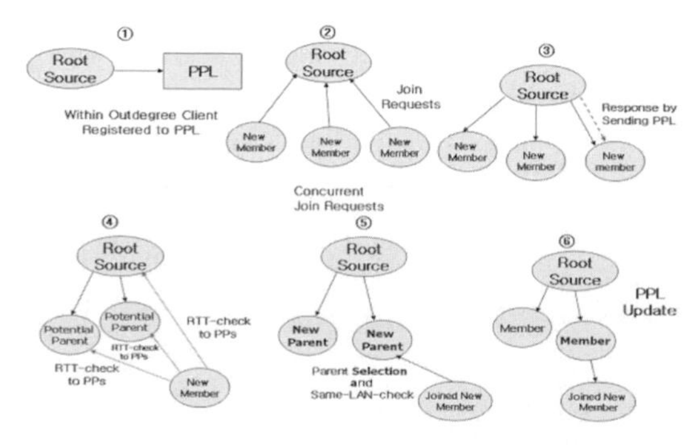

Fig. 4. A Simplified Architecture of *m*TBCP Mechanism and Concurrent Member Join

potential parent, the new member selects its parent and does same-LAN-check before it sends the update to the source root. The source root then updates the PPL, as in ⑤ and ⑥.

3.2 Concurrent Member Join under *m*TBCP

Under *m*TBCP, concurrent member join is treated well with negligible delay. When many new members want to join as in Fig. 4, each new requesting member is responded with appropriate PPL containing a proposal for possible potential parents for each of the new members to consider joining the session through them.

4 Simulation Setup and Results

4.1 Simulation Setup

A topology of 1,000 nodes was considered to simulate the algorithm with NS-2. Six group size - 10, 20, 40, 60, 80, and 100 nodes - categories were simulated for the traditional TBCP, the modified *m*TBCP, and the Random mechanisms. The overall performance measures were the Overlay Connection Throughput and the Latency.

4.2 Simulation Results

Three different mechanisms – *m*TBCP, TBCP, and Random - were simulated. Table 1 summarizes a few selected group sizes' results. It can be clearly noted that as the group size becomes bigger, the percentage overlay connection throughput gets better for all the three mechanisms under consideration. Nevertheless, the performance of the *m*TBCP-Based mechanism is impressively tremendous.

Table 1. The trend of Overlay Connection Throughput with respect to Group Size (nodes)

Size	*m*TBCP	TBCP	Random
10	7	5	8
20	11	8	8
40	35	19	19
60	56	38	38
80	80	58	58
100	97	73	73

The plots of the performance appraisal for the three mechanisms under discussion have been shown in Fig. 5 (a) and (b). In (a), it can be visualized that with a group size of 10 node, the Overlay Connection Throughput for the case of *m*TBCP is 7 with that of TBCP equal to 5 and 8 for the Random mechanism. This case does not show much discrepancy for the three mechanisms. However, as the group size gets larger the performance of the proposed *m*TBCP gets better defeating the other mechanism. In Fig. 5 (b), it is obvious that from medium group sizes to larger ones, the performance of the TBCP and the Random Mechanism almost over-write each other and they

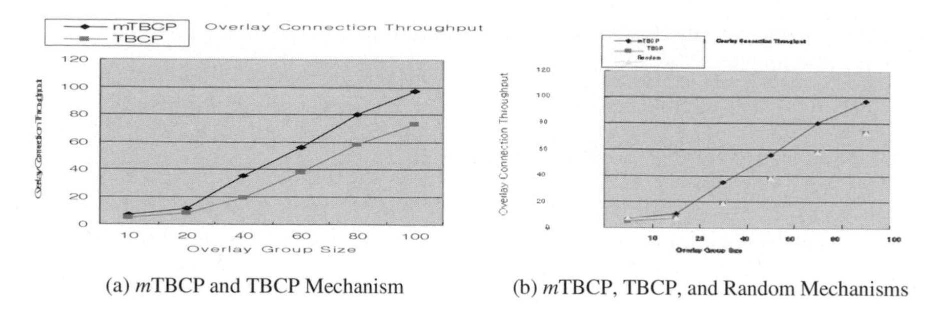

(a) *m*TBCP and TBCP Mechanism (b) *m*TBCP, TBCP, and Random Mechanisms

Fig. 5. Overlay Connection Throughput vs. Overlay Group Size

perform almost equally. For larger groups, the Overlay Connection Throughput of *m*TBCP is the best choice for Overlay Broadcasting Mini-System.

The Overlay Latency for the *m*TBCP has been proved to be the lowest among the three mechanisms. Table 2 shows that in the three cases, the latency starts to increase gradually from small to medium size groups and then it remains constant for the larger groups. Of the three mechanisms, the latency experienced by the *m*TBCP-Based Overlay is the least.

The *m*TBCP registers a latency of 5.94448*ms* for the group size of 10 nodes while it is more than 7*ms* and 8*ms*, respectively, for the TBCP and for the Random mechanisms.

For group sizes of 100, 80, and 60 nodes, *m*TBCP experience a constant overlay latency of a little greater than 7*ms* while for the same group sizes it is a little more than 9.7*ms* for the TBCP and the Random mechanisms.

Table 2. The trend of Overlay Latency in milliseconds with respect to Group Size (nodes)

Size	*m*TBCP	TBCP	Random
10	**5.94448**	7.029728	**8.506528**
20	**5.94448**	7.029728	**9.486016**
40	**6.66448**	9.702239	**9.766528**
60	**7.26639**	9.702239	**9.766528**
80	**7.26639**	9.702239	**9.766528**
100	**7.26639**	9.702239	**9.766528**

Fig. 6 shows that for the case of TBCP and Random mechanisms, we have a discrepancy in terms of overlay latency for small group sizes, with TBCP performing better up to group size of 40 nodes.

However, after that the latency performance for the two seems to overlap and to stay consistent. The *m*TBCP performs overall better, regardless the group size, as compared to the other two mechanisms.

The results confirm that the proposed *m*TBCP-Based Overlay Multicast Mechanism can lead to a well performing Overlay Broadcasting Mini-System where the PBSs can provide efficient and pleasant services to clients with just partial information about the hosts and partial knowledge of the topology.

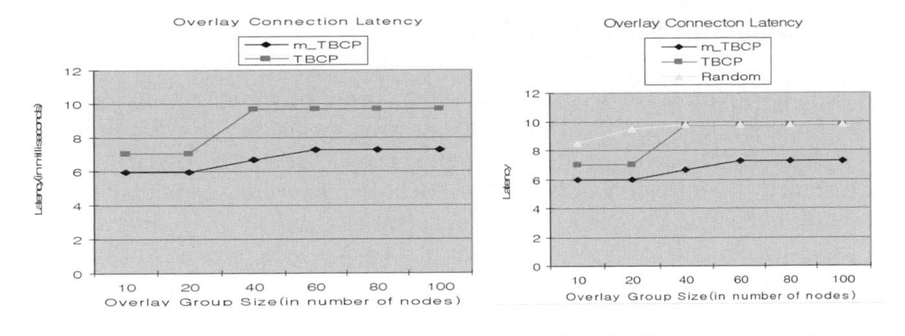

(a) *m*TBCP and TBCP Mechanisms (b) *m*TBCP, TBCP, and Random Mechanisms

Fig. 6. Overlay Latency vs. Overlay Group Size

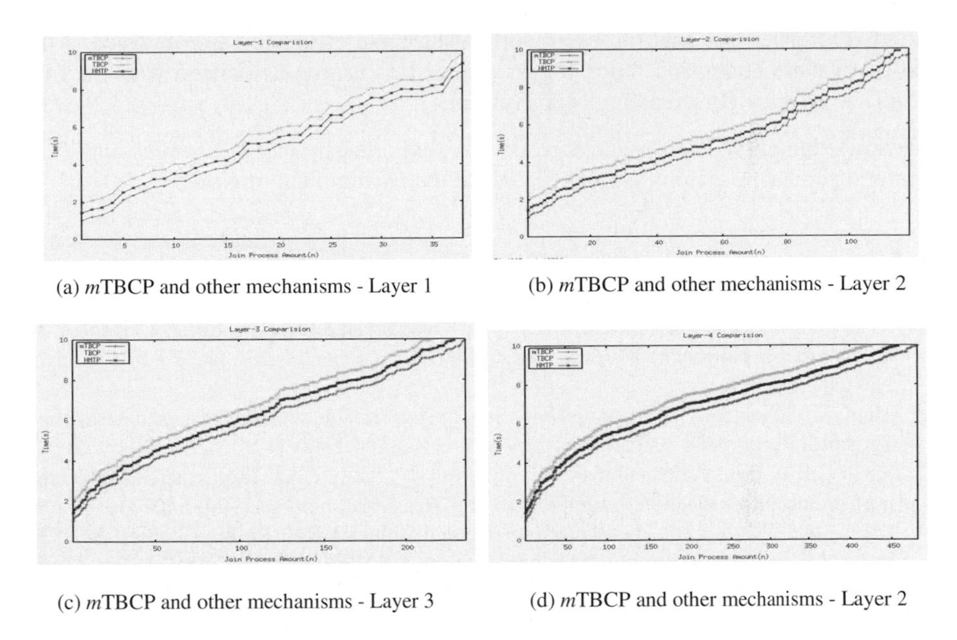

(a) *m*TBCP and other mechanisms - Layer 1 (b) *m*TBCP and other mechanisms - Layer 2

(c) *m*TBCP and other mechanisms - Layer 3 (d) *m*TBCP and other mechanisms - Layer 2

Fig. 7. Overlay Latency vs. Overlay Group Size

Figs. 7(a) to 7(d) show the results in terms of time required for the join process amount of 35 to 450 group members and for layers 1, 2, 3, and 4, respectively. The results illustrate the time comparison versus the amount (in number of nodes) that each of the traditional TBCP, HMTP, and the mTBCP protocols achieves. The mTBCP seems to be a favorable in terms of a short duration (in seconds) required to handle the same amount. The traditional TBCP seems to perform the worst among the three. It can be clearly visualized that at layer 1 and layer 2, the protocols obey the time complexity of the order $\boldsymbol{O(nlog_2n)}$ as opposed by the obedience of the order $\boldsymbol{O(log_2n)}$ for the cases of layers 3, 4, and beyond which means a, comparatively, better performance as the tree gets deeper from the source root.

5 Discussion and Further Scope

It has been proposed, described, and evaluated a better performing and more efficient mechanism, *m*TBCP-Based Overlay Multicast Mechanism, for Overlay Broadcasting Mini-System. The protocol operates with partial knowledge of the hosts and of the network topology. The special feature about *m*TBCP-Based Overlay Multicast mechanism is its power to be able to attend concurrent clients at a certain given time with a very remarkable latency and overlay connection throughput measures with respect to the group size. This mechanism has better results in terms of connection throughput and latency, especially when the group size grows bigger. High connection throughput makes the *m*TBCP-Based Overlay mechanism being the best candidate for membership management at the overlay multicast module in the Overlay Broadcasting Mini-System.

Since the simulation was not performed while including the actual Overlay Broadcasting Mini-System, in future the simulation including the Mini-System can be included to make sure that the exact performance and efficiency are observed. A topology of more nodes and much bigger group sizes can be associated with the field test of the Overlay Broadcasting Mini-System.

Acknowledgments. This research work was supported by the Electronics and Telecommunications Research Institute (ETRI) of the Government of Korea.

References

1. Mathy, L., Canonico, R., Hutchison, D.: An Overlay Tree Building Control Protocol. In: Crowcroft, J., Hofmann, M. (eds.) NGC 2001. LNCS, vol. 2233, pp. 76–87. Springer, Heidelberg (2001)
2. Shen, K.: Substrate-Aware Connectivity Support for Scalable Overlay Service Construction. Technical Report #800. Department of Computer Sc., University of Rochester (May 2003)
3. Farinacci, D., Lin, A., Speakman, T., Tweedly, A.: Pretty Good Multicast (PGM) Transport Protocol Specification. Internet Draft draft-speakman-pgm-spec-00. IETF (1998)
4. Levine, B., Garcia-Luna, J.: Improving Internet Multicast with Routing Labels. In: IEEE Intl. Conf. on Network Protocols (ICNP), Atlanta, USA, pp. 241–250 (1997)
5. Pendarakis, D., Shi, S., Verma, D., Waldvogel, M.: ALMI: an Application Level Multicast Infrastructure. In: 3rd USENIX Symposium on Internet Technologies, San Fransisco, CA, USA (March 2001)
6. Francis, P.: Yoid: Extending the Internet Multicast Arch. Tech. Report, ACIRI (April 2000)
7. Chu, Y.-H., Rao, S., Zhang, H.: A Case for End System Multicast. In: ACM SIGMETRICS 2000, Santa Clare, USA, pp. 1–12 (June 2000)
8. Jannotti, J., Gifford, D., Johnson, K., Kaashoek, F., O'Toole, J.: Overcast: Reliable Multicasting with an Overlay Network. In: USENIX OSDI 2000, San Diego, USA (October 2000)
9. NS-2 Network Simulator, `http://www.isi.edu/nsnam/ns`
10. Lee, M., Kang, S.: A FTTH network for integrated services of CATV, POTS and ISDN in Korea. In: Proceedings of the 1st International Workshop of Community Networking Integrated Multimedia Services to the Home, pp. 261–264, July 13-14 (1994)
11. Chan, S.-H.G., Yeung, S.-H.I.: Client buffering techniques for scalable video broadcasting over broadband networks with low user delay. IEEE Transactions on Broadcasting 48(1), 19–26 (2002)

A Network I/O Architecture for Terminal-Initiated Traffics in an Ubiquitous Service Server

Kyoung Park[1], Soocheol Oh[1], Seongwoon Kim[1], and Yongwha Chung[2]

[1] ETRI
161 Gajeong-dong, Yuseong-gu, Daejeon, 305-350 Korea
{kyoung,ponylife,ksw}@etri.re.kr
[2] Dept. of computer science, Korea University
208 Seochang-dong, Jochiwon-eup, Yonki-geun, Chungnam, 339-700 Korea
ychungy@korea.ac.kr

Abstract. In ubiquitous environment, terminal-initiated traffics(*e.g.*, sensor data streams) will become more popularized than human-initiated traffics(*e.g.*, web browsing, file transfer, and media streaming). The characteristics of the terminal-initiated traffics can be described as the streams of small sized packets with the large number of simultaneous connections. Thus, an ubiquitous service server will suffer from the burden of the terminal-initiated traffics since the per-packet-cost of a CPU will increase proportionally with the number of terminals. In this paper, we propose Latona architecture that offloads not only TCP/IP processing but also parts of socket processing from a host to minimize the per-packet-cost of a CPU in a Linux server. Based on the experimental results, the Latona kernel could save 50% and 79% of the kernel execution time of *send()* and *recv()* for 32 bytes transfer in the legacy TCP/IP stack. The packet-per-second were 21.9K and 18.3K for *send()* and *recv()*, respectively. The bandwidth increased as the size of payload increased. The profile of detail execution time showed that the bottleneck of the Latona was for handling the socket and the TCP in Latona hardware.

Keywords: Terminal-Initiated Traffics, Ubiquitous Service Server, TCP/IP Overhead, Offloading, Linux, Network I/O.

1 Introduction

Today's Internet consists of broadband connections mainly for PCs and servers. However, under the ubiquitous environment, new internet will consist of networks of thousands or millions of microchips, various electronic appliances, mobile devices as well as PCs and servers.

In the new internet(*i.e.*, future ubiquitous communication environment), *terminal-initiated traffics* will become more popularized than *human-initiated traffics*, like current internet communications[1]. The terminals can be characterized as various and huge amount of sensor nodes reporting simple presence data involved in a daily life to the servers[1,2,3,4]. Thus, the server will convert the received data into the meaningful context information which can be provided to ubiquitous service applications.

T. Vazão, M.M. Freire, and I. Chong (Eds.): ICOIN 2007, LNCS 5200, pp. 161–170, 2008.

The human-initiated traffics are mainly generated by PCs or mobile phone, so the number of maximum simultaneous connections to the server will be the same as the number of users. On the other hand, the terminal-initiated traffics are event-driven communications generated by themselves[2,3,4], so the number of maximum simultaneous connections to the server will be equal to the number of terminals. Furthermore, there will be much more terminals than users, because each user will be surrounded by several terminals reporting dynamic status of the personal environment[1].

The packet size of the terminal-initiated traffics is relatively small compared to that of the human-initiated traffics(*e.g.*, web browsing, file transfer, and media streaming), because the presence data is tiny information obtained from various sensors reporting location information, temperature, and so on. Each presence data has the size of less than 100 bytes and can be assumed to be one packet[1,4]. The terminals report the presence data to the server when new data is acquired or when an update-timer is expired[1,2,3]. With thousands of terminals, the server will receive thousands of small packets via thousands of simultaneous connections.

As the internet evolves into an ubiquitous network, the *ubiquitous service server* must manage not only the human-initiated traffics, but also a lot of the terminal-initiated traffics to provide ubiquitous services to users. Therefore, for network I/O performance, the server needs to be optimized to process bulk of the terminal-initiated traffics in the ubiquitous environment.

In this paper, we propose Latona architecture to lessen the burdens of network I/O from a low-end Linux server for the terminal-initiated traffics. The rest of this paper is organized as followings. Section 2 describes the motivation of this paper by examining previous works regarding TCP/IP processing. Section 3 discusses the design details of Latona architecture. In section 4, we discuss the experimental results, and finally section 5 presents our conclusion and future work.

2 Motivation

TCP/IP, the predominant protocol suite across the internet, traditionally has been implemented in software as a part of the operating system kernel. A frequently cited drawback of TCP/IP is that data copying and TCP/IP processing overhead[5,6] consumes a significant share of host CPU cycles and memory bandwidth, siphoning off system resources needed for application processing.

Generally, the TCP/IP processing overhead can be divided into two categories: *per-byte-cost*, primarily data-touching operations such as checksums and copies; and *per-packet-cost*, including TCP/IP protocol processing, interrupt overhead, buffer management overhead, socket handling, and kernel overhead[6].

To lessen some of the overhead, researchers developed several mechanisms that became common in today's TCP/IP processing. Today's advanced kernel and high performance NIC support zero-copy, segment offloading, checksum offloading, and interrupt coalescing to reduce the overhead of data touching overhead[6,9,10,11]. On the other hand, several industry players announced TCP Offloading Engine(TOE) devices that offload the TCP/IP processing from a host CPU[7,8].

The analysis by Jacobson, *et al.*[5] and Foong, *et al.*[6] shows that the overhead of protocol itself is small relative to the overhead of TCP/IP implementation and management. According to [6], the TCP/IP protocol processing takes up about 7% on receive, and 10~15% on transmit. On the other hand, the socket handling and corresponding libraries take up 34% for small transfers. Another large portion is the kernel overhead taking up 36% for small transfers and it consists of system_call routines that handle bookkeeping required to support system calls, parameter checking, and context switching. Finally, the driver code takes up 4~11%. Thus, the overhead of socket layer and kernel makes up large portion of TCP/IP processing for small transfers.

The generally accepted rule of thumb in TCP/IP processing is that 1bps of network link requires 1Hz of CPU processing[6]. However, the general rule can be accepted in case of bulk data transfers. For smaller packets, the processing requirement is 6~7 times higher and the per-packet-cost is high enough to be the bottleneck of the server since the overhead of socket layer and kernel increases with the number of packets[6].

In case of TOE[7,8], it increases the server's throughput while reducing CPU utilization by offloading the TCP/IP processing onto a specialized device. However, there is an ongoing debate[9,10] about whether TOE is suitable for general network applications. Mogul's research[9] shows the limitations of TOE technology, and Regnier, *et al.*[10] discuss that TOE is only suitable for bulk large transfers involving long-lived, few connections. The major reason of such limitation of TOE is that it focuses on offloading the TCP/IP protocol processing and data touching overhead(*i.e.*, per-byte-cost), while the major part of the per-packet-cost, the overhead of socket handling and kernel, still remains in a host system. Moreover, it requires another communication overhead between a TOE device and a host to lookup socket and TCP Control Block(TCB) [9,10]. Also, Its specialized API and stack code cause troubles in designing or porting general network applications[9].

As we learned from Matsumoto's study[1], the characteristics of the terminal-initiated traffics can be described as the streams of small sized packets with the large number of simultaneous connections. Then, the server will suffer from the burden of per-packet-cost, especially the overhead of socket handling and kernel in the ubiquitous environment.

In this paper, we propose Latona architecture that offloads the socket and TCP/IP layer to minimize the per-packet-cost of a CPU in a low-end Linux server for the terminal-initiated traffics.

3 Latona Architecture

3.1 Overview of Latona

Latona consists of Latona kernel and Latona hardware as shown in Fig. 1. The Latona kernel is a set of dedicated kernel modules which replace the traditional INET and TCP/IP protocol stack in a Linux server. It receives socket level commands from applications and delivers them to the Latona hardware through the doorbell interface. The Latona hardware is a kind of TOE which offloads the full TCP/IP stack and parts of the socket processing. The DMA engines eliminate per-byte-cost from a host CPU by performing direct data movement between the payload buffer and the memory locations of user address space. The doorbell interface provides simple and abstracted communications between the Latona kernel and the Latona hardware. The Latona hardware receives optimized socket level commands from the Latona kernel and also reports the results through the completion queue. The Latona is designed such that:

- eliminates the socket handling and TCP/IP processing overhead from a host to reduce per-packet-cost
- supports 10K of sockets via hardware socket resource pool in the Latona hardware
- supports a true zero-copy mechanism between user memory and the Latona hardware
- supports all standard socket API and file I/O API
- supports both of the legacy TCP/IP stack and the Latona hardware

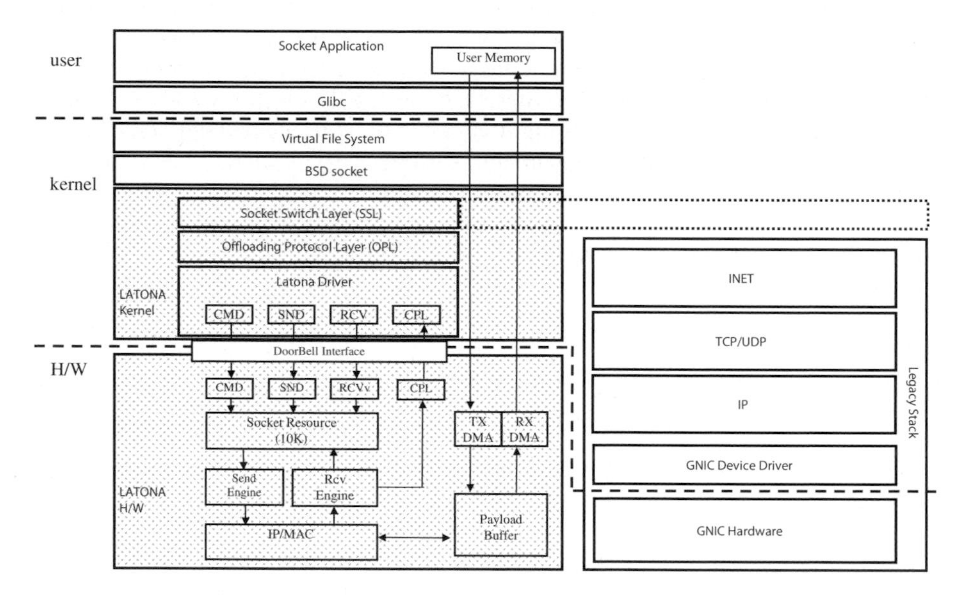

Fig. 1. The Overall Architecture of Latona

3.2 Design Details of Latona Kernel

The Latona kernel is composed of Socket Switch Layer(SSL), Offloading Protocol Layer(OPL) and the Latona driver as shown in Fig. 1. The socket level command generated by a network application is delivered to SSL through the BSD socket layer. Then, SSL determines whether it uses the Latona hardware or general NIC(GNIC) in serving this socket command, based on the result of bind and NIC selection rules. If SSL decides to use the Latona hardware, the command from the network application is delivered to the Latona hardware through OPL and the Latona driver. If GNIC is selected, SSL handles this command by calling the INET layer of the legacy TCP/IP stack.

By locating SSL above OPL and the INET layer of the legacy stack, the Latona kernel provides the binary compatibility with the BSD socket interface. It also supports both the legacy TCP/IP stack with GNIC and the Latona hardware. Thus, it is possible for traditional network applications to use the BSD socket interface without any modification or recompilation by using SSL that processes the standard BSD socket API.

3.2.1 Socket Switch Layer

SSL is the replacement of the INET layer. Fig. 2-(a) shows how SSL creates a socket briefly. When a network application calls *sys_socket()* function of Linux kernel, it internally calls *inet_create()* that is the socket creation function of the INET layer through *sock_create()* and *sock_alloc()*. The call, *inet_create()*, is replaced with *toe_create()* that is the socket creation function of SSL. *Sock_alloc()* of the BSD socket layer calls the *inet_create()* using a function pointer. The function pointer is stored in *structure net_families* that includes data about each protocol family and is maintained by the kernel. *Create*, one member of *net_families*, is a function pointer indicating a socket creation function of PF_INET that is the protocol family of TCP/IP. *Net_families[PF_INET]->create* indicates *inet_create()*, which is the socket creation function of the INET layer. In SSL, value of *net_families[PF_INET]->create* is replaced with *toe_create()* as shown in Fig. 2-(a).

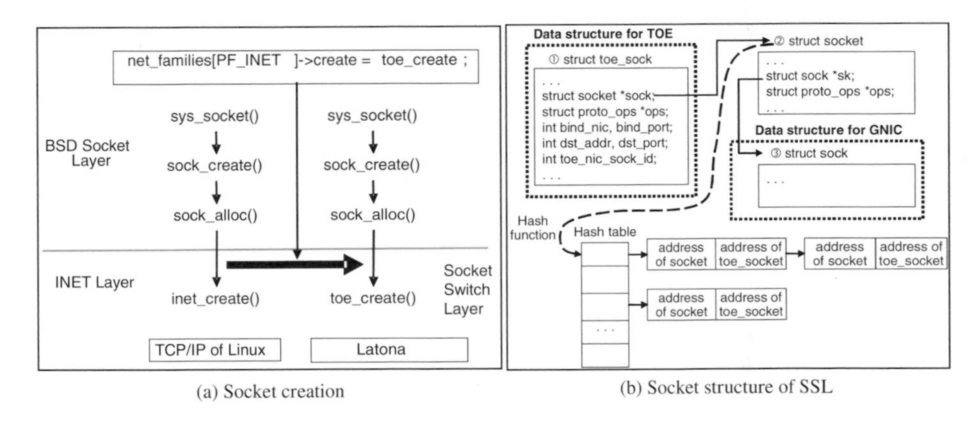

(a) Socket creation

(b) Socket structure of SSL

Fig. 2. Socket Handling in Socket Switch Layer

Note that, *toe_create()* creates a socket structure. In order to support both the Latona hardware and GNIC, SSL creates a socket structure supporting both of the Latona hardware and GNIC simultaneously. Fig. 2-(b) shows the socket structures for SSL. We expand the legacy socket structure by defining a new data structure, called *struct toe_sock,* and making this structure to point *struct socket.* The BSD socket layer creates *struct socket*(see ② in Fig. 2-(b)), and calls SSL with this structure. Then, SSL creates *struct toe_sock*(see ① in Fig. 2-(b)). Also, *struct sock*(see ③ in Fig. 2-(b)) is created by the INET layer of the legacy stack called by SSL. *Struct socket *sock*, one member of *struct toe_sock*, is the pointer to *struct socket. Struct proto_ops* has function pointers that indicate the INET layer functions called by SSL. *Int bind_nic* and *bind_port* store information about GNIC and port number that a socket is bound to. *Int dst_addr* and *dst_port* store the destination IP address and port number of remote peer. Finally, *int toe_nic_sock_id* is a socket ID returned by the Latona hardware in order to use the hardware socket resource in the Latona hardware.

Furthermore socket calls such as *bind()*, *send()* and *recv()* are served by SSL, instead of the INET layer. Fig. 3 shows the modification of *structure proto_ops* that is a member of *structure socket* in order to call SSL, instead of the INET layer.

TCP/IP of Linux	Latona
struct proto_ops ops = { .family = PF_INET, .bind = inet_bind, ...sendmsg = inet_sendmsg, .recvmsg = sock_common_recvmsg };	struct proto_ops ops = { .family = PF_INET, .bind = toe_bind, ...sendmsg = toe_sendmsg, .recvmsg = toe_recvmsg };

Fig. 3. Modification of *structure proto_ops*

3.2.2 Offloading Protocol Layer and Latona Driver

When the Latona hardware is selected for data transmission, SSL calls OPL. Data transmission in OPL is performed by a true zero-copy mechanism through DMA technique between user memory and the Latona hardware. For the true zero-copy, the user memory pages are pin-down and physical addresses of these pages are passed to the Latona hardware in the forms of scatter-gather list via the Latona driver.

The Latona driver provides the interface between the Latona hardware and OPL. The Latona driver exports functions which are used by OPL, and passes the request received from OPL to the Latona hardware via three kinds of doorbells; command (CMD), send(SND) and receive(RCV). Then, the Latona hardware processes the requests and reports the results in completion queue(CPL). It also issues an interrupt signal which is captured by the Latona driver and calls the interrupt service routine which is defined in OPL.

3.3 Design Details of Latona Hardware

The Latona hardware has a layered architecture as shown in Fig. 4-(a). The host interface adopts the industry standard PCIExpress fabric to interface a modern motherboard. It contains not only an endpoint protocol engine, but also dual DMA engines and doorbells. The transport layer processes the TCP and UDP protocols using both hardware and software. For software processing, two processors are embedded and each handles transmitting and receiving, respectively. The IP layer processes the IPv4 and ARP protocol, and the MAC/PHY layer transmits and receives bit stream to/from Gigabit Ethernet.

The Latona hardware has a socket pool in order to offload the socket handling overhead from a host CPU. The socket pool can store 10K of socket entries and also contain the socket search logic to reduce the searching time for one entry. Each entry maintains TCP connection status, transaction IDs, and scatter/gather list of each transaction for payload processing. Thus, the Latona hardware can process socket level transactions without the interface with kernel software.

We implemented the Latona hardware with FPGAs as shown in Fig. 4-(b), and its implementation specifications are summarized in Table 1.

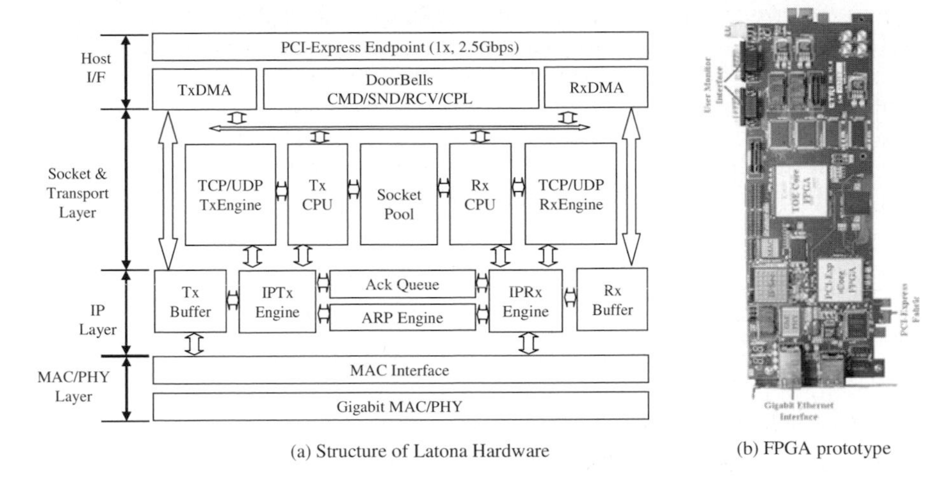

(a) Structure of Latona Hardware

(b) FPGA prototype

Fig. 4. Latona Hardware

Table 1. Implementation Specifications of Latona Hardware

Functions	Host Interface	Latona core	
		Scoket/TCP/IP	Datapath
Spec.	PCIExpress 1.0a Link speed : 2.5Gpbs Dual DMA channel Doorbells - 4 CMDs - 4 SNDs - 4 RCVs	Dual CPUs @250MHz - PPC405 Socket Resource Pool - 10K TCP/UDP Protocol IPv4 Protocol ARP protocol	Checksum offloading TCP segmentation Zero-copy Pipelined datapath - 32bits@125MHz
	- 2K of CPL queue Interrupt coalescing Datapath : 32bits@125MHz	Full offloading of TCP/UDP/IP/ARP Socket level transaction processing 10K of simultaneous connections	
Device	ALTERA Stratix-GX	Xilinx Virtex-II Pro	

4 Experimental Results

In this paper, we focus on the offloading effect which can be derived from the Latona architecture for small transfers such as the terminal-initiated traffics. We first measured the basic performance metrics including kernel execution time, packet-per-second, and bandwidth with small sized packet streams from multiple socket connections. Then, we measured the whole processing time in the Latona architecture and broke down it to find out the possible bottleneck. The experimental system was a PC having a Pentium 2GHz CPU, 512Mbytes of main memory. The operating system was Linux kernel 2.6.9.

4.1 Kernel Execution Time with Legacy TCP/IP Stack

Fig. 5 shows the execution times of *send()* and *recv()* of the legacy Linux kernel with National Semiconductor's DP83820 Gigabit Ethernet card and the Latona kernel with the Latona hardware. In order to measure the per-packet-cost only, we used 32 bytes of *send()* and *recv()*. In case of the legacy kernel, INET, TCP/IP, and device driver spent 22μs and 69μs for *send()* and *recv()*, respectively. However, the Latona kernel spent only 11μs for both *send()* and *recv()*. Thus, Latona kernel could save 50% and 79% of the kernel execution time for *send()* and *recv()*, respectively. Consequently, the system can utilize the saved time for application processing.

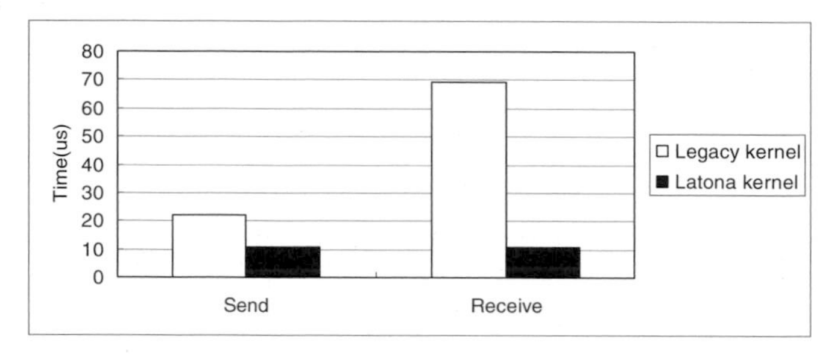

Fig. 5. Comparison of Kernel Execution Time(Legacy TCP/IP Stack vs. Latona Kernel)

4.2 Packets-Per-Second and Bandwidth of Latona Architecture

Fig. 6 shows the packet-per-second and bandwidth of the Latona architecture as increasing the size of payload from 32 bytes to 1024 bytes. In case of 32 bytes, *send()* needed 45.49μs(11μs for Latona kernel and 34.49μs for Latona hardware), and *recv()* required 54.45μs. Thus, the Latona could provide 21.9K and 18.3K of packet-per-second in *send()* and *recv()*, respectively. The bandwidth increased with the size of payload since the packet-per-second was not affected by the size of payload due to DMA between user memory and the Latona hardware.

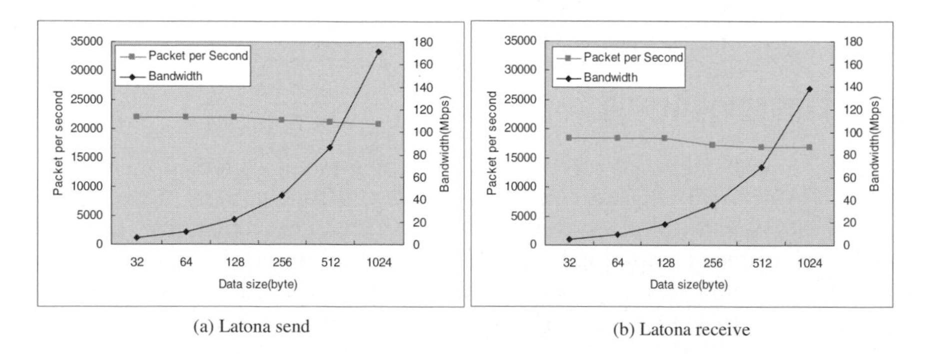

(a) Latona send (b) Latona receive

Fig. 6. Packet-Per-Second and Bandwidth of Latona

Note that, *send()* consumed relatively large amount of time, compared to the legacy TCP/IP stack(22μs)[1]. Though the Latona kernel offloaded some burdens in the legacy TCP/IP stack, we can predict that the packet-per-second and bandwidth are lower in case of *send()*. In case of *recv()*, however, the Latona could provide better performance than the legacy TCP/IP stack(69μs).

4.3 Profiling of Processing Distribution in Latona Architecture

The major functions in the Latona hardware were implemented with the combination of firmware and dedicated hardware logics. The firmware handles data structures that belong to socket, TCB, and the transaction requests from the host. It also controls internal hardware logics for processing the incoming and outgoing packets. We measured the execution time of each major function by tracing the firmware.

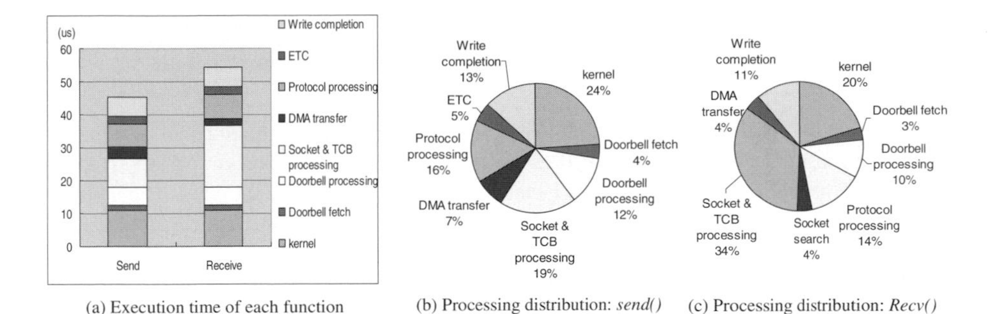

(a) Execution time of each function (b) Processing distribution: *send()* (c) Processing distribution: *Recv()*

Fig. 7. Processing Distribution of Latona Architecture

Fig. 7 shows the breakdown of processing time in the Latona that performs *send()* and *recv()* for 32 bytes of payload. The protocol processing took up 16% and 14% in *send()* and *recv()*, respectively. The major portion of execution time was for the socket and the TCB processing that took up 19% and 34% in *send()* and *recv()*, respectively. In case of *recv()*, the Latona hardware spent relatively large time to find out a socket entry, a request pended in the entry, and scatter/gather list of DMA transfer. Thus, the socket and the TCB processing in Latona hardware becomes a bottleneck and we need more efforts to reduce it.

5 Conclusion and Future Work

As the internet evolves into the ubiquitous network, the server will suffer from the burden of the terminal-initiated traffics that can be characterized as the streams of small sized packets with the large number of simultaneous connections.

In this paper, we proposed the Latona architecture that could offload not only the TCP/IP layer but also the parts of the socket layer to reduce the per-packet-cost of a CPU in a Linux server, whereas the server could provide ubiquitous services under

[1] The data transfer time of GNIC in the legacy TCP/IP stack is negligible for small sized packet transfers.

the ubiquitous environment. In order to prove the concept of the Latona architecture, we implemented the Latona hardware prototype with FPGAs and the Latona kernel on Linux kernel 2.6.9.

Based on the experimental results, the Latona kernel could save 50% and 79% of the kernel execution time for *send()* and *recv()* in the legacy TCP/IP stack. The packet-per-second of *send()* and *recv()* for 32 bytes packets were 21.9K and 18.3K, respectively. The bandwidth increased proportionally with the size of payload since the packet-per-second was not affected by the size of payload. Furthermore, the profile of detailed execution time showed that the bottleneck was for handling the socket and the TCP.

Currently, we are putting more efforts to optimize the Latona hardware to enhance the performance. For better understanding of the offloading effects and comparisons with other solutions, we will also perform various measurements with real workloads such as RFID and sensor data stream processing applications.

References

1. Matsumoto, M., Itho, T.: Study of Server Processing Load Evaluations in Ubiquitous Communication Environments. In: Proc. of the International Symposium on Applications and Internet Workshop, pp. 122–125 (January 2006)
2. IETF RFC2778: A Model for Presence and Instant Messaging (2000)
3. IETF RFC2779: Instant Messaging/Presence Protocol Requirements (2000)
4. Ararwal, P., Banerjee, A., Flammer, J.: RFID Technical Challenges and Reference Architecture, `http://dev2dev.bea.com/pub/a/2005/11/rfid-reference-architecture.html`
5. Clark, D., Jacobson, V., Romkey, J., Salwen, H.: An Analysis of TCP Processing Overhead. IEEE Communications Magazine 27(6), 23–29 (1989)
6. Foong, A.P., et al.: TCP Performance Re-Visited. In: Proc. of the International Symposium on Performance Analysis of Systems and Software, pp. 70–79 (March 2003)
7. Earls, A.: TCP Offload Engines Finally Arrive. Storage Magazine (March 2002)
8. Currid, A.: TCP Offload to the Rescue. Queue 2(3), 58–65 (2004)
9. Mogul, J.: TCP Offloading Is a Dumb Idea Whose Time Has Come. In: Proc. 9th Workshop on Hot Topics in Operating Systems. Usenix Assoc. (2003)
10. Regnier, G., et al.: TCP Onloading for Data Center Servers. IEEE Computer 37(11), 46–56 (2004)
11. Rangarajan, M., et al.: TCP servers: Offloading TCP Processing in Internet Servers. Design, Implementation and Performance, Tech. Rep. Rutgers University (2002)

Analyzing and Modeling Router–Level Internet Topology

Ryota Fukumoto, Shin'ichi Arakawa, Tetsuya Takine, and Masayuki Murata

Graduate School of Information Science and Technology, Osaka University, Japan
{r-fukumoto,arakawa,murata}@ist.osaka-u.ac.jp,
takine@comm.eng.osaka-u.ac.jp

Abstract. Measurement studies on the Internet topology show that connectivities of nodes exhibit power–law attribute, but it is apparent that only the degree distribution does not determine the network structure, and especially true when we study the network–related control like routing control. In this paper, we first reveal structures of the router–level topologies using the working ISP networks, which clearly indicates ISP topologies are highly clustered; a node connects two or more nodes that also connected each other, while not in the existing modeling approaches. Based on this observation, we develop a new realistic modeling method for generating router–level topologies. In our method, when a new node joins the network, the node likely connects to the nearest nodes. In addition, we add the new links based on the node utilization in the topology, which corresponds to an enhancement of network equipments in ISP networks. With appropriate parameters, important metrics, such as the a cluster coefficient and the number of node-pairs that pass through nodes, exhibit the similar value of the actual ISP topology while keeping the degree distribution of resulting topology to follow power–law.

Keywords: Power–law, Router–level topology, ISP topology, AS topology.

1 Introduction

Recent measurement studies on Internet topology show that the connectivities of nodes exhibit a power–law attribute (e.g., see [1]). That is, the probability $p(k)$ that a node is connected to k other nodes follows $p(k) \sim k^{-\gamma}$. In recent years, considerable numbers of studies have investigated power–law networks whose degree distributions follow the power–law [2]. Here, the degree is defined as the number of out–going links at a node. The theoretical foundation for the power–law network is introduced in Ref. [3] where they also presents the Barabashi–Albert (BA) model in which the topology increases incrementally and links are placed based on the connectivities of topologies in order to form power–law networks.

However, even if the degree distributions of some topologies are the same, more detailed characteristics are often quite different. A pioneering work by Li et al. [4] has enumerated various topologies with the same degree distributions, and has shown the relation between the characteristics and performances of these topologies. With the technology constraints imposed by routers, the degree of nodes limits the capacity of links that are

T. Vazão, M.M. Freire, and I. Chong (Eds.): ICOIN 2007, LNCS 5200, pp. 171–182, 2008.

connected to. Li et al. point out that higher–degree nodes tend to be located at the edges of a network. Their modeling method in [4] provides a new insight in that the location of higher–degree nodes are not always located at the core of networks. Actually, different to AS–level topology, each ISP constructs its own router–level topology based on strategies such as minimizing of the mileage of links, redundancies, and traffic demands.

Although Li et al.'s approach is significant, it is insufficient for ISP networks. As will be discussed in Sec. 2, the Sprint topology and Abilene–based topologies are quite different in terms of the cluster coefficient. The main difference may come from the fact that scientific networks like Abilene provide fewer opportunities to enhance their network equipment because of budgetary constraints, while ISPs make their efforts on enhancement of networks based on their strategies. The difference can be also seen from the graphs of the Abilene network (Fig. 6 (e) of Ref. [4]) and the Sprint network (Figs. 7 and 8 of Ref. [5]). More importantly, these differences greatly affect methods of network control. One typical example is routing control as we will demonstrate it in Sec. 4; the link utilization in the router–level topology is much far from the one in the conventional modeling method. The same argument could also be applied to the higher–layer protocols. That is, for vital network researches, a modeling method for a realistic router–level topology is urgently needs to be developed, which is our next concern.

In this paper, we develop a modeling method to construct ISP router–level topologies. To achieve this, we first reveal basic structures for the router–level topologies other than the power–law property of degree distribution. The results clearly reveal the ISP topologies had a much higher cluster coefficient than the AS topology [6], the topology examined by Li et al. [4], and the other topologies attained with conventional modeling methods. We therefore propose a modeling method for realistic router–level topologies. Our modeling method has two main features. When a new node joins the network, the ISP likely connects it to the nearest nodes, while the ISP add new links based on the utilization of nodes. With our modeling, important topology–related metrics such as the number of node-pairs passing through nodes have almost the same characteristics as the actual ISP topologies with appropriate parameter settings, while still keeping the degree distribution of the topology to follow the power–law. We also apply optimal routing method to the topology generated by our modeling method, in order to demonstrate that our modeling method constructs the realistic router–level topology, and can be actually used for evaluations on routing control. The results show that the characteristic of link utilization is similar to the actual ISP topology.

This paper is organized as follows. Section 2 discusses the basic structure of ISP's router–level topologies. We then discuss our development of a new modeling method in Section 3 to obtain realistic router–level topologies that can be applied to "traffic flow" level research, which will be demonstrated in Sec. 4. Finally, Sec. 5 concludes this paper.

2 Structural Properties of Router–Level Topology

In this section, we investigate the structure of router–level topologies as a first step to modeling a router–level topology, and discuss the differences between actual ISP's router–level topologies and topologies generated by existing modeling methods.

2.1 Network Motif

Milo et al. [7] have introduced the concept of *Network Motif*. The basic idea is to find several simple structures in complex networks. In this paper, we select four–node subgraphs as building blocks for router–level topologies following the Milo et al.'s approach, i.e., rectangular (Fig. 1(a)), tandem (Fig. 1(b)), sector (Fig. 1(c)), umbrella (Fig. 1(d)), and full–mesh. The case of a three–node subgraph, which has an exactly the same meaning as "cluster", will be discussed later. Figure 2 plots the frequency of four–node subgraphs appearing in each topology. The labels along the horizontal axis represent the ISP networks (from ISP1 to ISP7) that have been measured with Rocketfuel tools [5]. A topology generated by the BA model (Model1), such that the number of nodes and links is the same as that for the Sprint topology is also presented. The results from the Abilene–based topology used in Ref. [4] (Model2) is also plotted in the figure. We also show the results obtained by a AS–level topology from INET topology generator (Model5 in Fig. 2), and topologies generated by conventional modeling methods (Model3 by the BA model, and Model4 by the ER model [8] in which links are randomly placed between nodes) for comparison. Models 3, 4, 5 have the same number of nodes and links. We can see that: 1) there are many more "sectors" with the Sprint topology (ISP1) than with the BA topology (Model1), 2) "full-mesh" appears more often than model topologies in the router–level topologies of ISPs (Sprint, abovenet,

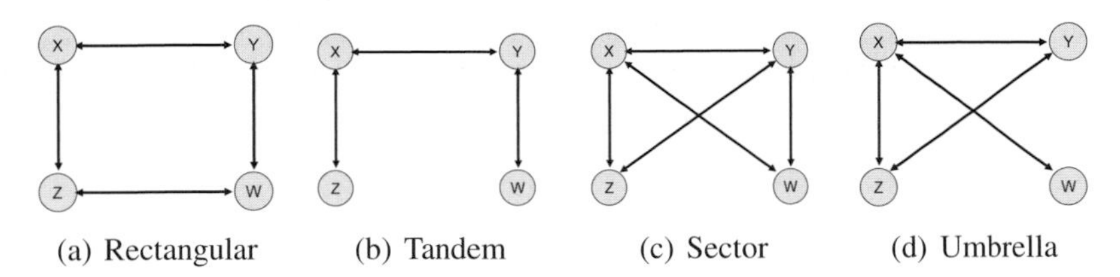

Fig. 1. Four–node subgraphs

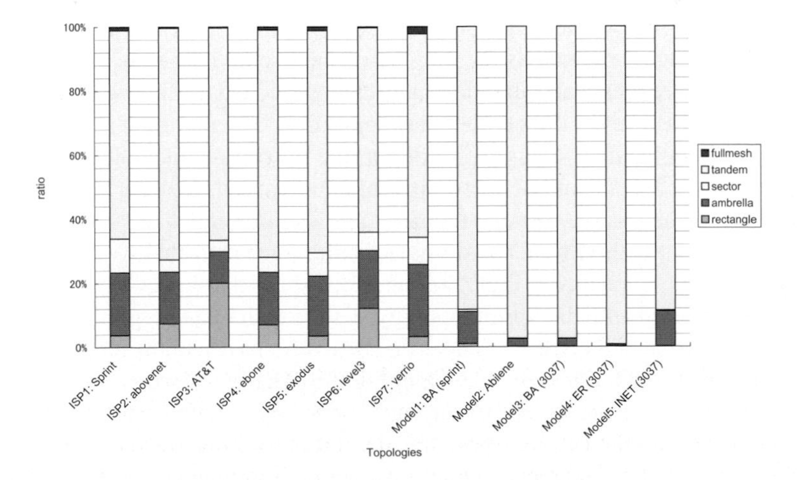

Fig. 2. Distribution of four–node subgraphs

AT&T, ebone exodus, level3, verrio), 3) the percentile sum for "rectangle", "umbrella", and "sector" is large (around 30%) for ISP topologies while not for model topologies.

From the figure, it is quite apparent that router–level topology is very different to the AS–level topology and the topologies generated with conventional modeling methods. Furthermore, ISP–level topologies (from ISP1 to ISP7) are highly clustered compared with the Abilene–based topology (Model2) presented by Li et al. [4]. We conjecture that the reason for differences derives from redundancy considerations in building the ISP networks. In what follows, we concentrate on the Sprint Topology (ISP1) and investigate the router–level topology in detail.

2.2 Detailed Analysis of Router–Level Topology

To compare how the previously–discussed structure for router–level topology affects the basic properties of networks, we prepare three topologies that have the same number of nodes and links. For the router–level topology, we use ISP1 (Sprint). Two topologies generated by the BA model (Model1 in Fig. 2) and the ER topology generated by the ER model are also used for purposes of comparison. The degree distributions for these three topologies follow the power–law, but not presented here due to space limitation.

We use the following metrics for node i to investigate the characteristics of topologies:

$A(i), D(i)$: Average and maximum number of hop–counts from node i to all other nodes. Hereafter, we will call the maximum hop–counts as diameters.

$C_e(i)$: Cluster coefficient [9] for a node, which is defined as

$$C_e(i) = \frac{2E_i}{d_i(d_i - 1)},$$

where d_i is the degree of node i, and E_i is the number of links connected between node i's neighbor nodes.

We also consider two centrality measures; degree centrality and betweenness centrality [10]. For each node i, degree centrality is defined as the degree of node i, and betweenness centrality is defined as the number of node–pairs that pass through node i. Note that the betweenness centrality does not represent actual traffic volume that passes through the node. However, the betweenness centrality is an important measure since it represents whether the node plays an important role for communicating nodes or not.

The cluster coefficient for each node is ranked in ascending order in Fig. 3(a). In the figure, the results of the Abilene topology are also presented. We can see that the cluster coefficient for the Sprint topology is much larger than that for the BA topology. Furthermore, the results in Figs. 3(a) and 3(d) show that lower–degree nodes are more highly clustered with the Sprint topology; a node with two out–going links always forms a cluster, while higher–degree nodes do not always have a high cluster coefficient. Other interesting observations can be seen in Figs. 3(b) and 3(c), which show the diameter $D(i)$ and average distance $A(i)$ from each node; both with the Sprint topology are larger than those with the BA topology. A node in the BA model tends to be connected to higher–degree nodes, and therefore any two nodes communicate with smaller hop–counts via the higher–degree nodes. However, the results for the router–level topology

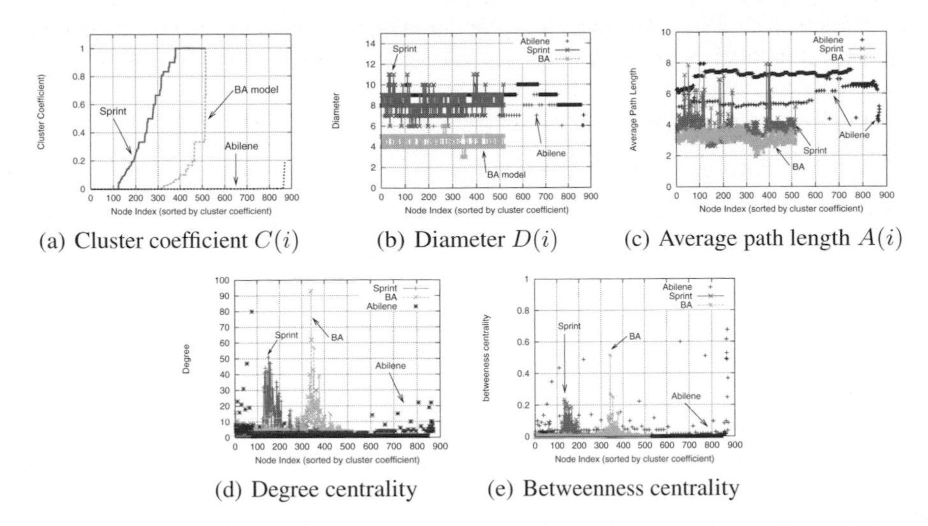

(a) Cluster coefficient $C(i)$ (b) Diameter $D(i)$ (c) Average path length $A(i)$

(d) Degree centrality (e) Betweenness centrality

Fig. 3. The basic properties of the router–level topology: Comparison among the Sprint, BA, and Abilene topologies

do not exhibit this effect. Since the average distance with the Sprint topology is larger than that with the BA topology, the small world property no longer hold with the router–level topology. Therefore, another attachment metric, rather than the degree–based metric, has to be considered to model the router–level topology, which we will discuss and propose in Section 3. The Abilene topology shows quite different characteristics in Fig. 3(a). With the Abilene topology, the cluster coefficient is even lower than the BA topology, and the average path length is much longer than the Sprint topology and the BA topology. The reason for this is apparent in that the Abilene topology is three–level hierarchical topology.

3 Modeling Methodology for Router–Level Topologies

The results in the previous section revealed that ISP–level topologies are very different to topologies using conventional modeling methods in that the cluster coefficient for lower–degree nodes is high. This indicates that ISP topologies are *locally* clustered networks, i.e., each node is connected to geographically closer nodes, and thus topologies attained by conventional models that do not use geographic information cannot appropriately evaluate for network control mechanisms, such as routing control.

Fabrikant et al.'s FKP model in Ref. [11] is a method that incorporates geographical information. However, they did not discuss in Ref. [11] whether the topologies resulting from the FKP model matches Internet topologies or not. The original FKP model, which adds one link for each node arrival, actually has numerous one–degree nodes, and is very different to the AS topology as shown in [12]. A question naturally arises as to whether the FKP model can actually predicts router–level topologies or not. In this section, we show that although topologies obtained with the FKP model are close to router–level topologies, they still have a lower cluster coefficient and do not match betweenness centrality. We therefore propose a new modeling method to generate router–level topologies in Sec. 3.2.

3.1 FKP Topology: Distance–Based Modeling

The FKP model proposed by Fabrikant et al. [11] revealed that the power–law property of degree distribution can still be obtained by minimizing "distance" metrics. This model does not use preferential attachment to add links, and instead uses minimization–based link attachment. More specifically, the FKP model works as follows. Each new node arrives at randomly in the Euclidean space $\{0,1\}^2$. After arriving at new node i, the FKP model calculates the following equation for each node, j, already existing in the network: $\alpha \cdot w_{ij} + l_{0j}$, where w_{ij} is the Euclidean distance (i.e., physical distance) between nodes i and j, and l_{0j} is the hop–counts distance between node j and a pre–specified "root" node (node 0). α is a parameter that weights the importance of physical distance. If α has a lower value, each node tries to connect to higher degree nodes; $\alpha = 0$ is an extreme scenario that creates a star–topology. If α has a higher value, each node tries to connect their nearest nodes. A topology with high a α is shown to behaves like an ER topology. The power–law property of the degree distribution appears at a moderate value of α value. Here, there are several hub–nodes in each region, and the hub–nodes form a power–law.

Figure 4 compares the ISP topology with the FKP model with regard to the same properties we previously discussed. In the figure, we do not use the actual Sprint topology (ISP1), but we modified the Sprint topology by eliminating one–degree nodes and their corresponding link since one–degree node has no impact on routing control. The resulting topology has 439 nodes / 1516 links, and the average degree is 3.46. In obtaining the results of the FKP topology, we add three links when each node arrived in order for setting the total number of links so that it is almost the same as for the modified Sprint topology. For the initial graph G_{init}, we use the 14–node NSFnet topology with

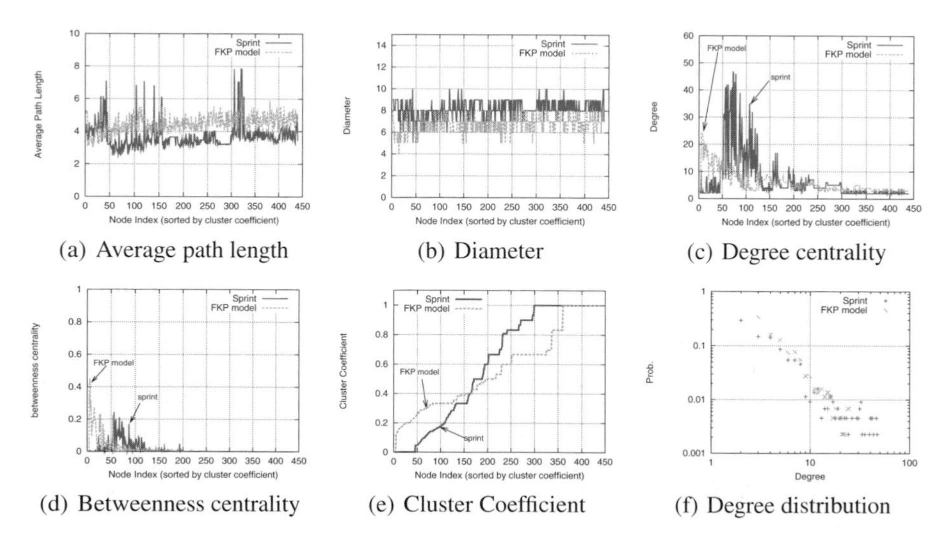

(a) Average path length (b) Diameter (c) Degree centrality

(d) Betweenness centrality (e) Cluster Coefficient (f) Degree distribution

Fig. 4. FKP model: $\alpha = 40$ (used in [11])

geographic latitudinal and longitudinal information. The value for α is set to 40 as used in Ref. [11].

A first impression of the results for the FKP topology is that the shape is closer than the results for the BA topology (see Figs. 3(a) through 3(e)). However, a clear difference appears again in the cluster coefficient; although the FKP model constructs a more highly–clustered network than the BA topology, the cluster coefficient is still smaller in lower–degree nodes. Another difference is that the maximum degree of the FKP topology is low. Note that the maximum degree depends on the parameter setting. As α gets smaller, the maximum degree can be increased. However, at the same time, a smaller value of α leads to a star–like topology and the betweenness centrality also becomes larger than the value in Fig. 3(e). Therefore, in the FKP model, fitting the degree distribution by appropriate α results in mismatches on the betweenness centrality of the modified Sprint topology.

3.2 New Modeling Method for Router–Level Topologies

The fact that the FKP model cannot construct router–level topologies because of much larger betweenness centrality drives us to develop a new modeling method by extending the FKP model. Our model incorporate the physical distance between nodes following the FKP model. However, unlike the FKP model, we also incorporate the enhancement of network equipments in ISP networks. For this, we add new links based on node utilization in the topology. However, the problem is where to place the new link. In this paper, we select a node that have the largest betweenness centrality in the network, and then attach a link between neighboring nodes. From the view point of graph theory, adding links to neighboring nodes increases to increase the cluster coefficient of the topology. From the view point of network design, on the other hand, this corresponds to improve reliability against network failures (e.g., link failures). It also corresponds to decreasing utilization of nodes in the topologies; some part of the traffic that has passed through the most utilized node is rerouted via added links.

More specifically, our algorithm works as follows. For a given initial network $G_{init}(V, E)$, when a new node joins the network, m links from that node are added (network growth). Besides, k links with no relation to m links are added based on node utilization of the network, which corresponds to network enhancements by ISPs (network enhancement). This procedure is continued until n nodes are added to the initial network. Since m links and k links are added to the network at each of node join, the resulting topology has $\|E\| + n \cdot m + k$ links, where $\|E\|$ is the number of links in the initial network. In the following, we explain the link attachment policy for network growth (m–link addition) and policy for network enhancement (k–link addition).

Network growth model

Step 0: Set the initial network.

Step 1: For each node i ($\in V$) already existing in the network, calculate the attachment cost to node i as

$$\alpha \cdot w_{ij} + \bar{h}_i, \quad (1)$$

where $\bar{h}_i$ is the average distance from node i to the other nodes.

Step 2: Select m nodes in an ascending order by Eq. (1). Then add one link to each of selected nodes.

Step 3: Go back to Step 1, until the number of nodes reaches n.

Network enhancement model

Add k links via the following steps.

Step 1: Calculate betweenness centrality for each node in the network, and then select a node, x, that has the largest betweenness centrality in the network.

Step 2: From the set of neighbor nodes from x, select two nodes y and z, that minimize,

$$\begin{aligned} &\beta \cdot w_{yz} + (1/D_z), \quad \text{if } D_z > D_y, \\ &\beta \cdot w_{yz} + (1/D_y), \quad \text{otherwise}, \end{aligned} \tag{2}$$

where β is the parameter for weighting importance to the physical distance, and D_p denotes the betweenness centrality of node p. Note that by using the equation $1/D_p$, more traffic on node x is rerouted via the link between node y and z.

3.3 Structural Properties of Our Modeling Method

We show the results with our modeling method in Fig. 5. Here, the number of joining nodes n is set to 425, and we use $m = 2$, i.e., when each node arrive, two links are prepared for newly arriving node. We set $k = 649$ so that the resulting topology has the same number of nodes (439) and links (1519) as the modified Sprint topology. If a one–degree node is necessary, the original FKP model that connects one link for node arrival can be applied. For the initial graph G_{init}, we use the NSFnet topology with

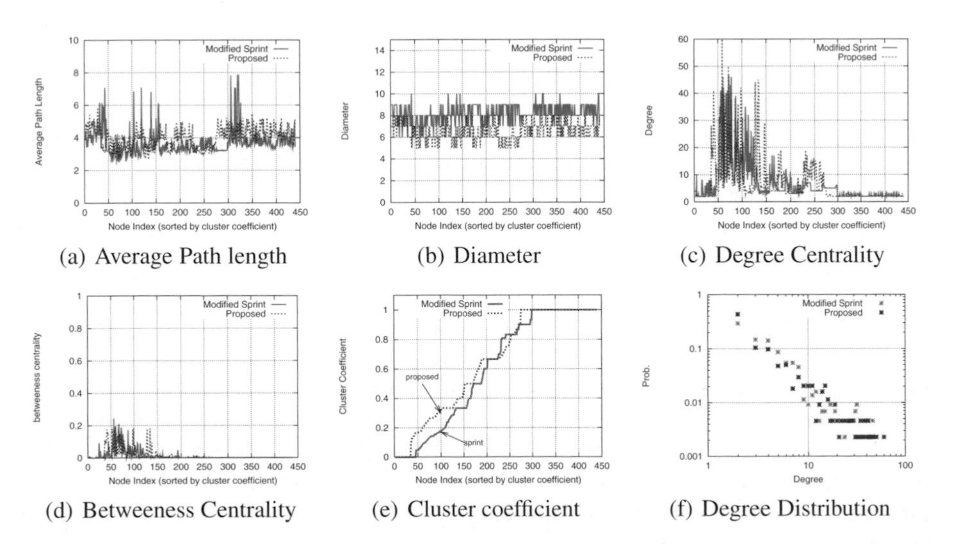

(a) Average Path length (b) Diameter (c) Degree Centrality

(d) Betweeness Centrality (e) Cluster coefficient (f) Degree Distribution

Fig. 5. Results with proposed modeling method: α is 25, and β is 200

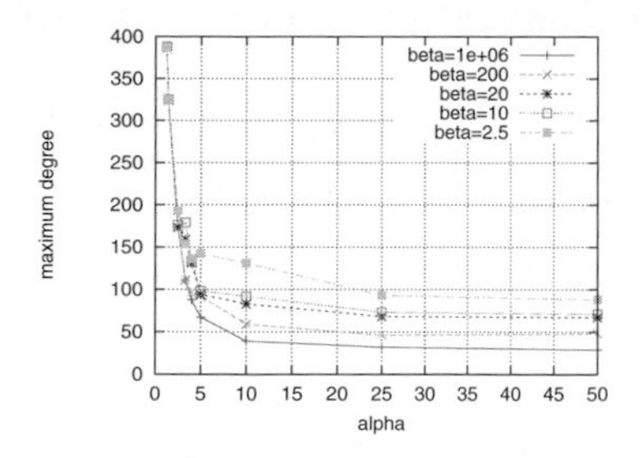

Fig. 6. Effect of α and β

geographic latitudinal and longitudinal information. By setting parameters α and β to be 25 and 200, the resulting topology is very close to the Sprint topology for both degree distribution and betweenness centrality. Note that we show the best parameter settings for the topology that looks like the modified Sprint topology in Fig. 5. Actually, depending on α and β, the topology differs from Fig. 5. To see the impact of parameter settings, we show the maximum degree dependent on α for each β in Fig. 6. Apparently, inherited parameter α from the FKP model shows the same tendency as presented in Ref. [11]; as α get smaller, the topology becomes a star–like topology. That is, if the maximum degree equals to n $(= 425)$, the topology becomes the star topology. β also impacts on the maximum degree in the topology; the maximum degree become larger as β gets smaller (i.e., weights on the physical distance becomes smaller). Considering that the maximum degree in the modified Sprint topology is 47, α should be greater than 20 and the β greater than 200, to generate a realistic ISP topology with a moderate maximum degree.

4 Application to the Evaluation on Routing Control

In this section, we demonstrate that our modeling method can be actually used for evaluations on routing control. For this purpose, we show the link utilization of the topology generated by our modeling method. In obtaining the link utilization, we have to determine 1) routing methods, 2) transmission capacities of links, and 3) amount of traffic between nodes.

For routing methods, we use minimum hop routing and optimal routing. Optimal routing is based on the flow deviation method (see Sec. 4.1). The transmission capacities are difficult to determine since there is no publicly available information. However, as Li et al. mentioned [4], constraints with router technology actually limit the degree (i.e., number of ports in the router) and line speed of a port. In this paper, we give a method to allocate the capacities of links based on router's specification. The details are described in Sec. 4.2. Note that link capacities obtained by Sec. 4.2 may not be actual link capacities. However, our primary purpose is to demonstrate that our modeling method can be applied to evaluations on routing control, and thus the exact information is not necessary here. By the same reason, we simply assume that traffic demand between nodes is identical for every node–pair.

4.1 Optimal Routing Method

In obtaining link utilization for each topology, we apply the optimal routing method that is based on the flow deviation method [13]. The flow deviation method incrementally changes the flow assignment along feasible and descent directions. Given objective function T, the method set w_l as a partial derivative with respect to F_l, where F_l is the amount of traffic that traverses link l. Then, the new flow assignment is solved by using the shortest path algorithm in terms of w_l. By incrementally changing from the old to the new flow assignment, optimal flow assignment is determined. In this study, we set objective function T to

$$T = \sum_{l} 1/(C_l - F_l), \tag{3}$$

where C_l is the capacity of link l and F_l is as defined above.

4.2 Method for Allocating Link Capacities

We allocate the capacities of links based on the technology constraints imposed by the Cisco 12416 router, which has 16 line card slots. When a router has 16 or less connected links, all the links can have 10Gbps capacity. If there are more than 16 links connected to the router, the capacity for one or more of the links have to be decreased.

However, it is difficult to determine in which link capacity should be decreased. Therefore, we allocate the capacities of links in a network so that the amount of traffic between a node–pair is maximized, while satisfying the following two technology constraints imposed by routers. See Ref.[14] for detailed algorithm.

4.3 Distribution of Link Utilization

In Fig. 7, we show the distribution of link utilization for the modified Sprint topology (Fig. 7(a)), BA topology (Fig. 7(b)), and the topology obtained by our modeling method (Fig. 7(c)). The vertical axis shows link utilization, and the horizontal axis represents link index. The link index is given in an ascending order of link utilization when the minimum hop routing method is used. Then, the link utilization of the optimal routing method in Sec. 4.1 is shown for each link index. Figure 7(a) shows that the optimal routing method gives smaller the maximum link utilization (about $1/3$) compared with

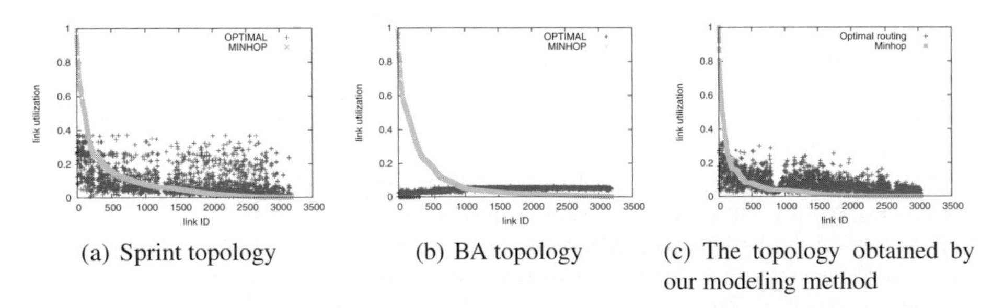

(a) Sprint topology (b) BA topology (c) The topology obtained by our modeling method

Fig. 7. Distribution of link utilizations by applying optimal routing method

minimum hop routing. However, it is observed from Fig. 7(b) that a topology by the BA model achieves much smaller of the maximum link utilization (about $1/10$). These results indicate that the link utilization in the router–level topology is much far from the one in the conventional modeling method. From Fig. 7(a) and Fig. 7(c), we observe that the distribution of link utilization in our topology is quite similar to that in the modified Sprint topology for both minimum hop routing and optimal routing method.

5 Concluding Remarks

For vital network researches, a method for modeling the realistic router–level topology urgently needs to be developed. However, we have shown that the structure of ISP topologies is quite different from that of topologies achieved with conventional modeling methods. Based on this, we have developed a new realistic modeling method for generation of router–level topologies. In our method, when a new node joins the network, it likely connects to the nearest nodes. In addition, we added new links based on node utilization in the topology, which corresponded to enhancing network equipments in ISP networks. The evaluation results have shown that our modeling method achieve a good compatibility with the Sprint topology with regards to degree distribution and the number of node-pairs passing through nodes.

References

1. Faloutsos, M., Faloutsos, P., Faloutsos, C.: On power–law relationships of the Internet topology. In: Proceedings of ACM SIGCOMM, pp. 251–262 (1999)
2. Gkantsidis, C., Mihail, M., Saberi, A.: Conductance and congestion in power law graphs. In: Proceedings of ACM SIGMETRICS, pp. 148–159 (2003)
3. Barabasi, A., Albert, R.: Emergence of scaling in random networks. Science 286, 509–512 (1999)
4. Li, L., Alderson, D., Willinger, W., Doyle, J.: A first–principles approach to understanding the Internet's router–level topology. ACM SIGCOMM Computer Communication Review 34(4), 3–14 (2004)
5. Sprint, N., Mahajan, R., Wetherall, D., Anderson, T.: Measuring ISP topologies with rocketfuel. IEEE/ACM Transactions on Networking 12(1), 2–16 (2004)
6. Bu, T., Towsley, D.: On distinguishing between Internet power law topology generators. In: Proceedings of INFOCOM, pp. 1587–1596 (2002)
7. Milo, R., Shen-Orr, S., Itzkovitz, S., Kashtan, N., Cheklovskii, D., Alon, U.: Network motifs: Simple building blocks of complex networks. Science, 824–827 (2002)
8. Erdös, P., Rényi, A.: On the evolution of random graphs. Publications of the Mathematical Institute of the Hungarian Academy of Sciences 5, 17–61 (1960)
9. Albert, R., Barabási, A.L.: Statistical mechanics of complex networks. Reviews of Modern Physics (2002)
10. Newman, M.E.J.: 2. In: Random graphs as models of networks, pp. 35–68. Wiley, Chichester (2002)
11. Fabrikant, A., Koutsoupias, E., Papadimitriou, C.H.: Heuristically optimized trade–offs: A new paradigm for power law in the Internet. In: Widmayer, P., Triguero, F., Morales, R., Hennessy, M., Eidenbenz, S., Conejo, R. (eds.) ICALP 2002. LNCS, vol. 2380, pp. 110–122. Springer, Heidelberg (2002)

12. Alvarez-Hamelin, J.I., Schabanel, N.: An Internet graph model based on trade–off optimization. European Physical Journal B 38(2), 231–237 (2004)
13. Fratta, L., Gerla, M., Kleinrock, L.: The flow deviation method: An approach to store-and-forward communication network design. Networks 3, 97–133 (1973)
14. Fukumoto, R., Arakawa, S., Murata, M.: On routing controls in ISP topologies: A structural perspective. In: Proceedings of Chinacom (2006)

Does the Average Path Length Grow in the Internet?

Jinjing Zhao, Peidong Zhu, Xicheng Lu, and Lei Xuan

School of Computer, National University of Defense Technology, Changsha 410073, China
misszhaojinjing@sina.com, {pdzhu,xclu,xuanlei}@nudt.edu.cn

Abstract. In this paper we focus on the trend of the average path length of the AS-level Internet, which is one of the most important parameters to measure the efficiency of the Internet. The conclusion drawn by the power-law and small world models is that the average path length of network scales as ln(n) or ln(ln(n)), n is the number of nodes in network. But through analyzing the data of BGP tables in recent 5 years, we find that the average path length of the Internet is descending and the descending rate is about 0.00025 which is much different from the result induced from the theories. We anatomize the reason and find many factors will affect the value of path length, like multi-homing, commercial relationships and so on. Besides, the trend of the average path length is also drawn with mathematics.

Keywords: As-level Internet, the average path length, power-law, small world, self-organized system.

1 Introduction

In recent years a considerable research effort has been focused on the field of complex networks. The main reason for this effort finds its rationale in the very pervasive presence of biological, social, or technological structures that can be described using the paradigm of complex networks.

The physical Internet is one of the most common examples of complex networks in the real society. In the absence of accurate Internet maps many research groups have started large scale projects aimed at the collection of data on the topology and structure of this network of networks [1–4].

In AS-level Internet, the path length between two ASs is defined as the number of edges along the shortest path connecting them. The average path length of the Internet, then, is defined as the mean path length between two ASs, averaged over all pairs of nodes. The average path length is one of the most important parameters to measure the efficiency of the Internet. And it is relevant in many fields, such as routing [9], searching [17], and transport of information [16]. All those processes become more efficient when the path length is smaller. The Internet, as a power-law and small world network, does display small path length property.

In this paper we care about the trend of the average path length of the AS-level Internet. In the previous studies [18-22, 10], the researches on the Internet path length can be divided into two aspects, one is the theoretical induction from the models of network, such as power-law and small world; the other is the analysis on the real data

T. Vazão, M.M. Freire, and I. Chong (Eds.): ICOIN 2007, LNCS 5200, pp. 183–190, 2008.

collected on network using traceroute or BGP data etc.. The conclusion drawn by the former is that the average path length of the Internet scales as ln(n) or ln(ln(n))[18-20], n is the number of nodes in network, But through analyzing the data of BGP tables in recent 5 years, we find that the average path length of the Internet is descending smoothly. What cause the difference between them? Does the path length of the AS-level Internet will keep reducing in future? And if so, what's the scope of its descending rate? These questions are the tasks for this paper to solve.

2 Background and Metrics

In this section, we introduce the topology and several topological properties of the Internet. Based on the power-law and small world nature of the Internet, the theoretical induction of the average path length is given.

2.1 Topology of the Internet

The Internet can be decomposed into sub-networks that are under separate administrative authorities. These sub-networks are called domains or autonomous systems.

The internet structure has been the subject of many recent works. Researchers have looked at various features of the Internet graph, and proposed theoretical models to describe its evolution. Faloutsos et al. [14] experimentally discovered that the degree distribution of the Internet AS and router level graphs obey a power law. Barab´asi and Albert [18] developed an evolutional model of preferential attachment, which can be used for generating topologies with power-law degree distributions. The Internet AS structure was shown to have a core in the middle and many tendrils connected to it. A more detailed description is that around the core there are several rings of nodes all have tendrils of varying length attached to them. The average node degree decreases as one node moves away from the core. We call these core nodes "rich" nodes and the set containing them the "rich club." The rich club consists of highly connected nodes, which are well interconnected between each other, and the average hop distance among the club members is very small. So the AS graph is also a small-world network with very small average path length compared with the network size.

2.2 Average Path Length

The average path length is one of the global metrics defined as the average of the closeness values for all nodes. Both in the power-law and small world networks, its value are smaller than the random networks. Now we analyze the theoretical value from these two models.

A. Small world model

Consider an undirected network, and let us define the average path length l to be the mean geodesic (i.e., shortest) distance between vertex pairs in a network:

$$l = \frac{1}{\frac{1}{2}n(n+1)} \sum_{i \geq j} d_{ij} \tag{1}$$

where dij is the geodesic distance from vertex i to vertex j and n is the number of nodes in the network.

The small-world effect has obvious implications for the dynamics of processes taking place on networks. For example, if one considers the spread of information, or indeed anything else, across a network, the small-world effect implies that spreading will be fast on most real world networks. On the other hand, the small-world effect is also mathematically obvious. If the number of vertices within a distance r of a typical central vertex grows exponentially with r—and this is true of many networks, including the random graph—then the value of l will increase as log n. In recent years the term "small-world effect" has thus taken on a more precise meaning: networks are said to show the small-world effect if the value of l scales logarithmically or slower with network size for fixed mean degree.

Logarithmic scaling can be proved for a variety of network models [18][29], including the Internet. So according to the small-world model, the average path length of the Internet will increase as log n.

B. Power-law model

Let us start from a given node and find the number of its nearest, next-nearest, … , m-th neighbors. Assuming that all nodes in the graph can be reached within l steps and Zm is the average number of m-th neighbors, we have

$$1 + \sum_{m=1}^{l} Z_m = N \tag{2}$$

For the power-law of networks, the degree distribution of nodes follows equation (3):

$$P(k) = Ck^{-\gamma}, \qquad \text{for } k \geq 1 \tag{3}$$

According to the equation (2) and (3), the average path length l can be induced as:

$$l = \frac{\ln(n) + \ln[\xi(\gamma)/\xi(\gamma-1)]}{\ln[\xi(\gamma-2)/\xi(\gamma-1)-1]} + 1 \tag{4}$$

Equation (4) indicates that its average path length scales logarithmically with its size n in power-law networks.

So in the networks of small world and power-law, the average path length would all increase equal or little less than logarithmically with its size. The Internet is both small world and power-law, so the theoretical value of its average path length l should also obey this rule. But is this really true? In the next section, we will verify the result with the statistic of the real data.

3 Statistic of Average Path Length

In order to observe the change of l, we collect the data of BGP tables over five years from Route-views project [27], and distill the AS_PATH attribution in them.

Fig 1 [27] is the average ASs path length sampled on 20000 points by AS1221 and AS 4637 during 2001-2006, the Average values are 3.4611 and 3.3352 separately. And we can also see that the value is reducing smoothly, which differs much from the theoretical induction in section 2. What causes this to happen?

In order to answer this question, we focus on the process of developing models of the Internet AS-level connectivity. Single ASs as the Internet's fundamental building blocks are designed largely in isolation and then connected according to both engineering and business considerations. So the true growth model of the Internet is not as simple as the famous BA model or WS model in power-law or small world networks. Just because the complexity of the network makes the real situation different from the result induced from the simple models. But the characteristics of the Internet are still obeying the basic principles as power-law and small world, as the degree correlations, small diameter, and small clustering coefficient etc.

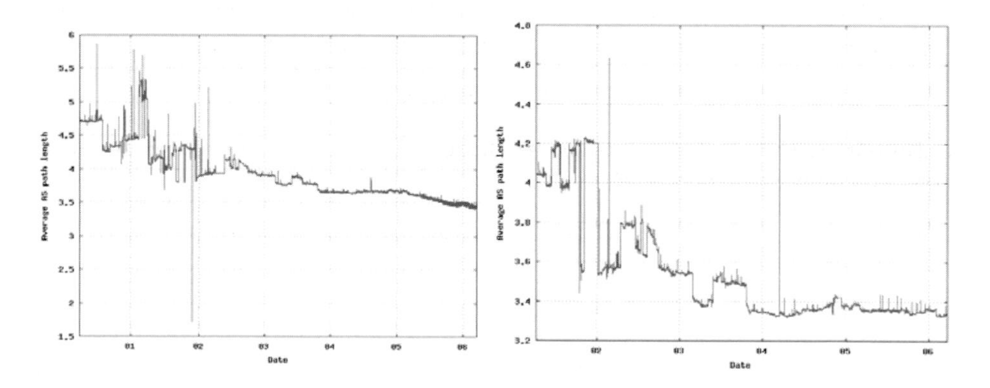

Fig. 1. The Average AS Path from 2001-2006

The reducing of the average path length of the AS-level Internet is because every AS has its own intentions (i.e. intelligence) and their games happened in the network. Concretely, some mechanisms like multi-home, commercial relationship, routing protocols, and even human beings can all change the trend of the average path length, and make it reduce.

1) Multi-home can make the path length shorter. Multi-home is the result of network evolution. The probability of the selecting another AS is descending exponentially. In this way, the distance between ASs can be shorter by these shortcuts, just as the long-range edges in the small-world WS model [26]. The probability of multi-home in the AS is increasing very quickly, which can reduce the forwarding cost greatly.
2) Commercial relationship makes the ASs together, especially the peering relationship. If two ASs attract each other with their forwarding ability, they make the peering relation. Sometimes, they build special tunnels between them to get the high speed or bandwidth. So the average path length of the network can reduce further.

The main spring of the Internet growth is the minimizing of forwarding cost. So in order to reach this ambition, the average path length must reduce and is reducing now.

4 Prediction on the Average Path Length

In section 3, we know the average path length of ASs is reducing because the special growth pattern of the Internet. This change looks random and out-of-order, but through the analysis of the previous data, we find that the trend of the average path length can be predicted. It also obeys a special rule. In this section we will analyze this trend and estimate its descending rate.

The Internet follows the power-law, so it has power-law parameters.

Definition 1. rank exponent: The out-degree, dv, of a node v, is proportional to the rank of the node, rv, to the power of a constant, R:

$$d_v \propto r_v^{\,R} \tag{5}$$

Definition 2. hop-plot exponent: The total number of pairs of nodes, P(h), within h hops, is proportional to the number of hops to the power of a constant, H:

$$P(h) \propto h^{\mathcal{H}}, \text{ h<Diameter(G)} \tag{6}$$

Definition 3. The average size of the neighborhood, NN(h), within h hops as a function of the hop-plot exponent, H, is

$$NN(h) = \frac{P(h)}{N} - 1 = \frac{N+2E}{N} h^{\mathcal{H}} - 1 \tag{7}$$

Here, NN(h) is the accumulation of the parameter Zm in equation (2). The average path length l can be induced by equation (2) and (7) as:

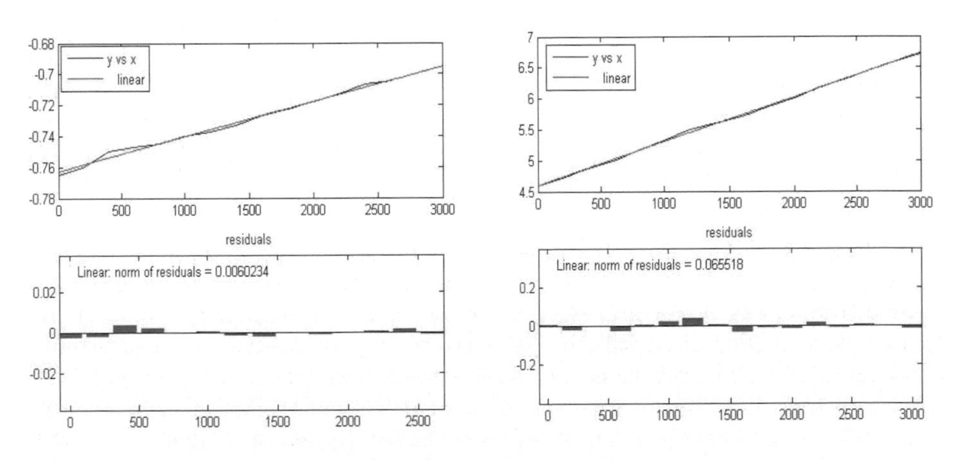

Fig. 2. The Curves of R and H and the Residuals of Linear Fitting

$$
\begin{aligned}
1 + NN(l) &= N \\
\frac{N+2E}{N} l^{H} &= N \\
l &= \left(\frac{N^2}{N+2E}\right)^{1/H}
\end{aligned}
\tag{8}
$$

According to the nature of power-law, the relation between the edges number E and the nodes number N is present in equation (9):

$$
E = \frac{N}{2(R+1)}\left(1 - \frac{1}{N^{R+1}}\right) \tag{9}
$$

The proof of formulation 9 can be found in [13][14]. So formulation 10 can be induced by formulation 8 and 9:

$$
\begin{aligned}
l &= \left(\frac{N^2}{N+2E}\right)^{1/H} \\
&= \left(\frac{N^2}{N + \frac{N}{R+1}\left(1 - \frac{1}{N^{R+1}}\right)}\right)^{1/H} \\
&= \left(\frac{(R+1)N^{R+2}}{(R+2)N^{R+1} - 1}\right)^{1/H}
\end{aligned}
\tag{10}
$$

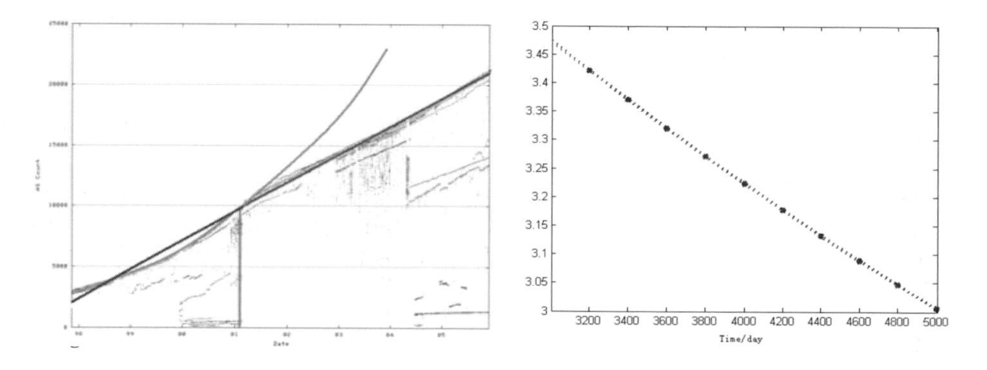

Fig. 3. The Growth of ASs Number

Fig. 4. Prediction of Average Path Length

So the average path length l can be calculated by R, N and H. The curves of R and H from 1997 to 2005 are printed in Fig.2. The number of ASs, N, was increasing in exponent rate by 2001, but the curve is nearly linear from then on, shown in Fig.3. In order to get the functions of time variations of R, N, and H, their curves are linear fitted by Matlab. Equations 11, 12, 13 were produced; the unit of t is day.

$$
N = 6t + 82 \tag{11}
$$

$$R = 2.26 \times 10^{-5} t - 0.76 \tag{12}$$

$$H = 7.12 \times 10^{-4} t + 4.6 \tag{13}$$

Fig.4. predicts the average path length of ASs from 2006 to 2012, which shows that the value is reducing smoothly. The slope of the linear is less than -0.00025. So we can see, the average path length of ASs will be 3.0 in 2012.

5 Conclusion

In this paper we focus on the trend of the average path length of the AS-level Internet. The conclusion drawn by the power-law and small world models is that the average path length of the Internet scales as ln(n) or ln(ln(n)), n is the number of nodes in network. But through analyzing the data of BGP tables in recent 5 years, we find that the average path length of the Internet is descending smoothly and the descending rate is less than 0.00025. We anatomize its reason and find many factors will affect the value of path length, like multi-home, commercial relationships and so on.

The physical Internet is one of the most common examples of complex networks in the real society. Its growing structure is the result of competitive and cooperative processes, in which individual choice, optimization criteria, and policy-driven strategies cooperate with the lack of any centralized control in determining the self-organized evolution of the system. All these factors lead to the differences from the regular growth models of power-law or small world. The Internet is a network which has the characteristics of power-law and small world, besides these, it is also an intelligence self-organized system. So in most cases, we should not analyze its properties just based on the regular models and simplex theories. This is the most important conclusion we draw from our work.

Acknowledgment

This research is supported by National Basic Research Program of China (Grant No. 2005CB321801), National Natural Science Foundation of China (Grant No. 60673169), and High-Tech Research and Development Program (Grant No. 20060101Z2134).

References

1. The National Laboratory for Applied Network Research (NLANR), sponsored by the National Science Foundation, `http://moat.nlanr.net/`
2. The Cooperative Association for Internet Data Analysis (CAIDA), located at the San Diego Supercomputer Center, `http://www.caida.org/home/`
3. Topology project, Electric Engineering and Computer Science Department, University of Michigan, `http://topology.eecs.umich.edu/`
4. SCAN project at the Information Sciences Institute, `http://www.isi.edu/div7/scan/`
5. Faloutsos, M., Faloutsos, P., Faloutsos, C.: Comput. Commun. Rev. 29, 251 (1999)

6. Govindan, R., Tangmunarunkit, H.: In: Proceedings of IEEE INFOCOM, Tel-Aviv, Israel, March 2000, pp. 1371–1380 (2000)
7. Broido, A., Claffy, K.C.: San Diego Proceedings of SPIE International Symposium on Convergence of IT and Communication, Denver, Colorado (2001)
8. Caldarelli, G., Marchetti, R., Pietronero, L.: Europhys. Lett. 52, 386 (2000)
9. Pastor-Satorras, R., Vázquez, A., Vespignani, A.: Phys. Rev. Lett., 258701 (2001); Vázquez, A., Pastor-Satorras, R., Vespignani, A.: Phys. Rev. E 65, 066130 (2002)
10. Chen, Q., Chang, H., Govindan, R., Jamin, S., Shenker, S.J., Willinger, W.: In: Proceedings of IEEE INFOCOM 2002, New York (2002)
11. Medina, A., Matta, I.: Boston University, Technical Report No. BU-CS-TR-2000-005 (2000)
12. Jin, C., Chen, Q., Jamin, S.: EECS Department, University of Michigan, Technical Report No. CSE-TR-433-00 (2000)
13. Siganos, G., Faloutsos, M., Faloutsos, P., Faloutsos, C.: Power-laws and the AS-level Internet topology. IEEE/ACM Trans. on Networking 11, 514–524 (2003)
14. Faloutsos, M., Faloutsos, P., Faloutsos, C.: On Power Law Relationships of the Internet Topology. In: ACM SIGCOMM, September 1-3, Cambridge MA, pp. 251–262 (1999)
15. http://bgp.potaroo.net/as1221/bgp-active.html
16. Goh, K.-I., Kahng, B., Kim, D.: Phys. Rev. Lett. 87, 278701 (2001)
17. Adamic, L.A., Lukose, R.M., Puniyani, A.R., Huberman, B.A.: Phys. Rev. E 64, 046135 (2001)
18. Albert, R., Barabasi, A.-L.: Rev. Mod. Phys. 74, 47 (2002)
19. Newman, M.E.J.: SIAM Review 45, 167 (2003)
20. Newman, M.E.J.: J. Stat. Phys. 101, 819 (2000)
21. Calvert, K., Zegura, E., Doar, M.: Modeling Internet topology. IEEE Trans. on Communication, 160–163 (December 1997)
22. Medina, A., Matta, I., Byers, J.: On the origin of powerlaws in Internet topologies. ACM SIGCOMM Computer Communication Review 30(2), 18–34 (2000)
23. Bollobás, B., Riordan, O.: The diameter of a scale-free random graph, Department of Mathematical Sciences, University of Memphis (preprint, 2002)
24. Barrat, A.: Comment on Small-world networks: Evidence for crossover picture, cond-mat/9903323 (preprint, 1999)
25. Newman, M.E.J., Moore, C., Watts, D.J.: Mean field solution of the small-world network model. Phys. Rev. Lett. 84, 3201–3204 (2000)
26. Barabási, A.L., Albert, R.: Science 286, 509 (1999)
27. Routeviews Project, http://bgp.potaroo.net/as1221/bgp-active.html
28. Albert, R., Barabasi, A.-L.: Statistical mechanics of complex networks. Reviews of Modern Physics 74 (January 2002)

An Implicit Cluster-Based Overlay Multicast Protocol Exploiting Tree Division for Mobile Ad Hoc Networks

Younghwan Choi, Soochang Park, Fucai Yu, and Sang-Ha Kim

Department of Computer Engineering, Chungnam National University,
220 Gung-dong, Yuseong-gu, Daejeon, 305-764, Republic of Korea
{yhchoi,winter,yufc}@cclab.cnu.ac.kr, shkim@cnu.ac.kr

Abstract. A characteristic of mobile ad hoc networks, mobility leads to frequently update multicast routing information based on network multicasting. For the reason, comparatively lots of control signaling messages are generated, so bandwidth of wireless media and limited battery power of mobile nodes are inefficiently consumed. To improve the shortcomings, overlay multicast protocols on application layer has been proposed; however, some of them have another shortcoming which is that the number of unnecessary data forwarding among multicast members increases. Additionally, they hardly support network scalability. Thus, this paper proposes *An Implicit Cluster-based Overlay Multicast Protocol Exploiting Tree Division (ICOM-TD)* for Mobile Ad hoc Networks to improve their shortcomings. Finally, we evaluate performance of *ICOM-TD* through numerical analysis about complexity of multicast routing and simulation in comparison with existing multicast mechanisms for ad hoc networks on end-to-end time delay, packet overhead, and delivery ratio.

Keywords: Mobile Ad hoc Networks, overlay ad hoc multicast, implicit-clustering.

1 Introduction

Mobile nodes comprising Mobile Ad hoc Networks (MANET)[1] have a function of host-specific routing. MANETs have two main considerations, limited battery power and dynamic topology. This paper proposes a protocol for group communications in MANETs. Multicast protocols in wired networks can exploit optimal routes to deliver multicast data in the absence of network failures. On the other hand, multicast protocols in MANETs must consider control overhead for maintenance of multicast routing trees owning to frequent changes of network topology. In other words, generation of lots of control messages for such maintenance inefficiently consumes bandwidth of wireless media and also promotes consumption of limited battery power of mobile nodes. That is, multicasting for MANETs needs to consider how to efficiently consume limited battery power and

T. Vazão, M.M. Freire, and I. Chong (Eds.): ICOIN 2007, LNCS 5200, pp. 191–201, 2008.

bandwidth of wireless media owning to topology changes than how to optimally establish multicast routes.

The rest of this paper is organized as follows. Section 2 shows analysis about related work, and then Section 3 fully presents *an Implicit Cluster-based Overlay Multicast Protocol Exploiting Tree Division (ICOM-TD)* for MANETs. The performance evaluation of the ICOM-TD is followed in Section 4. Finally, further works and concluding remarks are made in Section 5.

1.1 Analysis of Related Work

This paper significantly divides the previously proposed multicast protocols in MANETs into two categories. One of them is network layer multicast. This requires that all nodes should support multicast routing protocol on network layer although some of the nodes do not join multicast groups. The other is application layer multicast, which is called, overlay multicast. Application layer multicasting has more reliable transmission than network layer multicasting since application multicasting exploits unicast routing. Network layer multicasting enables to transmit multicast data to densely distributed nodes in the transmission range at one time, but application layer multicasting should transmit multicast data as many time as the number of multicast group members in the same situation because it exploits unicast media(see Fig. 1). Such a property could be a shortcoming for MANETs, which has mobility and limited battery power.

The network layer multicast protocols are divided into backbone-based multicasting and source-specific multicasting. Backbone-based multicasting exploits tree or mesh algorithm to establish multicast routing paths. They need control messages to maintain and update the routing information when network topology changes. It results in waste of wireless resources and limited battery power. A Multicast Protocol for Ad hoc Wireless Networks(AMRIS)[2] and Multicast Ad hoc On-demand Distance Vector Routing(MAODV)[3] are tree-based multicast protocols. They have a defect which is that children of a tree cannot receive data when their parents have a transmission fault. Mesh-based multicast protocols, such as On-Demand Multicast Routing Protocol(ODMRP)[4], establish all possible routing paths between multicast sources and receivers, so it is more

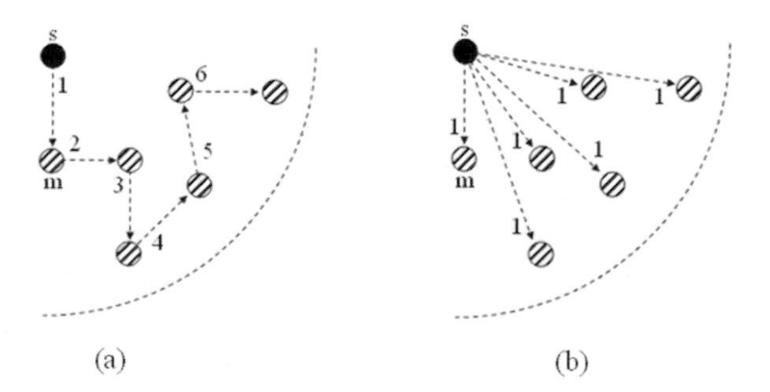

Fig. 1. A multicasting propagation model: (a) Overlay multicasting on application layer; (b) 1-hop multicasting on network layer

stable for link failures than trees. ODMRP would inefficiently consume bandwidth of wireless media since it exploits duplicate paths from a multicast source to receivers. Source-specific multicasting collects information of all nodes for multicasting and then delivers data through source-specific multicast routing. It, however, restricts to the number of data forwarding so only support small multicast groups. There is an example, Differential Destination Multicast-A MANET Multicast Routing Protocol for Small Groups(DDM)[5].

Application layer multicast protocols are divided into statefull and stateless multicasting. All members in statefull multicasting exploit their own multicast routing information to forward data (AMRoute: Ad hoc Multicast Routing Protocol[6]). To update multicast routing information due to topology changes, it also needs control messages like AMRIS, MAODV, and ODMRP. In addition to the shortcoming, its overlay multicast trees (OMTs) could increase the number of unnecessary data forwarding among multicast members on link layer. For instance, the number of data forwarding from S to m2 via m1 is 4 in terms of the OMT in Fig. 2(a). After network topology changes in Fig. 2(b), the number of data forwarding from S to m2 via m1 with the same OMT is 6, which is increased from 4. The control messages to optimize OMT results in the same problems as AMRIS, MAODV, and ODMRP.

Stateless multicasting, such as Effective Location-Guided Tree Construction Algorithms for Small Group Multicast (LGT)[7] and Efficient overlay multicast for mobile ad hoc networks (PAST-DM)[8], possibly improves the problems because only multicast sources collect information for OMTs and establish multicast routing paths. In similar to source-specific multicast routing, it also supports only small group communications since it piggybacks multicast routing information in data packets. Network scalability problems are possibly improved by Cluster-based hierarchical multicasting (Efficient End System Multicast for Mobile Ad Hoc Networks; NICE-MAN). It exploits broadcast media for multicasting to densely distributed nodes in the transmission range. For doing that, it establishes clusters and clusterhead-based hierarchical OMTs. NICE-MAN improves almost all shortcomings on multicasting in MANETs, but its clustering

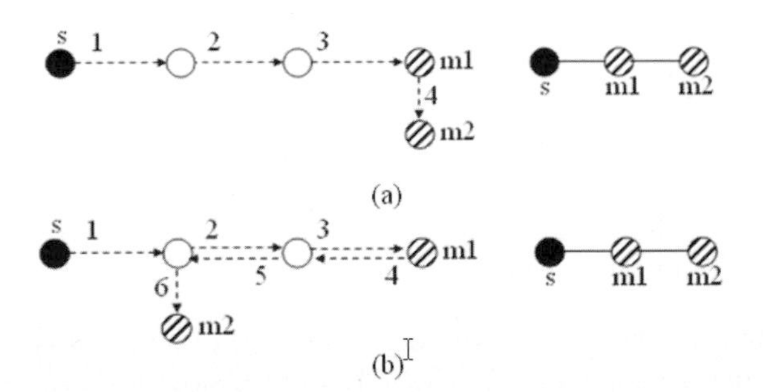

Fig. 2. Comparison of the number of propagation in terms of network topology changes: (a) Before network topology changes, data forwarding according to the overlay multicast tree, (b) After network topology changes, data forwarding according to the same overlay multicast tree

mechanism needs to periodically exchange explicit control messages for maintenance. Such explicit control messages cause link overhead and performance lowering in highly dynamic network topology.

This paper proposes a novel overlay multicast protocol to overcome the shortcomings which multicast protocols [2,3,4,5,6,7,8,9] have for MANETs.

2 An Implicit Cluster-Based Overlay Multicast Protocol Exploiting Tree Division (ICOM-TD)

The proposed protocol, named "An Implicit Cluster-based Overlay Multicast protocol Exploiting Tree Division (ICOM-TD)," in this paper belongs to application layer multicasting, stateless multicasting, and cluster-based multicasting. Following states protocol design principles and assumptions of ICOM-TD.

2.1 Protocol Design Principles and Assumptions

The ICOM-TD has four principles for protocol design. First of all, it is based on application layer multicasting for more reliable communications. Second, it exploits a single-hop clustering scheme. The single-hop clustering can help multicast data to densely distributed group members in the transmission range through broadcast media at one time. Explicit-clustering [10], however, exploits periodic beacon messages to establish and maintain their clustering structures. The ICOM-TD presents an implicit-clustering scheme by multicast sources. This is named, implicit-clustering. Third, it is also based on stateless multicasting in order to reduce control messages. Thus, all multicast members do not have to establish their own multicast routing information. It means they can save their limited battery power. Fourth, data packets in the ICOM-TD is delivered after piggybacking control information, such as multicast routing and clustering state information, so size of the control information cannot help being restricted. This is a cause not to support large number of multicast groups. To improve the problem, the ICOM-TD additionally exploits an overlay multicast tree division algorithm.

The ICOM-TD needs three assumptions for the principles like follows.

- All mobile nodes in MANETs should have homogeneous wireless network devices, so their transmission ranges are the same.
- All mobile nodes can sense their own geographical information.
- All multicast group members should have basic information for multicasting; for example, multicast addresses.

The ICOM-TD assumes that all mobile nodes communicate through homogeneous wireless network media and size of their transmission ranges are the same in order not to make it complicated. Abilities of mobile nodes to sense geographical information are required for construction of implicit-cluster.

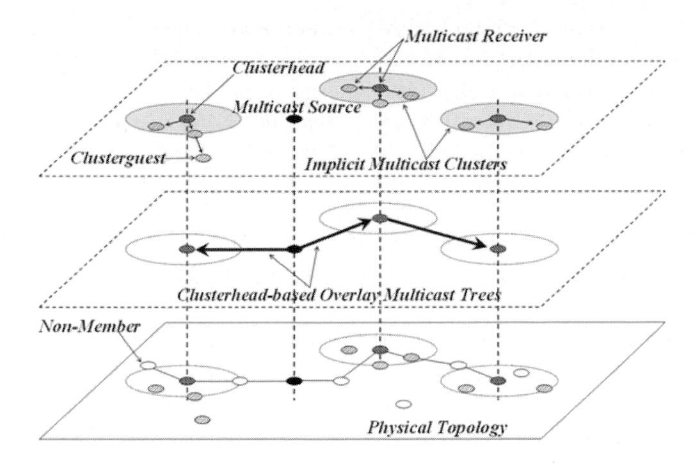

Fig. 3. Network architectures for Implicit Cluster-based Overlay Multicasting

2.2 Network Architectures

Mobile nodes joining multicast groups take the proposed protocol, ICOM-TD, on application layer. They are *Multicast Member (MM).* A MM which generates and sends multicast data is *Multicast Source (MS).* A MM which only receives the multicast data is *Multicast Receiver (MR).* Also, MRs neighboring in a transmission range are *Multicast Neighbors (MNs).* The Fig. 3 shows network architectures of ICOM-TD. Each MR belongs to a cluster, which is named *Multicast cluster.* A MR in a multicast-cluster is in charge of the other MRs as a clusterhead. Clusterheads has a function to receive data from a MS and multicast the data to its clustermembers through broadcast media. In addition, the clusterheads join overlay multicast trees (OMTs), which are created with their geographical information by MSs. The clusterhead-based OMTs are exploited as multicast paths from a MS to MRs.

2.3 Protocol Description

Operations of the ICOM-TD are divided into two procedures. They are construction of implicit multicast clusters, division of overlay multicast trees and propagation.

Construction of implicit multicast clusters. At first, MRs which want to join multicast groups send a *JoinRequest* message to a MS of the multicast group through unicasting. The JoinRequest message includes {coordinate #x, coordinate #y, node address ID} information, which is exploited for construction clusters by the MS. Whenever the MS receives the information, it adds the member information to its group member list. And then it selects clusterheads, which should have the most MNs in its transmission range to adjust the OMTs. Node density inside the transmission ranges of the selected clusterheads becomes the highest. Thus, a clusterhead and its MNs organize an implicit multicast cluster as shown in Fig. 3. After the MS constructing clusters, it sends *JoinReply* messages back to the MRs which wait to join groups. The *JoinReply* messages include {address ID of the MRs clusterhead} information, and each of the MRs recognizes its clusterhead ID after receiving the *JoinReply* message.

Division of overlay multicast trees and propagation. It is impossible that a data packet piggybacking unlimited size of multicast routing information. The unlimited size of routing information means scalable networks. The ICOM-TD supports such network scalability by using the algorithm of OMT division and propagation.

The divided OMTs construction algorithm is based on Prim's algorithm of the Minimum Cost Spanning Tree (MST) [11]. In the proposed algorithm, a constant, n, is the total number of clusterheads, and T is a set of links among clusterheads compris-ing a clusterhead-based OMT. TV is a set of clusterheads comprising a clusterhead-based OMT; however, it assumes that TV should include at least one clusterhead closest to a multicast source. Also, D(u, v) is distance between two clusterhead u and v, and r is a radio transmission range of ad hoc network nodes.

Algorithm I. *Construction of divided overlay multicast trees*

```
Begin
    T = { } ;
    TV = { ... } ; /* TV has at least one of the nearest
                   cluster-heads to multicast source */
    while ( T contains fewer than n-1 connected links )       (1)
    {
    let D(u, v) be a distance between u and v ;
    let (u, v) be the nearest to each other
        such that u TV, v TV, and r < D(u, v) < 3r ;           (2)
    if ( there is no such link to connect or                  (3)
    T has the limited number of connected links )
        break ;
    add v to TV ;                                             (4)
    add (u, v) to T ;                                         (5)
End
```

2.4 Group Communications of the ICOM-TD

Multicast Data Delivery from a MS to MRs is significantly divided into three stages. At the first stage, the MS transmits a data packet which piggybacks one of divided OMTs, to one of the clusterheads through unicasting. The divided OMTs mean overlay multicast routing information. The clusterhead which is sent the data packet to decides the next clusterhead to forward according to the divided OMT in its packet. And then it forwards the data and replies an *Acknowledgement* message to the MS. The *Acknowledgement* message includes information of clustermembers. All clusterheads do overlay routing in the same way. At the second stage, each clusterhead, which finishes forwarding the packet, broadcasts the received data to its own clustermembers. The clustermembers replies an *Acknowledgement* message to their clusterhead. Each clusterhead manages information their clustermembers via the *Acknowledgement* message. At the last stage, clustermembers which manage information of clusterguests forward the data to the clusterguests. The clusterguests replies *ForwardingAck* message to the clustermembers.

3 Performance Evaluation

Performance evaluation is established with one issue, complexity of multicast routing structures. Also, this paper compares performance of the ICOM-TD on time delay, packet overhead, and delivery ratio with existing multicast mechanisms through simulations. The simulation is established by the simulation tool, QualNet ver.3.8 [12], and parameters exploited in the simulations are given on a Table 1.

3.1 Analysis on Rates of Participation in Multicast Routing

Complexity of routing structures significantly affects performance of multicasting. The complexity depends on the number of nodes participating in multicast routing structures, and is presented by rates of participation in multicast routing, R which is rates of the number of nodes participating in multicast routing structures of total multicast group members as follows:

$$R = \frac{\#\ of\ nodes\ participating\ in\ multicast\ routing\ structures}{\#\ of\ total\ multicast\ group\ members} \quad (1)$$

Multicast routing protocols on network layer [2,3,4,5] should be loaded on not only multicast members but nonmembers. It is because both members and nonmembers should take part in multicast routing structures to forward multicast data. Hence, a rate of participation of multicasting on network layer, R_N, is given by:

$$\frac{m}{m} = 1 \leq R_N \leq \frac{m+n}{m}, for\ m \geq 1\ and\ n \geq 0, \quad (2)$$

where m is the number of multicast group members, and n is the number of non-members participating.

Table 1. Simulatioin Parameters

Simulation network space	1500m x 1000m
The total number of nodes	100 nodes
Multicast group size	10, 0, 30, 40, 50, 60, 70, 80, 90, and 100 nodes
Speed of nodes	uniform over [0, 20] m/s
Transmission range	uniform 250m
Simulation time	300 seconds
Mobility models	No mobility and Random way point

In the cases of nonhierarchical multicasting on application layer [6,7,8], only multicast group members take multicast routing protocols and join multicast routing structures. A rates of nonhierarchical multicasting on application layer, R_A, is figured out by:

$$R_A = \frac{m}{m} = 1, for\ m \geq 1 \quad (3)$$

Additionally, in the cases of hierarchical multicasting on application layer, based on clustering; for example ICOM-TD and [9], clusterheads, which are selected from their members, only participate in multicast routing structures. Its rate of participation, R_C, therefore, is given by:

$$R_C = \frac{r_{clusterhead}(m-1)+1}{m} \leq 1, for\ m \geq 1\ and\ 1 \geq r_{clusterhead} \geq 0, \quad (4)$$

where $r_{clusterhead}$ is a rate of the number of clusterheads and the number of members. A following result is derived from (2), (3), and (4) in comparison of their magnitudes:

$$R_C \leq R_A \leq R_N, for\ n \geq 1\ and\ r_{clusterhead} \neq 0, \quad (5)$$

At this point, if all clusters do not have any member ($r_{clusterhead}$=1) and there exist only multicast members in the networks ($n = 0$), magnitudes of their rates are the same as 1 ($R_N = R_A = R_C = 1$). Except the situation, hierarchical application-multicasting like the ICOM-TD has the fewest rate of participation of all. That means complexity of its multicast routing structure is also the lowest in comparison with network and nonhierarchical multicasting.

3.2 Simulation Experimental Results

The metrics of simulation for performance evaluation are time delay, packet overhead, and delivery ratio over changing the number of group members.

In addition, to evaluate performance of clustering and tree division schemes for application multicasting, this paper selects LGT[7] and NICE-MAN[9] as comparison targets. The LGT of the nonhierarchical application multicast protocols have comparatively stable operations for node mobility. On the other hand, The NICE-MAN stably supports scalability of MANETs by exploiting shared overlay trees in hierarchical application multicasting.

Simulation results for performance evaluation are analyzed with three stages. For the first stage, average end-to-end delay time between multicast sources and members in hierarchical application multicasting is a comparatively shorter than the others as shown in Fig. 4(a, d). In the LGT, the delay time increases in proportion to increment of group members, but increment rates of the delay time of ICOM-TD and NICE-MAN are not significant when the number of group members increase more than 50. The LGT should deliver multicast data to all group members via unicast relaying, while the ICOM-TD and NICE-MAN do

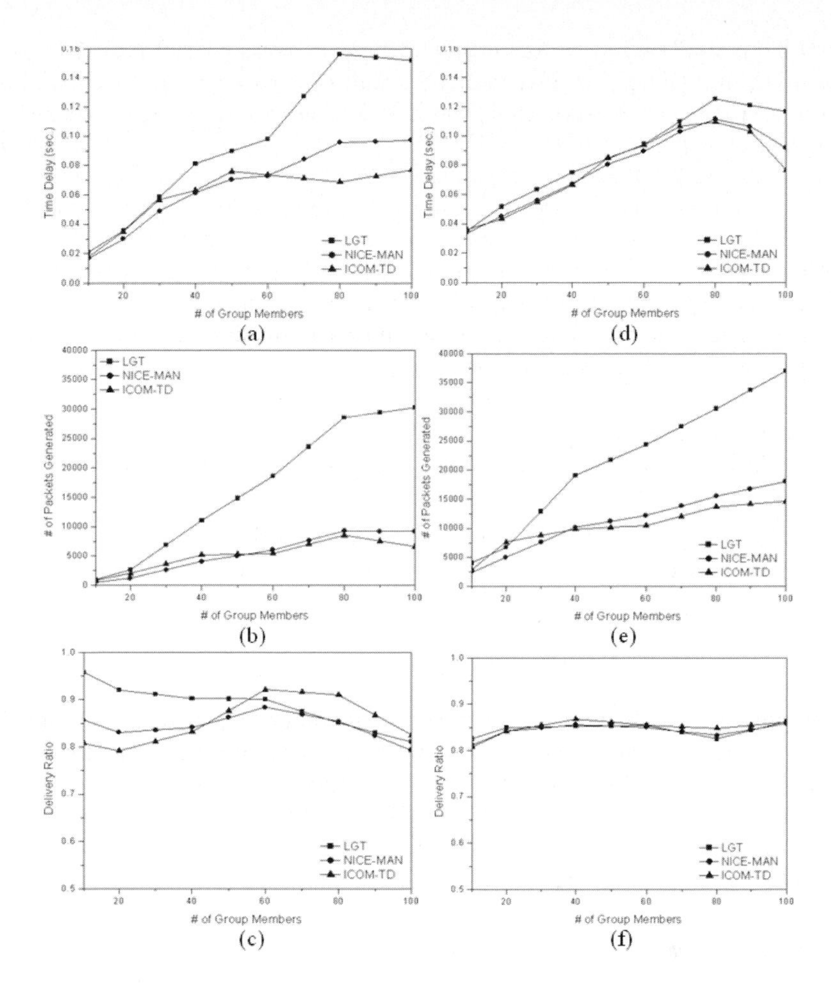

Fig. 4. Simulation results: (a) end-to-end time delay in static topology; (b) packet overhead in static topology; (c) delivery ratio in static topology; (d) end-to-end time delay in dynamic topology; (e)packet overhead in dynamic topology; (f) delivery ratio in dynamic topology

not need to deliver data to all group member via unicast relaying but only selected clusterheads, which are a part of all members. In addition, the number of clusterheads does not increase more than the threshold even though group members continuously increase. The other members except the clusterheads receive data via broadcast media; therefore, their end-to-end delay time is shorter. In Fig. 4(d), much longer delay time between 50 and 80 group members causes cluster reconstruction owing to topology change. Nonetheless, we can obtain the same simulation result as Eq. (5) in the subsection 4.1. It could be stated that The ICOM-TD and the NICE-MAN has similar results on end-to-end time delay, but performance differences between them is dependent on differences of their multicast routing structures.

For the second stage, the ICOM-TD generates lower signal and data packet overhead than the LGT and NICE-MAN. As shown in Fig. 4(b, e), the ICOM-TD generates around 51.3% signal and data packets of the LGT and around 65.4% signal and data packets of the NICE-MAN. In other words, the ICOM-TD reduces around 48.7% packet overhead of the LGT and 34.6% of the NICE-MAN. Likewise, there are the two reasons about the performance differences. First, it is periodic control messages. Second, clustering information in the ICOM-TD is piggybacked in *JoinRequest* messages and collected to a multicast source. Clusters are implicitly constructed and maintained by the multicast source. In addition, node's control information is replied to clusterheads and the multicast source with *ForwardingAck* messages. There is no explicit control message in the ICOM-TD. Second, as shown in Eq. (3), the LGT delivers data to all members via unicast relaying, but the ICOM-TD and NICE-MAN deliver data to only clusterheads via unicast relaying. The other members receive the data via broadcast media.

For the last stage, the ICOM-TD has similar performance to the LGT and NICE-MAN on delivery ratio as shown in Fig. 4(c, f). They are all application multicast routing protocols, so they exploit reliability of transport layer protocols and unicast routing protocols for data delivery. That is, if there are unicast routes from multicast sources to members, it is highly possible to deliver their multicast data; thus, they all have high delivery ratio.

Delivery ratios among them in static topology have differences while their delivery ratios are all similar in dynamic topology as Fig. 4(c, f). It is because multicast routing information can be repaired owing to node mobility but it is hard to repair routes to unreachable nodes once nodes are randomly distributed first in static topology.

4 Conclusion and Future Work

Efficient energy consumption, wireless bandwidth consumption, and network scalability are essential considerations for MANETs. This paper analyzed shortcomings of existing protocols and the ICOM-TD to improve the shortcomings. This paper proves that the ICOM-TD keeps high delivery ratio, supports network scalability via implicit clustering and tree division, and reduces inefficient energy consumption owing to high control and data packet overhead through simulation results. For doing that, we believe the ICOM-TD have better performance over improving such shortcomings of the others.

This research has two issues for future work. First, the ICOM-TD has a defeat for many-to-many multicast routing because it exploits source-specific multicast routing and clustering. On the other hand, the NICE-MAN has already considered many-to-many multicast routing through shared multicast routing tree schemes. Seocond, the algorithm to construct divided overlay multicast trees is also should be more improved. The divided overlay multicast trees must be optimized with regard to unicast routing information on network layer. It is necessary to reduce inefficient battery consumption and support network scalability.

References

1. Corson, S., Macker, J.: Mobile Ad hoc Networking (MANET): Routing Protocol Performance Issues and Evaluation Considerations. IETF RFC2501 (January 1999)
2. Wu, C.W., Tay, Y.C.: AMRIS: A Multicast Protocol for Ad hoc Wireless Networks. In: Proceedings of IEEE MILCOM, November 1999, pp. 25–29 (1999)
3. Royer, E.M., Perkins, C.E.: Multicast Ad hoc On-demand Distance Vector Routing (MAODV). IETF Internet-draft, Work in progress (July 2000)
4. Lee, S.J., Gerla, M., Chiang, C.: On-Demand Multicast Routing Protocol. In: Proceedings of IEEE WCNC, September 1999, pp. 1298–1302 (1999)
5. Ji, L., Corson, M.S.: Differential Destination Multicast-A MANET Multicast Routing Protocol for Small Groups. In: Proceedings of IEEE INFOCOM, April 2001, pp. 1192–1202 (2001)
6. Xie, J., Talpade, R., Mcauley, A., Liu, M.: AMRoute: Ad hoc Multicast Routing Protocol. In: Proceedings of Mobile Networks and Applications, vol. 7, pp. 429–439 (2002)
7. Chen, K., Nahrstedt, K.: Effective Location-Guided Tree Construction Algorithms for Small Group Multicast in MANET. In: Proceedings of IEEE INFOCOM, June 2002, pp. 1180–1189 (2002)
8. Chao, G., Mohapatra, P.: Efficient overlay multicast for mobile ad hoc networks. In: Proceedings of IEEE WCNC, March 2003, pp. 1118–1123 (2003)
9. Blodt, S.: Efficient End System Multicast for Mobile Ad Hoc Networks. In: Proceedings of IEEE PERCOMW, March 2004, pp. 75–80 (2004)
10. Yu, J.Y., Chong, P.H.J.: A Survey of Clustering Schemes for Mobile Ad Hoc Networks. IEEE Communications Surveys and Tutorials 7, 32–48 (2005)
11. Horowits, E., Sahani, S., Anderson-Freed, S.: Fundamentals of Data Structures in C, p. 289. Computer Science Press, New York (1993)
12. Scalable Network Technologies, QualNet, http://www.scalable-networks.com

An Efficient Address Assignment Mechanism for Mobile Ad-Hoc Networks*

Uhjin Joung[1], Dongkyun Kim[1,**], Nakjung Choi[2], and C.K. Toh[3,***]

[1] Department of Computer Engineering, Kyungpook National University, Korea
ujjoung@monet.knu.ac.kr, dongkyun@knu.ac.kr
[2] School of Computer Science and Engineering, Seoul National University, Korea
fomula@mmlab.snu.ac.kr
[3] EEE Department, University of Hong Kong, China

Abstract. In a mobile ad-hoc network (MANET), in order to route a packet between any nodes, nodes should have their unique IP address in the network. In our previous work, we introduced and compared three IP assignment mechanisms, namely RADA, LiA, and LiACR. In RADA, a randomly-selected IP address in a specified address space is assigned to a joining node, which results in poor utilization of the address space with a great deal of address conflict. LiA allows a joining node to be assigned to the current maximum address + 1, that is, linearly from the address space. Although LiA utilizes the address space better, it takes a long time to complete address assignments due to address conflict in case that several joining nodes require the same IP address. LiACR allows simultaneously joining nodes to have their IP addresses according to the node ID-based order. However, it relies on reliable exchange of control messages in the wireless network. Since broadcasting is inherently unreliable, we therefore propose an enhanced version of the LiACR protocol, called E-LiACR, which copes with the unreliable broadcasting through the help of neighbor nodes. Through ns-2 simulations, we show that E-LiACR performs better than LiACR in terms of IP address allocation time, number of address conflicts, and control message overhead.

1 Introduction

A mobile ad-hoc network (MANET) [1] is a wireless network consisting of mobile nodes without any infrastructure. In a MANET, as its network topology is so dynamic due to node mobility, routing packets in the network are very important, but difficult as compared to the fixed Internet. Hence, the Internet standard body like IETF (Internet Engineering Task Force) has standardized AODV [2] and OLSR [3] as reactive and proactive routing protocols, respectively.

In order to route a packet between any two nodes, they should be able to be identified uniquely in the network. Hence, for the purpose of the identification,

* This work was supported by the Korea Research Foundation Grant funded by the Korean Government (MOEHRD) (KRF-2005-003-D00295).
** Corresponding author.
*** C.K.Toh is supported by KNU BK21 Project, Daegu, Korea.

T. Vazão, M.M. Freire, and I. Chong (Eds.): ICOIN 2007, LNCS 5200, pp. 202–212, 2008.

a node is assigned its own unique IP address by considering a smooth future integration with the fixed Internet. Furthermore, an address assignment by a gateway node simplifies the solution. Since a MANET, however, is easily detachable from the fixed network, any two nodes in the detached MANET should also be able to communicate with each other and a new joining node should be able to obtain its own unique IP address in the detached network, which requires us not to rely on the gateway node. In addition, a centralized address assignment approach, where a selected server in the network is responsible for assigning a unique address to a new joining node, is not desirable due to the overhead of the server election, in the case that the server moves out of the network. Hence, many distributed IP address assignment mechanisms have been proposed. In particular, an effort to standardize so-called address auto-configuration protocol has been made for IETF Autoconf Working Group [4]. In the address auto-configuration technique, two main processes are considered: First, an efficient protocol should allow a new joining node to acquire its unique ID address in its joined network. Second, a MANET can be easily partitioned into several sub-networks, they can be merged into a MANET due to node mobility, and some different MANETs can be merged into a MANET. Since an address assignment can be performed independently of the different networks, the duplicate address detection and resolution mechanisms should be devised in the merged network.

In our previous work [5], we introduced and compared three IP assignment mechanisms, RADA, LiA and LiACR. In RADA, a random address from available IP addresses in the network is selected and assigned to a joining node. In particular, since a randomly-selected address can already be occupied by an existing node, it is possible that RADA spends much time in performing duplicate address resolution. To avoid the overhead, LiA enables an IP address of the maximum IP address + 1 to be assigned to a joining node with an additional goal of utilizing IP address space more efficiently. However, if several joining nodes try to get the same IP address simultaneously, only one node of them (called winner) will have the IP address after winning the contention and the rest of them (called losers) will continue to contend for the IP address of the new maximum IP address + 1. Such repeated procedures are needed until all contending nodes are assigned their unique IP addresses. Hence, LiACR allows the losers to select their addresses differently according to a precedence to avoid such a unnecessary repetition. However, LiACR assumes that during the contention resolution to decide a winner, control messages should be reliably broadcasted in the network. The violation of the strong assumption forces the LiACR protocol not to work properly. Since broadcasting is inherently unreliable we, therefore, propose an efficient address assignment to enhance the LiACR protocol.

In this paper, we focus on how to efficiently assign IP address to a joining node in a MANET. Since many protocols dealing with network partitioning and merging have been proposed, our protocol can adopt their approaches for the purpose. The rest of our paper is organized as follows. In Section 2, we will describe our previous work [5], namely LiA and LiACR mechanisms in detail, because our protocol is the enhanced version of the LiACR. Our protocol is presented in Section 3, which is followed by a simulation study using the ns-2 simulator in Section 4. Finally, concluding remarks are provided in Section 5.

2 Related Work

2.1 LiA Protocol

When a host is powered on or joins a MANET, it waits for a BEACON message from the network. If it does not receive the BEACON message within a certain time period, it concludes that it is the initial node of the network. Then it selects the first IP address from the usable IP address pool as its own IP address. However, if other nodes already exist, a node joining the network will receive a BEACON message containing the 'maximum IP address($Max.IP$)' used in the network during a specific period.

A new node joining the network selects its candidate IP address (i.e., $Max.IP+1$) and broadcasts the ANNOUNCE message to confirm the IP address that it will use. If the new node receives other control messages (i.e., an ANNOUNCE or a WINNER) with an identical IP address before the confirming timer expires, an address conflict is apparent. Hence, a winner and several losers are determined on the basis of the nodes' IDs, such as a MAC address. The winner with the smallest ID broadcasts the WINNER message so that other losers will re-attempt the address requisition process. In addition, the winner broadcasts its BEACON message with its new $Max.IP$.

Algorithm 1. LiACR Algorithm

1: **if** WAIT_BEACON_TIMER expires **then**
2: choose the initial address in address pool and broadcasts a beacon periodically.
3: **else if** a BEACON is received before WAIT_BEACON_TIMER expires **then**
4: sets a candidate IP address ($Max.IP+1$), broadcasts ANNOUNCE message, and starts ANNOUNCE_TIMER.
5: **if** ANNOUNCE_TIMER expires without any other control messages **then**
6: sets its local IP address using a candidate IP address and sends BEACON periodically as the initial node .
7: **else if** other ANNOUNCE message is coming with the same candidate IP address **then**
8: calculates the precedence value based on its ID.
9: **if** ANNOUNCE_TIMER expires and the nodes precedence value is 0 (the highest) **then**
10: sends WINNER and does the same process when ANNOUNCE_TIMER expires without any other control messages
11: **else if** ANNOUNCE_TIMER expires and the nodes precedence value is not 0 **then**
12: candidate IP Address += the precedence value and re-initiates the ANNOUNCE process.

2.2 LiACR Protocol

In LiA, if several joining nodes try to get the same IP address, only one node will acquire the IP address after winning and the rest of them will continue to contend for the IP address of the new maximum IP address + 1. Such repeated procedures are needed until all contending nodes are assigned their unique IP addresses.

In LiACR, therefore, a node which has sent an ANNOUNCE message waits for other ANNOUNCE messages during a pre-defined period, ANNOUNCE_TIMER. Using the IP address information collected during this period, it checks whether its ID is the smallest. If so, it becomes the winner, broadcasts a WINNER message, and then uses its candidate IP address as its local IP address. Otherwise, it is a loser and its candidate IP is increased, not by 1, but by a precedence value based on the collected node IDs. All losers retry to acquire their IP addresses by using the same procedure until they obtain their IP addresses (see Algorithm 1).

3 Our E-LiACR Protocol

3.1 Assumption and Notations

We assume that every node has a unique identifier, and that these identifiers are uniformly assigned throughout the field. For a description of our E-LiACR, we use the following notations.

- N_j = joining node.
- $\{N_j\}$ = set of joining nodes at the same time.
- N_{ADDR} = node's allocated address.
- N_{CAND} = joining node's candidate address.
- $Max.IP$ = the maximum IP address in the network.
- N_{MAX} = the node with the $Max.IP$ in the network, or the initial node in the network.
- N_{NEI} = joining node's neighbor node.
- N_{WIN} = the node with the smallest ID among the nodes which broadcasted ANNOUNCE messages (i.e., WINNER node).
- $\{N_{LOS}\}$ = other nodes with larger IDs than the WINNER node (i.e., LOSER nodes).
- N_{ID} = node ID (predetermined node value).
- RCV_{ID} = received message's node ID.
- REQ_{ADDR} = requested address included in the received message.
- P_{CNT} = the priority counter set after calculating RCV_{ID}. It is calculated with N_{ID}.
- MAX_{POOL} = maximum value in the address pool.
- $nextAddr$ = joining node's new candidate address from CANCEL message.

3.2 Control Messages and Timers

The four types of messages are the following:

- BEACON: N_{MAX} periodically broadcasts this message in order to notify other network nodes of $Max.IP$ (SEND_BEACON state). When a new joining node hears it, the node selects $Max.IP + 1$ as its candidate address.
- ANNOUNCE: N_j broadcasts this message in order to notify other network nodes of its candidate address (SEND_ANNOUNCE state). Other nodes like N_k (k != j) and N_{NEI} include their IP addresses and IDs in the message.
- WINNER: If a node which broadcasted its the ANNOUNCE message has not received any other the ANNOUNCE messages, or if it has the smallest ID among the nodes which broadcasted ANNOUNCE messages, the node broadcasts the WINNER message to notify other network nodes that it will have $Max.IP$ (SEND_WINNER state). On receiving the WINNER message, the N_{NEI} of the message checks if the IP address used by the node which sent the WINNER message (called winner node) was selected according to the ID-based order. If not, it will send the CANCEL message to the WINNER node.
- CANCEL: If a node broadcasted the BEACON or WINNER message, violating the address assignment rule according to node ID-based order, its neighbor node detects the violation and sends the CANCEL message to the node.

In addition, four timers are needed in our E-LiACR.

- WAIT_BEACON_TIMER: This timer is set as soon as N_j joins a MANET (WAIT_BEACON state) and is cancelled if a BEACON message is received from N_{MAX}.
- ANNOUNCE_TIMER: This timer set in order for a new joining node to notify its candidate address of other network nodes and to receive any other ANNOUNCE message(s) from them (SEND_ANNOUNCE state).
- SEND_BEACON_TIMER: This timer is set by a node with $Max.IP$. It is an interval that broadcasts a new BEACON message.
- WINNER_TIMER: If a node has not received any ANNOUNCE message during ANNOUNCE_TIMER, or it has the smallest ID among the nodes which broadcasted ANNOUNCE messages during the period, it will broadcast its WINNER message and wait for any CANCEL message during the WINNER_TIMER.

The operation of E-LiACR shown in Figure 1 is described according to each type of received message and the timer expiration.

3.3 E-LiACR

In order to overcome the unreliable broadcasting, E-LiACR allows the neighbor nodes of a joining node to provide the joining node with its IP address. E-LiACR operates as follows.

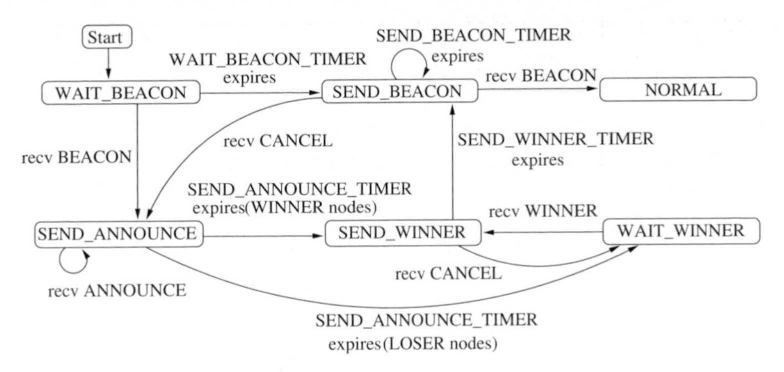

Fig. 1. State Transition

Initially, a joining node waits for a BEACON message during the WAIT_BEACON_TIMER period. If there is no BEACON message received before the timer expires, the node selects a random IP address and starts broadcasting the BEACON message with this selected address as the $Max.IP$. If its neighbor nodes, however, existed and the node failed to receive any other BEACON message due to unreliable broadcasting, they can send the CANCEL messages with the $Max.IP$ to the node in order to allow the node to cancel its selected address as well as to notify the node of the $Max.IP$ address. Thereafter, when the node becomes aware of the $Max.IP$ through the BEACON or CANCEL message, it broadcasts its ANNOUNCE message with $Max.IP + 1$ as its selected address. Through the ANNOUNCE messages from the nodes joining simultaneously, their IDs are exchanged. After an ANNOUNCE_TIMER expires, the node with the smallest ID will be assigned $Max.IP + 1$ and it will broadcast the WINNER message to notify the other nodes of the new $Max.IP$. In the case that a node broadcasts its WINNER message because it failed to know its position in the ID-based order among the joining nodes, its neighbor nodes will send the CANCEL message to the node in order to enable the node to know its candidate address, according to the ID-based order.

If a node which broadcasted its WINNER message, did not receive any other messages before a WINNER_TIMER expires, the node will start its BEACON message periodically and stop the address acquisition process. Thereafter, a node with the smallest ID which has not been assigned will broadcast its WINNER message, which is repeated until all N_j nodes are assigned addresses. Meanwhile, N_{MAX}, the node which broadcasts its BEACON message, will stop broadcasting when it receives any BEACON messages with an ID larger than its ID (see Algorithm 2).

Suppose that nodes A, B, C, D, and E join a MANET by a current $Max.IP$ is 10 and, at the same time, their node IDs are 5, 15, 17, 19, and 21, respectively. Initially, they will be in the WAIT_BEACON state where they will await any BEACON messages. LiACR spends more time on the address auto-configuration process than E-LiACR in two instances: Where a node fails to receive any BEACON messages and where an ANNOUNCE or WINNER message does not succeed in arriving at a node.

Algorithm 2. E-LiACR Algorithm

```
1: {N_j} start WAIT_BEACON_TIMER
2: P_CNT := 0
3:
4: if N_j's WAIT_BEACON_TIMER expires then
5:    N_j becomes N_MAX
6:    N_CAND := RAND(0, MAX_POOL)
7:    broadcasts BEACON & starts SEND_BEACON_TIMER
8: else if N_j's ANNOUNCE_TIMER expires then
9:    if P_CNT ≠ 0 then
10:      broadcasts WINNER & starts WINNER_TIMER
11: else if N_MAX's SEND_BEACON_TIMER expires then
12:    N_MAX sends BEACON & restart SEND_BEACON_TIMER
13: else if N_WIN's WINNER_TIMER expires then
14:    N_ADDR := N_CAND & N_MAX = N_j & starts SEND_BEACON_TIMER
15:
16: if N_j receives BEACON then
17:    N_CAND := Max.IP + 1
18:    broadcasts ANNOUNCE & starts ANNOUNCE_TIMER.
19: else if N_NEI receives BEACON then
20:    if myGroupID ≠ receivedGroupID then
21:       sends CANCEL to N_j
22: else if N_j receives ANNOUNCE then
23:    if N_CAND == REQ_ADDR then
24:       if RCV_ID ≤ N_ID then
25:          P_CNT++
26: else if {N_NEI} receive ANNOUNCE then
27:    save RCV_ID & REQ_ADDR in the table.
28: else if {N_NEI} receive WINNER then
29:    if REQ_ADDR in the message < (REQ_ADDR in the table + RCV_ID's
       precedence value based on ID) then
30:       sends CANCEL message to N_j.
31: else if {N_LOS} receive WINNER then
32:    if REQ_ADDR == N_CAND then
33:       N_CAND := N_CAND + P_CNT
34: else if N_WIN receives CANCEL then
35:    N_CAND := nextAddr & N_ADDR := NULL
```

In the first case, consider that node A has not received any BEACON messages successfully despite the existence of other nodes using $Max.IP$ in the network. Node A will think that node A is a unique node in the network, select its address and broadcast its BEACON message periodically. If the same procedure is applied to nodes B, C, and D, four independent networks will be created, which requires LiACR to perform a network merging process with a great deal of overhead. In E-LiACR, however, if node A starts broadcasting its BEACON message using its randomly selected IP address as $Max.IP$ because of the absence of other BEACON messages, neighbor nodes which receive the message will send CANCEL messages to node A. It enables node A to cancel its current IP address. The CANCEL message contains the current $Max.IP$, which allows node A to select its candidate IP address with 11 and broadcast its ANNOUNCE message, in order to obtain its new address. If other nodes from B to D do not receive any BEACON messages, each will receive the CANCEL message similarly and attempt to get its new address. This process is needed to assign IP addresses to joining nodes, not to resolve network merging. Hence, E-LiACR does not need much time to assign addresses to nodes, compared with LiACR.

In the other case, where all nodes from A to E receive a BEACON message, each node will select its candidate address with 11, broadcast the ANNOUNCE message, and wait for an ANNOUNCE or WINNER message from other nodes in the SEND_ANNOUNCE state. However, due to the absence of reliable broadcasting capability, we cannot expect that all ANNOUNCE messages will safely reach the nodes. Consider that the ANNOUNCE message sent by node A reachs nodes B and D, the ANNOUNCE message sent by node B did not reach any other nodes, the ANNOUNCE message sent by node C reached nodes D and E, and the ANNOUNCE messages sent by nodes D and E reached all other nodes. Nodes from A to E will select their candidate addresses with 11, 12, 11, 13, and 13, respectively. Figure 2 shows the candidate IP-ID mapping table for N_j at this time.

Thereafter, if nodes A and C broadcast their WINNER messages simultaneously and receive messages from each other, nodes A and C will attempt to broadcast their ANNOUNCE messages with new candidate address of 11 and 12, respectively. Hence, node A will be able to acquire its address with 11. However, node C detects the duplicate address with node B and will perform a new address contention with two remaining nodes for address 13. In the above-mentioned case, however, E-LiACR will process the same procedure until nodes from A to E select 11, 12, 11, 13, and 13, respectively for their IP addresses. If the ANNOUNCE_TIMERs expire, nodes A and C will broadcast WINNER

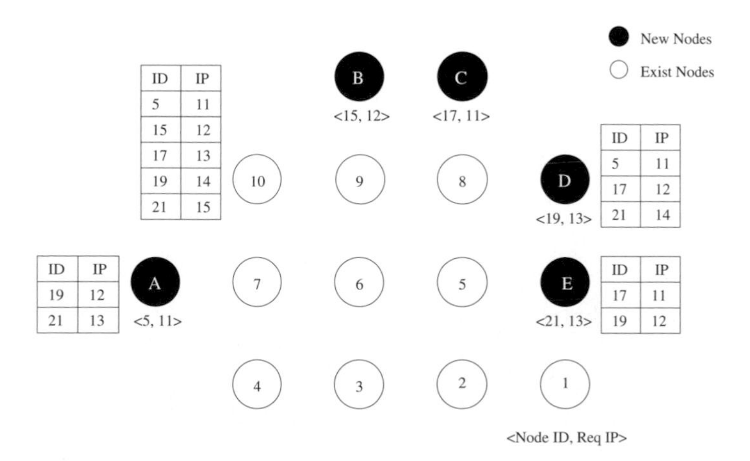

Fig. 2. E-LiACR Example

messages for the address, 11, at the same time. Node C's neighbor nodes receiving the WINNER message will become aware that node C selected the wrong address in the ID-based order. They will send a CANCEL message to node C and notify it of the correct position in ID-based order. Therefore, node C will be notified that it should select 13 as its candidate address through the CANCEL message. Next, after node B broadcasts its WINNER message saying that 12 was selected as $Max.IP$, nodes C, D and E will broadcast WINNER messages for the address, 13. Similarly, CANCEL messages will be sent to nodes D and E, which will allow them to select 14 and 15 as their candidate addresses, respectively.

As a result, E-LiACR reduces the number of address collisions because it has the number of retransmitted ANNOUNCE or WINNER messages, compared to LiACR. Therefore, E-LiACR does not need as much time as LiACR in order to complete IP address assignment.

4 Performance Evaluation

Since RADA requires a random address in an address pool to be assigned to a joining node, a large address pool is needed so that the probability of address conflicts is reduced. LiA and LiACR, however, work well even in an environment with a small address pool. In addition, LiACR performed better than LiA due to its efficient resolution of address conflicts. In this section, we compare our E-LiACR with the LiACR protocol in terms of three metrics: IP address allocation time (denoted by AAT); the number of address conflicts; and the number of control messages exchanged. AAT is the average time it takes for a joining node to obtain its IP address successfully. We implemented LiACR and E-LiACR on the NS-2 simulator [6].

Fifty nodes were initially spreaded over an area of 670 m x 670 m. Each node has a transmission range of 250 m and the propagation delay is assumed to be 200 ms.

Since we are focusing on the AAT evaluation, node mobility is not considered in this simulation. There are concerns that a node might move to other places during address allocation(considered as one of our future research issues). More simulation parameters are shown in Table 1.

In this simulation, we evaluated performances by varying the number of joining nodes during a short time period (10 x propagation delay) between 1 and 7, which is thought to be a reasonable amount in a realistic environment.

Table 1. Simulation Parameters

Parameter	Value
Total Number of Nodes	50 nodes
Simulation Area	670 m x 670 m
Simulation Time	500 seconds
MAC Layer	IEEE 802.11
Packet Size	512 bytes
Traffic Source Type	UDP

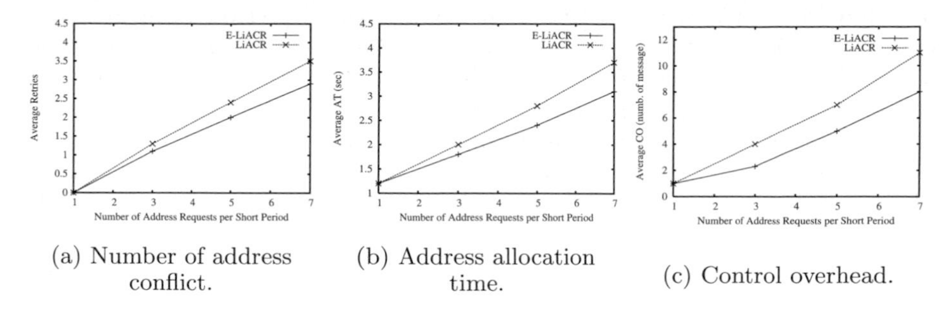

(a) Number of address conflict.

(b) Address allocation time.

(c) Control overhead.

Fig. 3. Performance Comparison

Figure 3(a) shows the average number of address conflict for LiACR and E-LiACR according to the number of nodes joining simultaneously. Accordingly, as more nodes request the same address at the same time, LiACR will have more address conflicts than E-LiACR. If the joining nodes don't receive the other nodes' ANNOUNCE messages, address conflicts will increase because more nodes request the same IP address using the WINNER message. In E-LiACR, however, joining node's neighbor nodes send the CANCEL message in order to enable the join node to know its candidate address, according to the ID-based order. This results in reducing the number of address conflicts.

Figure 3(b) shows how the AAT (Address Allocation Time) is affected by the number of joining nodes over a short period. LiACR allocates IP addresses sequentially during a beacon interval. Hence, if several nodes request IP addresses during such an interval, collisions will occur, which will cause AAT values to increase. The larger the number of address requests is, the larger the AAT will be. Some nodes which didn't receive a beacon message because of unreliable broadcasting, will select their own IP address randomly, and then they will start broadcasting BEACON messages. In addition, LiACR has an additional problem in that some nodes can't hear the other nodes' ANNOUNCE messages, which creates more address conflict and may cause some nodes to get the same IP address. In E-LiACR, however, neighbor nodes are able to detect these duplicate address requests and resolve them through their CANCEL messages. Hence, E-LiACR can reduce the AAT more than LiACR.

Figure 3(c) compares the overhead of control messages exchanged. In LiACR, ANNOUNCE messages from other joining nodes are gathered during the ANNOUNCE_TIMER and the nodes which have the smallest IDs will broadcast their WINNER messages. As mentioned before, however, some nodes may not receive other nodes' ANNOUNCE messages in LiACR. They could increase the number of contending nodes and produce collisions. In E-LiACR, however, the contending nodes can receive CANCEL messages from their neighbor nodes. It reduces the number of contending nodes with the Max.IP + 1 and also reduces the control message overhead.

Broadcasting over IEEE 802.11 is unreliable compared with unicasting because no response like ACK and no retransmissions for broadcast packets can be used.

Unfortunately, LiACR needs the broadcasting for the purpose of exchange important information in order to allocate the address. Since the unreliable broadcasting in the IEEE 802.11 does neither have collision detection nor recovery mechanism, neighbor nodes to send the CANCEL message to resolve the problems. Therefore, E-LiACR requires less time and less overhead than LiACR to complete IP address assignment (see Figure 3).

5 Conclusions

In this paper, we proposed an efficient IP address auto-configuration protocol called E-LiACR, an enhanced version of LiACR. Since LiACR relies on the reliable exchange of broadcasted control messages such as ANNOUNCE or WINNER, address conflicts occur excessively when applied to networks without reliable broadcasting protocol such as IEEE 802.11 WLANs. In case that some joining nodes, however, did not receive a message that is needed to perform address assignments successfully, such as the BEACON or ANNOUNCE message, E-LiACR allows their neighbor nodes to notify the joining nodes of their possible addresses through the CANCEL message.

Through the ns-2 simulation, we proved that regardless of the number of nodes joining a network at almost the same time, our E-LiACR performed better than LiACR. E-LiACR enabled the number of address conflicts to be reduced because it was able to reduce the number of retransmission of ANNOUNCE or WINNER messages, as compared to LiACR. Hence, E-LiACR requires lower IP address allocation time than LiACR and less overhead to complete IP address allocation.

References

1. Internet Engineering Task Force, Mobile Ad-hoc Networks (manet) Working Group Charter, `http://www.ietf.org/html.charters/manet-charter.html`
2. Perkins, C., Belding-Royer, E., Das, S.: Ad hoc On-demand Distance Vector (AODV) routing, RFC 3561, IETF (July 2003)
3. Clausen, T., Jaquet, P.: Optimized Link State Routing Protocol (OLSR), RFC 3626, IETF (October 2003)
4. Internet Engineering Task Force, Ad-hoc Network Autoconfiguration (autoconf) Working Group Charter, `http://www.ietf.org/html.charters/autoconf-charter.html`
5. Choi, N., Toh, C.-K., Seok, Y., Kim, D., Choi, Y.: Random and Linear Address Allocation for Mobile Ad Hoc Netwokrs. In: IEEE Wireless Communications and Networking Conference (WCNC 2005), New Orleans, USA (March 2005)
6. VINT Group, UCB/LBNL/VINT Network Simulator ns (version 2), `http://www.isi.edu/nsnam/ns`

Geocasting in Wireless Ad Hoc Networks with Guaranteed Delivery*

Farrukh Aslam Khan[1], Khi-Jung Ahn[2], and Wang-Cheol Song[2,**]

[1] Department of Computer Science, FAST National University of Computer and Emerging Sciences, Islamabad, Pakistan
farrukh.aslam@nu.edu.pk

[2] Department of Computer Engineering, Cheju National University, Jeju 690-756, South Korea
{kjahn,philo}@cheju.ac.kr

Abstract. In this paper, we address the problem of delivering the geocast packets to all nodes inside the geocast region in an ad hoc network, which are not directly connected to one another. We propose a geocast routing protocol that guarantees the delivery of geocast packets to all nodes inside a geocast region. In order to guarantee the delivery of packets to all nodes, we make use of the nodes outside the geocast region. We call the isolated group of nodes inside the geocast region as islands. There can be several nodes outside the geocast region that have direct connections with the islands, but we elect one node called Main Entry Point (MEP) which is responsible for delivering the packets to the nodes inside the geocast region. Our mechanism is quite efficient and guarantees the delivery of geocast packets to all nodes inside the geocast region.

1 Introduction

With the fast development and advancement of the Global Positioning System (GPS), we are now able to route packets inside a wireless network on the basis of physical locations of nodes. Several location-based unicast as well as multicast routing protocols for wireless ad hoc networks have been added into the literature during the past few years. Another concept called geocasting, which is a position-based variation of multicasting, has been seeking attention of researchers all over the world. In geocasting, a packet is supposed to be delivered to nodes inside a physical region.

Several geocasting protocols have been proposed by various researchers [4, 5, 6, 11, 12, and 13]. A detailed survey of geocasting protocols is presented in [7]. It is noted that most of the geocasting protocols at present are based on unicast routing protocols. In many cases, unicast protocols are enhanced to incorporate the geocasting features and then are transformed into a geocasting protocol. For instance, LAR [3] has been enhanced to make LBM [4], GRID has been modified to make GeoGRID [6], Geo-TORA [5] is the modified version of TORA, and AODV is modified to work for Geocasting [9]. Moreover, DSR, which a unicast protocol and ODMRP which is a multicast routing protocol, have been used as a basis for GAMER [1] which is a mesh-based geocast routing protocol.

* This work was supported by Cheju National University.

** Corresponding author.

T. Vazão, M.M. Freire, and I. Chong (Eds.): ICOIN 2007, LNCS 5200, pp. 213–222, 2008.

Mostly, geocasting protocols like LBM, Voronoi Diagram based geocasting [12], GeoGRID and GAMER, are all based on directed or limited flooding whereas Geo-TORA is a protocol without flooding. This directed flooding is carried out before the packet enters the geocast region. Inside the geocast region, all the protocols use simple flooding to deliver the packet to nodes inside the geocast region. Apart from that, some protocols like [10] and [13] use different strategies to make it possible to route the packets to all nodes inside the geocast region, even if they are positioned not in direct connection to one another. In this case, nodes outside the geocast region are also involved in order to guarantee the delivery of packets to all nodes in the region. Right hand rule traversal of nodes and face routing has been used in these algorithms. Although, these algorithms deliver packets to all nodes in a geocast region, they are quite expensive especially in terms of computation time as they are face traversal-based algorithms and therefore, spend more time in traversing faces in different manners.

We propose a geocasting protocol for wireless ad hoc networks which guarantees the delivery of packets in a geocast region. For our mechanism, we take a scenario where the geocast region is assumed to be fixed and the boundary co-ordinates of that region are known to all nodes in advance. This kind of scenario can be useful in military operations or in situations of physical disasters. Moreover, our mechanism can be applied to wireless sensor networks where a query message can be sent by a monitoring station (sink) to a group of sensor nodes in a fixed physical region. In the proposed mechanism, the source node sends packet to the geocast region by using the greedy forwarding, i.e. the packet is forwarded to that node which is physically nearest to the destination. In several geocasting protocols, when a packet arrives at the geocast region, the first node receiving the packet inside the region broadcasts it to all its neighbors, which is then flooded to all the connected nodes inside the region. The problem faced in this situation is that as in Fig.1 (a), the nodes in the upper left and right corners of the geocast region are unable to receive the flooded packets, as they are not in the radio range of any node that receives the geocast packet. Hence, even if we use simple flooding in order to deliver the packets to all the nodes, there are still certain nodes which are unable to receive geocast packets. We call these groups of nodes as islands. Since there is no direct connectivity among all nodes in the geocast region, packets cannot be delivered to all nodes through only the nodes present inside the geocast region. Nevertheless, by including some nodes from outside this region, the delivery of geocast packets can be guaranteed.

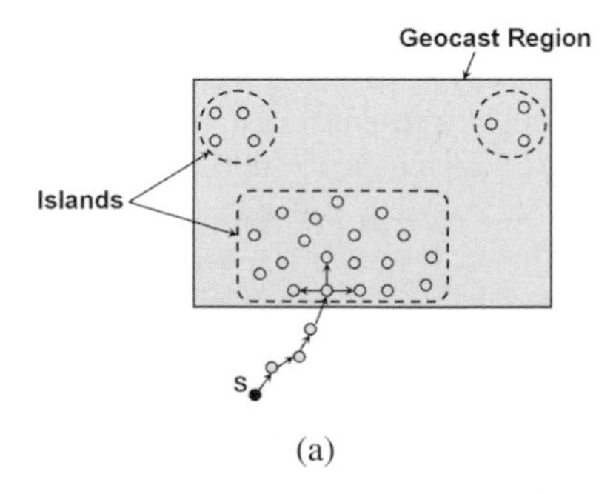

(a)

MEPs	Island	Position
M1	A	(x1,y1)
M2	A	(x2,y2)
M3	B	(x3,y3)
M4	C	(x4,y4)
⋮	⋮	⋮

(b)

Fig. 1. (a) Nodes at top right and left corners are unable to receive geocast packets (b) MEP table stored by the location server containing information about the MEPs

2 Proposed Mechanism

For our proposed mechanism having a fixed geocast region, we define a few terminologies that will be used in the following text. An *island* is a group of connected nodes that do not have direct connection with other nodes in the geocast region. Every island has a *leader* node which represents an island and it is elected in order to identify how many islands are there in the geocast region. A *gateway* is a node inside geocast region that has a direct connection with one or more nodes outside the geocast region. Furthermore, an *Entry Point* is a node which lies outside the geocast region and is directly connected with one or more gateways inside the geocast region. Out of several Entry Points, we elect one *Main Entry Point (MEP)* which is responsible for delivering geocast packets to nodes inside geocast region. MEPs are discussed in detail in section 2.2. We suppose that each node knows its own position with the help of a GPS receiver. A *location server* is a node that stores the location information of all MEPs present around the geocast region. There can be more than one location servers in the network. As mentioned earlier, a situation may arise when some nodes in the geocast region are unable to receive the geocast packet flooded by other nodes because of not being in the range of any node. In this case, we can make use of the nodes present outside the geocast region in order to deliver the geocast packet to all the nodes as they can have a path to the isolated islands in the geocast region.

2.1 Island Discovery and Leader Election

In order to know the topology information inside the geocast region and to figure out whether there are other islands in that region, we need to first elect the leader of an island. Any gateway node can be a candidate to become a leader. The leader should be that gateway node which is nearest to the boundary of the geocast region. A Gateway node that wants to become a leader sends a LEADER_ANNOUNCE packet to all reachable nodes in the geocast region. This packet contains the node-id and its position. If there is already a leader in the group, then it rejects its announcement by sending a REJECT packet. If the announcing node does not hear any other announcement from other gateway nodes, it becomes the leader and sends a LEADER_CONFIRM packet to all the nodes in the island. The leader then sends periodic LEADER_CONFIRM packets to tell other nodes about its existence. Once a node is elected as leader, the information is sent to the Main Entry Point (MEP) which then sends its own location and the leader information to the location server. This information helps the location server to decide which MEPs belong to which island and how many islands are there in total. In case of a leader failure, if a node does not hear any LEADER_CONFIRM packet from the leader for a certain predefined time, the leader election procedure is re-initiated in the same way. If there are some other islands in the geocast region, they can choose their own leader using the same procedure.

2.2 Main Entry Points (MEPs)

As we know, there can be several entry points in each island in the geocast region. We assume that the geocast region is in the form of a rectangle. In this case, each side of a geocast region should have one Main Entry Point (MEP) for each island. All entry points elect one MEP and then MEP sends its location information to the location server. The location server stores the id and location of each MEP on each side of the geocast region. The MEP table stored by the location server is shown in Fig 1(b).

Our mechanism works in the following manner: In order to send a packet to the geocast region, a source node S first sends a request to location server asking it for the location of nearest MEP outside the geocast region. The location server replies back with the location of the nearest MEP. The source then unicasts the packet to the nearest MEP based on greedy forwarding i.e. that node from the neighbors is selected for forwarding the packet which has shortest distance from the destination. The destination here is the nearest MEP of the nearest island from the source node. The location server also periodically sends the location information and id of all other MEPs to each MEP. However, location server does not send the path information. The path to other MEPs is found on-demand using any reactive routing protocol e.g., AODV or DSR. When a geocast packet arrives at one of the MEPs from the source node, it checks how many other islands are there in the geocast region by looking up the information received from the location server. If there are other islands, the MEP chooses one MEP from each island and sends packet to them. The route discovery process for finding a path for the MEPs is performed on-demand. There can be multiple MEPs belonging to one island, one on each side of the geocast region; therefore, only one MEP is selected per island based on the shortest distance from the sending MEP. Here, only MEPs are used to deliver geocast packets to the islands because we assume that MEPs have enough resources than other nodes and are more stable and have updated information about other MEPs. Moreover, there can be multiple location servers in the network and all the location servers collaborate with one another for keeping the updated information about MEPs. The procedure for routing the packet from Source S to the nodes in a geocast region is shown in the following steps:

Procedure

1. Source S contacts location server for MEP information.
2. Server replies back with the ID and position of MEP of the nearest island.
3. Source sends packet to the MEP based on greedy forwarding.
4. MEP checks MEP-Table received from location server and selects one MEP for each island based on shortest distance.
5. MEP sends packet to selected MEPs of each island using on-demand routing.
6. All MEPs then flood the packet to all nodes in their respective islands.

The procedure for electing an MEP node is as follows:

Procedure

1. An entry point that has a minimum distance from the geocast region is elected as the Main Entry Point (MEP).
2. Any entry point attached to an island in the geocast region can announce itself as the MEP. The announcement packet contains its node-id, position, the island it belongs, and the side of the geocast region.
3. As mentioned earlier, one MEP from an island is allowed on each side of the rectangular geocast region. If the island is connected with entry points from more than one side of the geocast region then each side will have one MEP. In Fig. 2 (b), the island D has two MEPs on each side of the geocast region.

4. If another entry point closer to the geocast region is present, it can reject the announcement and declares itself as an MEP.
5. If no other entry point claims to be closer to the geocast region for a predetermined period, the announcing node becomes the MEP of the island.

As shown in Fig. 2 (a), the source node S first contacts the location server to get the location information of the closest MEP. Then it sends the geocast packet to the nearest MEP based on greedy routing. When packet arrives at the MEP, it checks whether there are other islands in the geocast region by looking up the MEP table provided by the location server. It then selects one MEP from each island and sends packet to them. When packet reaches gateway node inside the geocast region, it floods the packet to all nodes inside its own island.

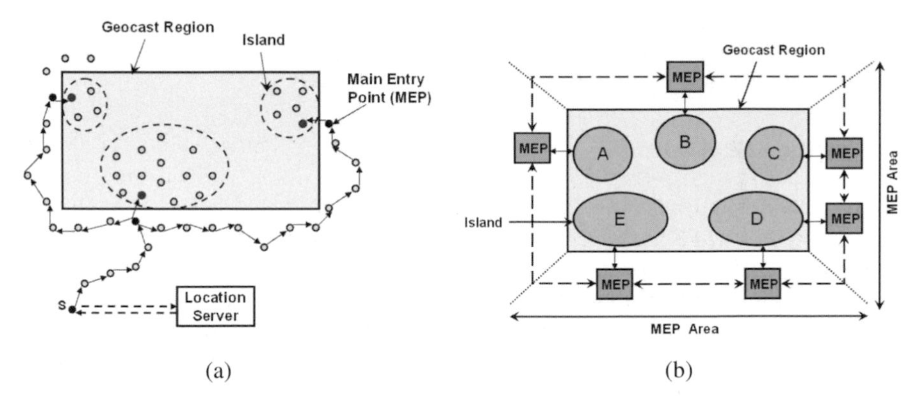

Fig. 2. (a) Working of the proposed method. Dark filled circles inside the geocast region are gateway nodes. (b) Each MEP around geocast region stores information about other MEPs. There are five islands inside geocast region i.e., A, B, C, D and E each having at least one MEP.

3 Maintenance of Geocast Region

3.1 Merging of Two Islands

When a node belonging to one island gets connected to a node in another island, the merger of two islands takes place. In this case, two islands combine to become one. Since every island has its own leader which represents the island, one leader has to voluntarily resign from serving the island as leader. As we know, every leader periodically broadcasts its presence in the form of LEADER_CONFIRM packet to all the nodes in its island. When the merger of two islands takes place, both leaders will also receive the LEADER_CONFIRM packet from each other. In the LEADER_CONFIRM packet, there is also the position of the leader and its id. The first receiving leader will compare its own position with the other leader. If it is nearer to the boundary of the geocast region, then it will reject the packet and keep on sending the LEADER_CONFIRM packet. If it is not, it will stop working as a leader. Since only gateway nodes can become leader, it will resume working as a gateway node. Similarly, other gateway nodes will also receive two conflicting leader-announcement packets. In this case, since the gateway nodes also have the position information of all other gateways as well as their leader, they will compare the new leader's position with the current leader and whoever is nearest to the boundary of geocast region, will be selected as leader. Merging process is shown in Fig. 3.

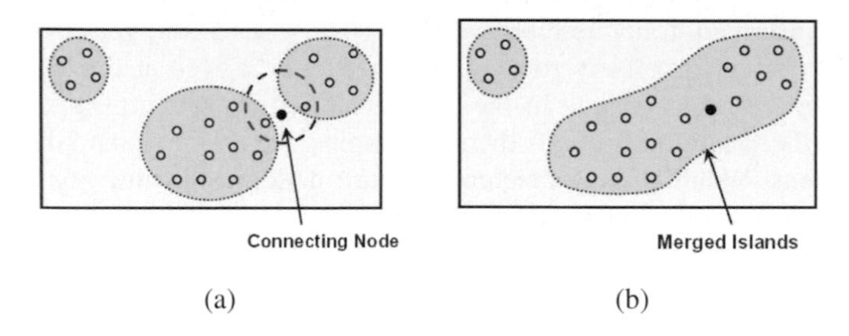

Fig. 3. Merging of two islands

3.2 Partitioning of Islands

When the connection between two or more nodes of an island is lost in such a way that it separates them into two or more groups, the island is said to be broken.

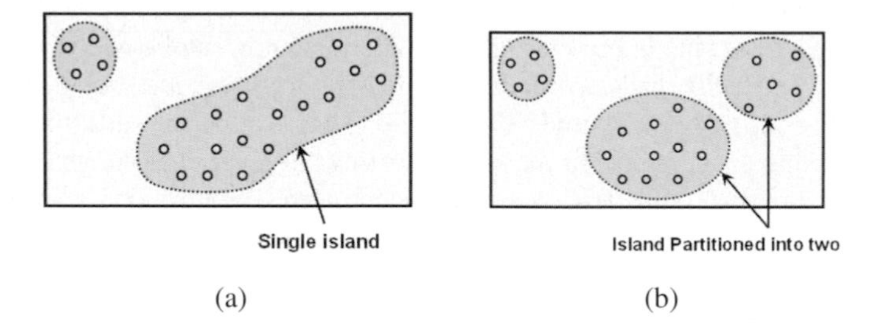

Fig. 4. Partitioning of an island into two

When an island breaks into two, one of them will have a leader whereas the other group will not have. In this case, the group which does not have a leader will elect their leader using the same leader election mechanism, as described in the previous section. Fig. 4 shows the partitioning mechanism.

4 Analysis and Discussion

In our proposed mechanism, every island can have one MEP on each side of the geocast region. As mentioned earlier, we assumed the geocast region to be in rectangular form. But for the purpose of evaluation and better generalization of our system, we consider the geocast region to be a square. We analyze the system from the very basic scenario of having one island in the geocast region to multiple islands. We discuss

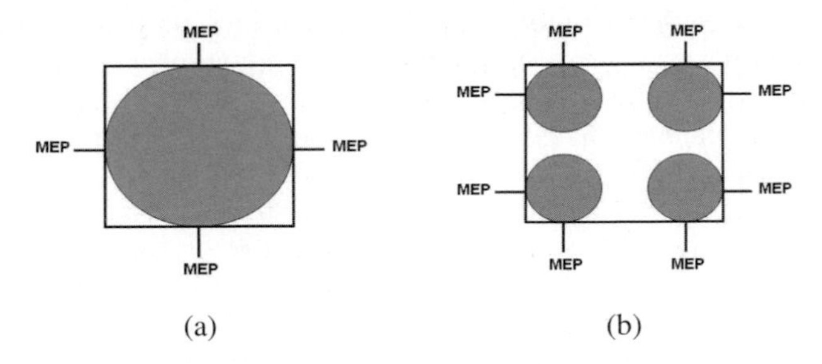

Fig. 5. (a) One island in the geocast region and each side of the region has one MEP (b) Four islands in the geocast region and each side of an island has one MEP

how our system is affected by increasing or decreasing the number of nodes, number of islands and other parameters in the geocast region. In order to analyze the system, consider the following scenarios as shown in Fig. 5.

In Fig. 5 (a), we see that there is one big island in the geocast region, which means that all nodes in the geocast region can receive geocast packets using simple flooding. Also, there are four MEPs, one on each side of the geocast region through which the geocast packets are delivered from outside the geocast region. Fig. 5 (b) shows four islands one at each corner of the geocast region each having two MEPs. If we increase the number of islands in the geocast region, the number of MEPs also increases. The maximum number of islands possible in a geocast region depends upon the size of the geocast region. If the size is big enough then more islands can be accommodated. For the purpose of generalization, we divide the islands into two main categories. One is corner-islands and the other is mid-islands. Fig. 6 shows the distribution of both corner and mid-islands. In Fig. 6, if we keep on increasing the number of islands in the geocast region, then at some point the islands will start merging with each other when one or more nodes from one island enter the radio range of another island. This situation has been analyzed by increasing the number of islands in the geocast region. As shown in Fig. 6, keeping the corner islands static and changing the number of mid-islands will affect the number of MEPs for each region.

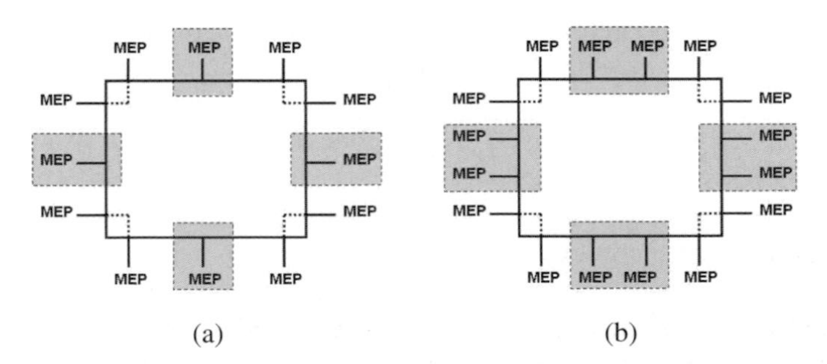

Fig. 6. (a) The shaded nodes are MEPs of single mid-island on each side of the geocast region (b) Two MEPs on each side of the geocast region representing two mid-islands

In Table 1, we see that on increasing the number of islands in the geocast region, the number of MEPs also increase. The increase in the number of MEPs is stopped at a certain point and then the number decreases with increase in the number of islands. At this point, the number of MEPs is maximum. For example, in Table 1(a), the maximum number of MEPs for 4 islands is 8. After this point, the value decreases with increase in number of islands. The reason for the decrease in number of MEPs after the maximum value is reached is that when the maximum MEP threshold is crossed, the nodes in an island start having direct connection with the nodes in other islands. When such situation arises, two or more islands start merging. This merging causes the decrease in the number of MEPs which can only be present one on each side of the geocast region. We generalize these scenarios by devising an algorithm which is shown in Fig. 7.

Table 1. (a) For one island, there are a maximum of 4 MEPs one on each side of the geocast region (b) Entries of maximum MEPs when there is one mid-island on each side of geocast region. Shaded area shows the maximum MEPs which in this case are 12.

Mid-Island = 0		
Actual Islands	No. of Islands (Iterations)	Max. MEPs
1	1	4
2	2	6
3	3	7
4	4	8
3	5	7
2	6	6
1	7	4

(a)

Mid-Island = 1		
Actual Islands	No. of Islands (Iterations)	Max. MEPs
1	1	4
2	2	6
3	3	8
4	4	9
5	5	10
6	6	11
7	7	12
8	8	12
7	9	11
6	10	10
5	11	9
4	12	8
3	13	7
2	14	6
1	15	4

(b)

5 Evaluation

Based on the analysis shown in section 4, we evaluate our system. We take several different values of the number of mid-islands and then figure out the effect on the number of Main Entry Points (MEPs) in the geocast region. Fig. 8 clearly shows that when there is only one island in the geocast region, there are a maximum of four MEPs in the region. By increasing the number of islands in the geocast region, the number of MEPs also increases until it reaches some maximum value. After that maximum threshold value, the number of actual islands starts decreasing by increasing the number of iterations until they become one island. Hence, our system performs better if there are large numbers of nodes in an island. In this case, the communication overhead decreases since for each island, the maximum numbers of MEPs are fixed, therefore, even if the numbers of nodes increase, the maximum number of MEPs would remain the same. But, if the number of islands increase, then after some threshold value, the maximum number of MEPs would decrease which means that we would have less communication overhead in terms of number of control packets generated to and from the location server.

```
Algorithm: Determining the maximum number of Main Entry Points (MEPs)
when the Mid-Islands are 0, 1, 2, 3 etc. up to some Mid_Range.

Inputs: Mid_island, Max_MEP, Max_Iterations; Output: Max_MEP

1.  For mid = 0 to Mid_Range
2.  If ( mid > 0 )
3.  Max_MEP = Max_MEP + 4;
4.  End if
5.  If ( mid > 0 )
6.  Max_Iteration=Max_Iteration + 8;
7.  End If
8.  Times=mid + 1;
9.  MEP(1) = 4;
10. For I = 2 to Max_Iterations
11. If I <= ( mid + 2 )
12. MEP(I) = MEP(I - 1) + 2;
13. Else
14. MEP(I) = MEP(I - 1) + 1;
15. End If
16. If I >= ( mid*3 + 4 ) AND ( times > 0 )
17. MEP(I) = MAX_MEP;
18. Times = Times - 1;
19. Else If I > ( mid*3 + 4 )
20. MEP(I) = MEP(I - 1) - 1;
21  End If
22  End If
23. End for
24. MEP(I) = value;
```

Fig. 7. Algorithm for determining the maximum number of Main Entry Points (MEPs) when the mid-islands are 0, 1, 2, 3 and so on up to some value called Mid_Range

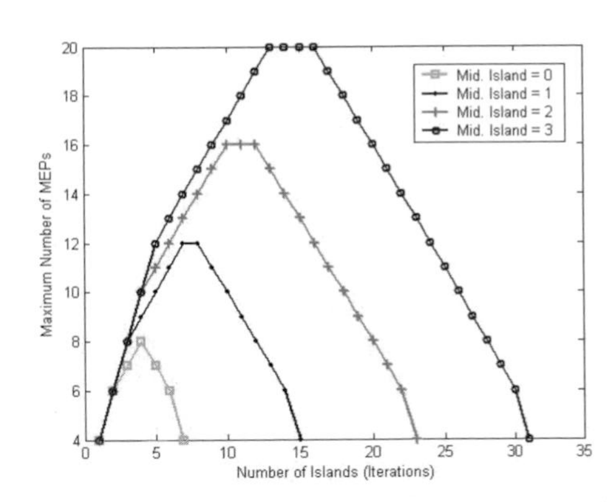

Fig. 8. Maximum number of MEPs increases by increasing the number of islands until the maximum threshold value is reached. After that, the number starts decreasing.

6 Conclusion

We have proposed a geocast routing mechanism in which we address the problem of guaranteeing the delivery of geocast packets to all nodes inside the geocast region in an ad hoc network. The nodes in the geocast region may not be connected directly to one another, so for this purpose we make use of the nodes outside the geocast region to guarantee the delivery of packets to all nodes inside the geocast region. We call the isolated group of nodes inside the geocast region as islands. There can be several nodes outside geocast region that have direct connections with nodes in the islands, but we elect one node called Main Entry Point (MEP) which is responsible for delivering packets to nodes inside the geocast region. We have shown various scenarios with examples which prove the significance of our algorithm. We also analyzed the impact of increasing the number of nodes as well as number of islands in the geocast region and conclude that using our mechanism; we can achieve less communication overhead among various MEPs and the location server. Our analysis and evaluation verify the suitability of our mechanism which guarantees the delivery of geocast packets to all nodes inside a geocast region.

References

1. Camp, T., Liu, Y.: An Adaptive Mesh-based Protocol for Geocast Routing. Journal of Parallel and Distributed Computing (2002)
2. Karp, B., Kung, H.T.: Greedy Perimeter Stateless Routing for Wireless Networks. In: Proceedings of the Sixth Annual ACM/IEEE International Conference on Mobile Computing and Networking (MobiCom 2000), Boston, MA, August 2000, pp. 243–254 (2000)
3. Ko, Y.-B., Vaidya, N.: Location-aided Routing (LAR) in Mobile Ad hoc Networks. Wireless Networks 6(4), 307–321 (2000)
4. Ko, Y.-B., Vaidya, N.: Geocasting in Mobile Ad-hoc Networks: Location-Based Multicast Algorithms. In: 2nd IEEE Workshop on Mobile Computing Systems and Applications, New Orleans, Louisiana (February 1999)
5. Ko, Y.-B., Vaidya, N.: GeoTORA: A Protocol for Geocasting in Mobile Ad Hoc Networks. In: IEEE International Conference on Network Protocols, Osaka, Japan (2000)
6. Liao, W.-H., et al.: GeoGRID: A Geocasting Protocol for Mobile Ad Hoc Networks Based on GRID. J. Internet Tech. 1(2), 23–32 (2000)
7. Maihofer, C.: A Survey of Geocast Routing Protocols. IEEE Communications Surveys & Tutorials 6(2), 32–42 (2004)
8. Mauve, M., Füßler, H., Widmer, J., Lang, T.: Position-Based Multicast Routing for Mobile Ad-Hoc Networks, University of Mannheim (2003)
9. Schwingenschlogl, C., Kosch, T.: Geocast Enhancements of AODV for Vehicular Networks. Mobile Computing and Communications Review 6(3) (July 2002)
10. Seada, K., et al.: On the Effect of Localization Errors on Geographic Face Routing in Sensor Networks. ACM IPSN (2004)
11. Seada, K., Helmy, A.: Efficient Geocasting with Perfect Delivery in Wireless Networks. In: IEEE Wireless Communications and Networking Conference WCNC (2004)
12. Stojmenovic, I., Ruhil, A.P., Lobiyal, D.K.: Voronoi Diagram and Convex Hull Based Geocasting and Routing in Wireless Networks, University of Ottawa, TR-99-11 (1999)
13. Stojmenovic, I.: Geocasting with Guaranteed Delivery in Sensor Networks. IEEE Wireless Communications 11(6) (2004)

Small-World Peer-to-Peer for Resource Discovery

Lu Liu[1], Nick Antonopoulos[2], and Stephen Mackin[3]

[1] Surrey Space Centre, University of Surrey, Surrey, U.K.
[2] Computing Department, University of Surrey, Surrey, U.K.
[3] Surrey Satellite Technology Limited, Surrey Research Park, Surrey, U.K.
{l.liu,n.antonopoulos}@surrey.ac.uk, s.mackin@sstl.co.uk

Abstract. Small-world phenomenon is potentially useful to improve the performance of resource discovery in decentralized peer-to-peer (P2P) networks. The theory of small-world networks can be adopted in the design of P2P networks: each peer node is connected to some neighbouring nodes, and a group of peer nodes keep a small number of long links to randomly chosen distant peer nodes. However, current unstructured search algorithms have difficulty distinguishing among these random long-range shortcuts and efficiently finding a set of proper long-range links located in itself or its local group for a specific resource search. This paper presents a semi-structured P2P model to efficiently create and find long-range shortcuts toward remote peer groups.

Keywords: Peer-to-peer, Small World, Information Search.

1 Introduction

Existing solutions for resource discovery over peer-to-peer (P2P) networks can be generally classified into two categories: structured and unstructured P2P systems. Distributed hash tables (DHTs) have become the dominant methodology for resource discovery in structured P2P networks [1]. Some current studies (e.g. [2, 3]) argued that the cost of maintaining a consistent distributed index is too high in the dynamic and unpredictable Internet environment. Some structured P2P protocols (e.g.[4, 5]) are beginning to seek ways to save the cost of maintaining a consistent index. In contrast, unstructured P2P systems (e.g. Gnutella) are more resilient in dynamic environments, but current unstructured P2P search techniques tend to either require high search overhead or generate massive network traffic.

Due to the similarity between P2P networks and social networks, where peer nodes are people and connections are relationships, social science theories can be potentially useful for improving the performance of object discovery over P2P networks. The small world phenomenon, postulated by Stanley Milgram in 1967, is the hypothesis that everyone in the world can be reached through a short chain of social acquaintances [6]. This phenomenon has also been observed in existing P2P networks (e.g. Gnutella, Freenet), which has proved useful in the design of P2P file-sharing systems on the Internet [7]. Duncan Watts proposed a mathematical model [8] to analyze the small world phenomenon with highly clustered sub-networks consisting of local nodes and random long-range shortcuts that help produce short paths to

T. Vazão, M.M. Freire, and I. Chong (Eds.): ICOIN 2007, LNCS 5200, pp. 223–233, 2008.

remote nodes. Duncan demonstrated that the path-length between any two nodes of his model graph is surprisingly small. This theory can be adopted in P2P networks: each peer node is connected to some neighbouring nodes, and a group of peer nodes keep a small number of long links to randomly chosen distant peer nodes. Jon Kleinberg discussed the problem of decentralized search in P2P networks with partial information about the underlying structure in [9]. However, current unstructured search algorithms have difficulty distinguishing among these random long-range shortcuts and efficiently finding a set of proper long-range links located in itself or its local group for a specific resource search. For this reason, the study [10] raised the open question about how to form and maintain inter-cluster connections and how to let nodes know which local nodes have external connections.

To address these problems, we present Small World Architecture for peer-to-peer Networks (SWAN) by combining techniques of both structured and unstructured search methods. The semi-structured P2P algorithm of SWAN is used to create and discover long-range shortcuts between different peer groups, which does not strictly rely on DHTs. It can still find the requested data inside and outside of peer groups with a high probability even though hash functions can not provide accurate information of data locations.

2 Related Work

Most studies of constructing small world behaviours on P2P are based on the group structure by clustering peer nodes into groups, communities, or clusters [11, 12, 13, 14, 15, 16]. PlantP is a content addressable publish/subscribe service for unstructured P2P, which uses gossiping to build content-addressable communities [11]. A study in [16] proposes an enhanced clustering cache replacement scheme for Freenet by forcing the routing tables to resemble neighbour relationships in a small-world acquaintance graph. Semantic Small World in [17] facilitates efficient semantic-based search in P2P systems where peers are clustered according to the semantics of their local data and self-organized as a small world overlay network. Despite the fact that unstructured P2P are more resilient in dynamic environments, the efficiency of these unstructured P2P approaches is still far lower than DHTs. Some hybrid P2P search methods (e.g. [18, 19]) are attempting to use combined techniques with both structured and unstructured search methods. However, it encounters the performance bottleneck of centralized super-peers and maintenance of the distributed index is disordered and redundant. In contrast, our model is built upon a flat P2P overlay network without the limit of super-peers, which is efficient and fault-tolerant for content discovery inside or outside of peer group.

3 Algorithm Descriptions

SWAN is built with the same group structures as Jon Kleinberg's model [19]. Studies like [20, 21, 22] have presented the methodologies of building an information sharing system by bootstrapping and grouping peer nodes that will not be discussed in this paper. This paper will focus on data publishing and searching algorithms of SWAN with generic group structures. By using a compact representation mechanism (e.g. Bloom Filters [23]), each peer node maintains an inconsistent list about members in the same group and regards other members as "acquaintances." A group of peer nodes keep a small number of long links to distant peer nodes. A simple example of SWAN topology is illustrated in Figure 1 that will be analysed in details in section 4 and 5. A semi-structured approach is presented in this section to create long-range links between groups as well as discover the local peer nodes that have specific external connections, which can satisfy the following requirements of design:

(1) Not every peer node needs to connect to other peer groups;
(2) Each peer node needs to know or can easily find which nodes have external connections to which peer groups;
(3) External links to other peer groups need to be distributed within the peer group and cannot be centralized in one or a few peer nodes.

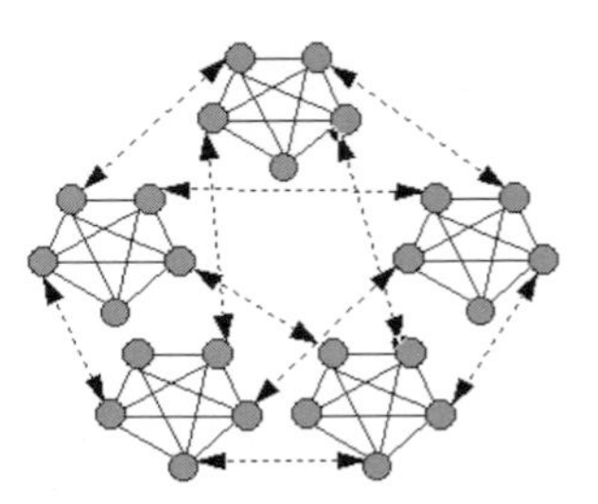

Fig. 1. Topology of SWAN

3.1 Intra-group Content Searching

Each shared file in SWAN is published by an associated content advertisement that provides the relevant meta-information of the file (e.g. name, node address, description), which is pushed to a target peer node according to the hash value of the name of file, as well as internal neighbours of the target peer node in the member list within a specific distance *d* to increase probability of discovery of the advertisement. The advertisement searching process involves two steps: a structured P2P search followed by an unstructured P2P search. The query originator firstly searches the target peer nodes generated from the same hash function (structured P2P search). If the requested advertisement cannot be found in the target peer node (e.g. the target peer node is offline at the moment), the query originator will continue to search the neighbours of the target peer node in the member list within distance *d* (unstructured P2P search). Figure 2 illustrates an example of content advertisement publishing and searching. *P1* shares a file with the name *K1*. The publication service on peer *P1* pushes the associated advertisement of *K1* to *P4* according to the hash value of *K1* and the neighbours of *P4* (*P3* and *P5*) within the distance *d=1*. Then other peer nodes in the same peer group can easily find the advertisement in a high probability. In this case, *P6* looks for the advertisement by generating the same hash value pointing to *P4* with the same hash function and sends a query to *P4* and find the advertisement with *K1* in *P4*. However, if the requested advertisement cannot be found in *P4*, the query originator will continue to search *P3* and *P5* that are neighbours of *P4* within distance $d = 1$. Publication and searching parameter *d* is defined based on users' requirements and present rate of peer nodes. Generally, a bigger *d* is required in a dynamic network with a lower peer present rate. The corresponding analysis and simulation results are shown in section 5 and 6.

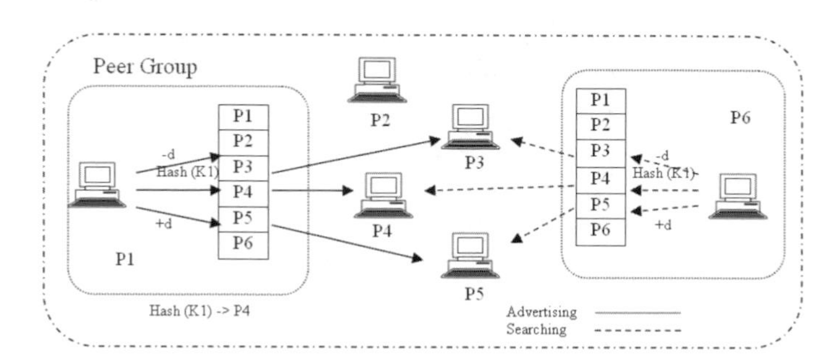

Fig. 2. Advertisement publishing and searching

3.2 Inter-group Content Searching

In SWAN, a new peer group is advertised by a group advertisement that provides the relevant meta-information about the peer group (e.g. ID, name, contact points, description), which will be multicast through the network. Not all the peer nodes in the network will receive the advertisement, but a large percentage of them will. When a peer node receives a peer group advertisement, it will push the advertisement to a target peer node in the peer group according to the hash value of the name of the peer group as well as the neighbours of the target peer node within a specific distance *d* to increase probability of discovery of the advertisement. Similar to the intra-group content searching, if the query originator cannot find the requested advertisement with the uniform hash function due to network churns, the requested advertisement will still be found in the neighbours of the target peer node with a high probability by using unstructured P2P searching. Therefore, even though only one peer node is informed, all the peer nodes in the same peer group potentially can find and pull the peer group advertisement. Figure 3 illustrates the process of inter-group link formation. When *P1* receives a group advertisement about group *G2* with contact point *P'3*, it will push the advertisement to the target peer node *P4* according to hash function as well as its neighbours (*P3* and *P5*) within distance $d = 1$. Then *P4* will inform a contact point of group *G2: P'3* with the advertisement of its peer group *G1*. When *P'3* gets the advertisement of *G1, P'3* will do the same as *P1* to forward the advertisement of *G1* toward the target peer node *P'5* according to hash function as well as its neighbours within distance *d*. When *P'5* receives the advertisement, *P'5* will do the same as *P4* to send the advertisement of its group *G2* back to *P4*. When *P4* receives it and sends the acknowledgement of inter-group link back to *P'5*, an inter-group link will be built between *G1* and *G2* and be maintained by *P4* and *P'5*. In the same way, more inter-group links will be created and maintained between *P3, P4, P5* and *P'4, P'5, P'6,* in case of $d = 1$. Each of them normally keeps 3 (*2d+1*) inter-group links as illustrated in Figure 5(a) which makes groups connected even in a highly dynamic environment. Inter-group search queries can be propagated toward the requested peer group efficiently via inter-group links and relevant shared files can be found with a high probability. The methods of resolving the pair of keyword of clusters and value of shared files have been discussed in [12, 14] and will not be described in detail in this paper.

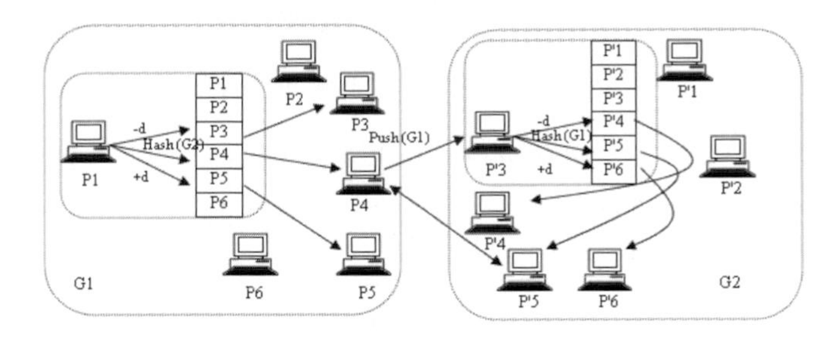

Fig. 3. Inter-Peer group link formation

4 Clustering Coefficient

In the Duncan's model [8], a small world network is a kind of network with a high clustering coefficient of nodes and short average path length. The clustering coefficient of a node is the proportion of the links between nodes within its neighbourhood divided by the number of links that could possibly exist between them. In this section, the clustering coefficient of SWAN is analyzed in a static environment. Publication and searching parameter d is set $d = 0$ in the static environment of peer present rate $p = 100\%$. There are n peer nodes in the network, each peer node has k neighbours in each peer group, and peer groups do not overlap and are connected by inter-group links as shown in Figure 1. Therefore, there are a total of $k+1$ peer nodes in each group and a total of $g = n/_{k+1}$ groups in the network.

If a peer node has i inter-group links, it has $k+i$ "neighbours" in the network (k internal neighbours and i external neighbours) as shown in Figure 4(a). Therefore, the possible links between its neighbours are $\frac{(k+i)(k+i-1)}{2}$. But in a static environment with $d = 0$, i external neighbours do not keep inter-group links to k internal neighbours. Moreover, in a large-scale network, the probability that two external

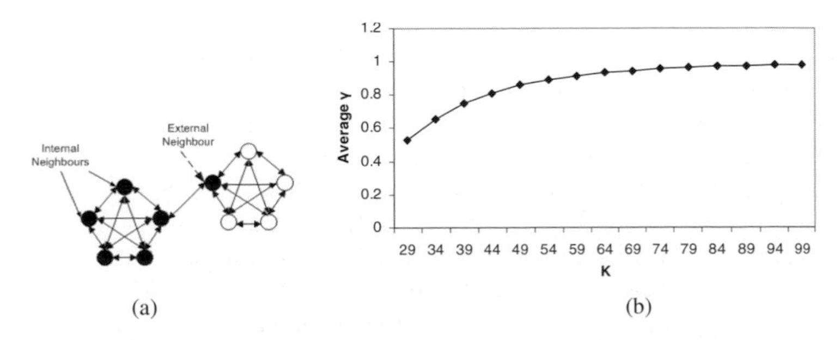

Fig. 4. (a) Neighbourhood of a peer node. (b) Clustering coefficient of SWAN.

neighbours are connected to each other by an inter-group link is very low (≈ 0). Therefore, the actual links in the neighbourhood of a peer node are the links among its k internal neighbours: $\frac{k(k-1)}{2}$. So the clustering coefficient of the peer node with i external neighbours is: $\gamma_i \approx \frac{2}{(k+i)(k+i-1)}\left(\frac{k(k-1)}{2}\right) = \frac{k(k-1)}{(k+i)(k+i-1)}$. The weighted average is $\bar{\gamma} = \sum_{i=0}^{l} p_i \gamma_i = \sum_{i=0}^{l} p_i \frac{k(k-1)}{(k+i)(k+i-1)}$, where p_i is the probability that a peer node keeps i external links and l is the maximum of inter-group links of a peer node. The distribution of inter-peer links can be regarded as a Poisson distribution: $p_i \approx \frac{\lambda^i \cdot e^{-\lambda}}{i!}$ where $\lambda = \frac{g-1}{k+1}$. So $\bar{\gamma} = \sum_{i=0}^{l} p_i \frac{k(k-1)}{(k+i)(k+i-1)} \approx \sum_{i=0}^{l} \frac{\left(\frac{n-k-1}{(k+1)^2}\right)^i}{i!} \cdot e^{-\frac{n-k-1}{(k+1)^2}} \cdot \frac{k(k-1)}{(k+i)(k+i-1)}$. Figure 4(b) shows the clustering coefficient in the networks with 10,000 peer nodes. The observed clustering coefficients are in a range of large values.

5 Performance Evaluation

We evaluated the effectiveness of SWAN in dynamic P2P environments with frequent peer nodes temporarily online and offline. In this section, we assume that the requested advertisements have been published successfully to the target peer node as well as its neighbours within a distance *d*. The success rate and the average number of messages per query will be evaluated with the present rate of peer nodes *p*.

5.1 Intra-group Search

A search for a content advertisement within a peer group will fail if the target peer node and its neighbours within *d* distance are all offline. Because advertisements are distributed, the query originator can possibly find the requested content advertisement in itself as well as in the other members. The probability of finding a requested advertisement in itself is $P(A) = \frac{2d+1}{k+1}$. The probability of finding an advertisement in other members is $P(B \mid \bar{A}) = \sum_{i=1}^{2d+1} p \cdot (1-p)^{i-1}$ and it requires i messages. The probability of failing to find an advertisement on other members is $P(\bar{B} \mid \bar{A}) = (1-p)^{2d+1}$ and it generates $(2d+1)$ messages. Therefore the success rate of finding a requested content advertisement within peer group is:

$$P_{\text{intra}} = 1 - P(\bar{A})P(\bar{B} \mid \bar{A}) = 1-(1-\frac{2d+1}{k+1}) \cdot (1-p)^{2d+1} = 1 - \frac{k-2d}{k+1} \cdot (1-p)^{2d+1}. \quad (1)$$

The average number of messages N_{intra} is calculated as follows:

$$N_{\text{intra}} = P(A) \cdot 0 + P(\bar{A})[P(B \mid \bar{A}) \cdot i + P(\bar{B} \mid \bar{A})(2d+1)] = \left(\frac{k-2d}{k+1}\right)\left\{\sum_{i=1}^{2d+1}\left[p(1-p)^{i-1} \cdot i\right] + (1-p)^{2d+1}(2d+1)\right\} \quad (2)$$

5.2 Inter-group Search

In SWAN, three conditions must be satisfied to find an advertisement in a different group as shown in Figure 5(a):

C = “succeed in finding an advertisement about the requested group”
D = “succeed in contacting the requested group”
E = “succeed in finding a requested content advertisement in the requested group”

The probability of failing in finding an advertisement about the requested peer group is: $P(\overline{C}) = \frac{k-2d}{k+1}(1-p)^{2d+1}$. As described in the section 3, the local peer group keeps inter-group links toward *2d+1* peer nodes in a remote peer group. We will fail to contact the request peer group, if all *2d+1* peer nodes are all offline. Therefore, the probability of failing in contacting the requested group is: $P(\overline{D} \mid C) = (1-p)^{2d+1}$. The probability of finding an advertisement about the requested peer group is $P(E \mid DC) = P(C)$. The success rate of finding a content advertisement in the requested peer group is:

$$P_s = P(EDC) = P(E \mid DC) \cdot P(D \mid C) \cdot P(C) = \left[1 - \frac{k-2d}{k+1} \cdot (1-p)^{2d+1}\right]^2 \left[1 - (1-p)^{2d+1}\right]. \quad (3)$$

If d is defined as a small value, the success rate is also very low in the network with a low present rate that is the situation needs to be avoided in practice. Figure 5(b) shows the minimal values of d to achieve different satisfactory success rates with different present rates of peer nodes.

The expected number of messages is:

$$\begin{aligned}
\mathrm{E}[\mathrm{N}] = {} & E[N(ADE)] + E[N(AD\overline{E})] + E[N(A\overline{D})] + \mathrm{E}[\mathrm{N}(\overline{A}BDE)] + E[N(\overline{A}BD\overline{E})] + E[N(\overline{A}B\overline{D})] + E[N(\overline{A}\overline{B})] = \\
& \left(1 - \frac{2d+1}{k+1}\right)\left\{\sum_{i=1}^{2d+1}\left\{p \cdot (1-p)^{i-1} \cdot \sum_{j=1}^{2d+1}\left[p \cdot (1-p)^{j-1}\left(\frac{2d+1}{k+1} \cdot (j+i) + \left(1 - \frac{2d+1}{k+1}\right)\sum_{m=1}^{2d+1} p \cdot (1-p)^{m-1} \cdot (m+j+i)\right)\right\} + \right. \\
& \sum_{i=1}^{2d+1}\left[p \cdot (1-p)^{i-1} \sum_{j=1}^{2d+1} p \cdot (1-p)^{j-1} \cdot \left(1 - \frac{2d+1}{k+1}\right) \cdot (1-p)^{2d+1} \cdot (2d+1+j+i)\right] + \sum_{i=1}^{2d+1} p \cdot (1-p)^{i-1}(1-p)^{2d+1} \cdot (2d+1+i) + \\
& (1-p)^{2d+1} \cdot (2d+1)\Big\} + \frac{2d+1}{k+1}\left\{\sum_{j=1}^{2d+1} p \cdot (1-p)^{j-1}\left[\frac{2d+1}{k+1} \cdot j + \left(1 - \frac{2d+1}{k+1}\right) \cdot \sum_{m=1}^{2d+1} p \cdot (1-p)^{m-1} \cdot (m+j)\right] + \right. \\
& \left. \sum_{j=1}^{2d+1} p \cdot (1-p)^{j-1} \cdot \left(1 - \frac{2d+1}{k+1}\right) \cdot (1-p)^{2d+1} \cdot (2d+1+j) + (1-p)^{2d+1} \cdot (2d+1)\right\}. \quad (4)
\end{aligned}$$

6 Simulation Results

6.1 Simulation in Dynamic Environments

We further evaluated the performance of SWAN by simulations in dynamic environments. In the simulations, we followed the same assumption as theoretical analysis that the requested advertisements had been published successfully to the target peer node as well as its neighbours within a distance *d*. Therefore, if either the

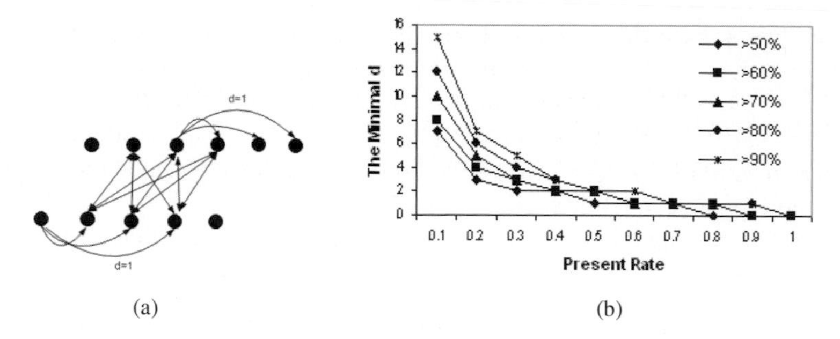

Fig. 5. (a) P2P searching in dynamic environments. (b)The minimal publishing distance for different required success rates.

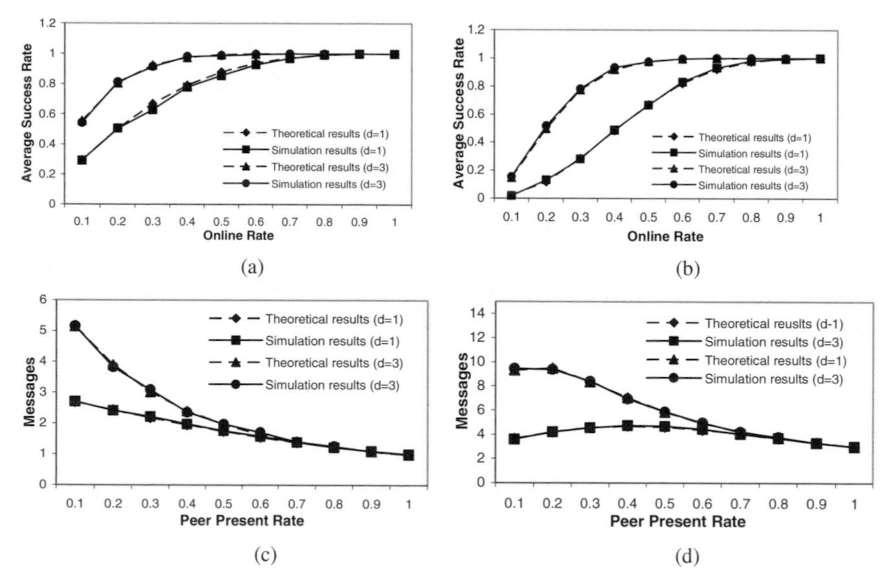

Fig. 6. (a) Success rate in intra-group search. (b) Success rate in inter-group search. (c) Average messages in intra-group search. (d) Average messages in inter-group search.

target peer node or one of its neighbours within distance *d* was visited successfully, this search succeeded. In the simulation, each peer group kept 500 peer nodes, all peer nodes in the same peer group were completely connected to each other, and all peer nodes were initialized as online in the network. At the beginning of each search, a set of peer nodes were randomly selected and set as offline according to the parameter of present rate of peer nodes, the query originator was randomly selected from the set of online peer nodes and the targeted peer node with its neighbours were randomly selected from the set of peer nodes regardless of their present situation. For each data search, the query originator initials a query that will be passed with the SWAN protocols.

In the simulations of intra-group search, the query originator and the target peer node were allocated in the same peer group. On the contrary, the query originator and the targeted peer node were separated into different groups in the simulation of inter-group search. Figure 6 (a)–(d) show the results of success rate and average number of messages per query in the intra-group search and inter-group search respectively (for 1000 queries), in which the theoretical results were generated from Equations (1)–(4). As shown in Figure 6 (a)–(d), the results of success rates from simulation results are very close to the theoretical results.

6.2 Performance Comparison

Unstructured P2P searching protocols are supposed to be more resilient in highly dynamic P2P environments. In this section, we compare the success rates of finding a shared file and traffic cost of SWAN to those of a Gnutella-like network in dynamic P2P environments. We simulated a SWAN network and a Gnutella-like network with blind flooding search. We assume the same number of neighbours (50 neighbours), the present rates of peer nodes are from 10% to 50%. Figure 7 (a) shows that the success rate of SWAN with $d = 5$ is much higher than that of the Gnutella-like network. Therefore, the performance and efficiency of the semi-structured network SWAN are much better than the unstructured Gnutella-like network in dynamic P2P environments due to hash functions directing search. The results in Figure 7 (b) show that the traffic cost of the Gnutella-like network increases super-linearly to a huge value as the present rate of peer nodes increases. However, the traffic cost is significantly reduced and remains stable in the dynamic environments by using SWAN because queries in SWAN are directed to relevant peer groups and relevant members of that peer group only.

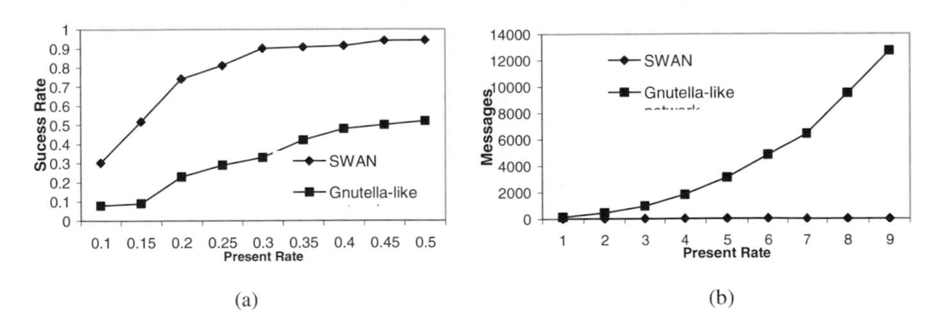

(a) (b)

Fig. 7. Performance comparison. (a) Success rates. (b) Traffic cost per query.

7 Conclusion

Small world phenomenon is a well-known hypothesis that greatly influences social and biological sciences. Due to the similarity between P2P networks and social networks, small-world phenomenon is useful for improving P2P resource search by building an artificial small-world environment. This paper presented small-world architecture for resource discovery in P2P networks. In SWAN, each node is connected to neighbouring nodes in the same peer group and peer groups are connected by a small number of inter-group links that can also be seen as long links to distant nodes in the network. Not every peer node needs to be connected to remote groups, but every peer node can easily find which peer nodes have external connections to a specific peer group in SWAN. A semi-structured P2P search method is introduced by combining techniques of both structured and unstructured search methods, which can find the requested data with a high probability even though hash functions can not provide accurate information of data locations. From our analysis and simulations, SWAN potentially has advantages of both structured and unstructured P2P networks and achieves good performance in dynamic environments with a high clustering coefficient and a short average path length.

References

1. Antonopoulos, N., Salter, J.: Efficient Resource Discovery in Grids and P2P Networks. Internet Research 14(5), 339–346 (2004)
2. Yang, B., Garcia-Molina, H.: Efficient Search in Peer-to-Peer networks. In: Proc. of International Conference on Distributed Computing Systems, Vienna, Austria (2002)
3. Rhea, S., Gells, D., Roscoe, T., Kubiatowicz, J.: Handling Churn in a DHT. In: Proc. the USENIX Annual Technical Conference, Boston, MA, USA (2004)
4. Maymounkov, P., Mazieres, D.: Kademlia: A Peer-to-Peer information system based on the XOR Metric. In: Proc. of International Workshop on Peer-to-Peer Systems (IPTPS), Berkeley, MA, USA (2002)
5. Castro, D., Costa, M., Rowstron, A.: Debunking some myths about structured and unstructured overlays. In: Proc. of the Symposium on Networked Systems Design and Implementation, Boston, MA, USA (2005)
6. Milgram, S.: The Small World Problem. Psychology Today, 60–67 (1967)
7. Hong, T.: Chapter Fourteen: Performance, Peer-to-Peer: Harnessing the Power of Disruptive Technologies, pp. 203–241. O'Reilly, Sebastopol (2001)
8. Watts, D., Strogatz, S.: Collective Dynamics of Small-World Networks. Nature 393, 440–442 (1998)
9. Kleinberg, J.: Navigation in a Small World. Nature 406, 845 (2000)
10. Iamnitchi, A., Ripeanu, M., Foster, I.: Locating Data in Peer-to-Peer Scientific Collaborations. In: Proc. of International Workshop on Peer-to-Peer Systems, Berkeley, MA, USA (2002)
11. Cuenca-acuna, F.M., et al.: PlanetP: Using Gossiping to Build Content Addressable Peer-to-Peer information Sharing Communities. In: Proc. of High Performance Distributed Computing, Seattle, Washington, USA (2003)
12. Hui, K.Y.K., et al.: Small World Overlay P2P Networks. In: Proc. of International Workshop on Quality of Service, Montreal, Canada (2004)
13. Triantafillou, P.: PLANES: The Next Step in Peer-to-Peer Network Architectures. In: Proc. of Workshop on Future Directions in Network Architectures Karlsruhe, Germany (2003)
14. Antonopoulos, N., Salter, J.: Improving Query Routing Efficiency in Peer-to-Peer Networks, University of Surrey Computing Sciences Report, CS-04-01 (2004)
15. Kleinberg, J.: Small-World Phenomena and the Dynamics of Information. In: Proc. of Advances in Neural Information Processing Systems, Vancouver, Canada (2001)
16. Zhang, H., Goel, A., Govindan, R.: Using the Small-World Model to Improve Freenet Performance. Computer Networks 46(4), 555–574 (2004)

17. Li, M., Lee, W., Sivasubramaniam, A.: Semantic Small World: An Overlay Network for Peer-to-Peer Search. In: Proc. of the International Conference on Network Protocols, Berlin, Germany (2004)
18. Loo, B.T., Huebsch, R., Stoica, I., Hellerstein, J.M.: The Case for a Hybrid P2P Search Infrastructure. In: Voelker, G.M., Shenker, S. (eds.) IPTPS 2004. LNCS, vol. 3279, pp. 141–150. Springer, Heidelberg (2005)
19. Traversat, B., Abdelaziz, M., Pouyoul, E.: Project JXTA: A Loosely-Consistent DHT Rendezvous Walker, Technical Report, Sun Microsystems, Inc. (2003)
20. Cuenca-Acuna, F.M., Nguyen, T.D.: Text-based Content Search and Retrieval in ad hoc P2P Communities. In: Gregori, E., Cherkasova, L., Cugola, G., Panzieri, F., Picco, G.P. (eds.) NETWORKING 2002. LNCS, vol. 2376, pp. 220–234. Springer, Heidelberg (2002)
21. Khambatti, M., Ryu, K.D., Dasgupta, P.: Structuring Peer-to-Peer Networks using Interest-Based Communities. In: Aberer, K., Koubarakis, M., Kalogeraki, V. (eds.) VLDB 2003. LNCS, vol. 2944, pp. 48–63. Springer, Heidelberg (2004)
22. Vassileva, J.: Motivating Participation in Peer-to-Peer Communities. In: Petta, P., Tolksdorf, R., Zambonelli, F. (eds.) ESAW 2002. LNCS, vol. 2577, pp. 18–23. Springer, Heidelberg (2003)
23. Bloom, B.: Space/time Trade-offs in Hash Coding with Allowable Errors. Communication of ACM 13(7), 422–426 (1970)

Proximity Based Peer-to-Peer Overlay Networks (P3ON) with Load Distribution

Kunwoo Park[1], Sangheon Pack[2], and Taekyoung Kwon[1]

[1] School of Computer Engineering, Seoul National University, Seoul, Korea
kwpark@mmlab.snu.ac.kr, tkkwon@snu.ac.kr
[2] School of Electrical Engineering, Korea University, Seoul, Korea
shpack@korea.ac.kr

Abstract. Construction of overlay networks without any consideration of real network topologies causes inefficient routing in peer-to-peer networks. This paper presents the design and evaluation of a *proximity* based peer-to-peer overlay network (P3ON). P3ON is composed of *two-tier* overlay rings. The high tier ring is a global overlay in which every node participates. Whereas, the low tier ring is a local overlay that consists of nodes in the same autonomous system (AS). Since the low tier ring consists of nearby nodes (in the same AS), the lookup latency can be significantly reduced if the first search within the low tier ring is successful. Also, to cope with skewness of load (of key lookup) distribution, P3ON effectively replicates the popular keys (and results) to neighbor nodes and neighbor ASs. Simulation results reveal that P3ON outperforms the existing ring-based P2P network in terms of lookup time and achieves relatively balanced load distribution.

Keywords: proximity, peer-to-peer, overlay network, load distribution.

1 Introduction

Recently several peer-to-peer (P2P) systems have been proposed to overcome the limitations of the traditional client-server model. P2P systems distribute functionality and share resources among peers. Depending on how to locate resources, P2P systems can be classified into two classes: unstructured and structured. Generally, in unstructured P2P systems, peers are unaware of how resources are located in the overlay networks. Therefore, lookup requests are typically resolved by flooding-like techniques. Gnutella [1] is a well-known unstructured P2P system. Due to the flooding technique, unstructured P2P systems incur a high volume of signaling traffic. On the contrary, in structured P2P systems, peers share the way in which resources are located. Thus, lookup requests can be directed to a specific peer and hence, much fewer lookup messages are needed. However, structured P2P systems require increased maintenance cost incurred by maintaining the overall structure. The most prominent approach in structured P2P systems is to use a distributed hash table (DHT) to locate resources.

In the literature, a number of DHT-based lookup algorithms have been proposed, [2][3][4]. The lookup time that of most of these algorithms is approximately bounded to log(N), where N is the number of nodes. However, the hop

T. Vazão, M.M. Freire, and I. Chong (Eds.): ICOIN 2007, LNCS 5200, pp. 234–243, 2008.

distance between two overlay nodes in the overlay network has nothing to do with the real distance between two nodes. To overcome this inefficient lookup problem, geographical proximity-based routing (i.e. proximity based neighbor selection) is proposed in P2P systems. Pastry [4] is a well-known proximity based algorithm. Since peers build their routing table entries depending on the proximity metric among all nodes in the network, a huge amount of control messages are required to measure proximity especially when a node joins the network.

There are a few attempts to achieve better lookup performance by adding an additional overlay in the system. Brocade [5] utilizes a new layer consists of supernodes, which are powerful nodes close to network access points such as routers. Each supernode manages a group of local nodes and every local nodes access resources via supernodes. The network traffic is reduced but a supernode may become the bottleneck. Plethora [6] organizes nodes into local overlays leveraging autonomous system (AS) information. Using cache in the local overlay significantly reduces the lookup latency. However, local overlay leaders, each of which uniquely exists in each local overlay are responsible for AS merge/split to keep the number of nodes in an AS appropriately.

In this paper, we propose a Proximity based P2P Overlay Network (P3ON). P3ON is composed of two overlays: high tier and low tier. The high tier ring is a global overlay, which includes every node participating in P3ON. In contrast to this ring, the low tier ring is a local overlay, which represents a single autonomous system (AS). That is, all nodes in an AS belong to the same low tier ring. We present a two-phase lookup algorithm to take advantage of local cache. Even if the input query is highly skewed, the overhead at a popular node is effectively distributed by a load distribution mechanism.

The rest of this paper is organized as follows. Section 2 details P3ON. Section 3 shows the numerical results of our system and Section 4 concludes this paper.

2 Proximity Based P2P Networks (P3ON)

If the hop distance in overlay is based on the proximity (e.g. geographical distance) between two nodes and there exists any semantic locality (the popular item will be looked up again by others) among the lookup queries, the lookup time in large-scale P2P networks will be substantially lowered [7]. To accomplish this, we design two decentralized algorithms: the proximity-based ID assignment algorithm and the two-phase lookup algorithm.

2.1 Proximity Based ID Assignment

We first assume that every node in P3ON possesses a unique IP address. By hashing the node's IP address in a collision-resistant manner (e.g. SHA-1, MD5), P2P systems obtain asymptotically almost a unique ID. However, the hash value does not reflect any proximity between the peer nodes. Therefore, P3ON proposes the following hierarchical ID assignment algorithm after mentioning our second assumption.

Our next assumption is that a node is feasible to figure out its AS number (e.g. [8][9]) and identical AS number (ASN) is assigned to all the nodes that belong to the same AS. The AS is usually a group of nodes governed by a single authority and in many cases, nodes in a same AS are closely located. IDs in P3ON are selected from a 176bit namespace. Since every node has a 16 bit AS number representing to which AS it belongs to, the first 16 bits of a node ID are adopted from its own ASN. The remaining 160 bits of the ID are determined by hashing the node's IP address with the SHA-1 algorithm. By concatenating those 16 bits (ASN) and 160 bits (SHA-1 value), we obtain a unique and uniformly distributed node ID, which namespace is 176 bits long. To utilize a distributed hash table (DHT), an item ID must be the same length as a node ID. Therefore an item ID must be also 176 bits long. The latter 160 bits of the item ID are derived from the hashing result of SHA-1 with its item name. As items do not have any similar concepts such as ASN, we prepend additional 16 bits by copying the last 16 bits of the latter 160 bits. Consequently, the node and item IDs are constructed as follows;

$$\text{Node ID} = (\text{ASN}) \mid\mid f(\text{node's IP address})$$

$$\text{Item ID} = (\text{last 16 bits of } f(\text{item name})) \mid\mid f(\text{node's IP address})$$

where f is a SHA-1 hash function

When constructing the item ID with 176 bits, we have two factors in mind. First, any ID constructed this way is unique. If two distinct items have different item names, the uniqueness is guaranteed by the property of SHA-1. Also, items with the same item name are mapped to an identical ID. Second, item IDs are well distributed over the 176 bit namespace. We perform a simple experiment to verify that item IDs in our scheme are evenly distributed. In our experiment, we first uniformly distributed 500 nodes in 176 bit namespace, and then distributed 1,000,000 items with IDs generated by our scheme. In an ideal case, if K items are uniformly distributed over uniformly distributed N nodes, K/N items will be located at each node.

2.2 Two Tier Ring

Figure 1 illustrates a two-tier ring in P3ON. The high tier ring is the main overlay in P3ON; therefore, every peer node is mapped to a position over the high tier ring. Owing to the proximity-based ID assignment algorithm, nodes, which are closely located in the same AS, are placed adjacently in the high tier ring. The high tier ring is partitioned into AS units. Note that unused ASNs will generate the empty node ID space. The keys corresponding to this empty space will be mapped to an immediate predecessor node. Since nodes in the same AS have the same ASN, the first 16 bits of those node's ID are identical. Therefore, those nodes are placed in the nearby area in the high tier ring naturally.

At the same time, a low tier ring is built by grouping the nodes in the same AS. As a result, the number of low tier rings equals to the number of participating ASs. To form a low tier ring, an additional link per AS is required. This link

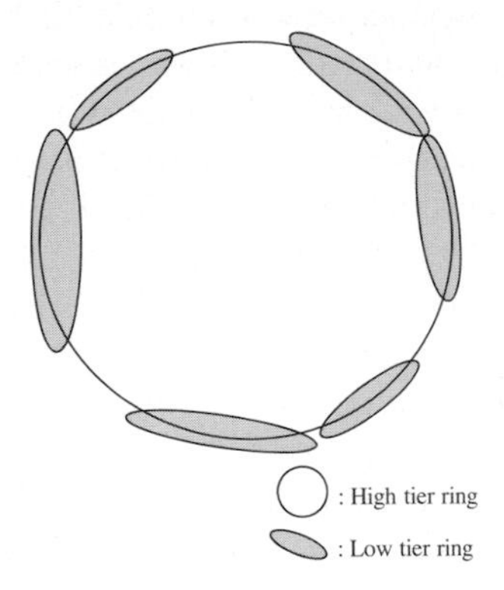

Fig. 1. Two-tier overlay

will connect two peer nodes with the highest and lowest IDs in the AS. A node whose ASN is different from the ASN of its predecessor realizes that it is the node with lowest ID within the AS and it has a responsibility to maintain the link to the node with the highest ID to form the low tier ring. This node queries a node with the highest ID and sets up a connection. As each node belongs to two different rings, high tier and low tier rings, finger tables must be maintained separately. Forming each finger table is identical with the process of Chord [2].

2.3 Two-Phase Lookup Algorithm

Phase I: Low Tier Ring Lookup. At first, a node tries to find the key within its low tier ring. To do this, the node creates a local item ID for the item. The local item ID is created by concatenating the 16 bit ASN of the node and the 160 bit result of SHA-1 with the item name. Initially, keys are located only in the high tier ring. Therefore, a cache miss will occur for the first lookup process searching the item in the low tier ring.

If a peer node that corresponds to the local item ID in the low tier ring (we call this node a local target node) stores the previously queried keys, the local target node can respond to the query from then on. To keep the previously queried keys in the low tier ring, each peer node maintains a local cache. A cache entry consists of item ID and the position of the item, the latest query originator. The latest query originator can give the location information about the item itself, since the originator will possess the item after lookup. The next node search for the same item can download it within the same AS. In the case of cache overflow, a famous replacement policy, such as least recently unused (LRU) can be used. If the local target node does not have any information about the query, the second phase lookup procedure is performed by the local target node.

Phase II: High Tier Ring Lookup. In the second phase lookup procedure, the local target node acts as a query originator. The local target node uses the original item ID described in Section 2.1 instead of the local item ID. By using this original item ID, the same lookup procedure as Chord is performed at the high tier ring. Since all items and nodes are located at the high tier-overlay ring, there is no possibility of lookup failure. The query response is delivered towards the local target node and the local target node relays the result to the original query initiator. When the local target node returns the query result to the query originator, it stores the local item ID and the IP address of the query originator in each local cache. Therefore, if any other node in the same AS tries to find the same item later, a query can be responded to the query originator with reduced delay.

2.4 Load Distribution

In an ideal P2P condition, all the nodes in the system should experience the same amount of lookup load. Here, the load is two-fold, i) the number of queries that the node receives in a unit time, ii) the number of keys that the node stores. Keys stored at a node occupy the storage space proportional to the number of keys. Incoming query consumes the node's computation power, network bandwidth, etc. In general, nodes with the relatively large number of keys tend to receive more queries than other nodes. Although these two aspects of the load have some correlations, it is feasible to decouple them. The reason is that in a real world, queries are not uniformly distributed among individual items, rather, the distribution of queries are highly skewed [10]. For example, a node with the most popular item can be overloaded by too many incoming queries.

Since recent machines have a sufficient storage space for small sized cache entry, the number of keys is not an issue. In P3ON, to cope with this problem, a dynamic load distribution using two thresholds (δ_1, δ_2) is proposed for load balance. Since each node has different power and link bandwidth, each node uses different thresholds. The central idea is that if the load (the number of queries per unit time) exceeds a certain threshold, the node triggers load distribution. Specifically, a node counts the number of times each key is referred. It is possible by simply modifying the structure of a cache entry by adding an additional field for counting. Keys with frequent access will make the node overloaded. When the number of incoming queries on all the keys of a node exceeds a pre-defined

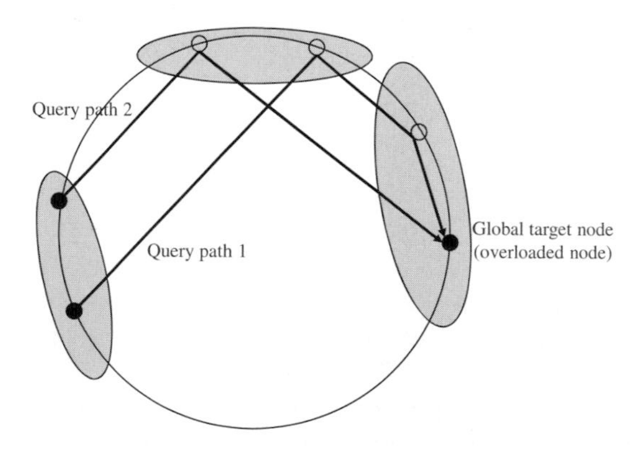

Fig. 2. Load distribution

threshold δ_1 of the node, the node sends out an intra-AS advertisement message and the message is delivered to all the nodes in the local ring (i.e. nodes in the same AS). The message contains the information on the most popular keys of the overloaded node.

When a new lookup process for the key in the above overloaded node is initiated, the key can be found in predecessor nodes before the query reaches the overloaded node. Intra-AS advertisement deals with queries routed over query path1 as shown in the Figure 2 which traverses via the low tier ring. If the size of low tier ring is small (i.e. the number of nodes in the AS is low), the ratio of queries routed over query path2 in Figure 2 will increase. However intra-AS advertisement cannot handle queries over query path2.

Although the overloaded node triggers intra-AS advertisement, the node can still be suffered from too many incoming queries. If the incoming query frequency exceeds the pre-defined threshold, δ_2, inter-AS advertisement is triggered. The intra-AS advertisement is relayed toward the predecessor AS. The message arriving at the predecessor AS will be delivered to all the nodes in that AS as if intra-AS advertisement is triggered. Both inter-AS and intra-AS advertisements distribute queries to mitigate the burden of the overloaded node.

3 Numerical Results

In this section, we evaluate the performance of P3ON in terms of the size of the network (total number of nodes in overlay) and the size of each AS (the number of nodes in each AS). The central idea of P3ON lies in exploiting the use of proximity between nodes. Therefore, the size of AS significantly affects lookup time and load distribution.

3.1 Simulation Parameters

We simulated a network that has up to 10,000 nodes, and in every network layout, 100,000 keys are distributed over the network. The low tier ring has a local cache with the size of 10 slots. That is, there are 10 keys and their locations in the cache of a node. Clearly, the more slots are provided in a cache, the better performance is achieved. However, for the purpose of emphasizing the impact of the local ring, we restrict the cache size to 10. Query initiators are

Table 1. One way delay between transit nodes

Continents	America	Latin	Europe	Asia	Africa	Oceania
Ameria						
Latin	222					
Europe	80	156				
Asia	125	237	159			
Africa	392	326	358	284		
Oceania	136	249	177	198	296	

chosen randomly among the entire nodes, and items to be queried are selected according to a Zipf-like distribution with an input parameter of $\alpha = 1.0$. We use an abstracted world topology. There are 6 continents and one representative transit node exists for each continent. As shown in Table 1, the latency between each transit node is based on the average of measure values during the period between August 2003 and June 2005 measured by IEPM [11]. Each AS belongs to one of the continents, and the delay to the transit node takes $[0 - 50]$ms. Up to 2000 nodes for each AS are deployed and an intra-AS delay of $[0 - 10]$ms is established in our experiments. For the parameters we refer to [11][12]. The stated parameters are used for our results, unless otherwise explicitly stated.

3.2 Lookup Latency

To verify the performance of P3ON, we measured the lookup latency with a virtual network, which varies in size by 250, 500, 1000, 2000, 4000, 8000 and 16000 nodes. Since P3ON is significantly affected by the AS size and the number of ASs, experiments with different numbers of ASs and different numbers of nodes in each AS are as important as experiments with the total number of the nodes being changed. Therefore, we increased the network size with two different ways. First, we fix the number of nodes in each AS to 50, and we increase the number of ASs as follows: 5(250), 10(500), 20(1000), 40(2000), 80(4000), 160(8000) and 320(16000). The values in the parentheses stand for the network size (total number of nodes in overlay). Second, we now fix the numbers of ASs to 50, and change the AS size as follows: 5(250), 10(500), 20(1000), 40(2000), 80(4000), 160(8000) and 320(16000).

Figure 3 shows the result of our experiments. Since P3ON is an enhanced version of the Chord's lookup algorithm, it is natural that P3ON outperforms Chord. We can see the first way of increasing the AS size with the fixed number of ASs maintains the lookup latency around $600 - 800$ ms. When the AS size is too small, i.e. 5 nodes, the total cache slots in the AS is 50, so that P3ON performs worse than Chord due to frequent cache miss. In this situation, the

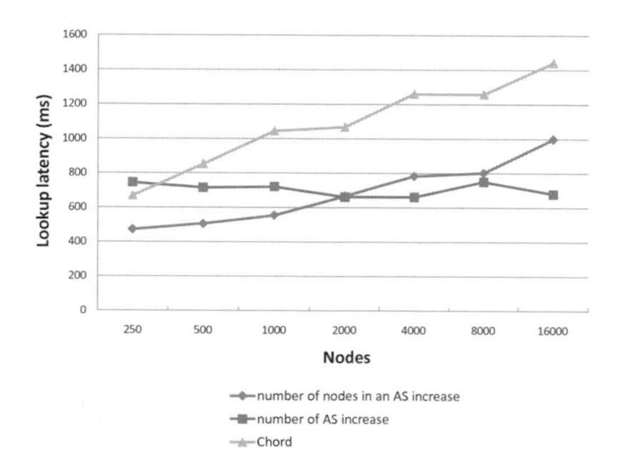

Fig. 3. Lookup latency versus network size (ms)

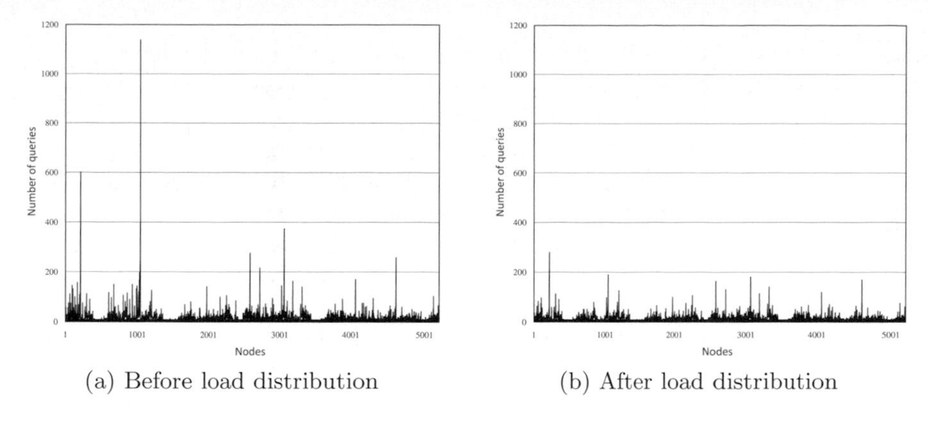

(a) Before load distribution (b) After load distribution

Fig. 4. Number of queries reached at each node before and after the load distribution

Table 2. Effect of load distribution

	before	after
Total number of query request	8999	6742
Average	1.73	1.30
Maximum	1133	271
Standard deviation	20.50	8.37

delay consumed by the first phase of two-phase lookup algorithm is only a burden to the system. The delay is beneficial when the cache size becomes bigger than 7. The lookup latency fixed around 600-800 is the counterbalance between effect of cache and network size increase. However, the lookup latency of the second way slowly increases as the number of ASs increases, with the fixed AS size. This can be explained as follows. Through the two-phase lookup algorithm, a node is able to utilize the whole information cached at all the nodes in the same low tier ring. Since a low tier ring corresponds to one AS, increasing the number of ASs does not increase the cache hit probability within a single AS.

3.3 Load Distribution

In this experiment, we reveal how the entire workload (the number of lookup requests) is distributed among all the nodes in the overlay. Figure 4 shows the amount of load at every node with the definition of the term 'load' at Section 2.4. Since queries are skewed, a small portion of the nodes dominates the target of the lookup requests before load distribution. The difference in the total number of query requests in Table 2 is due to the cache in each node. If the target key is in its own cache, the node does not initiate the lookup process at all.

Figure 4 shows the impact of load distribution in P3ON. Through the local cache, intra-AS, and inter-AS advertisement messages, the overloaded node's load is significantly reduced. Each local ring reduces a significant amount of load by facilitating the local cache. Query requests that take place due to cache misses are mostly filtered by the predecessor nodes of the overloaded node. After the load distribution process is in effect, the predecessor nodes experience slightly more load than before. If the predecessor node's capacity is insufficient to handle this additional load, it becomes an overloaded node and triggers its own advertisement to reduce the load. By that domino effect, the lookup load is efficiently distributed in a fully decentralized manner.

The performance of the proposed load distribution algorithm is affected by several factors as follows: the place where the overloaded node is located within the low tier ring, the size of the low tier ring, and the size of predecessor AS. If there are a lot of predecessors of the overloaded node in the low tier ring, the load will be distributed to the predecessors due to the intra-AS advertisement message. Likewise, if the predecessor AS has a number of nodes therein, they will also help mitigate the load of the overloaded node.

4 Conclusion

Peer-to-peer (P2P) overlay networks should be carefully constructed to reduce the lookup latency, especially taking into account real network topologies. Most of P2P protocols and systems have focused on how to reduce the hop distance in lookup operations. However, even a single hop in overlay networks can reach far more than 10 hops in real networking environments. In this paper, we propose a proximity based peer-to-peer overlay network (P3ON), which is a fast, scalable lookup algorithm. P3ON is composed of two overlays. The high tier ring is a global overlay in which every node participates. Whereas, the low tier ring is a local overlay that consists of nodes in the same autonomous system (AS). Since the low tier ring consists of nearby nodes (in the same AS), the lookup latency can be significantly reduced if the first search within the low tier ring is successful. To this end, previously queried keys and results (locations of keys) are stored in the corresponding node of the low tier ring. Also, to cope with skewness of load (of key lookup) distribution, P3ON effectively replicates the popular keys (and results) to neighbor nodes and neighbor ASs. This replication is realized by two advertisement messages: intra-AS advertisement and inter-AS advertisement. Simulation results reveal that P3ON outperforms the existing ring-based P2P network (i.e. Chord) in terms of lookup time and achieves relatively balanced load distribution. Since the real (underlying) network topology is a key issue in determining the performance of P3ON, we are currently working on more realistic experiments.

References

1. Gnutella, `http://www.gnutella.com`
2. Stoica, I., Morris, R., Liben-Nowell, D., Karger, D.R., Kaashoek, M.F., Dabek, F., Balakrishnan, H.: Chord: A Scalable Peer-to-peer Lookup Protocol for Internet Applications. IEEE/ACM Transactions on Networking 11(1), 17–32 (2003)
3. Ratnasamy, S., Francis, P., Handley, M., Karp, R., Shenker, S.: A scalable content-addressable network. In: Proc. ACM SIGCOMM 2001, pp. 161–172 (August 2001)

4. Rowstron, A., Druschel, P.: Pastry: Scalable, distributed object location and routing for large-scale peer-to-peer systems. In: Guerraoui, R. (ed.) Middleware 2001. LNCS, vol. 2218, pp. 329–350. Springer, Heidelberg (2001)
5. Zhao, B.Y., Duan, Y., Huang, L., Joseph, A.D., Kubiatowicz, J.D.: Brocade: landmark routing on overlay networks. In: Druschel, P., Kaashoek, M.F., Rowstron, A. (eds.) IPTPS 2002. LNCS, vol. 2429, pp. 34–44. Springer, Heidelberg (2002)
6. Ferreira, R.A., Grama, A., Jagannathan, S.: Enhancing Locality in Structured Peer-to-Peer Networks. In: Proceedings of Tenth IEEE International Conference on Parallel and Distributed Systems, Newport Beach, CA, July 2004, pp. 25–34 (2004)
7. Gummadi, K.P., Dunn, R.J., Saroiu, S., Gribble, S.D., Levy, H.M., Zahorjan, J.: Measurement, Modeling, and Analysis of a Peer-to-Peer File-Sharing Workload. In: Proc. of the 19th ACM Symposium on Operating Systems Principles, Bolton Landing, NY (October 2003)
8. Mao, Z.M., Rexford, J., Wang, J., Katz, R.H.: Towards an Accurate AS-Level Traceroute Tool. In: Proceedings of the 2003 ACM SIGCOMM Conference on Applications, Technologies, Architectures, and Protocols for Computer Communication, Karlsruhe, Germany (August 2003)
9. Exploiting Autonomous System Information in Structured Peer-to-Peer Networks. In: The 13th IEEE International Conference on Computer Communications and Networks (ICCCN 2004), Chicago, IL, October 11-13 (2004)
10. Ge, Z., Figueiredo, D.R., Jaiswal, S., Kurose, J., Towsley, D.: Modeling peer-peer file sharing systems. In: Proceedings of INFOCOM 2003, Santa Fe, NM (October 2003)
11. Internet End-to-end Performance Monitoring (IEPM), `http://www-iepm.slac.stanford.edu/`
12. Xu, Z., Mahalingam, M., Karlsson, M.: Turning Heterogeneity into an Advantage in Overlay Routing. In: Proceedings of the IEEE INFOCOM 2003, San Francisco, CA (April 2003)

Network Architecture and Protocols for BGP/MPLS Based Mobile VPN*

Haesun Byun and Meejeong Lee

Dept. of Computer Science and Engineering, Ewha Womans University, Korea
ladybhs@ewhain.net, lmj@ewha.ac.kr

Abstract. We propose a provider edge (PE)-based provider provisioned mobile VPN mechanism, which enables efficient communication between a mobile VPN user and one or more correspondents located in different VPN sites. The proposed mechanism not only reduces the IPSec tunnel overhead at the mobile user node to the minimum, but also enables the traffic to be delivered through optimized paths among the (mobile) VPN users without incurring significant extra IPSec tunnel overhead. The proposed architecture and protocols are based on the BGP/MPLS VPN. A service provider platform entity named PPVPN (Provider Provisioned VPN) Network Server (PNS) is defined in order to extend the BGP/MPLS VPN service to the mobile users. Compared to the existing mechanisms, the proposed mechanism requires less overhead with respect to the IPSec tunnel management. The simulation results also show that it outperforms the existing mobile VPN mechanisms with respect to the handoff latency and/or the end-to-end packet delay.

1 Introduction

VPNs have been attaining significant attention as a cost effective replacement of enterprise networks with private leased lines. Complexities in management, though, become one of the obstacles in the wider deployment of VPN services. PPVPN, for which the entire jobs related to the management as well as the establishment of a VPN are to be provided by the VPN service providers, is proposed so that the customers are relieved from the complexities of provisioning a VPN. PPVPN is under the process of standardization within the Layer 2 VPN (L2VPN) and Layer 3 VPN (L3VPN) working groups (WGs) of Internet Engineering Task Force (IETF). Our work is specifically related to supporting mobility within the layer 3 VPN.

For the layer 3 VPN, three mechanisms are proposed: BGP/MPLS VPN, Virtual Router VPN, and CE-based IPSec VPN [1]. In the BGP/MPLS VPN and the Virtual Router VPN, a customer edge (CE) device is connected to a PE device, and the PEs perform VPN establishment procedures. Since both the BGP/MPLS VPN and the Virtual Router VPN consist of PEs, they are

* This work was supported in part by MIC, Korea under the ITRC program(C1090-0603-0036) supervised by IITA.

T. Vazão, M.M. Freire, and I. Chong (Eds.): ICOIN 2007, LNCS 5200, pp. 244–254, 2008.

categorized as a PE-based VPN. In the CE-based IPSec VPN, on the other hand, CEs are directly connected with each other by establishing IPSec tunnels among themselves. In contrast to the PE-based VPN, the CE-based IPSec VPN is a CE-based VPN. In terms of the number of tunnels to provide VPN services and the complexities to add or to delete a site to or from a VPN, the PE-based VPNs are more scalable than the CE-based VPNs.

In addition to providng the management support for the fixed VPN users, supporting mobility the VPN services becomes important. Mobile VPN mechanisms to enable a mobile user to make a secure access to a server in the corporate network are proposed [2] [3] [4] [5]. In these mechanisms, a mobile user node is one of the end points of an IPSec tunnel when the user is in a foreign network, and hence the mobile user node is involved in the IPSec tunnel management overhead in this mechanism. Furthermore, it could be inefficient to provide a service for a mobile user which wants to communicate with one or more correspondent(s) residing in different VPN sites. Either the VPN traffic to or from the mobile user always has to detour the home network of the mobile user, or the mobile user has to build a separate IPSec tunnel with every gateway of the VPN sites in which its correspondents reside.

To reduce the IPSec tunnel management overhead at the mobile user node, the CE-based IPSec VPN mechanisms are extended so that the GW of the foreign network, on behalf of the mobile user, establishes IPSec tunnel with the VPN GW of the mobile user's home network [6] [7]. In [6], registration of a mobile VPN user to its HA is made via a service provider platform entity named Service Provisioning Platform (SPS), and SPS provisions a IPSec tunnel between the foreign GW and the VPN GW of the mobile user's home network. The inefficiency in supporting communications between a mobile user and one or more correspondents residing in different VPN sites still remains, though. For route optimization (RO), foreign GW serving the mobile VPN user has to establish a separate IPSec tunnel with each VPN GW of the VPN sites in which the correspondents of mobile user reside.

In this paper, hence, we propose a PE-based mobile VPN mechanism, which enables efficient communication between a mobile VPN user and one or more correspondents located in different VPN sites. The proposed mechanism not only reduces the IPSec tunnel overhead at the mobile user node to the minimum, but also enables the traffic to be delivered through optimized paths among the (mobile) VPN users, regardless of their locations, without incurring significant extra IPSec tunnel overhead. The proposed mechanism is based on the BGP/MPLS VPN, and an entity named PNS is defined within a service provider platform in order to extend the BGP/MPLS VPN service to the mobile users.

The rest of the paper is organized as follows. Section 2 presents the details of the proposed mobile BGP/MPLS VPN architecture and protocols. The management overhead of the existing mechanisms and the proposed mechanism are analyzed and compared in section 3. In section 4, simulation results on the performance of the existing mechanisms and the proposed mechanism are given. Finally, we conclude our work in Section 5.

2 Mobile BGP/MPLS VPN

This section explains the details of the proposed mechanism. First, the network architecture and protocol entities to provide the mobile VPN services based on the BGP/MPLS VPN are specified. The procedure for a mobile VPN user visiting a foreign network to attain the access to the VPN from the foreign network is then explained.

2.1 Network Architecture and Protocol Entities for Mobile BGP/MPLS VPN Services

Fig. 1 shows the network architecture and the tunnels to be established for the proposed mobile BGP/MPLS VPN services. For the proposed mechanism, the GW of foreign network has to assume the role of CE temporarily for the mobile users visiting its area. Differentiating it from the CE in the home network, let us call it as a foreign CE (FCE). A logical attachment circuit is dynamically established between the FCE and a PE. Similar to the static attachment circuit between the CE and PE of the BGP/MPLS VPN, IPSec tunnel should be established in order to provide the necessary security for a VPN over the dynamic attachment circuit. The PE to which the FCE is attached treats the VPN site behind the FCE in the same way as it does with a static VPN site of a BGP/MPLS VPN.

The PNS is an entity belonging to the service provider platform. It provides the mobility management for the mobile users of VPNs and provisions the VPNs accordingly. Specifically, if a mobile user moves out to a foreign network, the PNS provides the FCE and a selected PE with necessary information to establish a dynamic attachment circuit, i.e., an IPSec tunnel between them. It also provides the PE with the information such as the VPN topology, route target (RT), and route distinguisher (RD) so that the PE can participate in the BGP/MPLS VPN establishment operations for the mobile user [8].

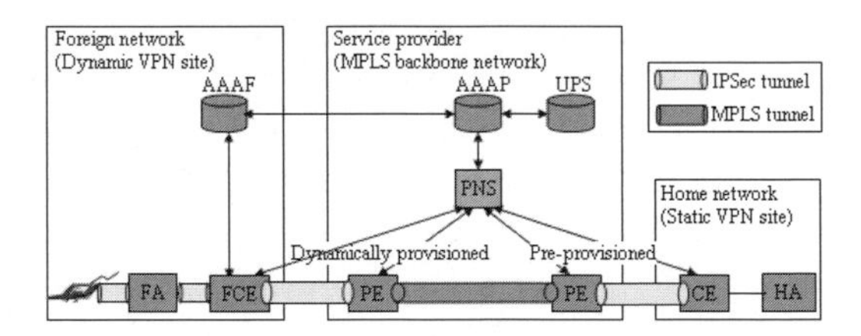

Fig. 1. Network architecture and the tunnels for the proposed mobile BGP/MPLS VPN

The accounting, authorization and authentication server for the foreign network (AAAF) and the AAA server for the service provider (AAAP) communicate with each other for the AAA of the mobile VPN users. The User Profile Server (UPS) maintains the user service profile information regarding the VPN.

2.2 Mobile BGP/MPLS VPN Establishment and Access

Fig. 2 illustrates the mobility management procedure for a mobile VPN user to obtain access to the VPN from a foreign network. In the proposed mechanism, Diameter MIPv4 protocol is used for the mobile user's registration and AAA service. Diameter MIPv4 is the protocol standardized by the AAA WG of IETF for the authentication, authorization, and accounting for a mobile user [9]. A mobile VPN user generates a MIP Registration Request message when it moves into a foreign network. Fig. 3 shows the MIP Registration Request message issued by a mobile VPN user. In order to have the registration request to be delivered to the PNS, the address of PNS instead of the address of HA is specified in the HA field of the MIP Registration Request message. It is assumed that the address of PNS is pre-configured in the mobile VPN user's node. A new field named FCE address is added to specify the address of FCE so that the PNS could find out the FCE serving the mobile VPN user. The address of FCE is assumed to be obtained from the Agent Advertisement. In the extension field of the MIP Registration Request message, the Network Address Identifier (NAI) of AAAP is specified instead of the AAAH NAI since the AAA service for the mobile VPN user is provided by the service provider instead of the user home network in the proposed mechanism.

Diameter MIPv4 uses two types of Care-of-Address (CoA): FA-CoA and Co-CoA. If FA-CoA is used, the MN sends the MIP Registration Request message to the FA (A2 in Fig. 2). Receiving the MIP Registration Request message,

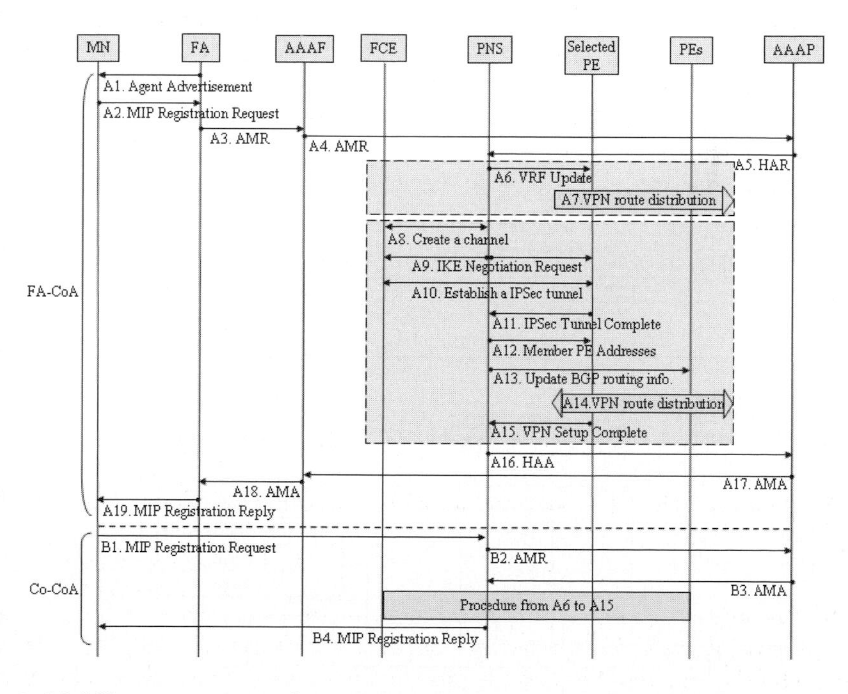

Fig. 2. Mobility management for a mobile VPN user in the Mobile BGP/MPLS VPN

the FA generates an AA-Mobile-Node-Request (AMR) message based on the information carried in the MIP Registration Request message and sends it to the AAAF (A3 in Fig. 2). The FCE address in one of the extension fields of the MIP Registration Request message is copied to AMR together with the other information in the extension fields. Fig. 4 shows the AMR message used for the proposed mechanism. It is extended with MIP-FCE-Address field specifying the FCE address.

Receiving the AMR message, the AAAF checks the NAI of the destination AAA server, which is the NAI of AAAP in the proposed mechanism, and determines that the message has to be relayed to AAAP (A4 in Fig. 2) [10]. The AAAP checks the MN-AAA authentication extension [11]. Upon the successful authorization, the AAAP sends Home-Agent-MIP-Request (HAR) message to the PNS (A5 in Fig. 2). Receiving the HAR message, the PNS starts to process the VPN provision and mobility management procedure for the mobile user. On the other hand, if Co-CoA is used, the MN sends the MIP Registration Request message to the PNS directly (B1 in Fig. 2). Receiving the MIP Registration Request message, the PNS generates the AMR message and sends it to the AAAP for the authentication and authorization of the mobile user and the VPN service (B2 in Fig. 2). The AAAP replies to the PNS with the AMA message (B3 in Fig. 2). Upon receiving the AMA message, the PNS starts to process the VPN access and mobility management procedure for the mobile user.

The PNS maintains the VPN Service Tunnel (VST) table to keep the necessary management information for each VPN as shown in Fig. 5. The VST table contains 5 entries for each VPN: an entry for the MN to FCE binding information (*MN_to_FCE_Binding list*), an entry to keep the information

Type	S	B	D	M	G	R	T	x	Lifetime
Home Address									
PNS Address									
Care-of-Address									
FCE Address									
Challenge and MN-AAA authentication extension									
...									

Fig. 3. Diameter MIPv4 Registration Request message for the mobile BGP/MPLS VPN

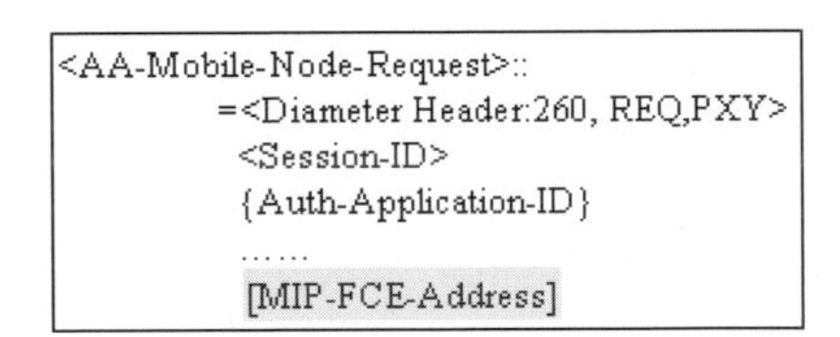
<AA-Mobile-Node-Request>::
=<Diameter Header:260, REQ,PXY>
<Session-ID>
{Auth-Application-ID}
......
[MIP-FCE-Address]

Fig. 4. Modified AMR message for the mobile BGP/MPLS VPN

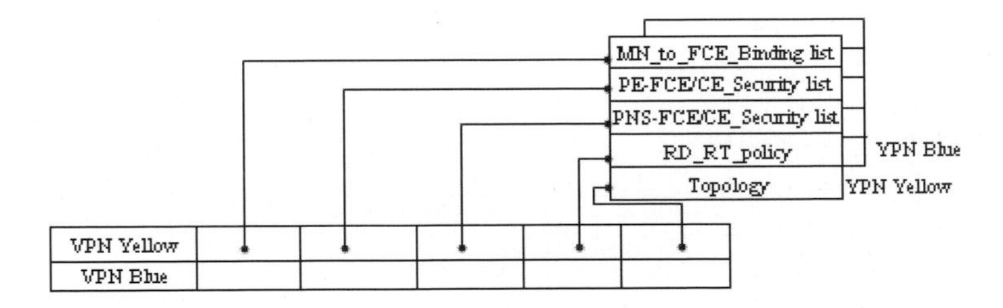

Fig. 5. VST table structure

for setting up an IPSec tunnel for each PE and an attached FCE/CE pair (*PE – FCE/CE_Security list*), an entry to keep the security information to be used between the PNS and a FCE/CE (*PNS – FCE/CE_Security list*), an entry for the RT/RD policy information (*RT/RD_policy*), and the entry to keep the topology information of the corresponding VPN (*Topology*). Except for the *RT/RD_policy*, which is assigned and fixed for each VPN, the above entries are updated as a mobile VPN user changes its location and is attached to a new FCE.

Receiving either the HAR message or the AMA message from the AAAP, the PNS determines the PE to provide the service of VPN access to the mobile user. The PE can be determined either in a static way based on the mobile user's VPN service profile and the management policy of the respective VPN, or in a dynamic way, for instance, among the PEs that can satisfy the mobile user's requirements, the nearest PE from the FCE can be selected.

The *PE – FCE/CE_Security list* of the VPN is then looked up from the VST table in order to check whether an IPSec tunnel already exists, between the selected PE and the FCE pair, for the VPN. If it does, the mobile user can be readily served by the existing one, and the PNS simply informs the PE of HoA of the new VPN user using the VRF Update message so that the VRF table in the PE can be updated (A6 in Fig. 2). Otherwise, the PNS first sends out the security and encryption related information to the FCE, which the FCE should use to communicate with the PNS through the secure remote configuration channel (A8 in Fig. 2). The security and encryption information for the FCE is then appended to the *PNS – FCE/CE_Security list* of the VST table. The PNS, then, sends out the IKE Negotiation Request message to the selected PE and the FCE through the secure remote configuration channel (A9 in Fig. 2). IKE Negotiation Request message is defined for the proposed mechanism in order for the PNS to request the PE and the FCE to establish an IPSec tunnel between them as an attachment circuit for a specific VPN.

The IKE Negotiation Request message sent to the PE also includes the HoA of the MN, the RD and the RT of the VPN, which are used for the PE to update its VRF table and to distribute the route information to the other PEs belonging to the same VPN. To the FCE, the PNS includes HoA and CoA of the MN in the IKE Negotiation Request message so that the FCE can maintain the mapping between the HoA, the CoA and the corresponding IPSec tunnels. This mapping information is used by the FCE for VPN data packet delivery. Upon receiving the IKE Negotiation Requests message, the PE and the FCE establishes the IPSec tunnel between themselves (A10 in Fig. 2). After the IPSec tunnel is established, the PE sends out IPSec Tunnel Complete message to the PNS (A11 in Fig. 2). Receiving the IPSec Tunnel Complete message, the PNS checks whether the PE is newly joining to the service of that particular VPN. Note that the PE had to establish a new IPSec tunnel with the FCE does not necessarily mean that the PE is new to providing service to the given VPN. The PE might have had another CE or an FCE of the corresponding VPN attached to itself already. If the newly established IPSec tunnel is the first one at the PE for the given VPN, then the PE is newly joining to the service of the given VPN. In that case, the PNS informs the address of the other PEs serving the same VPN to the new PE using the Member PE Addresses message (A12 in Fig. 2), and notifies the other PEs of that VPN to send the BGP routing update information of the corresponding VPN to the new PE (A13 in Fig. 2) so that the VRF table of the new PE can be completed.

After establishing the IPSec tunnel between the PE and the FCE, the PE inserts the mapping between the HoA of the MN and the IPSec tunnel into the VRF table of the corresponding VPN. Note that each entry of VRF table corresponds to a site for the static VPN whereas a single mobile VPN user takes a separate VRF entry. Upon updating the VRF table for the new mobile VPN user, the PE advertises the information to the other PEs serving the same VPN through the multi-protocol BGP (MP-BGP) sessions (A7 in Fig. 2). The other PEs update their VRF table with the received routing information, and forward it to the CEs of the VPN that are attached to themselves. Especially when the CE of the mobile user's home network receives the routing information for the mobile user, the CE forwards the information to the HA. The HA, then, updates the mobility management binding table entry so that the CoA of the MN is set to the address of the CE. If the PE is newly joining to the service of the given VPN, the other PEs serving the same VPN extract the set of routes that appear in their VRF table and send it to the new PE (A14 in Fig. 2).

When the PE determines that the VPN route distribution is completed, it sends out the VPN Setup Complete message to the PNS (A15 in Fig. 2). The VPN Setup Complete message includes the security information related to the IPSec tunnel established between the FCE and the PE. Receiving the VPN Setup Complete message, the PNS appends the MN to FCE mapping information to the $MN\_to\_FCE\_Binding\_list$, and the security information specified in the VPN Setup Complete message to the $PE - FCE/CE\_Security\ list$ in its VST table. If the PE is newly joined to the service of the given VPN, the $Topology$ entry also needs to be updated to include the PE. Finally, the rest of the the mobility management procedure for a mobile VPN user is proceeded as shown A16$\sim$A19 and B4 in Fig. 2.

3 Comparisons on Complexities and Overhead of Mobile VPN Access Mechanisms

Fig. 6 compares the existing and the proposed mobile VPN mechanisms with respect to the IPSec tunnels to be set up for optimal path data delivery. (a) and (b) correspond to the user- and CE-based mobile VPN respectively, and (c) corresponds to the proposed mechanism. In (a), if a mobile VPN user wants to communicate with multiple correspondents residing in one or more VPN sites, the MN should know the address of the CEs for those VPN sites, and has to establish separate IPSec tunnels toward each CE. The IPSec tunnel management overhead at a user node is, therefore significant.

In (b), for the end-to-end secure communication, IPSec tunnel is established between the MN and the FCE, and between the FCE to every CE under which the correspondents of the mobile VPN user reside. Even when a mobile VPN user wishes to communicate with multiple correspondents, only a single IPSec tunnel need to be established at a user node. Furthermore, if multiple mobile users belonging to the same VPN are under the service of the FCE and if they want to communicate with the same set of correspondents, the FCE can aggregate the requests onto the same set of IPSec tunnels, and hence reduces the IPSec tunnel establishment overhead compared to the mechanism in (a).

In (c), an IPSec tunnel needs to be established between the MN and the FCE, and then between the FCE to a PE for secure end-to-end communication of a mobile VPN user. Similar to (b), just a single IPSec tunnel needs to be established at a user node. The FCE also need to establish just a single IPSec tunnel toward a PE for a VPN. (c) incurs, though, route distribution overheads to update the routing table maintained in the PEs whenever a mobile VPN user changes its location. Both (b) and (c) have drawback of utilizing concatenated IPSec tunnel to provide an end-to-end secure communication.

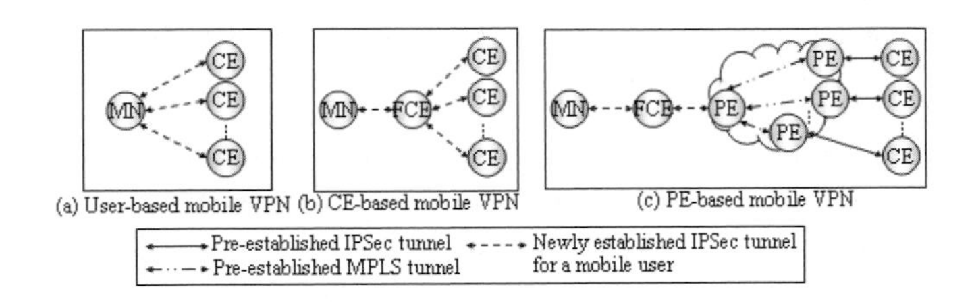

Fig. 6. IPSec tunnel overhead for different mobile VPN mechanisms

4 Simulation

The performance of the proposed mechanism (called as M-BGP/MPLS hereinafter) and the existing provider provisioned mobile VPN mechanisms are compared through simulation experiments. Specifically, the CE-based mobile VPN mechanism proposed in [6], and the variation of [6] for route optimization are compared, and they are called as M-CE IPSec without RO and M-CE IPSec

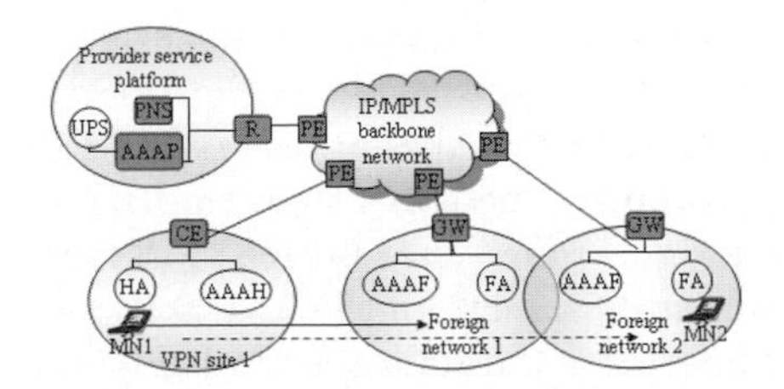

Fig. 7. Simulation network model

with RO hereinafter. The simulation experiments are done using the OPNET Modeler 11.0. The handoff delay, the average throughput over the handoff period, and the end-to-end packet delay are measured. Fig. 7 shows the simulation network model. The mobile VPN users MN1 and MN2 move out to the foreign network 1 and 2 respectively, and MN1 transmits 100Kbps of constant bit rate traffic toward MN2. The transit delay within a single network site is assumed to be 0.1msec, and the transit delay between different network sites are varied from 1msec to 0.5sec.

Fig. 8 shows the handoff delay that the MN1 experiences when it moves to the foreign network 1. For the M-CE IPSec with RO, the IPSec tunnel from the foreign network GW to the home network GW is first established and another IPSec tunnel between the GW of foreign network 1 and the GW of the foreign network 2 is then establish for optimal path data delivery. The moment that the direct IPSec tunnel is established between the GWs of the foreign networks is considered as the handoff completion moment for the M-CE IPSec with RO. The handoff delay is longest with the M-CE IPSec with RO since it has to go through the IPSec tunnel establishment procedure twice for handoff completion. The handoff delay for the M-CE IPSec without RO, which requires a single IPSec tunnel establishment, and the proposed M-BGP/MPLS, which requires the route distribution by BGP for the entire provider network to be completed, are similar.

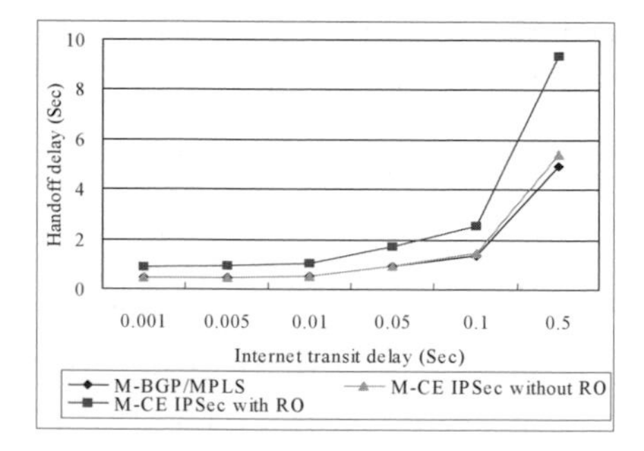

Fig. 8. Handoff delay

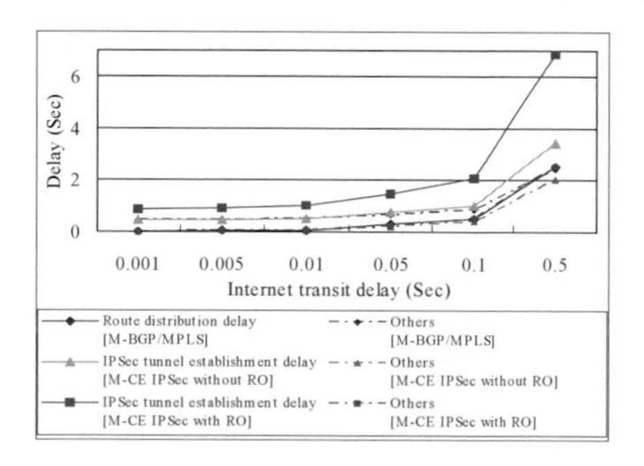

Fig. 9. Components of handoff delay

In Fig. 9, the solid lines identify the delay for IPSec tunnel establishment in the M-CE IPSec mechanisms, and the route distribution delay in the M-BGP/MPLS. The dotted lines identify the delay for the rest of the handoff procedure. Even though the route distribution in the M-BGP/MPLS incurs relatively larger volume of control messages than the IPSec tunnel establishment procedure, the delay for the route distribution does not exceed the delay incurred by IPSec tunnel establishment in the M-CE IPSec mechanisms since the transmission of route information from a certain PE is done in parallel with the other PEs' transmission. The delay for the other processes of handoff is a little longer in the M-BGP/MPLS.

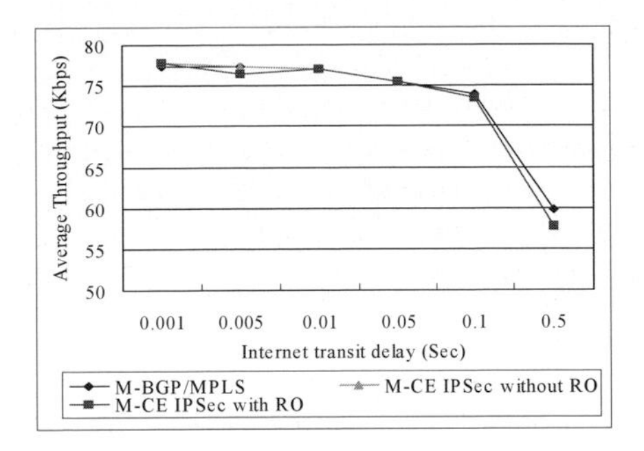

Fig. 10. Average throughput during handoff period

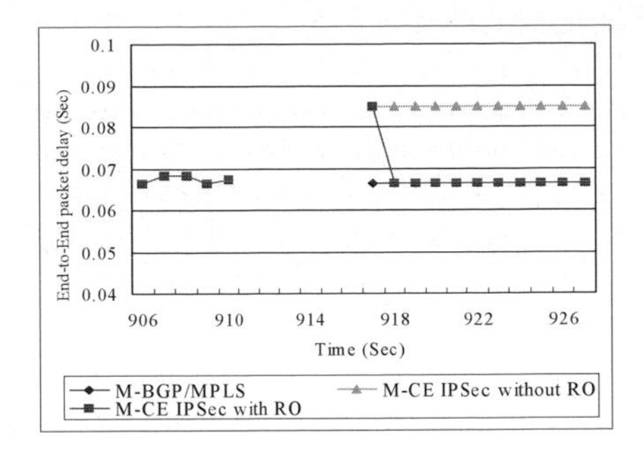

Fig. 11. End-to-end packet delay

Fig. 10 shows the average throughput during the handoff period. For these numerical results, the throughput is measured from the beginning of the handoff until the M-CE IPSec with RO, which incurs the longest handoff delay, completes the handoff procedure. All three mechanisms show similar performance with respect to the average throughput. Even though the M-CE IPSec with RO has longer handoff delay than the other two mechanisms, the average throughput is similar since the traffic transmission proceeds through the detouring path, i.e., through the home network, once the IPSec tunnel to the GW of home network is setup.

Fig. 11 shows the end-to-end packet delay during a certain period of simulation time. The Internet transit delay is set to 0.05sec for this experiment. While the traffic source MN1 is in the home network the three mechanisms show similar end-to-end packet delay. During the handoff, no packet is delivered, and after the MN1 moves to the foreign network 1, the M-CE IPSec without RO has longer end-to-end packet delay than the other two mechanisms since the traffic has to detour the home network to be delivered to MN2, which is in the foreign network 2.

5 Conclusions

A provider provisioned mobile VPN mechanism based on BGP/MPLS VPN is proposed. The proposed mechanism enables efficient communication between a mobile VPN user and one or more correspondents located in different VPN sites. The proposed mechanism not only reduces the IPSec tunnel overhead at the mobile user node to the minimum, but also enables the traffic to be delivered through optimized paths among the (mobile) VPN users without incurring significant extra IPSec tunnel overhead. Compared to the existing mechanisms, the proposed mechanism requires less overhead with respect to the IPSec tunnel management. The simulation results also show that it outperforms the existing mobile VPN mechanisms in terms of the handoff latency and/or the end-to-end packet delay.

References

1. Callon, R., et al.: A framework for Layer 3 PPVPNs, RFC4110 (July 2005)
2. Vaarala, S. (ed.): Mobile IPv4 traversal across IPsec-based VPN gateways (September 2003), http://draft-ietf-mobileip-vpn-problem-solution-03.txt
3. Vaarala, S. (ed.): Mobile IPv4 traversal across IPsec-based VPN gateways (September 2003), http://draft-ietf-mobileip-vpn-problem-solution-03.txt
4. Matteo, et al.: IP Mobility Support for IPsec-based Virtual Private Networks: an architectural solution. In: GLOBECOM 2003, vol. 3, pp. 1532–1536 (December 2003)
5. Barcelo, F., et al.: Design and Modelling of Internode: A Mobile Provider Provisioned VPN. Mobile Networks and Applications 8, 51–60 (2003)
6. Bhagavathula, R., et al.: Mobility: A VPN Perspective. IEEE MWSCAS, pp. 89–92 (August 2002)
7. Rosen, E., et al.: BGP/MPLS IP VPNs, RFC4364 (February 2006)
8. Calhoun, P., et al.: Diameter Mobile IPv4 Application, RFC4004 (August 2005)
9. Calhoun, et al.: Diameter Base Protocol, RFC 3588 (September 2003)
10. Perkins, et al.: Mobile IPv4 Challenge/Response Extensions, RFC 3012 (November 2000)

Migration toward DiffServ-Enabled Broadband Access Networks

Seungchul Park

School of Internet Media Engineering,
Korea University of Technology and Education,
307 Gajun-Ri, Byeongcheon-Myun, Cheonan, Chungnam, 330-708
Republic of Korea
scpark@kut.ac.kr

Abstract. In this paper, we propose several candidate models to support DiffServ QoS for multimedia applications in broadband access network environments, and discuss about smooth migration path from current best-effort access networks to DiffServ-enabled ones. Since broadband access networks are already widely deployed in the world, there are several important consideration factors when supporting DiffServ in broadband access networks. They are backward compatibility with DiffServ-unaware legacy systems, consistency with existing pricing infrastructure, effective QoS support for various applications, and so on. The DiffServ models proposed in this paper are divided into static and dynamic models. The static DiffServ models include Flat DiffServ providing per-subscriber DiffServ QoS and Structured DiffServ providing both per-service and per-subscriber DiffServ QoS. The dynamic DiffServ models include Direct DiffServ for peer to peer multimedia applications and Indirect DiffServ for applications of service providers. Based on the analysis of the pros and cons of the proposed models and the characteristics of current broadband access networks, smooth migration path toward QoS-enabled broadband access networks is also discussed.

1 Introduction

In the past decade there have been a lot of research and standardization activities for QoS architectures such as IntServ and DiffServ, traffic shaping and policing, queuing and scheduling, QoS signaling, and so on[1,2,3]. And in recent years there has been explosive growth in multimedia computing in Internet, which requires differentiated QoS support[4]. However, it is not easy to find QoS-enabled networks, particularly QoS-enabled broadband access networks, in real world because of huge investment to equip new QoS-enabled systems, difficulties in keeping backward compatibility with legacy QoS-unaware systems, complexity increase and performance degradation of QoS-enabled network systems, difficulties in keeping consistency with existing pricing infrastructure, and so on. Nevertheless, QoS support in broadband access network is indispensable because QoS-sensitive realtime multimedia applications such as voice of IP, teleconferencing, IP TV, and audio/video streaming are currently rapidly

T. Vazão, M.M. Freire, and I. Chong (Eds.): ICOIN 2007, LNCS 5200, pp. 255–264, 2008.

deployed in broadband access networks. We believe that it is very important to keep smooth migration from current best-effort networks toward QoS-enabled ones when supporting QoS in broadband access networks.

From both technical and economic viewpoints, DiffServ IP QoS architecture is accepted as a more practical solution in Internet world because the other IntServ architecture is more complex and has scalability problem. The DiffServ architecture can be differently applied according to the corresponding environments. In this paper, we propose several models to support DiffServ QoS in broadband access network environments. We also discuss about smooth migration path from current best-effort access networks to DiffServ-enabled ones after analyzing the characteristics of current broadband access networks and pros and cons of the proposed models. The DiffServ models proposed in this paper for broadband access networks are divided into static DiffServ models and dynamic DiffServ models. The static DiffServ models provided through static provisioning at subscription time, for example, include Flat DiffServ and Structured DiffServ. The Flat DiffServ provides per-subscriber DiffServ QoS, and the Structured DiffServ provides both per-service and per-subscriber DiffServ QoS. The dynamic DiffServ models, which provides dynamic mechanisms for changing DiffServ QoS parameters through the use of some QoS signaling protocols, include Direct DiffServ for peer to peer multimedia applications and Indirect DiffServ for applications of service providers.

2 Related Works

There were some proposals suggesting RSVP-based access networks under the assumption that RSVP would be widely deployed and most multimedia applications would be developed assuming RSVP as the resource reservation protocol[5,6]. But this does not seem to be the case right now, and it is believed that using RSVP even in the access network will introduce unneeded complexity. DSL-Forum suggested a two-phased approach based on DiffServ architecture for QoS-enabled DSL networks[7]. Phase 1 is a near-term solution and characterized by DiffServ provided through static provisioning, and phase 2 long-term solution adds a dynamic mechanism for changing the DiffServ QoS parameters through the use of a policy-based networking enhancement. [7] specifies only architectural requirements for the support of QoS-enabled IP services, but it does not present any idea on how to support phase 1 static DiffServ in current flat-rate pricing environments so as to keep backward compatibility with legacy QoS-unaware systems as much as possible. It also does not present any idea on what signaling interfaces are needed for phase 2 dynamic DiffServ QoS, on what signaling protocol can be used for the interfaces, and on how to migrate from phase 1 toward phase 2. Those issues will be mainly resolved in this paper. The issues related with static DiffServ are relatively simple. But signaling interface and protocol issues for dynamic DiffServ are different. European IST project AQUILA developed a dynamic signaling mechanism based on CORBA[8], [9] proposed COPS(Common Open Policy Service) protocol-based signaling mechanism for dynamic DiffServ. We propose dynamic DiffServ models in broadband access networks based on the COPS protocol since it is simple, application-independent, and standard protocol.

3 Static DiffServ Models

Static DiffServ is provided through static provisioning at subscription time or at the time when the SLA(Service Level Agreement) contracted at subscription time needs to be changed. If a SLA between a subscriber and an ISP is contracted, the subscriber always receives the QoS specified in the SLA and needs to be correspondingly charged. In this paper, we propose two different models to support static DiffServ in broadband access network environments. One is Flat DiffServ, the other is Structured DiffServ.

3.1 Flat DiffServ

Flat DiffServ is simplest model to support DiffServ in broadband access networks. A subscriber is provided with differentiated DiffServ QoS according to the SLA, but the QoS is flat in the sense that the traffic from a subscriber is equally treated. (a) of Fig. 1 shows the operational model of Flat DiffServ.

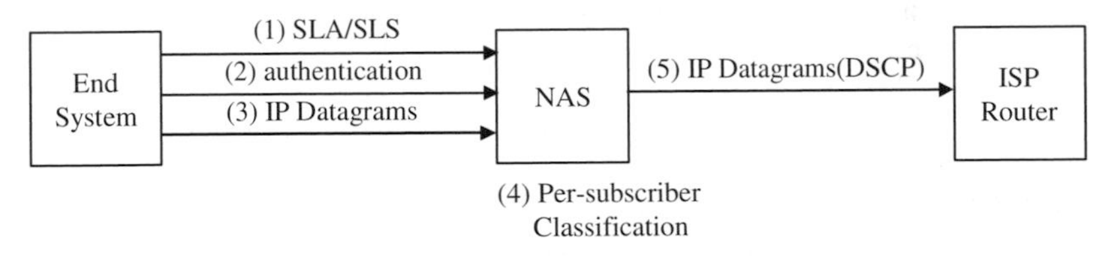

(a) Operational model of Flat DiffServ

Subscriber Class	QoS	PHB
Expedited	Dedicated bandwidth IP transfer service	EF
Assured	Statistical bandwidth IP transfer	AF
Best-effort	Best effort IP transfer service	BE

(b) An example of traffic classification for Flat DiffServ

Fig. 1. Flat DiffServ model

The SLA between a subscriber and an ISP is firstly established and corresponding SLS(Service Level Specification) is delivered to the NAS(Network Access Server), an equipment interconnecting an access network and an ISP backbone network, via off-line or on-line interface. The SLS of Flat DiffServ may include subscriber identification information, bandwidth allocated for the subscriber, service class, DSCP, pricing information, and so on. When a subscriber is trying to access Internet, the NAS will authenticate the subscriber based on the identification information of corresponding SLA and allocate IP address if necessary. Then the NAS will be ready to provide appropriate DiffServ for traffic from the subscriber based on the corresponding SLS. The NAS will appropriately mark and handle the IP datagrams from a subscriber based on the service class and DSCP of the SLS. In Flat DiffServ, classification of an IP datagram for DSCP marking will be done according to the source IP address because flat QoS is provided per-subscriber irrespective of application service. The marked datagrams will be appropriately treated in the DiffServ-enabled ISP backbone network.

In the Flat DiffServ model, subscribers of a broadband access network can be classified into several classes. (b) of Fig.1 shows an example classification of subscribers of a Flat DiffServ-enabled broadband access network. Expedited class provides dedicated bandwidth IP transfer service and is serviced through EF PHB(Per-Hob Behavior) defined in [10]. VPN(Virtual Private Network) subscribers may belong to this Expedited class. Subscribers who are using realtime multimedia application services can be classified into Assured class which provides statistical bandwidth IP transfer service. The Assured class may be divided into several sub-classes serviced through corresponding AF PHBs defined in [11]. Conventional Internet subscribers can be classified into Best-effort class serviced through BE PHB.

3.2 Structured DiffServ

Most significant problem of the Flat DiffServ model is that incoming traffic from a subscriber should be equally treated. That means that there is no ways to differently handle packets from a subscriber when congestion occurs, even though the subscriber are simultaneously using several application services of different QoS requirements. In the case of Expedited class where every packet is treated with highest priority in order to provide low delay, low jitter, and low error rate and in the case of Best-effort class where every packet is treated in FIFO-based, per-service differentiated QoS support for a subscriber will not be required. But it is different in the case of Assured class where statistical bandwidth IP transfer service is provided. In this case, packets of lower QoS requirement need to be discarded first by using WRED[12], for example.

Structured DiffServ allows an end-system to classify per-service traffic, if necessary, in addition to per-subscriber traffic classification of Flat DiffServ model at NAS, as shown in (a) of Fig. 2. (b) of Fig. 2 shows an example of per-service traffic classification for Structured DiffServ. Traffic of an Assured class subscriber can be further classified into CM(Conversational Multimedia) service traffic which has low drop precedence in case of congestion, SM(Streaming Multimedia) service traffic which has medium drop precedence, and CI(Conventional Internet) service traffic which has high drop precedence. If the Assured_x class is defined to be serviced through AFxy, the CM, SM, and CI service traffic can be serviced through AFx1, AFx2, and AFx3 respectively. Default per-service traffic class may be CI class. Traffic proportion of each per-service class within a per-subscriber class will be specified in SLA establishment, policed by NAS, and properly related with pricing policy.

As shown in (c) of Fig. 2, 6-bit DSCP field can be divided into 3-bit per-subscriber sub-DSCP field and 3-bit per-service sub-DSCP field to support Structured DiffServ. In Structured DiffServ-enabled broadband access networks, end-system is responsible for marking the per-service sub-DSCP and NAS is responsible for marking the per-subscriber sub-DSCP. If end-system does not mark the per-service sub-DSCP, NAS will mark default value for the sub-DSCP.

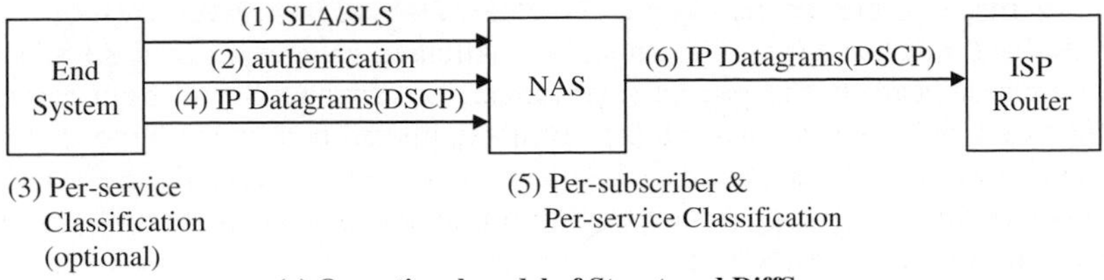

(a) Operational model of Structured DiffServ

Subscriber Class	Service Class	QoS		PHB
Expedited	-	Dedicated bandwidth IP transfer service		EF
Assured_x	CM	Low drop precedence	Statistical bandwidth IP transfer	AFx1
	SM	Medium drop precedence		AFx2
	CI	High drop precedence		AFx3
Best-effort	-	Best effort IP transfer service		BE

CM - Conversational Multimedia, SM – Streaming Multimedia, CI – Conventional Internet

(b) An example of traffic classification for Structured DiffServ

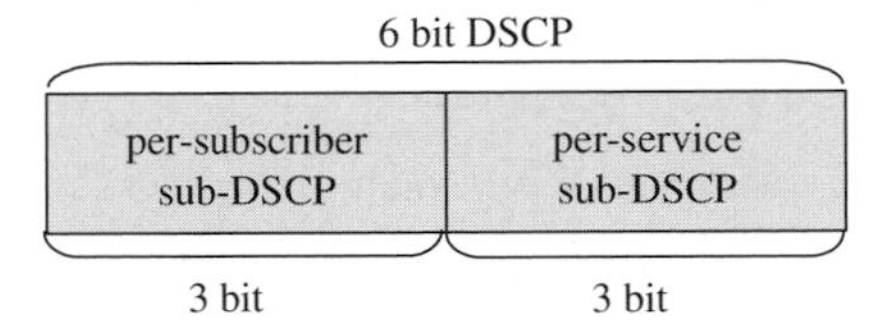

(c) An example of structured DSCP field

Fig. 2. Structured DiffServ model

4 Dynamic DiffServ Models

Since, in the static DiffServ models explained in the previous chapter, provisioning of network resources to a QoS client(a subscriber) is done on long-term scale(e.g., several months), it is difficult for the ISP to adapt to changes in traffic demand. Even when a realtime multimedia subscriber of Assured class does not use any realtime multimedia application, networks resources allocated to the client can not be released to support Best-effort class. This will lead to underutilization of network resources. There is no ways for a lower QoS subscriber to request higher QoS for a specific application(e.g., VoIP) in the static DiffServ models, too. Dynamic DiffServ models are proposed to overcome these problems.

Dynamic DiffServ allows a QoS client to dynamically request a QoS server to support and release the required QoS through a signaling mechanism. In the broadband access networks, NAS will become QoS sever. There may be two different QoS clients. In the case of P2P(Peer-to-Peer) applications, end-systems responsible for being charged will become QoS clients, and the DiffServ QoS will be directly requested by the application entities of end systems. It is different in the case of ASP(Application

Service Provider) applications where there are QoS proxy servers responsible for being charged. SIP Default proxy server, H.323 default gatekeeper, and RTSP streaming server may become QoS proxy servers, and DiffServ QoS will be indirectly requested by a QoS proxy server on behalf of the application entities of end systems. In this paper, two different dynamic DiffServ models, called Direct DiffServ and Indirect DiffServ, are presented to support P2P multimedia applications and ASP multimedia applications respectively.

4.1 Direct DiffServ

Fig. 3 shows an operational model of Direct DiffServ in broadband access networks. In order to request NAS to support DiffServ QoS, end system of an application collects identification information and QoS attributes for the constituent media streams when an application session is established((1) of Fig.3).

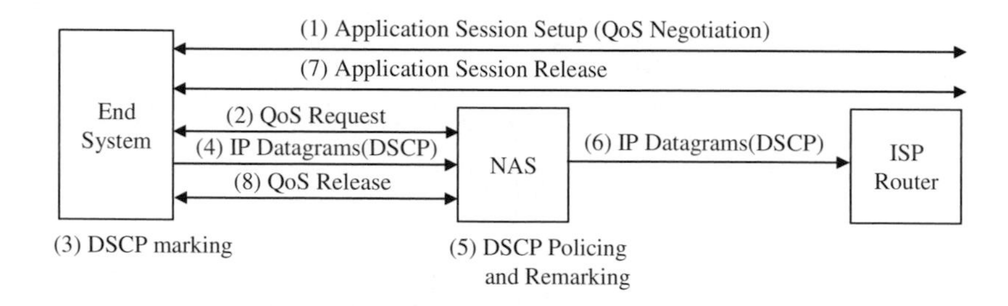

Fig. 3. Operational Model of Direct DiffServ

Typical stream identification information may include source IP address, destination IP address, transport protocol, source port, and destination port. And the QoS attributes may include media type(e.g., audio, video, data), codec type(e.g., PCMU, H.263), transmission type(e.g., sendrecv, sendonly, recvonly), and QoS support type(e.g., assured-QoS, enabled-QoS[6]).The end-system will determine the DiffServ class of each media stream based on the acquired QoS attributes, and then request NAS to admit the IP QoS required for the session. The request message will carry the QoS attributes and DiffServ class for each media stream of the session. The request will be admitted by NAS after checking pre-defined policy information of the user, resource allocation status of NAS, and result of query to bandwidth broker if necessary((2) of Fig.3).

End system authorized to use some DiffServ QoS for each media stream through the admission procedure will send corresponding DSCP-marked packets of the stream to NAS((3) and (4) of Fig.3), and the NAS will police whether the DSCPs are admitted ones or not. In case of invalid DSCP, the NAS will remark DSCP before queuing, prioritization, and forwarding toward an ISP router((5) and (6) of Fig.3). When the application session is released, end system will ask NAS to release QoS of the session((7) and (8) of Fig.3). Since, in this model, end system is performing the role of QoS PEP(Policy Enforcement Point) and NAS is acting as a QoS PDP(Policy Decision Point), the standard PEP-PDP COPS(Common Open Policy Service) protocol[13,14] can be used for the QoS signaling between end system and NAS. COPS REQ(Request), DEC(Decision), and DRQ(Delete Request) messages are correspondent with QoS request, QoS admission, and QoS release, respectively.

4.2 Indirect DiffServ

In Indirect DiffServ model, end systems are not directly involved in supporting QoS. Instead QoS proxy servers of a multimedia application are allowed to request and release DiffServ QoS on behalf of application entities of end systems. Fig. 4 shows an operational model of Indirect DiffServ in broadband access networks. When an application session is established, local QoS proxy server of the application(e.g., SIP

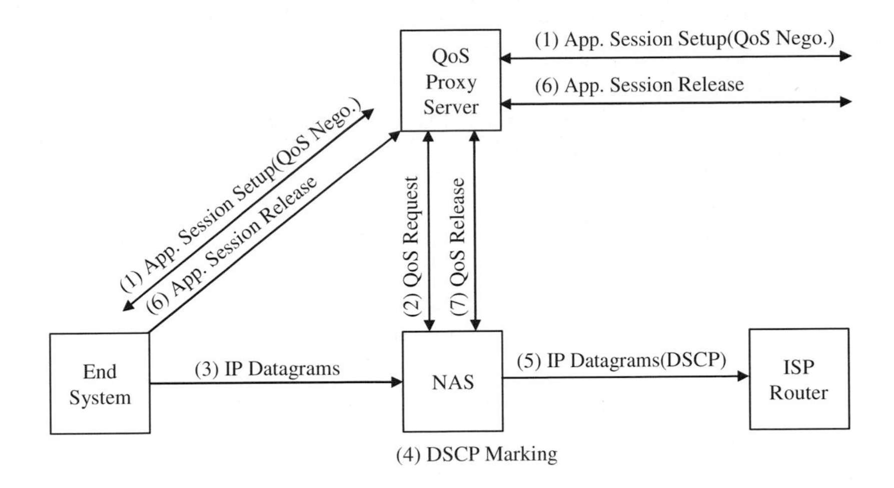

Fig. 4. Operational Model of Indirect DiffServ

default proxy server, H.323 default gatekeeper) collects the identification information and QoS attributes for the constituent media streams by capturing session establishment messages related with QoS negotiation, and determines the DiffServ class of each media stream((1) of Fig. 4). And then, the QoS proxy server requests NAS to admit the IP DiffServ QoS required for the session by sending a request message carrying the QoS attributes and DiffServ class for each constituent media stream. If the QoS request is successfully admitted, the QoS proxy server confirms a successful session establishment to the end system in application-dependent way((2) of Fig. 4).

The IP datagrams of a media stream arriving at NAS, after completion of the session establishment, will be appropriately classified, marked, processed, and forwarded toward an ISP router((3), (4), and (5) of Fig. 4). When the application session is released, the QoS proxy server will ask NAS to release QoS of the session((6) and(7) of Fig. 4). The QoS signaling between QoS proxy server and NAS can be based on the same COPS protocol as in Direct DiffServ model.

5 Comparison and Migration Path

Since broadband access networks are already widely deployed in the world, it is very important to keep backward compatibility with existing QoS-unaware end systems when supporting DiffServ in broadband access networks. Moreover, consistency with existing flat-rate pricing infrastructure needs to be importantly considered because flat-rate pricing seems to remain so long time. Fig. 5 shows a smooth migration path toward DiffServ-enabled broadband access networks. Since, in Flat DiffServ model, the QoS-related functions such as traffic classification and differentiated traffic handling are performed at NAS, end-systems(ES of Fig. 5) of Flat DiffServ model don't have to be QoS-enabled. This means that Flat DiffServ-enabled broadband access networks are backward compatible with legacy DiffServ-unaware end systems. Moreover, in the Flat DiffServ-enabled broadband access networks, existing flat-rate pricing infrastructure can be used with a slight modification to price based on both bandwidth and DiffServ class, not just on bandwidth. Therefore, it is relatively simple to migrate toward Flat DiffServ-enabled broadband access networks.

Structured DiffServ model inherits the advantages of Flat DiffServ model in the sense that per-subscriber DiffServ QoS is still provided. And the significant problem

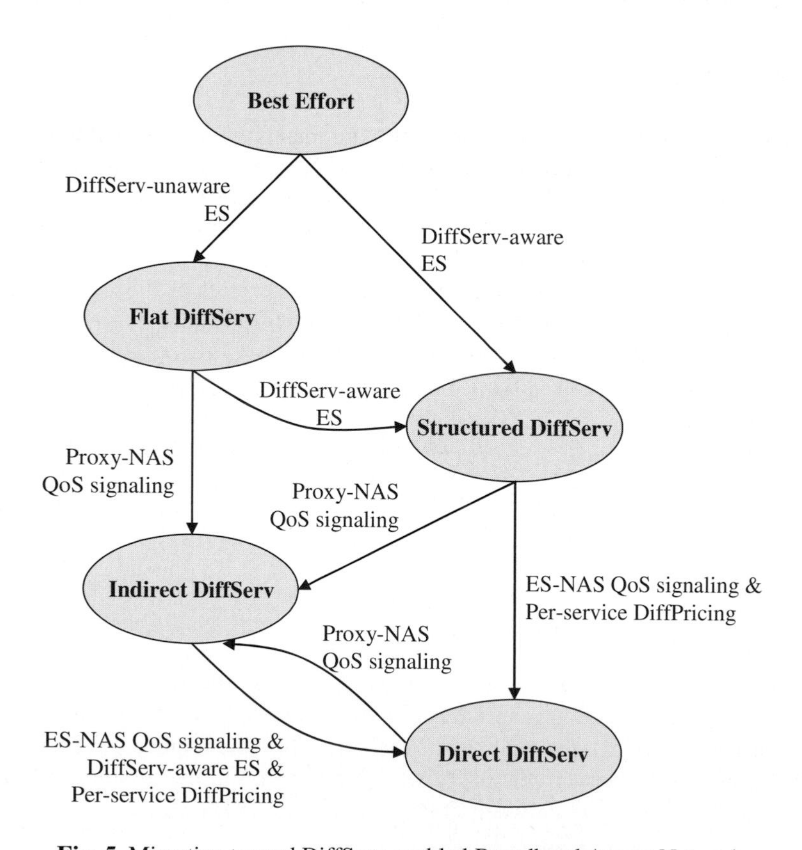

Fig. 5. Migration toward DiffServ-enabled Broadband Access Network

of Flat DiffServ model that incoming traffic from a subscriber should be equally treated is resolved in Structured DiffServ model, even though the per-service traffic classification is static and limited due to the constrained per-service sub-DSCP field. Flat DiffServ-enabled networks can migrate to support Structured DiffServ when end systems are able to support DiffServ. However, migration toward Structured DiffServ broadband access networks requires end-systems to be DiffServ-aware.

By using the QoS signaling mechanisms, dynamic DiffServ models can provide more flexible and effective DiffServ QoS for multimedia applications than static DiffServ models. However, in order to support Direct DiffServ in broadband access networks, several enhancements are needed in both end system and network sides. The end systems of Direct DiffServ-enabled networks should be DiffServ-aware and capable of supporting COPS protocol-based QoS signaling. NAS of a Direct DiffServ-enabled network needs to be enhanced to be capable of pricing based on per-service QoS usage for each subscriber as well as supporting DiffServ and COPS protocol-based QoS signaling. On the other hand, Indirect Diffserv model can be relatively easily deployed in the legacy DiffServ-unaware end system environments because end systems are not involved in supporting DiffServ QoS. However, Indirect DiffServ model can be supported only in ASP environments where there are QoS proxy servers, and requires a QoS signaling mechanism between QoS proxy server and NAS.

6 Conclusion

It is impractical to replace existing network equipments with QoS-enabled ones at a time in broadband access networks, even though QoS support is really necessary and related-technologies are readily available, because broadband access networks are already widely deployed in the world. Thus it is very important to keep smooth migration from current best-effort networks toward QoS-enabled ones when supporting QoS in broadband access networks. In this paper, we presented several DiffServ-based models to support QoS in broadband access networks, and discussed a migration path toward DiffServ-enabled broadband access networks. Though there might be different cases according to different requirements in different environments, it is believed that practical approach to support QoS in broadband access networks is to follow Flat DiffSrev model first, Structured DiffServ model second, Indirect DiffServ model third, and finally Direct DiffServ model. QoS support in broadband access network is indispensable because QoS-sensitive realtime multimedia applications such as voice of IP, teleconferencing, IP TV, and audio/video streaming are currently rapidly deployed in broadband access networks. We believe that the proposed DiffServ models and migration path discussed in this paper will contribute to rapid migration toward QoS-enabled broadband access networks.

References

1. Gozdecki, J., Jajszczyk, A., and Stankewicz, R. : Quality of Service Terminologies in IP Networks. IEEE Communications Magazine (March 2003)
2. Soldatos, J., Vayias, E., Kormentzas, G. : On The Building Blocks of Qualitity of Service in Heterogeneous IP Networks. IEEE Communications Surveys & Tutorials (First Quarter 2005)

3. Vali, D., Paskalis, S., Merakos, L., Kaloxylos, A.: A Survey of Internet QoS Signaling. IEEE Communications Surveys & Tutorials (Fourth Quarter 2004)
4. Bai, Y., Ito, M.R.: QoS Control for Video and Audio Communication in Conventional and Active Networks: Approaches and Comparsion. IEEE Communications Surveys & Tutorials (First Quarter 2004)
5. Bernet, Y. : The Complementary Roles of RSVP and Differentiated Services in the Full-Services QoS Network. IEEE Communications Magazine (February 2000)
6. Sargento, S., et al.: IP-Based Access Networks for Broadband Multimedia Services. IEEE Communications Magazine (February 2003)
7. DSL-Forum TR-059 : DSL Evolution - Architecture Requirements for the Support of QoS-Enabled IP Services (September 2003)
8. Engel, T., et al. : AQUILA: Adaptive Resource Control for QoS Using an IP-Based Layered Architecture. IEEE Communications Magazine (January 2003)
9. Salsano, S., Veltri, L.: QoS Control by Means of COPS to Support SIP-Based Applications. IEEE Network (March/April 2002)
10. Davie, B., et al.: An Expedited Forwarding PHB(Per-Hob Behavior). IETF RFC 2598 (March 2002)
11. Heinanen, J., et al.: Assured Forwarding PHB Group. IETF RFC 2597 (June 1999)
12. Floyd, S., Jacobson, V.: Random Early Detection Gateways for Congestion Avoidance. IEEE/ACM Trans. Net. 1(4) (August 1993)
13. Durham, D., et al. : The COPS(Common Open Policy Service) Protocol. RFC 2748 (January 2000)
14. Chan, K., et al. : COPS Usage for Policy Provisioning (COPS-PR). IETF RFC 3084 (March 2001)

Integration of Broadband and Broadcasting Wireless Technologies at the UMTS Radio Access Level

Natasa Vulic[1], Sonia Heemstra de Groot[1,2], and Ignas Niemegeers[1]

[1] Technical University of Delft, Faculty of Electrical Engineering, Mathematics and Computer Science, Department of Wireless and Mobile Communications, Mekelweg 4, 2628CD Delft, The Netherlands
{N.Vulic,S.M.HeemstradeGroot,I.Niemegeers}@ewi.tudelft.nl
[2] Twente Institute for Wireless and Mobile Communications, Institutenweg 30, 7521PK Enschede, The Netherlands

Abstract. The UMTS radio resources are insufficient to support services that demand high bandwidth. Integration of heterogeneous wireless access systems into UMTS may be a way to overcome this problem. In the market, there are different wireless technologies, ranging from those that provide high and cheap bandwidth to those specially tailored for mass multimedia services. In this paper, we address the integration of broadband and broadcasting technologies at the UMTS radio access level. More specifically, we discuss benefits of such integration for mobile operators and present necessary architectural and protocol modifications to the UMTS network.

Keywords: UMTS, integration, radio access, WLAN, WiMAX, DVB-H, resource management.

1 Introduction

The demands for accessing services at high data rates while on the move, anyplace and anytime, have resulted in the current research activities towards the integration of heterogeneous mobile and wireless systems. In the market, there are various types of wireless access technologies that cover different areas, support different levels of mobility, work in different frequency bands and are tailored for different services. The Universal Mobile Telecommunication System (UMTS) is a third-generation (3G) mobile network that initially offered voice and moderate data rates, while the recent enhancements have brought higher data rates and multicast services. Wireless Local Area Networks (WLANs), such as IEEE 802.11, provide at least one degree higher license-free capacity to slow-moving users at hotspots. Due to their complementary characteristics, 3G-WLAN interworking has drawn a lot of attention from the research community. When the Wireless Metropolitan Area Networks (WMANs) enhanced for mobility hit the market, like IEEE 802.16e that is also known as the Worldwide Interoperability for Microwave Access (WiMAX) system, research efforts have been redirected to the integration of this technology, on one side, with complementary WLANs, and on the other, with UMTS for extra capacity provision. Furthermore, although existing for a long time, Digital Broadcasting Systems (DBS) were given a high attention

T. Vazão, M.M. Freire, and I. Chong (Eds.): ICOIN 2007, LNCS 5200, pp. 265–274, 2008.

again with the emergence of new standards that provide one-to-many multimedia services to fast moving portable devices. Such technologies are the European Digital Video Broadcasting for handheld devices (DVB-H), the Korean Digital Multimedia Broadcast (DMB), the UK Digital Audio Broadcast (DAB-IP), the US Qualcomm's Media Forward Link Open Standard (MediaFLO), etc.

For interconnection of heterogeneous systems, mobility management mechanisms may be deployed at different protocol layers, from the Medium Access Control (MAC) to the application layer [1]. This may be accompanied with different solutions for the integrated QoS support and security as well as the coordinated radio resource management. Combinations of interworking mechanisms have resulted in a variety of interworking architectures. Basically, all of them may fit into two general interworking approaches: either the networks remain independent of each other, possibly belonging to different administrative domains or one network is embedded in another and run by the common operator. In this paper, we discuss the second approach, referred to here as the UMTS-based integration, where different broadband and broadcasting wireless systems may be integrated into the UMTS network. More specifically, we consider the UMTS-based integration at the radio access level. This interworking architecture promises good vertical handover performance and facilitates the optimized use of all available radio resources. In addition, most of the UMTS functionality related to QoS support and security may be reused in the whole integrated network without additional modifications.

This paper is organized as follows. In Section 2, we provide an overview of the UMTS network architecture and radio interface protocols as well as the UMTS-based integration at the radio access level. In Section 3, after briefly presenting candidate wireless technologies, we give a comparison and discuss how their integration may be beneficial for the mobile operator. Modifications to the UMTS architecture are described in Section 4. Finally, some conclusions are drawn in Section 5.

2 Integration at UMTS Radio Access Level

2.1 UMTS Network Architecture

The UMTS network architecture, as shown in Figure 1, consists of the Core Network (CN) and the Access Network (AN) [2]. The CN is responsible for higher layer functions, such as mobility management, session management, call control, connectivity to other data networks, etc. The UMTS Rel'6 CN includes the Circuit-Switched (CS) and Packet-Switched (PS) domains as well as the IP Multimedia System (IMS). We consider here only the CN PS domain that consists of the Serving GPRS Supporting Nodes (SGSNs) and the Gateway GPRS Support Nodes (GGSNs), interconnected via the Gn interface. These network entities are connected to the Home Location Register (HLR) over the Gr and Gc interfaces, respectively. For the support of Multimedia Broadcast Multicast Services (MBMS), the Broadcast Multicast Service Center (BM-SC) is attached to the GGSN via the Gmb and Gi interfaces [3].

The SGSN is further connected via the Iu interface to the ANs. Here, we present the Wideband Code Division Multiple Access (WCDMA)-based UMTS Terrestrial Radio Access Network (UTRAN). The UTRAN includes one or more Radio Network

Subsystems (RNS), which consist of a Radio Network Controller (RNC) and base stations NodeBs. The RNC is connected to NodeBs and neighboring RNCs via the Iub and the Iur interfaces respectively, while the Uu radio interface is defined between the User Equipment (UE) and a NodeB.

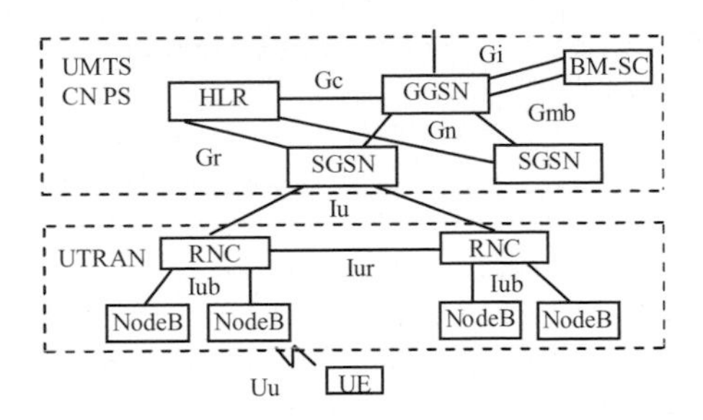

Fig. 1. UMTS Network Architecture

2.2 Radio Interface Protocol Architecture

The protocol architecture at the Uu radio interface consists of three layers. Above the physical layer (PHY), there is the link layer that includes the Medium Access Control (MAC), the Radio Link Control (RLC), the Packet Data Convergence Protocol (PDCP) and the Broadcast/Multicast Control (BMC) protocols. The key UTRAN control protocol is the Radio Resource Control (RRC) that belongs to the network layer and is responsible for the overall RNS resource management.

Several types of communication channels are defined: logical, transport and physical channels. The MAC data transfer service is offered to the RLC by means of logical channels, which indicate the type of information – signaling or data traffic - that is to be transferred. Mapping between the logical channels and transport channels is done at the MAC layer, according to the RRC decision. Transport channels denote how the transmission is actually performed and may be common for all users or dedicated to a single user. They are further mapped to the physical channels at the physical layer. The protocol architecture at the radio interface together with transport and logical channels specified for the UMTS Frequency Division Duplex (FDD) mode is depicted in Figure 2.

2.3 Radio Access Integration Approach

As shown in Figure 3, integration at the UMTS radio access level denotes that an additional wireless access technology is attached via a new UTRAN network interface (Iuw) at the RNC. Since the RNC is the highest common network element for all types of base stations, switching between them is expected to be seamless, similarly to pure UMTS handovers. In addition, the fact that most of the UTRAN Radio Resource Management (RRM) functionality is situated in the RNC and the information on radio resource utilization is available locally facilitates common management of all the available radio resources without specifying new network entities or considerable protocol modifications.

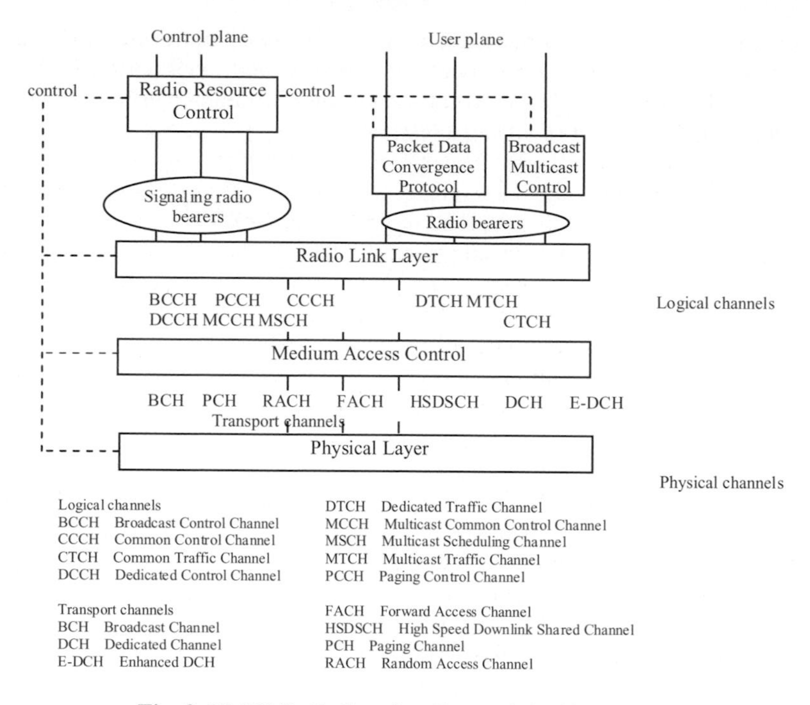

Fig. 2. UMTS Radio Interface Protocol Architecture

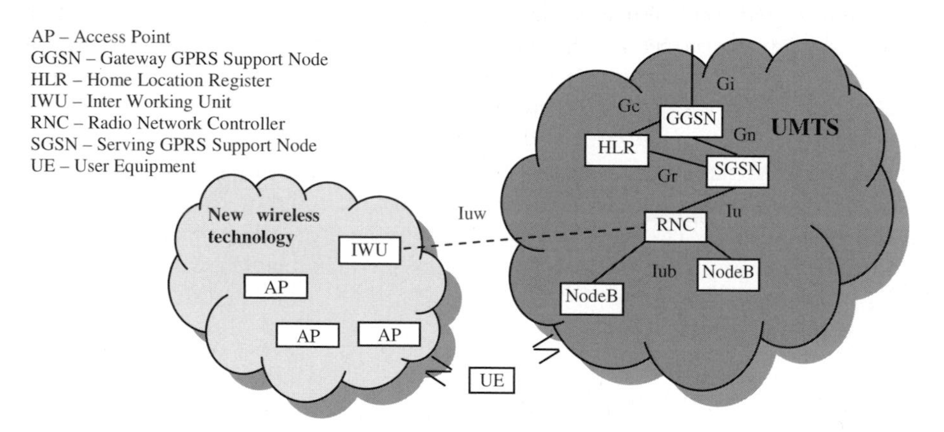

Fig. 3. Integration of an infrastructure-based wireless technology at the UMTS radio access network

The integration at the UMTS radio access network may be performed at different protocol layers [4]. Integration at the MAC level is chosen, as it is already responsible for scheduling different services on the air, in accordance with the decision made by the RRC. Besides, the upper data link layer protocols are not affected and the UMTS encryption and integrity protection (performed at the RLC or the MAC layer, depending on the RLC mode) can be applied in the whole integrated network without additional modifications. This may be important since security mechanisms specified for the embedded wireless technologies may considerably differ from UMTS. Furthermore, all UMTS higher layer control procedures, including those for the QoS negotiation and authentication may be reused without additional modifications.

The advantages of this integration approach come at the cost of modifications to the involved networks as well as the introduction of an Inter-Working Unit (IWU) for adjustment to the UMTS radio protocols. The modifications are discussed more in detail in Section 4, while here we only point out that they are technology specific and the integration mechanisms applied to one wireless technology cannot be straightforwardly reused for embedding another wireless access system. However, with a careful initial design, a set of common control procedures may be specified. The modifications also depend on the capabilities provided within the embedded wireless technology, referred to as architectural options. The first option denotes that only user traffic is transferred via the embedded technology, which requires the UMTS air interface to be always active for, at least, the provision of signaling. The second option allows transfer of both user and control information via any access technology, which makes sufficient the use of a single air interface at a time, but requires considerable modifications to various UMTS control mechanisms.

We consider here the first architectural option. The user connects to the integrated network via the WCDMA radio interface in the usual way, which is followed by the exchange of information on the device capability. Having information on the available UE air interfaces as well as the current heterogeneous environment, the RNC makes a decision on the most suitable type of base station for a particular session according to the session QoS requirements as well as current network load, cost, user preferences, velocity, etc. When needed, an ongoing session may be handed over to a base station of another type, which may be initiated and performed independently of other user sessions. Depending on the configuration, this interworking architecture may allow transfer of different sessions, different flows of the same session or temporarily a single session over different radio interfaces.

3 Candidate Wireless Technologies

In this section, we give a brief overview and a comparison of suitable wireless technologies for the integration.

WLAN The IEEE 802.11 standard specifies the PHY and MAC layers in license-free frequency bands, either 2.4GHz (802.11b and 802.11g) or 5GHz (802.11a). They provide up to 54 Mbps data rates, while the emerging 802.11n standard aims at data rates over 100 Mbps. The original standard specified best effort WLAN with limited mobility and weak security. Enhancements of the MAC layer resulted in the improved mobility management (802.11f), security mechanisms (802.11i) and QoS support (802.11e), while the work on resource management (802.11WGk), fast handover (802.11WGr) and other issues is underway [5]. A fundamental building block of the 802.11 infrastructure-based architecture is the Basic Service Set (BSS), consisting of an Access Point (AP) and mobile stations.

WMAN The original WiMAX IEEE 802.16 standard specified the physical and MAC layers for fixed wireless networks in the licensed 10-66GHz spectrum. The subsequent amendments extended the physical layer specification for non-line-of-sight communications in the licensed and license-exempt 2-11GHz frequencies. The support for mobility at vehicular speeds was specified by IEEE 802.16e [6]. The IEEE 802.16 MAC layer is connection-oriented and able to support different QoS requirements. The MAC layer consists of three sublayers. The service specific convergence sublayer maps different types of traffic, while the MAC common part sublayer functionality includes fragmentation and segmentation, scheduling and retransmission, QoS control, connection establishment and maintenance. The lowest privacy sublayer is responsible for authentication, secure key exchange and encryption. WiMAX may provide up to 70Mbps in the 20MHz channel bandwidth.

DBS As a represent of the emerging broadcasting technologies for fast moving mobile devices, we here consider DVB-H, an enhanced version of the Digital Video Broadcasting Terrestrial (DVB-T) standard. DVB-H [7] specifies the physical layer and the mandatory elements of the link layer. The link layer elements are time slicing and multiprotocol encapsulation-forward error correction (MPE-FEC). The time slicing technique transfers data in bursts, which enables significant power reduction. The optional MPE-FEC employs advanced channel coding and interleaving. The physical layer specifies the following additions: the upgrade of the transmitter parameter signaling (TPS), 4K-transmission mode, a new symbol interleaver and a new 5MHz channel bandwidth [8].

Characteristics of the presented wireless access technologies are summarized in Table 1. The main reason for the integration of different wireless technologies into UMTS is its insufficient capacity for the provision of high-bandwidth demanding multimedia services. Even with the newest enhancements for High Speed Packet Access (HSPA), the UMTS data rates (14.4 Mbps in the downlink and 5.7Mbps in the uplink) are still considerably lower than those provided by the existing broadband technologies, WLAN IEEE 802.11 (54Mbps) and WiMAX IEEE 802.16e (70Mbps).

802.11 may provide additional high and cheap bandwidth, but only to slowly moving users in hotspots. In contrast, WiMAX uses licensed spectrum as UMTS (although likely less expensive) and provides high data rates to fast moving users, which makes it an alternative to UMTS for wide area and a direct competitor. Nevertheless, the UMTS integration with the WiMAX technology as an overlay network may be beneficial to minimize the operator's investments or fill the capacity gap until the UMTS emerging enhancements enable similar data rates in the uplink and downlink. Due to fewer sites required, the WiMAX integration may provide less frequent handovers and thus improve system performance.

The UMTS enhancements for the MBMS services enable cost-efficient provision of one-to-many multimedia services over the existing infrastructure. Since DVB-H offers the same type of service and covers similar geographical areas, it can be also seen as a competitor to UMTS in this regard. According to [9], where the DVB-H and MBMS performance are compared, DVB-H is preferable because it can offer more simultaneous video channels within one cell with better resolution than UMTS MBMS. In addition, the larger DVB-H cell size also means less frequent handovers. However, the main problem with the UMTS MBMS service, especially when the number of UMTS subscribers increases, could be lack of resources for more profitable point-to-point services. The DVB-H integration into the UMTS MBMS architecture may be a solution. As a return channel for DVB-H, different UMTS transport channels may be used.

Different access technologies may be implemented in different environments, depending on the mobile operator's needs. For example, in rural areas, the selection of one wide-area technology that can support all types of services may be sufficient. In urban areas, with high demands for capacity, it may be that all the available bandwidth is needed. While the embedded wireless technologies bring new capacity for fast revenue generating services to mobile operators, the UMTS advantages include the reuse of well-defined and proven control mechanisms, such as resource management or billing mechanisms, which many of the technologies lack.

Table 1. Characteristics of wireless access technologies

	UMTS	IEEE 802.11	IEEE 802.16e	DVB-H
Area	Wide	Local	Wide	Wide
Frequency spectrum	Licensed 2GHz	License-free (2.4 GHz and 5GHz)	Licensed below 3.5GHz	Licensed 470-860 MHz (UHF) and 1.5GHz (L-band)
Cell radius	1km	300m	6km	30km
Velocity	High	Low	High	High
Max capacity [Mbps]	2 (rel'99) DL: 14.4 UL: 5.7 (HSPA)	54	70	5-11
Channel bandwidth [MHz]	5	20	1.25-20	5 (6,7,8)
Services	Voice, data, MBMS	Data	Services with different QoS requirements in point-to-multipoint and mesh modes	Broadcast
Direction	bidirectional			unidirectional
Billing mechanisms	yes	no		
Resource management	yes	under specification (802.11WGk)	no	

4 Modifications

Integration at the UMTS radio access level requires various modifications to the involved networks. We discuss here primarily the UMTS changes, aiming at designing them in such a way that every new wireless technology may be embedded with a little effort, introducing only unavoidable technology-specific modifications.

The UMTS modifications are mainly localized in the UTRAN, leaving the UMTS PS CN unaffected, except for an upgrade to support the additional traffic. The UTRAN modifications include modifications to the MAC and RRC protocols, specification of new protocols in the control and user plane at the Iuw interface, introduction of an IWU and the upgrade of the resource management algorithms.

Regarding the MAC protocol, the integration of the wireless broadband systems introduces modifications to the MAC-d entity, while the integration of a broadcasting technology affects the MAC-c/sh/m entity [10]. For each wireless technology, new transport channels are specified, which are depicted in Figure 4 together with potential mapping to the logical channels. For IEEE 802.11, the WLAN Downlink Shared Channel (WDSCH) and the WLAN Uplink Shared Channel (WUSCH) are introduced in the downlink and uplink. Integration of the WiMAX system adds the WiMAX Downlink Channel (MDCH) and the WiMAX Uplink Channel (MUCH), while a new DVB-H transport channel is named the DVB-H Downlink Channel (HDCH). The permitted mapping possibilities depend on the chosen architectural option. If both data and control information are to be transferred over a particular technology, then any type of logical channels may be mapped to the new transport channels. Otherwise, only traffic logical channels are allowed, while the related control logical channels should be mapped to an original UMTS transport channel.

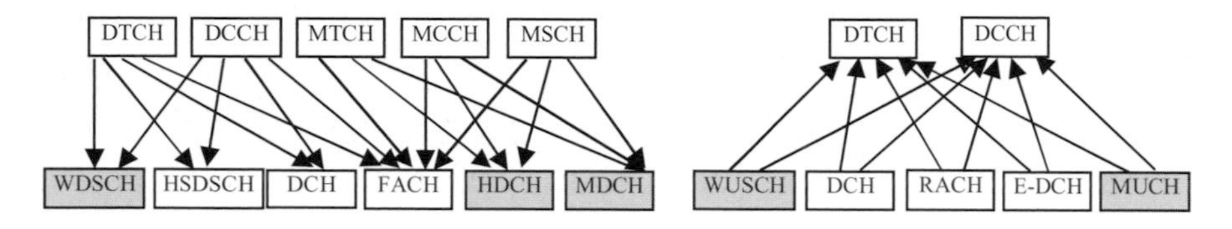

Fig. 4. An example of mapping between logical and transport channels in the downlink (left) and in the uplink (right)

Switching between the transport channels is made according to how the RRC configured the MAC over the related control interface. As an example, for multicast services, different transport channels may be used. For a large number of subscribers, UMTS Rel'6 standard specifies point-to-multipoint connections, where multicast specific logical channels (MTCH, MCCH, MSCH) are mapped to the common FACH transport channel. When the number of subscribers gets lower than the predefined threshold, the multimedia content may be provided over point-to-point connections (DTCH->DCH). The integration of DVB-H adds a new mapping in the downlink between the multicast logical channels and the HDCH transport channel. The multicast content can also be provided over other wireless technology with QoS and multicast support (here, also MDCH). The mapping in the uplink shows a diversity of transport channels that may be used as a DVB-H return channel.

In addition to the modifications of the existing MAC entities, a new MAC entity has to be added to the IWU. Its functionality depends on the characteristics of a particular wireless technology. For example, for the best effort 802.11, it may simply translate between the data formats and optionally prioritize between sessions and users. For 802.16e, it has to map between the frames received from the UTRAN and the WiMAX connections with an appropriate QoS translation. For DVB-H, it may multiplex streams sent from the RNC with location-specific information.

Switching among the transmission alternatives cannot be performed without modifications of the RRC protocol [11]. This complex UTRAN control protocol is responsible for setting the transport channel and the transmission parameters according to the negotiated QoS level, for handover to another base station, etc. The modifications are required for different RRC elementary procedures and the RRC state model. An example of vertical handover between a WiMAX IEEE 802.16e AP and a WLAN IEEE 802.11 AP is shown in Figure 5. After entering the WLAN coverage, the UE sends an updated measurement report to the RNC with the information on all types of neighboring base stations. If the WLAN interface is more suitable for an ongoing data session, the network may decide on vertical handover and send an RRC command for the association with a particular AP. The authentication and association requests are forwarded via the AP and the IWU to the RNC; first to check if the user is a registered subscriber, while the second message is sent for location information update. After the radio bearer reconfiguration to the WLAN transport channels, the session is transferred via WLAN, while the 802.16e resources are released.

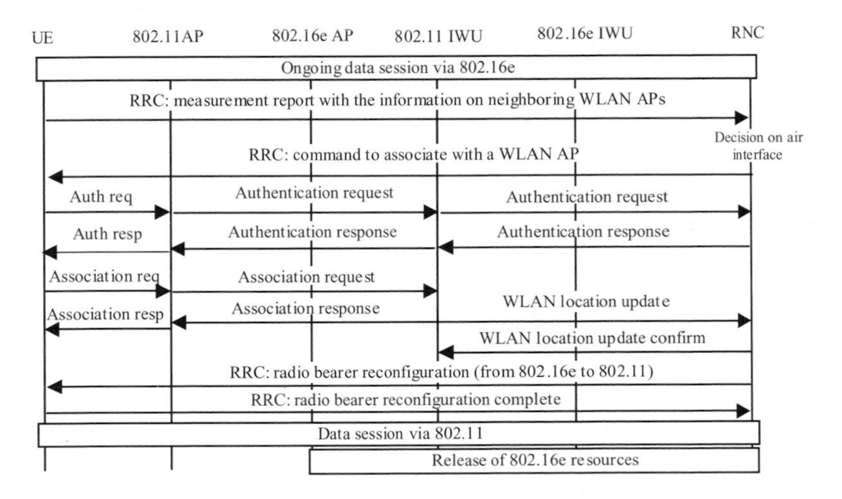

Fig. 5. Vertical handover between an IEEE 802.16e AP and an IEEE 802.11 AP, both integrated into UMTS at the radio access level

As it can be concluded from the previous example, such tight integration of heterogeneous networks introduces inevitable modifications to different resource management mechanisms and algorithms. The integration at the RNC level facilitates the addition of the Common Radio Resource Management (CRRM) functionality as an upgrade to the UTRAN RRM. Different degrees of CRRM-RRM cooperation are actually possible [12]. For this architecture, the degree primarily depends on the chosen architectural option in addition to the existence of the technology-specific RRM. Most wireless technologies do not specify resource management mechanisms, allowing the mobile device to select a point of attachment, usually according to signal strength. In such a case, the RRM of the embedded technology is likely to be added as part of the CRRM. Modifications are also needed to the mobile device; for the AP selection, the related algorithm has to give priority to the network decisions.

Finally, new protocols in the control and the user plane at the integration interface Iuw have to be specified. The Iuw control protocol has to allow communication between the IWU and RNC related to the user access via the embedded technology, handover issues and common radio resource management in general. The Iuw frame protocol has to support data transfer between the RNC and the IWU.

5 Conclusions

In this paper, we have addressed integration of heterogeneous wireless technologies at the UMTS radio access level. This integration approach provides flexibility when switching between the wireless access systems, promises seamless vertical handovers and facilitates common management of the overall radio resources. The advantages come at the cost of different modifications that are specific to the embedded technologies. With carefully designed initial modifications, this interworking architecture may enable integration of various wireless technologies in an easy manner, providing the users with an optimized heterogeneous environment.

References

1. Lampropoulos, G., Passas, N., Merakos, L., Kaloxylos, A.: Handover Management Architectures in Integrated WLAN/ Cellular Networks. IEEE Communications Surveys & Tutorials 7(4), 30–44 (2005)
2. Third Generation Project Partnership, http://www.3gpp.org
3. 3GPP TS 23.246 v6.10.0: Multimedia Broadcast / Multicast Service (MBMS); Architecture and Functional Description (2006)
4. Vulic, N., Heemstra de Groot, S., Niemegeers, I.: Architectural Options for UMTS-WLAN Integration at Radio Access Level. In: 1st International Workshop on Convergence of Heterogeneous Wireless Networks (2005)
5. IEEE 802.11 Wireless Local Area Networks, http://grouper.ieee.org/groups/802/11/
6. IEEE Std 802.16e-2005 IEEE Standard for Local and Metropolitan Area Networks Part 16: Air Interface for Fixed and Mobile Broadband Wireless Access Systems Amendment 2: Physical and Medium Access Control Layers for Combined Fixed and Mobile Operation in Licensed Bands and Corrigendum 1 (2006)
7. ETSI EN 302 304 V1.1.1 Digital Video Broadcasting (DVB): Transmission System for Handheld Terminals (DVB-H) (2004)
8. Faria, G., Henriksson, J., Stare, E., Talmola, P.: DVB-H: Digital Broadcast Services to Handheld Devices. Proceedings of the IEEE 94(1), 194–209 (2006)
9. Herrero, C., Vuorimma, P.: Delivery of Digital Television to Handheld Devices. In: 1st IEEE International Symposium on Wireless Communication Systems, pp. 240–244. IEEE Press, Los Alamitos (2004)
10. 3GPP TS 25.321 v6.8.0: Medium Access Control (MAC) Protocol Specification (2003-2006)
11. 3GPP TS 25.331 v6.10.0: Radio Resource Control (RRC) Protocol Specification (2006)
12. Perez-Romero, J., Sallent, O., Agusti, R., Diaz-Guerra, M.A.: Radio Resource Management Strategies in UMTS. Wiley, Chichester (2005)

A Mobility-Aware Resource Reservation Mechanism for the Future Generation Network Environment*

Seong-Ho Jeong and Ilyoung Chong

Department of Information and Communications Engineering
Hankuk University of Foreign Studies, Korea
{shjeong,iychong}@hufs.ac.kr

Abstract. This paper proposes a mobility-aware resource reservation mechanism to support quality of service in the future generation network environment where various wireless/wired access networks are converged to an IP-based core network. The proposed mechanism aims to provide real-time application-driven resource reservation in the mobile environment where mobile nodes move frequently between IP-based wireless access networks. The proposed mechanism uses localized path management to provide seamless QoS to the mobile nodes. We also present simulation and experimental results to show that the proposed/implemented mechanism works well.

1 Introduction

The Next Generation Network (NGN) is a popular phrase used to describe the future network that will replace the current network used to carry voice, fax, data, etc. The NGN is a managed packet-based network that enables a wide variety of services including VoIP, videoconferencing, instant messaging, e-mail, and other kinds of packet-switched communication services.

More specifically, according to ITU-T Recommendation Y.2001 [11], the NGN is defined as a packet-based network which is able to provide telecommunication services and to make use of multiple broadband, QoS-enabled transport technologies, and in which service-related functions are independent from underlying transport-related technologies. It also offers unrestricted access by users to different service providers and supports generalized mobility which will allow consistent and ubiquitous provision of services to users.

One of the important aspects of the NGN is to provide the ability for mobile entities to communicate and access services irrespective of changes of the location or technical environment. In particular, latency and data loss incurred during handover should be within a range acceptable to users (e.g., below a certain limit) for real-time services. To meet this requirement, there should be a way to provide fast resource reservation on the new path with mobility awareness.

RSVP [4] is a well-known resource reservation protocol to setup resources on the data path for real-time traffic. However, it is not suitable for resource reservation in

* This work was supported by Korea Research Foundation Grant funded by Korea Government (MOEHRD, Basic Research Promotion Fund) (KRF-2005-003-D00214).

T. Vazão, M.M. Freire, and I. Chong (Eds.): ICOIN 2007, LNCS 5200, pp. 275–284, 2008.

mobile networks. For example, a change in the location of a mobile node (MN) may make the reserved resources on the old path useless, and resources on the new path are reserved while maintaining the old reservation (double reservation problem). This results in the waste of network resources. In addition, signaling delay (end-to-end) is caused for resource re-reservation after handover, which may have a significant impact on the application.

To overcome such drawbacks of RSVP, many solutions have been proposed, which are mostly based on the modification or extension of RSVP [5, 6, 7]. However, most of RSVP-based solutions may not be useful for preventing service disruption during handover.

This paper proposes a mobility-aware resource reservation mechanism to support quality of service in the future generation network environment where various wireless/wired access networks are converged to an IP-based core network. The proposed mechanism aims to provide real-time application-driven resource reservation in mobile environments where mobile nodes move frequently between IP-based wireless access networks. The proposed mechanism uses localized path management to provide seamless QoS to the mobile nodes. The rest of the paper is organized as follows. Section 2 describes IP-based QoS signaling requirements for resource reservation, and Section 3 describes the proposed mechanism, called Mobility-aware Resource Reservation Mechanism (MRRM). Section 4 presents simulation and experimental results to show that the proposed/implemented mechanism works well. Finally Section 5 provides concluding remarks.

2 QoS Signaling Requirements

ITU-T Recommendation E.800 [12] defines QoS as the collective effect of service performance which determines the degree of satisfaction of a user of the service. Fig. 1 shows the relationship between a customer and a service provider with respect to QoS, and the next generation network environment where various wireless/wired access networks are converged to an IP-based core network.

To support QoS requested by a user, a certain amount of resources need to be reserved using QoS signaling. ITU-T Recommendation Q.Sup51 [13] provides requirements for signaling information on IP-based QoS at the interface between the user and the network (UNI), and at the interface between the different types of networks (NNI).

Some of the key requirements for UNI signaling include the derivation of attributes of user QoS request, flow control for user QoS requests/re-requests, network response to user QoS request, and user reaction to network QoS response.

Signaling requirements for NNI include the derivation of attributes of network QoS request, performance requirements for QoS requests and re-requests, response to network QoS request, and accumulating performance for additional request. Others include QoS release, performance optimization, support for the symmetry of information transfer capability, contention resolution, error reporting, parameters and values for transport connections, user-initiated QoS resource modification, support for emergency service, and reliability/priority attributes.

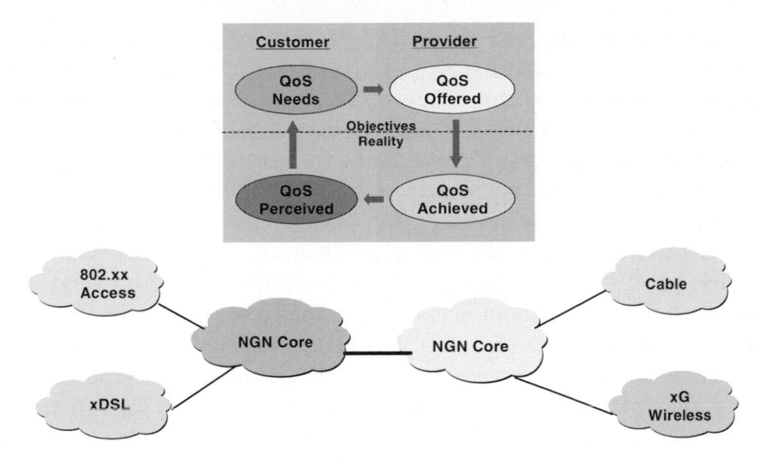

Fig. 1. QoS and Next Generation Network Environment

These signaling requirements can be used as a basis to enable the development of a signaling protocol capable of requesting, negotiating, and ultimately delivering IP QoS classes/parameters from UNI to UNI, spanning NNIs as required.

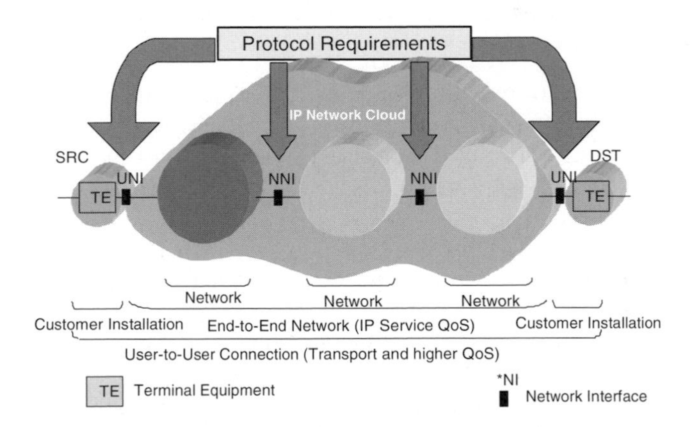

Fig. 2. Scope of ITU-T QoS Signaling Requirements

Terminal Equipment (TE) shown in Fig. 2 may be mobile and move frequently between different IP-based wireless access networks. IP-based mobility causes topological changes due to the change of network attachment point. Topological changes entail the change of routes for IP packets sent to or from an MN and may lead to the change of host IP addresses. These changes of route and IP addresses in mobile environments are typically much faster and more frequent than generic route changes in the wired network and may have some significant impact on resource reservation.

To develop a resource reservation mechanism which is suitable for mobile networks, it is first necessary to analyze the key differences between generic route changes and mobility. The generic route changes may occur due to load balancing, load sharing, or a link (or node) failure, but the mobility is associated with the change of the network attachment point. These will cause a merging point between the old path where resource reservation state has already been installed and the new path where data forwarding will actually happen. The flow identifier may not change after the route changes while the mobility may cause the change of the flow identifier by having a new network attachment point. Since the reservation session should remain the same even after a mobility event occurs, the variable flow identifier should not be used to identify the reservation session [1].

In general, a mobility event results in creating an old path, a new path, and a common/unchanged path. The old and new paths converge or diverge according to the direction of each flow. Such topological changes make the reservation on the old path useless, and thus it should be removed as quickly as possible. In addition, resources on the new path should be reserved in a fast manner, and the reservation state on the common path should be updated to reflect the change of the flow identifier.

3 A Mobility-Aware Resource Reservation Mechanism

In this section, we propose a fast resource reservation mechanism with mobility awareness, called Mobility-aware Resource Reservation Mechanism (MRRM) which operates in IP-based wireless access networks.

3.1 Localized Path Management

To minimize the impact of mobility, it is important to make the change of reservation state occur within the affected (local) path. The major issue in this case is to find a node which performs the localized path management in a fast manner. The most appropriate node is the crossover router (CR) because it is the merging point where the old and new paths meet.

In order to discover the CR, the following key fields are used: reservation identifier (RID), flow identifier (FID), virtual interface number (VIN), and handover object (HOB).

The RID is contained in the signaling messages for resource reservation and used to easily identify the involved reservation session. The RID remains the same while the FID may change after handover. Note that the RID is unique and used to solve the double reservation problem. On the other hand, the FID is used to specify the relationship between the address information and the reservation state update. In other words, the change of FID indicates topological changes, and therefore the reservation state along the common path should be updated immediately after the CR is discovered.

The VIN is a virtual interface number which identifies the logical incoming/outgoing interface and is used to recognize the change of the flow path. The HOB is used to inform of the handover event and therefore to initiate the CR discovery. The HOB contains two handover-related fields such as handover_sequence_number (HSN) and handover_start (HST) fields. The HSN field is used to detect the latest handover event and to handle the ping-pong type of handover. The HST field is used to explicitly inform that a handover is now initiated and fast resource reservation on the new path is needed. The value of the HST field can be obtained from a layer 2 trigger/event which indicates a forthcoming handover according to the link status, e.g., the signal strength received from neighboring base stations or access points.

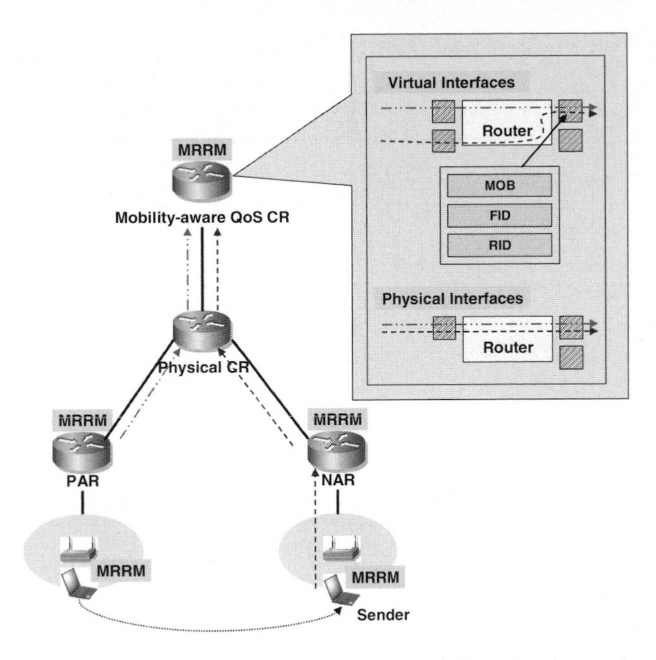

Fig. 3. Fast CR discovery using key identifiers

When handover occurs, the CR can be discovered by comparing the existing key identifiers (described above) with the identifiers included in the newly incoming signaling message sent by an MN or a correspondent node (CN). When a new signaling message arrives at a router, it should determine whether it is a CR or not. To do this, it first checks whether the CR has already been discovered. If not, it checks if the same RID and FID exist. It also checks whether the VIN of the particular flow has been changed. The HOB (in particular, HSN field) in the signaling message is also examined to find the latest handover event.

The CR discovery procedures can be divided further according to which node is a signaling initiator for reservation as shown in Fig. 4. Fig. 4 (a) shows the CR discovery procedure (when the MN is a sender), and Fig. 4 (b) shows the CR discovery procedure (when the MN is a receiver).

3.2 Fast Reservation and Teardown

It the CR is discovered, the reservation on the new path and the teardown on the old path can be performed in a fast manner. The CR discovery procedures are different according to the role of the MN. When the MN is a sender, it initiates resource reservation towards a CN along the new path (see Fig. 4 (a) step 1). The merging point where the old and new paths meet is a CR which is recognized using the key identifiers described in the previous section. The MQCR in Fig. 4 is a mobility-aware QoS CR which employs the MRRM mechanism. The MQCR sends a notification message to the MN to inform that it is the MQCR for the specific reservation session. The MQCR then sends a refresh message towards the CN to update the FID along the common/unchanged path (Fig. 4 (a) step 2), and it also sends a teardown message

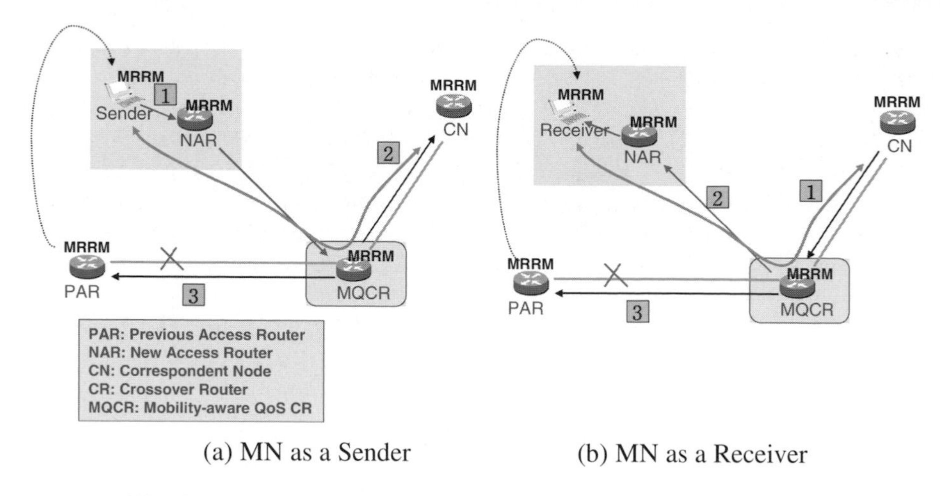

(a) MN as a Sender (b) MN as a Receiver

Fig. 4. Fast resource reservation/teardown in the mobile access network

towards the previous access router (PAR) to release the reserved resources on the old path quickly (Fig. 4 (a) step 3).

When the MN is a receiver, the CN sends a refresh message toward the MN to perform path management (Fig. 4 (b) step 1) after detecting the update of the binding cache in the CN. The binding cache is normally updated after receiving a BU from the MN. In this case, the MQCR is discovered by the CN-initiated refresh message along the common path, and the node from which the common path begins to diverge into the old and new logical paths realizes that it is the MQCR for the reservation session. After the MQCR is determined, it sends a resource reservation message to the MN along the new path (Fig. 4 (b) step 2), and afterward the MQCR sends a teardown message toward the PAR to release resources on the old path quickly (Fig. 4 (b) step 3).

One of the goals of the proposed mechanism is to avoid double reservations. The double reservation made along the common path can be torn down by maintaining a unique RID for each reservation session and the FID when a mobility event occurs. Therefore, the reservation session remains the same even when the MN is moved to a new access network. After resource reservation on the new path, the reservation state on the old path needs to be quickly removed to prevent the waste of resources. Although the release of the resources on the old path can be accomplished by the timeout of soft state (maintained using the refresh message), the refresh timer value may be long (e.g., default value of 30s in RSVP [4]). Therefore, the transmission of a teardown message along the old path is preferred to the use of a refresh timer. The release of resources on the old path is accomplished by comparing the existing and new VINs.

4 Experimental Results

In this section, we evaluate the performance of MRRM via simulations. The proposed mechanism is compared to RSVP and RSVP-MP [6]. Experimental results are also provided to demonstrate that the proposed/implemented MMRM works well in the real mobile environment.

We first used simulations to measure the performance of RSVP, RSVP-MP, and MRRM in terms of delay for resource reservation after handover. Fig. 5 illustrates a simulation topology where there are 8 MNs. The number of hops from the MN and the CN is 7, and every MN generates UDP traffic. It is assumed that the refresh period of RSVP and RSVP-MP is 30s. Initially, only one MN (e.g., MN1) which communicates with the CN generates UDP traffic. The traffic load increases when MNs other than the MN1 begin to generate UDP traffic. The amount of added traffic load is 0.1 when another MN starts to generate UDP traffic. Our simulation model is based on Marc Greis' RSVP model implemented in ns-2.1b3 and Rui Prior's RSVP model implemented in ns-2.26 which is an updated version of Marc Greis' model.

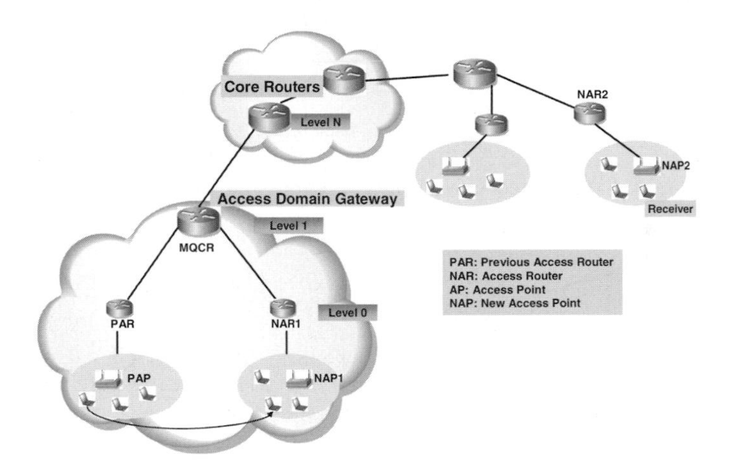

Fig. 5. A target network for simulation

With MRRM, the localized path management is performed after handover, and the existing reservation session on the common path is only updated without double reservation, which results in minimizing the resource reservation delay and avoiding the waste of resources. Therefore, MRRM shows better performance in terms of delay for resource reservation after handover, compared to RSVP and RSVP-MP (see Fig. 6).

To show that the proposed MRRM works well in the real mobile network environment, we configured a physical testbed which consists of four routers, a mobile node (MN), and a fixed node (CN) as shown in Fig. 7. The proposed MRRM is installed at each router and mobile/fixed node. All devices in Fig. 7 are based on Linux OS (Kernel version 2.4.26). Video LAN Client and MGEN6 were used to generate video traffic (which needs resource reservation) and best effort (BE) traffic, respectively. For traffic monitoring and measurement, we used Tele Traffic Tapper (TTT) software. We assume that the MN is a data sender, and the CN is not mobile.

In our experiment, the MN is initially attached to AR1 and moves to the new AR, AR2, at a certain time as depicted in Fig. 7. Before the MN moves to AR2, a certain amount of bandwidth is reserved on the current path for the video flows generated by the MN. The total amount of bandwidth available to video and BE traffic is 4.5 Mbps. As shown in Fig. 8, the video flow is using the reserved bandwidth (1.5 Mbps) before handover, while the BE traffic is allowed to use up to 3 Mbps (although MGEN6

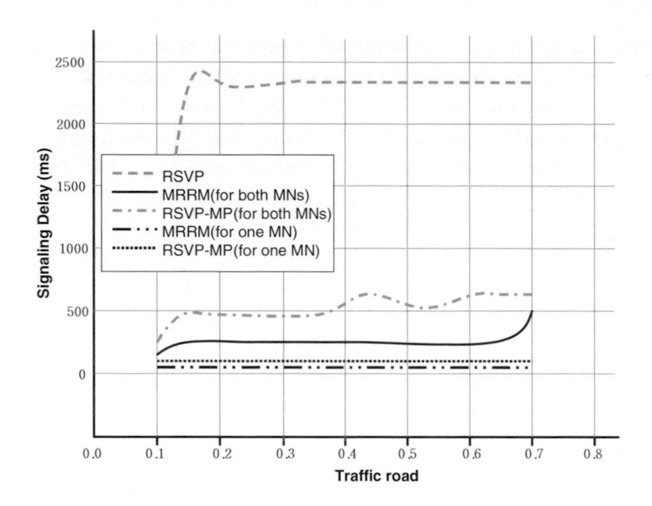

Fig. 6. Signaling delay performance for reservation after handover

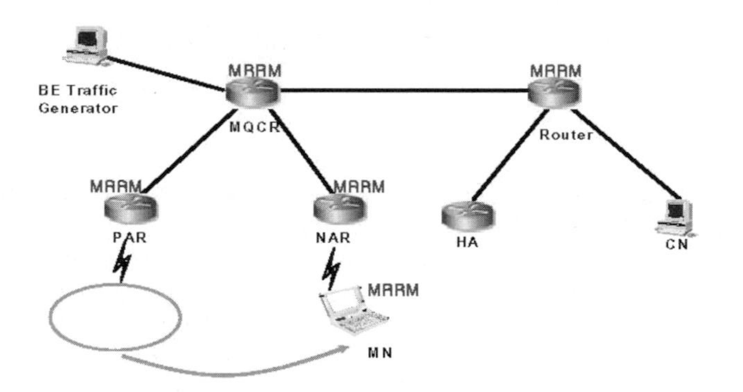

Fig. 7. A physical testbed for experiments

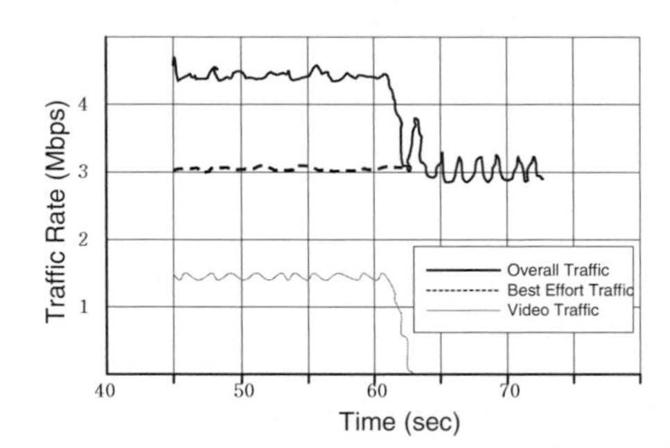

Fig. 8. Traffic rates for video and best-effort traffic before handover

generates BE traffic at the rate of more than 3 Mbps). At 61.309s, the MN moves to AR2, MIPv6 sends the BU message, and the MQCR initiates signaling messages to setup resource reservation on the new path. Fig. 9 shows that the video traffic uses the reserved bandwidth quickly after handover while the bandwidth consumption of BE traffic is limited. The low signaling delay was obtained using localized path management by the MQCR within the area affected by handover (end-to-end signaling for reservation is not performed), and the close coupling between MIPv6 BU and MRRM. In this experiment, the measured handover delay is approximately 500ms, and the measured signaling delay for re-reservation is about 100ms.

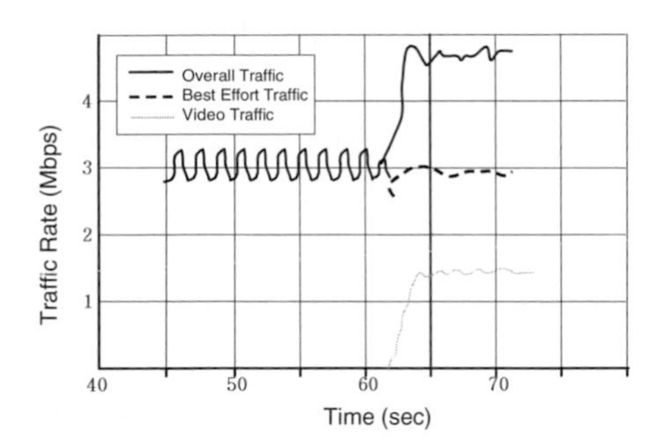

Fig. 9. Traffic rates for video and best-effort traffic after handover

5 Concluding Remarks

An important aspect of the future generation network is to provide the ability for other mobile entities to communicate and access services irrespective of changes of the location. In particular, latency and data loss incurred during handover should be within a range acceptable to users (e.g., below a certain limit) for real-time services. One of the key elements to meet this requirement is a fast resource reservation mechanism with mobility awareness. In this paper, we first presented IP QoS signaling requirements, and identified double reservation and end-to-end resource reservation problems in mobile environments. We then proposed a mobility-aware resource reservation mechanism for resolving the problems. We also demonstrated that the proposed mechanism works well by using localized path management and fast reservation/teardown.

References

[1] Hancock, R.: Next Steps in Signaling: Framework, IETF RFC 4080 (June 2005)
[2] Busropan, B., et al.: AN Scenarios, Requirements and Draft Concepts, IST-2002-507134-AN/WP1/D02 (October 2004)
[3] Lee, S., Jeong, S., Tschofenig, H., Fu, X., Manner, J.: Applicability Statement of NSIS Protocols in Mobile Environments, draft-ietf-nsis-applicability-mobility-signaling-05 (work in progress) (June 2006)
[4] Braden, B., Zhang, L., Berson, S., Herzog, S., Jamin, S.: Resource ReSerVation Protocol (RSVP) - Version 1 Functional Specification, RFC 2205 (September 1997)
[5] Talukdar, A., Badrinath, B., Acharya, A.: MRSVP: A Resource Reservation Protocol for an Integrated Services Packet Network with Mobile Hosts, Technical report DCS-TR-337. Rutgers University (1997)

[6] Chen, W., Huang, L.: RSVP Mobility Support: A Signaling Protocol for Integrated Services Internet with Mobile Hosts. In: Proc. IEEE Conference on Computer Communications, vol. 3, pp. 1283–1292 (2000)
[7] Mahadevan, I., Sivalingam, K.M.: Architecture and Experimental Results for Quality of Service in Mobile Networks using RSVP and CBQ. ACM Wireless Networks 6, 221–234 (2000)
[8] Schulzrinne, H., Hancock, R.: GIST: General Internet Messaging Protocol for Signaling, draft-ietf-nsis-ntlp-11 (work in progress) (August 2006)
[9] Van den Bosch, S.: NSLP for Quality-of-Service Signaling, draft-ietf-nsis-qos-nslp-11 (work in progress) (June 2006)
[10] Johnson, D., Perkins, C., Arkko, J.: Mobility Support in IPv6. RFC 3775 (June 2004)
[11] ITU-T Recommendation Y.2001, General overview of NGN (December 2004)
[12] ITU-T Recommendation E.800, Terms and definitions related to quality of service and network performance including dependability (August 1994)
[13] ITU-T Recommendation Q.Sup51, Signaling requirements for IP-QoS (December 2004)

Proactive Internet Gateway Discovery Mechanisms for Load-Balanced Internet Connectivity in MANET*

Youngmin Kim[1], Sanghyun Ahn[1,**], Hyun Yu[1], Jaehwoon Lee[2], and Yujin Lim[3]

[1] School of Computer Science
University of Seoul, Korea
{blhole,ahn,finalyu}@venus.uos.ac.kr

[2] Department of Information and Communications Engineering
Dongguk University, Korea
jaehwoon@dongguk.edu

[3] Department of Information Media
University of Suwon, Korea
yujin@suwon.ac.kr

Abstract. The mobile ad hoc network (MANET) is a rapidly configurable multi-hop wireless network without an infrastructure and originally proposed for the military use. In order to make the MANET a more commonly used network in our daily lives, it may be necessary to connect the MANET to the global Internet. A MANET node can communicate with an Internet node via the Internet gateway. To support fault tolerance and increase the bandwidth, multiple Internet gateways can be deployed within a MANET and, in this case, the network performance can be improved by balancing the load among Internet gateways. In this paper, we propose load-balancing Internet gateway discovery mechanisms for the MANET with highly mobile nodes and multiple stationary Internet gateways. Existing mechanisms are analyzed through simulations and, from the simulation results, problems of the existing mechanisms are figured out. Based on the analysis, several enhanced mechanisms are proposed and compared with the existing mechanisms by using the NS-2 simulator.

Keywords: Load-Balancing, Internet Gateway Discovery, Multiple Internet Gateways.

1 Introduction

The MANET is a multi-hop wireless network without any network infrastructure and, to increase the communication range and enhance the usability of the MANET, the Internet Gateway (IGW) is used for the connection of a MANET

* This work was supported by the University IT Research Center Project.

** Corresponding author.

T. Vazão, M.M. Freire, and I. Chong (Eds.): ICOIN 2007, LNCS 5200, pp. 285–294, 2008.

to the global Internet. We assume that stationary multiple IGWs connect the MANET to the Internet and broadcast their own prefix information to the MANET, and MANET nodes move freely within the communication range.

In order to connect a MANET to the Internet, mechanisms for default IGW (DIGW) discovery, path setup from a DIGW to a MANET node, address autoconfiguration, duplicate address detection and session management after changing the DIGW are needed. In this paper, we focus only on the DIGW discovery mechanism and the path setup mechanism from a DIGW to a MANET node, and assume that address conflicts do not occur and a node establishes a new session with its correspondent node after changing its DIGW.

With multiple IGWs, if any one of the IGWs fails, another IGW can take over the failed one. To increase the overall throughput of the MANET to the global Internet, balancing the load on IGWs is required. The existing load-balancing IGW discovery mechanisms for the MANET with multiple IGWs use the hop count between a MANET node and an IGW, the load of an IGW or the receiving interval of IGW advertisement (IGWADV) messages as the metric. From the performance analysis of the existing mechanisms, we figure out those factors which can affect the performance and propose enhanced mechanisms using the above-mentioned metrics complementary and verify the performance of our enhanced mechanisms by using the NS-2 simulator. In addition to this, we propose a scheme that reduces the routing control message overhead in the MANET. In this paper, we set the objective of the load-balancing to maximize the total throughput of MANET nodes communicating with Internet nodes.

The paper is organized as follows. Section 2 describes our simulation model used in evaluating the performance. The existing mechanisms providing load-balancing among IGWs and the performance evaluation of those mechanisms are given in section 3. In section 4, we figure out the problems of the existing mechanisms and propose enhanced mechanisms and carry out performance evaluation through simulations. Section 5 concludes this paper.

2 Simulation Model

For the simulation, we have used the NS-2.28 simulator [1] and AODV [2] as the MANET routing protocol. In the case when traffic generating nodes are deployed in a MANET uniformly, the possibility that the traffic is concentrated on an IGW becomes very low and, as a result, the load-balancing may not be required. Therefore, to get a benefit from the load-balancing, more traffic generating nodes are placed near an IGW and less near the other IGW (with assuming two IGWs in a MANET).

A wireless interface used in our simulation is IEEE 802.11, the transmission range of the interface is 150m and the bandwidth of the interface is 2 Mbps. 96 MANET nodes are deployed randomly in a rectangle, 1300m x 700m, and two IGWs are located at the center of the area A and the area B in Fig. 1. Each IGW broadcasts an IGWADV per second. The number of traffic generating nodes is 15 or 21. To see the effect of the load-balancing, in the network with 15 traffic

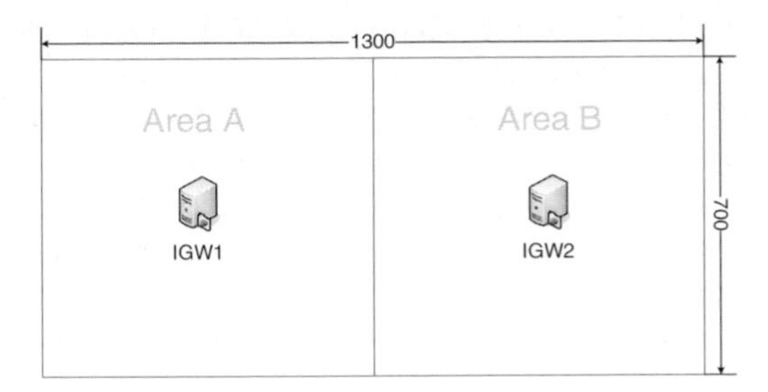

Fig. 1. Network for simulation

generating nodes, 10 nodes are deployed in the area A and 5 nodes in the area B and, in the network with 21 traffic generating nodes, 14 nodes in the area A and 7 nodes in the area B. The traffic generating rate of a node is 15 kbps in the constant bit rate (CBR) and the data packet size is 210 bytes. All traffic generating nodes communicate with Internet nodes through IGWs (i.e., no traffic between MANET nodes).

The random waypoint model is used as the mobility model and the node speed is chosen randomly from 20 m/s to 40 m/s. To see how the performance varies with the node mobility, various pause times (0, 20, 60, 120, 200, 300 and 500s) are used. The total time for each simulation is 500s and the number of simulation runs from which the average value is taken is 5. The performance comparison factors chosen for measuring the degree of the load-balancing are the packet delivery ratio and the throughput of IGWs:

- Packet delivery ratio (PDR): the ratio of the number of data packets received by Internet nodes from MANET nodes to the number of data packets sent by MANET nodes to Internet nodes.
- Throughput per IGW: the number of data bits successfully delivered through IGWs from MANET nodes to Internet nodes per second.

3 Performance Analysis of Existing Load-Balancing IGW Discovery Mechanisms

3.1 Existing Load-Balancing IGW Discovery Mechanisms

IGW discovery mechanisms can be categorized into proactive, reactive and hybrid approaches. In the proactive IGW discovery approach [3,4], each IGW broadcasts an IGWADV message containing the IGW information periodically. Each MANET node selects its DIGW based on the information in an IGWADV. In the reactive IGW discovery approach [3], a MANET node gets the IGW information after it broadcasts an IGW solicitation message. When an IGW receives a solicitation message from a MANET node, the IGW unicasts its IGW information to the MANET node. In the hybrid IGW discovery approach [5,6,7], each IGW broadcasts an IGWADV message periodically within a limited area and a

MANET node which has not received any IGWADV determines its DIGW using a reactive IGW discovery mechanism.

According to the performance evaluation of these three IGW discovery approaches in [6,7], the proactive IGW discovery approach produces the largest amount of control traffic due to periodic IGWADVs. However, the proactive approach shows the highest performance in terms of data throughput because MANET nodes always have the IGW information of all IGWs. On the other hand, the reactive IGW discovery approach produces the least amount of control traffic since a MANET node configures its DIGW only when it has data to send to a global Internet node. However, it gives the lowest data throughput due to the on-demand IGW discovery. The amount of control traffic and the data throughput of the hybrid IGW discovery approach are located in between the proactive approach and the reactive approach. Because the objective of this paper is to achieve higher throughput by performing the load-balancing, the proactive approach is used as the IGW discovery mechanism. In the case when MANET nodes move frequently, the reactive IGW discovery approach may cause the IGW solicitation message storm, so the proactive IGW discovery approach may suit well to this situation since it allows only IGWs to broadcast IGWADVs.

When all IGWs broadcast IGWADVs in a MANET with multiple IGWs, the number of IGWADVs increases as the number of IGWs increases. [8] proposes a scheme that limits the sending area of IGWADVs to fix the total number of IGWADVs even when the number of IGWs increases. When a MANET node having already configured its DIGW receives an IGWADV from another IGW, it does not forward the IGWADV any further. A MANET node receiving IGWADVs from only one IGW chooses the IGW as its DIGW. A MANET node receiving IGWADVs from two or more IGWs chooses the best IGW as its DIGW. An IGW and MANET nodes selecting the IGW as their DIGW form an area and, between two areas, a boundary is implicitly formed.

[8] proposes the shortest path (SP) scheme and the minimum load index (MLI) scheme as the DIGW discovery mechanism with the boundary concept. These schemes assume that IGWs broadcast IGWADVs periodically. In the SP scheme, a MANET node chooses as its DIGW an IGW with the minimum hop count from the node. The IGWADV message has a field having the hop count from a MANET node to an IGW. In the MLI scheme, the IGWADV includes the load of an IGW and a MANET node chooses the IGW with the minimum load as its DIGW. The operation of the MLI scheme is as follows:

- i) When a MANET node having not decided its DIGW yet receives an IGWADV from an IGW, it configures the IGW as its DIGW and forwards the IGWADV to its neighbors.
- ii) When a MANET node having already configured its DIGW receives an IGWADV from its DIGW, it updates the load metric of the IGWADV and forwards the IGWADV to its neighbors.
- iii) When a MANET node having already configured its DIGW (g) receives an IGWADV from an IGW (g') which is not the DIGW, if the following two conditions C1 and C2 are met, the DIGW of the MANET node is changed

to g'. In this case, regardless of the change of the DIGW, the MANET node does not forward the IGWADV to its neighbors.

- C1. The time using g as the DIGW is longer than the predetermined time t.
- C2. $L_{g'} + \frac{T_n}{C_{g'}} + \Delta \leq L_g - \frac{T_n}{C_g}$
  where L_g is the load of g, C_g the capacity of g and T_n the amount of traffic between a MANET node n and Internet nodes communicating with n.

By C1 and Δ in C2, we can prevent a MANET node from changing its DIGW frequently. Moreover, by not allowing the MANET node to forward the IGWADV to its neighbors, changing the DIGWs of MANET nodes at the same time is prevented.

Fig. 2 is an example showing MANET nodes deciding their DIGWs using the SP scheme and the MLI scheme. In the SP scheme, the boundary is decided by the hop distance from each IGW. On the other hand, in the MLI scheme, if we assume that all MANET nodes transmit the same amount of traffic, the number of MANET nodes selecting each IGW as their DIGWs becomes almost the same.

[9] proposes a proactive DIGW discovery mechanism using the running variance metric (RVM) as the metric for selecting the DIGW. RVM is the variance of the interval of IGWADVs. In the RVM scheme, IGWs broadcast IGWADVs periodically, but, since RVM scheme does not limit the forwarding range of the IGWADV, IGWADVs are forwarded to the entire MANET. A MANET node chooses the IGW with the minimum RVM (V_n) as its DIGW. The metric (V_n) of the RVM scheme is defined as follows:

$$\bar{t}_n = \alpha \cdot t_n + (1-\alpha) \cdot \bar{t}_{n-1}, \quad where \quad 0 < \alpha \leq 1 \tag{1}$$

$$V_n = \alpha \cdot (t_n - \bar{t}_n)^2 + (1-\alpha) \cdot V_{n-1}, \quad where \quad 0 < \alpha \leq 1 \tag{2}$$

Here, t_n is the difference between the time when the $n-th$ IGWADV arrives at the MANET node and the time when the $(n-1)-th$ IGWADV arrives at that

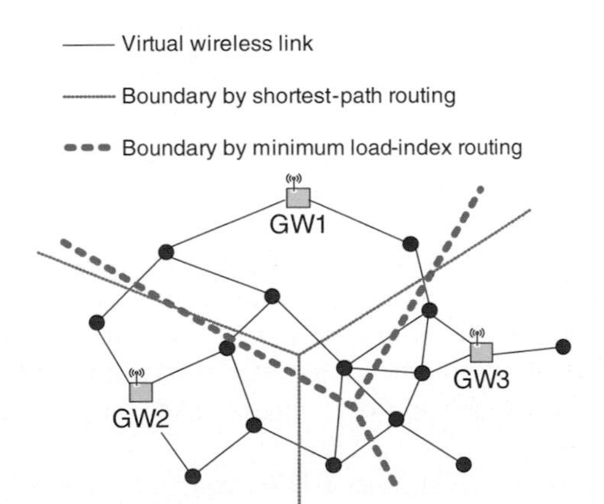

Fig. 2. Example of using the SP and the MLI schemes

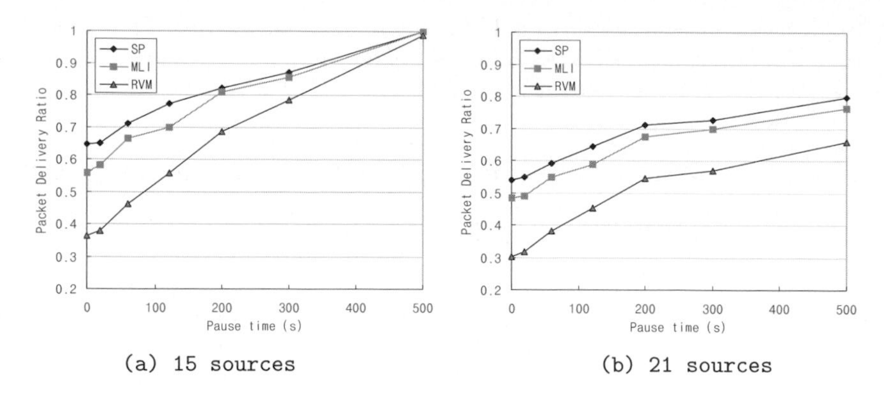

Fig. 3. Comparison of the existing load-balancing IGW discovery mechanisms in terms of the packet delivery ratio with varying the node mobility

node. $\bar{t}_n$ is the estimated average time of the interval of IGWADVs when the MANET node receives the $n-th$ IGWADV. Whenever it receives an IGWADV, each MANET node stores (IGW, upstream neighbor, RVM) or updates the RVM metric of (IGW, upstream neighbor) if it already has the entry. When a MANET node receives the same IGWADV from a different upstream neighbor, it forwards only the IGWADV from its upstream neighbor whose RVM is smaller.

3.2 Performance Evaluation of Existing IGW Discovery Mechanisms

The SP, the MLI and the RVM schemes are compared based on the simulation environment described in section 2. Fig. 3 shows that the PDR of the SP scheme outperforms that of the MLI and the RVM schemes for various pause times (Δ of the MLI scheme is set to $\frac{1}{30}$, an approximate value of the load generated by a sending node and α for the RVM scheme to 0.3).

In Fig. 3 (a), 15 sources communicate with Internet nodes and the fact that the PDR at pause time of 500s is almost 100% implies that the network is not saturated when nodes are static. Fig. 3 (b) shows the case when 21 sources communicate with Internet nodes. In this case, the PDR is about 80% even when all MANET nodes are static and this implies that the network is congested. In the case of using the MLI scheme which has a higher possibility of selecting an IGW located at a distance as a DIGW, if the amount of traffic from MANET nodes to Internet nodes is low, the MLI scheme performs pretty similar to the SP scheme (Fig. 3 (a)). In the congested situation such as Fig. 3 (b), however, the PDR of the MLI scheme is much lower than that of the SP scheme compared to the case of Fig. 3 (a) since the MLI scheme takes longer routes to IGWs than the SP scheme.

The RVM scheme shows the lowest PDR because IGWs broadcast IGWADVs to the entire MANET and, when MANET nodes change routes frequently, the RVM scheme may not get correct RVM values. On the other hand, the SP and the

MLI schemes may choose the best IGW quickly enough because these schemes can get correct metric values even when MANET nodes change routes frequently due to high mobility and high traffic load.

4 Enhancement of Existing IGW Discovery Mechanisms

4.1 Enhanced IGW Discovery Mechanisms

In this section, we enhance the existing schemes presented in section 3 and compare our enhanced schemes with the existing ones. The objective of the load-balancing in this paper is to maximize the total throughput of MANET nodes communicating with Internet nodes. The throughput of the MANET becomes low as the hop count between two nodes increases. Therefore, it may be good for MANET nodes to choose an IGW located as near as possible to the IGW to increase the throughput of the MANET. However, if traffic is concentrated on one IGW, it lowers the throughput due to interference. Consequently, we need a scheme considering both the hop count and the load of IGWs as the metric. Thus, we propose the SMN-HL (Selection of DIGW by a MANET node considering both the Hop count and the IGW's traffic Load) scheme. In addition to that, we propose the HRVM (considering both the Hop count and the Running Variance Metric) scheme.

In the SMN-HL scheme, a MANET node uses the following metric, M_g, to choose an IGW g with the minimum metric value as the DIGW:

$$M_g = W \cdot H_g + L_g \tag{3}$$

where W is the weight of H_g, H_g is the hop count from the IGW g to the MANET node and L_g is the traffic load of the IGW g. Each IGW measures the total amount of transmitting and receiving traffic for a predetermined period. And each IGW calculates a new L_g as follows:

$$L_g = \alpha \cdot \frac{T_{new}}{\lambda \cdot C_g} + (1 - \alpha) \cdot L'_g, \quad where \quad 0 < \alpha \leq 1 \tag{4}$$

where C_g is the capacity of the IGW g, T_{new} is the total amount of transmitting and receiving traffic for recent λ time and L'_g is the old load of the IGW g. An IGWADV having C_g and the measured L_g is broadcast to MANET nodes and each MANET node uses the information to choose its DIGW.

The SMN-HL scheme is similar to the MLI scheme in the forwarding of IGWADVs and the selection of the DIGW. The SMN-HL scheme shares i) and ii) of the MLI scheme mentioned in section 3.1. However, the second condition (C2) of iii) of the MLI scheme is changed as follows:

– $M_{g'} + \frac{T_n}{C_{g'}} + \Delta \leq M_g - \frac{T_n}{C_g}$

where M_g is the metric value of the IGW g, C_g is the capacity of the IGW g and T_n is the total amount of traffic of all sessions between the MANET node n and Internet nodes.

Since the RVM scheme broadcasts IGWADVs to the entire MANET, it generates more IGWADVs than the schemes with the boundary concept which restricts the forwarding area of IGWADVs. Besides, the RVM scheme can obtain correct RVM values only when MANET nodes receive IGWADVs properly, i.e., in a stable network, which implies that the RVM scheme is not appropriate for a highly mobile network. Therefore, to improve the performance of the RVM scheme, in our scheme, a MANET node having already configured with its DIGW is not allowed to forward IGWADVs with the information on IGWs different from its current DIGW. Moreover, our scheme uses the multiplication of the hop count from an IGW to a MANET node and the variance (V_n) of the intervals of IGWADVs as the metric so that more correct metric values can be obtained in a highly mobile network, and we call this the HRVM scheme:

$$M_g = H \cdot V_n \tag{5}$$

The metric for the HRVM scheme considers both the RVM which may fluctuate severely under highly mobile condition and the hop count H which may change slightly under highly mobile condition. This can prevent a highly mobile MANET node from changing its DIGW frequently and increase the probability of a MANET node selecting a closer IGW as its DIGW.

A route setup between a MANET node and its DIGW is necessary for the communication between the node and an Internet node. Current Internet connectivity mechanisms configure routes by broadcasting routing control messages to the entire MANET. However, if a network can be divided into areas like Fig. 2, we can limit the forwarding area of routing control messages to the area with the same prefix. Therefore, for the establishment of routes for the Internet connectivity, we propose the limited forwarding (LF) scheme in which only MANET nodes with the same prefix as that of the originator of a routing control message are allowed to forward the message. This scheme can be used in all IGW discovery mechanisms limiting the forwarding area of IGWADVs. We show that the LF scheme improves the performance of the load-balancing IGW discovery mechanism in section 4.2.

4.2 Performance Evaluation of Enhanced IGW Discovery Mechanisms

We use 0.8 as α and 1 second as λ for the load measurement of IGWs in the SMN-HL scheme. And the weight (W) for H_g is set to $\frac{1}{10}$ which gives relatively good results in terms of throughput for various simulation environment.

Fig. 4 (a) shows that all schemes give almost the same PDR because the traffic load is not saturated. As shown in Fig. 4 (b), the PDR of the SMN-HL scheme and that of the SP scheme are almost the same under highly mobile condition, but as the node mobility decreases the SMN-HL outperforms the SP scheme. If we compare the SMN-HL scheme and the SMN-HL with the LF (SMN-HL+LF) scheme, when the traffic load is low (Fig. 4 (a)), where path rerouting does not occur frequently, the effect of limiting the forwarding of routing control messages (i.e., the LF scheme) is so negligible that the performance of both schemes becomes almost the same. But, for the case when the traffic load is high

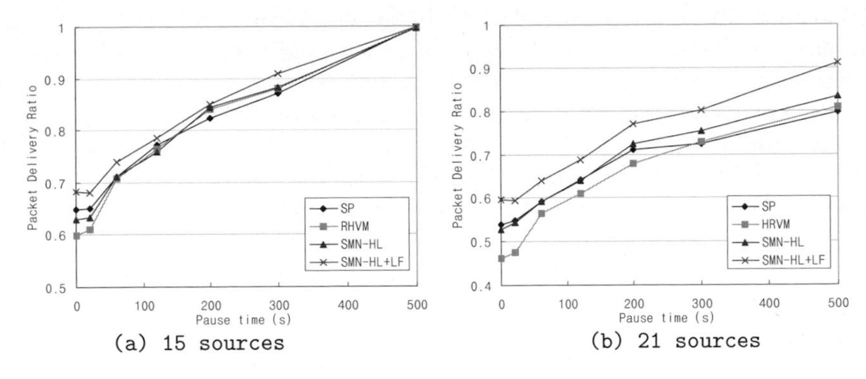

Fig. 4. Comparison of the load-balancing IGW discovery mechanisms in terms of the packet delivery ratio with varying the node mobility

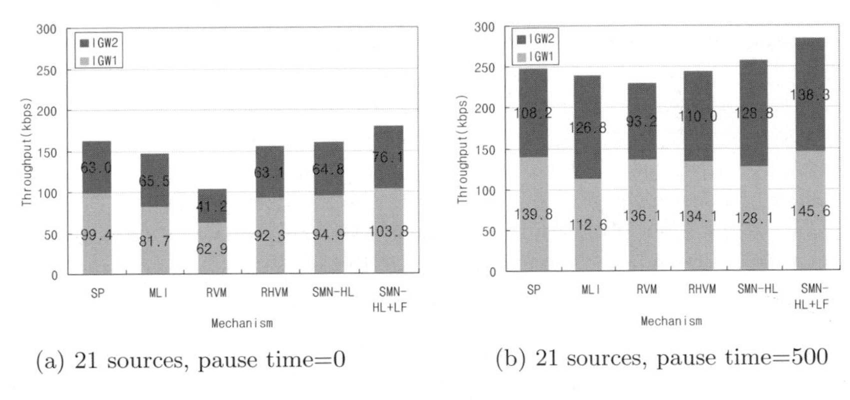

(a) 21 sources, pause time=0 (b) 21 sources, pause time=500

Fig. 5. Throughput of data successfully delivered by IGW1 and IGW2

(Fig. 4 (b)), the SMN-HL+LF scheme outperforms the SMN-HL by limiting the forwarding of routing control messages. From Fig. 3 (b) and 4 (b), we can observe that the HRVM scheme gives much better PDR than the RVM scheme. However, compared to the SP scheme, the HRVM scheme outperforms only when the node mobility is low. So using the RVM as a metric may not be good for balancing the traffic load among IGWs.

Fig. 5 shows the throughput of data successfully delivered by IGW1 and IGW2 when 21 sources transmit the CBR traffic. The MLI scheme balances the load well but the overall throughput is not high. The SMN-HL schemes with or without the LF scheme show good performance in terms of both the load-balancing and the throughput.

5 Conclusions

MANET nodes may communicate with Internet nodes through multiple IGWs. In this case, it is required for a MANET node to be able to control the selection of its default IGW so that the load on IGWs can be well balanced. According to our performance analysis among the existing load-balancing IGW discovery mechanisms, the SP scheme which considers only the hop count outperforms the MLI and the RVM schemes considering the load on an IGW and the variance of IGWADV intervals, respectively.

In this paper, we have proposed the SMN-HL and the HRVM schemes which overcome the problems of the MLI and the RVM schemes. The SMN-HL scheme uses both the load of an IGW and the hop count from an IGW to a MANET node as the metric for the DIGW selection, and the HRVM scheme uses the multiplication of the hop count from an IGW to a MANET node and the variance of IGWADV intervals. According to our simulation results, the SMN-HL scheme outperforms the SP and the MLI schemes. In addition to that, we have proposed the LF scheme which limits the forwarding area of routing control messages to improve the performance of the SMN-HL and the HRVM scheme.

References

1. The Network Simulator, NS-2, `http://www.isi.edu/nsnam/ns`
2. Perkins, C., Royer, E., Das, S.: Ad hoc on demand Distance Vector (AODV) routing, RFC 3561 (July 2003)
3. Wakikawa, R., Malinen, J.T., Perkins, C.E., Nilsson, A., Tuominen, A.J.: Global Connectivity for IPv6 Mobile Ad Hoc Networks. IETF Internet draft (March 2006), `http://draft-wakikawa-manet-globalv6-05.txt`
4. Jelger, C., Noel, T., Frey, A.: Gateway and Address Autoconfiguration for IPv6 Ad Hoc Networks. IETF Internet draft (April 2004), `http://draft-jelger-manet-gateway-autoconf-v6-02.txt`
5. Ratanchandani, P., Kravets, R.: A Hybrid Approach to Internet Connectivity for Mobile Ad Hoc Networks. IEEE WCNC (March 2003)
6. Ghassemian, M., et al.: Performance Analysis of Internet Gateway Discovery Protocols in Ad Hoc Networks. IEEE WCNC (March 2004)
7. Ruiz, P.M., Gomez-Skarmeta, A.: Adaptive Gateway Discovery Mechanisms to Enhance Internet Connectivity for Mobile Ad Hoc Networks. Ad Hoc and Sensor Wireless Networks (March 2005)
8. Huang, C., Lee, H., Tseng, Y.: A Two-Tier Heterogeneous Mobile Ad Hoc Network Architecture and its Load-Balance Routing Problem. IEEE VTC (October 2003)
9. Brännström, R., Åhlund C., Zaslavsky, A.: Maintaining Gateway Connectivity in Multi-hop Ad hoc Networks. IEEE WLN (November 2005)

An Effective Data Dissemination in Vehicular Ad-Hoc Network

Tae-Hwan Kim, Won-Kee Hong*, Hie-Cheol Kim, and Yong-Doo Lee

Department of Information and Communication Engineering,
Daegu University, Gyeong-San, Gyeong-Buk, Korea
{thkim76,wkhong,hckim,ydlee}@daegu.ac.kr

Abstract. Vehicular ad-hoc network (VANET) has several characteristics that are different from mobile ad-hoc network (MANET). Due to these characteristics, the network topology based protocol, often used in MANET, can not be applied to VANET. In this paper, we propose an emergency warning message (EWM) broadcast protocol using range-based relay node selecting algorithm, which determines the minimal waiting time spent by a given node before re-broadcasting the received warning message. Because the waiting time for a message transmission is randomly calculated based on the distance between the sender node and the receiver node, the chosen node as the relay node will have have a minimal waiting time. The results of experiment show the proposed algorithm performed better than the flooding and the distance-based relay nodes selecting algorithm in terms of the network load and the end-to-end delay time. Furthermore, the proposed algorithm can reduce message transmission latency under the circumstances of low node density and short transmission range in VANET.

1 Introduction

VANET (Vehicular Ad-hoc Network) is a temporarily established network through wireless connection between moving vehicles without infrastructure aid such as base stations or access points on the road[1]. The self-established ad-hoc network can facilitate smooth traffic flow as well as increase safety for drivers and pedestrians. Unlike MANET, however, the network topology is frequently changed in VANET because vehicles move faster that 100 kilometers per hour. Individual vehicle speed also differs, which changes topology. In addition, the vehicle density in a platoon within a given area is variable and irregular.

To prevent vehicle accidents, vehicles driving behind should be informed immediately to prepare for any emergencies. In general, the broadcast scheme is widely used for an emergency warning message (EWM) delivery to vehicles in the vicinity of accidents at the same time. Flooding is a representative type of broadcast schemes where data appear to flow into the network of nodes[1] like

* Corresponding author.
[1] Node stands for vehicles in this paper.

T. Vazão, M.M. Freire, and I. Chong (Eds.): ICOIN 2007, LNCS 5200, pp. 295–304, 2008.

water[2]. Lots of nodes except for root node and leaf nodes in the network play the role of sender as well as receiver. This occurs because they are not assured to be within a single hop distance of a root node. If all nodes can be reached within a single hop, the flooding can be seen as a very efficient broadcast scheme. However, in case of multi-hop communication, the network traffic increases dramatically because all the intermediate nodes are involved in delivering EWMs.

In order to resolve this problem, the distance-based broadcast schemes are proposed in the literature [5][6][7][8]. Unlike flooding, it minimizes the number of relay nodes. The relay node performs tasks to both receive and send EWMs to other nodes in the VANET. All nodes within transmission range of a relay node can be candidates for a next relay node. The Distance-Based Relay node Selecting (DBRS) algorithm is generally used to select relay nodes among intermediate nodes. All the intermediate nodes have their waiting time that they have to spend before re-broadcasting. The waiting time varies according to the node's position. That is, a node that is close to a relay node has a long waiting time while one that is far to the broadcasting node has a short waiting time. The waiting time is fixed depending on a node's position so that the chosen node as a relay node should be a border node. A border node is one that is the farthest from the broadcasting node. If not in the border of the communication range of the broadcasting node, it has to spend a given waiting time wastefully. Consequently, the overall EWM delivery time is increased due to the fixed waiting time.

In this paper, we propose a new relay node selecting algorithm, called a Range-Based Relay node Selecting (RBBS) algorithm to deliver EWMs to post-vehicles immediately. It allows an intermediate node to select its waiting time within a given time range so that a relay node has a minimal waiting time, although it is not a border node. The shortest waiting time in the range is equal to the waiting time of the border node. However, the longest waiting time in the range is determined by a node's position like DBRS. Therefore, the waiting time of an intermediate node in RBRS is always lower than or equal to the one in DBRS. This means that the RBRS can reduce unnecessary time spent by a relay node. It may happen that a node close to the previous broadcasting node has a shorter waiting time than the remote one. However, This is very low because nodes nearby have wider range of waiting time than remote ones. The near node's possibility of selecting shorter waiting time is relatively low compare to the remote one. The experimental results show that the RBRS has on average 18.5% higher network traffic than the DBRS due to a slight increase in the number of relay nodes. The end-to-end delay time of RBRS is 28% lower than that of DBRS. Especially, in the low node density and the short transmission range, RBRS has 39% lower end-to-end delay time than the DBRS. Consequently, the RBRS can improve overall performance from 19.5$\sim$40%, compared to the DBRS.

This paper is organized as follows. Section 2 introduces the related work. The RBRS algorithm is explained in section 3. The result of the experiments and performance evaluation is described in section 4. Finally, section 5 contains concluding remarks.

2 Related Work

Lots of EWM broadcast protocols for VANET are found in the literature as shown in Figure 1. They can be classified into four categories depending on the roles of nodes and the relay node selection methods such as flooding, distance-based, table-based and cluster-based broadcast scheme. Most of these protocols assume that vehicles can obtain their positions from the global positioning system (GPS).

The flooding scheme is quite simple and provides a high message reachability even in the mobile environment. It, however, suffers from contention and collision of the redundant packets because all the nodes in the network participate in broadcasting packets. Specially, as node density grows, network traffic and end-to-end delay time increase dramatically. The flooding scheme tries to suppress the number of re-broadcasting nodes but still has heavy network traffic and long delivery latency because of the broadcast storm problem[3]. The representative flooding schemes are I-IBA[2] and DOLPHIN[4].

The distance-based broadcast scheme allows only one node to be involved in relaying EWM at a time to decrease network traffic and delivery latency. The DBRS is a representative relay node selecting algorithm in the distance-based broadcast scheme. The relay node is determined by the distance from the prior node that broadcasts EWM. In other words, every intermediate node that is eligible to re-broadcast EWM has to hold the message for a given waiting time, which is computed based on the distance from the prior broadcasting node, before re-broadcasting it. Because the waiting time of a node is different from one another, only one node with the shortest waiting time is assured to re-broadcast the EWM. The farther the node from the prior broadcasting node is, the shorter the waiting time is. Therefore, the closest node to the border of communication range of the prior broadcasting node is selected as a next relay node. If a relay node is a border node, the shortest waiting time will be spent so that the lowest network traffic and the shortest end-to-end delay time can be guaranteed in the distance-based broadcast protocol. However, this is not always the case. If the relay node is not at the border of communication range of the previous relay node, it will hold back the message unnecessarily for a longer time because the waiting time is determined only by the distance. This case happens often in low node density. There are DDT[5], RBM[6], ODAM[7] and VCWC [8] in the distance-based broadcast protocol.

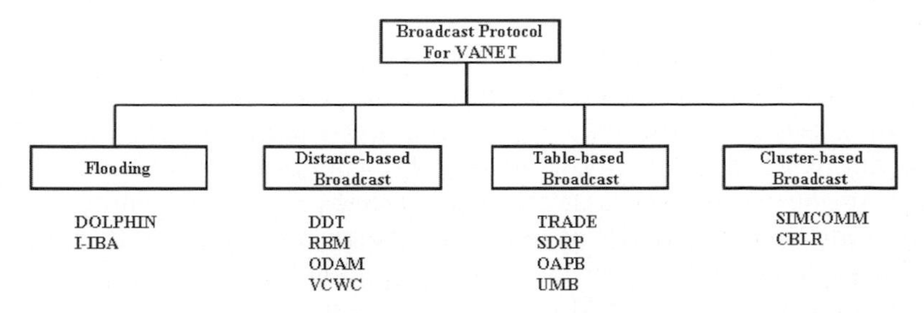

Fig. 1. Broadcasting protocols for VANET

In the table-based broadcast scheme, every node in the network maintains the list of neighbor nodes that is periodically updated through the query-reply mechanism. The relay node is determined by the prior relay node. This scheme provides better performance in terms of network traffic and delivery latency time than the flooding scheme. However. if the network topology is frequently changed, the performance sharply declines and thus, is not efficient for a network that shows high node mobility. Because as the node mobility increase, the period of control message exchange between nodes decrease and therefor the waste of network bandwidth increases. The table-based broadcast schemes are TRADE [9], SDRP [10], OAPB[11], and UMB [12].

The cluster-based broadcast scheme divides the road into several clusters and selects a cluster head among nodes within a cluster. Only the cluster head is eligible to broadcast EWM. This scheme shows good performance when the change of network topology is very low. However, it suffers from heavy network traffic and long latency time in high node mobility since it is required to reorganize the cluster members and reelect a cluster head more frequently. SIMCOMM[13] and CBLR[14] are the cluster-based broadcast schemes.

3 Range-Based Relay Node Selecting Algorithm

As mentioned in earlier section, the frequent change of network topology and node density, which is inherent in VANET, can adversely affect the EWM delivery in the flooding, the table-based broadcasting and the cluster-based broadcasting schemes. The distance-based broadcast scheme is an efficient broadcast scheme because it allows for changes in network topology and node density. However, it requires an efficient relay node selecting algorithm. The DBRS can offer an in-time EWM delivery only in a very high node density where most of relay nodes are located at the border of transmission range. Figure 2 illustrates the DBRS, where n_i stands for an intermediate node. If n_1, n_2 and n_3 are located at distance d_1, d_2 and d_3 from the previous broadcasting node (the EWM sender), each of them has to spend its waiting time proportional to the distance before re-broadcasting the EWM. The waiting time of an intermediate node, denoted by RWT_i, is represented by the following equation:

$$RWT_i = RWT_{max} \times (1 - d_i/R) \tag{1}$$

Where, d_i is the distance of n_i from EWM sender, RWT_{max} is the maximum waiting time and R is the range of the radio covered by the EWM sender. Thus, the n_3 that is the farthest from the broadcasting node has the shortest waiting time and is selected as the next relay node. According to the equation (1), the best performance is given only when all relay nodes are located at the border of transmission range. Every relay nodes, however, can not be guaranteed to be at the border in the VANET, if not in very high node density. If it is not at the border, a relay node has to consume the waiting for a given time wastefully. Consequently, the overall EWM delivery time is increased by the fixed waiting time.

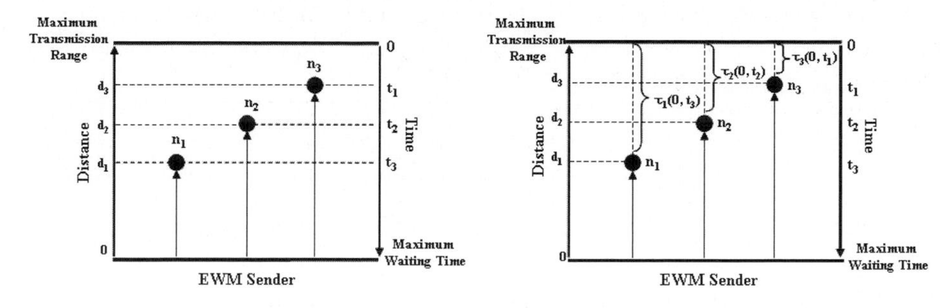

Fig. 2. Distance-based relay node selection **Fig. 3.** Range-based relay node selection

We propose a new relay node selecting algorithm, called RBRS, to decrease the waiting time of relay nodes, obtain low end-to-end delay time and obtain low network traffic during frequent change of network topology and node density. Our approach is similar to the DBRS in that the EWM re-broadcasting waiting time is based on the distance from the previous broadcasting node. In the RBRS, however, a relay node can choose its waiting time in a given time window. Furthermore, a relay node which is not at the border is allowed to have the waiting time of the border node. As shown in Figure 3, the intermediate nodes, n_1, n_2, and n_3 have their waiting time windows denoted by τ_1, τ_2, and τ_3, respectively. A lower bound of τ_i is equal to the waiting time of a border node, while an upper bound of τ_i is determined by n_i's position like DBRS. The intermediate node n_i takes its waiting time from the time window τ_i. Thus, the waiting time of an intermediate node in RBRS is always lower than or equal to the one in DBRS. This means that the RBRS can reduce the unnecessary time spent by a relay node. It may happen that a node close to the broadcasting node has a shorter waiting time than the remote node. However, it is very low because near nodes have a wider range of waiting time than remote ones. The near node's possibility of selecting shorter waiting time is relatively low compare to the remote one. In RBRS, the waiting time, RWT_i of a relay node n_i can be defined as follows:

$$RWT_i = \{\tau : \tau_{min} \leq \tau \leq \tau_{max}\} \tag{2}$$

$$\tau_{min} = RWT_{max} \times (1 - d_b/R) \, , \tau_{max} = RWT_{max} \times (1 - d_i/R)$$

Where, d_b is the distance of a border node from the previous relay node. The RBRS requires every node to know the position of the original broadcasting node as well as the previous relay node in order to compute its waiting time.

The information is contained in the EWM message. The position of the original broadcasting node is used to determine when the EWM relay should be finished. The EWM message format is presented in Figure 4. The length of EWM is 250 bytes and it consists of four fields such as original broadcast node position, relay node position, delivery range and emergency contents. As soon as it receives an EWM from the previous relay node, an intermediate node checks up if the message is new. If not, the message will be discarded. Otherwise, intermediate nodes pull positions information out of a new EWM message and

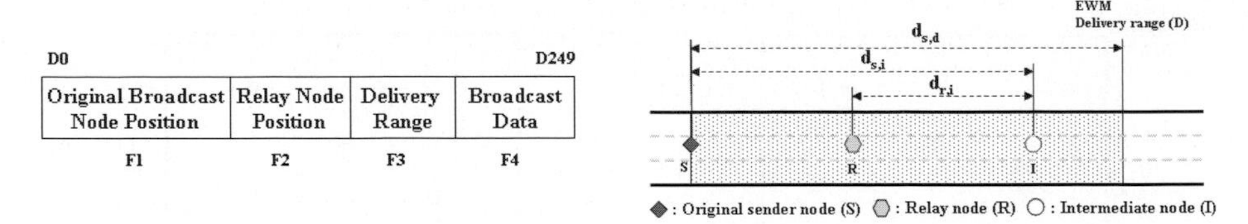

Fig. 4. EWM packet format

Fig. 5. Distance between nodes

obtain their positions from the GPS receiver. Then, an intermediate node calculates the distance $d_{s,i}$ from the original broadcast node, and the distance $d_{r,i}$ from the previous relay node as shown in Figure 5. An intermediate node with the shortest waiting time based on the equation (2) is selected as a new relay node. A new relay node changes the field of relay node position in the EWM with its position and then re-broadcasts it backward. Re-broadcasting the EWM ceases after $d_{s,i}$ exceeds the predetermined EWM delivery range $d_{s,d}$.

4 Performance Evaluation

In this section, we evaluate and analyze the performance of RBRS in VANET with changing node density and transmission range. The simulation parameters are presented in Table 1. It is assumed that all of nodes in the network can obtain a position of itself from the GPS and use IEEE 802.11 DCF(Distribution Control Function) MAC for inter-vehicle communication. Lane change and overtaking are not considered. The simulation has been performed one hundred times and simulation results are on averages.

Figure 6 shows the informed rate of EWM when the node density and the transmission range vary. The informed rate is the percentage of vehicles that receive EWM over all the vehicles within the VANET. As node density and transmission range increase, the informed rate dramatically increases before it is saturated. This is because the network fragmentation gets mitigated with the increment of node density and transmission range. Most broadcast protocols show a similar pattern change in the informed rate because the informed rate strongly depends on the node density and the transmission range. It is assumed

Table 1. Simulation paremeters

Network Environment		Road Environment	
Parameter	Value	Parameter	Value
Transmission range	$150m$	Length of road	$7\ km$
Packet length	$250byte$	Width of a lane	$3.6m$
Channel Bandwidth	$2Mbps$	Road Ditection	One way
propagation Delay of a packet	$0.125\ \mu s$	Number of lanes	3
Computation time	$1\ ms$	Average speed of node	$100\ km/h$
RWT_{max}	$10\ ms$	Traffic density	$13.33\ vehicles/lane/km$
EWM delivery range	$5\ km$	Length of a vehicles	4m

Table 2. EWM Informed Rate versus Transmission Range

	Node Density (Vehicles/lane/km)	
Transmission Rnage (m)	97% informed rate	100% informed rate
150	18.33	30
300	8.33	15
450	5	10

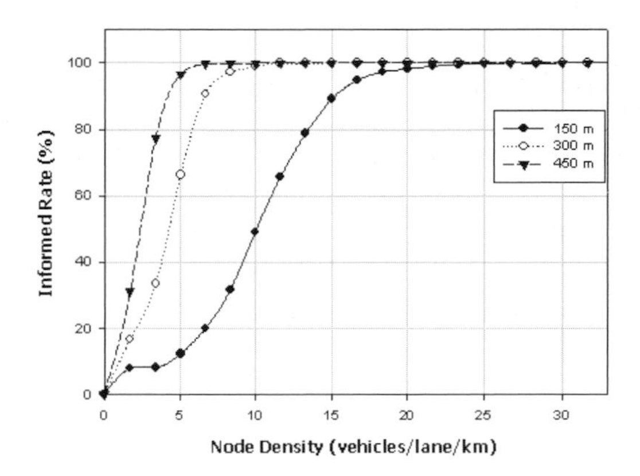

Fig. 6. Informed rate versus node density

that the network fragmentation[15] does not occur in this experiment. When the informed rate reaches over 97%, the informed rate scarcely increased. Thus, we made an experiment with node densities and transmission ranges that show 97% and 100% of informed rate respectively. The node density is presented in Table 2. 100% of informed rate requires almost two times node density compared to 97% of the informed rate.

Figure 7 shows the distribution of waiting time of relay nodes when the informed rate is 97% and the transmission range is $150m$. The RBRS has a shorter waiting time than DBRS though it needs about five more relay nodes on average. The average waiting time of a relay node is $0.27ms$ in RBRS, while it is $1.24ms$ in DBRS. The number of relay nodes used in RBRS is 40.44, while it is 34.8 in DBRS.

Figure 8 shows the change of network traffic when the node density and transmission range vary. The network traffic is measured by the sum of EWMs received by nodes. The flooding shows the highest network traffic because of the broadcast storm problem. While RBRS has 5.7∼10.3 times lower network traffic than flooding, it has 17∼20% higher network traffic than DBRS. This is because more relay nodes are needed in RBRS than in DBRS. However, the gap of network traffic between RBRS and DBRS is very insignificant, compared to the network traffic in the flooding.

As shown in Figure 9, the end-to-end delay time is more than $500ms$ in the flooding. On the other hand, it is shorter in RBRS than others. When the informed rate is 97% and 100%, the RBRS has 39% and 17% shorter end-to-end delay time than the DBRS respectively.

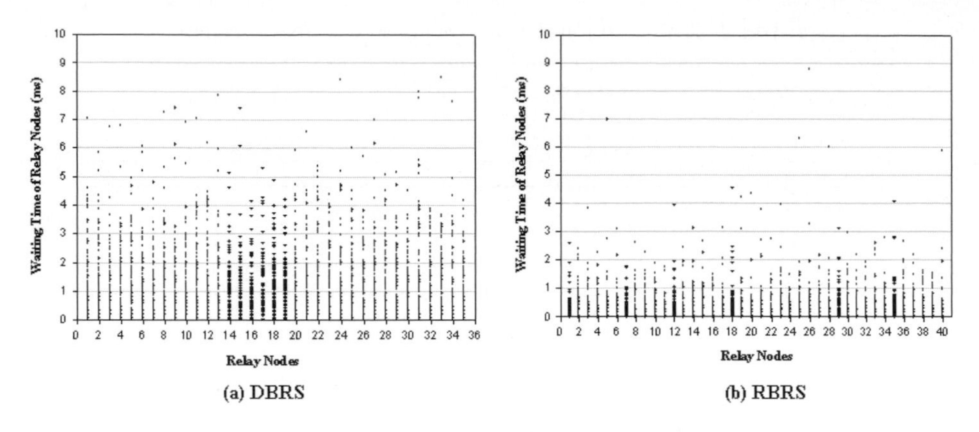

Fig. 7. Waiting time of the relay nodes

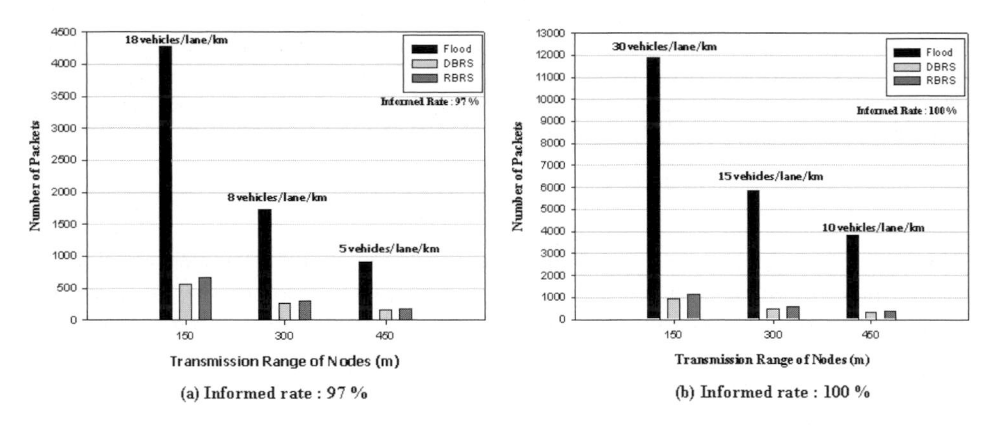

Fig. 8. Network traffic

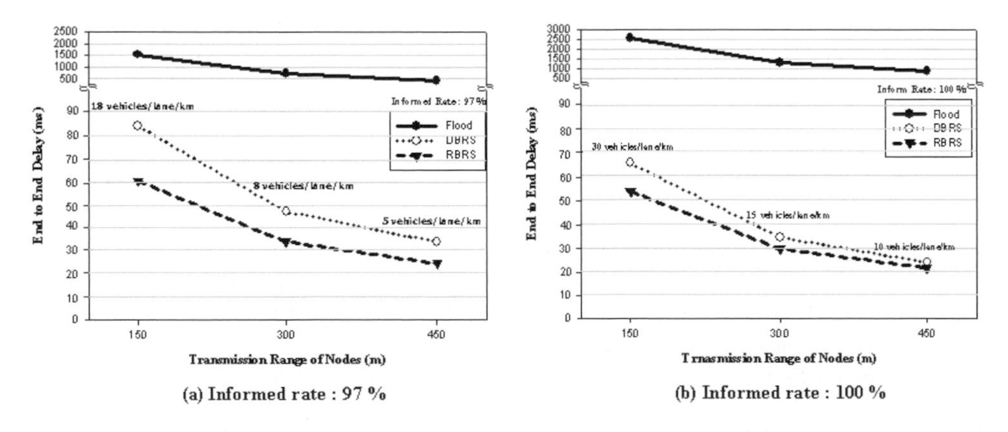

Fig. 9. The end-to-end delay

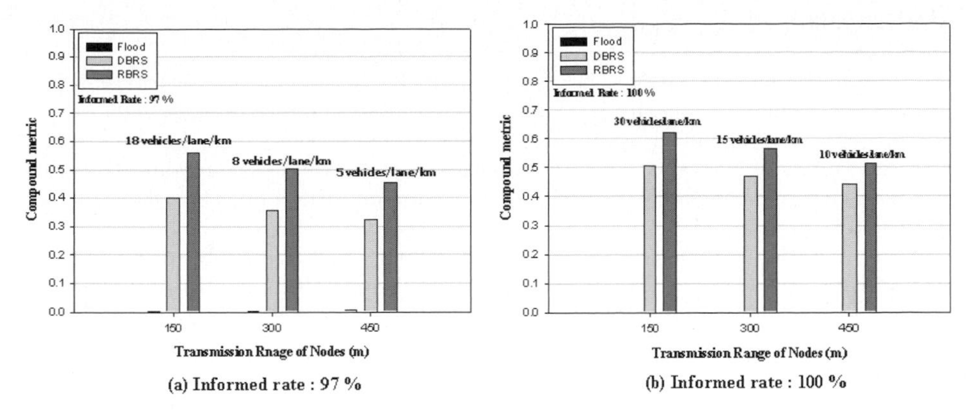

Fig. 10. Compound metric of the network traffic and the end-to-end delay time

The RBRS achieves a decrease in end-to-end delay time, with trading an increased network traffic. A new compound metric is used to measure the real performance improvement. It is computed with the end-to-end delay time and the network traffic. As shown in Figure 10, the overall performance of RBRS is 19.5%∼40.3% higher than that of DBRS.

5 Conclusion

VANET efficiently prevents vehicle accidents by promptly informing other vehicles about road emergencies. However, an efficient broadcast protocol is required to send EWM to neighboring vehicles promptly and reliably in VANET. Such broadcast protocol should be designed to function with inherent characteristics of VANET such as frequent change of network topology, node density, and high mobility.

The distance-based broadcast scheme suppresses the number of relay nodes to minimize network traffic. Only one node with the shortest waiting time is selected as the relay node for broadcast while other nodes abandon the message relay. In the DBRS, the waiting time is determined by the distance from the previous broadcasting node. However, it becomes another overhead to the overall end-to-end delay time since a relay node makes a slow delivery as much.

In this paper, the RBRS is proposed to minimize the waiting time of a relay node. An intermediate node is allowed to select its waiting time within a given time range that is at most as much as the waiting time in the DBRS. The experimental results show that the RBRS has 19.5∼40% better performance in terms of the compound metric of end-to-end delay time and network traffic when compared to the DBRS. Specially, when low node density and short transmission range are provided, RBRS has 39% shorter end-to-end delay time than the DBRS.

Acknowledgment

This work was supported by the Korea Research Foundation Grant Funded by the Korean Government (MOEHRD : KRF-2006-331-D00339).

References

1. Torrent-Moreno, M., Killat, M., Hartenstein, H.: The challenges of robust inter-vehicle communications. In: IEEE 62nd Conf. on Vehicular Technology 2005, vol. 1, pp. 319–323 (September 2005)
2. Biswas, S., Tatchikou, R., Dion, F.: Vehicle-to-vehicle wireless communication protocols for enhancing highway traffic safety. Communications Magazine 44(1), 74–82 (2006)
3. Ni, S., Tseng, Y., Chen, Y., Sheu, J.: The Broadcast Storm Problem in a Mobile Ad Hoc Network. In: ACM MOBICOM 1999, August 1999, pp. 151–162 (1999)
4. Tokuda, K., Akiyama, M., Fujii, H.: DOLPHIN for inter-vehicle communications system. In: Proc. of the IEEE on Intelligent Vehicles Symposium 2000, pp. 504–509 (October 2000)
5. Min-Te, S., Wu-Chi, F., Ten-Hwang, L., Yamada, K., Okada, H., Fujimura, K.: GPS-based message broadcast for adaptive inter-vehicle communications. In: IEEE 52nd Conference Vehicular Technology 2000, vol. 6, pp. 2685–2692 (September 2000)
6. Briesemeister, L., Hommel, G.: Role-based multicast in highly mobile but sparsely connected ad hoc networks. In: First Annual Workshop on Mobile and Ad Hoc Networking and Computing, MobiHOC 2000, pp. 45–50 (August 2000)
7. Abderrahim, B.: Optimized Dissemination of Alarm Messages in Vehicular Ad-Hoc Networks (VANET). In: Mammeri, Z., Lorenz, P. (eds.) HSNMC 2004. LNCS, vol. 3079, pp. 655–666. Springer, Heidelberg (2004)
8. Xue, Y., Jie, L., Feng, Z., Nitin, V.: Vehicle -to-vehicle Communication Protocol for Cooperative Collision Warning. In: The First Annual Int. Conf. on Mobile and Ubiquitous Systems: Networking and Services 2004, pp. 114–123 (August 2004)
9. Min-Te, S., Wu-Chi, F., Ten-Hwang, L., Yamada, K., Okada, H., Fujimura, K.: GPS-Based Message Broadcasting for Inter-vehicle Communication. In: Int. Conf. on Parallel Processing, 2000, pp. 279–286 (August 2000)
10. Massashi, S., Mayoko, F., Takaaki, U., Teruo, H.: Inter-Vehicle ad-hoc Communication Protocol for Acquiring Local Traffic Information. In: The 11th World Congress on ITS (November 2004)
11. Alshaer, H., Horlait, E.: An optimized adaptive broadcast scheme for inter-vehicle communication. In: IEEE 61st Conf. on Vehicular Technology 2005, vol. 5, pp. 2840–2844 (May 2005)
12. Gokhan, K., Eylem, E., Fusun, O., Umit, O.: Urban Multi-Hop Broadcast Protocol for Inter-Vehicle Communication Systems. In: Proceedings of First ACM Workshop on VANET 2004, pp. 76–85 (October 2004)
13. Durresi, M., Durresi, A., Barolli, L.: Sensor inter-vehicle communication for safer highways. In: 19th Int. Conf. on Advanced Information Networking and Applications, vol. 2, pp. 599–604 (March 2005)
14. Santos, R.A., Edwards, R.M., Edwards, A.: Cluster-based location routing algorithm for vehicle to vehicle communication. In: Radio and Wireless Conference, pp. 39–42. IEEE, Los Alamitos (2004)
15. Artimy, M.M., Robertson, W., Phillips, W.J.: Connectivity in inter-vehicle ad hoc networks. In: Canadian Conference on Electrical and Computer Engineering, 2004, vol. 1, pp. 293–298 (May 2004)

Packet Forwarding Based on Reachability Information for VANETs*

Woosin Lee[1], Hyukjoon Lee[1], Hyungkeun Lee[1], and Keecheon Kim[2]

[1] School of Computer Engineering, Kwangwoon University
{wlee,hlee,hklee}@kw.ac.kr
[2] School of Computer & Engineering, Konkuk University
kckim@konkuk.ac.kr

Abstract. In the VANET, where the duration of communication is extremely short, the large amount of control overheads associated with discovering and maintaining end-to-end path information may not be tolerable. This paper presents a new multi-hop forwarding protocol which does not use explicit path information, but instead, uses reachability information towards the destinations in determining next-hop nodes. The reachability information for a particular node merely indicates that the node is reachable. At each hop, one of the neighbor nodes which hold the reachability information towards the same destination is selected as the next-hop node by contention based on some priority values. The proposed protocol is designed to be integrated with the IEEE 802.11 MAC protocol in order to achieve higher efficiency and accuracy in its time-critical operations. It is shown through simulations that the proposed protocol outperforms the AODV in a realistic the VANET scenario in terms of both the end-to-end delay and packet delivery ratio.

1 Introduction

The VANET (Vehicular Ad-hoc Network) is drawing a significant amount of attention as one of the technical areas where the ad-hoc network technologies can be applied. The VANET can introduce a set of new services in a robust and cost-efficient manner. In the infrastructure-based systems, the radio coverage of a roadside unit (RSU) can be extended by having a node near the edge of the transmission range forward data to nodes outside the range. Imminent collision warning, rollover warning, work zone warning, platooning, cooperative route planning, and peer-to-peer entertainment are some of the public safety and non-safety related applications that can be enabled by the VANET.

Although there is a large body of work on mobile ad hoc network protocols [1-4], most of them are not suitable for the VANET. In general, topology-based unicast routing protocols — proactive, on-demand or hybrid of the two — such as DSDV, DSR and ZRP set up a path between two nodes before they exchange data. In the VANET scenarios, where network topologies change continuously and abruptly,

* This work was supported by Engineering Foundation and Research Grant of Kwangwoon University in 2007, Seoul R&BD Program and the ubiquitous Autonomic Computing and Network Project, the MIC 21st Century Frontier R&D Program in Korea.

T. Vazão, M.M. Freire, and I. Chong (Eds.): ICOIN 2007, LNCS 5200, pp. 305–314, 2008.

frequent route updates may be necessary. Route update operations, generally based on message flooding, generate an excessive amount of control message overhead which is one of the main sources of large end-to-end delay. The end-to-end delay is one of the most crucial protocol design parameters in the VANET scenarios, where the duration of communication may be extremely short. Moreover, the control message overhead may cause a significant media contention when communicating nodes are densely populated as in a crowded urban traffic environment [5]. Therefore, a routing protocol with a minimum amount of control overhead in path discovery is desired.

Position-based routing protocols can forward packets without path discovery or maintenance operation [6-9]. Forwarding decision at each node is made primarily based on the position of the destination and one-hop neighbor nodes. The position information of the destination node is carried in the packet header so that packets can be forwarded by intermediate nodes in the general direction of the destination node. However, unless a separate channel is available for the location service by which the source node to obtain the position of the destination, the position-based routing protocols can suffer from the overhead of location service that scales with $O(\sqrt{n})$, where n is the number of nodes [6]. This means the overhead of location service has approximately the same complexity as that of path discovery. Furthermore, the inaccuracy of position information caused by node mobility may lead to a significant decrease in terms of packet delivery ratio.

Our goal is to design a new multi-hop routing protocol for the VANET that does not perform path discovery or maintenance without using position information. Each node relies on reachability information collected from the packets received previously in making the forwarding decision. This new protocol called MMFP (Multi-hop MAC Forwarding Protocol) is designed as an extension to the IEEE 802.11 MAC layer [10] in order to ensure its functional accuracy in the time-critical operations.

The rest of this paper is organized as follows: In section 2 the MMFP is explained in detail. Simulation results are presented in section 3. Finally, some conclusions are drawn in section 4.

2 Multi-hop MAC Forwarding Protocol

2.1 The Main Operation

The operation of MMFP follows the principle of a MAC bridge that forwards a frame to a particular LAN segment, if the destination address of a frame has been registered to the filter table, and floods it to all LAN segments otherwise. Specifically, whenever a node receives a packet, the addresses of the transmitter, i.e., a 1-hop neighbor, and the source node are entered in the forward table as reachable nodes. Two modes of forwarding are defined:

- *Implicit unicast mode* is used to select a single forwarding node among the 1-hop neighbors by competition based on a priority value. This mode is used when the reachability information is available for the destination node.
- *Broadcast mode* is used to inform all its 1-hop neighbors to rebroadcast the received packet. This mode is used when the reachability information is not available.

A more detailed description on how to maintain the forward table is deferred to the next sub-section. The implicit unicast forwarding process is different from the conventional unicast forwarding process. Whereas each node forwards packets to the next-hop along the predetermined end-to-end path in the conventional unicast, each node broadcasts packets with the destination address specified in the implicit unicast. By allowing only one of the neighbor nodes receiving the broadcast frame to rebroadcast it, an operation similar to the unicast is achieved. This is in principle similar to the forwarding process of position-based routing.

The rebroadcast node is selected based on a priority value, which is determined by the effectiveness of forwarding by each neighbor node. The effective period of a forward table entry, Received Signal Strength Indicator (RSSI), the hop count, or the interface queue length are a few examples of possible metrics that can be used to determine the priority value. The position-based forwarding is achieved if the distance to the destination node is used as the priority value. The selected node sends an ACK so that the semantics of original IEEE 802.11 MAC is preserved. The black-burst method that allows a node sending the longest jamming signal to reserve the medium is used in order to have the highest-priority neighbor node send an ACK frame. Once the destination node receives a frame, it sends an ACK frame immediately after SIFS without sending the black-burst signal. If there are many nodes with the same priority, collisions may occur. The MMFP sends the black-burst signal of a random length once again to resolve the collision. Namely, our black-burst process consists of two black-burst phases; the priority-based first phase and the random backoff-based second phase. A more detailed discussion on the two black-burst phases is presented in section 3.3. The main algorithm of MMFP can be represented as follows:

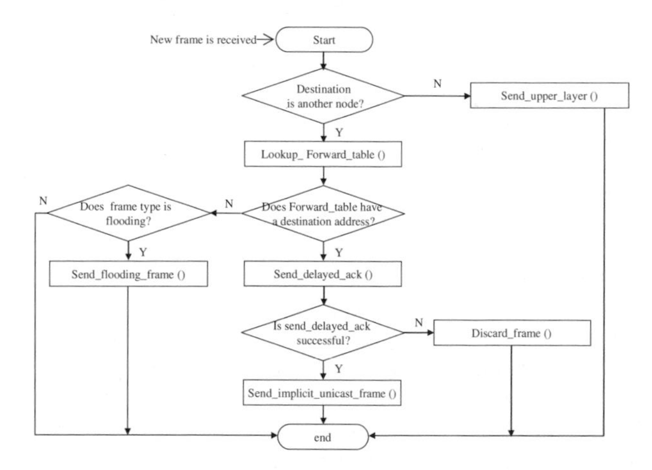

Fig. 1. The main algorithm of MMFP

2.2 Maintaining the Forward Table

The main propose of forward table is to provide information about all reachable nodes. Each entry of the forward table consists of two fields (*destination_address, refresh_timer*), of which *destination_address* represents the address of a node reachable and the *refresh_timer* indicates the effective period of an entry. An entry is automatically purged when the value of *refresh_timer* becomes zero.

Depending on the type of frames received, the forwarding table should be updated as follows:

1. *When a data frame is received:* Both the source node and transmitter node are reachable along the reverse path assuming all links are bidirectional. Hence, new entries for the source and transmitter nodes should be registered or the *refresh_timer* should be updated if the corresponding entries exist.
2. *When an ACK frame is received:* There are two sub-cases when an ACK frame is received:
   A. The received ACK frame acknowledges the data frame transmitted by the node itself. The destination node is reachable via a neighbor node. If the transmitted data frame is an implicit unicast frame, it means that the existing entry for the destination node is still valid. Hence, the *refresh_timer* should be reset. Otherwise, a new entry for the destination node should be registered.
   B. The received ACK frame acknowledges the data frame transmitted by a neighbor node. The destination of data frame transmitted by the neighbor node is reachable via the node from which the ACK has been received. Hence a new entry for the destination node should be registered.

In Fig. 2, creation of the implicit multi-paths is observed. Implicit multi-paths S-A-C-D and S-B-C-D between S and D are created as B adds D to the forward table, and the frame, therefore, can continue to be transferred even if either A or B node moves away. As a result, it is possible to reduce overheads significantly, compared to topology-based routing protocol that is subject to the path maintenance process.

If the destination is not registered in the forward table, a node should broadcast a flooding frame to all 1-hop neighbors. The flooding frames are repeatedly rebroadcast by subsequent nodes until they reach a node that has a forward table entry for the destination. From then on the frames are forwarded by the implicit unicast. Since the last nodes that rebroadcast a flooding frame receive an ACK from one of their 1-hop neighbor, i.e., case 2 above, they add a new entry for the destination to their forward tables. This type of forward table update is spread from the destination towards the source as more frames are sent by the same source to the same destination. As a result, the area of flooding is reduced quickly as communication between two nodes proceeds. An example is illustrated in Fig. 3, where none of node A and B initially

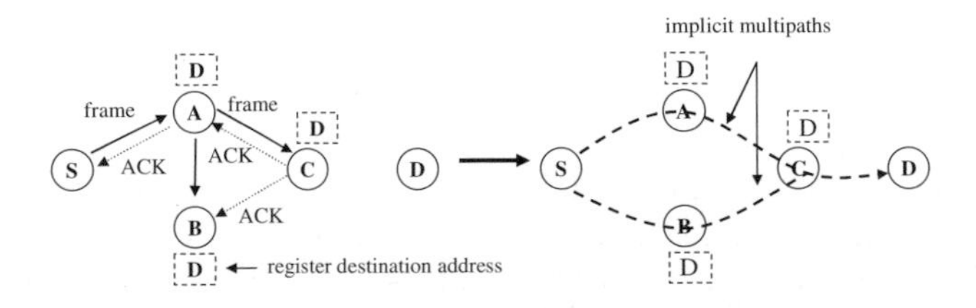

Fig. 2. An example of implicit multipath

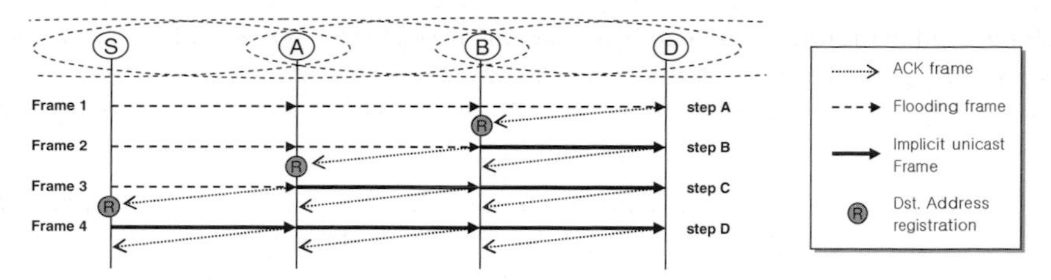

Fig. 3. An example of forward table update process

has a forward table entry for destination node D. The flooding frame sent by node S reaches destination node D via node B. Node D broadcasts an ACK which is received by B. Node B then adds a forward table entry for node D as explained above (Fig. 3 (step A)). When node B receives the next frame destined for node D from node A, since node B now has a forward table entry for node D, broadcast an ACK and sends an implicit unicast frame to node D. Upon receiving the ACK from node B, node A adds an entry for node D (Fig. 3 (step B)). Similar phases are taken when the next frame is sent by node S and now all of nodes S, A and B have an entry for node D (Fig. 3 (step C)), hence no more flooding frames are generated (Fig. 3 (step D)).

2.3 Forwarding Node Selection by Contention

As mentioned previously, all neighbor nodes that have the reachability information for the destination compete for a right to send an ACK using the black-burst method. The winner rebroadcast the frame (i.e., implicit unicast) whereas the losers discard the frame. This prevents uncontrolled rebroadcasting of the same frame. Since this ACK is delayed by black-burst, we call it a delayed_ACK.

Black-burst method was proposed in [11] and [12] in order to provide guaranteed access delays to rate-limited traffic. By allowing each node transmit a data frame only if the medium is free after sending out an energy burst (channel jamming signal) of which the length is determined independently based on a priority value, a node with the highest priority has the exclusive right to transmit the data frame.

All contending nodes send the black-bursts after they sense the medium is idle in SIFS+1 slot after receiving a data frame. Since it makes no sense to have the destination node contend with other nodes, the destination node is allowed to send an ACK in SIFS after receiving the frame as specified in the IEEE 802.11 standard. In other words, SIFS+1 slot of waiting by the other nodes ensures the priority access to the medium by the destination node taking into account the propagation delay of the ACK.

The length of black-burst is determined by:

$$\text{The length of black-burst} = \lfloor (priority\_value) \cdot D_r \rfloor \cdot slot\_time, \quad (1)$$

where *priority_value* is number in [0, 1] that increases as the effectiveness of forwarding by a node increases, D_r is the maximum number of slots allocated to the first phase black-burst, and *slot_time* is the length of a slot (i.e., 9 microseconds).

In our work, we use RSSI in calculating the priority value. The RSSI can be used to determine the distance between two communicating nodes based on the path-loss radio propagation model, namely, the ratio of the received signal strength P_{RX} at distance d from the transmitter, to the transmitted signal strength P_{TX}, is given by:

$$\frac{P_{RX}}{P_{TX}} = Cd^{-\alpha}, \tag{2}$$

where C is a constant that depends on the antenna gains, the wavelengths, and the antenna heights, α is the path loss factor ranging from 2 to 4 [13]. Using the distance, the farthest away node from the forwarding node among its contending neighbor nodes becomes the winner. Therefore, it is more likely that the closest nodes to the destination become the intermediate nodes in the forwarding path.

It is possible that more than one contending node have the same priority value and hence the same black-burst length. In this case, ACK's sent by these nodes can collide. In order to resolve the problem of colliding ACK's, all winning nodes perform the second phase black-burst one slot after the first-phase black-burst taking into account the propagation delay of the first-phase black-bursts. The length of the second phase black-burst is determined randomly from the range of allowed slots. Note that the per-hop transmission overhead generated by the two-phase black-burst would not be a significant loss compared to the overhead generated by the transmission of RTS/CTS pair that takes 13 slots in IEEE 802.11 a/g.

In Fig. 4, an example of the selection process of a forwarding node based on two-phase black-burst is illustrated. Three contending nodes (A, B and C) send the first phase black-bursts. In this example, node A and B send the black-bursts of the same length, and node C send a shorter black-burst since node A and B have the same priority values that is higher than node C. In the second-phase black-burst, node A sends a longer black-burst than B as determined randomly. Since A senses the idle channel for SIFS, it proceeds to send a delayed_ACK and rebroadcast the implicit unicast frame, and node B and C discard the frame.

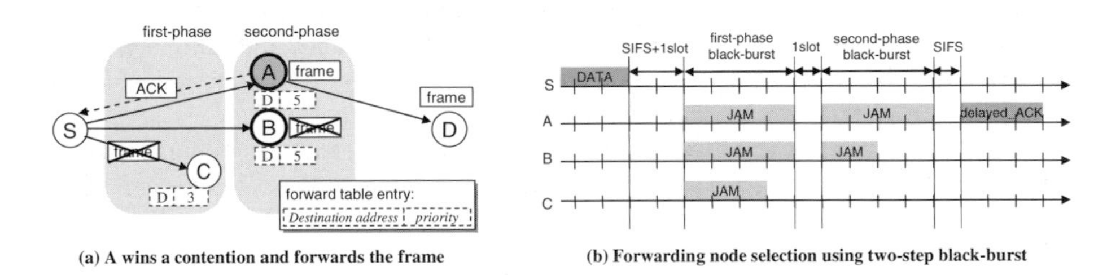

(a) A wins a contention and forwards the frame

(b) Forwarding node selection using two-step black-burst

Fig. 4. An example of contention-based forwarding node selection using two-phase black-burst

2.4 Maintaining the Sequence Number Table

In the MMFP, the routing loop is prevented by using the sequence number defined in the IEEE 802.11 MAC specification. The sequence number table consists of four fields including *source_address, sequence_number, forwarding_flag* and *refresh_timer.* When a node receives a frame whose source address matches that of a sequence number table entry with a sequence number equal to or smaller than the *sequence_number,* it discards the frame.

The *forwarding_flag* is used to resolve forward table errors due to the collision of delayed_ACK's that may occur because the two-phase black-burst works with a limited number of slots. If two forwarding nodes send the delayed_ACK's at the same time a collision occurs and the sender retransmits the frame for a specified number of times or until it finally receives an ACK. Because the sequence number of all retransmitted frames is the same, the forwarding nodes determine them as duplicate frames and discard them. In this case, the sender, deluding himself that the retransmission has failed, erroneously purges the corresponding entry. The default value of *forwarding_flag* is 0, and it is set to 1 if the frame is forwarded. If the value of *retry field* in the header of duplicated frame and *forwarding_flag* are both 1, the forwarding node recognizes that there has been a collision in sending the previous delayed_ACK, and it retransmits a delayed_ACK.

3 Simulation

In order to analyze the performance of MMFP, we performed the simulation using ns-2. The MMFP was implemented in a sublayer between the network and IEEE 802.11 MAC layer. The AODV was also implemented in the sublayer for a fair comparison. We set the values of *active_route_timeout* and *max_rreq_timeout* to 10 seconds, *local_repair_wait_time* to 0.15 seconds, and *rreq_retry* to 3 times as recommended by [14]. The physical layer of IEEE 802.11b was modified to operate as 802.11g by specifying the system parameters for the ERP-OFDM as shown in Table 1. For modeling a practical communication environment, two-ray ground model and ricean distribution [15] was chosen as the path-loss radio propagation and fading model.

A simulation scenario was designed to reflect the realistic inter-vehicle communication by 360 cars running on a two-way straight-line highway of four lanes with the occasional occurrences of entrances and exits (Fig. 5). Each node periodically makes random transitions with the probability varied from 0.0 to 0.4 between two states, i.e., 'on' and 'off' states, which represent entering and exiting the highway, respectively. Table 1 lists some of the simulation parameters.

The data rate was set to 54 Mbps with the transmission range of 200 meters. The distance between two nodes in the outer and inner lane was set to 90 and 88.95

Table 1. Simulation parameters

Parameter	Value
CWMin (slots)	15
SlotTime (microseconds)	9
Preamble length (bits)	120
PLCP Header Length (bits)	24
PLCP Data Rate (Mbps)	6
Data rate (Mbps)	54
Transmission range (m)	200
UDP payload size (bytes)	1024
Ricean K factor	6

meters, respectively. Two adjacent nodes in different lanes were initially separated by 5 meters. All nodes in each lane move at the difference speed and the difference in speed between two (passing and driving) lanes of different direction is maximum 40 m/s. Each node has nine 1-hop neighbor nodes within its transmission range. Each of the 10 randomly selected nodes sends data traffic at 5 pkts/s for 30 seconds to a destination node that is selected to be a specific distance apart at the beginning of a simulation session. Both the source and destination nodes remain in 'on' state during an entire simulation session. Half of the cars are randomly selected to be initially in 'on' state and the other half in 'off' state such that the network topology changes frequently. A series of simulations were run while changing the values of the distance between the source and destination nodes (720, 1080, 1440, 1800, 2160 m) and the on/off probability (0.0, 0.1, 0.2, 0.3, 0.4). Each simulation was repeated 20 times with different seed values for random numbers.

The performance of MMFP was measured with two priority values, based on the RSSI. Fig. 6 and 7 illustrate the performance of MMFP and AODV in terms of the end-to-end delay and delivery ratio, respectively, against the varying on/off probability values. Here, the distance between the source and destination nodes is fixed at 1440 m. Fig. 6 shows the end-to-end delay of MMFP is consistently lower than that of AODV regardless of the values of on/off probability: 27 ms and 34 ms for the MMFP and 207 ms and 306 ms for the AODV when the values of on/off probability are 0.1 and 0.4, respectively. We observed the AODV suffer from the frequent local repair of routes which increased the queuing delay and hence the end-to-end delay. By contrast, because the MMFP is able to forward the frames without the route repair via the implicit multi-paths, the end-to-end delay remains almost constant.

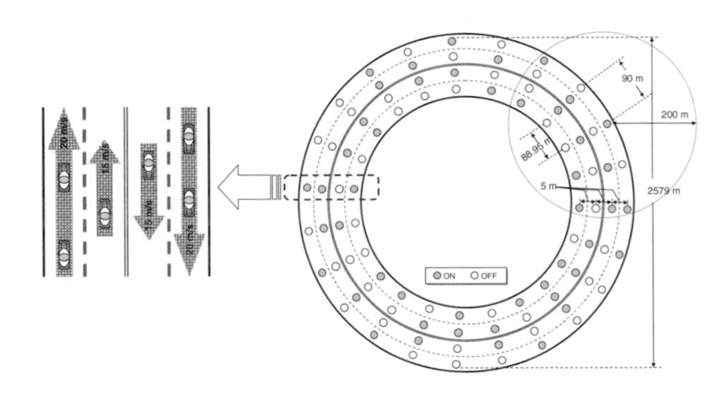

Fig. 5. Circular scenario

In Fig. 7, we can see that the MMFP outperform the AODV: 79.6 % and 76.5 % for the MMFP versus 62.5 % and 48.6 % for the AODV when the values of on/off probability are 0.1 and 0.4, respectively. We observed the AODV suffer from retransmission failure caused by frequent route failure and unstable wireless link. However, the MMFP achieves a lower packet loss ratio (average 33 %) than AODV (average 55 %) because of implicit multi-path.

Fig. 8 and 9 show the performance of MMFP and AODV in terms of the end-to-end delay and delivery ratio, respectively, against the different values of distance between the source and destination nodes with the fixed value of on/off probability

(0.3). In Fig. 8, it is shown the end-to-end delay of MMFP-RT and MMFP-RSSI is lower than that of AODV in all regions of the distance values: 29 ms and 58 ms for the MMFP versus 166 ms and 388 ms for the AODV when the values of distance are 720 m and 2160 m, respectively. The steep increase in the end-to-end delay of AODV is due to the increase in queuing delay caused by the route repairs as the probability of route failure increases with the distance. By contrast, for the MMFP, the end-to-end delay increases slowly as the queuing delay is barely affected by the increased distance. As shown in Fig. 9, the delivery ratios of MMFP and AODV both drops as the average speed of node increases: from 81.6% to 65.5% for the MMFP and from 61.8 % to 46 % for the AODV as the distance increases from 720 m to 2160 m. However, We observed that overall delivery ratio of MMFP is constantly higher than AODV (average 20%).

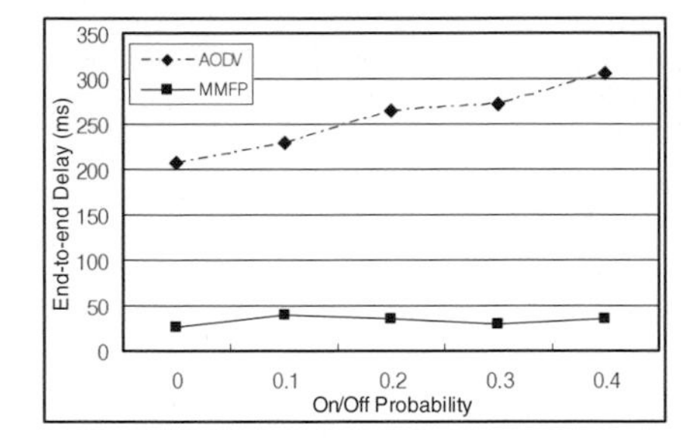

Fig. 6. End-to-end delay vs. on/off probability

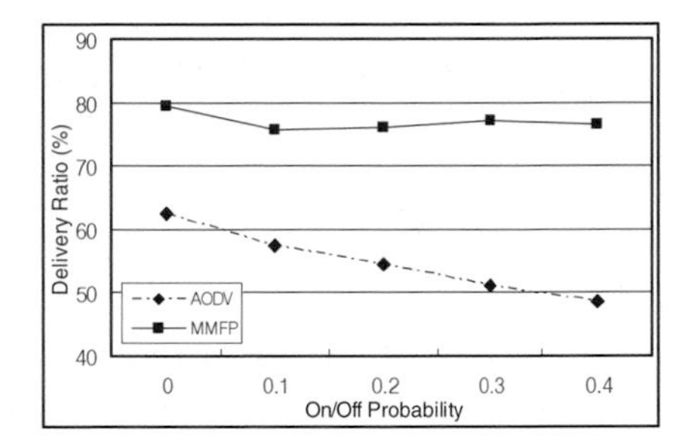

Fig. 7. Delivery ratio vs. on/off probability

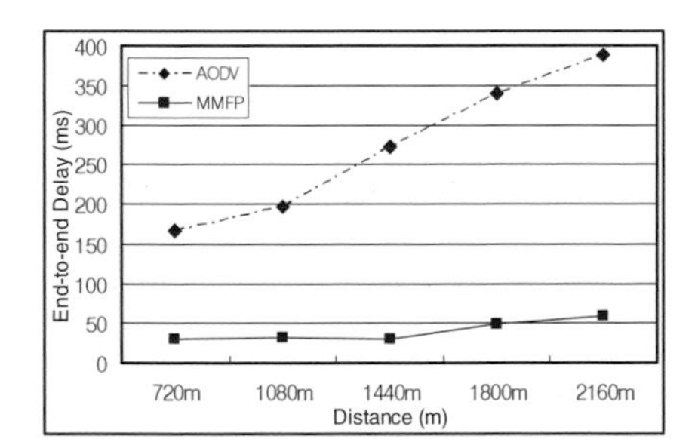

Fig. 8. End-to-end delay vs. intervehicular distance

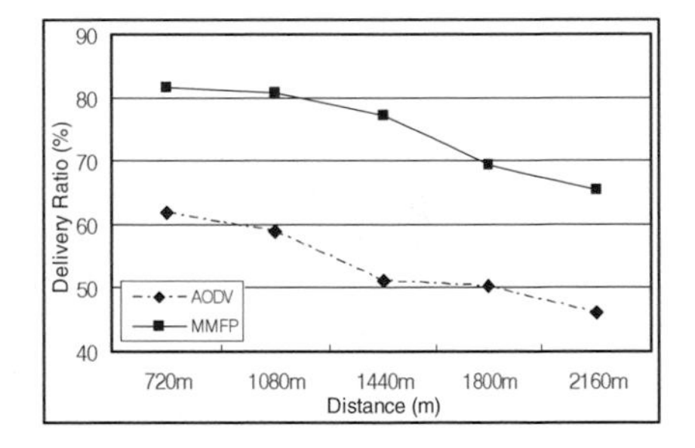

Fig. 9. Delivery ratio vs. intervehicular distance

4 Conclusions

In this paper, we propose a new multi-hop routing protocol for the VANET. The proposed protocol, MMFP, does not perform path discovery or use the position information of communicating nodes. Since no path discovery or maintenance is performed, the communicating nodes experience shorter delay which is critical in the high-mobility scenarios of the VANET. The fact that the MMFP is implemented as an extension to IEEE 802.11 MAC is a significant advantage in terms of reliable performance and rapid deployment. Additional simulations are being set out to evaluate the performance of MMFP in more realistic situations such as a two-way highway with multiple lanes in each direction and a blind intersection. Further investigations are also underway to improve the performance of the MMFP by integrating position information into the forward node selection procedure and by containing flooding frames within the general direction of the destination node.

References

1. Perkins, C.E., Bhagwat, P.: Highly Dynamic Destination Sequenced Distance-Vector Routing (DSDV) for Mobile Computers. In: Proc. of ACM SIGCOMM 1994, pp. 234–244 (1994)
2. Johnson, D.B., Maltz, D.A.: Dynamic source routing in ad hoc wireless networks in Mobile Computing. In: Imielinski, T., Korth, H. (eds.), ch. 5, pp. 153–181. Kluwer, Norwell (1996)
3. Perkins, C., Royer, E.: Ad-hoc On-Demand Distance Vector Routing. In: IEEE Workshop on Mobile Computing Systems and Applications, pp. 90–100 (1999)
4. Perlman, M.R., Haas, Z.J.: Determining the optimal configuration for the zone routing protocol. IEEE Journal on Selected Areas in Communications, 1395–1414 (1999)
5. Zhu, J., Roy, S.: MAC for Dedicated Short Range Communications in Intelligent Transport System. IEEE Communications Magazine, 61–67 (2003)
6. Mauve, M., Widmer, J., Hartenstein, H.: A survey on position-based routing in mobile ad hoc networks. IEEE Network, 15(6) (2001)
7. Karp, B., Kung, H.T.: GPSR: Greedy Perimeter Stateless Routing for Wireless Networks. In: MobiCom 2000, Boston, Massachusetts, pp. 243–254 (2000)
8. Basagni, S., et al.: A Distance Routing Effect Algorithm for Mobility (DREAM). In: MOBICOM 1998, Dallas, TX, USA, pp. 76–84 (1998)
9. Blazevic, L., et al.: Self-organization in mobile ad-hoc networks: the approach of terminodes. IEEE Communication Magazine (2001)
10. ANSI/IEEE: 802.11: Wireless LAN Medium Access Control (MAC) and Physical Layer (PHY) Specifications (1999)
11. Sobrinho, J.L., Krishnakumar, A.S.: Distributed multiple access procedures to provide voice communications over IEEE 802.11 wireless networks. In: GLOBECOM 1996 Communications: The Key to Global Prosperity, vol. 3, pp. 1689–1694 (1996)
12. Jacob, L., Xiang, L., Luying, Z.: A MAC protocol with QoS guarantees for real-time traffics in wireless LANs. In: ICICS-PCM 2003, vol. 3, pp. 1962–1966 (2003)
13. Rappaport, T.S.: Wireless communications, principles and practice. Prentice-Hall, Englewood Cliffs (1996)
14. The Network Simulator(ns-2), `http://www.isi.edu/nsnam/ns/`
15. Stüber, G.L.: Principles of Mobile Communication. Kluwer Academic, Dordrecht (1996)

Balanced Multipath Source Routing*

Shreyas Prasad, André Schumacher**, Harri Haanpää, and Pekka Orponen

Laboratory for Theoretical Computer Science, Helsinki University of Technology,
P.O. Box 5400, FI-02015 TKK, Finland
Shreyasp@tcs.hut.fi, Andre.Schumacher@tkk.fi,
Harri.Haanpaa@tkk.fi, Pekka.Orponen@tkk.fi

Abstract. We consider the problem of balancing the traffic load ideally over a wireless multihop network. In previous work, a systematic approach to this task was undertaken, starting with an approximate optimisation method that guarantees a provable congestion performance bound, and then designing a distributed implementation by modifying the DSR protocol. In this paper, the performance of the resulting Balanced Multipath Source Routing (BMSR) protocol is validated in a number of simulated networking scenarios. In particular, we study the effect of irregular network structure on the performance of the protocol, and compare it to the performance of DSR and an idealised shortest-path routing algorithm in setups with several source-destination pairs. For all network scenarios we consider, BMSR outperforms DSR significantly. BMSR is also shown to be more robust than the shortest-path algorithm, in that it can distribute the traffic load more evenly in cases where shortest-path routing is impeded by radio interference between proximate paths.

1 Introduction

Consider a scenario whereby a wireless multi-hop network is used to set up communications between a disaster recovery area and an operations centre, or is needed to replace a broken segment of a high-throughput fixed network. One issue that then arises is how to optimally allocate the total transmission bandwidth of the wireless network to carry the high volume of end-to-end traffic. Unless the traffic pattern and the design of the wireless network are fully predictable, some load balancing multipath routing scheme should be used to avoid congestion. Such schemes also improve reliability of the network.

A considerable amount of work exists on multipath routing in wireless networks. E.g. Nasipuri, Castañeda, and Das [1] extend the DSR [2] route finding process to consider alternate routes for a given destination. Wu and Harms [3] modify the procedure of forwarding route-reply messages back to the initiator of a route-request in order to discover alternative routes. See [4] for a discussion of different approaches to multipath routing in wireless networks.

* This research was partially supported by the Academy of Finland under Grant No 209300 (ACSENT).

** Corresponding author.

T. Vazão, M.M. Freire, and I. Chong (Eds.): ICOIN 2007, LNCS 5200, pp. 315–324, 2008.

Most of the existing literature on this topic takes as its starting point some natural heuristic for multipath routing and investigates its behaviour either analytically or by means of simulation studies. The recently introduced *Balanced Multipath Source Routing* (BMSR) protocol [5], however, takes a different approach. Here the starting point is a linear programming approximation algorithm [6,7] that *provably guarantees* a desired bound on the congestion performance of the routing scheme, and this algorithm is then given a *distributed implementation* by extending the DSR protocol. Since in DSR, alternate routes can easily be collected to the source node, it provides a simple and lightweight platform for multipath routing extensions, as also noted by other authors [1].

The present work validates the behaviour of the BMSR protocol in several networking scenarios, using the `ns2` network simulator [8]. In the original article [5], average throughput and packet delay provided by BMSR were compared to those of DSR on a simple square grid of 10×10 nodes, with one source-destination pair generating traffic left-to-right and another bottom-to-top on the grid. In this paper, we first investigate how the placement of nodes on a grid influences the behaviour of the algorithm, as opposed to a random placement with a similar node density. Observing that the effect is not significant, we continue with a grid placement in order to eliminate one source of random effects in the simulation results. The following sets of simulation experiments contrast the behaviour of BMSR to DSR and an idealised shortest-path routing scheme using various placements of source-destination pairs, and with attention to the scalability of the different algorithms as the number of source-destination pairs increases.

Overall, for all the network setups we considered, BMSR outperforms DSR significantly. BMSR also performs better than the shortest-path algorithm when source-destination pairs are placed densely or the shortest-paths are not disjoint and there are only few sources simultaneously active in the network.

The rest of the paper is organised as follows. In the next section, we first give a brief overview of DSR and describe its basic operation. Thereafter, we describe the basic linear programming congestion-control approximation algorithm [6,7] and its implementation as an extension to DSR. Section 3 presents the `ns2` simulations conducted and compares the performance of BMSR with DSR and a global shortest-path routing algorithm. Finally, Section 4 presents some conclusions and outlines future research directions.

2 The BMSR Protocol

The recently proposed BMSR protocol [5] extends the Dynamic Source Routing (DSR) [2] protocol with a distributed approximation algorithm to optimise network congestion.

In this section we first outline the basic operation of DSR and those aspects of it that we will modify. Thereafter, we describe how we obtain BMSR by integrating the approximation algorithm for load balancing into DSR.

2.1 DSR

DSR is an on-demand source routing protocol: the source includes the whole route in every packet sent, and routes are discovered only when required. The basic DSR protocol consists of two operations: *route discovery* and *route maintenance*. If a source node wishes to send to a destination to which it does not have a route in its route cache, it initiates the route discovery process by broadcasting a *route-request* (`RREQ`) message to its neighbours. Upon receiving the `RREQ`, if a node knows of a route to the destination it can send a *route-reply* (`RREP`) message back to the source; otherwise it will append its own address to the list of nodes in the `RREQ` and forward the request. Upon receiving the `RREQ`, the destination obtains a route from the source to the destination, and in the presence of bidirectional links it can simply reverse this route and use it for sending a `RREP` message along this route to the source. A sequence number mechanism ensures that a node only forwards each `RREQ` at most once. Since shorter routes require fewer hops, the first `RREQ` to reach the destination is likely to have taken a route that is minimum or close to minimum in terms of the hop count.

If a source route breaks, the source is notified by the intermediate node that detects the break. The source may then resend the packet using an alternative route in its route cache, or initiate a new route discovery. If the intermediate node has a different route to the destination in its own cache, it can initiate *packet salvaging* and forward the packet using this alternative route.

2.2 BMSR

In this section we describe a method for obtaining multiple source-destination routes for a given source-destination pair by a linear programming approximation algorithm that minimises flow congestion [7]. The algorithm relies on the computation of shortest paths determined by an adaptive cost metric using weights on the links. The weight updates are distributed to avoid dissemination of global information. Each shortest path computed becomes a source route for the BMSR protocol. The data flow is then equally distributed over these pre-computed routes.

We model balancing the traffic in the network as a multicommodity flow problem in a directed graph $G = (V, E)$ where the set of vertices V corresponds to the radio nodes in the network. There is a directed edge $(i, j) \in E$ from vertex i to vertex j if the radio node corresponding to j is within transmission range of the radio node corresponding to i; this is the well known *unit disk graph model.* Each source-destination pair is modelled as a commodity so that there is a supply of the commodity at the source node and a demand, modelled as a negative supply, at the destination node. With x^c_{ij} denoting the flow of commodity c along the edge from vertex i to vertex j and $s^c(i)$ representing the supply of commodity c at vertex i, to route the transmissions from the respective sources to destinations we must satisfy

$$s^c(i) + \sum_{(j,i)\in E} x^c_{ji} - \sum_{(i,j)\in E} x^c_{ij} = 0. \tag{1}$$

The version of the algorithm we use requires that each radio link (and thus every edge $(i,j) \in E$) has the same fixed capacity u_{ij}, but we do not limit the total flow $f_{ij} = \sum_c x_{ij}^c$ along an edge; rather our aim is to balance the flow in the network by minimising the maximum congestion in the network:

$$\min \max_{(i,j)\in E} \frac{f_{ij}}{u_{ij}} \ . \tag{2}$$

The optimisation method that we will implement as an extension to DSR is based on the work of Young [6], as summarised by Bienstock in [7]. To route a flow of rate $s^c(i)$ of commodity c from the source node i to a destination, the weight of each edge is initialised to 1. We then run I iterations: in each iteration, for each commodity c, the least-weight path from the source to the destination is determined, the flow of commodity c along each edge on the path is increased by $s^c(i)/I$, and the weight w_e of each edge e on the path is updated as

$$w_e \leftarrow (1 + \epsilon s^c(i))w_e. \tag{3}$$

where $0 < \epsilon \leq 1/2$ is a parameter of the algorithm. It can be shown that if the number of iterations I is sufficiently large, the result of this algorithm is optimal to a factor of $(1+\epsilon)$ [6,7].

Our routing method, BMSR, is an implementation of this algorithm on top of DSR. Each radio node maintains a record of the weights of its incoming links. Our balanced route request packet, `BREQ`, is a modified version of the standard route request packet (`RREQ`) of DSR that additionally contains a record of the total weight of the path the `BREQ` has taken so far. Contrary to DSR, if a node receives a `BREQ` packet related to a route request it has already seen, it may resend it if the second packet has a lower weight than the previously seen `BREQ` packets related to the same route request. This allows us to find routes of minimum weight.

After receiving a `BREQ` packet the intended destination waits for a while, aiming to make sure that lower weight routes represented by `BREQ` packets that arrive later are taken into account in choosing the route. The destination then sends a route reply packet back to the source, and all nodes on the path update the weights of the edges on the path according to (3).

To obtain a good selection of routes, a source node may run a large number $I$of route requests. The iteration number I also depends on the chosen approximation quality parameter ϵ: for larger ϵ fewer iterations are necessary in order to obtain an acceptable selection of routes, but the resulting flow may be further away from the optimum.

In BMSR, routes that are broken due to temporarily congested links are not invalidated. With each source having a balanced collection of routes to the destination the effect of a single link failure diminishes, as the source distributes the traffic equally among the routes in its the cache.

3 Simulation and Performance Evaluation

To experimentally validate the BMSR algorithm, we performed simulations for a number of different network topologies using the ns2 network simulator [8]. An initial set of experiments was used to investigate the algorithmic effect of arranging the simulated nodes in a grid structure, as opposed to the arguably more natural placement uniformly at random in a corresponding area. Thereafter, the number and position of source-destination pairs was varied to determine the effect of differing traffic patterns on the algorithm's performance, as well as to get an intuition how BMSR scales with the number of source-destination pairs.

As a performance metric in the evaluations, we have used the average throughput over the simulated time, taken over all source-destination pairs in use at each simulation. Using this metric, we compare BMSR to standard DSR and an idealised shortest-path routing (SPR) algorithm.

3.1 Review of Previous Experiments and Random Node Placement

As part of previous work [5] we tested BMSR on a simulated 2160 m $\times$ 2160 m grid network of 100 nodes. We placed two pairs of constant bit rate (CBR) traffic source and destination nodes at the boundary of the grid, so that the direct connections between both pairs would approximately form a cross shape. In this setup each node may communicate with the nodes beside, above or below it. The supply value s^c for both source nodes was chosen to be 1. However, one should note that due to the update rule given in (3), the particular choice for s^c has only a scaling effect on the approximation quality parameter ϵ. Therefore, s^c will be also fixed to 1 for all further experiments discussed in this paper.

We then compared DSR to BMSR for a range of CBR packet-sizes and BMSR parameters ϵ and I. BMSR clearly outperformed DSR, both in terms of packet delay and network throughput. The simulations showed a performance gain of 14% to 69%. We also analysed the effect of BMSR on network load and on collisions due to interference and interface queue (IFQ) overflows. BMSR led to a more balanced distribution of the load in the network, which effectively reduced packet loss resulting from IFQ overflows. A choice of $\epsilon = 0.05$ and $I = 160$ showed the best average throughput over both source-destination pairs.

In the present set of experiments, we first evaluate the impact of the regular network structure on the performance of BMSR and DSR, by considering nodes placed uniformly at random within a square area. The locations of the source and destination nodes, as well as the total number of nodes, are kept the same as in the experiments in [5]. In order to maintain connectivity, we roughly doubled the density of nodes within the network by scaling the network down by a factor of approximately $\sqrt{2}$, yielding a network size of 1530 m $\times$ 1530 m. We then compared results for CBR traffic throughput of a grid with the throughput of networks with randomly placed nodes using the same dimensions and location of sources and destinations. The simulation parameters are summarised in Tab. 1. The ns2 default value for transmission range, 250 m for our TwoRayGround propagation model, was used, enabling nodes to communicate with the closest nodes located in their vicinity, including diagonal neighbours.

Table 1. The parameters used in ns2 simulations

CBR packet size:	2048 B	MAC bandwidth:	1 Mbit
CBR data rate:	160 Kbit/s	MAC protocol:	802.11 with RTS/CTS
Antenna type:	OmniAntenna	Propagation model:	TwoRayGround
Transmission range:	≈ 250 m	Interference range:	≈ 550 m
Max. source route length:	22, 26	Max. IFQ length:	50
Node count:	100, 200	Network size:	1530 m × 1530 m, 2160 m × 4320 m
Simulation time:	1500 s	Balancing setup time:	500 s
BMSR parameters:	$I = 160,\ \epsilon = 0.05$		

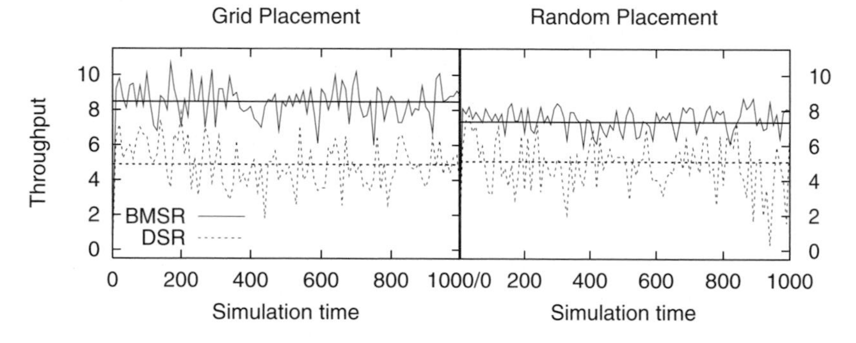

Fig. 1. Average throughput of both source-destination pairs in KB/s versus simulation time for a single run of DSR and BMSR, setup stage for BMSR is omitted from the plot. Performance over 15 runs is shown for each plot.

As can be seen in Fig. 1, random node placement does not affect the performance of either routing algorithm significantly. However, a slight performance decrease for BMSR is observable. This may be due to interference caused by the larger density of nodes in the network, as BMSR does not take into account interference between radio links.

Due to the only minor decrease and to prevent the introduction of an additional source of random effects, we focus further simulations on the grid topology.

3.2 Densely Placed CBR Pairs

In this section we describe experiments where the sources are located directly opposite to their respective destinations and the source-destination nodes are placed next to each other on the grid, as shown in Fig. 2. The width of the grid was doubled to determine the spread of routes over the network.

We also increased the bound for the number of hops in each route to facilitate the potential increase in route length. This value, which is a constant given in the ns2 implementation of DSR, determines the spreading of routing-control packets, such as RREQ, in the network as well as the connectivity between nodes.

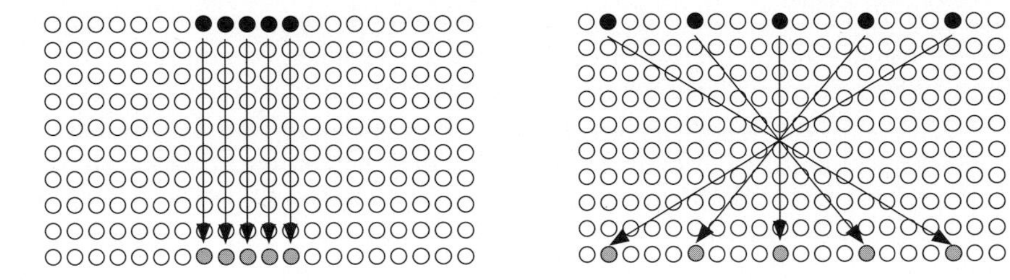

Fig. 2. Dense and twisted sparse network setup: Left unidirectional for all source and destination pairs, on the right side the row of destinations is twisted around

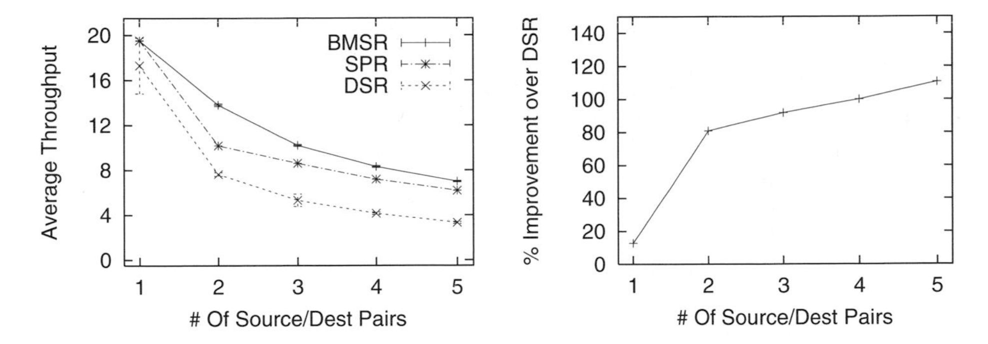

Fig. 3. Average throughput of multiple densely-placed source-destination pairs in KB/s using routing algorithms DSR, BMSR, and SPR. Errorbars represent the standard deviation over 15 repetitions.

However, ns2 resource consumption forced us to choose a rather conservative value of 26 hops. We then explored the performance of DSR, BMSR and SPR.

The SPR algorithm was initialised to use a route of minimum length from the source to the destination node for all packets during the simulation run, independent of route failures. In this sense it behaves similar to BMSR, which determines routes in the setup phase of the algorithm. The algorithm was chosen to evaluate the benefit from choosing multiple routes over a single shortest route for a given network setup.

Because the source and destination nodes are packed closely together, there exist a large number of nodes to the left and to the right, respectively, of the leftmost and rightmost source and destination nodes. In this setup, one can observe from Fig. 3 that BMSR outperforms both DSR and SPR. The latter two routing methods tend to use routes that are close to each other, so that up to three shortest paths can interfere with each other. Route interference, in turn, causes collisions and packet drops due to IFQ overflows.

Note that DSR buffers packets scheduled to be sent over the wireless interface in a queue. After several retransmissions fail, these packets are dropped and removed from the IFQ, causing a decrease in throughput. Hou and Tipper [9]

observed that one of the main reasons for the decline in throughput for congested networks using DSR is the overflow of the IFQ of congested nodes.

In the dense setup, as the number of source-destination pairs increases, shortest-path routes can be expected to be the most favourable in terms of causing less interference than any other choice of routes. Thus the comparative advantage of BMSR with respect to SPR decreases. This trend is observable in Fig. 3. However, as densely packed routes are subject to a higher rate of collisions and retransmissions, SPR also shows a decreasing performance for an increasing number of CBR pairs.

3.3 Sparsely Placed CBR Pairs

Subsequently, we separated the source and destination nodes by three intermediate nodes. Because of the spacing, adjacent shortest-path routes do not conflict with each other. As a consequence, one can see from Fig. 4 that the performance of SPR remains constant for an increasing number of source-destination pairs.

In this setup, BMSR is able to take advantage of the additional nodes between adjacent shortest-paths and shows an increased throughput compared to the dense placement of source-destination pairs, while DSR does not perform significantly better than for the dense setup. As DSR heavily relies on cached routing information, which is updated by routes overheard from neighbours or taken from forwarded packets, nearby sources tend to share parts of their routes over the long run.

In order to evaluate the performance of BMSR for a scenario with a large number of route-intersections, we created a worst-case scenario for route intersections by 'twisting' the aforementioned setup. Figure 2 depicts the resulting network. It is easy to see that the number of pairwise route intersections is $n(n-1)/2$, where n is the number of source-destination pairs.

Figure 5 shows performance results obtained for this network setup. It is interesting to see that BMSR performs very similarly to SPR. This can be

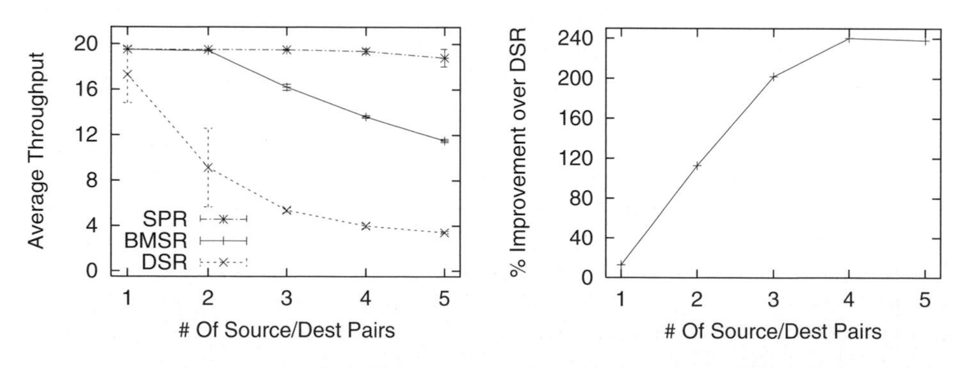

Fig. 4. Average throughput of multiple sparsely-placed source-destination pairs in KB/s using routing algorithms DSR, BMSR, and SPR. Errorbars represent the standard deviation over 15 repetitions.

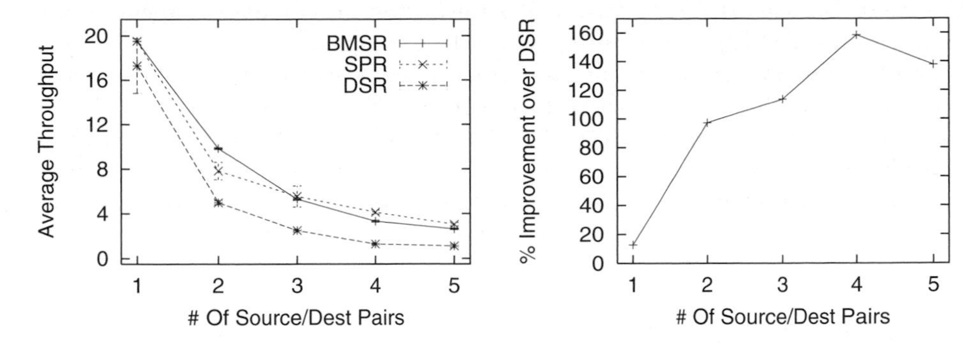

Fig. 5. Average throughput of multiple sparsely-placed source-destination pairs with twisted destination arrangement in KB/s using routing algorithms DSR, BMSR, and SPR. Errorbars represent the standard deviation over 15 repetitions.

explained by the fact that each route will necessarily cross any other route, which causes collisions and congestion at intermediate nodes. However, due to the spreading achieved especially for the inner source-destination pairs, BMSR still outperforms SPR for a smaller number of active sources. As the maximum source-route length restricts the choice of routes for the balancing algorithm, the degree of freedom for outer pairs is much smaller than for inner pairs. In fact, for five source-destination pairs, the outmost pair will always route over shortest paths, determined by the maximum source-route length of 26.

DSR again shows the least performance for this setup. Congested nodes and collisions due to the heavy load in the centre of the network cause routes to fail repeatedly. These have to be rediscovered regularly, causing additional overhead of routing control messages.

4 Conclusions

In this paper, we have applied a linear programming approximation algorithm to the problem of load balancing in ad hoc networks. When integrated into the DSR routing protocol, the resulting BMSR protocol relies on a mathematical formulation of the underlying problem. This is an advantage over routing protocols which aim to achieve load balancing by introducing heuristic rules, for which there is little mathematical justification like that provided here for BMSR.

A limiting factor of BMSR is its non-adaptivity towards node and permanent link failures, as well as changes in the demands of source-destination pairs and the network topology in general. An adaptive version of the proposed protocol could adjust edge flows by locally rerouting flow for example. The second limiting factor is the non-integration of edge interference into the model. This problem could be solved by considering clusters of interfering edges as units to be used for the load balancing procedure, rather than edges themselves.

BMSR is a lightweight, distributed implementation of a linear-programming approximation algorithm. Its integration into a standard routing protocol is enabled by the computation of shortest paths. In this paper, we have presented simulations that also indicate its robustness for different traffic patterns. For all simulations BMSR clearly outperforms DSR. BMSR also performs better than the idealised shortest-path algorithm when source-destination pairs are placed densely or the shortest-paths are not disjoint and there are only few sources simultaneously active in the network. Therefore, we conclude that load balancing based on source routing can provide benefits for stationary networks, e.g. mesh networks with a high throughput requirement.

References

1. Nasipuri, A., Castañeda, R., Das, S.R.: Performance of multipath routing for on-demand protocols in mobile ad hoc networks. Mobile Networks and Applications 6(4), 339–349 (2001)
2. Johnson, D.B., Maltz, D.A., Hu, Y.C.: The dynamic source routing protocol for mobile ad hoc networks (DSR). Technical report, IETF (2003), IETF Draft, work in progress (July 2004)
3. Wu, K., Harms, J.: Performance study of a multipath routing method for wireless mobile ad hoc networks. In: Proc.of 9th International Symposium in Modeling, Analysis and Simulation of Computer and Telecommunication Systems, Washington, DC, USA, pp. 99–107. IEEE Computer Society, Los Alamitos (2001)
4. Mueller, S., Tsang, R.P., Ghosal, D.: Multipath routing in mobile ad hoc networks: Issues and challenges. In: Calzarossa, M.C., Gelenbe, E. (eds.) MASCOTS 2003. LNCS, vol. 2965, pp. 209–234. Springer, Heidelberg (2004)
5. Schumacher, A., Haanpää, H., Schaeffer, S.E., Orponen, P.: Load balancing by distributed optimisation in ad hoc networks. In: Cao, J., Stojmenovic, I., Jia, X., Das, S.K. (eds.) MSN 2006. LNCS, vol. 4325, pp. 873–884. Springer, Heidelberg (2006)
6. Young, N.E.: Randomized rounding without solving the linear program. In: SODA 1995: Proc. of the Sixth Annual ACM-SIAM Symposium on Discrete Algorithms, Philadelphia, PA, USA. Society for Industrial and Applied Mathematics, pp. 170–178 (1995)
7. Bienstock, D.: Potential Function Methods for Approximately Solving Linear Programming Problems: Theory and Practice. International Series in Operations Research & Management Science, vol. 53. Kluwer Academic Publishers, Norwell (2002)
8. McCanne, S., Floyd, S., Fall, K., Varadhan, K.: The network simulator ns2, The VINT project (1995), `http://www.isi.edu/nsnam/ns/`
9. Hou, X., Tipper, D.: Impact of failures on routing in mobile ad hoc networks using DSR. In: Proc. of the Communication Networks and Distributed Systems Modeling and Simulation Conference (2003)

Design and Evaluation of a Multi-class Based Multicast Routing Protocol

Maria João Nicolau[1], António Costa[2], and Alexandre Santos[2]

[1] Departamento de Sistemas de Informação,
Universidade do Minho, Campus de Azurém,
4800 Guimarães, Portugal
joao@dsi.uminho.pt
[2] Departamento de Informática,
Universidade do Minho, Campus de Gualtar,
4710 Braga, Portugal
{costa,alex}@di.uminho.pt

Abstract. Most of current multicast QoS routing proposals are based on the principle that QoS routes must be computed for each request, where requests explicitly express their resource requirements. As a result, within this environment, the goal of QoS routing is to satisfy individual request requirements, resorting to resource reservation to maintain those requirements after a feasible path has been found. This type of strategy is suited within the IntServ model but does not seem adequate in presence of DiffServ networks. According to DiffServ model, traffic flows are aggregated into specific classes-of-service and each flow receives a specific treatment accordingly to its class-of-service. There are no per flow guarantees, only per class differentiation. In this environment instead of per flow path computation, per class path calculation should be made, and so, within multicast scenarios, multiple multicast trees must be computed in order to satisfy different QoS requirements of different traffic classes.

This paper presents a new multicast routing protocol enabling per class multicast tree computation. The proposed heuristics enable directed trees establishment, instead of reverse path ones, due to the importance of link asymmetry within an environment which is, essentially, unidirectional. The proposed protocol is implemented and simulated using Network Simulator. A set of simulation results are presented, analyzed and compared against PIM-SM, a widely deployed multicast routing protocol.

1 Introduction

The main goal of multicast routing protocols is to build a distribution tree or a set of trees in order to deliver data packets from sources to a set of receivers in an efficient manner, without incurring into network overloads. To minimize the resource usage in the network the multicast tree built should be the tree with minimum cost. The problem of finding such a tree is NP-complete and

T. Vazão, M.M. Freire, and I. Chong (Eds.): ICOIN 2007, LNCS 5200, pp. 325–334, 2008.

is called *Steiner Tree Problem*[1]. There are many heuristics in finding a sub-optimal Steiner tree [2]. When QoS is considered, besides the connectivity, the tree branches between the sources and each receiver should satisfy the QoS constraints which turns the problem of build a multicast tree even more complex.

Several strategies have been proposed to implement QoS Multicast Routing, most of them relying on flooding in order to find a feasible tree branch to connect a new member. QoSMIC[3] and QRMP [4] are examples of those strategies. The underlying idea is to obtain multiple paths where a new member may connect to the tree. Typically, multiple probe messages are sent over different possible routes collecting QoS information on the path. Among candidate paths the new member selects the one that is able to satisfy its QoS requirements. This type of strategy is better suited within the IntServ model[5]. The main strength of the IntServ model is its ability to provide service guarantees by means of (state-full) resource reservation. However it has several weaknesses too. Each router is required to maintain state information for each flow, thus, scalability problems do arise in operational environments. In addition a significant amount of processing overhead is required within each router, and the connection setup time may even sometimes be greater than the time required for the transmission of all the packets belonging to a specific flow. The alternative model is called DiffServ[6]. According to this new model, the traffic is aggregated into specific classes-of-service thus changing the scope from per flow guarantees to per class differentiation. Before entering a DiffServ domain packets are marked by border routers (or ingress routers) in one of the available classes-of-service. Inside domain, core routers just give them a specific treatment accordingly to its class-of-service. There are difficulties and challenges when trying to adequate multicast protocols to DiffServ model. The main assumption behind this work is that in presence of DiffServ networks, per flow computation is not adequate. Instead of that a per class path calculation must be made.

In this paper a new multicast routing protocol is proposed enabling per class multicast routing implementation. The proposed protocol takes link asymmetry into account as it defines a *shortest-path-tree* based routing strategy as opposite to a *reverse-path-tree* based one. This is an important feature because when routing constraints are introduced links become asymmetric in terms of the quality of service they may offer, thus link costs are likely to be different in each direction.

2 A Model for Multi-class Based Multicast Routing

The main objective of this work is to propose a new multicast routing protocol that enables per class-of-service multicast routing implementation. The key ideas of the protocol are:

- Build multiple trees, one per class of service. Within a DiffServ multicast scenario, multiple multicast forwarding trees may be found, one per Class of Service (CoS), in order to comply with different per-class Quality of Service

(QoS) requirements. The main objective of this work is to study the viability and efficiency of such an approach.
- Implementation of a directed-tree based routing strategy, instead of a reverse-path-tree one. We believe that reverse path routing is not adequate to address Quality of Service Routing. Links are asymmetric in terms of the quality of service they offer, which makes reverse path routing not suited to implement QoS routing.
- Use both shared trees and source based trees. In PIM-SM the use of both, shared and source based trees, is proposed. It allows nodes to initially join a shared tree and then commute to source based trees if necessary. The same idea is used in the proposed protocol.

Besides, the proposed protocol is aligned with current IP multicast model since it allows that sources and receivers may join or leave at any time and no previous group membership knowledge is assumed.

2.1 MCMRP Tree Construction Algorithm

First, a multiple shared tree mechanism is proposed in order to give receivers the ability to join the group without knowing where the sources are located.

The multiple shared tree mechanism proposed is inspired in Protocol Independent Multicast-Sparse Mode (PIM-SM)[7] with trees rooted at a Rendez-Vous Point (RP) router. A shared tree per class of service available is needed, in order to give sources the ability to start sending data in any class. It is assumed that the total number of classes of service "available" has a pre-established upper limit and is small when compared to the number of participants. Data packets originated by sources are sent towards the RP router, previously marked according to source defined QoS parameters. The RP router forwards data packets from sources through one of the shared trees, based on their class of service. Receivers must connect to all of the RP shared trees when joining the group. When a new receiver decides to join, the designated router sends an explicit join request towards the RP router. The routers along the way between the new receiver and the RP just forward the join request message and no sate information is introduced in these routers. When the RP receives a join request message from a new receiver it must send one join acknowledge message per class of service. These messages must travel towards the new receiver through the best unicast path per each class of service. Routers, along those paths, receiving such acknowledge message may then update their routing tables in order to build new multicast trees branches. Updating is done basically by registering with the multicast routing entry for that tree, the acknowledge message's incoming and outgoing router interfaces.

The process of joining the shared tree in MCMRP is detailed in Figure 1, where variables and flags have the same meaning as defined in PIM-SM[7]. In the illustrated scenario there are two different classes of service ($CoS = 1$ and $CoS = 2$) and router `A` (the designated router of the new receiver) issues a join request message. The routing table entries have the same fields as the PIM-SM

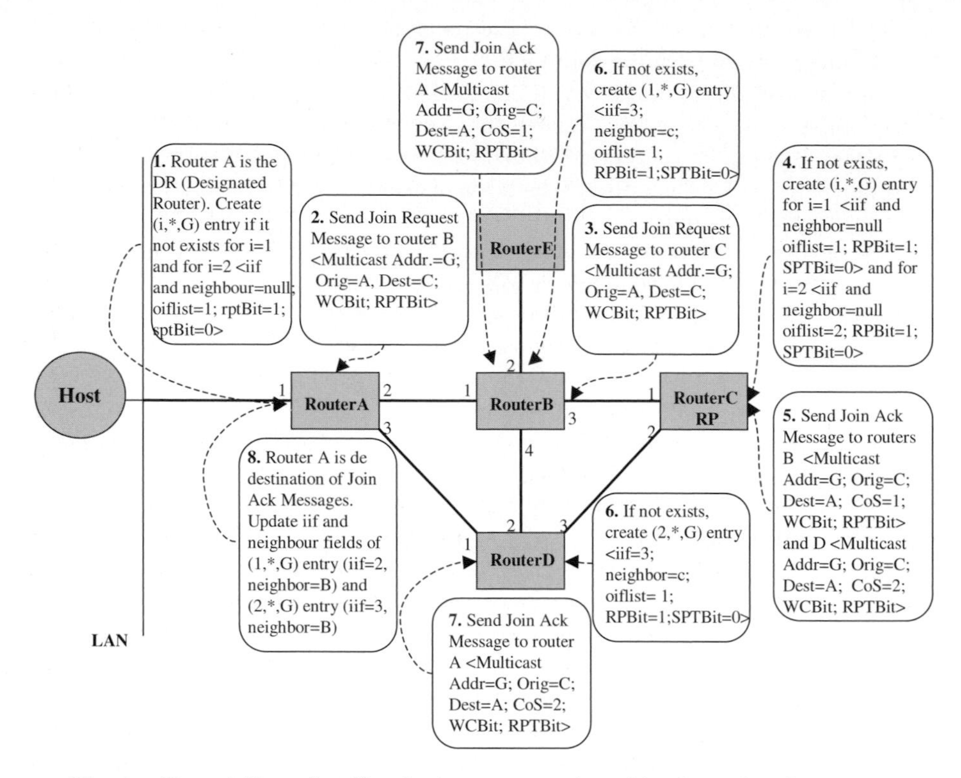

Fig. 1. Shared Trees Set Up. Actions are numbered in the order they occur.

ones, and an extra one: the upstream neighbor in the tree. This field has been introduced in order to implement the prune mechanism.

The multiple RP shared tree mechanism, presented so far, does not really allow receivers to specify their own QoS requirements. Traffic flows from sources to receivers through one of the shared trees, according only to the QoS parameters defined by sources. After a starting period a receiver may demand for a reclassification of source multicast traffic. This issue cannot be accomplished by a shared tree, but it may be met if the receiver joins a source-based tree. When initiating the join to source procedure, the receiver should include in the join request the desired Class of Service. It is up to the source to decide whether or not to accept the join, knowing that when accepting a join, traffic in the requested class of service must be generated. In this situation, each source may face several distinct requests of several distinct receivers for different classes of service within the same group. At the limit, for larger groups, there may be requests for all classes. Even with this worst case situation scalability problems do not arise because the total number of different classes will be much smaller than the total number of receivers. In practice this implies one source-based tree per class of service, unless some order relationship between the classes can be established.

When accepting a join for a new Class of Service, a source must generate an acknowledge message, addressed to the corresponding receiver. This procedure is similar to the one described for the construction of the shared trees. But in this situation only one join acknowledge message is generated per join request. Two different situations may occur. The receiver may decide to switch to a source based tree in the same class used by the source, or it may want to switch to a source based tree requesting a different class of service. In the first case, when a router in the path between the source and the receiver receives the join acknowledge message, if it is not already in the source based tree it must create a `(i,S,G)` entry and copy the outgoing interfaces list from the `(i,*,G)` entry to the outgoing interfaces list of the `(i,S,G)` new entry. This is because, in the future, packets from source `S` will be forward based on this new entry. Besides, when a router lying between the source and the receiver starts to receive data from that source, it must issue a prune of that source on the shared tree of that class. This prune indicates that packets of the class of service `i` from this source must not be forwarded down this branch of the shared tree, because they are being received by means of the source based tree. This mechanism is implemented by sending a special prune to the upstream neighbor in the shared tree of the class `i`. When a router at the shared tree of the class `i` receives this type of prunes, it creates a special type of entry (an `(i,S,G)RPT-bit` entry) closely like a PIM-SM router. In MCMRP the outgoing interface list of the new `(i,S,G)RPT-bit` entry is copied from the `(i,*,G)` entry and the interface deleted is the one being used to reach the node that had originated the prune, which may not be the arriving interface of the prune packet. This is because in MCMRP there are directed trees not reverse path ones. These `(i,S,G)RPT-bit` entries must be updated too when a join acknowledge message arrives in order to allow the join of a new receiver on a shared tree with source-specific prune state established.

When a receiver decides to join a source requesting a different class of service, the process is different. When a new `(i,S,G)` entry is created, the outgoing interface list should not be copied from the `(i,*,G)` entry, because in this case the other receivers connected through the corresponding shared tree still want to receive data packets in the source's default class of service. For the same reason these entries should not be updated when a posterior join to shared tree acknowledge message is received. In addition, the "prune of source in the shared three" mechanism must be triggered by the Designated Router when it receives the join acknowledge message. The prune messages must be sent to the shared trees of all classes except to the shared tree of the class for which the receiver commuted. This is because the receiver will start to receive the source's packets through the source tree in the desired class, so it can not continue to receive it by the shared tree of the source's default class of service.

The process of switching from the shared tree to a source based tree in MCMRP is detailed in Figure 2. In the illustrated situation the receiver decides to switch to a source based tree in class of service `1` ($CoS = 1$), supposing the source's default class of service is 2.

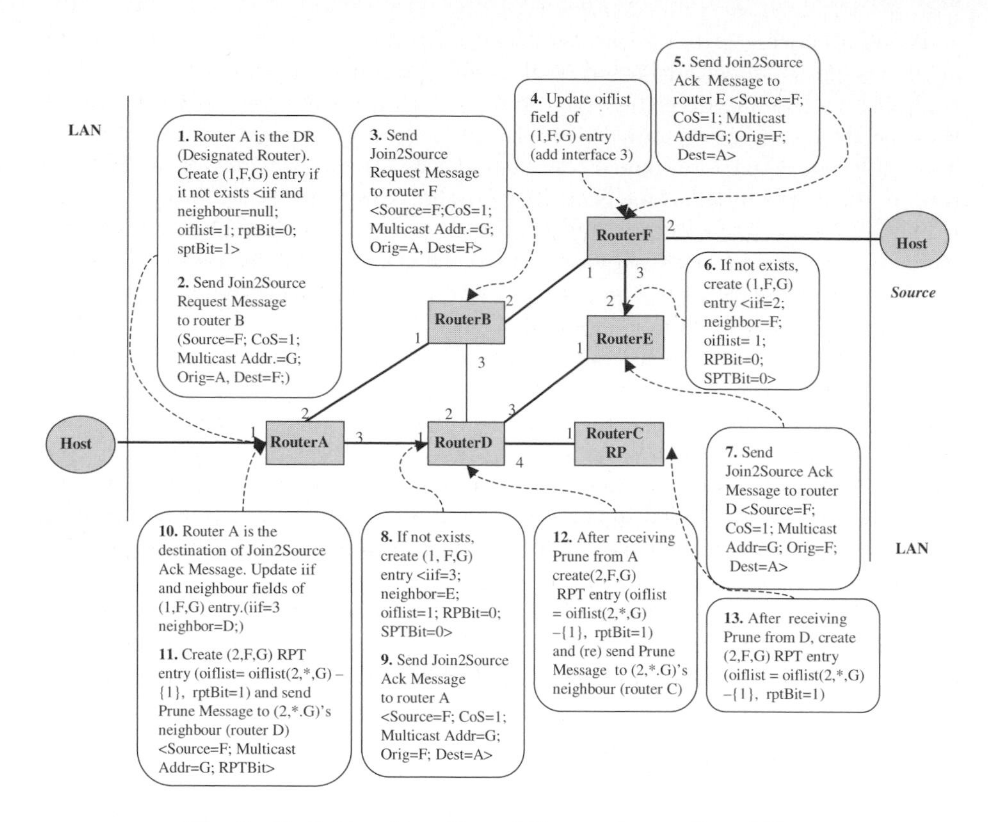

Fig. 2. Switching from Shared Tree to Source Based Tree

3 Multi-class Based Multicast Routing Protocol Implementation

MCMRP implementation is based on Directed Trees Multicast Protocol (DTMP). DTMP [8] is a multicast routing protocol that builds directed trees instead of reverse-path-ones. A complete description of this protocol, with implementation details and comparative results with PIM-SM may be found in [8]. DTMP uses both shared trees and source based trees, like PIM-SM, in order to get the advantages of the both strategies. It is suited for use in asymmetric networks where link costs between any two nodes are different in each direction.

Another major element of MCMRP is the unicast routing protocol in use. Although MCMRP is independent of the underlying unicast routing protocol, it must be a multi-class enabled unicast routing protocol. In other words, the unicast routing protocol must be able to find the unicast routes that can meet the QoS requirements of each Class of Service. In order to build a new tree branch for each Class of Service the multicast routing protocol will search the unicast routing table for the unicast path that is more adequate to satisfy the QoS requirements of each class. To accomplish this new feature, a new unicast routing protocol called CoSLSP (Class of Service Link State Protocol) was implemented.

CoSLSP aims to provide a class based unicast routing mechanism. The basic idea is to find one route per class-of-service, able to satisfy the QoS requirements of that class. Apart from the goal of satisfying the QoS requirements of each class, this protocol also addresses the problem of optimizing network utilization. Therefore, instead of computing just the routes that might meet the QoS requirements of each class, CoSLSP tries to find the shortest path that might satisfy those requirements. It is a unicast link-state protocol that uses a modified Dijkstra algorithm capable of finding the shortest path routes, if they exist at all, that can meet the QoS requirements of different classes of service. In few words: the path calculation algorithm starts by finding the shortest path, whose feasibility is then verified against the QoS requirements. If infeasible, the next shortest path is then iteratively verified, until a feasible path is found or a configured threshold is reached. In this way, a different route is found for each class of service and it is installed in the routing table. The packet forwarding process has been modified too in order to lookup for the appropriate route depending on the class of service of each packet.

CoSLSP has been implemented and evaluated with Network Simulator (NS-2)[9]. The simulations results show that CoSLSP in case of network congestion is able to find "better" routes in respect to the QoS metrics of each class of service.

4 Simulation Analysis

NS[9] has been used in order to simulate MCMRP and its results have been compared with a PIM-SM implementation. In our simulations we used MCMRP with CoSLSP, and an implementation of PIM-SM with LS (a link-state unicast routing protocol implementation included in NS-2 distribution).

To evaluate MCMRP we used two different metrics. To measure the quality of the multicast trees built by MCMRP we used a metric combining the number of data packet replicas with the cost associated to each link traversed by each packet. The second thing we intend to measure is the gain of using class-of-service multicast routing. For this purpose we used the average packet drops occurred in the flows of each class-of-service.

4.1 Simulation Scenarios

The topology used in a simulation scenario is a typical large ISP network[10]. It includes 36 nodes, 18 of them are core nodes, and the other 18 are edge nodes. Each of the core nodes is connected with one edge node by a symmetric link with the cost 1. The core nodes are inter-connected with each other by 30 asymmetric links. There are different alternative paths with different costs between any pair of core nodes. Link costs are integers randomly chosen from the interval [1, 10].

Three different classes of service with different QoS requirements in terms of losses were considered. Class 1 does not support any losses, class 2 supports well 25% of losses and finally class 3 can deal with 50% of losses. Each link has 3 physical queues (one per class) and two virtual queues corresponding to two different drop precedences. All queues are configured exactly in the same way in order to prevent inside node differentiation. Therefore the only class differentiation that can occur is caused by the action of the routing protocol.

For each simulation run, only one group is considered and the RP is randomly chosen within the set of all nodes. There are four fixed sources and each source generates traffic in a class-of-service randomly chosen. It is assumed that a single receiver is connected to each edge node in the topology and that all edge nodes have one potential receiver attached. At the beginning of the simulation there are no receivers joining the group. After an initial period, 9 receivers start to join the group building three shared trees rooted at RP, one per class of service. After all the receivers have joined 8 receivers randomly chosen, join the four different sources requesting a class-of-service randomly chosen too. This scenario is then kept till the end of the simulation. Before the simulation ends, all the receivers leave the group.

4.2 Simulation Results

Simulations results are presented in Figures 3 and 4, The results shown reflect the computed average after 100 independent simulations.

Figure 3 show the characteristics of the trees built with the two protocols (MCMRP and PIM-SM). The curves presented in Figure 3a, show the average tree cost in function of number of receivers. The tree cost is measured in term of number of replicas `times` the link cost. The curves presented in Figure 3b, show the total number of links in the topology that are involved in the multicast trees as a function of number of receivers.

The results shown in 3a bring to evidence that MCMRP constructs trees with lower costs than those created by PIM-SM. This is because MCMRP builds directed trees instead of reverse path trees. Note that CoSLSP does not choose the *best* unicast routing path, it chooses the *best* unicast path that can meet the QoS requirements of each class-of-service. Even with this characteristic the trees

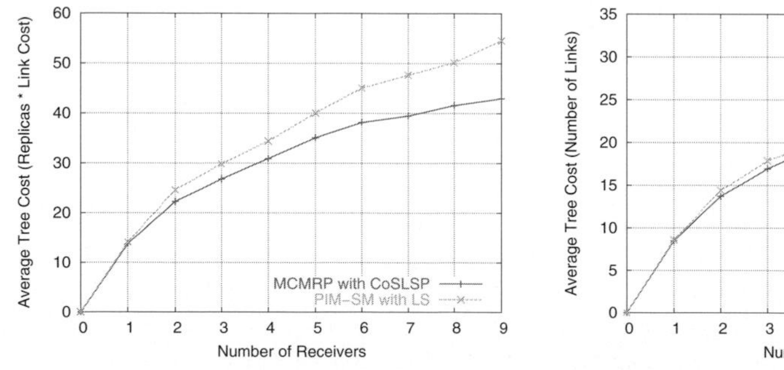

(a) *Tree Cost (packet replicas × link cost)*

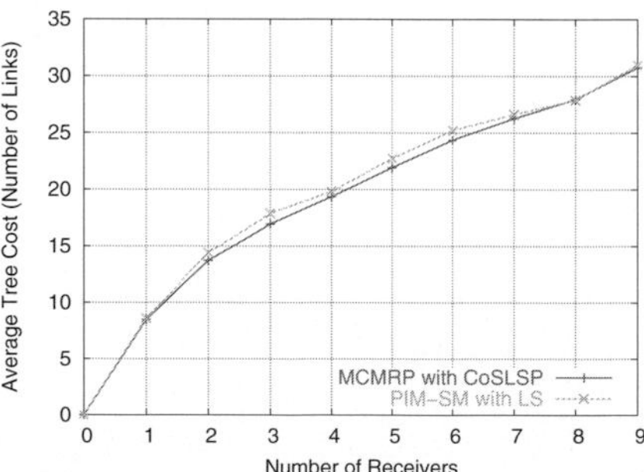

(b) *Number of Links in the multicast trees*

Fig. 3. Characteristics of the multicast trees

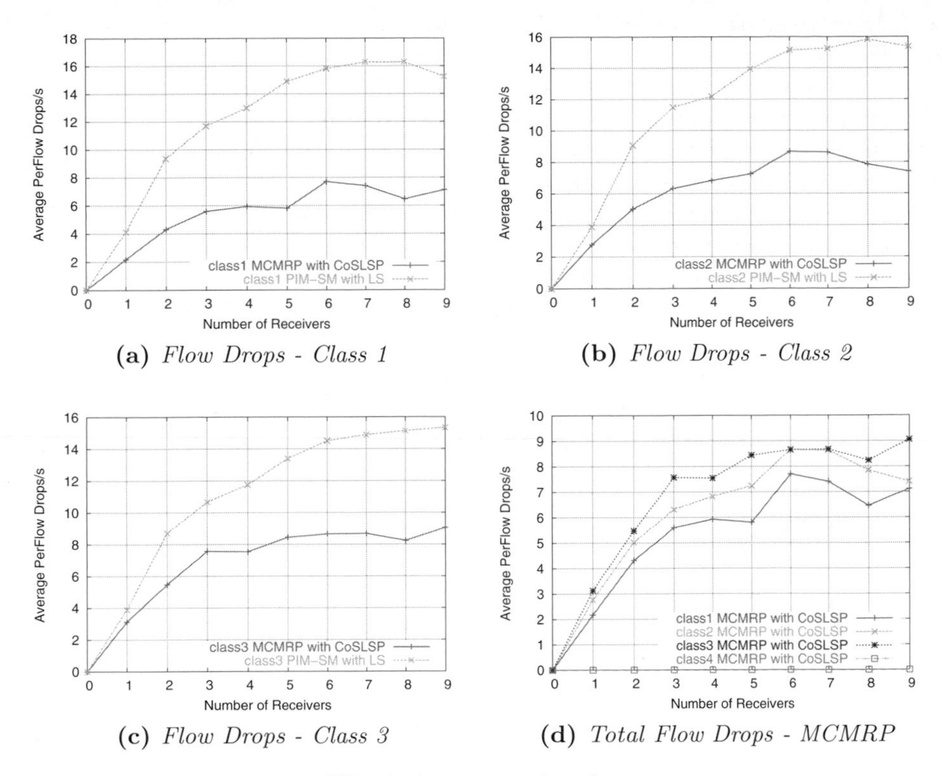

(a) *Flow Drops - Class 1* **(b)** *Flow Drops - Class 2*

(c) *Flow Drops - Class 3* **(d)** *Total Flow Drops - MCMRP*

Fig. 4. Average packet drops

built by MCMRP are *better* in terms of total costs than trees built by PIM-SM. In addition, observing figure 3b we conclude that MCMRP is able to build *better* trees than PIM-SM without enlarging their size.

Figure 4 shows the average packet drops suffered in function of number of receivers. Figures 4a, 4b, 4c show the average packet drops occurred in the flows of each class of service when using the two protocols (MCMRP and PIM-SM). Figure 4d shows the results obtained for all the three classes, in terms of packet drops per flow, when using MCMRP.

These results demonstrate that when MCMRP is used a considerable less amount of drops is verified. This is because MCMRP try to find routes less congested when links became bottlenecks. In addition results show that MCMRP routing strategy promotes the expected differentiation between classes. Observing figure 4d we conclude that the average packet drops suffered by class 3 is greater than the average packet drops suffered by class 2 and the average packet drops suffered by class 2 is greater than the average packet drops suffered by class 1. This is because class 1 have the highest QoS requirements, followed by class 2 and finally class 3 is the least demanding one.

5 Discussion

A new protocol is presented in this paper, MCMRP - a multicast routing protocol that implements multi-class based multicast routing, to be used in a DiffServ environment. Because class differentiation is inherently unidirectional, we propose the usage of source and shared directed trees instead of typical reverse path forwarding ones. The heuristic is based upon explicit join acknowledges sent by either source or RP routers in response to explicit join requests sent by receivers.

MCMRP has been implemented and tested with Network Simulator. The simulations results show that in presence of asymmetries within the network the MCMRP is a promising approach, enabling the establishment of directed multicast distribution trees with significant lower tree costs than either shared and source based trees created by PIM-SM. In addition, MCMRP, is able to find "better" trees in respect to the QoS metric of each class of service.

References

1. Winter, P.: Steiner problem in networks: A survey. Networks 17, 129–167 (1987)
2. Berman, P., Ramaiyer, V.: Improved approximations for the Steiner tree problem. In: Proceedings of the Third Symposium on Discrete Algorithms, pp. 325–334 (1992)
3. Faloutsos, M., Banerjea, A., Pankaj, R.: Qosmic: Quality of service sensitive multicast internet protocol. In: SIGCOMM, pp. 144–153 (1998)
4. Chen, S., Nahrstedt, K., Shavitt, Y.: A qos-aware multicast routing protocol. In: INFOCOM (3), pp. 1594–1603 (2000)
5. Mankin Ed., A., Baker, F., Braden, B., Bradner, S., O'Dell, M., Romanow, A., Weinrib, A., Zhang, L.: Resource ReSerVation protocol (RSVP) – version 1 applicability statement some guidelines on deployment. Request for Comments 2208, Internet Engineering Task Force (September 1997)
6. Blake, S., Black, D., Carlson, M., Davies, E., Wang, Z., Weiss, W.: An architecture for differentiated service. Request for Comments 2475, Internet Engineering Task Force (December 1998)
7. Estrin, D., Farinacci, D., Helmy, A., Thaler, D., Deering, S., Handley, M., Jacobson, V., Liu, C., Sharma, P., Wei, L.: Protocol independent multicast-sparse mode (PIM-SM): protocol specification. Request for Comments 2362, Internet Engineering Task Force (June 1998)
8. Nicolau, M.J., Costa, A., Santos, A., Freitas, V.: Directed Trees in Multicast Routing. In: Ajmone Marsan, M., Listanti, G.C.M., Roveri, A. (eds.) QoS-IP 2003. LNCS, vol. 2601, pp. 321–333. Springer, Heidelberg (2003)
9. Fall, K., Varadhan, K.: The NS Manual (January 2001), `http://www.isi.edu/nsnam/ns/ns-documentation.html`
10. Apostolopoulos, G., Guerin, R., Kamat, S., Tripathi, S.K.: Quality of service based routing: A performance perspective. In: SIGCOMM, pp. 17–28 (1998)

Design and Test of the Multicast Session Management Protocol*

Jung-Jin Park[1], Su-Jin Lee[1], Ilyoung Chong[2], and Hyun-Kook Kahng[1]

[1] Dept. of Electronics and Information Eng., Korea University, Seoul 136-701, Korea
{pjj,aza97,kahng}@korea.ac.kr
[2] Dept. of Information and communications Engineering,
Hankuk University of Foreign Studies, Seoul 130-790, Korea
iychong@hufs.ac.kr

Abstract. Conventional Multicast transport protocols do not include a dynamic mechanism for group management according to the join/leave of receivers and for the modification of membership information. Also, these protocols need the QoS management function that process various QoS requests from each user in the multicast group. In this paper, we propose a reliable multicast session management protocol called Multicast Session Management Protocol(MSMP). It provides efficient group management function and QoS management functions to support a reliable multicast group communication.The MSMP is an application-layer control protocol for managing QoS and security. The MSMP would be designed to provide the IP multicast-based multimedia applications with a QoS management required for the group multicasting such as QoS monitoring and reporting, key distribution, and key management for cryptographic groups, etc. The MSMP will operate over the conventional transport protocols and/or ECTP, and can be used as a control protocol together with the Group Management Protocol(GMP).

1 Introduction

The continuous development of Internet increases demands for efficient and secure network technologies which can cope with various problems introduced by latest emerging applications. Multicast protocols are emerging sets of technologies and standards that provide efficient delivery of data from a sender to receivers in a group. It reduces the transmission overhead of a sender, the network bandwidth usage, and the latency observed by the receivers. Even though multicast communications require a dynamic membership management function for the change of membership information, most existing multicast transport protocols do not have a strictly coupled mechanism for a group management. In this paper, we design the multicst session management protocol for a reliable

* "This research was supported by the MIC(Ministry of Information and Communication), Korea, under the ITRC(Information Technology Research Center) support program supervised by the IITA(Institute of Information Technology Advancement)" (IITA-2006-(C1090-0603-0005)).

T. Vazão, M.M. Freire, and I. Chong (Eds.): ICOIN 2007, LNCS 5200, pp. 335–344, 2008.

multicast group communication. The structure of the paper is as follows. In the section 2, we propose the Multicast Session Management Protocol(MSMP). In the section 3, we describe the design principles for MSMP. In the section 4, and the section 5, we describe the session management and the QoS management of MSMP. Lastly, we describe the traffic variation comparison and conclusion.

2 Proposal of Multicast Session Management Protocol

We propose the Multicst Session Management Protocol(MSMP) for a reliable multicast group communication. The MSMP needs the following motivations and the requirements for MSMP

2.1 Increasing Demand of QoS and Security for Multicast Applications/Services

The legacy multicast transport protocols do not include a dynamic mechanism for group management and QoS management according to status of receivers and for the modification of membership information. According to increasing demand of multicast applications/services, these applications/services need the QoS management that provides various QoS requests from each user in the session.

2.2 Requirements for MSMP

To provide the session management for multicast group communication, the MSMP shall be designed with the following requirements:

(1) Functionality of managing the sessions: Basically, the MSMP is purposed to provide the session management functions for the multicast applications/sessions. The session management function to be provided by MSMP includes the QoS management such as QoS monitoring, and reporting for maintenance. The MSMP could take information about the session creation and the membership status from GMP[1]. Based on the information from GMP, the MSMP could monitor the membership status and inform the members of the change of QoS level.

(2) Easy integration of legacy multicast applications with MSMP: The MSMP is a new control protocol in application layer to support QoS management function for managing the multicast session. Accordingly, all of the existing legacy multicast applications should be able to be used together with the MSMP, without any further modifications. That is, it should be guaranteed that the MSMP can be used along with any legacy multicast applications.

(3) Separation of MSMP control channel from the application data channel: To support the requirement of the easy integration of legacy multicast applications, the MSMP needs to operate as a control channel, separately from the application data channel. Accordingly, the MSMP may itself be implemented as a control module of library, which could be used by any application programs. When the multicast data application needs a QoS management functionality provided by the MSMP, the application could be implemented by the appropriately using the application programming interfaces (APIs) defined in the MSMP control.

3 Design Principles

The MSMP is an application-layer control protocol used for control and management of multicast sessions. The MSMP shall be designed with the following principles.

(1) Separation of control channel and data channel: The MSMP could operate and be implemented separately from the conventional multicast applications. That is, the MSMP will be implemented as a control protocol and thus it could be used by the multicast application programs.

(2) Support of QoS for a reliable session: The MSMP shall be designed to support the monitoring of the QoS level of the active session participants. If the change of QoS level is perceived, the MSMP server will inform the participants of the change of QoS level. The monitored information is also used in the QoS maintenance. QoS maintenance is performed to maintain the desired QoS level and to prevent the connection quality from being degraded below the negotiated QoS level. The MSMP will support the reliable session with QoS management such as QoS monitoring, reporting, and maintenance.

3.1 MSMP Overview

The MSMP is an application-layer control protocol for managing a quality of service for a group session. The MSMP would be designed to provide the IP multicast-based multimedia applications with a QoS management required for the group multicasting such as QoS monitoring and reporting. The MSMP will operate over the conventional transport protocols and/or ECTP, and can be used as a control protocol together with the GMP. Generally it is assumed that there are one MSMP server, one session creating client(or Session Creator), and one or more session participating clients(or Session Participants).

The MSMP supports a reliable integrated service for multicast group communication. This MSMP consists of a session management and a QoS management function. The Session management function is defined in the session management of Group Management Protocol, GMP. The Session management function in MSMP supports the QoS Values for QoS management. The QoS management function provides a stable management of QoS requirement for group members.

3.2 Protocol Model

Figure 1 shows the MSMP control protocol together with the legacy multicast application and the GMP. In the Figure 1, the GMP server provides the MSMP server with the session information and the membership information. Based on thr information, the MSMP server provides the session management and the QoS management to the participants in the session.

Figure 2 shows the protocol stack of the MSMP. As shown in the Figure 2, the MSMP could operate over ECTP[2]. The MSMP will give an API to the GMP to exchange information about the session and membership. Also, the MSMP

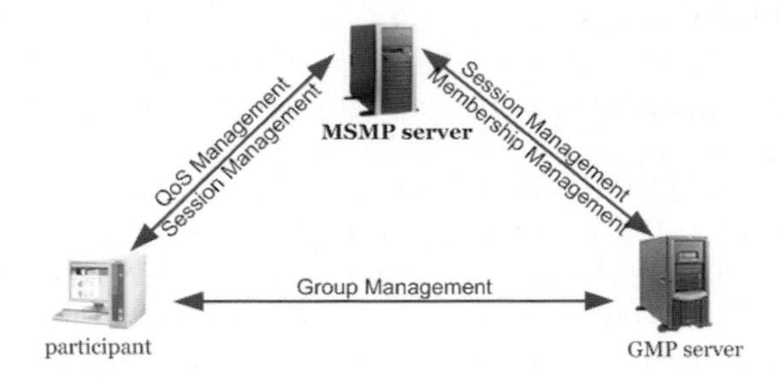

Fig. 1. MSMP, GMP, and Multicast Applications

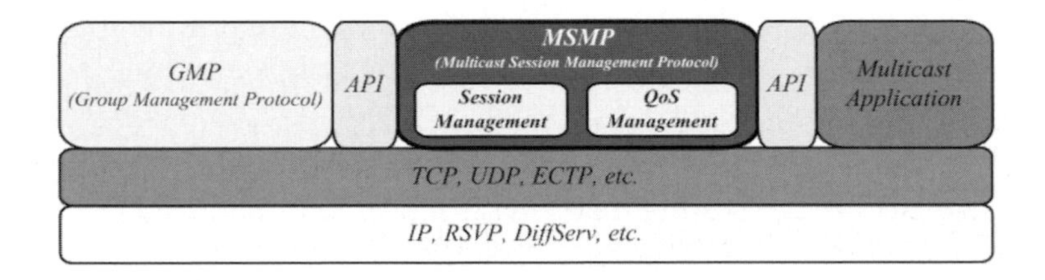

Fig. 2. MSMP Protocol Stack

will provide the MSMP functionality to the multicast applications using API. These APIs will be defined in the MSMP control.

The RSVP[3] supports the QoS for a multicast communication but the RSVP is limited for managing the group membership. Also, the GMP supports the group management but the GMP doesn't support the QoS for multicast group communications. Therefore, we need to manage the group management together with the QoS management for multicast group communications. The MSMP should coordinate the group management and the QoS management.

4 Session Management of MSMP

Session Management(SM) function may be achieved in eight distinct phases: creation, announcement, registration, enrollment, activation, de-registration, de-enrollment, and de-activation.[12] A particular client, called a session creator, creates a session. Then, SM updates the session list. The session creator will send a Session Creation Request message to the server with an initial QoS values. If accepted, the session creator will receive the Session Creation Accept message from the server. Then the session creator will send the detailed session information to the server and receive the confirmation message with a modified and more specified QoS values. If the session can not be created or the session creator does not have the necessary rights, then the Session Creation Reject message will be returned.

After successful session creation, the server will announce the new session to the clients with the more specified QoS parameter values. The announcement may be done by e-mail, web posting, and so on. From this point on, those clients may register in multicast groups. A client may register for the session, considering those QoS parameter values. After successful registration, the client belongs to the registered group.

When the session starts, the session's registered members will start a group application to send and receive session data. At this time, all preparations for the data transfer and group management are accomplished. The session's registered group member belongs to the enrolled group. When the session creator sends a connection request to the enrolled members with a proposed QoS parameter values, the QoS management is then activated.

4.1 Session Creation

Session creation is effected by a session creator, who will define and characterize the session with an initial QoS parameter values including media type, application type, additional information, and so on. A Session Creator defines and characterizes a session an initial QoS values and sends a Session Creation Request message to the session server with initial QoS parameter values such as throughput, delay, jitter, and loss. A Session Creation Request message is a more request asking whether a new creation is possible or not. Considering the multicast environment and its application, the server may allow a new session creation by replying with a Session Creation Accept message. Then, the Session Creator will send detailed session information with more specified QoS parameter values in Session Creation Information message, which may include media type, application type, etc. The server will acknowledge successful session creation with a Session Creation Confirm message and then update its session list and the QoS parameter values.

4.2 Session Registration

Session registration is to select a session and to let the server and creator know the intention of the participation. After successful the session creation, a session client may register for a session. A session client will select a session and send the Session Registration Request message considering the announced QoS parameter values to the server. The server will simply add the requesting client to the Registered Group Membership list and reply to the requestor with Session Request Accept message. After successful the registration, the client belongs to the registered group.

4.3 Session Enrollment

Session enrollment is the state where communication is possible among the Session participants and the session creator. Session participants, including the session creator, should send the Session Join Request message with own QoS parameter values. The server will arbitrate the QoS parameter values considering the number of the enrolled members. Those arbitrated QoS parameter values will be used for a session creator to decide the QoS parameter values for a transport connection.

5 QoS Management of MSMP

QoS Management may have four operation such as the QoS Arbitration, the QoS Reporting to maintained QoS, and the QoS Announcement for the late-join. The MSMP server sends the QoS parameter values to the session creator. The session creator will acknowledge with the arbitrated QoS parameter values to the MSMP server. The MSMP server sends the QoS Reporting request to the senders periodically to maintain a QoS status. In the late join case, a participant will send the QoS value request in order to get QoS information. The MSMP server will reply with the QoS value response witch includes QoS parameter values.

5.1 QoS Arbitration

The MSMP Server will send the QoS Setting Request message to the session creator to let the session creator to reserve the resource. The QoS Setting Request message includes QoS parameter values such as throughput, delay, delay jitter, and loss rate which are previously arbitrated. After receiving the QoS Setting Request from the server, the session creator arbitrates the QoS parameter values. The session creator will acknowledge with the final arbitrated QoS parameter values to the MSMP server via QoS Setting Response message.

5.2 QoS Reporting to Maintained QoS

The MSMP server maintains the QoS parameter values using the periodic QoS Reporting. The MSMP server sends the periodic QoS Reporting Request message to the session creator and participants who may send the multicast data along the multicast control tree. Each sender will acknowledge with own QoS parameter values via the QoS Reporting Response message. The MSMP server keeps updated the QoS parameter values.

5.3 QoS Announcement for the Late-Join

In the late join case, the late-joiner will send the QoS Value Request message in order to get QoS parameter values on-going session. The MSMP server will reply with the QoS parameter values via QoS Value Response message.

6 Test of the MSMP

In this paper, we experiment using 8 PCs on Linux kernel version 2.4.22, the kernel version provided by Red Hat Linux version 9.0. The pc-router has two network interface cards. one(eth0) is the 100mbit NIC and the other card(eth1) is the 10mbit NIC. Also, we use the "MpegTV Player version 1.0"[10] for the MSMP client's multicast application. Figure 3 shows the testbed for MSMP in our work.

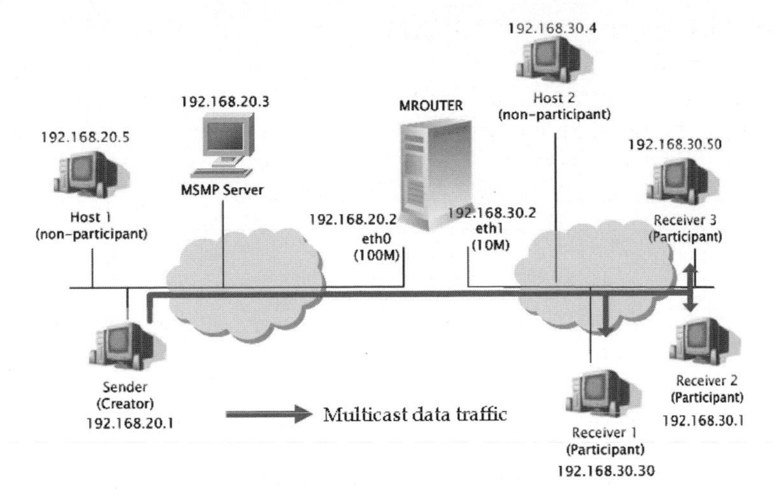

Fig. 3. Testbed for MSMP

6.1 Experimental Results

We experiment with two scenarios. In the first scenario, we used the "MpegTV Player" as the multicast application and in the second scenario, we used the "MGEN packet generator"[11] as the multicast application. In this scenario, the session creator is to send the multicast data traffic to group members, receiver1, receiver2, and receiver3. It means sender, receiver1, receiver2, and receiver3 are enrolled group members. Each of them is in enrolled state. In this case, there is no a background traffic. In the state of enrolled session, the MSMP server and the session creator finalize the arbitrated QoS values. After that the session creator will reserve the resource using PATH and RESV message in RSVP among enrolled group members.

Figure 4 shows the multicast data traffic originated from the sender to receivers using those QoS parameters which are arbitrated QoS values. The traffic source is the MpegTV Player. We monitored the multicast data traffic using the application, "Tele Traffic Tapper, TTT"[9]. In the Figure 4.(a), the multicast traffic is guaranteed by 1.6Mbps. However, Figure 4.(b) is the case that the QoS management is not applied. It shows that the QoS not guaranteed. Our implementation of MSMP with RSVP[3] works properly.

Figure 5 shows the throughput of the multicast data traffic at receiver1. Figure 5.(a) shows the throughput of the multicast data traffic using the QoS management with RSVP. Figure 5.(b) shows the throughput without the QoS management. From this, we can see that there is no big difference between two cases. That means there is no big difference under there is no background traffic.

In the second case, there is the background traffic. We used "MGEN packet generator" as the multicast application instead of "MpegTV Player". In this scenario, the session creator will send the multicast data traffic to group members, receiver1, receiver2, and receiver3, while the host1 sends a background traffic to host2.

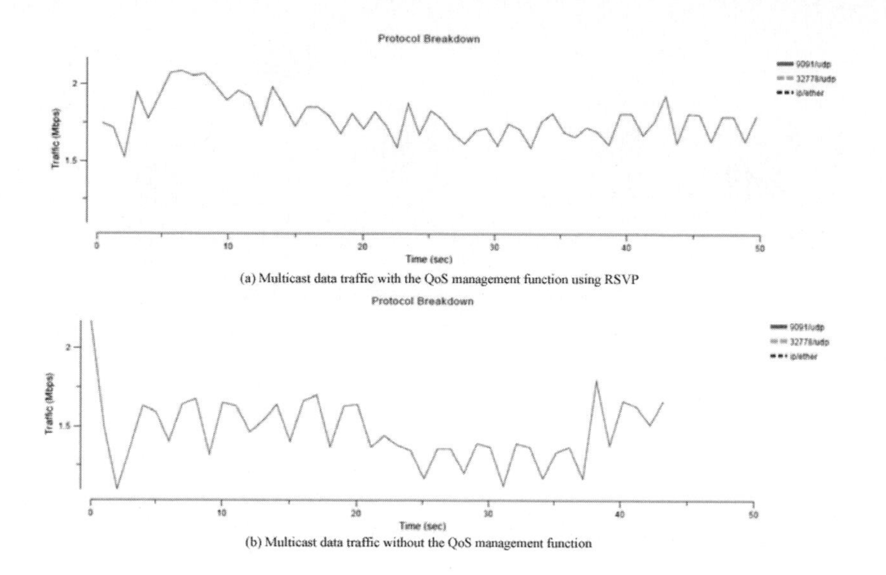

(a) Multicast data traffic with the QoS management function using RSVP

(b) Multicast data traffic without the QoS management function

Fig. 4. Multicast data Traffic with using RSVP

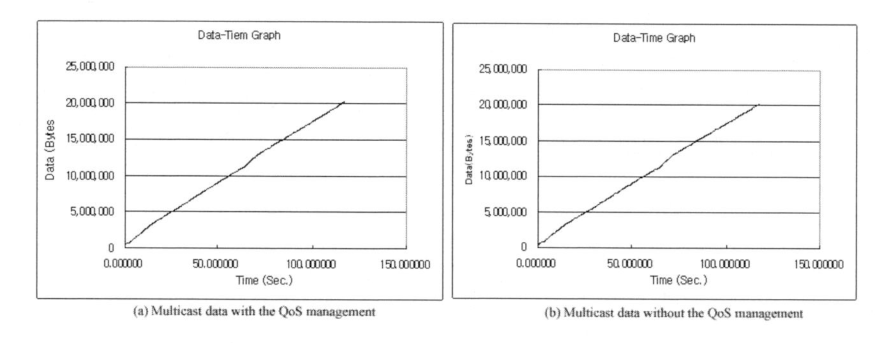

(a) Multicast data with the QoS management

(b) Multicast data without the QoS management

Fig. 5. Multicast data Traffic with using RSVP

Figure 6.(a) shows the multicast data traffic using the QoS management function with RSVP using those arbitrated QoS values. The red dot line is the multicast data traffic and the black dot line is the background traffic. In the Figure 6.(a) the multicast traffic is guaranteed by 5Mbps in spite of another obstruction traffic. Figure 6.(b) shows the multicast data traffic without the QoS management function. The green solid line is the multicast data traffic and the pink line is the background traffic. However, in the Figure 6.(b), the Multicast data traffic doesn't keep the data rate because of another background traffic.

Figure 7 shows the throughput of the multicast data traffic at receiver1. Figure 7.(a) shows the throughput of the multicast data traffic using the QoS management with RSVP. Figure 7.(b) shows the throughput without the QoS management. Figure 7.(a) shows that the Multicast data traffic keeps the token rate in the flowspec of RESV message while the Figure 7.(b) shows that multicast

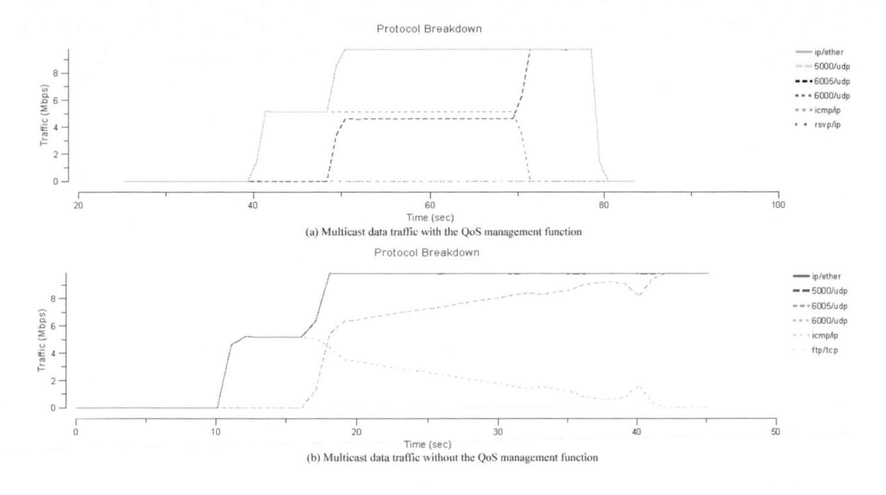

(a) Multicast data traffic with the QoS management function

(b) Multicast data traffic without the QoS management function

Fig. 6. Multicast data Traffic with using RSVP

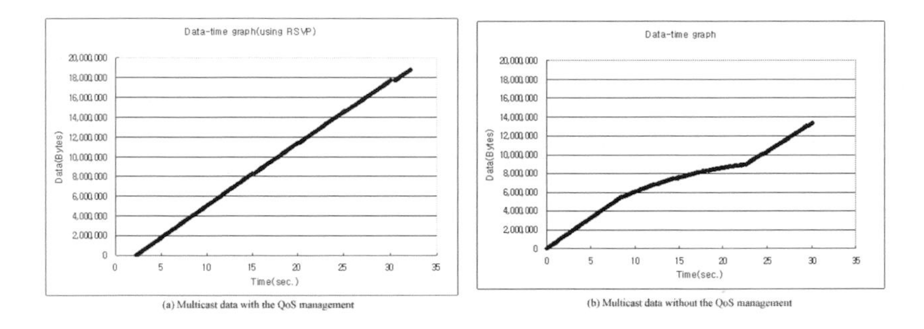

(a) Multicast data with the QoS management

(b) Multicast data without the QoS management

Fig. 7. Multicast data Traffic with using RSVP

data traffic degraded at receiver1. From this, the MSMP provides a reliable QoS service to the group members.

7 Conclusion

GMP provides a session management and membership management, but GMP does not support any QoS management function. In this paper, we proposed and designed the multicst session management protocol for a reliable multicast group management protocol. The MSMP provide the multicast session management and the QoS management for a multicast group session. We use RSVP to apply the QoS management for MSMP. In this paper, we show traffic variation in two senarios.

For the future works, We should evaluate the performance in the case of late-join and leave and measure the performance for the time such as delay and delay jitter. We should evaluate the performance according to the Group-Key management for a secure service for the multicast group communication.

References

1. ITU-T Rec.X.602—ISO/IEC 16513, Group Management Protocol (2004)
2. ITU-T Rec.X.606—ISO/IEC 14476-1, Enhanced Communications Transport Protocol: Specification of simplex multicast transport (2002)
3. Braden, R., Zhang, L., Berson, S., Herzog, S., Jamin, S.: Resource ReSerVation Protocol (RSVP) - Version 1 Functional Specification (1997)
4. Hubert, B., et al.: Linux Advanced Routing & Traffic Control HOWTO (2003)
5. Handley, M., Floyd, S., et al.: The Reliable Multicast Design Space for Bulk Data Transfer RFC 2887
6. McCloghrie, K., et al.: Internet Group Management Protocol MIB, RFC2933 (October 2000)
7. Fenner, W.: Internet Group Management Protocol, Version 2, RFC2236 (November 1997)
8. Ethereal, `http://www.ethereal.com`
9. Tele Traffic Tapper version, `http://www.csl.sony.co.jp/person/kjc/software.html`
10. MpegTV Player, `http://www.mpegtv.com`
11. Multi-Generator(MGEN), `http://cs.itd.nrl.navy.mil/work/mgen/index.php`
12. ITU-T Rec.X.601, Multi-Peer Communications Framework

Providing Full QoS with 2 VCs in High-Speed Switches*

A. Martínez[1], F.J. Alfaro[1], J.L. Sánchez[1], and J. Duato[2]

[1] Departamento de Sistemas Informáticos, Escuela Politécnica Superior
Universidad de Castilla-La Mancha, 02071 - Albacete, Spain
{alejandro,falfaro,jsanchez}@dsi.uclm.es
[2] Dept. de Informática de Sistemas y Computadores, Facultad de Informáica
Universidad Politécnica de Valencia, 46071 - Valencia, Spain
jduato@disca.upv.es

Abstract. Current interconnect standards propose 16 or even more virtual channels (VCs) for provision of quality of service (QoS). However, VCs increase the complexity of the switch and the scheduling delays. In a previous paper, we have shown how to use only two VCs for full QoS support at the switches. In this paper, we explore thoroughly two alternative switch designs that take advantage of this reduction. We analyze their feasibility in a single chip implementation and show that they get a noticeable performance while greatly reducing the cost and power consumption of the network.

1 Introduction

There are many proposals for quality of service (QoS) support in high-speed interconnects. Most of them incorporate 16 or even more virtual channels (VCs), devoting a different VC to each traffic class. In most of the recent switch designs, the buffers are the most silicon area consuming part.

To build the 64-port MIN, we need 48 switches and 192 links when using the 8-port switch designs (*Traditional 8 VCs*, *Traditional 2 VCs*, and *New 2 VCs-B* cases). The final cost of the interconnect greatly depends on the number of switches used, while the power consumption comes mostly from the transceivers needed to drive the links, and thus, depends on the actual number of links [1]. Note that with the *New 2 VCs-P* switch model, it takes just 16 switches (67% less than the traditional case) and 128 links (33% less than the traditional case) to build the 64-port MIN. This greatly reduces the cost and power-consumption of the network.

The buffers at the ports are usually implemented with a memory space organized in logical queues, which consist of linked lists of packets, with pointers to

* This work has been jointly supported by the Spanish MEC and European Comission FEDER funds under grants "Consolider Ingenio-2010 CSD2006-00046" and "TIN2006-15516-C04"; by Junta de Comunidades de Castilla-La Mancha under grant PCC08-0078-9856; and by the Spanish State Secretariat of Education and Universities under FPU grant.

T. Vazão, M.M. Freire, and I. Chong (Eds.): ICOIN 2007, LNCS 5200, pp. 345–354, 2008.

manage them. Therefore, the complexity and cost of the switch heavily depend on the number of queues at the ports. For instance, the crossbar scheduler has to consider 8 times the number of queues if 8 VCs are implemented (greatly increasing the area and power consumed by this scheduler). Then, a reduction in the number of VCs (and in the required buffer space) necessary to support QoS can be very helpful in the switch design and implementation.

In [2] we have proposed a strategy to use just two VCs at each switch port for the provision of QoS that obtains similar results as if we were using many more VCs. This is achieved by respecting at the switches some of the scheduling decisions made at network interfaces. Moreover, to provide guarantee even to the low-priority flows and to prevent starvation, a connection admission control (CAC) is used. In this way, at no link the load of regulated traffic will be higher than the available bandwidth.

In this paper, we throughly explore two alternative switch designs that take advantage of this reduction and we evaluate them with different traffic models. Simulation results show that our proposal provides a very similar performance compared with a traditional architecture with many more VCs both for the QoS traffic and for the best-effort traffic. Moreover, comparing our technique with a traditional architecture with 2 VCs, our proposal provides a significant improvement in performance for the QoS traffic, while for the best-effort traffic, the traditional model is unable to provide the slightest differentiation.

There are several differences between this study and the one at [2]. The first is that here we have made a thorough study of silicon area consumption by switch components taking into account the total silicon area available. Moreover, we also propose a chip design that takes advantage of the reduction of VCs. In addition to this, we offer quantitative results concerning the advantages of using our proposal, in terms of component count and power consumption. Finally, in [2] we considered a theoretical model for scheduling times, while here we take an approach based on the actual ASIC design of the switch.

The remainder of this paper is structured as follows. In the following section the related work is presented. In Section 3 we present our strategy to provide QoS support with only two VCs and we study its hardware implications. The details on the experimental platform are presented in Section 4 and the performance evaluation in Section 5. Finally, Section 6 concludes this study.

2 Related Work

During the last decade several switch designs with QoS support have been proposed. All of them incorporate VCs in order to provide QoS support. InfiniBand was proposed in 1999 both for communication between processing nodes and I/O devices and for interprocessor communication. InfiniBand Architecture (IBA) [3] proposes three main mechanisms to provide the applications with QoS, including the use of up to 16 VCs.

PCI Express Advanced Switching (AS) architecture is the natural evolution of the traditional PCI bus [4]. It defines a switch fabric architecture that supports high availability, performance, reliability, and QoS. AS ports incorporate up to 20 VCs that are scheduled according to some QoS criteria.

These proposals, therefore, use a significant number of VCs to provide QoS support. However, if a great number of VCs are implemented, it would require a significant fraction of silicon area and would make packet processing slower. Note that this paper deals with single-chip switches, where the buffers, the crossbar, and the scheduler are inside the same chip, in order to be able to provide the low latency necessary in current high-performance networking.

Traditional 2 VC proposals distinguish between just two broad categories (regular and premium) [5]. In contrast, the novelty of our proposal lies in the fact that, although we use only two VCs at the switches, the global behavior of the network is very similar as if the switches were using many more VCs. This is because we are respecting at the switch ports the scheduling decisions performed at the network interfaces, which have as many VCs as traffic classes. In the end, the network provides a differentiated service to all the traffic classes considered.

3 Providing Full QoS Support with Only Two VCs

In [2] we have proposed a new strategy to use only two VCs at each switch port to provide QoS. It achieves similar performance results to those using many more VCs. We review this proposal in this section.

Supporting a large number of queues in a switch is not easy. This affects the scheduling performed to configure the crossbar and the arbitration at the output ports. We are referring to classical unbuffered crossbars. However, a different switch architecture exists, that uses buffered crossbars [6]. In this case, the buffer space is neither at the inputs nor the outputs of the switch, but distributed at the crosspoints of the crossbar. When using this design, supporting many queues is even more difficult, since the space at the crosspoint buffers is very limited and to implement many queues is not possible. Moreover, in both crossbar types, the control data structures needed for managing the queues consume silicon area and switch control unit cycles. Therefore, proposals of 8 or 16 VCs for QoS support are very rarely implemented and, as we mentioned before, the trend is to increase the number of ports per switch, instead of the number of VCs.

Traditionally, network designers have overdimensioned the network in order to provide an acceptable QoS. However, this solution is becoming less and less interesting since the network is becoming the most expensive and power-consuming part of the system [1].

The key idea of our proposal is based in this observation: Assuming that the links are not oversubscribed, all the traffic flows through the switches seamlessly. Therefore, the basic idea of our proposal consists in using only two VCs at the switch ports. One of these VCs is used for QoS packets and the other for best-effort packets. Moreover, we propose to use a connection admission control (CAC) to guarantee that QoS traffic will not oversubscribe the links and we give QoS traffic absolute priority over best-effort traffic, which is not subject to the CAC.

Moreover, the network interfaces are responsible for injecting traffic of the different classes applying any desired algorithm. At the same time, the switches reuse this scheduling. This is the cornerstone of our proposal: To reuse at the switches the scheduling decisions taken at the host interfaces.

We assume that a static priority criterion exists to order packets. In this way, every packet would be stamped with a priority level (typically, 8 or 16 levels). This is necessary because packets arriving at the switches come in the order specified by the interfaces, and the switch has to merge these packet flows at the output ports. The way of performing this is very simple: The scheduler takes into account the service level of the packets (8 or 16 priorities), not just if they are at the QoS or best-effort VC. This is not very complex because very efficient priority encoder circuits have been proposed [7]. On the other hand, from the scope of these priority assignments, the packet ordering established at network interfaces does not need to be changed at any switch in the path because queuing delays for QoS traffic will be short.

Obviously, the network interface can only arbitrate among the packets it holds at a given moment. Therefore, when no more high-priority packets are available, a low-priority QoS packet can be transmitted. If this packet has to wait at a switch input queue, and other packets with higher priority are transmitted from the network interface, they would be stored in the same VC as the low-priority packet, and would be placed after it in the queue. Thus, the arbiter would penalize the high-priority packets, because they would have to wait until the low-priority packet is transmitted. But this situation has a small impact on performance because there is bandwidth reservation for QoS packets. This means that all the QoS packets will flow with short delay.

Remember that, although we assume that QoS traffic does not oversubscribe any link, no assumption is made about best-effort traffic. However, the network interfaces are still able to assign the available bandwidth (the one not consumed by QoS traffic) to the best-effort traffic in the configured proportions. In this way, they can still take into account the QoS requirements of this kind of traffic. Obviously, this is a coarse-grain QoS provision. If stricter guarantees were needed by a particular flow, it should be classified as QoS traffic.

Note that this proposal does not aim at achieving a higher performance but, instead, at drastically reducing buffer requirements while keeping the performance and behavior of systems with many more VCs. In this way, a sophisticated QoS support could be implemented at an affordable cost.

Summing up, our proposal consists in reducing the number of VCs at each switch port needed to provide flows with QoS. Instead of having a VC per traffic class, we propose to use only two VCs at switches: One for QoS packets and another for best-effort packets. Moreover, the scheduling decisions performed at network interfaces are reused at switches. In order for this strategy to work, we guarantee that there is no link oversubscription for QoS traffic by using a CAC strategy.

3.1 Implementation Considerations

In order to find out the advantages of using our approach, we have evaluated the cost of implementing the switch both with the 8 VCs (one per traffic class) and with just 2 VCs. In both cases, we have considered a combined

Table 1. Area consumption by components

	16 ports - 2 VCs		8 ports - 8 VCs	
Module	Tech. 0.18 μm	Tech. 0.13 μm	Tech. 0.18 μm	Tech. 0.13 μm
Buffers	64 mm^2	32 mm^2	64 mm^2	32 mm^2
Xbar and datapath	10 mm^2	5 mm^2	5 mm^2	3 mm^2
Scheduler	5 mm^2	3 mm^2	10 mm^2	5 mm^2
Total	79 mm^2	40 mm^2	79 mm^2	40 mm^2

input-output queuing architecture, because it is a common design for high-performance switches. Note that all the switch components must be implemented into a single chip, in order to be able to provide the low latency necessary in high-performance networking.

We have considered two switch designs in this section. The first benefits from our proposal and uses the saved silicon area to implement additional ports at the switch. In this case, the switch would have 16 ports. The other design would be a traditional 8 VCs switch, where only 8 ports would be possible.

The components we have considered for this study are the ports (mainly buffer space with some additional logic), the crossbar, and the scheduler. We have taken the design constraints like packet format, routing, and so on, from the PCI AS specification [4]. Table 1 shows area estimates for each module for two different technologies: 0.18 and 0.13 μm.

The internal clock of the system is 250 MHz and the datapath is 64 bits wide. This provides a speed of 16 Gbits/s, which is twice the speed of the external links. That means there is an internal speed-up of 2.0.

The buffers are 16 Kbytes per port with our proposal (which are shared between the two VCs, 8 Kbytes each) and 32 Kbytes per port in the 8 VCs case (4 Kbytes each VC), both as a compromise between silicon area and performance. The memory area estimates are based on datasheets of typical ASIC (Application-Specific Integrated Circuit) technologies available to European Universities. These numbers include the input and output buffers.

The crossbar and datapath estimates come from the actual numbers of the switch design in [8]. Finally, we base our estimates for the scheduler area on the data provided by McKeown at [9]. The increased area for the scheduler in the 8 VCs design takes into account the queues needed if 8 VCs are implemented (64 queues, 8 VCs multiplied by 8 output ports). The 2 VCs design needs half of the queues (32 queues, 2 VCs multiplied by 16 output ports).

Note that usually there is not a full utilization of the available area in an ASIC design and, therefore, the final chip would be larger. In order to find out more accurate estimates, all the design flow should be performed. However, these area estimates are very helpful to compare the alternative architectures.

Finally, we could also implement an 8-port switch design using our proposed 2 VCs, but with increased buffer space. It would be very similar to the 8 VCs design we have described, but just 16 queues per input port would be necessary

(2 per output port). In the performance evaluation section we will evaluate both alternatives for applying our proposal, compared with more traditional solutions.

4 Simulation Conditions

In this section, we will give details on the simulated network architecture and the load used for the evaluation.

4.1 Simulated Architecture

We have performed the tests considering four cases. First, we have tested the performance of our proposal, which uses 2 VCs at each switch port. It is referred to in the figures as *New 2 VCs*. Note that, with our proposal, the network interfaces still use 8 VCs. We have considered the two variants of this proposal presented in Section 3.1, which are noted *New 2 VCs-P* (16 ports) and *New 2 VCs-B* (8 ports, but larger buffers per VC).

We have also performed tests with switches using 8 VCs. In this case, it is referred to in the figures as *Traditional 8 VCs*. Finally, we have also tested a traditional approach with 2 VCs, noted in the figures as *Traditional 2 VCs*. In this case, the network interfaces also use 2 VCs. Therefore, we have two references to compare the performance of our proposals, one being the lower bound (*Traditional 2 VCs*) and the other the upper bound (*Traditional 8 VCs*). *Traditional 2 VCs* switches have the same number of ports than *Traditional 8 VCs*, but they have 4 times more buffer space per VC, allowing a good performance with bursty traffic.

The network used to test the proposals is a butterfly multi-stage interconnection network (MIN) with 64 end-points. The actual topology is a folded (bidirectional) perfect-shuffle. We have chosen a MIN because it is a usual topology for clusters. However, our proposals are valid for any network topology, including both direct networks and MINs. No packets are dropped because we use credit-based flow control between the switches at the VC level.

The CAC we have implemented is a simple one, based on average bandwidth. Each connection is assigned a path where enough resources are assured. It guarantees that less than 70% bandwith is used by QoS traffic at any link. We also use a load-balancing mechanism, which consists in assigning the least occupied route among those possible. The parameters of the network elements used in this performance study aretaken from PCI AS specification [4].

4.2 Traffic Model

In Table 2, the characteristics of the modeled traffic are included. We have considered the traffic classes (TCs) defined by the IEEE standard 802.1D-2004 [10] at the Annex G, which are generally accepted for interconnection networks. However, we have added an eighth TC, *Preferential Best-effort*, with a priority between *Excellent-effort* and *Best-effort*.

Table 2. Traffic injected per host

TC	Name	% BW	Packet size	Notes
7	Network Control	1	[64,512]	self-similar
6	Audio	16.333	128	CBR 64 KB/s conn.
5	Video	16.333	[64,2048]	750 KB/s MPEG-4 trc.
4	Controlled Load	16.333	[64,2048]	CBR 1 MB/s conn.
3	Excellent-effort	12.5	[64,2048]	self-similar
2	Pref. Best-effort	12.5	[64,2048]	self-similar
1	Best-effort	12.5	[64,2048]	self-similar
0	Background	12.5	[64,2048]	self-similar

The destination pattern of the traffic injected is based on Zipf's law [11], as recommended by the *network processing forum switch fabric benchmark specifications* [12]. In this way, there will be hot spots in the network.

The packets are generated according to different distributions, as can be seen in Table 2. *Audio*, *Video*, and *Controlled Load* traffic are composed of point-to-point connections of the given bandwidth. The self-similar traffic is composed of bursts of 60 packets heading to the same destination, with the packets' sizes governed by a Pareto distribution and the periods between bursts modelled with a Poisson distribution. With this distribution there is a lot of temporal and spatial locality and should show worst-case behavior.

5 Simulation Results

In this section, the performance of our proposals is shown. We have considered three traditional QoS indices for this performance evaluation: Throughput, latency, and jitter. Note that packet loss is not considered because no packets are dropped due to the use of credit-based flow control. We also show the cumulative distribution function (CDF) of latency and jitter, which represents the probability of a packet achieving a latency or jitter equal to or lower than a certain value.

Figure 1 shows the performance of QoS traffic. The average latency results are very similar for the four architectures. On the other hand, we can also see the CDF of latency and jitter at a normalized network load of 1.0. The *Traditional 8 VCs* case offers the best results. With our proposal, a small portion of the packets see their latency increased, and thus, the jitter. However, this is a small handicap compared with the benefits, as we will see later.

In Figure 2 we evaluate the best-effort traffic. In this case, the *Traditional 2 VCs* approach produces the same performance for all the TCs, which is an inadequate behavior, because excellent-effort and preferential best-effort traffic classes should have better performance. The reason for this inadequate behavior is that in the *Traditional 2 VCs* model, all the best-effort classes look the same for the schedulers at both the network interfaces and switches.

On the other hand, the arbiters using our technique take into account the priority of the packets, although they share the same VC. For that reason, our

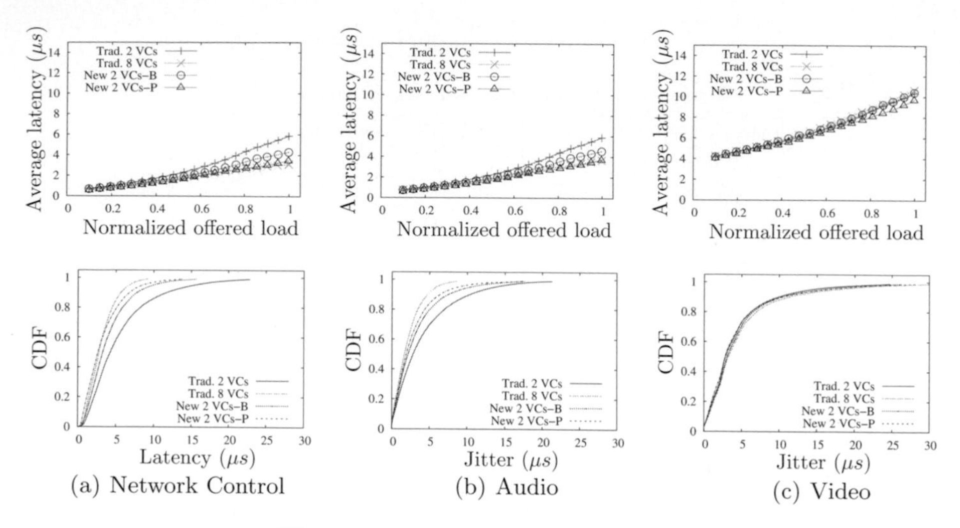

(a) Network Control (b) Audio (c) Video

Fig. 1. Results for QoS traffic classes

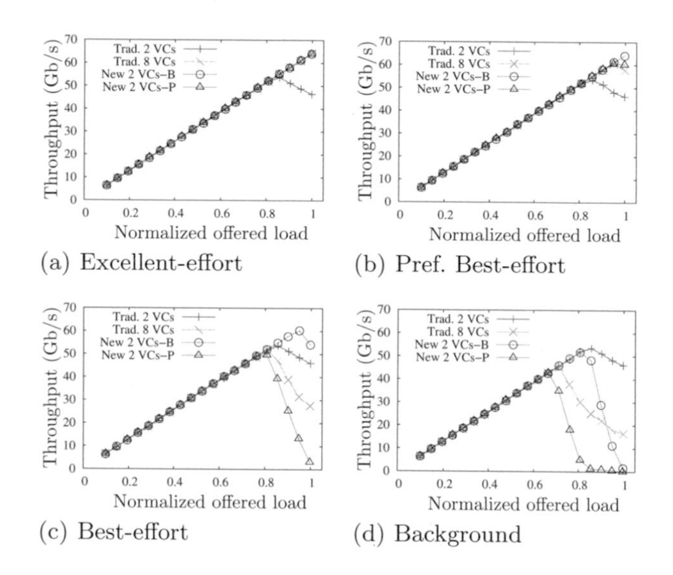

(a) Excellent-effort (b) Pref. Best-effort

(c) Best-effort (d) Background

Fig. 2. Throughput of best-effort traffic classes

proposals, which devote a single VC at the switches for all the best-effort TCs, can provide a behavior similar to that of the *Traditional 8 VCs* approach, which uses 4 VCs for the best-effort TCs.

The *New 2 VCs-B* offers the best performance because it provides the maximum flexibility in the use of the buffer space, while in the *Traditional 8 VCs* case the same amount of memory is statically partitioned. On the other hand, the *New 2 VCs-P* case offers a worse performance because the buffer space is smaller.

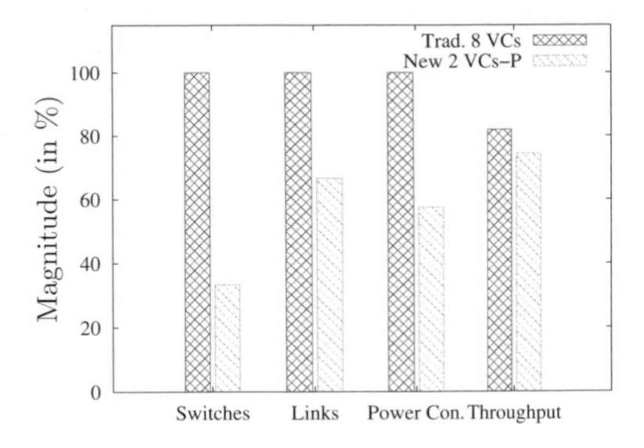

Fig. 3. Summary of the trade-offs of the *New 2 VCs-P* proposal, compared with the *Traditional 8 VCs* case

Figure 3 summarizes the trade-offs of using our *New 2 VCs-P* proposal. As can be seen, there is a very noticeable reduction in chip[1] and link counts and, therefore, in the associated power consumption of the interconnection network (calculated from estimates of the ASIC chip design). On the other hand, the global throughput achieved with bursty traffic is slightly lower (only a 10% reduction), but the reduction on the component count would lead to a much more cost-effective solution. Also note that the non-uniform traffic pattern for best-effort traffic is very bursty: Throughput under more uniform traffic is higher for both *Traditional 8 VCs* and *New 2 VCs-P* cases.

According to these results, we can conclude that our proposals can provide an adequate QoS performance. The reduction of the number of VCs to just two allows us to offer two alternative switch designs: The first keeps the expenses but produces a higher global throughput by using more flexible buffers, and the second greatly decreases the cost and power-consumption of the interconnection with a small degradation in performance (13% global throughput less than *Traditional 2 VCs* with non-uniform traffic).

6 Conclusions

In [2] we presented a proposal to use only two VCs at each switch port to provide QoS support. The first VC is used for QoS traffic and the other for best-effort traffic. In this way, we obtained a drastic reduction in the number of VCs required for QoS purposes at each switch port. In that paper, we showed preliminary results using multimedia traffic in a uniform scenario, without a clear study on how to apply this VC reduction.

This paper offers two novel designs for the switch that benefit from a reduction on the number of necessary VCs to provide QoS. The first design increases the global performance of the network against unbalanced traffic by using buffer

[1] Remember that both *New 2 VCs-P* and *Traditional 8 VCs* switches take equivalent sillicon area.

space in a flexible way. The second design greatly reduces the component count, and thus, the cost and power-consumption of the interconnection, at the cost of a small degradation of performance. However, even under worst-case traffic, this degradation of performance is worthy compared with the savings it brings.

References

1. Shang, L., Peh, L.S., Jha, N.K.: Dynamic voltage scaling with links for power optimization of interconnection networks. In: Proceedings of the 9th Symposium on High Performance Computer Architecture (HPCA), pp. 91–102 (2003)
2. Martínez, A., Alfaro, F.J., Sánchez, J.L., Duato, J.: Providing full QoS support in clusters using only two VCs at the switches. In: Bader, D.A., Parashar, M., Sridhar, V., Prasanna, V.K. (eds.) HiPC 2005. LNCS, vol. 3769, pp. 158–169. Springer, Heidelberg (2005)
3. InfiniBand Trade Association: InfiniBand architecture specification volume 1. Release 1.0 (2000)
4. Advanced Switching Interconnect Special Interest Group: Advanced Switching Core Architecture Specification. Revision 1.1 (2005)
5. Dally, W.J., Carvey, P., Dennison, L.: Architecture of the Avici terabit switch/router. In: Proceedings of the 6th Symposium on Hot Interconnects, pp. 41–50 (1998)
6. Duato, J., Yalamanchili, S., Ni, L.: Interconnection networks. An engineering approach. Morgan Kaufmann Publishers Inc., San Francisco (2002)
7. Huang, C., Wang, J., Huang, Y.: Design of high-performance CMOS priority encoders and incrementer/decrementers using multilevel lookahead and multilevel folding techniques. IEEE Journal of Solid-State Circuits 1, 63–76 (2002)
8. Simos, D.: Design of a 32x32 variable-packet-size buffered crossbar switch chip. Technical Report FORTH-ICS/TR-339, Inst. of Computer Science, FORTH (2004)
9. McKeown, N.W.: The iSLIP scheduling algorithm for input-queued switches. IEEE/ACM Transactions on Networking 7, 188–201 (1999)
10. IEEE: 802.1D-2004: Standard for local and metropolitan area networks (2004), http://grouper.ieee.org/groups/802/1/
11. Zipf, G.K.: The Psycho-Biology of Languages. Houghton-Miffin, MIT (1965)
12. Elhanany, I., Chiou, D., Tabatabaee, V., Noro, R., Poursepanj, A.: The network processing forum switch fabric benchmark specifications: An overview. IEEE Network, 5–9 (2005)

Fast Re-establishment of QoS with NSIS Protocols in Mobile Networks

Franco Tommasi, Simone Molendini, Andrea Tricco, and Elena Scialpi

Dept. of Innovation Engineering
University of Salento, Italy
{franco.tommasi,simone.molendini,andrea.tricco,elena.scialpi}@unile.it

Abstract. Re-establishment of the QoS after a Mobile Node handover must be done as quickly as possible in order to reduce degradation or interruption of QoS, especially when realtime applications are used. We propose a Semi-Proactive procedure for a faster QoS re-establishment in environments where there are Mobile Nodes. The basic functionality of this procedure is to perform as many operation as possible before the handover (in a proactive manner). Resources are reserved after the handover on the effective new data path in order to avoid their waste. Moreover, we propose to buffer at the Candidate CRossover Node - CCRN (an intermediate node on end-to-end data path) the packets directed to the Mobile Node during the handover. In this way the total QoS re-establishment time is reduced. This buffering also guarantees the same QoS treatment of both the buffered and the non-buffered packets.

Keywords: QoS, NSIS, mobility.

1 Introduction

Mobile IP [1] [2] is a protocol that allows a Mobile Node (MN) to communicate to other Internet nodes after changing its link-layer point of attachment without loosing its IP address. Thus, the Mobile IP provides the node to maintain a higher-layer connection while moving.

NSIS Working Group [3] [4] uses a two layer architecture for a signaling protocol:

- NTLP (GIST) [5] carries signaling messages between neighboring peers;
- QoS-NSLP [6] manages resource reservation.

In order to allow the signal packets to have a secure path, GIST provides the modality "connection" creating the GIST state so that there is an association between the adjacent GIST nodes. This association is known as Messaging Association.

As general rule, the re-establishment of the QoS occurs after the handover, the so called reactive way. This means that both GIST and QoS-NSLP act after the handover in order to re-establish the proper QoS on the New path. Resource reservation on the New path must be done as quickly as possible in order to

T. Vazão, M.M. Freire, and I. Chong (Eds.): ICOIN 2007, LNCS 5200, pp. 355–364, 2008.

reduce the degradation of QoS for the MN which is especially important when real-time applications are used.

A basic idea for reducing such delay could be to re-establish QoS before the handover on all the possible future paths, that is to act in a proactive way. Of course this means that there will be a waste of resources since they are assigned on all Candidate paths and we have no guarantee that the handover will occur. We therefore, propose to take the pros of both mechanism by creating the Messaging Associations in a proactive way and by assigning resources in the reactive way. We may say that we are acting in a Semi-Proactive way. Adopting this solution we anticipate the creation of the GIST state before the handover. The resource reservation is as usual performed after the handover but the total time will be minimal. Moreover, resources are assigned only to the effective path to avoid their waste.

In our solution the re-establishment is faster also because we introduce the possibility of buffering the packets directed to MN, in an intermediate node of the end-to-end path. This node is known as the Candidate CRossover Node (CCRN).

The procedures described in the following sections are applicable in networks served by Mobile IPv4 with Route Optimization [7] or by Mobile IPv6 (in which the Route Optimization is already integrated).

Other previous works are related to the same problems treated in this paper:

- Mobile resource reservation Protocol(MRSVP) was proposed by Talukdar [8] to provide QoS for Mobile IP through RSVP [9] [10]. MRSVP Protocol makes the resource reservations in advance at multiple locations where the mobile host may possibly visit during the service time. The mobile host can thus achieve the required service quality when it moves to a new location where the resources are reserved in advance.

  MRSVP wastes too much bandwidth in making excessive resource reservations in advance for an MN in all the neighboring locations of the MN.
- Route optimization strategies [7] is used to reduce the well known triangle routing problem of Mobile IP.

  An extension to the registration process, called smooth handoff [11], reduces data loss during a handoff. The foreign agent (FA) buffers any data packets it is forwarding to an MN. When a handoff occurs the new FA in turn requests that the previous FA hands over the buffered packets to the new location.

  Although Route Optimization with smooth handoff extension can, to some extent, reduces packet loss, this does not resolve an other problem: if the flow required a resource reservation, the packets that travel from the previous FA to the new will not (or could not) avail of the QoS present on the paths between the Correspondent node and the FAs.

This paper is organized as follow. In the next paragraph we describe the terminology used in this paper. In paragraph 3 and 4 we present our procedure from the NSLP and the GIST point of view respectively. In paragraph 5 we illustrate the new buffering proposed in order to reduce packets loss during the MN handover. The paragraph 6 shows the analysis and simulation results of the QoS re-establishment time when different procedures and types of buffering are used. We conclude in paragraph 7 mentioning future work.

2 Terminology

- NSIS-aware: a node is NSIS-aware when it supports NTLP and QoS-NSLP;
- Mobile Node (MN): a host that can change its point of connection to the network;
- Correspondent Node (CN): the network node with which the MN is communicating at that time;
- Access Router (AR): the access node to the network for an MN;
- Previous AR (PAR): the AR for the MN before the handover;
- Candidate AR (CAR): possible AR for a handover of the MN;
- New AR (NAR): the AR for the MN after the handover;
- New path: the path between CN and NAR;
- Old path: the path between CN and PAR;
- Candidate path: the path between CN and CAR;
- Crossover node (CRN): one of the NSIS-aware nodes of the intersection between the Old and New path;
- Candidate CRN (CCRN): one of the NSIS-aware nodes of the intersection between the Old and Candidate path.

3 Semi-proactive Procedure from the NSLP Point of View

In these paragraphs we will present a brief description of the traditional NSIS phases and a detailed description of the Semi-Proactive procedure.

Let us first recall that QoS-NSLP supports both the sender-initiated and the receiver-initiated reservation. The entity that sends the RESERVE message takes the name of QoS NSIS Initiator (QNI) and the entity that receives that message is called QoS NSIS Responder (QNR). The intermediary NSIS-aware takes the name of QoS NSIS Entity (QNE).

In the case of sender-initiated reservation (Fig. 1(a)), the RESERVE message (used to reserve the resources) travels in the same direction of the data flow. Therefore, the QNI is the NSIS-aware node that is closer to the sender of that flow. Moreover, going through the nodes, the same RESERVE message allows the GIST to create its state.

In the case of a receiver-initiated reservation (Fig. 1(b)), the RESERVE message travels in the opposite direction to the data flow. Therefore, the QNI is the NSIS-aware node that is closer to the afore mentioned flow receiver. In this case, due to asymmetric routing, the QNR must send a QUERY message to allow the GIST to create its state.

The procedure presented in this paper, the Semi-Proactive procedure, has the aim to re-establish the QoS as fast as possible.

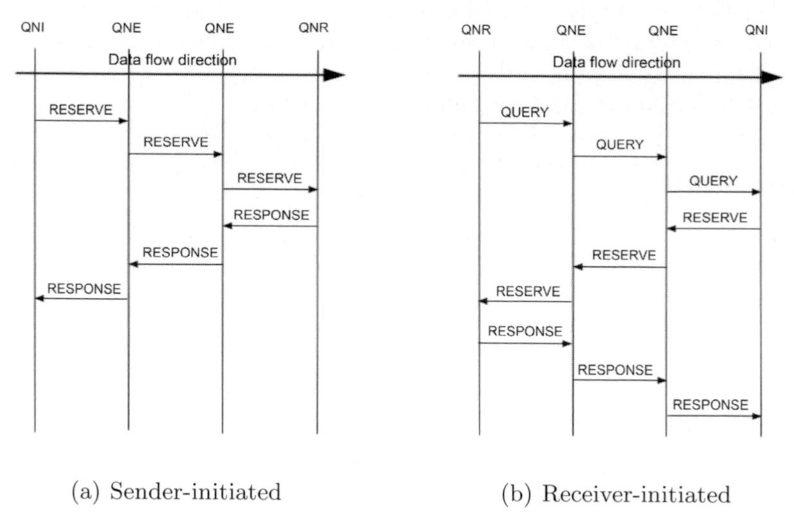

(a) Sender-initiated (b) Receiver-initiated

Fig. 1. Basic NSIS reservation

The best way is to initiate, in a proactive way, as many operations as possible but without the waste of the available resources. In order to do this we propose to create the GIST state before the handover and to reserve the resources on the New path after the moving of the MN. The RESERVE message therefore, can only be sent after the handover. In case of receiver-initiated reservation (Fig. 2(a)) it is easy to define a separation between the creation of the state and the reservation of the resources. In fact, it is only necessary that the QNR (close to the sender of data flow) sends the QUERY message before the handover and

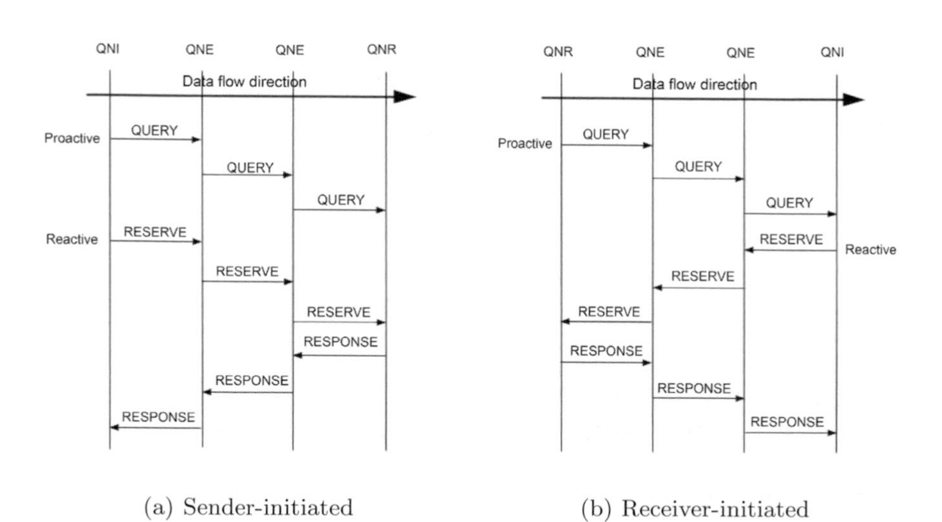

(a) Sender-initiated (b) Receiver-initiated

Fig. 2. Semi-Proactive QoS Re-establishment

the QNI waits the end of the handover to respond with a RESERVE message. The problem is the separation of the two phases when the reservation is sender-initiated. Therefore, also in this case, we could send a QUERY message in order to obtain the separation between the GIST state creation and the resource reservation (Fig. 2(b)). The QUERY message is sent in a downstream direction in the same way as the receiver-initiated, but in this case the message travels from the QNI to the QNR. After the handover, the QNR that has received a QUERY message will wait for the RESERVE message, without further action.

In addition, we would like to consider that in the proactive phase the PAR already has a list of CARs. Each one of these could act as a QNI or as a QNR depending on the type of reservation; it would send or receive the QUERY message for the GIST state creation.

In order to benefit from this procedure, the NAR must be one of the CARs, otherwise, there wouldn't be a GIST state with the CN and the procedure would change from being Semi-Proactive to totally Reactive.

The procedure consists of three main sections:

1. Triggering, in which the PAR communicates to the correct end-point so that it sends the QUERY message;
2. Proactive, in which the QUERY message allows the creation of the GIST state and the discovery of the CCRN;
3. Reactive, in which the resources are reserved.

4 Semi-proactive Procedure by the GIST Point of View

The fundamental function of the GIST protocol is to discover the path that signaled data will follow and to create a GIST state between the end-points of the data flow. As previously stated, in networks where there are MNs it is better to create a proactive GIST state on all of the Candidate paths. The GIST protocol discovers the CCRN during the creation of the GIST state. If the MN is the data flow sender the CCRN could be the first NSIS-aware node in which the Old and the Candidate path converge. If the MN and the receiver of the data flow the CCRN could be the last NSIS-aware entity before the Old and the Candidate path diverge. Consequently, the CCRN would be a node on the common trunk between the Old and the Candidate path.

At the end of the proactive phase there will be as many GIST states as the nodes contained in the CAR list. It is potentially possible to discover as many CCNRs as CARs.

5 The New Buffering

If the MN is the data flow receiver, it is necessary to have buffering mechanisms to avoid the loss of packets sent to MN during its handover. The buffered packets are sent to the MN across the NAR at the end of the handover.

This type of solution has already been defined, for example, Perkins [11] who proposed buffering at the PAR (in the document this node is called Foreign Agent), though this guarantees only a minimal loss of packets and not the proper QoS during the sending of the buffered packets to the MN. We have proposed the inclusion of buffering in the CCRN for various reasons. The most significant reason is that the CCRN is the closest node to the MN on the New path in which buffering is possible. This reduces the total time that the buffered packets use to reach the MN. The second reason for making this choice relates to the QoS. The buffered packets, sent to the MN after the reservation of resources are treated by the same QoS as the other packets from the same flow.

A CCRN starts to buffer packets as soon as it has been discovered.

The CRN ends the buffering of packets when it receives the RESERVE message from the QNI.

The other CCRNs stop the buffering of packets and discards the already buffered packets at the end of a timeout.

If the MN is the data flow sender it is not necessary to begin any buffering.

6 Evaluation of the QoS Re-establishment Time

In this paragraph we will evaluate the QoS re-establishment time when different procedure and type of buffering are used.

We reference the following definitions:

- N1: the number of node-to-node links between the NAR and the CRN;
- N2: the number of node-to-node links between the CN and the CRN;
- N3: the number of node-to-node links between the NAR and the PAR;
- Tstate: Time necessary to install GIST state on a node-to-node link;
- Tresv: Time necessary to reserve resources on a node-to-node link;
- Td: Transmission delay on a node-to-node link;
- TimeCRNsp/TimeCRNp/TimeCRNr: time necessary to re-establish the QoS after a handover with the Semi-Proactive/Proactive/Reactive mechanism and with the packets directed to the MN being buffered at the CRN. In the Proactive case we assume that the MN performs the handover at the node where the resources have been previously reserved;
- TimePARsp/TimePARp/TimePARr: time necessary to re-establish the QoS after a handover with the Semi-Proactive/Proactive/Reactive mechanism and with the packets directed to the MN being buffered at the PAR. In the Proactive case we assume that the MN performs the handover at the node where the resources have been previously reserved;
- TimeCRN/TimePAR: time necessary to re-establish the QoS after a handover with the Proactive mechanism and with the packets directed to the MN being buffered at the CRN/PAR. If the MN performs the handover at a different node from the one where the resources have been previously reserved, then the procedure used becomes Reactive.

6.1 Time Necessary to Re-establish the QoS with CRN and PAR Buffering

In this paragraph we describe the time necessary to re-establish the QoS after the handover when the packets directed to the MN are buffered at the CRN or at the PAR. It also includes the time necessary to send these packets to the MN after that the resources have been reserved. Of course, in both cases, the buffered packets are treated with the same QoS of the no-buffered ones.

Time necessary to re-establish the QoS with CRN buffering is calculated as:

$$TimeCRNsp = (N1 + N2) * Tresv + N1 * Td \tag{1}$$

$$TimeCRNp = N1 * Td \tag{2}$$

$$\begin{aligned} TimeCRNr = & (N1 + N2) * Tresv + \\ & +(N1 + N2) * Tstate + N1 * Td \end{aligned} \tag{3}$$

If PAR buffering is used, Route Optimization and the guarantee for the buffered packets of the same QoS of the non-buffered ones are required in order to match the same condition of the CRN buffering.

Then, the time necessary to re-establish the QoS is calculated as:

$$TimePARsp = (N1 + N2 + N3) * Tresv + N3 * Td \tag{4}$$

$$TimePARp = N3 * Td \tag{5}$$

$$\begin{aligned} TimePARr = & (N1 + N2 + N3) * Tresv + \\ & +(N1 + N2 + N3) * Tstate + N3 * Td \end{aligned} \tag{6}$$

6.2 Network Topology Used in the Simulation

We use a group of APs in a hexagonal cell model as shown in Fig. 3(a). The indexing (row, column) represents the position of an AP. We assume that if a MN moves from the AP at the row 0 upward it reaches the row 3 and if it moves from the row 3 downward it reaches the row 0. Moreover, if a MN moves from the column 0 to its left it reaches the column 3 and if it moves from the column 3 to its right it reaches the column 0. In a recurring cyclic. When a MN resides in a cell the probability to select one of the 6 neighbouring cells is 1/6 as displayed in Fig. 3(b).

When a MN performs a handover, N1, N2 and N3 change as shown in Fig. 3(c).

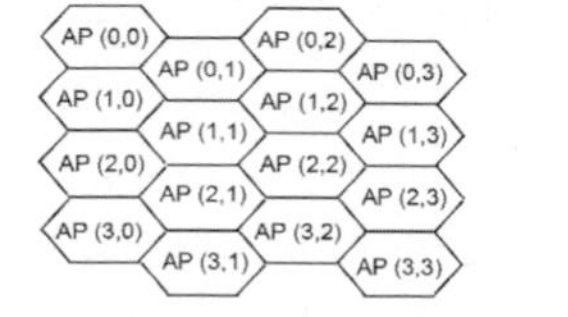

(a) Cell model

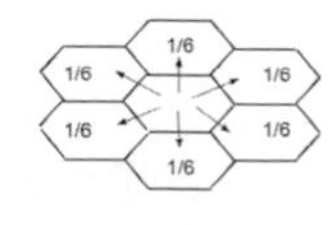

(b) Movement probability between the cells

		N1	N2	N3
column changes	0 <---> 1	3	2	6
	2 <---> 3	3	2	6
	0 <---> 3	4	1	8
	1 <---> 2	4	1	8
row changes	0 <---> 1	1	4	2
	2 <---> 3	1	4	2
	0 <---> 3	2	3	4
	1 <---> 2	2	3	4

(c) Number of the node-to-node links in the network topology

Fig. 3. Simulation topology

6.3 Analysis and Simulation Results

In both PAR or CRN buffering, the Reactive procedure uses much more time than the Proactive or the Semi-Proactive ones. Its clearly evident from the equations in the previous paragraphs that the Proactive procedure would seem the best solution. The equation (2) and the equation (5) are valid only in the case in which MN performs the handover at the node where the resources have been previously reserved. Since the probability that this occours is only 1/6, in the remaining 5/6 instances the Reactive procedure must be used and the equations (3) and (6) would be applied. Then, the equations that describe the Semi-Proactive procedure remain (1) and (4) while the ones that represent the Proactive procedure become:

$$TimeCRN = 1/6 * TimeCRNp + 5/6 * TimeCRNr \tag{7}$$

$$TimePAR = 1/6 * TimePARp + 5/6 * TimePARr \tag{8}$$

The values of N1, N2 and N3 at each handover are illustrated in Fig. 3(c). The values of N1, N2, N3 used in this time evaluation are obtained by the mean of 10000 handover.

The Semi-Proactive simulator is written in C language.

To study the equations (1), (4), (7) and (8), we fix one of the three variables (i.e Tstate, Tresv and Td). Tresv is chosen because it influences all the previous equations while Tstate would influence only the TimeCRN and the TimePAR. The variable Tresv is set to 2*Td because for reserving resources we require the RESERVE and RESPONSE messages.

Matching the equations we obtain three cross points that are:

- Td=5/2*Tstate between the equations TimeCRNsp and TimeCRN;
- Td=1/4*Tstate between the equations TimePARsp and TimeCRN;
- Td=5/2*Tstate between the equations TimePARsp and TimePAR.

These are shown in the Fig. 4 where Tstate has the value of 50 msec.

When increasing Tstate, we see that only the equations TimeCRNsp and TimePARsp are not influenced while TimeCRN and TimePAR will move up along the y-axis (demonstrating that the network delay during the installation

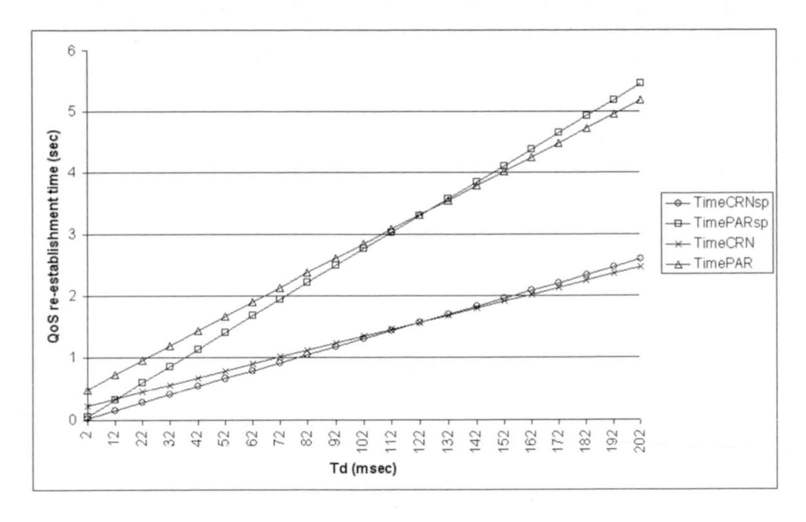

Fig. 4. QoS re-establishment time graph

of the GIST state does not influence the total QoS re-establishment time if the Semi-Proactive mechanism is used). Therefore, at a certain value of Tstate (approximately 80 msec) the cross points equal to 5/2*Tstate will fall out of the Td range. In these conditions the time for QoS re-establishment for each value of Td is minimum when using the Semi-Proactive procedure with CRN buffering (TimeCRNsp) and maximum when using the Proactive procedure with PAR buffering (TimePAR).

7 Conclusion and Future Work

In this paper we have proposed a procedure for the faster re-establishment of QoS with NSIS protocols in mobile networks. It has been named Semi-proactive procedure and its function is to perform as many operations as possible before the handover. We prepare in advance the multiple locations where the mobile host could visit during the service time, however, we reserve the resources only after the handover to minimize the waste of resources.

Moreover, we propose a new point of buffering for the packets directed to the MN during the handover. The buffered packets in this node are treated with the same QoS of the other packets of the same flow which are sent to the MN after the resources reservation on the New path.

Analysis and simulations evaluated the QoS re-establishment time when the Semi-Proactive, the Proactive and the Reactive procedures are used; we have also considered CRN and PAR buffering.

We conclude that CRN buffering is better than PAR buffering and that for low values of Td the QoS re-establishment time is minimized when using the Semi-Proactive procedure with CRN buffering and maximized when using the Proactive procedure with PAR buffering .

Future works will provide a detailed description of the algorithm for CCRN discovery. There will be other works to implement and test this Semi-Proactive procedure.

References

1. Perkins, C.: IP mobility support for IPv4. Request for Comments (Proposed Standard) RFC 3344, Internet Engineering Task Force (August 2002)
2. Johnson, D., Perkins, C., Arkko, J.: Mobility Support in IPv6. Request for Comments (Proposed Standard) RFC 3775, Internet Engineering Task Force (June 2004)
3. Hancock, R., Karagiannis, G., Loughney, J., Van den Bosch, S.: Next Steps in Signaling (NSIS): Framework. Request for Comments (Proposed Standard) RFC 4080, Internet Engineering Task Force (June 2005)
4. Fu, X., Schulzrinne, H., Bader, A., Hogrefe, D., Kappler, C., Karagiannis, G., Tschofenig, H., Van den Bosch, S.: NSIS: a new extensible IP signaling protocol suite. IEEE Communications Magazine, 133–141 (October 2005)
5. Schulzrinne, H., Hancock, R.: GIST: General Internet Signaling Transport. Internet draft (draft-ietf-nsis-ntlp-11), work in progress (August 2006)
6. Manner, J., Karagiannis, G., McDonald, A.: NSLP for Quality-of-Service signalling. Internet draft (draft-ietf-nsis-qos-nslp-12), work in progress (October 2006)
7. Johnson, D., Perkins, C.: Route Optimization in Mobile IP. Internet draft, work in progress (November 2001)
8. Talukdar, A., Badrinath, B., Acharya, A.: MRSVP: a resource reservation protocol for an integrated services network with mobile hosts. Wireless Networks 7(1), 5–19 (2001)
9. Zhang, L., Deering, S., Estrin, D., Shenker, S., Zappala, D.: RSVP: a new resource Reservation protocol. IEEE Network 7, 8–18 (1993)
10. Braden, R., Zhang, L., Berson, S., Herzog, S., Jamin, S.: Resource ReSerVation Protocol (RSVP) – Version 1 Functional Specification. Request for Comments (Proposed Standard) RFC 2205, Internet Engineering Task Force (September 1997)
11. Perkins, C., Wang, K.-Y.: Optimized smooth handoffs in Mobile IP. In: Proceedings of IEEE Symposium on Computers and Communications, pp. 340–346 (July 1999)

A Probabilistic Approach for Fully Decentralized Resource Management for Grid Systems

Imran Rao and Eui-Nam Huh

Department of Computer Engineering, KyungHee University
Yongin, Keyonggi, South Korea 446-701
imran@oslab.khu.ac.kr, johnhuh@khu.ac.kr

Abstract. The specific problem that underlies in collaborating Grids is scheduling of resources with no knowledge about availability of the resources due to the distributed and autonomous nature of the underlying Grid systems. In this paper*, we propose a fully decentralized and probabilistic resource management scheme for Grid systems collaborating based on peer-to-peer communication paradigm. The key idea we employ is to use benchmarked performance measures about the static resource information and calculate the job execution workload. Then this benchmarked job execution time is used to predict the job scheduling feasibility in the face of resource dynamism on the target system. We design our scheme as self adjusting to the actual resource behavior and performance. Simulation results validate the appropriateness of our scheme.

1 Introduction

The momentum of Grid adoption in business and scientific communities is growing rapidly to solve resources intensive problems which otherwise are not possible. Such Grid frameworks are needed to be highly adaptive to the user's requirements along with adjusting to the dynamically varying resource performance. That is, they must have the ability to dynamically reconfigure (a term also named as resource harvesting in some literatures [3], [4]) themselves on need basis. One of the reasons to allow participating entities to join and/or resign the Grid framework is load-balancing and load-sharing (invoking more resources on-peak hours to share load and revoking resources off-peak hours to reduce the system overhead). Resource reconfiguration also is a way for saving administrative cost by offloading un-utilized resources. In this study, we figure out the issues relating to resource scheduling in such a framework and propose a solution for that. Due to the independent and autonomous nature of the underlying resource management systems (RMSs) and need for dynamic reconfiguration, resource scheduling is one of the main research issues. Answer to the challenging scheduling questions, such as given below, is needed to be searched;

* This research was supported by MIC (Ministry of Information and Communication), Korea, under ITRC (Information Technology Research Center) support program supervised by the IITA (Institute of Information Technology Advancement). (IITA-2006-C1090-0603-0040).

T. Vazão, M.M. Freire, and I. Chong (Eds.): ICOIN 2007, LNCS 5200, pp. 365–374, 2008.

- How resource schedule can be calculated in a fully decentralized environment where no assumption about the availability of resources on remote sites can be made?
- Considering target resources being heterogeneous and reconfigurable, different machines will have different job execution time. How to conduct schedule-ability analysis of these resources when their performance characteristics are unknown or dynamic?

Ref. [6, 7, 8 and 9] suggests submitting jobs to all the systems in the framework and as soon as the job starts on any one of the machine, they revoke job from rest of the systems. This solution offers an optimal schedule but on a cost of job submissions to multiple sites. We present a novel solution to reduce this overhead by finding the feasibility of the schedule on a target machine without actually submitting a job to it. To find a good and appropriate schedule in such an environment, application specific information (such as job begin time, job completion time (deadline), job staging information, inputs, etc) and system specific information (such as resource type, resource capability, job arrival rate, job execution rate, job execution duration etc) needed to be integrated. We further classify this system specific information into static information (resource type, resource capacity, etc) and dynamic information (job arrival rate, job execution rate, etc). Due to the difficulty in consistently obtaining good dynamic-performance information about Grid machines, we implement our scheduling algorithm that is partially based on static information to calculate application workload, and partially based upon the probabilistic technique to capture dynamics of resources. Our simulation results demonstrate the effectiveness of our scheduling mechanism for long-term applications.

Our paper organization is as follows; Section 2 presents a detailed survey of related work. In Section 3, we introduce our proposed resource scheduling scheme. We in section 4 present the simulation model for our work and analyze the simulation results. We conclude our work by presenting summary and future directions in section 5.

2 Survey of Related Work

There are many approaches have been proposed to solve these issues [13]. Some dynamic online scheduling algorithms, such as those in [1] and [2], consider the case of resource reservation which is popular in Grid computing as a way to get a degree of certainty in resource performance. Algorithms in these two examples aim to minimize the makespan of incoming jobs which consist of a set of tasks. Authors in [5] present a resource planner system that uses performance prediction, based upon historical data, to identify the appropriate resources. Such systems don't take into account the resource reconfiguration and assume fixed resource performance which is not the case in our presented scenario. Authors in [4] propose a dynamic and self-adaptive task scheduling scheme based upon application-level and system-level performance prediction. In this scheme, authors assume a complete knowledge about the resources and number of tasks allocated on them and uses a statistical model to calculate the task execution time on the remote machine.

The most related work with that of us is K-Distributed Model [7] according to which a job is submitted simultaneously to the K lightly loaded sites instead of sending only to the most lightly loaded site. When a job is ready to start at any of the site, the site informs the scheduler at the job-originating site, which in turn contacts the other K-1 sites to cancel the jobs from their respective queues. In [6], authors propose a resource information free algorithm, which uses task-replication to cope with the heterogeneity of hosts and tasks, and also the dynamic variation of resource availability due to load generated by others users in the Grid. Tasks are replicated until a predefined maximum number of replicas are reached. Based on the assumption that the network transfer time is negligible is only existent for application with small input/output data. In [8], authors introduce Storage Affinity, which takes into account the fact that input data is frequently reused either by multiple tasks of the one application or by successive executions of the same application. We believe this assumption is only true for very limited scientific applications and will not hold for most of business and social scenarios. The rationale behind Suffrage [9] is that a task should be assigned to a certain host and if it does not go to that host, it will suffer the most. But when there is input and output data for the tasks and resources are clustered, conventional suffrage algorithms may have problems.

As far as the mechanization of the learning from previous experience in order to predict future events is concerned, it has led to vast amounts of research [11, 12] in the construction of predictive algorithms. Note that resource-availability predictions solely based upon the archived performance-data assume that a slow resource will always be slow and a busy resource will most likely be always busy on a particular time of the day. In our paper, we demonstrate how predictive algorithms enable us to anticipate the occurrence of events of interest related to system performance, such as CPU overload, bandwidth utilization, and low response time.

3 Proposed Scheduling Scheme

3.1 Application Model

We assume user application is composed of independent, coarsely grained and indivisible jobs and we define job to be anything which need a resource. Let job J_i be represented as $J_i = (B_i, C_i, S_i, T_i, \varepsilon_i)$. Where B_i is job begin time, C_i is job completion time (deadline), S_i is job staging information, T_i is the required resource type (for example, CPU, memory). The job required to be scheduled at begin time B_i and completed before deadline C_i with an expectable margin of error ε_i. Resource type T_i is used to select the benchmark Standard Performance Evaluation Corporation (SPEC) [10] to gauge the application workload.

3.2 System Model

Grid resource performance is variable and is a based upon static (e.g. total CPU, network, memory capacity) and dynamic information (current resource usage). This information is necessary to be known with a good degree of accuracy to predict the success of a schedule. We, therefore, model system using both of these parameters. Let there are n independent Grid systems offering resources to be shared and by resource we mean anything that can be scheduled. We assume each Grid resource to be reconfigurable and non-cooperative. Furthermore each Grid system has its own local resource scheduler to map jobs to the resources. We also assume the underlying schedulers are online-schedulers and map the jobs to the resources as soon as they arrive. We model Grid system G_j as follows; $G_j = (T_j, C_j, A_j, SPEC_j)$. Where T_j, C_j, A_j and $SPEC_j$ is resource type, capability, availability and benchmark factor of G_j respectively.

3.3 Scheduling Algorithm

In our proposed scheduling algorithm, instead of actually replicating the job to the participating Grid systems, we collect schedule feasibility value from prospected Grid systems by just sending the job information to save the data replication, communication and staging cost. After obtaining job execution estimate (which may not be correct) from a list of *preferred* collaborating Grid systems, we evaluated them on the basis of archived performance factor. Proposed scheme is described in Fig. 1.

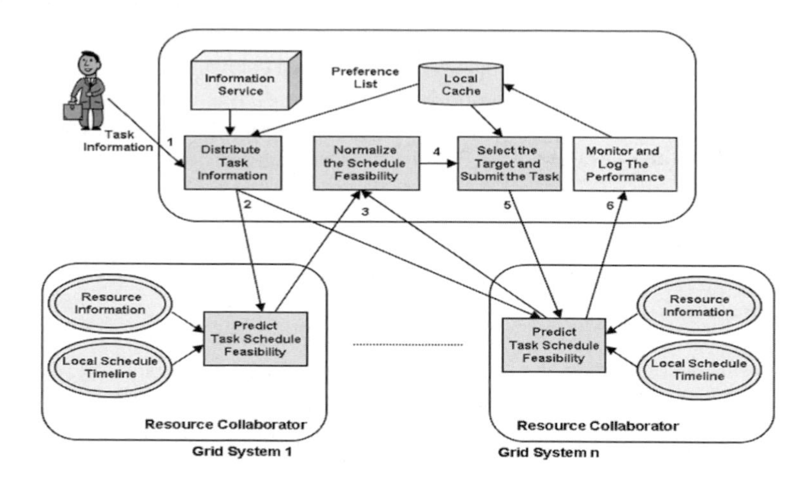

Fig. 1. Proposed resource scheduling scheme

Our scheduling objective is to find the target Grid system with optimal performance in terms of job execution duration and network latency. Let E_{ij} is the execution duration of job J_i on G_j and N_{ij} is the network latency between G_h and G_j then scheduling function to calculate normalized feasibility value Θ_i for J_i is defined as;

$$\Theta_i = Optimal(E_{ij}, N_{ij}) \forall j \in \{1,2,...,n\} \rightarrow \text{(Equation. A)}$$

In the above listed steps, scheduling algorithm, essentially the selection of Grid system with optimal performance, can be break down into following three main functions 1) Calculation of job expected execution duration using the benchmark information 2)

Prediction of schedule feasibility for the expected execution duration and 3) Normalization the feasibility value and selection of target Grid system. We calculate them one by one in rest of this section.

Job Benchmark Execution Duration E_i

The techniques for calculating job benchmark execution time are well known and well practiced. The only implication in using the benchmark scheme is that we need to determine the composition of the job in terms of the same code type, the process called as code profiling. The benchmarked job execution duration E_i using the $SPEC_i$ information is calculated as follows; $E_i = (J_i, SPEC_i)$. This benchmarked execution duration is utilized to calculate the schedule feasibility of job J_i at peer Grid systems.

Scheduling Feasibility F_{ij}

When a new job J_i with profiled workload E_i is submitted to G_j for scheduling, there are issues relating to heterogeneity of the underlying resources and dynamism of resource utilization at that particular time. Due to these factors, one job will have different job execution duration on different Grid systems. Assuming that job execution duration is directly proportional to the job workload, the job execution duration E_{ij} of the J_i on G_j is calculate as follows; $E_{ij} = SPEC_i * E_i / SPEC_j$.

The expected job completion time C_{ij} of job J_i on G_j is defined as the time at which job J_i completes, that is $C_{ij} = B_i + E_{ij}$

Based upon resource prediction scheme similar to [11], Grid system G_j will use this E_{ij} to calculate the feasibility value F_{ij} to predict the success of the schedule of this new job J_i. On G_j, let we have k randomly scheduled current jobs with begin times B_j and execution duration E_j. Then these jobs are shown on a single discrete time line ta as $ta = \{B_j, E_j\} \forall j \in \{1,2,...,k\}$ such that $B_j + E_j \le B_{j+1}$. We define tr as the time line to be scheduled and is calculated as $tr = ta - \sum_{j=0}^{k} E_j$. The new J_i cannot be scheduled successfully on G_j if its begin time conflicts with any previously scheduled job. That is; $(B_j \le B_i \le B_j + E_j)$. Or if a previously scheduled job has a begin time that conflicts with the J_i. That is; $(B_i \le B_j \le B_i + E_{ij})$.

Now the feasibility F_{ij} that new job J_i will be successfully scheduled on G_j is calculated as $F_{ij} = \Pr obability(X \cup Y \cap Z)$. Where X is the event that B_i doesn't conflict with the any previously scheduled job duration E_j, Y is the event of resolving a conflict in case B_i overlaps with E_j and Z is the event that no previously scheduled job has a begin time B_j that conflicts with E_{ij}. Since X and Y are mutually exclusive events; $F_{ij} = (P_s + P_\varepsilon) * P_d$. Where P_s, P_ε and P_d are probabilities of event X, Y and Z respectively. Certainly P_ε is the margin of error J_i enjoys on G_j because of the flexibility value ε_i in its deadline. Since begin time B_i is uncorrelated with any previously scheduled job, P_s is given by the fraction of the time line remaining to be unscheduled: $P_s = tr / ta$.

Now for each of the k previously scheduled jobs, the probability that begin time B_j will not conflict with E_{ij} is calculated as; $(1 - E_{ij} / tr)$. Hence P_d is obtained as follows; $P_d = \prod_1^k (1 - E_{ij} / tr) = (1 - E_{ij} / tr)^k$.

And finally we calculate P_ε. If there is an overlapping $(B_j \le B_i \le B_j + E_j)$ of job J_i with a previously scheduled job on G_j with begin time B_j and execution duration E_j, then the conflict can be resolved by adjusting the begin time of J_i, if $(\varepsilon_i \ge B_j + E_j - B_i)$ as shown in Fig. 2. Probability that conflict of J_i with a job execution duration E_j can be resolved as $= \varepsilon_i / E_j$ for $\varepsilon_i < E_j$ and = 1 otherwise.

The average probability, for resolving the conflict with k existing job is calculated as sum of the product of the probability that new job conflicts with previously scheduled job and probability that the conflicts can be resolved. Numerically;

$$P_\varepsilon = \sum_{j=0}^{k} (E_j / ta) * (\varepsilon_i / E_j) = k * \varepsilon_i / ta$$

As this feasibility F_{ij} is only an estimate weather J_i can be scheduled at time B_i on G_j and can complete its execution E_{ij} before C_i, the actual schedule may be different then what was predicted.

Normalized Feasibility Θ_i and Target Grid Selection

It is quite possible that a greater feasibility value may come from a Grid system with poor network bandwidth. Moreover, there may be situations when, due to the poor estimates, a job is assigned to slow or busy Grid system. To avoid such cases we need to normalize the schedule feasibility value. The normalized feasibility Θ_i is function

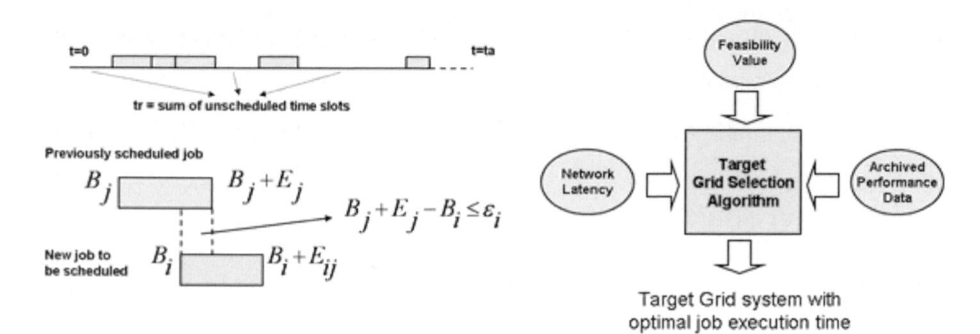

Fig. 2. Probability of scheduling in case of a conflict

Fig. 3. Feasibility normalization function and Target Grid selection

of schedule feasibility F_{ij}, network latency N_{hj} and archived resource performance P_{hj} as Fig. 3 shows.

Let N_{hj} is the time to transmit data S_i for job J_i to Grid G_j. If λ_j is the estimated bandwidth between G_h and G_j obtained through NWS [12] and is assumed to be consistent with expectable error latency then time taken to transmit S_i to G_j follows; $N_{hj} = S_i / \lambda_j$.

The performance factor P_{hj} of Grid system G_j for the host Grid G_h is calculated as shown in Fig. 4. The performance factor technique is self-adaptive against resource dynamism. To start with each Grid system is assigned an arbitrary uniform performance factor value. Performance threshold value is also to be adjusted according to the application scenario.

Let schedule feasibilities, network latencies, and archived performance counters for a set of n participating Grid systems $\overline{G}$ are $\overline{F}$, $\overline{N}$ and $\overline{P}$ respectively. Where $\overline{G} = \{G_1, G_2, ..., G_n\}$, $\overline{F} = \{F_{i1}, F_{i2}, ..., F_{in}\}$, $\overline{N} = \{N_{h1}, N_{h2}, ..., N_{hn}\}$ and $\overline{P} = \{P_{h1}, P_{h2}, ..., P_{hn}\}$.

```
current_job_status = JobMonitorAgent.getCurrentJobInfo(job[i]);

job_remaining_execution = current_job_status.jobRemainingExecution();
expected_remaining_execution = getBenchmarkValueOf(job[i], currentTime);

if ( ( job_remaining_execution - expected_remaining_execution) > threshhold )
        decrement_target_grid_preference ();
else
        increment_target_grid_preference ();
```

Fig. 4. Resource Performance Adaptive algorithm

The Equation A. to calculate normalized feasibility value Θ_i for target Grid system selection is reinterpreted as;

$$\Theta_i = Max(F_{ij} * P_{hj} / N_{hj}) \forall j \in \{1,2,..,n\} \rightarrow \text{(Equation B)}$$

4 Simulation Model and Results

We simulated our system for 10 Grid systems connected with varying values of N_{hj}. Each Grid system G_j has randomly generated k number of previously assigned jobs. Initially 0.50 value assigned to P_{hj} G_j. We calculated Θ_i by assigning equal weights to all three selection factors F_{ij}, N_{hj} and P_{hj} and presented the results in Fig. 5 (a). Results of Grid system 1, 2 and 9 of particular interest and require further analysis.

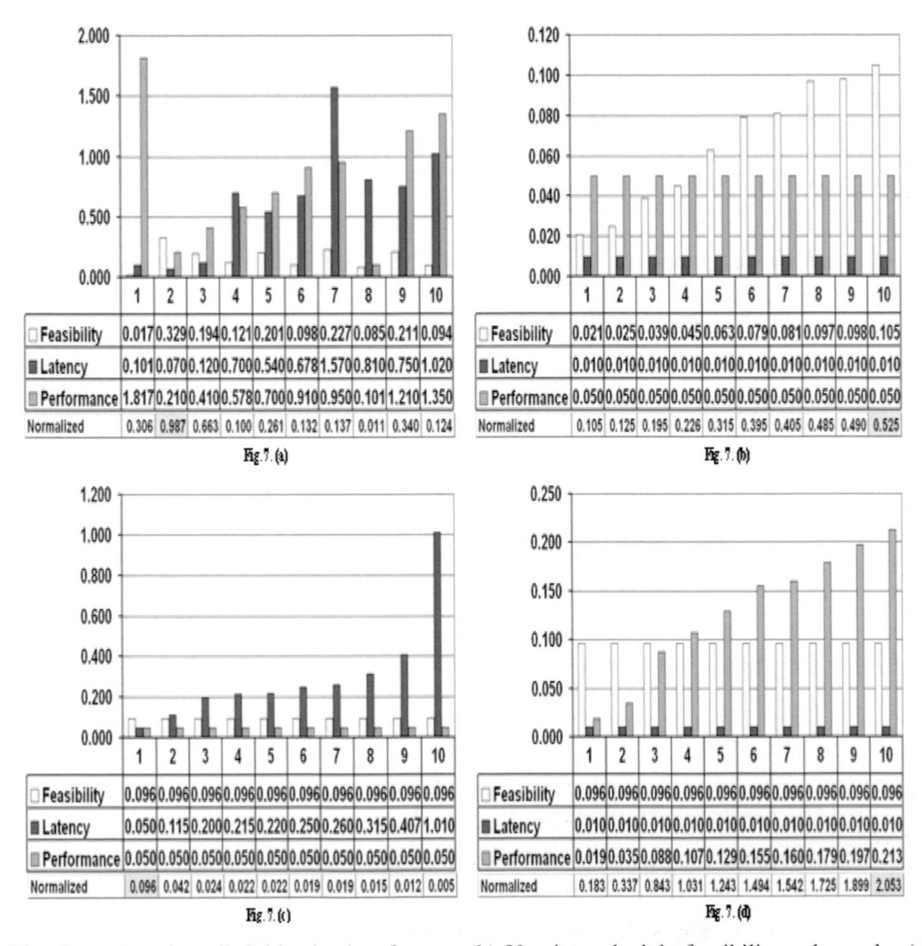

Fig. 5. (a) Varying all Grid selection factors, (b) Varying schedule feasibility value only, (c) Varying network latency value only and (d) Varying archived performance factor

Fig.7.(a)

	1	2	3	4	5	6	7	8	9	10
Feasibility	0.017	0.329	0.194	0.121	0.201	0.098	0.227	0.085	0.211	0.094
Latency	0.101	0.070	0.120	0.700	0.540	0.678	1.570	0.810	0.750	1.020
Performance	1.817	0.210	0.410	0.578	0.700	0.910	0.950	0.101	1.210	1.350
Normalized	0.306	0.987	0.663	0.100	0.261	0.132	0.137	0.011	0.340	0.124

Fig.7.(b)

	1	2	3	4	5	6	7	8	9	10
Feasibility	0.021	0.025	0.039	0.045	0.063	0.079	0.081	0.097	0.098	0.105
Latency	0.010	0.010	0.010	0.010	0.010	0.010	0.010	0.010	0.010	0.010
Performance	0.050	0.050	0.050	0.050	0.050	0.050	0.050	0.050	0.050	0.050
Normalized	0.105	0.125	0.195	0.226	0.315	0.395	0.405	0.485	0.490	0.525

Fig.7.(c)

	1	2	3	4	5	6	7	8	9	10
Feasibility	0.096	0.096	0.096	0.096	0.096	0.096	0.096	0.096	0.096	0.096
Latency	0.050	0.115	0.200	0.215	0.220	0.250	0.260	0.315	0.407	1.010
Performance	0.050	0.050	0.050	0.050	0.050	0.050	0.050	0.050	0.050	0.050
Normalized	0.096	0.042	0.024	0.022	0.022	0.019	0.019	0.015	0.012	0.005

Fig.7.(d)

	1	2	3	4	5	6	7	8	9	10
Feasibility	0.096	0.096	0.096	0.096	0.096	0.096	0.096	0.096	0.096	0.096
Latency	0.010	0.010	0.010	0.010	0.010	0.010	0.010	0.010	0.010	0.010
Performance	0.019	0.035	0.088	0.107	0.129	0.155	0.160	0.179	0.197	0.213
Normalized	0.183	0.337	0.843	1.031	1.243	1.494	1.542	1.725	1.899	2.053

On first Grid system the archived performance factor is highest (1.817) with second lowest N_{hj} (0.101). But the lowest F_{ij} (0.017) indicates poor chances of this job get scheduled on this Grid and hence overall Θ_i value is 0.306. Similarly, Grid 9 due to a high value of network latency, indicating its low network connection bandwidth with host Grid, results in a poor Θ_i value. Eventually Grid 2 with highest schedule feasibility value and low network latency, evaluated to be the optimal choice resulting normalized feasibility Θ_i value in (0.987). To analyze the affect of each F_{ij}, N_{hj} and P_{hj}, we further executed our simulation and results are shown in Fig. 5 (b), (c) and (d). It is clearly seen from the graph that our proposed formula to normalize these schedulability factors works well and choose the target Grid system with maximal F_{ij} and minimal N_{hj} value. We also compared our proposed scheme with the random scheduling scheme as shown in Fig. 6. Our proposed scheme gives better results as the resource performance data is collected and helps to select subsequent Grid system selection. The main distinction between our proposed scheme and random scheduling is due to the prediction of the future availability of the resources and thus avoiding the costly delegation of the job. The process of target system selection is further optimized by using the cached performance history.

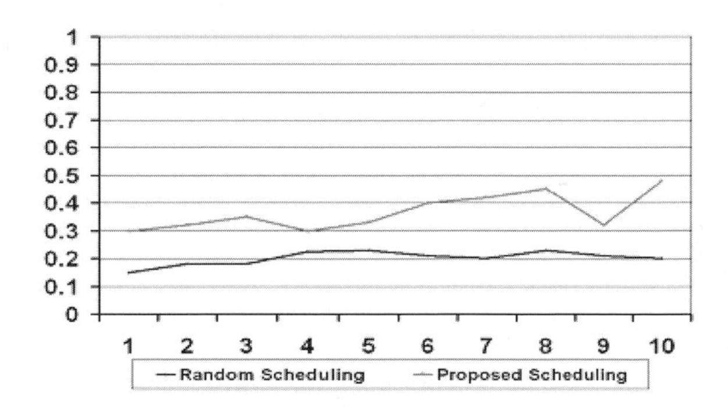

Fig. 6. Average resource selection throughput comparison with random scheduling

5 Conclusion and Future Work

In this paper, we propose a hybrid scheme for adaptive resource scheduling among distributed and reconfigurable Grid systems which are collaborating based on peer-to-peer communication paradigm. We based the metric of our resource scheduling scheme scheduling on optimal performance in terms of job execution and network latency. In future we plan to extend our work to autonomous resource reconfiguration based upon the application and resource adeptness.

References

1. Aggarwal, K., Kent, R.D.: An Adaptive Generalized Scheduler for Grid Applications. In: Proc. of the 19th Annual Int'l Symposium on High Performance Computing Systems and Applications (HPCS 2005), Guelph, May 2005, pp. 15–18. Ontario, Canada (2005)
2. Mateescu, G.: Quality of Service on the Grid via Meta-scheduling with Resource Co-Scheduling and Co-Reservation. Int'l Journal of High Performance Computing Applications 17(3), 209–218 (2003)
3. Lee, B.-D., Weissman, J.B.: Adaptive Resource Selection for Grid-Enabled Network Services. In: Second IEEE Int'l Symposium on Network Computing and Applications, pp. 75–81 (2003)
4. Sun, X.-H., Wu, M.: Grid Harvest Service: A System for Long-Term, Application-Level Task Scheduling. In: Int'l Parallel and Distributed Processing Symposium (IPDPS 2003), pp. 25–34 (2003)
5. Jang, S.-H., Wu, X., Taylor, V., et al.: Using Performance Prediction to Allocate Grid Resources, SC 2004 Posters, Pittsburgh, PA (November 2004)
6. Daniel, P., Crine, W., Brasileiro, F.: Trading Cycles for Information: Using Replication to Schedule Bag-of-Tasks Applications on Computational Grids. In: Kosch, H., Böszörményi, L., Hellwagner, H. (eds.) Euro-Par 2003. LNCS, vol. 2790. Springer, Heidelberg (2003)
7. Subramani, V., Kettimuthu, R., Srinivasan, S., Sadayappan, P.: Distributed Job Scheduling on Computational Grids Using Multiple Simultaneous Requests. In: 11th IEEE Int'l Symposium on High Performance Distributed Computing, pp. 359–364 (2002)
8. Santos-Neto, E., et al.: Exploiting Replication and Data Reuse to Efficiently Schedule Data-Intensive Applications on Grids. In: Proc. of 10th Int'l Workshop, JSSPP, New York, NY (June 2004)
9. Maheswaran, M., Ali, S., Siegel, H.J., Hensgen, D., Freund, R.F.: Dynamic Matching and Scheduling of a Class of Independent Tasks onto Heterogeneous Computing Systems. J. of Parallel and Distributed Computing 59, 107–131 (1999)
10. Standard Performance Evaluation Corporation (SPEC), http://www.spec.org/
11. Messing, F.: Predicting Scheduling Success. In: SpaceOps Symposium, Germany
12. Wolski, R., Spring, N., Hayes, J.: The Network Weather Service: A Distributed Resource Performance Forecasting Service for Metacomputing. Journal of Future Generation Computing Systems 15(5-6), 757–768 (1999)
13. Dong, F., Akl, S.G.: Scheduling Algorithms for Grid Computing: State of the Art and Open Problems. Technical Report, School of Computing, Queen's University, Ontario Canada (January 2006), http://www.cs.queensu.ca/TechReports/Reports/2006-504.pdf

ATS-DA: Adaptive Timeout Scheduling for Data Aggregation in Wireless Sensor Networks

Jang Woon Baek[1], Young Jin Nam[2,*], and Dae-Wha Seo[1]

[1] School of Electrical Eng. & Computer Science, Kyungpook National University
{kutc,dwseo}@ee.knu.ac.kr
[2] School of Computer & Information Technology, Daegu University
yjnam@daegu.ac.kr

Abstract. In wireless sensor networks, the timeout scheduling of data aggregation controls the time each sensor node has to wait to receive data from its child nodes. This paper proposes a new timeout scheduling scheme for data aggregation, the ATS-DA, which adaptively configures its length of timeout according to changing data patterns. The ATS-DA decreases the timeout when the variance of the received data (data variation) from children is lower than a pre-defined threshold because there are not any noticeable events, which reduces the consumed power and improves transmission latency. The ATS-DA, however, increases the timeout when data variation is more than the pre-defined threshold in order to fulfill more accurate data aggregation. Extensive simulation work under various workloads has revealed that ATS-DA not only enhances data accuracy by 33%, but also it improves power consumption and transmission latency by 5% and 58% respectively, as compared with the previous cascading timeout scheduling scheme.

Keywords: In-network data aggregation, timeout, wireless sensor network.

1 Introduction

According to advances in MEMS, wireless communication, and digital electronics technology, wireless sensor networks have been widely deployed to monitor and control physical environments [1]. Wireless sensor networks typically consist of a large number of sensor nodes which can observe physical phenomena, process sensed information, and communicate with other nodes. Since sensor nodes are equipped with limited battery power, low-power consumption is very crucial in wireless sensor networks [2]. It is regarded that power consumption is dominated by the costs of transmitting and receiving messages [3].

In-network aggregation can save a significant amount of energy by reducing the number of transmitted messages over wireless sensor networks. Sensor nodes with in-network aggregation can combine data from their child nodes and their locally-collected data before sending a message to their parent node. When a routing tree is

* Corresponding author.

T. Vazão, M.M. Freire, and I. Chong (Eds.): ICOIN 2007, LNCS 5200, pp. 375–384, 2008.

created, a sensor node sets up its *timeout* which is waiting time for messages from its children to arrive. If the timeout is lengthy, sensor nodes can receive more messages from their children, but transmission latency and power consumption increase. Various timeout scheduling schemes have been proposed [3-8]. In those schemes, the timeout is set by the depth of the routing tree and the maximum single-hop delay of the adjacent nodes. Those schemes assume that the single-hop delay is fixed. The single-hop delay, however, changes depending on the time to process, schedule packets, and reserve the channel [3]. Also, those schemes have not taken into consideration the data patterns of sensor readings. There is little change in sensor readings, from a wireless sensor network, as time passes. Sensor readings from physically-adjacent sensor nodes are similar. Significant changes exist for sensor readings when in the presence of noticeable events, such as earthquakes or forest fires.

This paper proposes a new timeout scheduling scheme for data aggregation, called ATS-DA, which adapts its timeout values according to changing data patterns. The ATS-DA decreases the timeout when the variance of the received data from children is lower than a pre-defined threshold. That improves power consumption and transmission latency. The ATS-DA, however, increases the timeout when the variance becomes higher than the pre-defined threshold, which can fulfill more accurate data aggregation. The remainder of this paper is organized as follows: Section 2 provides the background regarding in-network aggregation and timeout scheduling schemes. Section 3 gives a detailed description of the ATS-DA, and Section 4 compares the performance of the ATS-DA with the cascading timeout scheduling scheme. Finally, concluding remarks are presented in Section 5.

2 Background

2.1 In-Network Aggregation

A user broadcasts a query to an entire wireless sensor network through the base station (BS). Query propagation creates a routing tree whose root is the BS. If a leaf node receives the query, it sends its readings toward the BS. An intermediate node combines its own readings with the readings of its children via the aggregation function *f* and sends the aggregated data to its parent node. Fig. 1 shows an example of in-network aggregation. Node 2 sends only one message (z), as aggregation is performed by itself. Without aggregation, it has to send four messages (a, b, c, d).

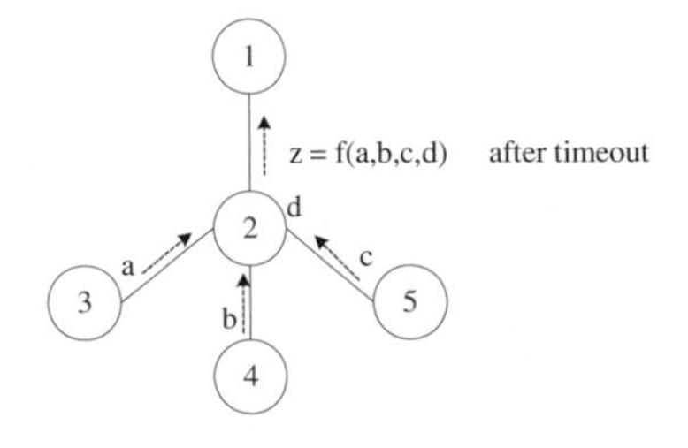

Fig. 1. An example of In-network Aggregation

2.2 Timeout Scheduling Schemes

Timeout scheduling schemes, with no clock synchronization, use hop counts (*hop-Count*), single-hop delays (*shd*), and periods (*T*) [3-5]. The cascading timeout scheduling scheme [5] sets up a timeout as in Fig. 2. In the cascading timeout scheduling scheme, a sensor node calculates its timeout by adding the *shd* to the child node's timeout. This scheme assumes that the *shd* is constant, but actually, it depends on network conditions. If the actual *shd* is larger than the pre-defined *shd* of the cascading timeout scheduling scheme, data loss occurs.

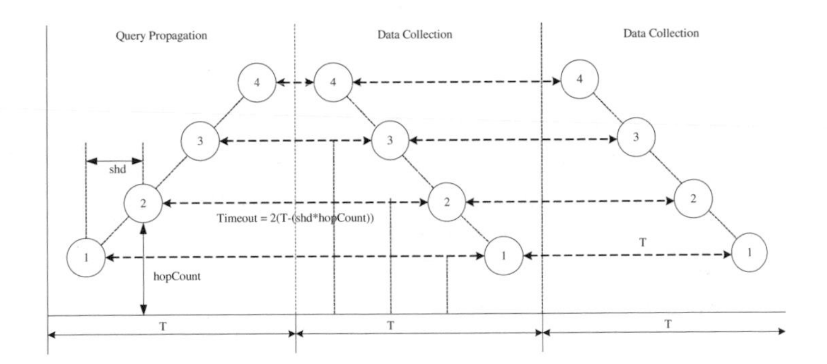

Fig. 2. The cascading timeout scheduling scheme

TAG [6] and Data Fusion [8] employ additional timing information for the synchronization of in-network aggregation. In the TAG, nodes send a query which includes timing information that describes the starting point of aggregation. In Data Fusion, the child node informs its parent node to be ready to receive a message, but these synchronization methods create the need for additional power.

The timeout is closely connected with the sensor node's power states. In general, power states consist of transmitting, receiving, listening and idle mode. Power dissipation is at its highest in the transmitting mode. The receiving and listening mode are much larger than the idle mode. A sensor node is in the listening/receiving mode during a timeout. Therefore, we should design a timeout which is as small as possible.

3 The Proposed Scheme

3.1 Adaptive Timeout Scheduling

We assume that data aggregation is performed periodically in a tree-structured wireless sensor network. It is also assumed that a single-hop delay between sensor nodes is time-variant, because the required times to process/schedule/retransmit packets and reserve channels can vary significantly due to link instability [3]. In general, the following characteristics can be observed in aggregating sensor data from a wireless sensor network.

[C1] As time passes, there is little change in sensor readings from a wireless sensor network.

[C2] There is little change in sensor readings from physically adjacent sensor nodes in a wireless sensor network.
[C3] In the presence of noticeable events, such as earthquakes or forest fires, there is significant change in the sensor readings from a wireless sensor network.

Adaptive timeout scheduling (ATS-DA) for data aggregation exploits the above characteristics of the sensor readings. First, by using the characteristics of C1 and C2, the ATS-DA decreases the timeout when the variance of the received data from child nodes becomes very low. That is, the timeout is set to the average length of the single-hop delay, as denoted by SHD_{avg}, when the data variation of the arrived messages is lower than a pre-defined threshold of data variation. The ATS-DA presumes that the messages which have not yet arrived are also within the threshold of data variation, and an aggregation is performed. Therefore, the ATS-DA can reduce the number of receiving messages and the amount of waiting time. As a result, the ATS-DA reduces the power consumption and transmission latency is improved. The ATS-DA can adjust the variation threshold according to the requirements of different applications which usually have different requirements from accuracy, power consumption to latency [3]. Small variation threshold leads to high accuracy, but power consumption and latency also increase. Large variation threshold leads to low power consumption and low latency, but the level of accuracy decreases.

Second, by exploiting the characteristics of C3, the ATS-DA increases the timeout when the variance becomes higher than a pre-defined threshold in order to perform more accurate data aggregation. That is, the ATS-DA sets its timeout as a maximum of the *shd* (SHD_{max}) in order to guarantee the arrival of messages from child nodes within its timeout, and next uses its parent node's timeout margin to receive as many messages as possible from the child nodes. The timeout margin at node *i*, denoted by MRG_i, is initially set to the difference between the SHD_{max} and SHD_{avg}. The SHD_{avg} is defined as an average of the *shd* to guarantee the arrival of messages from child nodes within its timeout. Usually, most of the messages arrive within the SHD_{avg}, which is much smaller than the SHD_{max}. If all messages from child nodes arrive within the SHD_{avg}, a timeout margin is created.

In sum, the ATS-DA scheme is given in Fig. 3. Basically, the ATA-DA operates at each sensor node, denoted as $node_i$. Also, the parent node and the child nodes of $node_i$ are denoted by $node_{pat(i)}$ and $node_{child(i)}$, respectively. A sensor node ($node_i$), which receives a query from a parent node ($node_{par(i)}$), initializes its hop counts ($hopCount_i$), timeout margin (MRG_i) and the data variation threshold (dv_i). Next, the ATS-DA commences its timeout timer (T_{cur}) and waits for messages from the child nodes. The sensor node fulfills partial aggregation and updates data variation (dv_i), which arrived for SHD_{avg}. If the data variation is within *DV*, the ATS-DA performs final aggregation. If the data variation is larger than *DV*, the ATS-DA waits for SHD_{max} considering the timeout margin that is used by the child nodes. If the ATS-DA receives all messages from the child nodes within this timeout, the ATS-DA performs a final aggregation. Otherwise, the ATS-DA adds the timeout margin of its parent node to its timeout and it receives messages during this additional time. If the additional timeout expires, the ATS-DA can no longer receive messages from children and it performs a final aggregation. Next, the sensor node sleeps until the next period.

```
procedure ATS-DA (Query from node_par(i))
  // running at node_i;
  // extract global parameters from Query & initialize local parameters
  extract hopCount_par(i), SHD_max, SHD_avg, and DV from Query;
  hopCount_i = hopCount_par(i) + 1;
  δ = (SHD_max - SHD_avg);

  while TRUE do;
    Timeout_i = SHD_avg; dv_i = MRG_i = 0;
    start TIMER with T_cur = 0;
    repeat
      receive data & MRG_child(i) from node_child(i);
      do partial data aggregation(curr_i); update dv_i;
      if (received from all child nodes) goto transmit;
    until T_cur ≤ Timeout_i;

    // if data variation is low enough, stop receiving data
    if (dv_i ≤ DV), then goto transmit; // [C1] & [C2]

    // now that an event occurred, set the timeout to SHD_max // [C3]
    // subtract the amount of time used by its child nodes
    Timeout_i = SHD_max - max{MRG_child(i)};
    repeat
      receive data from node_child(i);
      do partial data aggregation(curr_i);
      if (received from all child nodes) goto transmit;
    until T_cur ≤ Timeout_i;

    // unless data are received from all child nodes,
    // finally, use the timeout margin of its parent node // [C3]
    MRG_i = δ/2; Timeout_i = Timeout_i + MRG_i;
    repeat
      receive data from node_child(i);
      do partial data aggregation(curr_i);
      if (received from all child nodes) goto transmit;
    until T_cur ≤ Timeout_i;

  transmit:
    finally do aggregation with its sensing data;
    prev_i = curr_i;
    send curr_i & MRG_i to node_par(i);
    sleep until next period;
  done;
end ATS-DA
```

Fig. 3. The ATS-DA Scheme

3.2 Data Variation (dv_i)

Data variation of a sensor node is defined as a temporal variance rate between the last period data ($prev_i$) and the current period data ($curr_i$). Data variation is calculated as

$$dv_i = \left|\frac{curr_i - prev_i}{curr_i}\right| \times 100(\%). \tag{1}$$

The ATS-DA should have enough samples to determine whether the data variation is within the pre-defined threshold (*DV*). Thus, the ATS-DA waits for data from the child nodes during the SHD_{avg} to obtain samples, because most data are expected to arrive within the SHD_{avg}. By using data variation, the timeout of a node can be saved as much as the difference between the SHD_{max} and SHD_{avg}. If all nodes in the routing tree whose depth is n, satisfy a data variation condition, the total timeout saving is equal to n times the difference between the SHD_{max} and SHD_{avg}. Note that the amount of energy saved is directly associated with the amount of the reduced timeout.

Fig. 4 illustrates how the ATS-DA exploits data variation in order to reduce the timeout and power consumption in a wireless sensor network. In the cascading timeout scheduling scheme, as shown in Fig. 4(a), each sensor node, (i.e. node B), waits for messages from its child nodes, (i.e. node D, E, and F), during the SHD_{max} (2sec). Node B receives the last data from node E at 1.97sec, and an aggregation is performed. The operation of the ATS-DA with a SHD_{avg} of 1.5sec and a DV of 0.5% is given in Fig. 4(b). Node B performs an aggregation at 1.5sec. Next, node B no longer has to wait for data from node E because the variation of data which have arrived for SHD_{avg} is lower than DV. Therefore, the ATS-DA reduces the amount of data received and lowers the waiting period (power consumption) by 25% in the example.

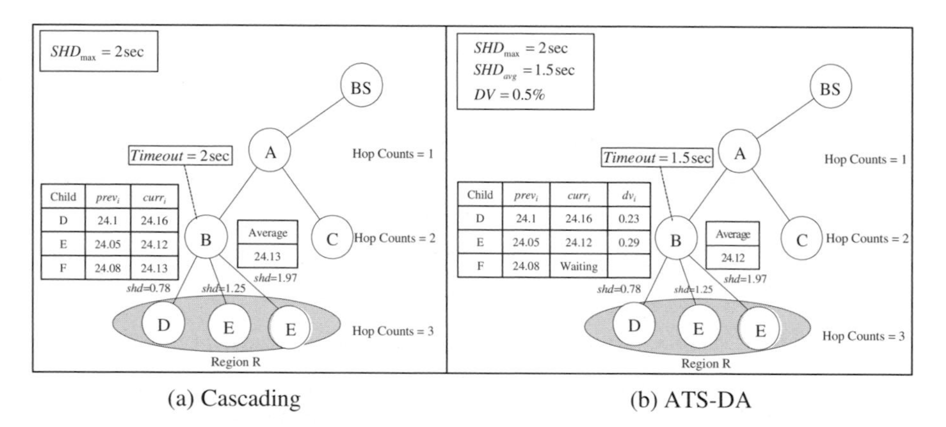

(a) Cascading (b) ATS-DA

Fig. 4. Illustrations of data aggregation (average) using data variation, where the SHD_{max} is 2sec, the SHD_{avg} is1.5sec, and DV is 0.5%: (a) The cascading timeout scheduling scheme and (b) the proposed scheme (ATS-DA)

3.3 Timeout Margin (MRG_i)

The timeout margin is additional time that is used to receive messages that did not arrive from children within the timeout when noticeable events occurred. As illustrated in Fig.3, each data aggregation period initializes the timeout margin at each node, as denoted by MRG_i as $SHD_{max} - SHD_{avg}$. The value of the SHD_{max} can be calculated as

$$SHD_{\max} = \max(sd_i) + \max(td_i + pd_i + qpd_i), \tag{2}$$

where the sd_i is the staggering delay of the packet at each node. The td_i and pd_i represent, respectively, the transmission and propagation delay of each node. The qpd_i accounts for queuing and processing delays of each node [5]. The value of SHD_{avg} is empirically obtained by averaging measured data during iterative experiments. While data is being aggregated, sensor nodes requiring additional time to their timeout employ half of its parent's timeout margin, $\delta = (SHD_{max} - SHD_{avg})$ in order to perform data aggregation more accurately. When using the timeout margin, the node has to inform the parent node of the amount of margin used. Then, the parent node waits for the SHD_{max} considering the timeout margin that is used by the child nodes; that is, the value of $Timeout_i = SHD_{max} - \max\{MRG_{child(i)}\}$.

Fig. 5 shows examples which illustrate how the ATS-DA exploits the timeout margin to increase its data accuracy. In the cascading timeout scheduling scheme, as shown in Fig. 5(a), node B waits for messages from its child nodes, say node D, E, and F, during the SHD_{max} (2sec). It, however, only receives data from node D. The data from node E and node F are lost. Given large data variation with noticeable events, the aggregation of lost data is very important. Thus, the ATS-DA uses the timeout margin of its parent node, (0.25 sec in the example). As a result, node B can receive data from all child nodes. Therefore, the ATS-DA provides more accurate data aggregation, as compared with cascading timeout scheduling.

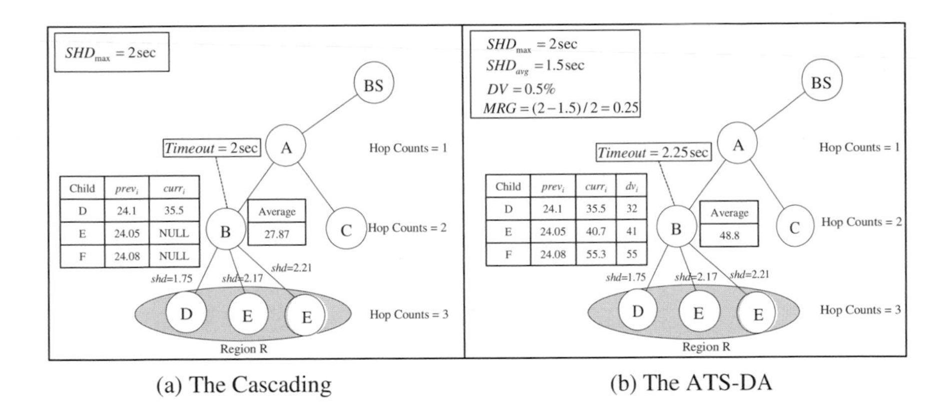

Fig. 5. Examples of data aggregation (average) using a timeout margin, where the SHD_{max} is 2sec, the SHD_{avg} is 1.5sec, and the DV is 0.5%: (a) The cascading timeout scheduling scheme and (b) the proposed scheme (ATS-DA)

4 Performance Evaluations

We perform extensive simulations using an ns-2 network simulator [11] to compare the proposed scheme (ATS-DA) with the cascading timeout scheduling scheme. We modified the AODV routing protocol and message passing agents in order to conduct in-network aggregation. In the simulations, a base station (BS) broadcasts a query to the entire wireless sensor network. Nodes that received the query initialize their timeout, margin and the threshold of data variation, as described in Fig. 3.

Each simulation begins by deploying 100 nodes randomly in a 1000-by-1000 grid, where the transmission range and the data rate of each node are set to 50 meters and 1Mbps, respectively. We employ the 802.11 MAC-layer protocol and AODV as a routing protocol. In the energy model, transmission and reception power consumption is set to 395 and 660mW, respectively [10]. The size of the packet is 30 bytes [6]. In addition to network traffic for data aggregation, heavy network workloads are randomly injected into the wireless sensor network to increase single-hop delays (longer than SHD_{max}). The SHD_{max} was set to 0.03 seconds, and the SHD_{avg} was set to 0.01 second. The SHD_{avg} was empirically obtained by running iterative experiments.

We employed the following metrics to compare the cascading timeout scheduling scheme and the ATS-DA scheme: Power consumption, data accuracy, and transmission latency.

4.1 Power Consumption

Power consumption is in proportion to the number of transmission and reception messages. Sensor nodes in the listening/reception mode consume more power than the idle mode. Table 1 summarizes the power consumption of the cascading timeout scheduling scheme and the ATS-DA scheme with different values of data variation (*DV*). Observe that the ATS-DA with large *DV* consumes less power. With larger *DV*, the listening time and the number of received messages decrease because the probability that data variation is within the pre-defined threshold *(DV)* increases. Our experiment reveals that the power consumption of the ATS-DA (0%), ATS-DA (5%) and ATS-DA (10%) is 1.3%, 4.8% and 7.5% lower than the cascading timeout scheduling scheme, respectively. Note that power consumption of the ATS-DA (0%) is similar to that of the cascading timeout scheduling scheme. This is because the actual timeout of the ATS-DA (0%) is not likely to be the SHD_{avg}, but mostly SHD_{max}.

Table 1. A comparison of power consumption via different timeout schemes: The cascading timeout scheduling scheme and the ATS-DA scheme with various data variation (*DV*)

Elapsed Time (sec)	Cascading	ATS-DA (0%)	ATS-DA (5%)	ATS-DA (10%)
2,000	0.114580	0.116512	0.112598	0.110619
5,000	0.290875	0.295228	0.276263	0.287145
10,000	**0.594700**	**0.587240**	**0.566339**	**0.550385**

4.2 Data Accuracy

Data accuracy can be measured as the relative distortion between the actual value (Xi) and the aggregated value (Xi'), where the relative distortion (RD) is defined as

$$RD = \sqrt{\frac{\sum_i |Xi - Xi'|^2}{N}}. \tag{3}$$

A large relative distortion means low data accuracy. On the other hand, small relative distortion means high data accuracy. Fig. 6 shows the results of relative distortion with different timeout schemes. The ATS-DA with a small *DV* has lower relative distortion on average, implying higher data accuracy. Note that the relative distortion of the cascading timeout scheduling scheme is much larger than that of the ATS-DA (0%) and ATS-DA (5%). The data accuracy of the ATS-DA, however, with a DV of 10% becomes worse than that of the cascading timeout scheduling scheme. Recall that heavy network workloads randomly generated into the wireless sensor network affect single-hop delay times between sensor nodes. Sometimes, the single-hop delays can be longer than the SHD_{max}. In such an event, the cascading timeout scheduling scheme continues to use the initially configured timeout value of the SHD_{max}. The ATS-DA can dynamically adapt to time-varying single-hop delays for higher data accuracy by employing the timeout margin of the parent node.

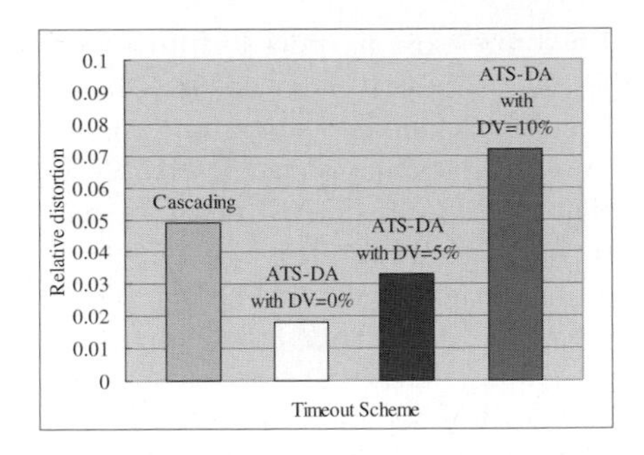

Fig. 6. A comparison of relative distortion according to different timeout schemes: The cascading timeout scheduling scheme and the ATS-DA scheme with various data variation (*DV*)

4.3 Transmission Latency

Fig. 7 shows the transmission latency results for the different timeout schemes. Note that the transmission latency of the cascading timeout scheduling scheme is much higher than that of the ATS-DA (5%) and ATS-DA (10%). Since average data variation is observed to be below 5%, small differences exist in the measured transmission latency of the ATS-DA (5%) and ATS-DA (10%). The transmission latency of ATS-DA (0%), however, slightly increases compared to that of the cascading timeout scheduling scheme. Recall that the ATS-DA (0%) not only uses the SHD_{max}, but it also partially employs the timeout margin of a parent node for better data accuracy.

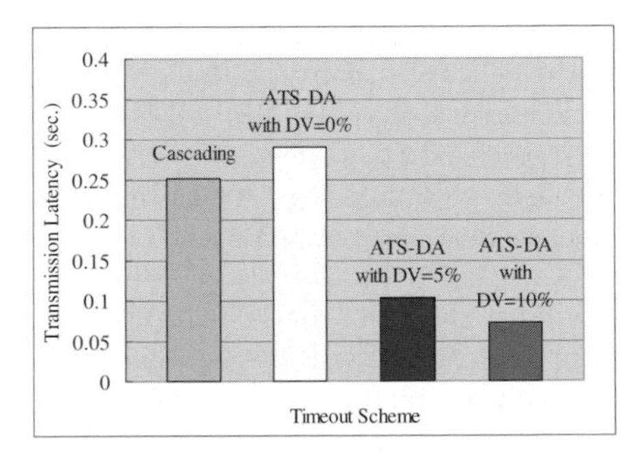

Fig. 7. A comparison of transmission latency according to different timeout schemes: The cascading timeout scheduling scheme and the ATS-DA scheme with various data variation (*DV*)

5 Concluding Remarks

This paper proposed the ATS-DA that a new timeout scheduling scheme for data aggregation in a wireless sensor network. The key to the ATS-DA is that it can adapt the timeout values according to changing data patterns (data variation); that is, the timeout decreases when the data variation from children becomes lower than a predefined threshold (*DV*), which reduces the amount of power consumed and improves transmission latency. On the other hand, the timeout increases when the data variation is more than the pre-defined threshold in order to fulfill more accurate data aggregation. Experiments in the ns-2 simulator showed that the ATS-DA scheme outperformed the cascading timeout scheduling scheme in terms of data accuracy by 33%, power consumption by 5%, and transmission latency by 58%.

In future work, we plan to devise a scheme that can automatically determine the optimal value of the *DV* for a given wireless sensor network. In addition, we will investigate the effect of sensor node mobility in the ATS-DA, and design a more robust ATS-DA scheme that can handle node failures in a wireless sensor network.

Acknowledgments. This research was in part supported by the MIC(Ministry of Information and Communication), Korea, under the ITRC(Information Technology Research Center) support program supervised by the IITA(Institute of Information Technology Assessment) (IITA-2006-C1090-0603-0045).

References

1. Levis, P., Madden, S., Gay, D., Polastre, J., Szewczyk, R., Woo, A., Brewer, E., Culler, D.: The Emergence of Networking Abstractions and Techniques in TinyOS. In: First USENIX/ACM Symposium on Networked Systems Design and Implementation (2004)
2. Akyildiz, I., Su, W., Sankarasubramaniam, Y., Cayirci, E.: Wireless sensor networks: A survey. Computer Networks 38, 393–422 (2002)
3. Yao, Y., Gehrke, J.: Query Processing for Sensor Networks. In: Proceedings of CIDR (2003)
4. Madden, S., Szewczyk, R., Franklin, M., Cullera, D.: Supporting Aggregate Queries Over Ad-Hoc Wireless Sensor Networks. In: Proceedings of WMCSA (2002)
5. Solis, I., Obraczka, K.: In-Network Aggregation Trade-offs for Data Collection in Wireless Sensor Networks. International Journal of Sensor Networks 1(2) (2006)
6. Madden, S., Franklin, M., Hellerstein, J., Hong, W.: TAG: a Tiny Aggregation Services for Ad-Hoc Sensor Networks. In: Proceedings of the Fifth Symposium on Operating Systems Design and Implementation (December 2002)
7. Motegi, S., Yoshihara, K., Horiuchi, H.: DAG based In-Network Aggregation for Sensor Network Monitoring. In: Proceedings of SAINT (January 2006)
8. Yuan, W., Krishnamurthy, S., Tripathi, S.: Synchronization of Multiple Levels of Data Fusion in Wireless Sensor Networks. In: Proceedings of GLOBECOM (2003)
9. Sharaf, M., Beaver, J., Labrinidis, A., Chrysanthis, P.: Balancing Energy Efficiency and Quality of Aggregate Data in Sensor Networks. The VLDB Journal (2004)
10. Intanagonwiwat, C., Govindan, R., Estrin, D., Heidemann, J., Silva, F.: Directed Diffusion for Wireless Sensor Networking. IEEE/ACM Trans. on Networking (February 2003)
11. VINT, The Network Simulator NS-2 (November 2005), http://www.isi.edu/nsnam

Sensor Network Deployment Using Circle Packings*

Miu-Ling Lam and Yun-Hui Liu

Department of Mechanical and Computer-Aided Engineering,
The Chinese University of Hong Kong, Shatin, Hong Kong
{mllam,yhliu}@mae.cuhk.edu.hk

Abstract. This paper presents a novel algorithm for autonomous deployment of active sensor networks. It enhances the sensing coverage based on an initial placement of sensor nodes. The problem of placing a number of circular discs (which model sensing coverage) of different radii to cover a field is intuitively transformed to the circle packing problem. Due to the fact that a unique *maximal packing* exists for a given set of combinatorics (triangulations) and boundary conditions, we can always find the minimum sensing range required for every interior node to satisfy these conditions. Though an extension from tangency packing to overlap packing, the interstices among triples (which represent coverage holes) can be eliminated. Based on a number of numerical simulations, we have verified that the proposed algorithm always yields sensor deployments of wide coverage and minimizes sensing range required for every interior sensing node to satisfy the packing and boundary conditions.

Keywords: Deployment, mobile sensor, robotics, sensing coverage, wireless sensor network.

1 Introduction

An active sensor network is a collection of wireless sensors mounted on spatially distributed mobile robots, which can provide better coverage of the environment, faster response to changes and superior mobility for active information gathering. Each sensor node is capable in communication, environmental sensing, data storage and processing and locomotion. Mobility enables a number of important functionality in sensor networks such as coverage maximization, adaptive sampling, network repair, localization and energy harvesting. This paper addresses the problem of autonomous deployment of a set of networked sensor nodes to enhance the coverage area of the sensors.

The work in [1] adopts a potential-field-based approach to spread sensor nodes throughout the target environment from a compact initial configuration. However, it does not consider some crucial problems like connectivity maintenance and topology control. The potential-field-based algorithm and the virtual force algorithm (VFA)

* This work is supported in part by Hong Kong RGC under grant 414505. It is also affiliated with the Microsoft-CUHK Joint Laboratory for Human-centric Computing and Interface Technologies.

T. Vazão, M.M. Freire, and I. Chong (Eds.): ICOIN 2007, LNCS 5200, pp. 385–395, 2008.

presented in [7] work in a similar fashion, that they increase sensor coverage by considering the virtual attractive and repulsive forces exerted on each sensor node by neighbor nodes and/or obstacles (if any). However, these works only consider homogeneous sensing model (i.e. sensors need to have an identical sensing capability), while in this paper, we address the problem of deploying heterogeneous sensor networks. Besides, VFA assumes all sensor nodes are able to communicate with their cluster head which is responsible for calculating sensor movement and the target location. In [6], three simple protocols for enlarging sensor coverage in a target area are presented. Unfortunately, the authors in [6] offer a computationally expensive algorithm as they fail to indicate that the minimum enclosing circle problem can be found in $O(n)$ time, where n is the number of points in the plane [2]. We have presented in [8] an improved version of the Minimax algorithm, called MEC (Minimum Enclosing Circle-based algorithm), by adopting a simple $O(n \log n)$ algorithm presented in [5] to compute the minimum enclosing circle instead. Besides, all the three protocols given in [6] disregard some primitive problems in wireless sensor network. In [8], we have presented an *ISOGRID* algorithm for autonomous deployment of mobile sensor networks. The principle is to redeploy the sensor nodes such that the communication graph approximates the layout of an isometric grid. Upon an initial random placement of sensor nodes, the algorithm iteratively computes node movements to enhance sensing coverage and avoid obstacles while ensuring sensor connectivity.

In this paper, the sensing regions are modeled as circular discs of variable sensing ranges. The problem of placing these circular discs on a field is intuitively transformed to the circle packing problem. A circle packing is a configuration of circles with specified patterns of tangency. [9] The central issues of the topic concern connections between the combinatorics of packings and their geometries, the variety among packings sharing combinatorics, computational methods, and connections with analytic function theory and conformal geometry. The study of circle packings was started by William Thurston in his famous notes. [10] Circle packings are computable, so they are introducing an experimental, and highly visual, component to research in conformal geometry and related areas. We adopt a circle packing algorithm to solve the sensor network coverage problem. For simplicity, we ignore the presence of obstacles and assumed all sensor nodes initially form a connected communication graph, i.e. there is no isolated node. We further assume that the communication graph can be transformed to some triangulations representing the geometric relation among the sensor nodes. Our problem is given a set of sensors and their maximum sensing ranges, find a deployment and the required sensing range of each associated sensor to give a large and connected coverage region. The central existence result derives from circle packings with certain extremal properties and these packings are called *maximal packings*. Given a set of combinatorics and boundary conditions, the maximal packing is univalent and essentially unique. Therefore, we can always find the minimum sensing range required for every interior node satisfying the boundary and packing conditions. Though an extension from tangency packing to overlap packing, the interstices of triples (which represent coverage holes) can be eliminated. As the circle packing algorithm utilizes only the local information about a sensor node and its neighbors, this module can be executed in a decentralized framework, and thus it is computationally efficient and scalable. Based on a number of simulation experiments, we have verified that the proposed algorithm always yields sensor deployments of wide coverage and desired topologies while setting all interior sensor nodes to their minimum required ranges.

2 Circle Packing

2.1 Preliminaries and Notations

We will adopt some notations in [9] to make the statements coherent. A hierarchy of circle packing structure consists of several levels of components, namely *circles*, *triples*, *flowers* and *packings*. The coordinates of circle centers refer to sensor node positions and radii refer to the corresponding sensing ranges. Here, all tangencies are referred as the external ones, each circle lying outside the disc bounded by the other. The fundamental units of the patterns are mutually tangent triples of circles (we will refer them as *triples*, for short), with each triple forming a triangular *interstice* (coverage hole). Triples are important to the rigidity associated with circle packings. The next level of circle packing structure is the *flower F*, consists of a central circle and some number of petal circles, the chain of successively tangent neighbors. The number of petals defines the *degree k* of the central circle. The condition that every circle has such a flower is a local planarity condition that we will enforce on all our packings.

Triangulation complex K: The tangency patterns for circle packings are encoded as abstract complexes *K*, which represent the triangulations of oriented topological surfaces. *K* is a combinatorial object, with no metric and no geometry. *K* has a finite number of vertices, edges, and faces. The vertices of *K* are of two types, interior and boundary. If u and v are neighboring vertices (i.e. $<u,v>$ is an edge of *K*) we write $u \sim v$. A vertex v and its neighbors form a combinatorial flower, $F_v = \{v; v_1, v_2, \cdots, v_k\}$: The petals v_j are listed in counterclockwise order about v with $v_{j+1} \sim v_j$; k is the degree of v, deg(v). When v is interior, the list of petals is closed; writing $v_{j+1} = v_1$, v belongs to the k faces $\{<v, v_j, v_{j+1}>: j = 1, 2, \cdots, k\}$. A sensor network topology is generally not a triangulated planar mesh. However, we adopt Delaunay triangulation to define a triangulation of communication graph. Compared to any other triangulation of the points, the smallest angle in the Delaunay triangulation is at least as large as the smallest angle in any other. As it is desirable to avoid narrow triangles, we consider it as the best choice among all conventional triangulation approaches. In Section 3, we will discuss an *Obtuse-Angle Pruning* to revise the complex *K* by trimming some boundary triangles of the Delaunay triangulation to improve the deployment result.

Circle packing P: A collection $P = \{c_v\}$ of circles is said to be a *circle packing* for a complex *K* if i) *P* has a circle c_v associated with each vertex v of *K*, *ii)* two circles c_u, c_v are externally tangent whenever $<u,v>$ is an edge of *K*, and *iii)* three circles c_u, c_v, c_w form a positively oriented triple whenever $<u,v,w>$ forms a positively oriented face of *K*.

Radius label R: R is a collection $\{R(v)\}$ of positive numbers associated with vertices v of K, where $R(v)$ represents the radius of circle c_v. It refers to the assigned values of sensing ranges over the set of sensor nodes.

Angle sums $\theta_R(v)$: For each triple of radii r_i, r_j and r_k, the Law of Cosines gives the angle α in a corresponding triple of circles.

$$\alpha(r;r_j,r_{j+1}) = \arccos\left[\frac{(r+r_j)^2+(r+r_{j+1})^2-(r_j+r_{j+1})^2}{2(r+r_j)(r+r_{j+1})}\right]$$

If we add these individual angles over the k triples involved, we get the angle sum $\theta_R(v)$ for this label at v. Suppose $F_v = \{v;v_1,v_2,\cdots,v_k\}$ is the flower for v in K. Vertex v belongs to m faces, where $m=k$ if v is interior and $m=k-1$ if v is boundary. In a flower having central label r and petal labels $\{r_1,r_2,\cdots,r_k\}$, the angle sum is given by the following summation formula, where $m=k$ and $r_{k+1}=r_1$ if the flower is closed, and $m=k-1$ otherwise: $\theta(r;r_1,r_2,\cdots,r_k)=\sum_{j=1}^{m}\alpha(r;r_j,r_{j+1})$. A label R is termed a *packing label* for K if the angle sum $\theta_R(v)$ equals to 2π for every interior vertex v. The packings we intend to compute are guaranteed by the fundamental existence and uniqueness result: *for a complex K with defined radius labels on the boundary vertices, there exists a unique packing label R for K with such that the labels on the boundary vertices conserve.* That is, given a set of fixed radii of all boundary vertices of K, there exists a unique circle packing (and the corresponding label R) such that the angle sum θ equals to 2π for every interior vertex of K. Then, we can define the circle packing problem as follows:

PROBLEM DEFINITION *Circle packing problem*: Given a complex K and the radii of all boundary vertices, compute the radii of interior vertices of the corresponding circle packing for K.

2.2 A Circle Packing Algorithm

The central issue of our circle packing problem is to adjust the radii of interior circles until all their angle sums approach 2π, which is the packing condition. One important observation inspires how we should adjust the radii of the central circles to achieve the packing condition: *the angle sum* $\theta(r;r_1,r_2,\cdots,r_k)$ *is strictly decreasing in r.* Suppose the radii of the petals are fixed. When the central circle radius increases from, the angle sum θ decreases. This monotonicity suggests that a strategy to adjust $R(v)$ to decrease the difference $|2\pi-\theta(v)|$ for circle c_v: decrease $R(v)$ if $\theta(v)<2\pi$; increase $R(v)$ if $\theta(v)>2\pi$. The strategy can be implemented in an iterative fashion. Given an interior vertex v we would ideally replace its current label r with that unique label $\bar{r}$ which gives angle sum 2π at v. However, we cannot yield a guaranteed stability without a proper choice of step size for the adjustment of $R(v)$. [9] gave another geometric monotonicity which suggests a very efficient estimation of unique label $\bar{r}$. When we compare the current k-flower F for v with a "uniform neighbor" model F', meaning a k-flower F' with label r for v but with petal labels set to a constant r'

chosen so that the angle sum θ is the same in *F'* as it is in *F*. The beauty of the model *F'* lies first in its simple computations. Then, $\frac{r'}{r+r'} = \sin\frac{\theta}{2k}$. Let $\beta = \sin(\frac{\theta}{2k})$ where k is the number of petals (i.e. degree of the central circle). Then, we have

$$r' = \left(\frac{\beta}{1-\beta}\right) r. \tag{1}$$

Now, the strategy is to replace r by a new central circle of radius ρ. The k-flower $\tilde{F}$ has uniform neighbors of radii r' and central circle of radius ρ, and angle sum θ precisely equals to 2π. Since $\frac{r'}{\rho+r'} = \sin\frac{2\pi}{2k} = \sin\frac{\pi}{k}$. Let $\delta = \sin(\frac{\pi}{k})$, then we have

$$\rho = \left(\frac{1-\delta}{\delta}\right) r'. \tag{2}$$

The new label always lies between r and the unique label $\bar{r}$. Therefore the strategy is conservative: it changes r in the correct direction yet never overshoots. In summary, a circle packing algorithm based on the uniform neighbor model is as follows:

Step 1. Initialize R: set boundary radii to assigned values, set interior radii to arbitrary values.

Step 2. For each interior circle c_v:

2.1 Compute the interior angle sum $\theta_R(v)$ using the law of cosines

2.2 Replace $R(v)$ by ρ as defined in equations (1) and (2).

Step 3. If $|2\pi - \theta_R(v)| < \varepsilon$ for all interior circles, where ε is a predefined threshold, then stop. Otherwise, go to Step 2.

3 Active Sensor Network Deployment Using Circle Packings

3.1 Obtuse-Angle Pruning

The uniqueness of a certain circle packing result depends on the associated triangulation of the network. However, in circle packing, boundary condition plays an important role in controlling to resultant coverage size and shape. The Delaunay triangulation always yields a convex mesh with the vertices of convex hull as boundary and this highly constrains the circle coverage size. Therefore, we propose an *Obtuse-Angle Pruning* method to reform the triangulation and increase the number of boundary vertices. The idea is simple but efficient: prune boundary edges if the associate triangle is obtuse at the opposite angle (angle opposite to the candidate boundary edge). Whenever an eligible edge is trimmed, the third vertex of the associated triangle turns to a boundary one and so the number of boundary vertices is increased by one. However, there are a few points to notice. First, the pruning process should be done one by one on the boundary vertices. Different choices of the starting vertex would sometimes lead to different resultant boundaries. However, it does not affect the performance of the algorithm because our aim is merely to enlarge the boundary set. Second, to make sure the resultant complex K is a simply connected triangulated mesh, we should not prune an edge with any of its both end vertices of degree<2. Moreover, we should not prune a candidate boundary edge if its third vertex in the associated triangle is already a boundary vertex. This can ensure the boundary edges always form a single enclosing loop with no crossover. The *Obtuse-Angle Pruning* is described as follows:

Start from any boundary vertex $v_{current}$.

Step 1. v_{next} denotes the next boundary vertex in anticlockwise sense and v_{middle} denotes the third vertex in the associated triangle of boundary edge $<v_{current}, v_{next}>$. If *i)* deg($v_{current}$)>2, *ii)* deg(v_{next})>2, *iii)* triangle $<v_{current}, v_{middle}, v_{next}>$ is obtuse at v_{middle} (i.e. $\angle v_{current} v_{middle} v_{next} \geq 90^\circ$) and *iv)* v_{middle} is not a boundary vertex, then prune boundary edge $<v_{current}, v_{next}>$ (i.e. delete triangle $<v_{current}, v_{middle}, v_{next}>$). Set v_{middle} as the next boundary vertex v_{next} and repeat this step again. Otherwise, go to Step 2

Step 2. Set v_{next} as the current boundary vertex $v_{current}$. If $v_{current}$ equals the starting vertex, then end. Otherwise, go to Step 1.

3.2 Overlap Packings

The central idea of circle packing problem is to adjust the radii of all interior circles to achieve the packing condition. Circle centers turn out to be secondary data. The geometric realization of a labelled complex *K(R)* can be done by fixing the coordinates of one circle and one of its neighbour, then the rest are consequently defined by the tangency relationships (Fig. 2(a)). The circle packing problem deals with tangency of triples and so interstices always exist. However, in sensing coverage problem, interstices are referred as coverage holes which are undesirable. Thus, we now discuss the overlap packing problem to eliminate the interstices. The notion of *overlap angle* $\phi_{ij} = \phi(c_i, c_j)$ suggests an index to measure the extent of overlap between two circles of euclidean centers z_i, z_j and radii r_i, r_j, and its formula is

$$\phi_{ij} = \cos^{-1} \frac{(r_i^2 + r_j^2) - |z_i - z_j|^2}{2 r_i r_j} . \tag{3}$$

In tangency case, the sides of triangle of any triple equals to the sum of two radii, i.e. $|z_i - z_j| = r_i + r_j$ for all $v_i \sim v_j$, and therefore the overlap angle always equals π. In overlap packing case, we reduce the relative distances among the centers (which in turns scale down the sizes of triangles) by certain scaling factor so that the interstices disappear. Let $\alpha \in (0,1)$ be a scaling factor for a triple $<c_1, c_2, c_3>$ with packing label r_1, r_2, r_3 to form its geometric realization, i.e.

$$|z_i - z_j| = \alpha(r_i + r_j) , \; i, j = 1,2,3 \text{ and } i \neq j . \tag{4}$$

Denote the three overlap angles of triple $< c_1, c_2, c_3 >$ as $\phi_{12} = \phi(c_1, c_2)$, $\phi_{23} = \phi(c_2, c_3)$ and $\phi_{13} = \phi(c_1, c_3)$. As shown in Fig. 1, the interstice vanishes if and only if the *summation* $\sum \phi$ of these three overlap angles is smaller than or equal to 2π, i.e.

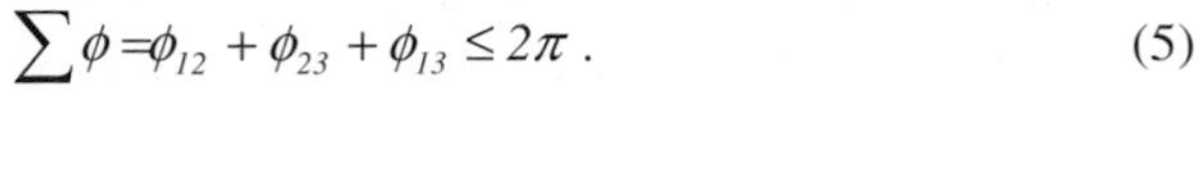

$$\sum \phi = \phi_{12} + \phi_{23} + \phi_{13} \le 2\pi \ . \tag{5}$$

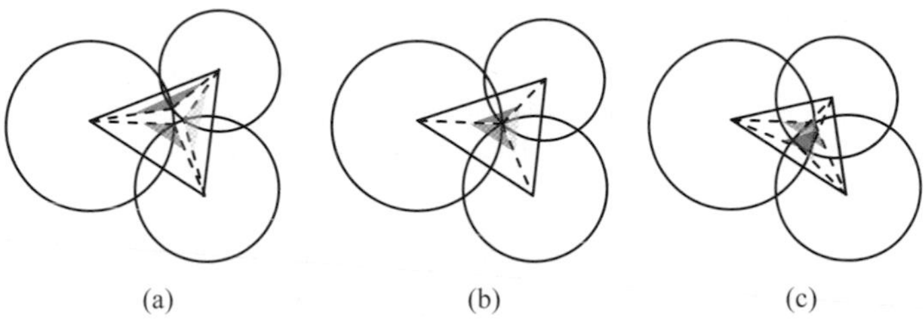

Fig. 1. Inversive distance triples: (a) Interstice exists $\sum\phi > 2\pi$ (b) Interstice just vanishes $\sum\phi = 2\pi$ (c) Interstice does not exist $\sum\phi < 2\pi$

The optimal scaling factor for a triple to precisely vanish its interstice ($\sum \phi = 2\pi$ Fig. 1(b)) depends on the ratios among the three radii. Moreover, it is impractical to impose different scaling factors for different triples, because that would cause incoherence in overall circle placement. Therefore, we should apply a proper scaling factor $\alpha \in (0,1)$ globally over the entire mesh and make sure the selected α is small enough to eliminate all interstices.

Theorem 3.1. The smallest scaling factor required to vanish the interstice of any triple is $\frac{\sqrt{3}}{2}$.

We have skipped the proof here due to the limited number of pages. Taking $\alpha = \frac{\sqrt{3}}{2}$ globally over the entire mesh can yield a set of sensor node positions for complex $K(R)$ while eliminating all interstices. Fig. 2 shows an example of geometric realizations obtained from conventional tangency packing and its overlap packing.

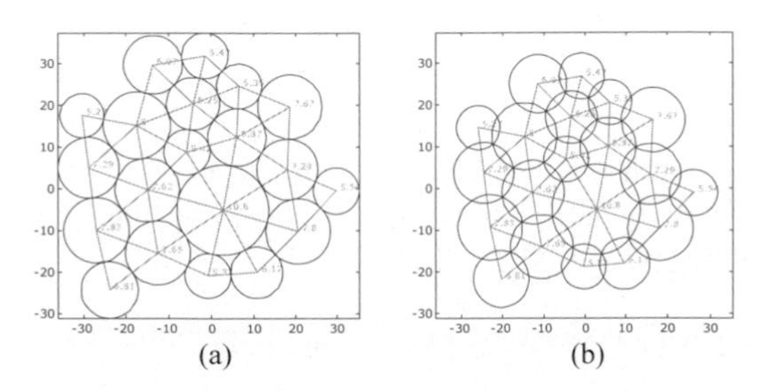

Fig. 2. (a) Tangency circle packing and (b) its overlap packing with $\alpha = \frac{\sqrt{3}}{2}$

3.3 Circle Packing Deployment Algorithm

In this subsection, we will summarize the circle packing and overlap packing strategies and describe the detail implementation of these approaches on the sensor network deployment problem. We first give the underlying assumptions:

1) At the beginning, the sensor nodes are randomly located (on a target region if applicable) while the communication graph is remained connected.
2) Each sensor node is capable in broadcasting its position and obtaining the relative distances and orientations and sensing range of its neighbors.
3) Each sensor node has a upper limit of sensing range and is capable to adjust its sensing range power.

Our central idea is to employ the circle packing result to calculate the required sensing ranges of each associated sensor node. A triangulated mesh describing the communication graph is generated upon a given initial deployment using Delaunay triangulation. Then, we use the *Obtuse-Angle Pruning* to increase the boundary size of the combinatorics complex *K*. The radii of the boundary circles will first be assigned as half of the maximum sensing ranges of the corresponding sensor nodes and gradually increase them to the maximum level. The radii of the interior sensing circles will be recursively calculated using the circle packing algorithm such that the packing condition is achieved while the communication graph is preserved as *K*. Note that if any calculated sensing circle radius exceeds the maximum range of the sensor, it means the required radius for that particular sensor cannot be achieved by its sensor node and the radius should be bounded to its maximum range. Next, the new position of each sensor node will be defined by the geometric realization of labeled complex *K(R)* with overlap scaling factor $\alpha = \frac{\sqrt{3}}{2}$. A new complex *K* will be defined by the newly obtained deployment and the whole process should be executed iterative until equilibrium state is reached. The *Circle Packing Deployment Algorithm* is summarized as follows:

Step 1: Set maximum number of iterations and threshold to determine if deployment has reached equilibrium.
Step 2. Define the triangulation complex K base on current sensor node positions.
Step 3. Increase the number of boundary vertices using the *Obtuse-Angle Pruning* method.
Step 4. Set the sensing range radii of boundary vertices to any arbitrary level in the first iteration. Then gradually increase them to their upper limits in successive iterations.
Step 5. Execute the *Uniform Neighbor Model Circle Packing* algorithm on each interior node in a distributed sense to find its sensing range radius. Bound all sensing ranges to their upper limits.
Step 6. Deploy all sensor nodes based on the circle packing result and take $\alpha = \frac{\sqrt{3}}{2}$ to ensure no coverage hole exists in the connected coverage region.

Step 7. Go to Step 2 if movement of sensor nodes is greater than the threshold (i.e. equilibrium not reached) and maximum number of iterations is not reached. Otherwise, end.

4 Simulation Examples

4.1 Example 1

We have implemented the proposed algorithm in Matlab to verify the approach and demonstrate its performance. In this example, we illustrate the capability of the proposed algorithm in deploying a large set of sensor nodes efficiently. 1000 sensor nodes are initially randomly located as shown in Fig. 3(a). The purple dots denote the boundary sensor nodes and the blue ones denote the interiors. Fig 3(b) shows the final deployment while taking the overlap scaling factor $\alpha = \frac{\sqrt{3}}{2}$.

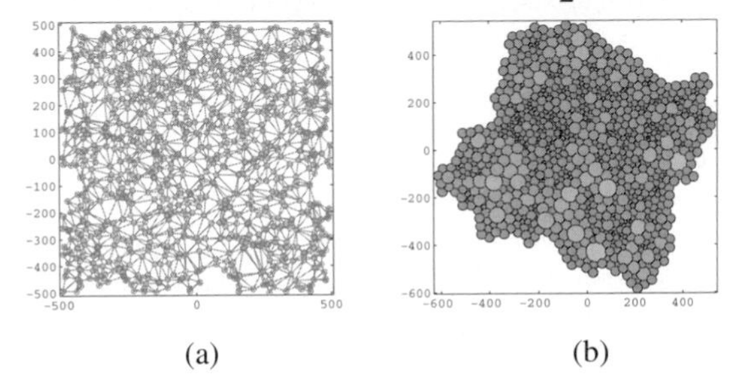

(a) (b)

Fig. 3. Example 1: (a) initial random placement (b) final deployment result

4.2 Example 2

In this example, we compare the deployment results obtained by two different boundary vertices definitions. In Fig. 4(a), the yellow boundary nodes are defined by the vertices of the associated convex hull. In Fig. 4(b), the boundary nodes are defined using the *Obtuse-Angle Pruning* method. The number of boundary vertices dramatically increases from *14* to *73*. The sensing radii of these boundary vertices are first set to be half of their corresponding maximum ranges, and then gradually increased to

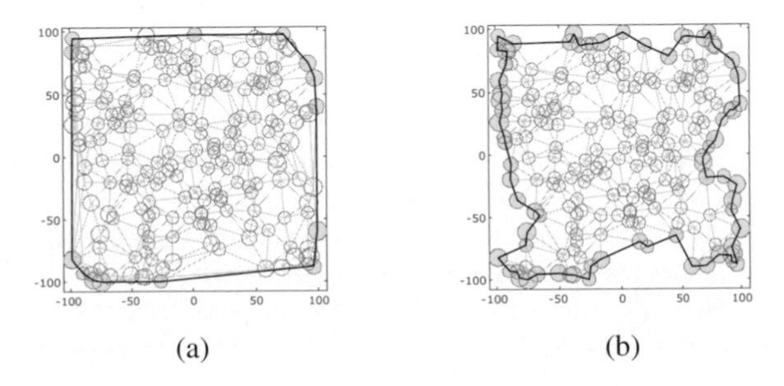

(a) (b)

Fig. 4. Initial placement and boundary using (a) convex hull and (b) *Obtuse-Angle Pruning*

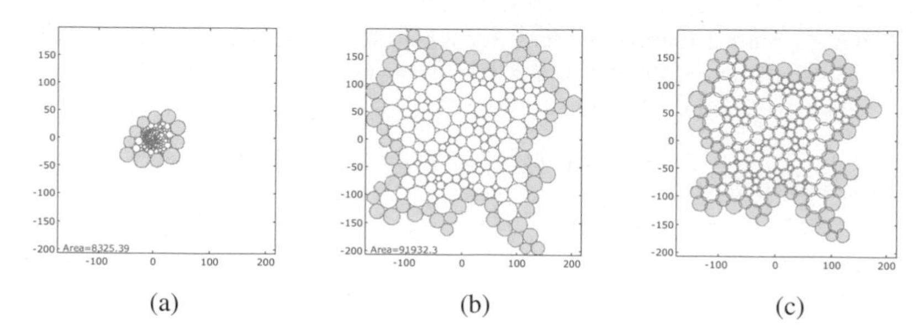

Fig. 5. Final deployments using (a) convex hull (b)-(c) *Obtuse-Angle Pruning* boundary

their maxima. After a few iterations, the circle packings reach the equilibrium and the final deployment results (with and without *Obtuse-Angle Pruning*) are shown in Fig. 5(a) and (b) respectively. The associated sensing area in Fig. 5(b) is over *11* times of that in Fig. 5(a). Fig. 5(c) is the overlap packing deployment result.

5 Concluding Remarks

This paper addresses the problem of autonomous deployment of active sensor networks. Upon an initial random placement, a triangulated mesh describing the neighboring relationship among the nodes is generated and refined by *Obtuse-Angle Pruning* to increase the number of boundary nodes. We have employed *Uniform Model Circle Packing* algorithm to find the radius of each interior sensing circle. Then, the geometric realization of the circle packing result is done by fixing one node and one of its neighbor via overlap packing. We have given a global scaling factor of $\alpha = \frac{\sqrt{3}}{2}$ which can always vanish interstices of any triple which represent coverage holes. We have implemented the algorithm in Matlab and demonstrate its performance with a number of simulation examples. We have verified that the proposed algorithm always yields sensor deployments of wide coverage and minimize the sensing range required for every interior sensing node.

References

1. Howard, A., Mataric, M.J., Sukhatme, G.S.: Mobile Sensor Network Deployment using Potential Fields: A Distributed, Scalable Solution to the Area Coverage Problem. In: Distributed Autonomous Robotic Systems, pp. 299–308. Springer, Heidelberg (2002)
2. Megiddo, N.: Linear-time Algorithms for Linear Programming in R^3 and Related Problems. SIAM J. Comput. 12(4), 759–776 (1983)
3. Okabe, A., Boots, B., Sugihara, K., Chiu, S.-N.: Spatial Tessellations: Concepts and Applications of Voronoi Diagrams. Wiley, Chichester (2000)
4. Shamos, M.I., Hoey, D.: Closest-Point Problems. In: Proc. 16th Annual IEEE Symposium on Foundations of Comput. Science, pp. 224–233. ACM, New York (1975)

5. Skyum, S.: A Simple Algorithm for Computing the Smallest Enclosing Circle. Information Processing Letters 37, 121–125 (1991)
6. Wang, G.-L., Cao, G., Porta, T.L.: Movement-Assisted Sensor Deployment. In: INFOCOM (March 2004)
7. Zou, Y., Chakrabarty, K.: Sensor Deployment and Target Localization based on Virtual Forces. In: INFOCOM (April 2003)
8. Lam, M.-L., Liu, Y.-H.: ISOGRID: an Efficient Algorithm for Coverage Enhancement in Mobile Sensor Networks. In: Proceedings of 2006 IEEE/RSJ International Conference on Intelligent Robots and Systems (IROS 2006), pp. 1458–1463 (2006)
9. Stephenson, K.: Introduction to Circle Packing: The Theory of Discrete Analytic Functions. Cambridge University Press, New York (2005)
10. Thurston, W.: The finite Riemann mapping theorem. In: An International Symposium at Purdue University in celebrations of de Branges' proof of the Bieberbach conjecture (March 1985)
11. Rodin, B., Sullivan, D.: The convergence of circle packings to Riemann mapping. Journal of Differential Geometry 26, 349–360 (1987)

TSD: Tiny Service Discovery

Pedro Fernandes and Rui M. Rocha

IT – Instituto de Telecomunicações, Instituto Superior Técnico, Technical University of Lisbon, Portugal
{paff,rui.rocha}@lx.it.pt

Abstract. Sensor Networks are heavily resource-constrained but nevertheless demand for generic flexible and efficient solutions to improve application development. By looking at the network as a collection of services, a user or a network node can use Service Discovery to browse locate and use the available services on the network. Moreover, Service Discovery besides enabling connection to external computer networks and interconnection between WSNs would simplify network self-organization and self-configuration. This paper proposes a simple solution for Service Support in WSNs based on Directed Diffusion. This paper addresses issues like efficiency, network interconnectivity, mobility and scalability. Preliminary simulation results show acceptable performance in distributed mode and point out the importance of good MAC and transport support.

Keywords: Wireless Sensor Networks, Service Discovery, Directed Diffusion.

1 Introduction

Recent technology advances in networking and embedded systems have enabled the research and development of new devices characterized for their communication ability, small size, sensor capabilities and programmability. Attracting much attention from the scientific community, a main goal to attain is to use these devices to build wireless sensor networks, long-lived, self-organized and disposable networks for a myriad of applications.

Sensor networks are data-centric, and thus, service discovery comes as a natural solution to this area. By looking at the network as a collection of services, a user or a network node can use Service Discovery to browse, locate and use the available services on the network. Moreover, Service Discovery easily enables connection to external computer networks as well as interconnection between WSNs. It can play a very important role in dynamic self-organization and self-configuration by allowing nodes to locate and use services to configure itself, as well as locate other services required for its normal operation.

Service discovery is a well studied topic in computer networks. However, the already proposed solutions cannot be directly applied in WSNs as they require memory, processor and bandwidth characteristics which sensor networks hardly can provide.

The protocol architecture and service information maintenance, the service access support, the service description schema, service browsing and discovery model can be

T. Vazão, M.M. Freire, and I. Chong (Eds.): ICOIN 2007, LNCS 5200, pp. 396–405, 2008.

identified as the major aspects that characterize a service discovery solution. The architecture defines the relation between the several protocol entities and can be classified as distributed, centralized or hybrid. Service information maintenance is closely tied to the architecture and is concerned to how and where service information is maintained in the network. Service access is related to the support offered by the protocol to the connection establishment between a service client and a given server. Service description concerns the naming schemes applied to the services and the structures that contain the services information. Service browsing is associated to the capability of discovering available services on the network with few or no service information known beforehand. Finally, the service discovery model refers to the way in which services are discovered. This can usually be a process initiated by the service announcer (push model) or by the requester (pull model).

NanoSLP [1] is one of the few proposals in this area. It is based on the IETF Service Location Protocol [2] and provides a very simplified version of it, adequate to WSN constraints. It is, however, a preliminary approach in the scope of the NanoIP [1] project, and still requires some amount of study concerning its functionalities and results. Two proposals for integrating Service Discovery are presented in the context of the EYES framework [3]. The first is a classical approach based on the description of services using simple identities and one-to-one communication. It is integrated with the routing protocol and lightweight but does not, however, address some issues, like service browsing. The second one is based on the publish/subscribe model and data-centric routing protocols and considers interfacing with external networks. Distributed service directories for increased performance and scalability are also proposed for both solutions.

Other projects [4], [5], [6], [7] are also of interest as they use the service paradigm to simplify interconnection between sensor networks and exterior networks based on IP overlay networks.

Application-related naming schemes are often employed, enabling functionalities similar to those provided by Service Discovery (e.g. network queries). In this context, sensor nodes are viewed as something which can provide or process information of a given type, which is virtually the same as considering sensor nodes as simple service providers. There are, some interesting developments in this area [8], [9], [10], [11], [12] which propose mechanisms for nodes to specify their interests on certain kinds of data, to which matching sensor nodes respond with the requested information. These solutions enable data fusion, mitigating problems like message implosion and overlapping.

Current service discovery solutions in WSNs tend to consider only subsets of the main SD aspects and are mainly focused on simplified versions of IP-based protocols. On the other hand, application related naming schemes are mainly concerned with data diffusion and imply a smaller abstraction level than the one provided by the services paradigm and do not really consider important issues like connection to the outside world and bidirectional data flow.

The goal of this work is to propose TSD, a Tiny Service Discovery solution for WSNs which allows nodes to easily discover, locate and announce services, providing a standardized abstraction to enable simple interoperability between heterogeneous nodes as well as interconnectivity with heterogeneous networks and the outside world.

The remainder of this paper is organized as follows: the main issues in Service Discovery are discussed in section 2, along with the proposed solutions. In section 3, an overview of the developed architecture is highlighted. Section 4 briefly covers preliminary protocol simulation and the main obtained results. Finally, section 5 concludes the paper and defines topics for further research.

2 Service Discovery Design in Sensor Networks

The various different aspects of service discovery need careful study in order to specify a suitable protocol for wireless sensor networks

The proposed approaches to the main issues related to the design of a suitable service discovery protocol for WSNs are discussed in the following subsections.

2.1 Service Discovery Protocol Architecture

Many sensor networks are characterized by flat topologies which offer simplicity, but in certain applications may hinder scalability and efficiency. Structured topologies can mitigate these issues by introducing central nodes organized in a hierarchical manner, forming clusters and backbones, but often imposing restrictions on mobility.

To handle both topologies types, TSD offers support by providing two different approaches. Flat topologies call for fully distributed solutions where service information is spread over all service providers in the network. On the other hand, in structured topologies, service directories are taken into account. Service directories can take advantage of the presence of backbone nodes and cluster heads by providing central service information repositories. This centralized approach may produce less traffic, but arises several other issues, such as self-organization of the central overlay and directory updates.

By caching service information from other service directories, traffic between central nodes can be reduced. Furthermore, cached information can also be used to identify paths to certain service types and forward queries to particular network areas or nodes instead of flooding it.

In addition to the services provided by individual sensor nodes, many services are associated with sensor networks as whole, or with regions within. These services are the ones more relevant to the outside. Taking advantage of the fact that nodes interfacing with the exterior are usually less restrained devices, in TSD these nodes act as gateways. The network services can thus be coherently announced to the outside as if they were provided by the gateway nodes themselves.

Services may also span multiple sensor networks. This functionality is assured by the usually called bridge node, which propagates service discovery messages between two connected neighbor networks according to rules defined by the protocol as to restrain uncontrolled message propagation.

2.2 Service Discovery Component Architecture

To avoid wasting resources, the protocol should provide independent modules, so that the various nodes in the network only to store and run the protocol components they actually need.

Integration with lower layers is also of importance. Using a data-centric routing protocol can have great impact in the protocol design, since it provides basic service discovery functionalities. In this context, Directed Diffusion [13] comes as a good candidate to support TSD. Since packet loss and packet retransmission imply additional energy consumption, providing congestion control and reliable transport can also be of importance.

2.3 Service Discovery Protocol

The protocol must be extremely streamlined, providing means to convey the necessary information with the least and smallest messages possible. Apart from efficiency, there are many issues the protocol has to address which can be divided in two categories: distributed and centralized.

Centralized Approach. Service directories maintain information about the services provided in their neighborhoods and can be contacted by local nodes and other service directories to provide requested service information. Periodic queries are the usual way to update service directories. However, they often imply the transmission of time-redundant data, and so, resource waste. It is preferable to have new nodes announcing their arrival to the network than waiting for the service directory to find them, letting the network react quickly to those arrivals. To detect node departure, every sensor node provides a special service: the heartbeat. Each node periodically sends a very small message indicating the service directory he is alive and well. In the same fashion, the service directory periodically broadcasts its presence, so that moving nodes may dynamically associate themselves to new service directories.

The use of the push model (requesting information) combined with the pull model (send announcements to the network), minimize the need for periodic messages.

Using the service directory in local queries has the advantage of producing a single response and the disadvantage of establishing non-optimal paths. As the number of service data packets is usually a lot larger than the SD protocol exchange, it is important that the path connecting client and server is optimized. Nonetheless, the service directory can answer in the place of nodes to which the request did not arrive. However, if the query is for service types (browsing), matters are different, since a node only wishes to know the types of the available services, for which the service directory has a prompt answer.

Depending on the query scope is, the local service directory may forward the request to other service directories and aggregate their responses.

Distributed Approach. In distributed mode, each sensor node announces its own services. A combination of the push and pull models is once more proposed. In this manner, a node joining the network announces its services so anyone interested can react accordingly, and a requester can send a query to the network when he sees fit. This allows good response to mobility while generating as little traffic as possible.

Common issues. As efficiency is a paramount requirement, the SD protocol must guarantee it by reducing protocol traffic as much as possible. Using attributes and operators in the queries and announcements can make them very flexible, optimizing the number of responses and their size. As responses converge on the requester node, data aggregation can be used to combine certain responses.

Since the connection to the outside is a requirement, there is usually a trade-off between interoperability and energy efficiency. IP-based protocols are usually unsuitable for resource-constrained WSNs and inefficient even when simplified. Interoperability can thus be assured at a higher level by the less resource-restricted gateways nodes.

To maintain scalability support, discoverability has to be sacrificed. Scopes have to be defined to avoid uncontrolled network floods. In this manner, queries and announcements will have a maximum number of hops they can travel, restraining network traffic, but also limiting the obtained results.

2.4 Service Browsing

A user may want at any time to check the available services in order to decide which services to use. Inside a sensor network, browsing may not be so important, as communication is machine to machine. However, if a node is a little more flexible and can operate according to the available services, then it can take advantage of service browsing in order to select which type to use.

2.5 Service Access

The common solution concerning service access is to simply provide all necessary information for the service client to connect to the server. Nevertheless if a node queries the network for a given service, then it will most likely want to access it. In this manner, when indicated in the query, the SD protocol can use the underlying routing protocol to set-up paths and connect the service providers to the requester.

Directed Diffusion sets paths only from the source to sink. Instead of one-way data flows, a service may present some interactivity, and so, messages may flow in both directions. Messages from the service user can be treated as regular queries, which may flood the network. However, as data travels to the requester, nodes in its path can temporarily cache information about the data and the sources from which they received it. Using this information, matching queries from the requester can be forwarded trough the inverse path.

2.6 Service Description

For maximum flexibility and simplicity, services are described as sets of attributes, which in turn are identifier/value pairs. Matching can thus be done with operators in an efficient and powerful manner. Service types organized in a semantic hierarchy provide the most expressive approach. It enables powerful service description and service browsing by allowing for nodes to lookup services of a given type or subtypes. Service types have a unique identifier in order to avoid the need to use the whole path in the hierarchy, and thus produce smaller protocol messages.

Finally, some services may have little interest to the outside of the network. On the other hand, other services may exist only to be exported to the outside of the network. In this context, the service description must contain an attribute which indicates if the gateways should export the service to the outside or not.

3 TSD Architecture

From the point of view of the proposed architecture, we consider four different entities: service providers; service requesters; service directories; gateways and bridges. As depicted in figure 1, service providers and requesters can be organized in groups, each group with its own service directory. The sensor network is connected to the outside through a gateway and may be connected to other sensor networks through bridges. Alternatively, in case the sensor network has a flat topology, service directories may not be present.

A *service provider*, as the name indicates, presents services that other nodes in the network can use. It announces its services when joining the network and it is able to respond to queries matching local services.

A node which uses services is a *service requester*. It is able to browse the network by sending queries for available service types, to obtain information on services matching a given description, and to access services. It can also use services in the network as part of services he provides himself. In this case, the node is also a service provider.

Service Directories keep track of the services provided in their vicinity by receiving the announcements of new nodes, and periodically check whether or not registered services are still available.

The *gateways* act as points from which users access the network, providing the functionalities of the usual sink nodes by collecting sensor data and issuing commands and request to nodes in the network.

Bridges interconnect independent sensor networks by simply forwarding service requests and announcements (as well as data) between them in a controlled manner.

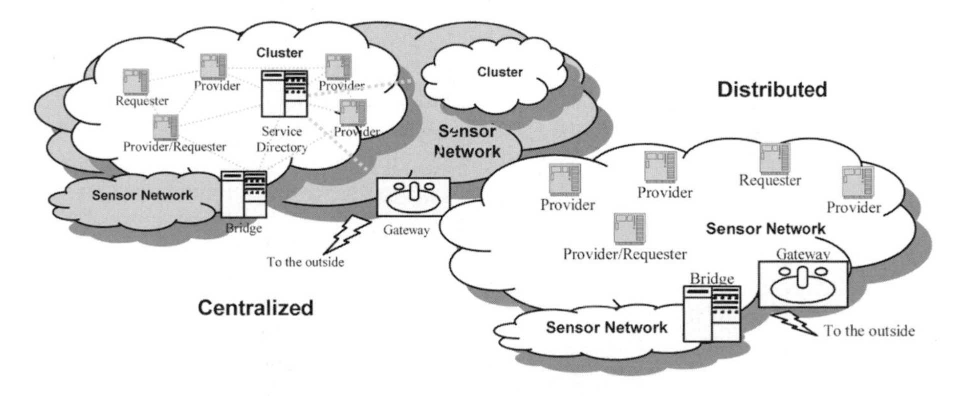

Fig. 1. TSD network architecture

In Directed Diffusion gradients are associated to a value which describes the data rate to each neighbor for a given data type. In our protocol, a message should not be forwarded by several different paths, since the underlying transport protocol should assure packet delivery. The gradient is thus used to select one preferred path and can be updated using feedback from the transport protocol. For equal gradients, paths will be chosen in an alternating manner, playing in favor of uniform energy consumption.

Protocol messages are sets of attribute type, operator, value tuples. Similarly to SCADDS [14], two types of attributes are considered: actual, which define actual values like the type of a service in a service response; and formal, which specify the values of attributes that define what we are looking for. Matching occurs when the formal attributes of a query match the correspondent actual attributes of a service description. A response to a query contains only the attributes present in the query, allowing in this manner selective access to information.

Actual attributes are identified by the use of the operator IS followed by the actual value of the attribute. Greater Than (GT), Less Than (LT), Greater and Less or Equal (GE and LE), Equal (EQ) and Different (NOT) are used to define formal attributes. ANY is also an operator which defines a formal attribute. It indicates that its value is not restricted, but should be present in the response

Attribute types have implicit data types. If a node receives a query containing an unknown attribute type, there is no possible match, and so the node acts accordingly.

The Service attribute identifies a service type. This is obviously an actual attribute of service descriptions and responses and a formal attribute in a query. HOPS is an optional actual attribute which can be used to specify the maximum scope of propagation for a given query or announcement.

Type	Attr.	Op.	Val.	Attr.	Op.	Val.	Attr.	Op.	Value	...
Query	Service	EQ	Temperature	Hops	IS	hop_number, max_hops	...	...	...	...
4 bit	8 bit	4 bit	16 bit	8 bit	4 bit	12 bit	8 bit / string size	4 bit	value type size	...

Fig. 2. Example protocol message structure

4 Simulation and Results

To perform a preliminary protocol analysis, its behavior was simulated using GloMoSim [15]. A simple scenario[1] was defined to study protocol traffic, effectiveness and efficiency in distributed, without node mobility, consisting of 10 to 50 nodes providing four different service types uniformly distributed. Once per hour, a node would then browse the network, and, according to the results, request access to a given randomly selected set of services based on node location.

Simulation parameters were based on 802.14.5 radios using TSMA [16] MAC and a simple upstream transport protocol based on implicit acknowledgements. MAC and transport overhead in terms of energy and protocol traffic are not considered in the results. For each set of parameters five simulation runs were performed to obtain statically correct results. Finally a simplified version of the protocol using flooding was considered for comparison purposes.

As expected, the protocol performs better than a simple flood based protocol in terms of energy and traffic efficiency, as illustrated in figures 3 and 4. The reference protocol produced from 637 to 830 bytes per node, while TSD generated traffic ranged from 551 to 637 bytes with little dependence of the number of nodes. As for delay, TSD presents considerable larger delays due to the transport protocol and its

[1] A sensor network providing temperature, luminosity, humidity and pressure measurements periodically accessed by a user to monitor climacteric conditions of a given area.

retransmissions. This, however, in not a major issue in WSNs. Success rate and discoverability (rate between existing matching services and effective query results) are not dissimilar in both protocols. The flood-based protocol presented a discoverability of 92.6% to 73.6% while TSD values vary between 95.5% and 70.3%. Even using TSMA, as the number of nodes increases, the number of packet collisions grows drastically, resulting in lost packets. As the TSD sends responses through single paths, packet loss affects severely its performance resulting in lower discoverability values than the flood-based protocol for larger networks.

Even though overall results are acceptable, they should be largely improved in future developments with the study of suitable MAC and transport protocols.

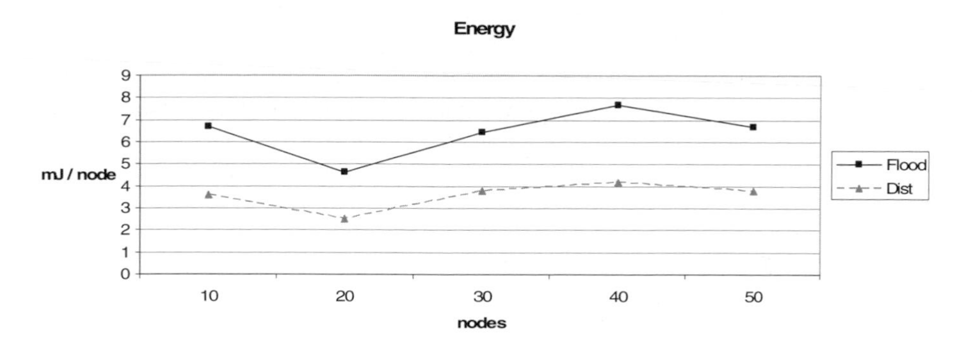

Fig. 3. Simulated energy results

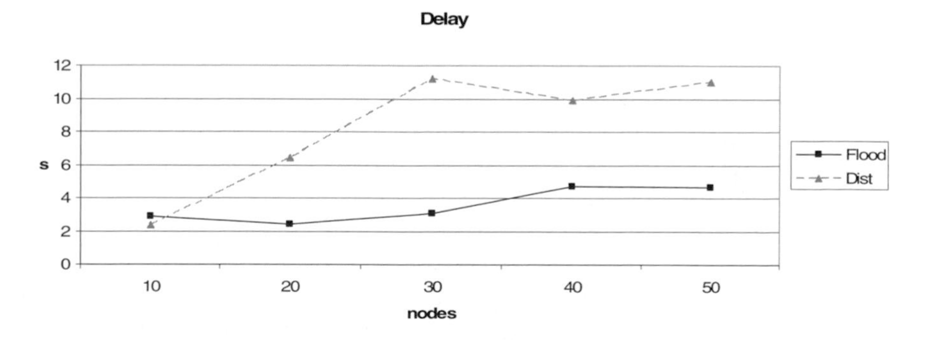

Fig. 4. Simulated delay results

5 Conclusions

This paper proposes the application of the service paradigm to wireless sensor networks. Service Discovery can play a very important role in self-organized and self-configured sensor networks. It allows new nodes to easily discover basic configuration services, to locate services required for their operation and to announce their own services to the network. Generic solutions bring important advantages, not only by simplifying application development, but also, by providing a standardized abstraction and enabling simple interoperability between heterogeneous nodes as well as interconnectivity with heterogeneous networks and the outside world. The main issues in designing a simple and efficient service discovery protocol, suitable for the constraints of sensor networks while maintaining all the functionalities were discussed and a solution was presented.

As a preliminary performance analysis, the protocol behavior was simulated in distributed mode and compared to a service discovery protocol based on flood messages. Overall results were acceptable and the energy efficiency was evident. The importance of efficient MAC and transport protocols on the service discovery protocol performance is a major conclusion to be drawn.

Finally, there is still much work left for future developments, namely the study of suitable MAC and transport protocols and synchronization mechanisms. Concerning result analysis, future work encompasses the study of the impact of aggregation; inverse paths and smart query forwarding; computation delays; service traffic; impact of scopes on discoverability; mobility; cluster formation and centralized operation.

References

1. Shelby, Z., Mähönen, P., Riihijärvi, J., Raivio, O., Huuskonen, P.: NanoIP: The Zen of Embedded Networking. In: Proceedings of ICC (2003)
2. Guttman, E., Perkins, C., Veizades, J., Day, M.: Service Location Protocol, Version 2. RFC 2608 (June 1999)
3. Handziski, V., Frank, Ch., Karl, H.:Service Discovery in Wireless Sensor Networks. Technical Report TKN-04-006, Telecommunication Networks Group. Technische Universität, Berlin (2004)
4. Shneidman, J., Pietzuch, P., Ledlie, J., Roussopoulos, M., Seltzer, M., Welsh, M.: Hourglass: Ann infrastructure for Connecting Sensor Networks and Applications. Technical Report TR-21-04, Harvard University (2004)
5. Tilak, S., Chiu, K., Abu-Ghazaleh, N., Fountain, T.: Dynamic Resource Discovery for Sensor Networks. In: EUC Workshops, pp. 785–796 (2005)
6. Isomura, M., Riedel, T., Decker, C., Beigl, M., Horiuchi, H.: Sharing sensor networks. In: Proceedings of the ICDCS (2006)
7. Song, H., Kim, D., Lee, K., Sung, J.: UPnP-Based Sensor Network Management Architecture. In: Proceedings of ICMU (2005)
8. Heidemann, J., Silva, F., Intanagonwiwat, C., Govindan, R., Estrin, D., Ganesan, D.: Building Efficient Wireless Sensor Networks with Low-Level Naming. In: Proceedings of the Symposium on Operating Systems Principles, pp. 146–159 (2001)
9. Bonnet, P., Gehrke, J., Mayr, T., Seshadri, P.: Query processing in a device database system. Technical Report TR99-1775. Cornell University (1999)
10. Madden, S., Franklin, M., Hellerstein, J., Hong, W.: TAG: a Tiny AGgregation Service for Ad-Hoc Sensor Networks. In: Proceedings of OSDI (2002)
11. Braginsky, D., Estrin, D.: Rumor routing algorithm for sensor networks. In: Proceedings of ICDCS-22 (2001)
12. Kulik, J., Heinzelman, W., Balakrishnan, H.: Negotiation-based protocols for disseminating information in wireless sensor networks. Wireless Networks 8(2/3), 169–185 (2002)
13. Intanagonwiwat, C., Govindan, R., Estrin, D.: Directed diffusion: A scalable and robust communication paradigm for sensor networks. In: Proceedings of the ACM/IEEE International 53rd Conference on Mobile Computing and Networking (2000)

14. Heidemann, J., Silva, F., Intanagonwiwat, C., Govindan, R., Estrin, D., Ganesan, D.: Building Efficient Wireless Sensor Networks with Low-Level Naming. In: Proceedings of the Symposium on Operating Systems Principles, pp. 146–159 (2001)
15. GloMoSim web page (visited on July 2006), `http://pcl.cs.ucla.edu/projects/glomosim/`
16. Chlamtac, I., Farago, A.: Making Transmission Schedules Immune to Topology Changes in Multi-Hop Packet Radio Networks. Proceedings of IEEE/ACM Transactions on Networking 2(1) (1994)

A Fault-Tolerant Event Boundary Detection Algorithm in Sensor Networks

Ci-Rong Li and Chiu-Kuo Liang

Department of Computer Science and Information Engineering
Chung Hua University, Hsinchu, Taiwan 30012, Republic of China
ckliang@chu.edu.tw

Abstract. This paper targets the detection of the reach of events in sensor networks with faulty sensors. Typical applications include the detection of the transportation front line of a contamination and estimation of the region in forest fire. We propose an algorithm for detection the boundary of such events. Our algorithm is purely localized and thus is suitable for large scale of sensor networks. The computational overhead is low since the detection algorithm is based on a simple clustering technique which only simple numerical operations are involved. Simulation results show that our algorithm can clearly detect the event boundary when as many as 20% sensors become faulty. Therefore, our algorithm achieves a great improvement over the previous algorithms. In addition, our proposed detection algorithm can accept any kind of scalar values as inputs. It can be applied as long as the "events" can be modeled by numerical numbers.

Keywords: Sensor networks, event boundary detection, fault tolerance.

1 Introduction

Wireless sensor networks are one of the most important technologies that will change the world [1] in that such networks can provide us with fine-granular observations about the physical world where we are living. Potential applications of wireless sensor networks include disaster rescue, energy management, medical monitoring, logistics and inventory management, and military reconnaissance, etc. With their capabilities for monitoring and control, the sensors are expected to be widely deployed. Such a network can provide a fine global picture through the collaboration of many sensors with each observing a coarse local view [7], [8].

One important task of a typical sensor network is to monitor, detect, and report the occurrences of interesting events (e.g. forest fire, chemical spills, etc.) with the presence of faulty sensor measurements. These events usually span some geographic region and in many application scenarios the detection of the event boundary may become more important than the detection of the entire event region. A good example is the timely estimation of the possible reach of the contamination in a surveillance network monitoring the transportation of chemical spills in soil. However, individual sensor reading is not reliable. Filtering out the faulty readings and transmitting only the boundary information to the base station can save energy and become crucial in

T. Vazão, M.M. Freire, and I. Chong (Eds.): ICOIN 2007, LNCS 5200, pp. 406–414, 2008.

sensor networks. In this paper we focus on the problem of detecting event boundaries in sensor networks with faulty sensors.

However, event boundary computation is not trivial at all. The most significant challenge task comes from the strict resource limitation (battery power, bandwidth, etc.). The sensors in a sensor network are powered by battery and once the sensors are deployed, the battery may not be recharged or replaced. It is very energy consuming to allow a base station collect all sensor measurements and identify the faulty sensors and compute event boundary in a centralized fashion [5], [8]. Therefore, we have to seek localized and computationally efficient algorithms for each node to determine whether it is faulty or whether it is close to the event boundary. The existence of faulty sensors constitutes another significant challenge for event boundary computation. Sensor readings may be faulty due to hardware crash, security attack, or environment disturbance. Thus a solid event boundary detection algorithm must be robust and fault-tolerant.

Our major contribution is to provide a localized algorithm for fault-tolerant event boundary detection. The algorithm is basically based on our previously proposed clustering technique [10] to identify the faulty sensors and then, by extending the clustering technique, we can efficiently determine whether the sensors are located in the event boundary or not.

This paper is organized as follows. We first briefly summarize the related work in Section 2 and some preliminaries, including the network model and useful notations, are presented in Section 3. A localized algorithm for fault-tolerant event boundary detection is proposed in Section 4. Performance metrics and analysis, and our simulation results are given in Section 5. We conclude our paper in Section 6 with future work discussion.

2 Related Works

Intuitively, when a remarkable change in sensor reading is detected, something must have happened. If the change is present with a single sensor only, the sensor is faulty. If most neighboring sensors observe the same phenomenon simultaneously, an event occurs. This observation is explored in [3], [7], and [11]. But all of the related algorithms in [3], [7], and [11] require only the most recent readings (within a sliding window) of individual sensors. No collaboration among neighboring sensors are exploited. The detector proposed in [7] computes a running average and compares it with a threshold, which can be adjusted by false alarm rate. In [11], the authors design kernel density estimators to check whether the number of "abnormal" readings is beyond an application-specific threshold. But none of these works can disambiguate faulty sensors and real event sensors since only observations from individual sensors are studied.

The faulty sensors can also be detected through route discovery and update. In [12], the authors propose to trace failed nodes in sensor networks at a base station, which assuming all sensor measurements will be directed to the base station along a routing tree. In their work, since the base station has a global view of the network topology, the failed nodes can be identified through route update messages. In [9], nodes can listen-in on the neighbor to detect failed or misbehaving neighbors. In [13], base stations initiate marked packets to probe sensors and rely on their responses to identify and isolate failed nodes. However, the above methods must rely on the routing or global topology information.

Recently, Ding, Chen, Xing, and Cheng [4] proposed a different approach, which is based upon the statistics, to identify the faulty sensors and also detect the event boundary. Their major contribution consists of one localized algorithm for faulty sensor identification and one localized algorithm for fault-tolerant event boundary detection. Their algorithms can identify the faulty sensors from event sensors and detect event boundary by exploring the collaboration among neighboring sensors. Their proposed algorithm can also detect many kinds of misbehaving nodes, as long as the "abnormal behavior" can be modeled by real numbers. One of the most important characteristics is that their algorithms do not need to rely on any routing or global topology information, thus provides better scalability and flexibility.

In this paper, we propose a better fault-tolerant event boundary algorithm than the method in [4] in the boundary detection accuracy. The basic idea of our approach is using the clustering technique which is based upon the maximum spanning trees [10], [14]. In our approach, the difference of readings between any two sensor nodes is represented as the "distance" between them. For a set of sensors, we can use the distances among them to classify the sensors into two clusters according to the properties of maximum spanning trees [14]. By doing some computations, the sensors which are located on the event boundary can be identified. Simulation results show that our clustering algorithm can achieve more accuracy in event boundary detection than the previous result in [4]. In addition, it is obvious to see that our algorithm is localized and collaborated with the neighboring sensors. This means that our algorithm does not rely on the network topology information to identify the event boundary sensors.

3 Network Model and Preliminaries

Let there be a $b \times b$ squared field, denoted as P, which is located on the 2-dimenstional Euclidean plane $\Re^2$. Throughout this paper, we assume that there is a set of n sensors, say $S = \{S_1, S_2, \ldots, S_n\}$, which is uniformly deployed on P. We say that, according to [2], a sensor's reading is *faulty* (abnormal) if it is different significantly from other readings of its neighboring sensors. Sensors with faulty readings are called faulty sensors. We use R to denote the radio range of sensors. Let x_i denote the reading of sensor S_i. Note that x_i is assumed to be the actual reading from sensor S_i which can reflect the status of the environment, such as temperature, light, sound, and so on. Therefore, x_i can be continuous or discrete. Let $N(S_i)$ denote a set of sensors that contains the sensor S_i and additional k sensors $S_{i1}, S_{i2}, \ldots, S_{ik}$ which are located in the closed disk area centered at S_i with radius R. We call $N(S_i)$ the neighborhood set of the sensor node S_i. Note that $N(S_i)$ represents a closed neighborhood of the sensor S_i.

In the following, we are going to explain the meaning of events. For consistence, we use the definitions in [4] for events. Informally, an event can be defined in terms of sensor readings. An event, denoted by E, is a subset of $\Re^2$ such that readings of the sensors in E are significantly different from those of sensors not in E. A faulty sensor can be viewed as a special event sensor which contains only one point, i.e., the sensor itself. A point $x \in \Re^2$ is said to be in the boundary of event E if and only if each closed disk area centered at x contains both points in E and points not in E. The boundary of the event E, denoted by $B(E)$, is the collection of all the points in the boundary of E. A circle, as an example, is the boundary of the region bounded by the circle if the region is an event. Fig. 1 shows the meanings of events and event boundary.

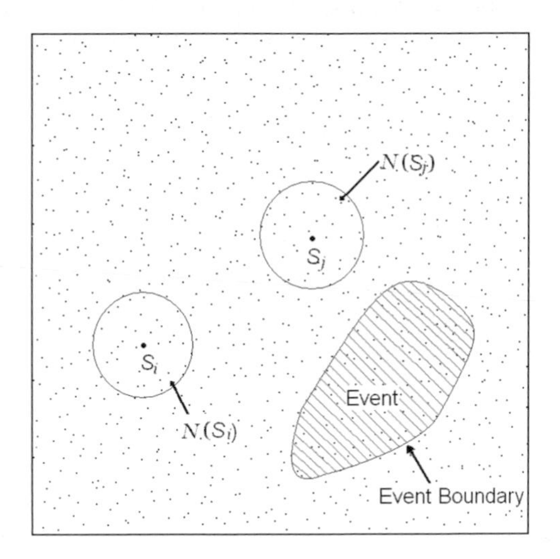

Fig. 1. The Event Boundary and $N(S_i)$

In order to detect the event boundary, a comparison between sensor S_i and part of its neighboring sensors is done by checking their difference, say d_i, between the reading of S_i and the median reading of part of its neighboring sensors. If d_i is large or large but negative, then it is very likely that S_i is close to the event boundary. Suppose that there are k sensor nodes (including S_i), say $S_1, \ldots, S_i, \ldots, S_k$, lie in the closed disk centered at S_i with radius R. In order to identify the event boundary sensors, the authors in [4] compute the mean μ and standard deviation σ for the set $D = \{d_1, \ldots, d_i, \ldots, d_k\}$. Then, they standardize the dataset D to obtain $\{y_1, \ldots, y_i, \ldots, y_k\}$, where $y_j = (d_j - \mu) / \sigma$ for $j = 1,2,\ldots k$. After that, they compare the value of y_j with a preselected threshold θ. If $|y_j| \geq \theta$, then S_j is treated as an event boundary sensor. In order to increase the accuracy of detection, some complicated procedures were proposed in [4] to verify the event boundary sensors.

In this paper, we propose a different approach to identify the event boundary sensors. The main idea of our approach is to partition a set of sensor nodes into homogeneous classes, called clusters. The homogeneity is expressed by some measure of dissimilarity between the readings of sensor nodes. Generally speaking, in a homogeneous partition, one wishes to minimize the maximum dissimilarity between elements of the same cluster. Therefore, we are looking to find a method to partition the sensor nodes into clusters such that the maximum dissimilarity of the readings of the same cluster is minimized. After the sensors are partitioned into clusters, the event boundary sensors can be identified in an easy manner. For example, suppose that there are 10 neighboring sensors around sensor S_i and all of them (include S_i) have been partitioned into two clusters according to their readings. If one cluster contains 6 sensor nodes and the other contains 5 sensors and the average readings of these two clusters are much different with each other, then we can say that sensor S_i is very much likely close to the event boundary. Since our main goal is to identify the event boundary sensors, only two different clusters should be identified. Therefore, we use the maximum spanning tree clustering technique to classify the sensors into two clusters according to their readings. In the following, we will briefly describe the clustering algorithm that based on the maximum spanning trees technique.

Consider a sensor node, say S_i, with reading x_i. Suppose that the area centered at S_i with radius R contains S_i and additional k - 1 sensors. That is, the set $N(S_i)$ consists of k sensors, say S_1, …, S_i, …, S_k. Also suppose that the readings at the sensors in $N(S_i)$ are x_1, …, x_i, …, x_k, respectively. Let $\alpha_{ij} = |x_i - x_j|$ denote the difference between x_i and x_j, where $1 \le i,j \le k$. Note that α_{ij} can be viewed as the distance between sensors S_i and S_j. Let $MXST(S_i)$ denote the maximum spanning tree constructed for $N(S_i)$ according to the distance measurement α. Since $MXST(S_i)$ is a bipartite graph [14], we can easily obtain two clusters of sensor nodes, say A and B. By comparing both the averaged readings of the sensors in clusters A and B, we can identify the sensor node S_i whether is faulty or not.

4 Localized Event Boundary Detection

In this section, we describe our algorithm for localized event region detection. To detect an event region, it is suffices to detect the sensor nodes near or on the boundary of the event. Thus, our boundary detection algorithm is to identify the sensors which are near or on the boundary of the event. Our algorithm needs two phases: faulty sensors detection and event boundary detection. Since the faulty readings may influence the accuracy of boundary detection, we need to eliminate the faulty readings. After eliminating the faulty readings, our event boundary detection algorithm will identify the sensors which are near or on the boundary of the event. We use our previous proposed algorithm [6] for detecting the faulty sensors, and followed by our event boundary detection procedure to implement our event boundary detection algorithm which is listed as follows. Note that our algorithm is localized and every sensor node in the field can perform the same steps individually.

Algorithm Event Boundary Detection
Phase 1: This phase is for faulty sensors detection. We only execute the faulty sensors detection algorithm in [6]. The faulty sensors' readings will be eliminated and the remaining readings are the input of phase 2.
Phase 2: This phase is for event boundary detection. The procedure is described as in the following steps:

Step 1: For each sensor S_i, construct the neighborhood set $N(S_i)$.
Step 2: Construct the distance matrix, α, where α_{jk} denote the distance between any two sensors S_j and S_k in $N(S_i)$.
Step 3: Construct the maximum spanning tree, $MXST(S_i)$, according to α.
Step 4: Partition the sensors into two clusters, say A and B, based on $MXST(S_i)$.
Step 5: Compute the average readings for both clusters A and B, namely ave_A and ave_B respectively.

Step 6: If $|ave_A - ave_B| \geq \theta$, $|A| \geq \lambda$ and $|B| \geq \lambda$, then assign S_i to F, where F denotes the sensors that are detected as the event boundary sensors by our detection procedure.

The output of our algorithm is the set F.

5 Simulation

In this section, we evaluate the performance of our algorithm. We compare the simulation results with the previous results done by [4]. The simulation environment is described as follows. The simulation program is written by *C* language. The sensor network contains 1024 nodes in a square region of size 32×32 units with one sensor randomly placed within each unit grid. Without loss of generality, we assume the square region resides in the first quadrant and the coordinates of sensors are defined accordingly. Normal sensor readings are drawn from $N(\mu_1,\sigma_1^2)$ while event sensor readings are drawn from $N(\mu_2,\sigma_2^2)$. In our simulation we choose $\mu_1 = 10$, $\mu_2 = 30$, and $\sigma_1 = \sigma_2 = 1$. Note that these means and variances can be picked arbitrarily as long as $|\mu_1 - \mu_2|$ is large enough compared with σ_1 and σ_2. In our simulation, we simply implement the algorithm proposed by [6] for the faulty sensors detection. For event boundary detection, our algorithm needs two thresholds, θ and λ. The usages of θ and λ are different: θ is for distinguishing the difference between two groups of readings while λ is for identifying whether the sensor is near or on the boundary of event. We use the settings in [4] as the settings of θ, which is listed in Table 1. In Table 1, the values of threshold depend on p, the probability that a sensor becomes faulty.

Table 1. The settings of threshold θ

p	5%	10%	15%	20%
θ	1.96	1.65	1.44	1.28

Here we present a brief examination on what value of λ is suitable for the event boundary detection. Let r denote the distance from the sensor S_i to the event boundary. If r is smaller than $R/2$, then the chance that sensor S_i will be near or on the boundary of the event is high. In such case, it is reasonable to identify S_i as an event boundary sensor. On the contrary, S_i can be said not an event boundary sensor if r is larger than $R/2$. However, how do we know the distance from sensor S_i to the event boundary? The answer is that we can measure the distance by knowing the number of sensors whose readings are different from the reading of S_i. This can be explained by using Fig. 2. Fig. 2 shows the area of the closed disk of radius R centered at S_i. Without loss of generality, we may assume that the portion of the boundary falling into the area is a line segment. Clearly, if $B(E)$ does not intersect the area or the portion of the intersection is less than the shaded area in Fig. 2, it is very likely that S_i may not be detected as a sensor on the boundary. Therefore, we only need to consider the number of sensors that falling into the shaded area in Fig. 2 to decide whether S_i is on the event boundary or not. Let m denote the expected number of sensors falling into the shaded area. It can be easily calculated that the shaded area is about 19.6% of the closed disk when $r = R/2$. The value of m depends on the average number of sensors falling into the closed disk centered at S_i, which is called the *density* of the sensor network. In our algorithm, we let $\lambda \geq m$ to be a threshold to distinguish whether a sensor is close to or far away from the event boundary. For a sensor S_i, if the number of sensors with different readings is larger than λ, we may say that S_i is likely close to the event boundary. Otherwise, S_i is likely far away from event boundary. Therefore, we can get a better performance by setting the value of λ. In the simulation, we use five different values of density with corresponding values of λ, as shown in Table 2.

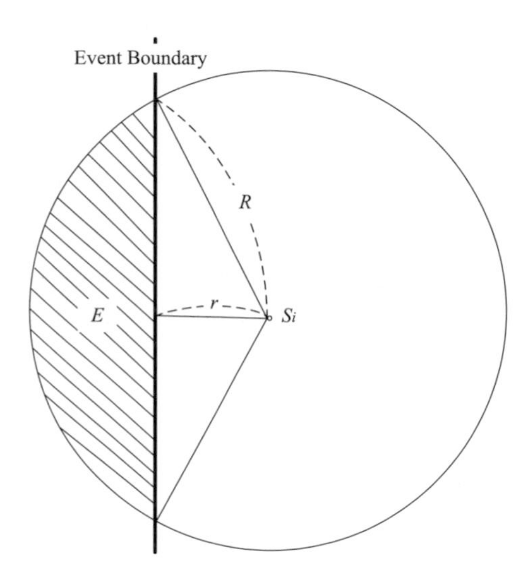

Fig. 2. The closed disk centered at S_i with distance r to event boundary

Table 2. The settings of λ

Density	10	20	30	40	50
m	1.96	3.92	5.88	7.84	9.8
λ	3	5	7	9	11

The simulation results are shown as in the following figures, each representing an averaged summary over 100 runs. To evaluate the performance, we compute the *degree of fitting* and *false detection rate* that are also used in [4]. Let $BA(E,r)$ denote the set of all points such that the distance of each point to the boundary $B(E)$ is at most r. Then, the degree of fitting is define to be $|BA(E,r) \cap F| / |BA(E,r) \cap S|$. Intuitively, the degree of fitting is to examine how many sensors that are close to the boundary are correctly detected as event boundary sensors. Let $A(E,R)$ denote the set of all points such that the distance of each point to the boundary $B(E)$ is at least $R/2$. Define the false detection rate of F to the following quantity: $|A(E,R) \cap F| / |A(E,R) \cap S|$. This means that the false detection rate is to examine how many sensors not close to the boundary are detected as event boundary sensors.

We report the degree of fitting and the false detection rate *vs.* network density and *vs.* p, respectively, in Fig. 3 and Fig. 4. Fig. 3 demonstrate the performance of F with variable density when $p = 0$. We observe that increasing density in general can increase both the degree of fitting and the false detection rate. This is due to that when we increase the density, we actually increase the transmission range R in our simulation. Thus, more sensors will fall into the closed disk centered at a sensor. Therefore, in our algorithm, it will be more likely to identify a sensor close to the event boundary. On the other hand, more sensors will also fall into the shaded area in Fig. 2 as density increased. Then, the chance a sensor which is far away from the event boundary may be detected as the event boundary sensor is high. Therefore, the false detection rate will be increased. However, our approach outperforms the previous approach both in degree of fitting and false detection rate.

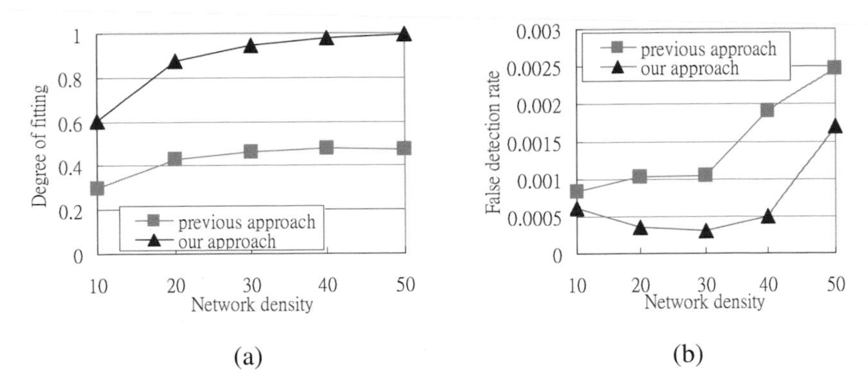

Fig. 3. (a) Degree of fitting, (b) False detection rate *vs.* network density when $p = 0$

When the sensor fault probability increases, the performance of F usually decreases, as shown in Fig. 4. Compared with Fig. 3, the degree of fitting is low when $p > 0$ for the same network density. This is obvious since faulty sensors may interfere with the identification of boundary sensors. Again, our approach still outperforms the previous approach both in degree of fitting and false detection rate.

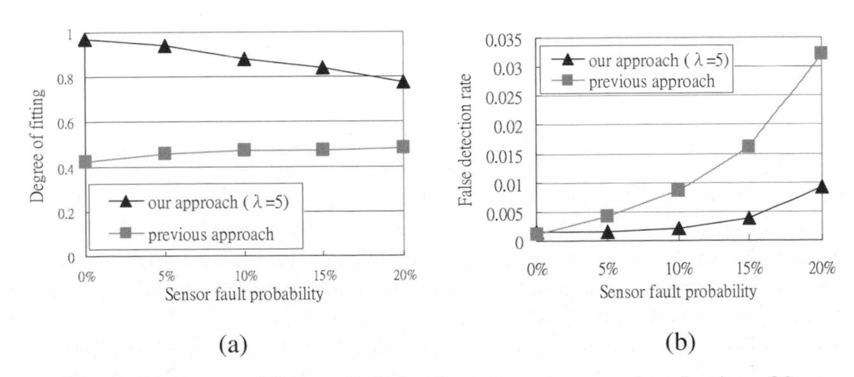

Fig. 4. (a) Degree of fitting, (b) False detection rate *vs.* p when density = 20

6 Conclusion

In this paper, we present a simple but more accurate algorithm for event boundary detection with faulty sensors in sensor network. Our approach is based on the idea of classifying two clusters with different characteristics (readings) from a set of sensor readings. The clustering algorithm is based on the technique of using properties of maximum spanning trees. Simulation results show that our approach can achieve a great improvement over the previous approach both in the degree of fitting and the false detection rate. The current algorithm is sensitive to the settings of thresholds, which are dependent on the sensor fault probability. By obtaining the sensor faulty probability from base stations, an adaptive threshold that better fits the application context can be designed.

References

1. 10 emerging technologies that will change the world. Technology Review 106(1), 33–49 (February 2003)
2. Barnet, V., Lewis, T.: Outliers in Statistical Data. John Wiley and Sons. Inc., Chichester (1994)
3. Chen, D., Cheng, X., Ding, M.: Localized Event Detection in Sensor Networks (manuscript, 2004)
4. Ding, M., Chen, D., Xing, K., Cheng, X.: Localized Fault-Tolerant Event Boundary Detection in Sensor Networks. In: IEEE INFOCOM 2005 (2005)
5. Hill, J., Szewczyk, R., Woo, A., Hollar, S., Heidemann, J.: System architecture directions for networked sensors. In: Proc. 9th International Conference on Architectural Suuport for Programming Languages and Operating Systems (November 2000)
6. Li, C.R., Liang, C.K.: A Localized Fault Detection Algorithm in Sensor Networks. In: IASTED International Conference on Wireless Sensor Networks WSN 2006, Canada (July 2006)
7. Li, D., Wong, K.D., Hu, Y.H., Sayeed, A.M.: Detection, Classification and Tracking of Targets. IEEE Signal Processing Magazine 19, 17–29 (2002)
8. Madden, S., Franklin, M.J., Hellerstein, J.M., Hong, W.: TAG: a tiny aggregation service for ad-hoc sensor networks. In: OSDI (December 2002)
9. Marti, S., Giuli, T.J., Lai, K., Baker, M.: Mitigating Routing Misbehavior in Mobile Ad Hoc Networks. In: ACM MOBICOM 2000, Boston, MA, August 2000, pp. 255–265 (2000)
10. Monma, C., Paterson, M., Suri, S., Yao, F.: Computing Euclidean maximum spanning trees. Algorithmica 5, 407–419 (1990)
11. Palpanas, T., Papadopoulos, D., Kalogeraki, V., Gunopulos, D.: Distributed Deviation Detection in Sensor Networks. SIGMOD Record 32(4), 77–82 (2003)
12. Staddon, J., Balfanz, D., Durfee, G.: Efficient Tracing of Failed Nodes in Sensor Networks. In: ACM WSNA 2002, Atlanta, GA, pp. 122–130 (September 2002)
13. Tanachaiwiwat, S., Dave, P., Bhindwale, R., Helmy, A.: Secure Locations: Routing on Trust and Isolating Compromised Sensors in Location-Aware Sensor Networks. In: ACM SenSys 2003, Los Angeles, California, pp. 324–325 (2003)
14. West, D.B.: Introduction to Graph Theory. Prentice-Hall, Englewood Cliffs (1996)

Collision-Free Downlink Scheduling in the IEEE 802.15.4 Network

Sangki Yun and Hyogon Kim

Korea University

Abstract. IEEE 802.15.4 is the Low-Rate Wireless Personal Area Network (LR-WPAN) standard that is suitable for wireless sensor networks and wireless home networks among others. The IEEE 802.15.4 is specifically designed for energy efficiency since many 802.15.4-compliant devices are expected to operate on battery. Because of inadequate design of the downlink frame transmission mechanism in the standard, however, 802.15.4 devices can waste their energy due to collisions even under modest downlink traffic in the network. In order to solve this problem, we propose a novel mechanism which evenly distributes the downlink frame transmissions and decimate collisions. It exploits the information already imbedded in the beacon frame, so it does not require modifications of the IEEE 802.15.4 standard. Our scheme significantly reduces the energy consumption of 802.15.4 WPAN devices under modest to heavy downlink traffic, while not adversely affecting the system performance under low utilization.

Keywords: 802.15.4, downlink collision, WPAN, ZigBee, energy consumption, battery.

1 Introduction

The IEEE 802.15.4 is a medium-access (MAC) and physical (PHY) layer standard for low-power, low-rate wireless communication [1], which is widely considered to serve the needs of sensor networks and home networks well. Many 802.15.4 devices are expected to operate on batteries, so energy efficiency is a prime concern in the design of 802.15.4 MAC. The 802.15.4 devices can turn off the radio transceiver when they do not have frames to send, only turning it on for periodic beacon frames. If the beacon frame notifies the device of a pending (downlink) data, it issues a Data Request frame and turns on the receiver until it receives the data. But if there are multiple devices to receive such downlink frames, according to the current 802.15.4 standard, they almost concurrently attempt the transmission of request frames, leading to high collision probability that leads to additional energy consumption and delay.

In this paper, we address the problem while not modifying the 802.15.4 standard specification. Instead, our approach simply utilizes the information already embedded in the beacon frame, so 802.15.4-conformant WPAN devices can implement it without causing the interoperability issues. In our scheme, each

T. Vazão, M.M. Freire, and I. Chong (Eds.): ICOIN 2007, LNCS 5200, pp. 415–424, 2008.

WPAN device is implicitly assigned its own superframe slot for the request frame transmission, which completely prevents the collisions among request frames. In contrast to the standard scheme that suffers higher collision and drop probabilities as the downlink traffic increases, our scheme has consistently low collision and drop probabilities and eventually low energy consumption regardless of the volume of the downlink traffic.

There is a dearth of prior work addressing this problem, because the IEEE 802.15.4 has been standardized only recently and this problem has not received much attention. Misic *etal.* [5] is the only one that notices the problem, but it simply suggests reducing the size of the Pending Addresses field to 3 or 4 to decrease the number of simultaneously requesting devices. Obviously, it would not only restrict the capacity and flexibility of the downlink traffic but also require the modification of the current standard. In contrast, our scheme proposed in this paper neither requires standard modification nor restricts the capacity of the downlink traffic.

The rest of the paper is organized as follows. We draw on the relevant parts of the IEEE 802.15.4 standard in Section 2. In Section 3 we characterize the downlink inefficiency problem of the IEEE 802.15.4, and introduce our solution to this problem. Section 4 presents the experimental evaluations. Section 5 concludes the paper.

2 Background

The IEEE 802.15.4 standard defines the PHY and MAC sublayer specifications for low data rate wireless connectivity with portable devices. It has data rates of 250kb/s, 40kb/s and 20kb/s. There are two types of devices in 802.15.4 WPAN networks: a full-function device (FFD) and a reduced-function device (RFD). The FFD has more computation capability and energy than the RFD and has a responsibility to be a PAN coordinator. An RFD is intended to be used in simple applications such as a light switch and a passive sensor, so it has minimal resources and memory capacity. An FFD can communicate with RFDs and other FFDs but an RFD can talk only to FFD. One FFD in a communication network is elected for a PAN coordinator. The coordinator has duties to transmit beacon frames, schedule a channel allocation, associate newly appearing devices. So usually the FFD which has consistent power supply and high computation capability (*e.g.* personal computer) becomes the coordinator. The standard specifies two network topologies: star topology and peer-to-peer topology.

2.1 Channel Structure

In a beacon-enabled network, an active period between a beacon interval is called superframe. It is divided into 16 equally sized superframe slots and organized into the Contention Access Period (CAP) and the Contention Free Period (CFP). The standard stipulates that a minimum of 9 slots should be used for the CAP. In the CAP, the device accesses the channel using slotted Carrier Sensing Multiple Access with Collision Avoidance (CSMA/CA). In the CFP, on the other hand, the coordinator assigns the superframe slot called Guaranteed Time Slot (GTS) to each device. The length of the superframe is determined by Superframe Order (SO), and the beacon interval by Beacon Order (BO). If the value of SO and BO is equal, the superframe length and the beacon interval becomes identical, so there is no inactive period between beacons. If the BO has a higher value than the SO, the gap between the superframe duration and the beacon interval becomes an inactive period in which all devices get into the power-down mode. One superframe slot is divided into several backoff periods, and the number of backoff periods in a superframe slot depends on the SO value.

2.2 Channel Access Mechanism

The channel access mechanism in the CAP is similar to that of the IEEE 802.11 DCF, with a few notable differences. The 802.11 device always turns on the receiver (barring power saving mode), so it performs Clear Channel Assessment (CCA) in every slot. But the 802.15.4 device turns on the receiver and performs the CCA only after the end of the random backoff, in order to save energy. Because 802.15.4 devices do not perform the CCA in every slot, they cannot freeze their backoff counter even when other nodes are transmitting, so keep decrementing it. When a device finishes the backoff countdown, it performs two CCAs in a row and transmits the frame if the channel is idle. If the channel is busy, it notches up the backoff stage unless it reaches the maximum. It is a major difference with 802.11 DCF: the backoff stage increases not upon the collision, but upon the busy channel. Collisions do not influence the backoff stage. Once the devices access channel and transmit frames, they reset their Backoff Exponent (BE) value regardless of the fate of the transmission.

2.3 Data Transfer Model

The data transfer mechanism is asymmetric between the coordinator and the device. The transmission from the device to the coordinator (*i.e.*, uplink) is straightforward. If the device has a frame to send to the coordinator, it simply transmits it using the CSMA/CA. If the coordinator successfully receives the frame, it sends an acknowledgement (ACK) frame to the device.

The transmission from the coordinator to the device (*i.e.*, downlink) on the other hand uses an indirect transmission mechanism. Fig. 1 illustrates the mechanism. The coordinator notifies the devices of the pending frames through the Pending Address field in the beacon frame. The length of the field is variable, but up to 7 devices can be addressed [1]. The notified devices must transmit the Data Request frame, and the instance of the transmission is dependent on the *macAutoRequest* parameter. If it is set true, which is the default value, the device should transmit the request frame in the immediately following superframe slot, within which backoffs are performed in the units of backoff periods (BPs). Upon successfully receiving the request frame, the coordinator transmits the downlink frame using the CSMA/CA. After receiving the frame, the device sends an ACK frame to the coordinator.

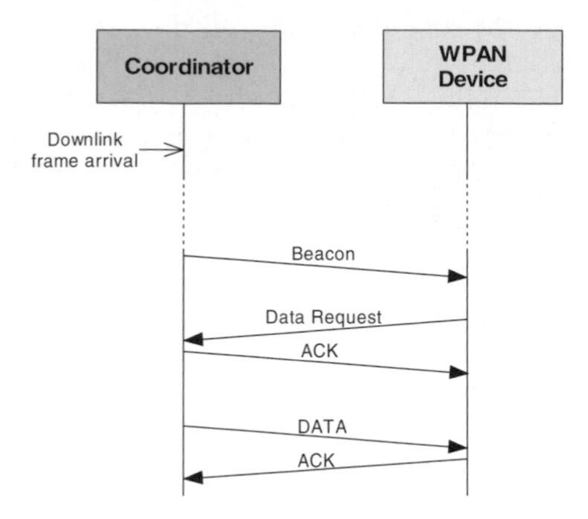

Fig. 1. Indirect transmission

In the indirect transmission, four frames are exchanged to transmit one data frame, so it looks inefficient. But there is a reason to use this indirect transmission: to turn off the device's radio receiver. If the downlink frame is transmitted by the direct transmission mechanism, the device must always turn on the receiver because it can not predict when the coordinator transmits the frame. On the other hand, in the indirect transmission, the device can turn on the receiver only for the beacon frame and immediately turn off if no pending frame exists.

3 Proposed Idea

3.1 The Problem of the Indirect Transmission

The indirect transmission achieves energy savings, but has a drawback. The collision probability easily reaches a high value for even a modest number of notified stations. Under the default macAutoRequest parameter setting, as soon as the beacon frame is transmitted, the involved devices jump into contention. The initial contention window is given by the macMinBE parameter, whose default value is upper bounded by 3 (it implies that initial maximum contention window size is 7). Thus if even 3 or 4 devices out of the maximum 7 contend, the collision and backoff probabilities become significant.

In addition to the high collision and backoff probabilities, the feature has one more negative impact: it elongates the duration for which the device's radio receiver remains on. As discussed earlier, after a device successfully transmits the request frame, it must turn on the receiver until the arrival of the data frame. Staying longer in the receiving mode is critical for battery life, because being in the receiving mode can consume comparable or even more energy than transmission [3]. So the 802.15.4 standard stipulates that the device can wait the downlink frame for only up to 61 backoff periods. If the downlink frame is

Table 1. Energy consumption in each state

State	Energy consumption (μJ/sec.)
Power Down (Sleep mode)	0.00064
IDLE	0.136
TRANSMISSION	5.57
RECEIVING	6.02

not received within this period, the device turns off the receiver and transmits the request frame in the next superframe again.

Table 1 shows the energy consumption when the device stays in each state for the duration of one backoff period, based on the popular TI Chipcon 2420 chipset data sheet [3]. The data indicates that receiving mode consumes slightly more energy than the transmission mode.

3.2 Time-Ordered Slot Appointment Rule (TSAR)

Our solution approach to the Data Request collision problem is to utilize the information about the number of pending devices that is already available in the beacon frame. Every device listens to the beacon frame so can notice how many devices will request the downlink transmission in the upcoming superframe by reading the Pending Addresses fields. By utilizing this information, we can distribute each device's contention period, and eventually lower the collision and drop probability. Below, we will call this scheme Time-ordered Slot Appointment Rule (TSAR) for convenience.

In TSAR, each WPAN device examines in the beacon frame to find where its address is positioned. If a device's address is located at n^{th} position in the Pending Address field, it starts its contention at the nth superframe slot from the left boundary. As discussed earlier, one superframe has 16 superframe slots of which the minimum of 9 slots are guaranteed for the CAP, and a beacon frame can convey up to 7 Pending Addresses. So all devices can be assigned their own superframe slots. This significantly reduces the collision probability for the Data Request frames.

Fig. 2 and Fig. 3 compares the implied behavior in the standard with that of TSAR, when 3 devices are notified of the pending frame. In the figures, B means backoff duration. In the standard scheme, every device starts the contention

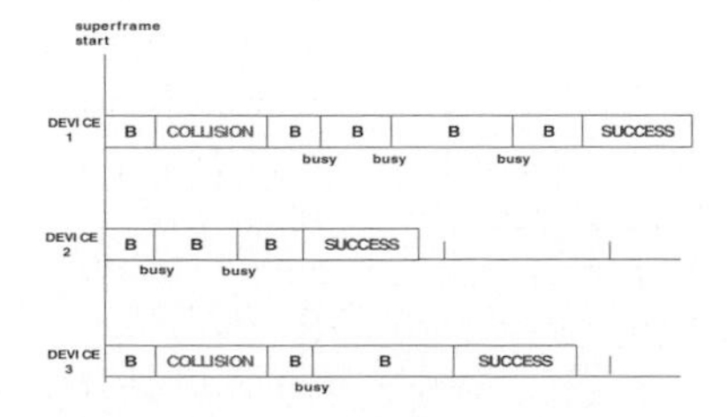

Fig. 2. Channel access in the standard (macAutoRequest)

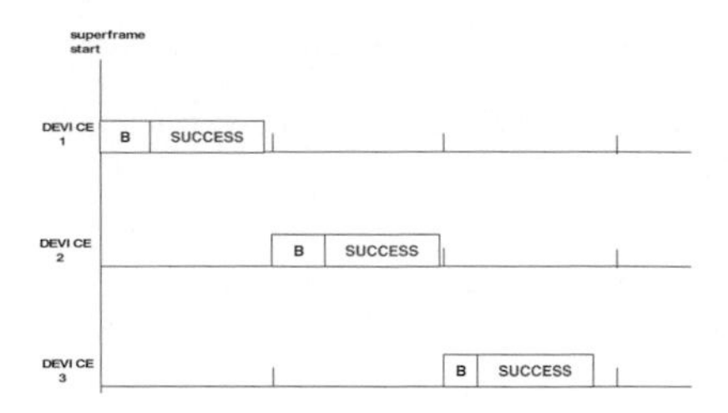

Fig. 3. Channel access in TSAR

from the first backoff period after the beacon frame. Therefore, they contend with each other and have high collision and backoff probabilities. On the other hand, in TSAR, each device starts its backoff in different superframe slot, so they are isolated in terms of contention. Although the uplink traffic can interfere, we will see later that such isolation is for the most part retained.

4 Performance Evaluation

In this section, we compare TSAR with the standard scheme. Although NS-2 [4] provides an environment for simulating 802.15.4, it poses difficulties for measuring the energy consumption and the delay per downlink frame. So for the experiments below, we use a home-brewed event-based simulator. Parameters used in simulations are summarized in Table 2. We assume that the WPAN has the star topology where a single coordinator communicates with 10 devices. The Superframe Order (SO) is set to 4, and the Beacon Order (BO) is set equal to the SO unless otherwise noted. We vary the downlink packet arrival rates while fixing the uplink packet arrival rate. In our experiments, the arrival rate represents the number of arriving packets over the entire system (not per device) in a given time.

4.1 Collision and Drop Probability

In this experiment, we measure two probabilities. First, P_c is the probability that the request or data frame collides with the transmission from other node(s). The

Table 2. Simulation settings

Parameters	Values
Simulation time	300 second
Transmission rate	250 Kbps
macMinBE	3
Topology	star-topology
Number of devices	10
Payload size	90 bytes
Superframe Order	4
Beacon Order	4, 5
Maximum backoff limit	5

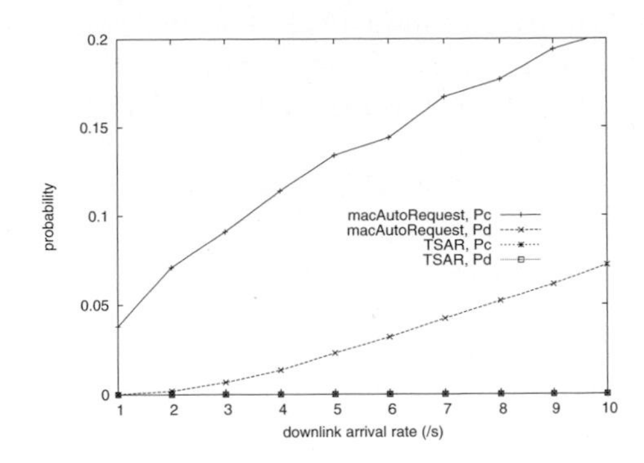

Fig. 4. Collision and drop probability, no uplink traffic

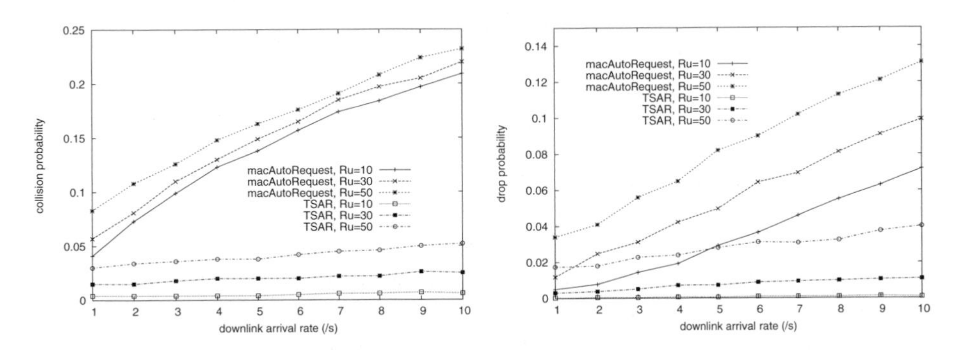

Fig. 5. Collision and drop probability with varying uplink traffic

second is P_d, the probability that a device gives up transmission due to repeated failures over limit.

Fig. 4 shows the P_c and P_d of the standard (labeled *macAutoRequest*) scheme and TSAR as functions of the downlink packet arrival rate, where there is no uplink traffic. The standard scheme suffers from the gradually increasing collision and blocking probability, whereas TSAR completely avoids either collisions or drops. It shows that TSAR successfully resolves channel accesses at least between downlink frames. Fig. 5 measures the probabilities in face of varying uplink traffic intensity. We set the uplink packet arrival rate R_u to 10, 30, and 50. Even when the uplink traffic interferes, TSAR maintains consistently low collision and drop probabilities compared with the standard scheme.

So far, we have set SO equal to BO. Here, we let SO be smaller than BO so there is an inactive period between superframes. Given the same packet arrival rate, it implies the traffic intensity and the contention level rises higher in this set of experiments since the traffic focuses on the active period. In particular, we set $SO = 4$ and $BO = 5$, *i.e.*, 50% duty cycle. So the superframes and the inactive periods each have 250ms duration. Fig. 6 shows the collision and drop probabilities.

The results are qualitatively similar to those without the inactive period, implying that the inactive period does not adversely affect TSAR. A noticeable phenomenon in the figure, however, is that TSAR has slightly decreasing collision probability, whereas the standard scheme has increasing collision probability with the downlink arrival rate. Note that with the idle period, the uplink frames arriving during the idle period accumulate, and they are launched as soon as the next superframe starts. In contrast, in the absence of the idle period they are spread over the superframe. The consequence is that with the idle period, the Data Request frames in the latter part of the superframe becomes less interfered by the uplink transmissions. Also note that the increase of the downlink traffic intensity means more and more superframe slots are utilized by TSAR, making the probability of encounter with the uplink transmissions lower as we go to the latter part of the superframe. So the average probabilities drop in Fig. 6. Although this effect can mitigate with intensified uplink traffic, it definitely contributes to the stability of the TSAR scheme.

4.2 Energy Consumption and Delay

One aspect of TSAR is that it can increase the delay for downlink data transmission as it intentionally moves all but one Data Request frame transmissions after the first superframe slot. But then again the backoff and collision probabilities are lower than the standard scheme, affecting the delay. So it is not straightforward to estimate the impact of TSAR on the delay performance. In this section, we weigh the benefit and cost of TSAR by comparing its delay and energy preservation performance with that of the standard scheme.

Fig. 7 compares the average downlink delay of the standard scheme with that of TSAR. Due to the indirect transmission, the downlink frame should wait for the next beacon frame for it to be notified to the device. Assuming uniform packet arrival, the mean of this delay is half the beacon interval. In this experiment, we set $BO = 4$ so it is 125ms in our setting. This delay is constant regardless of the arrival rate. After the beacon frame, the downlink frame waits for the request frame from the device and then it is transmitted. This delay is influenced by the backoff and collision probabilities and the instance that the

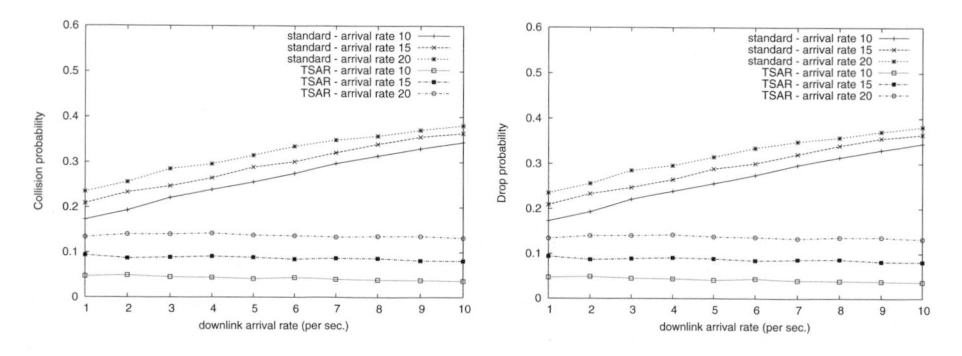

Fig. 6. Collision and drop probability with inactive period and uplink traffic

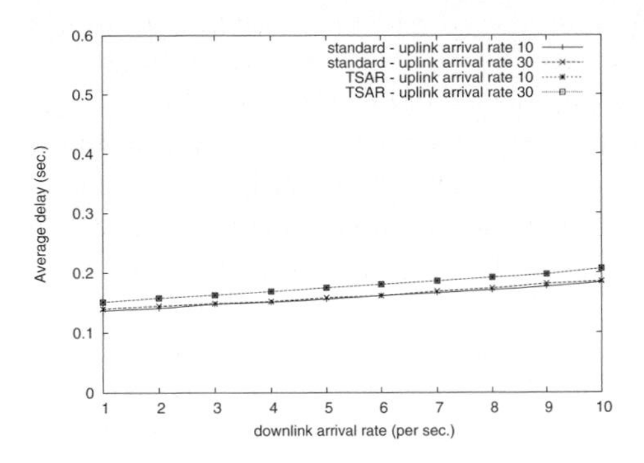

Fig. 7. Average delay comparison

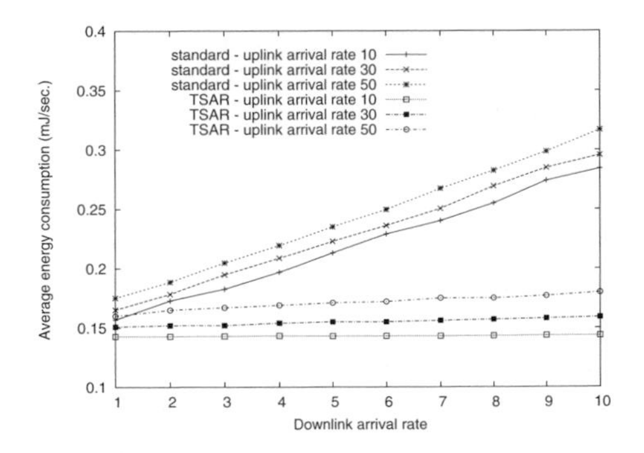

Fig. 8. Average energy consumption comparison

device jumps into contention for the request frame transmission, so depends on the channel access mechanism. From the simulation result, we observe that the former delay that TSAR does not affect contributes the majority of the total delay. TSAR adds only about 10ms to what the standard scheme already incurs. This delay difference is negligible, considering the delay requirements of most IEEE 802.15.4 WPAN applications [2].

Fig. 8 addresses the energy aspect. In this experiment, we measure the energy that the device consumes to receive one downlink data frame. The energy consumption in each state is based on Table 2 [3]. We observe that the energy consumption of the standard and TSAR is similar when the downlink traffic is low, but the gap grows as it intensifies. TSAR performance is about 50% superior when downlink arrival rate is 5 frames per second compared with the standard. It is mainly because the device with the standard scheme stays longer with its receiver on than the device with TSAR. In the standard scheme, requesting devices contend with each other with increasingly larger number of collisions with traffic intensity so the time to receive their downlink frames grows. But in TSAR, since the requesting devices do not contend with each other, the probability of such event happening is much lower even considering the interference from the uplink traffic.

5 Conclusion

The IEEE 802.15.4 is a WPAN PHY/MAC standard that regards the efficient use of the battery as a prime consideration. But the indirection transmission mechanism in the 802.15.4 turns out to harbor inadequacy that causes energy waste when moderate downlink traffic exists such as 3 or 4 frames per superframe. Our scheme, TSAR, solves this problem without the modification of the standard specifications. As such, our scheme can be implemented in the ZigBee-compliant devices. The experiment results show that TSAR always saves energy over the current standard with minimal delay cost. TSAR achieves this without adversely affecting the performance when the traffic is light.

Acknowledgement

This work was supported by grant No. R01-2006-000-10510-0 from the Basic Research Program of the Korea Science & Engineering Foundation.

References

1. IEEE Standard for Part 15.4: Wireless Medium Access Control Layer (MAC) and Physical Layer (PHY) Specifications for Low Rate Wireless Personal Area Networks (LR-WPANS) (October 2003)
2. ZigBee Alliance, `http://www.zigbee.org`
3. Chipcon data sheet for 2.4GHz 802.15.4, `http://www.chipcon.com/files/CC2420_Data_Sheet_1_3.pdf`
4. The ns-2 simulator, `http://www.isi.edu/nsnam/ns/`
5. Misic, J., Shafi, S., Misic, V.B.: Avoiding the Bottlenecks in the MAC Layer in 802.15.4 Low Rate WPAN. In: Proceedings of IEEE ICPADS (2005)

Successful Cooperation between Backoff Exponential Mechanisms to Enhance IEEE 802.11 Networks

Jose Moura[1], Valter Pereira[2], and Rui Marinheiro[1]

[1] ADETTI/ISCTE, Av. Forças Armadas,
1600-082 Lisboa, Portugal
{jose.moura,rui.marinheiro}@iscte.pt
[2] ADETTI/ISCTE, Av. Forças Armadas,
1649-026 Lisboa, Portugal
{vdc.pereira}@gmail.com

Abstract. The apparently never ending widespread of wireless networks requires an enhanced multimedia application support. In this paper, we show how to improve the performance of EDCA in terms of throughput, frame delay and station scalability, through the effective cooperation of two backoff mechanisms: EDM and CWD. Simulation results confirm our previous analytical study and enable the discussion about how to improve relevant Quality of Service metrics, considering different channel bit rate and network load.

Keywords: QoS, EDCA, Simulation, MAC, Throughput, Delay, Performance.

1 Introduction

In the last years, the use of multimedia applications has increased significantly. This fact imposes firm requirements on QoS metrics, such as throughput and delay. The IEEE 802.11e [2] amendment has appeared to support the QoS requisites in IEEE 802.11 networks [1]. Some previous works had studied several ways to manage the backoff algorithm in 802.11 networks, for improving the network performance [6][7][8][9][10][13].

In this work, we show that using traffic differentiated Exponential Decrease Mechanism (EDM) and a Contention Window increase during Defer periods (CWD) mechanism, it is possible to enhance EDCA's performance.

EDCA with EDM and CWD mechanisms, designated in this paper as Differentiated Exponential Collision Recover - DECR, as recently proposed [13], decrease Average Backoff Time (ABT), improving some QoS metrics, namely throughput, frame delay and multimedia station scalability, supporting provision for voice and video traffic.

The rest of this paper is as follows: the section 2 briefly describes related work; in section 3, we study how to enhance the performance of EDCA; in section 4, we validate the study with simulation results; in section 5, we point out our conclusions as well as future work.

T. Vazão, M.M. Freire, and I. Chong (Eds.): ICOIN 2007, LNCS 5200, pp. 425–434, 2008.

2 Background

The 802.11 only provides channel access with equal probability to all contending stations, not supporting different traffic QoS requisites [11][12]. The 802.11e amendment introduces a new function designated by Hybrid Coordination Function (HCF) that enhances 802.11. It introduces traffic differentiation based on priority, which can fulfil different QoS requisites. The HCF provides both a contention-based channel access function (Enhanced Distributed Channel Access - EDCA) and a controlled channel access one (HCF Controlled Channel Access - HCCA). The latter provides deterministic scheduling in both the Contention Periods (CPs) and Contention Free Periods (CFPs), and typically, it supports a large number of stations. The former manages the network access contention in a distributed way, and typically, it only supports a limited number of stations.

The EDCA maps eight user priority levels to four Access Categories (ACs). Therefore, each EDCA station has four frame queues and channel access functions, because there is one function that controls each queue. The EDCA uses new parameters, namely AIFS[AC], CWmin[AC] and CWmax[AC] for ensuring a different access service for each AC. The AIFS[AC] enables a distributed time division mechanism that reserves exclusive time intervals, one for each AC. In this way, there are a low number of queues in competition to channel access during each AIFS[AC] interval. In addition, the EDCA can differentiate the channel access between four traffic types: background, best effort, video and audio. On the other hand, the CWmin[AC] and CWmax[AC] parameters help to build a Contention Window (CW), i.e. [0, CW[AC]. This window enables the generation of a random BT, in order to ensure a collision avoidance mechanism between frames with equal priority but from different stations. When one AC queue receives the first frame, the AC function checks if the medium is idle during an AIFS[AC] time interval. Then, the EDCA function transmits the frame, if: i) The BT for that EDCA function has a value of zero; ii) The actual transmission function is not contending the channel access with another internal higher priority function. During contention, if the AC function senses the medium busy, it stops the running backoff process until the medium becomes again idle. This backoff mechanism of EDCA has a strong impact on throughput and frame delay, as we will see during this work. Our actual research tries to enhance it for improving QoS metrics, like throughput and delay.

One can find relevant survey work in [4][5], which classify many approaches to enhance wireless 802.11 networks and provide QoS. Reference [5] shows a hierarchical taxonomy of QoS mechanisms that enable service differentiation in 802.11 networks. Following their taxonomy, the proposals compared in this paper are priority-based methodologies, using backoff algorithms or CW differentiation.

We have seen some proposals that modify the backoff functionality. Some of them change the way to determine the CW [6][7][8]. Others propose new ways to determine the Backoff Timer (BT), introducing an exponential trend on their algorithms [9][10][13]. As the authors in [10] observed, EDCA performs badly for heavily loaded channels due to unavoidable collisions and idle slots created by contention access mechanism (i.e. backoff) and, for this reason, they presented a new proposal, the Adaptive Fair EDCF (AFEDCF). Their proposal was based in the Fast Collision Resolution (FCR) proposal [9] and the AFEDCF's novelty is a dynamic backoff threshold, distinct for each AC, which separates a linear backoff decrease stage from an exponential one, recalculated according to channel load. AFEDCF also adopted an FCR´s mechanism to improve access fairness: the CW is enlarged and a new BT is determined after each channel busy slot (CWD). This gives more transmission opportunities to high priority flows, because of their low CW_{max} value [10].

DECR - Differentiated Exponential Collision Recovery [13] proposal uses CWD but decreases the BT exponentially with a distinct exponent for each AC (EDM mechanism), instead of the EDCA linear BT decrease function. This difference and the contribution of CWD (not used by EDCA) enables DECR to enhance EDCA´s performance, namely in throughput, dropped frames and average delay [13]. When comparing DECR and AFEDCF, the former does not need to calculate a backoff threshold for each time slot and traffic class as the latter does. Therefore, DECR is easier to deploy than AFEDCF because the former introduces fewer changes to EDCA. In Section 4, we show that DECR decreases the channel idle time more efficiently than AFEDCF.

3 Enhancing EDCA´s Performance

Now, we study how to enhance throughput, frame delay and Average Backoff Time (ABT) in EDCA mode. Following [14], we assume that EDCA's throughput (i.e. S^h), for an AC of priority h and between two consecutive transmissions, is given by (1).

$$S^h = \frac{E[\mathit{Payload\ information\ successfully\ transmited}]}{E[\mathit{Length}\text{ of time }\mathit{between\ two\ consecutive\ transmissions}]} =$$

$$= \frac{\sum_t \sum_{h=0}^{3} \tau_t^h p_{succ,t}^h L^h}{\sum_t \tau_t^h \left(\sum_{h=0}^{3} T_s^h p_{succ,t}^h + T_c^h p_{coll,t} + aSlotTime \cdot p_{idle,t} \right)} \tag{1}$$

In (1), we have used the same symbols as in [14]. *E[x]* means the average value of x. τ_t^h is the probability that one station of priority h transmits at slot t. L^h is the packet's payload. T_s^h is the average time that the channel needs to perform a successful transmission. T_c^h is the average time that the channel deals with a collision. $p_{succ,t}^h$ (2), $p_{iddle,t}$ (3) and $p_{coll,t}$ (4) are respectively, in a slot t and priority h, the probability of a successful transmission, the channel being idle and a collision to occur.

$$p_{succ,t}^h = \binom{N^h}{1} \tau_t^h \left(1-\tau_t^h\right)^{N^h-1} \prod_{l \neq h} \left(1-\tau_t^l\right)^{N^l} \tag{2}$$

$$p_{idle,t} = \prod_{h=0}^{3} \left(1-\tau_t^h\right)^{N^h} \tag{3}$$

$$p_{coll,t} = 1 - p_{idle,t} - \sum_{h=0}^{3} p_{succ,t}^h \tag{4}$$

From (1) the throughput increases when one decreases either $p_{idle,t}$ or $p_{coll,t}$. For example, we could try to minimize $p_{idle,t}$ (i.e. using EDM mechanism) in a way that $p_{coll,t}$ does not increase so much (i.e. using CWD mechanism), and therefore the throughput could be increased. This effective interaction is behind the backoff process of DECR, which is represented in Fig. 1 through a Markov Chain.

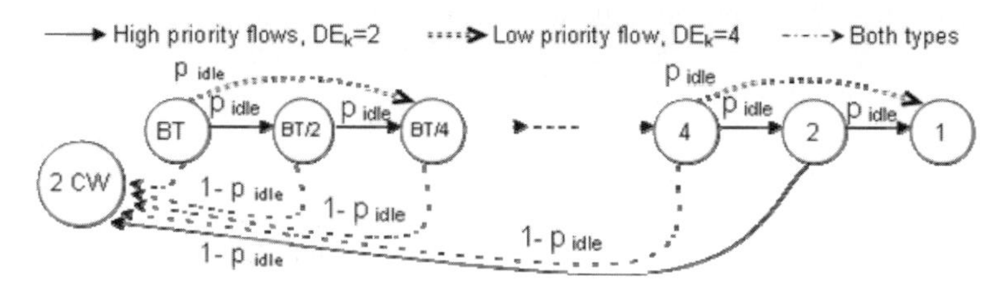

Fig. 1. Representation of DECR backoff process by a Markov chain

When a BT is created, its value is chosen randomly from the range [1, CW[AC]+1]. After that, whenever the medium is sensed idle (i.e. p_{idle}), for a time slot, the BT will be decreased exponentially (EDM) using the DE_K parameter. This action will obviously increase the transmission probability (τ_t^h) when compared to EDCA, since ABT will now be smaller (i.e. the channel idle time is decreased). As a disadvantage, this action will increase the collision probability. We believe that the solution for this problem is to, with a probability of $1 - p_{idle}$ (i.e. channel is busy), increase the CW (CWD) and restart the BT. By doing so, we force the BTs to be recalculated in a wider range. Then, in theory, we are probabilistically distributing the BT values far apart from each other, reducing the collision probability. One should note that the CW increase in Fig. 1 is noted by "2CW" state. One could preview a starvation phenomenon as a side effect of CWD, but the EDM mechanism works against this. Therefore, EDM and CWD can cooperate in order to enhance EDCA´s throughput.

The frame delay is the time elapsed between the start of the backoff process of a frame and its successful transmission. In order to decrease the frame delay, one could try to decrease the channel access delay. In this way, it is necessary to decrease the idle time. As we explained before, the EDM ensures idle time reduction, and the CWD mechanism is an effective solution to compensate the collision problem introduced by EDM. One could preview a CWD negative impact over the frame delay, essentially in traffic with higher CW_{max} value, but the EDM mechanism works against this. Therefore, one commitment between both mechanisms could enable bounded delay values.

As mentioned before, DECR decreases the channel idle time. The most prominent mechanism for this result is the existence of EDM in DECR. We will then perform a study in order to compare EDCA's ABT with DECR's ABT, without considering the existence of busy periods, ignoring the CWD mechanism in DECR and defer periods in EDCA. This will allow us to evaluate the influence of EDM on DECR´s idle backoff time and make a fair comparison with EDCA. In (5), we have calculated EDCA's ABT considering that the BT is chosen within the range [1, CW+1], keeping in mind that in EDCA, the BT is decremented linearly, and the channel is consecutively idle after each backoff process starts. This CW range is slightly different from the EDCA proposal for the sake of fair comparison among all proposals tested in our work. Also, note that w means CW_{max}. In (6) ABT has been calculated for the EDM mechanism, considering that, during each idle slot, one should divide the previous BT value by the DE_k value.

$$ABT^{EDCA}_{[1,cw+1]} = \frac{\sum_{i=0}^{w}(w+1-i)}{w+1} = \frac{(w+1)\times(w+2)}{2\times(w+1)} = \frac{w+2}{2} \tag{5}$$

$$ABT^{EDM}_{[1,cw+1]} = \frac{\sum_{i=0}^{w} \log_{DE_k}(w+1-i)}{w+1} = \frac{2\times(w+1)\times(\ln(w+1)-1)+\ln(w+1)+\ln(2\times\pi)}{2\times(w+1)\times\ln(DE_k)} \tag{6}$$

Table 1 clarifies the ABT value difference between EDCA and DECR, for each AC, when CW_{max} values suggested in [2] are used. DECR has lower ABT than EDCA, especially in background AC (i.e. CW_{max}=1023), as EDM uses a higher DE_k value for this AC.

Table 1. ABT values

AC	DE_K	CW_{max}	ABT^{EDCA}	ABT^{EDM}	$DE_{K,min}$
voice	2	15	8.5	2.765	1.253
video	2	31	16.5	3.667	1.167
Background	4	1023	512.5	4.282	1.012

Then, it is determined the minimum DE_k that gives the same ABT for both EDM and EDCA (7). Solving (7), we get the minimum value for DE_k - $DE_{k,min}$ for each AC as shown in (8). $DE_{k,min}$ is the value that should be used by an AC queue to exponentially decrease BT in order to obtain an ABT similar to the one achieved with the linear BT decrease of EDCA (see Table 1). This is a very important outcome since it implies that the linear decrease approach is a particular case of the exponential decrease approach. In addition, it should be expected a similar performance for both approaches, when $DE_k = DE_{k,min}$, as later results in section 4 will prove this.

$$ABT^{EDM}_{[1,cw+1]} \leq ABT^{EDCA}_{[1,cw+1]} \tag{7}$$

$$DE_{k,\min} = e^{\frac{2\times(w+1)\times(\ln(w+1)-1)+\ln(w+1)+\ln(2\times\pi)}{(w+1)\times(w+2)}} \tag{8}$$

4 Evaluation

We have implemented EDCA [2], AFEDCF [10] and DECR [13] in NS-2 [3][10] to simulate each one, using the wireless network topology shown in Fig. 2. We have used the topology of [10] to make a fair comparison of results. Each station sends three different flow types (e.g. voice, video, background) to one Access Point (Station 0), using the basic access mode, without RTS/CTS control messages, because the hidden problem is not a concern. We have also assumed no errors on the channel; all stations were stationary; power consumption was not relevant. The used traffic types will allow us to have a picture of throughput and delay behaviours for different application requirements and proposals. The voice traffic had the lowest throughput value, the video traffic had the highest throughput and the background traffic had almost the same throughput of video. All traffic was Constant Bit Rate and sent using UDP. We have followed the default EDCA suggestion for MAC parameters (see Table 2). Each scenario has run during a simulated time of 15 s. We have used for DECR simulation the DE_k values used in [13].

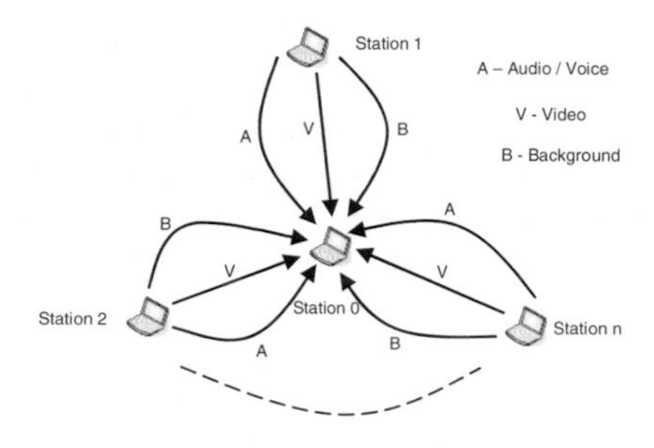

Fig. 2. Network Topology

Table 2. MAC and Flow Parameters

	Voice	Video	Back.
Transport	UDP	UDP	UDP
Priority	3	2	0
CWmin	7	15	31
CWmax	15	31	1023
AIFSN	2	2	7
Frame Size (Byte)	160	1280	1500
Frame Interval (ms)	20	10 ms	12.5
Flow rate (KByte/s)	8	128	120

We have studied how each proposal (i.e. EDCA, DECR and AFEDCF) reacts at different channel bit rates (e.g. 24, 36, 56 Mbps) and traffic loads. To help the comparison of results, we have considered the following three channel conditions perceived individually by each proposal: the Lightly Loaded (LL), Moderately Loaded (ML) and Heavily Loaded (HL) condition. For each proposal, the channel is in LL condition as the network is always capable of transmitting the entire additional traffic load (i.e. due to each new station incorporation) and the throughput increases at the same rate as the load increase. When a proposal that does not rise its throughput at the same rate of load amount, one can conclude that the channel is entering in a new condition, the ML (i.e. the channel only partially accepts the load increase). When the throughput decreases for load increase, the channel is in HL condition. In Fig. 3, all the studied proposals presented the above three load conditions, independently of channel bit rate, except for 56 Mbps where the HL condition is only visible in EDCA. Note that at least until 7 stations, each proposal forces the channel to be in LL condition (not shown in Fig. 3). We will focus the following explanation for the 36 Mbps scenario and with load variation. Typically, AFEDCF is the first proposal to change from LL to ML (9 stations). The first proposal to change from ML to HL is normally EDCA (11 stations), when also EDCA obtains its maximum throughput value. Furthermore, after 13 stations, EDCA's throughput becomes lower than AFEDCF's result and after this presents the worst throughput of all proposals. DECR is the proposal that enters in ML and HL for higher loads (respectively, 11 and 16 stations). Moreover, during the ML condition, DECR's total throughput has the highest value of all proposals. In addition, DECR keeps the total throughput close to its maximum (2500 KB/s) for a wider range of loads (12 – 16 stations) than others do.

One can confirm that during the LL channel condition, DECR's idle time (Fig. 4) decreases linearly as the load increases. After 11 stations (ML condition begins here), the decrease of idle time becomes less significant, while the collisions still have a marginal increase. This explains why throughput for DECR reaches and keeps a maximum value (see Fig. 3), starting from this load level. A side effect of EDM (explained in section 3) is a potential increase of collisions. However, as also previously explained, this increase is compensated by the CWD mechanism. This also justifies, in our opinion, why DECR keeps the total throughput near to its maximum for a wider range of channel loads. However, after a certain load (16 stations) the previous compensation mechanism is not sufficient and the number of collisions increases to more than 10% (HL condition begins here). As EDCA does not have the CWD mechanism to compensate collisions' increase, in Fig. 4 is visible that the HL condition for EDCA begins earlier (11 stations) than in DECR/AFEDCF (16 stations). It is also visible that using CWD has a drawback, which is a slight increase of idle time for DECR/AFEDCF relatively to EDCA (Fig. 4, between 11 and 16 stations). However, this last drawback is attenuated by exponential decrease backoff mechanisms, which justifies why DECR/AFEDCF perform better than EDCA. In addition, DECR has lower idle time values than AFEDCF (11 – 16 stations) because the former always uses EDM to decrease BT and the latter only partially uses it and with a lower DE_k value, which justifies why DECR has better throughput trend than AFEDCF. Finally, one can conclude that EDCA could improve its throughput as it adopts both CWD and EDM.

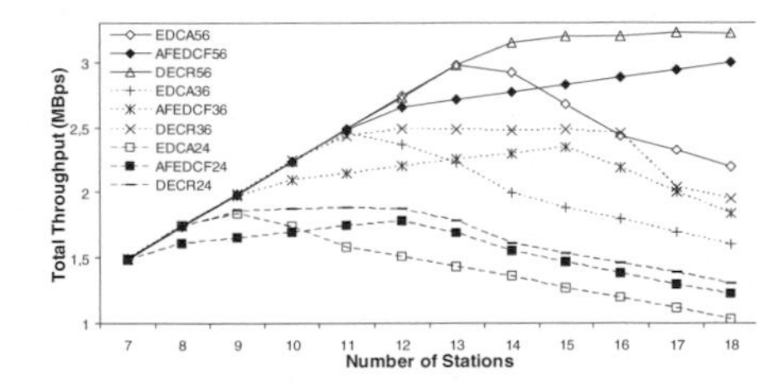

Fig. 3. Total Throughput 24 - 56 Mbps

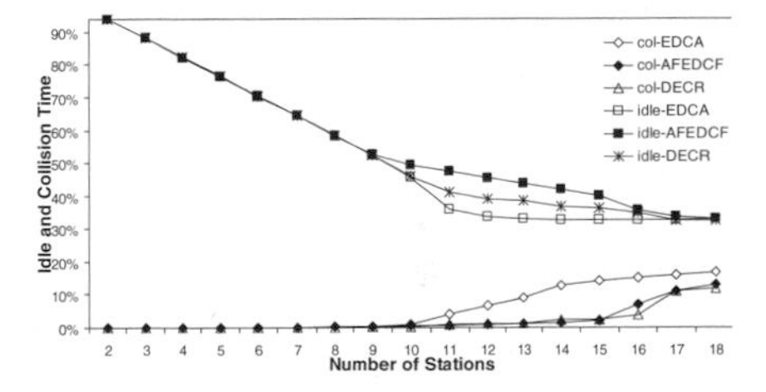

Fig. 4. Idle & Collision Time at 36 Mbps

Now, for each proposal, we have analyzed the frame delay results with 14 stations and a 36 Mbps bit rate, which according to [10] represents an 80% channel load. We have obtained the cumulative fraction function for delay. This function is a graphical presentation of how delay is distributed. For any delay value x, its cumulative fraction value is the fraction of delay values that is strictly smaller than x. Then, we have obtained the average (i.e. considering all stations) cumulative fraction delay for each AC. We have also decided, for clarity purposes, to show only the delay values corresponding to each 5% increment of cumulative fraction value. Fig. 5 presents the cumulative fraction delay for voice. EDCA is the proposal that has the highest delay values, e.g. 50% of EDCA frames suffer a delay value less than 3.6 ms, while AFEDCF and DECR frames suffer respectively less than 0.7 and 0.8 ms. These result differences happen because EDCA misses the CWD mechanism, and it has more collisions than others, as the network has more load. The previous EDCA increase on collisions reflects also its worst behaviour in frame delay. The EDM absence in EDCA is also partially responsible for its bad delay result because in this way the channel access delay is increased.

In video traffic (Fig. 6), the difference between EDCA and other proposals is even larger. Using EDCA, 50% of frames have less than 455 ms. While AFEDCF and DECR frames have approximately 1.4 ms. One can use the same CWD/EDM argumentation initially used for audio but now for video traffic. The worst delay results of EDCA are more visible now than previously, because the video traffic has a lower priority than audio. In this way, the CWD lacks effect in EDCA is more evident for video, as there are more collisions in video than in audio.

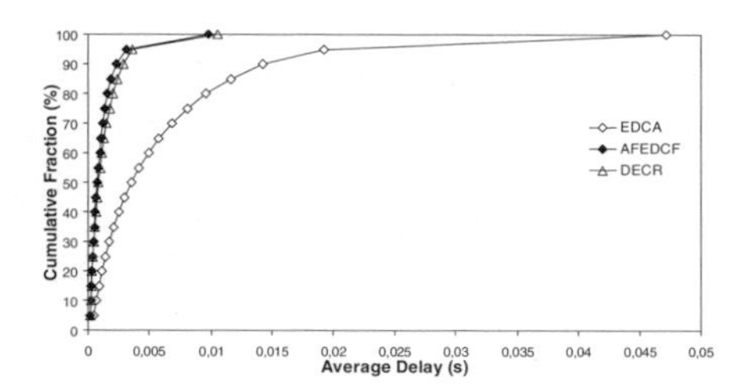

Fig. 5. Voice Cumulative Fraction Delay

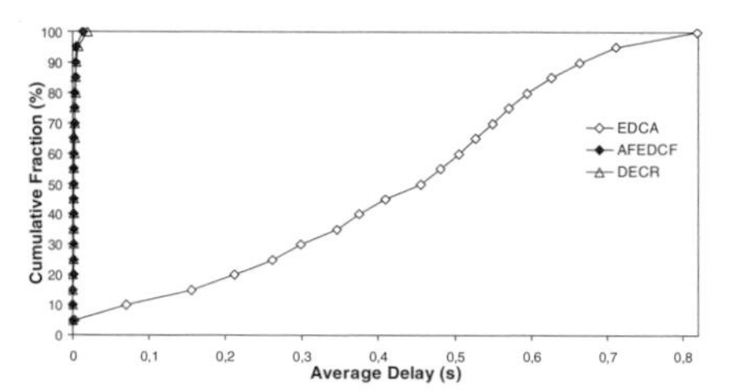

Fig. 6. Video Cumulative Fraction Delay

From Fig. 7, we see that until 20% of background frames, EDCA has delay values similar to DECR. After 20%, DECR becomes clearly the best proposal. AFEDCF by its turn becomes better than EDCA for more than 40% but presents always results worst than DECR does. This happens because the EDM mechanism (only partially used in AFEDCF and with a lower exponent decrease for background traffic) is an efficient mechanism to reduce the channel access delay (one component of frame delay). Here, it is also visible the drawback on EDCA behaviour, because as it does not use the EDM and neither the CWD, it presents the worst results of all.

In Table 3, we show the maximum average delay for 80% and 90% of sent frames. It is clear that in voice and video, AFEDCF has slightly lower delays than DECR. EDCA has higher delay values especially in video flows, confirming the results of Fig. 6. For Background flows, DECR is the best proposal, e.g. for 90% of transmitted frames, their delay is 50% lower than EDCA and 25% lower than AFEDCF. From this delay study, we can conclude that in terms of frame delay distribution, DECR and AFEDCF have good multimedia results. EDCA has the poorest results, especially in video traffic because it lacks the CWD and EDM mechanisms. In background data, DECR has clearly the lowest delay values for more than 20% of transmitted frames. From these results, one can conclude that EDCA could improve the frame delay as it adopts both the CWD and EDM mechanisms (i.e. DECR). In this way, we have validated the delay conclusions of section 3.

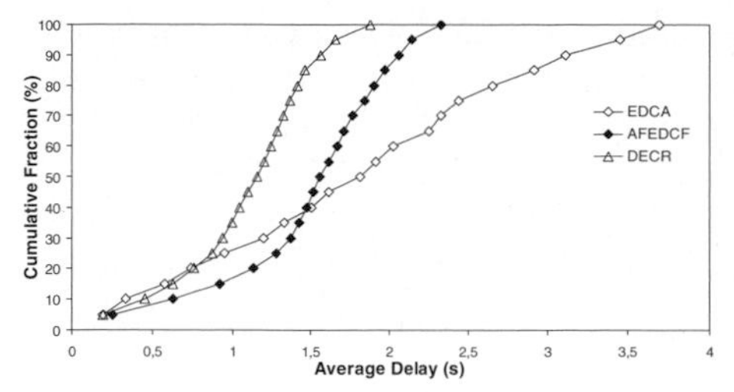

Fig. 7. Back. Cumulative Fraction Delay

Table 3. Max. Average Delay (ms) (DECR/EDCA/AFEDCF)

		Voice	Video	Back.
Maximum average delay for ***x*** **% of frames**	80%	2	3.5	1414
		9.6	594	2653
		1.6	2.6	1897
	90%	2.9	5.2	1561
		14.2	662	3116
		2.3	3.7	2057

We have made a couple of simulations intended to validate the ABT analysis made in Section 3. We have chosen the 6 Mbps channel bit rate and the following four proposals: EDCA, $DE_{k,min}$ (i.e. DECR without CWD and using $DE_{k,min}$), $DE_{k,min-CWD}$ (i.e. DECR using $DE_{k,min}$) and DECR with DE_k values used in [13], and shown in Table 1. Fig. 8 and 9 present results about multimedia average delay of all queues. Delay achieved by EDCA and $DE_{k,min}$ are identical, both in voice and video flows. CWD/EDM have a positive impact on mean delays, implying that, for instance lower values of delays (< 300ms), the channel supports more audio stations: the presence of CWD (EDM) enables approximately 2 (3) more stations than in EDCA. This is quite significant for a low channel bit rate. For video, one additional station is possible due to CWD.

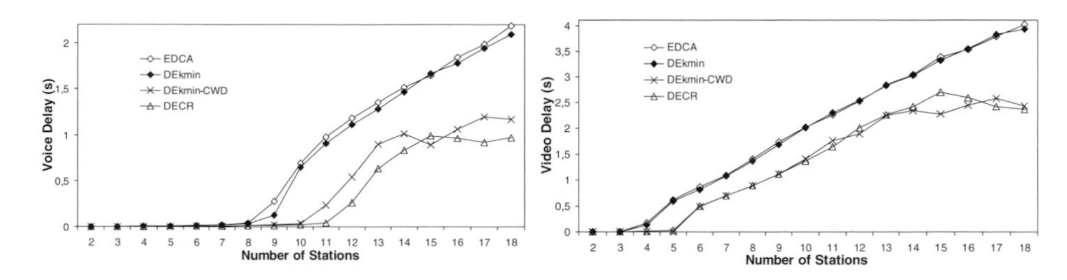

Fig. 8. Voice Delay at 6 Mbps

Fig. 9. Video Delay at 6 Mbps

These results prove our initial expectation, addressed in section 3, that EDCA trend can be replicated with an exponential function for BT decrease. This makes the linear BT decrease of EDCA a particular case of the exponential BT decrease, when $DE_k = DE_{kmin}$. We recall that these DE_{kmin} values were, one for each AC, previously calculated in section 3 considering DECR (without CWD) and EDCA with the same ABT. Another important conclusion is that EDCA's performance can be improved choosing proper DE_K values, i.e. with $DE_k > DE_{kmin}$. In addition, one should note that EDM/CWD seems to enable network's scalability.

5 Conclusions and Future Work

We have shown how to enhance the EDCA mode of IEEE 802.11e in terms of throughput, delay and station scalability, as it adopts both EDM and CWD mechanisms. Our results indicate that the linear Backoff Timer decrease of EDCA is a particular case of the exponential Backoff Timer decrease, as for each traffic type the following is valid: $DE_k = DE_{kmin}$. These results confirmed conclusions driven from our analytic study. Another conclusion is that EDCA's performance can be improved using EDM and proper DE_K values, i.e. with $DE_k > DE_{kmin}$. On the other hand, CWD seems to compensate the potential negative effect of increasing collisions due to EDM, principally in a network with a large number of stations. Finally, one should note that EDM/CWD seems to help the network's scalability.

Some preliminary work on proper DE_K value for each traffic type, in order to optimize EDCA throughput has been already done in [13]. One can envision a further investigation on other metrics, like frame delay and multimedia stations scalability.

References

1. IEEE Std 802.11: Local and metropolitan area networks. Specific Requirements Part 11: Wireless LAN MAC and PHY (1999)
2. IEEE 802.11e-2005: Local and metropolitan area networks. Specific requirements Part 11: Wireless LAN MAC and PHY. Amendment 8: MAC QoS Enhancements (2005)
3. NS-2 simulator, `http://www.isi.edu/nsnam/ns/`
4. Zhu, H., et al.: A Survey of Quality of Service in IEEE 802. 11. Networks. IEEE Wireless Communications 11, 6–14 (2004)
5. Ni, Q., Romdhani, L., Turletti, T.: A Survey of QoS Enhancements for IEEE 802.11 Wireless LAN. Journal of Wireless Comm. and Mobile Computing 4, 1–20 (2004)
6. Romdhani, L., Ni, Q., Turletti, T.: Adaptive EDCF: Enhanced Service Differentiation for IEEE 802.11 Wireless Ad Hoc Networks. In: IEEE WCNC, vol. 2, pp. 1373–1378 (2003)
7. Zhu, H., et al.: EDCF-DM: A Novel Enhanced Distributed Coordination Function for Wireless Ad Hoc Networks. IEEE Comm. Society 7, 3886–3890 (2004)
8. Wang, C., Li, B., Li, L.: A New Collision Resolution Mechanism to Enhance the Performance of IEEE 802.11 DCF. IEEE Trans. on Vehicular Technology 53, 1235–1246 (2004)
9. Kwon, Y., Fang, Y., Latchman, H.: A Novel MAC Protocol with Fast Collision Resolution for Wireless LANs. In: IEEE Infocom, vol. 2, pp. 853–862 (2003)
10. Malli, M., et al.: Adaptive Fair Channel Allocation for QoS Enhancement in IEEE 802.11 Wireless LANs. In: IEEE ICC, vol. 6, pp. 3470–3475 (2004)
11. Mangold, S., et al.: IEEE 802.11e wireless LAN for quality of service. European Wireless 1, 32–39 (2002)
12. Grilo, A., Nunes, M.: Performance evaluation of IEEE 802.11e. In: PIMRC, vol. 1, pp. 511–517 (2002)
13. Moura, J., Marinheiro, R.: MAC approaches for QoS Enhancement in Wireless LANs. In: Third Workshop in Electronics, Telecommunications, and Computer Engineering, JETC (2005)
14. Salkintzis, A., Passas, N.: Emerging Wireless Multimedia: Services and Technologies, pp. 156–167. Wiley, Chichester (2005)

Knowledge-Based Exponential Backoff Scheme in IEEE 802.15.4 MAC

Sinam Woo[1], Woojin Park[1], Sae Young Ahn[1], Sunshin An[1], and Dongho Kim[2]

[1] Department of Electronics and Computer Engineering, Korea University, Seoul, Korea
{snwoo,wjpark,syahn,sunshin}@dsys.korea.ac.kr
[2] Department of Computer Engineering, Halla University, Wonju, Kangwon, Korea
imi@hit.halla.ac.kr

Abstract. In wireless personal area network, a channel is shared by a number of nodes. Therefore packet collision may take place and it degrades the throughput performance. To improve the throughput of slotted CSMA/CA in IEEE 802.15.4, we propose knowledge-based exponential backoff (KEB) scheme for enhancing channel utilization solely based on the locally available channel state information. In addition, we propose an analytical model of slotted channel access mechanism with finite retry and derive the theoretical throughput limit. In simulation experiments, we show that the existing MAC scheme, binary exponential backoff (BEB) scheme, operates very far from the theoretical limits due to increased time for negotiating channel access. Also performance results indicate that KEB shows significant improvement in throughput performance over BEB with the knowledge of the network status.[1]

Keywords: Throughput, Backoff, Markov Model, IEEE 802.15.4, MAC.

1 Introduction

Wireless sensor networks are appealing to researchers due to their wide range of applications potential in areas such as target detection and tracking, environmental monitoring and industrial process monitoring. However, it is necessary to achieve an efficient medium access protocol for the optimum performance [9]. IEEE 802.15.4 standard [1] leads to the work in wireless sensor networks. In IEEE 802.15.4 networks, a central controller, called the PAN (personal area network) coordinator, builds the network in its personal operating space. The standard uses two types of channel access mechanism, depending on the network configuration. Non-beacon-enabled networks use an unslotted CSMA/CA channel access mechanism. Beacon-enabled networks use a slotted CSMA/CA channel access mechanism, where the backoff slots are aligned with the start of the beacon transmission.

[1] This research was supported by SK telecom in Korea.

T. Vazão, M.M. Freire, and I. Chong (Eds.): ICOIN 2007, LNCS 5200, pp. 435–444, 2008.

In a backoff algorithm, the duration of the backoff is usually selected randomly in the backoff interval. The backoff interval is dynamically controlled by the backoff algorithm. However, to determine the length of the backoff interval is not a trivial task. In the case of a fixed number of nodes, small backoff intervals do not reduce the collision among the contending nodes to a low enough level. It results in a still too high probability of collisions, lowering the channel throughput. On the other hand, large backoff intervals introduce unnecessary idle time on the channel and degrade the performance. The difficulty in designing a good backoff algorithm is to achieve the optimum performance. Many backoff algorithms have been proposed in the literature.

In [6], the authors proposed to tune the backoff window size on the number of stations, which is difficult to predict in the absence of the controller and derived the analytical formula for the IEEE 802.11 protocol capacity. An adaptive DCF that adopts its backoff procedure based on the number of collisions and freezes in IEEE 802.11 DCF was proposed in [4]. When a node succeeds in transmitting a data frame, it selects the preferred stage with high likelihood that the overall network throughput would increase. In [5], the authors proposed the sensing backoff algorithm (SBA) for supporting maximum channel throughput with fair access to active node on a shared medium. Nodes sensing successful packet transmission decrease the backoff intervals and every node that experience packet collisions increases the backoff intervals. Also the transmitter and the receiver of each successful transmission decrease the backoff interval. However, some problems still remain unresolved. In this paper, we study the backoff algorithm using the channel station information in IEEE 802.15.4 protocol. With the knowledge about the number of collisions and busy periods experienced in the recent past, the backoff procedure will be controlled for maximum network utilization.

The paper is organized in the following way. Section 2 discusses the exponential backoff scheme in IEEE 802.15.4. The KEB is introduced in Section 3. Section 4 analyzes the slotted CSMA/CA with finite retry and derives the theoretical throughput limit. Section 5 presents simulation results about KEB and BEB. Finally, we conclude the paper in section 6.

2 Exponential Backoff Scheme in IEEE 802.15.4

The standard uses two types of channel access mechanism: unslotted and slotted CSMA/CA channel access mechanism. The CSMA/CA algorithms shall be used before the transmission of data or MAC command frames transmitted with the contention access period (CAP), unless the frame can be quickly transmitted following the acknowledgement of a data request command.

The algorithm is implemented using units of time called backoff periods, where the backoff period shall be equal to *aUnitBackoffPeriod* symbols. In the slotted CSMA/CA, the backoff period boundaries of every device in the PAN shall be aligned with the superframe slot boundaries of the PAN coordinator, i.e. the start of the first backoff period of each device is aligned with the start of the beacon transmission.

Each device shall maintain three variables for each transmission attempt: NB, CW and BE. NB is the number of times the CSMA/CA algorithm was required to backoff while attempting the current transmission; this value shall be initialized to 0 before each new transmission attempt. CW is the contention window length, defining the number of backoff periods that need to be clear to channel activity (CCA) before transmission can commence; this value shall be 2 before each transmission attempt and reset to 2 each time the channel is assessed to be busy. CW is only used for the slotted CSMA/CA. BE is the backoff exponent, which is related to how many backoff periods a device shall wait before attempting to assess a channel. Fig.1 a) illustrates the steps of the slotted CSMA/CA algorithm. The MAC sublayer shall first initialize NB, CW, and BE and then locate the boundary of the next backoff period. The MAC sublayer shall delay for a random backoff periods and then request that the PHY perform a CCA on a backoff period boundary. The MAC sublayer shall ensure that, after the random backoff, the remaining CSMA/CA operations can be undertaken and the entire transaction can be transmitted before the end of the CAP. If the channel is assessed to be busy, the MAC sublayer shall increment both NB and BE by one, ensuring that BE shall be no more than *aMaxBE*. The MAC sublayer shall also reset CW to 2. If the value of NB is less than or equal to *macMaxCSMABackoffs*, the CSMA/CA algorithm shall return to backoff. If the value of NB is greater than *macMaxCSMABackoffs*, the CSMA/CA algorithm shall terminate with a *Channel Access Failure* status. If the channel is assessed to be idle, the MAC sublayer shall ensure that the contention window has expired before commencing transmission. To do this, the MAC sublayer shall first decrement CW by one and then determine whether is equal to 0. If it is not equal to 0, the CSMA/CA algorithm shall return to channel assessment. If it is equal to 0, the MAC sublayer shall begin transmission of the frame on the boundary of the next backoff period.

3 Knowledge-Based Exponential Backoff Scheme (KEB)

When collisions indicate congestion prevalent in the network, and once present, congestion is unlikely to drop sharply. When contention for the medium is low, the medium remains idle most of the time. Under such circumstances, it is desirable that overheads resulting from backoff procedure are minimum. Due to occasional collisions, it is possible that a node might have backed-off significantly even though the medium is underutilized. In BEB, after successfully transmitting a frame, it set BE equal to 0, as if congestion was non-existent. Therefore, it is desirable to maximize channel utilization solely based on the locally available channel state information. Let n_{tr}^{i} be the transmission counter at time i, which measures the number of times the packet transmits during two successful transmissions; T^{i} is the duration of consecutive packet transmission acknolwdged. Let n_{s}^{i} be the successes counter at time i. Let α be the collision threshold. Finally,

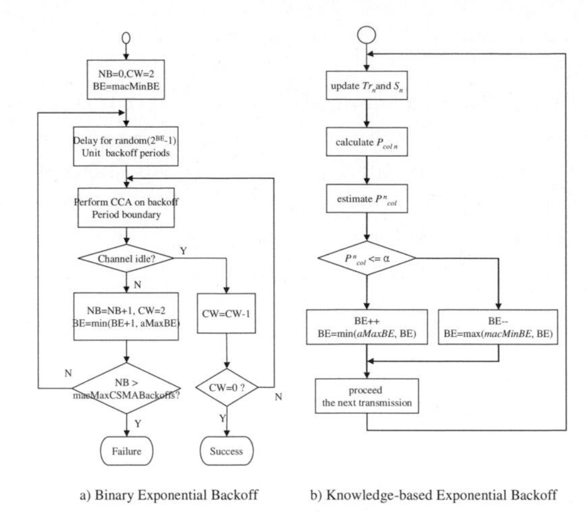

Fig. 1. Binary exponential backoff algorithm versus knowledge-based exponential backoff algorithm

p_c^i is the collision rate at time i, which measures the ratio between n_{tr}^i and n_s^i such as (1).

$$p_c^i = 1 - \frac{n_s^i}{n_{tr}^i} \tag{1}$$

We use exponential weighted moving average (EWMA) to compute the p_c^i at the successful transmission so that the bias against transient measurements of p_c^i can be minimized. Let β be the smoothing factor. Then

$$p_c^i = \beta p_c^{i-1} + (1-\beta)p_c^i, \qquad \beta \in [0,1] \tag{2}$$

The procedure of KEB scheme is presented in Fig.1 b). At each T^i, n_{tr}^i and n_s^i are updated and then p_c^i is calculated. With (2), p_c^i is recalculated to avoid harmful fluctuation. Finally BE is adjusted for degrading the collision probability and achieving the high throughput performance. The parameters (0.5, 0.8) that we choose in section 5 for (α, β) is simulated and compared with some other values. Noting the basic idea behind the KEB scheme, a node chooses the backoff duration with the locally available channel information. When the collision for the medium is low, it keeps the backoff duration to be low. Otherwise, the backoff duration will be increased. With the channel status, KEB can achieve the throughput performance higher than BEB.

4 Throughput Analysis with Finite Retry

In our analysis, we assume that network consists of a finite number of devices and each device always has a packet available for transmission in an ideal case as the characteristic behavior of a backoff algorithm is critical when the channel

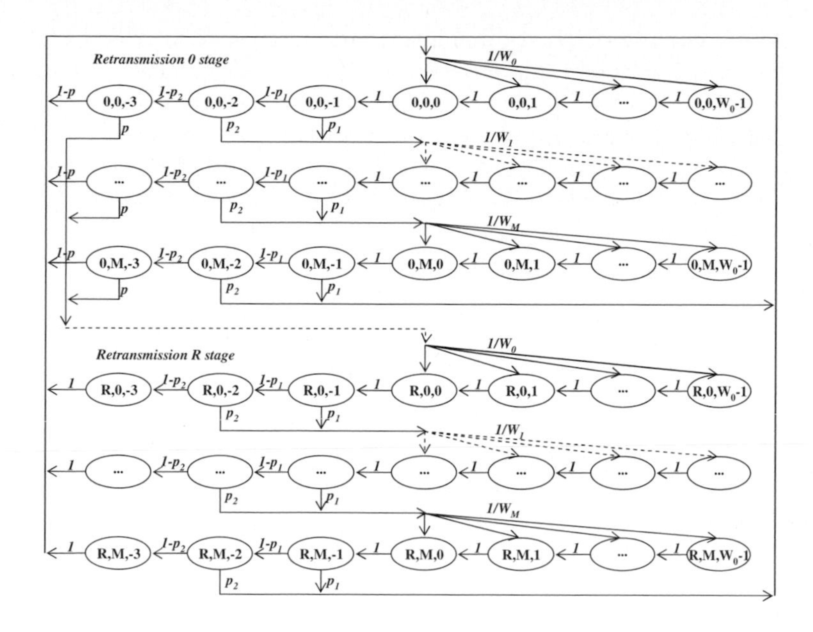

Fig. 2. Markov Chain Model

is heavily loaded. Our Markov model is based on [2] and [8] and is extended with finite retry.

We model the operation at an individual node using the state diagram in Fig.2, in which each node is in one of a number of backoff states in steady state. p denotes the probability that a transmission experiences a collision. p_1 and p_2 denote the probability that a channel is sensed busy at the first and the second CCA respectively. To analyze the behavior of IEEE 802.15.4 MAC, we introduce a number of random variables. Let $r(t)$ be a random process representing the number of retransmission at time t; the maximum value is defined as *aMaxFrameRetires*. Let $n(t)$ be a random process representing the number of backoffs at time t; it belong to the range 0 to *macMaxCSMABackoff*-1. Let $b(t)$ be a random process representing the backoff periods for a given node at time t which can take any value in the range 0 to $2^{BE}-1$.

Then, we develop three-dimensional Markov chain with $\{r(t), n(t), b(t)\}$. Let $P_{(i,j,k)(i',j',k')}$ be the number of retransmission going from i to i' , the backoff stage going from j to j', and the backoff period from k to k'. The transition probabilities are listed as follows. The value of R is defined by *aMaxFrameRetires*-1, and M represents the maximum value of NB, *macMaxCSMABackoff*-1.

$$\begin{array}{ll}
P_{(i,j,k)(i,j,k-1)}=1, & i\in(0,R), j\in(0,M), k\in(0,W_j-1) \\
P_{(i,j,0)(i,j+1,k)}=(p_1+(1-p_1)p_2)/W_{j+1}, & i\in(0,R), j\in(0,M-1), k\in(0,W_{j+1}-1) \\
P_{(i,j,0)(i+1,0,k)}=(1-p_1)(1-p_2)p/W_0, & i\in(0,R-1), j\in(0,M), k\in(0,W_0-1) \\
P_{(i,M,0)(0,0,k)}=(p_1+(1-p_1)p_2)/W_0, & i\in(0,R), k\in(0,W_0-1) \\
P_{(i,j,0)(0,0,k)}=(1-p_1)(1-p_2)(1-p)/W_0, & i\in(1,R-1), j\in(0,M), k\in(0,W_0-1) \\
P_{(R,j,0)(0,0,k)}=(1-p_1)(1-p_2)/W_0, & j\in(0,M), k\in(0,W_0-1)
\end{array} \quad (3)$$

The first equation shows the probability that the backoff time is decremented after each *aUnitBackoffPeriod*. The second equation denotes the probability to choose a random duration of the backoff period when the channel is sensed to be busy in both CCA backoff periods. The third equation determines the probability to choose a random duration of the backoff period when there is a collision and a device retransmits a packet. The fourth equation determines the probability to choose a random duration of the backoff period when the channel is sensed to be busy until maximum NB and a packet terminates with a *Channel Access Failure* status. The fifth equation shows the probability to choose a random duration of the backoff period when a transmission succeeds. The sixth equation shows the probability to choose a random duration of the backoff period regardless of channel states.

Let $b_{i,j,k} = \lim_{t\to\infty} P\{r(t) = i, n(t) = j, b(t) = k\}$ be the stationary distribution of the Markov chain for $i \in (0, R)$, $j \in (0, M)$, and $k \in (-3, W_j - 1)$. We now show that the stationary probabilities are calculated recursively through the following algorithm.

$$b_{i,j,0} = A^j b_{i,0,0}, \qquad for \quad i \in (0, R), j \in (0, M) \tag{4}$$

where A is the probability that the channel is sensed busy at either first or second CCA; $p_1 + p_2 - p_1 p_2$ and$B$ is the probability of transmitting the packet after two CCA; $(1 - p_1)(1 - p_2)$. Owing to the chain regularities and using (4), we have

$$\begin{aligned} b_{i,0,0} &= pB \textstyle\sum_{j=0}^{M} b_{i-1,j,0}, \qquad for \quad i \in (1, R) \\ b_{0,0,0} &= (1-p)B \textstyle\sum_{i=0}^{R-1} \sum_{j=0}^{M} b_{i,j,0} + B \sum_{j=0}^{M} b_{R,j,0} + A \sum_{i=0}^{R} b_{i,M,0} \\ b_{i,j,k} &= \tfrac{W_j - k}{W_j} b_{i,j,0}, \qquad for \quad i \in (0, R), j \in (0, M), k \in (0, W_j - 1) \end{aligned} \tag{5}$$

Thus, by equation (4) and (5), all $b_{i,j,k}$ values can be expressed as a function of $b_{0,0,0}$, p, p_1, and p_2. If the normalization condition is imposed such as (6), finally $b_{0,0,0}$ can be calculated.

$$1 = \sum_{i=0}^{R} \sum_{j=0}^{M} \sum_{k=-3}^{W_j - 1} b_{i,j,k} \tag{6}$$

Then the probability to start sensing the channel at a particular time is

$$\tau = \sum_{i=0}^{R} \sum_{j=0}^{M} b_{i,j,0} = \frac{1 - A^{M+1}}{1 - A} \sum_{i=0}^{R} b_{i,0,0} \tag{7}$$

Note that the ongoing packet transmission may have started one or more backoff periods earlier when the backoff counter reaches zero and CCA senses a busy channel. However, when the first CCA senses that the medium is idle but the second one finds it busy, the packet transmission must have started in the same backoff period. Therefore, the probability that the channel is busy at the end of backoff counting in the probability that other devices are in the process of transmitting or waiting for the acknowledgement. The probability is equal to

$$p_1 = (1 - (1 - \tau)^{n-1})(L_P + L_{LIFS} + L_{ACK})(1 - p_1)(1 - p_2) \tag{8}$$

Where L_p is the number of slots for packet size, L_{LIFS} is the is LIFS backoff slots and L_{ACK} is the number of slots waiting for acknowledgement packet. As noted in [8], the probability that the channel busy at the next backoff slot given that it is idle at the current backoff slot can be computed by noting that

$$p_2 = \left(1 - \frac{1-(1-\tau)^n}{2-(1-\tau)^n}\right)\left(1-(1-\tau)^{n-1}\right) \tag{9}$$

And the probability of a collision seen by a packet being transmitted on the channel is

$$p = 1-(1-\tau)^{n-1} \tag{10}$$

τ, p_1, p_2, and p are determined by solving the four simultaneous non-linear equations (7), (8), (9), and (10).

By means of the given model, we calculate the throughput S, which is expressed by the packet size P occupied for a successful packet transmission. We have

$$S = n\tau(1-\tau)^{n-1}P(1-p_1)(1-p_2) = \frac{P}{\frac{2+L-(1+L)(1-\tau)^n - L\tau(1-\tau)^{2n-2}}{(1-(1-\tau)^n+(1-\tau)^{n-1})n\tau(1-\tau)^{n-1}}} \tag{11}$$

Where L is the duration of a complete transmission measured in a backoff slot; $L_P + L_{LIFS} + L_{ACK}$. Finally, we study the theoretical bounds in the analytical model given above. The throughput S is maximized when the denominator is minimized in (11). We obtain the maximum achievable throughput by taking derivative of the denominator and by calculating the throughput in an iterative way. Fig.3 shows the achievable saturation throughput versus the probability to start sensing the channel in a backoff period for different number of nodes n and packet sizes P. Such as (11), it show that τ depends only on the network size and on the parameter P. We see that a small variation in the optimal value of τ leads to a greater decrease in the throughput. Intuitively, the network size has a substantial influence on the performance due to the increased number of collisions.

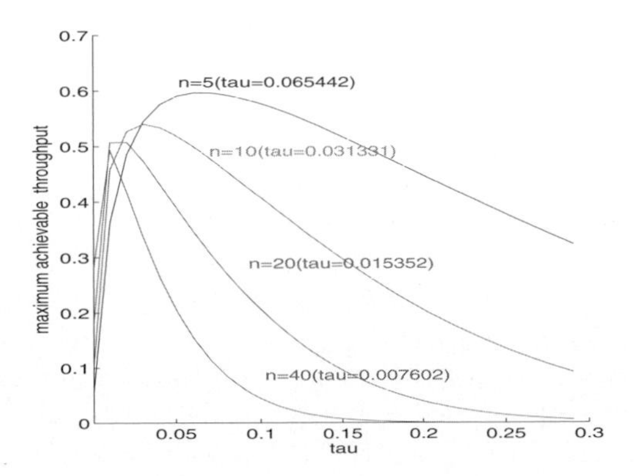

Fig. 3. Probability of starting sensing channel (P=12)

5 Performance Evaluation

We have run simulations to evaluate the performance of KEB scheme and compare KEB with the analytical throughput limit and BEB. The simulation models a network of n nodes that always have data to transmit and every transmitted packet has to be acknowledged, i.e. the packets which do not receive the positive acknowledgement have to be retransmitted.

Each node is fixed in the beacon-enabled star topology, where PAN coordinator is located at the center and the distance between PAN coordinator and nodes is $15m$. The simulation duration is $500s$, the application traffic is CBR, which is popularly used in most simulations, and the radio bandwidth is 250Kbps. We make beacon order (BO) and superframe order (SO) equal, which leads to the active cycle to be 100%; BO is 3. All simulations are run independently and their results are averaged under 5 different seeds. The system parameters used in performance evaluation are shown in Table 1. Note that packet size is represented by a backoff slot.

In Fig. 4, simulation results are presented as line and the analytical results are shown as discrete points. We observe that KEB operates closer to the theoretical limit and outperforms BEB scheme in saturation conditions. Thus, we infer that by using the readily available channel state information, it is possible to significantly improve the performance of slotted CSMA/CA. Note that the parameter set (α, β) in KEB is (0.5, 0.8).

Table 1. System parameters for performance evaluation

MaxPropagationDelay	$0.025ms$
aBaseSlotDuration/aUnitBackoffPeriod	60/20 symbol
macMaxCSMABackoffs	4
aMaxBE/macMinBE	5/3
aMinLIFSPeriod/macAckWaitDuration	40/50 symbol
aMaxFrameRetries	3

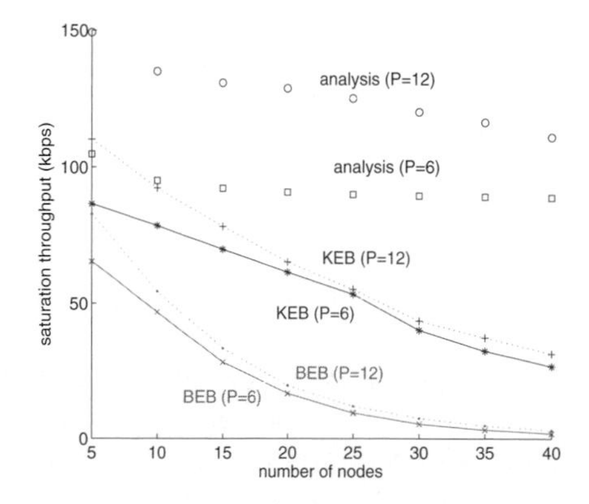

Fig. 4. Performance evaluation under saturation conditions ($\alpha = 0.5, \beta = 0.8$)

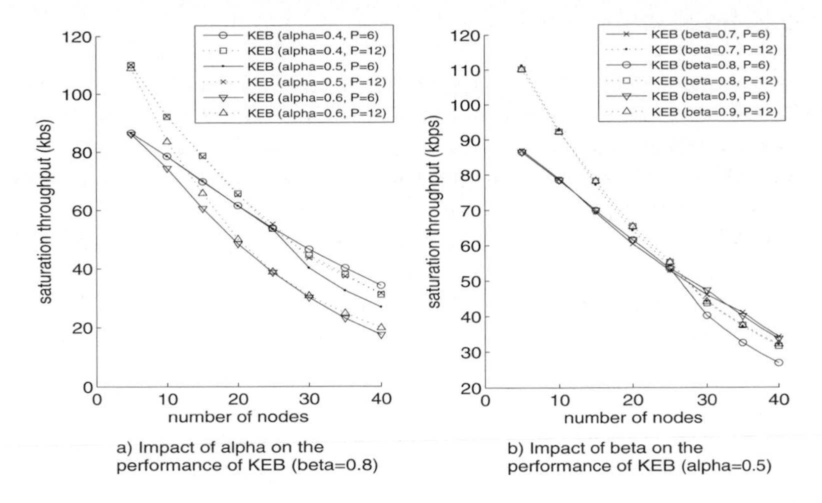

Fig. 5. Impact of parameters on the performance of KEB on the saturation conditions

Fig.5 a) shows the throughput of the KEB scheme with the different values of α. We find that as α increases over 0.5, the throughput performance degrades. Intuitively, when α increases, the frequency of the control for the backoff period becomes low and it results in more collisions. Therefore the throughput is degraded. Fig.5 b) presents the throughput with different β. β is used for biasing the transient network condition and avoiding the fluctuation. Operating with a parameter from 0.7 to 0.9, the throughput performance changes a little. When $\beta < 0.5$, transient condition influences the performance. When $\beta > 0.9$, network conditions in the past play a role in determining the collision probability. Thus, we use $\beta = 0.8$ for all simulation.

6 Conclusion

In wireless personal area network, a single channel is shared by a number of nodes. Packet collisions may take place as results of the random transmissions from active nodes. Therefore backoff algorithm needs to be controlled by the channel status for achieving high channel throughput. These channel information obtained locally can be productively used to predict the level for contention for the channel access.

In this paper, we proposed knowledge-based exponential backoff scheme for enhancing the throughput performance and derived the theoretical throughput limit using Markov model with finite retry. Also, we compared the performance of KEB with that of analytical maximum throughput and BEB through simulations. Under saturation conditions, we observed that KEB operates closer to the theoretical limit and outperformed BEB scheme. Thus, we infer that by using the readily available channel state information, it is possible to significantly improve the performance of slotted CSMA/CA scheme.

References

1. IEEE Standard for Wireless Medium Access Control (MAC) and Physical Layer (PHY) Specification for Low-Rate Wireless Personal Area Networks (LR-WPAN) (October 2003)
2. Bianchi, G.: Performance analysis of the IEEE 802.11 Distributed Coordination Function. IEEE Journal of Sel. Areas Comm. 18(3), 535–547 (2000)
3. Misic, J., Shafi, S., Misic, V.V.: Analysis of 802.15.4 beacon enabled PAN in saturation mode. In: SPECTS 2005, pp. 535–542 (2004)
4. Kuppa, S., Prakash, R.: Adaptive IEEE 802.11 DCF scheme with knowledge-based backoff. In: WCNC 2005, pp. 63–68 (March 2005)
5. Haas, Z.J., Deng, J.: On Optimizing the backoff interval for Random Access Schemes. IEEE Trans. on Comm. 51(12), 2081–2090
6. Cali, F., Conti, M., Gregori, E.: IEEE 802.11 Protocol: Design and Performance Evaluation of an Adaptive Backoff Mechanism. IEEE Journal of Sel. Areas Comm. 18(9), 1774–1786 (2000)
7. Park, T.R., et al.: Throughput and energy consumption analysis of IEEE 802.15.4 slotted CSMA/CA. Electronic letters 41(18) (September 2005)
8. Pollin, S., et al.: Performance analysis of slotted IEEE 802.15.4 medium access layer, Technical Report, DAWN Project (September 2005)
9. Demirkol, I., Ersoy, C., Alagoz, F.: MAC protocols for wireless sensor networks: a survey. IEEE Comm. Mag. 44(4), 115–121 (2006)
10. Samsung/CUNY, ns-2 simulator for 802.15.4, `http://www-ee.ccny.cuny.edu/zheng/pub`

Dynamic Backoff Time Adjustment with Considering Channel Condition for IEEE 802.11e EDCA

Yuan-Cheng Lai, Yi-Hsuan Yeh, and Che-Lin Wang

Department of Information Management
National Taiwan University of Science and Technology
laiyc@cs.ntust.edu.tw, amyyeh@mail.oit.edu.tw

Abstract. Many enhanced schemes of IEEE 802.11e have been proposed to meet the increasing demand for Quality of Service (QoS). Most of them adjust the parameters, such as Contention Window (CW), according to whether the transmission is successful or failed, but don't consider the channel condition (bandwidth) at transmitting frames. In a wireless network, the channel condition is time-varying due to some factors such as station mobility, time-varying interference, and location-dependent errors. In this paper, we propose the Enhanced Distributed Channel Access with Link Adaptation (EDCA-LA) algorithm by adapting with the time-varying channel condition. Our idea is to set the smaller backoff time for a station which has a better channel to achieve higher overall performance. Simulation results show that EDCA-LA outperforms the standard EDCA on throughput and average end-to-end delay.

1 Introduction

In recent years, the need for mobile communications increased rapidly, causing that Wireless Local Area Network (WLAN) gained strong popularity. However, the widespread use of emerging real-time multimedia applications over wireless networks made the QoS support be a key problem. Due to the importance of wireless QoS, IEEE 802.11e, an enhanced version of 802.11, has been proposed to provide the QoS extension in WLANs.

The architecture of IEEE 802.11 standard includes the definitions of the MAC sublayer and physical layer [1]. The IEEE 802.11 MAC sublayer has two access mechanisms: Distributed Coordination Function (DCF) and Point Coordination Function (PCF). The DCF mode is defined as the contention mode and adopts Carrier Sense Multiple Access/Collision Avoidance (CSMA/CA) for medium access. The PCF is known as the contention free mode and uses a centrally controlled polling method to support synchronous data transmission. To support QoS in WLAN, the new standard IEEE 802.11e has been proposed [2]. IEEE 802.11e introduces a new MAC access method named Hybrid Coordination Function (HCF), which consists of two parts: Enhanced Distributed Channel Access (EDCA) and HCF Controlled Channel Access (HCCA).

The existing researches focusing on 802.11e can be classified into two categories: to pursue the QoS achievement (service differentiation) [3-5] and to pursue the performance improvement [6-8]. In pursuing the QoS achievement, AEDCF dynamically

T. Vazão, M.M. Freire, and I. Chong (Eds.): ICOIN 2007, LNCS 5200, pp. 445–454, 2008.

adjusted contention window according to the transmission is successful or failed in order to provide service differentiation [3]. The SBB (Schedule Before Backoff), instead of SAB (Schedule After Backoff) adopted by EDCA, was proposed by using WRR (Weighted Round Robin) and varying CW for EDCA virtual contention [4]. The goal is avoiding the starvation of low priority traffic when load is high. In order to achieve service differentiation and good fairness, the proper AIFSs were calculated for different ACs in [5].

In pursuing the performance improvement, Malli et al. adjusted CW and backoff time according to the collision happening [6]. The main idea is fast increasing CW when the collision is serious and fast reducing CW when the channel is idle. Vollero et al. proposed an approach that QoS Access Point (QAP) uses a specific probability to skip transmitting ACK frame to the original data sender [7]. Chevillat et al. used the ACKs of transmitted frames as the measurement of channel condition [8]. The basic idea of this algorithm is that if the number of consecutive successful transmissions exceeds a threshold, the transmission rate is increased. On the other hand, if the number of consecutive failed transmission exceeds the other threshold, the transmission rate is decreased.

In reality, the channel conditions are time-varying in wireless networks. The 802.11a, 802.11b, and 802.11g provide several PHY modes that combine different coding rates with modulations to support multiple data rates. PHY modes can be dynamically selected through Link Adaptation to achieve a compromise between throughput and error rate [9]. Auto rate schemes proposed in [10] showed significant throughput gain by matching the data rate with the channel condition.

Although the dynamic link adaptation is an excellent approach, transmitting frames at a low data rate will damage the network throughput. Therefore, how to let the frames be transmitted at some higher-bandwidth links is worthily studied. A feasible way dealing with the channel variations in wireless networks is to combine EDCA with adaptive PHY mode selection. We have designed a new scheme called EDCA with Link Adaptation (EDCA-LA), which adaptively adjusts backoff time for each AC by taking the channel condition into account. The idea behind EDCA-LA is to increase the backoff time when the channel condition is bad and to decrease this time when the channel condition is good.

The paper is organized as follows. Section 2 describes an overview of the IEEE 802.11 MAC and IEEE 802.11e MAC for wireless communication. The EDCA-LA algorithm is presented in Section 3. Section 4 presents some simulation results and gives their implicit reasons. Finally, our conclusions are given in Section 5.

2 Background

2.1 IEEE 802.11 MAC

The main purpose of the 802.11 MAC layer is to provide reliable data services for higher layer protocols and to control fair access to the shared wireless medium. The basic medium access protocol is DCF that allows for fair medium sharing through the use of Carrier Sensing Multiple Access with Collision Avoidance (CSMA/CA). A station (STA) using DCF has to follow two medium access rules: One is that the STA shall be allowed to transmit the frames only if its carrier-sense mechanism determines that the medium has been idle for at least Distributed Interframe Space (DIFS) time; and the other is that the STA shall select a random backoff time, ranging between 0 and CW, after sensing the busy medium or detecting an unsuccessful transmission.

To provide reliable and high-performance data transmission, the 802.11 standard defines an optional mechanism, RTS/CTS. This mechanism can increase the sturdiness of the protocol and address the problem of "hidden node". Under this mechanism, a STA sends a RTS frame to the destination before transmitting any MAC Service Data Unit (MSDU). After the Short Interframe Space (SIFS) waiting, the destination then responds with a CTS frame once it has successfully received a RTS. The source can then send the MSDU after receiving the expected CTS response. All frames, including RTS and CTS frames, contain duration information about the length of the MSDU/ACK transmission. All ambient STAs will update an internal timer called Network Allocation Vector (NAV) according to the duration information and defer any transmission until the timer runs out.

2.2 IEEE 802.11e MAC

The IEEE 802.11e MAC is an emerging standard to support QoS. This standard introduces HCF, which combines functions from DCF and PCF with enhanced QoS-specific mechanisms and frame types. HCF has two modes of operation: EDCA and HCCA. HCF allocates QSTAs the right to transmit through Transmission Opportunity (TXOP), which defines the start time and the maximum duration during which a QSTA can transmit a series of frames. TXOPs are allocated via contention (EDCA-TXOP) or granted through HCCA (polled-TXOP).

EDCA provides four Access Categories (ACs), where each AC represents a specific QoS provision, as shown in Fig. 1. Each AC within a QSTA contends for accessing the medium and independently starts its backoff after sensing the medium idle for at least AIFS period in deferral. Differentiated ACs are achieved by differentiating the following three key parameters.

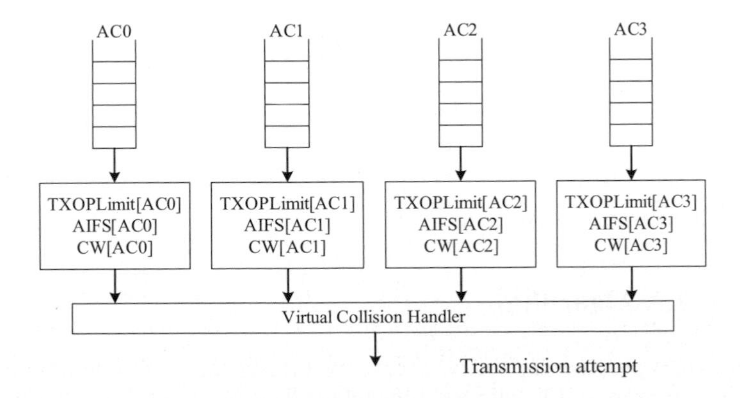

Fig. 1. ACs and virtual collision

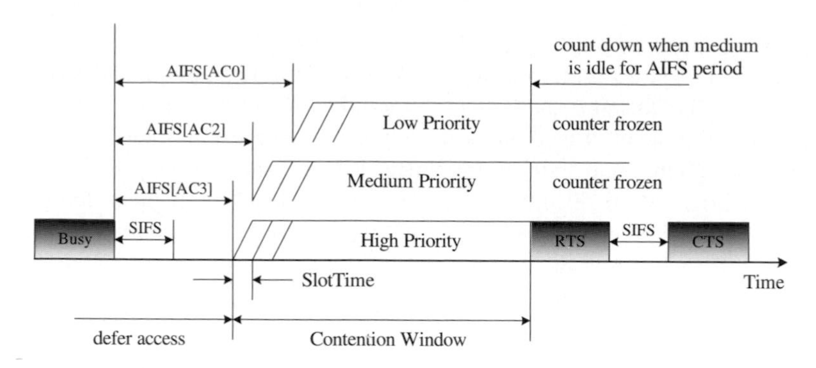

Fig. 2. AIFS parameter in IEEE 802.11e EDCA mechanism

- TXOP Limit: The maximum duration for which a QSTA can transmit after obtaining a TXOP. TXOP Limit can be used to ensure that high-priority traffic gets greater access to the medium. Each AC is assigned a specific $TXOPLimit[AC_i]$ $(i = 0,..,3)$ to provide QoS differentiation.
- Arbitration Inter-Frame Space (AIFS): The time interval between the wireless medium becoming idle and the start of channel access negotiation. Each AC is assigned a specific $AIFS[AC_i]$ $(i = 0,..,3)$ to provide QoS differentiation. Figure 2 shows the EDCA timing diagram, where three ACs, i.e., AC0, AC2 and AC3, are shown:. The AIFS for a given AC is determined by the equation $AIFS = AIFSN \times SlotTime + SIFS$, where $AIFSN$ is AIFS Number and determined by the AC and physical settings, and $SlotTime$ is the duration of a time slot.
- Contention Window (CW): A number used for generating the backoff time in the backoff mechanism. The backoff time, which is used to determine the time interval a QSTA has to wait before transmission after deferral, is a random number that lies between 0 and CW. The backoff time is computed as follows:

$$Backoff\ \ Time = Random(CW) \times SlotTime,$$

where Random(CW) is pseudorandom integer drawn from an uniform distribution over the interval [0, CW]. CW is an integer within the range of values CW_{min} and CW_{max} (i.e., $CW_{min} \le CW \le CW_{max}$). The initial CW is set as CW_{min}. This value is multiplied by a factor (usually double) after each failed transmission and is reset as CW_{min} after any successful transmission. In 802.11e, each AC is assigned a specific $CW[AC_i]$ (i.e., $CW_{min}[AC_i]$ and $CW_{max}[AC_i]$) to provide QoS differentiation. A smaller CW is assigned to a higher-priority AC to favor this AC.

3 EDCA-LA Algorithm

Although EDCA provides a priority scheme by differentiating TXOP limit, AIFS, and CW, the high priority traffic with low transmission rate (poor channel condition) leads long transmission time and decreases overall performance. Recently, many rate adaptation schemes have been proposed to utilize the multi-rate capability offered by the IEEE 802.11 wireless MAC protocol through automatically adjusting the transmission rate to best match the channel condition. The 802.11 physical layers provide multiple data transmission rates by employing different modulation and channel coding schemes. As shown in [11], the transmission rate should be chosen in an adaptive manner since the wireless channel condition varies over time due to such factors as station mobility, time-varying interference, and location-dependent errors. IEEE 802.11b PHY provides four PHY modes with data transmission rates 1/2/5.5/11Mbps.

In order to take the channel condition into account, we design an adaptive EDCA with Link Adaptation (EDCA-LA) scheme. In EDCA-LA, the QSTAs can adaptively adjust the backoff time based on the measured channel conditions. Our idea is to combine EDCA with adaptive PHY mode selection, so that the proper PHY rate as well as the better channel can be more easily adopted to achieve higher network performance.

In EDCA-LA, every AC adjusts its backoff time according to the measured transmission rate TR_{curr}^{j} at the *j*th update. The value of TR_{curr}^{j} is obtained from the current PHY rate. To avoid the effect from rapid fluctuation of channel condition, an estimator of Exponentially Weighted Moving Average is used to smoothen the measured channel condition. Let TR_{avg}^{j} be the average transmission rate at the *j*th update, and it is computed according to the following iterative relationship:

$$TR_{avg}^{j} = \alpha \times TR_{avg}^{j-1} + (1-\alpha) \times TR_{curr}^{j},$$

where α is the smoothing factor and determines the weight of history in the averaging process. After TR_{avg}^{j} is obtained, the *channel quality*, ρ, is derived from the following relationship:

$$\rho = \frac{TR_{avg}^{j}}{\frac{1}{2} \times TR_{\max}},$$

where $TR_{\max}$ represents the maximum transmission rate in the PHY mode (e.g. 11Mbps in 802.11b, 54Mbps in 802.11a, and 54Mbps in 802.11g).

Using ρ, after any successful transmission of frame in AC *i*, the backoff time is then updated as follows:

$$CW[AC_i] = CW_{\min}[AC_i],$$

$$Backoff\ Time = \frac{1}{\rho} \times Random(CW[AC_i]) \times SlotTime.$$

After each failed transmission of frame in AC *i*, the new CW of this AC is doubled and the backoff time is updated with the factor ρ as

$$CW[AC_i] = 2 \times CW[AC_i],$$

$$Backoff\ Time = \frac{1}{\rho} \times Random(CW[AC_i]) \times SlotTime.$$

A flow chart explaining the operation of EDCA-LA is shown in Fig. 3.

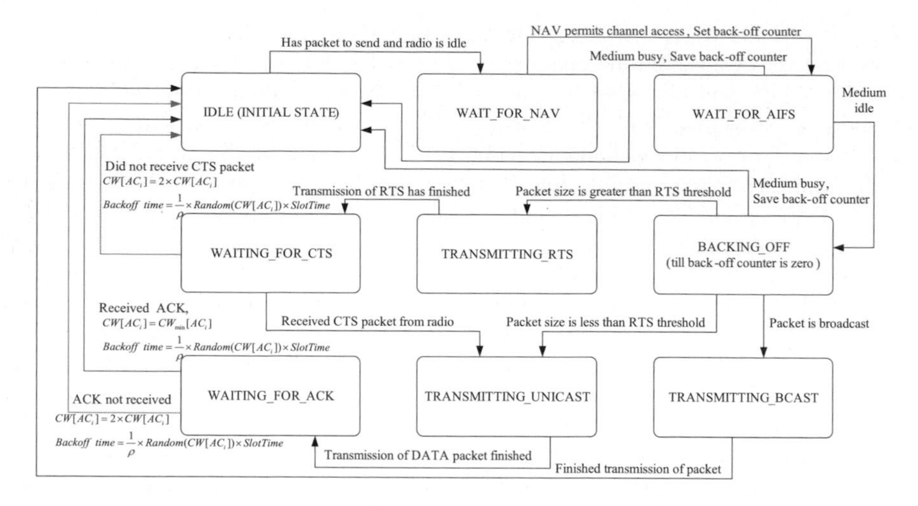

Fig. 3. The flow chart of the EDCA-LA algorithm

4 Simulation Results and Their Implications

In this section, we present the simulation results of EDCA-LA performance and compare it with the EDCA scheme.

4.1 Simulation Environment

In our simulation, the wireless topology consists of several QSTAs and a QAP. All QSTAs are located such that every QSTA is able to detect the transmission from any other QSTA. The number of QSTAs is varied from 1 to 15 to simulate the increase of network load.

For simplicity and without loss of generality, we assume that there are three active ACs in each QSTA, one is AC3 with highest priority traffic, and the others are AC2 and AC0 with lower priority traffic. We use 802.11b PHY, that is, four data transmission rates: 1/2/5.5/11Mbps. The values of parameters used in our simulation are listed in Table 1. The chosen traffic characteristics are shown in Table 2.

Table 1. Simulation parameters

SIFS	20 μs	**CW_{min}[AC3]**	7
SlotTime	20 μs	**CW_{max}[AC3]**	15
Number of QSTAs	1~15	**CW_{min}[AC2]**	15
AIFS[AC3]	*SIFS* + 1× *SlotTime*	**CW_{max}[AC2]**	31
AIFS[AC2]	*SIFS* + 1× *SlotTime*	**CW_{min}[AC0]**	31
AIFS[AC0]	*SIFS* +2× *SlotTime*	**CW_{max}[AC0]**	1023

Table 2. Type of traffic

Traffic type	Frame size	Interval
AC3	100 bytes	20 ms
AC2	250 bytes	40 ms
AC0	500 bytes	80 ms

4.2 Simulation Results

Figure 4 shows the throughput occupied by AC3, AC2 and AC0 for EDCA and EDCA-LA. Three points are easily observed. First, EDCA and EDCA-LA mechanism actually provide an effective way of throughput differentiation because of different AIFS and CW settings. When the number of QSTAs exceeds five, the lowest-priority traffic AC0 begins suffering from starvation because some bandwidth is grabbed by higher-priority traffic AC3 and AC2. The similar situation happens between AC3 and AC2 when the number of QSTAs grows up to eight. Second, as the number of QSTAs increases, the overall throughput first increase before the saturation (the number of QSTAs is about five), and then slightly decrease. The throughput drop causes from that more stations will generate more serious contention at accessing the wireless medium. Finally, the most important observation is that EDCA-LA has much more overall throughput than EDCA, observing from EDCA-LA Total>>EDCA Total. This is because EDCA uses the higher-bandwidth link to transmit the frames.

Deeply comparing the throughput of individual AC between EDCA-LA and EDCA, the throughput of EDCA-LA AC3 and EDCA AC3 are very similar, while EDCA-LA AC2 has larger throughput than EDCA AC2 and EDCA-LA AC0 has larger throughput than EDCA AC0 when the numbers of QSTAs exceed eight and five, respectively. The similarity of EDCA AC3 and EDCA-LA AC3 occurs because even when the number of QSTAs is 15, only AC3 traffic does not saturate the link, causing

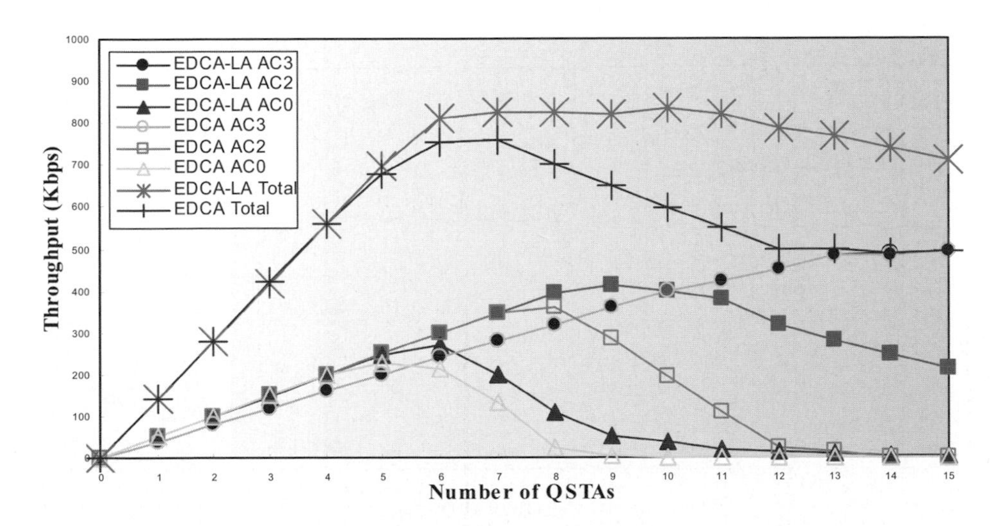

Fig. 4. Comparison between EDCA-LA and EDCA on throughput

that it can obtain the required bandwidth due to its highest priority. However, the bandwidth is actually grabbed from AC2 and AC0. As stated previously, EDCA-LA has more overall throughput than EDCA. Thus the grab in the former is less serious than the latter, implying that EDCA-LA AC2 will get more throughput than EDCA AC2 . The similar case also happens for AC0.

Figure 5 shows the average end-to-end delay of AC3, AC2 and AC0 for EDCA and EDCA-LA. The frame's end-to-end delay means the duration between it enters the AC queue and the original QSTA gets the corresponding ACK. Three points are also observed from this figure. First, EDCA and EDCA-LA mechanism actually provide an effective way to achieve end-to-end delay differentiation. The high-priority traffic AC3 always has much shorter delay than the lower-priority traffic AC2 and AC0 because the different AIFS and CW settings. The lower-priority traffic AC0 and AC2 begins suffering from starvation when the numbers of QSTAs are larger than five and eight, respectively. The starvation causes the rapid increase of the average delay of AC0 and AC2. Second, the delay of all ACs increases as the number of QSTAs increases. This is because the more stations there are, the more collisions there are, leading to a longer backoff time for each AC and then the increase in the average end-to-end delay of all ACs. Finally, the most important observation is that EDCA-LA has shorter delay than EDCA regardless of what AC. For AC2 and AC0, EDCA is saturated when the numbers of QSTAs are five and eight, which are smaller than the saturation values, six and ten, in EDCA-LA, respectively. This phenomenon causes that the end-to-end delay of EDCA also boost earlier than EDCA-LA. For AC3, the EDCA-LA has the shorter end-to-end delay than EDCA because the former uses the higher-bandwidth link to transmit its frames and obtains shorter transmission time, although they achieve the similar throughput.

Figure 6 shows the throughput in seven stations as the traffic load of each station increases. We adjust the interarrival time of AC traffic to control the traffic load. When traffic load increases, the throughput of the lower priority traffic, AC2 and AC0, decreases. The highest priority traffic AC3 grabs their bandwidth if the traffic load

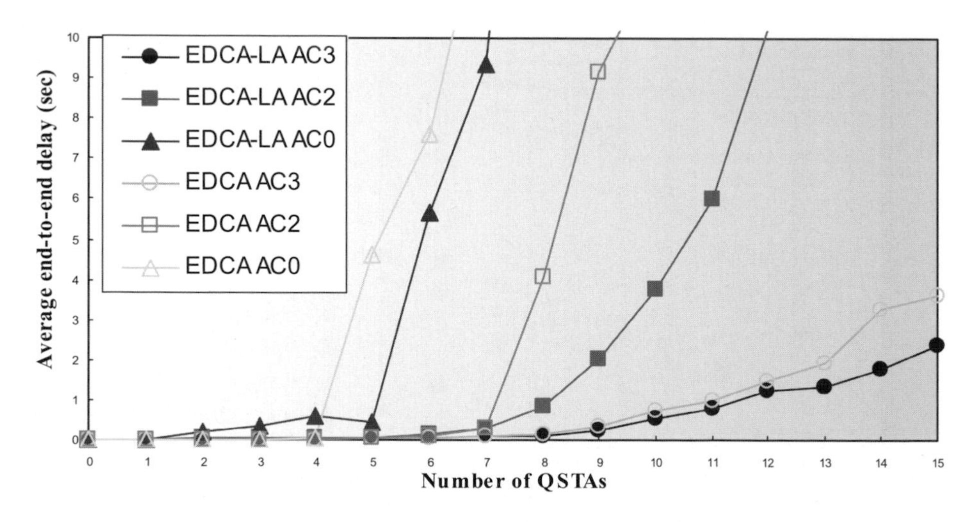

Fig. 5. Comparison between EDCA-LA and EDCA on average end-to-end delay

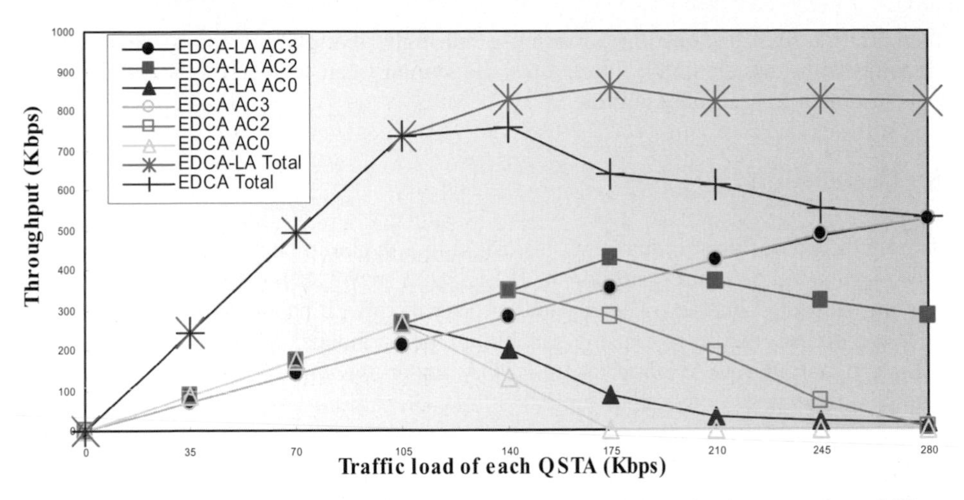

Fig. 6. The throughput in seven stations as the traffic load of each station increases from 0 Kbps to 280 Kbps

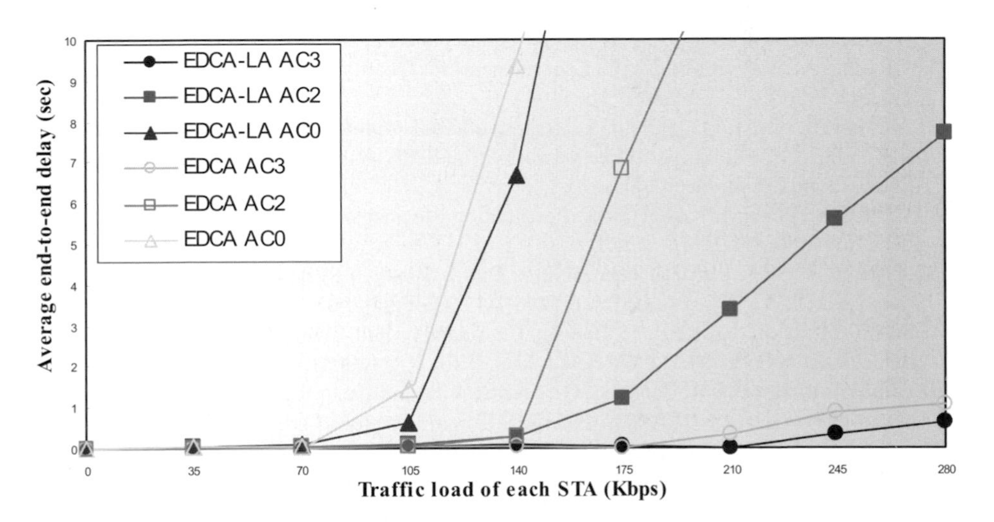

Fig. 7. The average end-to-end delay in seven stations as the traffic load of each station increases from 0 Kbps to 280 Kbps

exceeds 175Kbps. The total throughput of EDCA-LA has much more throughput than EDCA. Actually, figure 6 has the similar tread with figure 4, and the same reasons can interpret these curves.

Figure 7 shows the average end-to-end delay in seven stations as the traffic load of each station increases. The delay of the lower priority traffic, AC0 and AC2, rapidly increases after offered traffic load exceeds 105Kbps and 175Kbps respectively, and are significantly longer than that of the highest priority traffic AC3. For more than 105Kbps and 175Kbps, EDCA-LA AC0 and EDCA-LA AC2 has much shorter delay than EDCA AC0 and EDCA AC2, respectively. EDCA AC3 has slightly shorter delay than EDCA AC3 because the former use the higher-bandwidth link to decrease the transmission time. Actually, figure 7 has the similar tread with figure 5, and the same reasons can interpret these curves.

5 Conclusions

The main contribution in this paper is designing a dynamic backoff time scheme with considering the channel condition in IEEE 802.11e WLANs. By letting the station having the better channel owns the smaller backoff time, it has the larger probability to access wireless medium, causing the higher performance. Simulation results have shown that EDCA-LA outperforms EDCA under the environments with different numbers of stations and with different load in each station.

References

1. Wireless LAN Medium Access Control (MAC) and Physical Layer (PHY) Specifications. IEEE 802.11 Standard (1999)
2. Wireless LAN Medium Access Control (MAC) and Physical Layer (PHY) Specifications: Medium Access Control (MAC) Enhancements for Quality of Service (QoS). IEEE 802.11e Standard (2005)
3. Romdhani, L., Ni, Q., Turletti, T.: Adaptive EDCF: Enhanced Service Differentiation for IEEE 802.11 Wireless Ad-Hoc Networks. In: IEEE Wireless and Communications and Networking Conference (2003)
4. Kuppa, S., Prakash, R.: Service Differentiation Mechanisms for IEEE 802.11-based Wireless Networks. IEEE WCNC (2004)
5. Chou, C.T., Shin, K.G., Shankar, S.: Inter-Frame Space (IFS) Based Service Differentiation for IEEE 802.11 Wireless LANs. IEEE VTC 2003-Fall (2003)
6. Malli, M., Ni, Q., Turletti, T., Barakat, C.: Adaptive Fair Channel Allocation for QoS Enhancement in IEEE 802.11 Wireless LANs. IEEE ICC (2004)
7. Vollero, L., Banchs, A., Iannello, G.: ACKS: A Technique to Reduce the Impact of Legacy Stations in 802.11e EDCA WLANs. IEEE Communications Letters 9(4) (2005)
8. Chevillat, P., Jelitto, J., Barreto, A.N., Truong, H.L.: A Dynamic Link Adaptation Algorithm for IEEE 802.11a Wireless LANs. IEEE ICC (2003)
9. Chai, C.C., Tjeng, T.T., Leng, C.L.: Combined Power and Rate Adaptation for Wireless Cellular Systems. IEEE Transactions on Wireless Communications 4(1) (2005)
10. Wang, J., Zhai, H., Fang, Y., Yuang, M.C.: Opportunistic Media Access Control and Rate Adaptation for Wireless Ad Hoc Networks. IEEE ICC (2004)
11. Yamada, H., Morikawa, H., Aoyama, T.: Decentralized Control Mechanism Suppressing Delay Fluctuation in Wireless LANs. IEEE VTC 2003-Fall (2003)

A Framework for Detecting Internet Applications

António Nogueira, Paulo Salvador, and Rui Valadas

University of Aveiro / Institute of Telecommunications-Aveiro
Campus de Santiago, 3810-193 Aveiro, Portugal
{nogueira,salvador,rv}@det.ua.pt

Abstract. There are several network management and measurement tasks, including for example traffic engineering, service differentiation, performance or failure monitoring or security, that can greatly benefit with the ability to perform an accurate mapping of network traffic to IP applications. In the last years traditional mapping approaches have become increasingly inaccurate because many applications use non-default or ephemeral port numbers, use well-known port numbers associated with other applications, change application signatures or use traffic encryption. Thus, new solutions are needed for this problem and this paper presents a new approach, based on neural networks, that is able to solve the problem of application detection and at the same time can predict the traffic level associated with each application based on the overall aggregated traffic, while overcoming the limitations of the previous approaches. Results obtained show that the proposed framework constitutes a valuable tool to detect Internet applications and predict their traffic levels since it can achieve good performance results while, at the same time, avoid the most important disadvantages presented by the other detection methods.

Keywords: Port matching, protocol analysis, semantic and syntactic analysis, neural networks.

1 Introduction

The different techniques that are currently used to identify IP applications have important drawbacks that limit or dissuade their application. Port based analysis is the most basic and straightforward method to detect applications and users based on network traffic and is based on the simple concept that many applications have default ports on which they function. To perform port based analysis, administrators just need to observe the network traffic and check whether there are connection records using these ports: if a match is found, it may indicate a particular application activity. Port matching is very simple in practice, but its limitations are obvious: most applications allow users to change the default port numbers by manually selecting whatever port(s) they like; additionally, many newer applications are more inclined to use random ports, thus making ports unpredictable; there is also a trend for applications to begin masquerade their function ports within well-known application ports. Protocol analysis monitors traffic passing through the network and inspects the data payload of the packets according to some previously defined application signatures. However, this approach still has some shortcomings: IP applications are evolving continuously and

T. Vazão, M.M. Freire, and I. Chong (Eds.): ICOIN 2007, LNCS 5200, pp. 455–464, 2008.

therefore signatures can change; application developers can encrypt traffic making protocol analysis more difficult; signature-based identification can affect network stability because it has to read and process all network traffic. Syntactic and semantic analysis of the data flow avoids some of the disadvantages of port-based analysis and protocol analysis. This approach can perform protocol recognition regardless of any encapsulation and is able to extract data specific to each protocol. However, this analysis can be a burden to network stability due to its high processing requirements and is not appropriate when dealing with confidentiality requirements, because in these situations it is not possible to have access to the packet contents (the same drawback is found in the protocol analysis approach).

A new approach, based on Neural Networks (NNs), was developed to detect Internet applications based on the overall aggregated network traffic and estimate the amount of traffic (represented by the number of downloaded bytes or packets) corresponding to each application. The NN model is trained using a set of known traffic values associated with each application and the corresponding aggregate traffic; after the training phase, the trained NN model can identify the traffic level associated with each application based on new values of aggregate traffic that are presented as inputs. The correlation that exists between the temporal sequence of aggregate traffic values and the current distribution of traffic per application is taken into account by presenting the current and the last h (where h represents a configurable parameter) values of aggregate traffic as inputs of the NN model. The *a priori* identification of the applications that constitute the traffic values used in the training phase can rely on conventional application mapping approaches or can derive from known traffic generated in a controlled environment.

The ability to detect an IP application relies on the traffic characteristics or profile associated to that application. From Figure 2 we can see that typically the File Sharing service has a very high bandwidth usage pattern characterised by a large variability, which can be attributed to the large number of TCP sessions that are opened/closed. The HTTP service is characterised by a set of non-periodic high bandwidth utilisation peaks, with very short durations, that result from the user clicks. On the contrary, the Games service is characterised by periodic (high-frequency) short duration peaks and a small bandwidth usage, while the Streaming service is characterised by a constant medium bandwidth usage. In a real network scenario, what we have is a mixture of

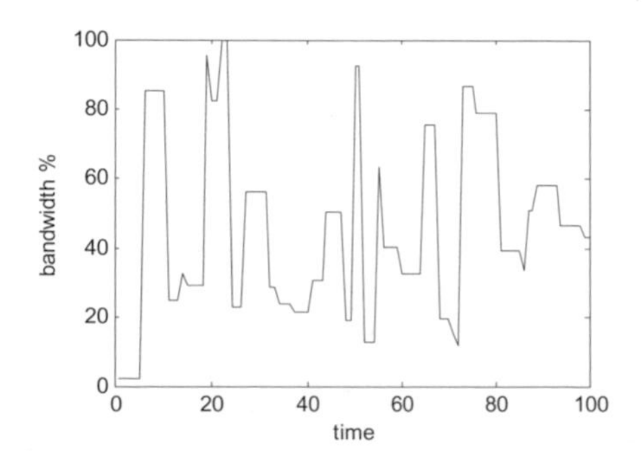

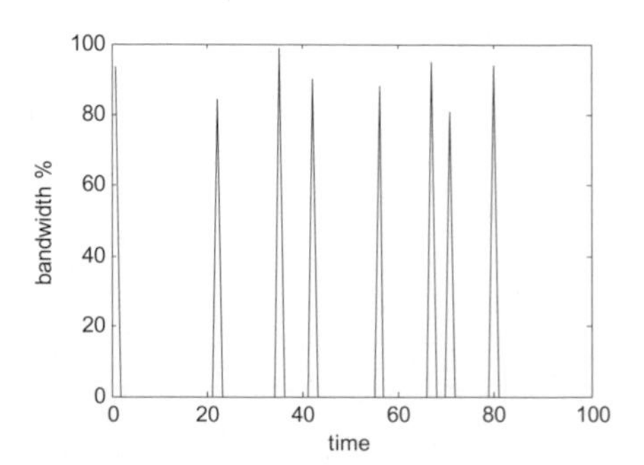

Fig. 1. Bandwidth utilisation (in percentage) per service: (left) File sharing; (right) HTTP

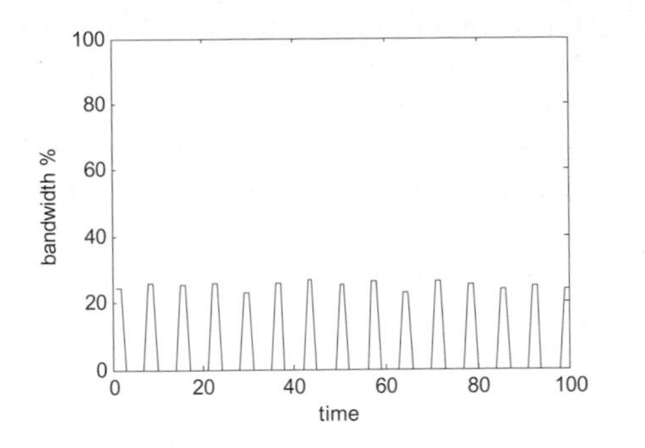

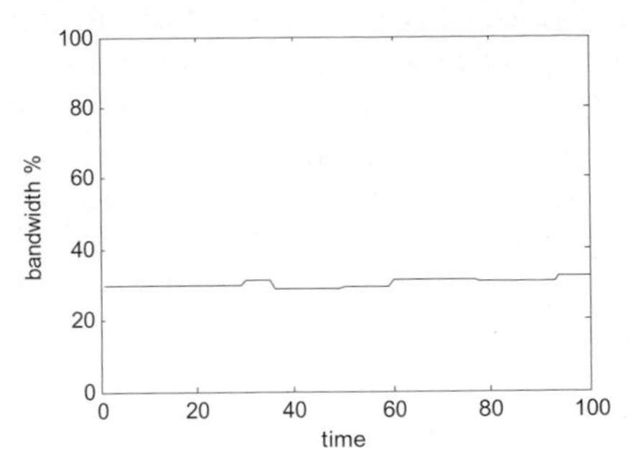

Fig. 2. Bandwidth utilisation (in percentage) per service: (left) Games; (right) Streaming

several services and the goal is to identify the different applications from the aggregated traffic. In the left part of Figure 3 we represent the percentage of bandwidth utilisation for a mixture of 80% of File Sharing and 20% of HTTP traffic: for this scenario, we can see high bandwidth usage and large variability for medium bandwidth values, typical characteristics of this level of File Sharing traffic, and non-periodic very short duration peaks, which is a typical behavior of the HTTP service. In the central part we represent the percentage of bandwidth utilisation for a mixture of 20% of File Sharing and 80% of HTTP traffic: in this case, it is easy to identify medium bandwidth usage and large variability for smaller bandwidth values, resulting from the presence of 20% of File Sharing traffic, and non-periodic very short duration peaks, which are due to the typical behavior of the HTTP service. In the right plot we represent the percentage of bandwidth utilisation for a mixture of 10% of File Sharing, 70% of HTTP traffic and 20% of Streaming traffic: in this case, we can identify a medium bandwidth comsumption, resulting from the File Sharing service, small variability for smaller bandwidth values, resulting from the presence of Streaming and File Sharing traffic, variability around a specific bandwidth value suggestting the presence of File Sharing and non-periodic very short duration peaks, which are due to the typical behavior of the HTTP service.

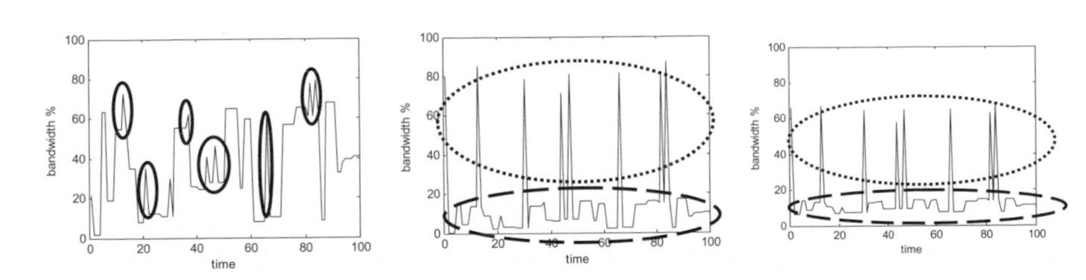

Fig. 3. (left) Bandwidth utilisation (in percentage) for a mixture of 80% of File sharing and 20% of HTTP; (center) Bandwidth utilisation (in percentage) for a mixture of 20% of File sharing and 80% of HTTP and (right) Bandwidth utilisation (in percentage) for a mixture of 80% of File sharing and 20% of HTTP

The capacity to detect and identify Internet applications can be used in Quality of Service (QoS) and security insurance mechanisms. In fact, once a service provider is able to associate traffic to its corresponding IP application it can use this information to group traffic with different or similar (depending on the most advantageous situation) statistical characteristics in order to optimise the bandwidth occupancy of the links. The ability to identify Internet applications can also be used to detect security attacks, like worms, zombies, among others, and consequently trigger the appropriate defence and/or repair actions.

Neural Networks have been successfully used in several applications, like pattern recognition [1], classification of Internet users [2,3], intrusion detection [4,5], among others, due to their advantageous properties like parallel processing of information, capacity to recognise patterns of information in the presence of noise and handle non-linearity, capacity to classify information and quick adaptability to system dynamics.

The results obtained in this paper show that NNs are able to detect Internet applications and their corresponding traffic levels, being at the same time immune to the most important drawbacks presented by traditional detection techniques. Once the NN model is conveniently trained it can be used to detect (without looking at the packets contents) the presence of Internet applications and predict their traffic values for new aggregate traffic values that are presented as inputs. The computational requirements of the training phase can be significant, but this is an off-line phase; the on-line simulation phase can be performed almost instantaneously.

The paper is organised as follows. Section 2 gives an overview of the traffic trace that is used in this study. Section 3 describes the way NNs are used to identify Internet applications and predict their traffic values: after identifying and describing the specific problem to solve, a suitable NN architecture is proposed and conveniently justified. Section 4 presents and discusses the results and, finally, section 5 presents the main conclusions.

2 Overview of the Traffic Trace

Our analysis resorts to a data trace measured in a Portuguese ISP that uses a CATV network and offers several types of services, characterised by the following maximum allowed transfer rates (in Kbit/s) in the downstream/upstream directions: 128/64, 256/128 and 512/256. The trace was measured during a whole week period, from 8h:13m AM of Wednesday July 2, 2003, to 7h:32m PM of Tuesday July 8, 2003. The measurements consisted in detailed packet level measurements, where the arrival instant and the first 57 bytes of each packet were recorded. This includes information on the packet size, the origin and destination IP addresses, the origin and destination port numbers, and the IP protocol type. Both upload and download traffic were measured, although this study only deals with download traffic. The traffic analyzer was a 1.2 GHz AMD Athlon PC, with 1.5 Gbytes of RAM and running WinDump. No packet drops were reported by WinDump in both measurements.

The specific data set used in this study was extracted from the whole data trace and includes 10 users (the ones that generate more aggregate traffic) characterised by their download traffic (in bytes) in 1 minute intervals. In order to train the NN and to evaluate

Table 1. Port numbers associated to each identified application group

Application group	Port numbers
File Sharing	1214,4662,412,413,1412,6699,20,21
HTTP	80,443
Games	2234,2344,2346,UDP 12203,UDP 27005
Streaming	1755

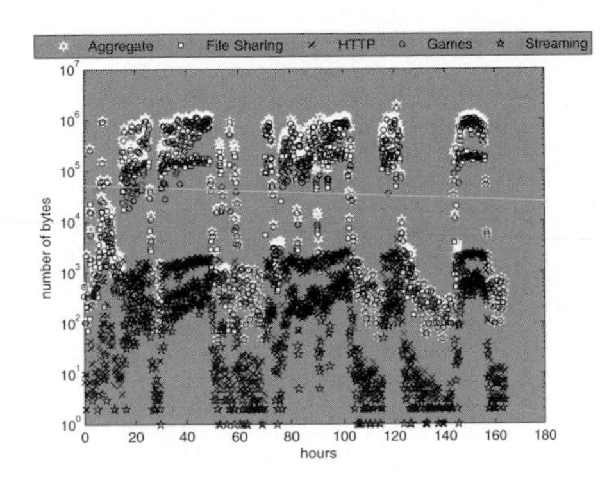

Fig. 4. Download traffic (in bytes) per application

its performance in identifying Internet applications and predicting their traffic levels, the different Internet applications that contributed to the measured data were identified using port-based analysis: four application groups were identified according to the port numbers presented in Table 1. Note that, according to some of the reasons that were mentioned in section 1, this classification process can be itself subject to some errors. Obviously, other IP applications have also contributed to the aggregate traffic, but the four identified applications are responsible for about 95% of the total measured traffic.

Figure 4 represents the download traffic (in bytes) corresponding to the 10 selected users, including either the aggregated traffic and the traffic corresponding to each application group. Looking at this graph, some remarks can be immediately pointed out: (i) the File Sharing and Games services contribute with the major part of the download traffic; (ii) the HTTP and Streaming services have small to medium download traffic values with non-periodic profiles. These characteristics will have obvious impacts on the detection and prediction results obtained by applying neural networks.

3 Neural Network Model to Detect Internet Applications

The main objective of this study is to detect the activity of different Internet applications based on the overall aggregate traffic (as represented by the number of download bytes) and, a step further, to estimate the amount of traffic (number of download bytes) corresponding to each application. Figure 5 schematically represents the main idea of this study: the neural network model uses the overall aggregate traffic to detect the

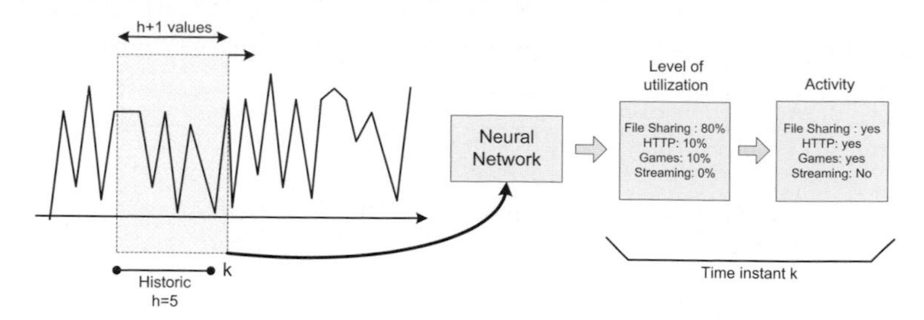

Fig. 5. Framework for detect IP applications and estimating their utilisation level

activity of each IP application and estimate its utilisation level. In order to incorporate some history in this detection/prediction process (that can take into account correlations that possibly exist between aggregate traffic and traffic distribution per application in adjacent time periods), the current and the last h values of the aggregate download traffic will be used to estimate the current distribution of traffic per application. This problem will be solved using a back propagation NN model.

Back propagation is a general purpose learning algorithm for training multilayer feed-forward networks that is powerful but expensive in terms of computational requirements for training. A back propagation NN model uses a feed-forward topology, supervised learning, and the back propagation learning algorithm. A back propagation network with a single hidden layer of processing elements can model any continuous function to any degree of accuracy (given enough processing elements in the hidden layer) [6].

For the dimension of our problem a conventional feed-forward back propagation network with three layers seems to be appropriate. The input layer will have $h + 1$ neurons, corresponding to the dimensionality of the input vectors: each input vector is composed by the current and the last h values of the aggregate traffic. We have tried different values of h, concluding that the performance gains obtained did not increase substantially from the ones given by a six neuron input layer NN ($h = 5$). Thus, we have considered a NN model having 6 neurons in the input layer. The output layer will have 7 neurons, since each output vector represents the distribution of traffic by each one of the 7 application groups previously detected. The number of nodes in the hidden layer is empirically selected such that the performance function (the mean square error, in this case) is minimised. Different NNs with variable number of neurons in the hidden layer were considered and their performance was calculated, leading to a choice of 5 hidden nodes. The final structure of the proposed NN model is presented in Figure 6.

In the proposed NN model, scalar $w^i_{n,j}$ represents the weight value corresponding to layer i, $i = 1, 2, 3$, that is multiplied by input n of neuron j, where n and j have different ranges depending on the network layer. Scalar b^i_j represents the bias associated with neuron j of layer i. For the input and hidden layers, a tan-sigmoid transfer function (represented by **f** in Figure 6) is used, generating outputs between -1 and 1 as the neuron´s input goes from negative to positive infinity. For the output layer, a linear transfer function (represented by **g** in Figure 6) is used in order to generate any output

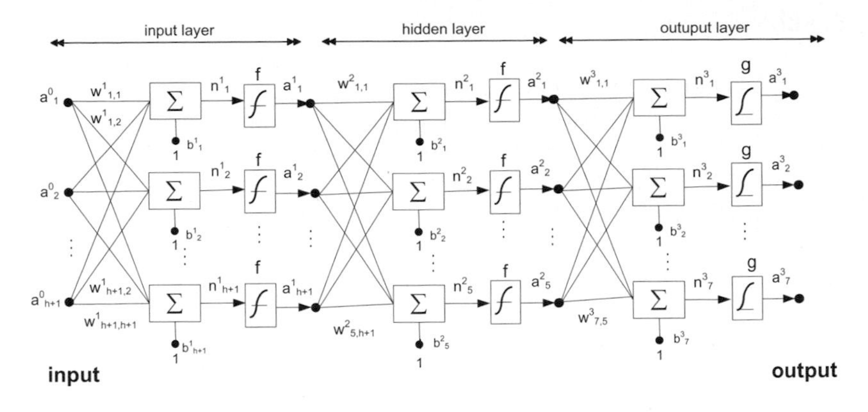

Fig. 6. Architecture of the Neural Network used to detect Internet applications

value. It is known that multiple layers of neurons with nonlinear transfer functions allow the network to learn nonlinear and linear relationships between input and output vectors [7,8]. Automated Bayesian regularisation is used to improve generalisation of the neural network, in order to avoid overfitting [9]. Using the above notation, the jth neuron in layer i produces an output a_j^i given by: $a_j^i = F(\sum_n (w_{n,j}^i a_n^{i-1}) + b_j^i), n = 1, 2, ...N, j = 1, 2, ..., J$, where N and J are the number of inputs and size of layer i, respectively, and F represents the transfer function of layer i (it can be **f** or **g**).

The application of a NN to solve a particular problem involves two phases: a training phase and a test phase. In the training phase, a training set is presented as input to the NN which iteratively adjusts network weights and biases in order to produce an output that matches, within a certain degree of accuracy, a previously known result (named target set). In the test phase, a new input is presented to the network and a new result is obtained based on the network parameters that were calculated during the training phase. There are two learning paradigms (supervised or non-supervised learning) and several learning algorithms that can be applied, depending essentially on the type of problem to be solved. For our problem the network was trained incrementally, that is, network weights and biases were updated each time an input was presented to the network. This option was mostly determined by the size of the training set: loading the complete training set at once and presenting it as input to the NN was very consuming in terms of computational memory. The training method used was the Levenberg-Marquardt algorithm combined with automated Bayesian regularisation, which basically constitutes a modification of the Levenberg-Marquardt training algorithm to produce networks which generalise well, thus reducing the difficulty of determining the optimum network architecture.

For solving our detection problem, a set including the 10 users that generated more aggregate traffic was selected from the original data set (containing 6655 users) and divided in two subsets of equal size, 5 users. The first subset was used to train the NN - it became the training set - and the second one was used to test the trained NN - the test set.

4 Results

The traffic prediction results given by the proposed framework are shown in the plots of Figures 7 and 8, where each plot corresponds to a specific application. Each one of these plots compares, for a given application, the target utilisation level and he utilisation level predicted by the NN. The utilisation level of a specific application corresponds to the fraction of overall aggregate traffic that belongs to that application.

From the obtained results we can see that the NN is conveniently trained (there are some discrepancies for the higher and lower target values) and this is reflected in the test phase that shows a good match between the target and predicted values, with a clear exception at the highest target value that is not conveniently predicted.

For the File Sharing service (left part of Figure 7) there are frequent peaks of utilisation level values (both in the training and test phases) and the NN model is conveniently trained being able to predict the peak values in the test phase. However, there are some errors (not very significant) in the prediction of the higher peaks, since the NN model has never seen target value equal to this ones in the training phase. A similar behaviour is obtained for the HTTP service (right part of Figure 7), where the prediction results are also very good.

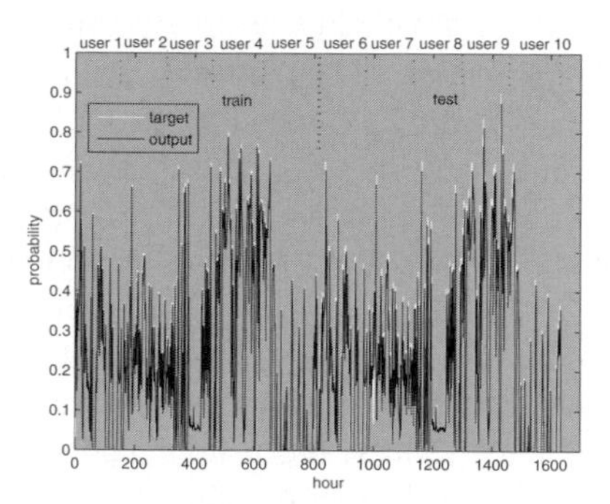

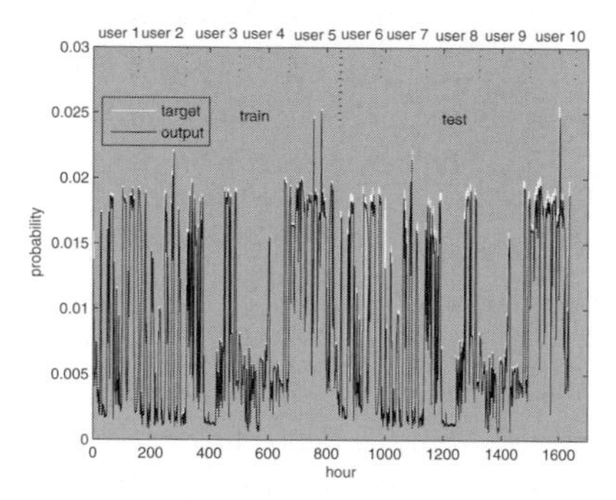

Fig. 7. Target and predicted utilisation levels: (left) File sharing service; (right) HTTP service

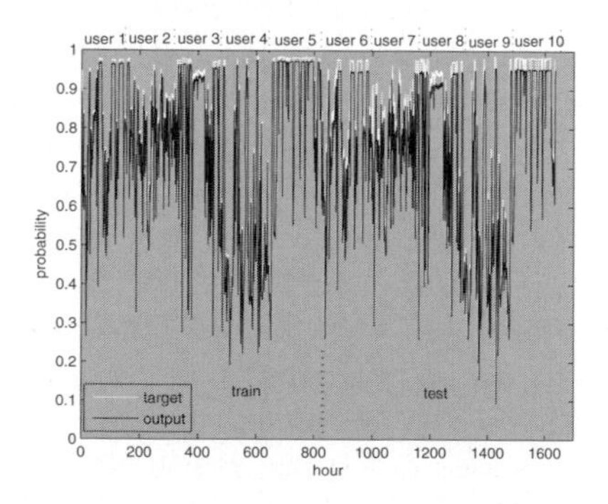

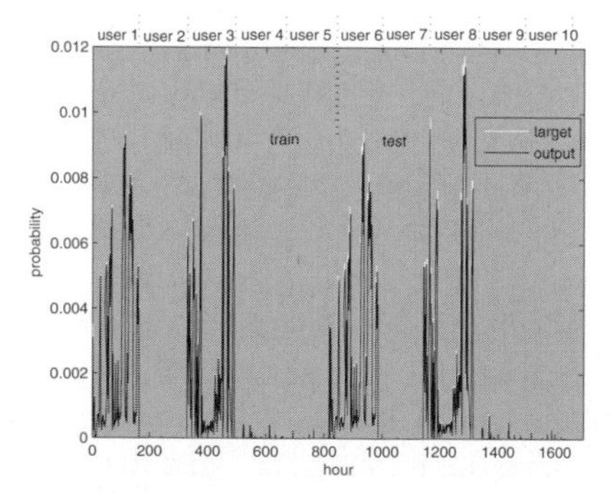

Fig. 8. Target and predicted utilisation levels: (left) Games service; (right) Streaming service

Table 2. Errors (in percentage) of estimating the download traffic and activity per application

	Downloaded traffic		Activity	
Service	Train	Test	Train	Test
File Sharing	4.62%	8.05%	2.20%	3.46%
HTTP	5.51%	10.47%	4.76%	6.46%
Games	7.31%	12.69%	5.46%	7.11%
Streaming	3.88%	5.61%	1.76%	3.28%

For the Games service (left part of Figure 8) the prediction errors are more significant, since this service is characterised by a highly varying pattern of probability values, that includes a lot of high values. In this case, the NN model has more difficulties in estimating the highest utilisation levels that are presented to the model in the test phase. The Streaming service (right part of Figure 8) presents an almost periodic pattern of utilisation levels, with very small probability values. In this case, the NN model is conveniently trained and is able to predict the target utilisation levels that are presented in the test phase. Note that, in this case, the utilisation level profiles of the train and test phases are very similar.

Table 2 presents the train and test errors (in percentage) per application, that is, the difference in percentage between the target and predicted utilisation levels for each service. These results give a more detailed insight into the results plotted in Figures 7 and 8. Table 2 also presents the train and test activity errors (in percentage) per application, that is, the ability of the NN model to detect the existence or absence of activity for each service. We consider that a given application is active if its number of downloaded bytes is higher than 0.001% of the download tranfer rate for each client (which is 512 Kbit/s), that is, 23.04 Kbyte/hour.

Concluding, we can say that the target utilisation levels for each application are well approximated by the NN model. There are, however, some discrepancies between the target and output values, specially for the higher target values, since the NN model is not generally able to predict very high target values that occur unfrequently.

5 Conclusions and Further Research

Network management and measurement tasks like traffic engineering, service differentiation, performance/failure monitoring, and security can greatly benefit from an accurate mapping of traffic to Internet applications. This paper proposes a new framework, based on neural networks, to detect Internet applications and predict their traffic levels. The obtained results are very promising, making this framework a potential tool to solve this problem. Besides, the proposed approach does not suffer from the most important drawbacks presented by other identification methodologies.

Acknowledgments

This work was part of project POSC/EIA/60061/2004 "Internet Traffic Measurements, Modeling and Statistical Analysis", funded by Fundação para a Ciência e Tecnologia, Portugal.

References

1. Looney, C.G.: Pattern recognition using neural networks: theory and algorithms for engineers and scientists. Oxford University Press, Inc., New York (1997)
2. Nogueira, A., Rosário de Oliveira, M., Salvador, P., Valadas, R., Pacheco, A.: Using neural networks to classify internet users. In: First Advanced International Conference on Telecommunications (July 2005)
3. Nogueira, A., Rosário de Oliveira, M., Salvador, P., Valadas, R., Pacheco, A.: Classification of internet users using discriminant analysis and neural networks. In: First Conference on Traffic Engineering for the Next Generation Internet (April 2005)
4. Ryan, J., Lin, M.-J., Miikkulainen, R.: Intrusion detection with neural networks. In: Advances in Neural Information Processing Systems, vol. 10. MIT Press, Cambridge (1998)
5. Mukkamala, S., Sung, A.H.: Identifying key features for intrusion detection using neural networks. In: ICCC 2002: Proceedings of the 15th international conference on Computer communication, Washington, DC, USA, pp. 1132–1138. International Council for Computer Communication (2002)
6. Demuth, H., Beale, M.: Neural Network Toolbox Users Guide. The MathWorks, Inc. (1998)
7. Fausett, L.: Fundamentals of Neural Networks. Prentice-Hall, Englewood Cliffs (1994)
8. Gurney, K.: An Introduction to Neural Networks. UCL Press (1997)
9. Foresee, F., Hagan, M.: Gauss-newton approximation to bayesian regularization. In: Proceedings of the 1997 International Joint Conference on Neural Networks, pp. 1930–1935 (1997)

Transport Layer Identification of Skype Traffic

Liang Lu, Jeffrey Horton, Reihaneh Safavi-Naini, and Willy Susilo

Center for Information Security
School of Information Technology and Computer Science
University of Wollongong, Australia
{ll97,jeffh,rei,wsusilo}@uow.edu.au

Abstract. The Internet telephony application *Skype* is well-known for its capability to intelligently tunnel through firewalls by selecting customized ports and encrypting its traffic to evade content based filtering. Although this capability may give some convenience to *Skype* users, it increases the difficulty of managing firewalls to filter out unwanted traffic. In this paper, we propose two different schemes, namely payload-based and non-payload based, for identification of *Skype* traffic. As payload based identification is not always practical due to legal, privacy, performance, protocol change and software upgrade issues, we focus on the non-payload based scheme, and use the payload based scheme mainly to verify its non-payload based counterpart. Our research results reveal that, at least to a certain extent, encryption by Skype to evade content analysis can be overcome.

1 Introduction

In recent years, Voice over IP (VoIP) - the transmission of voice over traditional packet-switched IP networks - has gained increasing popularity and has been widely deployed by many enterprises. Compared with conventional PSTN telephony service, it offers lower cost, greater flexibility and easier integration with other services such as call monitoring and auditing if traffic is not encrypted.

Among VoIP applications, Skype has a reputation for ease of use, superior sound quality and secure communication. Skype claims that it can work almost seamlessly across NATs and firewalls, and has better voice quality than other major free-to-use VoIP applications such as ICQ, AIM and MSN [7]. Skype calls are encrypted "end-to-end" and are not tappable by intermediate routing nodes. Additionally, Skype is built on a decentralized overlay peer-to-peer network. Calls can be relayed through an intermediate node if both ends are behind NATs or firewalls that prevent direct connections. Skype hence claims to offer a higher call completion rate than any other VoIP applications [7].

However, every coin has two sides - the aforementioned advantages of Skype may be undesirable under certain circumstances. There may be situations where phone calls are of a personal nature, not business-related, and it is desired to prevent if possible this abuse of network access. In such situations, the capability to intelligently tunnel through firewalls increases the overhead for network

T. Vazão, M.M. Freire, and I. Chong (Eds.): ICOIN 2007, LNCS 5200, pp. 465–481, 2008.

management and is not desirable by the organization. Some organizations, such as banks or government agencies, may be required to monitor communications between their employees and customers for training or auditing purposes. However, they are not able to do so with Skype communications because Skype uses a proprietary protocol and encrypts the communication end-to-end. Additionally, Skype running on a user's computer may route calls for other network users without the user's awareness. This can pose a problem on networks that have limited resources and Internet connectivity. Therefore, it may be seen as highly desirable to be able to detect or block individual Skype nodes running inside a managed network.

However, the ability to tunnel through firewalls by using customized ports increases the difficulty of detecting or blocking Skype traffic at the network layer without blocking some other applications that may be required by an organization. For example, blocking ports greater than 1024 to incoming data packets in order to frustrate use of these ports by Skype will also throttle some streaming protocols [3] [4]. Inspection of packet payloads at the application layer does not help much either, as Skype voice traffic is encrypted. These factors have combined to make detecting or blocking Skype a difficult task for firewalls or intrusion detection systems, and there has not been systematic approach described in the open literature to distinguish Skype voice traffic from other types of streaming traffic, such as real-audio or gaming traffic.

1.1 Related Work

Much research has been done on identifying peer-to-peer traffic. Sen *et al.* [15] presented a signature-based approach to identifying traffic generated by five popular P2P applications. They derived TCP signatures for each application mainly by examining packet-level traces and some available documents. Karagiannis *et al.* [17] extended this approach to nine P2P applications. Furthermore, they proposed two non-payload based heuristics to identify P2P applications, namely "TCP/UDP IP pairs" and "{IP, port} pairs". The "TCP/UDP IP pairs" heuristic identifies source-destination IP pairs that use both TCP and UDP transport protocols. It is based on the observation that most file-sharing P2P applications use UDP and TCP to transfer, respectively, query or query responses and actual data. The "{IP, port} pairs" heuristic utilizes the connection patterns of P2P applications. In P2P networks, a peer advertises its IP and port on the network so that other peers can connect. As a result, the number of distinct IPs connected to the advertised <IP, port> pair will be roughly equal to the number of distinct ports used to connect to it.

There is also considerable research interest in detecting Skype traffic [5] [19] [9]. When either caller or callee or both are behind NATs or firewalls that prevent direction connection, Skype traffic must be relayed via a third node. This situation is referred to as Skype traffic relaying. Suh *et al.* [9] proposed to detect relaying of Skype traffic by an end-host in a monitored network based on the correlation of packet sizes and bit rates between the incoming and outgoing (relayed) Skype traffic flows. They take the perspective of the operator of a large network, who is monitoring incoming and outgoing traffic at an access link. The goal is to determine whether Skype traffic is being relayed through an end-host belonging to the network. Their detection heuristic is based on the facts that 1) two flows carrying Skype-relayed traffic must have opposite directions (one entering, the other leaving the network), have the same end-host (same IP address) within the network being monitored, and have different end-hosts (different IP addresses) outside the monitored network; 2) Skype-relayed traffic is voice traffic and poses strict constraints on maximum delays and minimum bit rates; and 3) patterns of packet sizes and bit rates in Skype-relayed flows are well preserved during the relaying process by the relaying node. That being said, outgoing (relayed) Skype flow demonstrates similar patterns of packet sizes and bit rates to that of incoming Skype flow. Putting these together, two flows satisfying 1) with a relatively small delay in time are considered to be relaying Skype traffic if enough correlation can be found in their packet size and bit rate patterns.

In industry, although several commercial products [5] [19] claim the ability to detect or block Skype traffic, their details have not been made public and their source code is not available. Hence, we have not had the opportunity to evaluate the mechanisms used by these products.

1.2 Our Contribution

This paper focuses on detecting the use of Skype for phone call purposes in a managed network. That being said, we identify end-hosts that participate in phone calls using Skype. We study both payload and non-payload based schemes to identify Skype traffic. We start with a signature-based, i.e. payload based, scheme to detect hosts that are running Skype. By observing and comparing Skype traffic with other types of traffic, we extract signatures, i.e. bit strings, that are unique to Skype. Hosts generating traffic that contains such signatures are considered to be running the Skype client. Our non-payload based scheme is then presented afterwards. We characterize Skype traffic from various aspects including packet size pattern, byte rate pattern, inter-arrival time pattern, and connection pattern. The non-payload based scheme is then developed and implemented based on the characteristics observed. Accuracy and effectiveness of the non-payload based scheme is evaluated by comparing results of both schemes running against the same traffic trace. The experimental results reveal that, at least to a certain extent, encryption by Skype to evade content analysis can be overcome.

It is worth noting that payload based identification of Skype is not always practical. There may be legal, privacy, and performance issues on monitoring and analyzing VoIP communications. Moreover, the Skype protocol is proprietary and not publicly available. Signatures of packet payload are only empirically derived from our observations of a particular version of the Skype client, and may change as Skype evolves. These combine to render payload based identification a fragile detection scheme not resistant against change of protocol and software upgrade. This paper therefore focuses on the non-payload based scheme for Skype identification, and uses the payload based scheme mainly to verify its non-payload based counterpart. Our research results reveal that, at least to a certain extent, encryption by Skype to evade content analysis can be overcome, and our non-payload based identification scheme is more resistant to Skype version change than its payload-based counterpart.

Skype tends to use UDP for data transfer as much as possible [13]. However, there are circumstances where Skype has to use TCP. We leave this situation as our ongoing work.

The rest of this paper is organized as follows. Section 2 provides an overview of Skype. Section 3 introduces our payload based detection techniques. Section 4 analyzes the realtime characteristics of Skype voice traffic. Section 5 presents our non-payload based detection techniques of Skype traffic. Section 6 provides experimental results and Section 7 concludes the paper.

2 Skype Overview

Skype is a proprietary but free peer-to-peer VoIP application developed by N. Zennstrm and J. Friis, the creators of KaZaA. It allows users to place voice calls, send text messages and files to other users. Skype has the reputation for working seamlessly across firewalls and NATs, and providing better voice quality than Microsoft MSN and Yahoo IM [13]. Skype provides data confidentiality using end-to-end 256-bit AES encryption. Symmetric AES keys are negotiated using 1024 bit RSA [16].

Before a Skype client can place calls or send text messages to other users of Skype clients, it must first join the Skype peer-to-peer network. We refer to this process as *login* stage. It is during this process that a Skype client determines the type of NAT and firewall it is behind, authenticates its user name and password with the login server, advertises its presence to other peers and its buddies [13]. After login, users can place calls to other Skype clients through *call establishment* stage, during which a connection between caller and callee is created and a session key is established if one does not exist already [18]. Voice media is packetized and transferred between participating peers after that. Baset *et al.* [13] observed that call establishment was always signalled by TCP, while voice media was transferred over UDP as much as possible. In the following discussions, we refer to traffic regarding login, connectivity establishment, and call setup/teardown as *signalling messages*, and packetized voice data as *voice traffic*.

3 Payload Based Detection

We now describe our Skype traffic detection techniques. This section focuses on payload based techniques, and the section following presents non-payload based techniques. In the rest of this paper, all observations and experiments are performed for Skype version 1.3, unless otherwise stated.

Our payload-based identification of Skype traffic is based on characteristic signatures, i.e. bit strings, observed in packet payload, which potentially represent Skype signalling messages such as login or call establishment traffic. As Skype uses a proprietary protocol, we empirically derived a set of signatures by observing TCP and UDP traffic to and from Skype nodes. Traffic is captured and recorded using Ethereal [1] on end-point computers.

We define two signature types — simple signatures and composite signatures. A simple signature is certain characteristics that a single packet presents, and a composite signature represents characteristics that are presented collectively by a number of consecutive packets. We empirically derived a set of simple and composite signatures by observing traffic generated by Skype. We repetitively carry out independent experiments and observations over months by varying a number of factors, including caller ID, callee ID, caller IP type, callee IP type, date and time and duration of call, to ensure the effectiveness and stability of our signature set.

3.1 Notations and Preliminaries

Let [a,b] denote the set of numbers x such that $a \leq x \leq b$. Let *SrcIPAd*, *DestIPAd*, *SrcPort*, *DestPort*, *Protocol*, *Payload*, *Dir* denote source IP address, destination IP address, source port number, destination port number, transport protocol, packet payload, and direction of the packet respectively, where *SrcIPAd*, $DestIPAd \in [0, 2^{32} - 1]$, *SrcPort*, $DestPort \in [1, 2^{16} - 1]$, $Protocol \in \{TCP, UDP\}$, $Dir \in \{in, out\}$, and *Payload* represents a byte sequence of variable length. We use $|Payload|$ to denote the length of byte sequence represented by *Payload*. We also use $Payload[i - j]$, where i, $j \in N$, to denote the sub-sequence that begins with the ith and ends with the jth byte of *Payload*.

We assume that a packet is convertible to the following 7-tuple: {*SrcIPAd*, *DestIPAd*, *SrcPort*, *DestPort*, *Protocol*, *Payload*, *Dir*}. Packets can then be classified into flows, defined by the 5-tuple {*SrcIPAd*, *DestIPAd*, *SrcPort*, *DestPort*, *Protocol*}. A packet p belongs to a flow f if they have the same *SrcIPAd*, *DestIPAd*, *SrcPort*, *DestPort*, and *Protocol*. In the following, we use $p.X$ or $f.X$ to denote a particular field X that belongs to packet p or flow f. To simplify our notation, we also use $p[i - j]$ to denote $p.Payload[i - j]$.

3.2 Simple Signatures

For a packet $p = \{SrcIPAd, DestIPAd, SrcPort, DestPort, Protocol, Payload, Dir\}$, let *ByteVal* denote a sub-sequence of *Payload*, i.e. $\exists i, j \in N$, such that $ByteVal = Payload[i - j]$, and we use Idx to represent the pair $\{i, j\}$. Then we define a simple signature as a 5-tuple $\{ByteVal, Idx, |Payload|, Protocol, Dir\}$.

For example, assume outgoing UDP packets of length 18 having the third byte $0x02$ are observed frequently, then the simple signature for it can be denoted as $\{0x02$, $\{3, 3\}$, 18, UDP, out$\}$.

Table 1 summarizes the repetitively occurring simple signatures that we identified during the observation period on Skype traffic. It is worth noting that occurrences of these signature are independent of firewall and NAT configurations, as Skype always starts with the same signalling messages in any attempt to connect to the outside world.

Table 1. Simple Signatures

$ByteVal$	Idx	$\|Payload\|$	$Protocol$	Dir
$0x02$	{2, 2}	18	UDP	out
$0x01$	{3, 3}	23	UDP	out
$SrcIPAd$	{3, 6}	11	UDP	in

3.3 Composite Signatures

The rationale behind composite signatures is that, although encrypted by a proprietary protocol, Skype traffic still presents some characteristics of its protocol. For example, a query/response pair may contain the same cipher text that is encrypted from a common value shared by the query/response messages.

Let f denote a flow which consists of packet sequence $\{p_1, p_2, \ldots, p_N\}$ in flow f. Let $SigLength$ denote the number of packets in f, i.e. N. Let $PktLengths$ denote a sequence of natural numbers representing payload length of each packet, i.e. $\{|p_1.Payload|, |p_2.Payload|, \ldots, |p_N.Payload|\}$. Let $BoolCondition$ denote a boolean condition $p_a[i_a - j_a] <|=|> p_b[i_b - j_b]$, where a, $b \in [1, N]$, i_a, $j_a \in [1, |p_a.Payload|]$, and i_b, $j_b \in [1, |p_b.Payload|]$, meaning that the numeric value of byte sequence consisted of the i_ath to j_ath byte of $|p_a.Payload|$ is greater than, equal to, or less than that consisted of the i_bth to j_bth byte of $|p_b.Payload|$. For example, assume $|p_1.Payload| = 1d5002559528$ and $|p_2.Payload| = 1d5102559520$, then we say $p_1[0-1] < p_2[0-1]$, $p_1[2-3] = p_2[2-3]$, and $p_1[4-5] > p_2[4-5]$. Let $Condition$ denote a set of boolean conditions, i.e. $Condition = \{BoolCondition_1, BoolCondition_2, \ldots, BoolCondition_K\}$. We then define a composite signature as a 5-tuple $\{SigLength, PktLengths, Condition, Protocol, Dir\}$.

For example, assume that flows containing the sequence of consecutive packets as depicted in Figure 1 has been observed frequently,

UDP 1d500250ed4f9528ce2990... (18 bytes, out)
UDP 1d500782952893826e2052 (11 bytes, in)
UDP 1d501301826e205283952a... (23 bytes, out)

Fig. 1. An example of composite signature

Then the signature of the above pattern can be represented by a composite signature as {3, {18, 11, 23}, $\{p_1[0-1] = p_2[0-1] = p_3[0-1]$, $p_1[6-7] = p_2[4-5] < p_3[9-10]\}$, UDP, {out, in, out}}.

Table 2 summarizes the repetitively occurring simple signatures that we identified during the observation period on Skype traffic.

Table 2. Composite Signatures

$SigLength$	$PktLengths$	$Condition$	$Protocol$	Dir
4	{18, 11, 23}	$p_1[2] = p_4[2] = 0x02$ $p_1[0-1] = p_2[0-1] = p_3[0-1]$ $p_2[8-10] = p_3[5-7]$	UDP	{out, in, out, in}
2	{18, 26}	$p_1[2] = p_2[2] = 0x02$	UDP	{out, in}

4 Characterization of Skype Traffic

We now introduce a systematic approach to identifying Skype voice traffic at the transport layer, i.e. based on the IP and TCP/UDP header and packet arrival times, without relying on packet payload.

As an Internet telephony application, Skype traffic demonstrates realtime streaming characteristics, i.e. smaller packet sizes and short packet inter-arrival time. On the other hand, being a peer-to-peer software, Skype presents almost identical connection patterns to other peers as do other P2P applications. In the following, we characterize the behavior of Skype traffic from these two aspects, i.e. realtime characteristics and connection patterns.

4.1 Realtime Characteristics

Real-time applications need to emit packets at a relatively small interval for the simulation of continuous and non-delaying effect. This is not a major concern for traditional non-realtime applications that are not operating under strict time constraints. As a result, real-time applications tend to produce smaller packets than traditional client-server applications such as HTTP or FTP. Additionally, for VoIP applications that transfer real-time traffic, UDP is usually preferred over TCP for its timely delivery and smaller header overhead.

We study the realtime characteristics of Skype voice traffic from 3 aspects, i.e. packet size, packet inter-arrival time, and bandwidth burstiness. We study these realtime characteristics by analyzing traffic captured on end-point hosts running Skype. Skype traffic is generated with a number of independent and varying factors, including caller ID, callee ID, caller IP type, callee IP type, audio type, date and time, and duration of call. Experiments are repeated over months to ensure that the realtime characteristics observed are consistent and stable.

We illustrate the realtime characteristics of Skype by an example where calls are established between a host in a research lab and a host connected with shared residential ADSL and behind NAT. Call durations vary from 10 seconds to 30 minutes. In respect to the traffic capture point which is the host at the research lab in this example, voice traffic that is sent to/from the host behind residential ADSL is *outbound/inbound* traffic respectively.

Packet Size Characteristics. Figure 2 depicts the distribution of packet sizes when call durations vary in 10 seconds, 30 seconds, 1 minute, 3 minutes, 10 minutes and 30 minutes using *cumulative density function* (CDF). It can be seen that packet size distributions are self-similar over various call durations and time-scale independent. That being said, packet size distribution is consistent over varying durations, and mainly centers around 120 bytes. Packets that are longer than 50 bytes and shorter than 150 bytes constitute the major portion. We observed that packets falling outside this range are mainly captured at the call establishment stage, and thus believe that they are signalling messages instead of voice traffic. This can be justified by that cumulative density of packets with packet size around [50, 150] bytes increases as calls last longer. It can also be

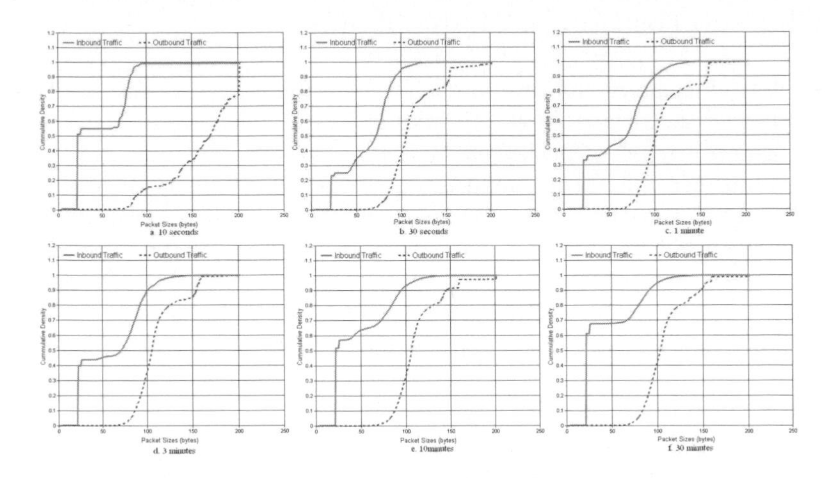

Fig. 2. Packet Size Cumulative Density Function

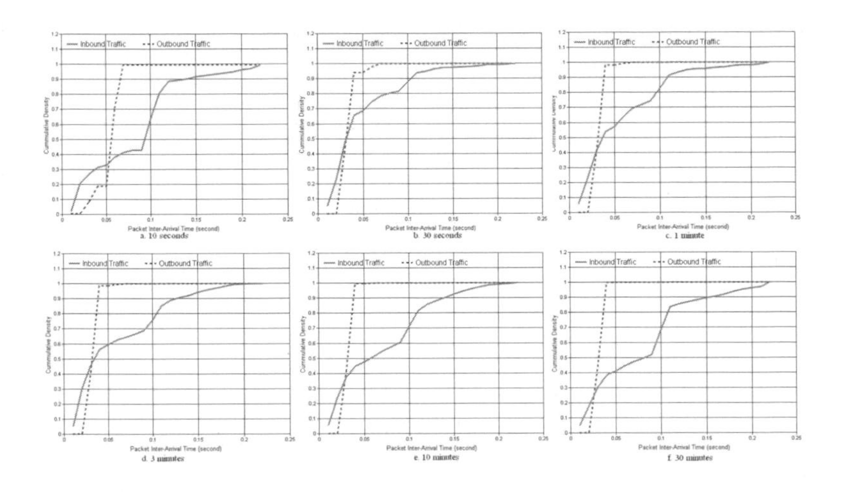

Fig. 3. Packet Inter-Arrival Time Cumulative Density Function

observed that the inbound traffic contains many packets between 20 to 30 bytes. We attribute this to signalling messages that are related to NAT, as they are only observed when the host that we communicate with is behind NAT.

Packet Inter-Arrival Characteristics. Figure 3 depicts packet inter-arrival time *cumulative density function* (CDF) of Skype voice traffic. It can be seen that most outbound packets, i.e. packets sent to the host behind residential ADSL, are sent in the range between 0.02 to 0.04 second. In addition, the percentage of packets arriving within 0.04 second increases as call duration becomes longer. We believe that packets with inter-arrival time greater than 0.04 second are signalling messages. For inbound traffic, packets arrive fairly uniformly over a

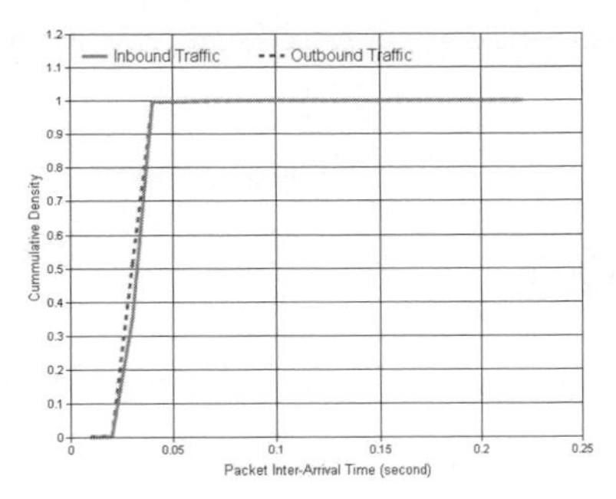

Fig. 4. Packet Inter-Arrival Time Cumulative Density Function Captured at Residential ADSL

range of 0.01 to 0.1 second. We attribute this spread of times to the delay imposed by the shared residential ADSL connection with very limited upload capability (128Kbps).

Bandwidth Burstiness Characteristics. We measure bandwidth burstiness by not only byte rate but also packet rate, i.e. the number of packets sent per time unit. Figure 5 and 6 depict byte rate and packet rate consumed by Skype voice traffic in the start-up 30 seconds and over a 30-minute period, respectively. It can be seen that, for outbound traffic, both packet rate and byte rate are fairly constant after a rise-up stage in the first few seconds. Packet rates stayed at 33 or 34 packets per second, and byte rate slightly fluctuated between 3 to 5 kilobytes per second. This observation is also supported by the empirical study at a larger scale in [14]. On the other hand, inbound traffic demonstrates strong fluctuation. We again attribute this fluctuation to delay and packet loss imposed by the shared residential ADSL connection which has very limited upload capability.

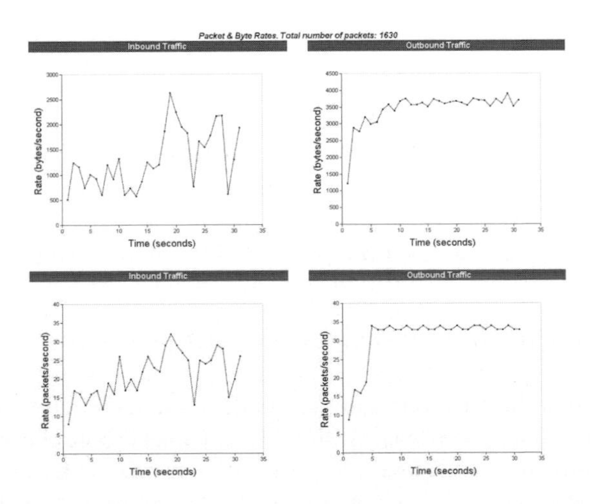

Fig. 5. Bandwidth Burstiness of the start-up 30 seconds

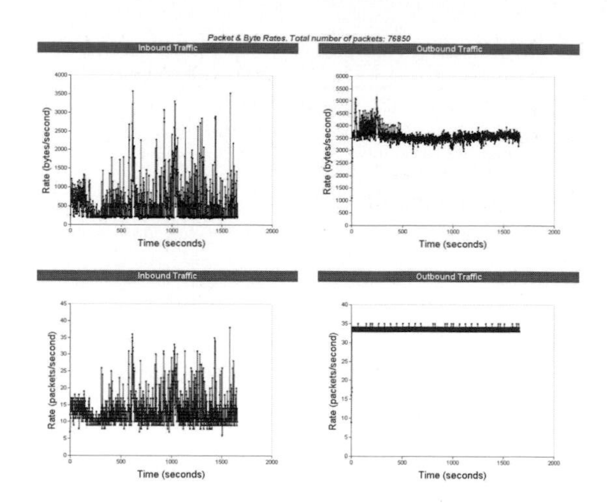

Fig. 6. Bandwidth Burstiness of 30 minutes

4.2 Connection Patterns

Peer-to-peer traffic demonstrates certain connection characteristics. Karagiannis *et al.* [17] proposed two non-payload based heuristics, namely "TCP/UDP IP pairs" and "{IP, port} pairs", to identify peer-to-peer traffic from other traffic. However, it is worth noting that they can only distinguish peer-to-peer traffic from other types of traffic, but cannot distinguish peer-to-peer traffic generated by one particular application from another.

"TCP/UDP IP pairs" identifies source-destination IP pairs that operate TCP and UDP on the same port. Unfortunately, we have observed frequent departure from this pattern in our experiments, i.e. Skype uses only one protocol for each source-destination IP pair. Hence, we are not going use this pattern in identifying Skype voice traffic.

"{IP, port} pair" utilizes the fact that for the advertised {IP, port} pair of host A, the number of distinct IPs connected to it will be equal to the number of distinct ports used to connect to it. As do most peer-to-peer applications, the advertised port used by Skype can be configured by users. Change to the port will be applied the next time Skype is started. In our experiments, we observed that this advertised port is not only used as a destination port for incoming connection attempts, but also as a source port for outbound voice traffic.

5 Non-payload Based Detection Technique

In this section we will introduce our non-payload based technique to identify Skype voice traffic from other types of traffic. Our identification heuristic combines the realtime characteristics of Skype voice traffic as discussed in Section 4 and the "{IP, port} pair" heuristic. That being said, we consider a host has had Skype conversation if a port is identified to be an advertised peer-to-peer port by the "{IP, port} pair" heuristic, and traffic associated with this peer-to-peer advertised port demonstrates realtime characteristics as discussed in section 4. We only consider outbound traffic with respect to the monitoring point which is relatively close to the point where outbound traffic is generated, as it can be seen from Figures 3, 5 and 6 that realtime characteristics of inbound traffic may not be well preserved due to transmission delay on Internet.

5.1 Conventional Client-Server Applications and Other Peer-to-Peer Applications

Conventional client-server applications, such as HTTP and FTP, demonstrate substantially distinct characteristics from Skype voice traffic. They usually use pre-defined, well-known source and/or destination port numbers; they exclusively rely on TCP, and exhibit large packet sizes. We considered a flow was generated by conventional client-server applications if its traffic conforms to such characteristics.

Many peer-to-peer applications are used for file and music sharing. In such applications, while reducing transmission errors and header overhead is considered important, timely delivery of data is usually trivial and not relevant. Hence, they typically use TCP for actual data transfer but may use UDP for signalling. On the other hand, Skype was designed from the beginning to deliver data in realtime, and prefers the use of UDP for voice transmission as much as possible [13]. We hence distinguish Skype from other conventional peer-to-peer applications.

5.2 Realtime Applications

We now focus on other types of realtime UDP-based applications. In the top 25 UDP application categories seen on Internet [11], Real Audio [2] accounts for the largest number of bytes among non-anonymous UDP applications. In addition to being a realtime application, Real Audio also demonstrates certain characteristics similar to peer-to-peer applications — it uses both TCP and UDP to transport data. To distinguish Real Audio from Skype traffic, we develop two heuristics as follows

- Real Audio traffic is dominated by specific packet sizes [12].
- Real Audio traffic is unidirectional. Volumes of inbound and outbound traffic are highly asymmetric.

Out of the top 25 UDP applications [11], 14 of them are Internet realtime strategic games, such as Starcraft and Half Life. Karagiannis's heuristic to identify gaming traffic is based on the viewpoint that Internet based realtime strategic games are inclined to employ packets dominated by specific packet sizes. This pattern is expected by Internet based games as each player sends out multiple copies of its current state to each other player [10] [6] [20]. However, games and gaming traffic differ a lot from one another, and some may employ packets not dominated by specific size. Due to different gaming types, packet sizes may vary in a wide range which is hard to predict. We hence extend Karagiannis's heuristic by taking into consideration the periodicity of gaming traffic, i.e. frequencies by which each player sends update of its status to other players. Our viewpoint is based on that gaming traffic demonstrates this periodicity with bandwidth consumption fluctuation and burstiness [20] [8]. On the other hand, packet rate of Skype traffic is relatively constant as demonstrated in Figures 5and 6.

Besides Real Audio and gaming traffic, other VoIP applications such as Microsoft MSN or GnomeMeeting may also potentially demonstrate realtime characteristics that are not distinguishable from Skype. However, they are mostly built on standard-based protocols such as H.323 or SIP, and transfer voice traffic through a dynamically negotiated port. As opposed to the advertised port that Skype uses to transfer voice traffic with the chatting peer and to signal other peers simultaneously, this negotiated port is dedicated to transferring voice traffic by the chatting peers of a particular session. We can therefore distinguish Skype from other standard-based VoIP applications.

5.3 Final Algorithm

We use x and y_1, y_2, y_3 to denote, respectively, the $\{IP, Port\}$ *pair* connection pattern and the realtime characteristics that Skype presents as follows,

- x <IP, Port> pair heuristics.
- y_1 Packet sizes are relatively small and the majority follows normal distribution.
- y_2 Relatively constant packet and byte rate.
- y_3 Packet inter-arrival time mainly resides in between [0.02, 0.04] second.

We use the matrix as shown in Table 3 to summarize distinct characteristics of various applications. It can be seen that Skype can be distinguished from each of other applications by at least one differentiating characteristic. Note that here we assume the worst case for other VoIP applications, i.e. they present the same realtime characteristics as do Skype.

Combining the techniques of all previous sections yields our final non-payload based identification method for Skype voice traffic. Note that our algorithm is designed for analysis of passive traffic traces, allowing multiple passes over the data if necessary. Adapting our algorithm to detect and block Skype traffic dynamically is part of our ongoing work. Our final algorithm is presented in Algorithm 1.

Table 3. characteristics matrix

Application	x	y_1	y_2	y_3
Skype	✓	✓	✓	✓
Conventional Web Apps				
Other P2P Apps	✓			
Real Audio			✓	
Online Games	✓			✓
Other VoIP Apps		✓	✓	✓

Algorithm 1: Nonpayload algorithm for Skype voice traffic identification

```
begin
    /* constants definition                                          */
    const FT = All Flows ;

    for each flow f in FT ;
    do
        if Use of <IP, Port> pair connection pattern then
            if Packet sizes are relatively small and the majority follows normal
            distribution then
                Packet_Size_Heuristic = true
            end
            if Relatively constant packet and byte rate then
                Packet_Rate_Heuristic = true
            end
            if Packet inter-arrival time mainly resides in between [0.02, 0.04]
            second then
                Inter_Arrival_Heuristic = true
            end
            if Packet_Size_Heuristic and Packet_Rate_Heuristic and
            Inter_Arrival_Heuristic then
                SkypeCall.insert(f) ;
            else
                OtherP2P.insert(f);
            end
        else
            NoneP2P.insert(f);
        end
    end
end
```

5.4 Discussions

Traffic classification techniques based on pattern recognition, including signature based techniques, can be invalidated by applying variations to the distinctive patterns. In particular, our non-payload based technique is built upon realtime characteristics of Skype traffic such as packet size and bandwidth consumption characteristics in addition to the peer-to-peer connection pattern, and hence can be circumvented by applying changes and variations to such realtime characteristics. For example, "junk bytes" can be added to Skype packets to change packet sizes. This also changes the bandwidth consumed by Skype. However, Skype is designed to deliver packets in realtime as much as possible. Although increase in packet size or bandwidth consumption can circumvent our detection scheme, it could also severely degrades call quality of Skype. On the other hand, unless voice compression techniques can be significantly improved in the near future, reducing packet size is desirable but not quite practical. Moreover, downward compatibility must be considered for any change made to a widely deployed application. This requires significant efforts and takes a relatively long time. Therefore, realtime characteristics presented by Skype as it is now is expected to remain at least in the near future. We empirically justified our viewpoint by experimenting our non-payload based technique, developed from Skype 1.3, with the latest Skype version 2.0 in Section 6.2.

6 Implementation and Experiments

We implemented the algorithm and techniques presented in all previous sections in Java. The experiments were performed on a Dell Optiplex GX280 with 1 GBytes of RAM and a 2.8GHz processor running Windows XP Professional.

We evaluate our schemes from two perspectives — false-positive and false-negative. False-positive evaluates accuracy of our schemes, i.e. likelihood that other non-Skype traffic are misclassified as Skype traffic. False-negative indicates the extent of misclassification where our schemes fail to identify Skype traffic.

6.1 False-Positive Evaluation

We first evaluate the false-positive, i.e. non-Skype traffic being identified as Skype traffic, of both payload and non-payload based schemes. We obtained from CAIDA[1] two OC-48 traffic traces that were captured in the early of 2003. They offered us large data sets that were expected to contain little or no Skype traffic because the initial version of Skype was released on August 29, 2003. All packets are truncated to 48 bytes, and hence UDP packets are preserved up to 16 bytes of data[2]. This is not an issue in our situation as the payload-based signatures examine up to the $10th$ byte in payloads, and our non-payload based scheme does not rely on any payload. Details of the traffic traces are described by trace file #1 and #2 in table 4.

On one hand, our non-payload based scheme identified 6 Skype flows in trace #1 and 1 Skype flow in trace #2. Therefore, it has resulted in at most 7 false-positives out of 2.1 million UDP flows, i.e. about 0.00003% flows are misidentified as Skype calls.

On the other hand, the payload based scheme identified 916 simple signature #1 and 4 simple signature #2 from trace #1, and 844 simple signature #1 from trace #2. Therefore, it has resulted in at most 1764 false-positives out of 172

Table 4. OC-48 Traffic Traces

Trace File No.	Packet No.	Bytes	Start Time	Dur.	Flow No.	UDP Flow No.
1	84 Mil.	43G	5:00pm, Apr 24, 2003	60 min.	8136 K	1498 K
2	88 Mil.	62G	4:59am, Jan 16 2003	26 min.	3176 K	653 K
3	14 Mil.	4.2G	4:00am, Mar 16, 2006	1 week	154 K	77 K

[1] http://www.caida.org

[2] Taking away 4 bytes Cisco HDLC header, 20 bytes IP header, and 8 bytes UDP header.

million packets, i.e. about 0.001% packets are mis-identified as simple signature and no packet sequence is mis-identified as composite signature.

6.2 False-Negative Evaluation

We next explore the extent of misclassification, where the non-payload based scheme fails to identify Skype traffic, by comparing the result from payload based scheme. We do not investigate the false-negative of our payload based scheme directly. Instead, we ensure that the signature set we derived is at least sufficient to the organization and environment where the experiments are conducted by repetitively carrying out independent experiments over months with a number of varying factors including caller ID, callee ID, caller IP type, callee IP type, date and time and duration of call.

We monitored a week's traffic on a boundary firewall in our organization. The experimental setup is illustrated in Figure 7. All hosts are Dell Optiplex series running Windows XP Professional and connected to Ethernet Switches. Trace file #3 in table 4 details the specification of traffic captured. We use Skype to call or receive calls from other users located on the other side of the firewall or on Internet, as indicated by dashed lines in Figure 7. As our non-payload based scheme heavily relies on packet inter-arrival timing, delay by traffic filtering may degrade its effectiveness and accuracy. Therefore, in order to study the effectiveness of our non-payload based scheme in the presence of filtering delay, we kept the firewall busy outside normal office hours by running random traffic generators. The randomly generated traffic traverses through the firewall as indicated by solid lines in Figure 7. During the experiment period, we also made a frequent use of other types of streaming applications when no Skype call is happening, including streaming video and audio, real time strategic online

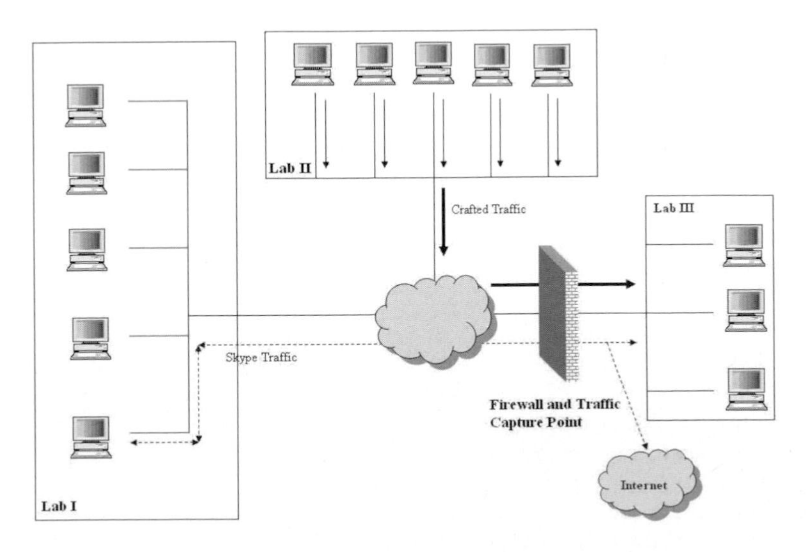

Fig. 7. Experimental Setup

games, and other types of VoIP applications such as MSN, Yahoo! Messenger and Google Talk, which including Skype combined to generate 4.2 GBytes streaming traffic.

During a week's time, we made in total 80 calls with varying caller ID, callee ID, caller IP type, callee IP type, date and time and durations of call which vary from 1 to 106 minutes averaging at 28 minutes. Our non-payload based scheme successfully identified all 80 calls. Within the duration of each identified call, payload based signatures are also identified. Therefore, although being encrypted, Skype traffic can still be effectively identified without looking at its encrypted payload, or at least as effective as its payload based counterpart.

Upgrade to Skype 2.0. All the experiments so far are performed with Skype 1.3. In order to test the resistance of our non-payload scheme against Skype version upgrade, we repeat the experiments in Section 6.2 with Skype 2.0. We made in total 60 calls with Skype 2.0 in a week's time. Call durations vary from 2 to 46 minutes, and average at 7 minutes. Our non-payload based scheme successfully identified all 60 calls with no false-positive. It is worth noting that our payload based scheme has not detected any occurrence of simple signature #2 and composite signature #1, i.e. these two signatures do not survive from version upgrade. This result shows that, although our non-payload based scheme is built on pattern recognition that can be invalidated by future changes to Skype protocol or variations to Skype traffic patterns, it is at least more resistant to such changes and variations than its payload (signature) based counterpart. The experimental result also empirically justifies our hypothesis that realtime characteristics of Skype will remain within at least the near future.

7 Conclusion

We presented both payload and non-payload based techniques for detecting Skype activities, especially Skype voice traffic. We believe that, with the presented techniques, institutions that do not permit use of Skype can detect breaches of networking policies and take relevant counter-measurements.

In terms of future work, we will focus on the behavior of Skype under UDP-restricted firewalls or proxies. Adapting the detection techniques to block Skype traffic dynamically will also be investigated. Additionally, opportunity to access ISP provided, large scale traffic traces will be highly regarded.

References

1. Ethereal, `http://www.ethereal.com`
2. `http://asia-en.real.com/guide/radio/list.html`
3. `http://service.real.com/firewall/rplay.html`
4. `http://www.microsoft.com/windows/windowsmedia/serve/firewall.aspx#portallocation`
5. `http://www.websense.com`

6. Pescape, A., Dainottiand, A., Ventre, G.: A Packet-level Traffic Model of Startcraft. In: Proceedings of the 2005 Second International Workshop on Hot Topics in Peer-to-Peer Systems, pp. 33–42. IEEE, Los Alamitos (2005)
7. Anonymous. Why is Skype better than Net2Phone, ICQ, AIM, MSN, etc?, `http://support.skype.com/index.php?_a=knowledgebase&_j=questiondetails&_i=70&nav2=General`
8. Chen, K., Huang, P., Huang, C., Lei, C.: Game Traffic Analysis: An MMORPG Perspective. In: Proceedings of NOSSDAV (2005)
9. Suh, K., Figueiredo, R., Kurose, J., Towsley, D.: Characterizing and Detecting Skype-Relayed Traffic. In: Proceedings of IEEE Infocom, Barcelona (April 2006)
10. Winslow, J., Claypool, M., LaPoint, D.: Network Analysis of Counter-strike and Starcraft. In: Proceedings of 22nd IEEE International Performance, Computering and Communication Conference (IPCCC), Phoenix, Arizona, USA, April 2003. IEEE, Los Alamitos (2003)
11. McCreary, S., Claffy, K.: Trends in Wide Area IP Traffic Pattern: A View from Ames Internet Exchange. In: Proceedings of ITC Specialist Seminar on Measurement and Modeling of IP Traffic, pp. 1–11. Cooperative Association for Internet Data Analysis (CAIDA) (September 2000)
12. Mena, A., Heidemann, J.: An Empirical Study of Real Audio Traffic. In: Proceedings of the IEEE Infocom, Tel-Aviv, Israel, March 2000, pp. 101–110. IEEE, Los Alamitos (2000)
13. Baset, S.A., Schulzrinne, H.: An Analysis of the Skype Peer-to-Peer Internet Telephony Protocol. Technical report, Department of Computer Science. Columbia University, New York (2004)
14. Guha, S., Daswani, N., Jain, R.: An Experimental Study of the Skype Peer-to-Peer VoIP System. In: Proceedings of IPTPS 2006, Santa Barbara, CA (February 2006)
15. Sen, S., Spatscheck, O., Wang, D.: Accurate, Scalable In-Network Identification of P2P Traffic Using Application Signatures. In: Proceedings International WWW Conference, New York, USA (2004)
16. Skype FAQ, `http://www.skype.com/help_faq.html`
17. Faloutsos, M., Claffy, K., Karagiannis, T., Broido, A.: Transport Layer Identification of P2P Traffic. In: Proceedings of the 4th ACM SIGCOMM conference on Internet measurement, Taormina, Sicily, Italy, pp. 121–134. ACM Press, New York (2004)
18. Berson, T.: Skype Security Evaluation. Technical report, Anagram Laboratories (October 2005)
19. Verso Technologies. Verso Netspective Enterprise, `http://www.verso.com/enterprise/netspective/netspective_brochure.pdf`
20. Feng, W., Chang, F., Feng, W., Walpole, J.: A Traffic Characterization of Popular On-Line Games. IEEE/ACM Transactions On Networking 13(3), 488–500 (2005)

Inter-Domain Access Volume Model: Ranking Autonomous Systems

Yixuan Wang, Ye Wang, Maoke Chen, and Xing Li

Network Research Center, Tsinghua University
{wangyx,sando}@ns.6test.edu.cn, {mk,xing}@cernet.edu.cn

Abstract. Exploring topological structure at Autonomous System (AS) level is indispensable for understanding most issues in Internet services. Previous models of AS graph involve address or connectivity information separately, and neither of these information can serve as a comprehensive metric for evaluating an AS's contribution to the global routing. In this paper, we propose a new model for AS ranking named IDAV (Inter-Domain Access Volume). IDAV introduces the quantity of routed addresses into AS graph, and enriches the methodology of Internet marco structure inference. In IDAV, the magnitude of AS is measured with the primary eigenvector of the access volume matrix (Carriage Matrix) for AS graph. We construct the AS graph by parsing the forwarding information bases in border gateways. The computation, compared with previous approaches, demonstrate that IDAV model results in more accurate AS rankings. We believe the IDAV model is promisingly useful for studying inter-domain routing and Internet service behavior.

Keywords: Internet topology, Inter-domain routing, Autonomous system ranking, IDAV model.

1 Introduction

Internet connects thousands of Autonomous Systems[1] (ASs) operated by many different administrative domains. Each domain maintains one or more ASs and hides intra-domain information from others. Macro observation of almost all the fundamental issues of Internet services, such as inter-ISP settlements, troubleshooting, and traffic engineering, strongly relies on the knowledge of Internet structure at AS granularity. Since AS level network topology is determined by inter-domain routing protocol, namely BGP[2] in today's Internet, study on BGP tables is the general way of exploring AS graph. Forerunners have made great efforts to gather BGP routing information. The RouteView project[3] periodically dumps BGP FIBs from multiple vantage ASs. Internet Routing Registries (IRR)[4] provide a database facility for sharing inter-domain routing policies.

With help of these resources, people are trying to get a comprehensive and accurate AS graph. Ge[5], Vazquez[6] and Siganos[7] examined the hierarchical Internet topology by observing the power-law[8] distribution of AS degrees. Policy-based AS graph was first studied by Gao[9]. She classified the AS relationship into 4 categories, and discovered the valley-free principle to infer

T. Vazão, M.M. Freire, and I. Chong (Eds.): ICOIN 2007, LNCS 5200, pp. 482–491, 2008.

AS relationships from BGP routing table. Subramanian[10], Battista[11] and Dimitropoulos[12][13] brought in more elaborate methods and extensive resources to solve the AS relationship problem. However, Internet structure is much more complicated than we have already captured. Besides degrees and edge-types, people continue adding other parameters to strengthen AS graph. There are many choices, such as address space issued by AS, customer cones affiliated to AS, etc. This initiated a new question: which parameter is better?

We believe AS ranking is a fundamental benchmark of modeling Internet topology with augmenting AS graph. By AS ranking, people characterize each AS's importance in Internet as a quantitive weight. There have been quite a few approaches to rank the major backbone ASs, such as CAIDA[14], FixedOrbit[15], Renesys[16], etc. Different metrics can result in different ranking criteria, any of which might be resonable. However, the common sense of AS ranking is that carrying more traffic equals ranking higher, because the main role of an AS is a packet-carrier. Unfortunately, the impossibility of measuring all the traffic between each pair of ASs has prevented such kind of precise ranking. Overwhelming majority of known AS rankings employ the degree-based or relationship-based AS graphs due to the lack of new theories.

In this paper, we originate a new model to evaluate the contribution of AS to Internet routing: Inter-Domain Access Volume (IDAV). Engineering experience tells us inter-domain routing policy is designed and implemented per prefix. We define *access volume* as the quantity of routed addresses propagated between neighbor ASs. This is an indicator of how much service one AS provides for the other. Theoretically, we rank the ASs by the primary eigenvector of the access volume matrix for AS graph. Then in practice, we construct a real AS graph, compute AS rankings, and verify the results from several different sources.

To the best knowledge of the authors, this is the first approach to model and measure Internet structure by investigating both connectivity and routed prefixes information. The rest part of this paper is organized as follows: section 2 presents the theories of IDAV model; section 3 tells the computation methods; results inferred by our model and verification of them is shown in section 4; finally we summarize the paper and briefly specify future works in progress.

2 IDAV Model

2.1 Definitions

Definition 1 (AS Graph). *An AS graph is a directed graph $G(V, E)$, where V is the set of ASs and E is the set of directed edges between two endpoint ASs. It may contain loops but must not contain multiple edges.*

Definition 2 (Carry & Transit). *In inter-domain routing, if ASy reaches prefix $N1$ through ASx, we say that ASx carries $N1$ for ASy, written as $N1 \in ASx \xrightarrow{c} ASy$. If multiple prefixes are carried, they could be bracketed into comma-separated lists, like $\{N1, N2\}$. Since there is no need for an AS to access itself through other ASs, we think every AS carries its own prefixes for itself.*

ASx transits prefix $N1$ for ASy, if and only if: (1) x, y and z are 3 different ASs; (2) ASy carries prefix $N1$ for ASx; (3) ASx carries prefix $N1$ or $N1$'s less-specific for ASz. Written as $N1 \in ASx \xrightarrow{t} ASy$.

The prefix $N1$ is presented in IPv4 (or IPv6) CIDR[17] format. In Internet routing, the definition of transit implies ASx redistributes prefixes received from ASy to ASz, so it complies with the common notion "transit AS".

Definition 3 (Access Volume). *The access volume of an edge $ASx \rightarrow ASy$ in the AS graph is a numeric parameter that equals the total number of unique addresses in the prefixes carried by ASx for ASy. Written as $C(ASx, ASy)$.*

As previously defined, ASx "carry" prefix $N1$ for ASy not only means ASx announces route $N1$ to ASy, but also requires ASy does reach network $N1$ through ASx, e.g. ASy must install route $N1$ into its forwarding table. We shall use each AS's FIB to calculate access volumes, which implies that $C(ASx, ASy)$ is much less than the total number of addresses that ASx announced. Duplicated addresses should be counted only once, e.g., if $AS1 \xrightarrow{c} AS2 = \{1.0.0.0/8, 1.1.0.0/16, 2.0.0.0/8\}$, then $C(AS1, AS2) = 2 \times 2^{24}$, not $2 \times 2^{24} + 2^{16}$.

Definition 4 (IDAV Model). *The Inter-domain Access Volume (IDAV) model is a weighted AS graph $G(V, E, W)$, where $G(V, E)$ is an AS graph as previously defined, and W is the set of numerical weight on each edge that $W(ASx \rightarrow ASy) = C(ASx, ASy)$.*

Definition 5 (CM). *The Carriage Matrix (CM) of a given IDAV model is an n-by-n square matrix M, where n equals the number of nodes in the IDAV model, and each element $m(i, j)$ equals $C(ASi, ASj)$, the access volume carried by corresponding edge. If edge $ASi \rightarrow ASj$ does not exist in the IDAV model, $m(i, j) = 0$.*

In the case that $i = j$, the diagonal element $m(i, i)$ equals the total addresses owned by ASi itself, complying with the definitions of carry and access volume. This mathematical description creates a bijective mapping between IDAV models and CMs: given an IDAV model, we can draw the CM, and vice versa.

2.2 AS Ranking

Unlike previous models, we use virtual traffic throughput as the ranking metric. The problem of AS ranking is: *given an IDAV model denoted by carriage matrix M, how to find an n-by-1 vector R, whose elements are virtual traffic flows $r_k, 1 \le k \le n$, sufficing that $\forall i, j, 1 \le i, j \le n$, if $r_i \le (\ge) r_j$, the real traffic throughput of ASi is not less (more) than that of ASj.* Moreover, besides relatively reflecting real traffic amount, could virtual traffic be proportional to it?

Theoretical Analysis. Ideally, the IDAV model and carriage matrix should suffice to the following conditions:

1. If $N1 \in ASi \xrightarrow{c} ASj$, then $N1 \notin ASk \xrightarrow{c} ASj$, $\forall k,\ 1 \leq k \leq n \, and \, k \neq i$.
2. $\sum_{i=1}^{n} m(i,j) = Const,\ \forall j,\ 1 \leq j \leq n$.

The first condition requires unique path selection, and the second condition means every AS should know routes to the whole Internet. While $Const$ in condition 2 equals the total address spaces that have been allocated by IANA[18] and utilized by ASs. In ideal situation, we have the following theorem:

Theorem 1. *Given an IDAV model denoted by carriage matrix M, the ranking R of the IDAV model suffices to the equation: $M \cdot R = \rho(M) \cdot R$, where $\rho(M)$ is the spectral radius (the maximum complex modulus of eigenvalues) of M.*

Proof

1. *Existence.* According to Perron-Frobenius theorem[19], the carriage matrix M derived from IDAV model is a non-negative matrix, so that $\rho(M)$ is the eigenvalue of M and there is a non-negative vector R, $R \neq 0$, holds the statement $M \cdot R = \rho(M) \cdot R$. Thus the ranking vector R shall always exist. Moreover, if $\sum_{i=1}^{n} m(i,j) = Const,\ \forall j,\ 1 \leq j \leq n$, the spectral radius $\rho(M) = Const$.

2. *Rationality.* First we normalize the carriage matrix M with $Const$: let $P(i,j) = m(i,j)/\sum_{i=1}^{n} m(i,j),\ \forall i,j,\ 1 \leq i,j \leq n$. $P(i,j)$ means the portion of access volume carried by ASi for ASj in the total address spaces. Let us consider example in figure 1. AS1 has 3 neighbors (left), each of which has some access volume carried by AS1. AS1 also carries its own addresses for itself. By adding a virtual intersection VI (right), the traffic incoming to VI equals the sum of traffic outgoing from VI's neighbors, including AS1 itself. For each neighbor, such as AS2, its egress traffic is most likely to be evenly distributed among all the destination addresses, so the traffic from AS2 to VI equals AS2's total egress traffic times $P(1,2)$. Let E_j and I_j be the egress and ingress traffic of ASj respectively, we have this formula: $I_j = \sum_{i=1}^{n} E_i \cdot P(i,j)$, e.g. $I = M/\rho(M) \cdot E$. Finally, because VI associated with AS1 is a pure exchange point, leaving no remaining traffic within itself, its egress and ingress traffic vector must be equal, thus we can get the ranking vector $R = E = I$.

Now let us illustrate the process of AS ranking using IDAV model. In figure 2, each AS owns only one address and provides a complete route forwarding for others. If we merely calculate the addresses, each AS would be viewed

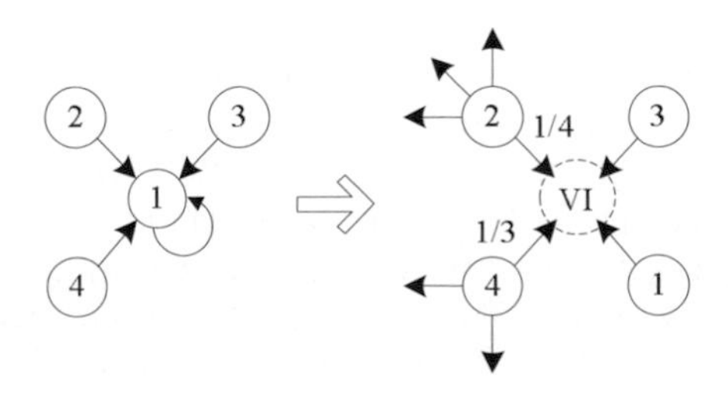

Fig. 1. VI: Proof of Theorem 2

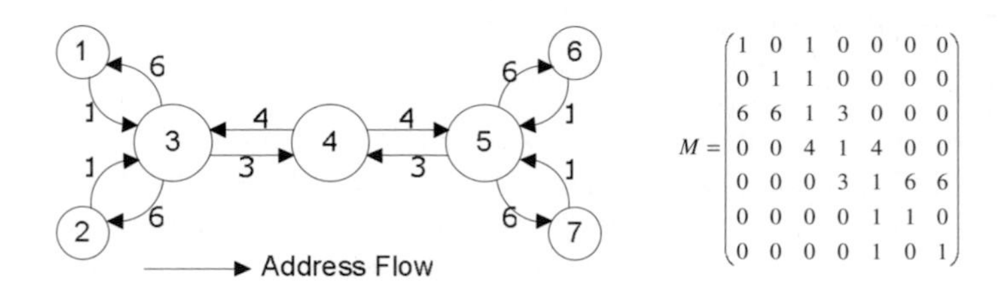

Fig. 2. IDAV-based AS Ranking

equally important. If we use the degree-based criterion, AS3 and AS5 are both bigger than AS4. Neither of the judgements is convincing. In real world, AS4 may be a global exchange point, while AS3 and AS5 are only national ISPs yet with more customers. Using IDAV, the CM of this AS graph is M. Solving equation $(M - 7 \cdot I) \cdot R = 0$, we get: $R = [0.08, 0.08, 0.51, 0.67, 0.51, 0.08, 0.08]$. We see although AS3 and AS5 have more customers, as long as AS4 is doing transit for both of them, AS4 ranks higher. This example shows that our method has obvious advantage because IDAV uses routing contribution for ranking.

Practical Considerations. In actual situation, the two conditions in previous section will no longer hold. Anyway, from theorem 1, we can still get R as the estimation of real traffic (albeit quite approximately): on premise of the two following rules, elements of R is at least positively related to potential traffic.

1. Capability Rule: The more access volume does an AS carry for other ASs, the higher it ranks.
2. Chain Rule: Suppose ASx carries some access volume for ASy, the higher does ASy rank, the higher ASx ranks.

Although we can not prove them, the two rules are compliant with people's common sense. The first rule says that capable ASs are always busy, while the second rule says that the contribution of an AS to the Internet depends on the weight of the node it serves. These rules ensured that even in non-ideal situation, we can employ theorem 1 to calculate AS ranking. For each ASj in the model, $r_j = \sum_{i=1}^{n} r_i \cdot m(i, j)$, if either the access volume $m(i, j)$ or another AS's ranking r_i goes bigger, the ranking r_j of ASj will get bigger consequently. It will nicely meet the two rules' requirements.

Admittedly, there are some limitations in applying our ranking theory, and the model would be refined with respect to the following factors: 1) It is almost impossible to gather all the route forwarding paths from every AS. 2) BGP signaling path is sometimes inconsistent with the actual packet forwarding path[20]. 3) Egress traffic of an AS is not evenly distributed among addresses[21][22]. The refinement to our model is another problem, which will be investigated in future works. Currently we use the no-correction form of IDAV model to demonstrate some results, and consider the influences of above phenomena negligible.

3 Computations

3.1 Construction of IDAV

To build up a comprehensive Internet AS graph in IDAV, we use the BGP dump data collected at 02:43 AM PDT on 2005-12-24 by university of Oregon's RouteViews project, many thanks to their generosity. The routes are presented in either binary zebra MRT[23] or human readable Cisco CLI[24] format, both of which can be parsed into BGP attribute tuples as shown in table 1. Of all these attributes of a BGP table entry, NETWORK and AS_PATH are the most important to our study. Our aim is to parse each route into some $N \in ASx \xrightarrow{c} ASy$ atoms. We use perl to analyze text, and do the construction and computation of IDAV model with C programming language.

In the AS_PATH attributes, each pair of neighbor AS numbers shows that the route is sent from the rightside AS to the leftside AS. As in tuple 1, we can get $8.8.9.0/24 \in AS27646 \xrightarrow{c} AS1239$ and $8.8.9.0/24 \in AS1239 \xrightarrow{c} AS1668$, etc. Sometimes the AS_PATH contains *repeated-prepending* or *aggregated AS_SETs* as in tuple 2 and 3, we just erase the duplicated AS numbers and decompose the aggregated ASs to count the routes separately. After parsing BGP routing tables into (network, AS-link) atoms, we create an IDAV model by scanning the Oregon data, then calculate each edge's access volume, and finally perform the following iterative algorithm on the CM to get the ranking vector.

3.2 Ranking Algorithm

In ideal situation, the IDAV model is a strong-connected graph, thus the CM is primitive, we have the following limit theorem[19]:

$$\lim_{m \to \infty} [\rho(M)^{-1} \cdot M]^m = L_{n \times n} > 0. \tag{1}$$

Thus the ranking vector R can be found through simple iteration starting from any initial non-zero vector $R(0) = l$, and $R(t) = M/\rho(M) \cdot R(t-1)$.

But in non-ideal situation, the limit theorem does no longer hold, for the CM is only a general non-negative matrix. Thanks to previous works like Pagerank[25] [26], under these conditions, we can still employ an iteration algorithm to solve the equation $M \cdot R = \rho(M) \cdot R$:

$$R(t) = (1 - \varepsilon) \cdot W \cdot R(t-1) + e(t-1) \cdot \ell_n, \tag{2}$$

Table 1. Sample BGP Table

NETWORK	NEXT_HOP	METRIC	LOCPRF	AS_PATH	ORIGIN
8.8.9.0/24	66.185.128.48	514	0	1668 1239 27646	?
166.111.0.0/16	66.185.128.48	514	0	1668 1239 4538 4538 4538	i
24.223.128.0/17	66.185.128.48	575	0	1668 10796 {11060,12262}	i

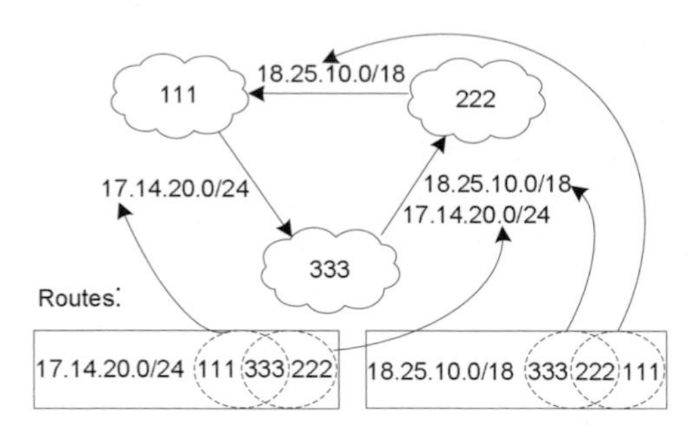

Fig. 3. Build IDAV model from BGP

where, $R(t)$ is the ranking vector after each iteration; W is the CM M normalized by column, e.g. $\sum_{i=1}^{n} w_{i,j} = 1,\ \forall j,\ 1 \leq j \leq n$; a small real number ε close to 0 and $\ell_n = [1, \ldots, 1]^T$ are used to guarantee convergency. Finally, for each t,

$$e(t-1) = \frac{1 - \|(1-\varepsilon) \cdot W \cdot R(t-1)\|_2^2}{2\|(1-\varepsilon) \cdot W \cdot R(t-1)\|_1}, \tag{3}$$

to force the condition $\|R(t)\| = 1$.

4 Results

4.1 Top 10 Ranking

Applying our ranking method to the final IDAV model, we calculated a numeric weight value for each of the 21,454 ASs appeared in the routing table. The weight vector is normalized to 1. As people are usually more interested in the bigger ones, in table 2 we show the top 10 ASs we found.

Table 2 contains most of the well-known global backbone ASs, such as AS1239 and AS3356, as well as those very large but not so famous ISPs, such as Cogent,

Table 2. Top 10 ASs Found

Rank	ASN	Weight	Degree	Organization
1	1239	0.500845	1755	Sprint
2	3356	0.443078	1281	Level 3
3	209	0.397985	1155	Qwest
4	7018	0.342186	2019	AT&T
5	701	0.314771	2420	MCI/UUNET
6	2914	0.207631	496	NTT
7	3549	0.140935	700	Global Crossing
8	3561	0.137122	583	SAVVIS
9	174	0.125386	1364	Cogent
10	1299	0.125087	370	TeliaNet

Table 3. Other Top 10 AS Rankings

	Caida	FO.	RS.	IDAV
1	701	3356	1239	1239
2	7018	6461	3356	3356
3	1239	1239	701	209
4	174	3303	7018	7018
5	3356	2914	2914	701
6	209	8075	3549	2914
7	3549	4637	209	3549
8	7132	209	3561	3561
9	4323	3549	1299	174
10	3303	2497	6453	1299

TeliaNet. From the degree values listed in the 4th column, we can see our result is quite different from degree-based rankings. A typical example is AS2914: based in America, peering with the world's top ISPs, it is doing global transit for most APNIC ASs, naturally its ranking is very high although its degree is much less than many lower-ranked ASs. The result emphasizes the significance of our IDAV-based ranking theory. Note that in ideal world the ranking weight value is the same as normalized traffic throughput, but our computation is in the non-ideal situation, so AS209's weight is twice of that of AS2914's doesn't necessarily mean the actual traffic also is. The weight is only a relative measure here.

To verify our results, we gathered some academic or commercial AS rankings from other sources in the Internet. These different versions of top 10 ASs including ours are listed in table 3 for comparison. Caida offers hybrid criteria, we choose to sort by degree first. AS701 is the biggest because it has the most (2,420) connections, refer to table 2. FixedOrbit uses average hops, the weight calculated to indicate the average number of AS hops that must be traversed from inside the network to any other IP addresses in Internet. In this sense, AS3356 is the largest because it reaches other networks through the shortest AS_PATHs. ReneSys, a commercial statistics that puts the customers' quantity and quality first, has the most in common with IDAV, which indicates the success of our model. The only one disagreement is that we replaced AS6453 with AS174 in the top-10. In our result, AS6453 ranks the 16th because, unlike AS174, it is not carrying much access volumes for the other top 10 ASs. As the two rules suggested in section 2.2, an AS can promote its ranking by carrying access volumes for other big ASs. It is the chain rule that works here.

Although all above ranking approaches are meaningful, our method is a little superior because, according to IDAV theories, the calculated weight reflects each AS's contribution to the Internet routing. Others may be more connected, nearer to reach, have greater number of customers or even earn larger portion of money, but without AS1239, the Internet will suffer the severest and most disastrous route failure, or redirection.

4.2 Validation

Due to the lack of authoritative source of AS ranking information, to validate our results, we look for some indirect approach. We employ the "Valley-Free" principle of inter-domain routing behavior. Introduced by Gao in [9], the general rule for inter-domain path selection is: one route should only take the paths first going from smaller ASs to bigger ASs, followed by some steps between equal-sized ASs, then from bigger ASs to smaller ASs, e.g. no part of the path could be a "valley".

Although originally, the relative magnitude of AS is defined by means of customer, provider, and peer relationships, we can slightly modify it using our ranking result without changing the thinking behind valley-free principle. A threshold number "2" is used to classify the AS relationship: if an AS's ranking weight is twice or more than the next-step's, identify the step is a downhill one, the opposite step is an uphill one accordingly. If neither of the two ASs is sufficiently larger than the other, identify the step as flat. We use this law to examine each AS_PATH to see whether it is compliant with the valley-free pattern. According to our examination, in the RouteViews data we used, 99.3% AS_PATHs obey the valley-free rule, only 0.7% (164,032/23,147,531) AS_PATHs have violations, which are mainly attributed to the following reasons:

1. Sibling ASs may transit arbitrary route for each other.
2. Some special address block have prefix-level traffic engineering.
3. Underestimation of certain ASs because of insufficient data.

Despite of these slight technical imperfections, this is an ultimate proof of our IDAV model and AS ranking theories. The wonderful consistency is a clue that both carriage matrix AS ranking and valley-free path selection are basic underlying characteristics of the Internet structure.

5 Conclusion

In this paper, we originate the Inter-Domain Access Volume (IDAV) model. This model investigates each individual AS's routing contributions to the Internet by adding access volume to the topological AS graph. Besides the formalization and construction of IDAV model, we also put forward a systematical AS ranking methodology using IDAV. The inferred AS magnitudes show more truthfulness than traditional approaches. And in return, the positive AS ranking results, as a benchmark for AS graph modeling, have validated the basic principles behind IDAV model. By leveraging information about both connectivity and the prefixes exchanged between neighbor ASs, the authors believe that this model will enable much more compelling applications, especially concerned with Internet routing architecture.

Future research topics, including visualization of the augmenting AS graph and applying the model to router-level intra-domain routing are ongoing. In addition, we are attempting to help traffic-engineering and smart-routing by employing this model, for example, we could make predictions of node congestions and link loads according to the network topology and routing solicitation, in order to adjust router configurations for better performance. The idea of traffic estimation upon IDAV model largely depends on establishing the relationship between real traffic flow and access volume. Some refinement to the model will be made and we believe such a mapping should be do-able.

References

1. Halabi, B., McPherson, D.: Internet Routing Architectures. Cisco Press (2000)
2. RFC 1771, A Border Gateway Protocol 4 (BGP-4)
3. University of Oregon RouteViews project, `http://www.routeviews.org/`
4. Merit Internet Routing Registry Services, `http://www.irr.net/`
5. Ge, Z., Figueiredo, D.R., Jaiswal, S., Gao, L.: On the Hierarchical Structure of the Logical Internet Graph. In: Proceeding of SPIE ITCOM (2001)

6. Vazquez, A., Pastor-Satorras, R., Vespignani, A.: Internet Topology at the Router and Autonomous System Level. Arxiv preprint cond-mat/0206084 (2002)
7. Siganos, G., Faloutsos, M., Faloutsos, P., Faloutsos, C.: Power Laws and the AS-Level Internet Topology. Transactions on Networking 11(4) (August 2003)
8. Barabasi, A.L., Albert, R.: Emergence of Scaling in Random Networks. Science 286(5439), 509–512 (1999)
9. Gao, L.: On inferring Autonomous System Relationships in the Internet. IEEE/ACM Transactions on Networking 9(6) (December 2001)
10. Subramanian, L., Agarwal, S., Rexford, J., Katz, R.H.: Characterizing the Internet Hierarchy from Multiple Vantage Points. IEEE INFOCOM (2002)
11. Di Battista, G., Patrignani, M., Pizzonia, M.: Computing the Types of the Relationships between Autonomous Systems. IEEE INFOCOM (2003)
12. Dimitropoulos, X., Krioukov, D., et al.: Inferring AS Relationships: Dead End or Lively Beginning? In: Nikoletseas, S.E. (ed.) WEA 2005. LNCS, vol. 3503, pp. 113–125. Springer, Heidelberg (2005)
13. Dimitropoulos, X., Krioukov, D., et al.: AS Relationships: Inference and Validation. ACM SIGCOMM Computer Communication Review 37(1) (January 2007)
14. AS ranking – Caida, http://as-rank.caida.org/
15. Knodes Index – FixedOrbit, http://www.fixedorbit.com/metrics.htm
16. Market Intelligence Rankings – Renesys, http://www.renesys.com/products_services/market_intel/rankings/
17. RFC 1519, Classless Inter-Domain Routing (CIDR): an Address Assignment and Aggregation Strategy
18. Internet Assigned Numbers Authority (IANA), http://www.iana.org/
19. Horn, R.A., Johnson, C.R.: Matrix Analysis. Cambridge University Press, Cambridge (1990)
20. Mao, Z.M., Johnson, D., Rexford, J., Wang, J., Katz, R.: Scalable and Accurate Identification Of AS-level Forwarding Paths. IEEE INFOCOM (2004)
21. Soule, A., Salamatian, K., Taft, N., Emilion, R., Papagiannaki, K.: Flow Classification by Histograms or How to Go on Safari in the Internet. Sigmetrics (2004)
22. Rexford, J., Wang, J., Xiao, Z., Zhang, Y.: BGP routing stability of popular destinations. In: Internet Measurement Conference (2002)
23. MRT format, http://www.zebra.org/zebra/Packet-Binary-Dump-Format.html
24. CLI format: Cisco IOS IP Command Reference, Volume 2 of 4: Routing Protocols, Release 12.3, http://www.cisco.com/
25. Page, L., Brin, S., Motwani, R., Winograd, T.: The PageRank Citation Ranking: Bringing Order to the Web. In: Proceedings of ASIS 1998 (1998)
26. Haveliwala, T.: Efficient Computation of PageRank. Stanford Technical Report (1999)

Achieving Network Efficient Stateful Anycast Communications

Tim Stevens, Filip De Turck, Bart Dhoedt, and Piet Demeester

Ghent University, Department of Information Technology (INTEC)
Gaston Crommenlaan 8 bus 201, 9050 Gent, Belgium
tim.stevens@intec.ugent.be

Abstract. Using IP anycast, connectionless services such as DNS can be decentralized to cope with high service request rates. Unfortunately, most Internet services cannot take direct advantage of anycast addressing because they rely on stateful communications. Subsequent packets of the same session may be routed to a different anycast group member, resulting in a broken service. Therefore, we introduce a transparent layer-3 architecture that provides anycast support for stateful services.

Taking into account user demands, available resources, extra network overhead and anycast infrastructure costs, we address optimal proxy router placement and infrastructure dimensioning, an issue that has not been tackled before for similar architectures. Simulation results indicate that even modest overlay infrastructures, consisting of a small number of proxy routers, often result in an effective stateful anycast solution where the detour via the proxy routers is negligible.

Keywords: anycast architecture, proxy router, service dimensioning.

1 Introduction

IP anycast enables communication between a source host and one member of a group of target hosts, usually the one nearest to the source [1]. As such, anycast is considered as a powerful mechanism to achieve flawless decentralization of connectionless network services, a prerequisite to scale such services to cope with high request rates. The use of replicated DNS root servers listening to a common—anycast—IP address is an example application where anycast has been proven useful [2].

At present, there are limitations that prevent widespread adoption of IP anycast for network service provisioning, however. First, session-oriented services (this includes all applications implemented on top of TCP) cannot take advantage of this addressing mode, because subsequent packets from the same source host (and session) may be routed towards a different target host. In a sense, application layer anycast [3] alleviates this issue, albeit at the expense of losing IP anycast transparency. Second, routes to anycast groups cannot be aggregated, leading to an explosive growth of network routing tables if anycast would be applied on a large scale. Possible solutions for this issue have been proposed by

T. Vazão, M.M. Freire, and I. Chong (Eds.): ICOIN 2007, LNCS 5200, pp. 492–502, 2008.

D. Katabi et al. [4] and H. Ballani et al. [5]. They do not directly address stateful anycast communications, however.

In this paper, we present a proxy-based architecture that enables IP anycast for session-oriented network services. Using this approach, advanced network services such as a grid computational service can be scaled to a large number of consumers, and this in a transparent way from an end-user perspective. After describing the anycast proxy architecture, we propose a dimensioning and placement optimization for the infrastructure components. Based on several criteria, including anycast proxy infrastructure costs, network operational costs, client demands and available resources, we solve the optimization problem by means of integer linear programming (ILP).

The proposed anycast proxy architecture is actually based on PIAS (Proxy IP Anycast Service) [5] and most PIAS features remain valid for our architecture. Whereas Ballani et al. focus on global routing scalability and justify part of the PIAS design by relying on BGP route stability, we tailor our anycast proxy architecture to the needs of regional network service providers, with explicit support for session-based communications. In addition to previous anycast architecture proposals, we also address infrastructure dimensioning and optimal placement of the proxy routers.

The remainder of this paper is structured as follows. Section 2 focuses on the anycast proxy architecture. We expose the ILP optimization problem for infrastructure dimensioning and proxy unit placement in Section 3. In Section 4, evaluation results from the implemented ILP are discussed and general findings are presented. Section 5 summarizes the main results of this paper.

2 Anycast Proxy Architecture

The design objectives for a regional anycast proxy infrastructure supporting session-based network services differ from a global anycast overlay infrastructure such as PIAS [5] or OASIS [6]. In addition to the PIAS objectives, we wish explicit session

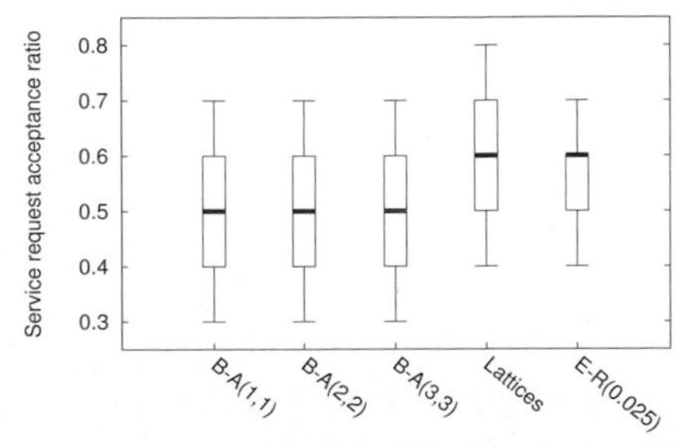

Fig. 1. Native anycast server selection efficiency percentiles (5-25-**50**-75-95) for different types of 100-node random graphs (for each graph 10 clients and 10 servers are selected at random and client demand is equal to server capacity). Requests arriving at an overloaded server are rejected.

support and a flexible, session-based bonding between client proxies and server proxies (and hence, between clients and servers). Contrary to OASIS, we wish that anycast network services are completely transparent from a client perspective, on each layer. Contrary to both PIAS and OASIS, we target a regional anycast solution (e.g., deployed in access and aggregation networks) and not a global one. As such, anycast infrastructure dynamics are easier to deal with.

Even if network sessions were supported by native anycast or some anycast proxy overlay, a semi-static coupling between clients and target servers (e.g., shortest paths) would be inefficient in terms of server resource utilization, potentially leading to service requests being rejected. This is especially true for long-lived, resource hungry network services. Fig. 1 depicts this inefficiency for different types of 100-node random graphs[1]. Therefore, we argue for a session-based coupling between client proxies and server proxies, even though this translates into an increase in the amount of state information in the proxy nodes.

Fig. 2 depicts the steps involved in setting up a session between a client and a target anycast server through the proxy system. Step R1 registers a server with unicast address **S** for the service offered by anycast address **A** and port **b**. Note that this registration uses native anycast to reach a server proxy (SP). Next, a client can initiate a session by sending a packet addressed to the anycast service of choice (step 1). When the packet arrives at the client proxy (CP), it is tunneled to a suitable SP (step 2), where it is tunneled again towards a target server (step 3). The return path (steps 4,5 and 6) is realized in the same way. *Stateful tunneling* occurs twice in each direction and is necessary to guarantee session continuity[2]. The IP tunnels cannot be avoided on the return path because both the CP and SP have to monitor the session state, for which packets have to traverse the system in both directions. In contrast with PIAS, we prefer tunneling to network address translation (NAT) for communication between SPs and anycast target servers, as this preserves end-to-end connectivity. This is important for IPsec support and application layer services that experience difficulties traversing NAT gateways. On the downside, the packet overhead increases on the path between the SP and target server (in both directions) due to the extra IP header. Because the second tunnel on the return path (step 5) is unavoidable, this is not an extra limitation on the return path.

Similar to PIAS, rendez-vous proxies can be introduced if the number of proxies gets too large and the signalling overhead significantly affects the system

[1] Throughout this paper **B-A**(x,x) stands for the class of Barabási-Albert random graphs [7] with x initial nodes, and during the growing process new nodes are connected by x edges. **Lattices** are square and they consist of 10×10 nodes. **E-R**(0.025) represents the class of connected Erdõs-Rényi random graphs [8] with connection probability 0.025. These three types of graphs cover a wide spectrum of random graphs, with regular lattices at one end of the spectrum and completely random E-R graphs residing at the opposite end. B-A graphs represent small world networks and are often used to model large networks (e.g., the Internet).

[2] Note that step 4 in Fig. 2 can be implemented as a stateless tunnel, as target servers always tunnel return packets towards the anycast address.

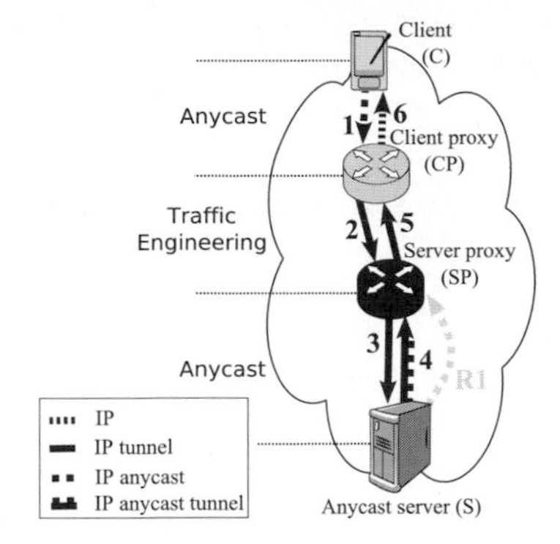

Step	Packet headers
R1	S:b → A:c
1	C:a → A:b
2	CP:C:a → SP:A:b
3	A:C:a → S:A:b
4	S:A:b → A:C:a
5	SP:A:b → CP:C:a
6	A:b → C:a

Fig. 2. Anycast communication through the proxy system. In the table capitals refer to IP addresses and lowercase characters point to the TCP/UDP port used.

performance. In this case, a CP queries a rendez-vous point for a suitable SP prior to initiating a new tunnel (towards a SP). Rendez-vous-based SP selection is not further elaborated in this paper.

The following section discusses how to dimension the proxy infrastructure and where to attach the proxies to the network.

3 Dimensioning the Anycast Infrastructure

Equipped with the anycast architecture outlined in Section 2, we wish to determine how many proxies are needed and where they should be attached to the network for a given client and server configuration. More formally, given a network $G(V, E)$, a set of client sites $C \subset V$ and their demands δ_c, a set of server sites $S \subset V$ and their capacities π_s, edge weights $l(u, v) : (u, v) \in E$ and edge capacities $c(u, v) : (u, v) \in E$, determine how many CP (resp. SP) with capacity κ_{CP} (resp. κ_{SP}) are needed and where they should be attached to the network. The optimization process should balance network operational costs (related to network resource utilization) and proxy infrastructure costs (determined by the fixed cost α (resp. β) associated with each CP (resp. SP)). The parameters for this optimization problem are summarized in Table 1.

We solve this optimization problem using ILP [9]. First *constraints* are specified, then we formulate the *objective function*. Where possible, constraints are grouped per connection step as depicted in Fig. 2. Subsequently, we discuss global constraints.

Constraints Related to Step 1 (see Fig. 2). Variable $P_{CP}(v)$ denotes whether $v \in V$ is selected as a CP. Therefore:

$$\forall v \in V : P_{CP}(v) \in \{0, 1\} \tag{1}$$

Table 1. Model parameters. The number of active sessions is our unit of measurement for capacities and client demands

Parameter	Description
$G(V, E)$	network topology, V and E denote the sets of vertices and edges
C	set of client sites ($C \subset V$)
S	set of server sites ($S \subset V$)
$l(u, v)$	edge weight (cost) ($\forall (u, v) \in E$)
Δ_{uv}	pre-computed shortest path distance between $u \in V$ and $v \in V$
$c(u, v)$	edge capacity ($\forall (u, v) \in E$)
δ_c	aggregated client demand for site c ($\forall c \in C$)
π_s	aggregated server capacity for site s ($\forall s \in S$)
κ_{CP}	capacity of a client proxy
κ_{SP}	capacity of a server proxy
α	cost for introducing one client proxy
β	cost for introducing one server proxy

We define the variable matrix $Y(|C| \times |V|)$ to express which node is selected as client proxy for every client site node. Each client site must select exactly 1 proxy.

$$\forall c \in C, \forall v \in V : Y_{cv} \in \{0, 1\} \wedge Y_{cv} \leq P_{CP}(v) \tag{2a}$$

$$\forall c \in C : \sum_{v \in V} Y_{cv} = 1 \tag{2b}$$

In the following constraints, $x_{cv}(p, q)$ denotes whether edge $(p, q) \in E$ is used for the path between $c \in C$ and $v \in V$. Additionally, constraint 3c enforces that the shortest possible path between c and v is chosen (native anycast).

$$\forall c \in C, \forall v \in V, \forall u \in V :$$

$$\sum_{p \in in(u)} x_{cv}(p, u) - \sum_{q \in out(u)} x_{cv}(u, q) = \begin{cases} -Y_{cv} & (u = c) \\ Y_{cv} & (u = v) \\ 0 & (u \neq c, v) \end{cases} \tag{3a}$$

$$\forall c \in C, \forall v \in V, \forall (p, q) \in E : x_{cv}(p, q) \in \{0, 1\} \tag{3b}$$

$$\forall c \in C, \forall v \in V : \sum_{(p,q) \in E} l(p, q) x_{cv}(p, q) = \Delta_{cv} \tag{3c}$$

Furthermore, the total demand for a CP should not exceed its capacity:

$$\forall v \in V : \sum_{c \in C} \delta_c Y_{cv} \leq \kappa_{CP} \tag{4}$$

Constraints Related to Step 2. $P_{SP}(v)$ is defined in the same way as $P_{CP}(v)$, but from a SP perspective. This yields:

$$\forall v \in V : P_{SP}(v) \in \{0,1\} \tag{5}$$

Variable matrix $N(|V| \times |V|)$ divides the demand arriving in a CP over all SPs. These variables obey the following rules.

$$\forall u, v \in V : N_{uv} \in \mathbb{N} \tag{6}$$

$$\forall u \in V : \sum_{v \in V} N_{uv} = \sum_{c \in C} \delta_c Y_{cu} \tag{7}$$

Next, $m_{uv}(p,q)$ denotes the amount of sessions that are forwarded over edge $(p,q) \in E$ on the path between $u \in V$ and $v \in V$. Note that a constraint similar to (3c) is omitted to allow for traffic engineering between CP and SP.

$$\forall u, v, w \in V :$$

$$\sum_{p \in in(w)} m_{uv}(p,w) - \sum_{q \in out(w)} m_{uv}(w,q) = \begin{cases} -N_{uv} & (w = u) \\ N_{uv} & (w = v) \\ 0 & (w \neq u, v) \end{cases} \tag{8a}$$

$$\forall u, v \in V, \forall (p,q) \in E : m_{uv}(p,q) = 0 \vee m_{uv}(p,q) = N_{uv} \tag{8b}$$

Constraints Related to Step 3. Analogous to constraint (2), $L(|S| \times |V|)$ establishes a similar relationship between a SP and a server.

$$\forall s \in S, \forall v \in V : L_{sv} \in \{0,1\} \wedge L_{sv} \leq P_{SP}(v) \tag{9a}$$

$$\forall s \in S : \sum_{v \in V} L_{sv} = 1 \tag{9b}$$

We introduce a last variable matrix $Q(|V| \times |S|)$ responsible for distributing the demand arriving in a SP (v) to connected servers (s).

$$\forall v \in V, \forall s \in S : Q_{vs} \in \mathbb{N} \tag{10}$$

The next constraints are the counterpart of (3). $k_{sv}(p,q)$ denotes the number of sessions forwarded over edge $(p,q) \in E$ from SP $v \in V$ to server $s \in S$.

$$\forall s \in S, \forall v \in V, \forall u \in V :$$

$$\sum_{p \in in(u)} k_{sv}(p,u) - \sum_{q \in out(u)} k_{sv}(u,q) = \begin{cases} -Q_{vs} & (u = s) \\ Q_{vs} & (u = v) \\ 0 & (u \neq s, v) \end{cases} \tag{11a}$$

$$\forall s \in S, \forall v \in V, \forall (p,q) \in E : k_{sv}(p,q) = 0 \vee k_{sv}(p,q) = Q_{vs} \tag{11b}$$

$$\forall s \in S, \forall v \in V : \sum_{(p,q) \in E} l(p,q) k_{sv}(p,q) = \Delta_{sv} Q_{vs} \tag{11c}$$

Each SP should balance its total demand over the connected servers:

$$\forall v \in V, \forall s \in S : Q_{vs} \leq \pi_s L_{sv} \tag{12}$$

$$\forall v \in V : \sum_{u \in V} N_{uv} = \sum_{s \in S} Q_{vs} \tag{13}$$

Constraints Related to Both Step 1 and 3. Because both clients and servers send packets to the same anycast address, the proxy node closest to a client should be its CP (14a) and it cannot be a SP (14b), unless CP and SP collide.

$$\forall c \in C, \forall x \in C : \sum_{v \in V} Y_{cv} \Delta_{cv} \leq \sum_{v \in V} Y_{xv} \Delta_{cv} \tag{14a}$$

$$\forall c \in C, \forall x \in S : \sum_{v \in V} Y_{cv} \Delta_{cv} \leq \sum_{v \in V} L_{xv} \Delta_{cv} \tag{14b}$$

Similarly, the proxy node closest to a server should be that server's SP.

$$\forall s \in S, \forall x \in S : \sum_{v \in V} L_{sv} \Delta_{sv} \leq \sum_{v \in V} L_{xv} \Delta_{sv} \tag{15a}$$

$$\forall s \in S, \forall x \in C : \sum_{v \in V} L_{sv} \Delta_{sv} \leq \sum_{v \in V} Y_{xv} \Delta_{sv} \tag{15b}$$

Constraints Related to Both Step 2 and 3. For every SP, the load should not exceed its capacity or the total capacity of the servers attached to that SP. Note that a server attaches itself to exactly one SP (the closest), due to the registration process by means of native anycast (see step R1 in Fig. 2).

$$\forall v \in V : \sum_{u \in V} N_{uv} \leq \kappa_{SP} P_{SP}(v) \tag{16}$$

$$\leq \sum_{s \in S} \pi_s L_{sv} \tag{17}$$

Global Constraints. Every edge $(p,q) \in E$ has a maximum capacity $c(p,q)$.

$$\sum_{c \in C} \sum_{v \in V} \delta_c x_{cv}(p,q) + \sum_{s \in S} \sum_{v \in V} k_{sv}(p,q) + \sum_{u,v \in V} m_{uv}(p,q) \leq c(p,q) \tag{18}$$

Objective Function. Based on the constraints above, minimizing Z yields the minimum cost solution.

$$\begin{aligned}\text{Minimize } Z = & \sum_{(p,q)\in E} \sum_{c\in C} \sum_{v\in V} \delta_c l(p,q) x_{cv}(p,q) + \alpha \sum_{v\in V} P_{CP}(v) + \\ & \sum_{(p,q)\in E} \sum_{s\in S} \sum_{v\in V} l(p,q) k_{sv}(p,q) + \beta \sum_{v\in V} P_{SP}(v) + \\ & \sum_{(p,q)\in E} \sum_{u\in V} \sum_{v\in V} l(p,q) m_{uv}(p,q)\end{aligned} \tag{19}$$

4 Evaluation

Due to the shortest path computations in the ILP of Section 3 (constraints (3),(8) and (11)), the number of variables in the system is bounded by $O(|V|^2|E|)$ or $O(|V|^4)$ for dense graphs. Starting from the assumption that network bandwidth is a less stringent resource than proxy router capacity, we propose a *relaxed version* of the ILP, neglecting link capacity constraints. This relaxation renders constraints (3),(8),(11) and (18) superfluous and essentially reduces the ILP to an extended version of the warehouse location problem (WLP) [10]. In this case, the objective function is rephrased using the pre-computed shortest distance matrix Δ, yielding an upper bound of $O(|V|^2)$ for the number of variables in the system. Due to the quadratic upper bound on the number of variables, simulation results for moderate to relatively large networks lie within reach. We implemented this relaxed ILP using CPLEX branch and bound software [11].

Fig. 3 depicts an example result of the ILP for a 6×6 lattice with three client sites and three server sites. If proxy unit costs are high (Fig. 3(a)), the number of proxies is small and the path stretch is large. Low proxy unit costs (Fig. 3(b)) result in more proxies being installed. Edge line width relates to the amount of traffic that edge is forwarding.

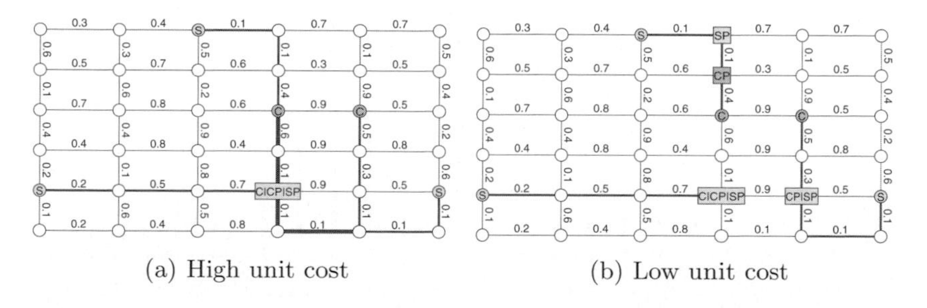

(a) High unit cost (b) Low unit cost

Fig. 3. An example result of the ILP for a 6×6 lattice. Nodes marked with a C (S) are client (server) sites. Boxed nodes are CPs or SPs. Note that one node (router) can fulfill multiple functions. Edge line width represents the traffic flow over this edge and indicates which proxy is selected by a client or server.

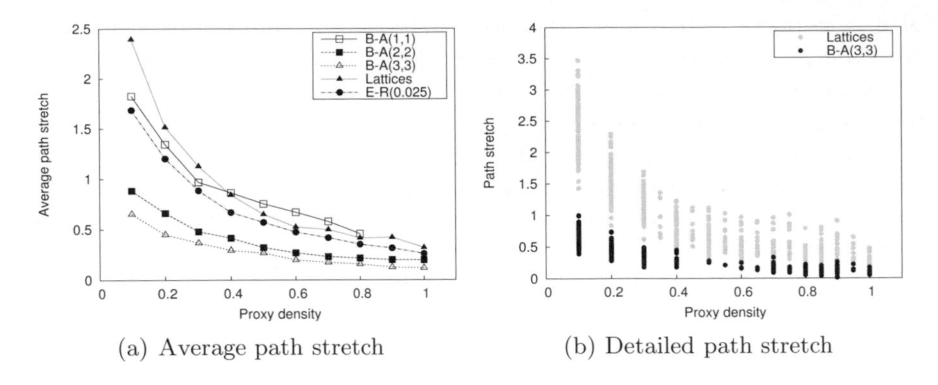

(a) Average path stretch (b) Detailed path stretch

Fig. 4. Path stretch induced by the proxy infrastructure in comparison with native anycast

We now wish to investigate differences in the proxy placement behavior and efficiency for different types of 100-node random graphs. Where possible, general findings are presented. Again, five types of random networks are generated: 10×10 lattices and networks from the classes B-A(1,1), B-A(2,2), B-A(3,3) and E-R(0.025)[1]. For each constructed network, edge weights $l(.)$ are selected randomly from the set $\{0.1, 0.2, \ldots, 0.9\}$ and 10 client and server sites are also selected at random. All random values are drawn from distinct Mersenne Twister instances (different seeds). From each class of graphs, 100 networks are generated, yielding a total of 500 evaluation networks. For the simulations, all client demands and server capacities are equal to 100 sessions. CPs and SPs are scaled in such a way they can handle the complete load (i.e., they can support 1000 simultaneous sessions or more). By varying the proxy unit costs, we influence the total number of proxies to be installed in the network.

Fig. 4(a) depicts the average path stretch related to the *proxy density* for each type of random graph. We define proxy density as $\frac{|\mathrm{CP}|+|\mathrm{SP}|}{|\mathrm{C}|+|\mathrm{S}|}$. As expected, the path stretch decreases as the proxy density increases. Because CPs can forward

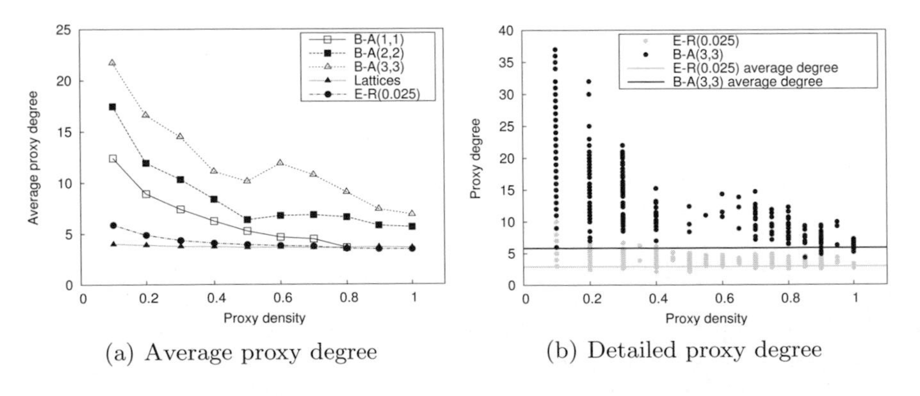

(a) Average proxy degree (b) Detailed proxy degree

Fig. 5. Degree of the network nodes elected to become a proxy

requests to the most suitable SPs (based on SP proximity and the available processing power in the servers behind each SP), the path stretch does not necessarily converge to zero. Fig. 4(a) and Fig. 4(b) show that both the path stretch and its spread is much smaller for small world graphs (B-A(x,x)) than for lattices. *Even for a small number of proxies, the anycast overlay does not impose a significant stretch in small world graphs.*

Another result that confirms intuition is depicted in Fig. 5. When proxy density is low, *highly connected nodes are elected to become an anycast proxy.* Due to their specific structure, B-A(x,x) graphs emphasize this property. The hiccup for B-A(x,x) graphs noticed in Fig. 5(a) can also be attributed to their structure: as shown in Fig. 5(b), a vast majority of ILP results for B-A(x,x) graphs are either low or high density solutions (without smooth transition), because the node degree is exponentially distributed.

5 Summary

In this paper, we presented an anycast proxy architecture tailored to the needs of a service provisioning platform (i.e., support for stateful communications, efficient use of available resources). This way, service providers can offer session-based network services in a scalable, robust way. From a client perspective, the proxy system is completely transparent, however. Moreover, end-to-end connectivity is preserved due to the double IP tunneling approach.

Once the anycast proxy architecture is being implemented, it is important to know how many proxy routers are needed and where they should be attached to the network. The second part of this paper addresses this issue by optimizing a balanced objective, combining fixed infrastructure investment costs with network operational costs related to the amount of extra traffic generated by the proxy system. We show that even with a small number of proxies the extra network operational cost can be small, especially in networks with small world properties.

References

1. Partridge, C., Mendez, T., Milliken, W.: RFC 1546: Host Anycasting Service (1993)
2. Sarat, S., Pappas, V., Terzis, A.: On the Use of Anycast in DNS. SIGMETRICS Performance Evaluation Review 33(1), 394–395 (2005)
3. Zegura, E., Ammar, M., Fei, Z., Bhattacharjee, S.: Application-Layer Anycasting: A Server Selection Architecture and Use in a Replicated Web Service. IEEE/ACM Transactions on Networking 8(4), 455–466 (2000)
4. Katabi, D., Wroclawski, J.: A Framework for Scalable Global IP-Anycast (GIA). ACM SIGCOMM Computer Communication Review 30(4), 3–15 (2000)
5. Ballani, H., Francis, P.: Towards a Global IP Anycast Service. ACM SIGCOMM Computer Communication Review 35(4), 301–312 (2005)
6. Freedman, M., Lakshminarayanan, K., Mazières, D.: OASIS: Anycast for Any Service. In: Proceedings of the 3rd Symposium on Networked Systems Design and Implementation (NSDI 2006), San Jose, California, United States (2006)

7. Barabási, A.-L., Albert, R.: Emergence of Scaling in Random Networks. Science 286, 509–512 (1999)
8. Bollobás, B.: Random Graphs. Academic Press, London (1985)
9. Nemhauser, G., Wolsey, L.: Integer and Combinatorial Optimization. Wiley-Interscience, New York (1988)
10. Khumawala, B.M.: An Efficient Branch and Bound Algorithm for the Warehouse Location Problem. Management Science 18(12), B718–B731 (1972)
11. ILOG CPLEX (2006), http://www.ilog.com/products/cplex/

Scalable RTLS: Design and Implementation of the Scalable Real Time Locating System Using Active RFID

Junghyo Kim, Dongho Jung, Yeonsu Jung, and Yunju Baek

Department of Computer Science and Engineering
Pusan National University Busan 609-735, South Korea
jhkim@juno.cs.pusan.ac.kr, yunju@pusan.ac.kr

Abstract. Interest in Real Time Locating Systems (RTLS), which is an RFID application, has been increasing recently. RTLS is used to locate and track object using RFID tags. Typically these objects are containers, pallets, and other commercially viable items. This paper presents the design and the implementation of an RTLS system using 433MHz active RFID tags considering scalability. Our system is developed using an RFID platform that takes RTLS standards into account. Also, in this paper a routing protocol is proposed to deliver data to the server via each reader. In order to evaluate the system's performance, some outdoor experiments are performed and the resulting errors reported in meters are discussed. Furthermore, simulation of the routing protocol is also included.

Keywords: RFID, RTLS, Localization, Tracking, Received signal strength, Wireless sensor networks.

1 Introduction

RFID transmits a variety of information using wireless communication technology about objects that have been tagged. Thus several routine tasks can be accomplished such as exchanging information, maintenance, and managing the tagged objects. The basic components of the RFID system are readers and tags. A reader communicates only with the tags and collects the information such as product ID, manufacturer, price, location, and other object-unique information [1].

Locating systems have become an important area of study and are applicable to many fields. Thus they have drawn wide-spread interest. These systems must utilize RFID technology and RTLS standards to locate the tagged objects.

International standards have been established for RTLS based on the ANSI/INCIT 371 standard [2][3][4]. The three parts of the standard have been defined as 2.4GHz RTLS, 433MHz RTLS and API. We will discuss the 433MHz RTLS standard using RFID in this paper.

A 433MHz RTLS is constructed to be used outdoors at distances of over 100m. This system has applications mainly in the distribution industry such as transporting containers in ports. Because ports cover large areas, this standard requires a number of RFID readers to be installed in order to have constant and uninterrupted coverage of the area. However, environmental restrictions have presented the installation of

T. Vazão, M.M. Freire, and I. Chong (Eds.): ICOIN 2007, LNCS 5200, pp. 503–512, 2008.

large numbers of hard-wired RFID readers. Therefore, wireless communication was used to solve this problem in RTLS.

A new data communication timing and message format for reader-to reader communication is presented, because the 433MHz RTLS system follows the principle of RFID that readers communicate only with tag. This paper presents the design and implementation of this RTLS using an active RFID.

The remainder of this paper is structured as follows. In Section 2, the RTLS standard is discussed as it relates to this work. Section 3 describes design issues of wireless communication - based RTLS using RFID readers. Routing protocols are also discussed. Next, an implementation of the RTLS system using an active RFID, and an analysis of its performance are presented in Section 4. Finally, Section 5 contains our conclusion.

2 Related Work

RTLS is the standard for locating objects with electronic tags in real time. This standard is stated in ISO/IEC 24730 based on the ANSI/INCIT 371 standard [5][6][7]. The ANSI/INCIT 371 standard has three parts. First, the 2.4GHz air interface protocol is defined in ANSI/INCIT 371-1. Second, the 433MHz air interface protocol is defined in ANSI/INCIT 371-2. Also, both explain the tag-to-reader communication architecture, message format and locating method. And last, ASNI/INCIT 371-3 defines Application Programming Interface (API) using server-client communication.

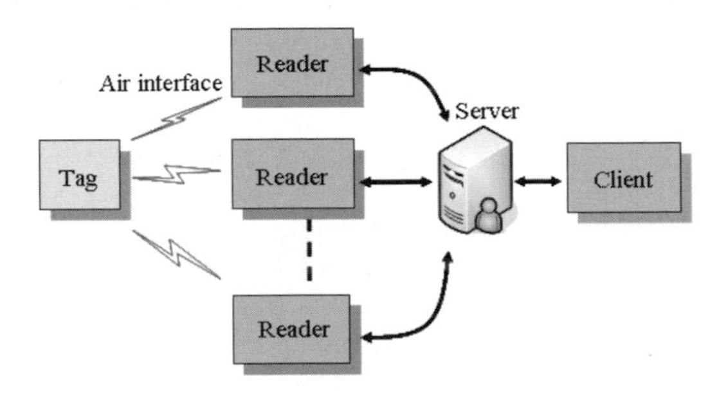

Fig. 1. General RTLS system

The general RTLS system is composed of tags, readers, a server and a client. Figure 1 illustrates the general RTLS system. The tag attached to the object has an original ID and sends information to the readers periodically. The readers communicate with the tag using wireless communication. This reader relays the server's requests to the tags or collects information from the tags. Then, the readers deliver the collected information to the server. The server gathers and analyzes this information. The client delivers the user's commands to the server and retrieves location information for application programs.

As mentioned in section 1, this paper focuses on the 433MHz RTLS system using active RFID. The characteristics of the 433MHz radio frequency are better than

2.4GHz in a port environment, because metal containers are stored, transported, and managed there. Two location methodologies are proposed in the RTLS standards [5][6]. They are Time Difference of Arrivals (TDOA) ranging and Received Signal Strength Indicator of Arrivals (ROA). The ROA ranging method has been used in this paper.

3 System Architecture

The design issues of the 433MHz RTLS system using an active RFID are as follows: (1) implementing the RTLS standards and (2) implementing scalability while maintaining RFID compatibility. In order to solve these problems, we suggest two methods which are a new data communication timing and message format for reader-to reader communication.

3.1 Design of Reader-to-Reader Communication

It is difficult to install a large number of RFID readers wired directly to the RTLS system. Consequently reader-to-reader wireless communication was suggested to maintain the scalability of the installation. In order to actualize our suggestion, new data communication timing for RFID reader is defined and reader-to-reader message format based on the RTLS standard is added.

3.1.1 Data Communication Timing

Data is transmitted in a packet format in RTLS data link layer. A packet is comprised of a preamble, data bytes, and a final logic low period. Data bytes are sent in the Manchester code format. The preamble is comprised of twenty pulses each with a duration of 60μs (30μs high and 30μs low), followed by a final sync pulse identifying the direction of the communication: 42μs high and 54μs low (tag-to-reader); and 54μs high and 54μs low (reader-to-tag) [3]. This is presented in figure 2.

Sync pulses have not been defined for reader-to-reader communication in the RTLS standard. Therefore, the tag-to-reader sync pulse is used to identify reader-to-reader communication. If the reader-to-tag sync pulse were used for reader-to-reader communication, a reaction by the tags would be induced. This reaction would cause compatibility problems with RFID. Consequently, a final sync pulse of 42μs high and 54μs low is used to identify two communication directions in this paper: tag-to-reader and reader-to-reader.

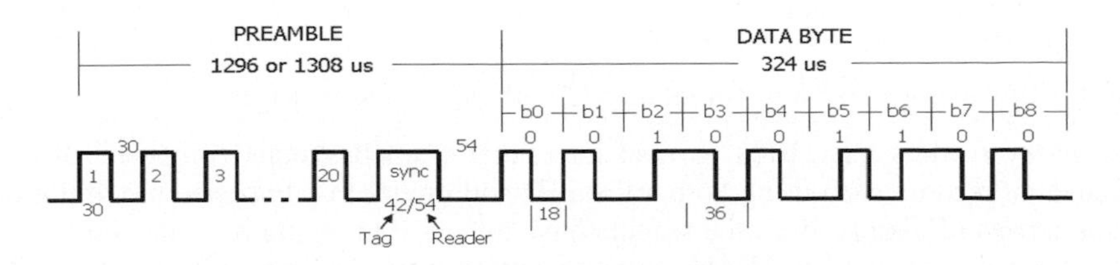

Fig. 2. Data communication timing in RTLS standard

3.1.2 Message Format

Tag-to-reader and reader-to-tag communication uses the RTLS standard message format shown in tables 1 and 2 [7]. A new message format has been added for reader-to-reader communication to these standard message formats as shown in table 3.

Table 1. Tag-to-reader message format

Protocol ID	Tag Status	Packet Length	Reader ID	Tag Mfr. ID	Tag ID	Command Code	Data	CRC
0x40	2 bytes	1 byte	2 bytes	2 bytes	4 bytes	1 byte	N bytes	2 bytes

Table 2. Reader-to-tag message format

Protocol ID	Packet Options	Tag Mfr. ID	Tag ID	Int ID	Command Code	Data	CRC
0x40	1 byte	2 bytes	4 bytes	2 bytes	1 byte	N bytes	2 bytes

Table 3. Reader-to-reader message format

Protocol ID	Command Type	Packet Length	Sender ID	Receiver ID	Destination ID	Data	CRC
0x42	1 byte	1 byte	2 bytes	2 bytes	2 bytes	N bytes	2 bytes

We define the Protocol ID using '0x42' for compatibility among the values '0x40' and '0x4F' in the reader-to-reader message format and use Sender ID, Receiver ID, and Destination ID to route information to its destination. The Destination ID specifies the sink ID or address.

3.2 Design of the Routing Protocol

In the proposed RTLS system, tags and readers communicate with each other using a radio frequency. If a number of readers and tags to communicate simultaneously, radio frequency interference happens frequently. This problem has been solved by using an efficient routing protocol system, Time Division Multiple Access (TDMA). The success of this approach depends on the three following assumptions, which are illustrated in figure 3.

- The readers which have unique IDs are deployed in a grid fashion at a regular distance, R, and synchronized.
- The transmission range of the readers and tags covers the distance defined by $\sqrt{2}$ R.
- It is possible for a sink to communicate with one or more readers.

All of the readers have been divided into several small groups composed of a set number of readers each using both a local ID and unique ID. For example, if a group is composed of nine readers, the readers have a local ID ranging in order from 1 to 9. We assign a time slot for TDMA communication using the local IDs of the readers. Therefore the readers that communicate at specific time all have the same local IDs in

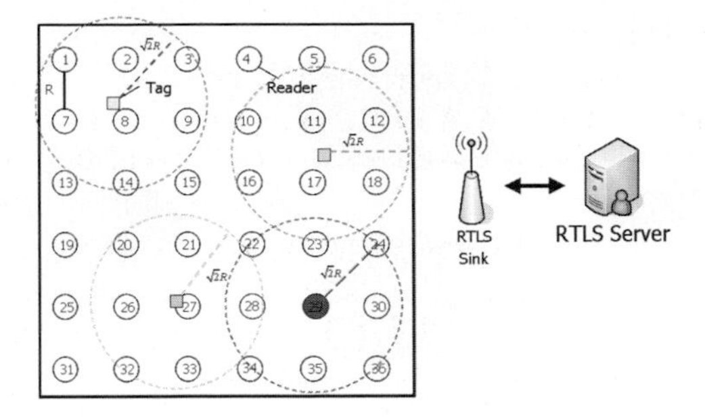

Fig. 3. Deployment of readers at regular intervals

the proposed RTLS system and they can communicate with the next local ID reader within the group without causing radio frequency interference with the readers of the other groups. Hence the reader with last local ID can collect information from the other readers within its group. It then transmits information to nearest reader of the next group in the neighborhood. According to this procedure, the sink gathers information from the readers and relays it to the RTLS server. The entire procedure is presented in figure 4.

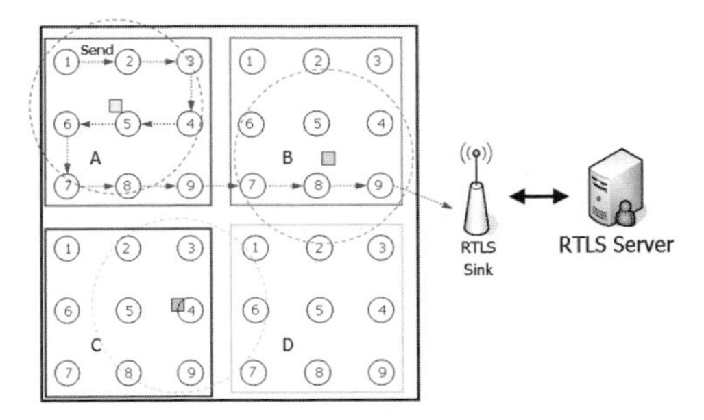

Fig. 4. Message routing from the readers using local ID to the sink

4 Implementation and Performance Evaluation

We implemented RTLS system based on the proposed architecture and routing protocol. In addition, we evaluated the accuracy with which objects were located with in field experiments and the performance of the routing protocol from a simulation results.

4.1 Implementation

The RTLS system was assembled using tags, readers, and a sink. The Atmel's Atmega128L was used as the processing unit for the RTLS platform. It has a maximum frequency of 8MHz providing reasonable processing power for a wide variety of

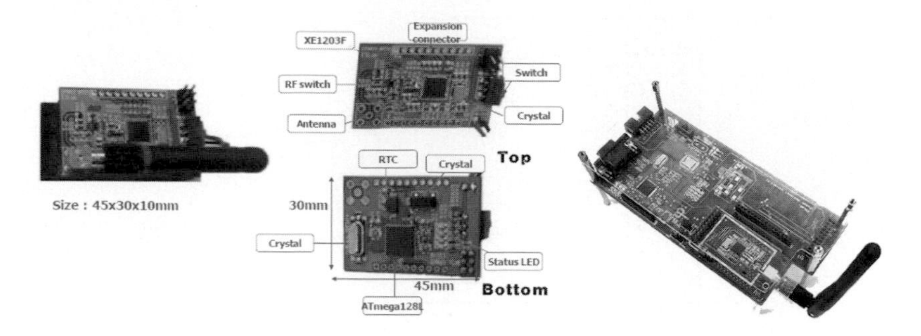

Fig. 5. Hardware component of RTLS system : Tag, Reader and Sink

applications. The communication module used the XEMICS's XE1203F radio chip. The XE1203F is a 433, 868, 915MHz compliant single-chip RF transceiver, which is designed to provide fully functional multi-channel FSK communication. Both the tags and readers were made from same hardware and the sink has some appended facility to connect with the RTLS server as shown in Figure 5 [8].

A general purpose PC is used as the RTLS server. It connects with the sink and estimates the location of the tags using RSS received from the readers.

4.2 Performance Evaluation

The following experiments were performed: (1) changes in signal strength were measured over distance in outdoor, (2) the accuracy of the locating system was assessed based on the result of the measurements and (3) the performance of the proposed routing protocol was evaluated through a simulation.

4.2.1 Field Experiments of RTLS System

Figure 6 depicts the environment where the signal strength experiment was held. The readers were arranged in a line, and then the tag emitted the signal. The results are the average of ten repetitions of the experiments. The results of the changes to the signal strength are shown in figure 7.

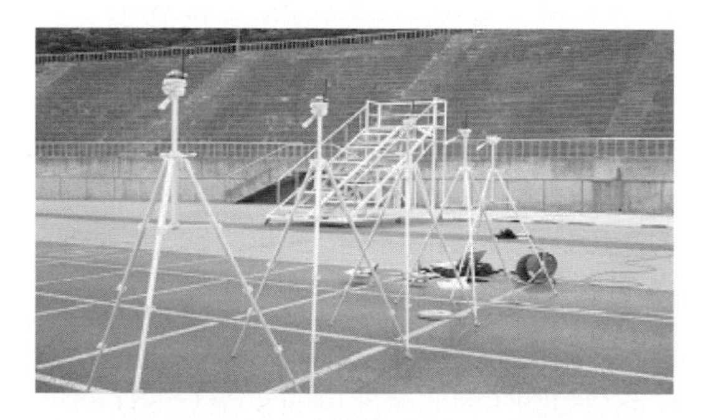

Fig. 6. Experiment environment of RSSI

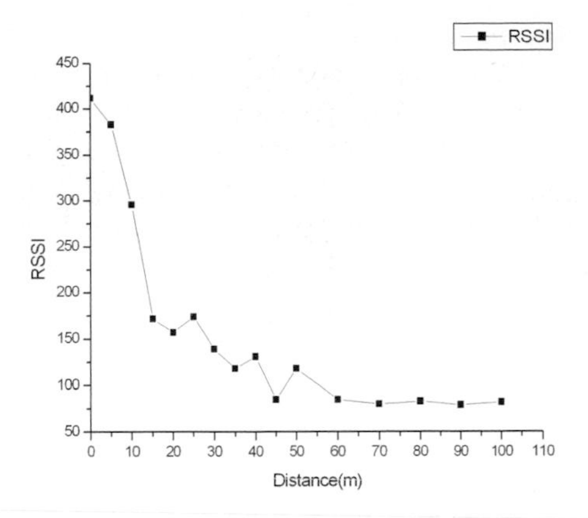

Fig. 7. Signal strength changes according to distance

The tag was located by either four or nine readers depending on the changes in its signal strength. The readers were deployed in a grid over a 30m x 30m flat area and the tag sent a blink message at 3 second intervals. Figure 8 shows the experimental set up.

Fig. 8. Fields experiment environment using 4 readers

The Results of the field experiment using four readers had an average location error of 3m and using nine readers, 2m as compared with the actual position. The distribution of location error is shown in Figure 9. As can be seen, RTLS having more readers increases the accuracy of the resulting location.

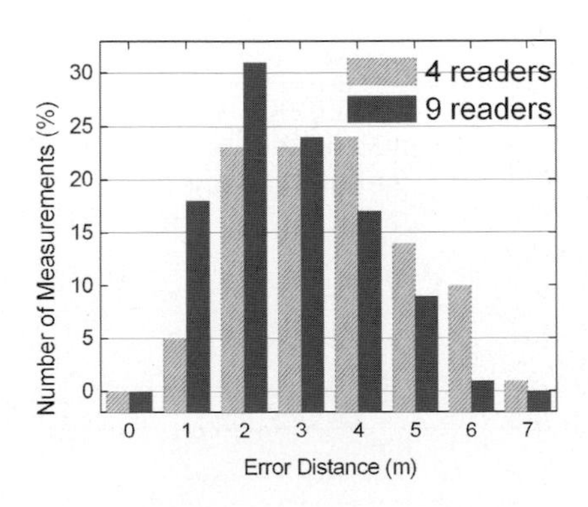

Fig. 9. Distance of location error

4.2.2 Simulation of Routing Protocol

An efficient routing protocol for communicating between readers was described in the previous chapter. The simulation environment was implemented using the NESLsim of PARSEC platform in order to evaluate the performance of the routing protocol [9]. Table 4 lists the values for different parameters used in the simulations. Each simulation scenario consisted of randomly placing unknown tags in a field.

Table 4. Simulation parameters

Parameter	Value
Field area	150×150 ~ 750×750
Number of readers	9, 36, 81, 144, 225
Number of tags	1, 5, 10, 15, 20, 25, 30, 35, 40, 45
Reader deployment	Grid
Tag deployment	Random
Transmission range	75
Interval of blink message	2, 5, 10, 30, 60 sec

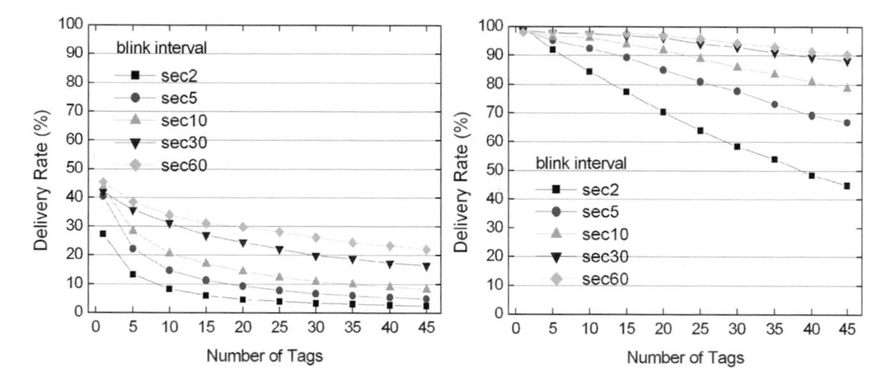

Fig. 10. Delivery rate comparison slotted aloha(left) and TDMA(right)

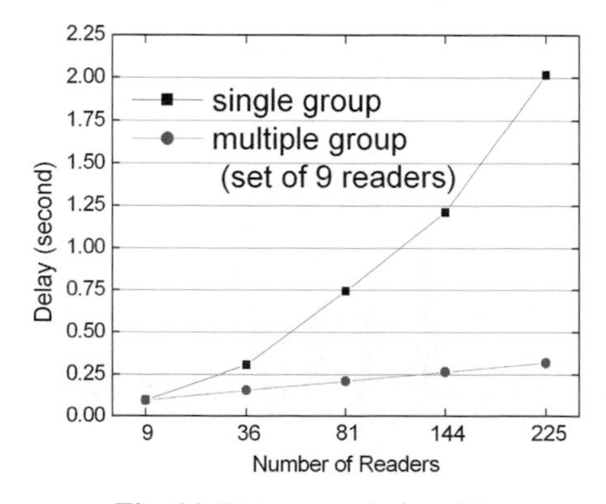

Fig. 11. Data transmission delay

The results are the average of over a hundred simulation runs. In first simulation, the performance of slotted aloha in the 18000-7 standard for active RFID was compared to that of TDMA using proposed routing protocol. The delivery rate represents an efficient use of radio frequency in figure 10.

In second simulation, we evaluated data transmission delays, because the property of real-time is important in RTLS. All of the readers are composed of one group were compared to all of them are divided into several small groups composed of a set number of nine readers. As can be seen in figure 11, the proposed routing protocol for RTLS has better performance in data transmission delays.

5 Conclusions

In this paper, we proposed a scalable RTLS system which makes use of reader-to-reader communication. First, new data communication timing for RFID readers was defined and a reader-to-reader message format was added to the RTLS standard that took RFID compatibility into account. Second, an efficient routing protocol for RTLS was designed and simulated. Finally, a scalable RTLS system was constructed and outdoor experiments were conducted.

In the future work, we would like to experiment with using containers in port logistics. In order to enhance the accuracy of locating objects, we would also like to explore localization techniques using radio interferometric positioning.

Acknowledgement

"This work was supported by the Korea Research Foundation Grant funded by the Korean Government(MOEHRD)" (The Regional Research Universities Program/Research Center for Logistics Information Technology).

References

1. ISO/IEC 18000-7, Information Technology - Radio frequency identification for item management - Part 7: Parameters for active air interface communications at 433 MHz (2004)
2. ANSI/INCITS 371-1, Real Time Locating Systems(RTLS) - Part 1: 2.4 GHz Air Interface Protocol (2003)
3. ANSI/INCITS 371-2, Real Time Locating Systems(RTLS) - Part 2: 433 MHz Air Interface Protocol (2003)
4. ANSI/INCITS 371-3, Real Time Locating Systems(RTLS) - Part 3: Application Programming Interface (API) (2003)
5. ISO/IEC 24730-1, Information technology - Automatic identification and data capture techniques - Real Time Locating Systems(RTLS) - Part 1: Application Programming Interface (API) (2003)
6. ISO/IEC 24730-2, Information technology - Automatic identification and data capture techniques - Real Time Locating Systems(RTLS) - Part 2: 2.4 GHz Air Interface Protocol (2003)

7. ISO/IEC 24730-3, Information technology - Automatic identification and data capture techniques - Real Time Locating Systems (RTLS) - Part 3: 433 MHz Air Interface Protocol (2003)
8. Cho, H., Baek, Y.: Design and Implementation of an Active RFID System Platform. In: Proceedings of the International Symposium on Applications and the Internet Workshops (SAINTW 2006) (IEEE CS) (January 2006)
9. PARSEC User Manual (1999), http://pcl.cs.ucla.edu/projects/parsec

Collision-Resilient Symbol Based Extension of Query Tree Protocol for Fast RFID Tag Identification

Jae-Min Seol and Seong-Whan Kim

Department of Computer Science,
University of Seoul, Jeon-Nong Dong, Seoul, South Korea
seoleda@gmail.com, swkim@uos.ac.kr

Abstract. RFID (RF based identification system) requires identification and collision avoidance schemes for tag singularization. To avoid the collision, there are two previous approaches: ALOHA based and binary tree algorithm.. They are essentially collision avoidance algorithms, and require much overhead in retransmission time. Previous research works on collision recovery protocol cannot distinguish tag collision from channel error. Because channel error significantly influences the overall performance of anti-collision protocols, we propose a robust and efficient tag collision recovery scheme using direct sequence spreading modulation; thereby we can reduce channel errors. Specifically, we propose MSQTP (multi-state query tree protocol) scheme, which is an extension of query tree protocol using modulated symbols. We experimented with two collision resilient symbols: orthogonal (Hadamard) and BIBD (balanced incomplete block design) code. MSQTP shows performance gain (decrease in iteration step for collision recovery) over previous query tree based collision recovery scheme, and shows graceful degradation in noisy environment with lower SNR.

1 Introduction

RFID (radio frequency identification) is a RF based identification system.. RFID system is easier to use than magnetic card and bar code. The RFID has high potential such as supply chain management, access control with identification card, and asset tracking system. As shown in Figure 1, RFID system is compose of a reader (transceiver) and tags (transponder), where RF reader reads and writes data from each entity (RF tag). The reader (transceiver) requests and receives information from tags using RF (radio frequency). Each tag has unique identification information, and tag responds to reader with its unique identification. Request signal also supplies energy for passive tags to make them respond to reader, and the strength of response signal sent by the tag is much smaller than the power of reader's request signal. To improve the signal to noise ratio of received signal from tags, we can use a direct sequence spreading, which spreads or repeats small energy, and increases the total received energy from tag to reader. As shown in Figure 1, all tags in reader's radio range can respond to reader's request signal simultaneously. If two or more tags are in a reader's radio range, the reader cannot uniquely identify tag without collision resolution

T. Vazão, M.M. Freire, and I. Chong (Eds.): ICOIN 2007, LNCS 5200, pp. 513–522, 2008.

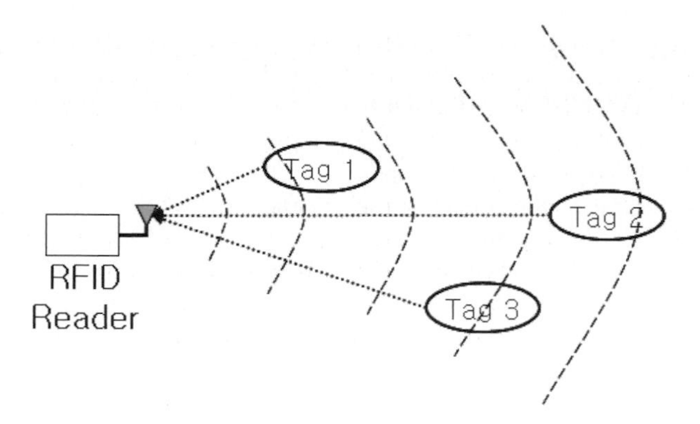

Fig. 1. Multiple tag identification in RFID system

scheme. In this paper, we propose a direct sequence spreading scheme based on collision resilient code symbols.

To prevent collision in RFID system, there are two previous researches: (1) multiple access protocol which is known to ALOHA from networking, and (2) binary tree algorithm, which is relatively simple mechanism [1]. The ALOHA is a probabilistic algorithm, which shows low throughput and low channel utilization. To increase the performance, slotted ALOHA (time slotted, frame slotted, or dynamic frame slotted) protocol is suggested. Binary tree algorithm and query tree protocol are deterministic algorithms, which detect the location of bit conflict among tags, and partitions tags into disjoint group recursively until there are no collision. It requires as many as the length of ID to identify one tag in worst case. To partition into tags, binary tree algorithm stores previous query at tag's register, but query tree protocol use prefix instead of the register. In this paper, we propose a variation of query tree algorithm with collision recovery vector symbol. When there is less than k responding symbols in reader's radio range, our protocol can identify the tags without any re-transmission. In section 2, we review previous approaches for tag collision, and propose our scheme with simulation results in section 3 and section 4. We conclude in section 5.

2 Related Works

To avoid collusion and share limited channel in communication system, there are many multiple access techniques - space division multiple access (SDMA), Frequency domain multiple access (FDMA), time domain multiple access (TDMA), code division multiple access (CDMA). But, these techniques assume that each user can use channel continuously, and are not suitable for RFID system. In RFID system, there two type of collision resolution scheme: (1) Probabilistic algorithm, which is based on ALOHA. (2) Deterministic algorithm which detects collided bits and splits disjoint subsets of tags. There are two open standards from ISO and EPC organizations. ISO 18000-6 family standard uses probabilistic algorithm which is based on ALOHA procedure, and EPC family standard uses deterministic algorithm.

ALOHA is very simple procedure, a reader requests ID, tags will randomly send their data. When collision occurs, they wait random time and retransmit. To enhance performance, they will uses switch off, slow down and carrier sense [2]. In slotted ALOHA, time is divided in discrete time slot, and a tag can send its data at the beginning of its pre-specified slot. Although the slotted ALOHA can enhance the channel utilization and throughput, it cannot guarantee the response time when there are many tags near reader.. To guarantee the response time, frame slotted ALOHA is proposed. In this scheme, all the tags response within frame size slots. As the frame size is bigger, the probability of collision gets lower, but the response time gets longer. When frame size equals to the number of tags, this scheme shows best high throughput [3]. In [3, 4], they suggest dynamic frame slotted ALOHA algorithm, which estimate the size of tags and dynamically change frame size. ALOHA based protocol, however, cannot perfectly prevent collisions. In addition, they have the tag starvation problem, where a tag may not be identified for a long time [6]. The starvation means that some tags have had no chance of transmissions for a long time, when they are collapsed repeatedly.

Deterministic algorithm, which has no starvation problem, is most suitable for passive tag applications. It is categorized into binary tree protocol and query tree protocol. Both of these protocols require all tags response at the same time and the reader identify corrupted bits [6]. In binary tree protocol, the tag has a register to save previous inquiring result. It has disadvantage of complicated tag implementation, and the tag in overlapped range of two readers will show incorrect operation. Query tree protocol does not require tag's own counter. Instead of using counter, the reader transmit prefix and tags are response their rest bits. The query tree protocol is memory-less protocol and tags has low functionality. However, it is slower than binary tree protocol for tag identification. In query tree protocol [6] as shown in Table 1, the reader requests their ID with no prefix, and all tags transmit their ID. As a result, received four bits are totally corrupted. Next, the reader requests it with prefix 0, 0001 and 0011 transmit their bits [0X1]. The reader can know third bit is in collision, it request ID with prefix 000 and only one tag whose ID is 0001 transmit fourth bit as one.

Table 1. Detailed Procedure of query tree protocol

Time	Reader request	Tag response	Note
t_0	null	Tag_1: 0001	All tags reply with their IDs, as a result, the reader knows that all bits are collusion.
		Tag_2: 0011	
		Tag_3: 1100	
t_1	0	Tag_1: 001	Tag 1 and tag 2 who match prefix 0 replies with their remaining IDs.
		Tag_2: 011	
		Tag_3: -	- : means not response
t_2	000	Tag_1: 1	Tag 1 who matches prefix 000 reply with its last bit. Tag 1 identified.
		Tag_2: -	
		Tag_3: -	

Although the prefix increase bits between tags and reader in query tree protocol, it makes tags low functionality, cost and robust to errors. In this paper, we suggest modified query tree protocol with collision resilient symbol.

3 Symbol Based Query Tree Protocol with Collision Recovery

The performances of query tree protocol and binary tree algorithm depend on how to detect collision bit position. However, if the majority of tags transmit 1 and a few tags transmit 0, the bit position will be decoded as 1. In this paper, we propose symbol based extension of query tree protocol to reduce identification speed, power consumption, and to increase robustness under low SNR regions. To identify tag, we propose multi-state extension of query tree protocol. Traditional query tree protocol uses round of bit queries and bit responses, and our scheme uses round of symbol prefix queries and responses. In each round, the reader broadcasts a message that tags whose ID contain a certain prefix should response to the reader with their remaining ID. If more than one tag answers, the reader knows that there are at least two tags that have the same prefix. The reader then appends additional symbol to the prefix, and continue to query with a longer prefix. When the reader gets unique response, which means that the prefix uniquely matches a tag, we can uniquely identify a tag.

Table 2 show the example of one tag identification procedure when there are four tags in reader's radio range. In this scenario, we can identify two tags (tag1 and tag2) at t2. When a reader requests tag identification signal with null at t0, each tag send their first symbols and received signal is $(\lambda_1 + \lambda_2)S_4 + \lambda_3 S_6 + \lambda_4 S_8$. The coefficient ($\lambda_i > 0$) depends on power, distance and channel state between tags and reader. With soft-decision (list) decoding algorithm [8], the reader may decode the symbol S4. After the first symbol is decoded, the reader requests with prefix S4, and, there are two prefix matched tags (tag1 and tag2) with S18 response. Finally, the reader requests with prefix S4S18, and we can identify both tags.

Table 2. Example of symbol based query tree protocol for one tag identification

Time	Reader request	Tag response	Note
t_0	Null	Tag_1: S_4	All tags reply with their first symbol. After this round, we can successfully identify maximum three symbols; in this case, we can identify S_4, S_6, S_8.
		Tag_2: S_4	
		Tag_3: S_6	
		Tag_4: S_8	
t_1	S_4	Tag_1: S_{18}	Using prefix selection strategy, we use the prefix S_4. We can extend our scheme to multiple prefix selection. - : means no response
		Tag_2: S_{18}	
		Tag_3: -	
		Tag_4: -	
t_2	S_4 S_{18}	Tag_1: S_5	Both of tag1 and tag 2 can be identified
		Tag_2: S_7	
		Tag_3: -	
		Tag_4: -	

To identify both of tag 1 and tag2 simultaneously as shown in Table 2, we can use special coding for symbols. Figure 2 shows the idea of collision recovery scheme. In error correction code, the distance of any symbols is at least D. and if the received

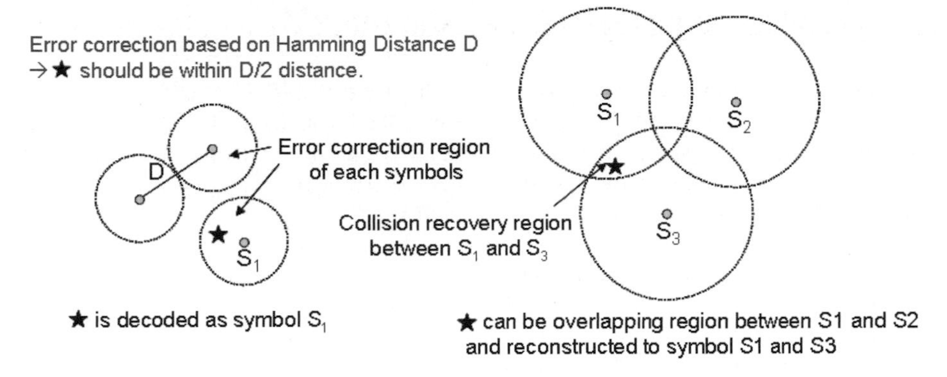

Fig. 2. Collision recovery vs. error correction code for collision resilience

signal is closer to one symbol then D/2, it will be corrected. However, the received symbol is far from any symbol, it will be error. The collision resilient symbol means that the distance of arbitrary two symbols uniform. Therefore, if the received symbol is same distance of any original symbols, we can reconstruct originally sent signals.

As a collision resilient symbol, we suggest two type of codebook. One comes from (v, k, λ)-BIBD and the other comes from orthogonal code. To make collision resilient symbol, we define frame-proof code [10].

Definition 1. *A (v, n)-code* Γ *is called a c-frameproof code if, for every* $W \subseteq \Gamma$ *such that* $|W| \leq c$, *we have* $F(W) \cap \Gamma = W$. *We will say that* Γ *is a c-frameproof code.*

The definition of (v, k, λ)-BIBD code is set of k-element subsets (blocks) of v-element set χ, such that each pair of elements of χ occurs together in exactly λ blocks. The (v, k, λ)-BIBD has total of n= λ $(v^2-v)/(k^2-k)$ blocks, and we can represent (v, k, λ)-BIBD code an v*n incident matrix, where C(i,j) is set to 1 when the i-th element belongs to the j-th block and set to 0 otherwise [9]. Suppose that the symbols are derived from (v, k, 1)-BIBD, all possible symbols are $n=(v^2-v)/(k^2-k)$ and it is one out of frame-proof codes [11]. Figure 3 shows the example of (7, 3, 1)-BIBD which can identify up to 2 symbols at one transmission. For example, when the 1-st and 2-nd symbols (column) collide, the first bit remains one and 5th and 7th bits are zero. On the contrary, if a reader receives that result, the reader knows that 1st and 2nd symbols really sent. If one or more bits are not corrupted, we can make partition into two disjoint subsets and the one has less than 3 tags and it has unique elements. e.g) when third bit is 1, the subset has first, sixth and seventh symbols. (7, 3, 1)-BIBD code can represent only 7 symbols and identify up to 2 symbols within one transmission, we can redesign the parameter (v, k). (16, 4, 1)-BIBD can support n = (16*15)/4*3=20 symbols.

Figure 4 compares the Hadamard code and BIBD code. As shown Figure 4 (a), the hamming distance of each others must be 8. As shown Figure 4 (b), the block represents as column and a first column has four elements, and its elements are 1, 2, 3

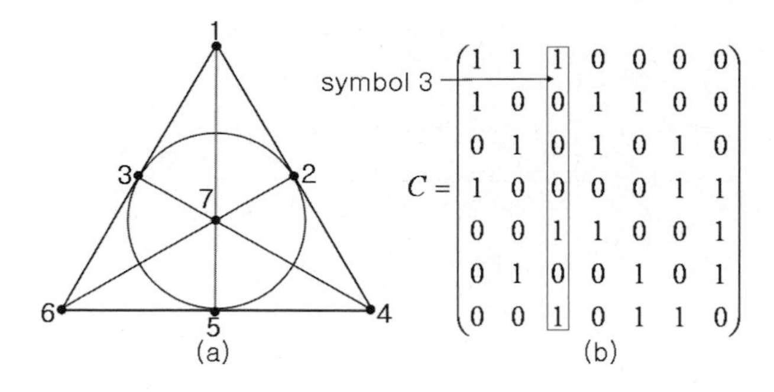

Fig. 3. Geometric (a) and incident matrix (b) representation of (7, 3, 1)-BIBD

and 4. In addition, when randomly select two elements, the block including them is only one. For example, third and fourth elements determinates first column. It stands-for the intersection between arbitrary two blocks is at most 1. The hamming distance of BIBD code can be calculated by this property. If two blocks share not any element, the distance will be 2*k. And, two blocks share only one element, the distance will be 2*(k-1). Although the BIBD code violates uniqueness in hamming distance, the BIBD code can support more symbol then same sized Hadamard code.

```
0000 0000 0000 0000      1111 1000 0000 0000 0000
0101 0101 0101 0101      1000 0111 1000 0000 0000
0011 0011 0011 0011      1000 0000 0111 1000 0000
0110 0110 0110 0110      1000 0000 0000 0111 1000
0000 1111 0000 1111      0100 0100 0100 0100 0100
0101 1010 0101 1010      0100 0010 0010 0010 0010
0011 1100 0011 1100      0100 0001 0001 0001 0001
0110 1001 0110 1001      0010 0100 0010 0000 1001
0000 0000 1111 1111      0010 0001 0000 1010 0100
0101 0101 1010 1010      0010 0000 1001 0100 0010
0011 0011 1100 1100      0001 0100 0000 1001 0010
0110 0110 1001 1001      0001 0010 0001 0000 1100
0000 1111 1111 0000      0001 0000 1100 0010 0001
0101 1010 1010 0101      0000 1010 0000 1100 0001
0011 1100 1100 0011      0000 1001 0100 0000 1010
0110 1001 1001 0110      0000 1000 1010 0001 0100
```

(a) Hadamard construction   (b) (16, 4,1)-BIBD construction

Fig. 4. The example of collision resilient symbols, with 16-bitHadamard matrix (a) and (16, 4, 1)-Balanced Incomplete Block design (b)

When a reader requests next symbol with a prefix, the tags who matches the prefix should response with their next 16-bit symbols. Table 3 shows the reader request and tag response scenario when we have four tags: [4 18 5], [4 18 7], [8 9 2], and [6 8 3]. At the 3rd iteration, although there are two tags whose prefix [4 18], the one tag whose closest from reader or strongest signal at reader will be identified with our protocol; or both of them can be identified simultaneously.

Table 3. The procedure for all tags identification using coded symbol based query tree algorithm

iteration	Reader request	Tags response
1	Null	[4]
2	[4]	[18]
3	[4 18]	[5]*1
3-1	[4 18 5]	Confirmed*2
4	[4 18]	[7]
4-1	[4 18 7]	Confirmed*2
5	[4 18]	no response
6	[4]	no response
7	Null	[8]
8	[8]	[9]
9	[8 9]	[2]
9-1	[8 9 2]	Confirmed*2
10	[8 9]	no response
11	[8]	no response
12	Null	[6]
13	[6]	[8]
14	[6 8]	[3]
14-1	[6 8 3]	Confirmed
15	[6 8]	no response
16	[6]	no response
17	Null	no response

*1: one of them response, *2: one tag is identified and muted

In the query tree protocol, a reader detects collision bit by bit. However, our scheme can detect collision with 16 bit vector symbols which can represent twenty tags. All tags that match the prefix transmit their remaining bits in traditional query tree protocol; whereas, our symbol based query tree protocol transmit their next symbols (each symbol takes 16 bits). The following procedure describes our protocol:

```
Algorithm: Coded Symbol based Query Tree Protocol

Set the prefix empty
do
    rx-signal = request (with the prefix)
        if (rx-signal is no response ) then
            if (the prefix is not empty) then
                delete last symbol in the prefix
            else
                terminate
            endif
         else
            Symbol = decode (the rx-signal)
            add symbol in to end of the prefix
         endif
         if (size of prefix == size of tags symbol) then
```

```
                        ensure that existence of the tag and
                        make it not response (mute)
                        delete last symbol in the prefix
                endif
        od
```

To solve the small number of tags and to be compatible with the electronic product code, we used 32-bit ID, which composed of two 16-bit BIBD codes. Each 16-bit BIBD code is based (16, 4, 1)-BIBD, and it can support 20*20 tags (users). If the RFID system uses 48 bits for IDs, we can use three symbols, we can support 8000 tags. Each tag has unique path in the query tree and its depth is 3. Therefore we can identify one tag at most 3 times transmission. To support 160000 tags, traditional query tree protocol requires 13 bits (8192 tags) and 13 iterations to identify one tag in worst case; however, our scheme requires only 4 iterations in worst case. Although our scheme sacrifices the number of supported tags, it has strong advantage in identification speed, low power consumptions and robustness under low SNR region. To increase the number of supported tag we can use hybrid scheme, where small part uses BIBD scheme to be compatible with EPC Global Code.

4 Experimental Results

To detect collision, traditional query tree protocol uses Manchester coding. To represent a message 'zero' and 'One', it sent 01 and 10 using an amplitude shift keying. Therefore, it can be described as (2, 2)-code. Table 4 shows the number of supported tags for collision resilient symbol based (BIBD), orthogonal code based (Hadamard) and bit based (query tree protocol). Table 4 summarizes our experiments, every code can be represented as (l, n)-code, l means the dimension of symbol, and n is the number of symbol. To generate tag ID, we permuted with each symbol. For example, the 80 (eighty) bits can divided into 5 segments. And each segment will be one of symbol. To support billions tags, In the Manchester encoding, the population of tags can be represented binary tree with depth 20. However, In BIBD or Hadamard matrix code, the maximum depth of tree is only 5.

Table 4. Comparison of collision resilient symbol based and bit based query tree protocol

Construction type	**Codebook parameter**	**Total Bits**	**Depth (segment)**	**Supported tags**
Binary Query Tree Protocol	(4,2)-code	80bits	20=80/4	2^20=1,048,576
MSQTP-Hadamard Matrix	(16,16)-code	80bits	5= 80/16	16^5=1,048,576
MSQTP - (16,4,1)-BIBD	(16,20)-code	80bits	5=80/16	20^5=3,200,000

In our experimentastion, we assume AWGN (additive white Gaussian noise) model without fading for radio channel, and tested three codebooks for collision recovery. To compare the performance of various codebooks, we count total iterations for all tag identification under various noisy environments. Figure 5 (b) and (c) shows the performance of 16 dimensional Hadamard code and BIBD code respectively. They

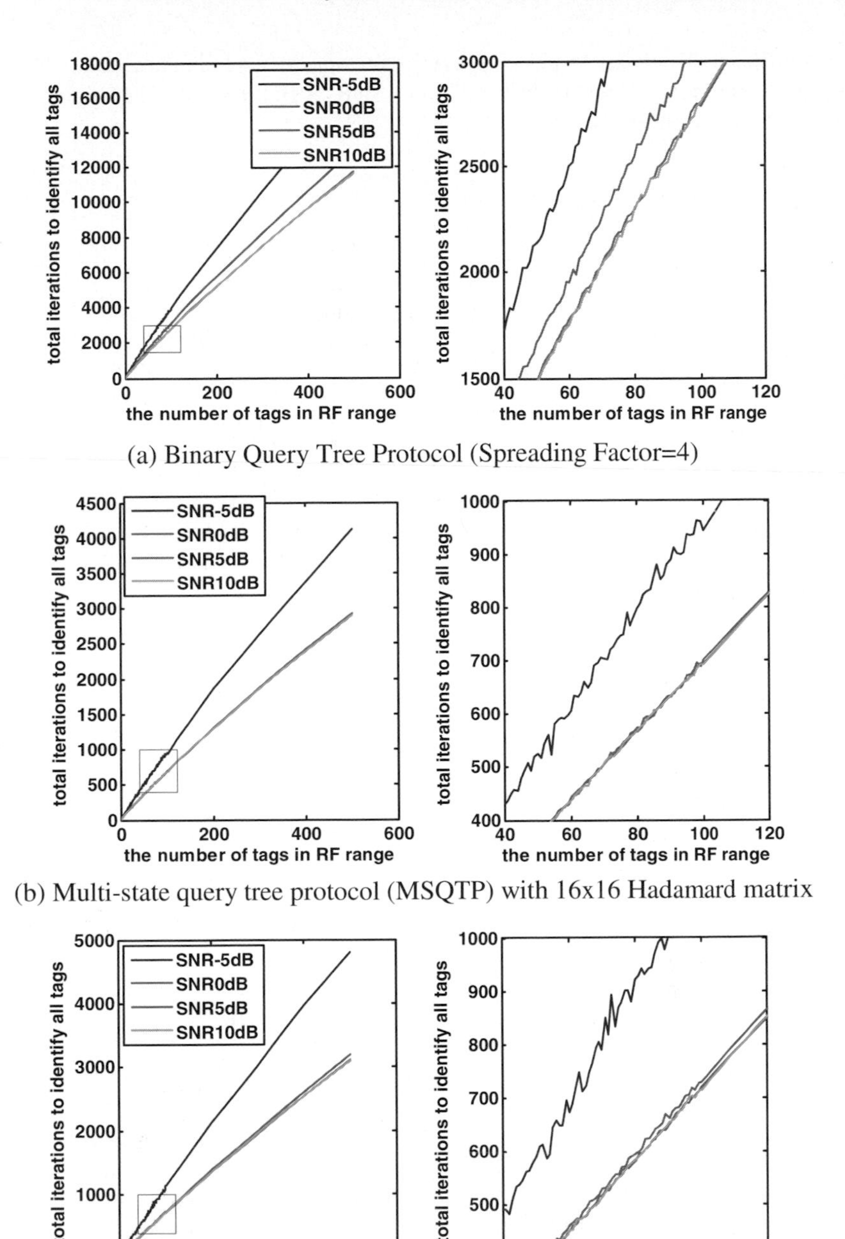

(a) Binary Query Tree Protocol (Spreading Factor=4)

(b) Multi-state query tree protocol (MSQTP) with 16x16 Hadamard matrix

(c) Multi-state query tree protocol (MSQTP) with (16, 4,1)-BIBD

Fig. 5. The performance comparison under various SNR

has no degradation of performance over SNR=0dB. And comparing iterations with Hadamard code, it shows more iterations than Hadamard code, however, the overhead is insignificant. Because, it has 3 times more tags then Hadamard code. However, there are difference between SNR=0dB and SNR 5dB. It means that binary symbol can not operate well in SNR=0dB.

5 Conclusion

In this paper, we proposed a collision detection and recovery algorithm for RFID tag collision cases. We designed the basic code using (v, k, λ) BIBD (balanced incomplete block design) code, and it can identify symbols when up to k symbols are collapsed. Our scheme does not require re-transmission, which costs power consumption. We simulated our scheme over various radio environments using AWGN channel model. Our scheme shows good collision detection and ID recovery (average k symbols for bad radio environments).

References

1. Finkenzeller, K.: RFID Handbook, Fundamentals and Application in Contact-less Smart Card and Identification, 2nd edn., pp. 195–219. John Wiley & Sons Ltd., Chichester (2003)
2. ISO/IEC 18000 Part 3- Parameters for Air Interface Communications at 13.56MHz, RFID Air Interface Standards
3. Cha, J., Kim, J.: Novel Anti-collision Algorithm for Fast Object Identification in RFID System. In: IEEE ICPADS, pp. 63–67 (2005)
4. Vogt, H.: Multiple object identification with passive RFID tags. In: IEEE Int. Conf. on System, Man and Cybernetics, vol. 3, pp. 6–9 (2002)
5. MIT Auto-ID Center, Draft protocol specification for a 900MHz Class 0 Radio Frequency Identification Tag (February 2003), http://www.epcglobalinc.org/
6. Myung, J., Lee, W.: An Adaptive Memoryless Tag Anti-Collision Protocol for RFID Netwroks. In: IEEE INFOCOM (2005)
7. Zhou, F., Chen, C., Jin, D., Huang, C., Min, H.: Evaluation and Optimizing Power Consumption of Anti-Collision Protocols for Applications in RFID System. In: ACM ISLPED 2004, pp. 357–362 (2004)
8. Guruswami, V., Sudan, M.: Improved Decoding of Reed-Solomon and Algebraic-Geometry Codes. IEEE Trans. on Information Theory 45, 1757–1767 (1999)
9. Colbourn, C.J., Dinitz, J.H.: The CRC Handbook of Combinatorial Design. CRC Press, Boca Raton (1996)
10. Stinson, D.R., Wei, R.: Combinatorial Properties and Construction of traceability Schemes and Frameproof Codes. SIAM Journal on Discrete Mathematics 11, 41–53 (1998)
11. Stinson, D.R., Trung, T.V., Wei, R.: Secure Frameproof Code, Key Distribution Patterns, Group Testing Algorithm and Related Structures. J. Statist. Planning Inference 86, 595–671 (2000)

Performance Study of Anti-collision Algorithms for EPC-C1 Gen2 RFID Protocol*

Joon Goo Lee, Seok Joong Hwang, and Seon Wook Kim

Compiler and Advanced Computer Systems Laboratory
School of Electrical Engineering
Korea University, Seoul, Korea
{nextia9,nzthing,seon}@korea.ac.kr

Abstract. Recently RFID systems have become a very attractive solution for a supply chain and distribution industry to trace a position or delivery status of goods. RFID has many advantages over barcode and vision recognition systems, but there are still many problems to be solved. One of the important issues is channel efficiency. To get higher efficiency, a reader uses an anti-collision algorithm. In this paper, we characterize a set of anti-collision algorithms based on a framed slotted ALOHA protocol when using EPC-C1 Gen2 protocol with different frame sizes. Additionally, we propose a simple and effective algorithm which called DDFSA. The proposed algorithm outperformed a conventional Dynamic Framed Slotted ALOHA with a threshold method by 14.6% on average. We also explain why Gen2 has a good channel efficiency in bad environment.

1 Introduction

Nowadays RFID (Radio Frequency IDentification) has become a very popular solution for an automatic recognition in a supply chain and a distribution industry as an alternative to bar-code and vision recognition systems. This RFID system exercises its influence over mobile and telematics services because of its advantages of contact-less recognition of tags, ease of maintenance and extendability over Internet services. However, there are still many problems such as slow standardization, radio regulation, and security issues. One of the obstacles in RFID is a low efficiency of tag identification in terms of recognition speed and channel efficiency.

To resolve a tag collision in passive RFID systems, there are two types of anti-collision protocols based on time division multiple access (TDMA). One is ALOHA and the other is Binary Tree scheme. The original ALOHA is the simplest probabilistic method, but it is inefficient. For this reason, RFID systems, such as Type A of ISO18000-6 [1], i-Code [8], and EPC Class-1 Generation2 [4] use dynamic framed slotted ALOHA [7]. The Binary Tree scheme, a deterministic method of anti-collision, has many varieties such as Bit-by-Bit in EPC Class-0 [2] and Bin slot in EPC Class-1 [3] for fast identification rate.

* This work was supported by LG Electronics and Brain Korea 21 Project in 2006.

T. Vazão, M.M. Freire, and I. Chong (Eds.): ICOIN 2007, LNCS 5200, pp. 523–532, 2008.

Amongst the protocols, Gen2 protocol has been widely accepted by all over the world very quickly due to high quality information flow— Gen2 became an ISO standard in 2006, and ISO/IEC named it ISO 18000-6 Type C. In general, RFID systems using DFSA (Dynamic Framed Slotted ALOHA) has a limited maximum frame size and varies a frame size using a power of two. If there are over 1000 tags to recognize and a protocol does not support enough frame size, like in case of Type A of ISO18000-6 and i-Code (both can vary between 1 and 256 frame size), the reader can hardly read tags' identification codes. However there is no necessity to solve this problem by adding an additional algorithm for the Gen2 protocol since it provides enough maximum frame size of 32768.

When we implement an RFID reader using DFSA anti-collision, it is very important to select an appropriate frame size for good performance. Even if we can improve speed of tag recognition by increasing transmission rate of a reader or a tag, there is a strict radio regulation allowed in each nation. With keeping this regulation, we should reduce tag collisions by choosing the best frame size for increasing channel efficiency.

Gen2 has a time constraint. A reader should fire one of `Query` series commands before T_2 time expired [4]. Consequently, a reader need to avoid complex algorithms and calculations to get the next frame size because of this constraint. Our algorithm, Dual-threshold DFSA, uses two-level thresholds, which allows to get the better frame size in a simple manner.

We implemented an emulator according to the Gen2 protocol and simulated some anti-collision algorithms that are based on DFSA. We applied Fixed Framed Slotted ALOHA, DFSA using a threshold method, DFSA using an increase method, and our own proposed DFSA using a dual threshold method algorithm. The proposed algorithm is simple, but showed the best performance in terms of channel efficiency. In addition, we analyzed the effect of receiver performance with the emulator since passive RFID systems often face bad environment because of poor tag performance.

The paper consists of the followings: In Section 2, we briefly review the feature of Gen2 protocol, and in Section 3 we present Algorithms based on Framed Slotted ALOHA for Gen2 and their performance. We introduce the Dual threshold Dynamic Framed Slotted ALOHA algorithm in Section 4 and the detail effect of receiver performance in Section 5. And finally the conclusion is made in Section 6.

2 EPC Class-1 Generation2

This protocol uses UHF (Ultra High Frequency), supports multi-reader environment with a session concept, and provides various Reader-to-Tag and Tag-to-Reader data rates using RTcal and TRcal calibration values. In this protocol, a tag communicates with a reader using backscatter modulation. A reader transmits data using PIE (Pulse Interval Encoding) and a tag encodes the backscattered data as either FM0 or Miller modulation of a subcarrier at the data rate while tag replying. A Gen2 tag has four kinds of memory banks including EPC

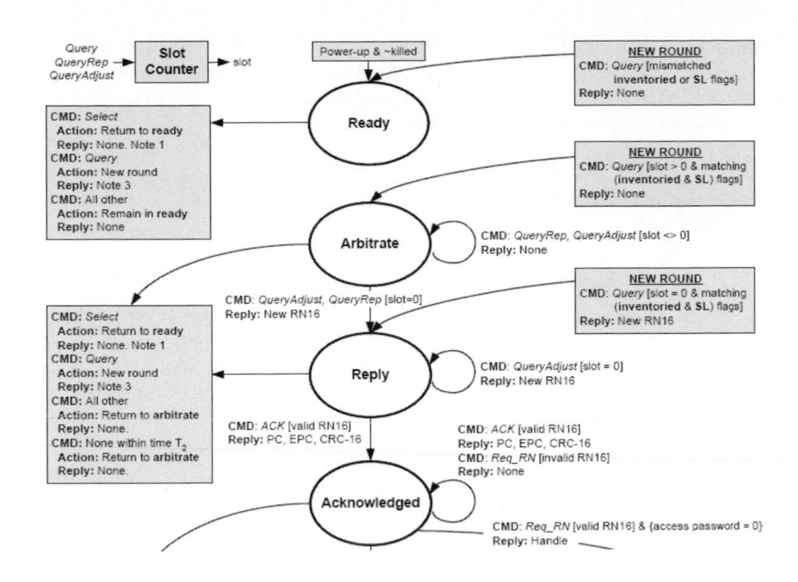

Fig. 1. State transition diagram for Gen2 protocol [4]

code. A tag and a reader can also detect error by using CRC5 or CRC16. A reader can use DSB-ASK, SSB-ASK or PR-ASK modulation for Reader-to-Tag RF envelope. Fig 1 shows the partial part of state transition diagram of Gen2. In the Gen2 protocol, a tag has 7 states (*Ready*, *Arbitrate*, *Reply*, *Acknowledged*, *Open*, *Secured*, and *Killed*) and an interrogator manages tag populations using three kinds of basic operations (Select, Inventory and Access). For tag inventory, we need to consider four states between *Ready* and *Acknowledged* and five kinds of reader commands (`Query` series, `ACK`, and `NAK`). By using the state transition, we can guess the tag's state and choose the proper next command.

3 Algorithms Based on Framed Slotted ALOHA

Framed Slotted ALOHA (FSA) is an evolved version of ALOHA for getting better performance. It consists of number of slots, a frame and a read cycle. A time slot is a time interval that tags transmit their reply, and a frame is a time interval between requests of a reader and consists of a number of slots. A read cycle is tag identifying process which comprises of a frame. Once frame size is fixed, frame size number of slots are prepared. In Gen2, this frame size can be defined as Q. If a reader commands Query with $Q = 4$, for example, the number of slots becomes $2^4 = 16$. Tags shall implement a 15-bit slot counter. Upon receiving a `Query` or a `QueryAdjust` command a tag shall select a value between 0 and $2^Q - 1$. This value becomes the time slot and tag transmits its reply RN16 which means 16-bit random number in its time slot only. In the reading zone of a reader, if there are many tags to recognize, a collision ratio will increase. If

```
Supply power to tags
end_cond_c = 0, end_cond_v = 0, end_inventory = FALSE, limit = 5
Send Select /*Select all tags in session_n, inventoried_flag -> A*/
Send Query /*Query with session_n and initial_Q(0~15)
DO
  calculate num_of_slots from Q, slot_cnt = 0
  DO
    Receive response of Query /* response is RN16 */
    IF no_error
      limit_cnt = 0
      DO
        limit_cnt = limit_cnt + 1
        Send ACK /*Ack with received RN16 */
        Receive response of ACK /*response is PC + EPC + CRC16 */
      UNTIL no_error .OR. (limit_cnt .EQ. limit)/*prevent infinite loop*/
    ELSE IF collision .OR. crc_error
      end_cond_c = end_cond_c + 1
      IF end_cond_c .EQ. slot_cnt
        end_inventory = TRUE break; ENDIF
    ELSE /*no response*/
      end_cond_v = end_cond_v + 1
      IF end_cond_v .EQ. slot_cnt
        end_inventory = TRUE break; ENDIF
    ENDIF
    IF slot_cnt .EQ. num_of_slots
      Send QueryRep /*QueryRep with session_n*/
    ELSE /*end of round*/
      Send QueryAdjust /*QueryAdjust with session_n and Q*/
    ENDIF
    slot_cnt = slot_cnt + 1
  UNTIL slot_cnt .GT. num_of_slots
UNTIL end_inventory .EQ. TRUE
```

Fig. 2. Applied FFSA algorithm on Gen2

there are few tags, empty slots increase and the collision ratio becomes lower. Using these response patterns, we can adapt to get an optimal Q value in Gen2.

In this section, we describe the existing FSA anti-collision algorithms and adapt each algorithm to Gen2, and we also briefly discuss of each performance result based on simulator results.

3.1 Fixed Framed Slotted ALOHA (FFSA) Algorithm

FFSA, what we called Basic Framed Slotted ALOHA, uses a fixed frame size and does not change the frame size during a tag identification process. This can be the simplest algorithm of DFSA. You can find the algorithm for FFSA on Gen2 at Fig 2. Theoretically, we can achieve maximum throughput of 36.8% [6], but when we using FFSA it is hard to obtain that throughput.

In the simulation results, if we choose proper Q value we can get relatively high channel efficiency around 22%~50%, especially when only a few tags is in the reading zone of the specific reader. However, when we chose only one level lower Q, the efficiency dramatically degraded. Similarly when we chose bigger Q than the effective Q, the efficiency degraded gradually. Also while we increase the number of tags, maximum throughput degraded to 22%. The reason is that if there are N tags in the reading zone of the specific reader, at first we can identify tags' EPC code well when we select the good Q gives the best efficiency, but even though we select it, after some read cycles the void reply ratio increased in compliance with the increase of inventoried tags. At result, we can not get higher performance than 22% if there are many tags. We simulated FFSA algorithm up to 4096 tags and observed that the maximum efficiency becomes stable to around 22% from over 1000 tags.

3.2 Dynamic Framed Slotted ALOHA Algorithm Using a Threshold Method

The concept of DFSA algorithm using a threshold method is that if there are too many tags in a reading zone, a collision ratio will increase, and in the reverse situation a void ratio will increase. From this pattern, we can decide thresholds to adjust the frame size. If we can find a good threshold, we can improve channel efficiency by issuing `Query/QueryAdjust` command with a proper Q value which gives better performance. We provide the algorithm of DFSA-t on Gen2 at Fig 3. When we simulated DFSA-t, we increased the frame size if the collision ratio over 75% and decreased the frame size if the void ratio greater than 30%.

From the simulated results, we could know that the maximum efficiency that can be obtained at when an initial frame size was near the number of unread tags. As the tag number increased, the maximum efficiency also increased near the theoretical best efficiency. When we varied an initial frame size from 1 to 2^{15} at the same number of tags, the efficiency started around 25.8%$\sim$29.0%, continually increased until the initial frame size approached the same value as an initial number of unread tags. After exceeding the optimal frame size, the performance degraded rapidly. If we use a large frame size as an initial frame size, it will degrade the performance when we identify small number of tags severely. If we choose a small frame size, for example 2 or 4, we can not get the best performance. However, if we do, we will get the sufficient performance at least 25.4% at almost cases.

3.3 Dynamic Framed Slotted ALOHA Algorithm Using a Increase Method

DFSA using a increase method, we called simply DFSA-i, is also simple algorithm to implement. It starts a read cycle with a small initial frame size which is either two or four. If no tag was identified during the previous read cycle, it simply increases the frame size and starts the next read cycle. It repeats this until at least one tag is inventoried. If a single tag was inventoried it stops the current read cycle immediately and starts another read cycle with the initial frame size. This algorithm can be found at Fig 4.

It shows very good performance result in terms of the channel efficiency when unread tag number is relative small. If we choose a larger initial frame size, we can get better efficiency than when we use a small one at a dense-tag environment. But the maximum efficiency was not good when we applied this algorithm at an dense-tag environment. We could only get 21.6% at 64 tags and 18.2% at 256 tags. As the number of tags increase, the maximum efficiency fall dramatically

```
Supply power to tags
end_cond_v = 0, end_inventory = FALSE, limit = 5 /*sample limit number*/
Ns = 0, Nv = 0, Nc = 0 /*Ns, Nv and Nc means number of success, void, */
                       /*and collision respectively */
Send Select /*Select all tags in session_n, inventoried_flag -> A */
Send Query /*Query with session_n and initial_Q(0~15) */
DO
  calculate num_of_slots from Q, slot_cnt = 0
  DO
    Receive response of Query /* response is RN16 */
    IF no_error
      Ns = Ns + 1
      limit_cnt = 0
      DO
        limit_cnt = limit_cnt + 1
        Send ACK /*Ack with received RN16 */
        Receive response of ACK /*response is PC + EPC + CRC16 */
      UNTIL no_error .OR. (limit_cnt .EQ. limit)/*prevent infinite loop*/
    ELSE IF collision .OR. crc_error
      Nc = Nc + 1
    ELSE /*no response*/
      Nv = Nv + 1
      IF Q .EQ. 0
        end_cond_v = end_cond_v + 1
        IF end_cond_v .EQ. slot_cnt
          end_inventory = TRUE  break; ENDIF
      ENDIF
    ENDIF

    slot_cnt = slot_cnt + 1
    IF slot_cnt .NE. num_of_slots
      Send QueryRep /*QueryRep with session_n*/
    ELSE /*end of round*/
      Up_threshold = Nc / (Ns + Nv + Nc)
      Dn_threshold = Nv / (Ns + Nv + Nc)
      IF Up_threshold .GT. Up_Threshold /*Up_Threshold to enlarge*/
        UpDn = 110b
        IF Q .LT. 15    Q = Q + 1  ENDIF
      ELSE IF Dn_threshold .GT. Dn_Threshold /*Dn_Threshold to shrink*/
        UpDn = 011b
        IF Q .GT. 0     Q = Q - 1  ENDIF
      ELSE
        UnDn = 000b
      ENDIF
      Send QueryAdjust /*QueryAdjust with session_n and UpDn*/
    ENDIF
  UNTIL slot_cnt .EQ. num_of_slots
  Ns = 0, Nv = 0, Nc = 0
UNTIL end_inventory .EQ. TRUE
```

Fig. 3. Applied DFSA algorithm using the threshold method on Gen2

nevertheless we chose the best initial frame size. But when a tag density was not high, below 16 tags, it showed better performance than FFSA and DFSA-t such as 17.1%~66.7% at Q=0, 20.2%~40.1% at Q=1, 21.8%~32.2% at Q=2.

3.4 Advanced Framed Slotted ALOHA Algorithm

AFSA algorithm estimates the number of tags to read and determines a proper frame size for the estimated number of tags by using a kind of estimation function [5,6]. In the AFSA algorithm, it was assumed that the tags that already been read respond during other read cycle. In Gen2, once a reader identified a

```
Supply power to tags
end_cond_v = 0, end_inventory = FALSE, limit = 5 /*example limit value*/
Ns = 0, Nv = 0, Nc = 0 /*Ns, Nv and Nc means number of success, void, */
                        /*and collision respectively*/
Send Select /*Select all tags in session_n, inventoried_flag -> A */
Send Query /*Query with session_n and initial_Q(0~15) */
DO
  calculate num_of_slots from Q, slot_cnt = 0
  DO
    Receive response of Query /* response is RN16 */
    IF no_error
      Ns = Ns + 1
      limit_cnt = 0
      DO
        limit_cnt = limit_cnt + 1
        Send ACK /*Ack with received RN16 */
        Receive response of ACK /*response is PC + EPC + CRC16 */
      UNTIL no_error .OR. (limit_cnt .EQ. limit)/*prevent infinite loop*/
    ELSE IF collision .OR. crc_error
      Nc = Nc + 1
    ELSE /*no response*/
      Nv = Nv + 1
      end_cond_v = end_cond_v + 1
      IF end_cond_v .EQ. slot_cnt
        end_inventory = TRUE  ENDIF
    ENDIF

    IF Ns .EQ. 1
      Q = initial_Q
      Send Query /*Query with session_n, Target = A, and Q */
      break;
    ELSE
      IF slot_cnt .NE. num_of_slots
        Send QueryRep /*QueryRep with session_n*/
      ELSE /*end of round*/
        UnDn = 110b /*Doubles frame size */
        IF Q .LT. 15
            Q = Q + 1
        ENDIF
        QueryAdjust /*QueryAdjust with session_n and Q*/
      ENDIF
    ENDIF

    slot_cnt = slot_cnt + 1
  UNTIL slot_cnt .EQ. num_of_slots
  Ns = 0, Nv = 0, Nc = 0
UNTIL end_inventory .EQ. TRUE
```

Fig. 4. Applied DFSA algorithm using the increase method on Gen2

single tag, this tag does not respond by inverting the inventoried flag [4]. It means this method can not be applied to the Gen2 protocol. Even though we can activate the tag that already been read using the `Select` command to do this, an overhead is very huge. We did not implement this algorithm because of these reasons.

4 Proposed Dual Threshold Dynamic Framed Slotted ALOHA (DDFSA) Algorithm

We thought that if there are lots of collisions, why doesn't a reader increase a frame size to four times not only two times? We applied this idea to DFSA-t,

```
Supply power to tags
end_cond_v = 0, end_inventory = FALSE, limit = 5 /*sample limit number*/
Ns = 0, Nv = 0, Nc = 0 /*Ns, Nv and Nc means number of success, void, */
                        /*and collision respectively */
Send Select /*Select all tags in session_n, inventoried_flag -> A */
Send Query /*Query with session_n and initial_Q(0~15)
DO
  calculate num_of_slots from Q, slot_cnt = 0
  DO
    Receive response of Query /* response is RN16 */
    IF no_error
      Ns = Ns + 1
      limit_cnt = 0
      DO
        limit_cnt = limit_cnt + 1
        Send ACK /*Ack with received RN16 */
        Receive response of ACK /*response is PC + EPC + CRC16 */
      UNTIL no_error .OR. (limit_cnt .EQ. limit)/*prevent infinite loop*/
    ELSE IF collision .OR. crc_error
      Nc = Nc + 1
    ELSE /*no response*/
      Nv = Nv + 1
      IF Q .EQ. 0
        end_cond_v = end_cond_v + 1
        IF end_cond_v .EQ. slot_cnt
          end_inventory = TRUE  break;    ENDIF
      ENDIF
    ENDIF

    slot_cnt = slot_cnt + 1
    IF slot_cnt .NE. num_of_slots
      Send QueryRep /*QueryRep with session_n*/
    ELSE /*end of round*/
      Up_threshold = Nc / (Ns + Nv + Nc)
      Dn_threshold = Nv / (Ns + Nv + Nc)
      IF Up_threshold .GT. Up_Threshold_q /*Threshold for quadruple*/
        IF       Q .LT. 14     Q = Q + 2
        ELSE IF  Q .LS. 15     Q = Q + 1
        ENDIF
        Send Query /*Query with session_n, Target = A, and Q*/
      ELSE IF Up_threshold .GT. Up_Threshold_d /*Threshold for double*/
        UpDn = 110b
        IF Q .LT. 15   Q = Q + 1  ENDIF
      ELSE IF Dn_Threshold .GT. Dn_Threshold_q /*Threshold for quarter*/
        IF      Q .GT. 1        Q = Q - 2
        ELSE IF Q .GT. 0   Q = Q - 1
        ENDIF
        Send Query /*Query with session_n, Target = A, and Q*/
      ELSE IF Dn_Threshold .GT. Dn_Threshold_h /*Threshold for half*/
        UnDn = 011b
        IF Q .GT. 0    Q = Q - 1  ENDIF
      ELSE
        UpDn = 000b
      ENDIF

      Send QueryAdjust /*QueryAdjust with session_n and UpDn*/
    ENDIF
  UNTIL slot_cnt .EQ. num_of_slots
  Ns = 0, Nv = 0, Nc = 0
UNTIL end_inventory .EQ. TRUE
```

Fig. 5. Applied DDFSA algorithm on Gen2

and we could get better channel efficiency. The channel efficiency of DDFSA when Q is 0, 1, and 2 and other simulation results are shown in Fig 6 and the pseudo code of DDFSA algorithm is in Fig 5.

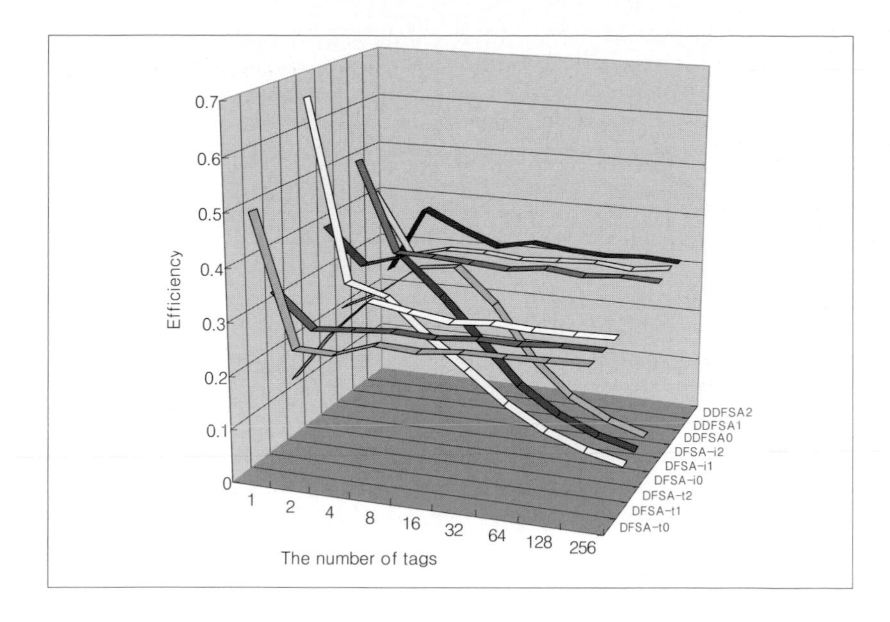

Fig. 6. Simulated channel efficiencies while tags increasing

The important thing is even if the number of tags increase the efficiency hold the value over 30%. Our algorithm is very simple but it outperformed a conventional DFSA algorithm using one level threshold which has the 75% upper threshold and the 35% lower threshold by 14.6% on average. We simulated DDFSA algorithm using the upper thresholds as 75% and 95% of the collision ratio, and the lower thresholds as 50% and 30% of the void ratio. It means a reader double the frame size if the collision ratio over 75% and quadruple it if the collision ratio over 95%. In case of the void ratio, if the void ratio over 30% we decrease the frame size to half, if the void ratio greater than 50% decrease it to one quarter.

5 Performance Analysis

In general, an RFID reader should have a robust receiver since tags have poor performance. In case of Gen2, a return frequency of a tag can vary ±4% to ±22%. If a reader has a humble receiver this variation gives negative influence to estimate responded data bits. For example, the receiver may misjudge a successful reply as a collide reply. A receiver performance is affected by many parameters such as the SNR (Signal Noise Ratio) of antenna, the gain of LNA (Low Noise Amplifier), filter performance, and so on. It is very hard to simulate with these all considerations. In this situation, we just consider BER (Bit Error Rate) as a receiver performance parameter. We applied this parameter to the tag transmitter, which can send wrong bits randomly according to the given BER.

We simulated DFSA-t with the initial frame size of two at BER of 1% and 0.5%. The result was that overall performance was degraded to 89.2∼95.5% at 0.5% BER and to 67.8∼91.5% at 1% BER, and the loss rate of tags was 10% at BER = 1% and 1% at BER = 0.5%. These overall performance degradation became stable rapidly when there are many tags to recognize. Note that if a reader is in a bad environment, it increases a collision ratio and drops overall performance. After a reader sent an `ACK` command, we could not accept a reply packet because of bit error. The reply of `ACK` was 12 bytes, and a tag sent this packet using FM0. At result, total length of the reply was 204 bits including a short preamble. If the BER is high, the reply of `ACK` has many error bits statistically, and this makes a reader can not detect the reply. Consequently, we should prevent infinite looping to detect the successful reply of `ACK`. Contrastively, when a reader scans slots of a frame, a tag reply only 16-bit random number. Including short preamble, this packet size is only 40 bits. So even if we face a bad environment, channel efficiency is not degraded so much. It means that we can apply anti-collision algorithms without regard to BER if a receiver guarantees the minimum performance in Gen2.

6 Conclusion

In this paper, we presented existing algorithms based on FSA, and applied these algorithms to Gen2 protocol. We could also find the performance characteristics of each algorithm from simulated data. We proposed DDFSA algorithm, which is extend version of DFSA-t, and showed that DDFSA has the best channel efficiency in most cases than the others. We could not find algorithm gives the optimal efficiency with low complexity of implementation. However we can get better system performance than other algorithms using very simple idea. This algorithm outperformed DFSA-t by 14.6% on average (65.7% on max) in the simulation result.

References

1. ISO/IEC FDIS 18000-6, ISO standard document, http://www.iso.org
2. EPC class 0, EPC global standard documents, http://www.epcglobalinc.org/standards_technology/specification.html
3. EPC class 1, EPC global standard documents, http://www.epcglobalinc.org/standards_technology/specification.html
4. EPC class 1 gen 2, EPC global standard documents, http://www.epcglobalinc.org/standards_technology/specification.html
5. Vogt, H.: Multiple object identification with passive RFID tags. In: International Conference on Systems, Mans, and Cybernetics, vol. 3 (October 2002)
6. Vogt, H.: Efficient Object Identification with Passive RFID Tags. In: Mattern, F., Naghshineh, M. (eds.) PERVASIVE 2002. LNCS, vol. 2414, pp. 98–113. Springer, Heidelberg (2002)
7. Finkenzeller, K.: RFID handbook, 2nd edn. John Wiley & Sons, Chichester (2003)
8. PHILIPS Semiconductor. I-CODE1 System Design Guide: Technical Report (May 2002)

A Lightweight Management System for a Military Ad Hoc Network

Jorma Jormakka[1], Henryka Jormakka[2], and Janne Väre[2]

[1] National Defence College, P.O.Box 07 00861 Helsinki, Finland
[2] Technical Research Centre of Finland, P.O. Box 1000 FI-02044 VTT, Finland
jorma.jormakka@mil.fi, henryka.jormakka@vtt.fi, janne.vare@vtt.fi

Abstract. Ad hoc networks are usually thought to be self-operating. Especially, they are not expected to need much management. This is not quite true: while the management system can be light, all networks need a management system if they are to stay operational for some time. This paper describes a network management solution to a military ad hoc network, but similar approaches could be applied also in non-military networks for distributing wireless services.

Keywords: policy based management, ad hoc networks, context-aware.

1 Introduction

Ad hoc networks are wireless networks where all network nodes are capable of forwarding data instead of using a central base station through which all data passes. Ad hoc networks use some form of auto-configuration so that network nodes can enter and leave the network without manually configuring network elements. Usually ad hoc networks are also dynamic. An ad hoc network with a connection to a fixed network is called a semi ad hoc network.

There are several potential applications of military ad hoc networks. Air force networks between airplanes and the ground station and navy networks between ships and base stations can gain from semi ad hoc operation mode. Ad hoc sensor networks are a much researched area, e.g. for replacement of infantry mines, active self-protection systems and surveillance. There is much research work on the army tactical ad hoc networks in several countries, the most famous being the USAs JTRS program [6]. Some militaries make a difference between the army tactical networks and wireless networks for brigade level headquarters and command posts. The management solution presented in this paper is basically intended for ad hoc type brigade headquarter networks, but with some reservation the concept may also be suitable to tactical ad hoc networks. The requirements for military ad hoc networks vary very much depending on the application. The IETF MANET working group is assuming battery powered devices in a very large and a very dynamic ad hoc network. These assumptions apply to the army tactical ad hoc network provided that no tactical core network is used. In most of the other applications either the energy constraint is not very critical and/or

T. Vazão, M.M. Freire, and I. Chong (Eds.): ICOIN 2007, LNCS 5200, pp. 533–543, 2008.

network is neither very large, nor very dynamic. The military ad hoc network for which this management solution is developed is neither energy-constrained, nor very large and the level of dynamicity is medium.

There exist rather few ad hoc management solutions. One ad hoc network management protocol ANMP is described in [4]. It is based on the management paradigm of CMISE and SNMP v3. Also the management solution Guerilla Management Architecture in [10] is based on SNMP. The manager-agent paradigm and Managed Information Base (MIB) approach are today often considered insufficient, especially since MIB does not include management of software configurations in the context of software updates and it does not allow easy addition of new actions, i.e., new procedures that can be called by the manager. Policy based network management (PBNM) protocols are better in this respect, since it is possible to distribute new policies from the policy database. However, in PBNM usually all nodes in the network follow the same policies. PBNM, joined with the other paradigm WBEM/CIM (Common Information Model) of the Distributed Management Task Force (DMTF) is a popular way for making management of autonomic systems, a typical example being PMAC [1]. CIM is too heavy for small ad hoc networks considered in this paper. One policy based network management approach for ad hoc networks called CAM (Context-Aware Management) was proposed in [3], but the RFC draft of CAM has already expired. Context-aware applications, services and management is a popular trend in mobile networks, e.g. [5] and [9]. Configuration management in the sense of adding new software is an active area in mobile networks [12], but not yet in ad hoc networks. In military ad hoc networks much attention has been paid to QoS management and ideas such as Service Level Agreements (SLA) have been proposed [11]. The setting in these developments seems to be an operationally difficult goal: obtaining radio bandwidth for supporting network-centric warfare in a situation where bandwidth is a limiting factor. In the brigade headquarter ad hoc network bandwidth is sufficient and sufficient service quality can be provisioned without SLAs. There is also much research on using policy-based approaches for alleviating problems caused by misbehaving nodes [2]. In the network studied in this paper, misbehaving nodes are not of major concern. The concept of mobile services does not apply easily to the intended military network: the services are either (group) voice and messaging services which do not need service mobility, or graphical database applications for group work, which contain lots of data and are more easily replicated than made mobile.

The use of ad hoc nodes in a brigade headquarter is rather similar to typical LAN usage. In office LANs there usually is an assigned person managing the network. The management tasks include the following. The users must be given access rights and addresses. There is firewall and antiviral software, which have settings and updates and thus must be managed. The users have email and access to some www-servers, and consequently can obtain malicious software. This means that fast updates to software are required. Network usage may be monitored and in some environments the manager has full remote management possibilities to all network nodes. Granting remote access is a too high security risk in the brigade headquarter network. Nodes have classified material and the managing person does not necessarily have the rights to access it. We can only grant limited access by using a management interface. Addressing is not assumed to require fast responses from management since the nodes will obtain temporary addresses by a self-configuration mechanism. Still, assigning home addresses and user rights is needed. It seems that the most relevant situations in office usage requiring faster responses than can be made by the management interface are software updates to mitigate attacks by malicious software and changing frequencies in a case of a jamming attack. Although the ad hoc network of the brigade headquarter is usually connected to a fixed network, we cannot always rely on a management centre situated in the fixed network. The network administrator should be able to reconfigure one of the ad hoc nodes to act as a replacement of the management centre.

2 Management System Overview

The management solution presented in this paper allows the ad hoc network to be managed from any node of the network. Every node supports a management interface from which the network administrator can do a small but sufficient set of management operations. The management module is a software module which logically has access to some operations of the management interface (MI), access to a set of policies and a protocol, by which it communicates with the management modules of other nodes. Instead of the operations of the management interface, the management module may use directly the same scripts which the MI evokes. Reading or assigning a parameter value from/to a network card or some other device is easiest done by running a command script. In this solution all management operations have corresponding command scripts (reading scripts and setting scripts) and the task of both the management interface and the management module is to run the scripts with suitable input parameters. Most of the complexity and all network/task specific issues are implemented in the scripts and the management solution can be simple and generic. The structure of the management module is presented in Figure 1.

The engine action is started by a timer trigger (like cron in Unix), by a sensor or by an input from the communication protocols of the distributed manager. The sensor concept here means an element receiving any input from an environment outside the distributed manager protocol. Sensors in Figure 1 do not usually mean sensing some analogue parameters (geographic position, temperature, time, humidity and so on), but more usually mean sensing some data in incoming connections. Naturally, sensors could be sensing analogue data as well if such hardware is available. Timing and power levels of incoming signals are useful analogue parameters to be sensed e.g. for location finding and transmission power control.

The engine controls that policy rules are followed. Each policy rule contains a condition and a setting script. Evaluating a condition requires reading configuration files and running reading scripts. If a condition is true, the engine runs

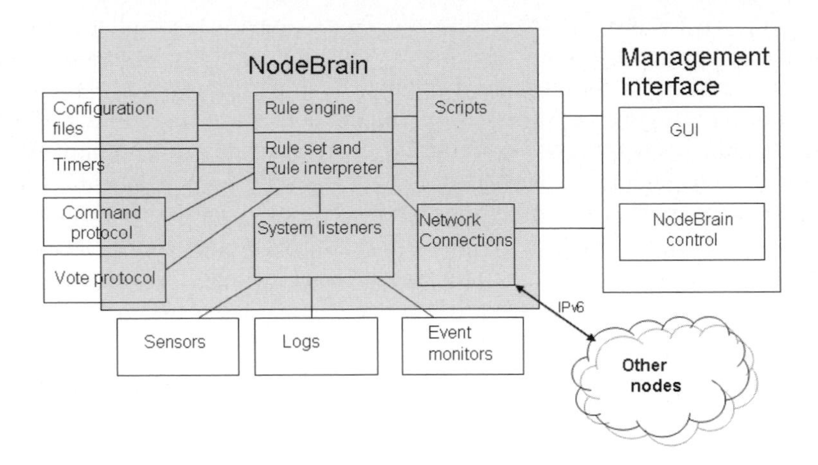

Fig. 1. Distributed management module structure

a setting script. A reading or setting script can run other reading and setting scripts, and it is possible to combine several read or set operations in one script if needed. The vote protocol is a protocol, where the nodes take part into a vote and agree to run some rules. Rules can be changed by the rule exchange protocol, but the decision of the change has to be made by the only node that has the rights to decide about the changes. As the vote protocol used for changing some common settings may be slow in some situations, the command protocol is used for fast changes, enforcing on the other nodes the change of rules. The command protocol additionally commands the engine to run a policy rule. The management interface (MI) uses the same reading and setting scripts but it does not use policies.

3 Services of the Management Interface (MI)

The services provided through the MI for the brigade headquarter network are listed below following the ISO division of managed areas into FCAPS.

The typical image of an ad hoc network is that the devices work perfectly or are replaced by new ones. In the military application replacements are not necessarily available and the management system must be capable of monitoring the state of the network nodes and receiving alarms from devices. The essential services are:

* Alarms of selected events, and
* Retrieval of the state of selected parameters.

Traditionally configuration management has been needed for configuring network devices. In an ad hoc network auto-configuration is used as much as possible. There is, however, still need for configuration management. Installing new software updates for provision of new services or for security patches is one of the main reasons why configuration management is still needed, and the manager sometimes needs to look at the settings. There is also a very specific need for file transfer in the army tactical networks: digital combat net radios have a number of configurable parameters, such as the frequency band, jump sequences for the selected spread spectrum solution, keys for security, power levels for the signals, modulation alternatives and so on. These settings are assigned by a communications officer and they are typically distributed offline, but it is too slow. The designed management system allows fast and secure transfer of files suitable for this management task. The following configuration management services are necessary:

* Installation of software modules,
* Retrieval of the state and setting values to configuration parameters,
* File transfer for software modules and external configuration files.

A traditional task of configuration management is to present the view of the managed network to the network administrator. In an ad hoc network this task would require polling the nodes and as some nodes are not always reachable, this would require constant monitoring. We have decided to leave this task to the applications. The main application in the intended usage is Command and Control (C2) application, which keeps track of active parties. This choice also means that the management solution does not generate much traffic.

Services for user account management are:

* Add/remove user entry. Adding a new user requires among other things provisioning him with private keys, a certificate issued by a Certification Authority (CA), and public keys of the CA. The user profile is created and stored at a database. When the user is removed from the database all the issued data is purged.
* Create and maintain user groups. User profiles and rights are updated accordingly.

The performance management services are optional, they comprise of:

* Assigning a proper QoS policy for nodes resources. In ad hoc networks some of the network resources may be provided by the users' terminals. In some cases the decision how much resources nodes provide to network functions and how much for their own usage should also belong to the management of network performance.
* Event reporting and alarm - to notify e.g. when the throughput is unsatisfactory, if any node has reached the maximum capacity and cannot service future requests, or when some performance thresholds have been surpassed.

The security management services are:

* Key management. The keys' generation and distribution to all the parties are provided, and the keys are stored in trusted repositories.

* Certificates issuing. In order to provide key management, certificate issuing authority has to be nominated. Taking into account the considerably small size of the network, the task can be assigned to the network manager.
* Maintenance of different security levels ranging from unclassified to the highest supported classification.
* Inspection of security logs - to protect the managed objects and prevent security breaches, maintenance of security logs is provided. The number of connections and disconnection to each node, information on types of management operations performed on the node, or statistics related to the usage of the node are stored. The logs are available for investigation by the administrator.
* Routing control is optional. It is for providing relay mechanisms to avoid specific networks or data communications link for purposes of security. Possibility for controlled switching of different routing protocols would also increase the security of the network.
* Handling security alarms. In case of security services violations e.g. unauthorized access attempt, security alarms are issued. However, sometimes security alarms may reveal too much information for an unauthorized listener. For that reason support of an adjustable alarm level that can be changed by the network administrator is necessary.

4 Protocols

The management module contains three protocols, which are shortly described below.

Vote Protocol. The vote protocol is an agreeing protocol where the nodes end up with common settings through a vote. Each node wishing to take part in a vote sends its common input parameters. These messages are flooded through the network, so every connected node gets them. Each node calculates an average value of each common input parameter using a parameter specific algorithm. Each node applies the same rule and as the input parameters are the same, each node makes the same decision which setting script to run. The script makes the same settings in all nodes. There are two special mechanisms: tunnelling and veto: If the network nodes decide on some parameter setting which would imply running a script which the node cannot or does not want to run, the node can establish a point to point connection with a nearby node so that all traffic is echoed through the node. A node can send a veto to a vote. If a given threshold number of vetoes are received, a vote is cancelled. This stops the network from selecting settings which are unacceptable to a too large number of nodes. Vote is an automatic protocol and does not need any user interaction.

Command Protocol. The vote protocol is slow and the command protocol is necessary for fast changes of settings. Any node can issue a command for moving to new common input parameter settings. Command is issued manually through the user interface. When a command is received, it optionally results in a prompt through the user interface. Before a command is accepted, the user on the targeted node may have to confirm it. This user interaction helps to avoid opposite commands which would be a result of issuing commands automatically. The command is a very simple protocol: the administrator orders running of a script.

Rule Exchange Protocol. Rules are not changed in the vote or command protocols. New rules must be occasionally added and old rules removed. For this reason the rule exchange protocol is necessary. It consists of three services: add-rule, disable-rule and enable-rule. All services require confirmation. The XML defined PDUs are mapped directly to SOAP/HTTP. The name of the service is given in SOAP ENVELOPE and sent in HTTP POST. The HTTP POST triggers a reply. Standard codes of HTTP POST reply can be used. XML/SOAP is verbose since all element and attribute names are typically encoded in plain text but SOAP allows also other encoding rules to be used. SOAP ENVELOPE contains the attribute and a name space. These two attributes are mandatory in a SOAP request, but as they unnecessary consume bandwidth, it is better to omit these attributes on the transmission and use default values. The SOAP specifications do not contain default values, but this is a small violation to the specification. More efficient coding of the XML documents is being studied. The mobile industry is moving towards the XML based SyncML Device Management mainly for management of software configurations in mobile phones [12]. SyncML uses a binary encoding of WAP Binary XML and this can be a quite efficient way to code the PDUs.

Assigning a Communication Mode. A network can be commanded into different modes. Five modes of communication have been defined in the management solution: Low-threat mode, Radio silence mode, Detection avoidance mode, Jamming tolerance mode and Deception mode. The low-threat mode, the detection avoidance mode and the jamming tolerance mode refer to different choices of signal strength, use of error correction codes and on the needed bandwidth. The radio silence mode is a mode where no communication is allowed in the network for a time spell starting from START RADIO SILENCE until STOP RADIO SILENCE. Deception mode means a scenario where nodes send traffic which is meant to deceive an enemy who is listening transmissions. Deception is always an operation ordered by a high level commander. Capability for deception means ability to generate traffic typical to some operation. This can be made by traffic generators. A command to move to a communication mode is ordered by an authorized commander using the command protocol or by an offline method. All nodes in the network set their policies to correspond to the communication mode. A communication mode command gives the network, the start time and the end time for the mode.

5 Implementation Overview

Since the goal was fast prototyping of the management solution NodeBrain software node was selected for basis of implementation. It is not necessarily the tool used in the final implementation. The NodeBrain program is an interpreter of a declarative rule-based language designed for construction of state and event monitoring applications. The software provides: rule engine, rule language and interpreter; message queue system (practical when nodes are out of network temporarily); authenticated and encrypted peer-to-peer communication (for remote management); API for external modules; remote command execution and file transfers; and system monitoring support. Although the NodeBrain software contained most of the required functionally, some features had to be added. One of the additions was a GUI that needed to be created to ease the use of the MI. The MI was designed with IPv6 in mind, while NodeBrain was implemented for IPv4. Thus, support for IPv6 was among the additions. All parts using network code were converted to IPv6 or being able to handle both IPv4 and IPv6. Some deviations from the design were made, for instance, the specified XML PDUs are replaced by NodeBrains own coding.

The implemented Management Interface offers an easy way to manage the ad hoc nodes and their policies in the prototype. Depending on the credentials of a user, the management can be done to all nodes, a group of nodes or a single node. The GUI was made in C++ and the graphics were built using QT Designer 3.3. Thus, running the management software requires the QT libraries. To some extent management can also be done with the command line interface. Supported actions in the GUI are: add and modify users (nodes); add and modify groups and group members; setup event monitoring and alarms in nodes; transfer files between nodes; logging of events; modify network parameters; change routing protocols; change security options and assign security levels to users (nodes); and automatic management by rules.

For example there is a monitoring case in a Linux node of the network. A NodeBrain command script checks syslog file every 30 seconds to see if the firewall has logged dropped packets in the file. The amount of dropped packets (unauthorized IPv6 traffic) is counted. After every ten dropped packets an alarm is sent to a group of nodes and when the amount of dropped packets increases to 100 the syslog file is copied to administrator node for further inspections. The log file is monitored by NodeBrain listener and interpreted with NodeBrain translator.

The performance of the designed system must be seen against the alternative solutions of the network manager physically accessing each node or some remote control application being used. The inherent higher vulnerability of an ad hoc network necessitates that the remote control system satisfies the requirements of a good security model. All aspects of the security model were not described; they include e.g. remotely sweeping the hard disc clean in case a node is abandoned. Additionally, the management application must interwork well with a C2 application and take advantage of the C2 functionalities. In the performance sense the designed solution can dispense with monitoring constantly the network

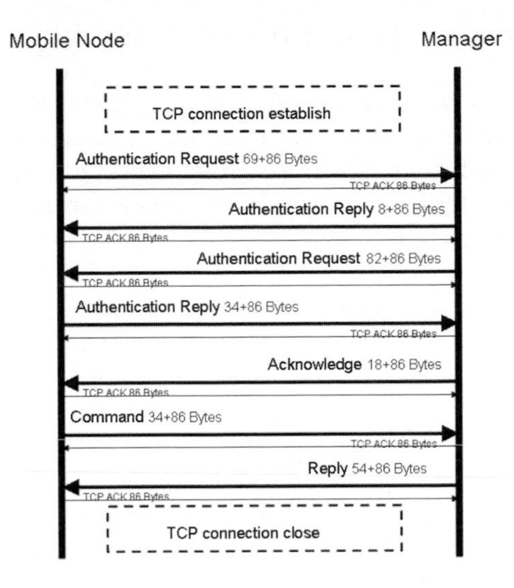

Fig. 2. Signaling chart of sending a vote

status because the C2 contains functionalities for this purpose, for instance blue force tracking systems that report changes of locations within some resolution. Alternative management solutions are therefore not compared mainly by network performance and quality of service issues but by functionality and security issues. A few remarks on performance can be made. The network is relatively small (max. 40 nodes) and has high bandwidth given by the MAC layers 802.11g and 802.16 series. Signalling chart in Figure 2 gives some data on packet sizes and supports the claim that if the number of hops is not high, manager initiated management operations do not create performance problems. The number of hops may be rather large, up to 30 in some cases. TCP is known to work poorly with 802.11 and 802.16 MACs and may result in low bandwidth. This issue can be helped by tuning TCP parameters and a better MAC is being designed, but this question is outside the scope of the paper. The Vote protocol may cause performance problems but it totally depends on the used scripts. The usefulness of context-aware management is still a research issue. The presented solution only enables context-aware operations and assumes that they have been carefully investigated.

6 Conclusions

The management system presented in this paper is made for a small ad hoc network of a brigade headquarter where bandwidth is sufficient, mobility is moderate and energy constraints are not hard. In most research papers of mobile ad hoc networks (MANETs) the assumptions are ambitious, but as [7] notices, there are neither commercial, nor military MANETs filling these more limiting conditions of a very large, very dynamic, high bandwidth, secure and survivable ad hoc network.

The management solution in Figure 1 and the services of the Management Interface (MI) are rather traditional, furthermore, an existing tool, NodeBrain, is used in the implementation. The research challenges are mainly in the selection of a suitable trade-off of functionalities for the particular application. The design choices are as follows:

* Full remote control was too insecure while the MI allows sufficient set of management services. The protection provided by the MI and the protocols is that of a well-designed wrapper: the management system allows an arbitrary file to be uploaded, installed and executed, thus full capability for misuse exists if an adversary can pass the protections of the security model (not described in this paper), but the intruder cannot use faults and needs much knowledge of the system.
* Network management usually monitors the network and because of this it may create too much traffic. We removed this monitoring. The reasons are: a C2 application gives the ability to see who is in the network; the network must support four communication modes; and as an ad hoc network, the topology is dynamic requiring too much polling for effective monitoring.
* We added the vote protocol enabling context-awareness. Initially CAM [3] was expected to be the management solution and as it was not implemented, the context-aware functionality of CAM was included to the design.
* We created a system where all complexity is hidden to scripts that can be easily changed during the operation. This provides a buzzword proof design. A military network for a conscript army cannot be changed every two years to follow a new fashion. For instance, the application does not use dynamic services, but should this be desired, suitable scripts and downloads give this ability through management services.

References

1. Agrawal, D., Lee, K.-W., Lobo, J.: Policy-Based Management of Networked Computing Systems. IEEE Comm. Magazine 43(10), 69–75 (2005)
2. Buchegger, S., Le Boudec, J.-Y.: Self-Policing of Mobile Ad Hoc Networks by Reputation Systems. IEEE Comm. Magazine 43(7), 101–107 (2005)
3. Candolin, C., Kari, H.: Dynamic management of core and ad hoc networks. In: Proceedings of InfoWarCon 2002, Perth, Australia (November 2002)
4. Chen, W., Jain, N., Singh, S.: ANMP: An Ad Hoc Management Protocol. IEEE Journal on Selected Areas in Communication 17(8), 1506–1529 (1999)
5. Gopal, H.: Resource-Aware Mobile Device Applications. Dr. Dobb's Journal, 10–14 (March 2006)
6. Joint Tactical Radio System (JTRS) Program, http://jtrs.army.mil
7. Nissen, C.A., Maseng, T.: Network-Centric Military Communications. IEEE Comm. Magazine 43(11), 102–104 (2005)
8. http://nodebrain.sourceforge.net
9. Pashtan, A., Heusser, A., Scheuermann, P.: Personal Service Areas for Mobile Web Applications. In: IEEE Internet Computing, pp. 34–39 (November/December 2004)

10. Shen, C.-C., Srisathopornphat, C., Jaikaeo, C.: An Adaptive Management Architecture for Ad Hoc Networks. IEEE Comm. Magazine 41(2), 108–115 (2003)
11. Sorteberg, I., Kure, O.: The Use of Service Level Agreements in Tactical Military Coalition Force Networks. IEEE Comm. Magazine 43(11), 107–114
12. Van Than, D., Hjems, A.M., Bjugrd, A., Loken, O.C.: Future management of mobile phones. Teletronikk, 143–154 (April 2005)

Efficient Partitioning Strategies for Distributed Web Crawling

José Exposto[1], Joaquim Macedo[2], António Pina[2], Albano Alves[1], and José Rufino[1]

[1] ESTiG - IPB, Bragança - Portugal
[2] DI - UM, Braga - Portugal

Abstract. This paper presents a multi-objective approach to Web space partitioning, aimed to improve distributed crawling efficiency. The investigation is supported by the construction of two different weighted graphs. The first is used to model the topological communication infrastructure between crawlers and Web servers and the second is used to represent the amount of link connections between servers' pages. The values of the graph edges represent, respectively, computed RTTs and pages links between nodes.

The two graphs are further combined, using a multi-objective partitioning algorithm, to support Web space partitioning and load allocation for an adaptable number of geographical distributed crawlers.

Partitioning strategies were evaluated by varying the number of partitions (crawlers) to obtain merit figures for: i) download time, ii) exchange time and iii) relocation time. Evaluation has showed that our partitioning schemes outperform traditional hostname hash based counterparts in all evaluated metric, achieving on average 18% reduction for download time, 78% reduction for exchange time and 46% reduction for relocation time.

1 Introduction

The importance of Web search services is undeniable. Its effectiveness and utility strongly depends on the efficiency and coverage of the underlying crawling mechanisms.

Crawling a dynamic and evolving Web is a complex task which requires a large amount of network bandwidth and processing power, thus suggesting crawling distribution as a suitable approach to Web information retrieval.

The quality of a distributed crawling system is largely dependent on: 1) the amount of computer nodes and storage capabilities to increase computing and data processing power and 2) the existence of multiple network connection points, to increase the total communication bandwidth and the dispersion of network overload. The deployment of these systems can also take advantage of the distributed infrastructure to obtain considerable gains when downloaded pages are fed to the information retrieval modules for indexing.

The slicing of the target Web space and the assignment of each part to the crawlers is a very well known partition problem, defined as it follows:

T. Vazão, M.M. Freire, and I. Chong (Eds.): ICOIN 2007, LNCS 5200, pp. 544–553, 2008.

For a given number of C distributed crawlers, residing in a pre-defined Web graph Internet topology, compute the C partitions that minimize page download time and crawler page exchange information, maintaining balance as possible.

Several different graphs may be obtained from the Web by using Internet communication metrics, the number of links between pages and other relevant data. When considering a significant period of time, graphs may evolve in conjunction with changes in: the amount of Web sites, the number of pages and links to pages and the Internet communication infrastructure, thus imposing a periodic update of the older computed partitions.

The information generated by the partitioning process may be spread among crawlers and used for routing the URLs. The routing process combines an IP aggregation mechanism similar to the Classless Inter-Domain Routing (CIDR), allowing a considerable reduction on the routing table's size.

To estimate download and exchange times metrics both average Round Trip Time (RTT) and bandwidth between each crawler and servers need to be computed. The available bandwidth for each crawler is bounded in each of the measurements of the RTT derived from real measures.

The total crawler download time is a function of the number of pages available at each server and the RTT between crawlers and servers. The number of existent links between servers is used to estimate the value for URLs exchange time, between two crawlers in different partitions connected by Web links.

Presently, our results have been only compared with the traditional hostname based hash assignment. Comparison with other authors' approaches is envisaged to future work.

2 Related Work

A taxonomy for crawl distribution has been presented by Cho and Molina [1]. They also proposed several partitioning strategies, evaluated using a set of defined metrics. The work included guidelines on the implementation of parallel crawler's architecture, using hostname and URL hashing as the base partitioning technique. In what follows, we will use a hostname hash based partitioning scheme to compare it with our own work results, discarding URL hash based distribution schemes, because it generates a large number of inter-partition links.

In [2] a hostname based hash assignment function is used as well, focusing a consistent hash mechanism.

The work in [3] proposes a redirection mechanism to allow a crawler to overcome balancing problems using a URL hash function after the hostname hash function is applied.

In [1] coordination is classified as: 1) independent, 2) static assignment and 3) dynamic assignment. Following this approach, the current proposal adopt an hybrid coordination strategy that uses both static and dynamic assignment. To accommodate Web evolution, dynamic assignment is reflected on the information collected in previous crawls where a central coordinator decides the initial partitions. Static assignment is reflected between partitioning stages, where each crawler decides on its own where to send the links found.

Another closely related research [4] presented a partitioning mechanism based on a hierarchical structure derived from the URL components. Coordination uses IBM TSpaces to provide a global communication infrastructure. In our work, Web space partitioning results from the application of a specific partitioning algorithm.

In a previous work [5], we evaluated a scalable distributed crawling supported by the Web link structure and a geographical partitioning.

3 Partitioning Strategies

Traditional hostname hash based partitioning strategies are considered good partitioning schemes, because they allow a crawler to extract all the URLs belonging to the same Web site thus reducing inter-partition communication. This scheme although robust, mainly because it allows a light scalable and decentralized URL routing mechanism, it does not take into account the real network and link infrastructures, thus reducing the chances to crawling optimizations.

Distributed crawling is aimed to: i) reduce Web page download times, ii) minimize the amount of exchanged information between partitions (crawlers); and iii) balance Web space load among crawlers. In our approach we create a simplified model of the Internet communication infrastructure and of the Web topology based on the relevant data collected by a first crawling run. Because there is no routing information available for the first URLs, a hash assignment mechanism is used to reach a first partitioning stage.

Subsequent partitioning stages are based on the previous graph configurations and partitions, thus preserving old graph values and updating only the new data, to be able reduce the time to calculate the new partitions. In this paper, the process of graph evolution is not described.

To optimize Web page download time, we use the communication distance between Web servers and crawler clients, which is computed as the Round Trip Times (RTT) between crawlers and servers, obtained by using a *traceroute* tool. The parsing the downloaded pages allows to calculate the amount of links pointing to pages allocated to other partitions, which will be used to compute the exchange time.

For each Web server, to balance the page download work assigned to each crawler, we also take into account the number of pages per server. Next, two separate graph representations are created using RTT and Web link data.

The partitioning of multiple graphs, whose vertices are shared, may be viewed as a single multi-objective graph with multi-edge weights. In what follows we focus on the RTT and link graphs. The study and exploitation of other sort of graphs, like geographic proximity and content awareness, are also under investigation.

3.1 Multi-level Partitioning

The graph partitioning problem aims to divide the vertices of a graph into a number of roughly equal parts, such that the sum of the weights of the edges is minimized between different parts. Given a graph $G = (V, E)$ where $|V| = n$, the partitioning of V into k subsets, $V_1, V_2, \ldots, V_k$, is such that $V_i \cap V_j = \emptyset, \forall_{i \neq j}$, $|V_i| = n/k$, $\bigcup_i V_i = V$, and $\min \sum_{j \in cut} w_j^e$, where *cut* is the set of edges of E whose incident vertices belong to different subsets, and w_j^e is the weight of edge j. $\sum_{j \in cut} w_j^e$ is also known as the edge-cut.

Partitioning problems are considered NP-complete, and several algorithms have been developed that find good partitions in reasonable time. Multilevel partitioning addresses the partitioning problem by successive coarsening phases, that transform the initial graph into smaller graphs.

Afterwards, a k-way partitioning algorithm is used to process the smaller graphs and produce other partitions that will be back projected (uncoarsed) to the initial graph. We use a k-way partitioning scheme adapted from the original Kernighan-Lin algorithm (KL) [6] which is similar to that described in [7]. Basically, the KL algorithm starts with an initial partition that iterates to find a set of vertices from each partition, such that moving that set to a different partition would yield a lower edge-cut. Whenever it finds a set that meets the required condition, the set is effectively moved to that partition and the algorithm restarts, until no further reduction of the edge-cut is achieved.

We started using Metis [8], which is a suitable tool for graph partitioning, however our approach imposes some particular constraints. First, we must ensure that each partition contains just one crawler. Second, the crawlers must be constrained to a single? partition. Then, moving a crawler using KL algorithm must be avoided. Finally, the initial phases of the coarsening process should benefit of the existent Web topology, to favour optimization and take advantages of the natural organization of the Web pages into Web hosts and IPs.

3.2 Multi-objective Partitioning

To achieve the multi-objective partitioning, we followed the procedure suggested in [9] and named it the Multi-objective Combiner. First, each graph representing each one of the objectives is partitioned separately. Afterwards, a third graph is created by assigning weights to the edges computed using the sum of the original weights normalized by the edge-cut of the associated objective.

Each normalized weight is affected by a preferential factor, to smooth the differences of magnitude existent between each of the initial graphs. The multi-objective partitioning will result from the application of the partitioning algorithm to the third graph. In order words, if we consider k objectives and k graphs, one for each objective, $G_1 = (V_1, E_1), G_2 = (V_2, E_2), \ldots, G_k = (V_k, E_k)$, after the partitioning of these graphs there is one edge-cut for each: $\mathrm{ec}_1, \mathrm{ec}_2, \ldots, \mathrm{ec}_k$. The combined graph $G_c = (V_c, E_c)$ where $V_c = \bigcup_{i=1}^k V_i$ and $E_c = \bigcup_{i=1}^k E_i$ and each E_c has a weight of $w_c^e = \sum_{i=1}^k \frac{p_i w_i^e}{\mathrm{ec}_i}$, where p_i is the preferential factor for each graph i. In what follows we will use constant values for p_i factors,

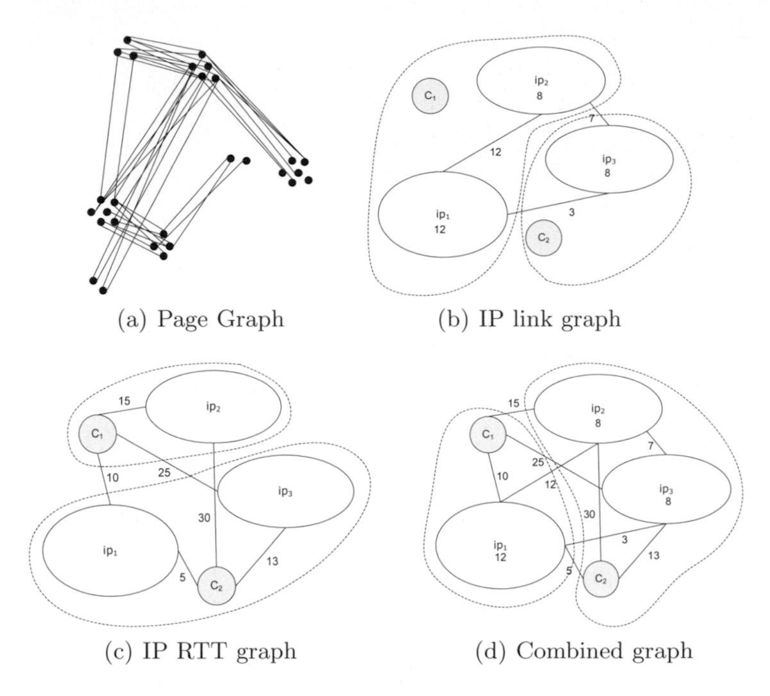

(a) Page Graph (b) IP link graph

(c) IP RTT graph (d) Combined graph

Fig. 1. Graph examples

however, preliminary experiments revealed promising results when considering p_i variation.

Two scenarios were considered for evaluation: 1) Graph vertices as Web hosts and 2) Graph vertices as IP hosts. Graph vertices as Web pages were not considered due to the additional complexity in the partitioning algorithms and to the difficulty to achieve a scalable representation of the routing tables.

Figure 1(a) depicts an example graph containing 28 pages (vertices) and 38 links (edges) among them. The partitioning algorithm starts by coarsening the graph into a IP graph like the one in Fig. 1(b). Hostname coarsening is not shown in the figure. In the coarser graph, pages are collapsed in the same vertex IP, resulting in a vertex weight calculated as the sum of the pages contained in the Web site at that vertex being the values of new edges computed as the sums of the link edges of the previous graph. Assuming that the depicted Web pages are hosted by some IPs and considering that there are two crawlers to download these pages, we may represent the RTT graph as in Fig. 1(c). Vertices weights in this graph are the same for all vertices. The dashed lines represent the resultant partitions.

Depending on the chosen scenario, we could have coarsened the page graph and we use the hostnames as vertices, or coarse one level deeper the hostname graph and we use IPs as vertices. In this example, we used IP graph coarsening. Each vertex has a weight which is the sum of weights of its hosts, being the edge weights the sums of the edge weights of the finer graph.

After each one of these graphs is partitioned (two partitions in this example) and their edge-cuts computed, a new combined graph is generated with their edges affected accordingly as mentioned before. The final partitions obtained from the partitioning of this combined graph are shown in Fig. 1(d). In these examples, the graphs weights do not correspond to actual RTT or link values.

4 Evaluation

To study and evaluate the proposed approach for multi-objective graph partition, we merged two independent Portuguese Web collections into a single one comprising 16,859,287 URLs (NetCensus [10] and WPT03 [11]).

Since then, the derived collection has supported the development and validation of the partitioning algorithms and produced statistics of the Portuguese Web. In particular, it is the source of the data information used to obtain the RTTs and the extracted link data, along with other sort of information related to Internet topology entities and their associated geographic entities.

As topological entities, we identified the Internet address block (Address Block), the address aggregate published by autonomous systems in BGP routing (Address Aggregate) and the autonomous system (AS). The identified geographical entities include cities and respective countries. The number of IPs for each of this entities is the following: 46,650 Hostnames, 6,394 IPs, 1,875 Address blocks, 620 Address aggregates, 363 ASs, 339 Cities and 24 Countries.

It is interesting to point out that a number of 4,627 additional IP routers, not previously included were discovered during *traceroute* operation. Another, surprising fact is that 52% of the IP servers actually reside outside of Portugal despite the fact that the evaluation refers to Portuguese Web space.

To obtain partitioning results, we used a variable number of up to 30 crawlers to process $1,903,336$ URLs, containing $26,472$ hostnames associated to a total of $1,700$ IPs. We initially constructed two different sets of graphs, based on RTT and Web link data obtained in previous experiments. The first set includes an IP graph and a hostname graph whose arcs are weighted by the computed RTT between nodes. The second set is made of two link graphs with IP and hostname nodes constructed using the method described in Sect. 3.2.

Partitioning strategies were evaluated by varying the number of partitions (crawlers) to obtain the following metrics: i) download time; ii) exchange time; and iii) relocation time. Although we believe that the maximum number of partitions used is acceptable, we plan to increase this number in future work.

The following metrics reflect a communication latency between crawler and servers, although we disregard server load, we assume this latency is far small than communication latency.

Download Time. For a set of partitions, the download time estimates the maximum time needed to download the totality of the pages included in the set. For a server i the download time needed by crawler j may be approximated by the formula: $dts_i = \frac{M_i}{L_j}(2RTT_{ij} + \frac{L_j \cdot ps_i}{BW_{ij}} + PT_i)$, where L_j is the number of

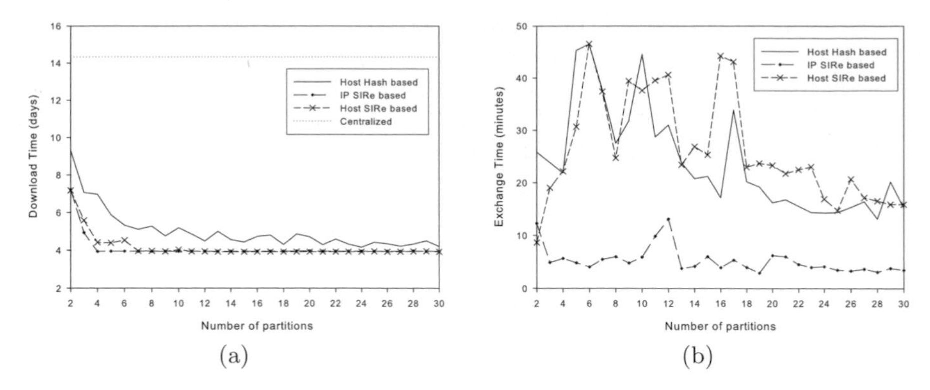

Fig. 2. Download and Exchange Time for SIRe and Hash Host based partitioning

pages downloaded in pipeline by `http` persistent connections between crawler j and server i; M_i is the number of pages of the server i; RTT_{ij} and BW_{ij} are the RTT and the available bandwidth between the crawler j and the server i, respectively; ps_i is the average page size of the server i. PT_i is a politeness wait time interval between consecutive connections to the same server. Note that, at each moment, a crawler can only have a single connection to a server.

Considering a total load of S_j servers for crawler j and N_j `http` simultaneous connections, the total download time for the crawler j is given by the equation: $dt_j = \frac{1}{N_j} \sum_{l=1}^{S_j} dts_l$.

The maximum total download time was defined by the time taken by the slowest crawler or heaviest partition of p partitions, given by the expression: $\max_{i=1}^{p} dt_i$.

Exchange Time. As each partition j has several links to other Web sites assigned to different partitions, the associated URLs have to be forwarded to the crawlers responsible for the associated partitions. The estimated time needed to exchange the foreign URLs is given by the equation: $et_j = \frac{1}{N_j} \sum_{l=1}^{P} 2RTT_{jl} + \frac{su \cdot nl_{jl}}{BW_{jl}}, l \neq j$, where RTT_{jl} is the RTT between crawlers l and j, nl_{jl} the total number of links from the partition j to the partition l, su is the average size of a single URL, BW_{jl} is the bandwidth between crawlers and N_j is the number of simultaneous links forwarded. The total time to process all the partitions is estimated as the maximum value over all the partitions.

Relocation Time. Relocation time estimates the amount of time taken to move the URLs from an original partition to a destination partition, whenever the addition of a crawler imposes a reassignment of the initial partition configuration.

It is equivalent to exchange time except for nl_{jl} that is the number of relocated links from the partition assigned to crawler j to the partition assigned to crawler l in the new configuration. A maximum for all partitions is calculated.

The results produced by the formulas above constitute the base for a series of experiments conduced to compare our results (which we call SIRe) and hostname

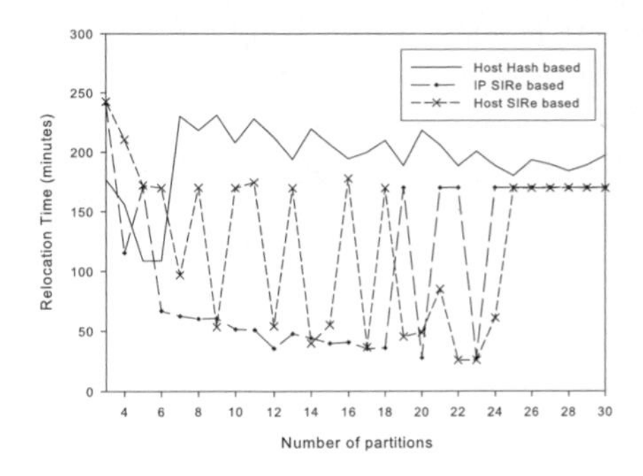

Fig. 3. Relocation Time for SIRe and Hash Host based partitioning

hash based partitioning schemes, using the following values: $L_j = 10$, $BW_{ij} = 16$ Kbps, $ps_i = 10$ KB, $PT_i = 15$, $su = 40$ bytes and $N_j = 10$. The time values obtained may seem high because of the low bandwidth used.

Figure 2(a) shows the estimated values for download time considering four schemes: i) SIRe with hostnames as vertices, ii) SIRe with IPs as vertices, iii) hostname hash based partitioning and iv) a centralized solution.

As expected, the increase in the number of partitions diminishes download time for all partitioning methods, effectively improving the centralized approach. Both SIRe based schemes outperform the hash based partitioning method, however no improvement is achieved by hostname SIRe based in comparison with IP SIRe based partitioning. In fact, there is a tendency for equality. On average an 18% reduction is achieved using IP SIRe based over hostname hash based partitioning.

Figure 2(b) presents the results obtained for the estimation of the exchange time. Host SIRe based partitioning behaves similar to hostname hash based counterpart. On the other hand, IP SIRe based partitioning clearly outperform the other methods, achieving on average 78% reduction over hostname hash based partitioning.

Figure 3 presents the results for the estimated relocation time of the URLs assigned to the partitions after a new crawler is added. SIRe based schemes outperform the hostname hash based scheme for the majority of partitioning configurations. However, IP SIRe based scheme tends to achieve better results, with a reduction of 46% on average of hostname hash based partitioning.

5 Conclusions and Future Work

This paper presents SIRe's (Scalable Information Retrieval environment) approach to Web partitioning, a project that seeks to optimize Web crawler download and exchange times. The approach is supported by the construction of two different weighted graphs. The first graph is used to model the topological communication infrastructure between crawlers and Web servers. The second graph allows to represent the amount of link connections between server's pages.

The values of the edges represent, respectively, computed RTTs and pages links between nodes. The two graphs are further combined to support Web space partitioning and load balancing for variable number of geographical distributed crawlers by means of a multi-objective partitioning algorithm.

The proposed approach differentiates instead of using a deterministic hash function to slice the target Web space, as most of the approaches referenced in the literature, is based on the possibility of using earlier knowledge of the servers and communication paths infrastructure to build an appropriate graph representation of the Web from where to derive a partitioning scheme.

To validate our work proposal, the partitioning algorithms were adapted to the specific requirements of the Web space under study, namely, crawler separation and compulsory assignment to at least one partition, and the need to provide additional control over the coarsening mechanism.

Evaluation showed that SIRe's based partitioning schemes outperforms hostname hash based counterparts for all the evaluated metrics. Justification for these results resides in the fact that traditional hostname hash based schemes do not consider the amount of pages in each server. SIRe's improvement is particularly relevant in situations where there is a significant difference in the total number of Web pages contained in each of the Web servers.

Differences between Host and IP SIRe approaches are not noteworthy for download times. We claim that IP SIRe is a better overall solution, even considering that Host SIRe partitioning strategy take longer time to run, because of the power-law distribution of pages on Web servers. In fact, a lot of servers with a small number of pages and few servers with a large number of pages, impose a maximum download time for the partition that hold heavy page load servers.

Host SIRe based partitioning exchange time was not able outperform Host Hash based scheme most likely due to the multi-objective preferential factor assigned to the link objective. IP SIRe based partitioning has considerably better performance because hostnames belonging to same IP have a larger number of links between them, thus decreasing the exchange time.

SIRe based partitioning relocation times are better than Host hash based scheme, although very unstable. Boundaries around 170 minutes for SIRe based schemes relocation time correspond to the relocation of heavy page load servers. In future work we will take into account this issue.

For all the 30 configurations considered, IP SIRe based partitioning strategy achieved a considerable time reduction, when compared to the hostname hash based scheme.

SIRe partitioning schemes seek to optimize multiple objectives which are optimized simultaneously, thus both download times and exchange times are counterbalanced. Changing the preferential factors for graph weight combination would certainly bias the results toward improved download times and worst exchange times, and vice versa. In the future we plan to vary preferential factors to check for better partitioning results.

Currently we are working to include other optimization objectives including i) content similarity between Web pages and ii) download optimization, taking into account the page size and frequency of change, when crawlers are working in incremental mode. The two graph edge weights used so far, might be extended with some content similarity between pages, thus allowing content aware partitioning to enable the creation of focused partitions; possible exchange overhead reduction may also be achieved based on page links affinity due to page similarity. Also, in incremental crawling mode, pages are visited several times based on its change frequency. Frequently visited pages induce an additional load to the crawler responsible for those pages, causing load unbalance among crawlers. The addition to the graph representation of the page change frequency estimation may result in the improvement of load balancing.

Finally, we think that graph evolution with real time updates and partitioning algorithm distribution is also a hot topic for further work.

References

1. Cho, J., Garcia-Molina, H.: Parallel crawlers. In: Proc. of the 11th International World–Wide Web Conference (2002)
2. Boldi, P., Codenotti, B., Santini, M., Vigna, S.: Ubicrawler: A scalable fully distributed web crawler. Software: Practice & Experience 34(8), 711–726 (2002)
3. Loo, B., Krishnamurthy, S., Cooper, O.: Distributed Web Crawling over DHTs. Technical report, EECS Department, University of California, Berkeley (2004)
4. Teng, S.H., Lu, Q., Eichstaedt, M., Ford, D., Lehman, T.: Collaborative Web Crawling: Information Gathering/Processing over Internet. In: Proceedings of the Thirty-second Annual Hawaii International Conference on System Sciences, vol. 5 (1999)
5. Exposto, J., Macedo, J., Pina, A., Alves, A., Rufino, J.: Geographical Partition for Distributed Web Crawling. In: 2nd International ACM Workshop on Geographic Information Retrieval (GIR 2005), Bremen, Germany, pp. 55–60. ACM Press, New York (2005)
6. Kernighan, B.W., Lin, S.: An efficient heuristic procedure for partitioning graphs. Bell Sys. Tech. J. 49(2), 291–308 (1970)
7. Fiduccia, C.M., Mattheyses, R.M.: A linear-time heuristic for improving network partitions. In: DAC 1982: Proceedings of the 19th conference on Design automation, Piscataway, NJ, USA, pp. 175–181. IEEE Press, Los Alamitos (1982)
8. Karypis, G., Kumar, V.: A Software Package for Partitioning Unstructured Graphs, Partitioning Meshes, and Computing Fill-Reducing Orderings of Sparse Matrices – Version 4.0. Technical report, University of Minnesota, Department of Computer Science / Army HPC Research Center (1998)
9. Schloegel, K., Karypis, G., Kumar, V.: A New Algorithm for Multi-objective Graph Partitioning. In: European Conference on Parallel Processing, pp. 322–331 (1999)
10. Macedo, J., Pina, A., Azevedo, P., Belo, O., Santos, M., Almeida, J.J., Silva, L.: NetCensus Project (2001), `http://marco.uminho.pt/~macedo/netcensus/`
11. XLDB Group: WPT 03. Linguateca (2003), `http://www.linguateca.pt`

A Secure Message Percolation Scheme for Wireless Sensor Network*

Md. Abdul Hamid and Choong Seon Hong**

Networking Lab, Department of Computer Engineering, Kyung Hee University
1 Seocheon, Giheung, Yongin, Gyeonggi, 449-701 South Korea
hamid@networking.khu.ac.kr, cshong@khu.ac.kr

Abstract. Wireless Sensor Network (WSN) deployed in hostile environments suffers from severe security threats. In this paper, we propose a Secure Message Percolation (SMP) scheme for WSN. We engineer a secure group management scheme for dealing with data gathered by groups of co-located sensors and analyze the robustness against increasing number of compromised sensor nodes. Key pre-distribution is performed with the concept of star key graph where one sensor node dominates other nodes. Analytical result shows that our protocol is robust against node compromise attack and scales well and requires a few pre-distributed shared keys per node.

Keywords: Secure Group, Node Compromise, Robustness, Star Key Graph.

1 Introduction

Lightweight secure protocol for resource-constrained WSN is challenging and much works are going on in designing storage and computationally inexpensive mechanism [2] [3] [16]. We consider few important issues in engineering our security protocol. Firstly, key storage for individual sensor node needs to be reasonably small. For example, if there are N nodes in the network, then we can not expect that a node can store $N - 1$ keys to share a secrete key with each of the other nodes. Secondly, in case where quite a good amount of sensor nodes are compromised by an adversary, the communications among other nodes should still be secure. Thirdly, it should be ensured that both local and global connectivity is maintained. A sensor node should be able to securely communicate with its local neighbors (i.e., sensor physically located within transmission range). Connectivity among local zones should provide global network connectivity [14]. Finally, asymmetric cryptography to WSN is too expensive, because they require expensive computations and long messages that might easily exhaust the sensor's resources [13]. That is why we take symmetric cryptographic operations and spread the load across the network in a distributed fashion. We take into consideration secure network formation (bootstrap), data aggregation by groups of locally co-located nodes and rekeying. Key management and rekeying are

* This work was supported by MIC and ITRC Project.
** Corresssponding author.

T. Vazão, M.M. Freire, and I. Chong (Eds.): ICOIN 2007, LNCS 5200, pp. 554–563, 2008.

basically performed in a distributed fashion to aiming at devising a more lightweight load for individual sensors. We analyze our scheme's robustness against node compromise attack when adversary compromises network nodes. We also show that our protocol scales well and requires a few pre-distributed shared keys per node.

Rest of the paper is organized as follows. Section 2 outlines the network assumptions and preliminaries. Section 3 presents our scheme in details. We analyze our SMP scheme in Section 4. Related works and comparisons are noted in Section 5 and Section 6 concludes the paper.

2 Network Assumptions and Preliminaries

Our network is a multi-hop in nature supporting node deletion/addition. We consider a heterogeneous network, where two types of sensors are deployed: ordinary sensor node (SN) and group dominator (GD) node. SN is simple, inexpensive and stringent in resources (power, memory and computation), while GD is rich is resources and more compromise-tolerant. We also assume that one GD can communicate with its neighbor GD to forward aggregated messages towards base station. There is no communication link among SNs within one group and between SNs in different group (Fig. 1b).

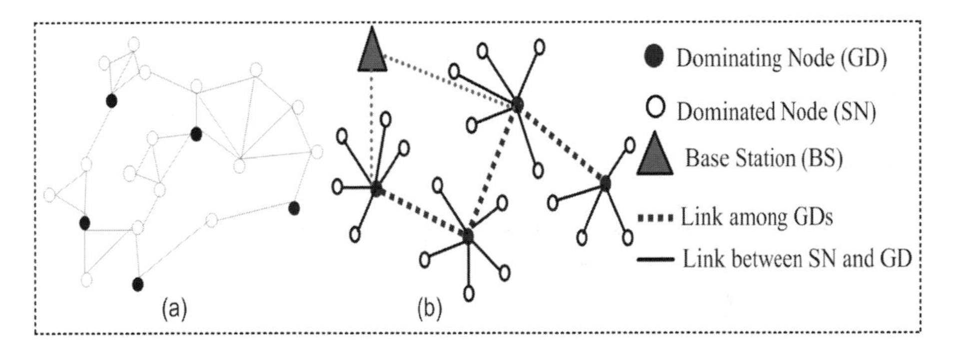

Fig. 1. (a) WCDS. (b) Star based-WCDS for proposed SMP scheme.

We assume that once the sensors are dispersed over the area of interest, they remain relatively static. We consider the sensors in the whole network as a graph G = (V, E), where V is the set of sensors in the network and E is the set of direct communication links. A direct communication link is present between SNs and its corresponding GDs if and only if they are within the transmission range of one another and share the same key. In such a setting, we apply our scheme to form a network-wide star based weakly connected dominating set (SWCDS) (Fig. 1b). A weakly connected dominating set (WCDS), S_W, is a dominating set where the graph induced by the stars of the vertices in S_W is connected. A star is comprised of the vertex itself and all the ordinary sensor nodes adjacent to it (all the black nodes in Fig. 1a). The underpinning of our proposed scheme is the star-based weakly connected dominating set. In fact, it is easy to see that each dominating node (or vertex) in the SWCDS is at the center of a star (Fig. 1b). Thus, for each dominating node in a SWCDS of the overall network, we have one star where all the other nodes in the star are just one hop apart. For the space constraint, we request the readers to look at references [1] and [14] for details about dominating set, connected dominating set and WCDS.

3 Proposed SMP Scheme

In our approach, key pre-distribution is performed using the concept of star key graph [17]. This is the special class of a secure group where each sensor node has only three keys to maintain: its individual key (shared between SN and GD), and a local group key that is shared by every user in the star graph with their corresponding GD and a pairwise key between SN and BS. BS stores all the keys of SNs and GDs. We use the notations in table 1 to describe our scheme.

Table 1. Notations used in SMP scheme

Notation	Definition
$i\ (0 \le i \le N)$	Ordinary sensor node i (SN_i)
$j\ (0 \le j \le Y)$	Group dominator j of ordinary sensor i (GD_j)
ID_i	ID of the ordinary sensor node i
ID_j	ID of the group dominator j
K_{Gj}	Group key shared by all sensors in a group j and GD_j
$K_{(SNi,GDj)}$	Pairwise key between a sensor i and GD_j
$K_{(SNi,B)}$	Pairwise key between sensor i and Sink/BS
$K_{(GDj,B)}$	Pairwise key between GD_j and Sink/BS
M_i	Event sensed by SN_i
M_{GDj}	Message aggregated by GD_j
MAC(K,M)	Computation of Message Authentication Code of message M using key K
E(K,M)	Encryption of message M using key K
X|Y	Concatenation of X and Y

3.1 Pre-deployment Key Pre-distribution and Rekeying

Key pre-distribution. In the offline key pre-distribution phase, we assign the group keys and individual keys to a group of nodes. For this, the key assignment is accomplished according to Fig. 2a. Each GD holds group key and all individual keys. Each SN holds group key and its individual key shared with GD. In this phase, all the SNs are also assigned unique ID_i $(1 \le i \le N)$ which are also stored by the respective GDs. Each GD is also assigned ID_j $(1 \le j \le Y)$, where Y is the total number of group dominators in the network.

Rekeying. During the offline key pre-distribution, all the nodes are assigned the keys but not all the nodes are deployed. When any of those remaining nodes is deployed, it sends the JOIN_REQ_NEW message using its own individual key. If authorized by the access list of GD, it joins the group. Otherwise, GD forwards this to BS. BS informs GD about the individual key of that SN. If authenticated by BS, GD generates a new group key and encrypts the new group key with the newly added node's individual key and sends it to the SN. All other nodes in the group know about the change by a multicasting by the GD of that group. For leaving a star graph, the node simply leaves a message to inform the GD which in turn generates a new group key and unicasts it within the group members (Fig. 2b). For example, let's say, SN_4 in Fig. 2b

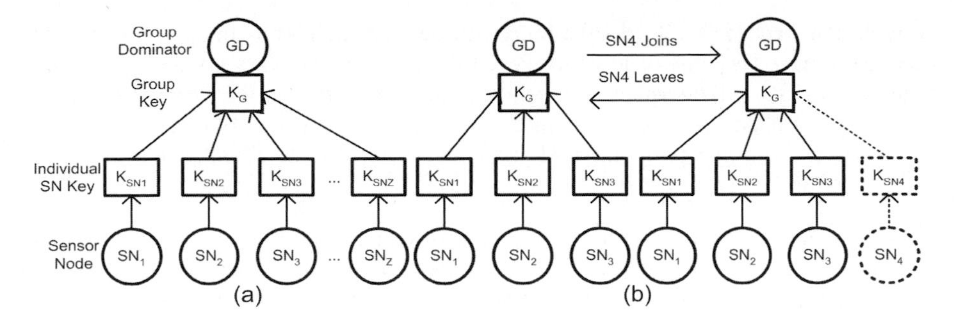

Fig. 2. (a) Pre-distribution of secret keys using star key graph. (b) Rekeying.

wants to join the existing group in the figure shown above. GD changes the group key K_G to a new key K_{Gnew}, and sends the following rekeying messages:

GD Encrypts new group key with the old group key: $GD_j \rightarrow all\ SN_i$**:** $E_{K_G}(K_{Gnew})$. GD also Encrypts new group key with the joining SN's individual key: $GD \rightarrow SN_4$**:** $E_{K_{SN4}}(K_{Gnew})$. Similarly, when any SN wants to leave the group, it just sends a leave message: $SN_4 \rightarrow GD$**:** E_{KSN4}(leave). GD deletes the leaving SN and updates the K_G to new K_{Gnew} and unicasts the following message: $GD \rightarrow SN_{i-1}$**:** $E_{KSNi-1}(K_{Gnew})$.

3.2 Post Deployment Phase

Ordinary sensor nodes and group dominators having the offline pre-distributed secret keys are deployed over the area of interest. In this section we describe secure bootstrap (network formation), data aggregation and rekeying. For convenience, we have described rekeying mechanism in section 3.1.

3.2.1 Secure Network Formation

It has been shown in [15] that sensors belonging to the same group can be deployed close to each other without the knowledge of sensor's expected location. Considering the fact in [15], we describe secure network formation with exception handling (e.g., a SN does not belong to same group).

After the deployment, each SN_i discovers its own GD_j. For this purpose, SN_i broadcasts an encrypted JOIN_REQ (using individual key K_{SNi}) message to all of the nodes within its transmission range. If the corresponding GD_j is within its transmission range (i.e., one hop distance), it gets the message and decrypts it as all the individual keys of the sub-ordinate nodes are known to the GD_j. Upon successful decryption of the message, the GD sends a JOIN_APRV message to the SN_i encrypting it with the group key. Thus the SN_i becomes a dominated node of a corresponding GD_j. If, for any SN_i, the GD_j assigned during pre-distribution of keys, is not within one hop distance; the SN_i needs to inform the BS for resolving the issue. We term this SN as the 'Orphan' (SN_{ORP}). On discovering itself as an Orphan the SN_i sends a GD_ERR message encrypting it with its individual key. This message is simply forwarded by other sensors in the network to reach to the BS. For resolving the unexpected issue of Orphans we consider two special cases.

Case I. The orphan has no dominator as its one-hop neighbor. In such a case, after getting the GD_ERR message from the Orphan, the BS issues a command to assign the role of a GD to the Orphan. For sending the command, the BS uses the individual key of the Orphan. This newly formed GD does not have any other dominated SN nevertheless; employing this method keeps the isolated node connected with the rest of the network.

Case II. The Orphan does not have its own GD within its one-hop neighborhood but a GD of another group is present in the vicinity. Failing to find out its own GD, SN sends the encrypted GD_ERR message to the BS. Now, as the GD of another group is present within its one hop distance, it eventually gets the encrypted GD_ERR message (only detects the type of message and just notes this incident) and informs this 'Orphan Information' using ORP_ERR (encrypted with its group key) to the BS. The BS eventually gets two separate but interrelated reports; one from SN_{ORP} and another from the neighboring GD. The BS checks whether both these reports tell about the same SN_{ORP} or not. If same SN_{ORP}, BS issues a command to that neighbor GD to be its adopter and also sends the individual key for the SN_{ORP}. The GD in turn uses this key to send its K_G to the SN_{ORP} to welcome it in its own group. Thus, all the stars could form a weakly connected network where the GDs of the logical groups (stars) are the dominating nodes and all other nodes in the network are dominated (Fig. 1b).

3.2.2 Secure Data Aggregation

Once the network is logically structured as a weakly connected dominating set, the sensory data from the sensors can be transmitted securely to the BS. GDs are responsible to aggregate data collected from different sensors. If there are *Z* number of ordinary sensors (SN) in a group, for fidelity and correctness of data, the GD waits for the same sensing event from at least q ($q \leq Z$) number of the SNs, where q is the threshold value set for a particular group. We consider any one group with ordinary SNs and its corresponding GD. Once an event occurs, q out of Z ($0 \leq q \leq Z$) ordinary sensors (ID_1, ID_2, . . ., ID_q) within the sensing area detect the event and send information to the GD.

$$SN_i \rightarrow GD_j : ID_i | E(M_i | MAC(M_i, K_{(SNi,GDj)}), K_{(SNi,GDj)})$$

Upon receiving the message sensed by SN_i (ordinary sensor), Dominator GD_j verifies every single MAC and generates an aggregated report (and discards the false packet if any). GD broadcasts the aggregated results M_{GDj} and MAC to all sensors.

$$GD_j \rightarrow all\ SN_i : ID_{GDj} |\ M_{GDj} | MAC(M_{GDj}, K_{Gj})$$

All SNs in a particular group *j* verifies the aggregated report whenever they receive it for the consistency with its own sensed event. It creates a MAC only to be verified by the Sink (BS) but to be relayed by its GD. The message format is

$$SN_i \rightarrow GD_j : ID_i | MAC(K_{(SNi,B)}, ID_i | M_{GDj})$$

Now, GD collects all the MACs from ordinary sensor nodes and sends q MACs, q IDs, ID_{GDj}, and M_{GDj} to the sink directly or via its neighboring GD (multi-hop path through consecutive GDs towards the Sink) as follows

$$GD_j \rightarrow Sink: ID_{GDj}, E(K_{(GDj,B)}, ID_{GDj} | M_{GDj} | ID_1 | MAC_{(SN1,B)} | \ldots . ID_q | MAC_{(SNq,B)})$$

The q MACs and aggregation report M_{GDj} are sent securely to the base station. Upon receiving an aggregation report, the sink first decrypts the message using the corresponding key $K_{(GDj,B)}$, then it checks whether the report carries at least q distinct

MACs from ordinary sensors and whether the carried MACs are the same as the MACs it computes via its locally stored keys. If no less than q MACs is correct, the event is accepted to be legitimate; otherwise it is discarded.

4 Analysis

In our scheme, we form a probable star based WCDS to cover almost all of the nodes in the network with minimum effort. As shown in Fig.1b, SWCDS requires less number (or equal to) of dominating nodes to cover the whole network than that of a connected dominating set (CDS) requires [14]. We use the distinct group keys for each of the GDs and distinct individual keys for each SN. So, the number of individual keys required for our network is much less than other probabilistic key management schemes. Table 2 shows the overhead of our scheme. We assume the length of ID, key and MAC is 2, 16 and 4 bytes respectively. We consider *20* ordinary sensor nodes in one group and an aggregated report travels 10 hops on an average. We calculate the overhead based on the message format described in section 3.2.2.

Table 2. Overhead of secure message percolation scheme (Excluding encryption)

Overhead Type	SN	GD
Storage	48 bytes (3 keys)	352 bytes (1+1+Z keys)
Communication	12 bytes (2 ID + 2 MAC)	1220 bytes ((1 ID_{GD} + (1 ID_{sn} + 1 MAC) Z) H hops)
Computation	3 MACs	21 MACs (1 MAC_{GD} + Z MAC_{SN})

4.1 Security Analysis

Our scheme ensures that, each of the GDs and the corresponding SNs can directly form the groups or stars maintaining the security of the network from the bootstrapping state. As encryption is used for message-transmission within the network from the very beginning of the network, our scheme can successfully defend Hello Flood Attack [2] and most of other attacks in wireless sensor networks. We contemplate security analysis of our scheme from various perspectives as described below.

Ordinary sensors (SNs) are captured. A compromised ordinary sensor in particular group may produce an invalid MAC by providing wrong guarantee for an aggregated report. The SMP scheme is robust against this kind of attack as long as no more than q sensors within a local group are compromised. Since we devise our scheme where each aggregated report carries q number of MACs from ordinary sensors. Only the base station checks the correctness and no less than q correct MACs from ordinary sensors is accepted by the base station. A compromised ordinary sensor may forge false event's information with valid MAC to its GD. We argue that a single faulty value will not hamper the aggregated result.

A GD is captured. When a GD is captured, it may fabricate a report. But to do that, at least q MACs need to be forged. The probability that at least q out of Z MACs is correct is given by $p_{GD} = \sum_{j=q}^{Z} \binom{Z}{q} p^{q}(1-p)^{Z-q}$, where, $p=1/2^L$ and L is the MAC size in bits. We claim that this probability is negligible; moreover, only one group out of entire network is in fact affected while other groups are not hampered.

Both GD and ordinary SNs are captured. We consider the situation where an adversary has compromised GD and some number x ($0 \leq x \leq q$) ordinary sensor nodes. To inject a false report, GD needs at least q valid MACs. Since GD has to forge $(q-x)$ more MACs, the probability that $(q-x)$ out of $(Z-x)$ is valid, is given by $p_{GD}^{x} = \sum_{j=q-x}^{Z-x} \binom{Z-x}{j} p^{j}(1-p)^{Z-x-j}$. Again, this probability is almost negligible. If an individual key is compromised, the attacker at best could send false report to the GD but when any message from the GD comes encrypted with the group key, it cannot decrypt it. So, for successful attack, it needs both the individual and group key at the same time. Moreover, compromising one key doesn't affect the rest of the keys used among other nodes and links in the network.

Network-wide Compromise of SNs and GDs. We analyze the robustness of our scheme when an adversary has randomly compromised Q ordinary SNs and j ($0 \leq j \leq Y$) GDs from the entire network. Let $p(j,q)$ be the probability that j^{th} GD (i.e., j^{th} group) having q ($0 \leq q \leq Z$) SNs compromised, we get

$$p(j,q) = \frac{\binom{Z}{q}\binom{N-Z}{Q-q}}{\binom{N}{Q}}$$

We define $g_{j,q}$ as: $g_{j,q} = 1$ if q ordinary SNs are compromised in j^{th} group and 0 otherwise. And let G_q denote the number of groups having q ordinary SNs compromised, we get $G_q = \sum_{j=1}^{Y} g_{j,q}$, and

$$E\left[\sum_{j=1}^{Y} g_{j,q}\right] = \sum_{j=1}^{Y} E\left[g_{j,q}\right] = Y.E\left[g_{j,q}\right] = Y.p(j,q) = Y.\frac{\binom{Z}{q}\binom{N-Z}{Q-q}}{\binom{N}{Q}}$$

Next, we assume that an adversary has compromised some groups having the z ($z \geq q$) SNs compromised and we call this situation as full group compromise. Let X be the number of fully compromised groups from entire networks. We can compute $E[X]$ by the following equation:

$$E[X] = \sum_{q=z}^{Z} G_q = \sum_{q=z}^{Z} Y.\frac{\binom{Z}{q}\binom{N-Z}{Q-q}}{\binom{N}{Q}}$$

For demonstration purpose, we take a simple example where total number of SNs is $N=170$, with $Z=10$ SNs in each group (i.e., $Y=17$ groups). Fig.3a shows the expected number of compromised groups against the entire network's compromised ordinary SNs.

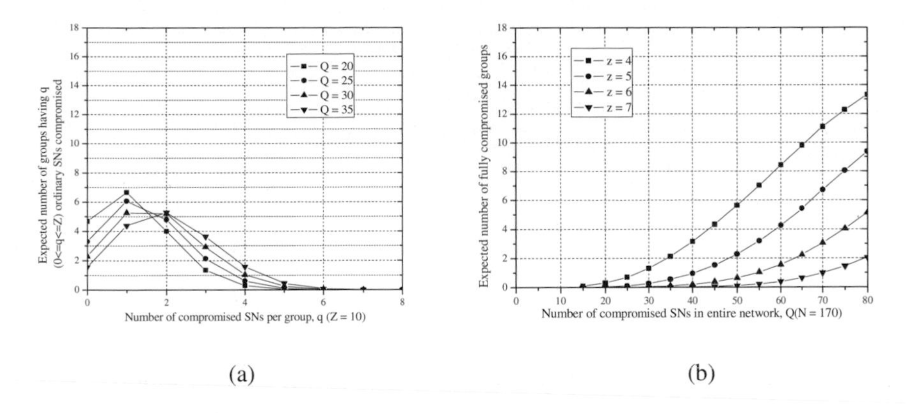

(a) (b)

Fig. 3. Analytical performance. (a) $E[G_q]$ and (b) $E[X]$.

When *20* SNs are captured, *5* groups having *0* SNs compromised and only *1* group having *3* SNs compromised. Fig.3b demonstrates the number of fully compromised groups that depends on the value *z*. Four cases are shown when *z* is *4*, *5*, *6* and *7*. When *z* is *4*, *3*(*16.66%* of entire network) groups are fully compromised against *40* (*23.5%*) ordinary compromised SNs. But, as the value *z* increases (*6* or *7*), number of fully compromised groups are much smaller. We observe that robustness can be improved significantly by increasing the value of *z*.

5 Related Works and Comparison

Extensive effort has been put so far on how to set up a pairwise shared secret between two sensors and how pre-deployment of secrets is performed to securely communicate among sensor nodes. In Table 3, we briefly compare the prior works and cite major features used including our proposed scheme.

Intuitively, WCDS will, in general, be smaller than connected dominating sets and the resulting induced graph will have smaller edges. This corresponds to fewer clusters and a sparser abstracted network. For comparison, Statistical En-Route Filtering [5] scheme, each intermediate forwarding node verifies one MAC and five hash computations (for Bloom filters) probabilistically if it has one of the keys in common, while in our scheme, only the GDs forward the aggregated report, but they don't perform this intermediate checking. SEF is constrained by sensor's storage since to increase one hop detection probability, the number of keys a sensor stores should be large. But in our scheme (SMP) only *3* keys are required.

SEF performs better when the number of hops a packet travels is very high, but SMP scheme has much smaller hops (SWCDS) and the overhead on forwarding aggregated reports gets to the powerful GDs only. Each report is about *15*

Table 3. Comparison of various security schemes

Schemes	Attacks Defended	Network Architecture	Key Management Scheme	Major Features
Statistical En-Route Filtering [5]	Information Spoofing	Large number of sensors, highly dense wireless sensor network	Partition global key pool and randomly assign m keys from one of the partitions	Detects and drops false reports during forwarding process
Radio Resource Testing, Random Key Pre-distribution etc. [6]	Sybil Attack	Traditional wireless sensor network	Random key pre-distribution	Random key pre-distribution, Registration procedure, Position verification and Code attestation for detecting sybil entity
TIK [7]	Wormhole Attack, Information or Data Spoofing	Traditional wireless sensor network	Used master key to generate other keys. Used Hash tree to generate HMAC for authentication	Requires accurate time synchronization between all communicating parties, implements geographical and temporal leashes
Random Key Pre-distribution [8][9][10]	Data and information spoofing, Attacks in information in Transit	Traditional wireless sensor network	Random key pre-distribution	Resilience, Protect the network even if part of the network is compromised, Provide authentication measures for sensor nodes
REWARD [11]	Black hole attacks	Traditional wireless sensor network	No cryptographic keys, identify malicious node and suspicious area by broadcast messages	Geographic routing, Takes advantage of the broadcast inter-radio behavior to watch neighbor transmissions and detect black hole attacks
SNEP & μTESLA [12]	Data and Information Spoofing, Message Replay Attacks	Traditional wireless sensor network	Each node is given a master key and all other keys are derived from this key	Semantic security, Data authentication, Replay protection, Weak freshness, Low communication overhead
Our Scheme	Insecure bootstrapping, False data injection, Node compromise, Hello Flood attack	Distributed Sensor Network	Star Key Graph based WCDS	Secure SWCDS network formation, Authenticated message delivery, Robustness against node compromise

bytes long in SEF scheme and communication overhead is $(15.h)$ bytes, where the number of hops h a report travels is very high (necessary for better performance).

6 Conclusions

In this paper, we have put an effort to devise a security mechanism that combines the star key graph to form a WCDS based distributed wireless sensor network to counteract the impact of compromised nodes. We have shown that the scheme has significantly less storage overhead for the individual constrained sensor node and the scheme is also scalable and efficient in storage, communication and computation. We evaluate our scheme through analysis to show that our SMP mechanism is resilient to an increasing number of compromised nodes. For further investigation, some important research issues are worth mentioning: a) To present experimental results on how often we have to use solutions described in Case I and/or Case II, b) how many messages have to be sent to the BS in order to fix the roles of the nodes after deployment, and c) to validate the analytical results through simulation.

References

1. Garey, M.L., Johnson, D.S.: Computers and Intractability: A Guide to the Theory of NP-Completeness. W. H. Freeman, San Francisco (1979)
2. Karlof, C., Wagner, D.: Secure routing in wireless sensor networks: Attacks and countermeasures. Elsevier's Ad Hoc Network Journal, Special Issue on Sensor Network Applications and Protocols, 293–315 (September 2003)
3. Akyildiz, I.F., Su, W., Sankarasubramaniam, Y., Cayirci, E.: Wireless sensor networks: a survey. Computer Networks 38, 393–422 (2002)
4. Erdos, Renyi: On Random Graphs. Publ. Math. Debrecen 6, 290–297 (1959)
5. Ye, F., Luo, H., Lu, S., Zhang, L.: Statistical en-route filtering of injected false data in sensor networks. IEEE Journal on Selected Areas in Communications 23(4), 839–850 (2005)
6. Newsome, J., Shi, E., Song, D., Perrig, A.: The sybil attack in sensor networks: analysis & defenses. In: Proc. of the third international symposium on Information processing in sensor networks, pp. 259–268. ACM, New York (2004)
7. Hu, Y.-C., Perrig, A., Johnson, D.B.: Packet leashes: a defense against wormhole attacks in wireless networks. In: Twenty-Second Annual Joint Conference of the IEEE Computer and Communications Societies. IEEE INFOCOM 2003, March 30- April 3, 2003, vol. 3, pp. 1976–1986 (2003)
8. Du, W., Deng, J., Han, Y.S., Varshney, P.K.: A pairwise key pre-distribution scheme for wireless sensor networks. In: Proc. of the 10th ACM conference on Computer and communications security, pp. 42–51 (2003)
9. Oniz, C.C., Tasci, S.E., Savas, E., Ercetin, O., and Levi, A.: SeFER: Secure, Flexible and Efficient Routing Protocol for Distributed Sensor Networks, `http://people.sabanciuniv.edu/~levi/SeFER_EWSN.pdf`
10. Chan, H., Perrig, A., Song, D.: Random key predistribution schemes for sensor networks. In: IEEE Symposium on Security and Privacy, Berkeley, California, May 11-14, 2003, pp. 197–213 (2003)
11. Karakehayov, Z.: Using REWARD to detect team black-hole attacks in wireless sensor networks. In: Workshop on Real-World Wireless Sensor Networks (REALWSN 2005), Stockholm, Sweden, June 20-21 (2005)
12. Perrig, A., Szewczyk, R., Wen, V., Culler, D., Tygar, J.D.: SPINS: Security Protocols for Sensor Networks. Wireless Networks 8(5), 521–534 (2002)
13. Menezes, A., Oorschot, P., Vanstone, S.: Handbook of Applied Cryptography. CRC Press, Boca Raton (1996)
14. Chen, Y.P., Liestman, A.L.: A Zonal Algorithm for Clustering Ad Hoc Networks. International Journal of Foundations of Computer Science 14(2), 305–322 (2003)
15. Liu, D., Ning, P., Du., W.: Group-Based Key Pre-Distribution in Wireless Sensor Networks. In: Proc. ACM WiSE 2005, September 2 (2005)
16. Walters, J.P., Liang, Z., Shi, W., Chaudhary, V.: Wireless Sensor Network Security: A Survey. Auerbach Publications, CRC Press (2006)
17. Wong, C.K., Gouda, M., Lam, S.S.: Secure Group Communications using Key Graphs. IEEE/ACM Transactions on Networking 8(1) (February 2000)

Prevention of Black-Hole Attack using One-Way Hash Chain Scheme in Ad-Hoc Networks

JongMin Jeong[1], GooYeon Lee[2,*], and Zygmunt J. Haas[3,**]

[1] School of Electrical and Computer Engineering,
Cornell University, Ithaca, NY 14853, USA
jj248@cornell.edu

[2] School of Information Technology, Kangwon National University,
192-1, Hyoja-dong, Chunchen, Kangwon 200-701, Korea
leegyeon@kangwon.ac.kr

[3] School of Electrical and Computer Engineering,
Cornell University, Ithaca, NY 14853, USA
haas@ece.cornell.edu

Abstract. In this paper, we consider the specific case of a black-hole attack in ad-hoc networks caused by an incorrect route reply from an intermediate node. As a solution to this problem, we propose a scheme based on a one-way hash chain mechanism and Further-Request/Further-Reply exchange. We prove the effectiveness of the scheme in terms of security and its efficiency in terms of network performance.

1 Introduction

There are various threats during the route establishment process in ad-hoc networks. Those can be divided into conventional threats and specific threats. Conventional threats occur in traditional general-purpose networks such as the Internet as well as ad-hoc networks. Unlike conventional threats, specific threats occur in ad-hoc networks due to their unique features. Examples include a black-hole attack that is brought about by an incorrect route message, a wormhole attack that is caused by tunnelling packets from one point to another point in the network [1], a rushing attack that is caused by duplicate suppression [2], and selfishness, in which users extend the lifetime of their devices by exhibiting selfish behavior. The black-hole attack can occur even in traditional networks. However, it is more likely to occur in an ad-hoc network, because of the lack of infrastructure and the mobility of ad hoc networks. Therefore we classify it as a specific attack of ad-hoc networks.

* The work of GooYeon Lee has been supported by Kangwon Institute of Telecommunications and Information(KITI).

** The work of Zygmunt J. Haas has been supported in part by the U.S. National Science Foundation under the grant number ANI-0329905 and by the DoD MURI (Multidisciplinary University Research Initiative) Program administered by the Air Force Office of Scientific Research (AFOSR) under the contract number F49620-02-1-0217.

T. Vazão, M.M. Freire, and I. Chong (Eds.): ICOIN 2007, LNCS 5200, pp. 564–573, 2008.

Motivation. Up until now, the concerns for security in ad-hoc networks were focused on secure routing protocols themselves. Because most of the works on securing ad hoc networks have only considered conventional threats, they are ineffective in defending against specific threats whose causes are more complex than those of conventional threats. Among the specific attacks, the black-hole attack is the most powerful, as there are various ways of generating an incorrect route during the route establishment process. In this paper, we propose a scheme to address the black-hole attack, especially in the specific case where the attack is caused by a polluted message from an intermediate node, as previously considered by Deng et al. [3]. Because most of the black-hole attacks happen during route establishment period, our solution resembles a secure routing protocol. However, our main goal is to solve the black-hole attack problem rather than proposing a secure routing protocol.

Contributions. The contributions of this work are as follows.

- **Introduction of a New Mechanism Based on a One-Way Hash Chain Scheme:** The conventional solution to solve the black-hole attack problem is to use digital signature (DS). However, DS requires significant processing and results in considerable transmission overhead. Hence, we adopt a one-way hash chain scheme which is quite lightweight compared to the DS scheme.
- **No Assumption on a One-Way Hash Chain Scheme:** In general, a one-way hash chain schemes assume that the last value of the chain is securely published for verification of the other hash chain values. In our study, we do not require this assumption.
- **Cost Effectiveness:** Simulation results show that a one-way hash chain is a more efficient scheme, as compared to the DS scheme.

Related Work. There are a number of previous works which proposed solutions against the black-hole attack [3][4][5][6][7][8]. However, all of these works have limitations.

- **DS Scheme:** Zapata et al. [4] proposed a Secure Ad-hoc On-Demand Distance Vector (SAODV) routing algorithm based on a DS scheme and a hash chain scheme to support the security functions of AODV. The DS mechanism can prevent a black-hole attack, however it requires a public-key based signature and hop-by-hop verification. These operations generate excessive overhead during the route establishment process.
- **Further-Request and Further-Reply Scheme:** Deng et al. [3] defined a specific type of black-hole attack that is caused by abnormal intermediate route replies, which is the problem that we address in this paper. They proposed a solution using alternative routes to confirm whether an intermediate node has adequate routes to reach the destination. However, when there are colluding nodes with the intermediate node, their scheme cannot definitely detect the black-hole attack.

In addition, Sequence Numbering scheme [5], Neighborhood-Based Detection scheme [6], and Collaborative Architecture scheme [7] have been proposed. However, these schemes have only considered the black-hole attack caused by incorrect route replies from the destination node. The Prevention of Cooperative Black-hole scheme [8] never examines the message integrity.

Section 2 defines the I^2black-hole attack considered in this paper, discusses the requirements to solve it, and explains a basic operation of the mechanisms used. Section 3 explains our proposed solution. Section 4 presents simulation results, and Section 5 analyzes several design factors. Section 6 concludes the paper.

2 Preliminaries

In this section, we define the I^2black-hole attack that we address in this paper. In addition, we discuss the requirements needed to solve the problem and we outline the underlying mechanisms used in our solution.

I^2Black-Hole Attack. We concentrates on a particular type of the black-hole attack, which is described by Deng et al. [3]. In general, the route establishment process in ad-hoc networks consists of route discovery and route maintenance operations. When a node requires a route to a destination, it initiates a route request (REQ) message. Once the REQ reaches the destination, the destination responds by unicasting destination a route reply (DREP) message. Many on-demand ad-hoc routing protocols, such as the dynamic source routing (DSR), allow an intermediate node (Inode) to reply the REQ packet if it has a fresh enough route to the destination. This mechanism is a useful strategy to decrease the routing delay. However, in the case that the intermediate route reply (IREP) is polluted by adversary, data packets will not be able to reach the destination. Here, we focus on this type of the black-hole attack, caused by a polluted intermediate route reply and refer to it as the I^2black-hole attack.

2.1 Requirements

To solve the I^2black-hole attack the following consideration should be taken into account:

- **R1. Integrity of IREP:** The reason that the I^2 black-hole attack occurs is an incorrect IREP from an Inode responding to an REQ by modification or fabrication of the normal DREP. Therefore, the key to detecting a polluted IREP is to verify the integrity of the IREP; i.e., whether or not it is derived from the DREP and is not changed during transmission from the Inode to the source.
- **R2. Freshness:** An IREP is based on a cached DREP. Yet, since a cached message often may not contain the most recent routing information, it can be useless, even if it has not been manipulated by a malicious node. Therefore, an IREP needs to be checked for its freshness.

- **R3. Destination Identification:** An IREP is derived from the DREP. Therefore, an Inode should guarantee that the IREP is based on information received from the destination.

One of the other controversial factors is the authentication of an Inode who sends an IREP back to the source. Most existing solutions against a black-hole attack focus only on the authentication of an Inode. However, even an authenticated intermediate node may have a compromised IREP. Furthermore, during the transmission from an Inode to the source, the IREP can be manipulated by a malicious node. Hence, the authentication of an Inode is not enough to confirm the correctness of an IREP. Rather, determining the integrity of an IREP is an essential issue to prevent the I^2black-hole attack.

2.2 The Network Environment

From a network-centric point of view, the following network model is considered.

- The underlying routing protocol is DSR, since it allows to cache a DREP. This allows any Inode to return the IREP to the source node on behalf of the destination.
- The destination is not compromised, and the DREP is normally stored in intermediate nodes. Although a black-hole attack can also be launched by a malicious destination or mutable DREP, this problem is outside the scope of this work. Our essential concern is to prevent the I^2black-hole attack which is defined as caused by a polluted IREP at an Inode.

2.3 The Basic Mechanisms

One-Way Hash Chain (OWHC). The idea of an OWHC was first proposed by Lamport [10] for a one-time password scheme and it consists of the following two processes:

- **Generation algorithm** of the hash chain returns hash chain values, $C = (c_0 \ldots c_n)$ for $n \in \mathbb{N}$ using hash function $h : (0,1)^m \rightarrow (0,1)^l$ for $m \geq l$, where m is the input length and l is the output length of the message. First, the initiator of the hash chain generates the first hash value $c_0 = H^1(M) = h(M)$ with an input message M, and calculates the other values by repeatedly applying the one-way hash function i-times $c_{i-1} = H^i(M) = h(h(\ldots h(M)))$.
- **Verification algorithm** of hash chain returns $True$ or $False$ according to whether $c_n \stackrel{?}{=} H^{n-i}(c_i)$. Conventional OWHC schemes assume that the last hash value c_n is securely distributed to all the nodes.

From a cryptographic point of view, we assume the following properties for a hash function to be practically and computationally secure:

- **Non-Inverseness(One-Wayness):** For a given c_{i+1} and c_i where $c_{i+1} = h(c_i)$, it is impossible to calculate $c_i = h'(c_{i+1})$.
- **Non-Collisioness:** For a given c_i and c_j that $c_i \neq c_j$, the possibility that $h(c_i) = h(c_j)$ is negligible.

Further-Request and Further-Reply Scheme. Here, we do not assume that the last value of the hash chain c_n is securely published. Instead, we use the Further-Request (FurREQ) and Further-Reply (FurREP) concept [3]. The FurREQ message requests *an additional information* from the next-hop node (NHN) which is an arbitrary node between the intermediate node and the destination node. The FurREP sends back the requested information that the NHN cached after receiving it from the source. For instance *an additional information*, in our scheme it is the hash chain value of the NHN.

3 Solution against I^2black-Hole Attack

As we assume that the DREP message is not compromised, the proposed solution builds upon the DREP procedure, since an IREP is based on a cached DREP message received from the destination. The process of replying with the IREP, the transmission of the FurREQ/FurREP, and the verification of the IREP are described next.

Phase 1 DREP. The DREP is divided into two steps: generation of the initial hash chain value and of the intermediate hash chain values.

Step 1. Initial hash value generation: The destination node DST generates the initial hash chain.

1. DST calculates the initial hash value $c_{DST} = c_0 = h(R \parallel REP \parallel LT_{DST} \parallel ID_{DREP})$, where R is the random seed $R \leftarrow \{0,1\}^r$ for $r \in \mathbb{N}$, REP is the conventional route reply message, LT_{DST} is the lifetime chosen by the DST, and ID_{DREP} is the sequence number of the $DREP$.
2. DST sends $DREP_0 = \{REP \parallel LT_{DST} \parallel ID_{DREP} \parallel h(hops) \parallel c_0\}$ to the previous node on the route. $h(hops)$ is a hashed hop count of the route.

To maintain the confidentiality of the DREP, a cryptographic mechanism can also be adopted during the route reply process. However, since the main focus of this paper is to prevent the I^2black-hole attack caused by a polluted IREP, it is assumed that there exists a method by which the DREP is transmitted to the intermediate route nodes. Thus, the confidentiality of the DREP is not explicitly addressed in this paper.

Step 2. Intermediate hash value generation: Each intermediate node N_i generates its hash value.

1. N_i extracts c_{i-1} from $DREP_{i-1}$.
2. N_i calculates its hash value $c_i \leftarrow h(c_{i-1})$.
3. N_i constructs $DREP_i = \{REP \parallel LT_{DST} \parallel LT_i \parallel ID_{DREP} \parallel h(hops) \parallel c_i\}$, where LT_i is the lifetime of $DREP_i$, and sends it to the previous node N_{i+1} on the route.

Phase 2 IREP. Each intermediate node N_i caches $DREP_{i-1}$, in case another REQ is generated by a different source with the same destination. If an Inode has a fresh enough route reply, it sends an IREP back to the source node.

1. Inode sends back $IREP = \{DREP_{Inode} \parallel ID_{Inode} \parallel c_{Inode}\}$, where c_{Inode} is the hash chain value of Inode, ID_{Inode} is the identity of Inode, and $DREP_{Inode}$ is the cached route reply received from the destination.

Phase 3 FurREQ/FurREP. After receiving the IREP, the source node SRC runs the FurREQ and FurREP process.

Step 1. FurREQ: The SRC sends a FurREQ to the next hop node (NHN) of the Inode.

1. SRC stores the hash chain value of Inode c_{Inode}.
2. SRC randomly selects $NHN \in \{\text{arbitrary node between Inode and DST}\}$.
3. SRC unicasts $FurREQ = \{ID_{SRC}, ID_{IREP}, ID_{NHN}\}$ via the Inode to the NHN, where ID_{SRC} and ID_{NHN} are the identification of the SRC and NHN, respectively, and ID_{IREP} is an identification number of the IREP.

Step 2. FurREP: The NHN that receives the FurREQ sends its chain value and hashed hop count to the SRC.

1. NHN sends c_{NHN} and $h(hops)$ to the SRC through alternative route which does not includes Inode.

Phase 4 Verification IREP: After receiving the FurREP, the SRC verifies the IREP.

1. SRC verifies whether $c_{Inode} \stackrel{?}{=} H^{dist}(c_{NHN})$, where $dist$ is the number of hops between the Inode and the NHN, and the SRC checks that the hop count was not changed. If both are valid, the SRC estimates that the IREP is neither modified nor fabricated.

4 Simulation Results

We use the ns-2 simulator with 50 nodes over a 670m x 670m network area. The initial positions of the nodes were random. The node pause time was set at 600 seconds and the mobility was set at 20 [m/s]. Three performance metrics are evaluated:

- **Path Reachability:** At step 2 of the FurREQ/FurREP phase, the FurREP uses an alternative route, as in Deng's scheme [3], to prevent it from being fabricated or modified by the Inode. We examine the possibility that there are alternative routes not passing through Inode from the source to the NHN.

- **Time Overhead Caused by FurREQ and FurREP:** The average time elapsed from when the source node sends the FurREQ to when it receives the FurREP.
- **Time Comparison between DS Scheme and Proposed Scheme:** The average time taken to complete the intermediate reply process.

Fig. 1(a) shows the average time from when the SRC sends the FurREP to the NHN to when the SRC receives the FurREP from the HNH; i.e. the additional time cost caused by the FurREQ and FurREP. The legend "Through Inode" is the case where the FurREP generated by the NHN reached the SRC via the Inode. The legend "Not through Inode" is the case where the FurREP reached the SRC through an alternative route that did not include the Inode as an intermediate node. The X-axis is the number of hops from the Inode to the DST.

In studying the path reachability, regardless of which node is selected as the NHN, there were always alternative paths that did not include the Inode from the source to the NHN. If the node density became more sparse, the possibility of no alternative paths increased. However, in the proposed configuration, there was always at least one alternative path. This means that the FurREQ and FurREP exchange using an alternative path is effective. The additional time overhead caused by using an alternative path was about 15%. In addition, the more hops between the Inode and the DST, the longer is the time overhead.

Fig. 1(b) shows a comparison of the DS and the proposed schemes. The X-axis refers to the number of hops between the source and the Inode and the Y-axis refers to the average time elapsed from when the Inode sent back the IREP to when the SRC completed the IREP process. The results shown as "DS scheme" include the time taken for the signing or verification operations by the SRC and other intermediate node located in between the SRC and the Inode. When the hop count between the source and the Inode is increased, the time for the source to receive the IREP is increased as well. The results labeled "Proposed scheme" incorporate the time taken by the FurREQ and FurREP exchange. Similarly, the time for the source to receive the FurREP increased as the distance increased. Nevertheless, the proposed scheme is still superior to the DS scheme.

5 Analysis

In this section, we analyze the proposed scheme from the following perspectives:

S1 Solution for I^2Black-Hole Attack. The I^2black-hole attack is caused by a fabrication or modification of a route reply message by an intermediate node. This vulnerability is solved by the proposed scheme, as demonstrated by the following scenarios.

- **Attack Scenario 1 - Message Attack.** Let us consider a message attack scenario in which a malicious Inode is trying to fake an IREP. The malicious Inode fabricates or modifies an IREP and generates hash chain values using

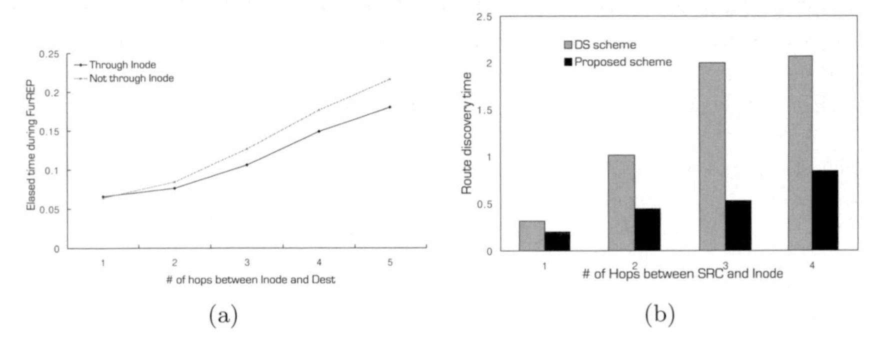

Fig. 1. Simulation results: (a) Delay and Reachability of FurREP (b) Comparison of time elapsed during IREP between the DS and the proposed schemes

the input with this polluted IREP. The Inode generates a forged or modified route L', chooses a random number $R' \leftarrow \{0,1\}^r$, and makes an abnormal hash chain list $C' = \{c'_0, c'_1, \ldots, c'_n\}$. Then, the forged IREP, $IREP' = \{L' \parallel \ldots \parallel c'_{Inode}\}$ is sent back, which includes a forged route and arbitrary hash chain value $c'_{Inode} \in C'$. After receiving $IREP'$, the source selects an NHN and verifies $IREP'$ through the hash value of NHN, c_{NHN}. To succeed, the malicious Inode should send back c'_{Inode}, which is the same as c_{NHN} generated by NHN. However, the Inode cannot know the NHN's hash value c_{NHN} due to the non-inverse feature of the hash chain. Furthermore the malicious Inode cannot know which node is selected as the NHN. Therefore, the scheme protects against fabrication or modification of IREP.

- **Attack Scenario 2 - Node Injection Attack.** Let us consider a node injection attack scenario that hides the true number of the nodes in the DREP route. The Inode injects several forged nodes into the route, but generates its hash chain value with the input of the previous hash value, which is based on the correct route. The Inode re-applies the hash function and advertises that there are more nodes on the route. That is, the Inode generates $c'_{Inode} = H^{\alpha}(c_{Inode})$ for $\alpha \geq 1$, changes the route to L', injecting α nodes into the correct route L, and then sends back $IREP' = \{L' \parallel \ldots \parallel c'_{Inode}\}$. In the verification phase, SRC uses hash values of both c'_{Inode} and c_{NHN}. Because c'_{Inode} is generated using the previous hash value, SRC cannot detect that c'_{Inode} is not an original hash value of the Inode. Therefore, SRC assumes that L' is a correct route. However, for L' to be a correct route, the Inode should also change the hop count in IREP to $h(hops + \alpha)$. Because our scheme includes a hashed hop count value in the FurREP message, the scheme will detect the illegitimate increased in the number of nodes on the route.

S2 Destination Identity. The proposed scheme supports implicit identification of the destination. That is, for identification of the destination, the proposed scheme relies on the values of c_{NHN} and c_{Inode}, rather than on direct information of the destination. If both, the Inode and the NHN, claim that they have received information from the same destination, and if this information is verified as correct through the one-way hash chain scheme, it is assumed that the IREP is indeed received from the destination.

S3 No Assumption on One-Way Hash Chain. If we were to assume that the last value of a hash chain could be securely communicated, there would be no need for the FurREQ and FurREP exchange. However, we do not require this assumption, as the secure distribution of the last hash value is a difficult problem by itself before a secure route is established. Therefore, the proposed approach is a more practical method, compared to other schemes addressing the same problem. Also, the simulation results showed that, although the FurREQ and FurREP exchange increases the transmission overhead, the proposed scheme is still efficient, as compared to the DS scheme.

S4 Freshness. For the freshness of IREP, the source node should verify whether or not the IREP has been sent sufficiently recently. In our scheme, the source node examines freshness of IREP through LT_{DST} generated by the destination.

S5 Colluding Attack. In Deng's scheme [3], a colluding attack is more probably, because the Inode itself selects the NHN node. If there are nodes which can potentially collude with the Inode, the Inode can simply choose one of such nodes as the NHN. Therefore, if there is at least one potentially colluding node between the malicious Inode and the destination, the probability of a colluding attack is $P(n, c, i) = 1$, where n and i are the number of nodes on the route and the number of hops between the source node and the intermediate node, respectively, and c is the number of possible malicious nodes on the route.

In contrast, in our scheme, the source node is the one which selects the NHN. Thus the Inode cannot know which node will be chosen as the NHN. Also, the Inode is unable to generate the hash value of NHN due to the feature of the one-way hash chain. To illustrate this point, the probability of colluding attack in our proposed scheme, $P(n, c, i)$, drastically decreases as a number of the nodes between Inode and the destination node:

$$P(n, c, i) = \frac{\sum_{r=0}^{n-i} \frac{r}{n-i} \binom{n-i}{r} \binom{i-1}{c-r}}{\binom{n}{c}}, \qquad (c - i \leq r \leq c)$$

where r is the actual number of colluding nodes between Inode and the destination. Fig. 2 shows an instance of probability of colluding attack in the case of $i = 4$.

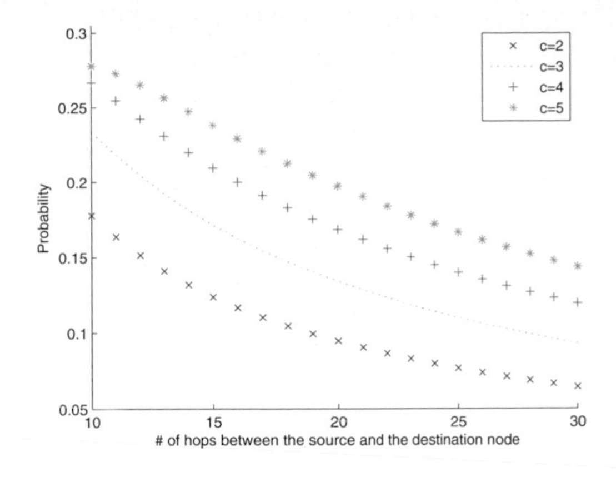

Fig. 2. Probability of colluding attack for $i = 4$

6 Conclusion

The study presented in this paper concentrates on the I^2black-hole attack, which results from a polluted intermediate route reply from an intermediate node. Although the I^2black-hole attack can be prevented with the DS scheme, we showed in this paper that it can also be prevented using the one-way hash chain scheme along with a modified further-message approach. The complexity of the proposed scheme is smaller, as compared with the complexity of the DS scheme.

References

1. Hu, Y.C., Johnson, D., Perrig, A.: Wormhole Attacks in Wireless Networks. IEEE Journal of Selected Area in Communication (2005)
2. Perrig, A., Hu, Y.C., Johnson, D.: Rushing Attacks and Defense in Wireless Ad Hoc Network Routing Protocols. In: ACM Workshop on Wireless Security (2003)
3. Deng, H., Li, W., Agrawal, D.P.: Routing Security in Wireless Ad hoc Networks. IEEE Communication 10(6), 70–75 (2002)
4. Zapata, M.G., Asokan, N.: Secure Ad-hoc Routing Protocols. In: WiSe 2002. ACM Press, New York (2002)
5. Al-Shurman, M., Yoo, S.M., Park, S.J.: Blackhole Attack in Mobile Ad Hoc Networks. In: ACMSE 2004 (2004)
6. Sun, B., Guan, Y., Chen, J., Pooch, U.W.: Detecting Black-hole Attack in Mobile Ad Hoc Networks. In: EPMCC 2003 (2003)
7. Animesh, P., Amitabh, M.: Collaborative Security Architecture for Black Hole Attack Prevention in Mobile Ad Hoc Networks (2003)
8. Ramaswamy, S., Fu, H., Sreekantaradhya, M., Dixon, J., Nygard, K.: Prevention of Cooperate Blackhole Attack in Wireless Ad Hoc Networks. In: ICWN 2003 (2003)
9. Marti, S., Giuli, T.J., Lai, K., Baker, M.: Mitigating Routing Misbehavior in Mobile Ad Hoc Networks. In: 6th MobiCom, Boston (2000)
10. Lamport, L.: Password Authentication with Insecure Communication. Communications of the ACM 24, 770–772 (1981)

On Modeling Counteraction against TCP SYN Flooding*

Vladimir V. Shakhov[1] and Hyunseung Choo[2,**]

[1] Institute of Computational Mathematics and Mathematical Geophysics of SB RAS, Prospect Akademika Lavrentjeva, 6, Novosibirsk 630090, Russia
shakhov@rav.sscc.ru
[2] School of Information and Communication Engineering, Sungkyunkwan Univ. Chunchun-dong 300, Jangan-gu, Suwon 440-746, South Korea
choo@ece.skku.ac.kr

Abstract. One of the main problems of network security is Distributed Denial of Service (DDoS) caused by TCP SYN packet flooding. To counteract SYN flooding attack, several defense methods have been proposed. In this paper we investigate a survivability of protected servers under SYN flooding. Analysis and comparison of some typical and well-known defense mechanisms are presented. We discuss critical parameters of the protection. Appropriated mathematical models based on stochastic processes are produced.

Keywords: DDoS attack, TCP SYN flooding, defense mechanism.

1 Introduction

One of the main problems of network security is Distributed Denial of Service (DDoS). The goal of DDoS attack is to exhaust victim server recourses so that legitimate users cannot access a service. Victims of DDoS attacks were a famous Web sites like Yahoo!, eBay, Amazon.com [1]. Computers with high-speed connections to the Internet were infected by computer viruses that, when activated, sent out a storm of requests and caused a denial of service at servers. The infected computers act like "zombies". So, a participation of DDoS initiator is not required on the last stage of attack. It is difficult to suppress all sources of spoofed traffic. But the time of DDoS attack is limited. Thus, an important question of network security is an endurance of DDoS defense schemes.

About 90 percent of DDoS attacks use vulnerabilities of TCP protocol [2]. A well-known example of that is TCP SYN Flooding. The theme of DDoS counteracting is popular and some defense methods against SYN flooding have been proposed. The most popular protections recommended by CERT [3] are SYN cookies and

* "This research was supported by MKE(Ministry of Knowledge Economy), Korea under ITRC(Information Technology Research Center) IITA-2008-(C1090-0801-0046) and ITFSIP(IT Foreign Specialist Inviting Program) IITA-2008-(C1012-0801-0006)."
** Corresponding author.

T. Vazão, M.M. Freire, and I. Chong (Eds.): ICOIN 2007, LNCS 5200, pp. 574–583, 2008.

SYN cache. Recently, novel approaches of anti SYN flooding protection via spoofed packets filtering have been proposed. But the question of protection quality is not closed. Marking shortcomings of SYN flooding defense schemes experts point a lack of clear rules for packets dropping policy. A modification of TCP protocol is also undesirable. Actually, if a filter is configured to block packets that arrive at the edge router of the source network with illegitimate source addresses then some existing protocols like Mobile IP are violated. Furthermore, a benefit of defense schemes is not felt directly. The necessity of a protection installation is non-obvious. To fill the gap, in this paper we offer mathematical models for a counteracting evaluation and comparison of protecting schemes.

The remainder of the paper is organized as follows. In the section 2 existing SYN flooding counteracting methods are considered. Corresponding papers are reviewed. In the section 3 we discuss a concept of network survivability and produce mathematical models for survivability estimation of an attacked server equipped by protection against SYN flooding. Section 4 evaluates the performance of the known SYN flooding counteracting schemes. Final conclusions are produced in section 5.

2 Preliminaries

2.1 TCP SYN Flooding Details

The idea of attacks named TCP SYN flooding is based on TCP's three-way handshake scheme and its limitation in keeping half-open connections. In usual situation, when a client computer tries to get a TCP connection to a server, the client and the server exchange series of messages which looks as follows [4]. The client requests a connection by sending a SYN (synchronize) message to the server. The server acknowledges this request by sending SYN-ACK back to the client, which, responds with an ACK, and the connection is established. If a server does not receive an ACK packet during given time (timeout) then the corresponded half-open connection is closed and a data structure describing pending connection is deleted from the server memory. The described scheme is named three-way handshake.

Under TCP Flooding a violator organizes a lot of TCP connections with target [3]. A victim host sends requests (SYN ASK packets) and waits confirmation or timeout expiration of SYN packets. But a violator does not complete a handshake. If an intensity of spoofed incoming flow is sufficiently large then a server buffer (also known as a backlog queue) will be overflowed. Thus, a legal user cannot overcome flooding and receive a service. A victim of this attack can be any network node supporting TCP-based services like mail, Web, or FTP servers.

2.2 Defense Methods

To defense against the exhaustion of resources in the system under attack, an obvious approach is to increase a configuration tolerance as follows. The timeout period can be reduced from default to a short time. A similar way is a timeout choice depending on the distance to a SYN packet source [5]. It helps to drop half-open connections quickly. But an essential part of legal requests is blocked. The reliability of victim system can be improved by the extension of backlog queue. In this case, the victim computer can keep more half-open connections than before. It increases a survivability of victim computer under DoS attacks. But in case of DDoS a victim computer quickly exhausts extra-recourses.

The protection schemes named SYN cookies and SYN cache are recommended by CERT [3]. SYN cookies mechanism counteracts to SYN floods by calculating cookies that are functions of the source address, source port, destination address, destination port, and a random secret seed [6], [7], [8]. When a SYN packet is received, the server calculates a SYN cookie and sends it to the requesting client as part of the SYN-ACK packets without allocating resources for that requests. When ASK packet is received, the connection is established if a valid cookie is included. In case of SYN cache [9], each SYN request is stored and a SYN-ACK response is sent. If a valid ACK comes back, a complete connection is created. If a waiting time of response is expired then the entry is deleted. When the SYN backlog queue overflows, the oldest entry is ejected by new incoming SYN packet.

Many of proposed novel methods for defending against TCP SYN flooding attacks are focused on a filtering policy [10]. The core of the mentioned protections is to accurately identify spoofed packets and filter them without dropping legitimate traffic. In this case a criterion of spoofed packet verification is an absence of ACK packet during some predefined time. A spoofed packet can be directly detected and blocked if the packet source has illegitimate addresses. Sometimes it is unrealistic to make distinguish between spoofed and legitimate packet or it requires an inadmissible long time. In this case some differences in behavior of spoofed and legitimate traffic is used. A fire-wall based protection is an example of filtering. Also a legal user can send some code for legitimacy recognition by victim computer. This way of SYN flood counteracting requires a TCP protocol extension. Let us remark the SYN cookies or SYN cache scheme is a kind of filtering policy.

The modern trend of SYN flooding investigations is an application of point-of-change technique based on copulative sum algorithm [11], [12]. Early detection of SYN flooding allows to activate a defense mechanism under attacks and deactivate it otherwise. It improve a network throughput since network recourses are not used for excessive verification of legal clients. Thus, change-point detection method can be implemented in SYN protections. But it should be used carefully. Actually, if a discard is omitted then a protection is not activated and a protected server is successfully attacked. On the other hand, an aggressive detection criterion results to high probability of false attack detection [13]. In this case, a network throughput is degraded.

3 Proposed Server Model in TCP SYN Flooding Counteraction

3.1 Server Survivability

Accordingly Federal Standard 1037C (USA), the survivability is the quantified ability of a system, subsystem, equipment, process, or procedure to continue to function during and after a natural or man-made disturbance. For a given application, survivability must be qualified by specifying the range of conditions over which the entity will survive, the minimum acceptable level or post-disturbance functionality, and the maximum acceptable outage duration [14].

Taking into account peculiarities of disturbance like TCP SYN flood attack, it is reasonable to focus on a survivability of legal SYN packets. Actually, an attacked server losses functionality but one does not receive damage. After completing the SYN flooding a server quickly renews a normal service. The real victim of TCP SYN flooding is a normal user which cannot get a service. Let us detail the factors of legal SYN packets acceptance by protected computer.

- **Blocking Probability** - A SYN packet is rejected if the SYN buffer is full. This factor is defined by the intensity of incoming SYN flow, the intensity of SYN packet treatment, and backlog queue size. This probability is also known as the packets drop rate or the probability of buffer overflow. Let p_{BLK} be the blocking probability.
- **Waiting Time for ASK Packets** - A SYN packet is deleted from the SYN buffer if corresponding SYN ASK packet does not arrive till the expiration of established timeout. Let p_{out} be the deletion probability for legal packet.
- **Filtering Criterion** - A SYN packet is rejected if it does not satisfy a criterion of valid packet. Let us remark, a filter can drop a legal packet. The reason of this can be a error of recognizing algorithm or a strategy to keep only repeated SYN packets. Let p_F be the probability of legal packets rejection by a server protection.
- **Repeated Requests for Connection** - If a SYN packet is rejected a source can repeatedly send new SYN packet. A number of trials n_t for connection affects on survivability. But this is not important for the protections comparison.

Now we can estimate survivability of an attacked server. The probability of connection success after n_t trials equals

$$Sur = 1 - (1 - (1 - p_{BLK})(1 - p_F)(1 - p_{out}))^{n_t};$$

if $n_t = 1$ then

$$Sur = (1 - p_{BLK})(1 - p_F)(1 - p_{out});$$

For efficient filtering policy the probabilities p_F and p_{out} are about zero. Hence, the blocking probability investigation is a main survivability factor for servers equipped by a filter. But under SYN cache protection $p_{BLK} = 0$. Thus, the separate model will be produced for this case.

3.2 Server Model Protected by Filter

Following the discussion in section above the server survivability is defined by the average rate of legal packets acceptance. The rate is defined by backlog queue status. So, for survivability estimation of an victim computer it needs to calculate the states probabilities of a SYN packets buffer. We assume that the incoming SYN packets flow is the Poisson process. In the case of DoS attack it is not completely correct because a vandal usually generates false SYN packets with maximal constant rate (deterministic flow). But in the case of DDoS attack a few servers are used for SYN flooding generating. The distances between an attacked computer and cracked servers are differ. The SYN packet path from source to destination can be randomly changed. So, a sum of independent stochastic flows is observed. This is a reason to use the Poisson process [15].

The SYN ASK packet waiting time is assumed to be exponential distributed. It is reasonable if the limited waiting time depends on distance to a SYN packet source [5]. Thus, outgoing flow of SYN packets from attacked buffer is also Poisson process. The intensity of legal incoming traffic we designate as λ_R, the intensity of spoofed SYN packets is S, the intensity of outgoing traffic is μ, and the backlog queue size of attacked computer is N. Let us remark an intensity of incoming SYN packets equals $\lambda = \lambda_R + S$.

In previous subsection we define the parameter named filtering criterion. Now we define other important parameters of the protection.

- **Quality of Filtering, QoF** - The probability of spoofed packets dropping, p_Q.
- **Coefficient of Deceleration, CoD** - A recognition of incoming packet requires some time. Hence, the intensity of SYN packets treatment is reduced. The coefficient C defines a deceleration of packets treatment through packets verification.

Generally, an attacked server behavior is described by M/M/N/N queue system. The blocking probability is calculated by B-formula of Erlang:

$$p_{BLK} = \frac{(\frac{\lambda^*}{\mu C})^N \frac{1}{N!}}{\sum_{i=0}^{N} (\frac{\lambda^*}{\mu C})^i \frac{1}{i!}},$$

where

$$\lambda^* = p_F \lambda_R + p_Q S.$$

Usually, $p_F \lambda_R << p_Q S$. As it was mentioned above, the survivability is defined by p_{BLK}.

It is clear that filtering worsens QoS under regular traffic. Thus, it is reasonable to use protection under DDoS and deactivate a packets verification under normal situation. According to the recommends for SYN cookies, if a number of SYN packets in buffer gets some threshold K then counteracting mechanism is activated. We offer to model the situation by Markov chains. The system behavior corresponds to a birth-death process with the state diagram shown in Figure 1. The state $i, i = 0 \ldots N$ means the number of entries in the backlog queue of attacked server equals i. The probability of state i is designated as π_i.

Steady-state balance equations are

$$p_0 \lambda = p_1 \mu,$$

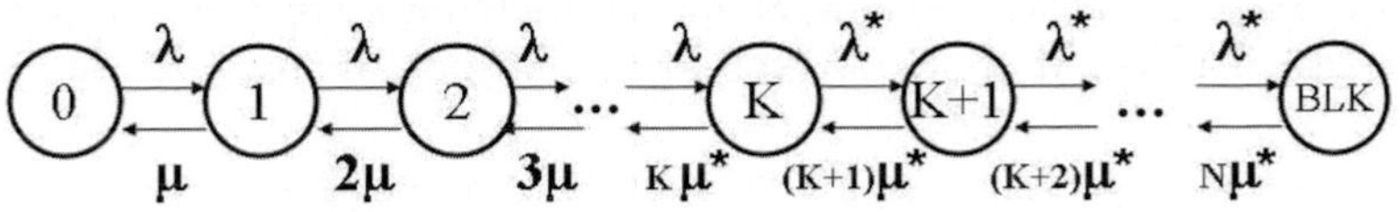

Fig. 1. Markov states diagram for protective filter

$$(\lambda + i\mu)p_i = \lambda p_{i-1} + (i+1)\mu p_{i+1}, i = 1 \ldots K-2,$$

$$(\lambda + (K-1)\mu)p_{K-1} = \lambda p_{K-2} + K\mu^* p_K,$$

$$(\lambda^* + K\mu^*)p_K = \lambda p_{K-1} + K\mu^* p_{K+1},$$

$$(\lambda^* + i\mu^*)p_i = \lambda^* p_{i-1} + (i+1)\mu^* p_{i+1}, i = K+1 \ldots N-1,$$

and the normalization condition

$$\sum_{i=0}^{N} p_i = 1$$

The steady-state probabilities $p_i, i = 1 \ldots N$ can be expressed as

$$p_i = \frac{1}{i!}(\frac{\lambda}{\mu})^i p_0, i = 1 \ldots K-1,$$

$$p_i = \frac{1}{i!}(\frac{\lambda}{\mu})^{K-1}\frac{\lambda}{\mu^*}(\frac{\lambda^*}{\mu^*})^{N-K} p_0, i = K \ldots N.$$

Taking into account the normalization condition the states probability is calculated

$$p_{BLK} = \frac{\frac{1}{N!}(\frac{\lambda}{\mu})^{K-1}\frac{\lambda}{\mu^*}(\frac{\lambda^*}{\mu^*})^{N-K}}{\sum_{i=0}^{K-1}\frac{(\frac{\lambda}{\mu})^i}{i!} + \sum_{i=K}^{N}\frac{1}{i!}(\frac{\lambda}{\mu})^{K-1}\frac{\lambda}{\mu^*}(\frac{\lambda^*}{\mu^*})^{N-K}}.$$

Now we can estimate the survivability of computers equipped by protective filters against SYN flooding.

3.3 SYN Cache Model

The SYN flooding protection scheme known as SYN cache does not satisfy the model above, although it is also a filtering based protection with criterion as following. The number of arrivals for the wafting time of a SYN ASK packet (hand-shake completion) should be less than $N-1$. Otherwise, the packet number N pushes out corresponding SYN packet form cache. The deleted entry can correspond to legitimate SYN packets. Thus, the quality of SYN cache scheme is defined by the probability of legal packets rejection. Let us calculate p_F.

Taking into account the model assumption, the number of arrival during time t is a variate distributed by Poisson with parameter $t\lambda$. Thus,

$$p_F = \sum_{n=N}^{\infty} \frac{(\lambda t)^n}{n!} e^{-\lambda t} = 1 - e^{-\lambda t} \sum_{n=0}^{N-1} \frac{(\lambda t)^n}{n!}.$$

The time t is a random value. Hence,

$$1 - p_F = \int_{-\infty}^{\infty} \sum_{n=0}^{N-1} \frac{(\lambda x)^n}{n!} e^{-\lambda x} f_t(x) dx,$$

here f_t is the distribution density of t. If t is distributed as exponential with parameter μ then

$$1 - p_F = \sum_{n=0}^{N-1} \frac{\lambda}{n!} \mu \int_0^{\infty} x^n e^{-(\lambda+\mu)x} dx = \frac{\mu}{\mu + \lambda} \sum_{n=0}^{N-1} (\frac{\lambda}{\lambda + \mu})^n.$$

Last expression is a geometrical progression. So, we have

$$\sum_{n=0}^{N-1} (\frac{\lambda}{\lambda + \mu})^n = \frac{1 - (\frac{\lambda}{\lambda+\mu})^N}{1 - \frac{\lambda}{\lambda+\mu}}$$

and

$$p_F = (\frac{\lambda}{\lambda + \mu})^N,$$

$$Sur = 1 - p_F = 1 - (\frac{\lambda}{\lambda + \mu})^N.$$

It is the formula for quality estimation of SYN cache scheme.

4 Performance Evaluation

In this section we estimate the quality of mentioned SYN flooding defense schemes. The legal packets drop rate is under consideration. Taking into account the results of previous section, we compare server survivability for differ DDoS defense scenarios defined by tuning of investigated SYN protection methods. A backlog queue size is selected as 32.

To demonstrate the effect of waiting time reduction, we carry out some calculations as well Fig.2.

This type of protection is recommended under SYN flooding. Keeping this assumption it is reasonable to use Poisson assumption for traffic model. Three

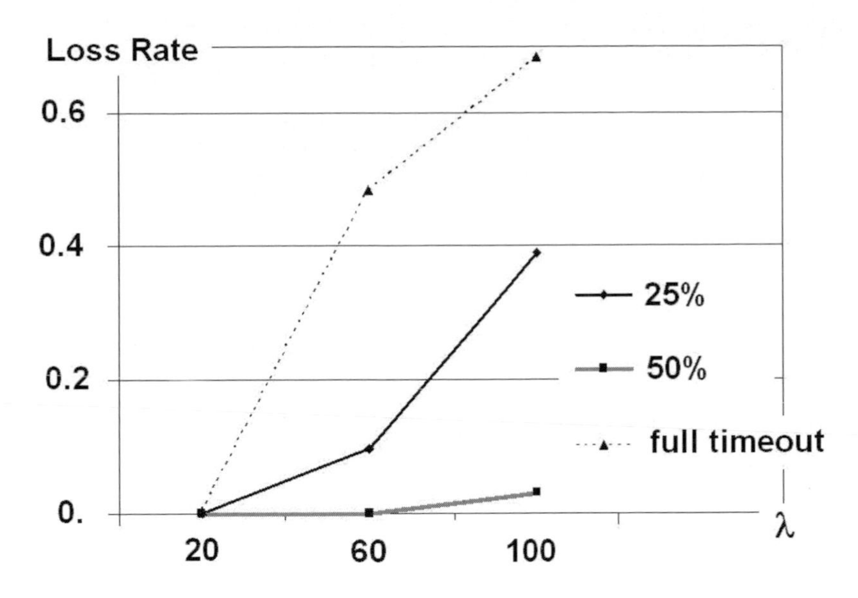

Fig. 2. Effect of timeout reduction

cases are considered: no timeout reduction, 50 percent reduction and 25 percent reduction. The calculations discover the following results. In the case of low attack intensity (DoS) a benefit of timeout reduction is absent. In the case of relatively average attack intensity (an intensive DoS or weak DDoS) any timeout reduction delivers some benefit. For high attack intensity (DDoS) drastically timeout increasing has got some reason. Victim computer has to communicate with local nodes only. But it is not enough for counteracting. More over, remote users have not any chance to get connection. The growth of probability p_{out} decreases a general survivability.

In case of filter based protections the parameter p_{out} is a secondary factor. The filtering policy seems more preferable technique. But it also has got shortcomings. For example, if the packet verification time and the legal packet treatment time are comparable then QoS drastically degrades under regular traffic. Hence, filtering algorithm with high CoD can not be used without mechanism of protection deactivation. Under real DDoS attack (thousandfold increasing of offered load) an admissible deviation of a recognizing algorithm (2 or 3 percent) reduces to zero an effect of protection filter using. Thus, QoF is very critical parameters. By the reason of low CoD and high QoF a defense strategy based on SYN cookies seems very attractive.

In Fig.3 we compare SYN cache and a filtering based defense scheme. The last approach is preferable for low attack intensity. For medium flooding intensity all methods deliver approximately same quality. In the case of high attack intensity an efficiency of SYN cache is higher than other SYN filtering. This is not surprising, since the intensity of unfiltered spoofed packets exceeds a node throughput.

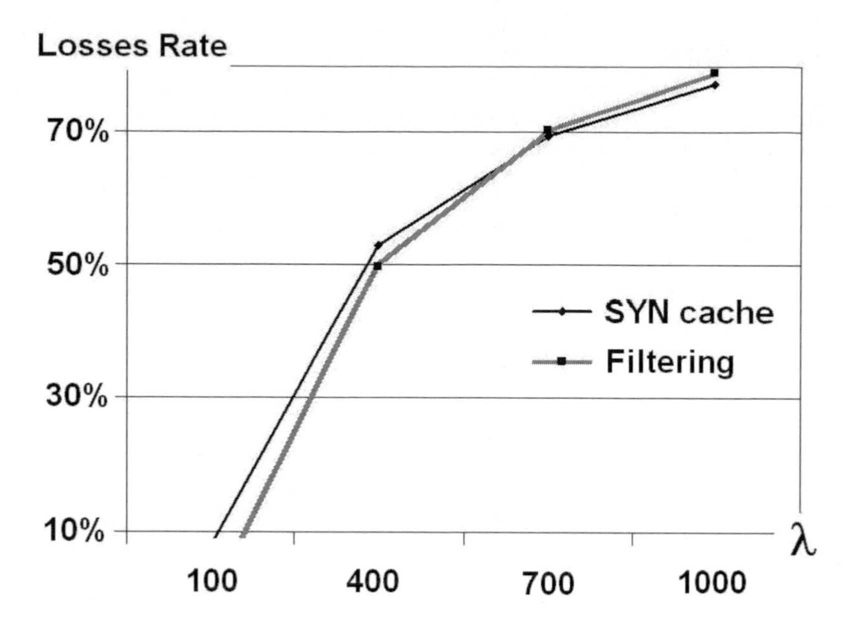

Fig. 3. Comparison of SYN cache and Filtering

5 Conclusion

In this paper we investigated a survivability of defense mechanisms against the DDoS attack named SYN flooding. A corresponding mathematical models based on stochastic processes have been developed. We defined main factors of a protection against SYN flooding. It is shown no defense method with pronounced advantages and disadvantages. A method, which is preferable for some fixed network environment, can demonstrate the worse results under insignificant deviation of network behavior. This is a reason for using multi-strategy for defense against SYN flooding. Hence, the detection methods for changing of network behavior and protection policy selection are useful. Thus, the future direction of SYN flooding counteracting can be an efficient point-of-change based technique for defense method selection in addition to SYN flooding counteracting.

Chiasserini, C.-F., Garetto, M.: An Analytical Model for Wireless Sensor Networks with Sleeping Nodes. Transactions on Mobile Computing, Vol. 5, Issue 12. IEEE (2006) 706 - 1718.

References

1. Garber, L.: Denial-of-Service Attack Rip the Internet, Computer (April 2000)
2. Moore, D., Voelker, G., Savage, S.: Inferring Internet Denial of Service Activity. In: Proceedings of USENIX Security Symposium (2001)
3. CERT Advisory CA-1996-21 TCP SYN Flooding and IP Spoofing Attacks (1996), `http://www.cert.org/advisories/CA-1996-21.html`

4. Stevens, W.: TCP/IP Illustrated, vol. I. Addison-Wesley Publishing Company, Reading (1994)
5. Fan, M., Jun-yan, Z., Wan-pei, L., Guo-wei, Y.: Tradeoffs of DDoS solutions. In: Proceedings of the Fourth International Conference on Parallel and Distributed Computing, Applications and Technologies, pp. 198–200 (2003)
6. Bernstein, D.J.: SYN cookies, http://cr.yp.to/syncookies.html
7. Schuba, C., Krsul, I., Kuhn, M., Spafford, G., Sundaram, A., Zamboni, D.: Analysis of Denial of Service Attack on TCP. In: Proceedings of IEEE Symp. Security and Privacy (1997)
8. Zuquete, A.: Improving the functionality of syn cookies. Communications and Multimedia Security, 57–77 (2002)
9. Lemon, J.: Resisting SYN flooding DoS attacks with a SYN cache. In: Proceedings of USENIX BSDCon. (2002)
10. Chen, W., Yeung, D.: Defending Against TCP SYN Flooding Attacks Under Different Types of IP Spoofing. In: ICN/ICONS/MCL (2006)
11. Wang, H., Zhang, D., Shin, K.G.: Change-point monitoring for the detection of DoS attacks. IEEE Trans. on Dependable Secur. Computing 1(4), 193–208 (2004)
12. Tartakovsky, A.G., Rozovskii, B.L., Blazek, R., Kim, H.: A novel approach to detection of intrusions in computer networks via adaptive sequential and batch-sequential change-point detection methods. IEEE Trans. on Signal Processing 54(9), 3372–3382 (2006)
13. Shakhov, V.V., Choo, H., Bang, Y.: Discord model for detecting unexpected demands in mobile networks. Future Generation Comp. Syst. 20(2), 181–188 (2004)
14. http://www.its.bldrdoc.gov/fs-1037/fs-1037c.htm
15. Trivedi, K.: Probability & Statistics with Reliability, Queueing, and Computer Science Applications. Prentice Hall, Englewood Cliffs (1982)

Energy-Efficient and Fault-Tolerant Positioning of Multiple Base Stations

Soo Kim, JeongGil Ko, Jongwon Yoon, and Heejo Lee*

Korea University
{sooo,jgko,yoonj,heejo}@korea.ac.kr

Abstract. As the nodes have limited battery power in Wireless Sensor Networks (WSNs), energy efficiency and fault tolerance should be the two major issues in designing WSNs. However, previous studies on positioning base stations (BSs) in WSNs are focused on energy efficiency only. Yet, mission-critical applications like emergency medical care systems should be guaranteed continuous services considering fault tolerance. In this paper we propose to place multiple BSs considering not only energy efficiency but also fault tolerance. We present two strategies to find the optimal position of BSs; (1) minimizing the average transmission energy for energy efficiency; and (2) minimizing additional energy consumption after a BS failure for fault tolerance. The optimal positions for multiple BSs are derived by the metric that considers both energy efficiency and fault tolerance, with a weight factor. Our simulation results show that fault tolerance is important and strongly related to elongation of network lifetime. In addition, we show that our proposed scheme is more energy-effective than previously suggested strategies on unexpected environmental changes which occur commonly in WSNs and sustain the network lifetime effectively under BS failures.

1 Introduction

Wireless sensor networks (WSNs) consist of a large number of sensors, which are small devices with sensing, processing and transmitting capabilities with limited power resources. Sensors in WSNs monitor a region and transmit information to the base stations (BSs) via wireless channels. The communication between sensors and BSs can be either direct (single-hop) or multi-hop. The energy consumption of each sensor is strongly related to the distance between the a sensor and a BS. Recent studies for energy efficiency in WSNs are mostly focused on minimizing the transmission distance by enhancing routing protocols or utilizing mobility of devices in WSNs. Recently, there have been several studies on BS placement for energy efficiency. Gandham *et al.* introduced the importance of energy efficient BS placement [1]. Vaas *et al.* presented the idea of moving the BS of a sensor network, in order to decrease the amount of energy required for communication [2].

Fault tolerance is also a critical issue in designing WSNs. Due to the limited battery and a hostile environment of WSNs, sensors and base stations are vulnerable and can be frequently inactive. Inactive devices increase the energy consumption of numerous

* Corresponding author.

T. Vazão, M.M. Freire, and I. Chong (Eds.): ICOIN 2007, LNCS 5200, pp. 584–593, 2008.

sensors and eventually decrease the network lifetime; it discontinues the services in a network at all. However mission-critical applications like the emergency medical care or national defense should be guaranteed with continuous services, minimizing failures of BSs and sensors. Especially, failures on BSs give much more impact to a network than failures on sensor nodes. Zimmermann *et al.* emphasized that a BS can be faulty and the BS failure can severely degrade the performance of WSNs [3].

In this paper we propose the positioning of multiple BSs considering both energy efficiency and fault tolerance. We use *minavg*, minimizing the average energy consumption, for the measurement of energy efficiency. This strategy is proven to be the best choice of three strategies by experiments in [2]. In addition, to consider fault tolerance of a WSN, we propose a new key-point to minimize additional energy consumption after a base station failure. Failure on a BS brings much more effects to a WSN than failure on a sensor, therefore we focus on BS failure to measure fault tolerance. The difference between our work and previous studies considering both energy efficiency and fault tolerance [4,5,6] is that we find the optimal placement of multiple BSs, while the authors of [4,5,6] try to enhance routing protocols of WSNs.

The main contribution of our proposed scheme is guaranteeing the elongation of the network lifetime in WSNs, while regarding the failure of BSs. Our simulation results show that using our scheme, which balances energy efficiency and fault tolerance, can sustain the network lifetime twice more than using the scheme considering only energy efficiency.

2 System Model

2.1 Assumptions

Before we describe the system model, we state our assumptions for this work.

- We assume that the sensors are randomly distributed, on a $n \times n$ region.
- There are multiple sink nodes in the network, which are called base stations (BSs). We assume that base stations do not have limited energy constraints.
- The sensors are static and have energy constraints.
- Only a single device may occupy a single x-y coordinate.
- The sensors operate in an event-driven way.
- The time is split into equal periods and we assume that an event can be reported only at the beginning of a time period.
- To calculate the transmission energy between a sensor and a BS, we assume that the sensors communicate with the BS directly, *i.e.* single-hop.
- The energy used for communication is proportional to d^{α}, where d is the transmission distance and α is the attenuation parameter, typically between 2 and 4. [2]
- Although sensing requires additional energy, this is far less than the energy used in communication; thus we neglect it. [2]

2.2 Minimizing Average Energy Consumption

Vass *et al.* proposed three strategies for positioning BSs: *minavg*, *minmax* and *minrel* [2]. The *minavg* strategy is to minimize the average (total) energy consumption using the distance between sensor nodes and BSs, the *minmax* strategy is to minimize the

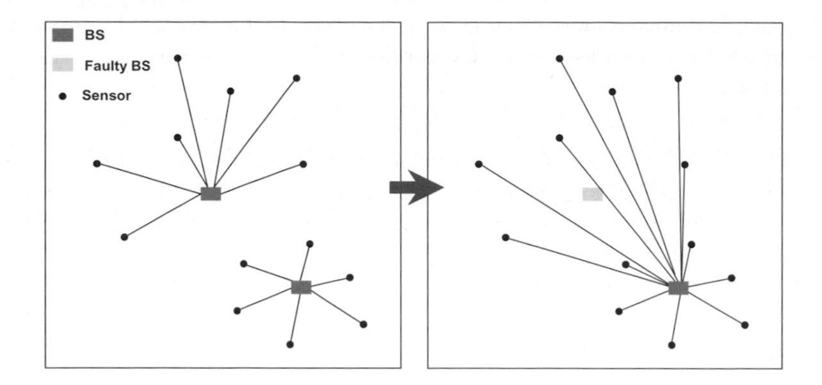

Fig. 1. Additional energy consumption caused by the failure of a BS

transmission energy for the most remote sensor in the network, and the *minrel* strategy is to minimize the maximum relative energy that a sensor node has to spend on transmission. Through the experimental results of these three strategies in [2], the *minavg* strategy is proved to be the best strategy for prolonging the network lifetime. Therefore we use the *minavg* strategy when considering energy efficiency. Let V denote the set of all sensors, and $A \in V$ the set of active sensors. Let (x, y) denote the coordinates of the BS, and (x_i, y_i) the coordinates of the ith sensor $(i \in V)$. The energy needed for node i to transmit data is

$$E_i = E_0\big((x - x_i)^2 + (y - y_i)^2\big)^{\alpha/2}, \tag{1}$$

where E_0 is a constant and α is the attenuation parameter. The energy consumed by all the active sensors is $E = \frac{1}{n(A)} \sum_{i \in A} E_i$, here $n(A)$ is the number of active sensors. It is clear that the optimal location which makes E the smallest is

$$(x_0, y_0) = \arg\min_{(x,y)} E. \tag{2}$$

In a WSN with the region of $n \times n$, where the sensors are uniformly deployed, the optimal location for the initial BS is $(\frac{1}{n}, \frac{1}{n})$. However, there is no closed solution for Eq. (2) since it depends on the network topology. Therefore, it should be solved using optimization methods.

2.3 Minimizing Additional Energy Consumption after BS Failure

We now measure the tolerance of a WSN against faults on BS. Although failure on a sensor node also effects the a network, this is far smaller than the effect by failure on a BS; thus, we only consider BS failure. There are various factors that make a BS faulty: energy depletion, attacks on purpose or natural disasters. Fig.1 shows the effect of BS failure in a WSN with two BSs. It is clear that the average distance between each sensor node and its nearest BS increases after a fault. It shows that failure on a BS increases total energy consumed in the network and eventually decreases the total network lifetime. Thus, a network is fault-tolerant if the difference between a network before the occurrence of faults and a network after the occurrence after faults is small. Additional energy consumption after faults on $f (0 \leq f \leq k)$ BSs is

$$E_f = \max E'(f) - E, \quad (3)$$

where $E'(f)$ is the average energy consumption of all active sensors when f BSs are faulty. Since E_f varies to the position of faulty BSs, we choose the maximum $E'(f)$ as the worst case. Eq. (3) is only possible when there are multiple base stations in a network, since with a single BS, E_f will be a value of 0 or ∞. Also, $E'(f) \geq E$ since routing distances of sensors which have been communicating with BSs that are faulty increases. A network is fault-tolerant if E_f is small, since it means the effect of faulty BSs to the energy consumption is low. Therefore, the optimal location for jth base station $(j > 1)$ is defined as

$$(x_j, y_j) = \arg \min_{(x,y)} E_f. \quad (4)$$

2.4 Synthesis of Energy Efficiency and Fault Tolerance

We proposed an energy-efficiency metric E and a fault-tolerance metric E_f in the previous subsections. The position that makes E and E_f to the minimum is the optimal position for BS positioning. However it is difficult to minimize both two metrics simultaneously. We show this tradeoff in the next section. To solve this problem we sum two metrics using a weight factor ω. $\Phi(f, \omega)$ is the function reflecting both energy efficiency and fault tolerance with f faulty nodes and weight factor ω. This is defined as,

$$\Phi(f, \omega) = (1 - \omega) \cdot \frac{E - E^{min}}{E^{max} - E^{min}} + \omega \cdot \frac{E_f - E_f^{min}}{E_f^{max} - E_f^{min}}, \quad (5)$$

where f is the number of faulty BSs as used in Eq. (3). E and E_f are all normalized to E^{max}, E^{min}, E_f^{max} and E_f^{min}. $\Phi(f, \omega)$ can be an energy-efficient function when $\omega = 0$, a fault-tolerant function if $\omega = 1$, or an equally balanced function if $\omega = 0.5$. The optimal position of the jth BS with f faulty nodes which is derived from Eq. (5) is

$$(x_j, y_j) = \arg \min_{(x,y)} \Phi(f, \omega). \quad (6)$$

2.5 Scheme for Optimal BS Positioning

With $\Phi(f, \omega)$ we can derive the optimal position to place a BS. Fig. 2 shows a brief scheme to find the optimal position for the $(k + 1)$th BS. When finding the optimal position of the initial BS, we use $\Phi(0, 0)$ since fault tolerance is not considered as previously mentioned. Otherwise we use $\Phi(f, \omega)$, which reflects both energy efficiency and fault tolerance. We search the position for BSs in a greedy method rather than a simultaneous method. This is because there is a big difference in complexity between the two methods. The complexity of the greedy method is $\mathcal{O}(n)$. Yet, the complexity of the simultaneous method is $\mathcal{O}(n^{\alpha})$ where α is the number of BSs. This is complexity is much higher compared to the complexity of the greedy method.

Optimal Positioning Scheme for the $(k+1)$th BS

---

k : number of previously deployed BSs
n : width/height of a WSN
(x, y) : optimal placement of the $(k+1)$th BS
f : number of faulty BSs $(0 \leq f \leq k+1)$
Φ_{ij} : $\Phi(f, \omega)$ when the $(k+1)$th BS is placed at (i, j)

$\Phi_{temp} = 1$, x = 0 , y = 0
if $k = 0$
  $for\ i = 1 : n$
    $for\ j = 1 : n$
      $\Phi_{ij} = \Phi(0, 0)$
      $if\ \Phi_{temp} > \Phi_{ij}\ then$
        $\Phi_{temp} = \Phi_{ij}, x = i, y = j$
    $endfor$
  $endfor$
else (if $k > 0$**)**
  $for\ i = 1 : n$
    $for\ j = 1 : n$
      $\Phi_{ij} = \Phi(f, \omega)$
      $if\ \Phi_{temp} > \Phi_{ij}\ then$
        $\Phi_{temp} = \Phi_{ij}, x = i, y = j$
    $endfor$
  $endfor$

Fig. 2. Pseudocode for the optimal position (x, y) of the $(k+1)$th BS using $\Phi(f, \omega)$

3 Simulation Results

In this section, using our simulations, we show the effectiveness of positioning multiple BSs while considering both energy efficiency and fault tolerance. First, we find the optimal position for the initial BS using E, the metric for energy efficiency. Second, we find the optimal position for the second BS using $\Phi(1, 0)$, $\Phi(1, 0.5)$ and $\Phi(1, 1)$, and compare the additional energy consumption by simulating a BS failure. Finally, we compare the energy consumption of a network with three BSs placed with the three different metrics. The network lifetime under different metrics ($\Phi(f, 0)$, $\Phi(f, 0.5)$ and $\Phi(f, 1)$) and the number of faulty BSs $(0 \leq f \leq 2)$ shows the necessity and effectiveness of considering both energy efficiency and fault tolerance, especially in vulnerable network environments.

3.1 The Optimal Position for the Initial BS

For simulations, we use a 20×20 region WSN with 100 sensor nodes deployed at random positions, as shown in Fig.3. As mentioned in the previous section, only energy-efficiency is considered in finding the optimal position for the initial BS; F is not capable if a single BS is deployed in a network. The optimal position of the initial BS in our

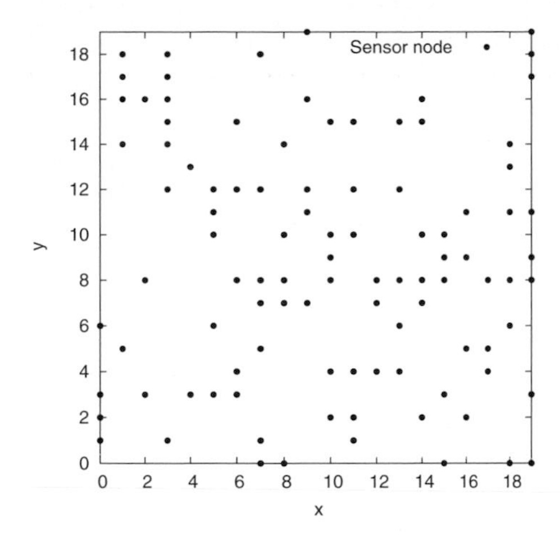

Fig. 3. Deployment of sensors for the simulation

testbed network is (9,9), where E is minimized to the smallest using Eq. (2). If sensor nodes are uniformly deployed in a network, the optimal location for the initial BS is center of a network.

3.2 The Optimal Position for the Additional BSs

Next, we find the optimal position for additional BSs. Unlike the initial BS, here we consider not only energy efficiency but also fault tolerance. We run simulations in three different conditions while varying the weight factor ω; (a) $\Phi(1,0)$ to consider energy efficiency only, (b) $\Phi(1,1)$ to consider fault tolerance only, and (c) $\Phi(1,0.5)$ to consider both energy efficiency and fault tolerant with equal proportion.

Fig.4 (a) shows that two additional BSs are located far from the first BS. This is because placing additional BSs near the initial BS hardly decreases E. However in cases like these, the energy consumption can be seriously increased if a failure occurs to any BS, especially the initial BS. In the latter part of this section, we measure and compare the energy consumption of each metric under the occurrence of a BS failure.

On the other hand, positioning the additional BSs near the initial BS makes a network tolerant to BS failure as shown in Fig.4 (b). The positions of two additional BSs that minimize $\Phi(1,1)$ are (9,8) and (9,10), marked as rectangles in Fig.4 (b). This is because the closer the additional BSs gets to the initial BS, the smaller the additional energy consumption gets when a BS faces a fault. However, here the average distance hardly decreases after the positioning of the additional BSs; the advantage of adding more BSs is technically undesirable.

$\Phi(1,0.5)$ reflects energy efficiency and fault tolerance with a balanced ratio of 50% each. As seen in Fig.4 (c), the distance between the three BSs are shorter than Fig.4 (a) but longer than Fig.4 (b). It shows that $\Phi(1,0.5)$ does not minimize both E and F_n, is more fault-tolerant than $\Phi(1,0)$ and more energy-efficient $\Phi(1,1)$.

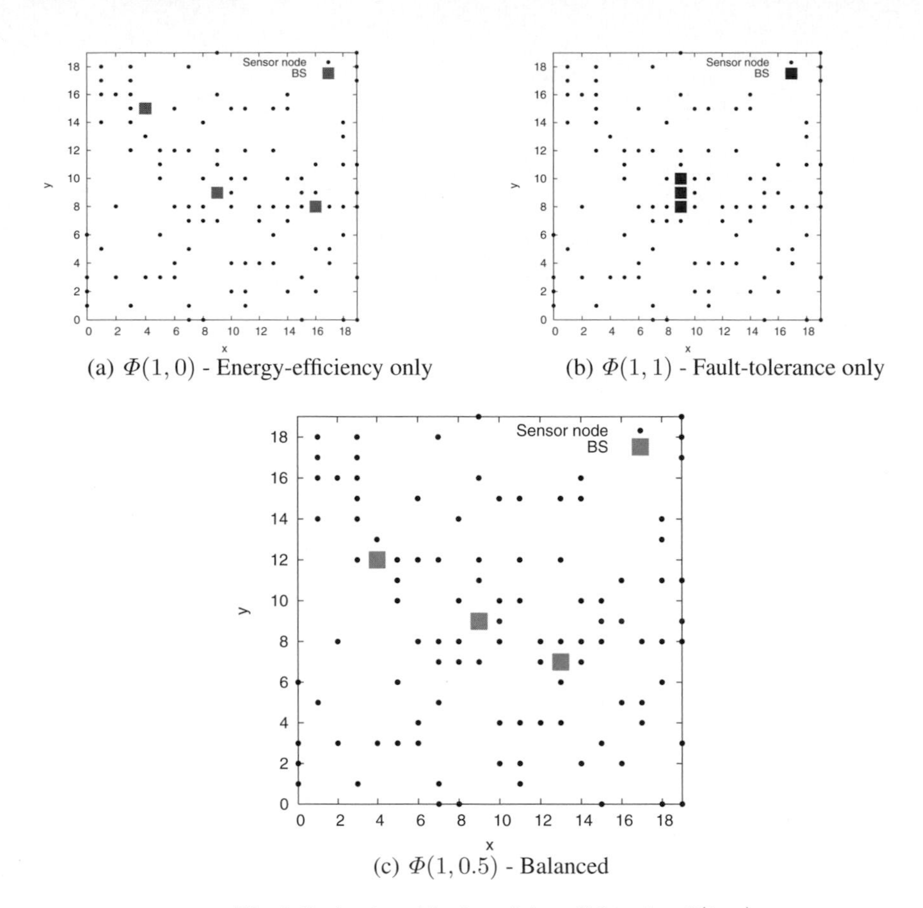

(a) $\Phi(1,0)$ - Energy-efficiency only

(b) $\Phi(1,1)$ - Fault-tolerance only

(c) $\Phi(1,0.5)$ - Balanced

Fig. 4. Optimal positioning of three BSs using $\Phi(1,\omega)$

3.3 Comparison of the Energy Consumption

We run simulations for comparing energy consumption for each base station placement algorithm. Each simulation is performed to check the number of sensor nodes that are still actuve in the network per each round. A round is the unit of time and consists of k events. The number of events within a round is modeled as a uniformly-distributed random variable between 0 and $n(A)$, where $n(A)$ is the number of active sensor nodes. A sensor node is randomly chosen to communicate with the nearest BS for every event. Every active sensor sends the same amount of data in a round, and communicate with the nearest BS directly, *i.e.* singlehop. The initial energy for each sensor is 300 J, E_0 is 0.25 J and the attenuation exponent α is set to 3.

First, we simulate the energy consumption of the network with a single BS and plot the number of sensor nodes that remain active for 100 rounds in Fig.5. The first sensor dies in the first round and the end of the 100th round 9 nodes remain alive when the BS is located at the worst position, *e.g.* where E is maximized (Fig.5 (a)). On the other hand, the first sensor dies in the sixth round and more than 28 nodes are alive after 100 rounds if the BS is located at the optimal position *e.g.* where E is minimized (Fig.5 (a)).

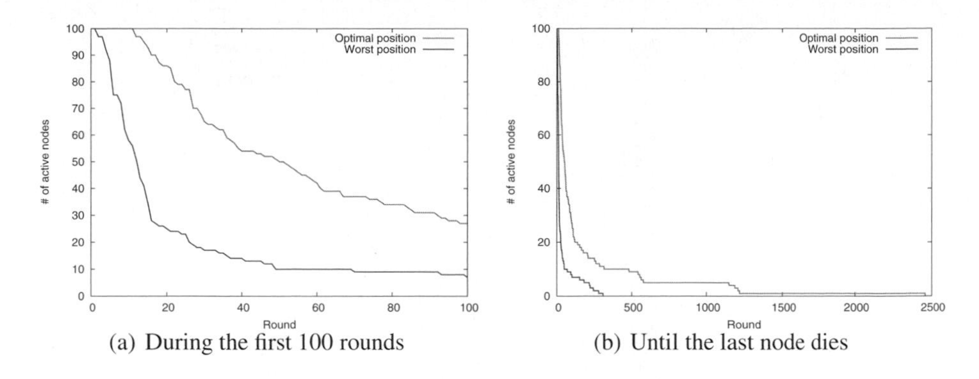

(a) During the first 100 rounds

(b) Until the last node dies

Fig. 5. Energy consumption with a single BS

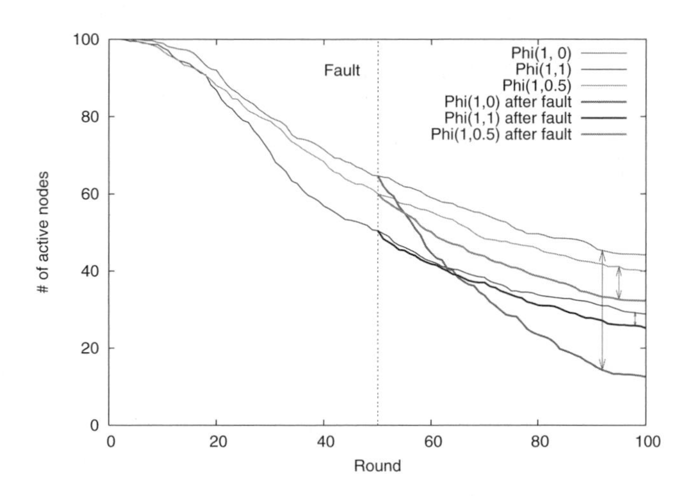

Fig. 6. Energy consumption with 2 BSs, a fault on one BS when round=50

Also when the BS is at its worst position, the last sensor dies within the 300th round. However, the final node dies near the 2500th round when the BS is place at a position with minimum E, showing performance of 800% than when E is maximized (Fig.5 (b)). As expected, placing the first BS to the position where E is minimized decreases the energy consumption of sensors and increases the network lifetime. Through Fig.5 it is seen that considering E as a metric is a necessity.

Second, we simulate the energy consumption of the network with two BSs and plot the number of sensor nodes that remain active for 100 rounds in Fig.6. The simulation environment is equivalent to the simulation done for the network with a single BS in Fig. 5. However, to measure fault tolerance, we assume there is a BS failure in the 50th round and only one BS can be active from that point. The network with two BSs placed at the $\Phi(1,0)$ position shows the best result among the three conditions when there is no faulty BS in the network (round 0 to 50). This result is what we expected, since the smaller the weight factor ω becomes, the more energy-efficient the network is.

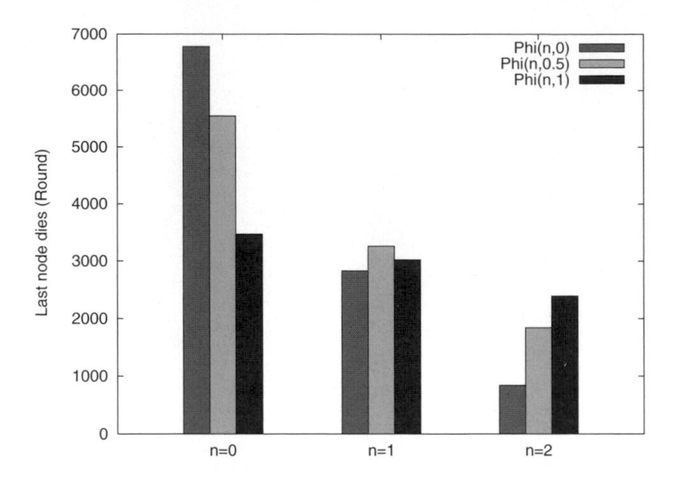

Fig. 7. Network lifetime with three BSs

However, the result becomes quite different when we start to consider faults in BSs. The arrows indicate the difference in the number of active nodes when a network consists of a BS with fault to a network with all BSs active. Since $\Phi(1,0)$ only considers energy efficiency and does not consider fault tolerance at all, $\Phi(1,0)$ shows the widest gap. Consequently when using $\Phi(1,0)$, we can imply that a BS failure in the network can rapidly decrease the network lifetime. The results of $\Phi(1,0)$ gets even worse than the network using $\Phi(1,1)$. On the contrary, the network using $\Phi(1,1)$ is the most fault-tolerant of all; since $\Phi(1,1)$ only considers fault tolerance, despite of the losses in energy efficiency. The network using $\Phi(1,0.5)$ shows moderate decrement in the number of nodes remaining active after a BS failure. The energy consumption without BS failure is not the best of the three, but approximates to the result of $\Phi(1,0)$. Also, the additional energy consumption due to a BS failure shows far better performance than the result of $\Phi(1,0)$. Although the numbers fall it still remains higher than that of $\Phi(1,1)$. Therefore, $\Phi(1,0.5)$ shows the most desired result between the three conditions in the network on a comprehensive basis.

Third, we perform the simulation of energy consumption in the network with three BSs and plot the round when the last node dies, showing the the network lifetime. Fig.7 shows the network lifetime for each metric: $\Phi(f,0)$, $\Phi(f,1)$ and $\Phi(f,2)$. We vary f, the number of faulty BSs, from 0 to 2, to identify the decrement of network lifetime by the number of faulty BSs. When using $\Phi(f,0)$, the network lifetime for the network without BS faults is the longest compared to the other two cases. Yet, the lifetime decreases rapidly by BS failures. On the contrary when using $\Phi(f,1)$, the network lifetime without fault is not desirable but fault tolerance is the best of the three metrics. This can be implied from the fact that the graph for $\Phi(f,1)$ does not change severely in any of the three cases (f=0, f=1, and f=2). The balanced metric $\Phi(f,0.5)$ reduces the additional energy consumption by BS failure and increases the network lifetime better than $\Phi(f,0)$. $\Phi(1,0.5)$ improves the network lifetime up to 15% than $\Phi(1,0)$, and the lifetime of $\Phi(2,0.5)$ is over 200% of $\Phi(2,0)$.

4 Conclusion

In this paper we propose an energy-efficient and fault-tolerant method in positioning multiple base stations for wireless sensor networks. Through our simulations, we show that by considering both energy-efficiency and fault-tolerance in selecting the optimal position for multiple base stations can increase the total network lifetime. This is compared to other work that have only considered energy-efficiency and fault-tolerance separately. In this work we have used ω as 0.5 to give equal balance to energy-efficiency and fault-tolerance. Yet, in our future work we plan to find the optimal value for ω. We expect the optimal value to vary in various situations, so many different situations will be considered in our future work. Also we plan to expand our work in a dynamic wireless sensor network where each sensor node has mobility.

Acknowledgement

This work was supported in part by the ITRC program of the Korea Ministry of Information & Communications, and grant No. R01-2006-000-10510-0 from the Basic Research Program of the Korea Science & Engineering Foundation.

References

1. Gandham, S.R., et al.: Energy efficient schemes for wireless sensor networks with multiple mobile base stations. In: Proc. IEEE GLOBECOM (2003)
2. Vaas, D., Vidacs, A.: Positioning mobile base station to prolong wireless sensor network lifetime. In: Proc. ACM CoNEXT (2005)
3. Zimmermann, K., et al.: Self-management of wireless base stations. In: Proc. IEEE MICMC (2005)
4. Ganesan, D., et al.: Highly resilient, energy efficient multipath routing in wireless sensor networks. In: Proc. ACM MobiHoc (2001)
5. Krishnamchari, B., et al.: The energy-robustness tradeoff for routing in wireless sensor networks. In: Proc. IEEE ICC (2003)
6. Sha, K., et al.: WEAR: A balanced, fault-tolerant, energy-efficient routing protocol for wireless sensor networks. International Journal of Sensor Networks 1(2) (2006)

Reducing Energy Consumption through the Union of Disjoint Set Forests Algorithm in Sensor Networks

Byoungyong Lee, Kyungseo Park, and Ramez Elmasri

Computer Science and Engineering,
The University of Texas at Arlington,
Arlington, TX 76019, USA
{bylee,kpark,elmasri}@uta.edu

Abstract. Recently, wireless sensor networks have improved for many applications aimed at collecting information. However wireless sensor networks have many challenges to be solved. One of the most critical problems is the energy restriction. Therefore in order to extend the lifetime of sensor nodes, we need to minimize the amount of energy consumption. In many cases, sensor networks use routing schemes based on the tree routing structure. But when we collect information from a restricted area within the sensor field using the tree routing structure, the information is often assembled by sensor nodes located on different tree branches. In this case unnecessary energy consumption happens in ancestor nodes located out of the target area. In this paper, we propose the Sensor Network Subtree Merge algorithm, called SNSM, which uses the union of disjoint set forest algorithm for preventing unnecessary energy consumption in ancestor nodes for routing. SNSM algorithm has 3-phases: first finding the disjoint set of the subtree in the sensor field; second connecting each disjoint subtree with the closest node; and third virtually disconnect the subtree connected to new tree branch from previous tree structure. In the simulation, we apply SNSM algorithm to a minimum spanning tree structure. Simulation results show that SNSM algorithm reduces the energy consumption. Especially, SNSM is more efficient as number of sensor nodes in a sensor field increases.

1 Introduction

Advances in MEMS (Micro-Electro Mechanical System) have led to the emergence of wireless sensor networks. Each sensor consists of a processor unit, storage unit, wireless transmission unit, power unit and sensing unit [1,12]. These sensor nodes are spread in a sensor field for measuring the environment. Each sensor node scattered in the sensor field is part of the network. When receiving a query from a user, the base station sends the query to nodes in the target area for collecting the information through the formed network. Because there is no wireless network infrastructure, each sensor node plays a role as either a routing or sensing node. The sensor node being of small size contains many restrictions

T. Vazão, M.M. Freire, and I. Chong (Eds.): ICOIN 2007, LNCS 5200, pp. 594–603, 2008.

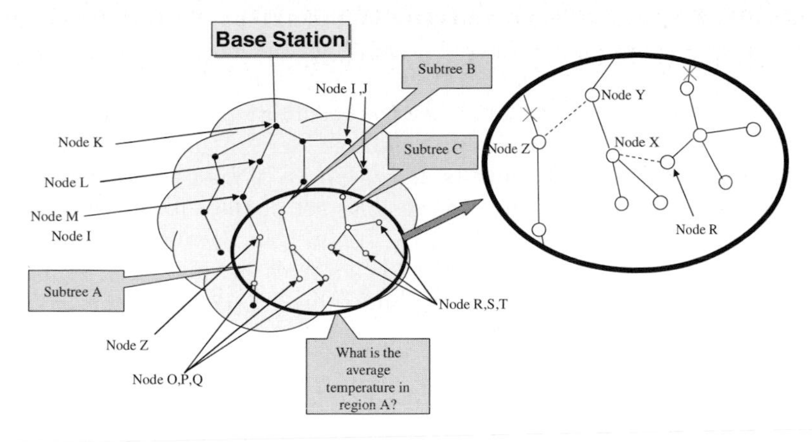

Fig. 1. The example of merging the subtrees

such as limited battery power, low capability of processor, short radio range, and limited storage [4,15]. The energy constraint is one of the most critical problems. It is almost impossible to replace the low level battery in many sensor nodes deployed in sensor fields. If we consider the aspect of energy consumption, we can observe that energy cost for transmitting is large when compared to the data processing cost [11]. Therefore in order to reduce the energy consumption for transmitting messages, many researchers try to find energy efficient techniques such as in-network aggregation [7], clustering [6,8,5,14], various multi-hop routing schemes [2,7,13], and so on.

In the in-network aggregation, there are several advantages for minimizing the communication cost. Especially, partial results that come from children nodes are combined in each intermediate node and then the aggregated results go up to the parent node. This saves considerable energy over the entire sensor network [9].

One method to save energy consumption is through clustering. It can also be used to reduce the energy consumption for sending the messages. One cluster head collects the sensing data from neighborhood nodes and then transmits to the parent node or base station.

Clustering methods and in-network aggregation work in tandem with the routing schemes for the wireless sensor network. Usually, single-hop routing schemes, in which each sensor is directly connected to the base station, needs more energy than multi-hop routing schemes, in which sensor nodes are connected to the base station through intermediate nodes [2]. This is because energy consumption of transmission is relative to distance. However, even if we use methods such as in-network aggregation and clustering and multi-hop routing schemes, unnecessary energy can be used for routing. For example, in Fig. 1, if a base station receives a spatial query like "What is the average temperature in region A", we have to use several routing subtrees to access the target area A. In this case, some ancestor nodes are used for routing unnecessarily. In the case of using the subtree B for routing, if we connect the subtrees A, B, and C to each other within the region

A, we do not need to use the nodes L and M and also ancestor nodes of subtree C such as nodes I and J for routing. Therefore we can reduce the overall energy consumption.

For preventing the unnecessary energy consumption in ancestor nodes for routing, we propose the Sensor Network Subtree Merge algorithm, called SNSM, using the union of disjoint set forests algorithm [3]. SNSM algorithm has three phases. In the first phase, we find the disjoint set of the subtrees in the target area. In Fig. 1, there are three subtrees A, B and C in the target area A. Hence through phase 1, we recognize the disjoint subtrees in the target area. In the second phase, we try to connect each disjoint subtree with its closest node in the target area. In the third phase, we disconnect any subtrees connected to a new tree branch from the previous tree structure.

The remainder of this paper is organized as follows. In section 2, we provide an overview of our network model. In section 3, we introduce the three phases of SNSM algorithm. The performance study is reported in section 4. Finally, section 5 presents concluding remarks.

2 Network Model

2.1 Radio Model

When transmitting and receiving the sensing data, each sensor node consumes energy. Therefore in order to measure the energy consumption in sensor network, we use the first order radio model presented in LEACH [5]. In this radio model, transmitter or receiver utilizes E_{elec}=$50nJ/bit$ and there is a transmit amplifier defined as ϵ_{amp}=$100pJ/bit/m^2$. It also assumes the radio channel to be symmetric, which means the cost of transmitting a message between node A and node B is same bidirectionally. This radio model calculates the energy used for k-bit message to be sent over a distance 'd' as:

$$E_{Tx}(k,d) = E_{elec} * k + \epsilon_{amp} * k * d^2 \tag{1}$$

$$E_{Rx}(k) = E_{elec} * k \tag{2}$$

In formulas (1) and (2), E_{Tx} is the energy used for transmission and E_{Rx} is the energy used for receiving k bits of data. The transmission energy is dependent on the distance parameter 'd'. The energy will be increasing at a high rate as the distance increase. Hence as distances increase, multi-hop routing structure consumes less energy than single-hop routing.

2.2 Preliminaries

In the sensor network, a routing structure corresponds to undirected graph G = (V , E). V is defined as the set of sensor nodes, $V = \{n_1, n_2, n_3, n_i\}$. E is the set of edges. Each sensor node n_i $\in$$V$ can send the sensing data within range of a radius denoted by R. If the distance between node n_i and n_j (denoted as d(n_i,n_j)) is within R, the edge between nodes n_i and n_j is defined as $e_{ij}$$\in$E. In

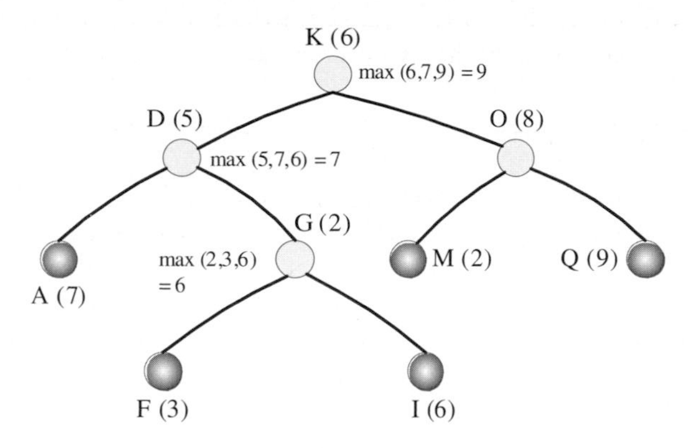

Fig. 2. Example of In-network aggregation subtrees

this case, the weight w(n_i,n_j) denotes the cost to connect n_i and n_j. In our sensor network, we apply SNSM algorithm to a minimum spanning tree structure used for routing. In this work, we assume the following properties for the suggested algorithm:

- The sensor nodes distributed over a geographical area are homogeneous and each has a unique node id.
- Each sensor node is aware of their position with the GPS.
- All sensor nodes can send a message to the base station via multi-hop routing and control the power of their radio range transmission depending on the distance

2.3 In-Network Aggregation

In wireless sensor network, constrained energy is one of the critical problems. The advantages of using the in-network aggregation are to minimize energy consumption and incur no approximation error [9]. In in-network aggregation, each node computes the query in its own place, and produces a local result. For example, if base station receives the MAX aggregation query, each node receives the max value of the sensing data from children nodes, then applies its sensing data to the max value and sends the result to the parent node. In Fig.2, sensor node G aggregates the data from the sensor nodes F and I. Also sensor node O aggregates from sensor nodes M and Q, and D from A and G. Finally the root node K aggregates data from nodes D and O. Hence in-network aggregation reduces the cost of transmission. Also, the value of max would be exact. In order to send and receive messages, all nodes are synchronized [7]. For example, when node G has two children, node G is allotted enough time period, called *epoch* in TAG [7], for receiving sensing data from nodes F and I. If node G is not allotted enough time, node G sends the result to its parent without receiving the sensed data from nodes F or I. Therefore query response time is affected by *epoch* duration. Also *epoch* duration is dominated by the depth of the routing tree.

Algorithm 1 : Finding the disjoint subtree in target area.

Input : position of target area.
1: *find the leaf node in the target area*
2: **for** *each* $n_i \in V[G]$
3: **if** $n_i \in$ *leaf node of target area*
4: **then** *leaf*$[n_i]$ ← leaf *node of target area*
5: **for** *each* $n_i \in$ *leaf*$[V]$
6: **while** (*leaf*$[n_i]$ *has parent* &&
P(*leaf*$[n_i]$) *is in the target area*)
7: **do** *leaf*$[n_i]$ ← *P*(*leaf*$[n_i]$
8: *root_node*$[n_i]$ ← *leaf*$[n_i]$
9: **return** *root_node*$[n_i]$
Output : root node of each subtree in target area.

3 The Sensor Network Subtree Merge Algorithm

In this section, we describe the sensor network subtree merge algorithm, called SNSM, using the union of disjoint set forest algorithm. The three phases of the algorithm are described in the following three subsections.

3.1 Phase-1: Finding the Disjoint Subtrees for a Given Range Query

In Phase-1, we describe Finding Disjoint Subtrees algorithm, called FDS algorithm, based on the union of disjoint-set forests algorithm. After the sensor nodes are distributed in the sensor field, an initial routing tree is formed for sending the initial information of each sensor node to the base station. The initial information of each sensor node is denoted by D = {position, unique id, id of neighborhood nodes}. The base station constructs the minimum spanning tree for routing from the initial information D received from all the sensor nodes. This minimum spanning tree becomes the basic tree structure for routing.

Once a spatial range query is submitted to the base station, the base station recognizes the disjoint subtrees within the target area of the query through the FDS Algorithm (Algorithm1). For example, in the FDS algorithm, if a base station receives the following spatial query from a user: "What is the average temperature in region A", the base station finds the sensor nodes within the target area A. leaf$[n_i]$Then the base station finds the leaf nodes within the target area (Line 1). Even if some sensor nodes have children nodes out of the target area, they become leaf nodes for this particular query. For example, in Fig.1, nodes O, P, Q, R, S, T are leaf nodes in target area A. In lines 2 - 4, we save the leaf nodes to array. In line 6, P(leaf$[n_i]$) means parent of leaf node n_i. Also root_node$[n_i]$ is the root node of a subtree (Line 8). After finding the leaf nodes, we find the root nodes of the subtrees (Lines 5-9). If a leaf node has a parent node, its parent node become a leaf node recursively until we find a root node of a subtree within the target area (Lines 6-8). In Fig. 1, the node Z become a root node of subtree

Algorithm 2 : Finding the closest node over different subtree branch.

Input : Set of node in each subtree
1: **for** $n_i \in S_m$
2: **for** $n_j \in S_n$
3: **do** *min_distance*$_i$ (n_i, n_j)
4: **if** *min_distanc*$_{i-1}$ > *min_distance*$_k$
5: **then** *subtree_connector1* ← n_i
6: *subtree_connector2* ← n_j
7: **connect** (*subtree_connector1, subtree_connector2*)
Output : Connection of the two nodes with the closest distance over different subtree

A within the target area. Therefore in Fig.1, we find three root nodes in the target area A through the Algorithm 1.

3.2 Phase-2: Finding the Closest Node over Different Subtree Branches

In Phase-2, we illustrate how to find the closest node in different subtree branches and connect the closest nodes to each other. Let all sensor nodes in subtree 'S_i' be the set $S_i = \{ n_1, n_2, n_3, ... , n_i \}$. For example, in Fig.1, there are three subtree sets S_a, S_b, and S_c. In this case, the distance of node Z in subtree A and node Y in subtree B is the closest. Also, the distance of node X in subtree B and node R in subtree C is the closest. Hence through Algorithm 2 we find the nodes which are connected with the closest distance in each pair of subtrees, such as node Z, Y, X, and R. Then, we connect node Z to node Y and node R to node Y. In Algorithm 2, n_i and n_j are nodes that belong to subtree S_m and S_n respectively (Line 1-2). In line 3, the min_distancei(n_i, n_j) means minimum distance between n_i and n_j. The minimum distance is defined as follows

$$min\_distance_i(n_i, n_j) = \{n_i \in S_m, n_j \in S_n \mid min\{distance(n_i, n_j)\}\} \quad (3)$$

In lines 5-6, subtree_connector1 contains a node of the subtree S_m and subtree_connector2 has a node of the subtree S_n. If we connect these two nodes, the distance of the two subtrees become the closest distance (Line 7).

3.3 Phase-3: Disconnecting a Subtree from the Previous Tree Structure

In this subsection, we show how to disconnect a subtree from the previous tree structure. The base station sends the changed routing information to subtree connectors such as Z, Y, X and R in Fig.1. Among the subtrees in the target area, we chose only one subtree for routing. Therefore other subtrees send the sensing data to subtree through the subtree connectors. Also we have several criteria to select the subtree for routing in the target area. For example, in

Fig.1, because subtree B has a smaller number of ancestor node than subtree C in its path to the base station as well as a longer depth in the target area A than subtree A, node Y becomes the root node in the target area A. For selecting the main subtree in a particular query target area, we check the number of ancestors and then if we have same number of ancestor, we check the depth of subtree. The subtrees A and C become the new branches of subtree B. Connection between node Z and node M is cut off for connecting between node Z and node Y.

4 Simulation

In this section, we present a simulation environment for SNSM algorithm and the results of the simulation. We are interested in studying the energy efficiency for spatial query routing. Thus, we evaluated the performance of the SNSM algorithm with the following metrics: 1) energy consumption of the spatial query in sensor fields, 2) effect of sensor density, and 3) effect of range for spatial query area. We compare SNSM algorithm with minimum spanning tree used by many researchers [10].

In the experiments, homogeneous sensors are deployed in a 500 X $500m^2$ sensor field area. We simulate 4 cases: 50 sensors/500X$500m^2$, 100 sensors/500X $500m^2$, 150 sensors/500X$500m^2$, 200 sensors/500X$500m^2$. For the radio range, all nodes can control the power of radio range. After connecting the disjoint set, each sensor in the target area transmits sensed data to the base station. Also for measuring the energy consumption for transmitting and receiving data, we used the LEACH energy model [5], using radio electronics energy $50nJ/bit$ and radio amplifier energy $100pJ/bit$. We assume that the amount of energy in each node is considered as 1 Joule and the packet size is 1500 bits.

4.1 Effect on Energy Consumption for Spatial Query

In a sensor network, it is important to reduce the energy consumption of each sensor node. If some sensor nodes die earlier than others, even if other sensor nodes have enough energy, the entire sensor network structure can collapse

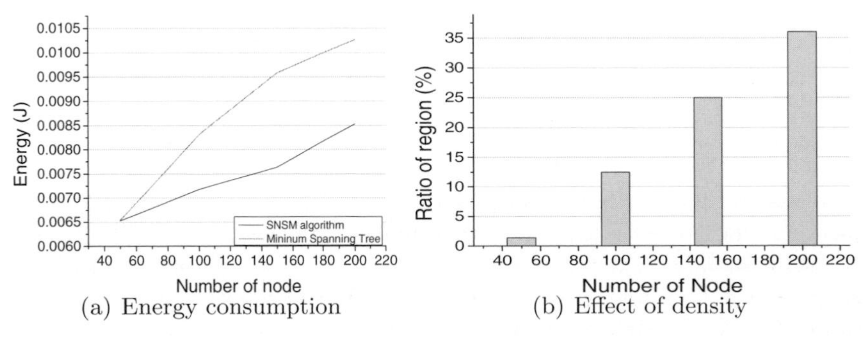

(a) Energy consumption (b) Effect of density

Fig. 3. These graphs show the energy (a), ratio of region on the number of node(b)

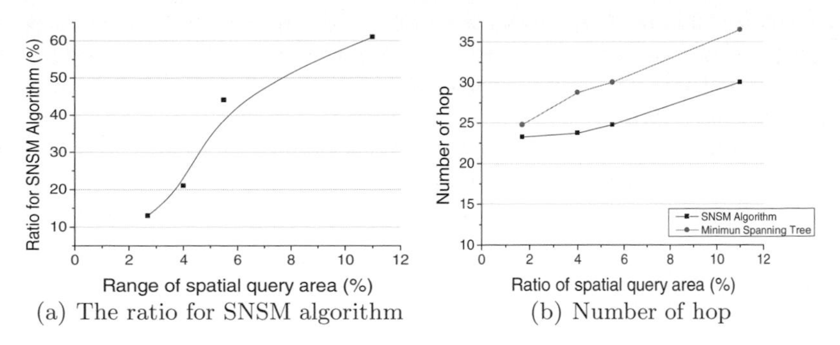

(a) The ratio for SNSM algorithm (b) Number of hop

Fig. 4. These graphs show the efficiency on the ratio of a special query area

rapidly. In this experiment, we compared SNSM algorithm with minimum spanning tree using the parameter of energy consumption of a spatial range query in the sensor field. For our experiment, we randomly chose the target area for the spatial query. The ratio of the query target area to the total sensor field area is 11%. In Fig.3 (a), as we increase the number of sensor nodes, energy consumption of total sensor nodes for spatial range query also increases. After we apply SNSM algorithm to minimum spanning tree, we obtain the result that energy efficiency of the merged tree is higher than minimum spanning tree structure. Especially, efficiency of energy consumption improves as the number of nodes increases. Therefore SNSM algorithm has better performance with a large number of sensor nodes than with a small number of sensor nodes.

4.2 Effect on the Density of Sensor Nodes

In this experiment, we measured percentage of query regions to which the SNSM algorithm changes the tree structure as the total number of nodes increases. In Fig 3. (b), when we spread 50 sensor nodes in the 500 X $500m^2$, SNSM algorithm improves performance in 1.38% out of the total sensor field area. But in the environment having 200 sensor nodes, the ratio of using SNSM increased to 36%. Therefore we can obtain the result that SNSM algorithm performs better in the environment that has high density of sensor nodes.

4.3 Efficiency on Ratio of Spatial Query Area

In the third experiment, we measured ratio for using SNSM algorithm and number of hops as we increased the range of the spatial query area. In Fig.4 (a), we show the result of this simulation. In this result, as we increase the region of spatial query area, this increases the benefits of SNSM algorithm. For example, consider a spatial query like "What is the average temperature in region A". When the area of region A is 2.7% of entire sensor area, the ratio of using SNSM algorithm is 13%. Then, as we increase the area of region A to 11%, the ratio of using SNSM algorithm increases to 61%. In the case of number of hops, as we increased the range of spatial query area, the number of hops also increases because of the larger target area, and the larger number of sensor nodes. In Fig.4 (b), the number of hops in SNSM algorithm is less than minimum spanning tree structure. Therefore SNSM has better performance than minimum spanning tree structure.

5 Conclusion

In this paper, we have described SNSM algorithm based on the union of disjoint set algorithm when applied to minimum spanning tree. Also SNSM algorithm works on the in-network aggregation schemes for sensor networks. We reduce the energy consumption for routing in sensor network for spatial range query through the SNSM algorithm. In the simulation, we applied SNSM algorithm to minimum spanning tree. If there are other kinds of routing tree structures, SNSM algorithm can also be applied to those. As we mention in the simulation section, SNSM algorithm improves energy consumption in sensor networks with tree structures because we remove the redundant energy consumption in ancestor nodes for routing.

References

1. Al-Karaki, J.N., Kamal, A.E.: Routing techniques in wireless sensor networks: a survey. Wireless Communications, IEEE [see also IEEE Personal Communications] 11(6), 6–28 (2004)
2. Boukerche, A., Cheng, X., Linus, J.: Energy-aware data-centric routing in microsensor networks. In: Proc. 6th ACM Intl. Workshop on Modeling, Analysis and Simulation of Wireless and Mobile Systmes (MSWiM), San Diego, CA, September 2003, ACM, New York (2003)
3. Cormen, T.H., Leiserson, C.E., Rivest, R.L., Stein, C.: Introduction to Algorithms, 2nd edn. The MIT Press, Cambridge (2001)
4. Deshpande, A., Guestrin, C., Madden, S., Hellerstein, J., Hong, W.: Model-driven data acquisition in sensor networks (2004)
5. Heinzelman, W.R., Chandrakasan, A., Balakrishnan, H.: Energy-efficient communication protocol for wireless microsensor networks (2000)
6. Li, Q., Aslam, J., Rus, D.: Hierarchical power-aware routing in sensor networks (May 2001)
7. Madden, S., Franklin, M.J., Hellerstein, J.M., Hong, W.: Tag: A tiny aggregation service for ad-hoc sensor networks (2002)
8. Manjeshwar, A., Agrawal, D.P.: Teen: Arouting protocol for enhanced efficiency in wireless sensor networks, p. 189 (2001)
9. Manjhi, A., Nath, S., Gibbons, P.B.: Tributaries and deltas: Efficient and robust aggregation in sensor network streams (June 2005)
10. Nath, S., Gibbons, P.B., Seshan, S., Anderson, Z.R.: Synopsis diffusion for robust aggregation in sensor networks, pp. 250–262 (November 2004)
11. Shnayder, V., Hempstead, M., Rong Chen, B., Allen, G.W., Welsh, M.: Simulating the power consumption of large-scale sensor network applications, pp. 188–200 (2004)

12. Wen, C.-Y., Sethares, W.A.: Automatic decentralized clustering for wireless sensor networks. EURASIP J. Wirel. Commun. Netw. 5(5), 686–697 (2005)
13. Yao, Y., Gehrke, J.: The cougar approach to in-network query processing in sensor networks. SIGMOD Record 31(3), 9–18 (2002)
14. Younis, M., Youssef, M., Arisha, K.: Energy-aware routing in cluster-based sensor networks, p. 129 (2002)
15. Younis, O., Fahmy, S.: Heed: A hybrid, energy-efficient, distributed clustering approach for ad hoc sensor networks. IEEE Transactions on Mobile Computing 3(4), 366–379 (2004)

Packet Delay and Energy Consumption Based on Markov Chain Model of the Frame-Based S-MAC Protocol under Unsaturated Conditions

Seokjin Sung, Seok Woo, and Kiseon Kim

Department of Information and Communications,
Gwangju Institute of Science and Technology (GIST),
1 Oryong-dong, Buk-gu, Gwangju, 500-712, Republic of Korea
{ssj75,swoo,kskim}@gist.ac.kr

Abstract. In this paper, we analyze packet delay and energy consumption of the sensor-medium access control (S-MAC) protocol under unsaturated conditions. Since the S-MAC protocol always behaves with a fixed frame length, the frame-based architecture should be reflected in the S-MAC analysis. In addition, the node contention for transmitting a packet is also considered for more practical analysis on packet delay than that in the original paper suggested by Ye et al. [2]. Hence, we employ a Markov chain model to express the S-MAC behavior for the consideration of back-off delay, including node contention. Numerical results show the average packet delay and energy consumption of the node according to offered load and duty cycle where a practical mote running S-MAC is used.

1 Introduction

Generally, because many sensor motes operate using limited-life batteries in wireless sensor fields, refined techniques to reduce energy consumption have been required, and many schemes have been proposed at the MAC-level [1]-[3]. In particular, the sensor-medium access control (S-MAC) protocol [1], one of the famous MAC protocols designed for wireless sensor networks (WSNs), is a classic. Nevertheless, it gives a good illustration for the frame-based MAC protocols applying sleep-scheduled techniques in WSNs.

The S-MAC protocol basically gains energy efficiency by using a periodic sleep interval determined by the duty cycle (DC), the ratio of listen interval to frame length [2]. Although the idea is to give reduced energy consumption of deployed nodes, the added sleep duration increases the time delay for transmitting a packet [1], [2]. In addition, because the S-MAC protocol uses contention-based channel access, the packet delay will also increase by contending nodes. Recently, analytic evaluations of S-MAC performance reflecting the behaviors of contending nodes have been presented [4], [5]. However, the analytic forms are expressed based on the analysis result of the IEEE 802.11 MAC protocol under saturated conditions [6]. That protocol has a slot-based architecture in which the distance between two consecutive distributed interframe spaces (DIFS) may vary

T. Vazão, M.M. Freire, and I. Chong (Eds.): ICOIN 2007, LNCS 5200, pp. 604–613, 2008.

depending on idle, collision, or successful transmission. In the frame-based S-MAC protocol, the frame is composed of a set of slots, and the distance between two consecutive DIFS is always fixed [8]. Because each node regularly repeats listen/sleep intervals, the nodes in one-hop should follow the same schedule, and they compose one virtual cluster to transmit their own packets to a relay or sink node. Therefore, its frame-based architecture is one significant feature in S-MAC modeling and performance analysis. Moreover, it is important to consider unsaturated conditions because the variation of offered load affects the energy consumption as well as the packet delay by the node contention.

In this paper, we analyze the packet delay and the energy consumption of the S-MAC protocol considering node contention under unsaturated conditions. For the analysis, we first employ the Markov chain model for one node running S-MAC [8]. Then, based on the S-MAC architecture with fixed frame length, we derive the delay and energy consumption for a successful packet transmission in one-hop under unsaturated conditions. Finally, the numerical results are shown according to the variation of DC and offered load where a practical mote is applied.

2 Architecture of the Frame-Based S-MAC Protocol

Although each node running the S-MAC can select its own schedule in multi-hop WSNs, the nodes within the same virtual cluster follow the same listen/sleep schedule to transmit their own packets to a relay or sink node. Thus, all nodes in one-hop always contend to access a shared channel, and the S-MAC protocol defines the access mechanism based on the 802.11 distributed coordination function, which includes physical and virtual carrier sensing, back-off (BO) and retransmission, as well as the request-to-send/clear-to-send (RTS/CTS) method [1].

Even though the channel access mechanism of the S-MAC protocol is based on that of the IEEE 802.11 MAC protocol, there exists a significant difference in architecture between the two protocols: a frame architecture with a fixed length is used in the S-MAC protocol whereas the IEEE 802.11 MAC protocol has a slot-based architecture [8]. That means that in the S-MAC architecture, the length of the slot including the transmitted RTS packet depends on how many idle slots occur before the RTS-transmission within the frame.

Fig. 1(a) and Fig. 1(b) show examples of frames where the collision of RTS packets occurs and exchanges of RTS/CTS and data/ackknowledgment (ACK) between a sender and a receiver are successfully accomplished, respectively. DIFS and short interframe spaces (SIFS) are also presented. From the figures, we note that the colliding slot length T_{col} or the successful slot length T_{suc} depends on the number of idle minimum slots that occur within each frame. Moreover, under unsaturated conditions, the number of idle minimum slots will also depend on the packet generation probability of each node. Therefore, this frame-based behavior, including the unsaturated conditions, should be reflected in the performance analysis of the S-MAC protocol.

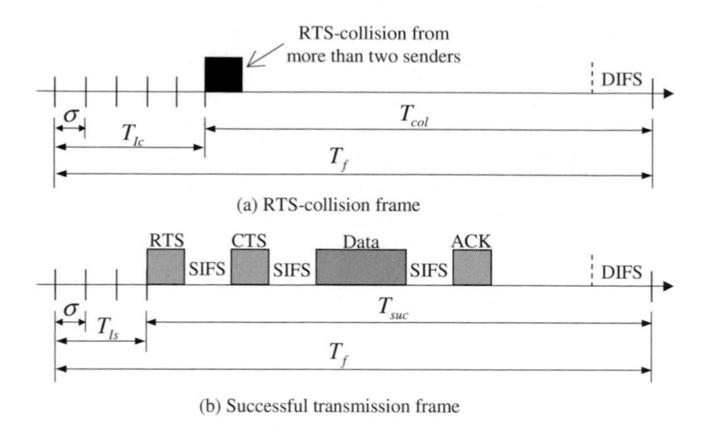

Fig. 1. Examples of colliding and successful frames in the S-MAC protocol

3 Analysis of Packet Delay and Energy Consumption

In this section, we analyze the time delay and energy consumption for transmitting one packet under unsaturated conditions. We consider a virtual cluster where M senders and one receiver follow the same schedule in one-hop. The key notations in this analysis are described in Table 1.

3.1 Preliminary Work

For this analysis, a discrete-time Markov chain for a sender running the S-MAC, as shown in Fig. 2, is used under the following assumptions [8]: (i) the channel is ideal without hidden terminals and capture [6] (ii) SYNC duration is omitted; (iii) each node can buffer only one packet for transmission in a queue; (iv) a data frame time T_{da} is fixed for all data packets; (v) propagation delay is negligible; (vi) packet arrivals follow a geometric distribution [7]; (vii) transmission error occurs only due to RTS packet collision [6], [7]; and (viii) the transmitted packet collision probability p_c is constant and independent of the BO procedure [6], [7]. In this model, $B(i,k)$ denotes each state in the BO procedure: the integers i and k indicate the BO stage and counter, where $i \in (0,m)$ and $k \in (0, W-1)$ for S-MAC, respectively. I presents the idle state.

Table 1. Notations for analysis

Notation	Description
σ	minimum slot length
T_f	frame length
T_{Ic}	idle length for carrier sensing within a colliding frame
T_{Is}	idle length for carrier sensing within a successful frame
$T_{R/A}$	length from RTS packet to ACK packet within a successful frame
W	fixed BO window size

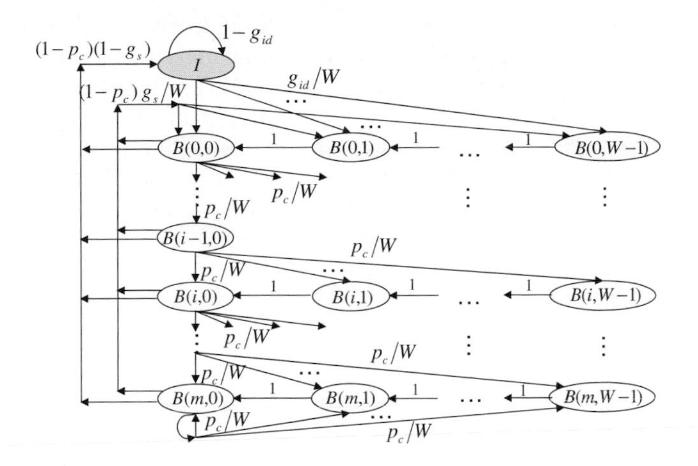

Fig. 2. Markov chain model for one sender running S-MAC

Although the Markov chain is the same model as that presented in [7], we should note that the probabilities that a node generates a new or rescheduled packet during an idle slot length, equal to T_f, and T_{suc} are given by $g_{id} = \min[1,\ T_f G/(T_{da}M)]$ and $g_s = \min[1,\ T_{suc}G/(T_{da}M)]$ where G is given by data packets per T_{da}, because every behavior in the S-MAC protocol is accomplished in the unit of the frame. Additionally, based on the frame-based S-MAC, we can express $T_{suc} = T_f - E[T_{Is}]$, where $E[T_{Is}]$ is given by [8]

$$E[T_{Is}] = \frac{\sigma\tau}{1-(P\{I\})^M}\sum_{z=1}^{M}\binom{M}{z}[1-P\{I\}]^z[P\{I\}]^{M-z}\frac{z(1-\tau)^{2z-1}}{[1-(1-\tau)^z]^2}, \quad (1)$$

where $P\{I\}$ is the stationary distribution of state I in steady state, given by

$$P\{I\} = \frac{2(1-p_c)(1-g_s)}{2(1-p_c)(1-g_s)+g_{id}(W+1)}, \quad (2)$$

and the conditional probability τ that a node transmits a packet in a slot, given that the BO procedure has been employed, is represented as $\tau = 2/(W+1)$. Thus, using p_c given by

$$p_c = 1-\left[1-\frac{2g_{id}}{2(1-p_c)(1-g_s)+g_{id}(W+1)}\right]^{M-1}, \quad (3)$$

we can obtain the unique solutions of the probabilities in steady state [8].

3.2 Average Packet Delay

We define the average packet delay between a sender and a receiver as the spent time interval from the time the BO counter of the sender is activated for the packet transmission, until an ACK message is received following successful transmission. When a target event occurs in a sensing area, the nodes detecting the event will send their own packets continuously, because the packets will be generated by the continuous event-detection. On the other hand, the nodes out of the event area will not attend the contention even though they are within the same virtual cluster. Hence, we assume that the number of nodes that attend the contention is maintained during the successful packet transmission of the observation node attempting to transmit its packet.

In order to analyze the average packet delay, careful consideration to two factors is needed: states of $M-1$ nodes except the observation node and the fixed frame length. The former means that the probability that $M-1$ nodes have the packets to be sent should be included, and the latter implies that the feature of the frame-based S-MAC protocol should be reflected. Now, under unsaturated conditions, the average packet delay D is expressed as

$$D = \Pr[I = M-1] \cdot E\big[D|I = M-1\big] + \Big(1 - \Pr[I = M-1]\Big) \cdot E\big[D|I \neq M-1\big], \quad (4)$$

where $\Pr[I = M-1]$ stands for the probability that all $M-1$ nodes are in the idle state, and $E\big[D|I = M-1\big]$ and $E\big[D|I \neq M-1\big]$ denote the conditional expectations of the delays given that $M-1$ nodes are in the idle state and at least one out of $M-1$ nodes has activated the BO procedure, respectively.

First, we can easily know

$$\Pr[I = M-1] = \big(P\{I\}\big)^{M-1}. \quad (5)$$

Second, given that the BO time is uniformly chosen in the range $(0,\, W-1)$, we obtain

$$E\big[D|I = M-1\big] = \sigma \cdot \frac{W-1}{2} + T_{R/A}, \quad (6)$$

because only the observation node will transmit its own packet within the given frame. Finally, considering the condition j_{r-TR}, given that the packet transmission of the observation node succeeds in the j-th retransmission, we have

$$E\big[D|I \neq M-1\big] = \sum_{j=0}^{\infty} E\big[D|I \neq M-1, j_{r-TR}\big] \cdot p_c^j \cdot (1-p_c), \quad (7)$$

where infinite retransmission of the packet is possible. To analyze equation (7), three components should be considered again: (i) the delay term for $j+1$ BO intervals of the observation node, including its initial transmission attempt; (ii) the delay term for j slots in collisions with the RTS packets, which are transmitted from the observation node; and (iii) the delay term for the final successful transmission of the observation node. From the components, we have the form

$$E\big[D\,|I \neq M-1, j_{r-TR}\big] = \sum_{n=1}^{M-1} \frac{\binom{M-1}{n}\big[1-P\{I\}\big]^n \big(P\{I\}\big)^{M-n-1}}{1-\big(P\{I\}\big)^{M-1}} \cdot \bigg[(j+1)E\big[T_{count-ob}|BO=n\big]\frac{W-1}{2} + jE\big[T_{col-ob}|BO=n\big] + T_{R/A}\bigg], \tag{8}$$

where $E\big[T_{count-ob}|BO=n\big]$ and $E\big[T_{col-ob}|BO=n\big]$ present the average length of a slot during the BO count of the observation node before it transmits its own RTS packet, and the average length of the colliding slot including the RTS packet of the observation node, respectively, given that n nodes except the observation node have activated their own BO procedures.

From the viewpoint of the frame-based S-MAC protocol, it is very significant that $E\big[T_{count-ob}|BO=n\big]$ and $E\big[T_{col-ob}|BO=n\big]$ also depend on how many minimum slots occur before the RTS-transmission within a frame. Consequently, they can be derived as

$$E\big[T_{count-ob}|BO=n\big] = T_f \frac{\big[1-(1-\tau)^n\big]}{\big[(1-\tau)^n\big]}\Big(-\ln\big[1-(1-\tau)^{n+1}\big]\Big), \tag{9}$$

and

$$E\big[T_{col-ob}|BO=n\big] = T_f - \sigma\tau\big[1-(1-\tau)^n\big]\frac{(1-\tau)^{n+1}}{\big[1-(1-\tau)^{n+1}\big]^2}. \tag{10}$$

The detailed derivations of equations (9) and (10) are given in Appendix.

Now, inserting equations (9) and (10) into equation (8), and then putting equation (8) into equation (7) again, we can obtain the closed form of the average packet delay by substituting the equations from (5) to (7) for each term in equation (4). Furthermore, simplifying the equation, we have the following final form:

$$D = T_{R/A} + \big(P\{I\}\big)^{M-1}\cdot\sigma\cdot\frac{W-1}{2} + \Big(1-\big(P\{I\}\big)^{M-1}\Big)\cdot\sum_{n=1}^{M-1}\frac{\binom{M-1}{n}\big[1-P\{I\}\big]^n\big(P\{I\}\big)^{M-n-1}}{1-\big(P\{I\}\big)^{M-1}}\cdot \bigg[\frac{E\big[T_{count-ob}|BO=n\big]}{1-p_c}\cdot\frac{W-1}{2} + \frac{p_c}{1-p_c}\cdot E\big[T_{col-ob}|BO=n\big]\bigg]. \tag{11}$$

3.3 Energy Consumption

Based on equation (11), we can express the total energy consumption $\mathcal{E}_{tot}$ of a node for successfully transmitting one packet. It is also derived from the fixed frame length architectural characteristic of the S-MAC protocol. Through a procedure similar to the one described in Subsection 3.2, we can finally obtain

$$\begin{aligned}\mathcal{E}_{tot} = \mathcal{E}[T_{R/A}] + (P\{I\})^{M-1} \cdot \mathcal{E}[\sigma] \cdot \frac{W-1}{2} + \\ \left(1-(P\{I\})^{M-1}\right) \cdot \sum_{n=1}^{M-1} \frac{\binom{M-1}{n}[1-P\{I\}]^n (P\{I\})^{M-n-1}}{1-(P\{I\})^{M-1}} \cdot \\ \left[\frac{\mathcal{E}[T_{count-ob}|BO=n]}{1-p_c} \cdot \frac{W-1}{2} + \frac{p_c}{1-p_c} \cdot \mathcal{E}[T_{col-ob}|BO=n]\right],\end{aligned} \quad (12)$$

where $\mathcal{E}[(\cdot)]$ stands for the average energy consumption during the interval $(\cdot)$. It is specified by the power consumption of the node in receiving, listening, transmitting, and sleeping.

4 Numerical Results

For numerical evaluation, we set $M = 5$ and $W = 20$. The parameters for the S-MAC are also included in Table 2, and power consumptions of sensor mote in receiving, listening, transmitting, and sleeping are given by 14.4 mW, 14.4 mW, 36 mW, and 15 μW, respectively, which follow the values for Mica Motes [2].

Fig. 3(a) and Fig. 3(b) illustrate the average packet delay D and the total energy consumption $\mathcal{E}_{tot}$ of a node according to the offered load G, respectively, for the successful transmission of one data packet. The DC values of 0.1, 0.3, 0.5, 0.7, and 1.0 are chosen. From the figures, it is seen that the packet delay and the total energy consumption increase as the offered load increases until reaching a saturated point, when the same DC value is given. In Fig. 3(a), we confirm that the selection of a lower DC incurs a longer packet delay at the same offered load. The reason is that a lower DC implies a longer frame length because the DC is determined by changing only the sleep duration [2]. In Fig. 3(b), it is noteworthy that the node with the lower DC consumes more total energy to successfully transmit a packet at the same offered load. We should note that it is not the figure of the energy consumption for the same duration. Actually, because the lower DC makes a longer frame length and the probability that

Table 2. Parameters for S-MAC [2]

Parameter	Value
Radio band width	20 kbps
Control packet length	10 bytes
Packet payload length	100 bytes
MAC header length	8 bytes
Duration of listen interval	115 ms
Duration of SIFS	5 ms
Duration of DIFS	10 ms
Duration of minimum slot	1 ms

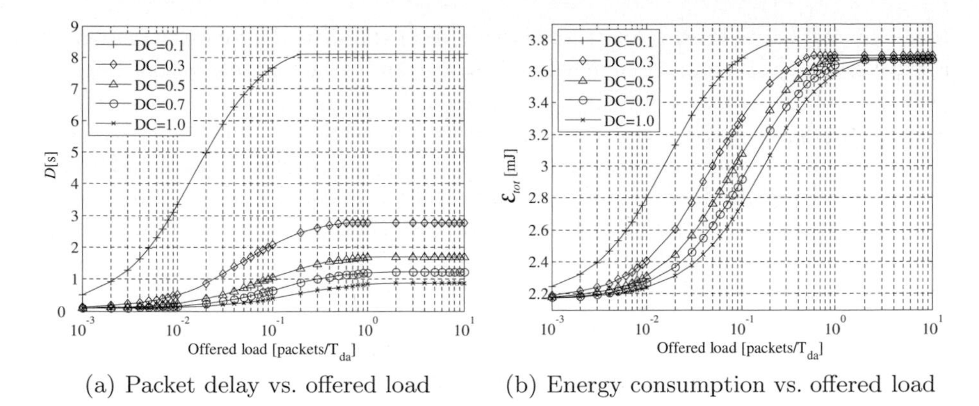

(a) Packet delay vs. offered load (b) Energy consumption vs. offered load

Fig. 3. Packet delay and total energy consumption for a successful packet transmission

each node has a packet to be sent also increases, the packet collision probability increases as the DC value decreases. Thus, $\mathcal{E}_{tot}$ somewhat increases as the DC value decreases. However, it is important that the energy consumption of a lower DC averaged on a fixed interval will be lower than that of a higher DC, because a lower DC induces a longer packet delay (Fig. 4(b)).

Fig. 4(a) and Fig. 4(b) show the relative packet delay D_{rel} and the relative energy consumption $\mathcal{E}_{rel}$ for a successful packet transmission with DC varying from 0.1 to 1.0, presented as $D_{rel} = D/D_{0.1}$ and $\mathcal{E}_{rel} = \mathcal{E}_{\sigma}/\mathcal{E}_{\sigma:0.1}$, respectively, where $D_{0.1}$ stands for the average packet delay in the case of $DC = 0.1$, and $\mathcal{E}_{\sigma}$ and $\mathcal{E}_{\sigma:0.1}$ denote the energy consumption averaged on the minimum slot length σ and its value in the case of $DC = 0.1$, respectively. In Fig. 4(b), the comparison of the relative values from the energy consumption normalized to σ is more suitable to observe network lifetime because total energy consumption $\mathcal{E}_{tot}$ is the value during packet delay varying according to DC. From Fig. 4(a)

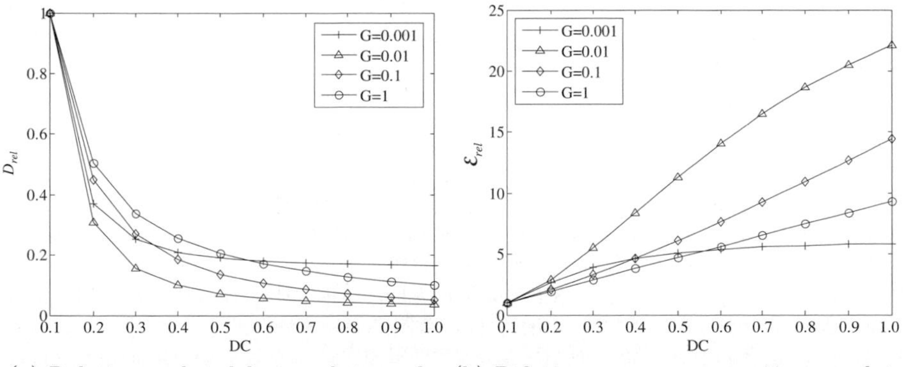

(a) Relative packet delay vs. duty cycle (b) Relative energy consumption vs. duty cycle

Fig. 4. Relative packet delay and relative energy consumption

and Fig. 4(b), we know that the relative packet delay decreases whereas the relative energy consumption increases as the DC value increases. The case of very low offered load (G=0.001) shows that the varying rates of the relative values rapidly become small after DC=0.6, although the DC increases. It is caused by the increase in the probability that only an observation node attempts the packet transmission, because the probability that other nodes generate the packets within one frame decreases due to the relatively short frame length at these DC values. Inversely, the result points out that the selection of a low DC in very low offered load will not guarantee very high energy efficiency, compared to the case of a higher offered load.

5 Conclusion

In this paper, we have analyzed the packet delay and energy consumption of the S-MAC protocol for a successful packet transmission in one-hop. In order to obtain the expressions, a Markov chain model for the S-MAC protocol is first used under unsaturated conditions. Then, we derive the packet delay and energy consumption considering the contending nodes in a shared channel.

The analysis is based on the architectural characteristic of the S-MAC protocol with a fixed frame length, and it reflects the probability that contending nodes attend the BO procedure, depending on the offered load. Hence, by using our analytic forms, the packet delay and the energy consumption of the S-MAC protocol can be numerically evaluated under various parameters, such as DC, offered load, etc.

For a practical mote running the S-MAC, we confirm that increased offered load induces increased packet delay and energy consumption. In addition, it is seen that a high DC leads to improved delay performance, but reduced energy efficiency, where the energy consumption normalized to the minimum slot is compared. In particular, under very low traffic loads, it is shown that the energy efficiency obtained from the choice of a low DC is low.

References

1. Ye, W., Heidemann, J., Estrin, D.: An Energy-Efficient MAC Protocol for Wireless Sensor Networks. IEEE INFOCOM 3, 1567–1576 (2002)
2. Ye, W., Heidemann, J., Estrin, D.: Medium Access Control with Coordinated Adaptive Sleeping for Wireless Sensor Networks. IEEE Trans. Networking 12, 493–506 (2004)
3. Dam, T.V., Langendoen, K.: An Adaptive Energy-Efficient MAC Protocol for Wireless Sensor Networks. ACM Sensys, 171–180 (November 2003)
4. Tseng, H., Yang, S., Chuang, P., Wu, H., Chen, G.: An Energy Consumption Analytic Model for A Wireless Sensor MAC Protocol. IEEE VTC Fall 6, 4533–4537 (2004)
5. Ramakrishnan, S., Huang, H., Balakrishnan, M., Mullen, J.: Impact of Sleep in a Wireless Sensor MAC Protocol. IEEE VTC Fall 7, 26–29 (2004)

6. Bianchi, G.: Performance Analysis of the IEEE 802.11 Distributed Coordination Function. IEEE J. Select. Areas Commun. 18, 535–547 (2000)
7. Ziouva, E., Antonakopoulos, T.: The IEEE 802.11 Distributed Coordination Function in Small-Scale Ad-Hoc Wireless LANs. Intl. J. Wireless Inform. Networks 10, 1–15 (2003)
8. Sung, S., Kang, H., Kim, E., Kim, K.: Energy Consumption Analysis of S-MAC Protocol in Single-Hop Wireless Sensor Networks. IEEE APCC (August 2006)
9. Chatzimisios, P., Boucouvalas, A.C., Vitsas, V.: Packet Delay Analysis of IEEE 802.11 MAC Protocol. IEE Electronics Lett. 39, 1358–1359 (2003)

Appendix

Let us derive the conditional expectations presented in equations (9) and (10).

First, to obtain $E\big[T_{count-ob}|BO = n\big]$, we note that $T_{count-ob}$ depends on the number of minimum slots within a frame. Because the duration of the minimum slot is geometrically distributed, by considering the probability that the RTS-transmission attempt occurs from at least one among n nodes except the observation node in the $(i+1)$th slot, after no transmission from $n+1$ nodes including the observation node for i minimum slots, we have

$$\begin{aligned} E\big[T_{count-ob}|BO = n\big] &= T_f \sum_{i=0}^{\infty} \frac{1}{i+1} \big[(1-\tau)^{n+1}\big]^i \big[1-(1-\tau)^n\big](1-\tau) \\ &= T_f \sum_{i=1}^{\infty} \frac{1}{i} \big[(1-\tau)^{n+1}\big]^{i-1} \big[1-(1-\tau)^n\big](1-\tau) \\ &= T_f \frac{\big[1-(1-\tau)^n\big]}{\big[(1-\tau)^n\big]} \sum_{i=1}^{\infty} \frac{1}{i} \big[(1-\tau)^{n+1}\big]^i. \end{aligned} \tag{13}$$

Using $\sum_{i=1}^{\infty} a^i/i = -\ln(1-a), \ -1 \le a < 1$, we get the form in equation (9):

$$E\big[T_{count-ob}|BO = n\big] = T_f \frac{\big[1-(1-\tau)^n\big]}{\big[(1-\tau)^n\big]} \Big(-\ln\big[1-(1-\tau)^{n+1}\big]\Big). \tag{14}$$

Second, $E\big[T_{col-ob}|BO = n\big]$ can be also derived by considering the frame-based S-MAC protocol, and it is given by $T_f - E[T_{Ic}]$ as illustrated in Fig. 1. In this case, because we should reflect the probability that the collision including the RTS packet transmitted from the observation node occurs in the $(k+1)$th slot after k minimum slots, it is represented as

$$E\big[T_{col-ob}|BO = n\big] = T_f - \sigma \sum_{k=0}^{\infty} k \big[(1-\tau)^{n+1}\big]^k \cdot \tau \sum_{v=1}^{n} \binom{n}{v} \tau^v (1-\tau)^{n-v}. \tag{15}$$

Now, using $\sum_{k=0}^{\infty} kb^k = b/(1-b)^2, \ |b| < 1$, we obtain the form in equation (10):

$$E\big[T_{col-ob}|BO = n\big] = T_f - \sigma\tau\big[1-(1-\tau)^n\big] \frac{(1-\tau)^{n+1}}{\big[1-(1-\tau)^{n+1}\big]^2}. \tag{16}$$

A Novel Local-Centric Mobility System

Nuno Ferreira[1], Rui L. Aguiar[1,2], and Susana Sargento[1,2]

[1] Instituto de Telecomunicações, Aveiro, Portugal
[2] Universidade de Aveiro, DETI, 3810 Aveiro, Portugal

Abstract. Efficient, fast and seamless mobility are key requirements for the ubiquitous access of mobile nodes in 4G networks. This paper presents a network architecture extension for IPv6 mobility scenarios able to provide fast and seamless handovers between intra and inter-domain networks. Our proposal follows a local-centric mobility architecture capable of reducing the handover times and packet losses, and also minimize the global mobility signaling in 4G scenarios. This paper shows some protocol details and describes how this method can significantly increase the network performance in IP mobility scenarios. Finally, test results obtained from a prototype test-bed are presented, showing negligible packet loss and handover timings.

Keywords: Fast and Seamless Mobility, localized architecture, global mobility signalling reduction, 4G, developing, tests evaluation, performance.

1 Introduction

The increasing complexity of next generation mobile networks (NGN), with multi-mode terminals always best connected, has brought mobility issues into a central role for the future networks. In this context, there are large initiatives, both industry and academia led, that address multiple aspects of mobility (e.g . [1, 10]). With the evolution of IP mobility protocols and architectures, one of the main objectives has become the optimization of the mobility protocols in order to achieve fast and seamless handovers over the cells, and reduce the mobility related signalling on the network infrastructure. Only with increased network performance will be possible for the next generation networks to support handling of real-time and multimedia traffic flows.

Recently, an IETF initiative has led to the acceptance of a hierarchical approach to this mobility problem. The *netlmm* working group [11] has developed a new set of specifications on mobility protocols, which has influenced the development of our work. We aim to develop a new system, optimizing the handover process, reducing mobility timings signalling over the global network. This paper is thus organized as follows. Section 2 discusses (necessarily briefly) previous work. In section 3, we describe our novel architecture, and in section 4 we present its evaluation with a developed prototype. Section 5 presents our conclusions.

T. Vazão, M.M. Freire, and I. Chong (Eds.): ICOIN 2007, LNCS 5200, pp. 614–628, 2008.

2 Previous Work

Many protocols were developed trying to overcome the early identified faults of the Mobile IP protocols, most specially the lack of performance during handovers. Note that with the advent of NGN, other shortcomings became apparent, such as QoS support, interaction with authentication, authorization, accounting and charging (AAAC) processes, and security support. With NGN, the reference mobile protocol became Mobile IPv6 (MIPv6) [1].

Protocols developed over MIPv6, like Hierarchical Mobile IPv6 (HMIPv6), Cellular IPv6 (CIPv6), and Fast Mobile IPv6 (FMIPv6), had substantial difficulties to solve all the identified problems.

HMIPv6 [2] is a mobility protocol based on hierarchical relations between mobility agents called Mobility Anchor Points (MAP). In this type of architecture the Mobile Host can acquire a Regional Care-of-Address (RCoA) that enables it to move on the same region without changing its global mobility Care-of-Address (CoA). This procedure addresses a better way to handle the mobility signalling problem providing better performances during the handover and reducing the global mobility signalling needs. This occurs because the Mobile Host does not need to send a Binding Update to its Home Agent after a handover between Access Routers (AR) in the same HMIP region. However, the HMIPv6 compels the Mobile Host to acquire a new RCoA after all the handovers, and a global mobility Care-of-Address after an inter-region handover. Still, HMIPv6 reduces the global signalling during the handovers, but harms the handover timings when the Mobile Host moves between different HMIP regions. This occurs because of the time wasted during the update of all the MAPs on the network.

FMIPv6 [3] is a mobility protocol that aims to accelerate the handover procedure allowing the reduction of handover time. Opposite to HMIPv6, the FMIPv6 architecture does not organize its agents in regions, but reduces the effective packet loss during handover. The FMIPv6 integrates a mechanism of predictive handover that enables the network to prepare the new AR to the Mobile Host arrival. Thus, the Mobile Hosts context is transferred between the old AR and new AR and the Mobile Hosts are prepared and pre-configured to be compliant with the new network configuration. This predictive mechanism allows the seamless handovers between ARs and avoids the Mobile Host configuration time right after the handover. However, FMIP is an enhancement of MIP with fast and predictive handovers but this is not enough to fulfil the localized mobility requirements: when a Mobile Host moves between ARs, it always need to acquire a new CoA, and therefore, send a Fast Binding Update (FBU) to its Home Agent. This fact compels the Mobile Host to reconfigure its IP configuration increasing the blackout time. Moreover, this also increases the signalling traffic in the global network and in the access network wasting core network resources and radio resources.

HMIPv6 with Fast Handovers [12] is an extension of classical HMIP with fast handover capabilities. This enhancement provides the ability of predict the handover inside and outside the HMIP regions. Thus, it is possible to prepared the new AR with the Mobile Host context and also pre-configure the Mobile Host to be compliant with the new IPv6 network configuration. These improvements will minimize the handover time between different AR and will almost avoid the blackout time. The combination of HMIP and FMIP is almost perfect; however, as the Mobile Host needs to change its RCoA whenever it moves between ARs, it will send a Fast Binding Update packet every time its IPv6 configuration changes. This procedure will waste network resources, specially radio network resources, and thus, it does not fulfill the "efficient use of radio resources" requirement.

CIPv6 [4] is an extension to the MIPv6 protocol, and its main objective is to provide better results in handover timings while the Mobile Host moves in nearby regions. In CIPv6 the Mobile Host does not change its IPv6 address while it moves between Base Stations (BS). This fact improves the handover procedure since the typical handover related signalling and timings do not exist. However, CIPv6 (as well as the Handoff-Aware Wireless Access Internet Infrastructure (HAVAII) [6]) has to solve the latency applied in the packet transition while travelling inside the core network, as well as substantial wireless resources wasted with micro-mobility related signalling. These two facts reduce significantly the performances of the access network and waste to much radio resources in 4G scenarios making it difficult to handle traffic flows like multimedia and real-time IP traffics with several users.

These protocols led to the overall idea that mobility could be divided in two areas, a micro-area in the access and a global area across the whole network. This type of division leads to less signalling needs for global mobility, better results in handover timings and less packet loss during the transitions. This is the approach that is being proposed in the *netlmm* working group, that set specifications for novel NGN mobility protocols [6].

As can be seen in the descriptions above, previous mobility protocols show some problems. This paper presents possible solutions with a new Local-Centric Mobility System (LMS) that will extend the classical global mobility scheme. The LMS is a protocol designed to be compliant with most *netlmm* requirements. It implements a localized mobility management architecture that aims for the: minimization of the handover timings and related signalling; reduction of packet losses during the handover; increased efficiency in wireless and core network resources usage; and integration of mechanisms for AAAC support.

3 LMS Mobility Architecture

LMS is an evolution to classical global IP mobility protocols. The LMS implements a protocol that enables a localized mobility environment in IPv6 networks, with retained wide area connectivity per itself. Our model uses an architecture organized in autonomous micro-domains with two hierarchical levels in each micro-domain.

This type of architecture optimizes the global mobility scenario improving handover procedures and reducing the signalling related traffic. In this type of architecture, the Mobile Host is enabled to move in the same micro-domain between Base Stations without changing its IP address (CoA). This fact results in an optimized process that increases the network and mobility performances.

Our protocol was influenced by previous analysis on advantages of the existing protocols [7, 8] to solve the problems of global IP mobility. The localized domains are based on the requirements of *netlmm* workgroup and the HMIP regions scheme. The LMS mobility micro-domains and Paging Areas were inspired on Cellular IP [9] and the predictive handovers were inspired on FMIP [3].

3.1 LMS Agents

As can be seen in Fig. 1, the LMS network architecture is divided in two regions: The Global Mobility and Local Mobility regions. The Global Mobility region will still run MIPv6, although its purpose is mostly for MIP backwards compatibility. The LMS mechanisms run over the Local Mobility region. The LMS is supported by four entities (in the figure):

1. MH – Mobile Host: this agent is responsible to enable the Mobile Terminal to get connected in the access network.
2. BS – Base Station: this agent is responsible for the packet filtering between the core access network and the edge access network. The core access network connects all access agents, all BS and the MAP; the edge access network connects all Mobile Hosts and BS.
3. MAP – Mobility Anchor Point: this agent is responsible for the management of all the intra-domain tasks. These tasks concern the access control during the intra-domain handovers and accounting procedures during the intra-domain hosting. This agent is also responsible for all the BS state-full auto-configuration.
4. MMP – Mobility Management Point: this agent is responsible for the management of all the authorization, inter-domain authentication, accounting aggregation and charging related tasks. This agent is also responsible for all MAPs state-full auto-configuration.

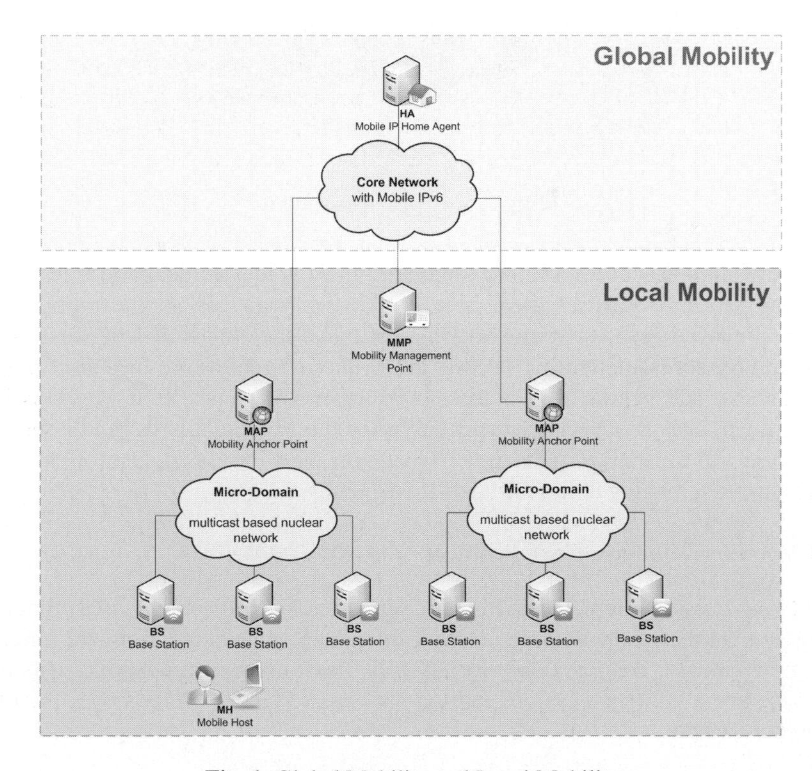

Fig. 1. Global Mobility and Local Mobility

Typically, the LMD region is also divided in the access (the connection between the BS and the MH) and the core (all the network from the BS upwards). We consider that a LMD region can be very large.

3.2 LMS Micro-domains

The LMS is based in a topology supported by (potentially) autonomous micro-domains (and as such can be seen as a global mobility architecture). Each micro-domain is constituted by a group of BS and one MAP Each micro-domain has a secret cryptographic key - used to derive a Personal ID (PID) for the Mobile Hosts in the micro-domain –, and one IPv6 network domain prefix (therefore when a MH moves on the same micro-domain it does not need to change its network configuration). For each micro-domain, it is also possible to associate it with a different security type of access, so it is possible to create restricted zones constituted with one or more micro-domains. The micro-domain cell groups can be mixed by restricted and unrestricted types of access (see the black zones in Fig. 2).

This strategy presents several improvements in the reduction of handover time and packet losses during the inter-micro-domain handover situation.

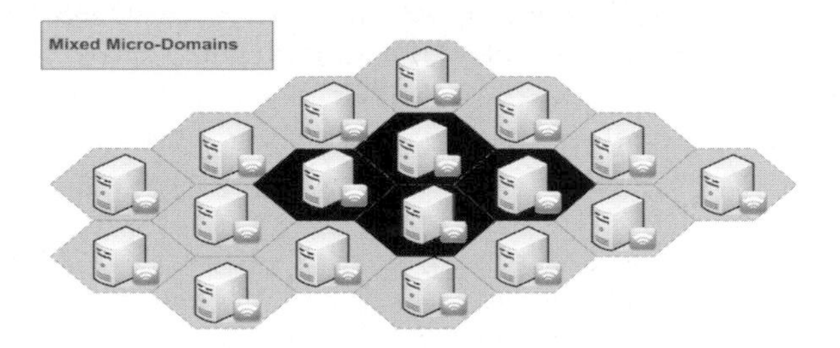

Fig. 2. Micro-Domain Cells

The network topology of a micro-domain (Fig. 3) is hierarchical and it is based on two levels. On the top level there is the MAP that manages all the micro-domain tasks. At the lower level there are all BSs that make the connection between Mobile Hosts and core access network. The BSs in the micro-domains are further grouped by paging areas. Each paging area is used to group mobile hosts in sub-regions in the micro-domains: this is especially important when the Mobile Host is in idle mode and it is necessary to route IP packets to it. We support the wireless concept of paging that can be mapped to paging at the L2 layer at each BS.

3.3 LMS Micro-domain Core Protocol

The network topology depicted in Fig. 3 is supported by three paging areas that represent the core of the micro-domain network. In this network the packets are forwarded over a new protocol designed only for LMS micro-domain core network. This protocol is based on a multicast system and the packets are label switched over the micro-domain core agents.

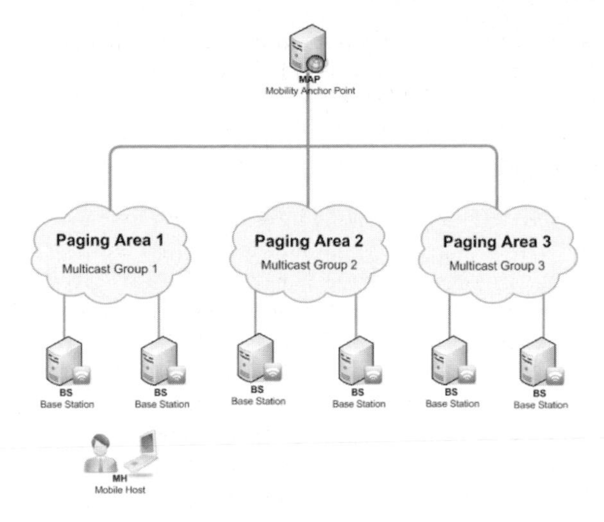

Fig. 3. Micro-Domain network topology

Multicast allows that several BS on the micro-domain can be reached without packet replication. This type of feature is very useful during the intra-micro-domain handovers of terminals running multimedia sessions. The architecture makes also possible the forwarding of packets of unicast sessions over this protocol reducing the use of the network resources, while reducing packet loss in handover. The packets that need to travel over the micro-domain network are encapsulated IP-over-IP in a multicast channel for the corresponding paging area. Each paging area is connected to MAP via a multicast channel that aggregates one set of BS. The multicast channels are managed by the MAP of each micro-domain that creates or removes the BS from the multicast groups dynamically. As seen in Fig. 4, the packets on the core of the micro-domain network can be sent for one specific BS or for a set of these without any packet replication.

With this protocol running on the LMS network it is possible to optimized the packet transmission during unicast and multicast sessions and improve the efficient use of network resources.

3.4 LMS AAAC Support

The LMS has mechanisms for AAAC integration. These are distributed between MAPs and MMPs.

The AAAC tasks are organized in two layers. The lower layer is constituted by all MAPs and manages the lower layer tasks like: Mobile Hosts flow control for accounting and intra-domain authentication control. Note that, as mentioned before, micro-domains can have access control, with a Mobile Host granularity. The higher layer is constituted by the MMP that merges all services reports from the MAPs, and manage all this information with any centralized agent.

LMS can be made compliant with protocols like Remote Authentication Dial in User Service (RADIUS), or can communicate directly with a LMS central data base.

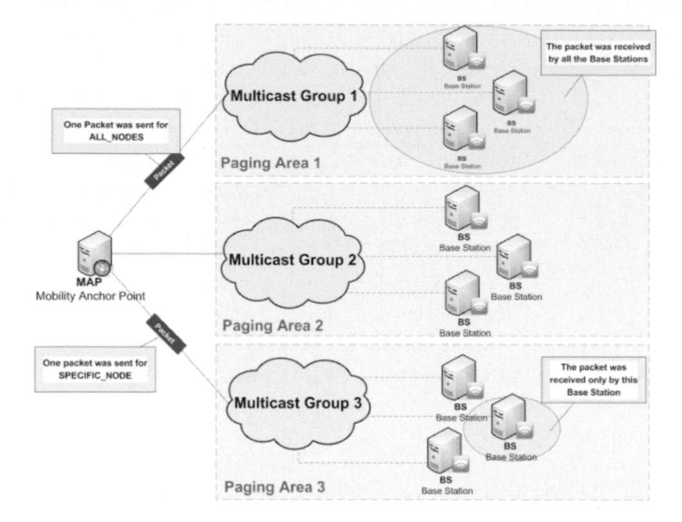

Fig. 4. Packets on the core of the micro-domain network

This data base can be a SQL based SGBD that contains information about the micro-domains (MAPs and BS) for auto-configuration services; it can also contain information about Mobile Host authentication control, authorization services, accounting and charging. The main advantage of the usage of a central data base is that it can also store information about network agents' information (auto-setup info, agent info).

3.5 LMS Internal Security

The LMS does not integrate any extra security for Mobile Host data transmission. As the LMS aims to be an extension protocol, it should not implement any extra security for data transmission, but rely on existing layers. Nevertheless, the LMS integrates cryptographic tasks for internal signalling.

The signaling packets of the protocol sent by Mobile Host are always authenticated with its PID. The PID is derived from the secret cryptographic key of the micro-domain network and it is generated by the MMP during the access registration on the micro-domain (see below). This PID is generated applying a MD5 hash function on a set of bits that represent the Mobile Host (128 bits) and the network key (128 bits).

The packets that contain confidential content are always encrypted and authenticated. This type of procedure ensures that most of the typical attacks over the access and control tasks of the micro-domain networks are avoided and the Mobile Hosts authenticity is secured.

3.6 Minimizing the Modifications on Mobile Hosts

The LMS was designed to be compliant with most of the *netlmm* requirements, and one of the most important is the "unmodified hosts" requirement. The handover mechanism of LMS is inspired in FMIPv6 predictive handovers in order to be fast enough to avoid blackout periods. The LMS design does not compels that Mobile Host needs any modification in order to guarantee that the main network functionalities work correctly.

However, in order to provide predictive fast handovers with authentication and authorization control, the LMS Mobile Host will need to integrate a simple software daemon that will guarantee these functionalities. Thus, using this enhancement on the Mobile Host, it is possible for the network to predict when the Mobile Hosts aims to move and where it will move to. Moreover, it is also possible to authenticate its movements along the network and guarantee the differentiated access control for different type of access areas. As an ultimate enhancement, as the LMS was designed to be a NGN protocol for operator networks, it is also possible to improve the protocol and guarantee some Quality of Service (QoS) reservation on the foreign local domain. Based on this aspects, we decided to give some intelligence to the Mobile Host in order to have all this increased functionalities mentioned previously; however, this intelligence capabilities only compel small modifications on the terminal.

3.7 LMS Mobile Host Registration

The registration of the Mobile Host occurs in two phases: operator database registration and network connection registration.

The first phase occurs before any attempt of network connection. The Mobile Host makes a registration on the central database of the operator, storing its NAI and a *Ticket_Key* (alternatively this *Ticket_Key* may be a credential delivered by the operator to the Mobile Host, e.g. by a SIM card). These two fields can univocally identify this Mobile Host and will be used during authentication and authorizations requests.

The second phase occurs when the Mobile Host connects to the network. For this purpose, the Mobile Host sends an authenticated *Registration Request* packet to the network with the confidential information encrypted. The packet is forwarded to the MMP because the BS and the MAP are not able to decrypt confidential information. For security reasons, the MMP is the only agent that can directly communicate with the operator database and, then, knows the *Ticket_Key* to decrypt the *Registration Request* packet information.

After decrypting the packet, the MAP verifies if the Mobile Host is authorized to entry in that specific micro-domain. In the negative case, the MMP sends a *Registration Response* with access denied. In the positive case, the MMP generates a PID for the Mobile Host and makes its registration on its caches and database. After this process, the MMP generates a new IPv6 for the Mobile Host based on the network-prefix IPv6 of the micro-domain and its MAC address. The MMP sends back the *Registration Response* with this information to the Mobile Host. While the packet travels the micro-domain network, the MAP will make the Mobile Host registration on the specific paging area that it aims to bind.

When the Mobile Host receives a positive registration response, it automatically sets up its configuration and starts to send heart beat packets to the network, necessary to avoid soft-state termination.

3.8 LMS Intra-micro-Domain Handover

The intra-micro-domain handover happens when a Mobile Host moves between two different BS on the same micro-domain (Fig. 5). When the Mobile Host intends to initiate the handover, it sends a *Handover Request* message for the network. The

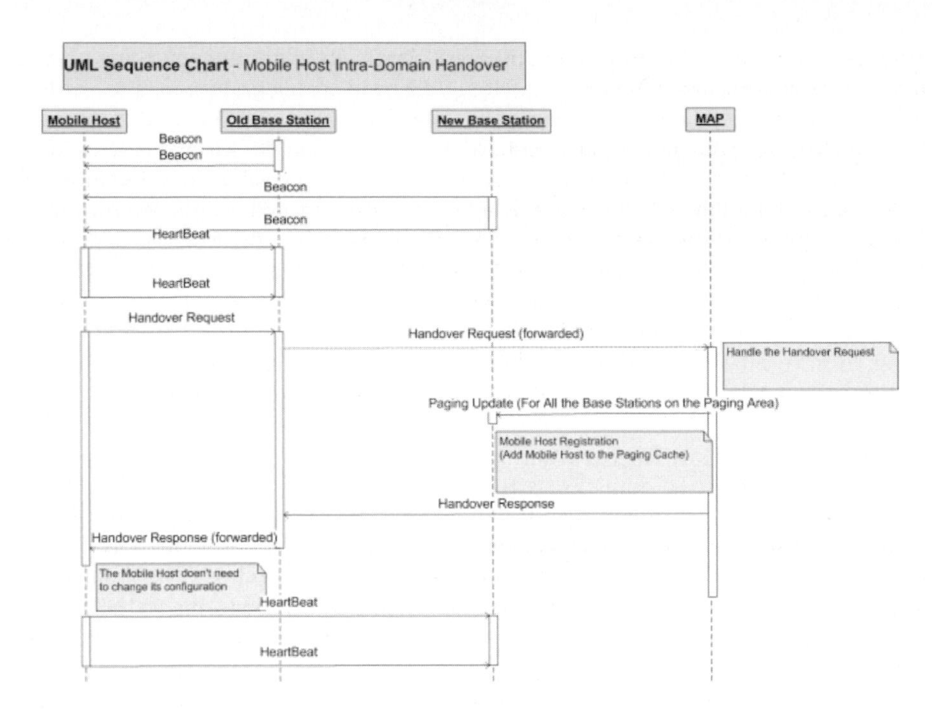

Fig. 5. Intra-micro-domain handover

Handover Request is an authenticated packet that informs the network about the new BS, new paging area and new network ID where the Mobile Host aims to move. The packet is sent for the old BS and it is forwarded to the MAP requesting a decision. When the MAP receives the packet, it knows that the Mobile Host intends to move on the same micro-domain network, because the new network ID in the *Handover Request* packet is the same of the current network ID.

In this type of handovers the MAP does not need any third-party authorization from MMP to process the *Handover Response*. Thus, the MAP processes a Mobile Host registration on the new paging area of the micro-domain and sends a *Paging Update* packet to it. When the BS in this paging area receives the *Paging Update*, it makes a registration in the Mobile Host caches allowing this Mobile Host to bind in any of them. The caches on the agents implement soft-states; after the Mobile Host chooses one, the others will be remove after some time.

After the *Paging Update*, the MAP sends a *Handover Response* allowing the Mobile Host to complete its handover to the new BS. The Mobile Host receives the packet, moves to the new BS, but does not need to change its network setup configuration (it only needs to change its routing table redirecting its traffic to the new BS). After this process, the Mobile Host sends periodically a heart beat packet to the new BS refreshing the soft-states.

If the Mobile Host is not able to predict its handover and send *Handover Request* related signalling, it can simply move to another BS and start its registration on the network again. In these cases, as the new BS does not know who the Mobile Host is, it needs to start a new registration. After its completion, the new registration on the network overlaps the previous one.

3.9 LMS Inter-micro-Domain Handover

The inter-micro-domain handover happens when a Mobile Host moves between BS in different micro-domains. When the Mobile Host intends to initiate the handover, it sends a *Handover Request* to the old BS, as before, which again forwards it to the MAP requesting a decision. The MAP, after receiving the *Handover Request*, knows that this handover is an inter-micro-domain handover through the new Network ID.

In this case, the MAP needs to delegate this decision task to the MMP agent and forwards the packet towards it. When the MMP receives the packet, it verifies if this Mobile Host can access the new micro-domain network. In negative case, the MMP generates a *Handover Response* with an access denied response and sends it back to old MAP. In positive case, the MMP notifies the new MAP that a new Mobile Host will move to its micro-domain. If the new MAP is able to register the Mobile Host on its micro-domain, then it will send a message with positive information to MMP notifying it that the registration was made successfully; otherwise, it will send a negative response to it. Based on the new MAP response, the MMP generates a new PID and a new IPv6 address based on IP network prefix of the new micro-domain, and sends a *Handover Response* back to the old MAP. When the old MAP receives the *Handover Response* from MMP, it forwards the packet to the Mobile Host. In the case of a positive *Handover Response*, the MAP also removes this Mobile Host from its caches.

After receiving the *Handover Response* packet, the Mobile Host knows if it can move or not to the new micro-domain. In a positive case, the Mobile Host changes its setup configuration and after all, it moves to the new BS (it also keeps sending heart beat packets periodically).

If this handover cannot be predicted, the Mobile Host needs to start a new registration on the network.

3.10 Summary of LMS Features

Our model presents the following advantages:

- Localized architecture – the LMS network is organized in semi-autonomous micro-domains providing very efficient local mobility support: the Mobile Host can move in the same micro-domain without changing the IPv6 address.
- Low handover-related signalling – in LMS, the handover-related signalling traffic is very low. As result, this technique especially improves the handover timings and network resources exploitation.
- Handover improvements – fast and seamless handover mechanisms with make-before-break techniques with very low (zero) packet losses are supported.
- Efficient use of access resources – the signalling packets across access network, shared between Mobile Hosts and BS, are small, improving the efficient use of wireless resources.
- Efficient use of core resources – LMS integrates a new packet-forwarding mechanism based on multicast services to improve core-network resources.

- Support for heterogeneous access link technologies - LMS is completely independent from the L2 technology, and can support multiple heterogeneous network link technologies.
- Secure mobility management – LMS supports security at the signalling level.
- Access control on mobility actions – LMS performs an explicit authorization before handovers to different micro-domains.

Furthermore, it has also two added features:

- No extra security between Mobile Host and network – LMS protocol does not implement any extra security for data flow between Mobile Host and network.
- Easy to support on the Mobile Hosts – LMS mechanisms do not change the Mobile Hosts significantly; LMS only needs a small mobility daemon running on the Mobile Hosts.

4 Evaluation Tests and Results

4.1 Testbed Description

The following figure (Fig. 6) presents the Linux-based testbed architecture used in the LMS tests. Code was developed for a *Gentoo* distribution with Linux kernel 2.6.11. Two micro-domains were used inside a single LMD. All network links were 10Mb/s Ethernet, in order to avoid wireless channel interference and L2-related handover issues, and handover is triggered by Mobile Hosts-initiated selection. All machines were set up in a laboratory, and as such link propagation times were small – which may bias slightly the results. All machines, with the exception of the Mobile Host, were PIII@600MHz. The Mobile Hosts was a Pentium Mobile@1.6GHz.

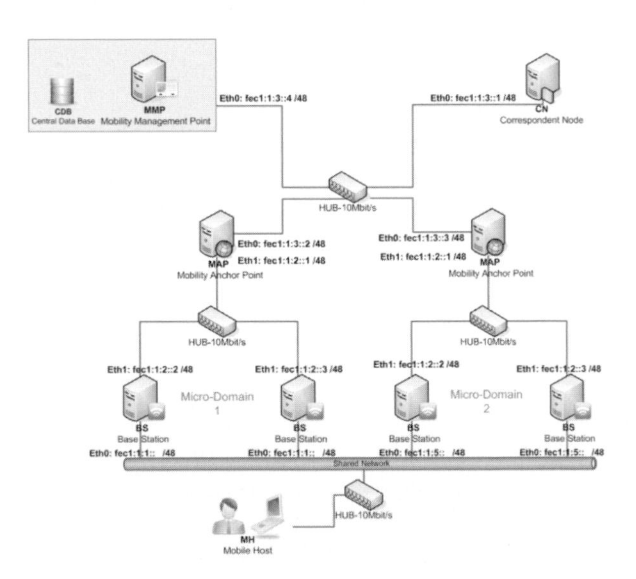

Fig. 6. Testbed architecture

4.2 Tests and Results

In the next sections it will be presented and described the tests that were made on the test-bed. For these tests, traffic was generally generated according to two different test sets (Test set A and B), as described in Table 1. Special traffic applications were used in some cases.

Table 1. Reference Test Sets

	Test set A	**Test set B**
Packet Size	64 Bytes	1024 Bytes (1KByte)
Packet Speed	1000 (packet per second)	1000 (packet per second)
Packet Type	ICMP	ICMP

Bandwidth Performance Tests

This type of performance tests were made to assess how LMS handles TCP and UDP traffic in a real scenario. The test consists in two special applications, one in the Mobile Hosts and other in a correspondent node, transferring data one to another. This two applications measure the real bandwidth available on the network. Table 2 shows the TCP and UDP bandwidth test results.

Table 2. Real Bandwidth results

	UDP	TCP
LMS Real Bandwidth (average)	8,61 Mbit/s	6,91 Mbit/s
Standard deviation (stddev)	0,34 Mbit/s	0.06 Mbit/s

As can be seen in the previous results the LMS makes an efficient use of network bandwidth providing almost 70% of physical bandwidth for TCP traffic and 86% for UDP.

Network Latency Tests

The network latency tests are used to know how the LMS reacts to network traffic. For efficient use of the network, the latency on packet propagation must be small. These types of tests were made using ICMP Echo Request / Response to measure the ICMP packet propagation time between Mobile Host and correspondent node. Table 3 shows the test results for different sets.

Table 3. Network Latency Results

	Test set A	Test set B
Packet Loss	0%	0%
Emitted Packets	35892 packets	9646 packets
Packet Delay	0,649 ms	3,06 ms

The LMS agents can react very fast to the network traffic compared with the typical latency on a 10 Mbit/s Ethernet network (~0,5ms). The latency increases with the packet size, but still, it is small even for 1KByte packet size (~3 ms). We also observe that the packet loss is zero percent (0%) for small and large packet sizes.

Intra-Micro-Domain Handover Performance Tests

The intra-micro-domain handover performance tests aim to assess how the LMS reacts in intra-domain handover situations. This test was made using a packet flow from Mobile Host to correspondent node during the handover procedure, and Test Set A was used, due to the reduced latency. The measured results are the packet losses, handover negotiation related signalling time and handover time (offline time). Table 4 shows these results.

Table 4. Intra-Micro-Domain Test Results

Packet Loss	0%
Negotiation Signalling Time (average)	6.48 ms
Negotiation Signalling Time (stddev)	2.20 ms
Handover Time (average) *offline time*	5.88 ms
Handover Time (stddev) *offline time*	0.86 ms

As can be seen in the previous results, the LMS intra-domain handover timings are extremely small compared with Mobile IP timings [8] and the packet losses are avoided. In this scenario the Mobile Host can move between two BS with zero percent (0%) packet loss in a fast and seamless intra-domain (micro-domain) handover. Note that, as the Mobile Host does not change its IPv6 address, it does not need to send Binding Updates to Home Agent and correspondent node. In this case, the blackout time for Mobile Host communications is below 6ms.

In these types of micro-domain handovers, as shown in [8], the Mobile IP handover latency (~50ms) can be avoided. Note that in [8] the results for Mobile IPv6 handover latency were simulated, and in real scenarios these times will be worse.

Inter-Micro-Domain Handover Performance Tests

The inter-domain handover performance tests are used to know how LMS reacts in inter-micro-domain handover situations. This test is made using Test Set A from Mobile Hosts to correspondent node during the handover procedure. The measured results are the same as in intra-domain case and are depicted in Table 5.

Table 5. Inter-Micro-Domain Test Results

Packet Loss	2%
Negotiation Signalling Time average	61.16 ms
Negotiation Signalling Time stddev	7.49 ms
Handover Time average *(offline tine)*	28.16 ms
Handover Time stddev *(offline time)*	9.49 ms

The LMS intra-micro-domain handover timings are still small, as well as the packet losses; however the standard deviation in offline time is substantial. Still in this scenario the Mobile Hosts can move between two different micro-domains with fast and seamless handover mechanisms. In this case, as the Mobile Host changes its IPv6 address it needs to send Binding Updates to Home Agent and correspondent node. Thus, the blackout time is the sum of the LMS offline time and Mobile IP Binding Update. Packet replication through the multicast groups can reduce this impact.

4.3 Protocol Comparison

The next table summarizes some characteristics of some mobility protocols and makes a comparison with the LMS – Local-Centric Mobility System. Note that the LMS can be arbitrarily large (such as a cellular operator network), and as thus can be seen as providing global mobility, while providing efficient localised mobility inside the micro-domains.

Table 6. Comparison between different mobility protocols

	CIP	HMIP	MIP	F-HMIP	FMIP	LMS
Local/Global	Local	Local	Global	Local	Local	Both
Fast Handover	Yes	No	No	Yes	Yes	Yes
Seamless Handover	Yes	Yes	No	Yes	Yes	Yes
Efficient Use of Core Resources	No	Yes	No	Yes	No	Yes
Efficient Use of Link Resources	No	Yes	No	Yes	No	Yes
Minimize CoA changes during Handover	Yes	No	No	No	No	Yes
Support of Paging Areas	Yes	No	No	No	No	Yes
Minimize Mobile Hosts Changes	No	Yes	Yes	Yes	Yes	Yes

5 Conclusions

This paper presents the design and performance evaluation of LMS (Local-Centric Mobility System): a localized-based approach for supporting mobility in wide-areas like in NGN (Next Generation Networks). The objectives of LMS were scalability, fast and seamless handovers, efficient exploitation of network resources and reliability. The LMS has been designed to solve IP mobility problems and to provide fast and seamless handovers with optimized network resources exploitation.

The LMS can integrate some AAAC services in a wide scenario with several micro-domains. Furthermore, it has a new mechanism for packet forwarding in the core network of micro-domains that increases the network performance especially during the handover situations. This protocol also presents a signalling mechanism that improves the exploitation of access network resources especially in wireless scenarios. Finally, it has some mechanisms to improve the security of network signalling without add extra security to data flows sent by Mobile Hosts.

The evaluation tests showed how LMS can optimize and solve some problematic aspects of the global mobility architecture and protocols. The LMS can minimize the handover timings from some hundreds of milliseconds (in a real mobility scenario) to 6ms in intra-micro-domain handover and 28ms in inter-micro-domain handover. Packet loss during handover was also mostly negligible, and the network overhead was also minimized improving the efficient use of the core and wireless network resources. Concluding, the LMS shows that it can be possible to optimize the mobility mechanisms to improve the timings, avoid packet losses and make efficient use of network resources in mobility scenarios.

As future work, we are optimizing wireless support in LMS (in order to avoid L2 issues) for future measurements in handover situations with integrated support for a global mobility protocol such a Mobile IPv6.

References

1. Johnson, D., Perkins, C.: Mobility Support in IPv6. RFC 3775 (June 2004)
2. Soliman, H., et al.: Hierarchical mobile IPv6 mobility management (hmipv6), IETF Internet RFC 4140 (August 2005)
3. Koodli, R.: Fast Handovers for mobile IPv6, IETF Internet RFC 4068 (July 2005)
4. Shelby, Z.D., et al.: Cellular IPv6, Internet draft, draft-shelby-seamoby-cellularipv6-00 (November 2000)
5. Ramjee, R., et al.: HAWAII: a domain-based approach for supporting mobility inwide-area wireless network. IEEE/ACM Transactions on Networking 10(3), 396–410 (2002)
6. Kempf, J., et al.: Requirements and Gap Analysis for IP Local Mobility, draft-ietf-netlmm-nohost-req-01 (February 2006)
7. Campbell, A., Gomez, J., Kim, S., Wan, C.: Comparison of IP micro-mobility protocols. IEEE Wireless Communications 9, 72–82 (2002)
8. Pérez-Costa, X., Torrent-Moreno, M., Hartenstein, H.: A Performance Comparison of Mobile IPv6, Hierarchical, Mobile IPv6. Fast Handovers for Mobile IPv6 and their Combination 7(4) (October 2003)
9. Gomez, J.: Design, Implementation and Evaluation of Cellular IP. IEEE Personal Communications 7(4), 42–49 (2000)
10. Aguiar, R.L., et al.: Scalable QoS-aware Mobility for Future Mobile Operators. IEEE Communications Magazine 44(6), 95–102 (2006)
11. IETF Netlmm Working Group, http://www.ietf.org/html.charters/netlmm-charter.html
12. Jung, H., et al.: Fast Handover for Hierarchichal MIPv6 (F-HMIPv6), draft-jung-mobileip-fastho-hmipv6-01.txt, IETF (June 2003)

An Efficient Proactive Key Distribution Scheme for Fast Handoff in IEEE 802.11 Wireless Networks

Junbeom Hur, Chanil Park, Youngjoo Shin, and Hyunsoo Yoon

Korea Advanced Institute of Science and Technology(KAIST)
{jbhur,chanil,yjshin,hyoon}@nslab.kaist.ac.kr

Abstract. Supporting user mobility is one of the most challenging issues in wireless networks. Recently, as the desires for the user mobility and high-quality multimedia services increase, fast handoff among base stations comes to a center of quality of connections. Therefore, minimizing re-authentication latency during handoff is crucial for supporting various promising real-time applications such as Voice over IP (VoIP) on public wireless networks. In this study, we propose an enhanced proactive key distribution scheme for fast and secure handoff based on IEEE 802.11i authentication mechanism. The proposed scheme reduces the handoff delay by reducing 4-way handshake to 2-way handshake between an access point and a mobile station during the re-authentication phase. Furthermore, the proposed scheme gives little burden over the proactive key pre-distribution scheme while satisfying 802.11i security requirements.

1 Introduction

Nowadays, wireless local area network (LAN) systems based on IEEE 802.11 standard are emerging as a competent technology to meet the requirements of users for high-speed wireless Internet connectivity. Due to the lack of mobility support of IEEE 802.11, however, seamless mobile services, particularly for real-time applications such as voice over IP (VoIP) are hard to be served in IEEE 802.11 networks when a mobile station (STA) moves from one access point (AP) to another. Especially, Authentication, Authorizing, and Accounting (AAA) servers are supposed to be located far away from each AP so that the full authentication delay requires about $1000ms$ [9]. This excessive latency of complete user authentication and security negotiations which should be performed at each AP during handoff can be a main obstacle to seamless services for real-time multimedia applications. Therefore, fast re-authentication and re-association schemes are essential during handoff between APs.

The current IEEE 802.11i [1] security architecture recommends an authentication process to follow EAP/TLS [4]. In addition, IEEE 802.11i makes use of IEEE 802.1x [3] model to authenticate the STA to the AAA server using AAA protocols to prohibit unauthorized access to the network. The complete EAP/TLS authentication, however, causes too large latency to support multimedia services whose overall latency should not exceed $50ms$ [8].

T. Vazão, M.M. Freire, and I. Chong (Eds.): ICOIN 2007, LNCS 5200, pp. 629–638, 2008.

To solve this problem, many previous studies proposed fast handoff schemes in diverse aspects [5]–[7]. A. Mishra et al. proposed the proactive key distribution (PKD) scheme using a mobility topology of the network, Neighbor Graph, which tracks potential APs to which an STA may handoff in near future [5]. Based on the PKD scheme that reduces the handoff delay by pre-authenticating an STA to next neighbor APs before handoff, M. Kassab et al. proposed two pre-authentication schemes to reduce the authentication exchange duration: PKD with anticipated 4-way handshake, and PKD with inter AP protocol (IAPP) caching [7]. However, these schemes not only heavily burden a current AP with excessive overheads but violate IEEE 802.11i trust assumptions.

In this study, an efficient pre-authentication scheme enhancing the proactive key distribution method is proposed. The proposed scheme reduces the number of exchanges for private session key generation between an STA and an AP of the re-association phase by exchanging key-generating materials in the pre-authentication phase before handoff. Therefore, the re-authentication delay of 4-way handshake during handoff can be reduced to that of 2-way handshake. In addition, the proposed scheme guarantees security requirements of IEEE 802.11i standard and secure communications between an STA and each AP.

2 Related Work

IEEE 802.11i uses IEEE 802.1x [3] framework to authenticate and authorize devices connected to the network. In the IEEE 802.1x framework, a supplicant authenticates to a central authentication server (AS) using an extensible authentication protocol (EAP) [4] like the EAP/TLS. After the mutual authentication, the authenticator and the supplicant establish keying materials, and then the AS directs the authenticator to allow the STA to access the network.

2.1 EAP/TLS Authentication

In the IEEE 802.11i authentication process using the EAP/TLS, the STA and the AAA server mutually authenticate each other based on a certificate from a common trusted certificate authority (CA). The mutual authentication process drives the STA and the AAA server to share a strong secret master key (MK) and to initialize pseudo-random functions (PRF) for generating further key materials. The STA and the AAA server generate a pairwise master key (PMK) separately.

$$\begin{aligned} PMK = PRF(&MK, \text{`}client\ EAP\ encryption\text{'} \mid \\ &ClientHello.random \mid ServerHello.random). \end{aligned} \tag{1}$$

Then, the AAA server sends the PMK to the associated AP. After that, the STA and the associated AP perform 4-way handshake through the EAPOL protocol [2] to confirm the PMK between them and to derive a session key, pairwise transient key (PTK). The 4-way handshake is described as follows:

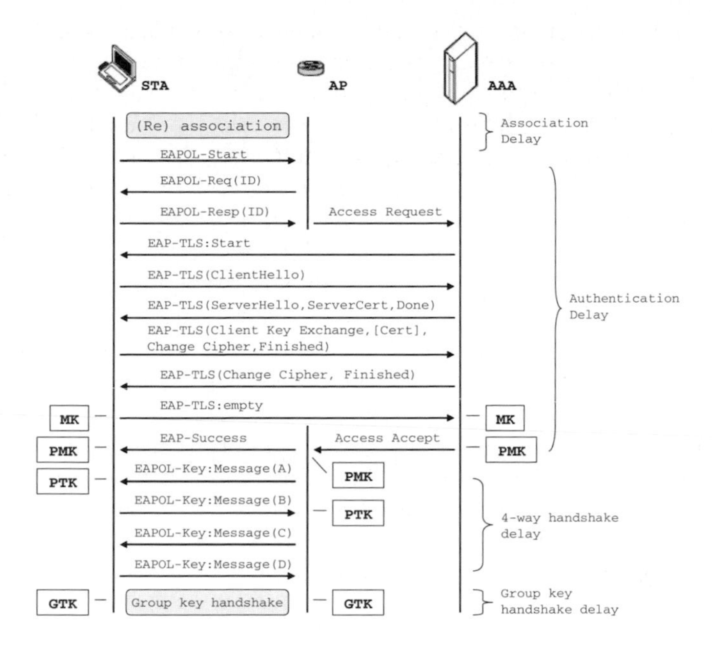

Fig. 1. Complete EAP/TLS authentication exchange

1. *Message(A): EAPOL-Key(ANonce, Unicast)*
   – This message contains ANonce, which is a nonce value generated by the AP. Once the STA has received this message, the STA can derive a PTK. This message is not encrypted or integrity-verified.
2. *Message(B): EAPOL-Key(SNonce, Unicast, MIC)*
   – This message contains SNonce, which is a nonce value generated by the STA, and a message integrity check (MIC) to protect its integrity. The AP derives the PTK using SNonce and verifies the MIC. If this step succeeds, the AP can confirm that the STA has the correct PMK and PTK, and that there is no man-in-the-middle attack.
3. *Message(C): EAPOL-Key(Install PTK, Unicast, MIC)*
   – This message tells the STA that the AP is ready to begin encryption using PTK. If this step succeeds, the STA can verify that the AP has the correct PMK and PTK, and that there is no man-in-the-middle attack.
4. *Message(D): EAPOL-Key(Unicast, MIC)*
   – After this message is sent, both sides install the PTK and begin data encryption using the PTK.

During the 4-way handshake, the STA and the AP generate PTK separately.

$$PTK = PRF(PMK, ANonce, SNonce, STA_{mac}, AP_{mac}), \qquad (2)$$

where STA_{mac} and AP_{mac} represent the MAC addresses of the STA and the AP, respectively. The PTK is shared only between the STA and the currently

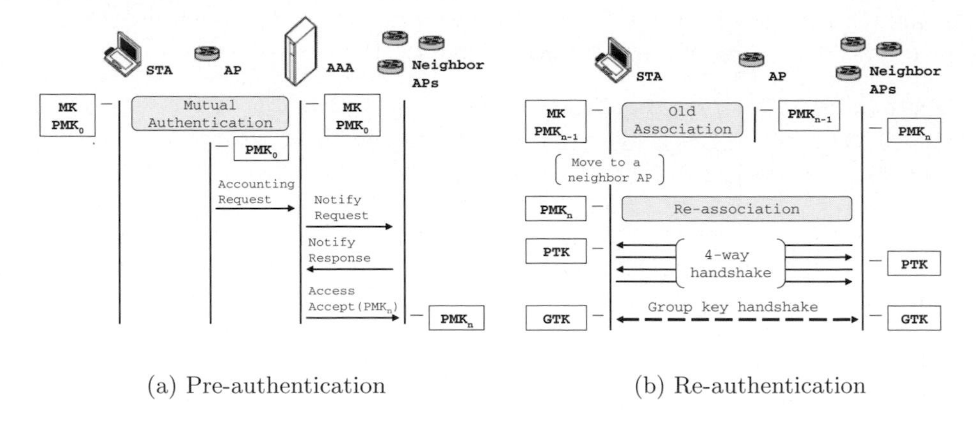

(a) Pre-authentication (b) Re-authentication

Fig. 2. Authentication exchange process with PKD

associated AP for secure communication between them. The confidentiality of the PTK is only based on a secrecy of the PMK because other key-generating materials are exposed. Fig. 1 describes the complete message exchanges and the point of each key generation time during a complete EAP/TLS authentication.

2.2 Proactive Key Distribution

The proactive key distribution (PKD) scheme [5] pre-authenticates an STA to next APs by generating and pre-distributing authentication keys, PMKs, to the neighbor APs of the currently associated AP before handoff.

$$\begin{aligned} PMK_0 &= PRF(MK, \text{`}client\ EAP\ encryption\text{'} \mid \\ &\qquad ClientHello.random \mid ServerHello.random), \\ PMK_n &= PRF(MK,\ PMK_{n-1} \mid AP_{mac} \mid STA_{mac}), \end{aligned} \tag{3}$$

where n represents the nth re-association. The PMK_0 is generated during a first mutual authentication between an STA and an AAA server. The AAA server pre-distributes the PMK_n to next neighbor APs for pre-authentication. This prevents other dissociated APs from generating the current PTK, which follows the IEEE Task Group I (TGi) trust assumption that the only associated AP and the AAA server are trusted [5]. Thus, the mutual authentication process between an STA and an AAA server after handoff is reduced to perform 4-way handshake and group key handshake as shown in Fig. 2. The PKD method reduces the full authentication delay of about $1000ms$ to the re-authentication delay of $60ms$ [9], but still exceeds the expected latency for real-time applications.

2.3 Other Approaches for Pre-authentication

Recently, M. Kassab et al. proposed two pre-authentication methods based on the PKD method [7]. The main idea of these methods is to reduce the re-authentication delay by performing 4-way handshake in the pre-authentication phase at the expense of additional loads and security degradation.

PKD with IAPP Caching. In the PKD with IAPP caching method, a current AP calculates all the PTK_x for its neighbor AP_x separately using the PMK, and pre-distributes the PTK_x and its valid time value to the corresponding neighbor AP_x through the inter access point protocol (IAPP) [2]. Upon handoff to a new AP_x, the STA derives the PTK_x and authenticates itself to the AP with the PTK_x through the group key handshake. Thus, the re-authentication phase is reduced to the group key handshake process without 4-way handshake. This re-authentication, however, is temporary authentication, which remains valid only within the time limit. After the time limit, the STA and the AP should authenticate each other and generate a permanent PTK for secure channel again.

PKD with Anticipated 4-way Handshake. In the PKD with anticipated 4-way handshake method, an STA and neighbor APs perform 4-way handshake through the current AP in the pre-authentication phase in advance. Thus, this method also reduces the re-authentication delay to the only group key handshake delay. To carry out 4-way handshake, the STA receives a list and MAC addresses of neighbor APs of the current AP from the AAA server. So, the STA can generate PTKs with the neighbor APs through its current AP using PMK_ns.

3 Efficient Proactive Key Distribution

In this section, a pre-authentication scheme based on the PKD method is proposed for fast handoff in the IEEE 802.11 network environment. The main idea of the proposed scheme is to perform 2-way handshake during a pre-authentication phase and perform remaining 2-way handshake during a re-authentication phase while satisfying security requirements of the IEEE 802.11i standard.

3.1 Modified EAP/TLS Authentication

To exchange the nonce values between an STA and APs in the pre-authentication phase, the STA transmits its nonce value to an AAA server through the following modified message exchange during the first full EAP/TLS authentication:

EAP-TLS:empty $\longrightarrow$ *EAP-TLS(SNonce).*

Then, the AAA server stores the nonce value received from the STA and delivers it with the PMK to the associated AP. Upon receiving the nonce value, the AP can generate the PTK for the STA. Then, the AP transmits its nonce value and MIC to the STA to verify that the AP has the correct PMK and PTK through the modified EAP-Success message exchange:

EAP-Success $\longrightarrow$ *EAP-Success(ANonce, MIC).*

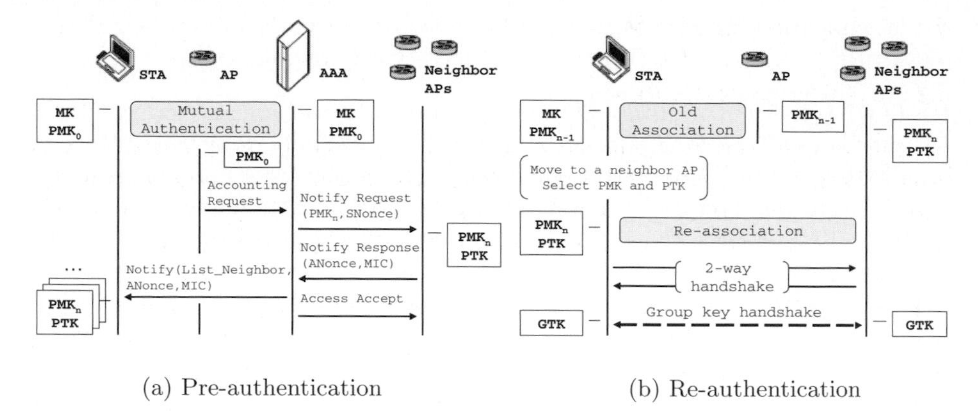

(a) Pre-authentication (b) Re-authentication

Fig. 3. Authentication exchange process with efficient PKD

Therefore, thereafter only 2-way handshake is required to establish the PTK between the STA and the AP, and check the integrity of the keying materials. The 2-way handshake process is described as follows:

1. *Message(A): EAPOL-Key(Install PTK, Unicast, MIC)*
   - This message tells the AP that the STA is ready to begin encryption using the PTK. If this step succeeds, the AP can verify that the STA has the correct PMK and PTK, and that there is no man-in-the-middle attack.
2. *Message(B): EAPOL-Key(Unicast, MIC)*
   - After this message is sent, both sides install the PTK and begin data encryption using the PTK.

3.2 Authentication with the Efficient PKD

After the first mutual authentication between an STA and an AAA server, the AAA server requests neighbor APs to pre-authenticate the STA by sending the corresponding PMK and SNonce of the STA. Upon receiving them from the AAA server, the neighbor APs generate their own PTK for the STA during a pre-authentication phase and respond to the AAA server with their own nonce values and MICs of the message. Upon receiving them, the AAA server transmits a list of neighbor APs, their nonce values, and MICs to the STA. After that, the STA generates PMKs and PTKs corresponding to each neighbor AP and verifies that each neighbor AP has the correct PMK and PTK. If these steps succeed, the AAA server completes the pre-authentication phase by transmitting an access accept message to the neighbor APs as described in Fig. 3(a). Upon handoff, the STA selects the corresponding PMK and PTK to the re-associated AP among the keys which were generated in the pre-authentication phase. Then, the STA and the AP check for the integrity of the keys and install the PTK by performing 2-way handshake as described in Fig. 3(b).

4 Protocol Analysis

4.1 Performance Evaluation

In this section, the performance of the four authentication schemes is analyzed: PKD, PKD with IAPP caching, PKD with anticipated 4-way handshake, and the proposed scheme. The overall results of the analysis are summarized in Table 1 in which m represents the average number of neighbor APs per each AP.

In Table 1, the communication factor represents the necessary number of message exchanges for the PMK and PTK establishment among the entities. The common exchanges of the first full EAP/TLS authentication exchanges, or group key handshake are not included in this analysis. The computation factor represents secret keys, which should be generated by each entity per handoff except the common key PMK_0 and MK. The memory requirement factor represents the memory consumption for a neighbor graph (NG) of the current AP, which should be maintained by each entity for key generation in the pre-authentication phase. The AP in the table is the current AP.

The IEEE 802.11i security factor represents whether the schemes satisfy the security requirements of the IEEE 802.11i standard: (1) There should be mutual authentication and fresh key derivation at each AP, (2) Mutual authentication should not cause man-in-the-middle attack. The PKD with IAPP caching method is vulnerable to the AP's compromise and the man-in-the-middle attack because each AP should participate in the process of other APs' secret key establishment. Thus, even a single AP's compromise can be a great threat to the security of the whole network.The total communication exchanges for authentication of the proposed scheme is the least compared to the other schemes. The PKD with anticipated 4-way handshake scheme has the shortest re-authentication delay; however, as the network size increases and the neighbor relationship of APs changes frequently, the total authentication efforts of the scheme may increase most greatly due to the overburdened pre-authentication process.

Compared to the PKD method, the proposed scheme requires one more communication exchange in the pre-authentication phase, but reduces the 4-way handshake to the 2-way handshake while keeping the other protocol exchanges intact and satisfying IEEE 802.11i security requirements. This can support the secure and seamless multimedia services in IEEE 802.11 network in that the re-authentication delay would be reduced from $60ms$ to the half.

4.2 Security Analysis

Key Freshness. To guarantee the freshness of a key derived at each AP, how to refresh the nonce value of an STA can be one of the considerable issues. Although the freshness of the PTK can be guaranteed by the freshness of the ANonce, a reuse of the SNonce may make a system vulnerable to the replay attack. An attacker who masqueraded as a participant in the system by forging a MAC address can eavesdrop on every message, remember nonces and MICs of each

Table 1. Performance analysis of authentication schemes

		PKD	PKD with IAPP caching	PKD with anticipated 4-way handshake	Proposed scheme
Communication	Pre-auth.	m (PMK)	m (PMK), m (PTK)	m (PMK) + 1 (list), $2m\times$4-way handshake	m (PMK) + 1 (list)
	Re-auth.	4-way handshake	<permanent> 4-way handshake, group key handshake	0	2-way handshake
Computation	STA	PMK_n, PTK, GTK	<temporary> PTK, GTK, <permanent> PMK_n, PTK, GTK	PMK_n, GTK, $m\times$ PTK	PMK_n, PTK, GTK
	AP	0	$m\times$ PTK	0	0
Memory Requirement	STA	0	0	local NG	local NG
	AP	0	local NG	0	0
IEEE 802.11i Security		Y	N	Y	Y

message, insert forged messages, and replay stored messages with a combination of known nonces and MICs. To refresh the nonce value of the STA, it can be one solution for the AAA server to regenerate the SNonce on behalf of the STA and distribute it to neighbor APs like the PMK_n pre-distribution. That is,

$$SNonce_n = PRF(MK, SNonce_{n-1}, STA_{mac}, AP_{mac}),$$

where n represents n^{th} re-association of the STA. This can achieve the freshness of the PTK. In addition, since the MK is securely shared between the STA and the AAA server, no other participants but they can generate or predict the appropriate SNonce per handoff.

DoS Attack. According to the security verification of 4-way handshake using Murφ model in [10], the 4-way handshake is analyzed to be vulnerable to a simple attack on $Message(A)$ that causes PTK inconsistency between the AP and the STA. An attacker who is impersonating the authenticator sends a forged $Message(A)$ to the STA after $Message(B)$. The STA will then calculate a new PTK corresponding to the nonce for the newly received $Message(A)$, leading to PTK inconsistency so that the subsequent handshakes to be blocked. The vulnerability of the 4-way handshake to DoS attack on the $Message(A)$ is actualized by the AP-initiated 4-way handshake in which the STA should must accept all messages to allow the handshake to proceed while the AP can initiate only one handshake instance and accept only the expected response within the expected time. So, the memory exhaustion attack on the STA always exists.

In the proposed scheme, however, the STA initiates the handshake, thus the STA needs not store all the unexpectedly received nonces and derived PTKs. This prevents the memory exhaustion attack on the typically resource-constrained STA. However, the STA-initiated 4-way handshake is still vulnerable to the DoS attack on the $Message(A)$. One possible solution is to add a MIC to the $Message(A)$ using a common secret such as a PMK to prevent an attacker from forging it, and to use a sequence counter to defend against a replay attack.

5 Conclusion

In this study, an efficient pre-authentication scheme based on the PKD method is proposed. The proposed scheme clearly improves the PKD method by reducing the re-authentication delay to 2-way handshake by transmitting nonce values between the STA and APs in the pre-authentication phase without security degradation. An efficient key distribution scheme for fast and secure handoff is an essential technology for secure and quality services in IEEE 802.11 networks. Since the proposed scheme is simple and does not require any impractical trust relationship among network entities, the scheme can be extensively adapted to the PKD-based pre-authentication methods for fast handoff.

References

1. IEEE 802.11i: Amendment 6: Medium Access Control (MAC) Security Enhancements. IEEE Computer Society (July 2004)
2. IEEE 802.11f: Recommnded Practice for Multi-Vendor Access Point Interoperability via an Inter-Access Point Protocol Across Distribution Systems Supporting IEEE 802.11 Operatoin. IEEE (July 2003)
3. IEEE 802.1x: IEEE Standards for Local and Metropolitan Area Networks: Port based Network Access Control. IEEE (June 2001)
4. Aboba, B., Simon, D.: PPP EAP TLS Authenticatoin Protocol. RFC 2716 (October 1999)
5. Mishra, A., Shin, M., Arbaugh, W.A.: Pro-active Key Distribution using Neighbor Graphs. IEEE Wireless Communications 11 (February 2004)
6. Pack, S., Jung, H., Kwon, T., Choi, Y.: SNC: A Selective Neighbor Caching Scheme for Fast Handoff in IEEE 802.11 Wireless Networks. ACM SIGMOBILE Mobile Computing and Communications Review (October 2005)
7. Kassab, M., Belghith, A., Bonnin, J., Sassi, S.: Fast Pre-Authentication Based on Proacitve Key Distribution for 802.11 Infrastructure Networks. In: ACM Workshop on Wireless Multimedia Networking and Performance Modeling (WMuNeP 2005), October 13 (2005)
8. International Telecommunication Union: General Characteristics of International Telephone Connections and International Telephone Circuits. ITU-TG.114 (1988)
9. Aboba, B.: Fast Handoff Issues. IEEE 802.11-03/155r0 (2003)
10. He, C., Mitchell, J.C.: Analysis of the 802.11i 4-Way Handshake. In: ACM Workshop on Wireless Security (WiSe 2004) (October 2004)
11. Burrows, M., Abadi, M., Needham, R.: A Logic of Authentication. ACM Transactions on Computer Systems 8(1), 18–36 (1990)

Appendix

Formal Analysis of the Proposed Protocol

Here, we analyze our pre/re-authentication scheme using a logic-based formal analysis tool [11] to ensure that our authentication protocol functions correctly.

We deem that authentication is complete between the STA and AP if there is a PTK such that both believe($\models$) the share of it($\overset{PTK}{\leftrightarrow}$):

$$STA \models STA \overset{PTK}{\leftrightarrow} AP, \qquad AP \models STA \overset{PTK}{\leftrightarrow} AP.$$

We idealize the protocol below, with STA and AP_n as the principals, AS as the AAA server, N_s and N_a as the nonce values, MAC_{AP_n} as the MAC addresses of AP_ns, and $MIC_k\{m\}$ as the MIC of the message m encrypted under the key k.

(Pre-authentication)
$Message$ 1. $AS \rightarrow AP_n : PMK,\ N_s$
$Message$ 2. $AP_n \rightarrow AS : N_a,\ MIC_{PTK}\{N_a\}$
$Message$ 3. $AS \rightarrow STA : MAC_{AP_n},\ N_a,\ MIC_{PTK}\{N_a\}$
(Re-authentication)
$Message$ 4. $STA \rightarrow AP_n : MIC_{PTK}\{\}$
$Message$ 5. $AP_n \rightarrow STA : MIC_{PTK}\{\}$

To analyze this protocol, we first give the following assumptions:

$STA \models STA \overset{PMK}{\leftrightarrow} AS,$ $\quad AS \models STA \overset{PMK}{\leftrightarrow} AS,$
$AS \models STA \overset{PMK}{\leftrightarrow} AP_n,$ $\quad STA \models AS| \sim N_s$(AS conveyed N_s),
$AP_n \models AS \triangleleft N_s$(AS is told N_s), $\quad STA \models \sharp(N_a)$($N_a$ is fresh),
$AP_n \models \sharp(N_s)$.

We analyze the idealized version of our authentication protocol by applying logical postulates of [11] to the assumptions; the analysis is straightforward. For brevity, we do not describe our deductions, and simply list the final results:

Analysis of $Message$ 1. $\quad AP_n \models STA \overset{PTK}{\leftrightarrow} AP_n$
Analysis of $Message$ 2,3. $\quad STA \models STA \overset{PTK}{\leftrightarrow} AP_n$
Analysis of $Message$ 4. $\quad AP_n \models STA \models STA \overset{PTK}{\leftrightarrow} AP_n$
Analysis of $Message$ 5. $\quad STA \models AP_n \models STA \overset{PTK}{\leftrightarrow} AP_n$

This state achieves more than the complete condition of the authentication. Each principal, STA and neighbor APs, knows a shared secret, PTK, with each other and has a knowledge of a shared secret that he believes the other will accept as being shared by the two principals. From this point, they can transfer data securely.

An Efficient Authentication Procedure for Fast Handoff in Mobile IPv6 Networks

Jun-Won Lee[1], Sang-Won Min[2], and Byung K. Choi[3]

[1] R&D team 2, SK Telesys, Suwon-shi, Kyungki-do, 441-230, Republic of Korea
jwlee@sktelesys.com

[2] Dept. of Electronics and Communications Eng. Kwangwoon University, Wolgye-dong, Nowon-gu, Seoul, 139-701, Republic of Korea
min@kw.ac.kr

[3] Dept. of Computer Science, Michigan Tech. University, Houghton, MI 49931, USA
bkchoi@mtu.edu

Abstract. Mobile IP (MIP) has been paid a lot of attention as a good candidate to provide such global mobility among heterogeneous networks. And, IPv6 incorporates the mobility into its extended function which is referred to as mobile IPv6 (MIPv6). On the other hand, authentication, authorization and accounting (AAA) service now plays an important role in many networks. Considering the future popularity of MIPv6 in many networks, investigating the MIPv6 architecture in a way that the fast handoff is well supported in the presence of AAA is desirable. In this paper, we propose a new authentication procedure for MIPv6-based networks where the original procedures of both fast handoff and AAA are still preserved to inter-operate with other networks. The proposed temporary authentication allows the roaming user to continue communication activities even before the user is fully authenticated by the corresponding AAA server in the home domain. The simulation results are very encouraging in that the proposed architecture shortens the handoff time significantly from that of the conventional approach up to 70%.

Keywords: temporary authentication, MIPv6, AAA, fast handoff and wireless mobile networks.

1 Introduction

The concept of mobility was originally designed for the mobile telecommunication networks such as in the global standard for mobile (GSM) for voice services. Recently, the mobile Internet Protocol (MIP) was designed to provide the mobile data service for IP-based networks. On the other hand, there is s trend that future networks such as third- and fourth-generation (3G/4G) mobile networks are being designed to support not only voice services but also data and video services even in heterogeneous wireless networking environment. To provide mobility in heterogeneous as well as homogeneous wireless networking environments, MIP has been paid a good amount of attention as it could be an excellent means to achieve seamless multimedia services in heterogeneous networking environments [1][2].

T. Vazão, M.M. Freire, and I. Chong (Eds.): ICOIN 2007, LNCS 5200, pp. 639–648, 2008.

Although MIP is designed to offer global mobility, fast handoff and seamless service were not a big concern. Hence, some micro-mobility schemes in support of fast handoff have been proposed to reduce the detection time of roaming, the frequency of registration, and the amount of registration traffic. Also, mobility has been incorporated into IPv6. Compared with mobile IPv4 (MIPv4), mobile IPv6 (MIPv6) is designed in a built-in fashion with the IPv6 extended header and Internet Control Message Protocol (ICMP)v6 [2]-[4].

AAA has become an important issue in mobile networks based on IP. When a mobile node (MN) moves to another administrative domain, it should be authenticated and authorized to access the visiting network. For example, when an MN with both wireless LAN (WLAN) and general packet radio service (GPRS) modules moves from a WLAN domain to a GPRS domain, the MN should be authenticated by the GPRS operator and there should be an agreement between WLAN and GPRS operators on roaming services.

In this paper, we propose a new authentication procedure of MIPv6 with AAA in support of fast handoff, and compare this with that of the conventional procedure. It is interesting to note that whereas the MIPv6 with AAA does not consider fast data forwarding, the MIPv6 with fast handoff such as hierarchical MIPv6 aims at achieving seamless data service with fast registration and short handoff latency [5]. The proposed authentication procedure in this paper is designed to provide AAA service in the context of fast handoff, meaning that it minimizes the time required for AAA service. The proposal adds a temporary authentication during handoff between the AAA servers in previous and new visiting domains to minimize AAA service time while the original AAA and fast handoff functions are preserved for inter-operability with other conventional networks. In the temporary state during handoff, the MN is permitted to transmit and receive data with some restrictions and the AAA server in the new visiting domain tries to obtain the full authentication from the AAA server in the home network. Since the fast handoff function is supported, the data delivered to the previous visiting domain are forwarded to the MN.

To show the efficiency of the proposed authentication procedure, we evaluate the performance by simulation which reveals the handoff latency and packet loss. As expected, the handoff latency of our procedure has much smaller values than that of the conventional procedure. The packet loss of the MN in the proposed procedure is ideally zero because fast handoff is supported. Our proposed authentication demonstrates a way that fast handoff service can be effectively supported even with AAA service in MIPv6-based networks.

Following this section, MIPv4 with AAA are explained in Section 2 where the conventional components of AAA and MIP are reviewed. The requirements on MIPv6 with AAA are briefly described in addition. In section 3, the architecture and the conventional authentication procedure of MIPv6 with AAA are described. In section 4, the proposed authentication procedure is explained with the temporary authentication, and many considerations are discussed compared with the conventional procedure. To show the effectiveness of our proposal, the simulation results and comparison study are presented in section 5. Finally, the conclusion is given in section 6.

2 Related Work

The MIP provides the mobility of an MN in the IP-based network without changing the IP address of the MN. There are two immediate concerns in MIP; one is the micro-mobility with fast handoff and the other is security. This has solicited various micro-mobility schemes. Also, the security issue of granting an access right to an MN to its visiting domain has become an important issue [6]. Whenever it roams to another administrative domain, an MN is required to be authenticated and/or authorized by both the home and visiting domains. Figure 1 shows the AAA architecture for an MIPv4 network where the AAA home (AAAH) and AAA foreign (AAAF) are the servers that authenticate and authorize access of an MN to the foreign domain. The AAA client functionalities are included in a home agent (HA) and a foreign agent (FA) which share security associations (SAs) with their AAA servers; SA-HA and SA-FA. And, the AAA servers share a SA-FH by which the credentials of the MN can be negotiated for authentication and integrity check purposes.

The MN sends a registration request to the FA with the network access identifier (NAI). In addition, it sends a response to the FA in response to the local challenge (LC) given in an agent advertisement message. The AAAF obtains MN's home network information form the registration request and delivers the request to the AAAH. Three session keys for the SAs, MHA-Key, MFA-Key and HFA-Key, which are used during the registration procedure, are shown in the figure.

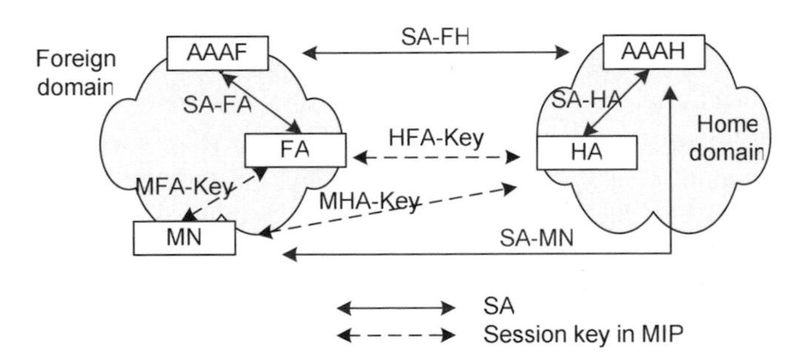

Fig. 1. Architecture of MIPv4 with AAA

Although MIPv4 and MIPv6 support the mobility at the same layer, there are some differences. The FA in MIPv4 does not exist in MIPv6 since the extended header of IPv6 and ICMPv6 support the function of the FA. In the collocated care-of address (CoA) mode, the FA informs this mode to MNs by setting a control bit in agent advertisement messages, meaning that it authorizes the MNs' access to the visiting network. The attendant node with FA will be responsible for authentication and authorization of the MN prior to its access to the resource. The attendant functionality can be implicitly supported in some other agents in wireless and mobile networks [7].

An access router (AR) is assumed in MIPv6 to provide the attendant function and to broadcast advent advertisement messages. In the architecture of MIPv4 with AAA support, the AAA servers should perform key generation and distributions for the SAs among MIPv4 entities.

3 Conventional Procedure

Following the MIPv4 architecture and the requirement on MIPv6 with AAA, we portray the architecture of MIPv6 with AAA in Figure 2, where an AR is assumed to carry out the attendant functionality of AAA. Although the key generation and agreements between AAAH and AAAF are accomplished, the AR should have the function of sharing a SA-AR with AAAF [8]. After receiving the advertisement and sending the information of credentials to the AR, an MN can be authenticated and can access the network through the AR. In order for the AAAH to generate and distribute the requested session key to the MN, a temporary session key (TSK) is needed between MN and AR. The TSK allows delegating functions such as entity authentication and key derivation to the visited domain.

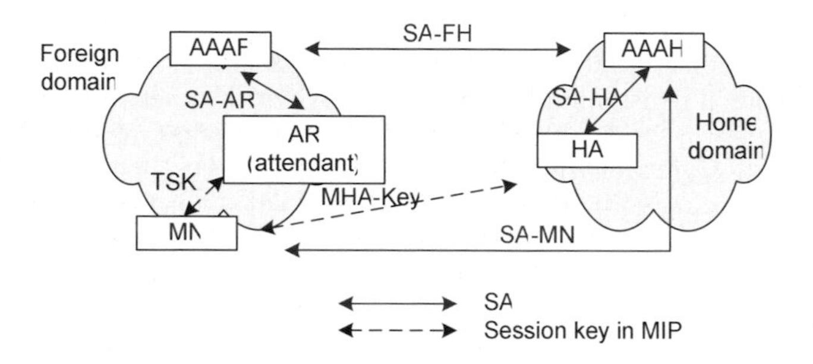

Fig. 2. Architecture of MIPv6 with AAA

In the MIPv6 with fast handoff, the important issue is to reduce the registration time and the amount of associated signaling traffic. As long as an MN stays in the authentication process, the fast handoff procedure is not taking place because the authentication has a higher priority. After it is authenticated and authorized, an MN resumes sending or receiving packets to/from a new visiting network. To recover normal operation from the transitional handoff processing state, the MN is first configured with a new CoA when it attaches to a visiting network, and then sends the credential of the MN to the AR in the visiting network, and then finally performs the binding update procedure with its HA and correspondent nodes (CNs).

The binding update procedure in the MIPv6 with AAA architecture is shown in Figure 3. After receiving an advertisement message including a LC, the MN sends a binding update (BU) message to the AR, where it contains the challenge response, the HA address, NAI, the MN's credential and the key-information. The MN's credential is used for the authentication of the MN by the AAAH. The key-information is used to establish SAs between the MN and the AR, and between the MH and the HA.

When there is no such a SA, an AAA client request (ACR) can be forwarded to a broker such as an AAA broker. The AAAH extracts the MN's credential and determines whether the MN can be authenticated based on the pre-registered database information. Before the binding update procedure, the AAAH randomly generates and delivers a session key to the MN, the HA, and the AR. A home agent MIPv6 request (HOR) with the MIPv6 BU is sent from the AAAH to the HA.

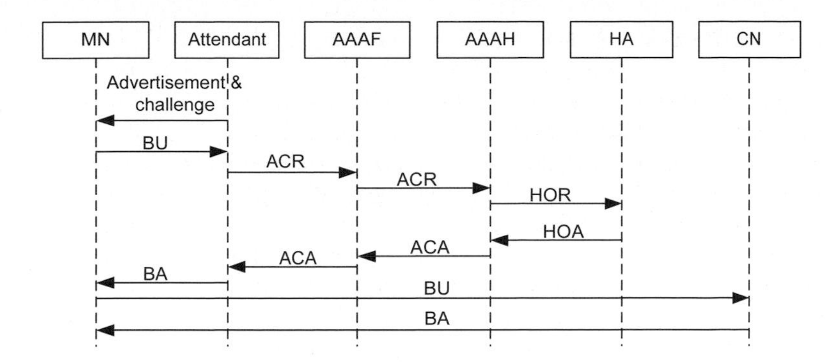

Fig. 3. Conventional Procedure of MIPv6-AAA registration and binding update

A home agent MIPv6 answer (HOA) with the MIPv6 binding acknowledgement (BA) is replied from the HA. After receiving the HOA, the AAAH updates the database about the MN, makes and sends an AAA client answer (ACA) to the AAAF, which contains the authorization information about the MN's roaming privilege, access scope and the related lifetime, as well as the MIPv6 BA and a TSK for the MN and the AR. The AAAF grants the access of the MN to the foreign domain based on the information of the ACA message, and forwards it to the MN. Finally, the MN can receive the MIPv6 BA in the ACA message. When the binding update procedure is successfully completed, the MN obtains the parameters and information about available service since the binding has been authenticated and authorized by the AAA servers.

4 Proposed Authentication Procedure for Fast Handoff in MIPv6

Since MIP has been designed to support data service, there might be long handoff latency and some packet could be lost accordingly. To solve these problems, there have been many efforts and proposals to support micro mobility and fast handoff, which reduce handoff latency, the frequency of registration, and the amount of associated signaling traffic. Hence, the packet loss can be decreased and seamless data service may be possible. In this section, we consider an MIPv6 architecture with fast handoff where AAA is supported in a similar manner as in the previous section.

We propose an authentication procedure of inter-domain handoff in MIPv6 and consider its related data flow form the ARs with the proposed authentication procedure. Figure 4 shows an MIPv6 network with AAA and fast handoff where the sequences of data traffic before and after handoff are indicated. Three data paths can be observed; the first is before handoff, the second is during temporary handoff, and the last is after handoff. The second path is one of the consequences of fast handoff. Without a fast-handoff scheme, there would be no second path.

In our proposed scheme, the MIPv6 fast handoff and the MIPv6-AAA procedure are executed simultaneously. To get the authentication from the AAAH, the MN executes the normal AAA procedure to the AAAH through the AAAF server in the new foreign domain. After MN's having the full authentication from the AAAH, there is no restriction to access to the visiting network under the pre-assigned contract of the MN.

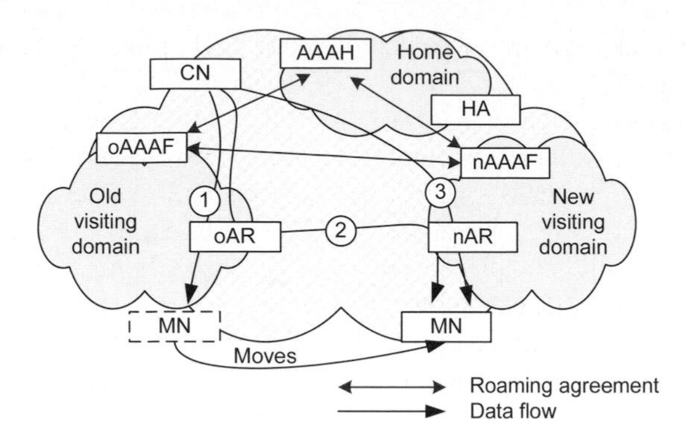

Fig. 4. Data flows in MIPv6 with AAA and fast handoff

Since an MIPv6 fast handoff scheme, which is defined as an extension of MIPv6, is needed, the layer-2 triggering scheme is assumed in our proposal and simulation. For MIPv6 fast handoff, both fast roaming detection and fast binding update to the HA and CNs are necessary. Solutions without layer-2 information could provide fast binding update but not fast roaming detection. Hence, the layer-2 triggering scheme, which is more appropriate in the case of wireless mobile network, is adopted for MIPv6 fast handoff in this paper.

The proposed procedure is in Figure 5. With layer-2 triggering information, an MN can detect its movement and sends a router solicitation message to a new AR (nAR). The nAR sends out its router advertisement message with the network prefix. On receiving the router advertisement message, the MN configures a new CoA (nCoA) with network prefix information and sends a fast binding update (FBU) message to the old AR (oAR) which contains a BU message to the HA. Binding of the nCoA and the oAR is performed to forward packets buffered during handoff. The FBU makes the oAR to start the packet buffering and configure a tunnel to either the MN with the nCoA information or the nAR if provided in the FBU message.

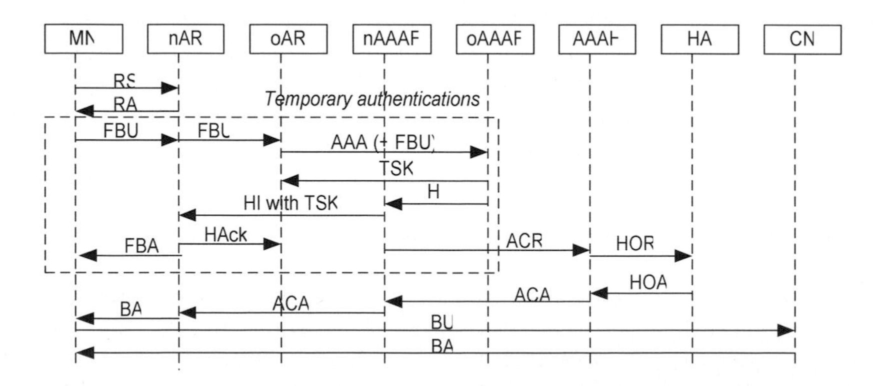

Fig. 5. Proposed Procedure of MIPv6-AAA registration and binding update for fast handoff

The oAR issues an AAA message including the FBU to its AAAF in the previous foreign domain (oAAAF) in which a handoff initiation (HI) message is generated to the AAAF in the new foreign domain (nAAAF) in order to execute the operation for fast handoff. Also, for a secure communication between the oAR and the nAR, a TSK is randomly generated by the oAAAF, which is transmitted directly from the oAAAF to the oAR, and to the nAR through the nAAAF with the HI message, respectively. When it is successfully completed, the nAR sends a handoff acknowledgement (HAck) message to the oAR as the response to the HI message. Also, the nAR generates a fast binding acknowledgement (FBA) message to the MN. After receiving the HAck and FBU messages, the oAR forwards the buffered packet. Upon receiving the FBA message, the MN can send or receive a packet through the nAR. Optionally, a fast neighbor advertisement (FNA) message could be used from the MN to the nAR in order to forward the buffered packet to the MN [8]. However, the MN may have a restriction because it is not authenticated by the AAAH.

During the temporary authentication procedure with the oAAAF and the MN, the nAAAF should execute the binding update procedure with the MN's HA simultaneously. Since the FBU message includes the BU message for MIPv6, the nAAAF can do the general AAA operation of the binding update procedure with the AAAH. Through the nAAAF and the AAAH servers, the MN can process the binding update procedure with the HA and CNs. This procedure is shown at the bottom part in Figure 5, which is the same as the conventional binding procedure without fast handoff.

With the proposed procedure, an efficient authentication during handoff can be performed to reduce the handoff latency and packet loss. While the fast handoff in mobile networks has been designed to shorten the delay of user traffic and the amount of registration signaling traffic, AAA has evolved from the basic network infrastructure independently of fast handoff features. Unless AAA is considered in the context of the fast handoff, the fast handoff and AAA would be independently applied to MIPv6 and the effectiveness of the fast handoff would significantly degrade. Although the fast handoff seems to provide the fast binding update, the AAA message exchanges would prolong the handoff processing latency and would be resulted in the source of performance bottleneck. In our proposal, the temporary authentication between the previous foreign domain and a new domain is adopted during handoff period before a full authentication is performed with the AAAH. The temporary authentication inherently supports the objective of fast handoff.

5 Performance Evaluation and Discussion

We use the *ns-2* extensions on Linux, *ns-2.1b*, to evaluate the performance of the proposed procedure. The *ns-2.1b* simulation platform, referred to as mobiwan, is widely used for MIPv6 in wide area networks [9]. In our simulation, we assume that there is a central router (CR) as a core network to interconnect four access networks, shown in Figure 8. Each access network is assumed to consist of its AAA server.

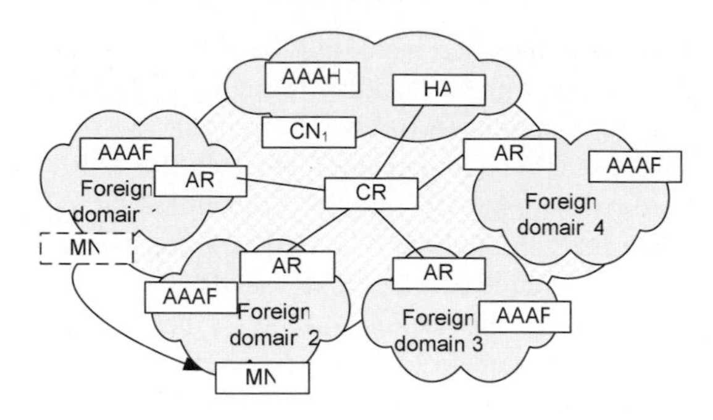

Fig. 6. Simulation model

We assume that there are four domains connected by a CR and the distance between ARs is 450 m and the transmission range is 250 m. Also, the wireless LAN of 2 Mbps bandwidth is assumed as a wireless domain, the wired connection is to be a duplex link with 5 Mbps, and the default link delay is 2 ms. Between the HA and the CR, a duplex link with 5Mpbs and the link delay of 10 ms are chosen in our simulation. The chosen mobility model is the random waypoint mobility model for the random movement with which an MN can move randomly within a wireless coverage area. Since an MN may generally compete with other MNs to send packets to the network through the AR, half of the MNs in a domain are assumed to receive data from the CR and the other half to send data to the CR. Here, the traffic source is assumed to be UDP constant bit rate (CBR) without acknowledgement.

The observed MN is assumed to have the maximum speed of 5 m/s and to be received 250 bytes at intervals of 10 ms of UDP CBR traffic while other MNs are assumed to generate background traffics with full-duplex at a rate of 32 Kbps. We assume that the number of MNs is twenty and the observed MN moves from domain 1 to domain 4 via domains 2 and 3 at 10 m/s and then reaches the starting point again. The characteristic of the link delay depends on the allocated bandwidth as well as the distance between network nodes. Hence, we choose the link delay as a simulation parameter in our simulation. As performance measures, we observe the handoff latency and packet loss as the value of the link delay changes. The handoff latency is defined as the elapse from the time when the last packet is received via the old router to the time when the first packet arrives through a new route after roaming.

In our simulation, we have compared the latency and the packet losses of the conventional MIPv6-AAA registration procedure and the proposed MIPv6-AAA with fast handoff procedure. Figure 7 shows the result of the handoff latency with varying link delay. Under 100 ms, there is no big difference between the conventional procedure and ours. However, above 100 ms where the distance between the foreign domain and the home domain becomes far, the latency difference between two schemes increases. Compared with the conventional procedure, the proposed procedure reduces the handoff latency between the roaming detection of the observed MN and the data redirection to its nCOA. Since the temporary authentication is accomplished quickly between the two foreign domains, it allows for low latencies. Without the

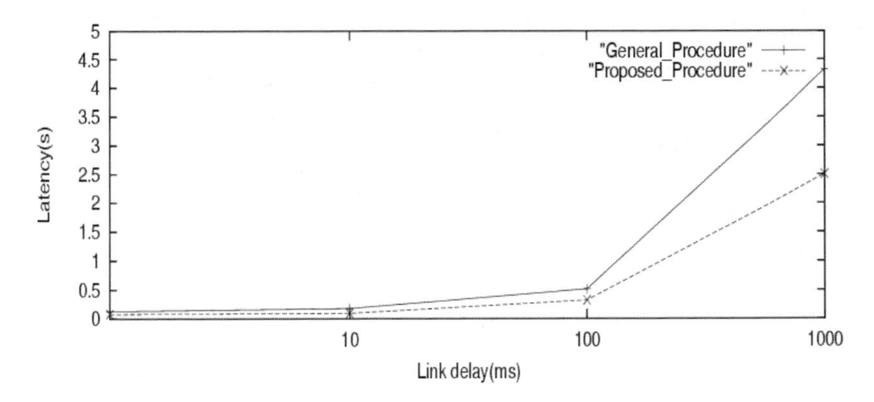

Fig. 7. Handoff latency vs. link delay

temporary authentication, the latency performance would be the same as the performance of the conventional procedure even with fast handoff.

Figure 8 presents the number of packets lost during handoff. Like the handoff latency performance, the conventional procedure has the similar pattern since long handoff latencies cause packet loss during handoff. After the link delay reaches 100 ms, the packet loss during handoff increases. If there is no need of authentication after handoff, it may be possible for the MN to send and receive with the nCoA. And, the packets transmitted through the oAR could be forwarded to the MN according to a fast handoff scheme. When authentication is required in the new visiting domain, however, the MN should wait until it is authenticated even though it is configured with the nCoA. In case that authentication is needed for the MIPv6 with fast handoff, our proposed procedure solves the packet loss problem within shorter packet delay than the conventional procedure with fast handoff does. This result may be similar to Figure 7 because the handoff latency can be considered as another consequence of the packet-forwarding delay during handoff.

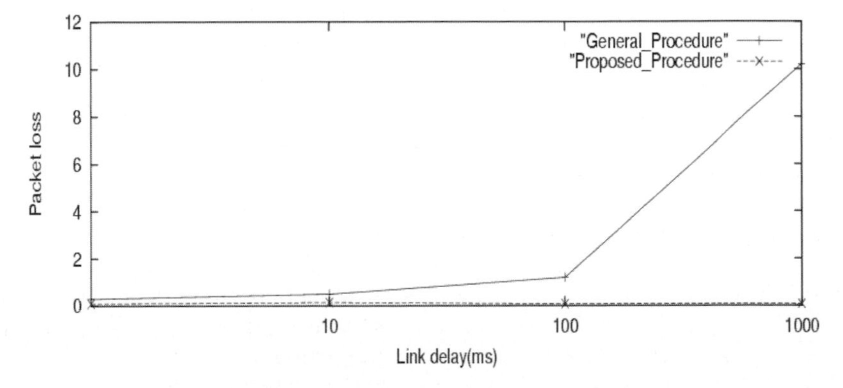

Fig. 8. Packet losses vs. link delay

6 Conclusion

In this paper, we considered the MIPv6 with fast handoff and AAA functions, and proposed an authentication procedure with temporary authentication. The full authentication state is still used without modification to the original AAA and the fast handoff functions. During the temporary state in handoff an MN can access the new visiting network with a restriction of the authentication between the previous and new visiting networks. And, the MN goes into the state of full authentication which is carried out by the original AAA procedure. Without an appropriate measure such as the proposed temporary authentication, the effectiveness of fast handoff is hard to be maintained since the authentication is performed at a higher priority than data delivery. Our solution in MIPv6 with both fast handoff and AAA works well as demonstrated by simulations, and is applicable to MIPv4 with those functions. Also, it is efficient in the case of hierarchical MIPv6 with AAA support, where a mobile anchor point will act as the AR. We expect our proposal be contributing to the all-IP environment in future networks.

References

1. Chen, J.-C., Zhang, T.: IP-Based Next Generation Wireless Networks. Wiley, Chichester (2004)
2. Lee, J.W., Min, S.W.: An Efficient Handoff Scheme for the Wireless LAN Based on the Mobile IPv6 (in Korean). In: The Proceeding of KICS (July 2003)
3. Yabusaki, M., Okagawa, T., Imai, K.: Mobility Management in All-IP Mobile Network: End-to-End Intelligence or Network Intelligence? IEEE Communication Magazine 43(12) (December 2005)
4. Koodli, R.: Fast Handoff for Mobile IPv6. IETF RFC 4068 (July 2005)
5. Calhoon, P., Johansson, T., Perkins, C., McCann, P.: Diameter Mobile IPv6 Application. IETF RFC 4004 (August 2005)
6. Johnson, D., Perkins, C., Arkko, J.: Mobility Support in IPv6. IETF RFC 3775 (June 2004)
7. Lee, C., Chen, L., Chen, M., Sun, Y.: A Framework of Handoff in Wireless Overlay Networks Based-on Mobile IPv6. IEEE Journal of Selected Areas on Communications 23(11) (November 2005)
8. Patel, A., Leung, K., Khalil, M., Arhtar, H., Chowdhury, K.: Authentication Protocol for Mobile IPv6. IETF RFC 4285 (January 2006)
9. Mobiwan: NS-2 Extensions to Study Mobility in Wide-Area IPv6 Networks, http://www.inrialpes.fr/planete/mobiwan/

Dynamic Distributed Authentication Scheme for Wireless LAN-Based Mesh Networks

Insun Lee[1], Jihoon Lee[1], William Arbaugh[2], and Daeyoung Kim[3]

[1] Comm.& Conncetivity Lab., Samsung Advanced Institute of Technology
P.O. Box 111, Suwon, 449-716, Korea
{insun,vincent.lee}@samsung.com
[2] University of Maryland, USA
waa@cs.umd.edu
[3] Information and Comminications University, Korea
kimd@icu.ac.kr

Abstract. Wireless LAN systems have been deployed for wireless internet services for hot spots, home, or offices. Recently, WLAN-Based Mesh Networking is developed with the benefit of easy deployment and easy configuration. Due to the characteristic of distributed environment, Wireless Mesh Networks(WMNs) need a new authentication scheme which allows multi-hop communication. In this paper, we propose a distributed authentication method which significantly eases the management burden and reduces the storage space on mesh points, thus enables the secure and easy deployment of WMNs.

Keywords: Wireless Mesh network, Distributed authentication, WLAN.

1 Introduction

Traditional single-hop network architecture dominates the current wireless and mobile communication technology. However, it imposes several limitations on its deployment in terms of coverage, cost, capacity and scalability. For example, widely deployed IEEE 802.11-based wireless LANs require a wired connection on each dedicated access point (AP). Since all mobile stations (STAs) need to be connected to a single AP, network throughput becomes poor with many stations operating simultaneously. Alternatively, multi-hop communication is known to overcome these limitations of single hop networking. Until recently, multi-hop techniques had been studied only in context of mobile ad hoc networks (MANETs).

Similar to MANETs, Wireless Mesh Networks (WMNs) is emerging as a multi-hop mobile network. Wireless mesh allows many individual nodes to be interconnected automatically with computers nearby, to create a large-scale, self-organizing network. Mesh is increasingly being adopted as an alternative to cable or DSL which are used for last-mile coverage[3]. In wireless mesh networks, intermediate nodes can function as a router to forward data transmitted from the source to a destination that is located in more than one hop away.

T. Vazão, M.M. Freire, and I. Chong (Eds.): ICOIN 2007, LNCS 5200, pp. 649–658, 2008.

Following list shows potential strengths of WMNs:

- Self-configuration: Mesh networks are self-organizing and self-configuring in the sense that they relieve service providers and/or IT departments from continuous administration. Deployment of mesh network also becomes easy and fast compared to other types of networks.
- Increased reliability: Redundant links in mesh networks provides added reliability even if some nodes fail to function.
- Increased capacity: Use of multiple channels and interfaces provides extra spatial reuse of available bandwidth, and it enables multiple data flows co-exist in the shared medium, thus the overall network capacity is increased.
- Coverage extension: Mesh network's multi-hop communication enables covering dead-zones. Any reachable mesh node can relay the traffic, so connectivity and network access can be provided.
- Energy conservation: Multi-hop communication, smart routing and medium access co-ordination contribute for saving power for energy-constraint devices.

In spite of its useful functions and characteristics, WMNs lack of efficient and scalable security solutions, because their security can more easily be compromised due to several factors: their distributed network architecture, the vulnerability of channels and nodes in the shared wireless medium, and the dynamic change of network topology [1]. Attacks on routing protocols and Medium Access Control(MAC) protocols are possible. For example, the backoff procedures and NAV for virtual carrier sense of IEEE 802.11 MAC may be misused by some attacking nodes, which causes the network to always be congested by these malicious nodes. WMNs wireless link is vulnerable to attack like other wireless medium, so a cryptographic protection has to be provided.

IEEE 802.11i standard[12] defines the security architectures for protecting the link layer between two entities- STA and AP. It provides the security architecture such as authentication, confidentiality, key management, data origin authenticity and replay protection. Authentication framework of this standard is for both infrastructure mode ad-hoc mode(IBSS mode). This authentication framework uses a combination of several protocols such as IEEE 802.1X and transport layer security(TLS). As shown in Fig.1, authentication is performed through the interaction of three entities- STA, AP, and Authentication server.

Authentication is performed to make only legitimate nodes can access the network. For infrastructure WLAN, this is performed through a centralized server such as RADIUS (Remote Authentication Dial-in User Service). Such a centralized scheme is not suitable for WMNs, where the network topologies are dynamic and distributed due to mobility and network failure that comes from their ad-hoc feature. Moreover, key management in WMNs is much more difficult than in infrastructure wireless LANs, because it's more complicated for a central authority to handle the distributed network, and the dynamic characteristic of WMNs makes the key management more complicated. Key management in WMNs needs to be performed in a distributed but secure manner. Therefore, a distributed authentication and authorization scheme with secure key management is needed for WMNs. Distributed Authentication with a public key infrastructure is straight forward for the implementers. It is, however, a major management and operational hurdle for end users. Historically, PKI's have only been

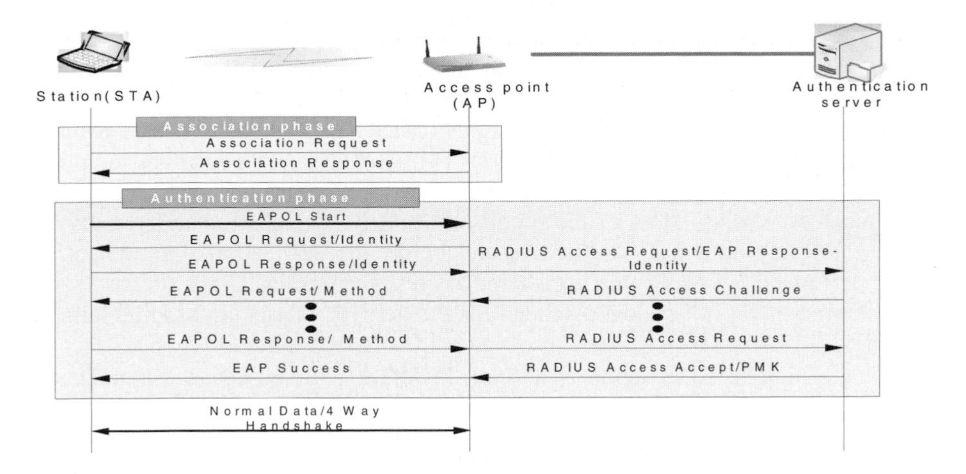

Fig. 1. IEEE802.11i Authentication model based on IEEE802.1X

attempted by large well-funded organizations and not by smaller organizations and home users. Yet these people (the SoHo users) are the very end users a distributed authentication mechanism is meant to target.

In this paper, we propose a novel distributed authentication algorithm which is suitable for WMNs. In our approach, the administrator does not have to provide each mesh point with the keys of all other mesh nodes, and a key for the authenticating station need only exist some where within the mesh. This significantly eases the management burden and reduces the storage space on mesh points.

The rest of the paper is organized as follows. Section 2 briefly describes some related works in WLAN and mesh networks. Section 3 presents the design and detailed description of Dynamic Distributed Authentication(DDA) scheme. Then, we present the evaluation results of analytical modeling and then finally we made a conclusion.

2 Related Works

Security issues on wireless mesh networks are very similar to those in sensor or ad-hoc networks, and these were well evaluated on the previous literatures[11]. Because of their large scale and dynamic topology change, key establishment based on public-key or symmetric key is one of the issues in wireless ad-hoc networks. Chan[9] and Jolly[8] introduced efficient key distribution schemes which have benefit for low power sensor nodes. TinySec[7] provides link layer encryption mechanism for access control, integrity, and confidentiality for TinyOS. This mechanism uses a symmetric key scheme for link layer data protection. TESLA[11] introduces an authenticated broadcast scheme which uses purely symmetric primitives and introduce asymmetry with delayed key distribution.

But these security problems and their solutions are not directly applicable to WMNs. The most important fact that impacts the performance for ad hoc sensor networks is power consumption, so important parameter that affects the performance of the security function is power consumption. The above schemes were designed for ad-hoc sensor networks where thousands of nodes are distributed over the large area, and the proposed schemes were based on the centralized access control under the assumption of existence of the central authority.

On the other hand, usage models for WLAN based WMNs includes office, residential, campus, public, safety/military networks[14]. As well as the power consumption, easier installation is one of the most important functionality that is necessary for rapid deployment of WMNs. Most of the proposed security schemes for ad-hoc network assume that every node has pre-shared secret before the installation, and authentication is performed by using this pre-shared secret. But pre-installing of the shared secret of every node is a critical overhead for the network consisting of large number of entities.

Authentication in IBSS mode is defined in 802.11i for ad-hoc communication, where one hop peer-to-peer communication is performed. This is a candidate authentication method for WMNs. But, to use this with either certificate or pre-shared key(PSK), all the certificates or keys of the nodes have to be pre-installed in each node. Also, authentication has to be performed twice between two nodes. i.e. one node plays the role of authenticator and supplicant in each authentication. Thus when there are N nodes in a WMN, N*(N-1) number of keys have to be maintained for the WMN, which results in high key management overhead. Thus an efficient authentication method for mesh needs to be provided.

3 Dynamic Distributed Authentication (DDA) Scheme

This section describes the detailed design of DDA scheme. First, we show the background and architecture of the proposed scheme. After that, we explain DDA scheme that performs the dynamic distributed authentication in detail.

3.1 Overview

As shown in the previous chapters, authentication and key management is one of the important hurdles for the deployment of Wireless Distributed Systems(WDS) which construct WMNs. In this section we introduce a novel distributed authentication mechanism that offers a high scalability. Especially, this algorithm allows small enterprises to use a shared secret mechanism and still allows a multi-hop environment to grow beyond a single AP in WDS. This method can enable the installation of the WDS easier, thus it enhance the easy deployment of WDS.

If an organization were to use the existing IBSS authentication for the Mesh, then the administrator would have to provide every mesh point with the key for every station. In a dynamic organization, this management burden would be intolerable. The only solution, currently, is to install a AAA server and perform centralized authentication. But, this management burden may be too heavy as well.

In our approach, the administrator need only establish a PSK with only one mesh point within the WMN and the station. The protocol, described below, will automatically find a shared key in the WMN if it exists, and establish a new fresh secret between the station and the mesh point to which this station is associating.

The typical usage scenario for this method will be that a small enterprise purchases a single AP to support one or more STAs. The usual form of security in this scenario will be to establish a shared secret between the AP and the STAs. This works without the need for a AAA server. But, it cannot grow easily beyond the single AP scenario without tedious manual key management. This algorithm is designed to allow the enterprise to add APs easily and still provide the same degree of security as the single AP case without additional work for the administrator.

A second scenario that this algorithm supports is when two isolated WDS join to form a single WMN. The system administrator need only establish a single shared secret between the two connecting APs, then STAs from each WDS will be able to roam freely between the two systems.

Our approach combines a modified Otway-Reese protocol with broadcasting for a novel distributed authentication algorithm within dynamic topologies that easily integrates into the extensible authentication protocol (EAP) and the IEEE 802.11i protocol. Otway-Reese Protocol is a proven security protocol for authentication and key exchange between 3 parties[10]. Our protocol uses the reactive routing protocol which is a mandatory routing protocol of IEEE802.11s standard which defines the WLAN based Mesh Networks.

Our notion is that all entities within a WDS are first class principals. That is both an AP and a STA authenticate in exactly the same method, i.e. both contain an IEEE 802.1X supplicant and authenticator module. In the rest of this paper, we identify the principals in an authentication as the STA for the MP(Mesh Point) that wants to join the WDS, and AP for the MP or Mesh Access Points. If a new AP wishes to join the WDS, then that AP acts as STA for the purposes of this protocol.

3.2 Operation Procedures of the Proposed Scheme

There are two cases we need to consider for authentication in WDS. The first is when the STA wishes to associate with an AP with which it shares a security association which may be either a shared secret or a public key certificate derived from a common authority, or the AP communicates directly with a central AAA server. The second is when the STA wishes to associate with an AP with which it doesn't share a secret, i.e. the AP can not authenticate the station directly nor through a AAA server. In the first case, the standard 802.11i protocol and EAP can be used. In the second case, our algorithm is used in conjunction with the 802.11i protocol to make the authentication of the STA possible in distributed manner.

Our approach enables the authentication without a common security association. When a STA wishes to associate with an AP, with which it does not share a secret, the AP will broadcast the identity of the STA to the WDS in the hopes of identifying an AP within the WDS with which the STA does share a secret. If no AP is found, then the STA cannot be authenticated. If an AP is found, then there are two cases. The first is when the Trusted AP (T) is one hop away from the requesting AP (i.e. T is a neighbor of the requesting AP), and the second is when the T is more than one hop away from the requesting AP. In both cases, reactive routing protocol is used to find T as described in the following section. Fig. 2 depicts the multi hop case.

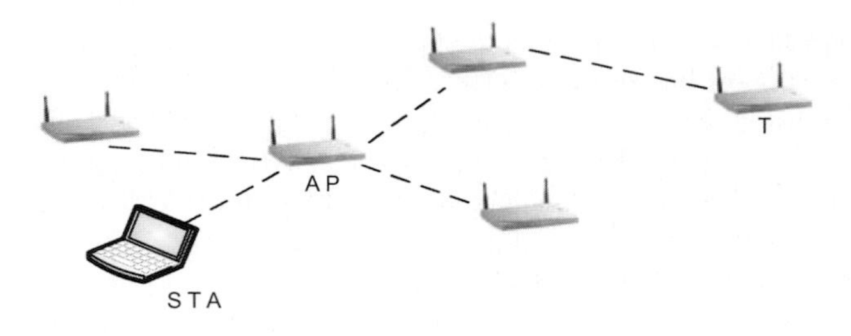

Fig. 2. Multi hop authentication

The protocol uses the following terminologies and has the following assumption for the compatibility with the current 802.11i;

Terminology	Description
E	Encryption E is AES-CCM that is the mandatory encryption method for 802.11i
$E_{K_{A\leftrightarrow B}}(C)$	Means that C is encrypted with AES-CCM with the secret key between A and B
K	128 bit session key that T generates for STA and AP to share
N_{STA}, N_{AP}, N_T	256 bit nonces chosen by STA, AP, and T respectively
M	256 bit transaction identifier chosen by STA

The authentication protocol is as follows. It consists of four phases :

i. Initializing the authentication
ii. Finding T(i.e. finding the Trusted AP)
iii. Authentication and key distribution
iv. Session key distribution

The detailed operation procedure of the proposed scheme is as follows:

Phase 1: Initializing the authentication
As shown in Fig.3, STA starts Traditional EAP to start the authentication procedure. On receiving EAP-START message from STA, AP request the STA's identity,then STA sends its identity to AP. This makes our method compatible with 802.11i.

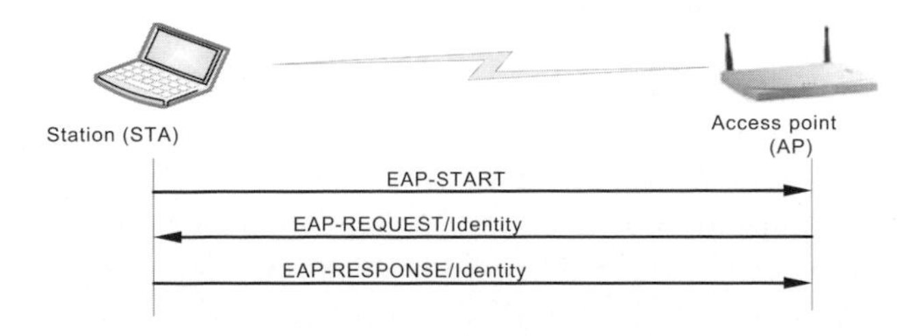

Fig. 3. Initialization of authentication

Phase 2 : Finding T .
In this phase, AP needs to find T. When the STA send the EAP-Response/Identity message to AP in Phase1, STA may or may not know the identity of T. If it knows the identity of T, it can send AP the identity of T. On receiving the identity of T, AP need to check the MeshID of T. Then there are two cases- The Mesh ID of T is the same as that of AP or not.

When the Mesh ID of T is the same as that of AP, AP will check its routing table to check if T is in its routing table. If T is in its routing table, then phase 2 is over. If T is not its routing table, then it sends the "T lookup request message" with the action frame in the RREQ message of IEEE802.11. On receiving this RREQ message, T sends RREP message to AP, and now the route between AP and T is established. If the Mesh ID of T is not the same as that of AP's, T and AP are not in the same WDS. In this case, sending RREQ message and RREP massage procedure is the same, but we should follow the interworking protocol that is used in the given mesh network.

If the STA doesn't know the identity of T, then on receiving EAP-Response/identity message from the STA, AP broadcast the "T lookup request message" using the action frame of the RREQ in the hope of finding T. In this case, AP sends STA's Identity in the action frame to ask the WDS if any node knows this STA. When T, who has pre-shared secret with STA, receives this RREQ, it'll send the RREP message to AP with the routing information. If more than one T responds to the requesting AP, then AP may use any algorithm such as thresh-hold scheme to select one T.

On receiving the RREP, the requesting AP will pick a T with which it already shares a secret. If the requesting AP does not share a secret with any of the responding T, then the closest responder is selected and a shared secret between AP and this responder has to be established prior to responding to the STA. For this, our algorithm works for the requesting AP, and this requesting AP plays the role of the STA of our algorithm at this time.

Using our algorithm, each node in WDS only need to maintain the keys which are authenticated to it, so the number of keys that is to be maintain for each nodes are the number of nodes that are authenticated to that nodes.

Phase 3: Authentication and Key Distribution
In this phase, modified Otway-Reese Protocol is used to distribute the key, k, between AP and STA. Through this step, AP and STA perform the mutual authentication, and share k at the end of the authentication phase.

T, which was found as the result of phase 2, sends nonce N_T to AP for generating k. N_T is encrypted with the key $K_{T\leftrightarrow AP}$ and $K_{T\leftrightarrow STA}$ for AP and STA respectively. On receiving this nonce from T, AP relays N_T which is encrypted with $K_{T\leftrightarrow STA}$ to STA. Now STA received N_T and it generates its nonce N_{STA} and encrypts N_{STA} and N_T with $K_{T\leftrightarrow STA}$. At this time STA may generate the transaction ID, M, for anonymity purpose. Then STA sends this packet with its ID to AP and AP relays this packet to T. By doing this, STA can show its authenticity to T. At the time AP relaying this packet to T, AP also generates its nonce N_{AP} and sends it to T encrypted with the key $K_{T\leftrightarrow AP}$, so that AP can show its authenticity to T. On receiving both N_{STA} and N_{AP}, T can

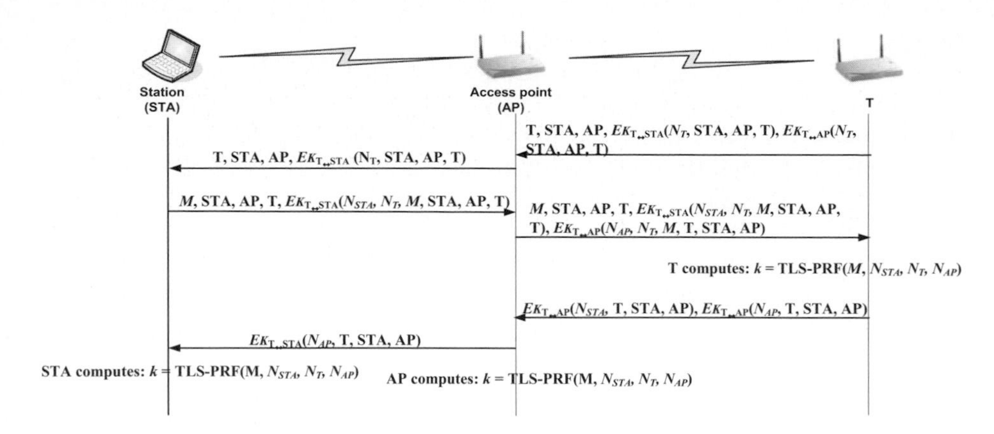

Fig. 4. Authentication & Key distribution phase

compute the master key k using the TLS-Pseudo Random function (TLS-PRF). Then T sends N_{STA} and N_{AP} to AP and then AP relays N_{AP} which is encrypted with $K_{T\leftrightarrow STA}$. Now both AP and STA have N_T, N_{AP}, and N_{STA} so that they can generate the master key k. Now authentication and key distribution is accomplished. Fig. 4 summarizes this procedure.

Phase 4: The last phase is for achieving the session key. This ensures that only the AP and STA know the derived session key as well as proving freshness. Following the completion of the above protocol, AP and STA can complete the IEEE 802.11i 4-way handshake for compatibility with the standards.

In summary, DDA scheme eases the key management burden of WMNs. Also, it reduces the control message overhead induced by single point failure problem by using the distributed authentication scheme.

4 Analytic Evaluation

In this section, we analyze how much our scheme reduces the expected authentication latency. In our simulation, we used the NS-2 implementation of the IEEE 802.11b physical layer and MAC protocols. The radio model was modified to have a nominal bit rate of 11Mb/sec while supporting a transmission range of 250 meter. In addition, we chose as a routing protocol model the Ad Hoc On-demand Distance Vector (AODV) protocol [6]. A peer relationship can be established when two nodes are located in close proximity. We use the random walk mobility model with various pause times and maximum speed [13].

Fig. 5 shows that the authentication delay increases when the load over the network increases regardless of the authentication type (i.e., existing scheme or proposed scheme). However, the proposed scheme has lower authentication delay compared to the existing scheme with the centralized architecture because of the distribution of authentication traffic.

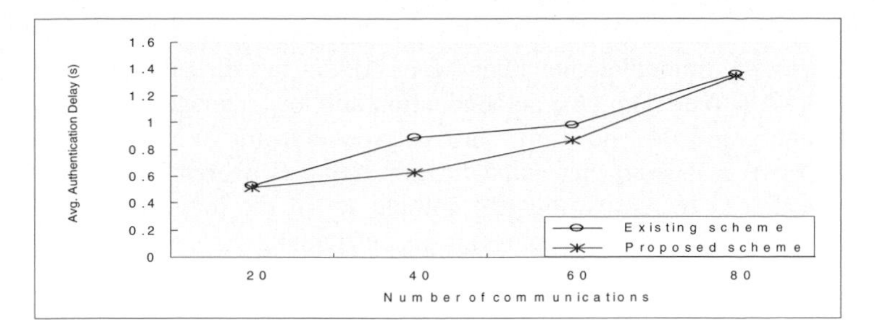

Fig. 5. Effect of distributed authentication in terms of authentication delay

Generally, the objective of increasing the number of node with the authentication functionality is to distribute the load over the nodes, hence, to enhance the performance of the authentication operation and the performance of the network accordingly.

Our simulation (Fig. 6) shows that as the number of trusted nodes (T) is increased, the authentication delay is decreased for multiple communications. This is expected since the distribution of authentication nodes should reduce the authentication overhead, which is expected to positively affect the performance of the authentication operation and hence the network performance. Note that the backoff effect of authentication is decreased by increasing the number of authentication nodes. Therefore, while the increase of the number of authentication nodes tend to decrease the authentication delay due to load distribution, the load on the network increases as a result of having faster flow start. Consequently, this leads to more packets in the network, which may lead to increasing the authentication delay due to contention/interference characteristics of IEEE 802.11 interface.

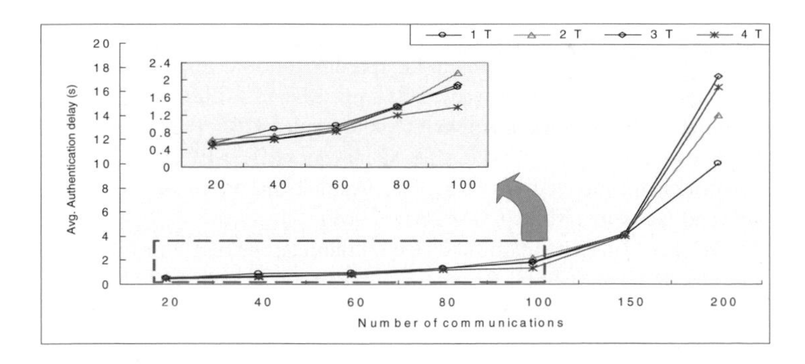

Fig. 6. Authentication delay as the number of communications increases for variable authenticator nodes

5 Conclusion

In this paper, we proposed a Dynamic Distributed Authentication (DDA) method for WMN. Instead of using a central authentication server, we distributed the trusted nodes (T) which can play the role of authentication server over the WMN. This enables the authentication of the new STA possible even though it doesn't have pre-shared key with the immediate neighbor AP or AP doesn't directly communicate with the central AAA server. Thus the authentication and key management of the network becomes simpler, and the end users can easily develop the Wireless Mesh Network using WLANs. We showed that authentication delay is decreased as the number of the trusted node (T) increases, thus the complexity of the network overhead is also decreased, and the network management is also efficient.

References

1. Akyildiz, I.F., Wang, X.: A Survey on Wireless Mesh Networks. IEEE Radio Comm. Mag., 23–30 (September 2005)
2. Yang, H., Luo, H., Ye, F., Lu, S., Zhang, L.: Security in mobile ad hoc networks: challenges and solutions. IEEE Wireless Communications 11(1), 38–47 (2004)
3. Perkins, C., Belding-Royer, E.: Ad-hoc on demand distance vector (AODV) routing. In: IEEE workshop in Mobile computing Systems and Applications (February 1999)
4. Buttyan, L., Hubaux, J.-P.: Report on a working session on security in wireless ad hoc networks. ACM Mobile Computing and Communications Review 7(1), 74–94 (2002)
5. Gong, L.: Increasing Availability and Security of an Authentication Service. IEEE Journal on Selected Areas in Communications 11(5) (June 1993)
6. Pabst, R., et al.: Relay-based deployment concepts for wireless and mobile broadband radio. IEEE Communications Magazine 42(9), 80–89 (2004)
7. Karof, C., Sastry, N., Wagner, D.: TinySec: A Link Layer Security Architecture for Wireless Sensor Networks. In: Proceedings of the Second ACM Conference on Embedded Networked Sensor Systems (SenSys 2004) (November 2004)
8. Jolly, G., Kuscu, M.C., Kokate, P., Younis, M.: A Low-Energy Key Management Protocol for Wireless Sensor Networks. In: IEEE Symposium on Computers and Communications (2003)
9. Chan, H., Perrig, A., Song, D.: Random key predistribution schemes for sensor networks. In: Proceedings of Security and Privacy, 2003, pp. 197–213 (2003)
10. Menezes, A., et al.: Handbook of Applied Cryptography. CRC Press, Boca Raton (1996)
11. Perrig, A., Szewczyk, R., Wen, V., Culler, D., Tygar, J.D.: SPINS: Security Protocols for Sensor Networks. In: Proceedings of Seventh Annual International Conference on Mobile Computing and Networks MOBICOM 2001 (2001)
12. IEEE Std. 802.11i, IEEE Standard for Telecommunications and Information Exchange between Systems-lan/man Specific Requirements, Part 11: Wireless Medium Access Control and Physical Layer (phy) Specifications Amendment 6: Medium Access Control (MAC) Security Enhancements (2004)
13. Capkin, S., Hubaux, J.P., Buttyan, L.: Mobility Helps Security in Ad hoc Networks. In: ACM MobiHoc (2003)
14. IEEE 802.11 TGs document IEEE P.802.11-04/662r14, http://www.ieee802.org/11

A MAP Changing Scheme Using Virtual Domain for Mobile IPv6 Networks*

Jae-Kwon Seo, Tai-Rang Eom, and Kyung-Geun Lee**

Department of Information and Communication Engineering
Sejong University, Seoul, Korea
{jaekwon,tairang,kglee}@nrl.sejong.ac.kr

Abstract. Hierarchical Mobile IPv6 (HMIPv6) supports micro-mobility within a domain and introduces a new entity, namely mobility anchor point (MAP) as a local home agent. However, HMIPv6 has been found to cause load concentration at a particular MAP and longer handover latency when inter-domain handover occurs. In order to solve such problems, this paper establishes a virtual domain (VD) of a higher layer MAP and proposes a MAP changing scheme. The mobile node (MN) changes the routing path between a MN and a correspondent node (CN) according to the mobile position and the direction of the MN before inter-domain handover occurs. The MAP changing scheme enables complete handover by using binding-update of the on-link care of address (LCoA) when inter-domain handover occurs. In addition, the concentrated load of a particular MAP is distributed as well. We simulate the performance of the MAP changing scheme and compare it with HMIPv6.

1 Introduction

The rapid growth of Internet and wireless networks drives the need for IP mobility. The Internet Engineering Task Force (IETF) proposed Mobile IPv6 (MIPv6) which allows a mobile node (MN) to maintain its IP connectivity regardless of its point of attachment. MIPv6 defines the home agent (HA) for mobility and location management. Although an MN might moves to a foreign network, MIPv6 supports mobility by specifying temporary addresses, that is one or more care of addresses (CoA)[1]. However, when an MN changes its point of attachment, the MN has to register both the HA and the correspondent node (CN). Authenticating binding-update requires an approximate 1.5 round-trip time between the MN and each CN. In addition, one round-trip time is needed to update the HA. Therefore, as the number of nodes increases and the MN moves frequently, more traffic is generated and packet loss or longer delay may then occur during the registration process. Hierarchical Mobile IPv6 (HMIPv6) has been proposed to compensate for the problems in employing MIPv6 such as handover latency

* This work was supported by the Korea Research Foundation Grant, KRF-2004-013-D00038, and in part by the Brain Korea 21 project of the Ministry of Education.
** Corresponding author.

T. Vazão, M.M. Freire, and I. Chong (Eds.): ICOIN 2007, LNCS 5200, pp. 659–668, 2008.

and signaling overhead. HMIPv6 supports micro-mobility within a domain and introduces a new entity, namely mobility anchor point (MAP) as a local HA. In HMIPv6, if an MN moves off the home network to a foreign network, the MN generates two temporary addresses, an on-link CoA (LCoA) and a regional CoA (RCoA)[2].

Although HMIPv6 performs efficiently in supporting micro-mobility, HMIPv6 is not appropriate in supporting macro-mobility. HMIPv6 leads to longer handover latency with more packet loss than MIPv6 when inter-domain handover occurs because the MN has to register two CoAs. Owing to these problems, an MN uses a distance-based selection scheme in choosing a MAP by default. The distance-based selection scheme involves the MN selecting the furthest of the MAPs to register with using the DISTANCE field within the MAP option included in the router advertisement (RA) message. This is because the handover occurs frequently if the MN selects the lower layer MAP with a narrow domain area. However, if the number of MNs is increased, all MNs select the furthest MAP by a distance-based selection scheme. In other words, the selected MAP may become the focal point of a performance bottleneck prompting a longer processing latency. Consequently, if an MN selects a higher layer MAP (HMAP), the load is concentrated at the MAP and if an MN selects a lower layer MAP (LMAP), more frequent inter-domain handover occurs. Thus, various MAP selection schemes considering mobile speed and range of an MN and inter-domain handover schemes have been proposed. In this paper, we propose an efficient MAP changing scheme that changes MAP before the MN moves off the domain of the LMAP or the virtual domain (VD) of the HMAP. We evaluated the performance of the MAP changing scheme in comparison to HMIPv6 through simulation experiments.

The remainder of this paper is organized as follows. Section 2 introduces related works about MAP selection schemes. The proposed MAP changing scheme is presented in Section 3. A simulation model and the results for performance evaluation are presented in Section 4. Finally, we conclude the paper in Section 5.

2 MAP Selection Schemes

In multilevel HMIPv6, the load is concentrated at a HMAP because a MN uses a distance-based MAP selection scheme (or furthest MAP selection scheme) without considering the MN's mobile range and location. This is because a MAP performs encapsulation and decapsulation for every packet destined for MNs. Therefore, various MAP selection schemes that take the characteristics and states of MNs into account have been proposed for the load distribution of a MAP.

One of these which limit the number of MNs to be registered at a MAP is introduced for the purpose of load control[3,11]. In this scheme, if a MAP receives a binding-update of a MN, the MAP checks the maximum number of MNs to be registered and decides whether to receive or to reject the registration request of an MN. The MN which is rejected by a certain MAP then selects the next candidate MAP, and hence the load concentration of a specific MAP is restrained.

In [4,5,6] velocity-based MAP selection schemes are introduced. In these velocity-based schemes, a MAP is selected depending on the estimated velocity of the MN. In [4], the MN's binding-update interval is employed to estimate the velocity of the MN. Then, a fast MN selects an HMAP because inter-domain handover occurs frequently. Meanwhile, a slow MN selects an LMAP. Thus, the load is distributed between the HMAP and LMAPs. In [5], an MN selects a MAP depending on the estimated velocity of the MN and if the number of MNs to be registered to the MAP are exceeded the load is distributed by selecting another MAP. In [6], this scheme considers both the velocity of an MN and the moving range of the MN using MN's 'mobile history'. Most of these MAP selection schemes limit the maximum number of MNs in order to distribute the load, or select the MAP considering the range of moving velocity of the MN.

However, estimating the velocity of the MN using binding-update interval may not reflect the current velocity of the MN. In other words, MNs do not always move with constant velocity and direction. Therefore, this paper proposes an efficient MAP changing scheme using virtual domain (VD).

3 MAP Changing Scheme Using Virtual Domain

The MAP changing scheme satisfies the fast inter-domain handover and load distribution of MAPs. The MAP changing scheme is a simple method which predictably changes the MAP by considering only current moving direction and position of the MN. However, our prediction system reduces the significant overhead generated by complex computation procedures. The basic idea of the MAP changing scheme is as follows: if an MN moves to an edge access router (AR) of the current MAP, the MN performs MAP changing to the HMAP while receiving packets before an inter-domain handover. This is because that there is

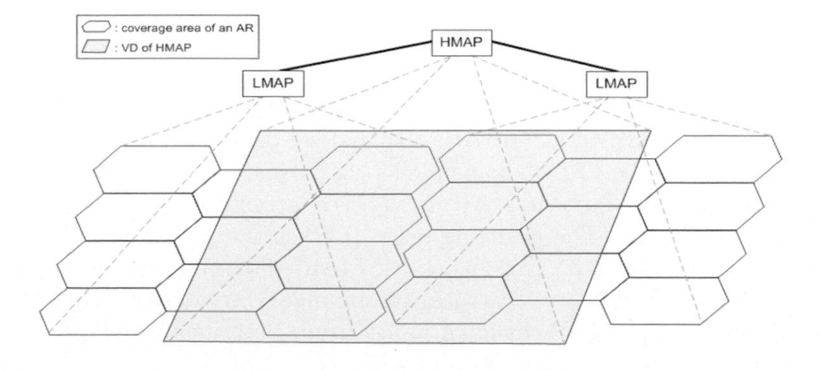

Fig. 1. The virtual domain of HMAP

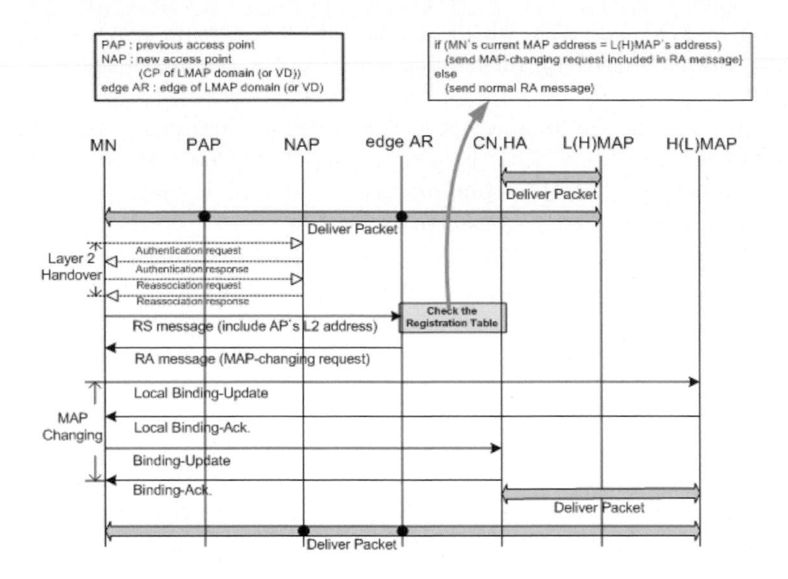

Fig. 2. Message flow of the MAP changing procedure

a higher probability of the inter-domain handover. Thus, the MN completes the inter-domain handover with binding-update of LCoA only. However, such a predictive method generates significant cost when the prediction is incorrect. If this happens the MN performs MAP changing again to the previous LMAP. As a result, the MAP changing operation generates unnecessary traffic.

The MAP changing scheme reduces the costs of such prediction errors by assigning a VD. The VD is to assign ARs which have higher probability of the inter-domain (LMAPs) handover among the ARs that belong to the HMAP domain. Fig. 1 shows the VD of HMAP. HMAP is a high prefix router of LMAPs. By assigning the VD, even though the MN changes its direction right after performing MAP changing operation and moves back into the domain of the previous LMAP, the probability of performing frequent MAP changing operation is reduced. This is because the current AR is far from the AR that has to perform MAP changing operation.

Fig. 2 depicts the major steps of the MAP changing procedure. As mentioned above, the VD is assigned among ARs that belong to the HMAP domain. Edge ARs of the VD or the LMAP domain keep the *registration table*. The address of both the MN and current MAP of the MN are stored in the *registration table* of the edge ARs. The changing point (CP) is the area where the layer 3 interacts with the layer 2. This paper uses IEEE 802.11 based networks and access point(AP)s which are located in the boundary area of the edge ARs are operated as CPs. The MAP changing procedure is described as follows:

– Step 1: If an MN moves from the previous access point (PAP) to the new access point (NAP) which operates as CP, the MN sends the router

solicitation (RS) message including the NAP's link layer address to an edge AR of the VD (or the LMAP domain).
- Step 2: The edge AR which received the RS message checks one's *registration table* because the MN locates in the CP. If the address of current MAP of the MN in the *registration table* is the address of the L(H)MAP, the AR sends a MAP-changing request message included in the router advertisement (RA) message to the MN's LCoA.
- Step 3: The MN receives the MAP-changing request message, then the MN generates RCoA and performs local binding-update to the H(L)MAP.
- Step 4: The MAP changing procedure is completed when the MN performs the binding-update to the HA. The MN then receives packet through new MAP.

The MAP changing is operated by sending binding-update messages to MAP and HA while the MN receives packets. If the MAP changing operation is completed, the MN enters into the VD of HMAP. HMAP intercepts packets destined to the RCoA and tunnels them to the LCoA. Then, once the inter-domain handover has occurred, between LMAPs, the MN completes the handover using binding-update with LCoA only to HMAP. Although the MN moves back to the previous LMAP again after the MAP changing operation, the MN does not need another MAP changing operation since the MN has been within the VD of HMAP. The criterion of MAP changing operation is only the current MAP of the MN at the CP of edge ARs. Thus, the overheads provided by complex computation procedures are avoided. In addition, The MAP changing procedure does not generate packet loss because the procedure operates during the communication[10,12].

Fig. 3 presents the ping-pong movement of the MN. If the MN moves as in Fig. 3, the MN has to perform global binding-update in the nearest MAP selection scheme. Then, critical packet loss and delay may occur by inter-domain handover at every handover. If the MN selects the HMAP, the ping-pong problem does not occur. However, if all MNs in each domain select the HMAP, the load is concentrated at the HMAP. Consequently, if MNs select an HMAP for solving the ping-pong problem, the load is concentrated at the HMAP. On the other hand, if MNs select a LMAP for the load distribution at the HMAP, inter-domain handovers occur frequently and also the ping-pong problem may occur. In the velocity based selection scheme, fast MNs select the HMAP for avoiding frequent inter-domain handover and slow MNs select LMAP for load distribution. However, slow MNs can also move as illustrated in Fig.3. In addition, the velocity measured using binding-update interval time may not be the current velocity of the MN. More effort is required towards for increasing the accuracy of measurements. Computational overheads, as expected, are increased.

In the MAP changing scheme, if the MN registered with the LMAP performs MAP changing operation at the CP of an LMAP domain, the MN is registered with the HMAP. After this, the ping-pong problem does not occur at the boundary of the LMAP domains since the MN can move using local binding-update only, in area A (Fig. 3). If the MN continually moves back and forth beyond area B (Fig. 3), frequent MAP changing occurs. However, the frequency of MAP

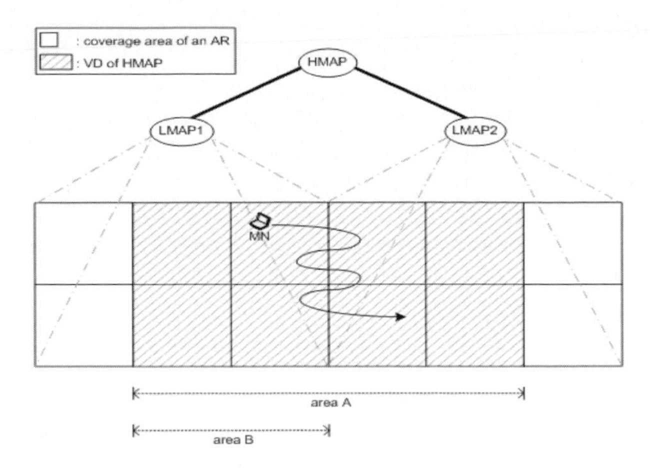

Fig. 3. Effect of the ping-pong movement

changing is decreased in comparison with the ping-pong problem. This is because the area B is always broader than the area in which ping-pong problems occur.

In addition, area B in Fig. 3 is within the domain of LMAP1 and included in the VD of HMAP as well. Then, the MNs registered with HMAP and LMAP1 coexist in this area. Thus, the MNs registered with the MAPs are distributed meaning the load concentration of a particular MAP is distributed. If we assume that the MN moves the same distance, in the proposed algorithm, the average number of registrations with the HA and the CN is increased compare to HMIPv6. However, the entire load added to the MAP is reduced since the MAPs need to encapsulate all the packets toward the MNs and then the MNs linked by the MAP are distributed.

4 Performance Evaluation

Simulation Setup. In this section the performance of the MAP changing scheme is evaluated using NS-2[7,8]. Simulations were carried out so as to examine both the inter-domain handover performance and the degree of load distribution as the number of MNs.

First, we built a simulation environment for comparing the MAP changing scheme with HMIPv6 (furthest and nearest) as in Fig. 4. The VD of HMAP is established from AR3 to AR6 and then the MN changes the MAP from LMAP1 (or LMAP2) to HMAP at the CP of AR4 (or AR5). If the MN is located in AR3 (or AR6), the MN performs the MAP changing operation from HMAP to LMAP1 (or LMAP2) at the appropriate CP.

The CN and the MN operate as a TCP source and a TCP sink and both use an FTP as an application. Initially, a home address is assigned to the MN and the MN locates at AR1. In 5 seconds, the CN connects the TCP with

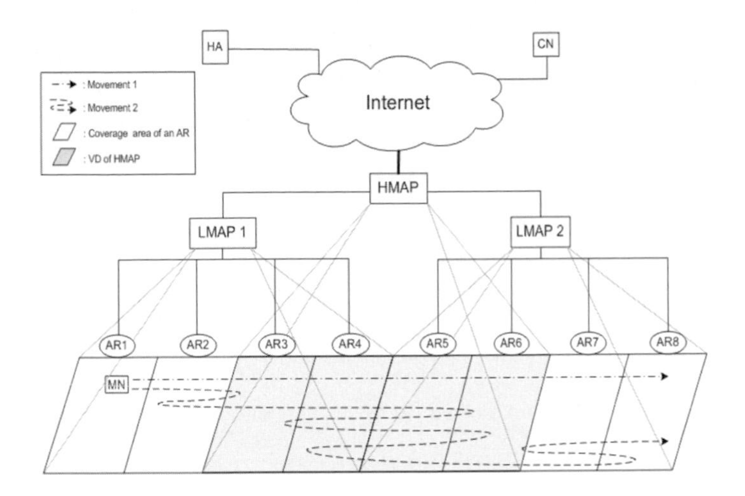

Fig. 4. The network environment used in simulation

the MN and sends FTP traffic. After 10 seconds, the MN moves according to movement 1 in the figure for comparing the handover latency of each scheme in the first experiment. In the second experiment, the MN moves according to movement 2 for comparing the effect of ping-pong movement in each scheme. This simulation scenario includes measuring load distribution by increasing the number of MNs and the start of movements from different positions and mobile directions. We measured the number of encapsulated packets from each MAP to the MNs respectively over 120 seconds.

Simulation Results. We measured the handover delay in order to compare the MAP changing scheme with HMIPv6 at simulation. As previously stated, with HMIPv6 the MN selects the HMAP by the furthest MAP selection scheme. Fig. 5 shows the TCP sequence number received by MN when the inter-domain handover between LMAPs occurs. Every handover is completed with only local binding-update in the furthest MAP selection scheme because the MN selects the HMAP. Thus, the packet delay between the MN and the CN equal to the packet delay of the intra-domain handover.

In nearest MAP selection scheme, we observed the longer packet delay when inter-domain handover occurred between the domain of LMAP1 and LMAP2. This is because the MN has to register both RCoA and LCoA to LMAP2 and HA when the MN moves from the AR4 to the AR5. The MAP changing scheme completes handover by the binding-update with LCoA only because the MAP changing scheme performs the MAP changing operation before the inter-domain handover and then the MN enters into the VD of HMAP. Therefore, the packet delay is approximately equal to the furthest MAP selection scheme.

Fig .6 shows the TCP sequence number when the MN moves according to movement 2 as shown in Fig.4. In the nearest MAP selection scheme, packet

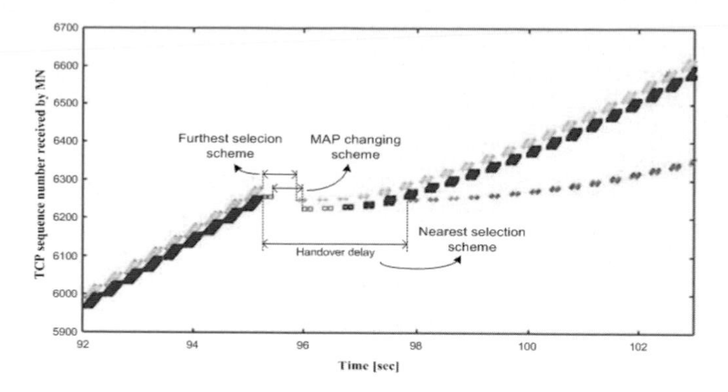

Fig. 5. TCP sequence number received by MN when the inter-domain handover occurs

throughput is decreased since handover delay is increased when the MN moves back and forth at the boundary of LMAP domains. We can observe from Fig. 6 that the packet throughput is decreased when the MN starts it's ping-pong movement. On the other hand, the MAP changing scheme completes handovers using binding-update of LCoA only, during the simulation. This is because the MAP changing scheme performs the MAP changing operation at the boundary of the LMAP domains and the MN moves toward the VD of HMAP. Thus, the MAP changing scheme shows approximately equivalent performance with the furthest MAP selection scheme in terms of inter-domain handover. In the velocity based selection scheme, a fast MN selects the HMAP and a slow MN selects the LMAP. The handover performance of the fast MN is equivalent to the furthest MAP selection scheme and the slow MN is equivalent to the nearest MAP selection scheme. The handover performance of the MAP changing scheme is always equivalent to the furthest MAP selection scheme regardless of the MN's speed. In addition, the MAP changing scheme does not generate computational overheads by measuring the velocity of MNs.

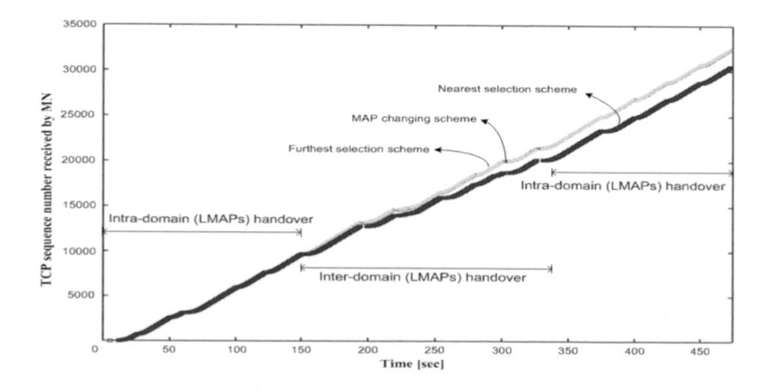

Fig. 6. TCP sequence number received by MN when the ping-pong movement occurs

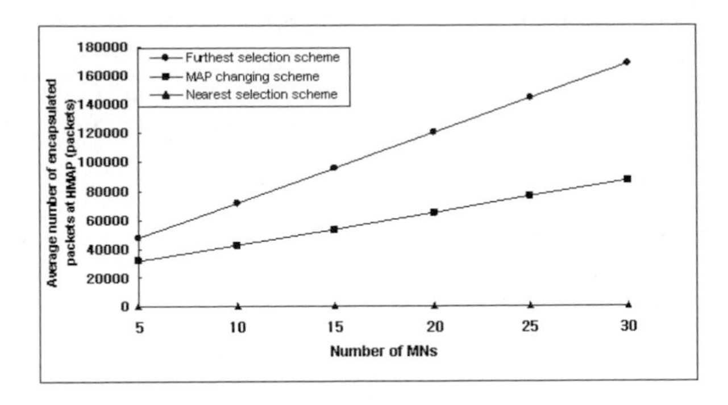

Fig. 7. The load of HMAP

Fig. 7 presents the average number of the encapsulated packets at HMAP when the MN selects HMAP using the furthest MAP selection scheme in HMIPv6 and changes the MAP according to the MN's mobile direction and location in the MAP changing scheme. In the nearest MAP selection scheme, tunneling overheads at the HMAP are zero because MNs do not select the HMAP. The load is concentrated at HMAP when the MNs select HMAP since all packets which are sent to the all MNs' RCoA are encapsulated at the HMAP. The results of the MAP changing scheme show the number of the encapsulated packets to be distributed. The reason is that the MNs change the MAP as they move back and forth between the domain of LMAP1 and LMAP2 and the VD of HMAP.

Summarizing the simulation results, longer latency is not generated by registering the RCoA when the MN selects HMAP. This is because the inter-domain handover does not occur when the MN moves between AR4 and AR5. However, a slightly different scenario generates load concentration at HMAP when it is selected by the MN since the domain of HMAP is enlarged and it intercepts and forwards all the packets sent to the MN's RCoA. In contrast, when the MNs select LMAP1 and LMAP2, the packet encapsulation process is distributed between LMAP1 and LMAP2 according to the MNs' location. However, longer latency is generated in case of inter-domain handover by registering the RCoA with the HA and the CN. It has been shown that the MAP changing scheme solves the longer latency problem in case of inter-domain handover and the load concentration at the HMAP by employing the furthest MAP selection scheme. This is because the MAP changing scheme predictably performs the MAP changing operation according to the information of the MN's mobile direction and location using the VD.

5 Conclusion

HMIPv6 has been proposed to compensate for the problems in employing MIPv6 such as handover latency and signaling overhead. However, HMIPv6 causes longer handover latency than MIPv6 when inter-domain handover occurs and also the load concentration is found at a particular MAP. In order to solve such problems, we propose a MAP changing scheme using the VD of the HMAP whereby the routing path is changed between the MN and CN before an inter-domain handover occurs. In order to evaluate the performance of the MAP changing scheme, we performed simulation experiments. Our results show that the MAP changing scheme reduces the inter-domain handover latency. In addition, the load concentration of a particular MAP is distributed because MNs registered with the HMAP and the LMAP coexist in the VD. In the future work, we will decide the optimal range of a VD among the ARs which are connected to the MAP.

References

1. Johnson, D., Perkins, C., Arkko, J.: Mobility Support in IPv6, RFC3775 (June 2004)
2. Soliman, H., Castelluccia, C., El Malki, K., Bellier, L.: Hierarchical Mobile IPv6 mobility management, RFC4140 (August 2005)
3. Pack, S.H., Lee, B.W., Choi, Y.H.: Proactive Load Control Scheme at Mobility Anchor Point in Hierarchical Mobile IPv6 Networks. IEICE Trans. Inf.& Syst. E87-D(12), 2578–2585 (2004)
4. Kawano, K., Kinoshita, K., Murakami, K.: A Multilevel Hierarchical Distributed IP Mobility Management Scheme for Wide Area Networks. In: Proc. IEEE ICCCN 2002, Florida, USA, pp. 480–484 (October 2002)
5. Bandi, M., Sasase, I.: A Load Balancing Mobility Management for Multilevel Hierarchical Mobile IPv6 Networks. In: Proc. IEEE PIMRC 2003, Beijing, China, vol. 1, pp. 460–464 (September 2003)
6. Kumagai, T., Asaka, T., Takahashi, T.: Location Management Using Mobile History for Hierarchical Mobile IPv6 Networks. In: Proc. IEEE GLOBECOM 2004, Texas, USA, vol. 3, pp. 1585–1589 (December 2004)
7. Network Simulator 2, http://www.isi.edu/nsnam/ns
8. Hsieh, R., Seneviratne, A., Soliman, H., El-Malki, K.: Performance Analysis on Hierarchical Mobile Ipv6 with Fast-handoff over End-to-End TCP. In: Proc. IEEE GLOBECOM 2002, Taipei, Taiwan, vol. 3, pp. 2488–2492 (November 2002)
9. Omae, K., Okajima, I., Umea, N.: Mobility Anchor Point Discovery Protocol for Hierarchical Mobile IPv6. In: Proc. IEEE WCNC 2004, Atlanta, USA, vol. 4, pp. 2365–2370 (March 2004)
10. Park, H.K.: An Inter-MAP Handover Algorithm Dynamic MAP Switching, Master's thesis, Dept. of Electrical and Electronic Engineering, Yonsei University (2003)
11. Pack, S.H., Choi, Y.H.: A Study on Performance of Hierarchical Mobile IPv6 in IP-based Cellular Networks. IEICE Trans. Commun. E87-B(3), 462–469 (2004)
12. Seo, J.K., Lee, K.G.: Efficient MAP Changing Algorithm for Macro-Mobility and Load Distribution in Multilevel Hierarchical Mobile IPv6 Networks. In: Proc. WMSCI 2006, Florida, USA, vol. 2, pp. 380–385 (July 2006)

Proposal and Evaluation of a Network Construction Method for a Scalable P2P Video Conferencing System

Hideto Horiuchi, Naoki Wakamiya, and Masayuki Murata

Graduate School of Information Science and Technology, Osaka University
1–5 Yamadaoka, Suita-shi, Osaka 565–0871, Japan
{h-horiuti,wakamiya,murata}@ist.osaka-u.ac.jp
http://www.anarg.jp

Abstract. Recently, video conferencing systems based on peer-to-peer (P2P) networking technology have been widely deployed. However, most of them can only support up to a dozen of participants. In this paper, we propose a novel method to construct and manage a P2P network for a scalable video conferencing system. Our method consists of three parts: a network construction mechanism, a tree reorganization mechanism, and a failure recovery mechanism. First, the network is constructed as new peers join a conference. Then, the tree topology is reorganized taking into account the heterogeneity of the available bandwidth among peers and their degree of participation so that, those participants, i.e., peers that can have many children peers and/or often speak are located near the root of the tree. As a result, the delay from speakers to the other participants is reduced. Through simulation experiments, we verify that our tree reorganization mechanism can offer smooth video conferencing.

Keywords: video conferencing, P2P, scalability, tree construction.

1 Introduction

With the proliferation of the Internet, video conferencing systems are getting widely accepted making it possible to have a meeting or a discussion among people at different and distant places. Especially, video conferencing systems using an application level multicast (ALM) technology based on P2P communication have been introduced due to their ease of deployment and low cost of operation [1] [2]. However, they still have the scalability problem and most of them can only support at most a dozen of participants. For example, a company with worldwide branches and convenience chain stores may involve hundreds of managers in a business meeting. There has been a number of research works in scalable ALM algorithms, but they mainly consider distribution type of applications [3] [4]. Therefore, we need a video conferencing system that can accommodate hundreds or thousands of interactive participants.

In this paper, we propose a novel method for constructing and managing a P2P network for a scalable video conferencing system. We assume that

T. Vazão, M.M. Freire, and I. Chong (Eds.): ICOIN 2007, LNCS 5200, pp. 669–678, 2008.

participants, i.e., peers, dynamically join and leave a conference. Peers are heterogeneous in terms of the network capacity available for video conferencing. Our proposed system consists of three parts: a *network construction mechanism*, a *tree reorganization mechanism*, and a *failure recovery mechanism*. The network construction mechanism sets up a hierarchical distribution network, which consists of distribution trees consisting of tens or hundreds of peers, and a core network which interconnects these trees with each other. To have a smooth conference, it is necessary to keep the delay from speakers to other participants small. To accomplish this goal, we focus on the fact that the number of simultaneous speakers is limited whereas speakers dynamically change in accordance with the agenda. The tree reorganization mechanism dynamically reorganizes a distribution tree so that speakers are located near the root in a distribution tree. In addition, to reduce the height of a distribution tree, the tree reorganization mechanism dynamically moves peers with higher available bandwidth toward the root. Furthermore, in the case of failure in distribution of conference data due to a halt or disappearance of a peer, the failure recovery mechanism reconfigures the distribution network through local interactions among peers using local information acquired during network construction.

The rest of this paper is organized as follows. We describe our proposal in Section 2. Then, we present some simulation results in Section 3. Finally, we summarize the paper and describe some future work in Section 4.

2 Network Construction Method for Scalable P2P Video Conferencing

In this section, we give an overview of the scalable P2P video conferencing system consisting of the network construction mechanism, the tree reorganization mechanism, and the failure recovery mechanism. In the following, we use the terms peer and participants interchangeably.

2.1 Overview of Scalable P2P Video Conferencing System

Our system consists of a login server, peers, and a distribution network. Delivery and exchange of streaming data, i.e., video and audio are done through the distribution network. For low bandwidth requirement and management cost, we adopt a shared-tree architecture to the distribution network. The distribution network consists of the core network and the distribution trees in which the root is connected to the core network. In this paper, we call a peer which belongs to the core network *leader peer*, and all other peers *general peers*. A leader peer manages the IP addresses of neighboring leader peers and all children peers that are directly connected to it. A general peer keeps the IP addresses of its parent and children, and the list of the IP addresses, which it knows, in its *ancestor list*. Peers have a limitation on the number of acceptable children called *fanout* in accordance to their available bandwidth. The login server is responsible for registration and management of the conference, and the authentication of

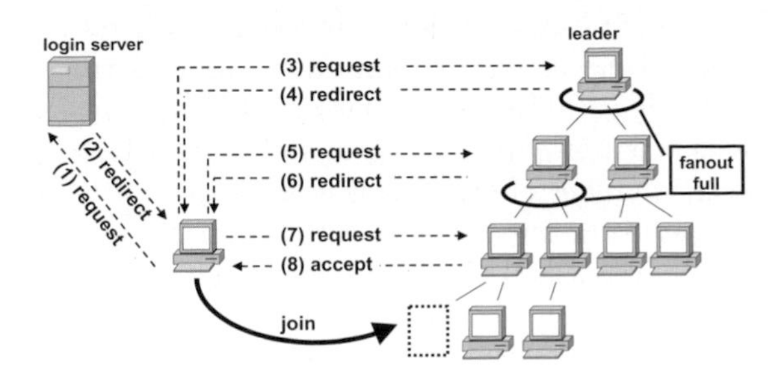

Fig. 1. Participation to a tree through sequential introduction

participants. It manages only information of leader peers and the number of general peers in each tree, and not the structure of each tree.

The overview of the system behavior is as follows. First, a newly participating peer requests the login server for authentication. At this time, the participating peer is notified whether it should become a leader peer or general peer. Then, it connects to either the core network or a distribution tree to join the conference. Then, the participant is involved in the conference as a speaker or an audience in accordance with the agenda. Since we do not consider any management of speech coordination in this proposal, all participants can speak freely. Streaming data from a speaker is once transmitted to the root of the tree to which it belongs, and then broadcasted to the other peers in the tree and to peers in the other trees via the core network. Our method makes peers with high fanout, i.e., high bandwidth, and active speaking move to the root of tree. We call this *promotion*. The promotion reduces the tree height and delay between active peers and others. In video conferencing systems, peers may leave because of failures in routers or links. So our method dynamically recovers from the failure in the distribution network so that peers can continuously receive streaming data.

2.2 Network Construction Mechanism

In our method, a participating peer first gets authenticated by the login server and then connects to the distribution network. With consideration of the fanout of the peer and the number of peers in each tree, the login server determines the role of the peer. If a participating peer is determined as a leader peer, the peer gets the IP address list of other leader peers, measures delay to them, and connects to the neighbor peers. If the participating peer is specified as a general peer, it connects to the designated distribution tree by sequential introduction of a temporary parent as shown in Fig. 1 [5]. First, the login server notifies the participating peer of the IP address of an appropriate leader peer as a temporary parent (Fig. 1:1-2). In our mechanism, the leader peer to be introduced is selected in a round-robin fashion. Therefore, without any peer leaving, the number of peers is equal among trees. The participating peer deposits the notified IP address in its ancestor list and sends a participation request message to the temporary parent (Fig. 1:3). The temporary parent which receives the participation request message compares its fanout with the number of children. If its number of children is less than $fanout - 1$, the temporary parent accepts the request and connects to the peer. The reason for comparing with $fanout - 1$ is that the tree reorganization mechanism requires one spare link as will be explained later. On the other hand, if the number of children is equal to $fanout - 1$, the temporary parent introduces one of its direct children to the participating peer as a new temporary parent. We call this procedure *redirect* (Fig. 1:4). If the temporary parent has information about the topology of its descendants, by introducing a child with the lowest or smallest subtree, we can build a balanced tree. However, for this purpose, peers have to maintain the up-to-date information by exchanging control messages very often. Therefore, in our mechanism, we consider that a new temporary parent is selected among children in a round-robin fashion. We can expect that a distribution tree is constructed in breadth-first order and the delay from the leader peer can be reduced. The participating peer adds the IP address of the introduced temporary parent to its ancestor list and sends a participation request message (Fig. 1:5). By repeating these procedures, the participating peer can eventually connect with the temporary parent, which has an available link, and join the distribution tree (Fig. 1:6-8). The participating peer has all IP addresses of its ancestors in the ancestor list when connecting to the tree.

In this mechanism, there is only small overhead at the login server and peers, because no centralized unit manages the distribution tree topology and the additional load of processing messages occurs only at temporary parents. An additional advantage is that a peer can reconfigure the distribution tree during failure with only local interaction because a peer has the knowledge of the complete ancestor list.

2.3 Tree Reorganization Mechanism

In our method, a peer with high activity and high fanout moves to the root of the tree for low delay and smooth conferencing. We call it *promotion*. In addition, to reduce tree height by completing the fanout, a peer which has less than $fanout - 1$ children invites its grandchild as a direct child. In this section, we describe the details of this mechanism.

Peer Promotion. A peer starts the promotion process if the participant speaks continuously. Additionally, a peer compares its fanout with that of its parent periodically. If the fanout is more than its parent, the peer starts the promotion process. However, the promotion process does not occur if the peer is involved in other tree reorganization or failure recovery. The promotion means that a peer becomes a child of its grandparent as shown in Fig. 2. Firstly, peer A which starts the promotion sends a promotion request message to parent peer B and its children (Fig. 2:1). If a peer receiving the request is involved in other tree reorganization or failure recovery, it rejects the request, otherwise it sends back an accept message. The accept message from peer B has the IP address of its parent peer C. On receiving the accept message from all peers, peer A sends a

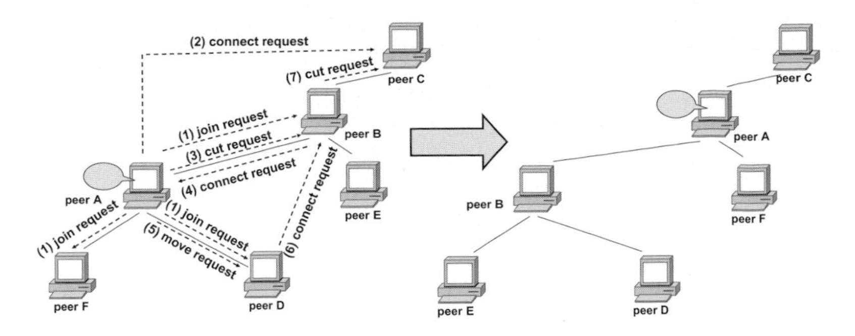

Fig. 2. Promotion of peer A for speaking

connection request message to peer C (Fig. 2:2). If peer C is involved in other tree reorganization or failure recovery, peer C sends back a reject message, otherwise it makes a connection to peer A and sends back an accept message. If the number of children becomes equal to the fanout on peer C, the accept message from peer C includes information indicating that the spare link is used.

After connecting with peer C, peer A sends a disconnection request message to its previous parent B (Fig. 2:3). This request includes information whether the spare link of peer C is used. After receiving the request, peer B terminates the connection with peer A. If the spare link is not used on peer C, the promotion is completed at this time. Now, both peer A and B are children of peer C.

If the spare link is used on peer C, peer B, the previous parent of the promoted peer A, becomes a child of peer A to make one link free on peer C. First, peer B sends an adoption request message to peer A (Fig. 2:4). If the number of children is less than $fanout - 1$ on peer A, peer A accepts peer B as its child. On the other hand if equal, peer A sends a moving request message to peer D which is selected from peer A's children in a round-robin fashion to make a room for peer B (Fig. 2:5). The request message includes the IP address of peer B. Then peer D sends a connection request message to peer B (Fig. 2:6). Peer B accepts peer D as its child, peer D terminates the connection with peer A, and peer A becomes a parent of peer B. Then, peer B sends a disconnection request message to peer C (Fig. 2:7) and peer C terminates the connection. As a result, peer C obtains a new spare link. In this way, the promotion is completed.

Completing the Fanout. Peers periodically compare the number of their children with the fanout. If the number is less than $fanout - 1$, a peer starts completing the fanout. However, if a peer is involved in other reorganization or failure recovery, the process does not occur. Peer A, which can accommodate more children, sends an introduction request message to peer B which is selected in a round-robin fashion from its children (Fig. 3:1). If peer B does not have any children, peer B sends back a reject message to peer A. Otherwise, peer B sends a moving request message to peer C which is selected in a round-robin fashion from its children (Fig. 3:2). The moving request includes the IP address of peer A. Peer C sends a connection request to peer A (Fig. 3:3) and makes the connection.

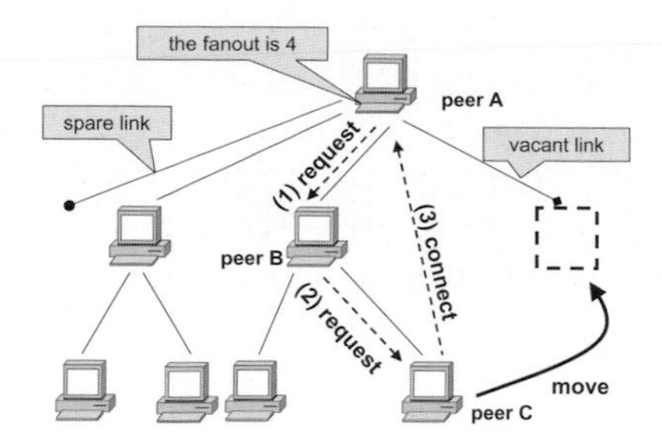

Fig. 3. Completing the fanout of peer A

After establishing the connection with peer A, peer C terminates the connection with peer B and this process is completed. If either of peer B and C or both are involved in other tree reorganization or failure recovery, peer A receives a reject message and the process is canceled.

2.4 Failure Recovery Mechanism

A peer may become to be unable to receive data due to not being able of accessing its temporary parent in tree construction/reorganization, a halt of links or routers, or a parent peer leaving the conference. We define this event as *failure*. In the failure recovery mechanism, a peer detecting a failure tries to make a new connection with another peer in its ancestor list [5].

If a peer fails in sending a message to a temporary parent, it sends a re-connection request message to the previous temporary parent, which introduced the missing temporary parent. On the other hand, if a peer detects the leaving or a fault of its parent, it sends a re-connection request message to its grandparent in the ancestor list (Fig. 4:1,3). In both cases, the IP address of the missing parent is removed from the ancestor list. If the recovering peer fails in sending the re-connection request message due to departure of the new temporary parent, it first removes the corresponding IP address from the ancestor list and then moves to the next ancestor at the bottom of the list. If the list becomes empty, the recovering peer goes to the login server and joins the distribution tree again

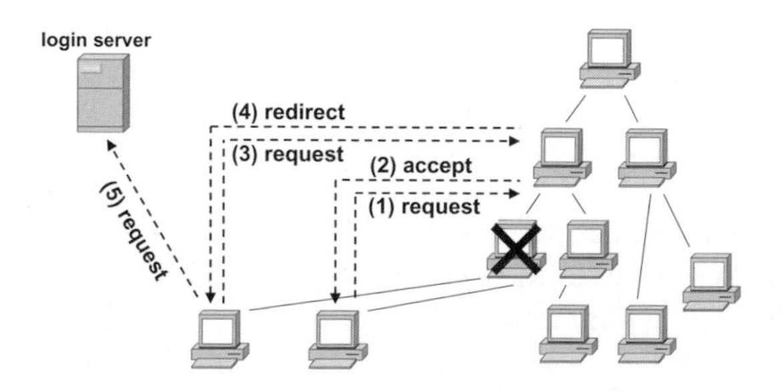

Fig. 4. Failure recovery

as a new peer. (Fig. 4:5). On receiving the re-connection request message, the temporary parent establishes a connection with the recovering peer if the number of children is less than $fanout - 1$ (Fig. 4:2), or introduces a child to the recovering peer as a new temporary parent otherwise (Fig. 4:4). In the latter case, the requesting peer eventually joins the tree and reorganizes its ancestor list by the same process as the initial join.

If a child of a leader peer detects the failure of the leader peer, the peer notifies the login server of the failure. The login server appoints the peer which first sends the notification as a new leader peer and updates the information of leader peers. The other children of the missing leader peer also report the failure and are redirected to the new leader peer.

3 Simulation Experiments

In this section, we show simulation results to evaluate the tree reorganization mechanism which contributes to smooth video conferencing.

3.1 Simulation Conditions

In this paper, we focus on the performance and effectiveness of the tree reorganization mechanism and thus consider a single distribution tree in simulation experiments. Evaluation of the whole proposal is left as a future work. First, we create a physical network, which follows the power-law principle based on BA model [6] using BRITE [7]. The average degree of this network is 2, and it consists of 101 nodes, i.e., routers. Each router has one peer to participate in the conference. One peer of these serves as the login server. The fanout is fixed during the simulation and is equal to the degree of the designated router plus 1. The delay among peers is computed by physical hops over the shortest path by the Dijkstra method ignoring the access link between a peer and a router, and the propagation delay over one physical link is 1 msec. We do not consider transmission delay and processing delay. Peers participate in the conference at random time with uniform distribution from 0 to 10 seconds. The first participating peer becomes the leader peer. After 100 peers participate in the conference and construct the tree, no further peer joins and leaves the conference.

After all peers participate in the conference, a peer begins to speak and the tree reorganization is conducted. We call peers to speak as *candidates*. Ten candidates are randomly chosen at the start and are fixed during the simulation. The duration of each speech is exponentially distributed with a mean value of 6 seconds [8] and the minimum duration is 1 msec. Any one of the candidates is always speaking during the simulation. In other words, when a candidate stops speaking, the next speaker is randomly chosen among candidates and starts speaking immediately. The same candidates would be chosen as the next speaker, but only one candidate speaks at the same time. In the following figures, time zero corresponds to the instant when the first speech starts. A speaking peer starts the promotion when it continuously speaks for more than 5 seconds, and as long

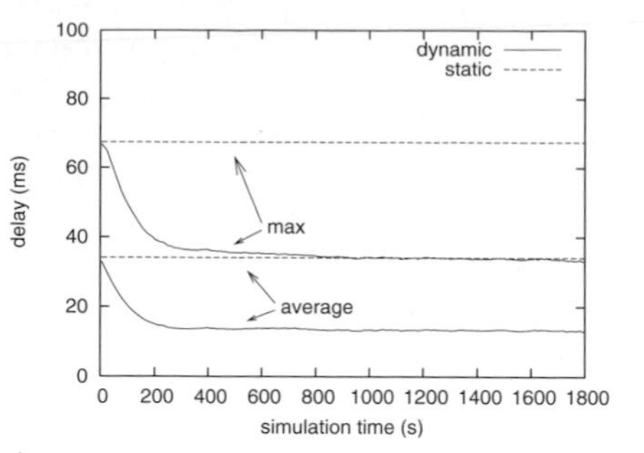

(a) Delay between the leader peer and all peers

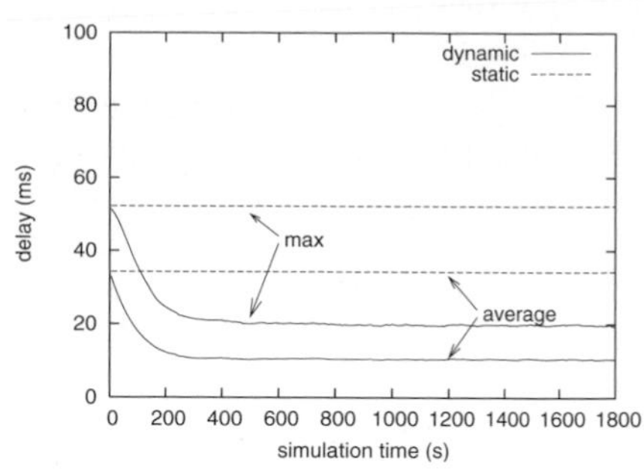

(b) Delay between the leader peer and candidates

Fig. 5. The transition of delay

as it is speaking, it tries the promotion every 5 seconds. However, as described in Section 2.3, if the preceding promotion is not completed, the next promotion is not triggered. All peers compare its fanout with its parent every 24 seconds and they may start the promotion depending on the result. To distribute the timing of the promotion among peers, the first comparison occurs at a random time with uniform distribution from 0 to 24 seconds. Peers compare the number of their children with their fanout every 7 seconds and they may start completing the fanout depending on the result. To distribute the timing, the first comparison occurs at random time with uniform distribution from 0 to 7 seconds.

We evaluate our method from the viewpoint of the average and maximum of delay from all peers to the leader peer and from all candidates to the leader peer, and the average and maximum number of received messages per peer. In the figures, we also show results of the case that a distribution tree does not change during a simulation experiment, denoted as *static*, to compare with results of the case with the tree reorganization mechanism, denoted as *dynamic*. Following results are the average over 1000 simulation experiments, each of which lasts for 30 minutes in simulation time unit after the first speaker begins to speak.

3.2 Simulation Results

Figure 5(a) illustrates the average and maximum delay between the leader peer and all peers. The figure shows that the tree reorganization mechanism can effectively reduce both of the average and maximum delay. Among promotions, those invoked by fanout comparison and completion mainly contribute to the initial reduction of delay. When the maximum delay between the leader peer and all peers in a distribution tree is D and that between leader peers in the core network is L, the maximum end-to-end delay among all peers can be derived as $D \times 2 + L$. Except for the initial stage, D is about 35 msecs in Fig. 5(a). Therefore, if we can construct a core network in which the delay among leader peers is less than 30 msecs, we can offer video conferencing with the end-to-end

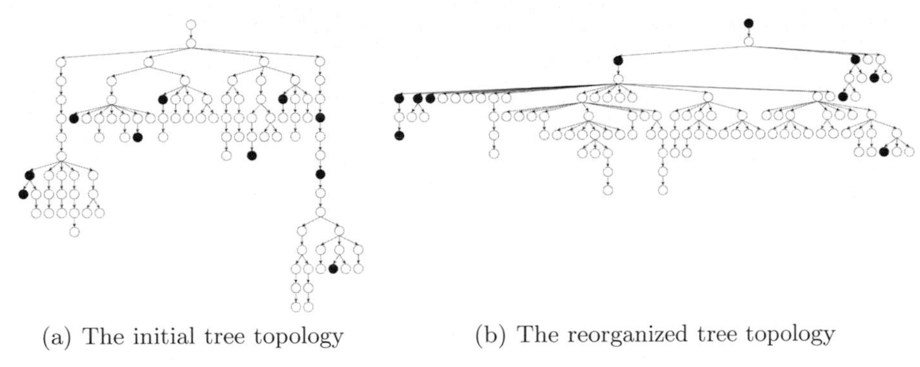

(a) The initial tree topology (b) The reorganized tree topology

Fig. 6. Result of tree reorganization

delay less than 100 msecs, which is smaller than the recommended one way delay for voice communication [9].

Figure 5(b) shows the average and maximum delay between the leader peer and candidates. When comparing to Fig. 5(a), the delay for candidates is less than that for all peers. It means that speakers have better and smoother conversation. We should note here that the delay for candidates remains constant after 300 seconds. This is because that the candidates have moved near the root by 50 speeches before 300 seconds.

Figures 6(a) and 6(b) illustrate how a tree was reorganized in a certain simulation run. In these figures, filled circles correspond to the candidates and open circles indicate other peers. The figures show that the tree reorganization mechanism reduces the height of the tree. With the 1000 simulation experiments, the average hop distance from the leader peer to all peers is reduced from about 7 hops to about 4 hops, and the maximum hops changes from about 14 hops to 10 hops by promotion related to fanout comparison and completion. Furthermore, we can see that the candidates have moved near the root of the tree. With 1000 simulation experiments, the average hop distance from the leader peer and candidates decreases from about 7 hops to about 3 hops, and the maximum hops changes from about 10 hops to 6 hops by promotion for speaking. However, all candidates are not necessarily located near the root depending on the timing of speaking or the duration of speaking, as shown in Fig. 6(b).

The number of messages received per second of a single peer is 0.0839 on average and 1.95 at maximum. By assuming the message size to be 5 Bytes and the packet size including the header to be 33 Bytes, the bandwidth consumed by control messages for a peer is 22 bps on average and 515 bps at maximum. This is very small compared to the rate of the data streaming in video conferencing which ranges from 64 kbps to 8 Mbps.

4 Conclusion

In this paper, we proposed a network construction method for a scalable P2P conferencing system consisting of the network construction mechanism, the tree reorganization mechanism, and the tree recovery mechanism. We evaluated the tree reorganization mechanism through the simulation experiments. We showed that the tree reorganization mechanism can offer smooth conferencing with low delay by moving peers which have high bandwidth or/and are actively speaking to the top of the distribution tree. In addition, we showed that the load of the control messages for the mechanism is very low.

However, we conducted the simulation under the assumption that no failure occurs. As one of future works, we will evaluate our method with peers dynamically joining/leaving and extend our experiments to several trees.

References

1. SmoothCom, http://www.zetta.co.jp/ecom/smoothcom/
2. WarpVision, http://www.ocn.ne.jp/business/infra/warpvision/
3. Jin, X., Cheng, K.L., Chan, S.H.: Sim: Scalable island multicast for peer-to-peer media streaming. In: Proceedings of IEEE International Conference on Multimedia Expo. (ICME 2006), pp. 913–916 (2006)
4. Zhang, R., Hu, Y.C.: Borg: a hybrid protocol for scalable application-level multicast in peer-to-peer networks. In: Proceedings of the 13th International Workshop on Network and Operating Systems Support for Digital Audio and Video (NOSSDAV 2003), pp. 172–179 (2003)
5. Suetsugu, S., Wakamiya, N., Murata, M.: A hybrid video streaming scheme on hierarchical P2P networks. In: Proceedings of The IASTED European Conference on Internet and Multimedia Systems and Applications (EuroIMSA 2005), pp. 240–245 (2005)
6. Barabasi, A., Albert, R.: Emergence of scaling in random networks. Science 286, 509–512 (1999)
7. Medina, A., Lakhina, A., Matta, I., Byers, J.: BRITE: An approach to universal topology generation. In: Proceedings of the Ninth International Symposium in Modeling, Analysis and Simulation of Computer and Telecommunication Systems (MASCOTS 2001), pp. 346–353. IEEE Computer Society, Los Alamitos (2001)
8. Kawahara, T.: Recognition and understanding of voice communication among humans (in japanese). Technical report, Kyoto University (2005), http://www.ar.media.kyoto-u.ac.jp/lab/project/
9. ITU-T: Recommendation G.114 - one-way transmission time. Switzerland (2003)

H.264 Video Broadcast Scheme Using Feedback Information over IEEE 802.11e WLAN

Myungwhan Choi, Jun Hyung Cho, and Jung Min Kim

Dept. of Computer Science and Eng., Sogang Univ., Seoul 121-742, Korea
{mchoi,agic,lakki78}@sogang.ac.kr

Abstract. We propose a new H.264 video broadcasting scheme with retransmission feature using feedback information. Our scheme efficiently maps the video data partitioning technique of H.264 onto the QoS control features of the 802.11e based MAC layer. The proposed scheme effectively suppresses the feedback messages which can potentially cause significant increase of network traffic and the processing overhead at the access point. Simulation results show that the significant overall video service quality enhancement can be achieved. Another advantage of the proposed scheme is that it uses the application level retransmission requests and therefore it can be applied without any modification to the existing MAC protocol.

Keywords: H.264, broadcast, 802.11e WLAN.

1 Introduction

As the speed of wireless LAN increases as demonstrated by IEEE 802.11g (54 Mbps) and IEEE 802.11n (100 Mbps) along with the development of QoS-based medium access control mechanism [1], the multimedia services such as voice-over-IP and video multicast/broadcast over WLAN will be a reality in the near future. The recently developed H.264 video coding scheme further accelerates this trend: H.264 coding scheme achieves a significant improvement in compression efficiency over the previous coding schemes. Additionally, H.264 standard introduces a set of error-resilience tools such as parameter set structure, NAL unit syntax structure, data partitioning, and flexible macroblock ordering [2].

Recently Ksentini et al. [3] proposed a cross-layer architecture to efficiently map the specific characteristics of the H.264 video streams onto the QoS control features of the 802.11e based MAC layer. More specifically, it relies on a data partitioning technique at the application layer and associates each partition with an access category provided by 802.11e enhanced distributed channel access (EDCA) mechanism. The proposed architecture is however limited to the unicast service.

For the multicast/broadcast service in the cellular network, forward error correction (FEC) codes are designed to protect against channel erasures [4]. If the number of the erased packets is less than the decoding threshold for the FEC code,

T. Vazão, M.M. Freire, and I. Chong (Eds.): ICOIN 2007, LNCS 5200, pp. 679–688, 2008.

the original data can be perfectly recovered. Both acknowledged and unacknowledged modes are proposed. This scheme however represents a lack of efficiency since FEC does not adapt to varying channel conditions. Also notice that if this scheme is to be applied to the WLAN, the WLAN MAC should be redesigned.

In this paper, we propose a new broadcasting scheme with retransmission feature. Our scheme relies on the data partitioning technique to provide the error-resilience capability and the cross-layer architecture between the application layer and the 802.11e MAC layer. It also effectively suppresses the duplicate retransmission requests. It does not require any changes to the 802.11e MAC mechanism.

2 H.264 Standard Overview

H.264 consists of the video coding layer and the network abstraction layer. In this H.264 overview, we particularly focus on the data partitioning and the NAL layer features. For more details of H.264, please refer to [2], [5], [6].

2.1 Video Coding Layer

A picture is partitioned into fixed-size macroblocks and slices are a sequence of macroblocks. And a picture may be split into one or more slices. In H.264, each slice can be coded using different coding types to create I, P, or B slice [2]. While I slice contains the coded information using only intra prediction, P and B slices contain the coded information using inter prediction.

Data partitioning is one of the many H.264 error-resilience tools and it creates three bit strings (called partitions) per slice and each partition contains different levels of importance for decoding [5].

- Partition A: contains the most important information which is crucial to the decoding of the other partitions.
- Partition B (intra partition): carries intra coded block pattern and intra coefficient. This partition is more important than Partition C because this information can stop further drift.
- Partition C (Inter partition): carries only inter coded block pattern and inter coefficient.

The VCL layer also generates a coded slice of an instantaneous decoding refresh (IDR) picture. An IDR access unit contains an intra picture only and can be used to limit the drifting effects due to the transmission error. An IDR access unit may be transmitted every transmission of a group of pictures.

2.2 Network Abstraction Layer

The NAL is designed to facilitate the transport of the H.264 VCL data using various communication systems and H.264 VCL data can be packetized for UDP/IP transport over wired/wireless Internet services among many others.

NAL units are classified into VCL and non-VCL units. VCL NAL units contain the coded video data and non-VCL units contain any additional information such as parameter sets and supplemental enhancement information.

A parameter set is supposed to contain information that is used by a large number of VCL NAL units. There are two types of parameter sets: sequence parameter sets and picture parameter sets. A sequence parameter set contains parameters such as the number of reference frames, the decoded picture size, and the choice of progressive or interlaced coding, to be applied to a sequence of pictures. A picture parameter set contains parameters such as the entropy coding algorithm and initial quantizer parameters, to be applied to one or more decoded pictures within a sequence.

Each NAL unit contains a NAL header and a Raw Byte Sequence Payload (RBSP), a set of data corresponding to coded video data or header information. The NAL header contains three fields. The forbidden_zero_bit is specified to be zero in H.264 encoding. The Nal_ref_idc can be used to signal the importance of a corresponding RBSP for the reconstruction process. A value of 0 indicates that the corresponding RBSP is not used for prediction, and hence its loss does not cause drifting effects. The higher the Nal_ref_idc is, the higher the impact of loss of the corresponding RBSP unit is on the drifting effects.

Nal_unit_type is 5-bit long and indicates the type of the RBSP. The types includes coded slice data partition A, coded slice data partition B, coded slice data partition C, coded slice of an IDR picture, sequence parameter set, and picture parameter set.

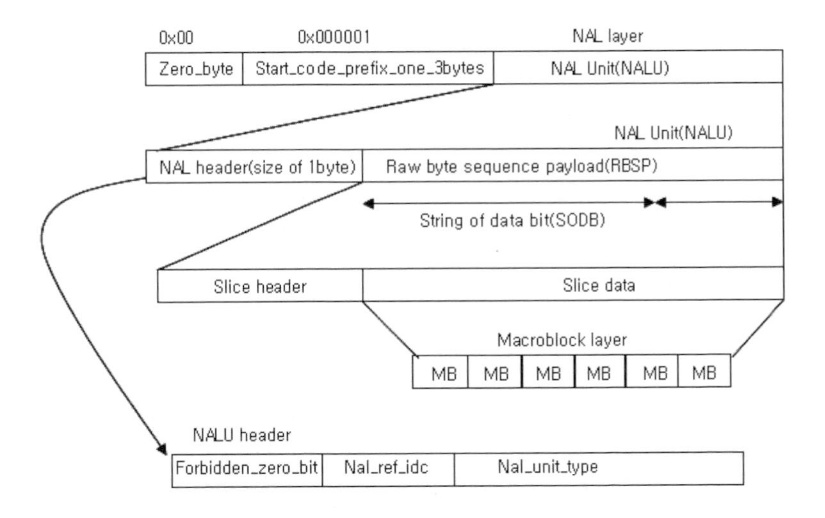

Fig. 1. The structure of NAL unit

3 Wireless Broadcasting Network Model: Cross-Layer Architecture

The wireless network architecture under consideration consists of one access point and multiple mobile receivers. The broadcast service is provided through the access point. Due to the time-varying nature of the wireless channel characteristics, packets of different importance will be lost for different nodes. Due to the real-time requirement for the video packet delivery, more important packets need to be transmitted earlier than less important packets to enhance the overall quality of the video packet broadcast service with feedback feature.

The EDCA mechanism of the IEEE 802.11e WLAN is utilized to provide different wireless network access priorities. EDCA supports four access categories (ACs) representing four priorities, with four independent backoff entities. The backoff entities are prioritized using AC-specific contention parameters, that is, $CW_{min}[AC]$, $CW_{max}[AC]$, $AIFS[AC]$, and $TXOP-limit[AC]$. Generally, AC3 corresponds to the highest access priority and AC0 to the lowest.

Based on the importance of the H.264 NAL units, each NAL unit can be assigned to the corresponding access category at the MAC layer. The importance of the NAL unit is based on the value of the Nal_unit_type field of the NAL header. Table 1 shows an example mapping between the Nal_unit_type and the access category.

Table 1. Mapping between the Nal_unit_type and the access category

Slice type	Nal_unit_type	Access category
Parameter set	7, 8	4
IDR picture, Partition A	5, 2	3
Partition B, Partition C	3, 4	2

4 The Proposed Broadcasting Algorithm

When a receiver receives the in-order broadcast packet, it uses the received packet for the decoding. If a receiver detects a missing packet by means of the RTP sequence number (because of the transmission error), it will send a NACK packet to the access point for the retransmission request of the missing packet. The transmission time of the NACK packet is delayed for the random amount of time between 0 and T msec. The parameter T will be dependent on the video packet transmission time deadline.

The access point receiving a NACK retransmits the corresponding video packet if it can arrive at the receivers before its deadline expires. Receiving the successfully retransmitted packet, the receivers which have pending NACKs for the retransmission request of the corresponding packet will withdraw them. This will greatly reduce the potentially large number of NACK message transmissions. The maximum number of NACK message transmissions for each of the missing packets allowed for each terminal is a system parameter and will affect the network overhead due the NACK message transmissions. The optimal value of it will be determined by the wireless channel characteristics and the wireless network load characteristics.

Some notations for the description of the algorithm are as follows.

Notations

- $P_{k,l}$: Packet k transmitted by access point for the l-th time.
- $t(P_{k,l})$: The time packet $P_{k,l}$ is transmitted.

- $NACK_i(d_j, P_{k,l})$: NACK packet asking the retransmission of the packet i, generated by node j with the transmission delay time of d_j after its reception of $P_{k,l}$.
- $d_j(P_i, P_{k,l})$: Random amount of transmission delay time computed by the receiving node j when it receives $P_{k,l}$ to ask the retransmission of packet i.
- $d(P_i, P_{k,l})$: transmission delay time of the $NACK_i(d_j, P_{k,l})$ which arrives in first at the access point requesting the retransmission of the packet i based on the $P_{k,l}$ reception.
- RTT: round trip time
- t_c: the current time
- T_{avg}: mean interarrival time of $P_{k,1}$, $k = 1, 2, \cdots$.
- deadline(P_i): deadline for packet i to be acceptable by the receiver.
- $P_{k'}$: the packet with the largest index k' among packets arrived in sequence at a receiver
- $count(P_k)$: number of retransmission requests for packet k. Initially zero.
- $ReTx_{max}(k)$: maximum number of retransmission requests allowable to each node for packet k.

4.1 Broadcast Algorithm at the Access Point

The access point broadcasts $P_{k,l}$ with $l = 1$ with regular interval which is a system parameter. When the access point receives $NACK_i(d_j, P_{k,l})$, it checks if this is the first message among the messages arriving from the receivers which have received the packet $P_{k,l}$. If this is the case, it saves the value $d_j(P_i, P_{k,l})$ in $d(P_i, P_{k,l})$ and then retransmits packet i if it can be delivered to the receivers within its deadline.

If $NACK_i(d_j, P_{k,l})$ is not the first message among the messages arriving from the receivers which have received the packet $P_{k,l}$, we check if $d_j(P_i, P_{k,l})$ - $d(P_i, P_{k,l}) >$ RTT. If this is the case and the retransmitted packet can be delivered to the receivers within its deadline, the access point retransmits the packet i. Notice that the packet retransmitted after receiving the first $NACK_i(d_j, P_{k,l})$ may not be successfully transmitted to some receivers which did not successfully receive the previous packet.

Pseudo-code for Access Point Processing

AP receives $NACK_i(d_j, P_{k,l})$:
    if first $NACK_i(d_j, P_{k,l})$ for the transmission of $P_{k,l}$ then
        $d(P_i, P_{k,l}) \longleftarrow d_j$.
        if $t_c - t(P_{i,1}) + \frac{1}{2}RTT < deadline(P_i)$ then
            re-broadcast packet i
        end if
    else if $d_j - d(P_i, P_{k,l}) > RTT$, then
        if $t_c - t(P_{i,1}) + \frac{1}{2}RTT < deadline(P_i)$ then
            re-broadcast packet i
        end if
    end if

4.2 Retransmission Requests by the Receivers

When a node receives a packet, it checks whether it is a duplicate packet or not. If it is a duplicate packet, discard it. Otherwise, the node stores it and discards the NACK message, if any, which is scheduled to be transmitted sometime later for the retransmission request for the received packet. It also checks whether there are some missing packets or not. If it turns out that there are some missing packets, it forms a NACK message to be used to ask the retransmission of the packet. The packet to request for the retransmission is the one with the smallest packet sequence number among those which can be delivered before their respective transmission deadline expires. The actual NACK message transmission is delayed by an amount which is chosen in random between 0 and T msec, which contributes to the suppression of the NACK messages actually transmitted.

Pseudo-code for Receiver Processing

A packet $P_{k,l}$ arrives successfully at a node j:
  if the arrived packet is the duplicate, then
    discard it.
  else
    store P_k
    if there are some $NACK_k(\cdot,\cdot)$ to be scheduled, then
      discard $NACK_k(\cdot,\cdot)$.
    end if
  end if
  if there are some missing packets (if $k > k' + 1$), then
    find the smallest K such that K < k, $count(P_K) < ReTx_{max}(K)$, and
      t_c - $T_{avg} \times$ K < deadline(P_K) - RTT.
    compute $d_j(P_K, P_{k,l})$
    $count(K) \longleftarrow count(K) + 1$
    schedule $NACK_K(d_j, P_{k,l})$ to be transmitted $d_j(P_K, P_{k,l})$ later.
  end if

5 Simulation

5.1 Model

To demonstrate the potential broadcast performance enhancements compared with the conventional broadcasting scheme which does not allow any lost packet retransmissions, we conducted simulation tests using a wireless LAN configuration with one access point and 10 nodes. These nodes are placed randomly in a 300 m $\times$ 300 m area and the access point is placed at a corner of the area.

For the network simulation, ns-2 version 2.28 [7] and the IEEE 802.11e EDCA and CFB Simulation Model for ns-2 [8] are modified to simulate our proposed algorithm. For the wireless channel model, path loss exponent of 4 is assumed and Orinoco 802.11b LAN card specifications are used to simulate the receive power. To simulate the BER for a given receive power, the Intersil HFA3861B empirical

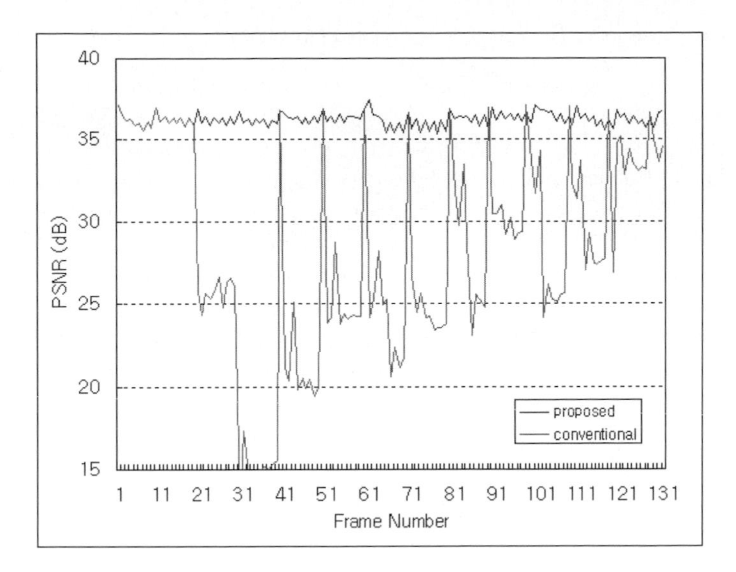

Fig. 2. PSNRs for a selected mobile node using the proposed and conventional schemes

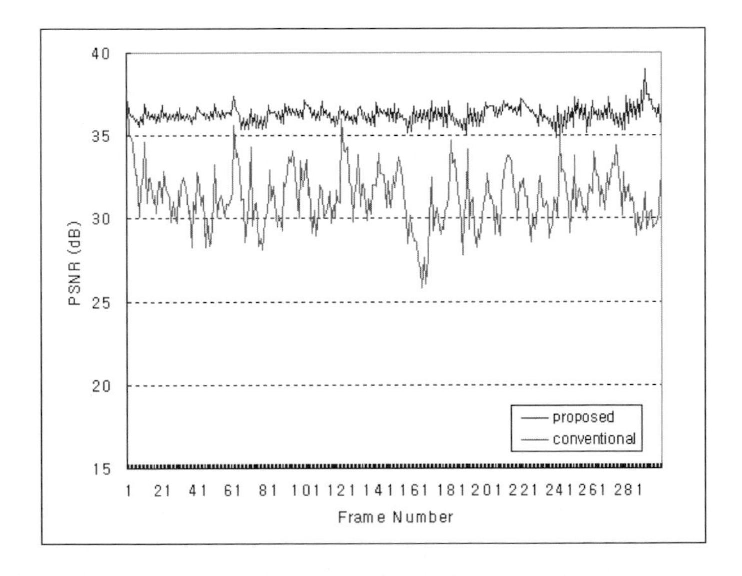

Fig. 3. Average PSNRs for proposed and conventional schemes

BER vs. SNR curve is used [9]. The frame error rate is computed under the assumption that the bit error is randomly distributed for a given BER. Video data are assumed to be broadcast at the rate of 2 Mbps and we set the parameter T to 10 *msec*.

For the video source traffic generation, we used H.264 JMware 10.2 [10] for video encoding. Foreman.YUV of QCIF format is used as the source video.

Fig. 4. Comparison of the 108th decoded frames using the proposed and conventional schemes

Table 2. Retransmission overheads and NACK suppression

	Partition A	Partition B	Partition C
Number of original packets to transmit	299	299	299
Number of candidate NACK packets for suppression	69	44	55
Number of suppressed NACK packets	62	40	49
Suppression ratio for NACK packets	0.89	0.91	0.89
Number of retransmitted packets	182	169	171
Portion of the retransmissions	0.609	0.565	0.572

Frames are generated 30 frames per second for 10 seconds. Three partitions (A, B, and C) are generated per slice. The GOP encoding scheme is IBPBPBPBPB type to limit the effect of the transmission errors to a certain degree.

5.2 Results

From this experiment, we observed that the video quality perceived at all of the mobile nodes using the proposed scheme is almost identical to the original video quality while only five mobile nodes can obtain good quality of video service using the conventional scheme. We also observed that the video quality using the conventional scheme severely deteriorates especially when the partition A of the video packets are lost. Obviously using the proposed scheme, the lost packets including partition A video packets are recovered by retransmissions.

To show the received video performance enhancement of the proposed scheme, we measured the peak signal-to-noise ratio (PSNR). Figure 2 shows the PSNR performance of the proposed and conventional schemes. In this figure, the PSNR curves for a mobile node not in that good channel condition are shown to show the performance differences between two schemes more clearly.

The average PSNRs for the proposed and conventional schemes are also shown in Fig. 3 which shows the superiority of the proposed scheme. Realizing that the significant portion of the PSNR curves for the mobile nodes using the conventional scheme severely fluctuate as shown in Fig. 2, we can see that the average

PSNR performance of the conventional scheme fluctuating near 30 dB does not mean its performance is good. It means that only some limited number of nodes receive high quality of video service.

Figure 4 shows the decoded sample frames at the mobile node whose PSNR performances are shown in Fig. 2. This performance enhancement could be achieved by virtue of the retransmission overheads. In the proposed scheme, transmission error detections by at least one receiving nodes for each new packet transmission invoke the retransmission requests. So as the number of receiving nodes increases, the retransmission overhead may increase. The embedded NACK suppression algorithm mitigates this problem. Table 2 shows the experimental data for the retransmission overhead and the effectiveness of the NACK suppression mechanism.

6 Conclusions

In this paper, we proposed a new broadcasting algorithm suitable for the IEEE 802.11e WLAN and showed that the significant video service quality enhancement can be achieved. The proposed scheme effectively suppresses the feedback messages which can potentially cause significant increase of network traffic and the processing overhead at the access point. However, the amount of feedback messages generated will be dependent on the distributions and the number of mobile users. The detailed study on the effect of the increased number of mobile users on the overall video broadcast service performance and the way to effectively reduce the retransmission overhead is reserved for future work.

Acknowledgments. This research was supported by the MIC (Ministry of Information and Communication), Korea, under the ITRC (Information Technology Research Center) support program supervised by the IITA (Institute of Information Technology Assessment) (IITA-2006-(C1090-0603-0011)).

References

1. Mangold, S., Choi, S., Hiertz, G.R., Klein, O., Walke, B.: Analysis of IEEE 802.11e for QoS Support in Wireless LANs. IEEE Wireless Communications 10, 40–50 (2003)
2. Wiegand, T., Sullivan, G.J., Bjontegaard, G., Luthra, A.: Overview of the H.264/AVC Video Coding Standard. IEEE Trans. Circuits and Systems for Video Tech. 13, 560–576 (2003)
3. Ksentini, A., Naimi, M., Gueroui, A.: Toward an Improvement of H.264 Video Transmission over IEEE 802.11e through a Cross-Layer Architecture. IEEE Communications Magazine 44, 107–114 (2006)
4. Jenkac, H., Stockhammer, T., Liebl, G.: H.264/AVC Video Transmission over MBMS in GERAN. In: IEEE International Workshop on Multimedia Signal Processing (2004)
5. Wenger, S.: H.264/AVC over IP. IEEE Trans. Circuits and Systems for Video Tech. 13, 645–656 (2003)

6. Richardson, I.E.G.: H.264 and MPEG-4: Video Compression. Wiley, Chichester (2003)
7. The Network Simulator - ns-2, http://www.isi.edu/nsnam/ns/
8. Wiethölter, S., Hoene, C.: An IEEE 802.11e EDCA and CFB Simulation Model for ns-2, http://www.tkn.tu-berlin.de/research/802.11e_ns2/
9. Xiuchao, W.: Simulate 802.11b Channel within NS2, http://www.comp.nus.edu.sg/~wuciucha/research/reactive/release.html
10. http://iphome.hhi.de/suehring/tml/download/jm10.2.zip

Quality Adaptation with Temporal Scalability for Adaptive Video Streaming

Sunhun Lee and Kwangsue Chung

School of Electronics Engineering, Kwangwoon University, Korea
sunlee@adams.kw.ac.kr and kchung@kw.ac.kr

Abstract. In video streaming applications over the Internet, TCP-friendly rate control schemes are useful for improving network stability and inter-protocol fairness. However it does not always guarantee a smooth quality for video streaming. To simultaneously satisfy both the network and application requirements, video streaming applications should be quality-adaptive. In this paper, we propose a new quality adaptation mechanism to adjust the quality of congestion controlled video stream by controlling the frame rate. Based on the current network condition, it controls the frame rate and sending rate of video stream. Through the simulation, we prove that our adaptation mechanism appropriately adjusts the quality of video stream while improving network stability.

1 Introduction

The Internet has recently been experiencing an explosive growth in the use of audio and video streaming applications. Loss-tolerant real-time multimedia applications such as video conferencing or video streaming prefer UDP (User Datagram Protocol) to avoid unacceptable delay introduced by packet retransmissions. As the use of real-time multimedia applications increases, a considerable amount of UDP traffic would dominate network bandwidth because UDP does not have congestion control mechanism. As a result, the available bandwidth to TCP (Transmission Control Protocol) connections is oppressed and their performance extremely deteriorates because the current Internet does not attempt to guarantee an upper bound on end-to-end delay and a lower bound on available bandwidth [1].

Researches on the TCP-friendly rate controlled streaming protocol has been increasingly done since the 1990s [2,3,4,5]. A TCP-friendly rate control mechanism regulates its data sending rate according to the network condition, typically expressed in terms of the RTT (Round Trip Time) and the packet loss probability. Therefore they improve the network stability and fairness with competing TCP connections. However, by considering only the network requirements, they ignore the quality of video stream which is delivered to the end user. To satisfy the network requirement and application requirement simultaneously, video streaming applications should be quality-adaptive. That is, the application should adjust the quality of the delivered video stream such that the required bandwidth matches congestion controlled rate-limit.

T. Vazão, M.M. Freire, and I. Chong (Eds.): ICOIN 2007, LNCS 5200, pp. 689–698, 2008.

In this paper, we propose a new mechanism to adjust the quality of video stream on-the-fly. To design an efficient quality adaptation mechanism, we need to know the properties of the deployed congestion control mechanism. Previously we have proposed the TF-RTP (TCP-Friendly Real-time Transport Protocol) which is TCP-friendly rate control scheme based on RTP [5]. Similar to previous congestion controlled streaming protocols it adjusts the sending rate of video stream in a TCP-friendly manner. With TF-RTP, our quality adaptation mechanism controls the quality of video stream based on the client's buffer occupancy. Therefore, not only does our mechanism maintain the network stability, but it also achieves the smoothed playback by preventing the buffer underflow or overflow.

The rest of this paper is organized as follows. In Section 2, we simply review several previous quality adaptation mechanisms. In Section 3, we present the concept and algorithms introduced in our new quality adaptation mechanism. Our simulation results are described in Section 4. Finally, Section 5 concludes the paper and discusses some of our future works.

2 Related Works

Previously, much attention has focused on developing congestion control algorithms for video streaming applications. These were some variants of TCP with no in-order and reliable delivery of TCP. More recently, some TCP-friendly congestion control algorithms, like RAP [2], SQRT [3], and TFRC [4], have been proposed. They have focused on reducing large oscillations associated with TCP's congestion control and improving the TCP-friendliness. As does TCP, TCP-friendly algorithms react to indications of network congestion by reducing the transmission rate. The transmission rate of TCP-friendly algorithms is typically smoother than that of TCP. Nevertheless, because network congestion occurs at multiple time scales, the bandwidth available to TCP-friendly streams typically fluctuates over several time scales. Rate smoothing is not useful for a best effort network, since the Internet does not provide any information about the bandwidth evolution in advance. Moreover, a smooth data rate does not always guarantee a smooth quality for video streaming [6]. In video streaming applications over the Internet, quality adaptation mechanism is required to improve the quality because a TCP-friendly rate control algorithm is insufficient for providing a better video streaming service.

Several researches have been conducted in adaptive video streaming and various approaches have been proposed [6,7,8,9,10,11,12]. With hierarchical encoding [7], the server maintains a layered encoded version of each stream. As more bandwidth becomes available, more layers of the encoding are delivered. If the average bandwidth decreases, the server may then drop some of the layers being transmitted. Layered approaches usually have the decoding constraint that a particular enhancement layer can only be decoded if all the lower quality layers have been received. The design of a layered approach for quality adaptation primarily entails the design of an efficient add and drop mechanism that maximizes quality while minimizing the probability of base-layer buffer underflow. Most of all quality adaptation mechanisms have adopted a layered approach.

Rejaie et al. propose a quality adaptation mechanism using receiver buffering for AIMD(Additive Increase Multiplicative Decrease)-controlled transmission and playback of hierarchically-encoded video [8]. Long-term coarse-grained adaptation is performed by adding and dropping layers of the video stream, while using AIMD to react to congestion. A new layer will be added only if, at any point, the total amount of buffering at the receiver is sufficient to survive an immediate backoff and continue playing all of the existing layers plus the new layer, and the instantaneous available bandwidth is greater than the consumption rate of the existing layers, plus the new layer. When the total amount of buffering falls below the amount required for a drop from a particular rate, then the highest layer is dropped. Additionally, buffer space is allocated between layers so as to place a greater importance on lower layers, thereby protecting these layers upon a reduction in the transmission rate.

Nick et al. extends the results of Rejaie et al. and finds that the combination of a non-AIMD algorithm that has smaller oscillations than AIMD and a suitable receiver buffer allocation [9]. For more smoothed quality control and efficient buffer usage, they use a SQRT scheme as a congestion control module. Naoki et al. proposes TCP-friendly MPEG-4(Moving Picture Experts Group) video transfer methods to fairly share the bandwidth with conventional TCP data applications [10]. For the video quality control, they assume that video stream is encoded by FGS (Fine Granular Scalability) video coding algorithm [11] of MPEG-4 standards.

Existing quality adaptation mechanisms use the hierarchical encoding scheme, specifically FGS algorithm, to match the quality of video stream with its determined sending rate. It provides an effective way for a video streaming server to coarsely adjust the quality of a video stream without transcoding the stored data. However, in the real world, the hierarchical encoding scheme is not deployed to streaming applications because an encoder with this capability is small. Moreover, unlike MPEG-4, where the FGS mechanism is specified to use hierarchical coding, the H.264/AVC standard [13], a recently proposed video compression scheme, does not contain any specification for this feature. While most of all the previous works done until now concern essentially hierarchical encoded video streaming, our work is different from those by using the new concept of temporal scalability introduced in the MPEG family of standards.

3 Quality Adaptation Mechanism

This section briefly introduces our quality adaptation mechanisms to adjust encoded video rate to the desired sending rate determined by TCP-friendly rate control scheme. Firstly, we describe the end-to-end architecture of proposed quality adaptation mechanism. After that, the detail algorithms on adaptive video streaming are described.

3.1 Overall Architecture

Fig. 1 depicts our end-to-end system architecture. Stored video stream is encoded by MPEG family of standards with a constant frame rate. The congestion control module can effectively estimate the TCP-friendly rate. All active frames are multiplexed into a single transport session by the server. At the client side, incoming frames from the network are demultiplexed and each one goes to its corresponding buffer. The decoder drains data from buffers and feeds the display.

Normally a video stream consists of three types of frame they are I, P, and B-frame. Frame is the basic unit of video data and is equivalent to the picture. A sequence of frames beginning from an I-frame is called GOP (Group Of Picture) and defined by two parameters, number of P-frames between two I-frames and number of B-frames between two P-frames. In this paper we assume each frame types have the same data size. This is unlikely in a real codec, but it simplifies the analysis.

Each RTCP (Real-time Transport Control Protocol) packet reports the most recent network state and data delivery information to the server. Having an estimate of RTT and a history of transmitted packets for each frame, the server can estimate the TCP-friendly rate and the amount of prefetched buffers for each frame. After estimating the current network state, the congestion control module informs the TCP-friendly rate to the quality adaptation module. Then the quality adaptation module efficiently adjusts the number of active frames and fairly allocates available bandwidth to each active frame. In our architecture, TF-RTP, previously proposed in [5], is deployed as the transport and congestion control module.

The quality adaptation mechanism is required when a data rate of originally encoded video stream is higher than the current available bandwidth. When the quality adaptation mechanism is applied, some frames of a GOP are activated and transmitted but the others are not. In Fig. 2, the number of frames and bytes of originally encoded and adapted video stream in a GOP are defined to explain our quality adaptation mechanism. In case of originally encoded video stream, number of frames and bytes in a GOP are determined when raw data is compressed by using the video compression scheme. In case of adapted video stream, those are dynamically changed because available bandwidth is time-variant. The number of active frames of adapted video stream is always equal

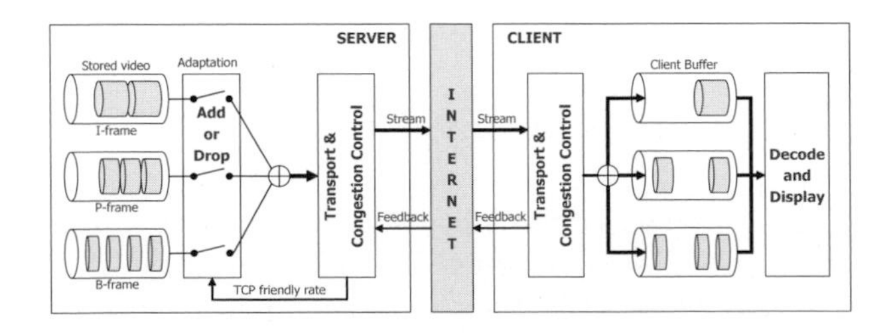

Fig. 1. End-to-end architecture

Common definition:

B_K : number of bytes of K-frame, where $K \in \{I, P, B\}$

Originally encoded video stream:

NF_K: number of K-frame

Number of frames in a GOP: $NF_{GOP} = NF_I + NF_P + NF_B$

Number of bytes in a GOP: $B_{GOP} = NF_I * B_I + NF_P * B_P + NF_B * B_B$

Encoded frame rate per 1 sec is FR_{GOP}

Adapted video stream:

α, β, γ: numbers of active I, P, and B-frame

Number of frames in a GOP: $NF_{Active} = \alpha + \beta + \gamma$

Number of bytes in a GOP: $B_{Active} = \alpha * B_I + \beta * B_P + \gamma * B_B$

Active frame rate per 1 sec is FR_{Active}

Fig. 2. Definition of video stream characteristics

to or lower than the number of frames of originally encoded video stream ($\alpha \leqq NF_I$, $\beta \leqq NF_P$, $\gamma \leqq NF_B$).

3.2 Estimation of Client Buffer Occupancy

Let $X(t)$ denote the TCP-friendly rate at time t. By permitting prefetching data in client buffers, our server sends the video data at the maximum rate $X(t)$. The stored video is divided into three frame types, I, P, and B-frame. Under congested network state, the available bandwidth is not enough for admitting an originally encoded video stream as it is. In this case, the quality adaptation mechanism should adjust the number of active frames in a GOP and their bandwidth shares. Therefore, the allocated actual bandwidth to a GOP is calculated by Eq. (1). This is the same with the sum of allocated bandwidth to each frame type.

$$X_{GOP} = \frac{X(t)}{FR_{Active}/NF_{Active}} = X_I + X_P + X_B \tag{1}$$

The client stores I, P, and B-frames data coming from the network in prefetched buffers. Let $Y_I(t)$, $Y_P(t)$, and $Y_B(t)$ denote the amount of data in the client prefetched buffers at time t. They are periodically reported to the sender, by RTCP packets, to estimate the client buffer occupancy. We denote $\Delta(t)$ for the number of each prefetched frame contained in the client buffers. Since the allocated bandwidth to a GOP is re-allocated to I, P, and B-frame in proportion to their number of active frames and bytes, the number of prefetched frames can be calculated as shown in Eq. (2). The number of prefetched GOPs can be estimated by Eq. (3) because three types of frame organize a GOP and they are synchronized by each other within a GOP.

$$\Delta_I(t) = \frac{Y_I(t)}{B_I}, \Delta_P(t) = \frac{Y_P(t)}{B_P}, \Delta_B(t) = \frac{Y_B(t)}{B_B} \tag{2}$$

$$\Delta_{GOP}(t) \approx \frac{\Delta_I(t)}{\alpha} \approx \frac{\Delta_P(t)}{\beta} \approx \frac{\Delta_B(t)}{\gamma} \tag{3}$$

3.3 Conditions for Preventing Underflow or Overflow

Without prefetching at the decoder, the perceived video quality may be degraded considerably. Therefore buffer underflow and overflow have to be avoided to improve the quality of video stream. Depending on the number of prefetched GOPs, condition for preventing underflow or overflow can be regulated. If the sum of the number of transmitted GOPs (θ) and the number of prefetched GOPs (Δ) per unit time is higher than the number of consumed GOPs (ϕ) by decoder, buffer underflow is avoided. After decoding some prefetched data, if the number of prefetched GOPs is lower than the client maximum buffer size, overflow can be avoided. In our works, we apply a more conservative prevention method by using the minimum (Δ_{MIN}) and maximum (Δ_{MAX}) thresholds because buffer underflow and overflow can be more disturbing for the overall quality.

$$\phi = \frac{FR_{Active}}{NF_{Active}} \text{ and } \theta = \frac{X(t)}{X_{GOP}} \tag{4}$$

Condition for preventing underflow:
(Number of consumed GOPs + Δ_{MIN})
≤ (Number of transmitted GOPs + Number of prefetched GOPs)
➜ $(\phi + \Delta_{MIN}) \leq (\theta + \Delta_{GOP})$
Condition for preventing overflow:
Number of prefetched GOPs ≤ Δ_{MAX}
➜ $\Delta_{GOP} \leq \Delta_{MAX}$

Fig. 3. Conditions for preventing underflow and overflow

3.4 Quality Adaptation Algorithms

A new frame can be added as soon as the instantaneous available bandwidth exceeds the consumption rate of the active frames. A more practical approach is to start sending a new frame when the instantaneous bandwidth exceeds the consumption rate of the active frames plus the new added frame. A new added frame is selected according to the type of current active frame and amount of available bandwidth. There is some excess bandwidth from the time the available bandwidth exceeds the consumption rate of the current active frames until the new frame is added. This excess bandwidth can be used to transmit data for current active frames at the client.

- Condition 1: $({X_{GOP}}^{n} - {X_{GOP}}^{n-1}) \geqq \frac{X_i}{j}, where\ i \in \{I, P, B\}, j \in \{\alpha, \beta, \gamma\}$

This bandwidth constraint for adding frame is still not sufficiently conservative. Some excess bandwidth makes more prefetched data at the client and is possible to induce buffer overflow. If prefetched buffers are expected to overflow ($\Delta_{GOP} \geqq \Delta_{MAX}$), the server improves the video quality by increasing the frame rate. Then, the client buffer occupancy may be lowered because prefetched active frames are more rapidly consumed by the decoder. Although network condition is more congested transiently, buffer overflow is efficiently prevented nevertheless.

– Condition 2: $\Delta_{GOP} \geqq \Delta_{MAX}$

An active frame can be dropped when the consumption rate of the active frames exceeds the estimated TCP-friendly rate. However, to provide higher quality as long as possible, a frame should be slowly dropped within the limits of the possible. If client buffer is expected to underflow ($\Delta_{GOP} \leqq \Delta_{MIN}$), the server degrades the video quality by decreasing the frame rate with no sending rate control. Dropping an active frame makes lower consumption rate at the client and delivers more GOPs per unit time because the sending rate is remained the same.

– Condition 3: $\Delta_{GOP} \leqq \Delta_{MIN}$

If these three conditions are not satisfied, the server keeps up the current active frames and transmits them in by order of frame importance. Under the congested network state, server adjusts the sending rate to the estimated TCP-friendly rate. In case of stable network state, the sending rate is increased to improve the TCP-friendliness.

4 Performance Evaluations

In this section, we present our simulation results. Using the ns-2 simulator, the performance of the proposed quality adaptation mechanism has been measured [14]. To emulate the congested network conditions, background TCP traffic is introduced. Fig. 4 shows the topology for our simulations. All of our experiments use a single bottleneck topology and the bottleneck queue is managed with the drop-tail mechanism. Each simulation consists of about 150 sec of the network lifetime. From 0 to 30 sec, only one adaptive video stream is introduced and occupies all available bandwidth. Beginning at 30 sec, one competing TCP connection is started and begins competing with existing connection for the bandwidth share. The RTCP feedback packets are periodically reported to the server with 1 sec interval.

To evaluate performance, we use a sample video data. The characteristics of video sample are described in Table 1. We assume that the video stream has encoded at 1.4 Mbps with 25 frames per second. It also has three types of frame with different data size.

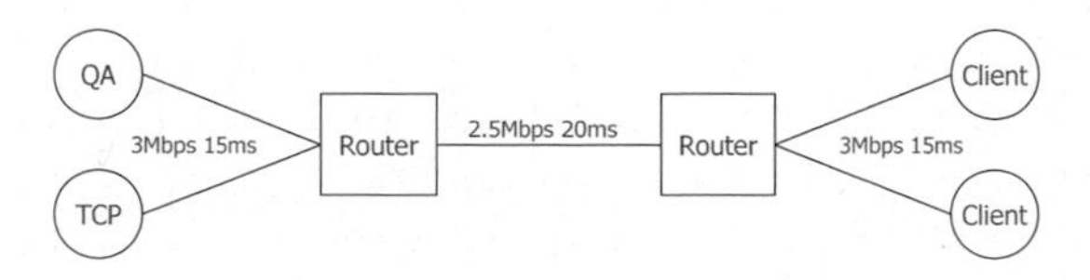

Fig. 4. Network topology

Table 1. Characteristics of video sample

GOP Sequence	I-B-B-P-B-B-P-B-B-P		
Frame Rate	25 fps		
Encoded Rate	1.4 Mbps		
Frame Types	I-frame	P-frame	B-frame
Number of Frame in a GOP	1	3	6
Size of Each Frame(bytes)	25 K	10 K	2.5 K

4.1 TCP-Friendly Rate Control

In [5], we have proposed the RTP-based rate control scheme for video streaming. Our quality adaptation mechanism applies this RTP-based rate control scheme as a congestion control module for calculating the TCP-friendly rate. Fig. 5(a) shows that our congestion control module can estimate the TCP-friendly rate and dynamically control the sending rate of video stream on the basis of the current network state. Congestion control module periodically estimates the available bandwidth. If the estimated available bandwidth is lower than the originally encoded rate of video stream, then the quality adaptation module adjusts the sending rate of video stream and number of active frames. Approximately, TCP connection shares on average 1.55 Mbps and adaptive video stream shares on average 1.25 Mbps. In Fig. 5 (b), packet loss occurrences are depicted. Our quality adaptation mechanism can decrease the packet loss probability as compared with no adaptation because it adjusts the sending rate of video stream based on the current network state. Approximately, our adaptation mechanism decreases the packet loss probability by about 23 %.

4.2 Quality Adaptation

Fig. 6 (a) shows that our quality adaptation mechanism appropriately controls the number of active frames. From 80 to 145 sec, the quality adaptation mechanism drops active frames and adds new frames. Dropping an active frame is

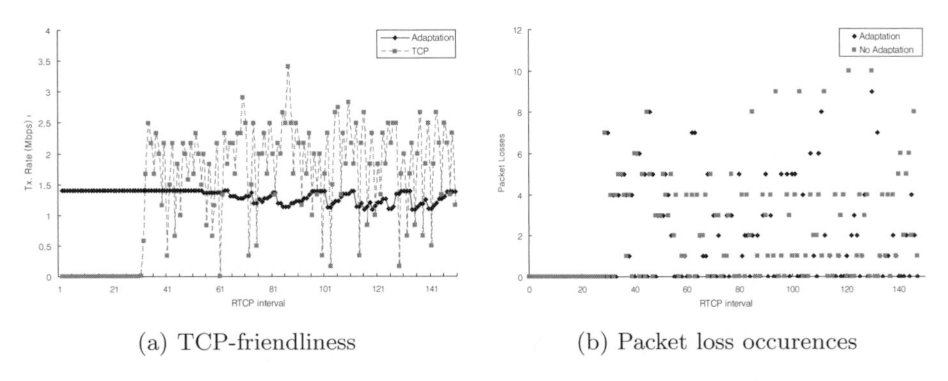

(a) TCP-friendliness (b) Packet loss occurences

Fig. 5. TCP-friendly rate control of quality adaptation

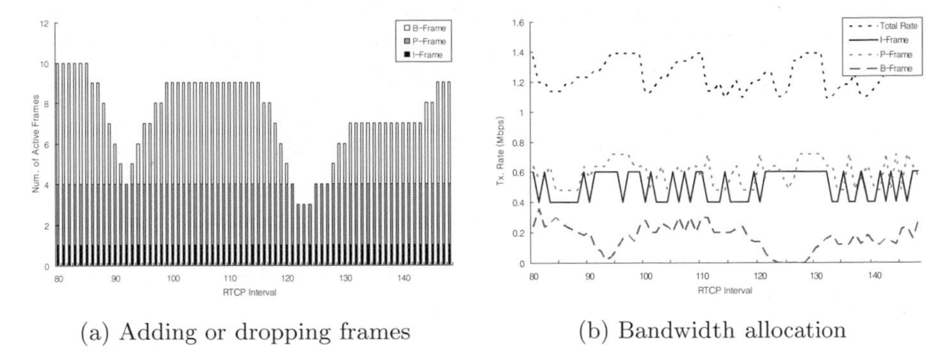

(a) Adding or dropping frames (b) Bandwidth allocation

Fig. 6. Adaptive frame rate control

done to prevent buffer underflow. On the other hand, a new frame is added when current network condition is available to admit a new frame. Allocated available bandwidths to each type of frame are shown in Fig. 6 (b). The sending rate of I-frame is regularly maintained because it has no frame rate control. However, the sending rates of P and B-frame are dynamically changed according to the estimated TCP-friendly rate. Specifically, B-frame is early dropped and lately added than P-frame because of their lower data importance. Fig. 6 shows that our quality adaptation mechanism with congestion control module efficiently controls the number of active frames and successfully transmits them by allocating available bandwidth.

5 Conclusion

We have presented a quality adaptation mechanism to improve the quality of video streaming applications over the Internet. Existing quality adaptation mechanisms mainly exploited the flexibility of hierarchical encoding. However, our new mechanism is based on temporal scalability of video stream to overcome limitations of hierarchical encoding schemes. The key issue is to select appropriate active frames among the frames in a GOP. We have described an efficient mechanism that dynamically adjusts the number of active frames while maintaining the network stability and inter-protocol fairness.

Our simulation results reveal that proposed mechanism can appropriately adjust the quality of video stream based on current network state. It also can provide a smoothed playback to the user by preventing buffer overflow and underflow. In the future, we will further enhance the stability of proposed mechanism and extend our quality adaptation mechanism to wireless network environments.

Acknowledgement

This research was supported by the MIC(Ministry of Information and Communication), Korea, under the ITRC(Information Technology Research Center) support program supervised by the IITA(Institute of Information Technology Assessment). This research also has been conducted by the Research Grant of Kwangwoon University in 2006.

References

1. Floyd, S., Fall, K.: Promoting the use of end-to-end congestion control in the Internet. IEEE/ACM Transaction (1999)
2. Rejaie, R., Handley, M., Estrin, D.: RAP: An end-to-end rate based congestion control mechanism for real-time streams in the Internet. IEEE INFOCOMM (1999)
3. Bansal, D., Balakrishnan, H.: Binomial congestion control algorithms. IEEE INFOCOM (2001)
4. Floyd, S., Handley, M., Padhye, J., Widmer, J.: Equation-based congestion control for unicast applications. ACM SIGCOMM (2000)
5. Lee, S., Chung, K.: TCP-friendly rate control scheme based on RTP. In: Chong, I., Kawahara, K. (eds.) ICOIN 2006. LNCS, vol. 3961, pp. 660–669. Springer, Heidelberg (2006)
6. Kim, T., Ammar, M.H.: Optimal quality adaptation for MPEG-4 fine-grained scalable video. IEEE INFOCOM (2003)
7. Lee, J., Kim, T., Ko, S.: Motion prediction based on temporal layering for layered video coding. ITC-CSCC (1998)
8. Rejaie, R., Handley, M., Estrin, D.: Layered quality adaptation for Internet video streaming. IEEE JSAC (2000)
9. Feamster, N., Bansal, D., Balakrishnan, H.: On the interactions between layered quality adaptation and congestion control for streaming video. In: PV Workshop (2001)
10. Wakamiya, N., Miyabayashi, M., Murata, M., Miyahara, H.: MPEG-4 video transfer with TCP-friendly rate control. In: Al-Shaer, E.S., Pacifici, G. (eds.) MMNS 2001. LNCS, vol. 2216. Springer, Heidelberg (2001)
11. Li, S., Wu, F., Zhang, Y.Q.: Study of a new approach to improve FGS video coding efficiency. ISO/IEC JTC1/SC29/WG11, MPEG99/M5583 (1999)
12. Kim, T., Ammar, M.H.: Optimal quality adaptation for scalable encoded video. IEEE JSAC (2005)
13. ITU-T Rec. and final draft international standard on joint video specification: H.264/ISO/IEC 14496-10 AVC, JVT-G050 (2003)
14. UCB LBNL VINT: Network Simulator ns (Version 2), `http://www.isi.edu/nanam/ns/`

Simple and Efficient Fast Staggered Data Broadcasting Scheme for Popular Video Services

Hong-Ik Kim[1] and Sung-Kwon Park[2]

[1] CJ CableNet, 1254 Sinjeong 3-dong, Yangchon-gu,
158-073 Seoul, Korea
hongik@cj.net

[2] Hanyang University, 17 Haengdang-dong, Seongdong-gu,
133-791 Seoul, Korea
sp2996@hanyang.ac.kr

Abstract. In a popular video broadcasting scheme, one of the major challenges is how to reduce a viewer's waiting time maintaining a given bandwidth allocation and how to reduce a client's buffer requirement. In order to solve these problems, many significant broadcasting schemes had been proposed. Among them, the harmonic broadcasting scheme has the best performance for viewer's waiting time and staircase broadcasting scheme has the best performance for client's buffer requirement. However, these previously proposed broadcasting schemes required managing many segments of a video and frequency of channel hopping as well as using many channels at the same time. These complexities make it difficult to implement. In this paper, we propose a fast staggered broadcasting scheme which has simple structure and channel efficiency in a given bandwidth. By comparing with other previously proposed broadcasting schemes, the proposed scheme is substantially reduced dividing segments of a video, a number of managing channels and frequency of channel hopping. Moreover, the proposed scheme is achieved significant bandwidth efficiency in viewer's waiting time and buffer requirement. The numerical results demonstrate that viewer's waiting time of the fast staggered broadcasting scheme converge to that of harmonic scheme as the bandwidth is increased and buffer requirements of this can be decreased below that of staircase scheme by adjusting a short front part of a video sizes.

Keywords: broadcasting, waiting time, buffer requirement, fast staggered broadcasting scheme, simple structure, channel efficiency.

1 Introduction

The multimedia services through the broadband networks are developed along with high-speed communication technology developments. Especially, video-on-demand (VoD) service, as the most important part of multimedia service, is already used in practical area. VoD allows clients to select any given video from a large on-line video library and to watch it through their set-top box (STB) at home. Unlike conventional television broadcasting services in which clients can't select videos, clients have been

T. Vazão, M.M. Freire, and I. Chong (Eds.): ICOIN 2007, LNCS 5200, pp. 699–708, 2008.

given more choices of videos than ever. To access a VoD program, the customer only requires a STB connected to the television set. The STB allows the client to navigate the electrical program guide, to handle the reception and to display of the video once the client has made a choice.

Despite of the attractiveness of VoD services, it does not seem practical to provide VoD services for a great number of clients. One of the reasons is the high cost in providing VoD services. Because a video consumes wide bandwidth even after significant amount of video compression, communication networks must have enough transmission channels to maintain multiple customers concurrently. Therefore, one of the major challenges of designing a VoD system is how to reduce bandwidth consumption with maintaining viewer's waiting time and buffer requirement.

Generally, VoD systems can be categorized into True-VoD (TVoD) systems [1] and Near-VoD (NVoD) systems [2]-[10] by the way that video is delivered. In TVoD, the system must reserve dedicated transmission channels from server resources to each client. Clients can receive video data without any delay via dedicated transmission channels as if they use their own Video Cassette Recorders (VCRs). Such systems, however, may easily run out of the channels because the channels can never keep up with the growth in the number of clients. Therefore, many researchers have investigated how to reduce the channels concurrently used. In NVoD, clients have to wait by some delay time because a video program is broadcasted over several channels with a periodical cycle. The number of broadcasting channels is due to the allowable viewer's waiting time, not the number of requests. Thus, it is more suitable for distributing hot video programs [11].

Many NVoD broadcasting protocols, including the fast [3], harmonic [4],[5], staircase [6], pyramid [7], [8], skyscraper [9] and pagoda [12], [13] broadcasting schemes have been proposed. All these broadcasting schemes can be roughly classified into three categories pyramid-based, harmonic-based, and pagoda-based protocols [10]. Among of these broadcasting schemes, the harmonic scheme [4] has best performance for viewers' waiting time and the staircase scheme [6] has best performance for client buffer requirement [15], [16], [17]. The broadcasting schemes above divide the video into a series of segments and broadcast each segment periodically on dedicated server channels. While a client is playing the current video segment, it is guaranteed that the next segment is downloaded on time and the whole video can be played out continuously. A client will have to wait for the occurrence of the first segment before they can start playing the video.

A drawback of these various broadcasting schemes approach is that a client in all of these schemes are managing many segments of a video, a frequency of channel hopping, using many channels for watching and joining many channel at the same time. These leads to complexity in a VoD system design.

In order to overcome these problems mentioned above, this paper proposes a new VoD scheme named as a fast staggered broadcasting scheme. The proposed scheme divides a video into a short front part and a long rear part. And then, the short front part of a video uses the fast broadcasting scheme [3] and the long rear part of a video uses the staggered broadcasting [14]. In this scheme, the advantage of short viewer's waiting time in the fast scheme is preserved and advantage of simplicity in the staggered scheme is also maintained. Moreover, the short front part leads to reduce buffer requirement significantly. These reasons make it possible to implement easily and more practical than previous proposed VoD schemes.

The simulation results show that the fast staggered broadcasting scheme has outstanding performance in aspects of both the viewer's waiting time and buffer requirement with simplicity of scheme. For a given bandwidth allocation, viewer's waiting time of the fast staggered broadcasting scheme is close to that of the fast scheme and converges into that of harmonic scheme as the bandwidth is increased. The buffer requirement of the fast staggered broadcasting scheme can be reduced below that of the pyramid, harmonic, fast and staircase broadcasting scheme by adjusting the short front part of a video sizes.

2 Fast Staggered Data Broadcasting and Receiving Scheme

The fast scheme has bandwidth efficiency in a video delivery and the staggered scheme has simple system structures. In the fast staggered scheme, these features are combined. The fast staggered scheme is divided a video into a short front part and a long rear part, then the short front part is broadcasted in the fast scheme while long rear part is broadcasted in the staggered scheme. This framework leads to bandwidth efficiency in transmission with simplicity of VoD scheme.

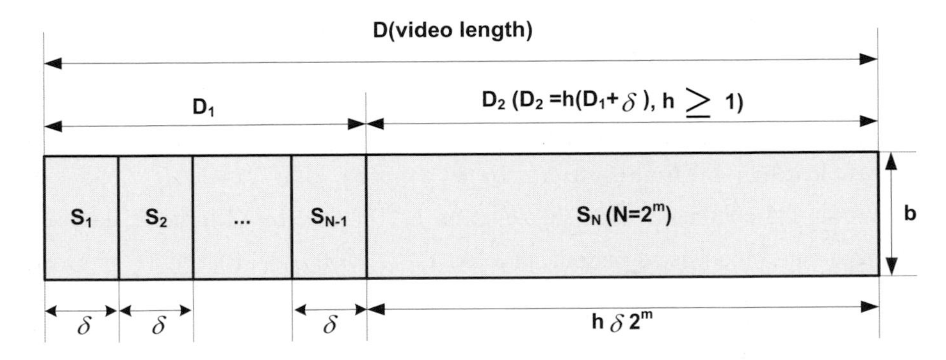

Fig. 1. Basic partition operating of the fast staggered broadcasting scheme for a video when $k = \beta$

Referring to Fig.1, on the server side, suppose there is a video with length D . The consumption rate of the video is b . Thus, the size of the video is $V = D \times b$. Suppose the bandwidth that we can allocate for the video is B , $B = \beta \times b$, $\beta \geq 2$. On the server side, the fast staggered broadcasting scheme involves the following steps,

1. The clients wait to download the video data until the start of segment S_1 occurs first on channel C_1^{Da} The video length D is divided into a short length of front part D_1 and a long length of rear part D_2 . Thus, the size of part D_1 is $V_1 = D_1 \times b$, and the size of part D_2 is $V_2 = D_2 \times b$. The each part relation is

$$D_2 = h(D_1 + \delta),\ h \geq 1 \tag{1}$$

where, h is a video dividing coefficient, δ is a divided data segment S's length in D_1 part. Here, h has only a positive integer and this is equal to assigned number of channel in D_2 part.

2. The bandwidth B is equally divided into k logical channels, where

$$k = \left\lfloor \frac{B}{b} \right\rfloor = \lfloor \beta \rfloor,\ k \geq 2 \tag{2}$$

Consider a number of assigned channels in D_1 part as m ($m \geq 1$), and a number of assigned channels in D_2 part as n ($n = h$, $n \geq 1$). Thus, the number of logical channels is $k = m + n$. Let $\{C_0^{D1}, C_1^{D1}, ..., C_{m-1}^{D1}\}$ represent the m channels of assigned D_1 part and $\{C_0^{D2}, C_1^{D2}, ..., C_{n-1}^{D2}\}$ represent the n channels of assigned D_2 part. The notation C_l^{Dp} (p=1,2) represents the l th channel with Dp's part.

3. D is divided into N segments, D_1 is equally divided into $N-1$ ($= 2^m - 1$) segments and D_2 is a single segment, where

$$N = \sum_{i=0}^{m-1} 2^i + 1 = 2^m \tag{3}$$

Suppose S_i is the i th segment of a video. Every segments concatenated in the order of increasing segment numbers constitutes the whole video. A each segment length in D_1, from S_1 to S_{N-1}, is δ.

4. Within the continuous data segments on C_i^{D1} ($i = 0, ..., m-1$), the 2^i data segments are broadcasted periodically. On C_j^{D2} ($j = 0, ..., n-1$), the D_2 is broadcasted using the staggered scheme with the n channels. The time of staggered channel interval is T_s, $T_s = D_1 + \delta$. Therefore, the number of staggered channels allocated for the staggered scheme is given by $n = D_2 / T_s$, and also number of assigned channel in D_1 part obtains as $m = k - h$.

At the client side, suppose there is enough buffer space at client site for storing the video data, we may store extra portion of the video into local buffer. For watching a video, the following steps are involved.

1. Begin to download the first data segment S_1 of the required video at the first occurrence on C_0^{D1} and to download other related data segments from C_1^{D1} to C_{m-1}^{D1} concurrently. If the first data segment play point is equal to the start of a staggered point, the last data segment S_N simultaneously buffers the ongoing staggered channel. Otherwise, the staggered S_N buffering waits till the start of the staggered point, only if buffer is necessary to the client.
2. Right after beginning to download the data segments, the client can start to consume the video with its consumption rate in the order of $S_1, S_2, ... S_N$.

3. Stop loading from channel C_i^{D1} ($i = 0,...,m-1$) when the client has received 2^i data segment from that channel. Stop loading from channel C_j^{D2} ($j = 0,...,n-1$) when the client has received completely S_N data segments from that channel.

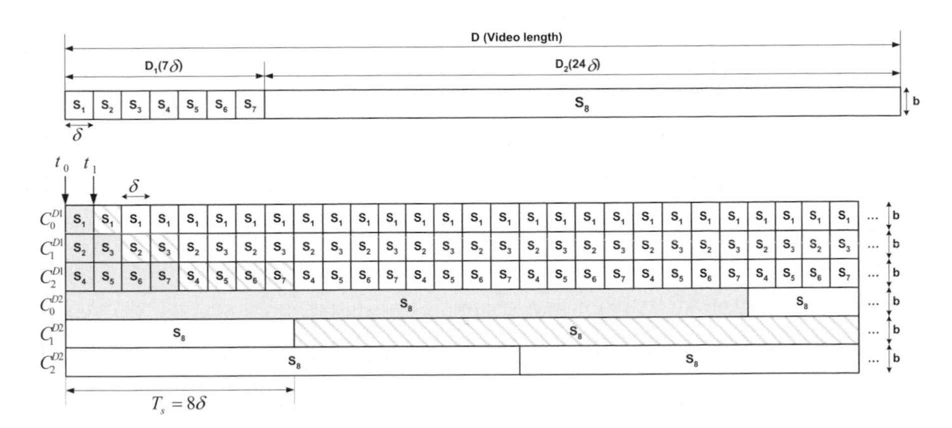

Fig. 2. Example of the fast staggered broadcasting when k =6, β =6, h =3, m =3 and n =3

Fig. 2 depicts the operation of the fast staggered broadcasting scheme for a video when k =6, β =6, m =3, n =3 and h =3. The video length D is divided into D_1 and D_2 by the coefficient h =3. The D_1 is partitioned into 7 segments and these segments are broadcasted over 3 channels (C_0^{D1}, C_1^{D1}, and C_2^{D1}). The last segment S_8 is broadcasted using 3 channels using staggered scheme (C_0^{D2}, C_1^{D2}, and C_2^{D2}). The staggered interval T_s of D_2 is 8δ, where δ is segment length in D_1. In this example, δ is $D/31$, which is also the maximum viewer's waiting time.

3 Analysis and Performance Comparison

3.1 Viewer's Waiting Time

The waiting time for a viewer is defined as the duration for which one has to wait for a video to watch after sending a request to a server. When a client just miss a S_1 of a requested video, the maximum waiting time will be equal to the access time of S_1 from the first channel C_0^{D1}. The size of S_1 is $D_1 b/(N-1) = D_1 b/(2^m - 1)$. The bandwidth of C_0^{D1} is B/k. Therefore, for a given bandwidth allocation in the fast staggered broadcasting and receiving scheme, the maximum waiting time to access a broadcast video is

$$\delta = \frac{D_1 b/(N-1)}{B/k} = \frac{D_1}{2^m - 1} \times \frac{k}{\beta} \tag{4}$$

As β is integer, the client will have $k=\beta$ and $\delta = D_1/(2^m-1)$. Here, the video length is $D = D_1 + D_2$ and $D_2 = h(D_1+\delta)$, $h \geq 1$. Thus, the maximum waiting time to access a broadcast video is represented differently as

$$\delta = \frac{D}{h(2^m-1)\frac{\beta}{k} + (2^m-1)\frac{\beta}{k} + h} \tag{5}$$

As β is integer, we will have $k=\beta$ and $\delta = D/(h2^m + 2^m - 1)$.

Fig. 3 shows the relationship between viewer's waiting time δ and bandwidth requirement β in each scheme for the length of video D =100 minutes and fast staggered scheme with h =3,5. Distinctively, fast staggered scheme results appear from $\beta = h+1$ because long rear part channel in fast staggered scheme n is assigned h channels ($\beta = h$) and short front part channel in fast staggered scheme m has allocated at least one channel ($\beta \geq 1$). In fig. 3, the viewer's waiting time of fast staggered scheme (h =3) was much shorter than that of pyramid scheme, and increases slightly as comparing that of fast scheme and staircase scheme. In case that if 6.9 channels is allocated for the bandwidth β, the viewer's waiting time for fast staggered scheme (h =3) is about 1.49 minutes whereas for pyramid scheme it is 5.03minutes, for fast and staircase scheme it is 1.38 minutes and for harmonic scheme it is 0.17minutes. Thus, the performance of viewer's waiting time for fast staggered scheme (h = 3) is almost same that of fast and staircase scheme, about 70% better than pyramid scheme but slightly higher then that of harmonic scheme. However, as the bandwidth is increased the waiting time δ also decreases sharply in fast staggered scheme, so the waiting time for fast staggered scheme gradually becomes almost same as that of harmonic scheme. It is also found that as h is decreased the waiting time δ also decreases on the same bandwidth allocation in fast staggered scheme.

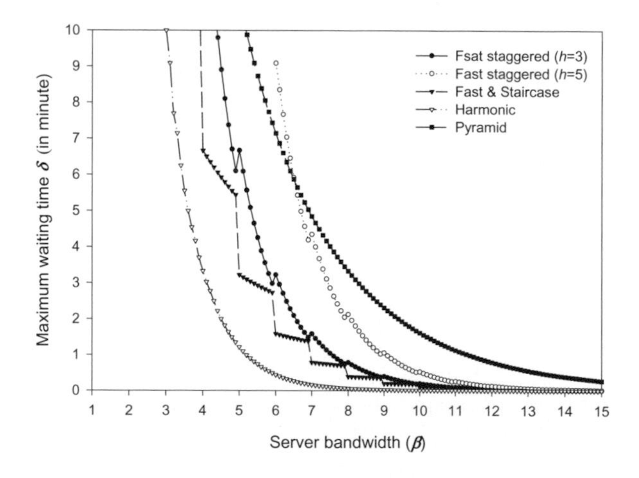

Fig. 3. Comparison of maximum waiting time of different broadcasting schemes when a video length D is 100 minutes and fast staggered scheme has h =3, 5

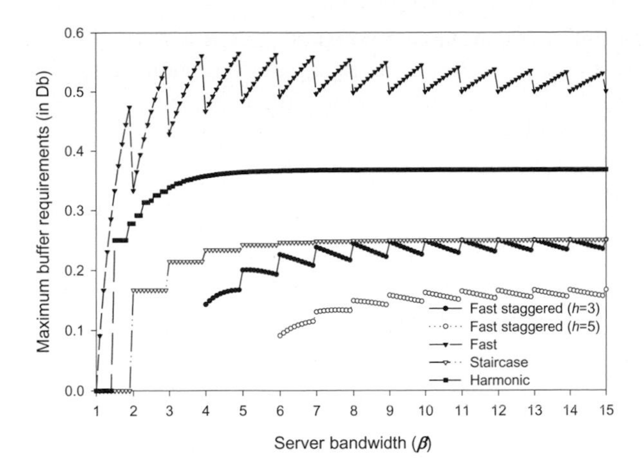

Fig. 4. Comparison of maximum buffer requirement of different broadcasting schemes when fast staggered scheme has h =3, 5

3.2 Buffer Requirement

At the client end, the arriving rate of the video data is greater than its consumption rate of the video. Therefore, a client needs to have local buffer to store the extra video being downloaded. In the fast staggered scheme, the total buffer requirement increases until the D_1's last channel C_{m-1}^{D1}.

Considering the data segments on the D_1's last channel C_{m-1}^{D1}, the client is received $(2^m - 1)\delta b$ of a video data within $2^{m-1}\delta$ but the data size which will be consumed within this period is $2^{m-1}\delta b$. Therefore, the last data segment S_N is buffered until $(2^m - 1)\delta b$. The maximum buffer size requirement at the client end will be given as

$$Z = (2^m - 1)\delta b = D_1 b \frac{k}{\beta} \tag{6}$$

Equation (6) shows that the client buffer requirement of the fast staggered scheme is decided by the short front part V_1 size which is controlled by the video dividing coefficient h. In Fig.2, the best case of buffer requirement happens when the client starts to receive the video at t_1 and no buffer is required. The worst case occurs when the client starts to receive the video at t_0. In worst case, the client buffer must reserve 7 δb amounts of data.

Fig. 4 shows relationship between maximum buffer requirement Z and bandwidth requirement β in each scheme when fast staggered scheme has h=3, 5. Buffer requirement Z of fast staggered scheme (h=3, 5) was less than that of the fast, staircase and harmonic schemes. It is found that if 10 channel is employed for bandwidth β, buffer requirement is obtained as 0.162 Db for fast staggered (h = 5), 0.499 Db for fast, 0.367 Db for harmonic and 0.249 Db for staircase scheme. Thus, the performance of buffer requirement for fast staggered scheme ($h = 5$) is found about 68% better then that of fast, 56% better then that of harmonic and 35% better then that of staircase scheme. It is also found that buffer requirement Z decreases as h increases in fast staggered scheme.

3.3 Complexity

The complexity of the scheme is an essential factor. The proposed scheme has lower complexity than other VoD schemes because it reduces a number of segments of a video, using channels at the same time, managing channels and hopping channels for a VoD services.

The comparison of complexity among different schemes is shown in Table 1 for a video when N_H is a number of using channel in the harmonic scheme which is obtained by $\beta = \sum_{i=1}^{N_H}(1/i)$ and β is a positive integer. From Table 1, it is observed that fast staggered scheme is most simple structures. For example, if $\beta = k = 5$ and assigned channels of short front part channels in fast staggered scheme m are 2. In this case, a number of segments of a video are obtained as 4 for fast staggered, 31 for fast, 341 for staircase and about 3403 for harmonic scheme. A number of managing channels and channel hopping for watching a video are obtained as 3 for fast staggered, 5 for fast, 31 for staircase and about 82 for harmonic scheme. A maximum number of using channels at the same time for watching a video are obtained as 3 for fast staggered, 5 for fast, 16 for staircase, about 82 for harmonic scheme. These results show that fast staggered scheme achieves much lower complexity in a number of segments, managing channels, using channels at the same time and hopping channels for a VoD service on a given bandwidth allocation.

Table 1. Comparison of complexity among different broadcasting schemes for a video, when N_H is a number of using channel in the harmonic scheme which is obtained by $\beta = \sum_{i=1}^{N_H}(1/i)$

Scheme	Segments	Managing channels	Using channels at the same time	Hopping channel
Fast	$2^k - 1$	k	$1 \sim k$	k
Staircase	$\sum_{i=1}^{k}(2^{i-1})^2$	$2^k - 1$	2^{k-1}	$2^k - 1$
Harmonic	$N_H(N_H + 1)/2$	N_H	N_H	N_H
Fast Staggered	2^m	$m+1$	$1 \sim m+1$	$m+1$

4 Conclusions

In VoD system, a large number of channels are needed if many video programs are to be transmitted without long client's waiting time. Many significant broadcasting schemes have been proposed to reduce the viewer's waiting time and buffer requirement but they would seem to be difficult to implement because of complexity. In this paper, we have proposed a new broadcasting scheme called the fast staggered broadcasting scheme which is combined the fast scheme with the staggered scheme. This scheme has both significant simplicity and bandwidth efficiency, comparing with the existing VoD schemes.

We compared the proposed scheme performance with the staggered, pyramid, fast, staircase and harmonic scheme. The results showed that that viewer's waiting time of the fast staggered broadcasting scheme converge to that of harmonic scheme as the bandwidth is increased. The buffer requirement can be reduced less than the staircase scheme by adjusting a video dividing coefficient. Furthermore, our proposed scheme achieves much lower complexity in a number of segments in a video, channels hopping, managing channels and using channels at the same time. Therefore, it is achieved much lower complexity with bandwidth efficiency against previously proposed VoD schemes. The proposed fast staggered scheme could provide a practical and simple implement solution for VoD services with only minor modifications in a server and STBs.

References

1. Tseng, Y.-C., Yang, M.-H., Chang, C.-H.: A recursive frequency-splitting scheme for broadcasting hot videos in VoD service. IEEE Transactions on Communications 50, 1348–1355 (2002)
2. Viswanathan, S., Imielinski, T.: Pyramid Broadcasting for video on demand service. In: IEEE Multimedia Computing and Networking Conference, vol. 2417, pp. 66–77 (1995)
3. Juhn, L.-S., Tseng, L.-M.: Fast data broadcasting and receiving scheme for popular video service. IEEE Transactions on Broadcasting 44, 100–105 (1998)
4. Juhn, L.S., Tseng, L.M.: Harmonic broadcasting for video-on-demand service. IEEE Trans. Broadcasting 43(3), 268–271 (1997)
5. Juhn, L.S., Tseng, L.M.: Enhanced harmonic data broadcasting and receiving scheme for popular video service. IEEE Trans. Consumer Electron. 44, 343–346 (1998)
6. Juhn, L.S., Tseng, L.M.: Staircase data broadcasting and receiving scheme for hot video service. IEEE Trans. Consumer Electron. 43(4), 1110–1117 (1997)
7. Viswanathan, S., Imielinski, T.: Metropolitan area video-on demand service using pyramid broadcasting. IEEE Multimedia Syst. 4, 197–208 (1996)
8. Aggarwal, C.C., Wolf, J.L., Yu, P.S.: A permutation-based pyramid broadcasting scheme for video-on-demand system. In: Proc. IEEE Int. Conf. Multimedia Computing and Systems, pp. 118–126 (1996)
9. Hua, K.A., Sheu, S.: Skyscraper Broadcasting: a new broadcasting scheme for metropolitan video-on-demand systems. In: SIGCOMM 1997, pp. 89–100 (1997)
10. Sul, H.K., Kim, H., Chon, K.: A hybrid pagoda broadcasting protocol with partial preloading. In: Proc. Int. Conf. Multimedia and Expo., pp. I-801–804 (2003)
11. Chien, W.-D., Yeh, Y.-S., Wang, J.-S.: Practical Channel Transition for Near-VOD Services. IEEE Transactions on Broadcasting 51(3), 360–365 (2005)
12. Paris, J.F., Carter, S.W., Long, D.D.E.: A hybrid broadcasting protocol for video on demand. In: Proc. 1999 Multimedia Computing and Networking Conf., pp. 317–326 (1999)
13. Paris, J.F.: A simple low-bandwidth broadcasting protocol for video-on-demand. In: Proc. 8th Int. Conf. Computer Communications and Networks, pp. 118–123 (1999)
14. Almeroth, K.C., Ammar, M.H.: The use of multicast delivery to provide a scalable and interactive video-on-demand service. IEEE Journal on Selected Area in Communications 14(5), 1110–1122 (1996)

15. Yang, Z.-Y., Juhn, L.-S., Tseng, L.-M.: On optimal broadcasting scheme for popular video service. IEEE Trans. Broadcasting 45, 318–322 (1999)
16. Chand, S., Om, H.: Modified Staircase Data Broadcasting Scheme for Popular Videos. IEEE Transactions on Broadcasting 48(4), 274–280 (2002)
17. Yang, H.-C., Yu, H.-F., Tseng, L.-M., Chen, Y.-M.: An efficient staircase-harmonic scheme for broadcasting popular videos. In: Consumer Communications and Networking Conference, pp. 122–127 (2005)
18. Chan, S., Yeung, S.: Client buffering techniques for scalable video broadcasting over broadband networks with low user delay. IEEE Trans. Broadcast. 48(1), 19–26 (2002)

An Enhanced Bandwidth Allocation Algorithms for QoS Provision in IEEE 802.16 BWA

Ehsan A. Aghdaee, Nallasamy Mani, and Bala Srinivasan

Monash University, Melbourne, Australia
{ehsan.aghdaee,Nallasamy.Manii}@eng.monash.edu.au,
{bala.srinivasan}@infotech.monash.edu.au

Abstract. A pair of QoS oriented bandwidth allocation algorithms for both BS and SS in broadband wireless access (BWA) networks, are presented. These algorithms offer soft QoS provisioning by letting each class of service exploiting the available bandwidth considering the delay bounds of the exisiting connection's queues so that lower priority classes can get a reasonable minimum share of bandwidth. This method is taking advantage of a bonus allocation scheme while keeping the connection priorities in focus. The simulation results show that the proposed scheduler provides QoS support for different types of traffic classes whilst maintaining the standards of IEEE 802.16.

1 Introduction

In recent years, there has been a considerable growth in demand for high-speed wireless Internet access and this has caused the emergence of new short-range wireless technologies (viz. IEEE 802.11) and also long-range wireless technologies (viz. IEEE 802.16).

Long-range wireless technologies, in particular IEEE 802.16, offer an alternative to the current wired access networks such as cable modem and digital subscriber line (DSL) links. The IEEE 802.16 has become an attractive alternative, as it can be deployed rapidly even in areas difficult for wired infrastructures to reach and also, it covers broad geographical area in more economical and time efficient manner than traditional wired systems.

At the same time, the growth in adopting a broadband wireless access (BWA) network has significantly increased the customer demand for guaranteed quality of service (QoS). The provision of QoS has become a critical area of concern for BWA providers. The IEEE 802.16 standard appear to offer a solution for this problem, by establishing a number of unique and guaranteed QoS parameters in terms of delay, jitter and throughput. This enables service providers to offer flexible and enforceable QoS guarantees, a benefit that has never been available with other fixed broadband wireless standards [1].

The IEEE 802.16 standard focuses on "First-mile/last-mile" connection in wireless MANs. The IEEE 802.16 working group is developing a standard for the physical and MAC layers. The IEEE 802.16 is utilizing a time division and polling based approach at its MAC layer, which is more deterministic as compared to

T. Vazão, M.M. Freire, and I. Chong (Eds.): ICOIN 2007, LNCS 5200, pp. 709–718, 2008.

contention-based MAC schemes. The 802.16 MAC layer classifies the application flows based upon their QoS requirements and maps them into four basic classes to one of the connections with distinct scheduling services(UGS, rtPS, nrtPS and BE)[2].

While the standard has defined the differentiated treatment for each class of service, the details of packet scheduling algorithms have been left undefined [3]. In this paper, we present a new bandwidth allocation schemes for the BS, which is based on an idea of reserving a minimum amount of bandwidth for each class of service during each frame-time, and sharing the reserved bandwidth among the connections with the same class of service according to their immediate bandwidth requirements. The excess bandwidth is distributed among all the connections according to their priority. We also propose a new scheduling algorithm for the SS which consider the severity of the packet expiry at each connection's queue and this will improve the fairness over the previous research.

The rest of the paper is organized as follows. Section 2 describes the Scheduling framework of the IEEE 802.16. The proposed base station (BS) bandwidth allocation architecture is explained in section 2.1. In section 2.2, our proposed subscriber station (SS) packet scheduler is illustrated. Section 3 is concerned with the simulations of the proposed scheduling algorithms and results. Conclusion is presented in section 4.

2 Scheduling in IEEE 802.16

In the IEEE 802.16 standard, the BS and SS must reserve resources to meet their QoS requirements. Bandwidth is the primary resource that needs to be reserved. The BS controls and can allocate the required bandwidth to downlink and uplink connections according to their traffic profile. The IEEE 802.16 supports UGS, rtPS, nrtPS and BE service classes, each of these service classes has its own QoS requirements. Each connection in the uplink direction is mapped into one of these existing four types of uplink scheduling service. Some of these scheduling services prioritize the uplink access to the medium for a given service, for example by using unsolicited bandwidth grants. Others use polling mechanisms for real time and non-real time services or even pure random access for non real time Best Effort (BE) services[3][5].

Scheduling in IEEE 802.16 is divided into two tasks. The first task, performed at the BS is the scheduling of the airlink resources (uplink subfrarme) among the SSs according to the information provided with the bandwidth requests (BW-request), which is received from the SSs. The second scheduler task is the scheduling of individual packets at SSs and BS. The SS scheduler is responsible for the selection of the appropriate packets from all its queues, and sends them through the transmission opportunities allocated to the SS within each subframe.

As specified in the current IEEE 802.16 standard[2], BW-requests from SSs reference individual connections while each bandwidth grant from the BS is addressed to the SS, not to its individual connections. As a result, SS scheduler is responsible for sharing the bandwidth among its connections while maintaining

QoS. The importance of building a QoS aware scheduler is undeniable, since different connections must be provided with individual QoS levels.

Over the past few years, an enormous amount of research has been conducted in the area of traffic scheduling. With the advent of the wireline integrated service networks, many packet-level traffic scheduling policies have been proposed for switches/routers in order to meet the requirements of different services[4]. However, the objective of most of these algorithms is to provide strict QoS guarantees (i.e., hard QoS) to traffic streams by requiring them to strictly conform to predetermined traffic profiles. Non-conforming traffic is not guaranteed, and so could suffer substantial performance degradation, even though over a period of time the traffic stream may have under-utilized its allocated bandwidth. One of the biggest challenges for all the schedulers is the reality of unpredictable workloads.

In wireless multimedia networks it is difficult to predetermine profiles of real-time traffics; this can lead to degradation of the expected QoS level of the non-conforming traffic, which may be detrimental to the overall QoS experienced by the end user. Soft QoS provisioning can be defined as graceful acceptance of traffic profile violation when excess bandwidth is available, provided that the session does not exceed its overall reserved bandwidth in the long term. In wireless networks, soft QoS provisioning can be ideal for scalable multimedia applications which can tolerate occasional degradation in network performance due to channel errors[6]. In the following two sections the details of the scheduling algorithms at the BS and SS will be illustrated.

2.1 Base Station Bandwidth Allocation Architecture

The BS bandwidth allocation architecture consists of connection classifier, scheduling database module, bandwidth scheduler and MAP generation module.

The connection begins to send out the bandwidth requests after being admitted into the system by the admission control. All connections bandwidth requests are first classified by the connection classifier according to their connection identifier (CID) and service type, they are then forwarded to the appropriate queues at the scheduling database. The scheduler allocates bandwidth to the connections according to the bandwidth requests retrieved from the queues. Based on the result of the bandwidth allocation the uplink-MAP (UL-MAP) will be generated and broadcasted to the SSs.

The following information are kept in our BS scheduling database module for each active connections in the system: Connection Identifier (CID), traffic classe (UGS, rtPS, nrtPS, BE), queue length status and time of expiry (TOE) value. In our proposed BS scheduler architecture, we assume that every BW-request message transmitted by the SS to the BS, carries the reservation request information which consist of traffic queue-length status and also time of expiry (TOE) value of the first packet in the connection traffic queue at the SS. We take the TOE value of the first packet of each queue into account, as a means of prioritizing a more urgent BW-requests. A TOE value of the first packet in the queue represents the amount of time that the scheduler has before allocating a bandwidth to this connection. If the scheduler does not allocate a bandwidth to this connection before a TOE deadline, at least the first packet of this connection and probably all packets belonging to the same traffic burst, will expire. The BS updates the TOE and queue length information for each of the connections whenever it receives a new BW-request from the connections.

In the current draft of the IEEE 802.16 standard [2], BW-request header quantifies the bandwidth request in the number of bytes that are required by the corresponding connection. This is a quantity that merely represents the queue length or changes in queue length. As the queue length is just an indirect measure of indicating the current traffic demand or load, the sole measure cannot be used to deal with the QoS requirements, especially for delay-sensitive or loss-sensitive applications. It needs to be used in combination with some other parameters. A new QoS management message was proposed in [7], that does not change any part of the existing IEEE 802.16 standard. A proposed message can carry various dynamic traffic parameters, thus providing an opportunity for the SS to transmit more information about the current status of each of its traffic queues. We have adopted this new management message to improve the QoS support of the current standard. The followings are the notations that we use throughout this section:

B_{uplink}: total bandwidth (bps) allocated for uplink transmission
$P(t)$: the remainder of bandwidth at time t
$EB(t)$: the excess bandwidth at time t
R_{total}: aggregated reserved bandwidth for all service classes
$L(t)$: the sum of leftover (bit) bandwidth at time t

Base Station Scheduler. As it is stated before the goal of our BS scheduler is to provide a soft QoS support for the connections which are admitted into the system, while maximizing the channel utilization. The proposed scheduling algorithm reserves a minimum amount of bandwidth (which is changed dynamically) for each class of service during each frame-time, and then distributing the reserved bandwidth ($R_{\text{rtps}}, R_{\text{nrtps}} and R_{\text{be}}$) for each class of service among the corresponding connections of that class. The excess bandwidth ($EB(t)$) is distributed among all the connections according to their instantaneous bandwidth requirements. The distribution of the excess bandwidth among different classes follows priority logic, from highest to lowest: rtPS, nrtPS and BE. In another word the procedure boils down into the following two stages:

- Reserved bandwidth of each priority class is distributed among the admitted connections of that class, according to the earliest deadline first with bandwidth reservation (EDF-BR) scheduling discipline.
- The excess Bandwidth would be allocated to those connections that have not been granted a part or total of their requested bandwidth.

  The excess bandwidth is defined as follows:

$$R_{\text{total}} = \sum_{c=1}^{N} R_c \quad , \quad c = \{ugs, rtps, nrtps, be\} \quad , \quad N = 4 \tag{1}$$

$$P(t) = B_{\text{uplink}} - R_{\text{total}} \tag{2}$$

$$EB(t) = P(t) + L(t) \tag{3}$$

The $EB(t)$ is comprised of a left over $(L(t))$ of the reserved bandwidth of each traffic class, if any, and also the remainder of bandwidth $(P(t))$. The main purpose of the second round of scheduling is to maximize the system utilization and to provide soft QoS, by allocating the left over bandwidth of one class of service to another class of service which currently requiring more bandwidth than its reserved bandwidth, due to the arrival of traffic in burst.

The BS scheduler follows the IEEE 802.16 policy mechanism in allocating bandwidth to the UGS connections, which requires the allocation of a fixed size data grants at periodic intervals. However, the bandwidth allocation to connections of other classes will be based on the scheduling algorithm EDF-BR initially proposed as EDF in [8].

The EDF-BR algorithm is carried out in the following four stages:

1. The scheduler first visits each non-empty priority (rtPS, nrtPS, BE) queue and sorts the BW-requests in the ascending order of their TOE.
2. The scheduler checks the requested data size of the first BW-request packet in the rtPS queue, if it is less than or equal to R_{rtps}, the reserved bandwidth is reduced by the size of the BW-request, and all the connection's requested bandwidth is allocated to it. Otherwise, the BW-request is reduced by the size of R_{rtps} and the connection's bandwidth requirement would be partially allocated to it. This process will be repeated until either the R_{rtps} is no more greater than zero or the request queue is empty. In the case that the request queue is empty but R_{rtps} is greater than zero, R_{rtps} will be added to $L(t)$. When any of the above conditions occurs, the scheduler moves on to serve the nrtPS request's queue. After the nrtPS connections being serviced the scheduler moves on to service the BE request's queue.
3. The scheduler allocates the excess bandwidth $EB(t)$ to the connections which have not been serviced in the first round of bandwidth allocation, starting from the highest priority queue. The scheduler moves to serve the next priority queue if the $EB(t)$ is greater than zero and also the request's queue is not empty.
4. When the bandwidth size to be allocated to each connection is determined, the allocations in the frame are made such that each SS gets contiguous allocation opportunities in order to follow the IEEE 802.16 standard that the bandwidth allocation is per SS not per connection.

While our proposed airlink scheduler follows priority discipline to meet delay and loss requirements of different classes, it also tries to maintain its fairness by reserving a minimum amount of bandwidth for each class of service during each frame time. The values of R_{rtps}, R_{nrtps} and R_{be} can be adjusted dynamically by the policy of the admission controller. The admission control policy determines the maximum and minimum values of these parameters based on the current traffic load of each class of service.

2.2 Subscriber Station Scheduler

In the IEEE 802.16 BWA standard, an uplink packet scheduler is located at the SS, which schedules packets from the connection queues into transmission opportunities allocated to the SS within each frame. The packet scheduling at the SS occurs just after the BS allocates the bandwidth to the SS.

At the SS each class of service (UGS, rtPS, nrtPS, BE) is associated with a delay bound of (D_{ugs}, D_{rtps}, D_{nrtps}, D_{be}), whenever an incoming packet of class i arriving at its queue at time t, it is stamped with a TOE of $t + D_i$, and packets are getting served in an increasing order of their TOE.

The first algorithm, which was developed, employs priority queuing (PQ) algorithm. The algorithm allocates all allocated bandwidth ($B_{allocated}$)to the first UGS connection. The remaining bandwidth would then pass on to second, and so on. After UGS connections were looped through, it would move on to rtPS connections in a similar fashion. After rtPS, then nrtPS and finally BE connections. The main problem with this scheduler was that it could starve lower priority connections of bandwidth.

Therefore, more intuitive policing method was introduced. In this method the scheduler first calculate the packet's expiry level of each connection's queue and accordingly a severity multiplier (β)would be determined for each of them. Following is the method that we have applied to calculate the β.

Improved Priority Queuing (IPQ): In this method, the scheduler first compute the weighted average waiting time (T_k) of the existing packets in each connection's queue from (4). Weighted average takes into account the overall criticality of packet's delay bounds together with the length of queues by giving higher priority/weight to packets that are pending to expire and in the same situation the packets in the longer queue are prioritized.

Since we have different priority classes in our system and each of them are associated with different delay tolerance, in order to compare a T_k of different classes, we normalize them into scale of 0 to 1 by dividing the average delay to the maximum delay tolerance of that class. The smaller the normalized weighted average waiting time (δ_k) is, the lower the severity of the packet expiry is at that queue.

d_{i_k} denotes the amount of time that packet i waited in the queue k, w_{i_k} represent the weight of packet i in queue k, and also n_k denotes the total number of packets waiting to be served at the queue k. D_k indicate the maximum delay tolerance of queue k.

$$w_{i_k} = \frac{n_k - i}{n_k} \tag{4}$$

$$T_k = \frac{\sum_{i=1}^{n_k} d_{i_k} w_{i_k}}{\sum_{i=1}^{n_k} w_{i_k}} \tag{5}$$

$$\delta_k = \frac{T_k}{D_k} \tag{6}$$

(δ_k) determines the severity of the packet expiration and accordingly a β will be assigned against the connection's service contract rate (r_i). Table 1 shows, the

Table 1. Severity of traffic expiration with bonus multiplier

	UGS & rtPS				
δ_k	< 0.20	< 0.30	< 0.40	< 0.50	> 0.50
β	0.6	0.65	0.70	0.85	1
	BE & nrtPS				
δ_k	< 0.25	< 0.50	< 0.6	< 0.75	> 0.75
β	0.4	0.55	0.60	0.70	1

details of the severity multiplier (β) against the δ_k. We have selected this values experimentally by running the simulation for number of times, though they can be determined more deliberately depending on the policy of the service provider.

Once the multiplier is determined, the scheduler would allocate bandwidth to each connection's queues first staring with UGS. The connection i would be allocated bandwidth up to its service contract rate(r_i) multiplied by the determined multiplier (β_i). The scheduler would move on to rtPS connections when all UGS connections were satisfied, and then nrtPS and finally BE connections. After the first round of bandwidth allocation, if the remaining bandwidth would be greater than zero the scheduler would allocate the remaining bandwidth to the queues in the order of their priority like the classic PQ algorithm.

3 Simulation Results

A C-coded event-driven simulator is used to evaluate the performance of the proposed packet scheduling schemes. Specifically, the MAC layer functionalities of the SS and the BS were implemented. Since the performance of the scheduling algorithms are of major concern, we disregard the contention resolution process among the reservation requests during the contention mode and assume that the contending SSs can successfully send their reservation requests during this period. Sufficient amount of buffer is also assumed to be available for each of the SS's queues so that no data packet is droped due to buffer overflow. The uplink and downlink subframes are assumed to be equal. The rtPS traffic is modeled according to 200Mbps over 1 hour of MPEG-2 VBR stream of Jurassic Park. For the other classes packet arrival follows Poisson process distribution with rate different rate λ. packet sizes are drawn from a negative exponential distribution. The delay requirements and traffic descriptions of different each classes of service, used in simulation are outlined in Table 2. The performance

Table 2. Traffic Sources Description

Service class	UGS	rtPS	nrtPS	BE
Maximum delay (ms)	30	40	1000	3000
Mean PDU size (byte)	52	NA	100	150
Average bandwidth (KBps)	48.8	NA	39.06	36.62

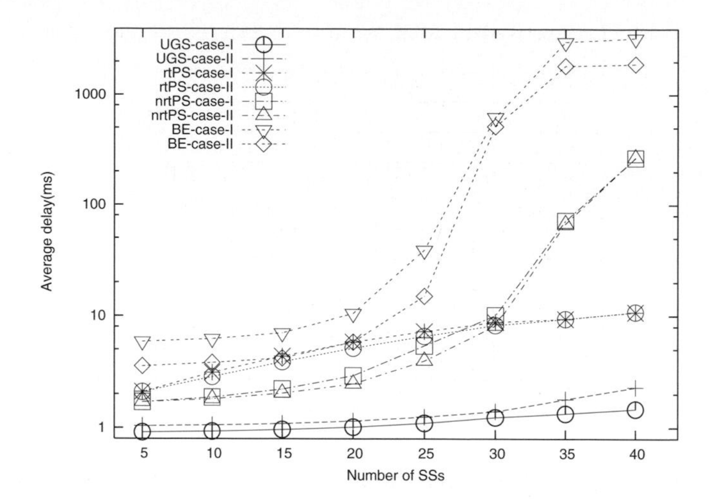

Fig. 1. Average delay

of the proposed BS and SS schedulers are studied under different traffic load. Two cases were considered, in the first case the PQ was applied as a SS packet scheduler , and in the second one the IPQ was used as a SS packet scheduler. The BS uplink scheduler was EDF-BR for both cases.

Figure 1 shows the average delay of each traffic class when the number of SSs, and hence the offered load increases. As can be seen, the average delay curves of UGS class for both case-I and case-II increase smoothly. When the number of SSs increases to more than 30 (the system becomes overloaded), a slightly higher delay is provided by the case-II than the case-I, this behavior can be explained by the way that IPQ calculates the severity multiplier (β) for each connection. The UGS connections are getting lower (β) than other classes since their normalized average delay is fairly low. It should be mentioned here that the provided UGS delay by both case-I and case-II are less than 30ms, which is the delay tolerance of UGS class. This clearly shows that the BS scheduler allocates enough bandwidth to the UGS connections in order to meet their QoS requirement.

When the number of SSs increases from 5 to 40 SSs, the rtPS delay curves of both case-I and case-II rises from 2.2 ms to almost 12 ms. According to Figure 1, the nrtPS delay curves increase smoothly when the number of SSs is less that 30. However, when the system becomes overloaded (e.g., the number of SSs becomes greater than 30), the delay curves increase sharply. Case-I and case-II provide almost the same average delay for nrtPS connections and also rtPS connections. As can be seen in the Figure 1, when the number of SSs is less than 20 the BE delay curves increase smoothly. The average delay considerably rises when the number of SSs increases from 20 to 25 but it is still below 40 ms. When the traffic load increases to more than 25 SSs, the delay rises sharply. When the number of SSs increases to more than 35 there is not much increase in the BE average delay curves since most of the incoming packets would be expired before being

Table 3. Simulation result

Num SSs	5	10	15	20	25	30	35	40
	case-I							
drop ugs	0	0	0	0	0	0	0	0
drop rtps	0	0	0	0	0.007	0.01	0.02	0.05
drop nrtps	0	0	0	0	0	0	0	0.02
drop be	0	0	0	0	0	0.07	0.33	0.72
	case-II							
drop ugs	0	0	0	0	0	0	0	0
drop rtps	0	0	0	0	0.002	0.008	0.03	0.07
drop nrtps	0	0	0	0	0	0	0	0.02
drop be	0	0	0	0	0	0.02	0.31	0.70

serviced. It can be clearly seen that the IPQ algorithm significantly reduces the BE average delay while keeping the other classes average delay almost the same as PQ algorithms. For instance when the number of SSs is equal to 35 SSs, the delay of 1530 ms and 2542 ms are obtained for case-II and case-I, respectively. This is because the IPQ algorithm does not let the BE connection queues to grow indefinitely since it allocates the bandwidth to each connection's queue according to its weighted average waiting time, which is more fair than the PQ which only considers the priority of the connection.

Table 3 shows the drop rate of different service classes obtained from case-I and case-II. In both cases and under different traffic loads the UGS drop rate is zero. This is because the UGS class has the highest priority and BS allocates the regular transmission opportunity to it. When the system is not overloaded (the number of SSs less than or equal to 30), IPQ algorithm either reduces the drop rate of rtPS, nrtPS and BE or keeps it at the same rate as the PQ algorithm does. This is because the IPQ algorithm accurately measures the packet expiry level of each connection's queue and accordingly distribute the bandwidth within them. However, when the system becomes overloaded the IPQ algorithm provides slightly higher drop for rtPS connections than PQ, but at the same time it reduces the BE drop rate. This is because when the system becomes overloaded all connections queue size grow. Since all the connections have a critical packet expiry conditions, each of them would be allocated a fair share of bandwidth by the IPQ algorithm. As a result in the overloaded conditions it prevents the higher priority classes to starve the lower ones.

4 Conclusion

In this paper, we have presented a new bandwidth allocation algorithm for IEEE 802.16 BWA. We have also introduced an enhancement to the original PQ, namely IPQ and characterized its delay and loss rate in comparison with original PQ. It was established from the simulation that the IPQ algorithm outperforms the PQ performance by preventing a bandwidth starvation to happen. We have also shown the service differentiation, in terms of delay and loss rate between real-time (served via UGS and rtPS) and non real-time (served via nrtPS and BE) traffic. The simulation results showed that the proposed solution could be used to provide QoS support for all types of traffic classes.

References

1. Pidutti, M.: 802.16 Tackles Broadband Wireless QoS Issues. Comms Design (December 2004)
2. IEEE 802.16 Working Group on Broadband Wireless Access. IEEE Standard for Local and Metropolitan Area Networks. Part 16: Air Interface for Fixed Broadband Wireless Access Systems. IEEE (2004)
3. Eklund, C., Marks, R.B., Stanwood, K.L., Wang, S.: IEEE standard 802.16: A Technical Overview of the WirelessMAN^{TM}Air Interface for Broadband Wireless Access. IEEE Communications Magazine 40(6), 98–107 (2002)
4. Guerin, R., Peris, V.: Quality-of-service in packet networks: Basic Mechanisms and directions. Comput. Networks, special Issue on Internet Telephony 31(3), 169–189 (1999)
5. Cicconetti, C., Lenzini, L., Mingozzi, E.: Quality Of Service Support in IEEE 802.16 Networks. IEEE Network (April 2006)
6. Wong, W.K., Zhu, H., Leung, V.C.M.: Soft QoS Provisioning Using The Token Bank Fair Queuing Scheduling Algorithm. IEEE Wireless Communications (June 2003)
7. Tcha, K., Kim, Y.B., Lee, S.C., Lee, K.J.H., Kang, C.G.: IEEE P802.16e/D4-2004 QoS Managment for Facilitation of Uplink Scheduling
8. Sivaraman, V., Chuissi, F.: Providing End-to-End statistical Delay Guarantees with Earliest Deadline First Scheduling and Per-Hop Traffic Shaping. In: Proc IEEE INFOCOM 2000 (2000)

Robust Transmission Power and Position Estimation in Cognitive Radio*

Sunghun Kim, Hyoungsuk Jeon, Hyuckjae Lee, and Joong Soo Ma

School of Engineering
Information and Communications University
119 Munjiro, Yuseong-gu Daejeon, 305-732, Korea

Abstract. The transmission power and position of the primary user in cognitive radio(CR) is very precious information because these information of the primary user determines the spatial resource. This opportunistic spatial resource is available to secondary users to exploit it. To find position of the primary user, we try to use existing positioning or localization schemes based on ranging techniques but those require the primary user's transmission. Since most primary users in CR are legacy system, and there are no beacon protocol to advertise useful information such as transmission power. Some of existing localization schemes don't require the transmission power, but those don't work in *Outer case* that the primary user is out of convex hull of secondary users' coordinates. We propose the constrained optimization method to estimate transmission power and position without the prior information of the transmission power. Also, we do extensive simulations on two major cases of network deployments to prove that the proposed constrained optimization method increases performance in mean square error(MSE).

1 Introduction

From the spectrum allocation aspect, the conventional static spectrum allocation results in low spectrum efficiency considering the increased demand for bandwidth. Recently, this problem has encouraged many researchers to investigate a cognitive radio (CR) as a promising technology to maximize the spectrum utilization. Fundamentally, the CR utilizes radio resources by the following rules: 1) sensing the current spectrum environment and 2) making smart decisions. Joseph Mitola III[1], a pioneer of CR, introduced the concept of cognition cycle which consists of *radio scene analysis*, *channel state estimation and prediction*, and *action* to transmit signal. All of actions in CR are based on the result of *radio scene analysis* which analyzes RF stimuli from radio environment and then identifies white space, position and transmission parameters of primary users or other systems. From this, radio scene analysis is very important task to apply

* This research was supported by the MIC(Ministry of Information and Communication), Korea, under the ITRC(Information Technology Research Center) support program supervised by the IITA(Institute of Information Technology Advancement)" (IITA-2006-C1090-0603-0015).

T. Vazão, M.M. Freire, and I. Chong (Eds.): ICOIN 2007, LNCS 5200, pp. 719–728, 2008.

CR into the wireless communication systems. Moreover, user's location and its transmission power for CR can be considered as important parameters to enable the agile transmission[2]. Thus, one of the challenging issues on CR is how to identify and analyze a primary user and other systems that coexists with CR.

In this paper, we propose robust and accuracy estimation scheme to capture the position of users and to find their transmission power. There are many localization schemes in previous works, however we know that most primary users are users of legacy systems that are not aware of secondary users, so we can't know the primary users' transmission power. This limitation is very critical to localize primary users, because most previous schemes, which use RSS measurements, require transmission power of the radio frequency signal. Therefore we proposed estimation scheme to find both transmission power and its position at the same time.

The rest of the paper is consisted of follows. Section 2 introduces the related works about localization and positioning schemes. Section 3 discusses the system model and propagation models to use in this paper, and then mentions measurement models to reduce disturbance of measurements based on propagation models. Section 4 gives the estimation problem to solve, methodologies to approach, and description of the proposed scheme. In Section 5, we evaluate the performances and show the improvements of the proposed scheme. Finally we summarize our work and conclude in Section 6.

2 Related Works

Many existing localization schemes attempt to solve the problem for estimating the position of the user. In this section, we classify previous works into two categories where we present range-based and range-free localization schemes.

Range-based localization schemes are using ranging metric to measure the distance between users with certain signal type. Schemes using range-based localization are divided by signal metrics (AOA: angle of arrival, TOA: time of arrival, TDOA: time difference of arrival, and RSS: received signal strength). TOA and TDOA use the propagation time to estimate the distance. AOA is measuring the angle of signal arrivals to estimate positions. Range-based localization schemes use above metrics to range between users, and then apply the triangulation method to estimate the position[6]. We consider estimations using RSS metric, but in CR, most primary users are legacy systems so we cannot know their transmission powers. This is the reason to be hard to range in CR system with RSS metric.

Range-free localization schemes doesn't use signal metric to range, but use protocol oriented metric such as the number of listened beacons, the hop count. [3] measures the number of listened beacons and use the implication of the proximity to estimate positions of users. [4] exploits the relationship between the hop count and the distance, and then uses the triangulation method. These approaches have the problem or don't work when the user to be estimated is out of the convex hull of anchors(see [7]). Designing positioning method, we must consider this problem.

3 System Model

In CR, there are 2-type of users: 1) primary user: an user that has priority to access spectrum. 2) secondary user: an user that has no or lower priority than primary users. Figure 1 represents the network configuration for position and transmission power estimation in CR. In figure 1, primary users are emitting the signal through air, and the secondary users are receiving the signal from primary users. A bold dotted line denotes the primary uesr's signal. Secondary users share the information of measured RSS values at each user and position of users. Dotted lines are secondary user's communications to share the information.

3.1 Assumptions

We devise the proposed scheme under several assumptions. The following items represent assumptions of this system model.

- Primary users' transmission powers are unknown.
- Secondary users' positions are known.
- Secondary users measure the RSS values from primary users.
- There are at least 4 secondary users receiving the signal from the primary user.
- A shadowing effect to each secondary user is independent.

Under those assumptions, the unknown primary users' position and transmission power can be estimated by secondary users.

3.2 Propagation Models

The proposed method uses RSS which is modeled by propagation model, so we touch on the related propagation model. Firstly, let the real position of the primary user or object to estimate be $[x, y]$ and the coordinate of the i_{th} secondary

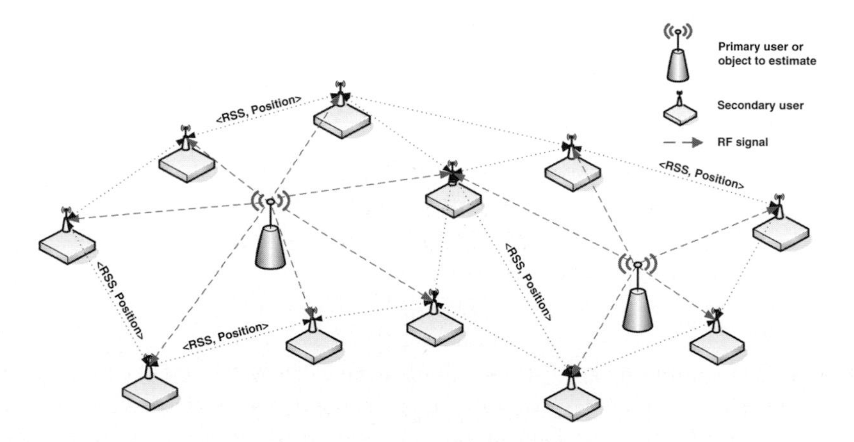

Fig. 1. Network Configuration

user be $[x_i, y_i]$, $i = 1, 2, ..., N$, where N is the total number of secondary users receiving primary user's signal. We denote d_i, the distance between the primary user and the i_{th} secondary user, represented as the following equation.

$$d_i = \sqrt{(x - x_i)^2 + (y - y_i)^2}, i = 1, 2, ..., N \tag{1}$$

Secondly, we discuss simplified propagation model. The ideal RSS or received power at the i_{th} secondary user is denoted by P_i^{rx}, is expressed as

$$P_i^{rx} = RSS_i = K_i \frac{P^{tx}}{d_i^{\alpha}}, i = 1, 2, ..., N \tag{2}$$

where P^{tx} is the transmission power of signal emitter, K_i is the rest of all other factors that affect the received signal power including the antenna gain, antenna height, and α is the the path-loss exponent. We also deal with the shadowing effect using log-normal path loss model.

$$P_i^{rx} = RSS_i = K_i \frac{P^{tx}}{d_i^{\alpha} S_i}, i = 1, 2, ..., N \tag{3}$$

where $S_i = 10^{0.1 X_i}$ is a Lognormal Random Variable and X_i is a Gaussian Random Variable with mean $= 0$, and variance σ^2.

3.3 Measurement Model

In this subsection, we describe the measurement model to estimate position and transmission power of the primary user based on propagation models. The proposed scheme deals with shadowing effect in propagation. This shadowing effect causes wide variation in RSS values, thus we model this effect using log-normal path loss model(3).

RSS values at each users have severe disturbances caused by the shadowing effect, so we measure a raw RSS values and refine samples data by using sample mean.

$$m_i(M)[dBm] = \frac{\sum_{j=1}^{M} RSS_{i,j}[dBm]}{M} \tag{4}$$

where $RSS_{i,j}$ is j_{th} sample RSS value at the i_{th} secondary user in dBm and M is total number of samples for the unit measurement.

Then, we use an unit measurement model defined as the below equation

$$r_i(M) = \frac{K_i}{m_i(M)} \tag{5}$$

where $m_i(M)$ sample mean of RSS values at the i-th node and K_i is the propagation factor in (2) and (3). The unit measurement model is based on the sample mean of RSS values at each node. Those two preprocesses reduce shadow disturbances of RSS values in (3).

When the number of samples for the unit measurement is M, the relationship between the unit measurement and the propagation model(3) is modeled as the following equation.

$$r_i(M) = \frac{K_i}{m_i(M)} \approx \frac{K_i}{RSS_i} = \frac{d_i^\alpha}{P^{tx}} S_i, i = 1, 2, ..., N \tag{6}$$

where $S_i = 10^{0.1X_i}$ is a Lognormal Random Variable and X_i is a Gaussian Random Variablethat that denote shadowing effect error at the i-the secondary user. We use (2) and (3) to analyze and formulate the transmission power and position estimating problem.

4 Problem Formulation and Methodology

This section is to formulate our estimation problem to solve, and proposed two schemes to find the solution of that problem. If there are no disturbance in the measurement, then (3) is expressed by

$$\frac{K_i}{RSS_i} = \frac{d_i^\alpha}{P^{tx}} = \frac{[(x - x_i)^2 + (y - y_i)^2]^{\frac{\alpha}{2}}}{P^{tx}}, i = 1, 2, ..., N \tag{7}$$

That means there is no shadowing effect on the RSS measurement. After taking power $\frac{2}{\alpha}$ on (7), we get the below derivations

$$\begin{aligned} (\frac{K_i}{RSS_i})^{\frac{2}{\alpha}} &= \frac{(x - x_i)^2 + (y - y_i)^2}{(P^{tx})^{\frac{2}{\alpha}}} \\ &= \frac{x^2 + y^2 - 2xx_i - 2yy_i + x_i^2 + y_i^2}{(P^{tx})^{\frac{2}{\alpha}}} \\ \Rightarrow x_i^2 + y_i^2 &= 2xx_i + 2yy_i + (\frac{K_i}{RSS_i})^{\frac{2}{\alpha}} p - R^2, i = 1, 2, ..., N \end{aligned} \tag{8}$$

where $R^2 = x^2 + y^2$ and $p = (P^{tx})^{\frac{2}{\alpha}}$ are intermediate variables in order to simplify in terms of x, y, R^2, and p. (8) hold at each node secondary users, which are receiving the primary user's signal. Hence, we can express (8) in a matrix form

$$\mathbf{A}\theta = \mathbf{b} \tag{9}$$

$$\text{where} \quad \mathbf{A} = \begin{pmatrix} 2x_1 & 2y_1 & \frac{K_1}{RSS_1}^{\frac{2}{\alpha}} & -1 \\ 2x_2 & 2y_2 & \frac{K_2}{RSS_2}^{\frac{2}{\alpha}} & -1 \\ 2x_3 & 2y_3 & \frac{K_3}{RSS_3}^{\frac{2}{\alpha}} & -1 \\ 2x_4 & 2y_4 & \frac{K_4}{RSS_4}^{\frac{2}{\alpha}} & -1 \end{pmatrix}, \theta = \begin{pmatrix} x \\ y \\ p \\ R^2 \end{pmatrix}, \text{and} \quad \mathbf{b} = \begin{pmatrix} x_1^2 + y_1^2 \\ x_2^2 + y_2^2 \\ x_3^2 + y_3^2 \\ x_4^2 + y_4^2 \end{pmatrix}$$

Finally, we formulate the estimation problem and transform the problem into the matrix form.

4.1 Least Square Method (LS)

The formulated estimation problem in the matrix form can be solved easily using least square method to minimize the disturbance caused by shadowing effect in measurements. Let $\hat{\theta}$ be estimated position and transmission power, then the solution is computed as the following equation.

$$\begin{aligned} \hat{\theta} &= \arg\min(\mathbf{A}\theta - \mathbf{b})^{\mathbf{T}}(\mathbf{A}\theta - \mathbf{b}) \\ &= (\mathbf{A}^{\mathbf{T}}\mathbf{A})^{-1}\mathbf{A}^{\mathbf{T}}\mathbf{b} \end{aligned} \tag{10}$$

Also we can get the solution to better performance with respect to mean square error (MSE).

4.2 Constrained Optimization Method (RoTPE)

In this subsection, we introduce the constrained and weighted optimization method to solve (9) with considering disturbances in measurements. We know that each secondary user has different disturbances caused by different shadowing effects in measurements, but least square method does not consider it. To consider these differences of measurements at each secondary user, we define the objective function to minimize mean square error(MSE) plugging a weighting factor into (10); the proposed scheme use the constraint using the relationship of the intermediate variable R in (8) and (9).

$$R^2 = x^2 + y^2 \tag{11}$$

Thus, all relevant components, objective function and constraints, are collected to build the constrained optimization problem with weighting factor. The following optimization problem is for estimating transmission power and position of the primary user.

$$\hat{\theta} = \arg\min(\mathbf{A}\theta - \mathbf{b})^{\mathbf{T}}\mathbf{W}(\mathbf{A}\theta - \mathbf{b}) \tag{12}$$

subject to

$$\mathbf{q}^{\mathbf{T}}\theta + \theta^{\mathbf{T}}\mathbf{P}\theta = 0, \quad \text{where} \quad \mathbf{P} = \begin{pmatrix} 1\,0\,0\,0 \\ 0\,1\,0\,0 \\ 0\,0\,0\,0 \\ 0\,0\,0\,0 \end{pmatrix}, \quad \mathbf{q} = \begin{pmatrix} 0 \\ 0 \\ 0 \\ -1 \end{pmatrix} \tag{13}$$

We translate (11) to (13) for building the constraint with the matrix form($-R^2 + x^2 + y^2 = 0 \Rightarrow \mathbf{q}^{\mathbf{T}}\theta + \theta^{\mathbf{T}}\mathbf{P}\theta = 0$). In order to set the weighting matrix $\mathbf{W}$, the disturbance in the matrix $\mathbf{A}$ containing measurements of RSSs at each secondary user should be studied. First of all, the measurement $\frac{K_i}{RSS_i}$ of (6) with disturbance can be linearlized using Taylor series as

$$\begin{aligned} \left(\frac{K_i}{RSS_i}\right)^{\frac{2}{\alpha}} &= \left(\frac{d_i^{\alpha}}{P^{tx}}S_i\right)^{\frac{2}{\alpha}} = f(X) = (XS_i)^{\frac{2}{\alpha}} \\ &= f(a) + f'(a)(X-a) + \frac{1}{2!}f''(a)(X-a)^2 + ... \end{aligned} \tag{14}$$

$$\approx (X)^{\frac{2}{\alpha}} + \frac{2}{\alpha}(X)^{\frac{2}{\alpha}} S_i - \frac{2}{\alpha}(X)^{\frac{2}{\alpha}}$$
$$\approx (\frac{d_i^\alpha}{P^{tx}})^{\frac{2}{\alpha}} + \frac{2}{\alpha}(\frac{d_i^\alpha}{P^{tx}})^{\frac{2}{\alpha}} S_i - \frac{2}{\alpha}(\frac{d_i^\alpha}{P^{tx}})^{\frac{2}{\alpha}}$$

where $X = \frac{d_i^\alpha}{P^{tx}}$, $a = \frac{1}{S_i}X$. Secondly, we define the disturbance of measurements as difference between true value($\frac{d_i^\alpha}{P^{tx}}$) and estimated value($\frac{K_i}{RSS_i}$). This difference implies the disturbance of the shadowing effect.

$$\varepsilon_i = (\frac{K_i}{RSS_i})^{\frac{2}{\alpha}} - (\frac{d_i^\alpha}{P^{tx}})^{\frac{2}{\alpha}} \approx \frac{2}{\alpha}(\frac{d_i^\alpha}{P^{tx}})^{\frac{2}{\alpha}}(S_i - 1) \tag{15}$$

The vector form of the disturbance is represented as $\varepsilon = \{\varepsilon_i\}$. Then the covariance matrix of the disturbance is expressed as

$$\Psi = E\{\varepsilon\varepsilon^{\mathbf{T}}\} = \mathbf{BQB} \tag{16}$$

where $\mathbf{B} = \text{diag}\{\frac{2}{\alpha}(\frac{d_1^\alpha}{P^{tx}})^{\frac{2}{\alpha}}, \frac{2}{\alpha}(\frac{d_2^\alpha}{P^{tx}})^{\frac{2}{\alpha}}, ..., \frac{2}{\alpha}(\frac{d_N^\alpha}{P^{tx}})^{\frac{2}{\alpha}}\}$ and $\mathbf{Q} = \text{diag}\{E[(S_1-1)^2], E[(S_2-1)^2], ..., E[(S_N-1)^2]\}$.

We think that a low value of error covariance implies a high reliability of measurements. In this implication, the appropriate weighting matrix for (13) is $\mathbf{W} = \Psi^{-1}$ that is determined by the unknown d_i and P^{tx} since we can not measure the unknown d_i and P^{tx} directly. However measurements of RSSs at each secondary user imply these unknown variables. Applying the relationship between RSS values and unknown variables, we get the following equation.

$$\Psi \approx \hat{\mathbf{B}}\mathbf{Q}\hat{\mathbf{B}} \tag{17}$$

where $\hat{\mathbf{B}} = \text{diag}\{\frac{2}{\alpha}(\frac{K_1}{RSS_1})^{\frac{2}{\alpha}}, \frac{2}{\alpha}(\frac{K_2}{RSS_2})^{\frac{2}{\alpha}}, .., \frac{2}{\alpha}(\frac{K_N}{RSS_N})^{\frac{2}{\alpha}}\}$,$\mathbf{Q} = \text{diag}\{E[(S_1-1)^2], E[(S_2-1)^2], ..., E[(S_N-1)^2]\}$

(17) is a result plugging (2) into (16). To set $\mathbf{Q}$, an analysis of S_i is a required step. First of all, we find probability density function(pdf) of S_i; after that, the analysis of pdf for a random variable $(S_i-1)^2$; finally, we find mean value of the random variable $(S_i-1)^2$. Next, we solve the constrained optimization problem by transforming the original problem into Lagrange dual problem.

$$L(\theta,\lambda) = (\mathbf{A}\theta - \mathbf{b})^{\mathbf{T}}\Psi^{-1}(\mathbf{A}\theta - \mathbf{b}) - \lambda(\mathbf{q}^{\mathbf{T}}\theta + \theta^{\mathbf{T}}\mathbf{P}\theta) \tag{18}$$

where λ is Lagrange multiplier. (18) is Lagrangian of the original problem. After differentiating (18) with respect to θ and λ, the results are the following equations

$$\frac{dL(\theta,\lambda)}{d\theta} = 2(\mathbf{A}^{\mathbf{T}}\Psi^{-1}\mathbf{A} - \lambda\mathbf{P})\theta - 2\mathbf{A}\Psi^{-1}\mathbf{b} + \lambda\mathbf{b} \tag{19}$$
$$\frac{dL(\theta,\lambda)}{d\lambda} = \mathbf{q}^{\mathbf{T}}\theta + \theta^{\mathbf{T}}\mathbf{P}\theta \tag{20}$$

To satisfy optimal condition, (19) and (20) are zero. Then θ is extracted from (19) and expressed as the function of λ.

$$\hat{\theta}(\lambda) = (\mathbf{A}^{\mathbf{T}}\Psi^{-1}\mathbf{A} - \lambda\mathbf{P})^{-1}(\mathbf{A}^{\mathbf{T}}\Psi^{-1}\mathbf{b} - \frac{\lambda}{2}\mathbf{q}) \tag{21}$$

Since λ is not yet defined, we plug (21) into (20), and we diagonalize the matrix $(\mathbf{A}^{\mathbf{T}}\Psi^{-1}\mathbf{A})^{-1}\mathbf{P}$ as

$$(\mathbf{A}^{\mathbf{T}}\Psi^{-1}\mathbf{A})^{-1}\mathbf{P} = \mathbf{U}\mathbf{\Lambda}\mathbf{U}^{-\mathbf{1}} \tag{22}$$

where $\Lambda = \text{diag}(\gamma_1, \gamma_2, \gamma_3, \gamma_4)$ and $\gamma_i, i = 1, 2, 3, 4$ are eigenvalues of $(\mathbf{A}^{\mathbf{T}}\Psi^{-1}\mathbf{A})^{-1}\mathbf{P}$. Applying (22), we finally find the below equation.

$$\begin{aligned}\mathbf{c}^{\mathbf{T}}(\mathbf{I}+\lambda\mathbf{\Lambda})^{-\mathbf{1}}\mathbf{f} - \frac{\lambda}{\mathbf{2}}\mathbf{c}^{\mathbf{T}}(\mathbf{I}+\lambda\mathbf{\Lambda})^{-\mathbf{1}}\mathbf{g} + \mathbf{e}^{\mathbf{T}}(\mathbf{I}+\lambda\mathbf{\Lambda})^{-\mathbf{1}}\mathbf{\Lambda}(\mathbf{I}+\lambda\mathbf{\Lambda})^{-\mathbf{1}}\mathbf{f} \\ -\frac{\lambda}{2}\mathbf{e}^{\mathbf{T}}(\mathbf{I}+\lambda\mathbf{\Lambda})^{-\mathbf{1}}\mathbf{\Lambda}(\mathbf{I}+\lambda\mathbf{\Lambda})^{-\mathbf{1}}\mathbf{g} - \frac{\lambda}{\mathbf{2}}\mathbf{c}^{\mathbf{T}}(\mathbf{I}+\lambda\mathbf{\Lambda})^{-\mathbf{1}}\mathbf{\Lambda}(\mathbf{I}+\lambda\mathbf{\Lambda})^{-\mathbf{1}}\mathbf{f} \\ +\frac{\lambda^2}{4}\mathbf{c}^{\mathbf{T}}(\mathbf{I}+\lambda\mathbf{\Lambda})^{-\mathbf{1}}\mathbf{\Lambda}(\mathbf{I}+\lambda\mathbf{\Lambda})^{-\mathbf{1}}\mathbf{g} = \mathbf{0}\end{aligned} \tag{23}$$

where $\mathbf{c}^{\mathbf{T}} = \mathbf{q}^{\mathbf{T}}\mathbf{U} = [c_1, c_2, c_3, c_4]$, $\mathbf{g} = \mathbf{U}^{-1}(\mathbf{A}^{\mathbf{T}}\mathbf{\Psi}^{-1}\mathbf{A})^{-1}\mathbf{q} = [q_1, q_2, q_3, q_4]^{\mathbf{T}}$, $\mathbf{f} = \mathbf{U}^{-1}(\mathbf{A}^{\mathbf{T}}\mathbf{\Psi}^{-1}\mathbf{A})^{-1}\mathbf{A}\mathbf{\Psi}^{-1}\mathbf{b} = [f_1, f_2, f_3, f_4]^{\mathbf{T}}$ Since the rank of the matrix $(\mathbf{A}^{\mathbf{T}}\Psi^{-1}\mathbf{A})^{-1}\mathbf{P}$ is 2 and this means that two of its eigenvalues are zero, we can simplify (23) into the equation that has 5-root with respect to λ. We find λ using the standard root finding method and then check KKT optimality conditions. These optimality conditions filter infeasible root of the optimization problem. The estimated θ is computed by applying the feasible λ into (21). Finally, p in the desired θ must be translated into P^{tx} using $P^{tx} = p^{\frac{\alpha}{2}}$.

5 Performance Evaluation

In this section, we do simulations to evaluate the least square(LS) and the constrained optimization method(RoTPE) on MSE. Main factors to effect the performance of schemes are the number of samplings for an unit measurement(M) and the number of nodes(N) in the localization system. To evaluate robustness of the proposed scheme, simulations are used the log-normal path loss model with $\alpha = 3$, $\sigma_X = 4$, $P^{tx} = 1$, and $K_i = 1, i = 1, 2, ..., N$. Network deployments of secondary users and the primary user are 2 cases whether the primary user in the convex hull of secondary users or not: 1) *Inner case*, 2) *Outer case*. This because some related works[3][4][7] can estimate the primary user without RSS-based ranging technique in *Inner case* of the network deployment, but that scheme doesn't work in *Outer case* of the network deployment. Our proposed method works well in *Inner/Outer cases*.

5.1 Inner Case

Inner case of the network deployments has the primary user in the convex hull of secondary users. We set up positions of secondary users and the primary user at [110, 260]m, [270, 90]m, [400, 245]m, [260, 400]m, [145, 15]m, [350, 155]m, [130, 380], [365, 345]m, and [240, 210]m respectively. MSEs are computed based on independent 1000 times of simulations. The left graph of Figure 2 represents MSE as increasing the number of sampling(M) for an unit measurement with the first

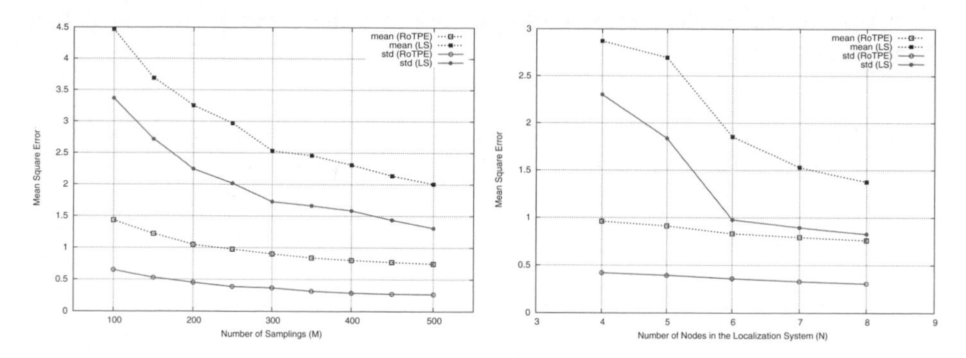

Fig. 2. Inner Case: MSE over Number of Samplings(Left) and MSE over Number of Nodes(Right)

4-secondary user in the system(N=4). Our proposed constrained optimization method (RoTPE) outperforms over standard least square method (LS) with respect to the mean and standard deviation. The right graph in Figure 2 shows the performance affected by the number of nodes(N) in the system when the number of samplings for an unit measurement is 200($M = 200$). It also shows the proposed constrained optimization scheme (RoTPE) has better performance in accuracy. These results, because the constrained optimization method (RoTPE) considers the differences in levels of the disturbance at each secondary user.

5.2 Outer Case

In this case, the primary user is out of the convex hull of secondary users' positions. We set up positions of secondary users and the primary user at [90, 260]m, [240, 110]m, [190, 110]m, [110, 190]m, [140, 280]m, [160, 145]m, [100, 225]m, and [180, 290]m, and [240, 210]m respectively. All the rest configurations of simulations are the same as *Inner case*. As results in Figure 3, the improvement of the constrained optimization method over of the standard method is apparent

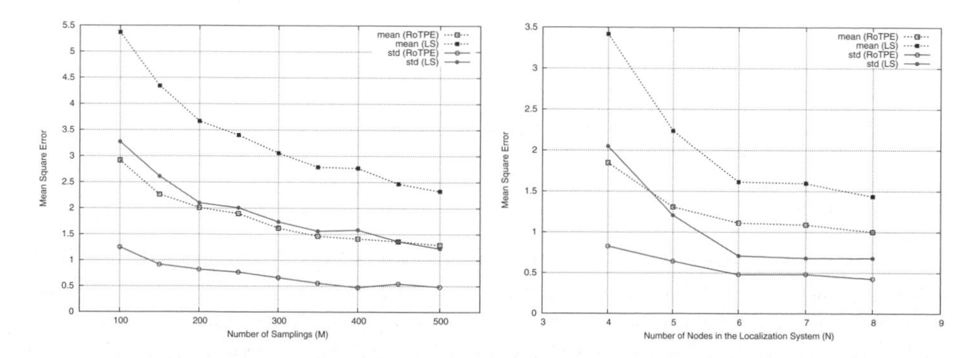

Fig. 3. Outer Case: MSE over Number of Samplings(Left) and MSE over Number of Nodes(Right)

in MSE, and also we know that our proposed scheme (RoTPE) is worked well in *Outer case.*

6 Conclusion

The transmission power and position of the primary user in CR is very precious information because that information of the primary user determines the spatial resource. To find position of the primary user, we try to use existing positioning or localization schemes based on ranging techniques but those schemes require the primary user's transmission. Since most primary users in CR are legacy system, and there are no beacon protocol to advertise useful information such as transmission power. Some of existing localization schemes need not the transmission power, but those don't work in *Outer case* that the primary user is in out of convex hull of secondary users' coordinates. We propose the constrained optimization method to estimate transmission power and position without the prior information of the transmission power. The proposed scheme use the linearlization technique to approximate relationship between RSS measurements and unknown power and coordinates of the primary user to set weighting factor that considers the differences of the quality of measurements, and then apply constrained optimization method containing appropriate weighting factor. We also do extensive simulations on different network deployments to prove improvement of performance in MSE. As a result, our proposed scheme(RoTPE) outperforms over standard least square(LS) method both in *Inner* and *Outer Case.*

References

1. Mitola, J., et al.: Cognitive radio: Making software radios more personal. IEEE Personal Communnications 6(4), 1318 (1999)
2. Haykin, S.: Cognitive Radio: Brain-Empowered Wireless Communications. IEEE Journal on Selected Areas in Communications 23(2) (February 2005)
3. Bulusu, N., et al.: GPS-less Low Cost Outdoor Localization for Very Small Devices. IEEE Personal Communications Magazine 7(5) (October 2000)
4. Niculescu, D., Nath, B.: DV Based Positioning in Ad hoc Networks. Journal of Telecommunication Systems (2003)
5. Cheung, K.W., et al.: Received Signal Strength Based Mobile Positioning via Contrained Wighted Least Square. IEEE ICASSP (2003)
6. Lim, H., et al.: Zero-Configuration, Robust Indoor Localization: Theory and Experimentation. In: Proceeding of IEEE INFOCOM (2006)
7. He, T., et al.: Range-Free Localization Schemes for Large Scale Sensor Networks. In: Proceeding of ACM MOBICOM (2003)

On the Use of Manchester Violation Test in Detecting Collision

Yee-Loo Foo and Hiroyuki Morikawa

Dept. of Info. and Comm. Eng., The University of Tokyo, Tokyo 113-8656, Japan
{ylfoo,mori}@mlab.t.u-tokyo.ac.jp

Abstract. Smart devices in the ubiquitous computing environment implement service/device discovery protocol that helps discovering each other and the services provided. As client device may receive multiple service description messages, it implements at its MAC layer a collision resolution mechanism to resolve the collision of messages. Effectiveness of collision resolution relies on the accuracy of detecting collision at the PHY layer. In this paper, we question the reliability of the conventional collision detection technique, which inaccuracy will affect the completeness of service/device discovery. Our analysis shows that capture effect and packet reception failure can cause failure in collision detection when the conventional technique is used. We suggest a detection technique that makes use of Manchester violation test. Implementation and evaluation of the proposed technique on some smart devices show its superiority over the conventional approach.

Keywords: Wireless radio, service/device discovery, concurrent transmissions, collision detection and resolution, capture effect, Manchester coding and violation.

1 Introduction

Ubiquitous computing envisions a world where various computing services run on a wide range of devices in our surroundings. Automated discovery of these devices and the services they offer will certainly enhance our quality of life and change the conventional concept of how service is found and delivered to us. This service/device discovery mechanism automates the process of identifying a device and describing the service it provides, and if necessary, setting up a connection with it. It involves passing *discovery* message that contains the device identity (ID), description of the service it provides, device configuration and connection setup information.

We are particularly interested in the possible implementation of discovery protocol on low-power 'smart' devices, which have only limited computing resources, and use only low-power radio to communicate over short distance at low data rate. Running service/device discovery protocol on these resource-limited devices is challenging. Nevertheless, these smart devices are 'powerful' in the sense that they are cheap and tiny, and thus can be easily tagged with other devices or even daily objects. The presence of a large number of smart devices in the surroundings provides us an unimaginably intelligent environment which we will benefit from.

T. Vazão, M.M. Freire, and I. Chong (Eds.): ICOIN 2007, LNCS 5200, pp. 729–740, 2008.

As a device may receive multiple discovery messages simultaneously, it implements at its MAC layer a collision resolution mechanism to resolve the collision of messages. Effectiveness of collision resolution relies on the accuracy of detecting collision at the PHY layer. In this paper, we raise a question on the reliability of the conventional collision detection technique, which inaccuracy will affect the completeness of service/device discovery. Our analysis shows that capture effect and packet reception failure can cause failure in collision detection when the conventional technique is used. We thus suggest a collision detection technique that makes use of Manchester violation test, which is different from the conventional one. Implementation and evaluation of the proposed technique on some smart devices show its superiority over the conventional approach.

The rest of this paper is organized as follows. Section 2 provides the problem analysis which also includes the background study. Section 3 describes how we approach the problems. Section 4 describes the implementation of our algorithm, experiment setup and procedures, followed by evaluation results. Related works are given in Section 5. Finally Section 6 concludes this paper.

2 Problem Analysis

2.1 Background

2.1.1 Service/Device Discovery Methods

There are two types of discovery method: *push*-type (or announcement-based) and *pull*-type (or on-demand based) [1]. Following the push-type approach, a service provisioning device advertises its service by repeatedly sending out discovery message through broadcast announcement. Interested parties gather all kinds of advertisements, picking up those that provide desired services for further actions, while filtering the unwanted ones. It is the service provider that bears the responsibility of getting its message delivered to its targets. A targeted device does not put effort in making sure that it receives the discovery message directed to it. On the other hand, in a pull-type discovery model, a client device proactively queries for desired services. Prospective service providers respond with their respective discovery message. However, it is the client's responsibility to make sure that it correctly receives all discovery messages targeted to it. A discovery protocol can be designed to support both discovery methods.

2.1.2 Collision Resolution

The choice of discovery method affects the design of Medium Access Control (MAC) protocol that runs at the lower layer, which governs the access to a shared channel from multiple devices. When push-type method is chosen, the MAC protocol implemented at the service provisioning devices is mainly a *collision avoidance* mechanism that aims to avoid simultaneous sending of discovery messages which will cause message collision at a listening device. Collision causes loss of discovery messages, implicating incomplete service/device discovery. When the pull-type discovery model is to be followed, in addition to the implementation of collision avoidance mechanism at service providers, clients should implement *collision resolution* mechanism that acts to resolve collision when it occurs. The resolution mechanism aims to recover the information lost due to collision, by principally requesting those involved in the collision to retransmit their messages. The ultimate goal of collision resolution is to allow every conflicting sender to successfully deliver its message to its target receiver. Collision resolution scheme, for example, is deployed in Radio Frequency Identification (RFID) readers (where it is called *anti-collision* scheme) to resolve collisions arise from RFID tags identification process [2, 3].

2.1.3 Collision Detection

Collision resolution mechanism starts operating whenever collision is detected, and it ends after resolving all collisions and no new collision is detected. The activation and deactivation of collision resolution mechanism closely relies on the detection of collision at the physical (PHY) layer. PHY layer must have function that is capable of detecting collision and provides the correct information to the collision resolution mechanism. Accuracy of collision detection is very important and becomes our concern here.

How does a low-power radio receiver detect collision? Collision refers to the situation where a receiver fails to receive the packet (that contains a message) sent by a transmitter due to interference from other simultaneous packet transmissions. The receiver can identify such situation (i.e. collision) when it senses the presence of a packet transmission, but detects incompleteness in the information that it receives (e.g. incorrect packet checksum). In short, conventionally the presence of collision is identified when erroneous packet is received. However, the completeness of this logic has not been previously questioned. The assumption is that concurrent transmissions always result in reception of erroneous packet (at a listening radio). In this paper, we reveal that this is not true.

2.2 Issues

2.2.1 Capture Effect

Instead of collision, concurrent transmissions can result in *capture*, i.e. successful reception of the packet that is the strongest among all transmitted packets. The 'captured' packet is received with full integrity i.e. correct checksum.

While collision is the result of interference induced by other concurrent transmissions, capture is the result of receiver's tolerance against the interference. When capture effect prevails, collision goes undetected. Based on this wrong indication of the absence of collision, the collision resolution mechanism fails to take action to resolve the occurred collision, resulting in incomplete service/device discovery.

Some early works that described about capture effect are [4, 5]. Commonly used in the literature to quantify capture effect [4 - 8], *capture ratio* refers to the minimum required signal-to-interference ratio (*SIR*) for a signal to be successfully received despite the presence of other transmissions. When capture ratio is low (i.e. close to 0 dB), concurrent transmissions result in collision more likely than capture. As the capture ratio increases, the trend is reversed that it is more likely to have capture.

2.2.2 Packet Reception Failure

There are collision conditions where receiver fails to receive a packet, not even an erroneous one. As a result, it does not sense the collision or even the presence of transmissions itself. When this happens, collision resolution mechanism does not activate, resulting in incomplete discovery.

A generic packet reception process is explained as follows. A radio packet is preceded by preamble, followed by synchronization (SYNC) bytes, and then data [9]. Preamble is usually a series of alternating bit 0 and 1. It informs the presence of a packet and allows a listening receiver to achieve bit synchronization with this packet transmission. Without achieving bit synchronization, the receiver will not be able to correctly identify the first bit of the SYNC byte that follows. Hence, a receiver that is yet to receive preamble would keep itself busy in searching for one. After successfully receiving preamble, the receiver starts tracing for SYNC bytes in order to reach byte synchronization with the packet transmission. After byte synchronization is achieved, the receiver is able to correctly identify the first byte of data that follows. Without receiving preamble and SYNC bytes, the receiver assumes that whatever that has been received is noise. Only after receiving SYNC bytes, the receiver starts buffering data bits that follow. There is no point of start buffering without first receiving both preamble and SYNC bytes, because synchronization would not have been achieved, or there simply is not any packet to be received. After all data bits have been buffered, the receiver examines the checksum. Incorrect checksum implies collision.

The problem is that preamble and SYNC bytes can be corrupted because of collision or noise. If we follow the conventional collision detection principle, we find that only collision that corrupts packet data but not preamble and SYNC bytes can be detected. Collision that corrupts preamble and SYNC bytes cannot be detected because the receiver cannot even confirm the presence of transmission.

Noise is another cause of corruption, although it is not related to collision or capture. In the case of single packet transmission, even when there are no concurrent transmissions that can possibly cause collision, the packet preamble and SYNC bytes can be corrupted by noise. As a result, the transmitted packet will not be received, giving the same impact of incomplete service/device discovery.

3 Approach

This section describes how we approach the presented challenges. An effective approach should be able to track down capture and packet reception failure, which cause the problems. When either capture or packet reception failure is identified, from the perspective of collision resolution scheme it should be counted as a collision. In fact, the conventional view on 'collision detection' and 'collision resolution' should be renewed to as follows: *Instead of detecting collision, we should make effort to detect the presence of concurrent transmissions. Instead of resolving collision, we should resolve the conflict arises from concurrent transmissions.* Here we introduce the use of Manchester violation (MV) test as our approach.

Manchester coding embeds transmitter's clock information into the bit stream, making synchronous transmission possible [10]. Manchester code requires a signal level transition in the middle of a bit. If such transition is not observed, the Manchester coding format is said to be violated. Violation detection is determined by how 'balanced' a bit looks. A distorted bit is likely to cause a violation. And bit distortion is usually a result of noise or interference contributed by other transmission. Therefore a violation could serve as a good indication of the presence of concurrent transmissions.

We suggest the following methodology when applying MV test for the purpose of detecting concurrent transmissions that result in collision or capture. When there is a positive detection, the collision resolution scheme is notified and it will activate. The algorithm suggested in the following is to be implemented on the radio receivers of client devices that are in search of services following the pull-type discovery approach.

A client broadcasts a query, and allocates a fixed duration to collect the discovery messages responded by potential service provisioning devices. Within the duration, if preamble and SYNC bytes have been received, all the bits that are received after them will be buffered and examined for data integrity through checksum function e.g. Cyclic Redundancy Code (CRC) check. Following the conventional collision detection technique, if checksum is incorrect, collision is assumed to be the cause and the collision resolution scheme will be alerted and activated. However, if the checksum is correct, collision is assumed to be absent because a packet has been received with complete integrity. No consideration is given to the possibility of capture.

Our approach is different, and it is capable of detecting both collision and capture. It requires a listening radio to record the MV status of every data bit (i.e. all bits that are received after the preamble and SYNC bytes). Let say the number of violation bits in the data part of a packet is N_{data}. $N_{data} = 0$ means the received data has not experienced distortion i.e. collision is regarded to be absent. On the other hand, $N_{data} > 0$ means certain degree of distortion has been experienced due to either interference resulted from concurrent transmissions or noise. Regardless of the cause, the correctness of the received data is now in doubt. The radio receiver simply assumes that collision is present and thus activates the collision resolution scheme.

As we will show in the Results subsection (Fig. 3), our technique is capable of detecting all collisions that are detectable through conventional detection technique. On top of that, it can also detect to some extent the capture cases that are not detectable through conventional technique. This is one of the reasons the suggested technique is considered more superior to the conventional one that relies on checksum algorithm.

On the other hand, if preamble and SYNC bytes have not been received within the listening duration, no special action is taken in the conventional packet reception process. No consideration has been given to the possibility that collision has actually occurred and corrupted the preamble and SYNC bytes. Here we add a mechanism to help detecting this. When the radio receiver is listening within a specified duration, it examines each received bit for its MV status and records the result. Let say the number of violation bits observed within the duration is $N_{duration}$. Now when the duration is over, and if preamble and SYNC bytes have not been received, $N_{duration}$ is examined. If no transmission has been made within the listening range of a radio receiver, not a single Manchester encoded bit would have appeared at the input of

Manchester decoder in the receiver, and thus there should be a large number of violations (i.e. large $N_{duration}$) observed. A reasonable guess is that at least half of the total number of received bits could have committed violation. On the other hand, when at least one transmission has been made, a series of Manchester encoded bits (though some could have been distorted) must have been available at the decoder input, and thus a relatively smaller number of violations (i.e. smaller $N_{duration}$) are expected. If we can determine a threshold level, $N_{threshold}$, that can differentiate between $N_{duration}$ of these two conditions, the algorithm simply includes a comparison between $N_{duration}$ with $N_{threshold}$. If $N_{duration} > N_{threshold}$, it is assumed that there has been no transmission and only noise has been received. However, if $N_{duration} < N_{threshold}$, the algorithm regards the corruption of preamble and SYNC bytes has been due to collision or noise. Regardless of the cause, the collision resolution scheme is to be activated.

The following pseudo code represents the methodology we suggest to implement on the radio receiver of client device. The lines not in bold letter are the pseudo code of a generic packet reception mechanism that is constructed based on recommendations given in [11]. Conventional collision detection algorithm is also described in these lines. The lines in bold letter represent the algorithm we suggest.

```
  Initial state = PREAMBLE
  begin
  Receive 1 bit
  if (MV is present)
        N_duration ++
  Shift the bit value into a shift register
  goto current state

state: PREAMBLE
  if (shift register value = PREAMBLE)
        Preamble_count ++
  if (Preamble_count > Requirement)
        goto SYNC state

state: SYNC
  if (shift register value = SYNC)
        goto DATA state
  else
        Error ++
  if (Error > Tolerance)
        goto PREAMBLE state

state: DATA
  if (MV is present)
        N_data ++
  if (Packet length is reached)
        // Packet reception completes
        Examine N_data
        if (N_data > 0)
             Detect possible collision
             Report to collision resolution scheme
        else if (N_data = 0)
             No collision
        Perform CRC check
        if (checksum correct)
             No collision
```

```
        else if (checksum incorrect)
            Detect collision
            Report to collision resolution scheme
// end of all states

   if (End of listening duration is reached)
        if (Preamble & SYNC not received)
              Examine N_duration
              if (N_duration > N_threshold)
                   No collision
              else if (N_duration < N_threshold)
                   Detect possible collision
                   Report to collision resolution scheme
```

4 Implementation and Evaluations

4.1 Implementation

For the purpose of evaluating our approach, we prepare three units of smart device. Each consists of the following main components: a Chipcon CC1000 low-power radio transceiver chip [12], which is controlled by PIC18LF4620 low-power microcontroller from Microchip [13]. In our following experiments, one of the devices plays the role of client which is in search of service, while the other two act as service provisioning devices that respond to the client. In order to replicate collision and capture effect, the latter two are made to reply simultaneously to every query issued by the client.

A generic packet transmission and reception mechanism is implemented in the form of software that runs on the microcontrollers of these devices. A CRC-16 checksum algorithm is also implemented in the software, following the guide given in [14]. On top of that, we implement our collision detection algorithm on the client device.

The CC1000 transceiver chip comes with MV test feature, where the violation status of each received bit is available at one of its pins. In receive mode, CC1000's demodulator passes each received bit and its accompanied MV status to the microcontroller for further processing. CC1000 use FSK modulation, and they are set to communicate at the rate of 4800 bps.

A packet is preceded by 4 bytes of preamble, followed by 2 bytes of SYNC, 28 bytes of data, and finally 2 bytes of CRC checksum, giving a total length of 36 bytes. When a packet is received perfectly without its bits experiencing distortion, all the received bits will be free from Manchester violation, except the initial few bits of the preamble. This is because the synchronization process takes place at preamble and before a receiver achieves synchronization with a transmission, the received bits most likely violate the Manchester coding format. We are interested in the part of a packet that can possibly give $N_{duration} = 0$ when an undistorted packet is received, since $N_{duration} = 0$ clearly indicates the absence of interference. As a result, we choose to skip monitoring the MV status of the preamble. In order to do this, we synchronize the transmitter and receiver to start transmitting and receiving at the same time. The receiver then skips recording the MV status of the first 4 received bytes (which are likely to be preamble if there is a transmission). Only the following 32 received bytes (or 256 bits) their MV status are recorded. From these 256 bits, the total number of Manchester violating bits gives $N_{duration}$.

After a receiver successfully receives preamble and SYNC bytes, it proceeds buffering the 28 bytes of data and 2 bytes of CRC checksum that follow. All these 30 bytes or 240 bits that are buffered will be examined for their MV status. The total number of Manchester violating bits gives N_{data}.

4.2 Experiments

To replicate collision and capture effect, we do the followings. Upon receiving a query message from the client device, the other two devices simultaneously send to the client a packet (which should contain the discovery message). Depending on the capture ratio, which describes the power relationships between these devices, the client device either captures one of the two packets, or the two collide and destroy each other and thus the client receives none of them. When capture ratio is high, capture effect becomes dominant, and vice versa.

To produce collision and capture effect in our experiments, we vary the capture ratio accordingly. This is done by first fixing the distance between client device and one of the other two devices. We then vary the distance of the remaining device from the client. As all devices transmit at equal and constant power, the client device receives different signal strength from the two devices that are located at different distance. To quantify capture ratio, it requires the client's radio to measure the received power from each device's transmission. The received power can be read out from the Received Signal Strength Indicator (RSSI) pin of the CC1000 radio. The ratio between the powers received from the two devices gives capture ratio. The received power values that have been taken into the calculation of capture ratio are the averages over a period of 1 second. Finally, we get a set of locations of all three devices, where each set of locations gives different capture ratio. Our experiment is repeated over these different sets of locations. For each set of locations, the client sends out 500 queries one after another, and thus triggering 500 concurrent transmissions from the other two devices, which end up in either collision or capture, allowing us to evaluate the effectiveness of our algorithm in detecting them. When the receiver in the client fails to receive any packet and $N_{duration} < N_{threshold}$, it is considered as a collision. And if the receiver receives a packet and $N_{data} > 0$, it is also counted as a collision.

4.3 Results

We are interested to find out a suitable value for $N_{threshold}$. We examine the $N_{duration}$ results of the cases where packet reception has failed due to corrupted preamble and SYNC bytes. We have observed 389 such cases for a particular set of devices locations that gives capture ratio of 0 dB. From the results we gather, we find that $N_{duration}$ is distributed between 10 and 160 (refer Fig. 1), and it gives a mean value of 58, out of 256 received bits which we have examined their MV status. Next, we examine the $N_{duration}$ results of the cases where no packet has been transmitted and all that the receiver receives is noise. We find that in this condition where only noise is received, $N_{duration}$ is distributed between 130 and 200 (refer Fig. 1), and it gives a

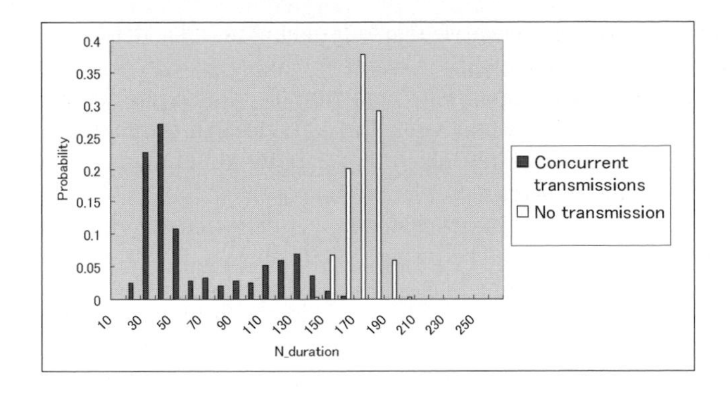

Fig. 1. Probability density function of $N_{duration}$ that belong to two different conditions

mean value of 165, out of 256 received bits which we have examined their MV status. There is therefore an overlapping region (from 130 to 160) between the two distributions of $N_{duration}$. This implies that if a receiver gets a $N_{duration}$ value that falls within this region, it cannot conclude that whether there has been a collision or not. Fortunately, the likeliness of overlapping is actually very small. From Fig. 1, we find that a good choice for $N_{threshold}$ is 140, i.e. collision is assumed to be present when $N_{duration} < 140$, and vice versa. By taking $N_{threshold}$ as 140, the probability of mistaking collision as being absent is only 1.80%, while the probability of mistaking collision as being present is as small as 0.26%.

Next, we are interested to find out the effectiveness of our technique in detecting concurrent transmissions that result in collision or capture. Fig. 2 shows the detection rate. The cases of packet reception failure with $N_{duration} < N_{threshold}$, and successful packet reception with $N_{data} > 0$, are both counted as 'Successful detection'. The cases of packet reception failure with $N_{duration} > N_{threshold}$, and successful packet reception with $N_{data} = 0$, are both counted as 'Unsuccessful detection'. With reference to Fig. 2, when the capture ratio is low (0 to 1.5 dB), the detection rate is rather high, ranging from 76% to 100% (giving an average of 84%). When the capture ratio is high (2 to 4 dB), the detection rate drops to near zero as the number of undetected captures grows. The explanation is that at low capture ratio, concurrent transmissions most likely

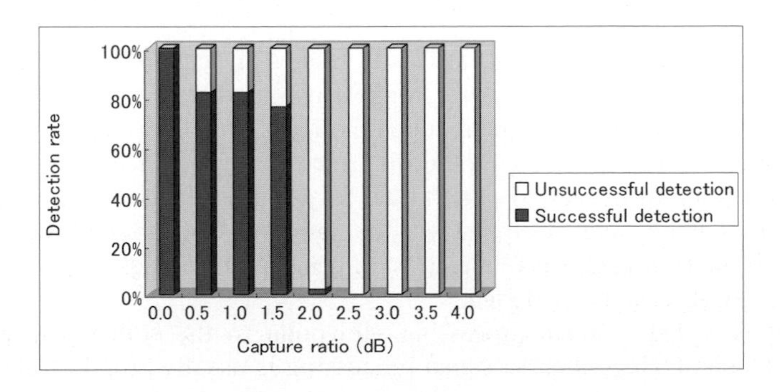

Fig. 2. Collision detection rate of proposed technique

result in collision. And we observe that our technique can detect all the collision cases. However, as the capture ratio increases, capture is more likely to become the end result of concurrent transmissions, and our technique can detect only cases of weak capture but not the severe ones (cf. the conventional technique can detect none of capture case). The above results show that capture effect gives a severe impact on collision detection, and remains a problem to be resolved.

Nevertheless, in Fig. 3 we show that our technique is still more superior to the conventional one. In the region where capture ratio is low (0 to 1.5 dB), our technique is at least twice more effective than the conventional one in terms of collision detection rate.

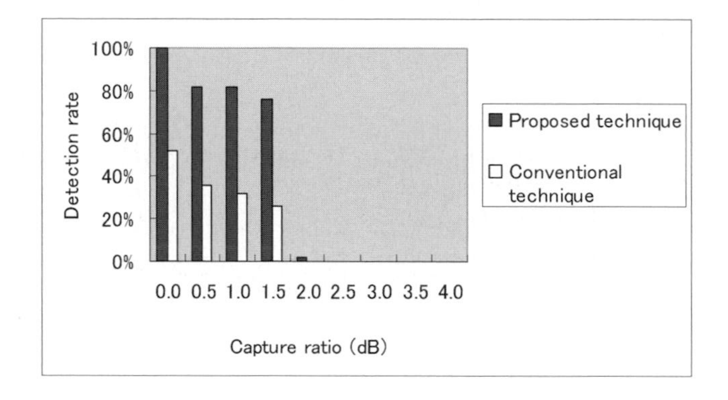

Fig. 3. Performance comparison between proposed technique and conventional technique

5 Related Works

In the literature, there has been no comprehensive investigation on the reliability of collision detection in a radio receiver. The possible problems posed by capture effect and packet reception failure have not been previously addressed. Nevertheless, in the context of RFID system, [15] did report that capture effect could actually influence the number of detectable RFID tags. On the other hand, we have not found any experimental work that evaluates the effectiveness of MV test as a means of collision detection.

The work by Whitehouse et al. [16] was related to capture effect and collision detection, which is close to ours. They suggest making use of capture effect to help detecting collision. Their technique requires the receiver to keep tracing for a second set of preamble even after receiving the first one. If a second set exists, the receiver drops the reception of the earlier packet, and resynchronizes to the later one. Instead of both packets collide and corrupt, at least the later one can be saved. This technique is effective if the later packet is relatively stronger than the earlier one, so that capture effect is present and can be exploited.

[17, 18] provided a signal processing viewpoint to the collision detection and resolution problem. Nevertheless, signal processing is usually prohibitively high cost to low-power radios.

6 Conclusion

In this paper, we raise a question on the reliability of the conventional collision detection technique that is deployed on radio receiver. We then elaborate on the issues that originate from capture effect and packet reception failure. We suggest a different detection technique that is based on the use of MV test. The performance of the proposed technique has been evaluated on some radio devices. The results are promising, and they show that the proposed technique outperforms the conventional one.

Acknowledgments. This work is supported by Ministry of Public Management, Home Affairs, Posts and Telecommunications, Japan.

References

1. Zhu, F., Mutka, M.W., Ni, L.M.: Service Discovery in Pervasive Computing Environments. IEEE Pervasive Computing, 81–90 (2005)
2. EPCglobal. 13.56 MHz ISM Band Class 1 Radio Frequency (RF) Identification Tag Interface Specification, `http://www.epcglobalinc.org/standards_technology/Secure/v1.0/HF-Class1.pdf`
3. EPCglobal. 900 MHz Class 0 Radio Frequency (RF) Identification Tag Specification, `http://www.epcglobalinc.org/standards_technology/Secure/v1.0/UHF-class0.pdf`
4. Roberts, L.G.: ALOHA Packet System with and without Slots and Capture. ACM SIGCOMM Computer Comm. Review 5(2) (1975)
5. Leentvaar, K., Flint, J.: The Capture Effect in FM Receivers. IEEE Trans. Commun. 24(5), 531–539 (1976)
6. Krauss, H.L., Bostian, C.W.: Solid State Radio Engineering. John Wiley & Sons, New York (1980)
7. Ash, D.: A Comparison between OOK/ASK and FSK Modulation Techniques for Radio Links. Technical report, RF Monolithics Inc. (1992)
8. Nelson, R., Kleinrock, L.: The Spatial Capacity of a Slotted ALOHA Multihop Packet Radio Network with Capture. IEEE Trans. Commun. 32(6), 684–694 (1984)
9. Torvmark, K.H.: Short-Range Wireless Design, `http://www.embedded.com/story/OEG20020926S0055`
10. Halsall, F.: Data Communications, Computer Networks and Open Systems. Addison-Wesley, Reading (1996)
11. Torvmark, K.H.: Application Note AN009: CC1000/CC1050 Microcontroller Interfacing, `http://focus.ti.com/lit/an/swra082/swra082.pdf`
12. Chipcon. CC1000: Single Chip Very Low Power RF Transceiver, `http://www.chipcon.com/files/CC1000_Data_Sheet_2_3.pdf`
13. PIC18F2525/2620/4525/4620 Data Sheet, `http://ww1.microchip.com/downloads/en/DeviceDoc/39626b.pdf`
14. Williams, R.: A Painless Guide to CRC Error Detection Algorithms, `http://www.ross.net/crc/download/crc_v3.txt`
15. Floerkemeier, C., Lampe, M.: Issues with RFID Usage in Ubiquitous Computing Applications. In: Ferscha, A., Mattern, F. (eds.) PERVASIVE 2004. LNCS, vol. 3001, pp. 188–193. Springer, Heidelberg (2004)

16. Whitehouse, K., et al.: Exploiting the Capture Effect for Collision Detection and Recovery. In: IEEE EmNetS-II, Sydney (2005)
17. Tsatsanis, M.K., Zhang, R., Banerjee, S.: Network-Assisted Diversity for Random Access Wireless Networks. IEEE Trans. Signal Processing 48(3), 702–711 (2000)
18. Zhang, R., Sidiropoulos, N.D., Tsatsanis, M.K.: Collision Resolution in Packet Radio Networks Using Rotational Invariance Techniques. IEEE Trans. Commun. 50(1), 146–155 (2002)

Dynamic Routing Algorithm for Asymmetric Link in Mobile Ad Hoc Networks*

Jongoh Choi[1], Si-Ho Cha[2,**], GunWoo Park[3], Minho Jo[4], and JooSeok Song[3]

[1] Defense Agency for Technology and Quality, Seoul, Korea
cjo31@naver.com
[2] Dept. of Information and Communication Engineering, Sejong University, Seoul, Korea
sihoc@sejong.ac.kr
[3] Dept. of Computer Science, Yonsei University, Seoul, Korea
{gwpark,jssong}@emerald.yonsei.ac.kr
[4] Graduate School of Information Management and Security, Korea University
minhojo@korea.ac.kr

Abstract. This paper proposes an improved algorithm that can replace the existing algorithms that use the unidirectional link. The proposed algorithm allows a first sink node of the unidirectional link to detect the unidirectional link and its reverse route and provide information regarding the reverse route to other sink nodes, thereby minimizing unnecessary flows and effectively using the unidirectional link. This paper also proposes a redirection process that actively supports a unidirectional link generated during data transmission to deal with a topology change due to the movement of node. Further, this paper proves the superiority of the proposed algorithm to the existing algorithms in terms of a delay in data transmission and throughput through a simulation in which the performance of the proposed algorithm is compared with those of the existing algorithms.

1 Introduction

Most of Ad Hoc protocols are designed to effectively operate over a bidirectional wireless link consisting of homogeneous nodes. However, a unidirectional link arises in a real network due to various causes such as the difference in transmission power between nodes, collision, and interference (or noise). Effective use of the unidirectional link reduces a transmission path, thereby lessening time required for data transmission, a delay in data transmission, and load on intermediate nodes during a relay. However, the unidirectional link increases overhead in a routing protocol designed for a bidirectional link, and discontinuity in a network may occur when the unidirectional link is not considered during path setup [1] [2]. The various algorithms

* This research was supported by the MIC (Ministry of Information and Communication), Korea, under the (Information Technology Research Center) support program supervised by the IITA (Institute of Information Technology Assessment).

** Corresponding author.

T. Vazão, M.M. Freire, and I. Chong (Eds.): ICOIN 2007, LNCS 5200, pp. 741–750, 2008.

for a unidirectional link have been suggested but they have many defects. For instance, reverse route discovery is not performed on the assumption that a unidirectional link always has a reverse route, or route discovery is performed irrespective of whether a reverse route exists. Second, a predetermined detour route is maintained through route discovery after a unidirectional link is detected, thereby wasting a lot of control packets such as a route request (RREQ) message and a route reply (RREP). Lastly, the existing algorithms are applicable to only a unidirectional link detected before data transmission node. Therefore, it is difficult to use a unidirectional link generated caused by a topology change due to the movement of node [3]. To solve these problems and effectively use the unidirectional link, this paper applies the notion of order to the sink nodes that have a unidirectional link relationship with a source node so as to allow a first sink node to provide other sink nodes with information regarding the reverse route of the unidirectional link. The application of the notion of order draws many advantages. First, routing for a reverse route from each sink node to the source node can be skipped. Second, it is possible to prevent unnecessary data from being transmitted when a reverse route is not detected. Therefore, it is possible to allowing a node with the longest battery life to act as a relay node and minimize transmission of control packets of other nodes lastly, it is possible to use a unidirectional link generated due to a topology change during data transmission, using redirection.

Section 2 discusses the characteristics, advantages, and disadvantages of the existing algorithms designed for a unidirectional link. Section 3 proposes an algorithm that detects a unidirectional link, maintains a reverse route of the unidirectional link, and solves problems due to a topology change. Section 4 compares the proposed algorithms with the existing algorithms through a simulation. Finally, section 5 provides the conclusion of this paper.

2 Related Works

A unidirectional link may be detected using a hello message, GPS-based detection, or information regarding transmission power. To make a unidirectional link available at a data link layer, a reply message, e.g., an acknowledgment, to data sent from a transmitter must be transmitted in the reverse route of the unidirectional link. In this connection, IP Tunneling [4], replying to the data using a GPS-based variable power value, and relaying of a MAC frame in the reverse route have been suggested.

Information regarding the reverse route of the unidirectional link may be provided by using reverse routing to set a predetermined detour route, separately setting a bidirectional route (forward and reverse routes), or using a Reverse Distributed Bellman-Ford (RDBF) algorithm.

A sink node detecting a unidirectional link performs routing to set a detour reverses to a source node of the unidirectional link, thereby maintaining the reverse route. The unidirectional link will be used during subsequent routing for all routes. Forward and reverse routes are separately determined. The forward route is detected through route discovery from a starting node to a destination node, and then, the reverse route is detected through route discovery from the destination route to the starting node. The shortest route from each node to other nodes is computed using the RDBF algorithm obtained by changing the existing DBF. Next, each node transmits information regarding the shortest route and the address of a first node in the shortest route, thereby maintaining the reverse route. Each node periodically transmits an update message specifying the shortest route to the other nodes, and guarantees loop free using the address of the first node in the shortest route.

The conventional algorithms have some problems. First, reverse route discovery is not considered on the assumption that the unidirectional link always has a reverse route. Second, route discovery is performed irrespective of whether the reverse route exists. Third, a predetermined detour route is maintained by performing route discovery after the unidirectional link is detected, thereby wasting control packets such as a RREQ message and a RREP. Lastly, since the conventional algorithms are applicable to only a unidirectional link detected before data transmission, it is difficult to use a unidirectional link generated by a topology change due to the movement of node.

3 Proposed Algorithm

The unidirectional link is likely to be generated when nodes have different ranges of transmission due to the difference in transmission power. Although nodes have similar transmission power, a unidirectional link may also be generated due to collision or noise. However, effects of the unidirectional link are regional or temporary in this case, and thus, a network is not greatly affected by such a unidirectional link. Accordingly, this paper will study a unidirectional link caused due to the difference in transmission power between nodes.

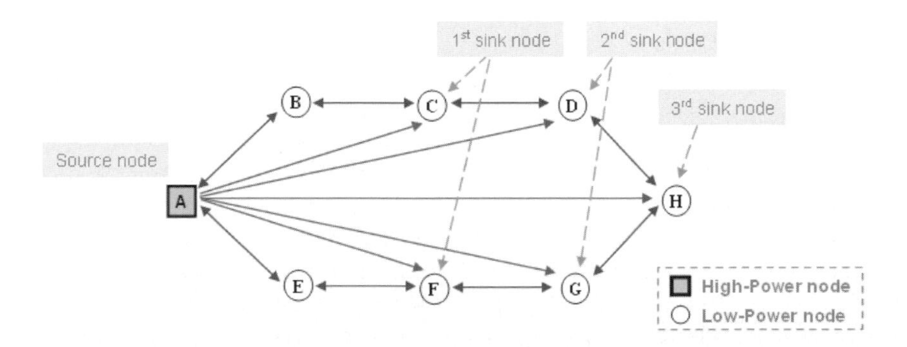

Fig. 1. A unidirectional link

Fig.1 illustrates a unidirectional link generated due to the difference in transmission power between nodes. A plurality of sink nodes may be present with respect to a source node of a unidirectional link, which has a high transmission power level.

3.1 DRAMA Algorithm

The existing algorithms require unidirectional link discovery to be performed in stages, i.e., an idle stage, before a data transmission stage. Thus, the existing algorithms are difficult to be applied to a unidirectional link detected during data transmission. To solve this problem, this paper proposes a DRAMA (Dynamic Routing for Asymmetric link in Mobile Ad hoc network) algorithm that actively supports unidirectional link discovery in the idle stage, and a unidirectional link generated during data transmission Stage. Before data transmission, a first sink node detects a reverse route of the unidirectional link and provides information regarding the reverse route to other sink nodes. Therefore, it is possible to effectively use the unidirectional link without routing the reverse route. A topology change is immediately reflected through redirection to use a unidirectional link generated even during data trans-mission.

Detection of Unidirectional Link. During exchange of a hello message, each node compares its neighbor list with the neighbor list of a neighbor node to determine whether a unidirectional link exists. If a node is not listed in a neighbor's neighbor list, it indicates that there is a unidirectional link that uses the neighbor node as a unidirectional source node and the node as a sink node. As shown in Fig. 2, each node compares its neighbor list with a neighbor list given from a neighbor. The neighbor list of each node is shown in Table 1. Referring to Table 1, nodes C, D, and E are not listed in the neighbor list of node A. Thus, nodes C, D and E have a unidirectional link relationship with node A.

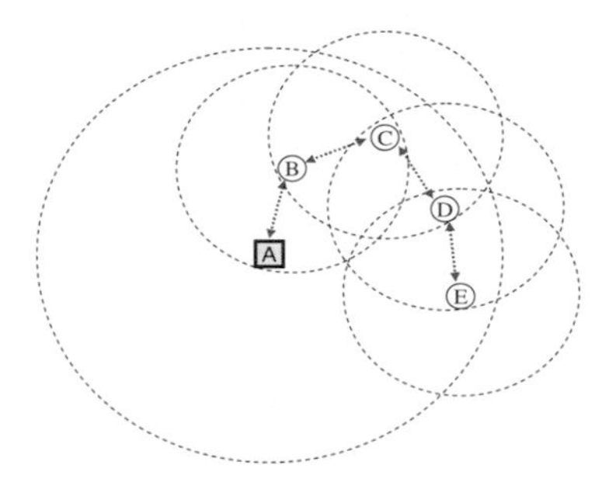

Fig. 2. Topology of unidirectional link

Table 1. Neighbor List of each node

Node \ NL	NL_A	NL_B	NL_C	NL_D	NL_E
A	{B}	{A,C}			
B	{B}	{A,C}	{A,B,D}		
C	{B}	{A,C}	{A,B,D}	{A,C,E}	
D	{B}		{A,B,D}	{A,C,E}	{A,D}
E	{B}			{A,C,E}	{A,D}

NL_i : NL of node i　　　[shaded] : NL of itself

Reverse Route Setup. When a sink node detects a unidirectional link using a hello message, the sink node compares its neighbor list with a neighbor list of a unidirectional source node to determine whether a common node exists. The presence of the common node proves that the unidirectional link has a reverse route. The sink node is referred to as a first sink node. The first sink node broadcasts a Link_info message specifying the source node of the unidirectional link and the order of the first sink to inform the source node and other sink nodes of the reverse route. Table 1 shows that node B, which is a common node, is listed in both the neighbor list of node C and the neighbor list of source node A of the unidirectional link, and node C is a first sink node. Node C creates a Link_info message and broadcasts it to neighbor nodes.

Processing of Link_info Message. When an intermediate node adjacent to a source node receives the Link_info message, the intermediate node recognizes a reverse route of a unidirectional link and rebroadcasts the Link_info message to act as a relay node. In this case, priority for transmitting the Link_info message is given to only one node to prevent all intermediate nodes from broadcasting the Link_info message. A sink node receiving the Link_info message updates information regarding the reverse node and compares a neighbor list of a source node with a neighbor list of each of neighbor nodes. When a first node is listed in the neighbor list of a second node but the second node is not listed in the neighbor list of the first node, the first and second nodes have a unidirectional link relationship. When detecting a sink node that is a next hop neighbor having the unidirectional link relationship with the sink node, the Link_info message is broadcasted again. In this case, information regarding the order of a sink node transmitting the Link_info message is included within the Link_info message. Each of the other sink nodes receives the Link_info message only when the order of the sink node transmitting this message is smaller than its order. The above process is repeated until a final sink node receives the Link_info message. In this way, all sink nodes having the unidirectional link relationship to the source node can recognize the reverse route.

Link_info Message with Priority. A plurality of intermediate nodes or sink nodes, each transmitting the Link_info message to an adjacent sink node, may be present over a unidirectional link. In this case, the performance of a network is lowered when all the nodes relay the Link_info message, and broadcasting flooding may occur when the Link_info message is transmitted using multi-hop path.

Accordingly, the back-off time is changed using formula (1) to reduce overall overhead by minimizing unnecessary flows and allow a node with a high battery life to act as an intermediate node. That is, the existing back-off time is determined according to the battery cost of each node, thereby increasing a probability that a node with a high battery life would transmit the Link_info message.

$$Back\_offTime_{Uni} = R_i \times Back\_off(\,)$$

$$* \; R_i = 1 / C_i \tag{1}$$

R_i: battery cost at node I

C_i: node i's residual battery capacity

$$* \; Backoff(\,) = Random(\,) \times Slot\ Time$$

An intermediate node with the highest battery life over unidirectional link transmits the Link_info message, and intermediate nodes that transmit the Link_info message determine the reverse route.

The process of detecting a unidirectional link and a reverse route thereof over an Ad hoc network and processing a Link_info message is illustrated in the flowchart of Fig. 3. As shown in Fig. 3, ① indicates a process of detecting the unidirectional link. ② indicates a process in which a first sink node detects the reverse route of the unidirectional link from a sink node to a source node. ③ indicates a process in which when a node being not located over the unidirectional link receives the Link_info message, the node acts as a relay node in the reverse route, as an intermediate node between a sink node and the source node. ④ indicates a case where a sink node, which is not a first sink, receives the Link_info message. ⑤ indicates a process in which the node illustrated in ④ determines whether there is an additional sink node and transmits the Link_info message to the detected additional sink node if any.

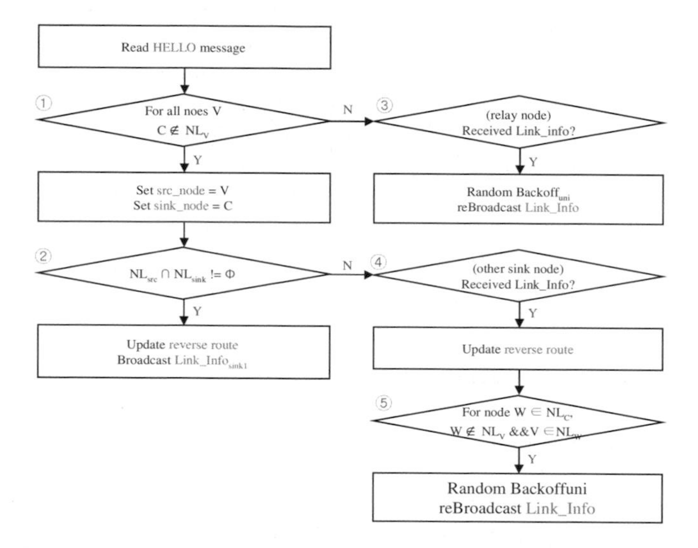

Fig. 3. Flowchart of detecting a unidirectional link and a reverse route

Redirection. The existing algorithms that support a unidirectional link are used to detect the unidirectional link and perform reverse route discovery before data transmission, but are applied to only the detected unidirectional links during data transmission. Thus, the existing algorithms are improper to deal with a dynamic topology change since they do not consider a unidirectional link being not detected before data transmission or being generated due to the movement of node. To actively deal with the dynamic topology change, it is required to detect and use a newly generated unidirectional link even during data transmission. Therefore, this paper proposes redirection to detect and use a unidirectional link even during data transmission.

If a high-power node moves around a node that is transmitting data, the high-power node overhears transmission of data of neighbor nodes. When transmission of data of first and last nodes is overheard, the high-power node determines them as target nodes, and broadcasts a redirect message for redirection to the target nodes. The first node receiving the redirect message determines the high-power node as a next node for subsequent data transmission. A sink node overhearing the redirect message transmits the Link_info message to use a unidirectional link.

Fig. 4 illustrates a redirection process. Node G with a high power level moves around node A that is transmitting data to node F. It overhears transmission of data of first node B and last node D and transmits the redirect message to first and last nodes B and D. During the transmission of the redirect message, node E also overhears the redirect message, and realizes that it is also a sink node over a unidirectional link. If node E was over a bidirectional link, a last node that node G detected should have been node E, not node D. Node E sends the Link_info message to node G, which is a source node of the unidirectional link. Node G uses the unidirectional link for subsequent transmission of data.

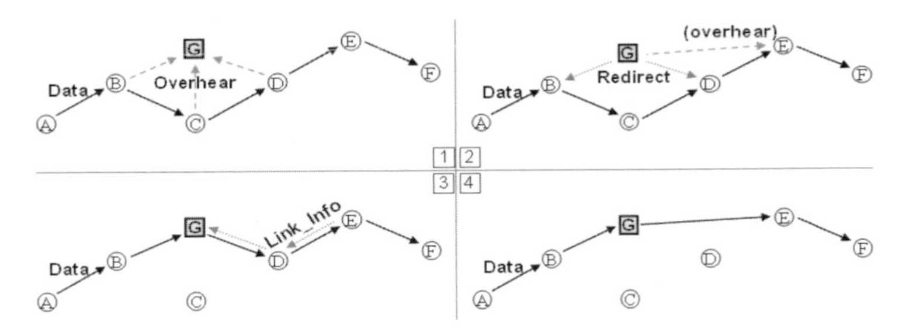

Fig. 4. A redirection process

4 Performance Evaluation and Analysis

4.1 Simulation Environment

We carry out simulation for the performance evaluation in which proposed DRAMA is applied to the AODV protocol [4], which is a representative routing protocol. Note that DRAMA is independent of underlying routing protocols. In the AODV protocol, a desired route is obtained through route discovery and maintained while the route maintenance procedure is performed. If the route is not further used, use of the AODV protocol is completed. Whether nodes are connected to a network is determined by broadcasting a control message. It is possible to detect neighbor nodes by broadcasting the control message, similarly to when broadcasting the hello message. When no control message is broadcast for a predetermined time, a connection to the network is maintained by broadcasting the hello message to the neighbor nodes. To transmit data to the destination node, the transmitting node begins route discovery by broadcasting an RREQ message. An intermediate node receiving the RREQ message records the reverse route toward the transmitting node, and rebroadcasts the RREQ message. An intermediate node, which contains the latest information regarding a route to a target node or the destination node, receives the RREQ message, produces a RREP, and unicasts the RREP to the originator following a reverse route determined based on the RREQ message. While the RREP is transmitted, a forward route to the destination node is determined. After sending the RREP to the originator, a routing table is updated and data transmission begins. If two or more RREPs are sent to the originator, a shorter one of two routes to the destination node is chosen.

In the simulation, QualNet 3.7 manufactured by Scalable Network Technologies (SNT) was used as a simulator. 100 nodes were uniformly distributed in an area of 1500×1500m, low-power nodes were moved at a maximum speed of 10m/s, and high-power nodes were moved at a maximum speed of 20m/s. Also, it was assumed that the high-power nodes assume 30% of the 100 nodes, i.e., 30 high-power nodes are present in the area. The simulation was performed for 400 seconds while changing an originator and a target node at intervals of 100 seconds to generate 512-byte CBR traffic per second.

Table 2. Simulation parameters

Radio propagation	Two ray Ground ($1/r^4$)
MAC layer	802.11b DCF mode
Network layer	AODV, Mobile IP
Mobility model	Random way-point model
Traffic	TCP, CBR
Antenna	Omni direction
Protocol	AODV
Simulation time	400s

4.2 Evaluation of Performance

Throughput and End-to-End Delay. The throughputs of the existing and proposed algorithms were measured for 400 seconds while changing the transmitting and destination nodes at intervals of 100 seconds to generate 512-byte CBR traffic per second. The graph of Fig. 5 shows that the throughputs of the existing and proposed algorithms supporting the unidirectional link are greater than that of an algorithm for a bidirectional link. In particular, the throughput of the proposed algorithm is the highest. This is because the existing algorithms can be applied to only unidirectional links detected before data transmission, whereas the proposed algorithm can be applied to even a unidirectional link generated due to the movement of node during data transmission.

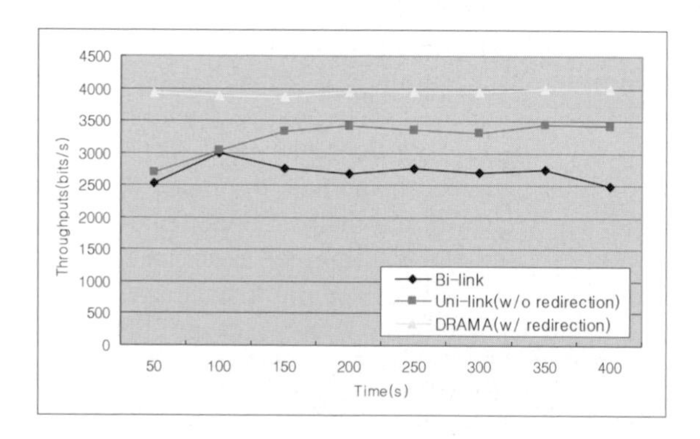

Fig. 5. Throughputs

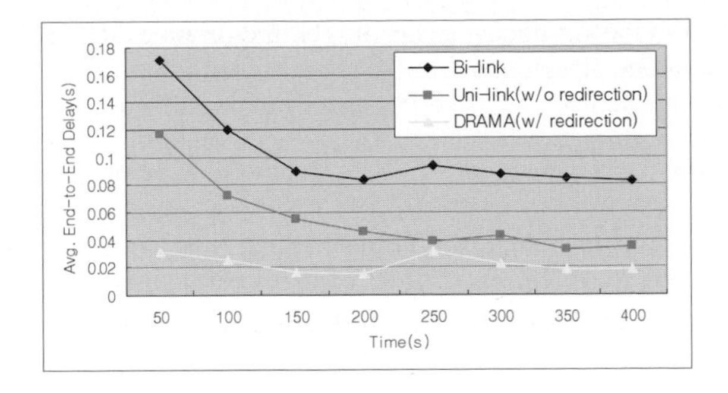

Fig. 6. Average end-to-end delay

As shown in Fig. 6, end-to-end delays in the algorithms for the unidirectional link are less than that in the algorithm for the bidirectional link. Further, as described above, the proposed algorithm can reduce and constantly maintain an end-to-end delay through redirection, compared to the existing algorithms.

Energy Consumption. In general, a high-power node requires more energy than a low-power node, but it can be consistently supplied with sufficient power from a vehicle or a portable device. In contrast, the low-power node receives the restricted amount of power from a portable battery. Therefore, much attention has been paid to effectively operating the low-power node with less energy. Accordingly, the amounts of battery power consumed by only low-power nodes for the respective algorithms were measured to evaluate the performances of the algorithms.

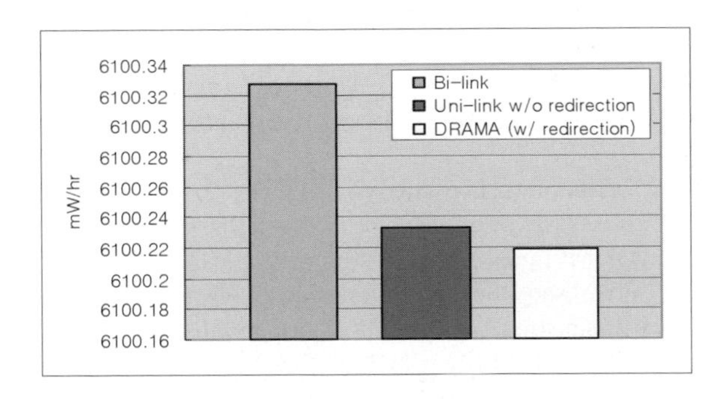

Fig. 7. Energy Consumption

Fig. 7 illustrates that a route (hop counts) required for data transmission over a network for a bidirectional link is greater than over a network for a unidirectional link on the average. Also, load on intermediate nodes over the network for the bidirectional link is more than over the network for the unidirectional link. Therefore, energy consumption of the low-power node over the bidirectional link is greater than over the unidirectional link. Also, DRAMA suggested in the proposed algorithm performs redirection to use an additional unidirectional link during data transmission, thereby lowering load on low-power nodes. Accordingly, the proposed algorithm requires less energy than the existing algorithms.

5 Conclusions

The existing algorithms that support a unidirectional link over a mobile Ad Hoc network perform routing to set a reverse route of the unidirectional link, thereby causing excessive flooding of control packets. Further, the existing algorithms can be applied to only stages to be performed prior to data transmission. That is, the existing algorithms have been designed not in consideration of a unidirectional link generated during data transmission, and thus, it is impossible to immediately deal with problems due to a dynamic topology change. To solve this problem, this paper has proposed methods of effectively detecting a unidirectional link before data transmission and using a unidirectional link generated during data transmission.

The above simulation revealed that the throughput of the proposed algorithm was about 1.2 times higher than those of the existing algorithms and an end-to-end delay in the proposed algorithm was about twice lower than in the existing algorithms. Also, the proposed algorithm reduces energy consumption of a low-power node that experiences a restriction to battery power or a storage capacity, thereby increasing the lifetime of node and the connectivity of network.

However, both the existing and proposed algorithms require control packets to be transmitted periodically for detection of a unidirectional link. Therefore, a method of reducing overhead caused by periodical transmission of control packets must be investigated further.

References

1. Kumar, A.: Ad Hoc Wireless Networks, Tutorial session on Wireless Networks. In: ICCC 2002, India (2002)
2. Kumar, A., Karnik, A.: Performance Analysis of Wireless Ad-Hoc Networks. The Handbook of Ad-Hoc Wireless Networks. CRC Press, Boca Raton (2003)
3. Prakash, R.: Unidirectional Links Prove Costly in Wireless Ad Hoc Networks. In: Proc. of ACM DIAL M 1999 Workshop, Seattle, WA (August 1999)
4. Nesargi, S., et al.: A Tunneling Approach to Routing with Unidirectional Links in Mobile ad hoc Networks. In: Proc. Of IEEE ICCN 2000, Las Vegas, Nevada, October 16-18 (2000)
5. Perkins, C., Belding-Royer, E., Das, S.: Ad-hoc On-Demand Distance Vector (AODV) Routing, IETF RFC 3561 (July 2003)

Dynamic Power Efficient QoS Routing Algorithm for Multimedia Services over Mobile Ad Hoc Networks*

Zae-Kwun Lee and Hwangjun Song

Department of Computer Science and Engineering,
Pohang University of Science and Technology (POSTECH),
San 31 Hyoja-dong, Nam-gu, Pohang, Gyungbuk, 790-784, Republic of Korea
{zklee,hwangjun}@postech.ac.kr
http://mcsl.postech.ac.kr

Abstract. In this work, we present a power efficient QoS routing algorithm for multimedia services over mobile ad hoc networks. Generally, multimedia services need various quality of service over the network according to their characteristics and applications but it is not easy to guarantee quality of service over mobile ad hoc networks since the resources are very limited and time-varying. Furthermore only a limited power is available at mobile nodes, which makes the problem more challenging. Now, we propose an effective routing algorithm over mobile ad hoc networks that provide the stable end-to-end quality of service with the minimum total transmission power consumption.

Keywords: Mobile ad hoc networks, Power-efficient, Multimedia, QoS Routing, IEEE 802.11.

1 Introduction

Mobile ad hoc networking technology has recently gained a large amount of interest and demand. However, there are many problems to be successfully deployed in the field, e.g. the network topology must be dynamically maintained because links may be unstable due to the random movement of mobile nodes, and the resources such as battery power, radio propagation range and bandwidth etc. must be efficiently managed since they are very limited. Especially, it is a challenging problem to support multimedia services over mobile ad hoc networks since the stringent QoS (quality of service) is required.

In general, multimedia data requires diverse QoS (e.g. delay, jitter, loss rate and bandwidth) considering their characteristics and applications. So far, some effective QoS routing algorithms have been proposed in [1, 2, 3]. But, most of existing QoS

* This research was supported by the MIC(Ministry of Information and Communication), Korea, under the ITRC(Center for Mobile Embedded Software Technology) support program supervised by the IITA(Institute of Information Technology Assessment) (IITA-2008-C1090-0801-0045) and grant No. R01-2006-000-11112-0 from the Basic Research Program of the Korea Science & Engineering Foundation.

T. Vazão, M.M. Freire, and I. Chong (Eds.): ICOIN 2007, LNCS 5200, pp. 751–760, 2008.

routing algorithms suffer from disadvantages of routing overhead and power efficiency because each node periodically sends a hello packet to know network state information. And power-aware QoS routing protocols [4] have been proposed, but these protocols don't consider both the delay-constraint and the transmission power-level control at the same time.

In this work, we propose a power efficient QoS routing algorithm for multimedia services over mobile ad hoc networks. The goal of the proposed algorithm is to provide more stable end-to-end QoS required for media services with the minimum total battery power consumption at mobile nodes. There are several unique features of the proposed algorithm to achieve the goal. First, we provide the required bandwidth by guaranteeing the received power level over every hop and fast prune links not to promise the power level to avoid the unnecessary computational complexity. Second, the proposed *on-demand algorithm* is designed to work in a distributed way with a small amount of flooding packets. Third, route maintenance mechanism is presented to reduce delay jitter to support stable QoS. And, the load-balancing problem is implicitly considered. The remaining of this work is organized as follows. IEEE 802.11 is briefly reviewed in Section 2, the proposed routing algorithm is presented in Section 3, experimental results are provided in Section 4, and the concluding remarks are given in Section 5.

2 Review of IEEE 802.11 b/g

Lately, WLAN (wireless local area networks) is becoming increasingly popular and IEEE 802.11 standards are widely employed as the wireless MAC protocol. In the case of IEEE 802.11, multi-rate service is provided by changing the modulation method based on the received power strength at the receiver [5, 6]. For example, IEEE 802.11 b/g changes its modulation scheme based on the received power strength to support multi-rates. The 802.11b standard provides data rates of up to 11Mbps and the 802.11g provides optional data rates of up to 54Mbps. The delay at a node is approximately the sum of queuing delay and transmission delay under the assumption that the processing delay is negligible since the routing table size of the proposed algorithm is relatively very small, i.e.

$$D_{N_k}(j) = D_{N_k}^T(j) + D_{N_k}^Q(j), \tag{1}$$

where $D_{N_k}^Q(j)$ and $D_{N_k}^T(j)$ are the queuing delay and the transmission delay of the j_{th} packet at node N_k, respectively. Model-based approach is adopted for queuing delay $D_{N_k}^Q(j)$ to reduce the control overhead. The MAC model of IEEE 802.11 is *M/G/1* queuing system using stop-and-wait ARQ [5, 7]. Now, the average transmission delay at node N_k is calculated by

$$D_{N_k}^T(j) = \frac{RTT_j}{1-p}, \tag{2}$$

where RTT_j is the round-trip time between a sender and a receiver and determined by [5, 6, 8].

$$RTT_j = (DIFS + 3SIFS + BO + T_{RTS} + T_{CTS} + T_{DATA} + T_{ACK}) \times 10^{-6} s$$
$$= (1542 + \frac{8L_{DATA\_MAC}}{R_{DATA}}) \times 10^{-6} s, \tag{3}$$

where *DIFS* is the distributed inter-frame space, *SIFS* is the short inter-frame space, *BO* is the average of back off time, T_{RTS}, T_{CTS}, T_{DATA}, and T_{ACK} are the transmission time of RTS, CTS, DATA, and ACK packet, respectively, L_{DATA_MAC} is the data length of the MAC layer's packet, and R_{DATA} is the transmission rate. And p is the probability that the packet is lost, which is calculated by

$$p = 1 - (1 - p_{ACC}) * (1 - p_{CTS}) * (1 - p_{ACK}), \tag{4}$$

where p_{ACC} is the probability of channel access failure, p_{CTS} is the probability of event that CTS packet has not returned after sending RTS packet, and p_{ACK} is the probability of event that ACK packet has not arrived after sending data packet. And the queuing delay at node N_k is determined by Pollazcek-Khinchin formula [7] under the assumption that packets arrives according to Poisson process with rate μ. That is,

$$D_{N_k}^{Q}(j) = \frac{\mu \cdot (1+p) \cdot RTT_j^2}{2(1-p)(1-p-\mu \cdot RTT_j)}. \tag{5}$$

3 The Proposed Power Efficient QoS Routing Algorithm

Now, we present a routing algorithm over mobile ad hoc networks to provide the QoS required for multimedia services with the minimum total transmission power consumption. First of all, we make the following assumptions.

1. The multi-rate service is supported at every node and each node can dynamically control its transmission power.
2. Neighborhood is commutative, i.e. if node A can hear B, then node B can hear A.
3. Packets in the queue are serviced in FCFS (first-come-first-service) at each node.

For the stable multimedia service, the channel bandwidth at every hop must be larger than the minimum bandwidth (BW_{min}) required for the media service as a necessary condition and end-to-end delay must be less than the tolerable maximum delay (D_{max}) at the same time. Actually, BW_{min} is provided over a hop along the route if the received power strength is guaranteed and modulation method is automatically determined based on the received power strength at the physical layer. However, BW_{min} may not be guaranteed for a connection because it is shared among multiple connections. Thus we need to consider both bandwidth and delay constraints over the link simultaneously. In the work, *route includes both the intermediate nodes and their transmission power level information.* Now, we formulate the given problem as follows.

Problem Formulation: Determine the route between a source and a destination to minimize the following cost function

$$\sum_{N_k \in R_i} \left\{ P_{N_{(k-1)}\_tx} + \alpha \cdot \max\left\{ P_{media\_\min} - P_{N_k\_rx}, 0 \right\} \right\}, \tag{6}$$

$$\text{subject to } D_{R_i} \le D_{\max},$$

where R_i is a route between a source and a destination, N_k is the k_{th} node along the route R_i (N_0 denotes the source), αis a constant, $P_{N_{(k-1)}\_tx}$ is the transmission power strength at the $(k-1)_{th}$ node, $P_{N_k\_rx}$ is the received power strength at the k_{th} node, P is the minimum received power strength at the receiver to guarantee BW, $media\_\min$ min $D_{\max}$ is the tolerable maximum delay for the media, and D_{R_i}, the end-to-end delay along the route R_i, is defined by

$$D_{R_i} = \frac{1}{M} \sum_{j=n-M+1}^{n} D_{R_i}(j), \tag{7}$$

$$D_{R_i}(j) = \sum_{N_k \in R_i} D_{N_k}(j), \tag{8}$$

where n is the last received packet number, M is the window length for calculating end-to-end delay (M is related to buffering delay of media, i.e. real-time media generally requires smaller M than non-real-time media), $D_{R_i}(j)$ and $D_{N_k}(j)$ are the end-to-end delay along the route R_i and the delay at the node N_k for the j_{th} packet. When α is set to a large value, links that do not promise the required $P_{media\_\min}$ are excluded from the route candidate, which means that the bandwidth constraint is used to fast prune routes that do not satisfy the bandwidth requirement while searching the optimal route. And, this problem implicitly considers the load balancing because it searches for another route to meet the delay requirement if a node is overloaded and thus the delay of packets passing the node is increased.

3.1 Fast Route Discovery Mechanism

To obtain the optimal solution of the above constrained problem, Lagrange multiplier method is adopted, that is, penalty function is defined by combining the cost function and the delay constraint, and the optimal route is obtained by minimizing this penalty function as follows.

$$R^* = \arg\min_{R_i} C(R_i), \tag{9}$$

$$C(R_i) = \sum_{N_k \in R_i} \left\{ P_{N_{(k-1)}\_tx} + \alpha \cdot \max\left\{ P_{media\_\min} - P_{N_k\_rx}, 0 \right\} \right\} + \lambda \cdot D_{R_i},$$

where λ is the Lagrange multiplier. In this case, full search is needed to get the optimal solution and a huge number of flooding packets are required because every link with the power larger than $P_{media\_\min}$ must be broadcasted. Thus, the above equation is simplified as follows to reduce the amount of flooding packets. Each intermediate node calculates the penalty function of interim routes between the source and itself (actually, it is the sub-vector of the route between the source and the destination), and determines the interim route with the minimum cost. That is,

$$\tilde{R}^*_{sub} = \arg\min_{\tilde{R}_{sub,j}} C\left(\tilde{R}_{sub,j}\right), \tag{10}$$

where $\tilde{R}_{sub,j}$ is the j_{th} interim route between the source and an intermediate node. Now, only an interim route $\tilde{R}^*_{sub}$ is broadcasted to the next nodes. Hence, a route between the source and the destination is obtained by minimizing $C\left(\tilde{R}_{sub,j}\right)$ at each node with a small amount of flooding packets at the cost of performance. Now, the detail procedure of the route discovery mechanism is presented with the current λ, and then λ is adjusted by bisection method for the fast convergence.

***Step 1**: Broadcast RREQ packet*

A source inserts its address, λ, P_{init_tx} and P_{media_min} to RREQ packet as well as the traditional routing information for loop-free such as a source address, a destination address, a broadcast ID and etc. and broadcasts it to neighbors with the maximum power level P_{init_tx}. Every node received the RREQ packet calculates the distance using P_{init_tx}, the received power strength at the link layer, and the power model. And then the node decides the transmission power level range of the previous node to guarantee P_{media_min}, and the corresponding delay values at a node are estimated based the delay model. Now, the optimal power level $P_{N_{(k-1)}\_tx}$ is determined to minimize $C\left(\tilde{R}_{sub,j}\right)$ in the transmission power level range. The node remembers the address and the optimal power level of the previous node and updates RREQ packet. The updated RREQ packet with $\tilde{R}^*_{sub}$ is broadcasted with its own initial maximum power level and the other RREQ packets are discarded.

***Step 2**: Destination-driven Route Determining Process*

When the destination node decides the optimal route, it sends back RREP packet to the source along the determined route. RREP packet contains the power level of the previous node. When RREP packet arrives at an intermediate node, each node updates its routing table by investigating RREP packet. By the way, an intermediate node may be included in the different routes simultaneously and thus a routing table must keep the connection information (i.e. source address, source's port number, destination address, and destination's port number) of data packets.

***Step 3**: Transmit Data Packet by using Node's Route Table*

When a data packet arrives, each node recognizes its connection information and searches for the transmission power level in the routing table. Then, data packets are forwarded to the next hop by the power level.

***Step 4**: Adjust λ by Bisection Method*

We consider how to dynamically determine λ to satisfy the required bandwidth and the delay constraint with the minimum power consumption over mobile ad hoc networks. In this work, bisection method is adopted for the fast convergence of λ. First of all, λ is set to a large value to make sure that the delay is less than D_{max}, and λ is updated based on the bisection method. By the way, it is difficult to measure the end-to-end delay when UDP is employed for multimedia data delivery because ACK packets are not used. Thus, the source inserts time information into data packets (i.e. IP Header Extension) and the destination puts time information of the received packets into a control packet (or a RTT packet) and periodically sends back the RTT packet to the source. Consequently, the source can estimate the end-to-end delay using the RTT packet. If the observed end-to-end delay is between $D_{\max}$ and D_{under_th}, then the route is retained. Otherwise the source performs the route discovery process again with the updated λ. As λ becomes larger, the delay constraint and load balancing becomes more stringent.

3.2 Predictive Route Maintenance Reducing Delay Jitter

Although the above fast route discovery mechanism works well for non-real-time application, it is observed during the experiment that it is not sufficient for real-time media since delay jitter caused by link failure may seriously degrade the media service. When link failure occurs, a large delay is inevitable to find a new route. Even though several alternative routes are stored in the routing table, it is not guaranteed that they are still alive. It may sometimes cause even larger delay. Thus, we need to predict node's mobility to reduce the link failures. Some effective algorithms have been proposed to search for the stable route and predict link failures [9, 10]. Most of them use hello packets, which can provide the stable route at the cost of the increased routing overhead. For example, RABR (route-lifetime assessment based routing) [9] sends a hello packet periodically to check link stability. Preemptive route maintenance algorithms have been proposed to avoid actual link failure in [10].

In this work, we consider a simple but effective measurement-based predictive route maintenance mechanism to reduce the delay jitter. In the proposed algorithm, each node monitors the received power continuously during data packet transmission to predict link failures without additional overhead. It is reasonable for multimedia service because multimedia data packets are in general transmitted continuously and periodically. Now, we adopt the linear prediction method to estimate the power level as follows.

$$\log P(\hat{t}) = \frac{\log P_{t_{curr}} - \log P_{t_{prev}}}{t_{curr} - t_{prev}} \left(\hat{t} - t_{curr}\right) + \log P_{t_{curr}}, \tag{11}$$

where $\hat{t}$ is the estimated arriving time of the next packet, $P_{t_{prev}}$ and $P_{t_{curr}}$ are power levels when receiving packets at t_{prev} and t_{curr}, respectively ($t_{prev} < t_{curr} < \hat{t}$). When $P(\hat{t})$ is less than a threshold value (i.e. the link failure is expected soon), the intermediate receiver node notifies it to the source and a new route is established before the actual link failure occurs. As a result, we can decrease the delay jitter. That is, If $P(\hat{t}) < P_{th}$, then the intermediate receiver node sends a warning message packet to the source and then another route is searched, where $P_{th} = (1+\Delta) \cdot P_{media\_min}$ and Δ is a positive real number. As Δbecomes larger, the more stable route is found at the expense of power efficiency.

4 Experiment Results

During the experiment, NS-2 is employed and Orinoco datasheet [11] is used to determine multi-rate throughput. Two-ray ground reflection power model is adopted for wireless channel since it has been found to be reasonably accurate for predicting the large scale signal strength over distances of several kilometers for mobile radio systems and NS-2 supports. And log-distance path loss model is employed to determine the required transmission power at each node so that the uncertainty of wireless channel are somewhat considered.

The shortest path algorithms such as AODV [12] and DSR [13] are used for the performance comparison, and they are also implemented to select their multi-rate by using RBAR (receiver-based auto-rate) protocol [14] for more fair performance comparison. End-to-end delay (average delay and delay jitter), total power, routing overhead and throughput are employed as performance measures. Routing overhead is defined by the number of routing packets transmitted per data packet delivered at the destination and throughput is a ratio of the data packets delivered to the destination to those generated by the sources. During the experiment, $BW_{\min}$, D_{under_th}, M and D_{max} *are set to 1Mbps, 17 ms, 1 and 20 ms, respectively.*

4.1 Proposed Fast Route Discovery Mechanism

Now, 30 nodes are randomly located in 500m*150m square and the simulation is performed for 300 seconds. We examine general case that nodes are moving [0, 5m/s] by random-way point model. Multiple source-destination pairs want to communicate each other: 15 connection pairs are tested. It is assumed that every source sends four packets per second whose size is 512 bytes. Almost the same phenomena are observed in the Figure 1, that is, delay decreases as λ becomes larger, vice versa. In terms of power, the proposed algorithm show better performance than DSR and AODV for all λ, but the gain decreases as λ becomes larger. One reason is that the path with a low end-to-end delay is selected when some of nodes is overloaded if λ becomes larger.

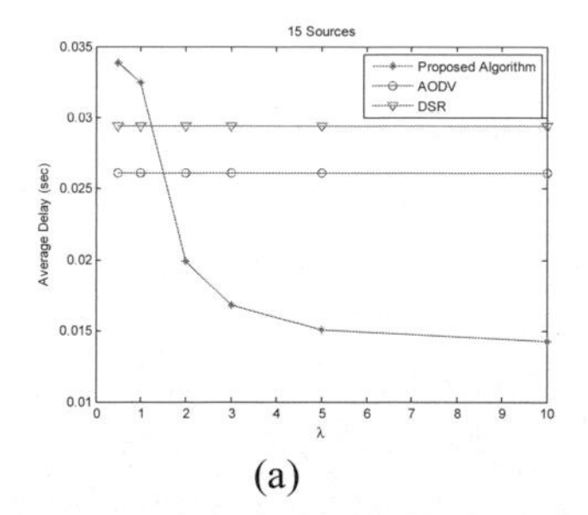

(a)

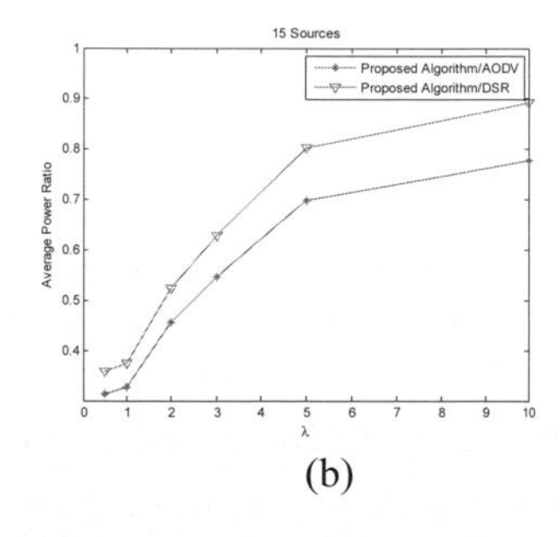

(b)

Fig. 1. Performance comparison with DSR and AODV in the case of multiple moving nodes when number of sources is 15: (a) average delay and (b) power ratio

4.2 Predictive Route Maintenance Mechanism

The performance of the proposed algorithm without predictive route maintenance mechanism is presented in Figure 2. In this figure, nodes start to move after 30 second to show the λ adaptation process. The first large delay is due to the initial route discovery process. As shown in (a) of Figure 2, the delay jitter is relatively small in the case of no mobility, which is caused by queuing delay and transmission delay at the node. However, the delay jitter becomes extremely larger because of link failures when nodes dynamically moving as shown in (b) of Figure 2. In fact, it is almost impossible to support the smooth real-time media through the channel. Now, two examples of the proposed predictive route maintenance mechanism with different Δvalues are given in Figure 3 and the overall performance is summarized in Figure 4. Delay

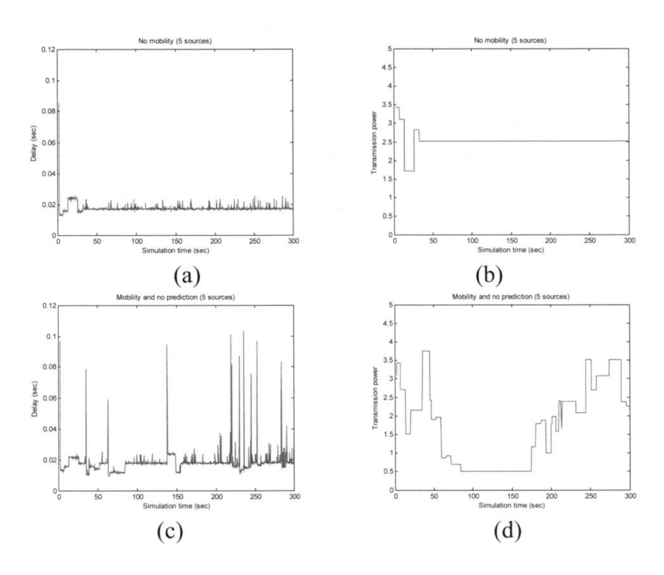

Fig. 2. Performance of the proposed algorithm without predictive route maintenance when number of sources is 5: (a) delay of no mobility case, (b) power of no mobility, (c) delay of moving node case, and (d) power of moving node case

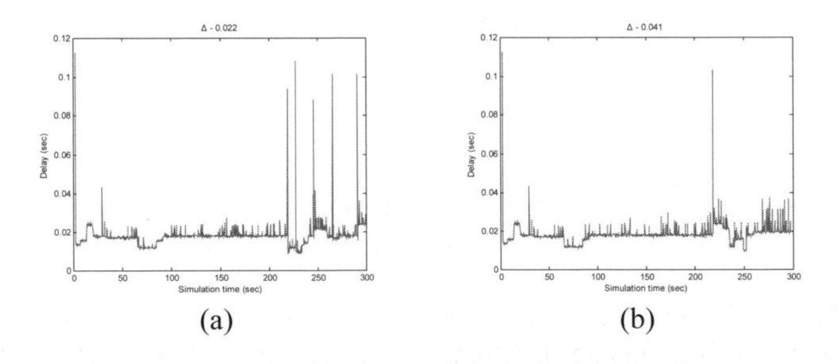

Fig. 3. Performance of the proposed algorithm with predictive route maintenance when number of source is 5: (a) delay when $\Delta = 0.022$, and (b) delay when $\Delta = 0.041$

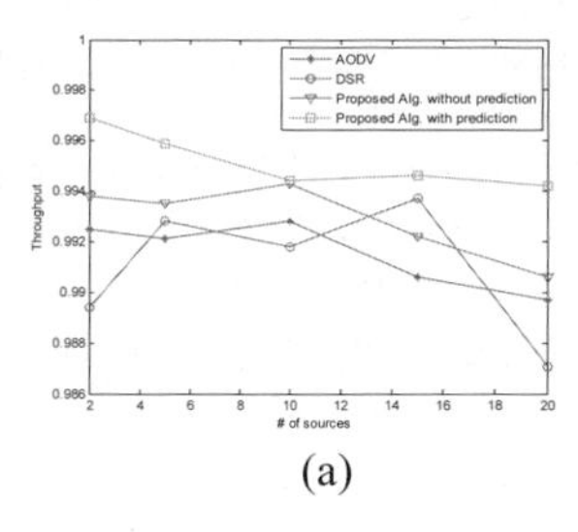

(a)

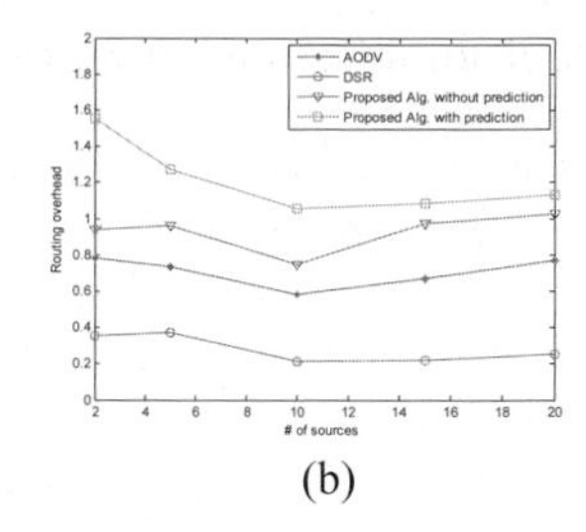

(b)

Fig. 4. Throughput and routing overhead comparison with DSR and AODV in the case of multiple moving sources: (a) throughput and (b) routing overhead

jitter is obviously reduced for the larger Δ as shown (a) and (b) in Figure 3. The reason is that the route discovery process is executed before the actual link failure occurs. In terms of routing overhead and throughput, the performance comparison is provided in Figure 4. As shown in (a) and (b) of the figure, the proposed algorithm presents better throughput at the cost of higher routing overhead compared with DSR and AODV. As a result, the proposed algorithm with prediction provides much more stable route for real-time media delivery at the price of higher routing overhead. We need intelligently adjusting Δ based on the node's mobility to improve the performance.

5 Conclusion

We have presented an effective routing algorithm to support various QoS over mobile ad hoc networks in terms of bandwidth, delay and delay jitter with the minimum total power consumption. By the experimental result, we have shown that the proposed algorithm can provide an effective route for the stable multimedia service over mobile ad hoc networks. In addition, it implicitly takes into account a load-balancing problem.

References

1. Sinha, P., Sivakumar, R., Bharghavan, V.: CEDAR: A Core-Extraction Distributed Ad hoc Routing Algorithm. IEEE Journal on Selected Area in Communications 17(8), 1454–1465 (1999)
2. Chen, S., Nahrstedt, K.: Distributed Quality-of-Service Routing in Ad Hoc Networks. IEEE Journal on Selected Area in Communications 17(8), 1488–1505 (1999)
3. Chen, L., Heinzelman, W.B.: QoS-Aware Routing Based on Bandwidth Estimation for Mobile Ad Hoc Networks. IEEE Journal on Selected Area in Communications 23(3), 561–572 (2005)
4. Zhang, W., Li, J., Wang, X.: Cost-Efficient QoS Routing Protocol for Mobile Ad Hoc Networks. In: 19th International Conference on Advanced Information Networking and Applications (AINA) (2005)

5. IEEE 802.11, Part 11: Wireless LAN Medium Access Control (MAC) and Physical Layer (PHY) Specifications (1999)
6. IEEE 802.11b, Part 11: Wireless LAN Medium Access Control (MAC) and Physical Layer (PHY) specifications: Higher-Speed Physical Layer Extension in the 2.4 GHz Band (1999)
7. Bersekas, D., Gallager, R.: Data Networks, 2nd edn. Prentice Hall, Englewood Cliffs
8. Jun, J., Peddabachagari, P., Sichitiu, M.: Theoretical Maximum Throughput of IEEE 802.11 and its Applications. In: Proc. of the 2nd IEEE International Symposium on Network Computing and Applications (NCA 2003) (2003)
9. Agarwal, S., Ahuja, A., Singh, J.P., Shorey, R.: Route-Lifetime Assessment Based Routing (RABR) Protocol for Mobile Ad-Hoc Networks. In: IEEE International Conference on Communications (2000)
10. Goff, T., et al.: Preemptive Routing in Ad Hoc Networks. ACM SIGMOIBLE (2004)
11. Datasheet for ORiNOCO 11b PC Card, http://www.orinocowireless.com
12. Perkins, C., Belding-Royer, E., Das, S.: Ad hoc on-demand distance vector (AODV) routing. IETF RFC 3561, http://www.ietf.org/rfc/rfc3561.txt
13. Johnson, D.B., Maltz, D.A., Broch, J.: DSR: the dynamic source routing protocol for multi-hop wireless Ad Hoc networks. In: Perkins, C.E. (ed.) Ad Hoc networking, pp. 139–172. Addison-Wesley, Reading (2001)
14. Holland, G., Vaidya, N., Bahl, P.: A Rate-Adaptive MAC Protocol for Multi-Hop Wireless Networks. In: ACM/IEEE International Conference on Mobile Computing and Networking (MOBICOM 2001) (2001)

Two-Hops Neighbor-Aware Routing Protocol in Mobile Ad Hoc Networks

N.W. Lo and Hsiao-Yi Kuo

Department of Information Management
National Taiwan University of Science and Technology
nwlo@cs.ntust.edu.tw,
M9309111@mail.ntust.edu.tw

Abstract. Routing Protocols for mobile ad hoc networks (MANETs) have been discussed extensively in recent years. Because of node mobility and energy conservation consideration in MANET, reactive routing protocols get more attention than others with their on-demand packet transmission philosophy. AODV as one of the most discussed reactive routing protocols establishes only single routing path after handshaking process. In this paper, we explore the effectiveness of limited multipath routing by utilizing beacon packets to collect routing information of neighbor nodes in two-hops distance. We present a two-hops neighbor-aware reactive routing protocol against AODV to increase the packet delivery ratio up to 10% and reduce the normalized routing load from 1.5 to 3 times in MANET environment.

1 Introduction

With no existing infrastructure every node in MANET acts as both end client and relay point through its wireless communication module. In addition, there are several interesting characteristics in MANET such as dynamic topology, constrained bandwidth, variable capacity links, constrained energy and limited physical security. Therefore, how to design an effective routing protocol for MANET environment has become an interesting challenge. Based on routing protocols proposed in recent years they can be generally classified into three categories: proactive, reactive (on-demand), and hybrid class. Proactive routing protocols such as DSDV and OLSR attempt to maintain consistent, up-to-date routing information from each node to every other node in the network. They use one or more tables to store routing information, and update their tables by exchanging control packets. The advantage of proactive routing protocols is that they know current topology situation immediately even if the topology changes frequently. On the contrary, reactive routing protocols such as AODV [1], DSR and ABR establish routes only when it is desired by the source node. A source node initiates a route discovery process when it requires a route to a destination. Once a route is established, the source node keeps maintaining this route information as long as it needs to send data through this route. In contrast with proactive routing protocols, on-demand routing reduces a large number of control overhead, but

T. Vazão, M.M. Freire, and I. Chong (Eds.): ICOIN 2007, LNCS 5200, pp. 761–770, 2008.

at the same time this class of protocols increase the delay time while propagating data packets. Hybrid routing protocols such as ZRP include proactive and reactive routing characteristics. They usually take advantage of node grouping concept to achieve routing operation.

Among proposed routing protocols, AODV has become a very popular topic to study and compare with new on-demand routing protocols in academic research because of its simple design and quite effective routing capability. We notice that in AODV after route discovery phase in which flooding scheme is invoked to broadcast route request (*RREQ*) control packets, between each pair of the source node and the destination node only one valid main route is established and maintained with the best effort. Therefore, the source node needs to look for valid route again when current valid route is broken. However, this mechanism will produce a large number of *RREQ* control packets and route reply (*RREP*) control packets correspondingly in order to find a new route. This observation motivates us to explore the possibility of acquiring routing information of neighbor nodes in limited range and the effectiveness of new protocol derived from acquired neighbor information. In this paper, we present the two-hops neighbor-aware routing protocol which maintain multiple routes by adopting enhanced beacon (*HELLO*) packets. The enhanced *HELLO* packet includes information of one-hop neighbors of the source node so that each node could know two-hops neighbors when a node receives a *HELLO* packet sent by its one-hop neighbor. Our protocol utilizes these information to automatically switch data packets to a backup route when the original established route is broken.

The rest of this paper is organized as follows. Previous research work is presented in Section 2. The proposed routing protocol is introduced in Section 3. Section 4 analyzes simulation results and discusses related issues. Finally, conclusion and future work are addressed in Section 5.

2 Related Work

In this section we review on-demand routing schemes proposed in recent years with multipath design in mind.

In [4], Sengul and Kravetsthe proposed bypass routing concept by taking advantage of cache to store alternative routes. When the original route breaks, the source node searches route cache and patches the broken route with alternative one if it has valid backup route in its route cache. If there are no backup routes in the cache, the source node queries its one-hop neighbors by broadcasting *RREQ* packet to wait for reply packets, i.e., *RREP* packets, sent from those neighbors with valid routes to the destination node.

AODV-BR in [2] and AODV-2HBR in [3] were similar to each other. They both take advantage of overhearing *RREP* packets to identify backup routes. AODV-BR maintains backup routes through nodes are one-hop away from established main route. In comparison with AODV-BR, AODV-2HBR protocol maintains backup routes through nodes are two-hops away from established main route. Since AODV-2HBR increases the search range for backup routes, it outperforms AODV-BR in terms of packet delivery ratio and average end-to-end delay in high mobility cases. However, both protocols did not update identified backup routes after source node starts to send data through established main route. When network topology changes frequently, both protocols failed most of time to find a valid backup route.

In [5], Tang and Zhang proposed a Robust-AODV protocol in which *HELLO* control packet is substituted with route update message. All nodes on established main route broadcast this control message to one-hop neighbors. With alternative route information inside this control message, nodes along active main route in one-hop distance can utilize this route information and switch data packets to backup route when main route is broken.

MRAODV routing protocol introduced in [6] modifies MNH routing protocol to discover more routes by adopting a number assignment scheme to a routing table parameter called branch identifier from which backward link direction of each mobile node that received forwarding *RREQ* from source node can be decided. However, MRAODV require two new control packets, namely, *RDEL* and *FORWARD*. In addition, time synchronization is required while receiving *RREP* packets in order to determine a branch route *ID*.

3 Two-Hops Neighbor-Aware Routing Protocol

Based on the review work in Section 2, we notice that most of multipath-oriented on-demand routing protocols tend to wait for the establishment of main route before invoking their multipath discovery and maintenance mechanisms. Since people try to avoid the adoption of traditional flooding scheme as much as they can to provide cheaper route repair, complex control messages and extra fields in routing table are developed to build and utilize local multipath information along the main route. Unfortunately, identified backup routes are usually invalid after short amount of time when nodes move quickly or network topology changes frequently. In reactive routing protocols, *HELLO* packets are used to check if wireless connection between two neighboring nodes has broken. Since each node in ad hoc network sends *HELLO* packets periodically, we can use *HELLO* packets to carry one-hop neighbor information conveniently. Specifically, a node appends all identifications (*ID*s) of its one-hop neighbors to its *HELLO* packet before broadcasting. When a node receives *HELLO* packets from its neighboring nodes, it retrieves and stores these *ID*s from *HELLO* packets. With this simple mechanism every node in an ad hoc network will possess an *ID* list of its corresponding two-hops neighbors dynamically in a short period of time. Our proposed two-hops neighbor-aware routing mechanism (2HNa) uses the two-hops neighbor list in each node as the resources for constructing alternative routes against an established main route. Because *HELLO* packets are broadcasted periodically, 2HNa should be able to get more up-to-date neighbor information than other on-demand routing protocols with a tolerable overhead cost against the enlarged *HELLO* packet size. In Section 4 we will discuss the performance and overhead of 2HNa protocol in details.

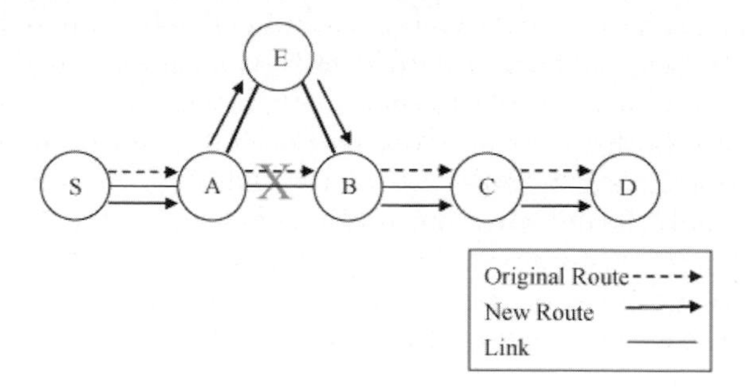

Fig. 1. 2HNa route repair pattern – one-hop neighbor node E of main route located between the upstream node A and the downstream node B of the broken link $A - B$

In this paper we investigate the effectiveness of 2HNa mechanism by implementing 2HNa onto AODV routing protocol. To simplify the description, we only address the difference between AODV-2HNa and AODV in the following.

AODV-2HNa adopts the same route discovery phase as AODV. When an established main route is broken, before original AODV route maintenance phase is invoked, 2HNa mechanism is activated to find alternative route for data packets. If the main route cannot be repaired and no other route is found by 2HNa mechanism, original AODV route maintenance process will be triggered. There are two route repair patterns in 2HNa mechanism. Figure 1 shows the first pattern in which a broken link $A - B$ on established main route is identified by node A; by using stored two-hops neighbor information node A can find a repair route through one-hop neighbor E to node B. The second pattern is described in Figure 2 in which node A finds the repair route through its one-hop neighbor E to downstream node C after identifying the broken link $A - B$.

To implement AODV-2HNa protocol, one new *neighbor table* is required for each node in MANET to store the *ID*s of neighbor nodes. When a node receives a

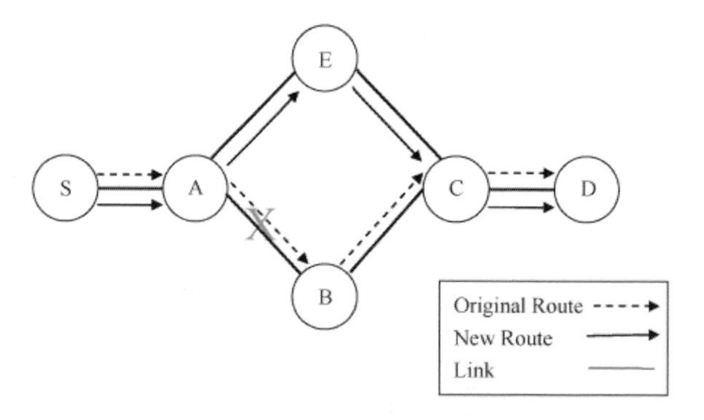

Fig. 2. 2HNa route repair pattern – one-hop neighbor node E of main route with connections to both the upstream node A and the next downstream node C of the broken link $A - B$

HELLO packet, its corresponding neighbor table will be updated to fill in triple column fields, namely, *one-hop neighbor*, *two-hops neighbor* and *timestamp*. The value of field *two-hops neighbor* contains an *ID* of the node's two-hops neighbors. The field *timestamp* records the arrival time of the corresponding *HELLO* packet. When a link connection is broken, the upstream node of this broken link can search the *neighbor table* to find an up-to-date repair route by sorting records on *timestamp* field. In order to maintain the entry correctness of *neighbor table*, an entry in this table will be deleted if this entry holds a two-hops neighbor *ID* which does not exist in the arrived *HELLO* packet.

4 Simulations and Discussions

4.1 Simulation Environment

Network simulator ns-2 is used to evaluate the overall performance of AODV-2HNa. In this paper we modified the AODV implementation of ns-2.27 version which supports multi-hop wireless networks with physical, data-link and MAC layer models to construct AODV-2HNa environment. The radio model operates at 2 Mb/s bit rate with radio range 250 m. The capacity of sending buffer of each node is 64 packets. Traffic model adopts continuous bit rate (CBR). Mobility model uses random way point. The interface queue has a maximum size of 50 packets and is maintained as priority queue served in FIFO order. Control packets get higher routing priority than data packets.

In ns-2.27 AODV adopts MAC layer notification mechanism as the default link-broken detection scheme instead of utilizing *HELLO* packet. Because AODV-2HNa needs *HELLO* packets to deliver neighbor information, in order to observe the influence of link detection mechanism three protocol models, AODV, AODV(HELLO)

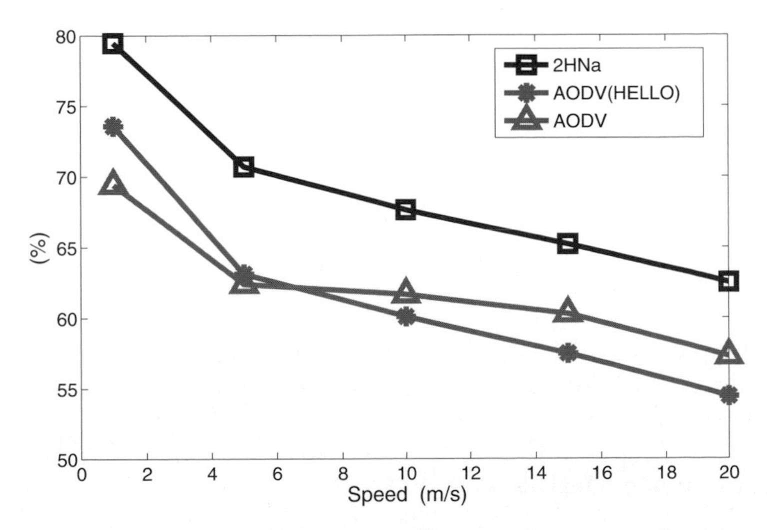

Fig. 3. Packet delivery ratio for nodes at different maximum speeds with pause time 0 second

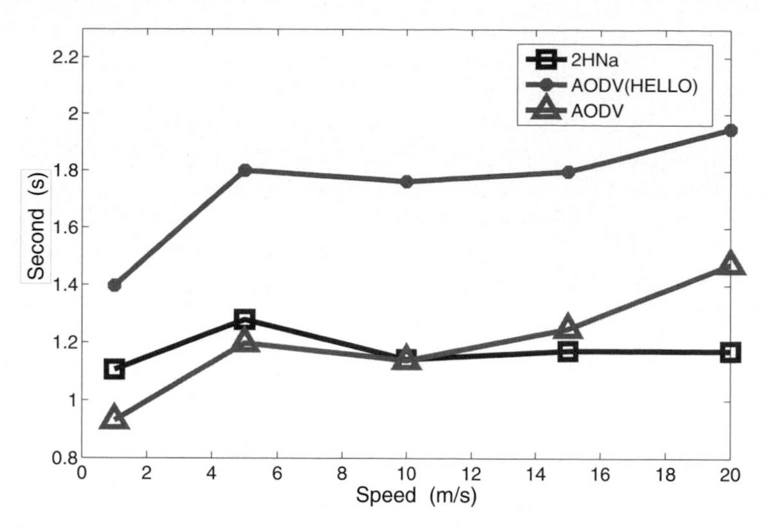

Fig. 4. Average end-to-end delay for nodes at different maximum speeds with pause time 0 second

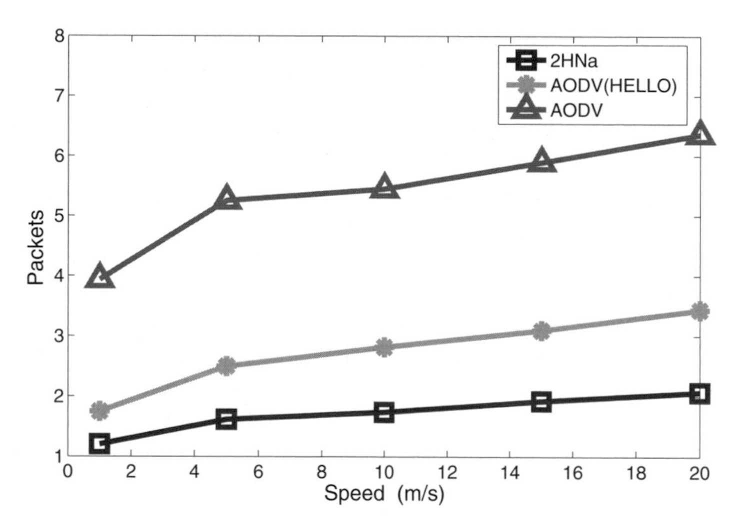

Fig. 5. Normalized routing load for nodes at different maximum speeds with pause time 0 second

and AODV-2HNa, are investigated in our simulations. Simulation results are generated after 1000 simulation seconds every time and each data record represents an average of ten runs.

4.2 Performance Metrics

The following metrics are used to evaluate the performance of targeted routing protocols.

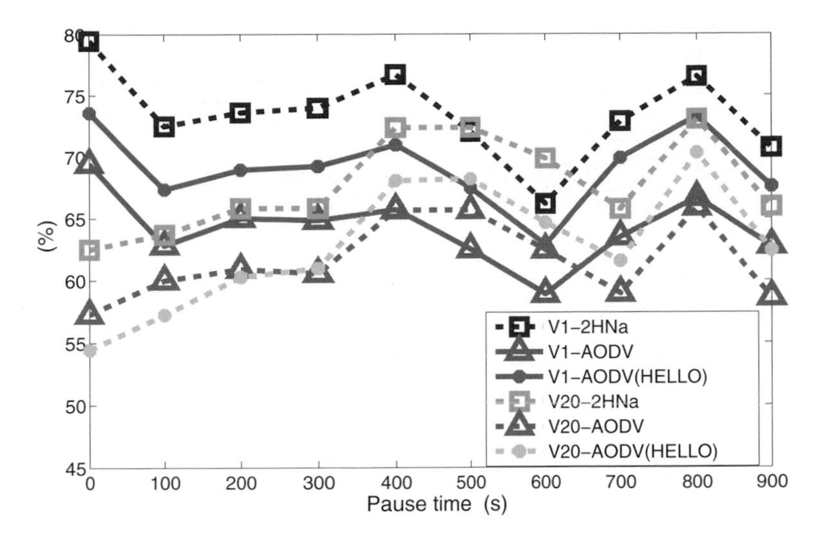

Fig. 6. Packet delivery ratio for nodes with different pause times while the maximum speed of each node is at 1 m/s and 20 m/s, respectively

- Packet delivery ratio: The ratio of the data packets delivered to the destination nodes to those generated by the CBR source nodes. In other words, packet delivery ratio indicates the effectiveness of data transmission of a routing protocol.
- Average end-to-end delay: Time latency between a data packet sent from source node and the same data packet arrived at the destination node in average. Time delay can be caused by node buffering, queuing at the interface queue, retransmission at the MAC layer and propagation. Average end-to-end delay denotes the efficiency of data transmission of a routing protocol.
- Normalized routing load: The number of routing control packets transmitted per data packet delivered at the destination node. Each hop-wise transmission of a routing control packet is counted as one transmission. Normalized routing load describes the overhead spent to complete a data packet transmission.

4.3 Results and Analysis

Because 2HNa mechanism depends on the neighbor awareness of each node in two-hops range to identify alternative routes, the most devastating environment to protocols with 2HNa mechanism is nodes in ad hoc network move constantly. Under this situation an established route can be easily broken. If a route repair mechanism is not robust enough, it will perform poorly in this case. In addition, the chance to invoke resource-consuming route discovery process increases proportionally with the increase of moving speed of each node. In Figure 3 we show that AODV-2HNa outperforms AODV and AODV(HELLO) from 6% to 10% in terms of packet delivery ratio while nodes move constantly at different

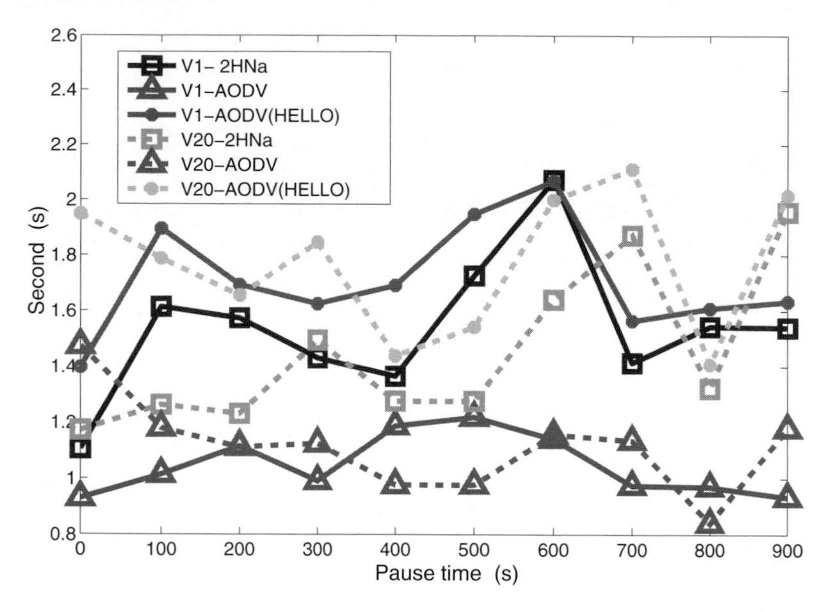

Fig. 7. Average end-to-end delay for nodes with different pause times while the maximum speed of each node is at 1 m/s and 20 m/s, respectively

speeds. Regarding to protocol efficiency, Figure 4 depicts that AODV-2HNa has similar end-to-end delay pattern as AODV does but runs over AODV(HELLO) up to 1.6 times. Notice that AODV-2HNa is more stable than AODV when node speed exceeds 10 m/s. Normalized routing load among the three routing protocols is drawn in Figure 5. AODV-2HNa reduces the routing overhead over AODV 3 times and AODV(HELLO) up to 1.75 times, accordingly. The reason for AODV-2HNa issuing much less routing packets than AODV per data packet delivered is because the route repair scheme of AODV-2HNa reduces the need of route rediscovery. In addition, the default link detection mechanism of AODV is too sensitive to stop sending unnecessary link-broken signals which consequently invoke route rediscovery process. In summary, AODV-2HNa routing protocol is very suitable for ever changing network topology with high speed node movement.

Another interesting question is how the dynamics between the pause time of a node and the speed of node movement influences the performance of a routing protocol. Therefore, we performed simulations with various pause time periods for nodes at two speeds of node movement, namely, 1 m/s and 20 m/s. In Figure 6 we show that in general, regardless of pause time of node, all protocols have better performance while nodes move slowly in terms of packet delivery ratio. The AODV-2HNa scheme outperforms AODV scheme about 10% at node speed 1 m/s and 5 $\sim$ 8% at node speed 20 m/s, accordingly. It indicates that AODV-2HNa can still take advantage of its local route repair mechanism to deliver more data than AODV even nodes move very slowly. We also notice that when node's pause time get longer, ad hoc network with nodes moving at high

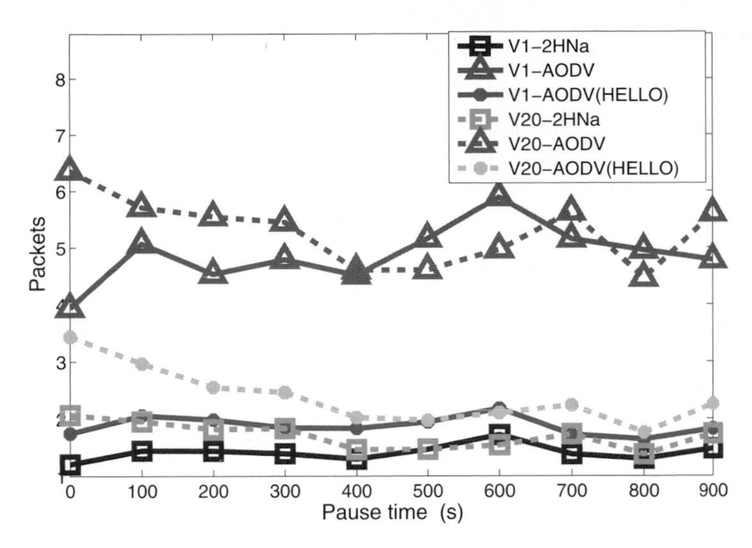

Fig. 8. Normalized routing load for nodes with different pause times while the maximum speed of each node is at 1 m/s and 20 m/s, respectively

speed (20 m/s) gains more performance lift in comparison with short pause time situation than the one with nodes walking at low speed (1 m/s).

Regarding to average end-to-end delay metrics shown in Figure 7, we learn that AODV-2HNa spends more time to deliver a data packet than AODV and it just gets better a little when compared with AODV(HELLO). By utilizing route repair scheme AODV-2HNa pays the price of longer end-to-end routing in average than original AODV to get better effectiveness.

Normalized routing load for nodes in various pause times and moving speeds are drawn in Figure 8. AODV-2HNa has the lowest overhead in comparison with AODV and AODV(HELLO) regardless of the moving speeds of nodes. Nevertheless, moving speed of node does impact the amount of normalized routing load. The faster node moves the higher overhead goes.

In summary, based on simulation results AODV-2HNa reveals better effectiveness and lower overhead than AODV and AODV(HELLO) with a moderate efficiency.

5 Conclusion

In this paper, we proposed the 2HNa routing mechanism which takes advantage of enhanced *HELLO* packet to establish alternative routes by utilizing the awareness information of two-hops away neighbors in each node. Simulation results show that AODV-2HNa, the AODV implementation plus 2HNa module, has better performance against different moving speeds and pause times when compares with AODV and AODV(HELLO). The packet delivery ratio increases up to 10% and the normalized routing load decreases from 1.5 to 3 times in MANET environment.

References

1. Perkins, C.E., Royer, E.M., Das, S.R.: Ad Hoc On Demand Distance Vector (AODV) Routing, IETF Draft (February 2000), http://www.ietf.org/ID.html
2. Lee, S.J., Gerla, M.: AODV-BR: Backup Routing in Ad Hoc Networks. In: Proc. IEEE WNMC, September 2000, vol. 3, pp. 1311–1316 (2000)
3. Chen, H.-L., Lee, C.-H.: Two Hops Backup Routing Protocol in Mobile Ad Hoc Networks. In: Proc. IEEE ICPADS, July 2005, pp. 600–604 (2005)
4. Sengul, C., Kravets, R.: Bypass Routing: An On-demand Local Recovery Protocol for Ad Hoc Networks. Ad Hoc Networks 4(3), 380–397 (2006)
5. Tang, S., Zhang, B.: A Robust AODV Protocol with Local Update. In: Proceedings of the 2004 Joint Conference of the 10th Asia-Pacific Conference on Communications and the 5th International Symposium on Multi-Dimensional Mobile Communications, August 2004, vol. 1, pp. 418–422 (2004)
6. Higaki, H., Umeshima, S.: Multiple-route Ad Hoc On-demand Distance Vector (MRAODV) Routing Protocol. In: Proceedings of the 18th International Parallel and Distributed Processing Symposium (IPDPS 2004), April 2004, p. 237 (2004)

H_2O: Hierarchically Optimized Hybrid Routing Strategies for Multihop Wireless Networks

Namhi Kang[1], Younghan Kim[1,⋆], and Jungnam Kwak[2]

[1] Ubiquitous Network Research Center, Soongsil University
Sangdo 5-Dong 1-1, Dongjak-Ku, Seoul 156-743 Korea
nalnal@dcn.ssu.ac.kr, yhkim@dcn.ssu.ac.kr
[2] Access Solution Team, Xener Systems, Inc.
Hyundai Bldg. 261, Nonhyeon-dong, Gangnam-gu, Seoul 135-832 Korea
muxxc1@xener.com

Abstract. This paper presents a semi-infrastructured architecture for multihop wireless networks and its routing strategies called H_2O (Hierarchically Optimized Hybrid). We show that the semi-infrastructured network architecture is very efficient in terms of scalability and reliability. Such advantages mainly come from the hierarchical network architecture combined with a hybrid routing protocol. We evaluate the proposed approach by using simulation tests with several scenarios. In the simulation, we observed that no specific routing strategy (e.g. a single hybrid way to combine a proactive and reactive routing protocol adapted in previously proposed ZRP or ZHLS) can always satisfy various requirements of applications. Therefore, it is required to make a hybrid routing strategy differently from application to application.

1 Introduction

These days most of information societies look forward to the ubiquitous world, where a set of smart objects that are any type of objects equipped with processing modules and storage communicate with each other without any restriction of time and space. Mobile ad hoc networks (MANETs) technology is one of the most promising approaches since it introduces a good feature of performing self-organizing an arbitrary network in a cost efficient manner. In particular, the most important property of MANET we consider is the capability of multi-hop relaying data from a source to a destination in the absence of fixed infrastructure.

To be capable of multi-hop relaying, it is necessary to develop an appropriate routing protocol for MANET. It is regarded as a difficult challenge to discover and maintain a stable route to the destination in MANET. Moreover, these challenges become harder to solve in large scale and/or densely populated MANET [1]. This is mainly due to the fact that the communication only relies on mutual and cooperative routing functionalities of ordinary nodes without any help of specific relaying devices such as a router in wired network or a BS in cellular

⋆ Corresponding author.

T. Vazão, M.M. Freire, and I. Chong (Eds.): ICOIN 2007, LNCS 5200, pp. 771–780, 2008.

network. Therefore, routing protocols used in conventional networks can not be employed into MANET directly. In addition, there exist inherent difficulties to deploy MANET into the real world. These include a highly dynamic topology change due to the movement of nodes and lack of resources in both nodes and wireless links. During decades, lots of routing protocols have been proposed in the literature to solve these problems [2]. However, most of such routing protocols still introduce several limitations including reliability, scalability, load balancing, security and others in order to make such networks practical for various applications (i.e. commercial applications rather than specific-purpose oriented applications such as a military operation or disaster discovery). Among these challenges, we focus on the scalability and reliability problems that result in severe performance degradation when the network size increases.

This paper presents a semi-infrastructured MANET architecture called *"u-Zone based network architecture"* and its routing strategies called *"H_2O (Hierarchically Optimized Hybrid)"*. To solve the scalability problem, the proposed approach is intended to avoid network-wide flooding by using the hierarchical network architecture combined with a hybrid routing protocol. The reliability can be enhanced by utilizing the wireless back bone directly connected between *u-Zone masters* that are the main component of the u-Zone based network (see section 2). Fig. 1 gives the conceptual model of the proposed approach.

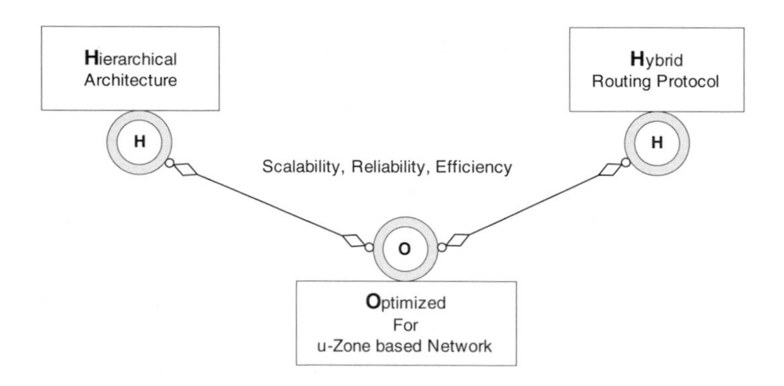

Fig. 1. Conceptual model of H_2O routing stratagies

Several results in the literature (e.g. [3], [4] or [5]) have shown that clustering can enhance scalability. The authors of [6] considered the impact of clustering on the scalability of routing protocols. The results said that clustering reduces the overhead by a factor of $O(1/M)$, where M is the cluster size. For this reason, we borrow the basic concept of cluster based routing protocols but we remove the highly expensive processing to elect a cluster head. Instead, each cluster (referred to as a u-Zone in our approach) has a u-Zone master. That is, a MANET of large size is divided into several u-Zones and one u-Zone master is assigned to each of u-Zones. The u-Zone master assists its member nodes in gathering and/or relaying routing information so that the nodes can discover and maintain a route at a low cost. In addition, a direct wireless link between u-ZMs, referred to as wireless back bone (WBB), offers reliable communication and reduction in hop counts to the destination (i.e. path length). In MANET, larger hop counts results in worse performance. As a result, the proposed approach can reduce lots of overburden of ordinary nodes and the amount of control packets necessary to maintain both route and cluster.

Based on the clustering, researchers typically used a hybrid routing approach (i.e. a combination of proactive and reactive routing protocol) in their protocols such as HARP [7], ZRP [8] and ZHLS [9]. In particular, they utilized a proactive approach inside a zone (say intra zone routing) and a reactive approach beyond the zone (say inter zone routing). However, such a single hybrid way can not meet with all the requirements of diverse applications. In this paper, we show that each of hybrid routing combinations (four different combinations as discussed in section 3) introduces its own properties. Thus, to deploy an optimized routing strategy, it is required to consider several factors such as the zone radius, mobility of nodes, size of the network, nodes density, traffic pattern and the most importantly the primary requirements of applications.

The remainder of this paper is organized as follows. The proposed architecture is given in section 2. In section 3, we address H_2O routing strategies for the proposed network architecture. In section 4, we show the simulation results of each of routing strategies. Finally, we conclude this paper in section 5.

2 u-Zone Based Hierarchical Network Architecture

The proposed u-Zone based network architecture is described in Fig. 2, where a large scale of ad hoc network is divided into a set of sub-regions. A sub-region is referred to as a u-Zone (ubiquitous-Zone) in which a u-Zone Master (u-ZM) is placed. The u-ZM is the central component to form a hierarchical architecture. We suppose that a service provider installs u-ZMs to build a temporary ad hoc network. For instance, a host of a conference or travel agency in a cultural place builds a network where any fixed infrastructure is not available. The u-ZM is portable but operates as a stationary node. In addition, unlike ordinary nodes, u-ZM has high computing power and robust electrical power as a super node. In some scenarios, a sink node of sensor network may be a member node of a u-Zone. Also a mobile node is able to access Internet or different kind of networks through the gateway (GW denoted in Fig. 2.).

The u-ZM manages most of routing functionalities such as monitoring consistent and up-to-date routing information so that member nodes do not need to follow frequently changes in the network topology. That is, highly expensive computing necessary to generate and maintain a routing table is performed by the u-ZM. Further, as an additional benefit from using u-ZM, a node can recognize some location information of the target node approximately since the u-ZM, which covers the target node, has its own identifier and its location information can be obtained by a network administrator. Such a property can be used very efficiently in some scenarios such as emergencies and natural disasters without any support of GPS or satellite based systems. Fig. 3 shows our first prototype

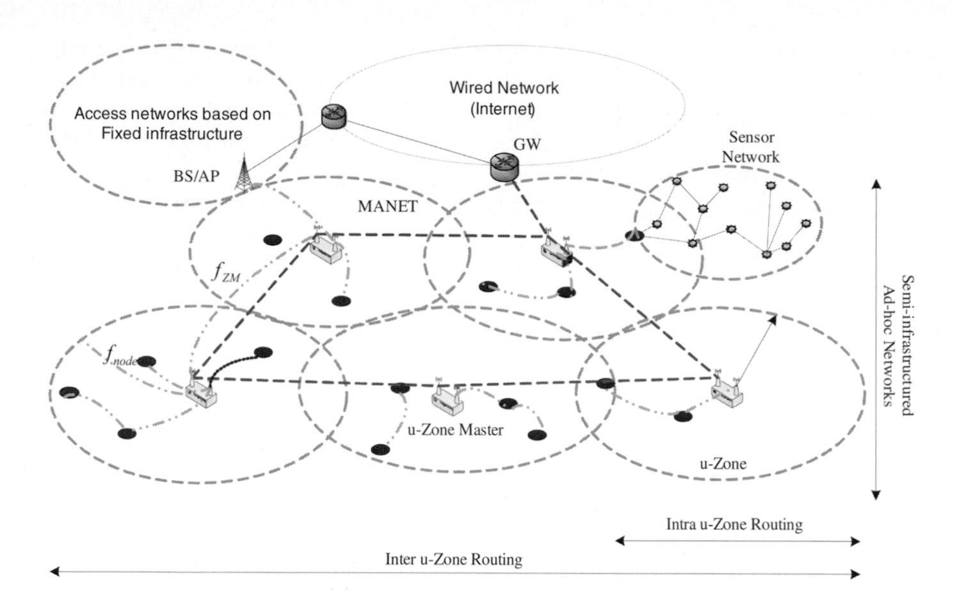

Fig. 2. u-Zone based hybrid MANET Architecture

of the u-ZM in which HRPC (our hybrid routing protocol) has been installed. To evaluate the HRPC module, we have implemented DYMO (Dynamic MANET On-demand Routing Protocol) [10] as a reactive routing protocol and ported OLSR (Optimized Link State Routing Protocol) [11] as a proactive routing protocol (see [12] in detail).

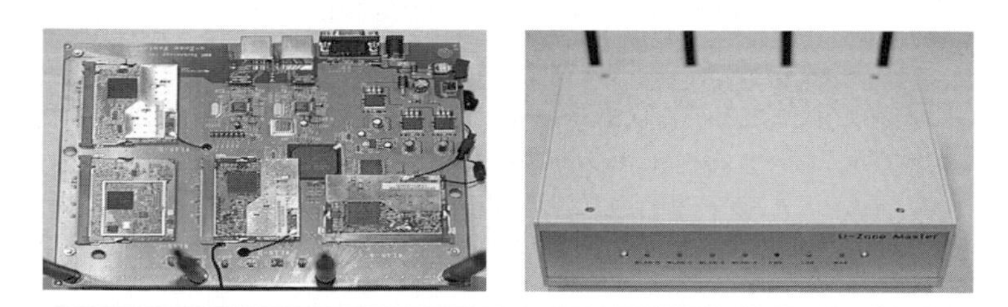

Samsung S3C2510 MCU based on ARM940T
Four different mini-PCI interfaces
: four 2.4Ghz/5Ghz dual-band wireless LAN interfaces (Atheros chipset)

Fig. 3. u-Zone Mater

We note that the proposed architecture is not a fixed infrastructure network and the u-ZM is different from a fixed network device such as AP or BS in other wireless networks. The u-ZM is an assistant device to reduce routing overheads of mobile nodes as does a cluster head in cluster-based approaches. But, the AP or BS is the main device; therefore, failure of such a device results in breakdown of the network. Besides, in AP/BS based network, all data must go through

Intra zone	Inter zone	WBB	Note
Proactive Routing Protocol	Proactive Routing Protocol	use	• Routing information of a u -Zone is periodically updated . • u-ZM manages inter u -Zone routing information . • Scalability problem
		disuse	• Similar to pure proactive routing protocol • No hierarchical u -Zone network gains • Scalability problem
Proactive Routing Protocol	Reactive Routing Protocol	use	• Best combination for delay sensitive applications • Insensitive to any changes in topology of other u -Zones
		disuse	• Same to clustering based routing protocols (ZRP, ZHLS,..) • Lots of hop counts in large network (bad performance) • Sensitive to changes in topology of other u -Zones
Reactive Routing Protocol	Proactive Routing Protocol	use	• Difficult to employ (because of high memory requirements) • u-ZM must figure out all routing information of nodes in the network proactively .
		disuse	• More of less infeasible to employ • Worst combination
Reactive Routing Protocol	Reactive Routing Protocol	use	• Least memory requirement for routing • Less control overhead than pure reactive routing protocol thanks to the hierarchical architecture
		disuse	• Similar to pure reactive routing protocol • No hierarchical u -Zone network gains • Less memory requirement for routing

Fig. 4. Hybrid Routing Strategies

the AP/BS even though a receiver is placed within the transmission coverage of a source. In the proposed network, however, if a source has routing information to the receiver then he can send data directly. Hence the proposed architecture is regarded as a semi-infrastructured MANET.

To enhance reliability, the u-ZM may be allowed to adjust his transmission power differently (i.e. different radio level). As shown in Fig. 3, u-ZM uses four different interfaces with various power level (e.g. f_{node} and f_{ZM} in Fig. 2). The signal power for transmitting data between u-ZMs should be higher than power for sending data from u-ZM to member nodes. In addition to the reliability gain, such an approach offers a way to achieve spatial reuse thus to enhance the performance of overall network. The wireless backbone (WBB) formed by linked u-ZMs provides robustness against a scenario where no boarder nodes, that play a role in relaying a packet between zones in ZRP, exist in an overlapping region. Hence, packets can be transmitted via the wireless backbone without aid of boarder nodes. If the wireless backbone is not used, traffic may concentrate on just a few nodes resulting in a long end-to-end delay due to congestion at the nodes and a high energy consumption of the nodes. Therefore, it may lead to network isolation and degradation of lifetime of networks.

3 Hybrid Routing Strategies

In this section, we introduce the hybrid routing strategies suited for the hierarchical u-Zone based network. The routing methodology is also hierarchically constituted by two levels: intra u-Zone routing and inter u-Zone routing according to the destination's location corresponding to the source (i.e. within the u-Zone or beyond the u-Zone).

Hybrid routing approach exploits the advantages of both proactive routing protocol and reactive routing protocol. Most of previously proposed hybrid routing protocols took only a single way into account. That is, they utilize a proactive approach inside a zone (i.e. intra zone routing) and a reactive approach beyond the zone (i.e. inter zone routing). In some cases, such a fixed combination may be inefficient to use.

On the other hand, we have more opportunities to select an appropriate combination of routing methodologies according to the requirements of applications or network environments. In our approach, four different combinations of flat routing protocols can be applied for building a hierarchical hybrid routing protocol. Fig. 4 presents such combinations and their properties in short. Here WBB is a direct link interconnected between u-ZMs to offer the back bone channel between adjacent u-Zones. The hierarchical architecture formed by the u-ZM and WBB enables an ordinary node to reduce computing power and communication overhead in both proactive and reactive routing protocol.

To maximize the using of u-ZM and WBB, u-ZM and nodes in the proposed approach perform both proactive and reactive routing functionalities in a different way.

- *Proactive routing protocol:* The u-ZM manages routing information of his u-Zone and then broadcasts it to member nodes periodically. Member nodes in the u-Zone are thus capable of communicating with each other directly by using such information. The information includes a list of the u-Zone member nodes and TTL value which is used to restrict a broadcasting region of control packets within a u-Zone. The main difference from the pure proactive routing protocol, only u-ZM updates link information of the u-Zone in our scheme.
- *Reactive routing protocol:* When a source has a packet to send, routing discovery procedure is activated. In general, the source broadcasts route request message toward the corresponding destination. In our architecture, on the other hand, such control overheads are less than others since the architecture restricts control packets broadcasting within a u-Zone and uses the WBB link to deliver the packets beyond the u-Zone.

4 Performance Evaluation

We evaluate H_2O routing strategies by using the Qualnet v3.8 simulator. Different routing combinations are compared in terms of the control overhead and end-to-end delay as the performance parameter. OLSR routing module of the

Qualnet is used as a proactive routing protocol of H_2O. We have implemented DYMO routing protocol into the simulator to use as a reactive routing protocol.

In simulation, we suppose that all nodes and u-ZM are equipped with IEEE 802.11 MAC protocol as the f_{node}. In addition, we assume that u-ZM has one more powerful wireless interface (i.e. f_{ZM}) to build WBB. The transmission coverage for mobile nodes is 150m in maximum distance. The overall simulation time is 50 seconds. During this time, there are 100 randomly established sessions; a selected source repeats 10 times of transmitting four UDP packets of 512 bytes in size per second.

Now, the simulation results of three different scenarios are given. We consider the u-Zone radius (i.e. the size of a u-Zone), network size and number of nodes as the parameters that impacts on the scalability. Additionally, we show the hierarchical gains of the proposed architecture and the efficiency of the H_2O routing strategies in comparison with both ZRP and flat routing protocols. In the descriptions, we commonly use the term of *'A-B'* routing combination, where *'A'* and *'B'* denote the type of routing protocol used for the intra u-Zone routing and the inter u-Zone routing protocol respectively. We exclude *Re-Pro* combination from the simulation due to the impractical property described in Fig. 4, where *Re-Pro* denotes the reactive-proactive routing combination.

4.1 Effects of the u-Zone Radius

The u-zone radius is defined as the number of hops from the u-ZM to a border node within a u-Zone. In this simulation, 100 nodes are uniformly placed at the network of 1000m×1000m in size. The size of overall network is fixed so that the increasing radius of u-Zone (i.e. from 1 to 5) results in the decreasing number of u-Zones (i.e. 9, 4, 3, 2, 1 respectively) and the increasing number of member nodes of a u-Zone.

Fig. 5 shows that the amount of control overhead and delay of hybrid routing approaches including ZRP are increased as the u-Zone radius is increased. In Fig. 5, we omitted the delay results of ZRP because it is too high to compare with ours. Also, both pure flat routing protocols are implicitly included in the results. When the radius meets 5, there is no inter u-Zone routing protocol since overall network is covered by a single u-Zone. Namely, either a proactive (in case of *Pro-Pro* and *Pro-Re* combinations) or a reactive routing protocol (in case of *Re-Re* combination) is only activated.

In H_2O routing strategies, the reactive-reactive (*Re-Re*) routing strategy is the most scalable combination with respect to the control overhead. But such a combination is the worst case in terms of delay, thus such a combination is not suited for time-sensitive applications. The proactive-proactive (*Pro-Pro*) and proactive-reactive (*Pro-Re*) combinations give the similar results in both cases. This is because the proposed hierarchical architecture (i.e. using u-ZM and WBB) allows member nodes to reduce inter u-Zone routing overheads and end-to-end delay. Namely, most of overheads are generated within a u-Zone. In particular, increasing u-Zone radius results in more retransmission of TC messages of OLSR hence smaller u-Zone radius gives more gains.

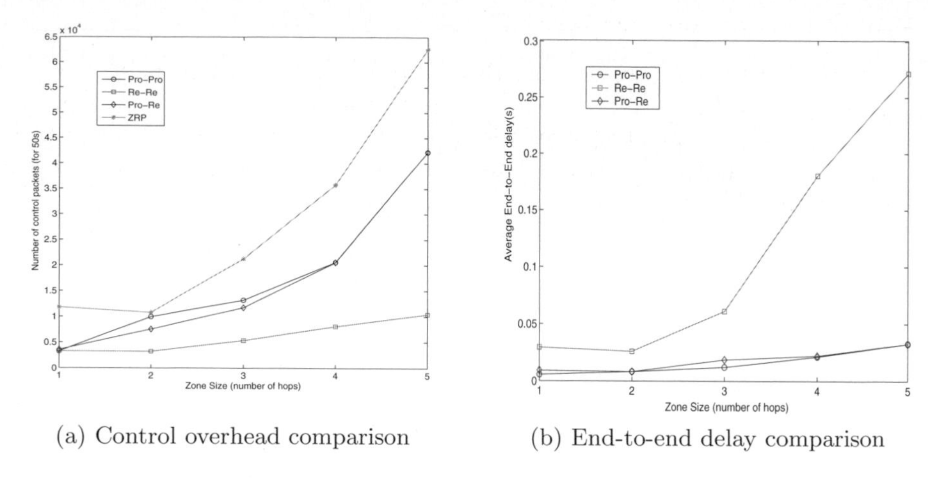

(a) Control overhead comparison (b) End-to-end delay comparison

Fig. 5. Simulation results in increasing u-Zone sizes

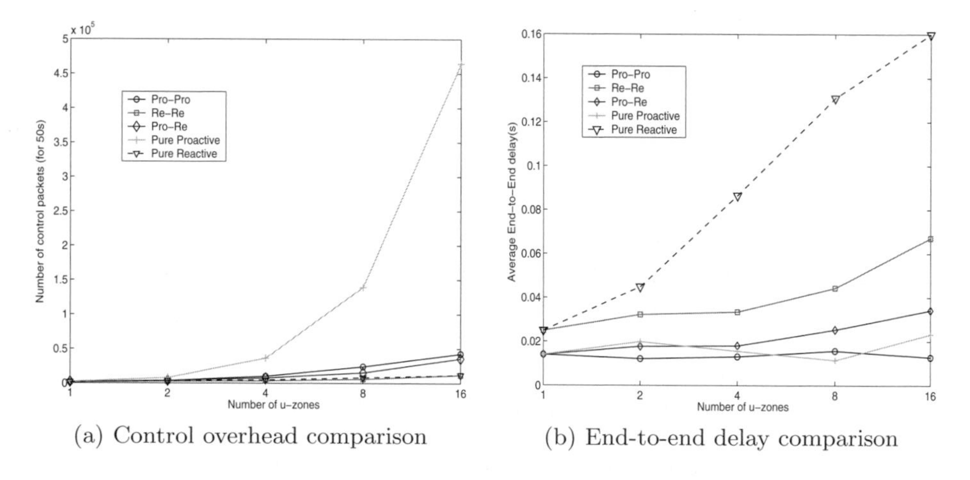

(a) Control overhead comparison (b) End-to-end delay comparison

Fig. 6. Simulation results in increasing network size

4.2 Effects of the Increasing Number of u-Zones

In this simulation, we evaluate the effects of the increasing number of u-Zones (i.e. growing size of overall network). All of u-Zones include twenty five member nodes (i.e. the same node density) and the radius is commonly two hops. Like the first simulation scenario, we compare each of routing combinations with respect to the control overhead and delay. The results are given in Fig. 6. We verified again that the proactive approach is better than the reactive approach from the aspect of delay, but worse in terms of the control overhead. Also, we observed that the hierarchical architecture gives us exponential gains as the scale goes up. Such results and their reasons are similar to those of the first scenario.

4.3 Effects of the Increasing Number of Nodes

In the final evaluation, we increase the number of nodes within the fixed size of network (i.e. 800m × 800m in size). The number of u-ZM in the network and its radius are four and two hops respectively as illustrated in Fig. 7, (a). We increase the number of nodes per u-Zone from 5 to 80 resulting in 20 to 320 nodes in the whole network.

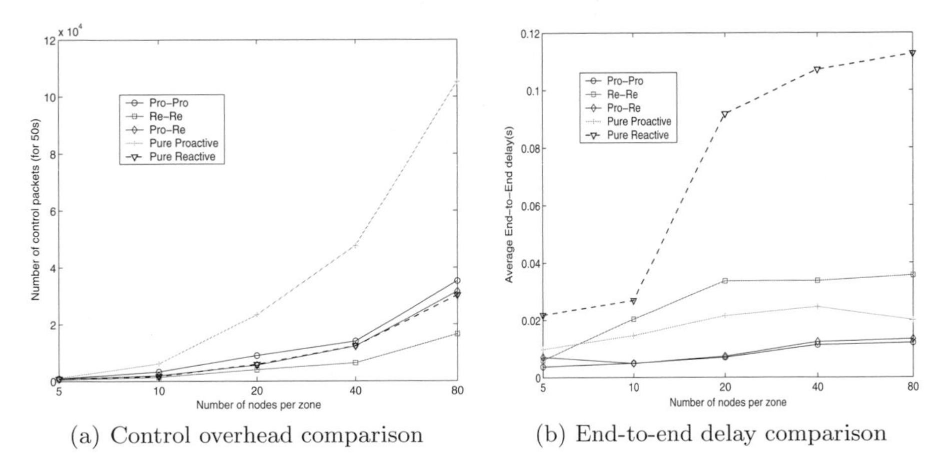

(a) Control overhead comparison (b) End-to-end delay comparison

Fig. 7. Simulation results in increasing number of nodes

In this simulation, we also observed that the performance gains come from the hierarchical architecture along with hybrid routing protocol. We have evaluated the packet delivery ratio to show the reliability of our approach. The results have shown that reactive-reactive routing combination is the best (no figure appeared because of length limit). This is due to the fact that the proactive routing protocol requires a bit more convergence time it takes to build routing table at all nodes. Packet losses may be occurred during the period.

5 Conclusion and Future Work

The fundamental question of this paper was how to enhance the scalability and reliability in wireless networks. Under the context, we proposed the u-Zone based network architecture and hybrid routing approaches. In particular, we showed that it is highly coupled with various requirements of the application (e.g. time sensitivity) to decide an optimal combination of routing protocols to build a hybrid routing protocol. In the simulation test, we have less considered the support of mobility under the assumption that MANET routing protocols can provide limited mobility (i.e. pedestrian nodes) with mobile nodes. For future work, to build a large MANET, we are required to find a solution to support high mobility. Also, we plan to enhance both radio level and packet level of u-Zone Master, thereby building a real test-bed in our campus.

Acknowledgement

This research is supported by the foundation of Ubiquitous Computing and Networking (UCN) Project, the Ministry of Information and Communication (MIC) 21st Century Frontier R&D Program in Korea.

References

1. Hong, X., Xu, K., Gerla, M.: Scalable Routing Protocols for Mobile Ad Hoc Networks. IEEE Network Magazine 16 (2002)
2. Royer, E.M., Barbara, S., Toh, C.K.: A Review of Current Routing Protocols for Ad hoc Mobile Wireless Networks. IEEE Personal Communications (1999)
3. Iwata, A., Chiang, C., Pei, G., Gerla, M.: Scalable Routing Strategies for Ad Hoc Wireless Networks. IEEE Journal on Selected Areas in Communications 17(8) (August 1999)
4. Lim, M.L., Yu, C.: Does Cluster Architecture Enhance Performance Scalability of Clustered Mobile Ad Hoc Networks? In: Proc. of the Int'l. Conf. on Wireless Networks (June 2004)
5. Lee, B., Yu, C., Moh, S.: Issues in Scalable Clustered Network Architecture for Mobile Ad Hoc Networks. The Mobile Computing Handbook. CRC Press, Boca Raton (2005)
6. Wu, H., Abouzeid, A.: Cluster-based Routing Overhead in Networks with Unreliable Nodes. In: Proc. of IEEE WCNC 2004 (March 2004)
7. Navid, N., Shiyi, W., Bonnet, C.: HARP: Hybrid Ad hoc Routing Protocol. In: Proc. of IST (2001)
8. Haas, Z.J., Pearlman, M.R.: The Zone Routing Protocol (ZRP) for Ad Hoc Networks. IETF Internet Draft (1998)
9. Joa-Ng, M., Lu, I.: A Peer-to-Peer Zone-Based Two-Level Link State Routing Protocol for Ad Hoc Wireless Networks. Journal of Communications and Networks 4(1) (March 2002)
10. Chakeres, I., Belding-Royer, E., Perkins, C.: Dynamic MANET On-demand (DYMO) Routing. IETF Internet Draft (Work in Progress) (2006)
11. OLSR daemon(olsrd), `http://www.olsr.org`
12. Kang, N., Yoo, S., Kim, Y., Jung, S., Hong, K.: Heterogeneous Routing Protocol Coordinator for Mobile Ad Hoc Networks. In: Youn, H.Y., Kim, M., Morikawa, H. (eds.) UCS 2006. LNCS, vol. 4239, pp. 384–397. Springer, Heidelberg (2006)

Anomaly Detection of Hostile Traffic Based on Network Traffic Distributions

Koohong Kang

Department of Information and Communications Engineering, Seowon University, 231, Mochung-dong, Cheongju, Chungbuk 361-742, South Korea
khkang@seowon.ac.kr

Abstract. Protecting network systems against novel attacks is a pressing problem. In this paper, we propose a new anomaly detection method based on inbound network traffic distributions. For this purpose, we first present the diverse distributions of TCP/IP protocol header fields at the border router of a real campus network, and then characterize the distributions when well-known denial-of-service (DoS) attacks are present. We show that the distributions give promising baselines for detecting network traffic anomalies. Moreover we introduce the concept of entropy to transform the obtained distribution into a metric of declaring anomaly. Our preliminary explorations indicate that the proposed method is effective at detecting several DoS attacks on the real network.

Keywords: Network Traffic Anomaly, Network Traffic Distribution, DoS, Entropy.

1 Introduction

1.1 Background and Related Work

Seamless and secure traffic streams through the Internet are becoming increasingly important today as corporations, public organizations, and even individuals are more heavily depending on the Internet for their daily activities. An important component of the Internet security is intrusion detection system (IDS). The main purpose of IDS is to identify suspicious or malicious activity, note activity that deviates from normal behavior, catalog and classify the activity, and if possible, respond to the activity[1].

Early IDS models were designed to support a single host such as a mail server or web server - host based IDSs. However more recent models examine activity on the network itself by monitoring the traffic crossing the network link - network based IDSs. Network based IDSs are classified as signature based or anomaly based. For example, Snort[2] uses rule files to detect signatures of known attacks, such as a specific string in the application payload. However anomaly based systems model normal network behavior, and then identifies the abrupt changes of network behavior. Most recent research topics mainly focus on anomaly detection because it has the advantage that no rules need to be written, and that it

T. Vazão, M.M. Freire, and I. Chong (Eds.): ICOIN 2007, LNCS 5200, pp. 781–790, 2008.

can detect novel attacks. Moreover, its position on computer security markets is getting top sooner or late because of zero time for the vendor to release a path and zero time for users to download and install the patch - zero day attack; that is, zero time between the discovery of the vulnerability and public knowledge of that vulnerability.

Network anomalies typically refer to circumstances when network operations deviate from normal network behavior. Network anomalies can arise due to various causes such as malfunctioning network devices, network overload, malicious DoS attacks, and network intrusions that disrupt the normal delivery of network service[3,4]. In this paper, we only consider security-related problems, such as DoS and network intrusions.

Thottan *et al.*[3] briefly review the commonly used methods for anomaly detection. But we simply classify them as two categories; one is the time series modeling approach, and the other is the statistical modeling approach.

The first approach attempts to build diverse traffic profiles in time series, such as symptom-specific feature vectors or just traffic volume[3,4,5]. These profiles are then categorized by time of day, day of week, and special days. When newly acquired data fails to fit within some confidence interval of the developed profiles then an anomaly is declared. In [5], the Holt-Winters forecasting algorithm is used to predicting the values of a time series one time step into the future. When an observed value of sequence of observed values is too deviant from the predicted values then an anomaly is declared.

The second approach learns a statistical model of normal network traffic, and flags deviations from this model [6,7]. Models are usually based on the distribution of source and destination addresses and ports per transaction. In [6], packet header anomaly detector (PHAD) learns the normal range of values for 33 fields of the Ethernet, IP, TCP, UDP, and ICMP protocols. PHAD uses the rate of anomalies during training to estimate the probability of an anomaly while in detection mode. If a packet field is observed n times with r distinct values, there must have been r anomalies during the training period. If this rate continues, the probability that the next observation will be anomalous is approximated by r/n.

Although there are a number of variations of these two approaches, we can note some general and important shortcomings of them. In time series modeling approach, we have to big tune for many parameters to characterize the statistical behavior of abnormal traffic patterns. For example, we have to choose the values of six parameters in [5]. In statistical modeling approach, we have to train attack-free training data which will not always be available in a practical system [6].

1.2 Motivation and Contribution

Figure 1 shows a campus network used throughout as an experimental network in this paper, where the network consists of a duplicated backbone network and multiple access networks. This style of network infra-structure is also very typical for modern enterprise networks. We note that there are 17 cass C (/24) scale sub-networks in Fig. 1.

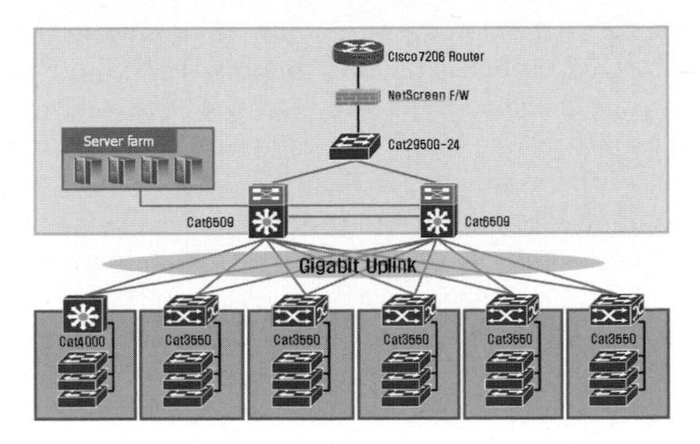

Fig. 1. A typical infra-structure of campus network

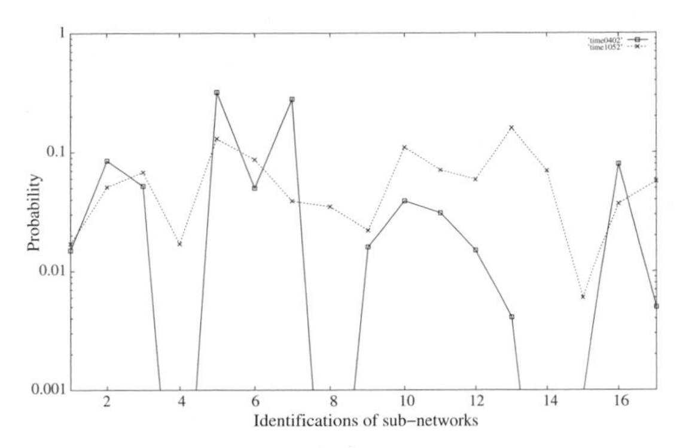

Fig. 2. Histogram of the destination sub-networks of ingress IP packets at the border router (solid line: 04:02AM, dotted line: 10:52AM)

Figure 2 shows the histograms of destination sub-networks of ingress IP packets at the border router at 04:02AM and 10:52AM respectively, where probabilities are determined based on counting IP packets for 5 minutes. Just as expected, we observe that the histogram at working hour 10:51AM is obviously different from the one at 04:02AM. Thus the knowledge of histogram at a particular time instant can be an important clue to detect network traffic anomaly. Moreover we also presume that the distributions of other values in the protocol header fields might be very different between these time instants. Although some published methods [6,7] use the values of protocol header fields, no one uses the distributions to detect anomalies directly. In order to declare the anomalies we fully rely on the diverse traffic distributions which can be determined in real time.

Our contribution is two-fold. First, we show the diverse traffic distributions about 5 fields - packet length, flags, time to live (TTL), protocol, and destination IP address - of the IP protocol, and destination port address of the TCP/UDP protocols, and flags of the TCP protocol by monitoring the ingress traffic of a real campus network border router. This kind of distributions gives some insights to the security and traffic engineering research areas. Second, we demonstrate the potential to apply network traffic distributions to the problem of network anomaly detection. In particular, we introduce the concept of entropy because it is difficult to use the distribution as an anomaly detection metric directly. We show through comprehensive on-line tests that the proposed anomaly detection method can detect 100% of the well known DoS attacks in a real network.

The paper is organized as follows. In the next section, we describe the experimental scenarios, and discuss characteristics of the ingress traffic distributions of campus network, and then show the substantially different traffic distributions compare to the normal ones when hostile traffic is mixed. Section 3 describes the entropy concept of traffic distribution and illustrates some results showing that the application of entropy concept is in effective detecting network traffic anomalies. In Section 4, we conclude with future work.

2 Ingress Traffic Distributions of Campus Network

2.1 Characteristic of Campus Network Traffic

Table 1 explains 14 different distribution categories under study in this paper. In order to avoid the traffic filtering function of firewall, all of the distributions are obtained at the border router whilst the virtual probing points of local networks are located at the corresponding sub-networks.

In Table 1 (1) and (2), we monitor the protocol field of ingress IP packets to get the IP protocol distributions for global and each sub-network. We also monitor destination IP address of the packets to get the destination sub-network distribution for Table 1 (3). In case of port address distribution of Table 1 (4), we choose three events; (i) well-known ports (ftp, telnet, SMTP, DNS, HTTP, RPC, and SNMP), (ii) less than 1024 except for the well-known ports, and (iii) the other port addresses. In Table 1 (5) and (6), we obtain conditional distributions given well-known services (ftp, telnet, and HTTP) and IP protocols (tcp, udp, and icmp). In Table 1 (7) - (14), we obtain the distributions of packet length (events: less than 64 bytes, 64 - 128 bytes, 129 - 512 bytes, greater than

Table 1. Distributions for characterizing campus network traffic (Global: the whole ingress traffic, Local: the traffic bound for each sub-network)

No	Distribution	Scope
1,2	IP protocol	Global, Local
3	Destination sub-network	Global
4	Port address	Global
5	Destination sub-network given each well-known service	Global
6	Destination sub-network given each IP protocol	GLobal
7,8	Packet length	Global, Local
9,10	TTL value	Global, Local
11,12	Fragmented packet	Global, Local
13,14	TCP SYN packet	Global, Local

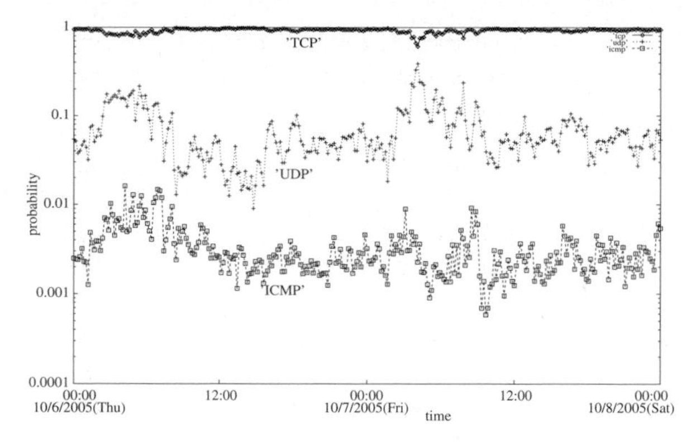

Fig. 3. The IP protocol distributions for two days (from Oct.6 to Oct.7, 2005)

512 bytes), TTL value (events: less than 32, 32 - 64, 65 - 128, greater than 128), fragment packet, and TCP SYN packet for global and each sub-network, respectively. We note that all observations are made at fixed 5 minutes duration.

Due to the space limitation, we only show two result diagrams, but you can refer to [9] for more additional results of Table 1. Figure 3 shows the IP protocol distributions for two days from Oct.6 (Thu.) to Oct.7 (Fri.) 2005. The graph illustrates a slight daily pattern, where occupied with most TCP protocol. Although most people think of TCP protocol occupies greater than 85% of the total traffic, it is only valid for working time. Figure 4 shows the destination sub-network packet distributions of incoming packets. Like presuming it in Fig. 2, all of sub-networks are busy for working hours from 08:00 - 20:00.

2.2 Traffic Distribution When Malicious Traffic Is Mixed

In this subsection, we observe the changes of traffic distributions when we attack our campus network with the real well-known DoS attacks - Jolt, WinNuke, Synk4, Teardrop and UDP flooder attacks shown in Table 2 - using datapool

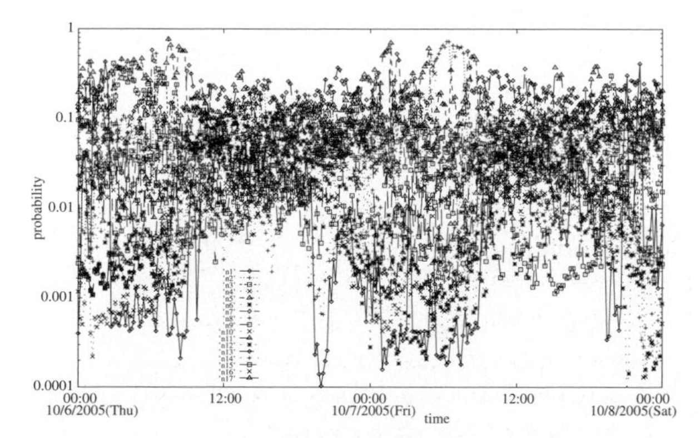

Fig. 4. The destination sub-network packet distributions for two days (from Oct.6 to Oct.7, 2005)

Table 2. DoS attacks

DoS Attack	Feature
Ping of Death(Jolt)[12]	Send very large, fragmented ICMP packets
WinNuke[14]	Send "junk" information to TCP port 139(NetBIOS)
Synk4[16]	TCP SYN flooder attack
Teardrop[13]	Buffer overflow vulnerability of overlapping IP fragments
UDP flooder[11]	UDP flooder attack

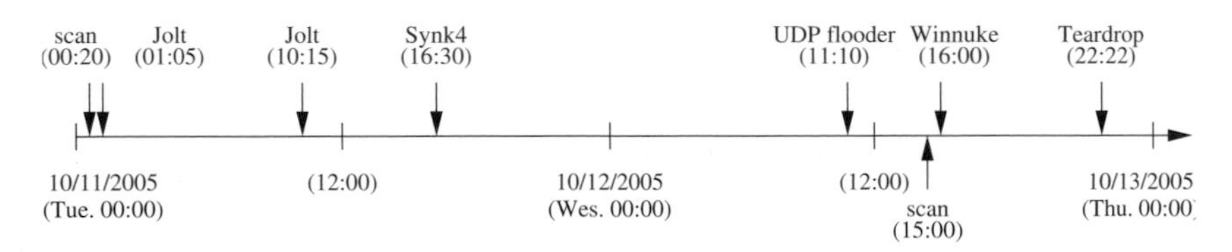

Fig. 5. Generation times of well-known DoS Attacks

version 3.3[10] from external network. These DoS attacks are generated as shown in Fig. 5.

As shown in Fig. 6, the IP protocol distributions changed abrruptly at the instants of Jolt, Synk4, and UDP flooder attacks. However it is difficult to notice any remarkable sign of changes at instants of WinNuke and Teardrop attacks because they do not generate a large volume of malicious traffic compared to the previous attacks.

Figure 7 illustrates the TTL distributions under the attack scenario of Fig. 5, where we can observe the symptoms of WinNuke and Teardrop attacks. Actually we can find these symptoms from the other different distributions explained in Table 1. In particular, the fragmented packet distribution gives the certain indication of Teardrop attack.

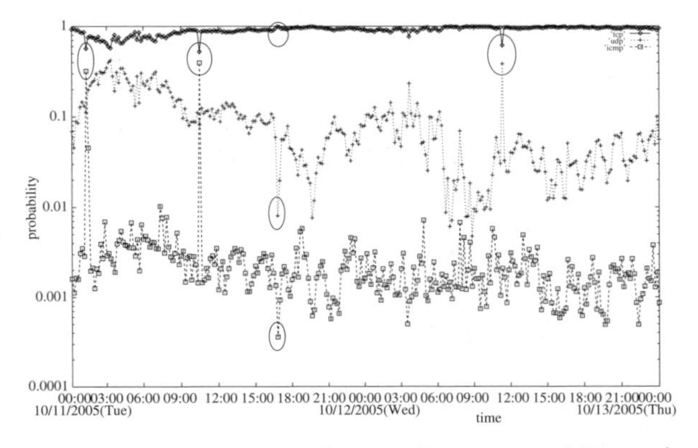

Fig. 6. IP protocol distributions under the attack scenario of Fig. 5 (cross points line: UDP, circle points line: TCP, and rectangular points line: ICMP)

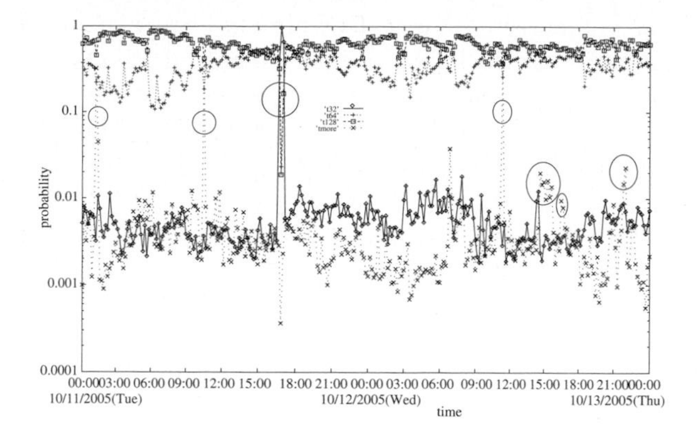

Fig. 7. TTL value distributions under the attack scenario of Fig. 5 (t32: less than 33 bytes, t64: 33 - 64 bytes, t128: 64 - 128 bytes, tmore: greater than 128 bytes)

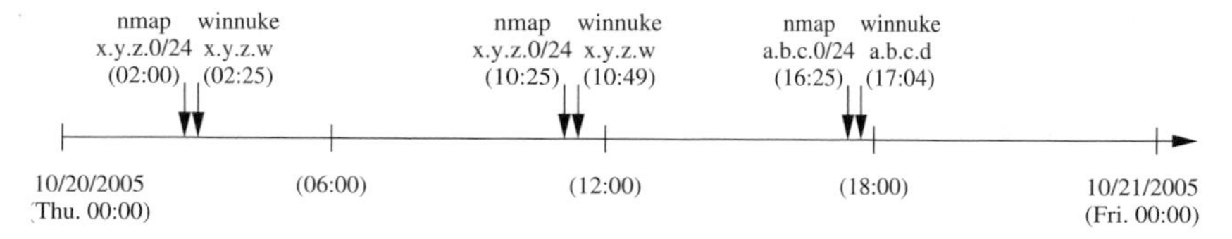

Fig. 8. Generation times of WinNuke attack for a day (Oct.20, 2005)

In order to identify the effect of WinNuke attack, we deliver only WinNuke attack against two different sub-networks for a day as shown in Fig. 8, where sub-network x.y.z.0/24 is for server farm.

3 Anomaly Detection Based on Entropy

Although we get and scrutinize more than 28 traffic distributions for the previous experiments, we show only some limited results. However we believe the presented figures are enough to explain that the traffic distributions are really good base for detecting network traffic anomalies. Now we need a way to transform the distributions into a decision metric for declaring anomaly. For this purpose, we introduce and apply the entropy concept.

Entropy is an information-theoretic measure of the amount of uncertainty in a variable[8]: that is, the entropy of x is

$$H(x) = -\sum_{i=1}^{n} p(X = x_i) \lg p(X = x_i) \ , \quad (1)$$

where, "lg x" is the base 2 logarithm of x. The conditional entropy of X given Y.

$$H(X|Y) = -\sum_{i=1}^{n} p(X = x_i|Y = y_i) \lg p(X = x_i|Y = y_i) \ . \quad (2)$$

Figure 9 shows the entropy of Fig. 6, where we can find abrupt changes of entropy at the instants of attack generations except for WinNuke and Teardrop attacks. If we consider the conditional entropy, we can identify more detail properties of the attacks. For example, it is obvious that the third attack "synk4" is using TCP protocol from Fig. 10. From Fig. 10 we also slightly detect scan attacks at 00:20 AM on Tue. and 15:00 on Wed. respectively, where the entropy is very flat because the nmap[15] rapidly scans the whole hosts using TCP protocol. Hence we can conclude that it is more effective for declaring anomalies to use the conditional entropies of different kinds instead of single entropy.

Figure 11 shows the entropies of port address distributions of two sub-networks under the scenario explained in Fig. 8, where we can detect the WinNuke attacks even if there are some unidentified anomalies.

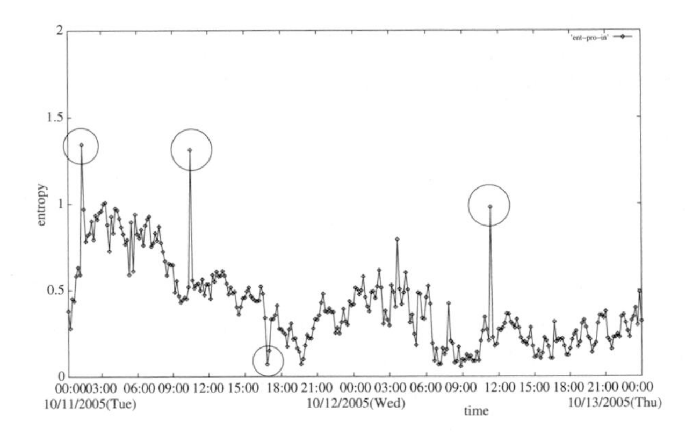

Fig. 9. Entropy of the protocol distributions of Fig. 6

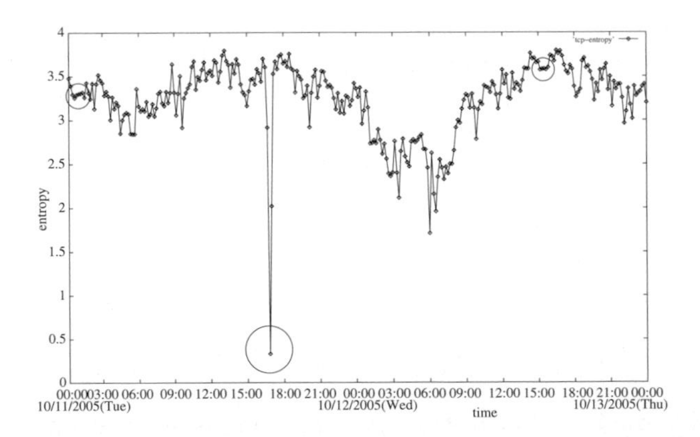

Fig. 10. Conditional entropy of destination sub-networks given TCP protocol under the attack scenario of Fig. 5

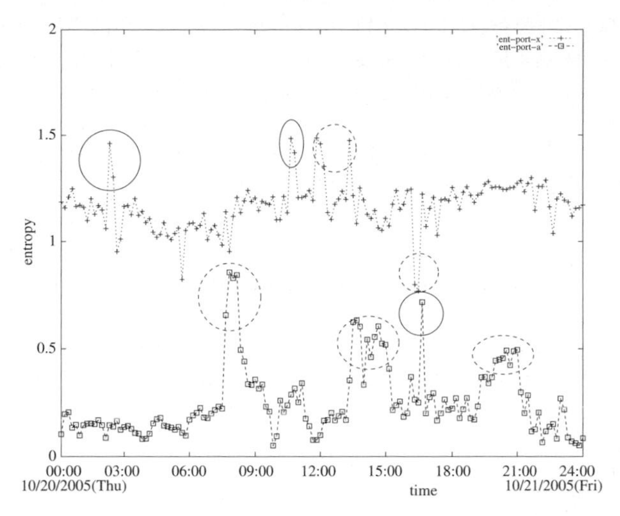

Fig. 11. Entropy of IP port address distributions of sub-networks under the attack scenario of Fig. 8 (cross points line: x.y.z/24, rectangular points line: a.b.c.0/24, solid circles: WinNuke attack, and dotted line circle: unidentified anomalies)

4 Conclusion and Future Work

In this paper, we presented the traffic distributions when well-known DoS attacks are delivered into a real campus network, and discussed that the distributions gave promising baselines for detecting network traffic anomalies. Moreover we demonstrated the potential to apply the entropies of distributions to the problem of network traffic anomaly detection. Our preliminary explorations indicated that our entropy-based detection method is highly accurate to detect the well-known DoS attacks on the real network.

Two related issues for future research are how to model the entropy into a time series with trend and seasonal components as in [5], and how to integrate the diverse entropies and distributions to improve the performance of anomaly detection such as false positives and negatives.

References

1. Conklin, W.A., Williams, D., White, G.B., Davis, R.L., Cothren, C.: Principles of Computer Security: Security+ and Beyond. McGraw-Hill, Burr Ridge Illinois (2004)
2. Rosech, M.: Snort Lightweight Intrusion Detection for Networks. In: Proc. USENIX LISA 1999 (1999)
3. Thottan, M., Ji, C.: Anomaly Detection in IP Networks. IEEE Trans. on Signal Processing 51(8) (2003)
4. Barford, P., Plonka, D.: Characteristics of Network Traffic Flow Anomalies. In: Proc. Of the ACM Internet Measurement Workshop (2001)

5. Brutlag, J.D.: Aberrant Behavior Detection in Time Series for Network Monitoring. In: Proc. USENIX LISA XIV (2000)
6. Mahoney, M.V.: Network Traffic Anomaly Detection Based on Packet Bytes. In: SAC 2003, Melbourne, Florida (2003)
7. Anderson, D., Terea, F.L., Harold, J., Ann, T., Alfonso, V.: Detecting unusual program behavior using the statistical component of the Next-generation Intrusion Detection Expert System (NIDES), Computer Science Laboratory SRI-CSL 95-06 (1995)
8. Bishop, M.: Computer Security: Art and Science. Addison-Wesley, Reading (2003)
9. Kang, K.: A Study on Network Anomaly Detections Based on Baseline and Anomaly Traffic Modeling, ETRI Final Report of Collaborative Research (2004)
10. Spender: datapool3.3, `http://packetstorm.linuxsecurity.com/DoS/indexsize.html`
11. www.cert.org: CERT Advisory CA-1996-01 UDP Port Denial-of-Service Attack, `http://www.cert.org/advisories/CA-1996-01.html`
12. www.cert.org: CERT Advisory CA-1996-26 Denial-of-Service Attack via ping, `http://www.cert.org/advisories/CA-1996-01.html`
13. www.cert.org: CERT Advisory CA-1996-01 IP Denial-of-Service Attacks, `http://www.cert.org/advisories/CA-1996-01.html`
14. www.nac.net: The WinNuke Relief Page, `http://www.users.nac.net/splat/winnuke/`
15. Wolfgang, M.: Hot discovery with nmap, `http://www.rootsecure.net/content/downloads/pdf/nmap_host_discovery.pdf`
16. Zakath: Syn Flooder, `http://packetstorm.linuxsecurity.com/Exploit_code_Archive/synk4.c`

Investigation of Secure Media Streaming over Wireless Access Network

Binod Vaidya[1], SangDuck Lee[2], JongWoo Kim[2], and SeungJo Han[2,*]

[1] Dept. of Electronics & Computer Eng., Tribhuvan Univ., Nepal
bnvaidya@gmail.com
[2] Dept. of Information & Communication Eng., Chosun Univ., Korea
dandylsd@hanmail.net, mmm@7.co.kr, sjbhan@chosun.ac.kr

Abstract. With the popularity of wireless networks, demand for multimedia services also rises rapidly. However, efficient multimedia services are still challenging research problem due to user mobility, limited resources in wireless devices and expensive radio bandwidth. To implement multimedia services over wireless network, IP header compression scheme can be used for saving bandwidth. In this paper, we present an efficient solution for header compression, which is modified form of ECRTP. It shows an architectural framework adopting modified ECRTP along with SRTP. We have conducted simulation to analyze the effects of different header compression techniques along with SRTP while delivering real-time services to wireless access networks.

1 Introduction

Wireless IP networks are becoming more popular and the demand for multimedia services in these networks rises with the number of their implementations. However, efficient services in these multimedia systems are open and challenging research problem due to user mobility, limited resources in wireless devices and expensive radio bandwidth.

Provisioning of high network capacity is a critical point for wireless access networks. With continuously growing device capabilities, users demand services similar to those accessible on wired networks and the services requested by users are extending and become more sophisticated. For wireless networks with high bit error rates (BER) and high latency, it is difficult to attain those high bandwidths required. Since in many multimedia applications, payload of IP packet is almost of same size or even smaller than the header, it is possible to compress those headers and thus save bandwidth and use expensive resources efficiently.

In this paper, we present an efficient solution for header compression. Modified enhanced header compression scheme is used to carry IP header compressed packets in wireless access network while securing the voice information at the transport layer by using the secure RTP.

* Corresponding author.

T. Vazão, M.M. Freire, and I. Chong (Eds.): ICOIN 2007, LNCS 5200, pp. 791–800, 2008.

2 Securing Real-Time Transport Protocol

Multimedia applications often use RTP, UDP, and IP as protocols. Real-time transport protocol (RTP) [1] is an IP-based protocol providing support for the transport of real-time data such as video and audio streams. RTP provides end-to-end delivery services for data with real-time characteristics. However RTP itself does not provide all of the functionality required for the transport of data and, therefore, applications usually run it on top of a transport protocol such as UDP. RTP is designed to work in conjunction with a control protocol, Real Time Control Protocol (RTCP), to get feedback on quality of data transmission and information about participants in the on-going and to provide minimal control over the delivery of the data. [2]

The Secure Real time Transport Protocol (SRTP) [3] has been designed to create an efficient security solution for the RTP, which would work in constrained environments. SRTP provides confidentiality, authenticity, integrity, and replay protection for the RTP and RTCP packets, providing all the important elements to secure a media stream.

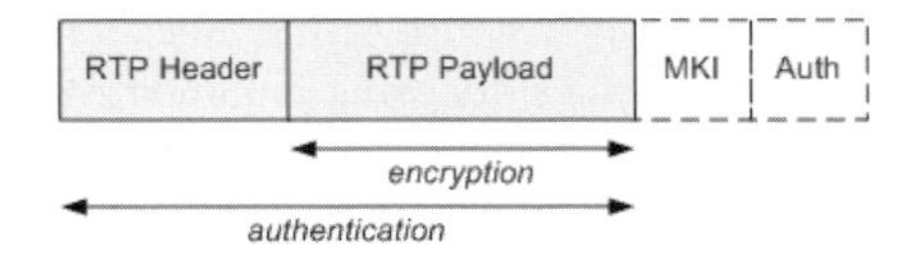

Fig. 1. Overview of the SRTP packet

Fig. 1 shows overview of the SRTP packet. SRTP supports the AES [4] algorithm in a stream cipher mode for encryption and HMAC-SHA1 [5] for the message authentication. SRTP encrypts only the data component (payload) of a voice packet and does not use additional encryption headers. SRTP can be used in conjunction with IP header compression and compressed RTP without packet manipulation with little or no effect on quality of service (QoS).

3 Compression Techniques

In many multimedia applications such as Voice over IP (VoIP), messaging etc, the payload of the IP packet is almost of the same size or even smaller than the header. Over the end-to-end connection, comprised of multiple hops, these protocol headers are extremely important but over just one link (hop-to-hop) these headers serve no useful purpose. It is possible to compress those headers, thus save the bandwidth and use the expensive resources efficiently. IP header compression [6] also provides other important benefits, such as reduction in packet loss and improved interactive response time. IP header compression schemes have always been an important part of saving bandwidth over bandwidth limited links. Header compression techniques need to be robustness with aspect to packet loss, and to misidentified streams.

3.1 Compressed Real-Time Transport Protocol

RTP header compression (CRTP) [7] was designed to reduce the header overhead of IP/UDP/RTP datagram by compressing the three headers. IP/UDP/RTP headers are compressed to 2-4 bytes most of the time. CRTP was designed for reliable point to point links with short delays. For lossy links and long round trip delays, CRTP does not perform well. After a single lost packet several sequential packets are lost within the round trip time. Thus, CRTP is not suitable for wireless links, which have typically a very high and variable BER.

CRTP is designed to compress IP/UDP/RTP flows. CRTP uses four packets formats: full header, Compressed UDP, Compressed RTP and context state.

After a full header packet has been sent to establish the context, the transition to Compressed RTP packets may occur. Each Compressed RTP packet indicates that the decompressor may predict the headers of the next packet on the basis of the stored context. Compressed RTP packets may update that context, allowing for common changes in the headers to be communicated without full header packet being sent. The format of a Compressed RTP packet is shown in Fig 2.

3.2 Enhanced CRTP

The Enhanced CRTP (ECRTP) [8] for links with high delay, packet loss and reordering feature includes modifications and enhancements to CRTP to achieve robust operation over unreliable point-to-point links. Thus ECRTP was developed to overcome the problems of CRTP as CRTP does not perform well over links with long round trip time that lose and reorder packets.

ECRTP extends CRTP by repeating context updates and by sending absolute values along with delta values when encoding monotonically increasing header fields to increase robustness. It inserts a header checksum when UDP checksum is missing, to improve error recovery and fail checks for the compression.

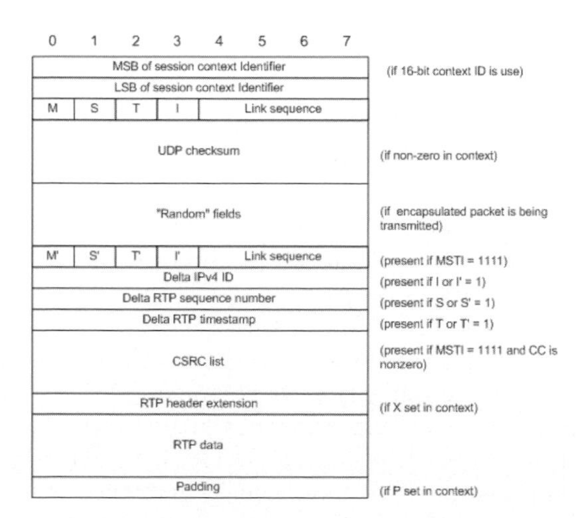

Fig. 2. Compressed RTP Packet

The packet format Full Header had some changes. The first two length-fields in the IP/UDP/RTP header and possible encapsulating headers are changed in the ECRTP compressor. The fields are used to send information about the flow to the decompressor.

In CRTP, the compressor sends a context identifier, a sequence number and a generation number. In ECRTP, the same fields are sent but the formats of the fields are changed. The C bit is included in the length fields indicating the new header checksum, replacing the missing UDP checksum over the compressed link. A check for a zero UDP checksum when parsing through the whole header must be done.

3.3 Modification in ECRTP

In order to reduce header overhead, the packet format of ECRTP header compression scheme is modified.

In all Compressed RTP, ECRTP packets have 2 bytes of either the UDP checksum or the compressor inserted Header Checksum. The average header size can be reduced if we send these checksums in some packets only. Robustness will be achieved if there is some way of conveying a correct decompression of these packets to the compressor.

For instance, if it is chosen to send the checksum only thrice for every 16 packets, then, assuming at least one of these packets is acknowledged, it can be expected the average header size to drop by about 1.6 bytes, given the fact that Compressed RTP packets are sent most often. Furthermore, this clearly implies a reduction in implementation complexity.

The Compressed RTP header includes the whole IP/UDP/RTP. Compressed RTP with individual RTP fields is shown in Fig. 3. With respect to the original packet structure of Compressed RTP, the T bit is replaced by the C bit, which

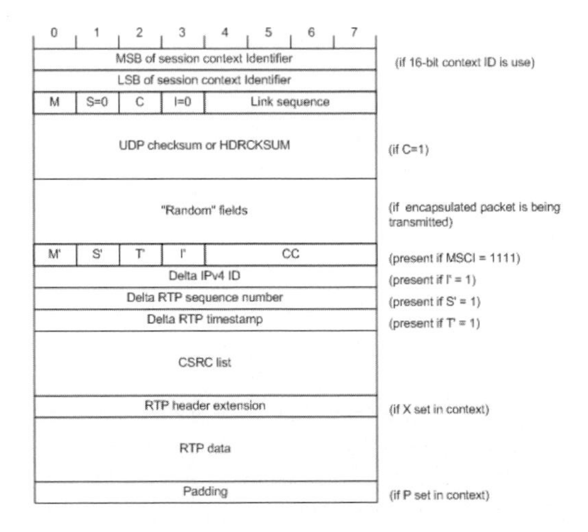

Fig. 3. Compressed RTP Packet for modified ECRTP

represents whether or not a Header Checksum is included in this packet. The S and I bits are set to 0. Hence, it is distinguishable from the Compressed UDP F=1 packet, where the corresponding bits are I and T, at least one of which, is 1. It is distinguishable from Compressed UDP F=0, because the corresponding bits are I and dI, both of which are 1, since the packet is a refresh packet.

4 Related Works

Voice communication in wireless mobile ad hoc network is challenging. It has been analyzed the impact of CRTP performance over cellular environment using real-time traffic such IP telephony and shown that CRTP does not cope with packet loss very well. [9] Enhanced CRTP mechanism improves the header compression performance, especially for highly error?prone links and long round trip times. [10]

The authors showed that ECRTP uses local retransmissions to more efficiently recover from wireless link errors compared with RTP and CRTP in an error-prone and bandwidth limited wireless network. [11]

The authors have shown that the packet error rate performance of ECRTP and robust header compression (ROHC) is similar while the local repair mechanism in ROHC is not considered. [12]

The authors presented a proposition to secure VoIP packets drawing inspiration from the existing voice security solutions. [13]

In this paper, we have presented the efficient IP header compression technique to deliver multimedia traffic in wireless access network and investigated performance of different header compression schemes while the real-time traffic is secured by using SRTP.

5 Performance Evaluations

5.1 Overviews

SRTP has been designed for the network, where bandwidth, delay, needed computational resources, and transmission errors have been critical factors taken into account. To allow improvement in bandwidth saving, SRTP does not obstruct header compression of the RTP packet; that is, it is possible to apply header compression of the whole IP/UDP/RTP header while using SRTP. In order not to obstruct header compression, SRTP does not encrypt the IP/UDP/RTP headers. Thus SRTP does not provide any protection of the IP/UDP/ RTP headers, and protects only the RTP payload.

As mentioned earlier, a voice payload is in the order of 33 bytes and the headers (IPv4, UDP, and RTP) are together 40 bytes. SRTP adds no new header overhead above the existing 40 bytes (IPv4) of an ordinary RTP packet. This overhead can usually be compressed to some degree, depending on the compression technique in use. Hence, applying header compression technique such as ECRTP is important to reduce the bandwidth consumption over the wireless links.

In this paper, the effective use of modified ECRTP in securely delivering voice traffic is as a main goal. Thus, main intention of this investigation is to show the performance evaluation of different header compressions in wireless access network while using SRTP to protect voice packets at transport layer.

5.2 Simulation Setup

In order to validate the conceived network architectural model, we have simulated a scenario that includes IPv4 audio streaming to the wireless access network over public IP network while using SRTP for protection of real-time traffic at transport layer. We have used OPNET Modeler, [14] which a discrete event-driven simulator tool capable of modeling both wireless and wireline network.

Fig. 4 shows the conceived system which comprises of a service provider, IP backbone network and wireless access networks. The service provider consists of audio streaming servers. And the wireless access network consists of 10 wireless clients using wireless IEEE 802.11b devices. In our experimental analysis, we have selected G.729 as the voice codec. And certain assumptions were also made about the voice characteristics. Most fields were assumed to remain constant. Silence suppression was assumed - so, the RTP Timestamp will jump by a constant amount, except at the end of a silence interval. Speech is bursty, with talk-spurts followed by silence periods. No packets are sent during these silence intervals. We have considered the IP public network ie. Internet with 5% packet discard ratio and average packet latency of 0.1 sec.

As per IETF RFC specification, SRTP was implemented in OPNET Modeler. The encryption and authentication throughputs were obtained from Crypto++ 5.2.1 Benchmarks [15] with computer configuration: Pentium 4 with processor speed - 2.1GHz and operating system used Window XP SP1, which were used within the model to compute authentication and encryption delay. For SRTP,

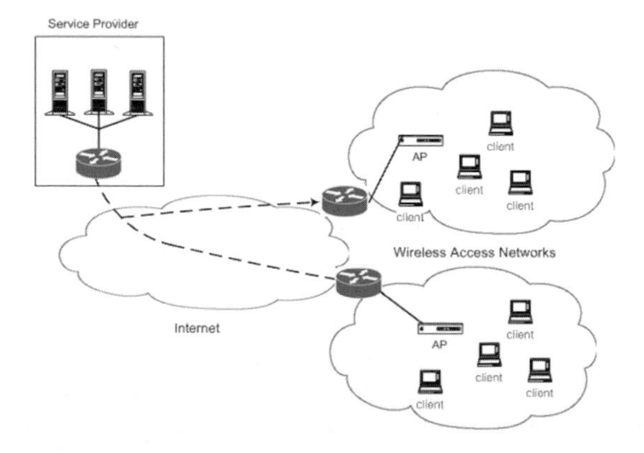

Fig. 4. System architecture

AES (128) CBC as encryption algorithm and HMAC-SHA1 as authentication algorithm were considered. Accordingly the throughputs of AES-CBC and SHA1 are 6.93 Mbps and 8.49 Mbps respectively.

In the simulation, we have conducted series of experiments in turn, the results obtained, in terms of performance metrics such as packet loss rate (PLR), average header size and probability loss in context as a function of BER and decompression (reconstruction) error rate as a function of PLR have been investigated.

5.3 Simulation Results

In order to analyze simulation results for various header compression schemes, we have conducted performance comparisons between CRTP, ECRTP and modified ECRTP (mECRTP) header compression techniques in terms of packet loss rate and decompression error rate whereas between ECRTP and mECRTP schemes in terms of average header size and probability loss in context.

Fig. 5 shows the packet loss rate as a function of BER for CRTP, ECRTP and mECRTP header compression schemes while using SRTP. As shown in Fig. 5, the packet loss rate in CRTP scheme is very high and the packet loss rates in ECRTP scheme as well as in mECRTP scheme have been significantly reduced in compare to CRTP scheme. The lowest packet loss rate can be observed in case of mECRTP scheme. It can be seen that at a BER of 10^{-3}, the packet loss rate for ECRTP scheme is slightly higher than that for mECRTP scheme.

Fig. 6 shows the decompression (reconstruction) error rate during packet loss for CRTP, ECRTP and mECRTP header compression techniques while using SRTP. It can be seen that decompression error rate in CRTP scheme significantly increases with the increase in packet loss rate. Whereas for header compression schemes ECRTP and mECRTP, the error rate is considerably low for the all the packet loss rates.

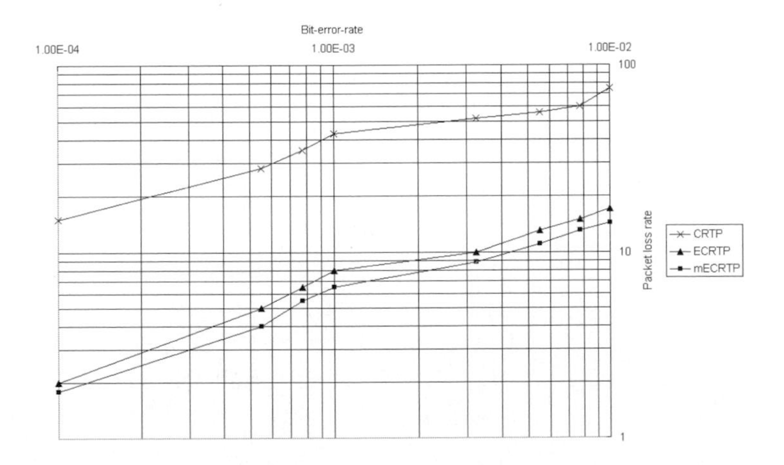

Fig. 5. Packet loss rate vs BER

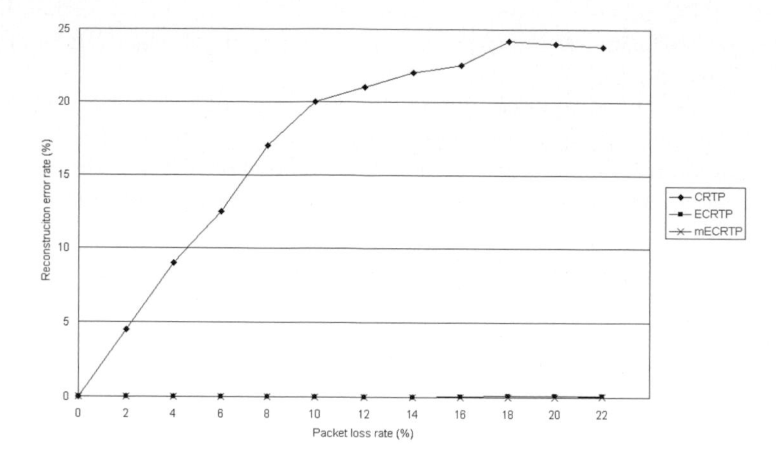

Fig. 6. Error rate vs Packet loss rate

Fig. 7 shows the average header size plotted against BER for ECRTP and mECRTP header compression schemes while using SRTP. The difference between ECRTP and mECRTP is mainly that the in latter case, checksums are sent only in certain packets. Hence, mECRTP introduces significantly less header overhead than ECRTP in a given environment. For BER of 10^{-3}, the average header size of mECRTP drops by about one byte than that of ECRTP.

Fig. 8 shows the probability of loss in context as a function of BER with ECRTP and mECRTP schemes while using SRTP. The result shows that in case of mECRTP compression scheme, it can be seen that there is a slight increase in the context loss rate.

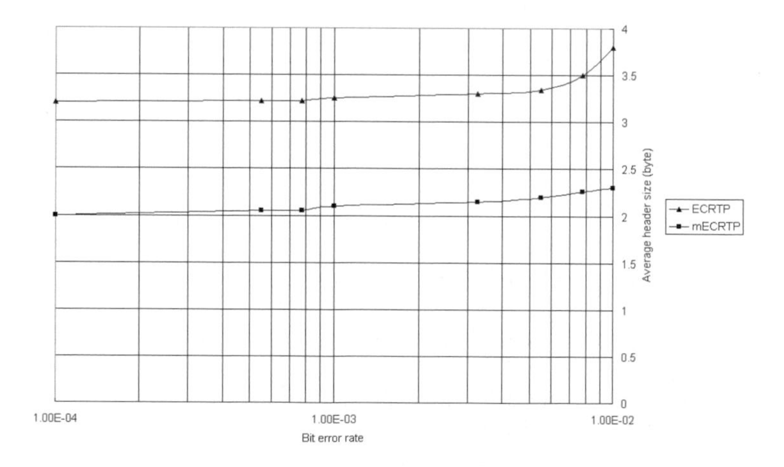

Fig. 7. Average header size vs BER

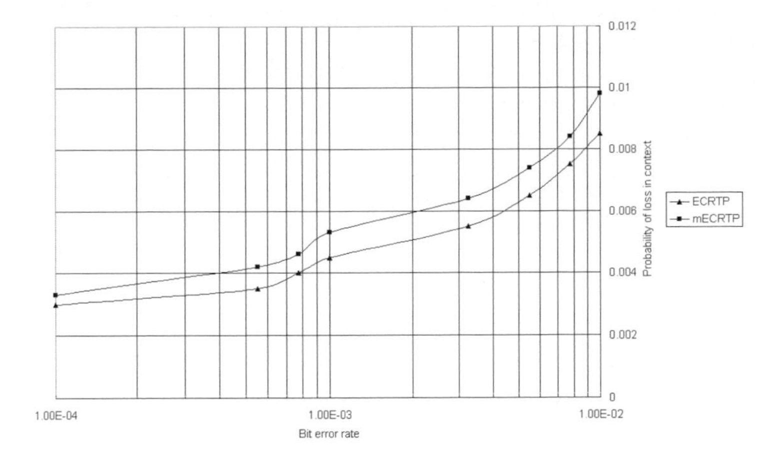

Fig. 8. Probability of loss in context vs BER

6 Conclusions and Future Work

This paper provides a framework for multimedia services using SRTP to protect real-time traffic through the public IP backbone network to the wireless access network. The simulation scenario shows that mECRTP scheme has better overall performance over wireless access network in compare to CRTP and ECRTP compression schemes. Based on the results, in mECRTP scheme, the average header size can be significantly dropped although there is slight increase in probability of loss in context while comparing with ECRTP scheme. Simulation results show that the mECRTP scheme significantly reduces the overhead of packet headers.

Furthermore, in order to investigate the performance of mECRTP compression scheme, a comparative study shall be carried out with a robust header compression scheme - Robust Header Compression (ROHC). [16]

References

1. Schulzrinne, H., et al.: RTP: A Transport Protocol for Real-Time Applications. IETF RFC 3550 (July 2003)
2. Perkins, C.: RTP: Audio and Video for the Internet. Addison Wesley, Reading (2003)
3. Baugher, M., et al.: The Secure Real-time Transport Protocol (SRTP). RFC 3711 (March 2004)
4. National Institute of Standards and Technology (NIST): Advanced Encryption Standard (AES) FIPS PUB 197
5. Krawczyk, H., Bellare, M., Canetti, R.: HMAC: Keyed-Hashing for Message Authentication. RFC 2104 (February 1997)
6. Degermark, M., Nordgren, B., Pink, S.: IP Header Compression. RFC 2507 (February 1999)

7. Casner, S., Jacobson, V.: Compressing IP/UDP/RTP Headers for Low-Speed Serial Links. IETF RFC 2508 (February 1999)
8. Koren, T., et al.: Enhanced Compressed RTP (CRTP) for Links with High Delay, Packet Loss and Reordering. RFC 3545 (July 2003)
9. Degermark, M., et al.: Evaluation of CRTP Performance over Cellular Radio Links. IEEE Personal Communications 7(4), 20–25 (2000)
10. Svanbro, K., et al.: Wireless Real-time IP Services Enabled by Header Compression. In: Proc. of the IEEE Vehicular Technology Conference (VTC), Tokyo, Japan, vol. 2, pp. 1150–1154 (2000)
11. Chen, W.T., Chuang, D.W., Hsiao, H.C.: Enhancing CRTP by retransmission for wireless networks. In: Proc. of the Tenth International Conference on Computer Communications and Networks, pp. 426–431 (2001)
12. Jin, H., Hsu, R., Wang, J.: Performance comparison of header compression schemes for RTP/UDP/IP packet. In: Proc. of IEEE Wireless Communications and Networking Conference (WCNC 2004), March 2004, vol. 3, pp. 1691–1696 (2004)
13. Bassil, C., Serhrouchni, A., Rouhana, N.: Critical Analysis and New Perspective for Securing Voice Networks. In: Lorenz, P., Dini, P. (eds.) ICN 2005. LNCS, vol. 3421, pp. 810–818. Springer, Heidelberg (2005)
14. OPNET Modeler Simulation Software, http://www.opnet.com
15. Crypto++ 5.2.1 Benchmarks, http://www.eskimo.com/~weidai/benchmarks.html
16. Bormann, C., et al.: Robust Header Compression (ROHC): Framework and four profiles: RTP, UDP, ESP, and uncompressed. RFC 3095 (July 2001)

A Reliable Multicast Transport Protocol for Communicating Real-Time Distributed Objects*

Jin Sub Ahn[1], Ilwoo Paik[2], Baek Dong Seong[2], Jin Pyo Hong[2], Sunyoung Han[3], and Wonjun Lee[4]

[1] Mobile Telecommunication Research Division, ETRI, Daejeon, Korea
jsahn@etri.re.kr
[2] Dept. of Information & Communications Engineering, Hankuk University of Foreign Studies, Kyongki-do, Korea
{steigensonne,iceboy98,jphong}@hufs.ac.kr
[3] Dept. of Computer Science and Engineering, Konkuk University, Seoul, Korea
syhan@cclab.konkuk.ac.kr
[4] Dept. of Computer Science and Engineering, Korea University, Seoul, Korea
wlee@korea.ac.kr

Abstract. The TMO (Time-triggered Message-triggered Object) model is a well-known real-time object model for distributed and timeliness-guaranteed computing. A distributed environment of the model may be configured by a number of the TMO nodes as a private network. It requires high reliability due to the feature of a distributed IPC message. The TCP seems to be suitable to this model. However, if a message needs to be broadcasted or multicasted to the other objects, the more the number of the nodes increases, the less efficient the repetitive unicast delivery of the message is. A multicast transport protocol can be considered to overcome this problem. In this paper, we propose a reliable multicast transport protocol suitable for supporting a distributed environment of the TMO model and discuss its performance with respects to the real-time delivery and throughput comparing with the alternative protocols. Results from the extensive performance measurement demonstrate that the proposed protocol outperforms the conventional TCP and existing RMT protocols.

1 Introduction

In the past, research on real-time computing focused on the functionality of a kernel, but recent focus moves to the development of real-time system using a real-time object model. The TMO [1] model has been greatly issued as a new programming paradigm. It integrates all merits of real-time programming, distributed system programming, concurrent programming, and object oriented programming. Thereupon a TMO model is proposed as a real-time object model for the concept of timeliness-guaranteed computing. The distributed TMOs communicate with each other by using the TMO methods on a logical unicast or multicast channel [2]. They may pass

* This research was supported by the MIC (Ministry of Information and Communication), Korea, under the ITRC (Information Technology Research Center) support program supervised by the IITA (Institute of Information Technology Assessment).

T. Vazão, M.M. Freire, and I. Chong (Eds.): ICOIN 2007, LNCS 5200, pp. 801–810, 2008.

messages to the other objects, which require high reliability and very short response time. If the TCP is applied, it is needed to establish connections as many as the number of distributed objects, moreover, the repetitive copy and delivery of a message is required whenever broadcasting or multicasting the message is needed. Therefore, the response time will undoubtedly increase in proportion to the number of distributed objects, and the real-time property might be inherently violated.

The IP multicast has been considered as an efficient way to deliver a message to a group of the distributed objects at a time, while it does not guarantee perfect reliability. In this paper, we propose a reliable multicast transport protocol suitable for supporting real-time reliable communication among distributed TMO objects and present the design and implementation of the protocol. Also, we discuss its performance with respects to the real-time delivery and throughput comparing with the alternative transport protocols.

2 Backgrounds

2.1 Communicating TMO Objects

The TMO channel is identified by a channel ID, which should be assigned to be unique value within a distributed environment of objects. The TMO objects communicating with a channel may define their channel access modes: write, read, or read/write mode.

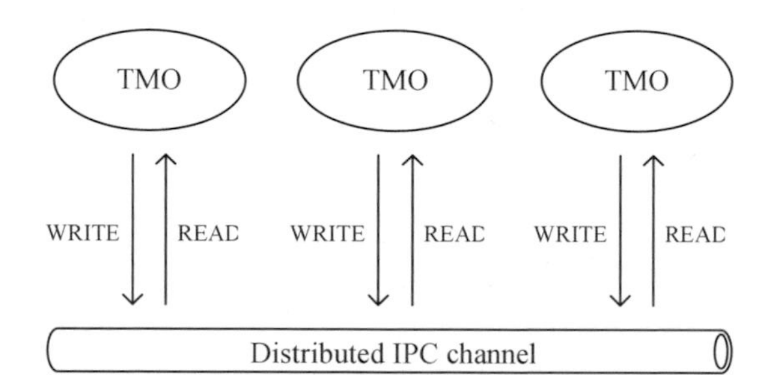

Fig. 1. TMO IPC Model with Channel

The read-mode or read/write-mode objects are allowed to receive all the messages passed through the channel. It means that each receiving object receives a copy of the same message sent by the sending object. A sender with the write or read/write access mode to the channel may send a message through the channel, but this message should be delivered to all the receiving objects waiting for reading the channel. This is why the distributed TMO model inherently requires multicasting capability.

Features like above are defined as distributed IPC interface [3] which has functions for channel assignment, synchronization, message transfer, close, etc. TMO engine based on Linux, TMO-Linux [4], recently restrict the maximum number of distributed IPC channels to 32 and the length of a message to 56Kbytes.

2.2 Reliable Multicast Transport Protocol Standards

To support reliability in one-to-many multicast transport, IETF RMT WG processes the standards for error and flow control of reliable multicast transport protocol. There are representative multicast transport protocols such as NORM (Nack-Oriented Reliable Multicast) [5][6] and ALC (Asynchronous Layered Coding) [7] based on LCT (Layered Coding Transport) [8].

The NORM protocol is designed for one-to-many multicast transport of bulk data. The basic behavior is that a receiver request retransmission to a sender polling NACK only when a packet is lost in order to prevent ACK implosion. When a receiver polls NACK, random backoff timer works to prevent NACK implosion. At this time, its duplicated NACK is checked by other receivers, unnecessary NACK is suppressed. A sender transfers data after FEC [9] encoding process. It can enable a receiver to recovery errors with FEC decoding. Consequently it brings the number of retransmission to decrease. The NORM protocol offers relatively high reliability but it is inferior to reliability of TCP. Another weak point is that response time takes long due to complexity of inner structure and congestion control (TFMCC [10]).

Another multicast protocol, ALC, is designed for massively scalable multicast distribution on wireless/satellite environment. As compared with NORM, the major difference is that the ALC can support multiple transfer rates for various heterogeneous receivers by using congestion control such as WEBRC [11]. However, there is no feedback between a sender and receivers, and thus the sender cannot confirm the status of packet reception. The ALC can rely upon FEC codec for reliability. Therefore, existing RMT protocols such as NORM and ALC are not suitable to distributed environment of TMO model in terms of real-time response and perfect reliability.

3 RM-IPC: A RMT Protocol for Communicating Distributed Objects

3.1 RM-IPC Message Format

We can classify RM-IPC messages into three types of CMD, DATA and ACK. DATA message is used for application data delivery, and the other messages are for control and session management. The message types are defined as fixed header and

Table 1. RM-IPC Message Types

Packet	Acronym	Transport Type	From	To
Timestamp	CMD(TS)	Multicast	Sender	Receiver
ACK for Timestamp	ACK(TS)	Unicast	Receiver	Sender
Receiver List for Advertisement	CMD(ADV)	Multicast	Sender	Receiver
Data	DATA	Multicast	Sender	Receiver
Flush	CMD(FLUSH)	Multicast	Sender	Receiver
ACK for Flush	ACK(FLUSH)	Unicast	Receiver	Sender
End of Transmission	CMD(EOT)	Multicast	Sender	Receiver
ACK for EOT	ACK(EOT)	Unicast	Receiver	Sender

header extension as each message should have different header values depending on its usage. The size of each message is variable.

Table 1 summarizes the characters of messages used in RM-IPC. And Fig. 2 presents the major RM-IPC header formats as below.

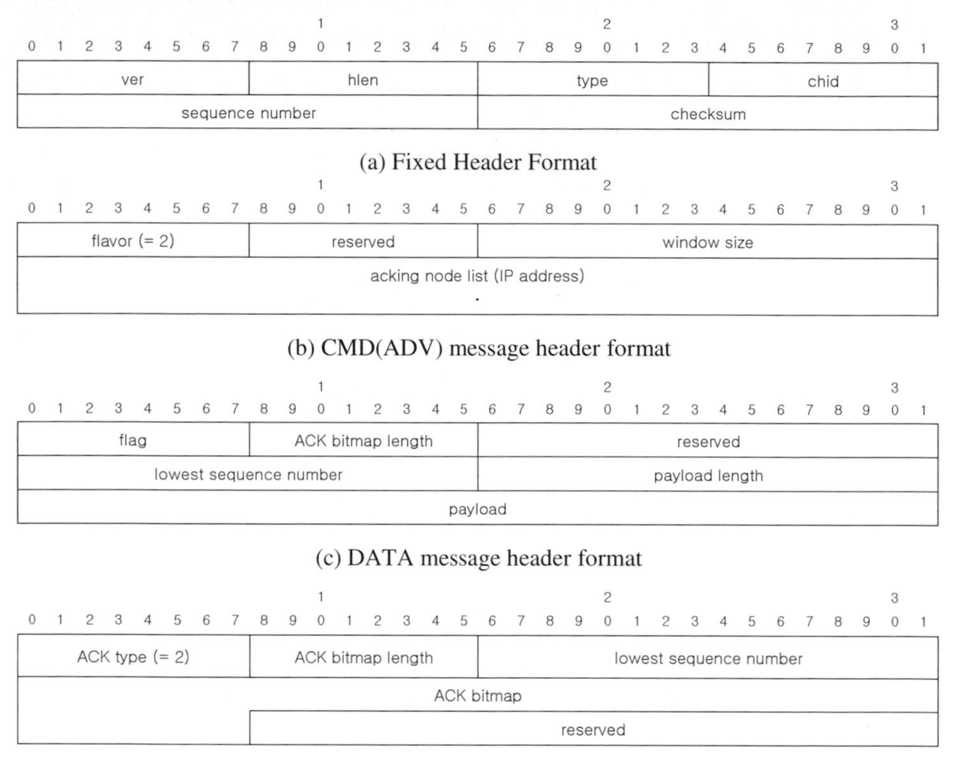

(a) Fixed Header Format

(b) CMD(ADV) message header format

(c) DATA message header format

(d) ACK(FLUSH) message header format

Fig. 2. RM-IPC message header formats

3.2 Session Creation and Termination

The important role of RM-IPC is that a sender should maintain a receiver list and measure GRTT(Group RTT). The message exchange between a sender and receivers for session creation/termination is similar to 3-way handshake in TCP, but the only difference is to enable to measure GRTT and to manage receivers of the session. Fig. 3 presents the state transition diagram.

3.3 Data Transmission and Reception

After session creation, the RM-IPC sender can transfer data, delivered from an application, to receivers by multicast. The sender copies the data to inner buffer by the API function `rmipc_send()` invoked from an application. And then, it processes segmentation procedure in order to fit the maximum segment size(1456 bytes) of RM-IPC. Then the sender configures header of the data, sets sequence number, calculates

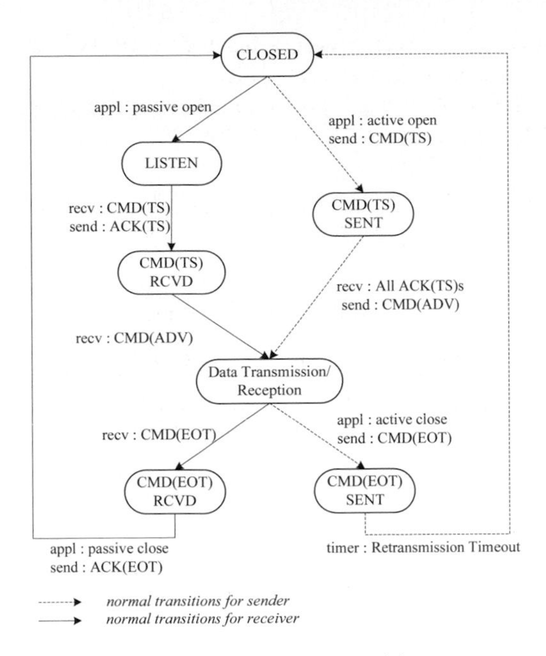

Fig. 3. RM-IPC state transition

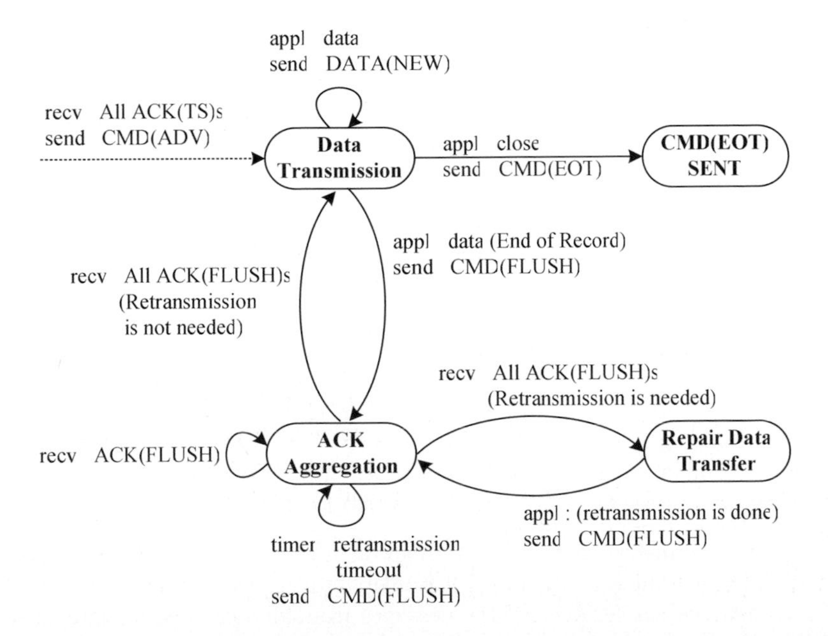

Fig. 4. 'Data Transmission' state transition (sender)

checksum, and transfer data. The RM-IPC receiver checks the sequence number and evaluates header checksum. Then, a cumulative feedback is sent to the sender by

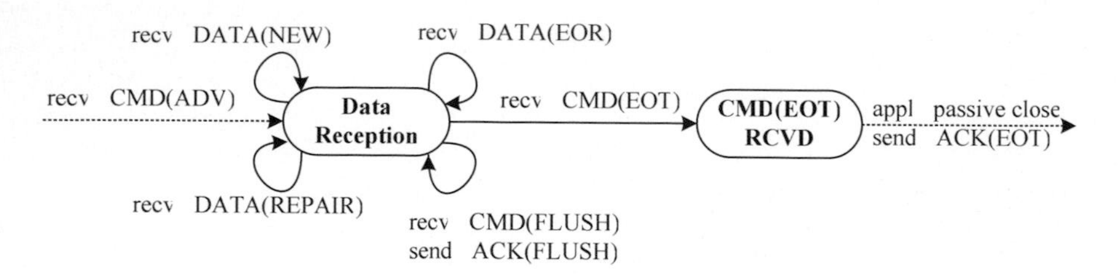

Fig. 5. 'Data Reception' state transition (receiver)

ACK(FLUSH) bitmap. All data, received from the sender, are assembled in an order that the sender has sent and delivered to the application. Fig. 4 and Fig. 5 present the state transition in Data Transmission/Reception state according to the role of a sender and a receiver.

3.4 Repairing Data

In this section, we describe error detection, retransmission by using ACK bitmap and error recovery mechanism related to retransmission of a sender.

If the checksum is not valid, the packet is regarded as incorrect and dropped. The sender can detect packet loss by comparing sequence number with the very next one. Fig. 6 shows the sequence of message exchanges from session creation to termination between a sender and receivers. RM-IPC does not provide the status of packet reception by type of range such as selective ACK in TCP or NORM protocol. If the receiver gets a message successfully, it sets ACK bitmap 1, which corresponds to the sequence number. If not, set to 0. This is to let the sender know the total reception status efficiently by delivering an ACK bitmap once. The sender and receivers maintain the following values to manage an ACK bitmap.

- Low Sequence Number (LSN): the lowest sequence number of message which a sender sent. This corresponds to the first bit in ACK bitmap that a receiver transfers through ACK(FLUSH) message.
- Highest Sequence Number (HSN): the highest sequence number of a message which a sender sent.
- ACK bitmap length: the total number of messages which a sender sent. The valid ACK bitmap length is equal to the expression, HSN-LSN+1. Considering the size of a distributed IPC message and MSS (1456 bytes) of RM-IPC in TMO-Linux, the maximum length of ACK bitmap is 40. In RM-IPC, a sender makes a receiver poll ACK bitmap by sending CMD(FLUSH) or DATA(EOR).

Each receiver transfers the ACK bitmap as type of ACK(FLUSH) message. ACK(FLUSH) is used not only for ACK bitmap deliver, but also for GRTT update.

If a sender receives ACK(FLUSH) messages from all receivers, it aggregates them to decide whether it requests retransmission or not. If the result of AND operation of all bitmap is not equal to 1, the sender retransmits. The sender repeats the cycle from 'ACK Aggregation' to 'Repair Data transfer' until error recovery is finished. Even if ACK(FLUSH) message is received in the retransmission, GRTT is not updated.

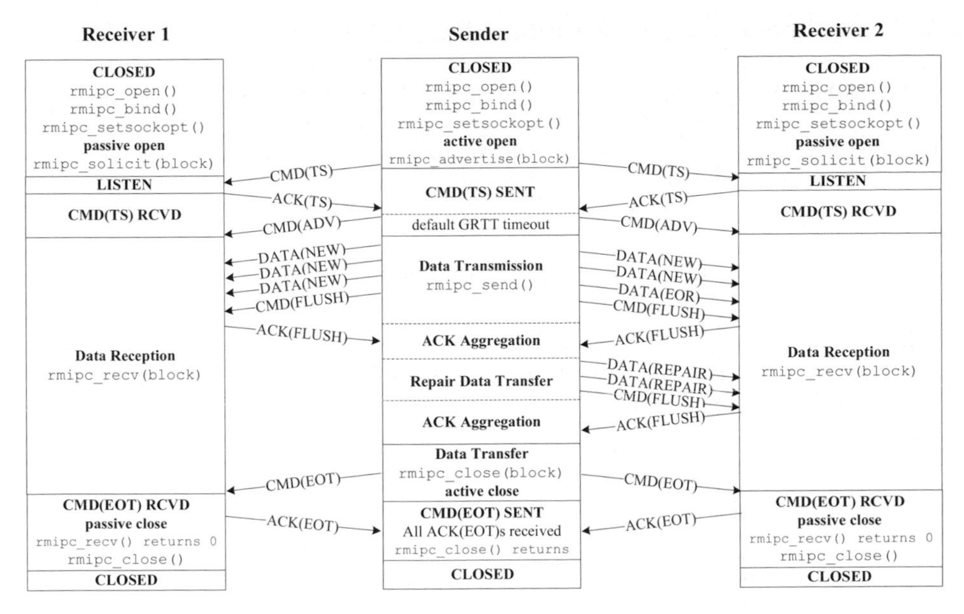

Fig. 6. Packet exchanges in RM-IPC session

3.5 API

Table 2 explains API functions of RM-IPC. RM-IPC API is designed at the view of following.

- Include the general RMT function in distributed IPC interfaces
- Prefix, 'rmipc' is to present RM-IPC protocol
- Design RM-IPC API on the foundation of the Berkeley Socket API.

Table 2. RM-IPC API

Function name	Description
rmipc_open()	create socket (with setup distributed IPC channel) and decide a role (sender or receive)
rmipc_bind()	assign a local protocol address and port to a socket.
rmipc_setsockopt()	set socket options (for interface, loopback).
rmipc_getsockopt()	get socket option info.
rmipc_advertise()	transfer CMD(TS) message for RTT (for only sender)
rmipc_solicit()	transfer ACK(TS) message in response to RTT (for only re ceiver)
rmipc_send()	send Application data
rmipc_recv()	delivery received data to Application
rmipc_close()	terminate session and close a socket.

4 Experiments and Performance Evaluation

4.1 RM-IPC Experiments with TMO-Linux

The RM-IPC has been implemented under Linux environment and has reflected not only requirements of TMO based distributed IPC interface, but also those of TMO-Linux. It has been integrated and tested with the TMO-Linux of TMO engines under the environment as follows:

- OS : Over the version of Linux Kernel 2.6
- Compiler: g++
- Timer resolution: 1ms
- Distributed IPC Channel: 32 channels (1~32)

4.2 Performance Evaluation

Considering a distributed environment of the TMO model, we evaluate the performance of the TCP, MCL-NORM, UDP/IP multicast and RM-IPC. Fig. 7 shows the

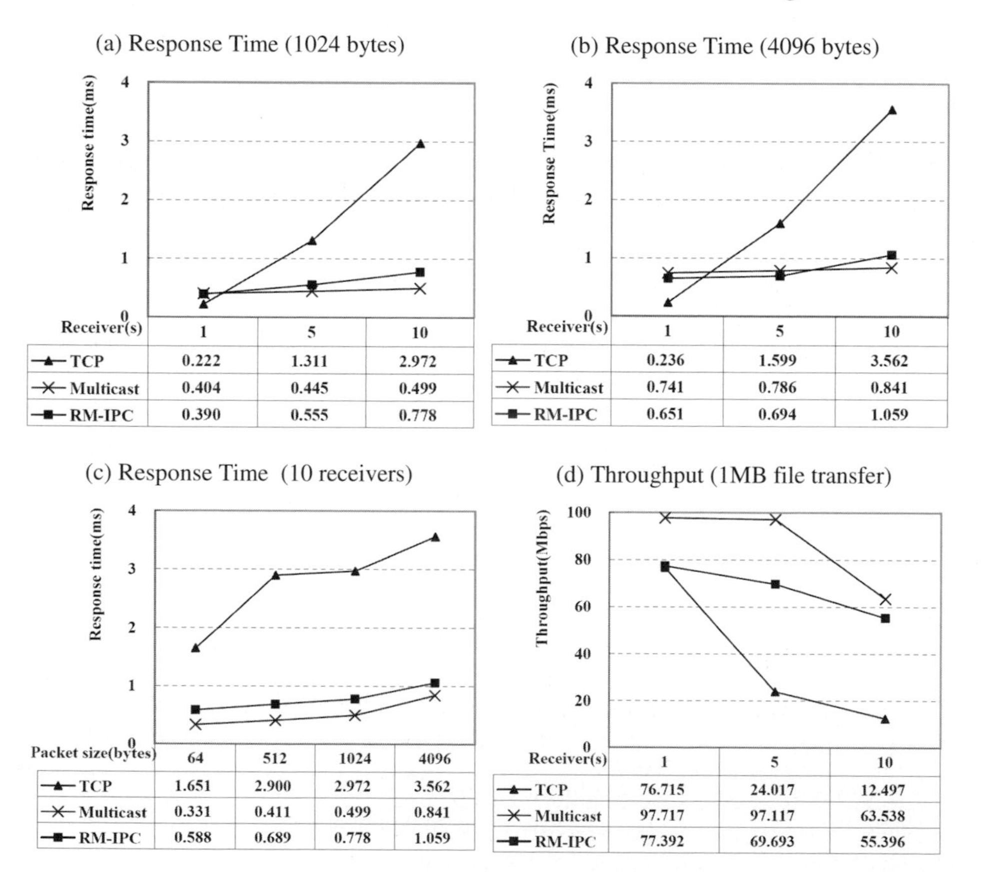

Receiver(s)	1	5	10
TCP	0.222	1.311	2.972
Multicast	0.404	0.445	0.499
RM-IPC	0.390	0.555	0.778

Receiver(s)	1	5	10
TCP	0.236	1.599	3.562
Multicast	0.741	0.786	0.841
RM-IPC	0.651	0.694	1.059

Packet size(bytes)	64	512	1024	4096
TCP	1.651	2.900	2.972	3.562
Multicast	0.331	0.411	0.499	0.841
RM-IPC	0.588	0.689	0.778	1.059

Receiver(s)	1	5	10
TCP	76.715	24.017	12.497
Multicast	97.717	97.117	63.538
RM-IPC	77.392	69.693	55.396

Fig. 7. Performance comparison of RM-IPC, UDP multicast, and TCP

results of measuring response time and throughput in UDP/IP multicast, TCP, and RM-IPC. We have measured the NORM protocol using MCLv3 [12] open source and MCL-NORM library. The response time and throughput of the MCL-NORM has been shown as abnormally bad performance regardless of the number of receivers, since it excludes congestion control such as TFMCC. In practice, the response time of the MCL-NORM has been measured from 3 to 7 seconds while the other protocols could be measured by the unit of millisecond, and thereupon the result is so far and meaningless that it is excluded in those graphs.

UDP/IP multicast may not support reliability, but we have evaluated it for comparing the speed of the RM-IPC. The result in comparison between TCP and RM-IPC is meaningful. The TCP should establish connections as many as the number of TMO nodes to support the distributed IPC interface. Total response time of TCP excluding connection establishment phase is also increasing in proportion to the number of receivers because it should transfer the same message repeatedly.

Although the number of receivers is increasing, it only takes the time for polling ACK bitmap in the RM-IPC. Compared with TCP, the RM-IPC does not increase the response time too long unlike the case of TCP. In addition, the RM-IPC is able to transfer a message to all receivers by one shot of multicast without the consideration of the number of receivers.

5 Conclusions

The transport layer protocol, which is the most suitable to a distributed environment of a TMO model, should meet the conditions of reliability and short response time. One of the representative multicast protocols NORM inherently takes a long response time. Therefore, it is unfit to a distributed interface of a TMO model. A reliable unicast protocol such as the TCP meets the requirement for reliability, but the more the number of TMO node increases, the longer response time takes. In order to overcome the existing protocols, we have analyzed the requirements for the distributed IPC interface of the TMO model, designed and implemented a new RMT protocol named RM-IPC. We have also evaluated and compared the performance of the RM-IPC with the TCP and the other protocols. The proposed RM-IPC protocol has been shown to be fit to not only TMO, but also to various distributed environments.

References

1. Kim, K.H., Kopetz, H.: A Real-Time Object Model RTO.k and an Experimental Investigation of Its Potentials. In: Proc. 18th IEEE Computer Software and Applications Conference, November 1994, pp. 392–402 (1994)
2. Kim, K.H.: Realization of Autonomous Decentralized Computing with the RTO.k Object Structuring Scheme and the HU-DF Inter-Process-Group Communication Scheme. In: Proc. ISADS 1995 (April 1995)
3. Kim, J.G., Kim, M.H., Kim, K., Heu, S.: TMO-eCOS: An eCos-based Real-time Micro Operating System Supporting Execution of a TMO Structured Program. In: Proc. ISORC 2005 (May 2005)

4. Kim, J.G., et al.: TMO-Linux: A Linux-based Real-time Operating System Supporting Execution of TMO's. In: Proc. ISORC 2002 (April 2002)
5. Adamson, B., Bormann, C., Handley, M., Macker, J.: Negative-Acknowledgment (NACK)-Oriented Reliable Multicast (NORM) Building Blocks, IETF RMT WG, RFC 3941 (November 2004)
6. Adamson, B., Bormann, C., Handley, M., Macker, J.: Negative-Acknowledgment (NACK)-Oriented Reliable Multicast (NORM) Protocol, IETF RMT WG, RFC 3940 (November 2004)
7. Luby, M., Gemmell, J., Vicisano, L., Rizzo, L., Crowcroft, J.: Asynchronous Layered Coding (ALC) Protocol Instantiation, IETF RMT WG, draft-ietf-rmt-pi-alc-revised-01.txt (October 2005)
8. Luby, M., Vicisano, L., Gemmell, J., Rizzo, L., Handley, M., Crowcroft, J.: Layered Coding Transport (LCT) Building Block, IETF RMT WG, draft-ietf-rmt-bb-lct-revised-01.txt (October 2005)
9. Luby, M., Vicisano, L., Gemmell, J., Rizzo, L., Handley, M., Crowcroft, J.: Forward Error Correction (FEC) Building Block, IETF RMT WG, RFC 3452 (December 2002)
10. Widmer, J., Handley, M.: TCP-Friendly Multicast Congestion Control (TFMCC): Protocol Specification, IETF RMT WG, RFC 4654 (August 2006)
11. Luby, M., Goyal, V.: Wave and Equation Based Rate Control (WEBRC) Building Block, IETF RMT WG, RFC 3738 (April 2004)
12. MCLv3 (Multicast Library v3), http://www.inrialpes.fr/planete/people/roca/mcl/mcl.html

Achieving Proportional Delay and Loss Differentiation in a Wireless Network with a Multi-state Link

Yuan-Cheng Lai and Yu-Chin Szu

Dept. of Information Management, National Taiwan University of Science and Technology
laiyc@cs.ntust.edu.tw, D9109102@mail.ntust.edu.tw

Abstract. Many algorithms for providing proportional delay differentiation and proportional loss differentiation have been proposed under wired networks. However, these algorithms suffer from low performance at encountering some distinct characteristics, such as location-dependent and time-varying channel capacity, which exist in wireless networks. This paper proposes a novel algorithm, Wireless Proportional Delay and Loss differentiation (WPDL) including a capacity-aware scheduler and a debt-aware dropper, to provide the proportional delay differentiation and proportional loss differentiation in a wireless network with a multi-state link. WPDL considers the channel state and debt information in order to improve the performance of scheduling and dropping. From simulation results, WPDL actually achieves proportional delay differentiation, proportional loss differentiation, lower queueing delay and loss, and higher throughput, compared with other methods in the wireless environment.

1 Introduction

The proportional differentiation model has received a lot of attention because it can perform the controllable and predictable relative service differentiation. That is, the proportional differentiation model offers the network manager a means of varying quality spacing between service classes according to the given pricing or policy criteria, and ensures that the differentiation between classes is consistent in any measured timescale. The proportional services can be differentiated according to different performance metrics, such as throughput, delay, loss, or jitter. When adopting queueing delay and loss as the performance metric, the proportional differentiation model are referred as the proportional delay differentiation model and proportional loss differentiation model, respectively.

To provide proportional delay differentiation in wired networks, some algorithms, such as Waiting Time Priority (WTP) [1], Proportional Average Delay (PAD) [1], Advanced Waiting Time Priority (AWTP) [2], Hybrid Proportional Delay (HPD) [3], and VirtualLength [4], have been proposed. To provide proportional loss differentiation in wired networks, some methods, such as Proportional Loss Rate (PLR) [5], Average Drop Distance (ADD) [6], Debt-aware [7], and DRED [8], have been proposed.

As wireless technology rapidly advances and lightweight portable computing devices become popular, wireless networks have become more pervasive. Accordingly, proportional delay differentiation and proportional loss differentiation are urgently

T. Vazão, M.M. Freire, and I. Chong (Eds.): ICOIN 2007, LNCS 5200, pp. 811–820, 2008.

required for wireless environments, just as it was for wired networks. However, above approaches designed in a wired network are not applicable in a wireless environment, which has some specific characteristics, such as high error rate and burst errors, location-dependent and time-varying capacity, and scarce bandwidth [9].

This paper proposes a novel algorithm, named Wireless Proportional Delay and Loss differentiation (WPDL), including a scheduler and a dropper. By these mechanisms, WPDL can provide proportional delay differentiation and proportional loss differentiation in a wireless with a multi-state channel. Additionally, WPDL considers the dynamic channel capacity of a wireless link to operate scheduling and dropping, so it can achieve higher throughput and lower queueing delay and loss.

The organization of this paper is as follows. Section 2 outlines the background, including the proportional differentiation model, waiting time priority scheduler and proportional loss rate dropper. In section 3, we describe our proposed algorithm WPDL in details. In section 4, we evaluate the effectiveness of our algorithm by simulation. Finally, some conclusions are given in section 5.

2 Background

2.1 Proportional Differentiation Model

The structure of the proportional differentiation model is shown as Figure 1. The arrival traffic is classified into *N* service class where each class has a dedicated queue. Let q_i denote the measured performance of class *i*. For proportional differentiation model, the following equation should be satisfied for all pairs of classes.

$$\frac{q_i}{q_j} = \frac{c_i}{c_j} \qquad (1 \le i, j \le N) \tag{1}$$

where $c_1 < c_2 < \cdots < c_N$ are the generic quality differentiation parameters (QDPs).

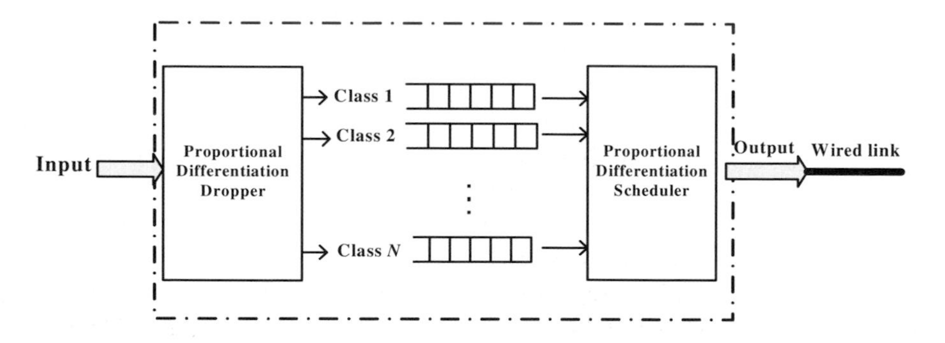

Fig. 1. The proportional differentiation model

The queuing delay, packet loss, or jitter could be as the performance metric in this model. Adopting queueing delay and packet loss, this model is called the proportional delay differentiation model and proportional loss differentiation model. Let $\overline{W}_i$ be the average queueing delay of class-*i* packets, and δ_i be the delay differentiation parameter (DDP) of class *i*. The proportional delay differentiation model has the following constraint for any pair of classes:

$$\frac{\overline{W}_i}{\overline{W}_j} = \frac{\delta_i}{\delta_j} \qquad (1 \le i, j \le N) \tag{2}$$

Let $\overline{L}_i$ and σ_i be the average loss rate and the loss differentiation parameter (LDP) of class *i*, respectively. For all pairs of service classes, *i* and *j*, the proportional loss differentiation model is specified by

$$\frac{\overline{L}_i}{\overline{L}_j} = \frac{\sigma_i}{\sigma_j} \qquad (1 \le i, j \le N) \tag{3}$$

To support proportional differentiated services, a proportional differentiation dropper and a proportional differentiation scheduler are necessary components. The dropper determines which packet should be dropped in case of buffer overflow to control the loss rate of each class, while the scheduler decides the packet serving order to control the queueing delay of each class.

2.2 Waiting Time Priority Scheduler

The waiting time priority (WTP) scheduler is a priority scheduler in which the priority of a packet increases in proportion to its waiting time [10]. According to the waiting time of the packet in queue, WTP adjusts the priority of its service. Let $P^k_i(t)$ denote the priority of the *k*-th packet of class *i* at time *t*, and $W_i^k(t)$ be its waiting time. The packet priority is calculated as its normalized waiting time in the following,

$$P_i^k(t) = \tilde{W}_i^k(t) = \frac{W_i^k(t)}{\delta_i} \qquad (1 \le i \le N) \tag{4}$$

Because the head-of-line (HOL) packet is the earliest arrival among all packets currently queuing in the buffer, so it has the longest waiting time. Therefore, a HOL packet has the highest priority within its queue, implying that only the HOL packet of every class needs to be considered when comparing priorities. For simplicity, the index *k* is skipped when *k*=1. The WTP scheduler chooses the HOL packet of class $J = \arg\ \max P_i(t)$, and actually transmits it. The WTP scheduler can successfully approach the targeted delay proportion when the network traffic load is heavy, but can't achieve it under light load [3].

2.3 Proportional Loss Rate Dropper

C. Dovrolis proposed the Proportional Loss Rate Dropper (PLR) to offer the proportional loss differentiated services [5]. In order to determine which packet should be dropped, PLR uses a Loss History Table (LHT) to record the loss rate of each class at present. Let $\bar{L}_i(t)$ be the average loss rate, $\tilde{L}_i(t)$ be the normalized average loss rate, $L_i(t)$ be the number of dropped packets, and $A_i(t)$ be the number of arrived packets, of class i. PLR chooses class J to drop its tail packet as follows,

$$\bar{L}_i(t) = \frac{L_i(t)}{A_i(t)} \qquad (1 \le i \le N) \tag{5}$$

$$\tilde{L}_i(t) = \frac{\bar{L}_i(t)}{\sigma_i} \qquad (1 \le i \le N) \tag{6}$$

$$J = \arg\ \min\ \tilde{L}_i(t) \tag{7}$$

PLR aims to maintain an unanimous normalized average loss rate among all classes. Depending on the number of packets that PLR estimates, the calculated average loss rate is different, so there are two kinds of algorithms, namely PLR with infinite memory (PLR (∞)) and PLR with memory M (PLR (M)). When calculating the average loss rate, PLR(∞) counts packets from initial to present, while PLR(M) observes the last M packets of every class at present. From the long-term observation, the result of PLR(∞) is closer to targeted proportion than that of PLR(M). Thus when the class load distribution is stationary, adopting PLR(∞) is suitable. However, when the class load distribution is non-stationary, adopting PLR(M) is preferred because of its adaptation, but determining an optimal M is difficult.

2.4 Proportional Differentiation Model in a Wireless Network

Figure 2 depicts the proportional differentiation model in a wireless network. In this model, all mobile hosts share one wireless link. Since each host could be located at different places, different capacities exist when this scheduler transmits data to different mobile hosts via this wireless link. Also as the mobile host moves, the destined capacity to this host varies. Thus a wireless link has a location-dependent and time-varying capacity, and it is called as a multi-state link herein. For simplicity, the term channel j means the wireless link at transmitting the packet to the mobile host j. Let

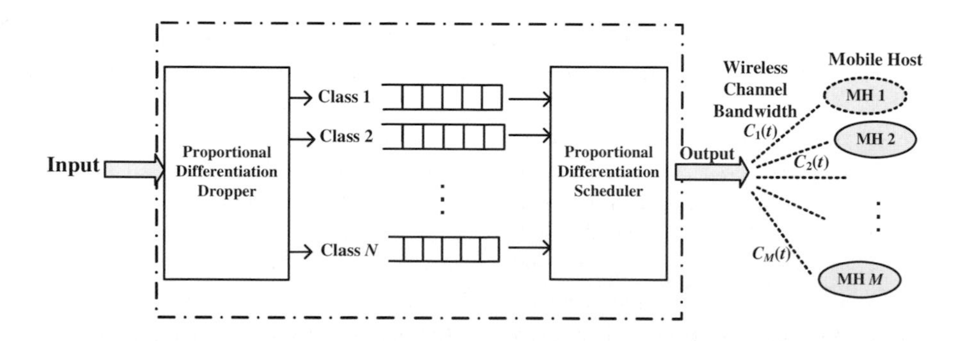

Fig. 2. The proportional differentiation model in a wireless network

$C_j(t)$ denote the encountered capacity when the scheduler transmits packets via channel *j* at time *t*.

3 WPDL Algorithm

Our proposed WPDL aims at the following goals in a multi-state wireless network: 1) to provide proportional delay differentiation, 2) to provide proportional loss differentiation, 3) to offer lower queuing delay and loss, and 4) to provide higher throughput.

WPDL uses a proportional differentiation scheduler for performing delay differentiation, and uses a proportional differentiation dropper for achieving loss differentiation. The scheduler is to decide when and which packet to be transmitted or dropped. The dropper is to decide when and which packet to be dropped. Note that the scheduler in WPDL may drop a packet if it feels necessary. The detailed WPDL algorithm is shown in Figure 3.

3.1 Proportional Differentiation Dropper

The upper part in figure 3 presents the pseudo code for the dropper of WPDL. When the buffer has empty space, the dropper simply inserts the packet into a proper queue. When buffer overflow occurs, the dropper selects an appropriate packet to drop. Let $L_i(t)$, $A_i(t)$, and $S_i(t)$ be the number of dropped packets, the number of arrived packets, and the loss debt of class *i* at time *t*, respectively. $S_i(t)$ having the positive value represents that it has the debt, implying that the number of loss happening is less than the expectation in class *i*. Restately, some other classes instead of class *i* drop the packet. Similarly, the value of $S_i(t)$ being negative and zero mean that class *i* has more and equal losses than its expectation, respectively.

When an arriving packet encounters a full buffer, the dropper selects a proper packet, which may be this arriving packet or a packet in the buffer, to keep the proportional loss among classes. At determining which packet to be dropped, the dropper first calculates the normalized average loss rate $\tilde{L}_i(t) = L_i(t) / A_i(t)\sigma_i$ for each class *i*. Then the class with the smallest normalized average loss rate, that is $J = \arg\ \min \tilde{L}_i(t)$, is regarded as the candidate class. Let $H_J(t)$ be the HOL packet of class *J* and $C(H_J(t))$ be the capacity of the destined channel of the packet $H_J(t)$ at time *t*. Also $maxC(t)$ denotes the maximum encountered capacity of all packets before time *t*.

The dropper considers two important factors to determine which packet to be dropped. One is whether the candidate class has the debt, and the other is how good of the channel. If the candidate class has the debt, the packet belonging to this class will be dropped because this class should return the previous debt immediately. On the other hand, a probability P_J is randomly generated for candidate class *J* to compare with the bandwidth percentage of the destined channel of its HOL packet to determine whether dropping this packet. The concept is the better capacity the channel owns, the less probability its corresponding packet is dropped. Thus the condition of dropping the packet of the candidate class J is as follows.

$$\text{If } (S_J > 0 \text{ or } P_J > \frac{C(H_J(t))}{maxC(t)}) \tag{8}$$

Note the dropper drops the HOL packet, rather than the tail packet or arriving packet, of candidate class J because using this method can reduce the queuing delay of queued packets.

If the candidate class J does not satisfy the above condition, the dropper will choose the candidate class K, which has the next smallest normalized average loss rate. Judging whether the HOL packet of candidate class K will be dropped is similar to equation 8, that is, $S_K > 0$ or $P_K > C(H_K(t)) / maxC(t)$. However, if the packet of class K is dropped, the debt S_K and S_J should be updated as S_K -1 and S_J +1, implying that class K instead of class J drops a packet, that is, class K borrows class J once.

3.2 Proportional Differentiation Scheduler

The lower part in figure 3 presents the pseudo code for the scheduler of WPDL. The scheduler judges two conditions as described in equation 9. When these two conditions are all satisfied, the scheduler drops the HOL packet of this candidate class.

$$\text{If } (S_J > 0 \text{ and } P_J > \frac{C(H_J(t))}{\max C(t)}) \tag{9}$$

Because dropping the packet needs to meet two conditions at the same time, WPDL only drops a few packets at this stage. Also WPDL does not have more packet losses than other algorithms because these packet drops generate the extra space in the buffer, leading to fewer losses made by the dropper later. That is, WPDL creates an early loss in the scheduler to replace a later loss in the dropper. The early drop of packets encountering a poor-capacity channel not only causes the shorter queuing delay, but also generates higher throughput because packets are usually transmitted on a high-capacity channel.

4 Simulations

The simulation evaluates WPDL and WPLR, which combines the WTP scheduler with the PLR(∞) dropper over a wireless link. The model we simulated is depicted as in figure 2. In all simulations, three service classes (N=3) are assumed, the corresponding DDPs are set as $\delta_1 = 1$, $\delta_2 = 2$, $\delta_3 = 4$, and the corresponding LDPs are set as $\sigma_1 = 1$, $\sigma_2 = 2$, and $\sigma_3 = 4$. Packet arrival follows a Poisson process and its mean arrival rate is $\lambda = 0.9$ packet/sec. The packet size is fixed at 441 bytes for all classes, and the full wireless link capacity is 2646 bytes/sec. The total buffer size is 20 packets and the number of hosts is five, i.e., M=5. The wireless channel is simulated by a multi-state Markov process, which has five states with the value of capacity

L_i : the number of dropped packets of class i
A_i : the number of arrived packets of class i
σ_i : the loss differentiation parameter of class i
δ_i : the delay differentiation parameter of class i
S_i : the loss debt of class i at time t
$C(H_i)$: the channel capacity encountered when the scheduler transmits the HOL packet of class i
$B(t)$: the set of backlogged classes at present t

Proportional Differentiation Dropper

{
  A class-*i* packet arrives, A_i++;
  If (buffer overflow){
    calculate $\tilde{L}_i(t) = L_i(t) / A_i(t)\sigma_i$, i=1,2,..,N
    J= arg $\min_{i \in B(t)} \tilde{L}_i(t)$;
    P_J =*random*();
    If ($S_J > 0$ or $P_J > \frac{C(H_J(t))}{maxC(t)}$) {
      drop the HOL packet from class *J*;
      $L_J ++$;}
    Else{
      do {
        find $K = \arg$ next min $\tilde{L}_i$;
        P_K =*random*();
        If ($S_K > 0$ or $P_K > \frac{C(H_K(t))}{maxC(t)}$) {
          drop the HOL packet from class *K*;
          $L_k ++$;
          $S_K --$;
          $S_J ++$;}
      }while(one packet is dropped or all classes have been visited)
      if (no packet is dropped){
        drop the HOL packet from class *J*;
        $L_J ++$;}
    }
  Accept the incoming packet;
}

Proportional Differentiation Scheduler

{
  calculate $P_i(t) = \tilde{W}_i(t) = \frac{W_i(t)}{\delta_i}$, i=1,2,..,N
  $J = \arg \max_{i \in B(t)} P_i(t)$;
  P_J =*random*();
  If ($S_J > 0$ and $P_J > \frac{C(H_J(t))}{maxC(t)}$) {
    drop the HOL packet from class *J*;
    $L_J ++$;
    $S_J --$;}
  Else transmit the HOL packet of class *J*;
}

Fig. 3. The WPDL algorithm

varying among 0% (purely bad), 25%, 50%, 75%, and 100% (purely good) of the full capacity. The transition matrix of channel capacity is set as

$$\begin{array}{c} \\ 100\% \\ 75\% \\ 50\% \\ 25\% \\ 0\% \end{array} \begin{array}{c} \begin{array}{ccccc} 100\% & 75\% & 50\% & 25\% & 0\% \end{array} \\ \begin{bmatrix} 1-a & ap_1 & ap_1^2 & ap_1^3 & ap_1^4 \\ ap_2 & 1-a & ap_2 & ap_2^2 & ap_2^3 \\ ap_3^2 & ap_3 & 1-a & ap_3 & ap_3^2 \\ ap_4^3 & ap_4^2 & ap_4 & 1-a & ap_4 \\ ap_5^4 & ap_5^3 & ap_5^2 & ap_5 & 1-a \end{bmatrix} \end{array}$$

where a is the state transition rate to other states and p_i is the probability of state i being translated to its neighbor states when the transition occurs. The default value of a is 0.2, and the values of p_1, p_2, p_3, p_4, and p_5 can be calculated by letting the sum of each row equal to 0. In each simulation, at least 1,000,000 packets for each class are generated for the sake of converge.

To exhibit the throughput improved by WPDL, the throughput improvement is defined as $(T^W - T^P)/T^P$, where T^W and T^P are the throughput obtained by WPDL and WPLR, respectively.

4.1 Absolute Performance

Figure 4 shows absolute delay, absolute loss rates of WPDL and WPLR, and throughput improvement. Figure 4(a) shows that the absolute queuing delay of three classes in WPDL is less than those in WPLR. There are two reasons. First, because the packet with a poor channel is easier to be dropped than that with a good-channel, non-dropped packets usually encounter a good channel, causing that they has short transmission time, and thus short queuing delay. Second, WPDL dropping the HOL packet, rather than dropping the tail packet in WPLR, reduces the queuing delay of the queued packets. Observed from figure 4(b), WPDL has the less drop rates of three classes than WPLR. This phenomenon is caused from that WPDL enjoys the high throughput, making fewer packet losses.

Figure 4(c) reveals that the throughput improvement of three classes. The throughput improvement of each class achieved by WPDL is in the order class 3 > class 2 > class 1, and the total throughput improvement is around 18%. For class 3 which is the lowest-priority class, more packets are dropped and fewer packets are serviced, so that its throughput improvement is up to 65%.

4.2 Loss Ratio and Delay Ratio

In this simulation, the delay and loss ratios between successive classes are measured over five time intervals - 100, 500, 1000, 5000, and 10000 p-units, where a p-unit is the average packet transmission time, i.e., 1/6 sec. During each time interval, the delay ratios and loss ratios of class 2/class 1 and class 3/class 2, are averaged.

Figure 5 shows five percentiles, 5%, 25%, 50% (median), 75%, and 95%, of the average delay ratio, average loss ratio and throughput improvement. In Fig. 5(a), both schedulers have broad ranges in a short timescale and condensed ranges in a long timescale. Also, WPDL has the more concentrated ranges than WPLR, resulting in more predictable behavior. Also WPDL has delay ratio nearer the targeted proportion than WPLR.

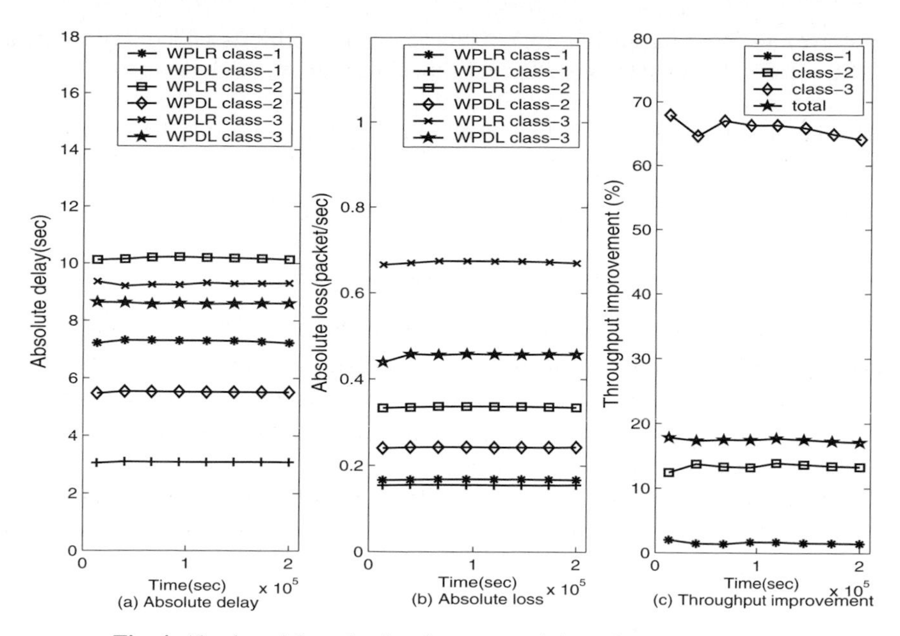

Fig. 4. Absolute delay, absolute loss rate and throughput improvement

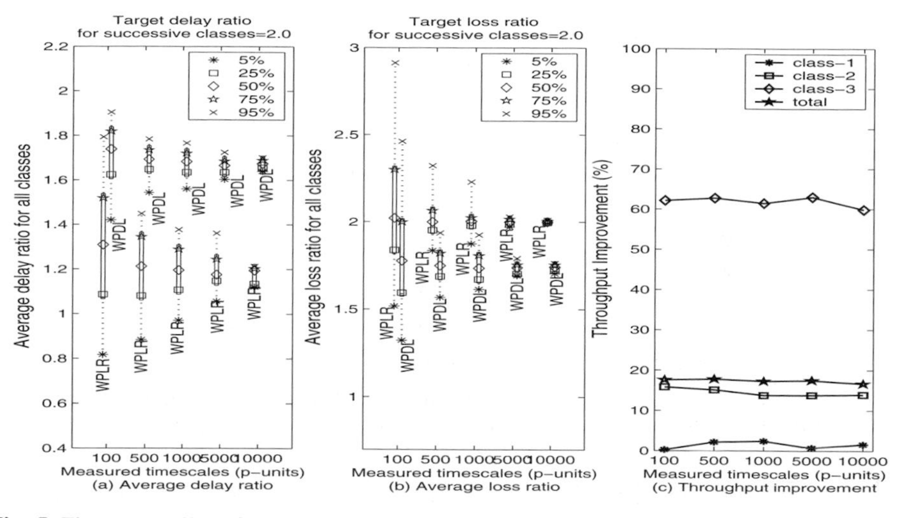

Fig. 5. Five percentiles of average delay ratio, average loss ratio, and throughput improvement on various measured timescales

Fig. 5(b) reveals that five percentiles of the average loss ratio. The average loss ratio of WPLR is nearer the target than WPDL. WPDL considers the debt S_i to achieve the loss differentiation because the class having $S_i > 0$ should return the debt quickly. However, WPDL does not design a specific mechanism to compensate the extra drop for the class with $S_i < 0$, leading to the slight bias of loss ratio from the targeted proportion.

Fig. 5(c) plots the throughput improvement for three classes under various measured timescales. Observed from different timescales, the throughput improvement is very stable. The throughput improvement of class 3 is better than the others because the absolute loss rates between WPDL and WPLR have larger difference in class 3.

5 Conclusions

The characteristic of time-varying and location-dependent channel capacity exhibited in wireless communication makes packet scheduling and dropping be challenges. We presented WPDL to achieve proportional delay differentiation and proportional loss differentiation in the wireless network with a multi-state link. WPDL not only considers the normalized waiting time and normalized average loss rate, but also considers the channel state and debt information, in order to improve the performance of scheduling and dropping.

From the simulation results, WPDL actually achieves the following goals in a wireless network with a multi-state link: 1) to provide proportional delay differentiation, 2) to provide proportional loss differentiation, 3) to offer lower queueing delay and loss, and 4) to provide higher throughput.

References

1. Dovrolis, C., Stiliadis, D., Ramanathan, P.: Proportional Differentiated Services: Delay Differentiation and Packet Scheduling. ACM SIGCOMM Computer Communication Review 29(4), 109–120 (1999)
2. Lai, Y.C., Li, W.H.: A Novel Scheduler for the Proportional Differentiation Model by Considering Packet Transmission Time. IIEEE Communications Letters 7(4), 189–191 (2003)
3. Dovrolis, C., Stiliadis, D., Ramanathan, P.: Proportional Differentiated Services: Delay Differentiation and Packet Scheduling. IEEE/ACM Transactions on Networking 10(1), 12–26 (2002)
4. Wei, J., Li, Q., Xu, C.Z.: VirtualLength: A New Packet Scheduling Algorithm for Proportional Delay Differentiation. In: The 12th International Conference on Computer Communication and Network (ICCCN), October, pp. 331–336 (2003)
5. Dovrolis, C., Ramanathan, P.: Proportional Differentiated Services, Part II: Loss Rate Differentiation and Packet Dropping. In: IEEE/IFIP Int. Workshop Quality of Service (IWQoS), June, pp. 52–61 (2000)
6. Bobin, U., Jonsson, A., Schelen, O.: On Creating Proportional Loss-Rate Differentiation: Predictability and Performance. In: Wolf, L., Hutchinson, D.A., Steinmetz, R. (eds.) IWQoS 2001. LNCS, vol. 2092, pp. 372–379. Springer, Heidelberg (2001)
7. Zeng, J., Ansari, N.: An Enhanced Dropping Scheme for Proportional Differentiated Services. In: IEEE International Conference on Communications (ICC), vol. 3, pp. 11–15 (2003)
8. Aweya, J., Ouellette, M., Montuno, D.Y.: Weighted Proportional Loss Rate Differentiation of TCP Traffic. International Journal of Network Management 14, 257–272 (2004)
9. Cao, Y., Li, V.O.K.: Scheduling Algorithm in Broad-Band Wireless Networks. Proceedings of the IEEE 89(1), 76–87 (2001)
10. Kleinrock, L.: Queueing Systems, vol. 2. Wiley-Interscience, New York (1976)

Directional Flooding Scheme with Data Aggregation for Energy-Efficient Wireless Sensor Networks

Sung-Hyup Lee[1], Kang-Won Lee[2], and You-Ze Cho[2]

[1] Industry Supprot Division, Korea Radio Promotion Agency, 78 Garak-dong, Songpa-gu, Seoul, 138-803, Korea
sunghyup.lee@gmail.com
[2] School of Electrical Engineering & Computer Science, Kyungpook National University, Daegu, 702-701, Korea
{kw0314,yzcho}@ee.knu.ac.kr

Abstract. We propose a directional flooding with data aggregation (DFDA) using hop-count values of sensor nodes for energy-efficient wireless sensor networks. In DFDA, the hop-counts of sensor nodes are used to discover the route to the sink. The packets of sensor nodes are aggregated in intermediate nodes and delivered to the sink or neighbors with direction to the sink. In addition, we develop a constrained data aggregation mechanism that discards the redundant data in intermediate nodes. Data aggregation is widely accepted as an essential paradigm for energy-efficient routing in wireless sensor networks. The proposed scheme, DFDA, achieves energy savings by reducing the redundant transmissions using directional flooding based on hop-count and data aggregation. Our simulation results show that DFDA achieves significant improvement in energy consumption.

1 Introduction

Recent advances in wireless communication and micro-mechanical systems have enabled the development of extremely small, low-cost sensors that possess sensing, signal processing, and wireless communication in short distances. These tiny sensor nodes, which consist of sensing, data processing, and communication components, affect the idea of sensor networks based on collaborative effort of a large number of nodes. A wireless sensor network of large numbers of inexpensive but less reliable and accurate sensors can be used in a wide variety of commercial and military applications such as target tracking, security, environmental monitoring, and system control. In wireless sensor networks, it is critically important to save energy. Battery-power is scarce and expensive in wireless sensor devices. Hence, energy-efficient communication techniques are essential to increase the network lifetime [1], [2]. Flooding is clearly a straightforward and simple solution, but it is very costly. In addition, it is a robust, fault tolerant, and scalable data delivery protocol. However, in wireless sensor networks, packets are not necessary to reach all nodes. Especially, for case when sensing data packets

T. Vazão, M.M. Freire, and I. Chong (Eds.): ICOIN 2007, LNCS 5200, pp. 821–830, 2008.

are directed towards a single destination (i.e., a sink) from other sensor nodes, the packets are required only to reach the sink [3]. Therefore, some researchers have proposed directional flooding protocols using directionality information for wireless sensor networks [3]-[5]. These protocols can reduce the number of transmission/reception and number of intermediate nodes unnecessarily forwarding packets. Hence, they reduce the total energy consumption and achieve greater efficiency in data delivery.

Data aggregation is widely accepted as an essential paradigm for energy-efficient routing in wireless sensor networks consisting of sensor nodes with severe energy constraints. The efficacy of data aggregation in wireless sensor networks is a function of the degree of spatial correlation in the sensed phenomenon [6], [7]. In order to conserve energy, many of the routing protocols proposed for wireless sensor networks reduce the number of transmitted packets by pursuing in-network data aggregation [8]. Therefore, we propose a directional flooding scheme using a constrained data aggregation mechanism.

The remainder of the paper is organized as follows. In the next section, we discuss related works. In section 3, we explain the details of DFDA. A performance evaluation of DFDA is presented in section 4, which illustrates the simulation environment, results, and a comparison with existing schemes in terms of energy consumption and network lifetime. Finally, our conclusions are presented in section 5.

2 Related Works

In the classical flooding protocol, a sensor node floods its packet to all of its neighboring nodes. Each receiving node stores a copy of the packet and rebroadcasts the packet once. This mechanism keeps going until all sensor nodes in a network that are connected the sensor node with sending packets have received the packet [6]. The classical flooding protocol is robust to node failure and radio collision. It requires that sensor nodes have the source ID and the sequence number of the packet. Therefore, the sensor nodes uniquely identify each packet and prevent the rebroadcast of the same packet more than once [7]. However, classical flooding can cause severe problems such as radio collision and redundant forwarding.

Therefore, some researchers have proposed several methods to solve the problems of the classical flooding protocol [3]-[5], [7]. In [3], authors have proposed a new directional flooding protocol for utilizing directional information to achieve the efficiency in data delivery. This directional flooding protocol can lead flooded packets to flow in the "right direction" towards their destination or the sink, hence eliminating unnecessary packet forwarding and reducing the total energy consumption. However, authors assume that all sensor nodes know their own location information and sink's location information using location system like the GPS because the flooding decision is made with considering the directionality information towards the sink. In [3], a sensor node calculates an estimated minimum hop-count between itself and the sink. This is based on the knowledge about the two location information and a sensor node's transmission range. To solve the assumption in [3], authors in [5] propose a novel gradient approach that utilizes for gradient setup at each node. The gradient usually means a direction state set toward the neighbors through which a sink is reached. In wireless sensor networks, the majority of packet transmissions are delivered in the direction to a sink from distributed sensor nodes. Thus, each sensor node can be provided with the direction to forward sensing data to the sink [5].

Our proposed scheme is similar to the gradient approach of the above explained scheme. In particular, the main concept of hop-count discovery (HD) process is identical, but the detailed operation of the mechanism is somewhat different. In the gradient-based flooding scheme in [5], to setup the initial gradient values of the sensor nodes, a sink floods its neighbors with a short initiation message (INIT). When receiving a first INIT packet, a sensor node sets its hop-count value to INIT's hop count plus 1. Then, the sensor node resends the INIT packet to all its one-hop neighbors after replacing the INIT's hop count value by its new hop-count value. An important energy saving mechanism for sensor nodes is to exploit in-network data aggregation. In wireless sensor networks, the raw sensed data are typically forwarded to a sink node for processing. The main idea of in-network data aggregation is to eliminate unnecessary packet transmission by filtering out redundant data and/or by performing incremental assessment the semantic of data (i.e., picking the maximum temperature reading) [8]-[11]. The data aggregation mechanism in DFDA promotes filtering and discarding-based data aggregation as much as possible and therefore tradeoffs increased per node waiting and processing delay, and finally raises the overall delivery latency. We show this through simulation in terms of delay. Similar to the data aggregation mechanism of [12], the data aggregation mechanism of DFDA is an in-network aggregation technique intended to save communication-related energy consumption.

3 DFDA: Directional Flooding with Data Aggregation

Since various sensor nodes often detect common phenomena, there is some redundancy in the data which the various sources send to a sink. Hence, in-network discarding and processing techniques that eliminate redundancy can help to conserve scarce energy resources [13]. Therefore, in this paper, we propose a constrained data aggregation mechanism. In DFDA, after the HD packet propagation phase has been complete, all sensor nodes indicate their hop-count values through the neighbor discovery phase. In the data delivery phase, sensor nodes perform data dissemination using the hop-count based data delivery mechanism toward the sink and the constrained data aggregation algorithm. In the following subsections, we explain the details of DFDA.

3.1 Operation of DFDA

A gradient-based data dissemination scheme [5] uses the hop count information for the gradient setup. However, DFDA only uses the hop count information to drop a redundant and duplicate packet, but the main concept of the two

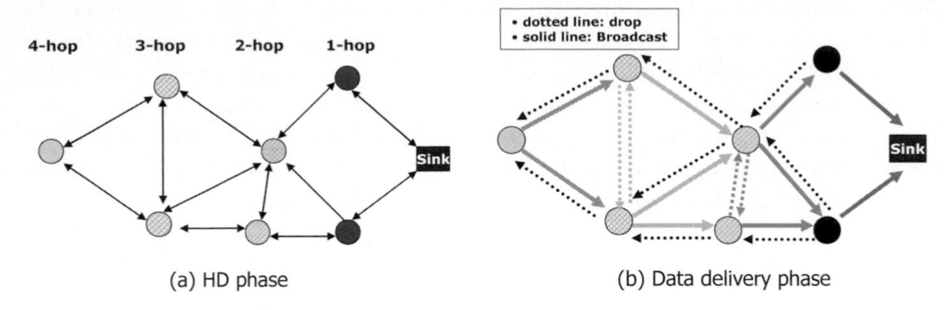

Fig. 1. The operations of each phase in DFDA

schemes are identical. DFDA has two phases: (i) the HD phase, (ii) the data delivery phase.

HD Phase: Fig. 1(a) shows the HD phase of DFDA. The sink broadcasts HD packets to the network with the hop-count set to 1. When a sensor node receives the HD packet, it caches this hop-count value as its hops to the sink in memory and increases the hop-count by 1, and then re-broadcasts HD packet with hop-count value modified. Also, whenever a sensor node receives the HD packet, it compares its hop-count value with that of the received HD packet. If its hop-count value is smaller than that of the received HD packet, a sensor node discards the received HD packet. Conversely, if its hop-count value is bigger than that of the received HD packet, a sensor node replaces its hop-count value with that of the received HD packet, and then re-broadcasts the received HD packet with hop-count value plus 1. This process continues until all sensor nodes receive the HD packets at least once within the HD phase.

Data Delivery Phase: As shown in Fig. 1(b), the data delivery mechanism of DFDA is fundamentally similar to that of the new directional flooding scheme in [3] and gradient-based data dissemination in [5]. However, the data delivery mechanism of DFDA is based on the packet discarding and data aggregation. When a sensor node receives packets from its neighbors, it compares its hop-count value with that in the received packet. If the hop-count value of the received packet is smaller than or equal to its hop-count value, the sensor node discards the received packet to avoid duplicate transmission of the same packets as much as possible. If the hop-count value of the received packet is bigger than its own hop-count value, a sensor node replaces the hop-count value of the received packet with its hop-count value and re-broadcasts it to the network. Thus, the received packet in a sensor node is delivered toward the sink through multi-hop data dissemination without unnecessary transmission in the intermediate nodes.

3.2 Data Aggregation Mechanism

We used a constrained data aggregation mechanism for an energy-efficient and directional flooding scheme. The number of inserting packets in the input queue of an intermediate node is limited by the size of queue and the length of the packet

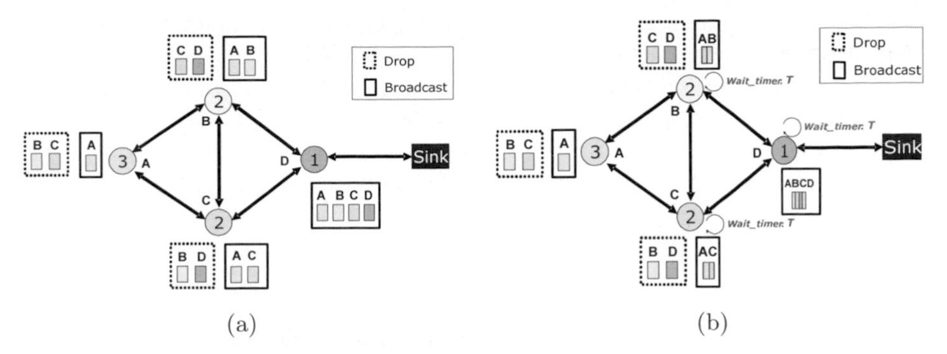

Fig. 2. The overview of DFDA. (a) Directional flooding without data aggregation (b) Directional flooding with data aggregation.

conserving the unique characteristics of incoming packets received by a sensor node. Fig. 2 shows that the overview of DFDA; directional flooding scheme without data aggregation(a) and directional flooding scheme with data aggregation(b). Constrained data aggregation can avoid that the information and unique characteristics of data are distorted and lost due to excessive data aggregation and compression in the intermediate nodes. For the above reasons, we illustrate an example of the constrained data aggregation in which two incoming packets are aggregated into one packet in an intermediate node. Additionally, intermediate nodes can aggregate during wait_timeout T to reduce the processing delay and overhead.

Fig. 3 shows the constrained data aggregation mechanism. To enable data aggregation, intermediate nodes will process or delay the received data for T. When packets of node 1 and 2 are incoming to node 3 during T, node 3 starts the aggregation process after it checks their data for duplication or similarity. If the data of node 1 and 2 are the same, node 3 floods the aggregated data after it aggregates its data(C) and the data (A) of node 1 and 2 into one unit data as depicted in Fig. 3(a). If the data of node 1 and 2 are different, node 3 floods one unit of data that aggregated into the packets of node 1 and 2. Then node 3 floods its packet after it floods the aggregated data as shown in Fig. 3(b). DFDA uses the data aggregation mechanism with wait_timer for energy-efficient data dissemination.

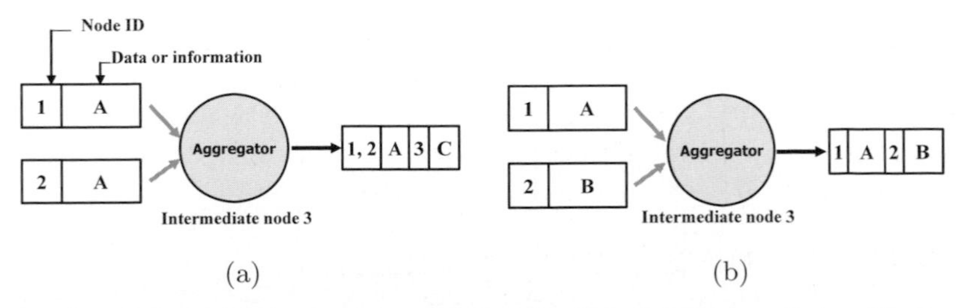

Fig. 3. The constrained data aggregation mechanism. (a) Two packets with the same data into an aggregator (b) Two packets with the different data into an aggregator.

If the intermediate node does not receive the packets from its neighbors during T in Fig. 3(b), it only broadcasts its own packet in the memory to the network without any data aggregation. The constrained data aggregation mechanism of DFDA eliminates unnecessary packet transmission by filtering out redundant data and conserves the data or information characteristics of incoming packets. Therefore, this mechanism can reduce energy consumption of wireless sensor networks.

4 Performance Evaluation

In this section, we compare the performance of DFDA with classical flooding and directional flooding in terms of the number of transmission/reception, energy

Table 1. Parameter values in simulation

Item	Value
Radio propagation speed	$3 \times 10^3\ m/s$
Radio link bandwidth	1 Mbps
Data size	500 byte
Radio mode	Power consumption (mW)
Transmit	14.88
Receive	12.50

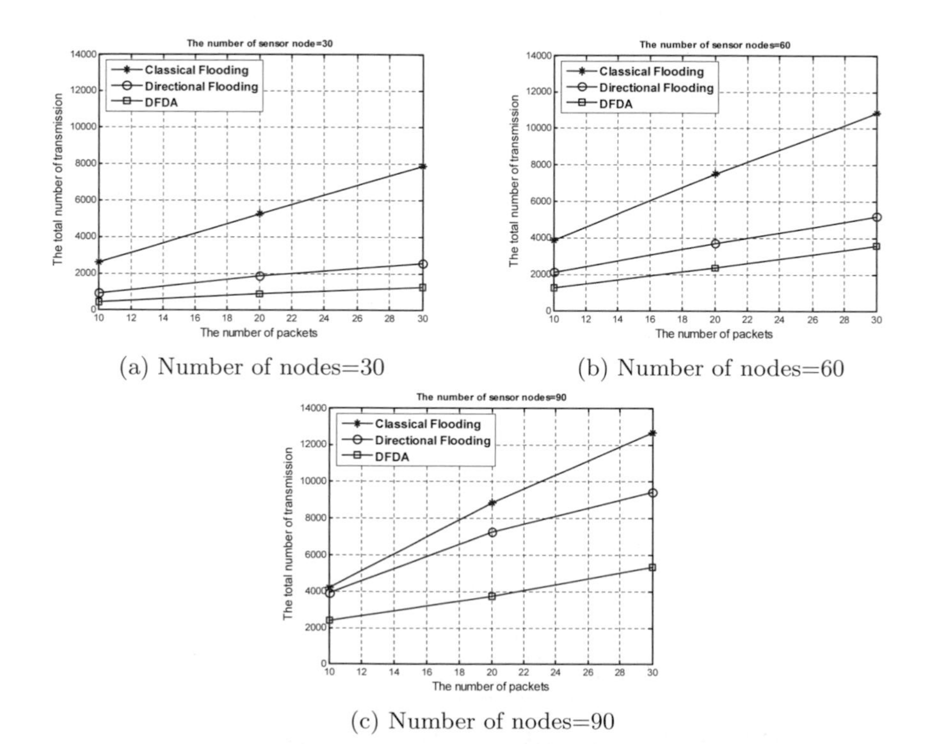

(a) Number of nodes=30

(b) Number of nodes=60

(c) Number of nodes=90

Fig. 4. Total number of transmissions versus the number of sensor nodes

consumption, and network delays through simulation using TOSSIM [14]. We use an omni-antenna and several simulation parameter variations in order to analyze our scheme. The network dimensions are assumed to be $100m \times 100m$. A sink is located in (200m, 200m) at a remote location. The sensor characteristics are given in Table 1 [15]. The transmission range of a sensor node is 40m [16]. We studied the performance of DFDA according to two factors: (1) when the number of packets generated by sensor nodes was 10, 20, and 30; and (2) when the number of sensor nodes was 30, 60, and 90. We ran the simulation 20 times for each scheme and averaged the results.

4.1 The Number of Transmissions

Fig. 4 shows the number of transmissions for classical flooding, directional flooding, and DFDA. This means the transmission trials need to deliver all packets generated by sensor nodes to the sink. With an increase in the number of sensor nodes, the total number of transmissions increases for all flooding schemes. However, the number of transmissions of DFDA is much less than those of other schemes because it discards redundant transmissions and uses the constrained data aggregation mechanism. Therefore, the declination of DFDA is the most efficient of all schemes.

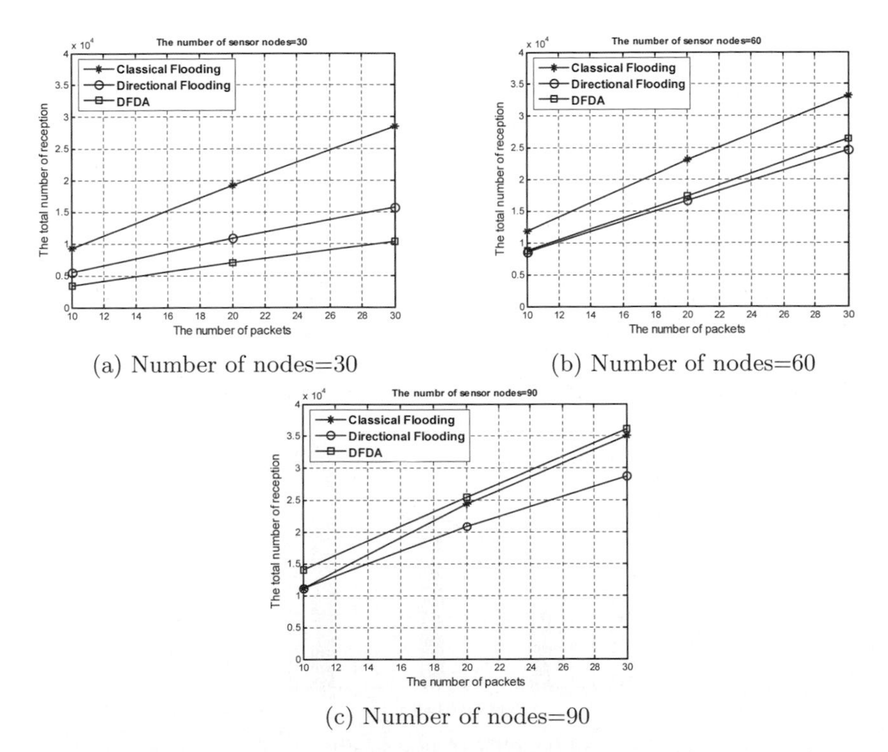

(a) Number of nodes=30

(b) Number of nodes=60

(c) Number of nodes=90

Fig. 5. Total number of receptions versus the number of sensor nodes

4.2 The Number of Receptions

Fig. 5 shows the number of receptions for classical flooding, directional flooding, and DFDA. This means the reception trials need to deliver all packets generated by sensor nodes to the sink. As shown in Fig. 5(a), the number of receptions of DFDA is much less than those of other schemes. Conversely, as shown in Fig. 5(b) and (c), the number of receptions of DFDA is more than those of other schemes because it can reduce the dropped packets by interference such as radio collision and overflow in memory using the directional flooding approach and the constrained data aggregation mechanism.However, the number of transmissions of DFDA does not increase because the unnecessary and redundant packets are discarded by the directional flooding using hop-count and some packets is aggregated by the constrained data aggregation mechanism in intermediate nodes.

4.3 Energy Consumption

Fig. 6 shows the total energy consumption of the three flooding schemes when varying the number of sensor nodes and packets when the network size is fixed. As expected, in all cases DFDA consumed much less energy than directional flooding and classical flooding. DFDA and directional flooding, both variants of the gradient approach, achieve improved energy-efficiency by reducing the many of redundant transmissions. In addition, the total energy consumption of

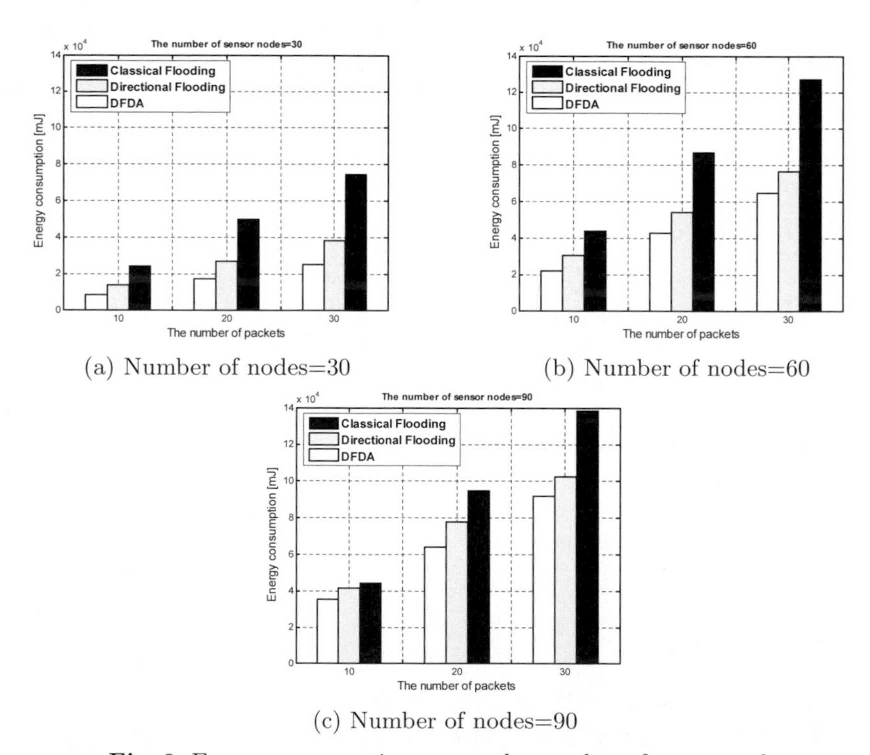

(a) Number of nodes=30

(b) Number of nodes=60

(c) Number of nodes=90

Fig. 6. Energy consumption versus the number of sensor nodes

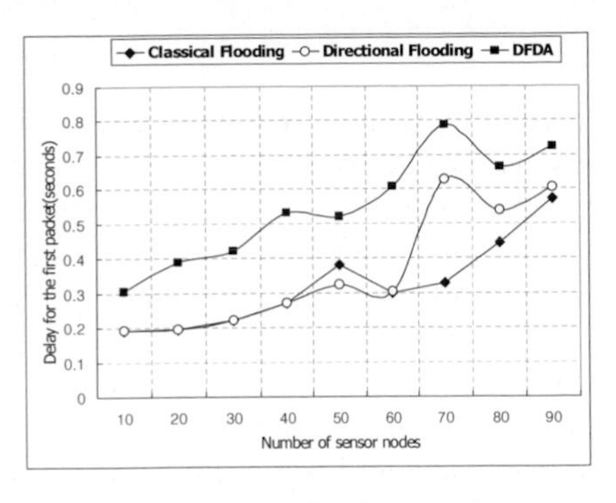

(a) Delay for the first packet.

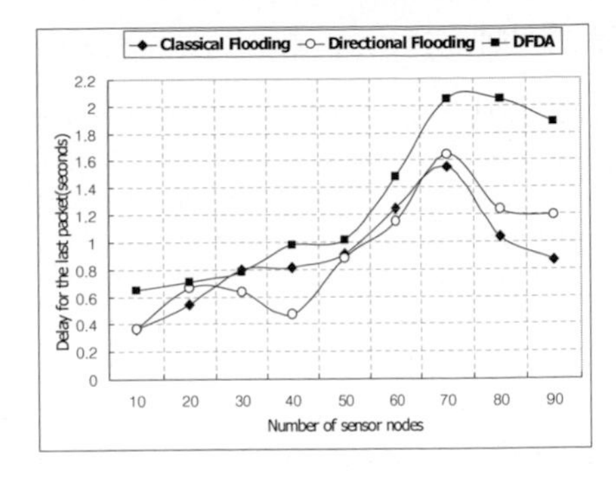

(b) Delay for the last packet.

Fig. 7. Delay versus the number of sensor nodes

directional flooding scheme is low compared to classical flooding scheme, but not as low as DFDA. DFDA is the most energy-efficient scheme because it uses both directional information and data aggregation mechanism.

4.4 Network Delay

Fig. 7 plots network delay as the number of sensor nodes. We measured the delay for the first packet received and the last packet received. It is the time at which all nodes in the network have transmitted the first and last flooded packet, respectively. DFDA introduces some delay during the data delivery phase because it uses wait_timer for the constrained data aggregation in intermediate nodes. However, energy efficiency is much more important than network delays because energy awareness is an essential design issue in wireless sensor networks.

5 Conclusion

We proposed a directional flooding scheme with data aggregation (DFDA) using hop-count values of sensor nodes for energy-efficient wireless sensor networks. In our scheme, packets transmitted by sensor nodes are aggregated in intermediate nodes and delivered to the sink or neighbors with direction to the sink. DAFA involves data discarding-based data dissemination toward the sink. Data aggregation is an essential paradigm for energy-efficient routing in wireless sensor networks consisting of sensor nodes with severe energy constraints. DFDA achieves energy savings by reducing redundant transmissions using directional flooding based on hop-count and data aggregation. Simulation results show that DFDA achieves the significant improvement in energy consumption.

Acknowledgement

This work was supported by the Korea Science and Engineering Foundation (KOSEF) grant funded by the Korea government (MOST)(No.: R01-2006-000-10753-0(2006)), Korea.

References

1. Akyldiz, I.F., Su, W., Sankarusubramaniam, Y., Cyirci, E.: Wireless sensor networks: a survey. Computer Networks 38(4), 393–422 (2002)
2. Durresi, A., Paruchuri, V.K., Lyengar, S.S., Kannan, R.: Optimized broadcast protocol for sensor networks. IEEE Trans. on Computers 54(8), 1013–1024 (2005)
3. Ko, Y.-B., Choi, J.-M., Kim, J.-H.: A new directional flooding protocol for wireless sensor networks. In: Kahng, H.-K., Goto, S. (eds.) ICOIN 2004. LNCS, vol. 3090, pp. 93–102. Springer, Heidelberg (2004)
4. Shim, H.-K., Lee, W.-J., Kim, J.-H., Lee, C.-H.: A restricted flooding scheme for stateless greedy forwarding in wireless sensor networks. In: Chong, I., Kawahara, K. (eds.) ICOIN 2006. LNCS, vol. 3961. Springer, Heidelberg (2006)
5. Han, K.-H., Ko, Y.-B., Kim, J.-H.: A novel gradient approach for efficient data dissemination in wireless sensor networks. In: Proc. of IEEE VTC (2004)
6. Su, W., Akyildiz, I.F.: A stream enabled routing (SER) protocol for sensor networks. In: Proc. of Med-hoc-Net (2002)
7. Lim, H., Kim, C.: Flooding in wireless ad hoc networks. Computer Communications 24, 353–363 (2001)
8. Sabbineni, H., Chakrabarty, K.: Location-aided flooding: an energy-efficient data dissemination protocol for wireless sensor networks. IEEE Trans. on Computer 54(1), 36–46 (2005)
9. Pattem, S., Krishnamachari, B., Govindan, R.: The impact of spatial correlation on routing with compression in wireless sensor networks. In: Proc. of IEEE IPSN (2004)
10. Xue, Y., Cui, Y., Nahrstedt, K.: Maximizing lifetime for data aggregation in wireless sensor networks. ACM Klwer Mobile Networking & Application (2004)
11. Akkaya, K., Younis, M., Youssef, M.: Efficient aggregation of delay-constrained data in wireless sensor networks. In: Proc. of ICQAWN (2005)
12. He, T., Blum, B.M., Stankovic, J.A., Abdelzaher, T.F.: AIDA: Adaptive application independent aggregation in sensor networks. ACM Trans. on Embedded Computing System (2003)
13. Krishnamachari, B., Estrin, D., Wicker, S.: The impact of data aggregation in wireless sensor networks. In: Proc. of IWDES (2001)
14. TOSSIM, http://www.cs.berkeley.edu/~pal/research/tossim.html
15. Akkaya, K., Younis, M.: A Survey of routing protocols in wireless sensor networks. Ad Hoc Network Journal 3/3, 325–349 (2005)
16. Zuniga, M., Krishnamachari, B.: Optimal transmission radius for flooding in large scale sensor networks. In: Proc. of IEEE ICDCSW (2003)

TCP with Explicit Handoff Notification for a Seamless Vertical Handoff

Young-Soo Choi[1], Kang-Won Lee[2], and You-Ze Cho[2]

[1] User Terminal Development Department, BcN Business Unit, KT, 17, Woomyeon-dong, Seocho-gu, Seoul, 137-792, Korea
cys@kt.co.kr

[2] School of Electrical Engineering and Computer Science, Kyungpook National University, Korea
{kw0314,yzcho}@ee.knu.ac.kr

Abstract. This paper proposes a new variant of TCP for seamless vertical handoffs in heterogeneous wireless/mobile networks, which exploits explicit handoff notifications to alert handoff types and packet losses due to handoffs for efficient congestion control and packet error recovery mechanisms. Through ns-2 simulation, we show that the proposed scheme performs better than the existing TCP protocols in terms of throughput during vertical handoffs.

1 Introduction

It is well known that a wireless link and user mobility can significantly affect TCP performance. Existing wireless networks offer mobile users a tradeoff between mobility support, coverage area, network capacity, power consumption, and costs. The user can choose the most suitable wireless network at a given time and location, for example, by switching between Wireless LAN (WLAN) and a 3G cellular network such as Universal Mobile Telecommunications System (UMTS), while keeping ongoing data transfers. To integrate heterogeneous wireless networks, it is crucial that a seamless vertical handoff is ensured. Vertical handoffs are divided into two types; upward vertical handoffs and downward vertical handoffs. An upward vertical handoff is a handoff to an overlay with a larger coverage area and a lower bandwidth/area. A downward vertical handoff is a handoff to an overlay with a smaller cell size and a higher bandwidth/area. When a vertical handoff occurs, there are sudden changes in terms of round-trip delays and bandwidth. The availability of heterogeneous interfaces has given rise to new challenges to TCP.

Various modifications for a regular TCP have been proposed to remedy the deficiency, which can be classified as (1) end-to-end approaches (e.g., Freeze TCP [1]), (2) local recovery approaches (e.g., Snoop TCP [2]), and (3) split connection approaches (e.g., Indirect TCP [3]). Recent research has focused on preventing or hiding handoff-induced packet losses from the transport layer. Wireless/mobile TCP modifications have attempted to avoid reducing the sending rate in response

T. Vazão, M.M. Freire, and I. Chong (Eds.): ICOIN 2007, LNCS 5200, pp. 831–840, 2008.

to handoff losses and to restart transmissions at a full rate with the old window size upon entering a new network.

Mobile IP/IPv6 can handle a vertical handoff without breaking the ongoing connection. However, packets often get lost, delayed, or are out of order during a handoff, which in turn can trigger unwanted TCP congestion control, thereby degrading TCP performance. This is because TCP congestion control is based on the assumption that the end-to-end path of a connection remains largely unchanged after a connection is established. This assumption of a *constant path* can hold in the case of a horizontal handoff, making it safe to resume transmissions with the same window size as prior to the horizontal handoff. For a vertical handoff, however, TCP cannot continue with the old data rate which only reflects the congestion state of the old path, rather than that of the new, unknown environment. Therefore, existing schemes, which mainly consider horizontal handoffs, do not work well because they overshoot or underutilize the available bandwidth after vertical handoffs.

In this paper, we propose a new variant of TCP, called ECP (Explicit Handoff Notification TCP). The proposed scheme ensures a seamless vertical handoff. In the proposed scheme, the TCP sender and receiver use a reserved field in the TCP header in order to recognize an impending handoff, types of handoff, and handoff-induced packet losses. By using explicit handoff notification, ECP readjusts transmission rate and round trip time (RTT) related information. In addition, it recovers handoff-induced packet losses efficiently.

The remainder of this paper is organized as follows: Section 2 briefly discusses related works. Section 3 describes the ECP protocol. Section 4 presents the simulation results, and conclusions are given in Section 5.

2 Related Works

TCP congestion control is composed of two major algorithms: slow start and congestion avoidance algorithms which allow TCP to increase the data transmission rate without overwhelming the network. TCP uses a variable called a congestion window (*cwnd*) and cannot inject more than *cwnd* segments of unacknowledged data into the network. In the slow start phase, TCP quickly determines the amount of available capacity in a network path by doubling *cwnd* for every RTT until the *cwnd* reaches the slow-start threshold (*ssthresh*). This allows an upper bound to be quickly reached, when the first packet loss is detected. In the congestion avoidance phase, TCP increases *cwnd* by one packet for each RTT and halves *cwnd* in the event of a packet loss. The TCP congestion avoidance algorithm is called the Additive Increase Multiplicative Decrease (AIMD), and it is the basis for steady state congestion control.

Many schemes have been proposed to enhance TCP over wireless links. The proposed approaches can be classified as end-to-end, local recovery, or split connection. Local recovery schemes modify the link layer protocol that operates over the wireless link to hide losses using local recovery mechanisms such as FEC and ARQ. In split connection approaches, the TCP connection is split at the base station (BS). TCP is used from the sender to the BS (wired part), while either TCP or a special purpose transport protocol is chosen over a wireless network. The TCP sender is only affected by the congestion in the wired network; hence, the sender is shielded from wireless losses. End-to-end schemes solve problems at the TCP sender, receiver, or both sides. Although local recovery and split connection schemes provide performance improvements, both approaches have some limitations: (1) Most protocols fail if encrypted mechanisms are used; (2) several schemes require modifications at the BS, making it difficult for these schemes to be deployed; and, (3) there is a possibility that an intermediate node will become the performance bottleneck itself. The buffering overhead is not negligible. In addition, to support handoffs, the connection state needs to be handed over to the new BS, and this becomes significant amount of overhead. All things considered, the current study focuses on the end-to-end schemes.

In Freeze TCP, the mobile node monitors the signal strength and sends Zero Window Advertisements (ZWA) if it detects an impending handoff. Upon receiving ZWA, the TCP sender enters the persist mode and keeps all TCP variables, such as the *cwnd* size. Upon reconnection detection, the mobile node (MN) retransmits 3 duplicate ACKs for the last data segment it received with a non-ZWA. Through these mechanisms, the Freeze TCP sender can resume its transmission at the full rate using the old *cwnd* size upon entering a new network. The Freeze TCP is a promising approach that does not require the involvement of intermediates.

Several recent papers have discussed problems concerning TCP, which are associated with a vertical handoff. [4] study TCP performance with a vertical handoff between GPRS and WLAN in real environments, where handoff delays were found to cause TCP to timeout, thereby degrading the TCP performance. In addition, the high buffering in GPRS deteriorated TCP performance, due to inflated RTT and RTO values.

[5] compared the effect of a vertical handoff regarding transport protocols, such as TCP and TCP Friendly Rate Control (TFRC). It was revealed that TFRC has significant difficulties in adapting to the network after a vertical handoff, due to the change in the link bandwidth. While [5] suggested an over-buffering scheme, this scheme required knowledge of the BDP (Bandwidth*Delay Product), which is impractical.

[6] proposed a mechanism, called STCP (Slow start TCP) in this paper, to improve TCP during vertical handoffs by using the physical layer information. MN measures the received signal strength and its velocity to determine impending handoff. In the case of a horizontal handoff, the sender operates in the same way as the Freeze TCP. For a vertical handoff, the TCP sender resumes data transmission at the slow start state in order to adjust quickly to the new network bandwidth.

Although Freeze TCP and STCP, which avoid retransmission timeouts and packet losses, can provide significant improvements, the choice of when to send a ZWA to freeze the sender is difficult. Since both schemes do not have the support of an intermediate node that can start buffering when a disconnection is predicted, they are susceptible to in-flight packet loss. In addition, it is difficult to send the ZWAs soon enough to avoid any in-flight packet loss, while, at the same time, making sure that the sender is not frozen too soon. [1] suggests that the round trip time can act as a reasonable *warning period* (WP). In practice, however, the TCP receiver cannot estimate the RTT and the assumption that the MN can detect precisely future disconnections is not practical.

3 ECP: TCP with Explicit Handoff Notification for Heterogeneous Mobile Networks

When a horizontal handoff is completed, the MN is reconnected to a similar cell where the traffic pattern is likely to be the same as before. Therefore, in this case, transmission can be resumed with the same window size prior to the horizontal handoff. However, with a vertical handoff, the path characteristics are drastically changed, and the MN enters a new network with an unknown congestion state and a different traffic pattern. The main idea of the proposed scheme is to freeze TCP during a handoff. After the handoff, the MN transmits an Explicit Handoff Notification (EHN) packet, with which the TCP sender adjusts the *cwnd*, sends the sequence number and RTT-related information appropriately. The proposed scheme requires minimal modification at a TCP sender. If the sender does not support the EHN packet, the proposed scheme will operate in a similar way as Freeze TCP. The proposed scheme assumes the existence of a link layer assist, such as IEEE 802.21 Media Independent Handover (MIH) function.

3.1 Overall Mechanism

Since the latency of a vertical handoff can be significant, the ECP sender temporarily halts data transmission and stops the retransmission timer, similar to Freeze TCP. However, the proposed scheme differs from the Freeze-TCP in initiating a slow start phase for a vertical handoff, and using explicit handoff-induced loss information to differentiate handoff-induced losses from congestion-induced packet losses.

If the receiver detects an impending handoff through a MIH function, it sends a few (at least one) ZWA. As soon as the handoff is completed, the receiver then sends three EHN packets to the TCP sender to resume transmission. The MIH information service is also used by the receiver to notify the sender of the type of handoff, which is encoded using two EHN bits. For a horizontal handoff, only the EHN bit is marked, whereas for a vertical handoff, the Explicit Vertical Handoff Notification (EVHN) bit is also marked. The EHN packet is a specially marked TCP ACK packet containing the sequence number of the last data packet successfully received by the MN. In the proposed scheme, the MN uses two bits in a reserved field in the TCP header to indicate the type of handoff. Based on this encoded feedback, the TCP sender then resumes transmission. The use of EHN bits allows the TCP sender to distinguish between packet loss which is due to either congestion or a handoff. In other words, the TCP sender assumes that

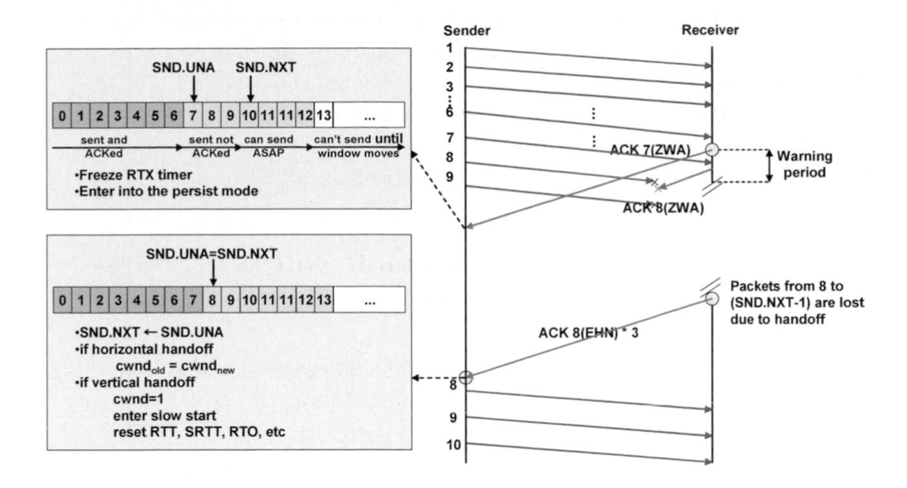

Fig. 1. Example of ECP operation

outstanding packets are lost due to a handoff when it receives an EHN packet. Hence, ECP assumes that the packets from the SND.UNA to the SND.NXT are consecutively lost by a handoff, where SND.UNA and SND.NXT represent the "oldest unacknowledged sequence number" and "next sequence number to be sent," respectively and updates SND.NXT. Since handoff-induced lost packets are removed from the network and packet loss does not indicate congestion, the TCP sender does not change *cwnd* in case of a horizontal handoff. Meanwhile, for a vertical handoff, TCP enters a modified slow start phase. This congestion window management mechanism will be discussed later. An example of ECP is shown in Figure 1.

3.2 Congestion Window and RTO Management

After a vertical handoff, TCP must choose the initial *ssthresh*. If the *ssthresh* is set too high relative to the network BDP, the exponential increase of *cwnd* generates multiple consecutive packet losses and retransmission timeouts. If the initial *ssthresh* is set too low, TCP exits the slow start phase and enters the congestion avoidance phase, where *cwnd* grows linearly, resulting in poor utilization especially when BDP is large. It may be possible to "jump-start" a TCP connection from that rate (with appropriate pacing), rather than using slow start. However, for this, it is necessary to estimate the number of nodes sharing the wireless access links or intermediate node supporting. Additionally, these approaches usually cannot adjust to changing network conditions. For example, if there are multiple TCP flow handoffs at approximately the same time, these approaches will have set the initial *ssthresh* too high, resulting in multiple packet losses.

To probe the available bandwidth of a new network, the proposed scheme begins a modified slow start only after a vertical handoff. The modified slow start

```
/* At sender */
On arrival of ACK:
if ACK with ZWA /* impending handoff */
    cancel RTX timer;
    stop transmission;
else if ACK with EHN /* handoff is completed */
    set SND.NXT = ACK;
    restart RTX timer;
    cwnd = cwnd;
    if EVHN flag is set /* Handoff type = vertical handoff */
        cwnd = 1;
        initialize RTO, SRTT, RTTVAR;
        enter modified slow start;
    resume transmission;
else /* Normal ACK */
    if(modified slow start phase)
        if RTT is increasing /* congestion avoidance mode */
            cwnd += 1/cwnd;
    else /* slow start mode */
        cwnd += 1;
else /* Regular TCP */
    if cwnd < ssthresh /* slow start */
        cwnd += 1;
    else /* Congestion avoidance */
        cwnd += 1/cwnd;
```

```
/* At receiver */
if impending handoff /* notify impending handoff */
    set ZWA;
if handoff is completed
    set EHN flag;
    if(vertical handoff)
        set EVHN flag;
send ACK;
```

Fig. 2. Pseudo code for ECP

algorithm has two modes; congestion avoidance mode and slow start mode. In the modified slow start phase, *ssthresh* is initially set as an arbitrary large value after a vertical handoff. Thereafter, the TCP sender measures whether the network bandwidth is fully utilized, and it determines whether to shift to the congestion avoidance mode or remain in the slow start mode. When the bottleneck link utilization is below 100%, the sender uses the slow start mode, where *cwnd* increases exponentially. Conversely, if the bottleneck link is fully utilized, the proposed scheme moves to the congestion avoidance mode, where the *cwnd* increases linearly. By repeating linear and exponential increase, *cwnd* adapts to the desired window in a timely manner. The proposed scheme switches to the regular TCP as soon as the first packet loss is experienced. To estimate the bottleneck link utilization, the proposed scheme uses the RTT. When the measured RTTs indicate an increasing trend, full utilization of the link bandwidth is assumed, and vice versa.

After a vertical handoff, the RTO increases or decreases rapidly, and it takes some time to converge into the proper value. For a downward handoff, the TCP sender has a large RTO value until it converges, despite a much smaller RTT. If packet loss occurs during this period, the TCP sender waits a long period of time for this timeout to occur so that normal transmission can restart. Meanwhile, for an upward handoff, the TCP sender can experience some spurious retransmissions

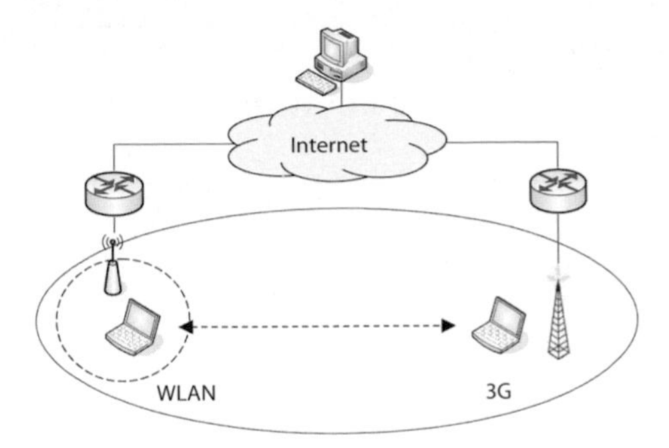

Fig. 3. Network topology for simulation

before the RTO value converges. Therefore, the proposed scheme simply resets the RTO-related parameters, such as the RTT, SRTT, RTTVAR, and RTO, after a vertical handoff. The pseudo code for the ECP is presented in Figure 2.

4 Simulation Results and Discussion

In this section, we compare the performance of the proposed TCP scheme in the case of a handoff between a WLAN and a 3G cellular network (e.g. UMTS). The network topology for the simulation is shown in Figure 3. A 3G link has 384 kbps bandwidth and 150 ms latency in downlink and 64 kbps, 150ms in uplink. Within the WLAN, the data rate and RTT were 5Mbps and 10ms, respectively. Each wired link has a bandwidth of 10Mbps and a delay of 20ms. The packet size was 1000 bytes. The ECP, STCP, and Freeze TCP were implemented based on the TCP Newreno. To investigate the impact of the WP for Freeze TCP, the WP was varied, equal to RTT (ideal WP) and RTT/2. The buffer size at the router was adjusted to 50KB, for WLAN and 20KB, for 3G. The handoff occurred at 10 sec (17 sec) and was completed by 13 sec (20 sec) for upward (downward) vertical handoff, respectively. Due to space limitations, representative simulation results are presented.

Figure 4 and 5 show the *cwnd* and packet sequence number when a vertical handoff occurs, respectively. After a downward vertical handoff, the available bandwidth drastically increases. In the case of an ideal WP, since there is no method for a TCP sender to know the bandwidth increase, the Freeze TCP cannot utilize the full bandwidth for an extended time because in the congestion avoidance mode, the *cwnd* increases just by one for one RTT. If the WP is smaller than the RTT, the Freeze TCP and STCP experience retransmission timeout and will enter the slow start phase. After an upward vertical handoff, the available bandwidth drastically decreases. As shown in the figures, Freeze TCP resumes its transmission and sends many packets back-to-back. The problem is that the Freeze TCP delays the point at which “forthcoming” congestion

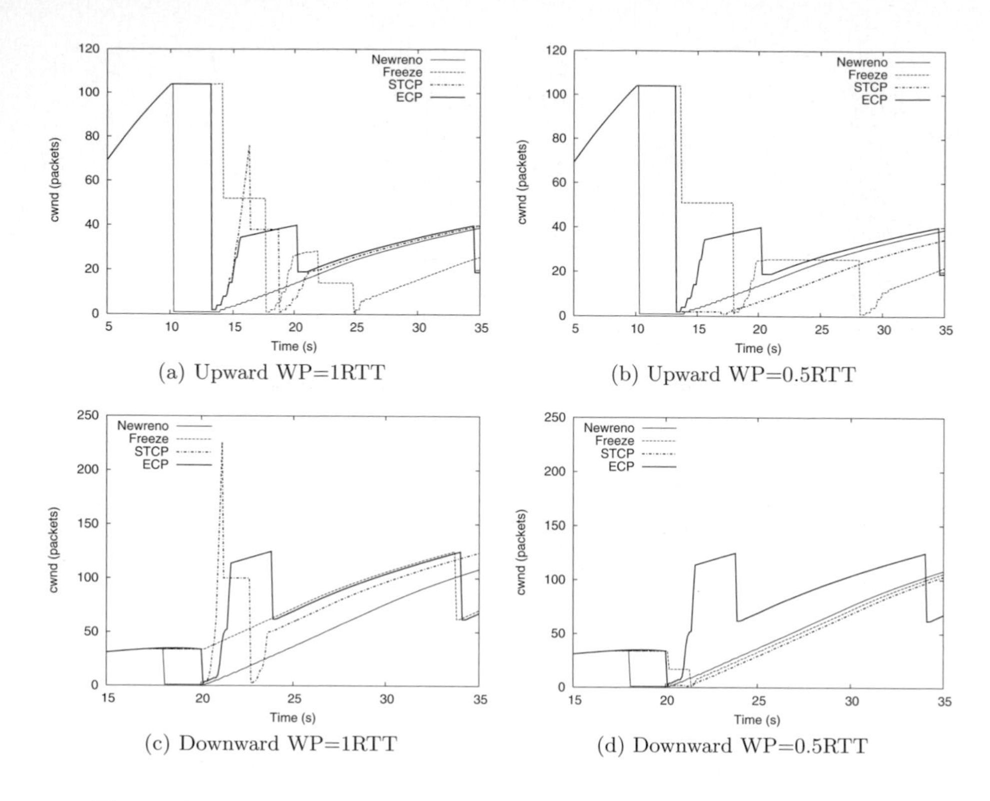

(a) Upward WP=1RTT (b) Upward WP=0.5RTT

(c) Downward WP=1RTT (d) Downward WP=0.5RTT

Fig. 4. Comparison of congestion windows (*cwnd*) during a vertical handoff

Table 1. Comparison of goodput after vertical handoff

		Newreno	Freeze TCP	STCP	ECP
Upward	WP=RTT	81.6 kbps	57.6 kbps	137.6 kbps	249.6 kbps
WLAN→3G	WP=0.5RTT	81.6 kbps	25.6 kbps	16.0 kbps	249.6 kbps
Downward	WP=RTT	1.1 Mbps	3.2 Mbps	1.5 Mbps	3.5 Mbps
3G→WLAN	WP=0.5RTT	1.1 Mbps	1.0 Mbps	0.7 Mbps	3.5 Mbps

is detected. This causes senders to congest the new network and to produce much more congestion-induced packet loss. STCP results in multiple losses and retransmission timeouts because *ssthresh* is set too high after a vertical handoff. Also, STCP does not perform well if the WP is smaller than RTT. This means that choosing the time to send ZWA is critical in STCP and Freeze TCP. ECP probes the available bandwidth quickly and gently. Additionally, ECP is robust against the wrong choice of time to send ZWA. Table 1 compares the average goodput during five-second period after a vertical handoff. The throughput of ECP outperformed other TCP variants regardless of WP or handoff types.

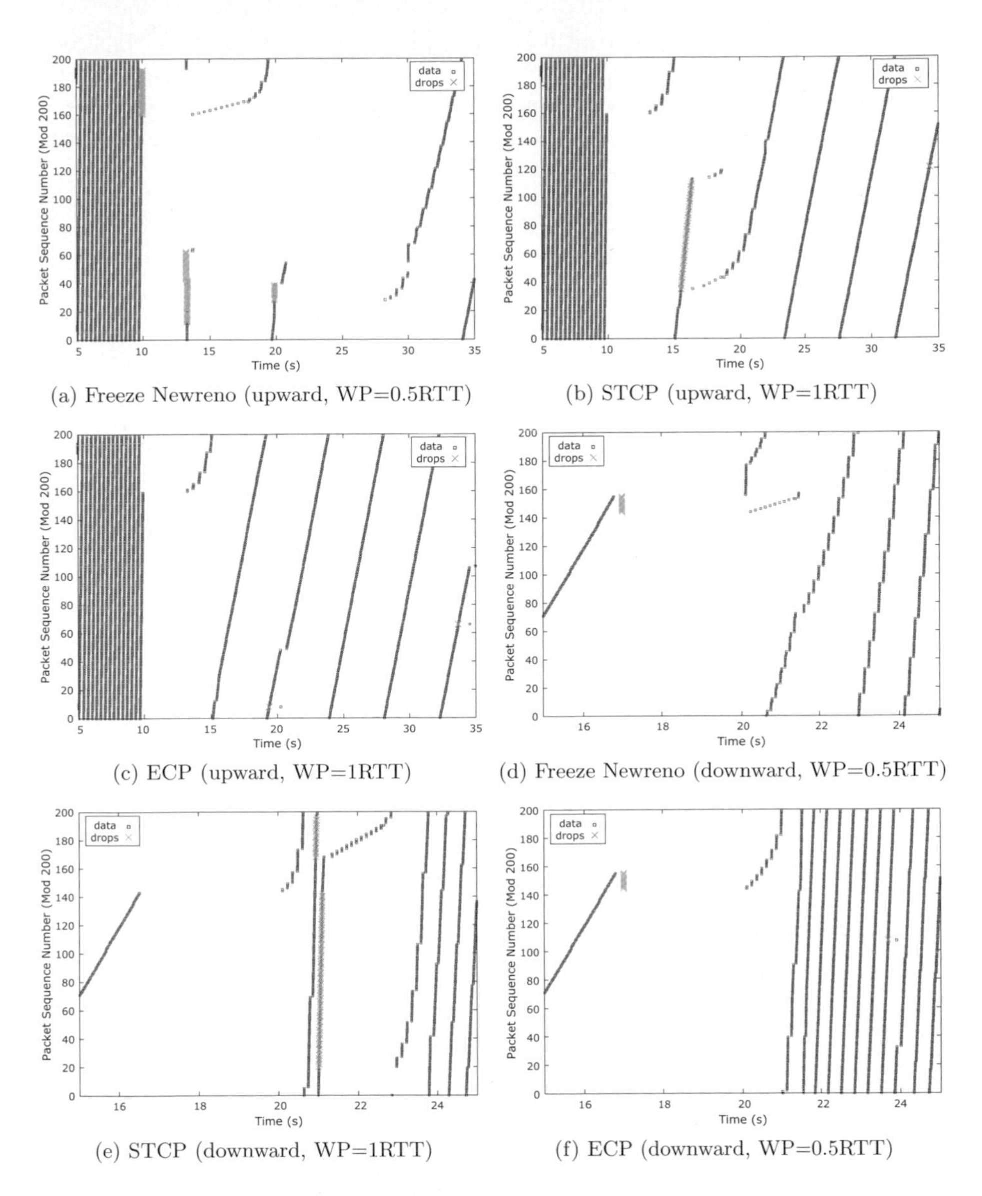

(a) Freeze Newreno (upward, WP=0.5RTT) (b) STCP (upward, WP=1RTT)

(c) ECP (upward, WP=1RTT) (d) Freeze Newreno (downward, WP=0.5RTT)

(e) STCP (downward, WP=1RTT) (f) ECP (downward, WP=0.5RTT)

Fig. 5. Comparison of packet sequence number during a vertical handoff

5 Conclusions

This paper proposed a new variant of TCP for vertical handoffs using explicit handoff notification. In the proposed scheme, layer 2 monitors the physical link qualities and triggers impending handoff events when appropriate. In addition, a MN sent an explicit handoff notification to the sender when a handoff was performed. Since the network environment is drastically changed after a vertical handoff, the proposed TCP readjusted its transmission rate and RTT. Freeze TCP and STCP, a previous end-to-end approach, have to predict precise future disconnections and suffer from severe performance degradation when the prediction fails. The proposed scheme also provided efficient packet loss recovery against handoffs by using explicit handoff notification.

Acknowledgement

This work was supported by the MIC under the ITRC program (C1090-0603-0036) supervised by the IITA(Institute of Information Technology Assessment), Korea.

References

1. Goff, T., Moronski, J., Phatak, D., Gupta, V.: Freeze-TCP: A True End-to-End TCP Enhancement Mechanism for Mobile Environments. In: Proc. of IEEE INFOCOM, vol. 3, pp. 1537–1545 (2000)
2. Balakrishnan, H., Padmanabhan, V., Katz, R.: Improving Reliable Transport and Handoff Performance in Cellular Wireless Networks. Wireless Networks 1(4), 469–481 (1995)
3. Bakre, A., Badrinath, B.: I-TCP: Indirect TCP for Mobile Hosts. In: Proc. of ICDCS, pp. 136–143 (1995)
4. Chakravorty, R., Vidales, P., Subramanian, K., Pratt, I., Crowcroft, J.: Performance Issues with Vertical Handovers - Experiences from GPRS Cellular and WLAN Hot-spots Integration. In: Proc. of PerCom, pp. 155–164 (2004)
5. Gurtov, A., Korhonen, J.: Effect of Vertical Handovers on Performance of TCP-Friendly Rate Control. ACM MC^2R 8(3), 73–87 (2004)
6. Kim, S., Copeland, J.: TCP for Seamless Vertical Handoff in Hybrid Mobile Data Networks. In: Proc. of IEEE GLOBECOM, vol. 2, pp. 661–665 (2003)

Author Index

Printing: Mercedes-Druck, Berlin
Binding: Stein+Lehmann, Berlin